W9-AQV-011

PRENTICE HALL

BIOLOGY

The Living Science

Kenneth R. Miller, Ph.D.
Professor of Biology
Brown University
Providence, Rhode Island

Joseph Levine, Ph.D.
Science Writer and Producer
Adjunct Assistant Professor of Biology
Boston College
Boston, Massachusetts

PRENTICE HALL
Upper Saddle River, New Jersey
Needham, Massachusetts

Components

- Student Edition
- Teacher's Edition
- Laboratory Manual and Annotated Teacher's Edition
- Teaching Resources
- BioLog and Annotated Teacher's Edition
- Transparency Box with Teacher's Guide
- Computer Test Bank
- BioVue and BioVue Plus CD-ROMs

The photograph on the cover shows female lions hunting at night on the African savanna—a brief glimpse into the amazing diversity of life on our planet.

Credits begin on page 971.

ISBN 0-13-415563-7

4 5 6 7 8 9 10 01 00 99 98

PRENTICE HALL
Simon & Schuster Education Group
A VIACOM COMPANY

Staff Credits for Prentice Hall *Biology: The Living Science*

Advertising and Promotion: Judy Goldstein, Carol Leslie, Rip Odell
Business Office: Emily Heins
Design: Laura Jane Bird, Kerri Folb, Kathryn Foot, AnnMarie Roselli, Gerry Schrenk
Manufacturing and Inventory Planning: Katherine Clarke, Rhett Conklin
Market Research: Eileen Friend, Gail Stark
Media Resources: Martha Conway, Libby Forsyth, Vickie Menanteaux, Maureen Raymond, Emily Rose, Melissa Shustyk
National Science Consultants: Charles Balko, Patricia Cominsky, Jeannie Dennard, Kathleen French, Brenda Underwood
Pre-Press Production: Carol Barbara, Kathryn Dix, Paula Massenaro
Production: Christina Burghard, Elizabeth O'Brien, Marilyn Stearns, Elizabeth Torjussen
Science Department
Director: Julie Levin Alexander
Editorial: Laura Baselice, Joseph Berman, Christine Caputo, Maureen Grassi, Rekha Sheorey, Lorraine Smith-Phelan
Marketing: Arthur Germano, Andrew Socha, Kathleen Ventura, Jane Walker Neff, Victoria Willows
Technology Development: Matthew Hart
Electronic Services: Greg Myers, Cleasta Wilburn

Acknowledgments

Teacher Advisory Panel

Leslie Ferry Bettencourt
Lincoln High School
Lincoln, Rhode Island

Jean T. (Caye) Boone
Hume-Fogg Academic High School
Nashville, Tennessee

David A. Dowell
Carmel High School
Carmel, Indiana

Deborah H. Fabrizio
Seminole High School
Seminole, Florida

Yvonne Favaro
Fort Lee High School
Fort Lee, New Jersey

Steve Ferguson
Lee's Summit High School
Lee's Summit, Missouri

Patricia Anne Johnson
Ridgewood High School
Ridgewood, New Jersey

Mamie Lew
George W. Brackenridge High School
San Antonio, Texas

Ned C. Owings
Florence School District One
Florence, South Carolina

Kathey A. Roberts
Lakeside High School
Hot Springs, Arkansas

College Reviewers

Brian Alters, Ph.D.
Harvard University
Cambridge, Massachusetts

Lauren Brown, Ph.D.
Illinois State University
Normal, Illinois

Maura Flannery, Ph.D.
St. John's University
Jamaica, New York

Ann Lumsden, Ph.D.
Florida State University
Tallahassee, Florida

Gerry Madrazo, Ph.D.
University of North Carolina
Chapel Hill, North Carolina

Cynthia Moore, Ph.D.
Washington University
St. Louis, Missouri

Laurence D. Mueller, Ph.D.
University of California
Irvine, California

Carl Thurman, Ph.D.
University of Northern Iowa
Cedar Falls, Iowa

High School Reviewers

Louise Ables
A & M Consolidated High School
College Station, Texas

Bernard Adkins
Wayne High School
Wayne, West Virginia

Tony Beasley
Science Consultant
Nashville, Tennessee

Victor Choy
John Oliver Secondary School
Vancouver, British Columbia, Canada

Gary Davis
Albuquerque Academy
Albuquerque, New Mexico

Barbara Foots
Houston Independent School District
Houston, Texas

Elaine Frank
Durant High School
Plant City, Florida

Truman Holtzclaw
Sacramento High School
Sacramento, California

Michael Horn
Centennial High School
Boise, Idaho

Jerry Lasnik
Agoura High School
Agoura, California

Marva Moore
Hamilton Southeast High School
Fishers, Indiana

Michael O'Hare
New Trier High School
Winnetka, Illinois

Susan Plati
Brookline High School
Brookline, Massachusetts

James Pulley
Science Consultant
Independence, Missouri

Eddie Rodriguez
Math Science Academy
San Antonio, Texas

Beverly St. John
Milton High School
Milton, Florida

John Young, Ph.D.
Council Rock School District
Newtown, Pennsylvania

Student Reviewers

Emma Greig
Home School
Rochester, Michigan

Rebecca Irizarry
Columbia High School
Maplewood, New Jersey

Laura Stearns
Park Ridge High School
Park Ridge, New Jersey

Adam D. Stuble
Green River High School
Green River, Wyoming

Keri Ann Wolfe
Clarkstown Senior High School South
West Nyack, New York

Elizabeth Ashley Wolgemuth
Santa Margarita Catholic High School
Rancho Santa Margarita, California

Laboratory Teacher's Panel

Judith Dayner
Northern Highlands Regional High School
Allendale, New Jersey

Paul Fimbel
Northern Valley Regional High School
Old Tappan, New Jersey

Deidre Galvin
Ridgewood High School
Ridgewood, New Jersey

Patricia Anne Johnson
Ridgewood High School
Ridgewood, New Jersey

Carole Linkiewicz
Academic High School
Jersey City, New Jersey

Joan Picarelli
Leonia High School
Leonia, New Jersey

Robert Richard
Hillsborough High School
Belle Mead, New Jersey

Tom Russo
Weehawken High School
Weehawken, New Jersey

Sandy Shortt
Ridgewood High School
Ridgewood, New Jersey

Contributing Writers

Sandra Alters, Ph.D.
Salem State College
Salem, Massachusetts

John C. Kay
Iolani School
Honolulu, Hawaii

LaMoine Motz, Ph.D.
Oakland Schools
Waterford, Michigan

Sue Whitsett
L. P. Goodrich High School
Fond du Lac, Wisconsin

Ronnee Yashon, Ph.D.
Tufts University
Medford, Massachusetts

Reading Consultant

Laurence J. Swinburne, President
Swinburne Readability Laboratory

C O N T E N T S

UNIT 1 The World of Life 1

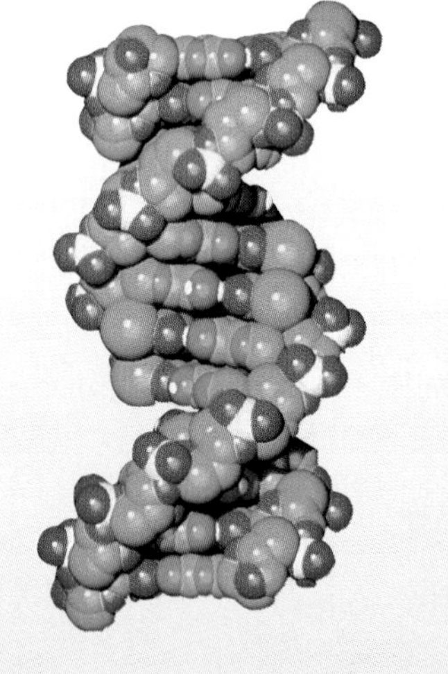

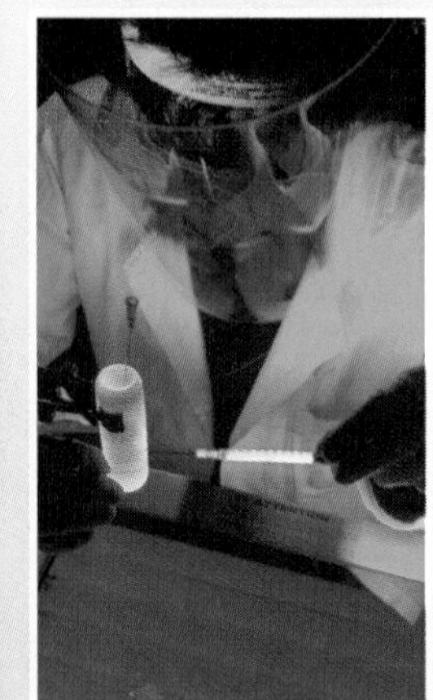

5
4
ATLANTA GP '96

CHAPTER 11 THE MECHANISMS OF EVOLUTION 240

CHAPTER 12 THE ORIGINS OF BIODIVERSITY 262

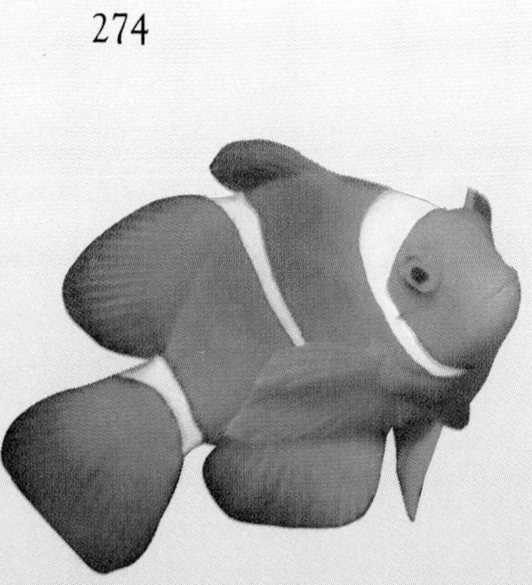

UNIT 4 Ecology 280

UNIT 5 Life on Earth: An Overview 384

CHAPTER 20 ANIMALS: VERTEBRATES 468

CHAPTER 21 HUMAN HISTORY 494

From Bacteria to Plants 512

UNIT 7 Animals 626

UNIT 8 The Human Body 762

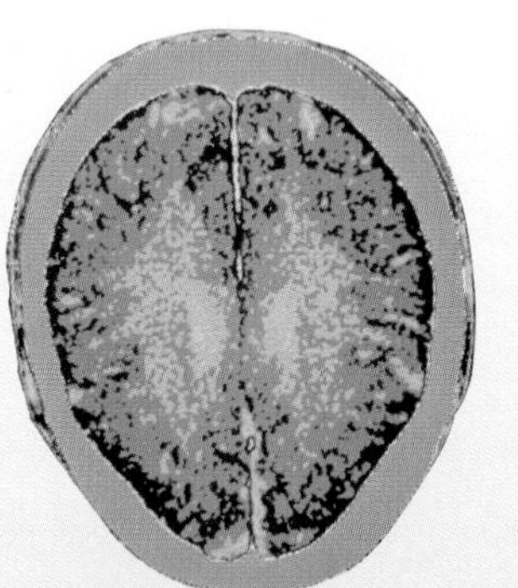

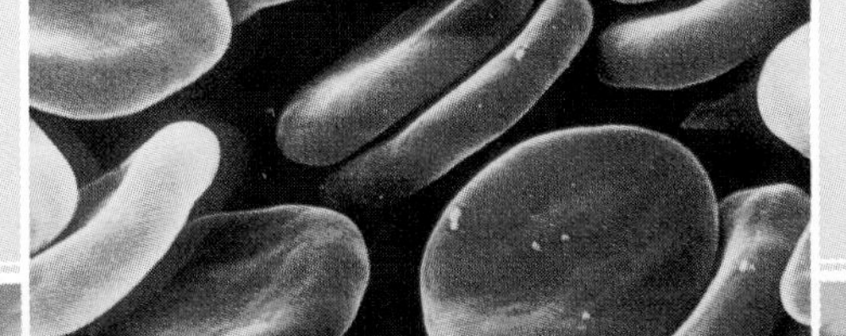

Reference Section

Features

Career Track

Connections

Laboratory Investigations

Mini Labs

Mini Labs *continued*

Mini Labs *continued*

Problem Solving

Visualizing . . .

WELCOME TO BIOLOGY!

YOU MAY HAVE HEARD PEOPLE SAY: "The future of our planet lies in the hands of our children." That might not mean much to you right now, but eventually the reins of power in our society, including the right to vote, will be passed to you and to your generation. As new discoveries in science and technology occur at an incredible pace, virtually every choice you make will require an understanding of science. Whether you become a scientist or an educated citizen in another career, the quality of your life, as well as the life of generations to come, will depend on how wisely you make those choices.

We wrote this book to inform, and maybe even to inspire, you about the living science of biology. We are all part of a great web of life that covers this planet, and your future depends on the survival and success of that life. The discoveries of science have greatly improved your quality of life over that of your parents and grandparents. But biology is still wide open to new discoveries that can make our world an even better place for you and your children.

The words, photographs, and illustrations in this book have been selected to help you master the basic concepts of biology in an intriguing and enjoyable manner. Many features, some of which are highlighted on these pages, have been included to help you learn and develop skills that can be used as you study biology, as well as other disciplines. We invite you to read the story of biology, to wonder and question, and to appreciate the beauty of this truly fascinating science.

The **Chapter Opener** helps you to focus on the themes and concepts that will be presented in the forthcoming pages. The **chapter-opener photograph** was selected as an interesting representation of the main ideas of the chapter.

Focusing the Chapter can be a very helpful organizational tool to use as you begin your study. All the major divisions, or sections, of the chapter are listed to help you familiarize yourself with the content of the chapter. The **theme** of the chapter helps you to make connections among the concepts presented.

You may be surprised to discover that you are already familiar with many of the concepts in the chapter. By using the **BioJournal,** you can determine what you already know and start thinking about what you would like to know.

Every section begins with a **Guide for Reading,** which highlights several of the more important ideas presented. As you read the section, you will also see **key terms** and **key ideas** presented in boldface type. This should alert you that these are important and may, indeed, be items that you will be tested on later.

Use the **visuals** and **captions** to help clarify the concepts. They have been carefully selected to visually enhance the content, as well as to relate the topic to real-world situations. Notice that the letters in the caption match the letters that identify the pictures.

In most of the chapters, you will find a feature called **Visualizing. . . .** This example shows **Visualizing Aquatic Biomes.** A lot of information is conveyed through pictures and illustrations, so you should pay close attention to these pages to ensure a complete understanding of the big ideas being presented.

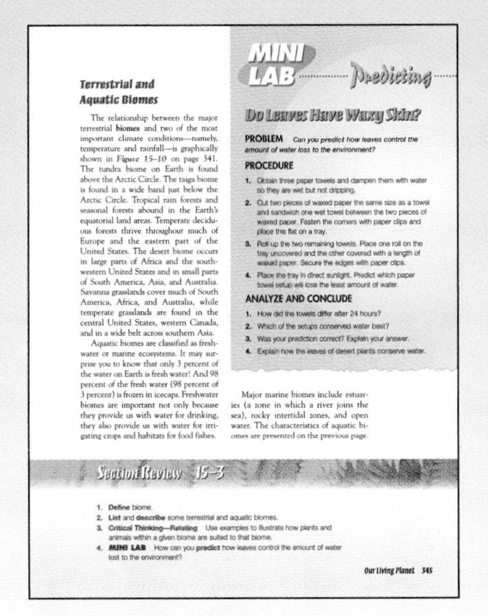

Terrestrial and Aquatic Biomes

The relationship between the major terrestrial **biomes** and two of the most important climate conditions—namely, temperature and rainfall—is graphically shown in *Figure 15–10* on page 341. The tundra biome on Earth is found above the Arctic Circle. The taiga biome is found in a wide band just below the Arctic Circle. Tropical rain forests and seasonal forests abound in the Earth's equatorial land areas. Temperate deciduous forests thrive throughout much of Europe and the eastern part of the United States. The desert biome occurs in large parts of Africa and the southwestern United States and in small parts of South America, Asia, and Australia. Savanna grasslands cover much of South America, Africa, and Australia, while temperate grasslands are found in the central United States, western Canada, and in a wide belt across southern Asia.

Aquatic biomes are classified as freshwater or marine ecosystems. It may surprise you to know that only 3 percent of the water on Earth is fresh water! And 98 percent of the fresh water (98 percent of 3 percent) is frozen in icecaps. Freshwater biomes are important not only because they provide us with water for drinking, they also provide us with water for irrigating crops and habitats for food fishes.

MINI LAB Predicting

Do Leaves Have Waxy Skin?

PROBLEM *Can you predict how leaves control the amount of water loss to the environment?*

PROCEDURE

1. Obtain three paper towels and dampen them with water so they are wet but not dripping.
2. Cut two pieces of waxed paper the same size as a towel and sandwich one wet towel between the two pieces of waxed paper. Fasten the corners with paper clips and place this flat on a tray.
3. Roll up the two remaining towels. Place one roll on the tray uncovered and the other covered with a length of waxed paper. Secure the edges with paper clips.
4. Place the tray in direct sunlight. Predict which paper towel setup will lose the least amount of water.

ANALYZE AND CONCLUDE

1. How did the towels differ after 24 hours?
2. Which of the setups conserved water best?
3. Was your prediction correct? Explain your answer.
4. Explain how the leaves of desert plants conserve water.

Major marine biomes include estuaries (a zone in which a river joins the sea), rocky intertidal zones, and open water. The characteristics of aquatic biomes are presented on the previous page.

Section Review 15–3

1. **Define** biome.
2. **List** and **describe** some terrestrial and aquatic biomes.
3. **Critical Thinking—Relating** Use examples to illustrate how plants and animals within a given biome are suited to that biome.
4. **MINI LAB** How can you **predict** how leaves control the amount of water lost to the environment?

Our Living Planet 345

No study of biology would be complete without some kind of hands-on experience. The **Mini Labs,** as well as the **Laboratory Investigations** (not pictured), give you that opportunity—often by using materials as simple as those found in your own kitchen! They also give you the opportunity to exercise your creative thought processes as you put on a scientific hat and **design your own experiments.**

The way to an A is in large part your responsibility. By evaluating yourself and your study habits, you can discover how much you know and where you need more review. To help you assess your own progress, **Checkpoint** questions (not pictured) are integrated throughout the content. The **Section Review** questions are also designed to help you do this.

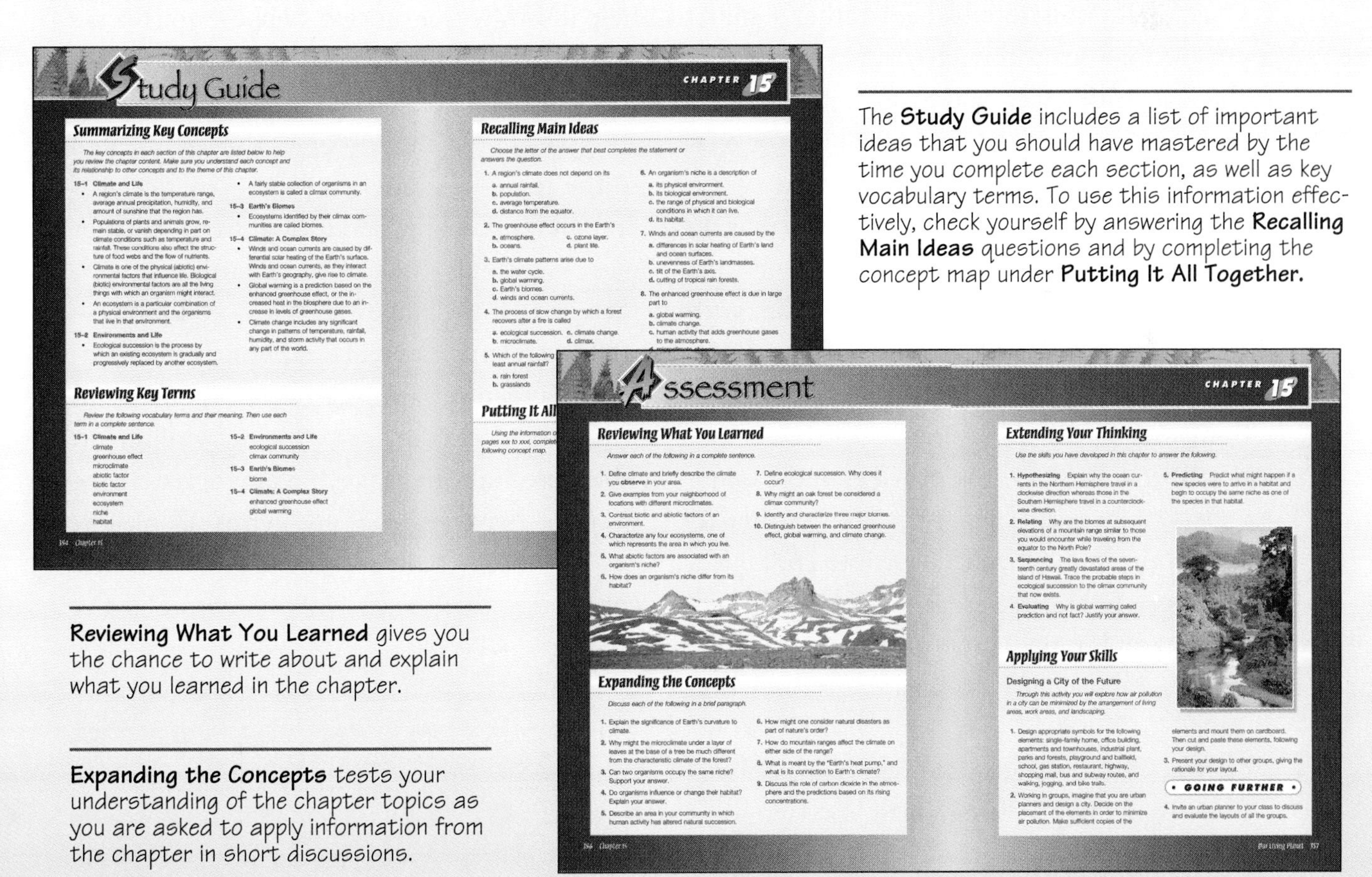

Study Guide — CHAPTER 15

Summarizing Key Concepts

The key concepts in each section of this chapter are listed below to help you review the chapter content. Make sure you understand each concept and its relationship to other concepts and to the theme of this chapter.

15–1 Climate and Life

- A region's climate is the temperature range, average annual precipitation, humidity, and amount of sunshine that the region has.
- Populations of plants and animals grow, remain stable, or vanish depending in part on climate conditions such as temperature and rainfall. These conditions also affect the structure of food webs and the flow of nutrients.
- Climate is one of the physical (abiotic) environmental factors that influence life. Biological (biotic) environmental factors are all the living things with which an organism might interact.
- An ecosystem is a particular combination of a physical environment and the organisms that live in that environment.

15–2 Environments and Life

- Ecological succession is the process by which an existing ecosystem is gradually and progressively replaced by another ecosystem.
- A fairly stable collection of organisms in an ecosystem is called a climax community.

15–3 Earth's Biomes

- Ecosystems identified by their climax communities are called biomes.

15–4 Climate: A Complex Story

- Winds and ocean currents are caused by differential solar heating of the Earth's surface. Winds and ocean currents, as they interact with Earth's geography, give rise to climate.
- Global warming is a prediction based on the enhanced greenhouse effect, or the increased heat in the biosphere due to an increase in levels of greenhouse gases.
- Climate change includes any significant change in patterns of temperature, rainfall, humidity, and storm activity that occurs in any part of the world.

Reviewing Key Terms

Review the following vocabulary terms and their meaning. Then use each term in a complete sentence.

15–1 Climate and Life: climate, greenhouse effect, microclimate, abiotic factor, biotic factor, environment, ecosystem, niche, habitat

15–2 Environments and Life: ecological succession, climax community

15–3 Earth's Biomes: biome

15–4 Climate: A Complex Story: enhanced greenhouse effect, global warming

354 Chapter 15

Recalling Main Ideas

Choose the letter of the answer that best completes the statement or answers the question.

1. A region's climate does not depend on its
   a. annual rainfall.
   b. population.
   c. average temperature.
   d. distance from the equator.
2. The greenhouse effect occurs in the Earth's
   a. atmosphere. c. ozone layer.
   b. oceans. d. plant life.
3. Earth's climate patterns arise due to
   a. the water cycle.
   b. global warming.
   c. Earth's biomes.
   d. winds and ocean currents.
4. The process of slow change by which a forest recovers after a fire is called
   a. ecological succession. c. climate change.
   b. microclimate. d. climax.
5. Which of the following ... least annual rainfall?
   a. rain forest
   b. grasslands
6. An organism's niche is a description of
   a. its physical environment.
   b. its biological environment.
   c. the range of physical and biological conditions in which it can live.
   d. its habitat.
7. Winds and ocean currents are caused by the
   a. differences in solar heating of Earth's land and ocean surfaces.
   b. unevenness of Earth's landmasses.
   c. tilt of the Earth's axis.
   d. cutting of tropical rain forests.
8. The enhanced greenhouse effect is due in large part to
   a. global warming.
   b. climate change.
   c. human activity that adds greenhouse gases to the atmosphere.

Putting It All ...

Using the information o... pages xxx to xxxi, complet... following concept map.

Assessment — CHAPTER 15

Reviewing What You Learned

Answer each of the following in a complete sentence.

1. Define climate and briefly describe the climate you **observe** in your area.
2. Give examples from your neighborhood of locations with different microclimates.
3. Contrast biotic and abiotic factors of an environment.
4. Characterize any four ecosystems, one of which represents the area in which you live.
5. What abiotic factors are associated with an organism's niche?
6. How does an organism's niche differ from its habitat?
7. Define ecological succession. Why does it occur?
8. Why might an oak forest be considered a climax community?
9. Identify and characterize three major biomes.
10. Distinguish between the enhanced greenhouse effect, global warming, and climate change.

Expanding the Concepts

Discuss each of the following in a brief paragraph.

1. Explain the significance of Earth's curvature to climate.
2. Why might the microclimate under a layer of leaves at the base of a tree be much different from the characteristic climate of the forest?
3. Can two organisms occupy the same niche? Support your answer.
4. Do organisms influence or change their habitat? Explain your answer.
5. Describe an area in your community in which human activity has altered natural succession.
6. How might one consider natural disasters as part of nature's order?
7. How do mountain ranges affect the climate on either side of the range?
8. What is meant by the "Earth's heat pump," and what is its connection to Earth's climate?
9. Discuss the role of carbon dioxide in the atmosphere and the predictions based on its rising concentrations.

356 Chapter 15

Extending Your Thinking

Use the skills you have developed in this chapter to answer the following.

1. **Hypothesizing** Explain why the ocean currents in the Northern Hemisphere travel in a clockwise direction whereas those in the Southern Hemisphere travel in a counterclockwise direction.
2. **Relating** Why are the biomes at subsequent elevations of a mountain range similar to those you would encounter while traveling from the equator to the North Pole?
3. **Sequencing** The lava flows of the seventeenth century greatly devastated areas of the island of Hawaii. Trace the probable steps in ecological succession to the climax community that now exists.
4. **Evaluating** Why is global warming called prediction and not fact? Justify your answer.
5. **Predicting** Predict what might happen if a new species were to arrive in a habitat and begin to occupy the same niche as one of the species in that habitat.

Applying Your Skills

Designing a City of the Future

Through this activity you will explore how air pollution in a city can be minimized by the arrangement of living areas, work areas, and landscaping.

1. Design appropriate symbols for the following elements: single-family home, office building, apartments and townhouses, industrial plant, parks and forests, playground and ballfield, school, gas station, restaurant, highway, shopping mall, bus and subway routes, and walking, jogging, and bike trails.
2. Working in groups, imagine that you are urban planners and design a city. Decide on the placement of the elements in order to minimize air pollution. Make sufficient copies of the elements and mount them on cardboard. Then cut and paste these elements, following your design.
3. Present your design to other groups, giving the rationale for your layout.

• GOING FURTHER •

4. Invite an urban planner to your class to discuss and evaluate the layouts of all the groups.

Our Living Planet 357

The **Study Guide** includes a list of important ideas that you should have mastered by the time you complete each section, as well as key vocabulary terms. To use this information effectively, check yourself by answering the **Recalling Main Ideas** questions and by completing the concept map under **Putting It All Together.**

Reviewing What You Learned gives you the chance to write about and explain what you learned in the chapter.

Expanding the Concepts tests your understanding of the chapter topics as you are asked to apply information from the chapter in short discussions.

Extending Your Thinking challenges your knowledge and skills as you solve the creative problems and the case studies and apply your knowledge to new situations.

Applying Your Skills allows you to put your skills to the test as you complete challenging tasks involving one or more skills.

CONCEPT MAPPING

THROUGHOUT YOUR STUDY OF science, you will learn a variety of terms, facts, and concepts. Each new topic you encounter will provide its own collection of words and ideas—which, at times, you may think seems endless. But each of the ideas within a particular topic is related in some way to the others. No concept in science is isolated. Thus, it will help you to understand the topic if you see the whole picture; that is, the interconnectedness of all the individual terms and ideas. This is a much more effective and satisfying way of learning than memorizing separate facts.

Actually, this should be a rather familiar process for you. Although you may not think about it in this way, you analyze many of the elements in your daily life by looking for relationships or connections. For example, when you look at a collection of flowers, you may divide them into groups: roses, carnations, and daisies. You may then associate colors with these flowers: red, pink, and white. The general topic is flowers. The subtopic is types of flowers. And the colors are specific terms that describe flowers. A topic makes more sense and is more easily understood if you understand how it is broken down into individual ideas and how these ideas are related to one another and to the entire topic.

It is often helpful to organize information visually so that you can see how it all fits together. One technique for describing related ideas is called a **concept map.** In a concept map, an idea is represented by a word or phrase enclosed in a box. There are several ideas in any concept map. A connection between two ideas is made with a line. A word or two that describes the connection is written on or near the line. The general topic is located at the top of the map. That topic is then broken down into subtopics, or more specific ideas, by branching lines. The most specific topics are located at the bottom of the map.

To construct a concept map, first identify the important ideas or key terms in the chapter or section. Do not try to include too much information. Use your judgment as to what is really important. Write the general topic at the top of your map. Let's use an example to help illustrate this process. Suppose you decide that the key term is Biology. Write and enclose this word in a box at the top of your map.

Now choose the subtopics that are related to the topic—Botany, Zoology, Genetics, Microbiology, Ecology. Add these words to your map. Continue this procedure until you have included all the important ideas and terms—study of plants, animals, inheritance, microscopic organisms, interactions of organisms with one another and with their environment. Then use lines to make the appropriate connections between ideas and terms. Don't forget to write a word or two on or near the connecting line to describe the nature of the connection.

Do not be concerned if you have to redraw your map before you show all the important connections clearly. If, for example, you rely on observation and experimentation as well as analysis, you may want to place these subjects next to each other so that the lines do not overlap.

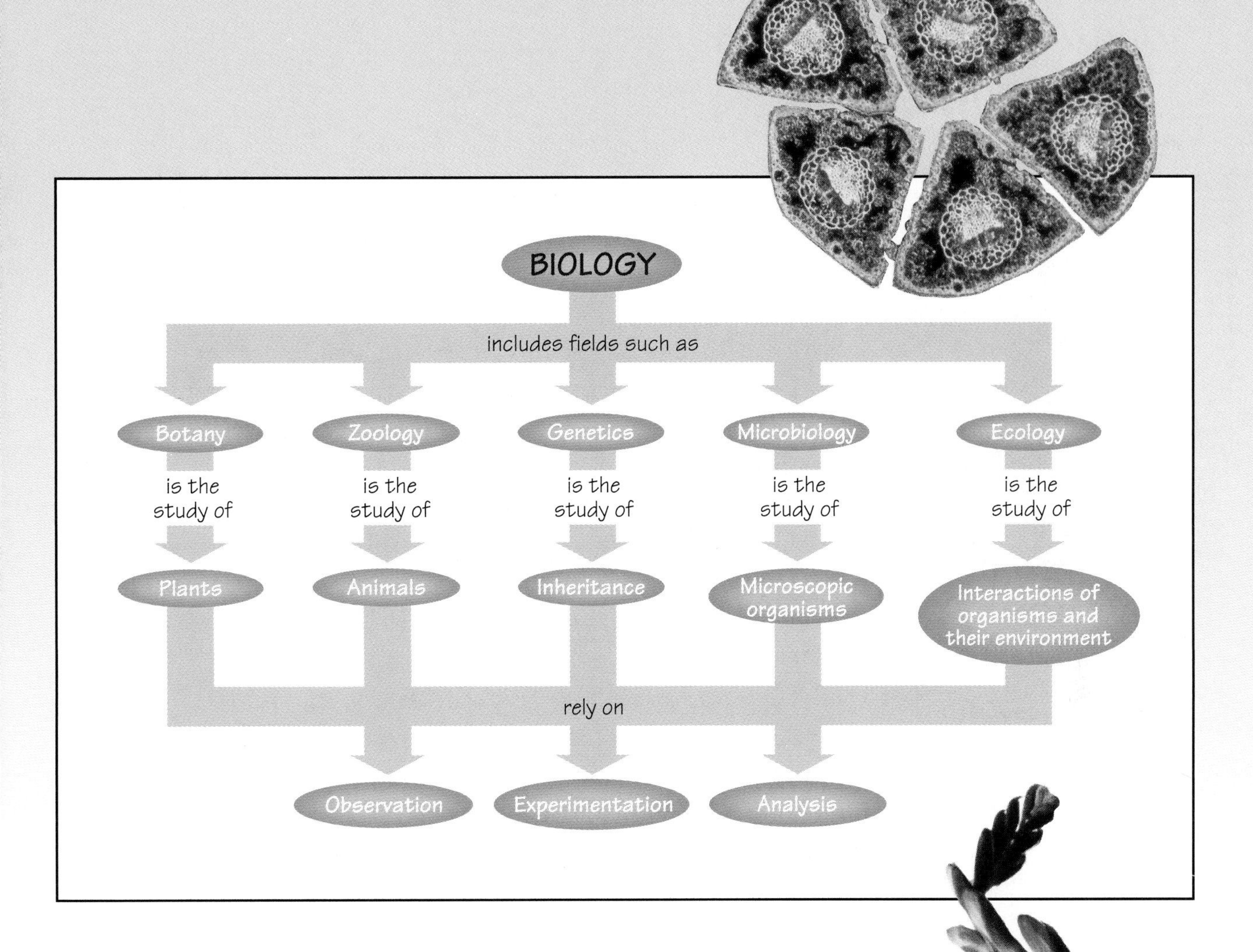

One more thing you should know about concept mapping: Concepts can be correctly mapped in many different ways. In fact, it is unlikely that any two people will draw identical concept maps for a complex topic. Thus, there is no one correct concept map for any topic. Even though your concept map may not match those of your classmates, it will be correct as long as it shows the most important concepts and the clear relationships among them. Your concept map will also be correct if it has meaning to you and if it helps you understand the material you are reading. A concept map should be so clear that if some of the terms are erased, the missing terms could easily be filled in by following the logic of the concept map.

TO THE STUDENT

WHEN I WAS 9 OR 10, I WANTED TO BE AN explorer. I imagined myself hiking through wild country, a hunting knife strapped to my ankle, fearlessly stepping into adventures at every turn. Growing up in suburban New Jersey, my friends and I "invented" a wilderness around the creek in a little woods at the end of our street. Almost every day we pretended to discover the place for the first time. On cool days we blazed new trails in the imaginary forest and on hot days we built dams and canals in the clay bed of the creek.

Four or five years later, when I walked into Mr. Zong's ninth-grade biology class in Rahway Junior High, I had become way too grown up to play at the creek anymore. Like most of the boys in my class, I was more interested in acting cool and learning how to do the latest dance—they called it "the twist"!

Our teacher had filled his classroom with specimens—stuffed animals, mounted bugs, pressed flowers, bones, leaves, cocoons. Everywhere you looked there was something new, something unknown, something mysterious. I began to get interested. Then one day, he asked for a volunteer. On my way to school the next morning I made a trip to the old creek. All grown up, the woods now seemed small and ordinary. I filled a jar of creek water to bring to school.

In lab that day we put a drop of the water under the microscope, and I could hardly believe my eyes. The water was teeming with life! After my last class that day, Mr. Zong let me spend another session at the microscope, and this time he helped me identify the tiny critters. We found rotifers and ciliates and a half dozen different kinds of algae. There were things that even he couldn't name for sure.

When I walked home from school that day my view of the creek had changed—and so had my view of the world. For the first time, I understood that the neighborhood I thought I knew was filled with hidden secrets, and I wanted to learn every one of them. To me, that's what biology still is like—a chance, every day, to understand a little more about the mystery of being alive.

With all my heart, I hope that this textbook, every now and then, will give you a little bit of the same feeling I had when I looked at that water under the microscope for the very first time. Life is the single most amazing thing about this remarkable little planet. I hope you will enjoy reading this textbook, but I also hope it will encourage you to study the living things outside your classroom. Who knows? Depending on what you find in that drop of water, you just might grow up to be an explorer, after all!

Ken Miller

"AH, THOSE WONDERFUL ZOO SMELLS!"

That was my reaction as a child whenever I walked into the elephant house at the Bronx Zoo. The grand old building was well-stocked with awesome animals, but it didn't have much fresh air. While my parents held their noses, I took deep breaths, enjoying the "organic" odors. My father reminded me of this a year ago, when he and I took my two-year-old son to the zoo. We had a great time. I was glad that the elephant house now has better ventilation, because my little boy thinks zoos can be "stinky."

Just weeks later, Dad lost his lifelong struggle with a disease called familial hypercholesterolemia. That jawbreaking medical term is the scientific way of saying that a defective gene left Dad's liver unable to control the amount of cholesterol in his blood. Despite surgery and medication, clogged arteries damaged his heart beyond repair while he was still a young man.

Why do I tell you this? Because finishing this textbook has made me stop to think about how biology has affected me and what it will mean to my son—and to you—in the future. As you'll learn, my brother Daniel and I each had a 50-percent chance of inheriting that bad cholesterol gene from Dad. Unfortunately, we both got it. Happily, we were diagnosed in early childhood and received treatments that Dad never had as a boy. In time, Daniel joined a biomedical team to battle heart disease with his own research. Their findings prolonged Dad's life, offered more years to many other patients, and we hope will keep us around for a while longer, too. Meanwhile, researchers elsewhere are studying a host of other disease-causing genes, including some that may affect you or your family.

But biological links between past and future involve more than genes. Why? Because some human genes enable us to learn and to teach. My father—by encouraging his sons to think for themselves, to ask questions, and to search for answers—planted seeds that grew into a lifelong love of learning. By wandering with us through zoos, in and out of museums, and along hiking trails till his legs ached, he gave us our love of nature. Whenever I take my son to the zoo, read him a book about animals, or write a book myself, I remember that love and respect for life aren't passed from generation to generation by chromosomes.

That's why my part of this textbook has a dual dedication. It is dedicated to you, in the hope that it will help to open your eyes to the joy of a science that can better all our lives, as it helps humanity survive into the future that belongs to my son and to you. And it is dedicated to the best father and grandfather anyone could have—the man who laid the foundation for everything worthwhile that I will ever do.

Joe Levine

UNIT 1

The World of Life

CHAPTERS

"The most incomprehensible thing about the world is that it is comprehensible."

— Albert Einstein

CAREER TRACK

As you explore the topics in this unit, you will discover many different types of careers associated with biology. Here are a few of these careers:

- Biologist
- Respiratory Therapist
- Cytotechnologist
- Biological Technician
- Histologic Technician

Zebras grazing in Africa

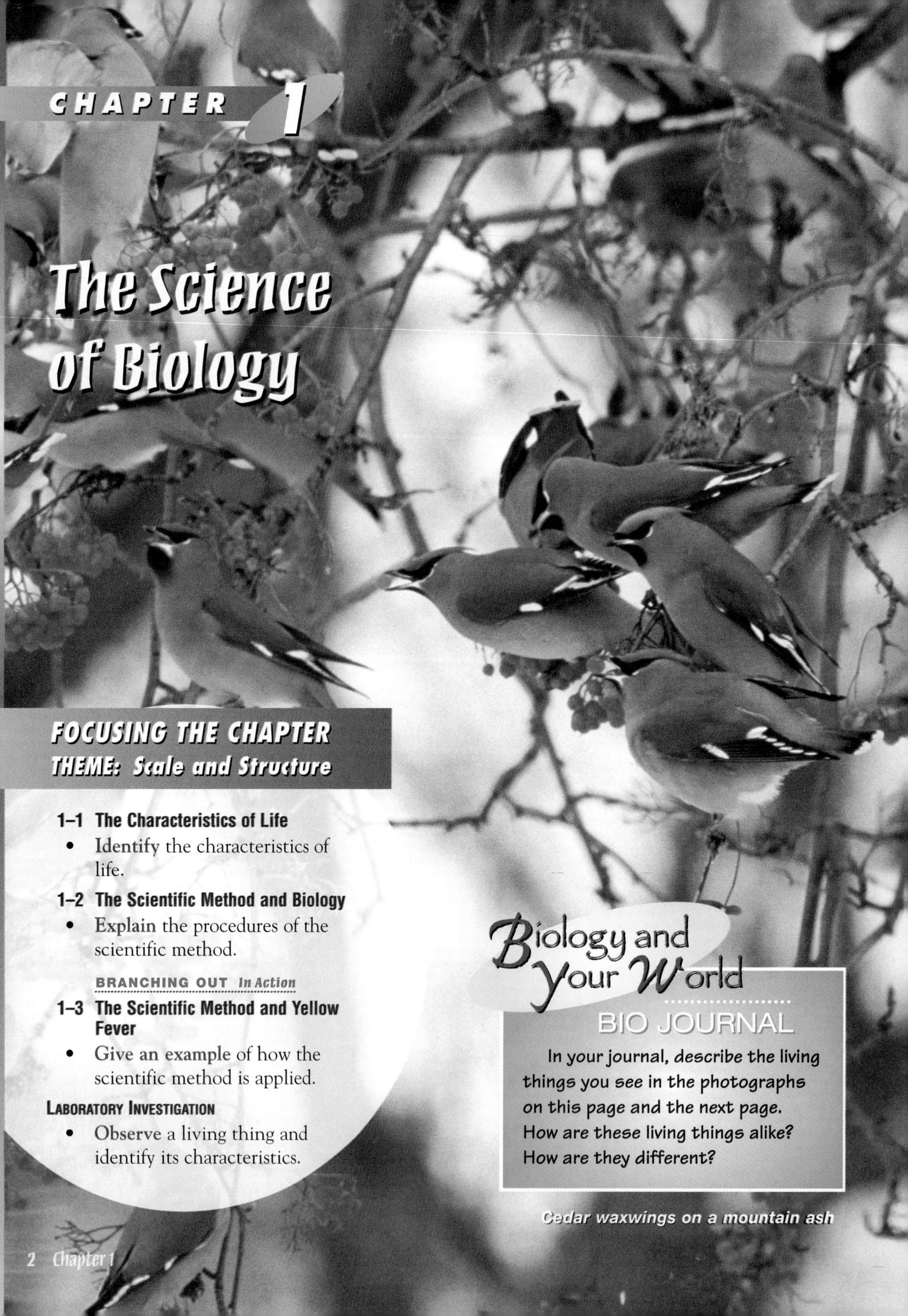

CHAPTER 1

The Science of Biology

FOCUSING THE CHAPTER THEME: *Scale and Structure*

1–1 The Characteristics of Life
- **Identify** the characteristics of life.

1–2 The Scientific Method and Biology
- **Explain** the procedures of the scientific method.

BRANCHING OUT *In Action*

1–3 The Scientific Method and Yellow Fever
- **Give an example** of how the scientific method is applied.

LABORATORY INVESTIGATION
- **Observe** a living thing and identify its characteristics.

Biology and Your World

BIO JOURNAL

In your journal, describe the living things you see in the photographs on this page and the next page. How are these living things alike? How are they different?

Cedar waxwings on a mountain ash

The Characteristics of Life

Guide for Reading

- **Identify** five characteristics of living things.
- **Describe** the different levels of questions that biologists ask.

MINI LAB

- **Predict** whether a sample is living or nonliving.

We live in a remarkable place, and we live in remarkable times. Our home, the Earth, is unlike any other planet that orbits the sun. The Earth is the only planet with great oceans of liquid water. It is the only planet surrounded by an atmosphere that is rich in oxygen. And it has something else that may not exist on any other planet: The Earth has life.

No matter where you live—city, farm, small town, or suburb—life is all around you. In fact, it is almost impossible to think of a single place on Earth that is not a home to living things. Life is found in golden grasslands, scorching deserts, deep oceans, and even the frozen wastes of the Antarctic.

It's often easy to think that life is the most ordinary thing in the world. However, life is far from simple or ordinary. As you study the living things of the world, you will come to appreciate their special qualities all the more.

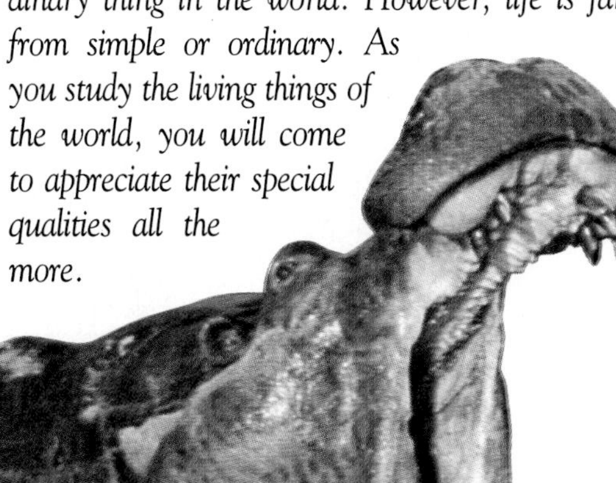

Questions About Life

Why do many trees and other plants sprout leaves in spring and lose them in autumn? How does a bird learn to fly and to build a nest? Why are there so many kinds of insects, and why are they so different from one another?

If every now and then you have asked a question about something you have seen in the world of living things, you have taken the first step toward understanding that world. What steps should you take next? The curiosity that makes you human really leaves you with only one choice—to try to answer the questions you ask.

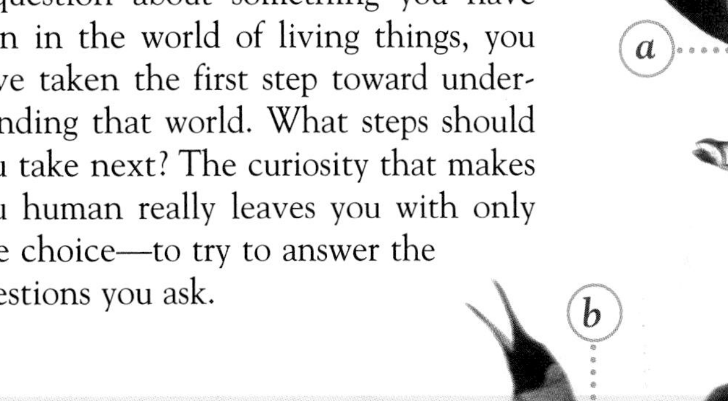

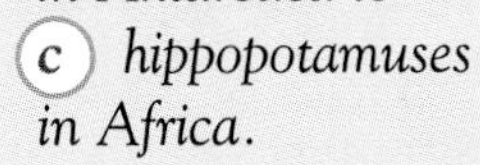

Figure 1–1 *The Earth is home to an enormous number of living things—from (a) fishes in the oceans to (b) penguins in Antarctica to (c) hippopotamuses in Africa.*

Biology as a Science

For most of human history, the laws and forces of nature were great mysteries. In the last few centuries, however, much of that has changed. Thanks to human curiosity and intelligence, we have developed a remarkable process of thinking and learning about the world around us. This process is called **science.** As you know, there are many fields of science, each of which tries to explain one aspect of our world. **Biology,** the science of life, is the subject of this textbook.

What Is Life?

Is something "alive" simply because it can move? The answer is not so simple! Cars move, but you would not consider a car to be alive. And plants are not as mobile as animals, yet plants are alive. So just what are the characteristics of life?

The answer to this question is not an easy one—even for the experts. Especially because of viruses and certain other things found in nature, biologists find it very difficult to draw an exact line between living and nonliving things. However, we can list five basic characteristics that are common to all living things. **Living things are made of cells, grow and develop, obtain and use energy, respond to their environment, and are able to reproduce.** These characteristics are described in detail in the illustrations on the next page.

You may be able to think of a few other characteristics that describe living things. In fact, you may even be able to think of nonliving things that have some of the characteristics of life just listed. For example, how many of these characteristics does a copying machine have?

☑ **Checkpoint** What are five characteristics of living things?

Organisms

An individual living thing is called an **organism.** You are an organism, as is each animal and plant. Many organisms, such as redwood trees, are so big you just can't miss seeing them. However, we are also surrounded by billions of organisms that are so small their existence wasn't even suspected until microscopes were invented less than 400 years ago.

As you study one type of organism after another in your biology course, try to see how each one meets the five characteristics of life that we just presented. Let's take a close look at one interesting organism to better understand these characteristics.

☑ **Checkpoint** What is an organism?

Figure 1–2
(a) This small organism, a dust mite, is magnified 150 times larger than its actual size. (b) Organisms can also be as large as these sequoia trees.

Visualizing the Characteristics of Living Things

How do living things differ from nonliving things? In some cases, answering this question is not so easy! However, we can list five basic characteristics of all living things.

1 Living things are made up of cells.

Cells are small, self-contained units that are the building blocks of organisms.

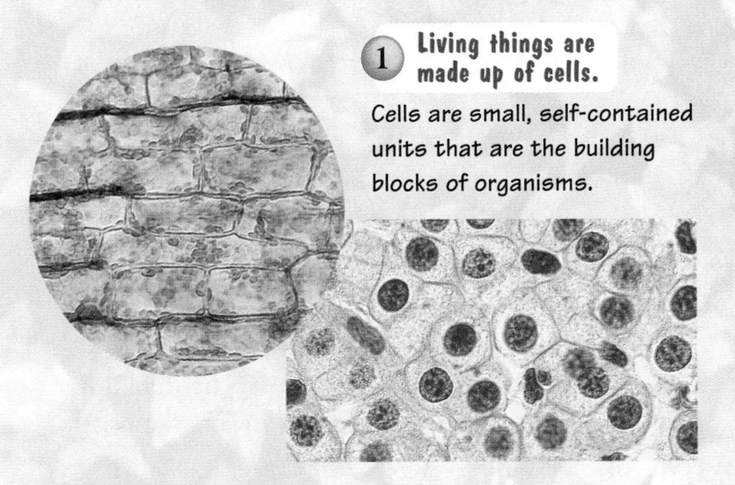

2 Living things obtain and use energy.

Most plants obtain energy directly from the sun, whereas most animals get energy from chemical compounds in their food.

3 Living things respond to their environment.

Anything in an organism's environment that causes it to react is called a stimulus. Organisms typically react in ways that keep their bodies suitable for life—a process called homeostasis.

4 Living things grow and develop.

An organism grows and develops by absorbing raw materials and processing them into new tissues and structures.

5 Living things can reproduce.

Living things produce new living things of the same type. Although a living thing does not need to reproduce in order to survive, life would quickly die out without reproduction.

Figure 1–3
Corals are found in shallow ocean waters—as here off the coast of Florida. Coral is alive because it exhibits the five characteristics of living things.

Coral

The Florida Keys are a chain of small islands that reach into the waters of the Caribbean Sea off the southern coast of Florida. The waters around the islands are warm and inviting, and many visitors enjoy swimming in the shallow areas near the shoreline. It is in these warm, shallow waters that divers encounter a stunning sight—great branches and colorful mounds that look like underwater forests from a fantasy world.

What are these branches and mounds? They are small organisms called coral, and they inhabit warm, shallow waters in oceans around the world. Corals form great clusters called coral reefs. In fact, some coral reefs are so large that they form whole islands.

Why Coral Is Alive

How do we know that coral is alive? First, look very closely at the surface of a coral, as shown in ***Figure 1–4.*** In this photograph, you can see tiny tentacles waving in the passing currents. What makes up these tentacles? Through a high-powered microscope, you could see that they are made up of individual units called cells—an important characteristic of living things.

Next, as anyone who lives near a coral reef will tell you, a coral reef grows from year to year as its branches and mounds become larger. Corals also obtain and use energy, and they respond to

Figure 1–4
If you could magnify this photograph of orange clump coral, you would discover that it is made of cells—evidence that orange clump coral is alive.

their environment. A coral's tentacles react to the presence of food particles in the passing water by capturing them. Then the coral uses the energy from the food to move and maintain its body.

The last characteristic on our list—the ability to reproduce—is also true of corals. From all these observations, then, we can state without a doubt that coral is alive.

☑ **Checkpoint** Why is coral alive?

A Living Colony

Coral is not just a single organism but exists as a colony of thousands of individual organisms. Some of these coral organisms produce an intricate supporting skeleton of calcium carbonate, a material also found in bones. Throughout its life, coral builds up larger and larger amounts of calcium carbonate. When it dies, a new coral grows directly on the remains. In this way, the coral colony gets larger and larger over time.

Are They Alive?

PROBLEM *How can you **predict** whether something is living or nonliving?*

PROCEDURE

1. Obtain a set of five unidentified samples from your teacher.
2. Carefully inspect each sample—both with the unaided eye and with a magnifying glass. You may also touch the samples. **CAUTION:** *Do not taste them.* Record your observations.
3. For each sample, predict whether the sample is composed of living or nonliving things. Give reasons for each prediction.

ANALYZE AND CONCLUDE

1. From your results, is it possible to determine whether each sample is living or nonliving? Explain.
2. What further questions do you have about each sample? What answers would you expect if the sample was alive? If it was not alive?
3. Ask your teacher to identify the samples. How accurate were your predictions? Which characteristic of living things proved most helpful in classifying your samples?

Biology at Different Levels

Deciding whether something is alive is only the first step in trying to understand it. As a biologist, you must also decide where to begin to ask questions. Living things can be studied on many levels, and a biologist will ask different types of questions, depending on the size and scale of what is being studied. Let's take a look at the different levels of questions that might be asked about coral.

Questions at the Chemical Level

What are the chemical compounds that make up the coral and its skeleton? What other chemical compounds are found in a coral's body? Questions like these are asked in the field of biochemistry, an important branch of biology. A biochemist might seek to learn how the coral animals obtain calcium from sea water or how they process calcium into calcium carbonate. ●

INTEGRATING CAREERS

Visit the library to discover more about biochemists and their work.

Questions at the Molecular Level

Every organism contains a variety of molecules, which are fundamental units of matter. Like other animals, a coral contains specialized molecules that regulate its growth and development, that help it to break down food and to eliminate waste, and that give shape and texture to its body.

Figure 1–5
The size of chemical compounds and molecules such as calcium carbonate ($CaCO_3$) and DNA is measured in tiny units called nanometers. The period at the end of this sentence is about 500,000 nanometers in diameter.

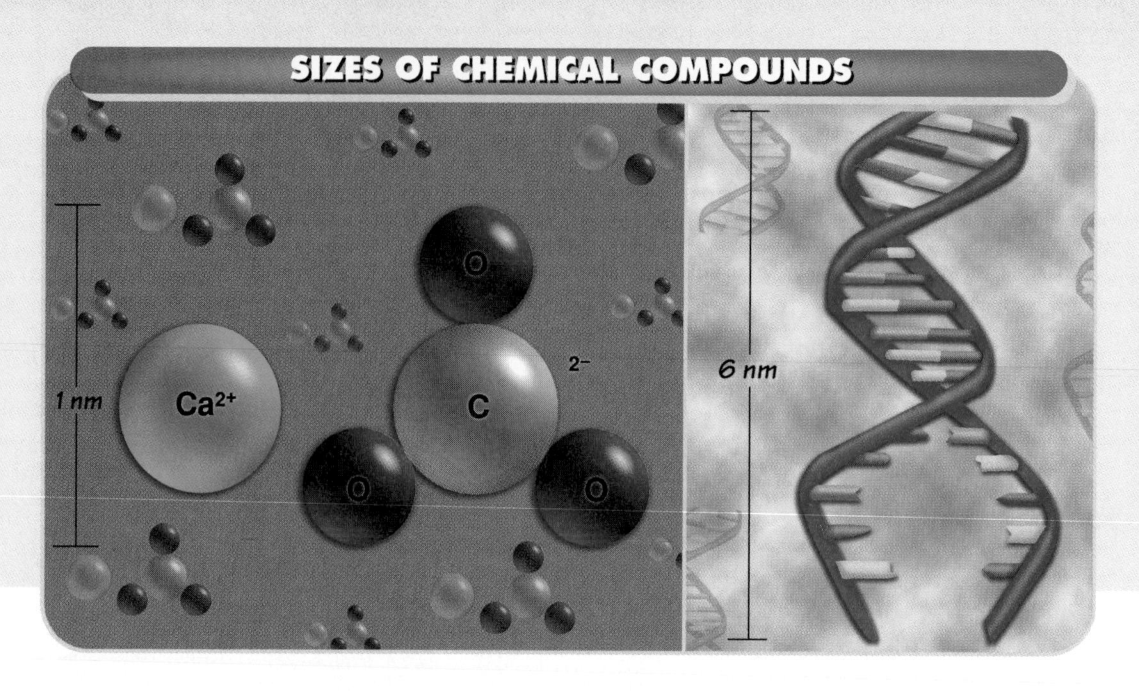

What are these molecules? How are they made? How do they work? A molecular biologist asks questions such as these.

Questions at the Cellular Level

How are coral cells organized, and how do they interact? Like all organisms with many cells, coral grows when cells divide to produce new cells. How does this process take place? Why are new cells sometimes different from their parents? Cell biologists specialize in studying questions at the cellular level.

Questions at the Organism Level

What controls the rate at which coral reproduces? How does one type of coral differ from another? Did coral exist hundreds of millions of years ago? Most species of coral contain single-celled green organisms called algae. What are algae doing in coral?

Many types of biologists would be interested in questions at this level. Paleontologists, who study ancient life, might wonder how coral has changed over time. Ecologists, who study the relationships among different types of organisms, would be interested in the relationship between coral and algae.

Questions at the Population Level

When thousands of coral animals form a coral reef, they begin to interact with their environment in interesting and important ways. Coral reefs provide homes for hundreds of other organisms, including sponges, seaweeds, and an enormous variety of fishes.

How do coral reefs affect these and other marine organisms? When the population of one kind of animal rises, is the population of another kind of animal affected? Ecologists and population biologists study questions such as these.

Figure 1–6
A coral colony may provide food or protection—or both—for other marine organisms, including this small fish called a coral goby.

Biology AND EARTH SCIENCE Connections

As Lifeless as a Grain of Sand?

Spread on the sunscreen and break out the blankets and the beach balls—it's time for a day at the beach! On a hot summer afternoon, a clean, sandy beach is a great place to relax and have fun. It's also a good place to ask questions. Where does beach sand come from? Why does beach sand come in different colors and textures? What keeps many beaches clean and attractive? As you will see, there's more to beaches than meets the eye!

Beach Sand

In many cases, beach sand comes from the silt and other materials that rivers carry with them to an ocean or large lake. But sand comes from other places as well. White sand comes from the remnants of countless numbers of marine organisms. Black sand, which can be found on islands in the Pacific Ocean, was originally lava flows from volcanoes.

Yet whatever its origin, beach sand acquires its texture and consistency from the waves that repeatedly pound and weather it. Given enough time, waves will grind even the rockiest coastline into stretches of finely ground particles.

Biology at the Beach

Certainly, an earth scientist can find much to investigate in the sand at the beach. But is the sand also of interest to a biologist? You may be surprised to learn that the answer is yes.

A handful of sand may contain as many as 10,000 creatures living on the surfaces of the grains! These creatures include tiny worms, mites, miniature shellfish, and even organisms so strange and unique that biologists have had to place them in their own phylum (a special category) of the animal kingdom. Biologists are discovering dozens of new creatures in beach sand every year.

What do these creatures do? The same things that animals do everywhere: They search for food, fight battles with predators, and reproduce their own kind—only on what you might consider to be a very small scale!

Any Dangers?

Does the presence of so many strange, tiny creatures mean that you should stay away from the beach? No, not at all. In fact, the cleanest and most attractive beaches seem to have the greatest numbers of these tiny animals. They cruise the surfaces of sand grains, scouring for algae and bacteria and removing organic wastes. You may not realize they are there, but their secret kingdom is the key to the glistening beaches that you enjoy.

A nematode (top, magnification: 650X) and a kinarhynch (bottom, magnification: 360X)—both tiny organisms found on a beach

Making the Connection

Why might an earth scientist want to work with a biologist to investigate the sand at the beach? For what other reasons might an earth scientist and a biologist work together?

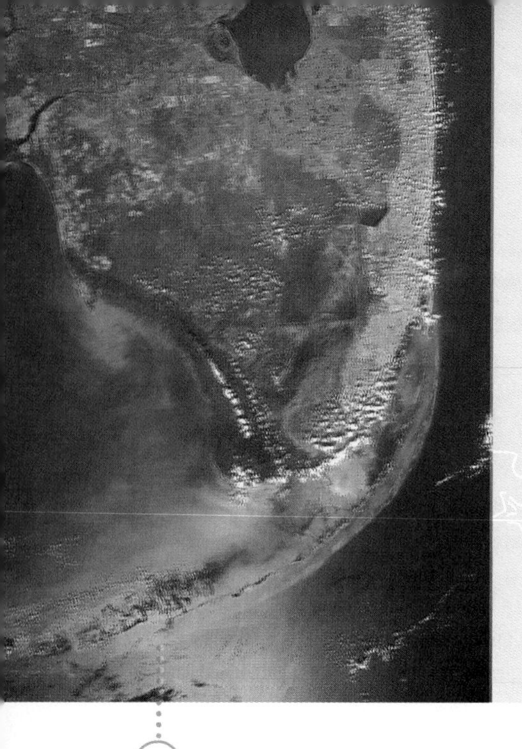

Figure 1–7
(a) *Although you cannot see any individual coral organisms in this photograph of southern Florida, the organisms are there nevertheless. Coral, like every other organism, can be studied at the global level.* (b) *A population biologist might study how the numbers of coral organisms change over time.*

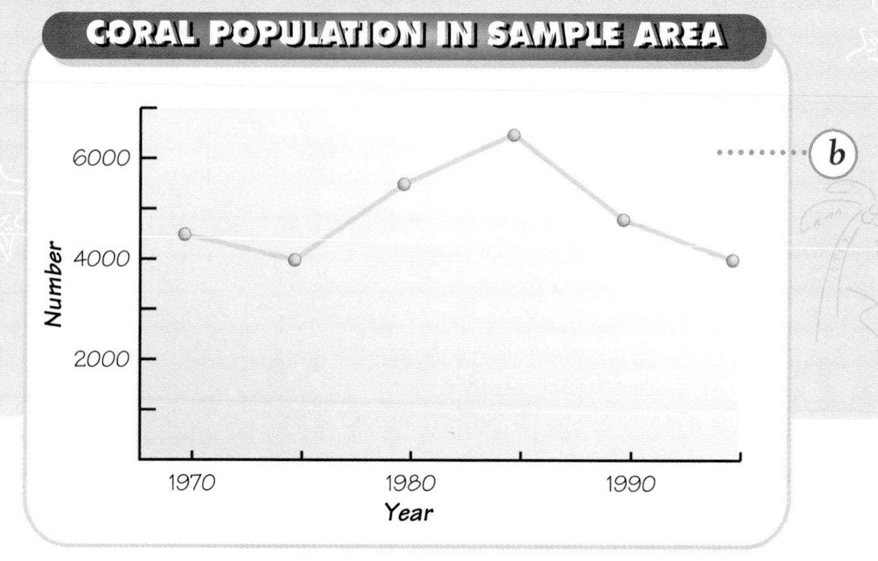

Questions at the Global Level

Many biologists take a worldwide view of biology, asking questions about organisms and their environment on a global scale. Is coral a global concern? Indeed, it is.

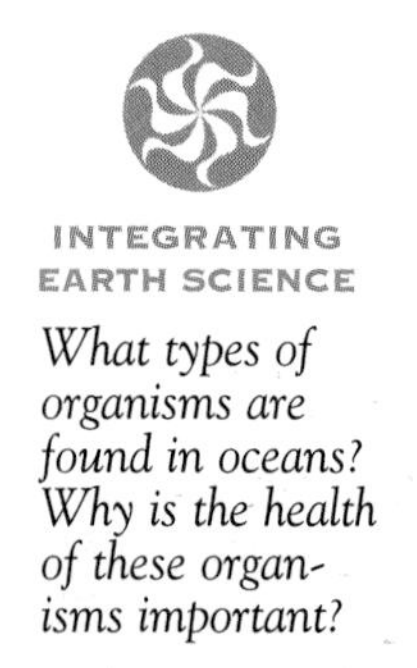

What types of organisms are found in oceans? Why is the health of these organisms important?

In fact, a coral reef is the largest and most spectacular structure on the whole planet that living things assemble. The Great Barrier Reef, which is found along the coast of northern Australia, is more than 2000 kilometers long and more than 100 kilometers wide in some areas. Because coral is affected by a delicate balance of temperature, minerals, and chemicals in ocean water, it may serve as an indicator of the overall health of Earth's oceans—home to so much life on our planet. ●

Asking questions on so many levels makes biology a very broad subject. A single living thing—a coral, for example, or a fish or a human—can interest any number of different biologists asking all sorts of different questions. Taken together, the results of their efforts provide us with as complete a view as possible of life on Earth.

Section Review 1–1

1. **Identify** the five characteristics of living things.
2. **Describe** the different levels of organization at which biologists ask questions.
3. Is coral alive? **Explain** your answer.
4. **Critical Thinking—Making Judgments** You are a biochemist studying chemicals in frogs and toads. You just received an invitation to a conference on pollution, the damaged ozone layer, and other global concerns. Could attending this conference help your research? Explain your answer.
5. **MINI LAB** How can you **predict** whether something is living or nonliving?

The Scientific Method and Biology

SECTION 1–2

GUIDE FOR READING

- **Define** the scientific method.
- **Describe** the steps of the scientific method.
- **Compare** a hypothesis and a theory.

MINI LAB

- **Formulate a hypothesis** based on observations.

HUMANS ARE ALWAYS ASKING questions about the world around them. But once we pose a question, the next step is to decide on how to go about answering it. Although this may sound difficult, it is something that people do automatically every day.

People often answer their questions about life by thinking about their everyday experiences. However, science demands that questions be answered by the use of a precise method. In this section, you'll discover just what that method is.

The Scientific Method 

The precise method used by scientists is called the **scientific method,** and it separates science from other ways of studying and learning. **The scientific method is a system of asking questions, developing explanations, and testing those explanations against the reality of the natural world.**

Other fields—such as art, music, history, and philosophy—all have a great deal to tell us about the world. But although each is important and deserving of careful study, none of them uses the scientific method, which means that none of them is a field of science.

Figure 1–8
The scientific method has been applied to answer all sorts of different questions—and to solve a huge variety of problems. (a) *In 1928, Scottish biologist Alexander Fleming used the scientific method to discover penicillin, the first drug to fight infections effectively.* (b) CAREER TRACK *This biologist is measuring an unusually large flower on the island of Borneo.*

MINI LAB ········ Hypothesizing ····

The Mystery Box

PROBLEM *What is inside the mystery box?* ***Formulate a hypothesis.***

PROCEDURE

1. Obtain a mystery box from your instructor. The box contains a unique arrangement of partitions and one or more marbles.
2. Tilt, turn, and tap the box to move the marbles inside. The sounds and sensations provide clues to the arrangement of partitions inside the box.
3. On a sheet of paper, sketch your hypothesis of how the partitions are arranged inside the mystery box.

ANALYZE AND CONCLUDE

1. How certain can you be of your hypothesis? Explain your answer.
2. Without opening the mystery box, what further tests might you perform to verify your hypothesis?

The Steps of the Scientific Method

The biologist Claude Villée once described the scientific method as "organized common sense," and that's a good description. No matter how complex science may seem, at its heart is an organized system of asking and answering questions.

In practice, scientists apply the scientific method in a number of different ways. However, a formal version of the scientific method can be described as a series of steps:

- First, state a specific problem or question based on observations of the natural world. Sometimes the most important questions become clear only after you have learned a great deal of information about the problem you are investigating.
- With that problem in mind, propose a **hypothesis.** A hypothesis is a possible explanation, a preliminary conclusion, or even a guess at the solution to your problem.
- Next, test the hypothesis. Sometimes the test involves gathering more observations to see if they are consistent with the hypothesis. In other cases, it is possible to set up a controlled test, or **experiment,** to check the hypothesis. The best hypotheses allow scientists to make clear predictions about the results of experiments.
- Finally, analyze the experimental results and draw conclusions. Sometimes the results are so clear-cut that you can say the hypothesis is either correct or incorrect. Other times the results suggest new hypotheses that, in turn, require new experiments.

When you are sure of your results, there is another step that you may need to take. This final step is to repeat the experiment to make sure that your results are reliable. Experiments are often complicated, and careful scientists want to make sure that their results hold true whenever the experiment is performed. If your experiment gives different results on repeated trials, then you should try to find out what is influencing your results.

☑ *Checkpoint* What is the scientific method?

In the Garden

Is the scientific method a complex technique that only scientists can use? In fact, people use the scientific method every day. No special training is required—just a bit of common sense.

For example, suppose you grow tomatoes in a small garden. Each spring you plant 20 seedlings, and by autumn you have usually harvested 50 to 60 tomatoes. But last spring, following your neighbor's advice, you mixed 5 kilograms of fertilizer into the soil. Instead of the usual crop, your 20 plants produced more than 100 tomatoes—a great yield!

You may be tempted to believe that you have proven that the fertilizer increased your tomato crop. But did it really? What other factors might have influenced your results?

Last year's weather might have been very good for tomatoes. Could the weather have been partly responsible? You also remember that you used a different type of tomato seed. Could the seed type have caused your success? Maybe your garden had fewer insects last year. Suddenly, things don't seem so simple. In fact, many factors could explain the good harvest.

Figure 1–9
The growth of tomatoes is influenced by many factors, including sunlight, soil, temperature, and rainfall. Performing a controlled experiment is the best way to test the effects of any factor.

A Logical Experiment

The best way to find out what caused your results is to use the scientific method. Start with a hypothesis: Using fertilizer increases the tomato crop. Then design a logical experiment to test the hypothesis.

This year, divide your garden into two sections, planting 12 seedlings grown from the same seed supply in each section. Each section of the garden should receive the same amount of rainfall and sunlight, be subjected to the same insects and other pests, and contain similar soil. However, add fertilizer to one section and not to the other.

By managing the garden this way, you make the two sets of plants as similar as possible—except for the presence of fertilizer. These two sections provide an accurate, scientific test for your hypothesis.

The Results

What will happen? The result might be as shown in ***Figure 1–10*** on the next page—the fertilized plot produces nearly 50 percent more tomatoes than the unfertilized plot. This result is much more meaningful than your observation from the year before because this time you have carried out a controlled, logical experiment.

What role did the section of unfertilized plants serve in this experiment? These plants were treated just like the plants in the other section except for the use of fertilizer. Therefore, you can reasonably conclude that the fertilizer was responsible for any differences between the two sections.

Control and Variable

The group of unfertilized plants is called the **control** group. The control group provided a benchmark that allowed you to measure the fertilizer's effect. The fertilizer is called the **variable** of the experiment. A variable is the factor that differs among the test groups.

Figure 1–10 (a) *The only difference between these two plots of a tomato garden is the presence of fertilizer.* (b) *At the end of the growing season, the fertilized plot produced far more tomatoes than the unfertilized plot.*

Every scientific experiment has a variable that is measured against a control.

Does this experiment on tomato plants seem almost too simple? Remember that the scientific method is nothing more than organized common sense—and a controlled experiment.

☑ **Checkpoint** What is a control group? A variable?

Science and Truth

Although the scientific method is a powerful way to learn about the world around us, one of the most important lessons to be learned about scientific knowledge is how changeable it can be. Many "scientific facts" of the past are now known to be incorrect.

A Human Activity

For example, years ago high school and college students were taught that genetic information is carried only in a molecule called DNA. Scientists now know that this is not quite true. Another molecule, called RNA, carries genetic information in certain viruses.

It is also worth noting that scientists, like other people, are influenced by their own prejudices, beliefs, and intuitions. They sometimes close their minds to results that "don't make sense," and they often jump to conclusions that the facts may not actually support. After all, science is a human activity, and it is as likely to go wrong as anything else that people do.

This does not mean, however, that you should disregard scientific results. The fact that science is open to new ideas and the constant testing of old ones should give us confidence that science eventually will arrive at the correct answers for the important questions that we put to it. Indeed, what makes science special is the demand that scientific conclusions be constantly tested against the reality of the natural world.

Theories

When repeated experiments consistently confirm a hypothesis, the scientific

Figure 1–11
The world of living things holds all sorts of wonders for you to study—from (a) the spiny underbelly of a starfish to (b) a fruit bat (which you should handle only with proper supervision) to (c) the delicate leaves of a fern.

community may come to accept the hypothesis as valid. In this book, the authors have tried to rely on generally accepted hypotheses and to point out those issues over which the scientific community disagrees.

In many cases, however, a series of closely related hypotheses have been confirmed so many times that they can properly be described as a **theory.** A theory is a logical explanation that explains a broad range of observations. The theory that certain diseases are caused by germs is one such example.

Are theories always true? No theory is ever beyond dispute. Therefore, it would be a serious mistake to believe that science ever produces absolute truth. Nonetheless, in science, theories represent solid, logical explanations of the natural world—explanations that have stood up under repeated analysis. Put another way, a theory is the best explanation that the process of science has produced to date.

Biology and You

Both authors of this textbook have a simple message to tell you about biology. Biology is, as far as we're concerned, the most interesting subject in the world!

No matter where you live and no matter what your school is like, you are surrounded by living organisms. Each organism has its own story to tell, and each is fascinating in its own right. Don't think for a second that you need a license or a degree to call yourself a biologist. All you need is an interest in living things. You will find plenty of them to investigate.

Throughout your biology course, take the time to observe the living things that share the world with you—the birds that feed in your yard, the insects that hover near lights on a summer night, the trees that line the streets of your town or city. You will come to understand and appreciate biology even more.

Section Review 1–2

1. **Define** the scientific method.
2. **Describe** the steps of the scientific method.
3. How do a hypothesis and a theory **compare**?
4. **Critical Thinking—Applying Concepts** Your friend argues that there are no facts in science—only theories that could be proved false. Do you agree or disagree with this argument? Explain your answer.
5. **MINI LAB** After you **formulate a hypothesis,** how can you prove or disprove the hypothesis?

SECTION 1–3

BRANCHING OUT In Action

The Scientific Method and Yellow Fever

Guide for Reading

- **Explain** the role of testing in science.
- **Describe** Walter Reed's experiment on yellow fever.

MINI LAB

- **Design an experiment** to test the effects of water and sunlight on the growth of bread mold.

INTEGRATING SOCIAL STUDIES

Why did the United States and Spain fight the Spanish-American War?

AS YOU HAVE READ, GENUINE science requires much more than observation. Scientists must develop explanations for their observations, then use the scientific method to test those explanations.

Why has science become so important in our lives? In part, because the scientific method has proved to be effective in solving one problem after another.

The Fight Against Yellow Fever

During the Spanish-American War of the late 1800s, United States soldiers fighting in Cuba faced an especially deadly problem—an enemy far more terrifying than the Spanish soldiers they were defeating on the battlefields. This enemy was yellow fever, a disease well known in tropical Central and South America. Before the smoke of battle had cleared, United States troops were suffering from fever and nausea, vomiting black liquid, and their skin had turned a ghastly shade of yellow, which gave the disease its name.

Checkpoint What is yellow fever?

Walter Reed

When United States military leaders realized that yellow fever was causing more deaths than enemy bullets, they appealed to their government in Washington, DC. Help came in the form of a

Figure 1–12
(a) The fight against yellow fever was led by Dr. Walter Reed. (b) Because so many soldiers were dying from yellow fever, several soldiers volunteered to take part in the experiments of Reed (left center) and Findlay (far left).

commission headed by an army research doctor named Walter Reed.

Reed's first step was to analyze the old ways of fighting yellow fever. Because the disease moved from one section of a town to another, most people believed that yellow fever must be spread from person to person. To fight the disease, they isolated the sick, boiled their bedsheets and clothes, and sterilized their plates, cups, and forks.

These precautions would make good sense if yellow fever was transmitted by personal contact. Reed quickly discovered, however, that none of these measures stopped the spread of the disease.

Findlay's Hypothesis

Reed's commission listened to many physicians, including a Cuban doctor named Carlos Findlay. Findlay hypothesized that the disease was spread by mosquitoes, of which Cuba had more than its share.

Most people believed that better sanitation was the key to controlling yellow fever, and they did not believe Findlay's hypothesis. But Reed believed that there was not enough evidence to draw any logical conclusions. **To test Findlay's hypotheses, Reed decided he needed to conduct a controlled experiment.**

It was thought at the time that yellow fever affected people but did not affect animals. So Reed performed his experiment on brave but frightened groups of human volunteers.

☑ *Checkpoint* What was the goal of Reed's experiment?

Reed's Experiment

Reed's commission assembled two groups of volunteers for a terrifying experiment. One group spent 20 nerve-wracking days wearing the filthy clothing of yellow fever patients, sleeping on their bedsheets, and eating from plates they had used. During this time, however, they lived behind screens that protected them from being bitten by mosquitoes.

The other group of volunteers used only fresh clothing, slept in clean beds, and remained totally isolated from yellow fever patients. These volunteers, however, were not protected by mosquito netting and so were bitten by mosquitoes. Unwilling to let soldiers take a risk that they would not take themselves, three doctors on the commission joined this group. (Reed also wanted to take part, but his associates refused his request to do so.)

A Moldy Question

PROBLEM *Why does bread turn moldy? **Design an experiment** to help answer this question.*

PROCEDURE

1. Mold will grow on bread that is exposed to air at room temperature. Design an experiment to test the effects of water and sunlight on the growth of bread mold. You may use up to four slices of bread and any materials available in your classroom.
2. Under your teacher's supervision, perform the experiment you designed.

ANALYZE AND CONCLUDE

1. What were the variables in your experiment? What were the controls?
2. Could factors other than water and sunlight have influenced the results? Explain.
3. What conclusions can you draw from this experiment? Explain your answer.

(a) REED'S EXPERIMENT ON YELLOW FEVER

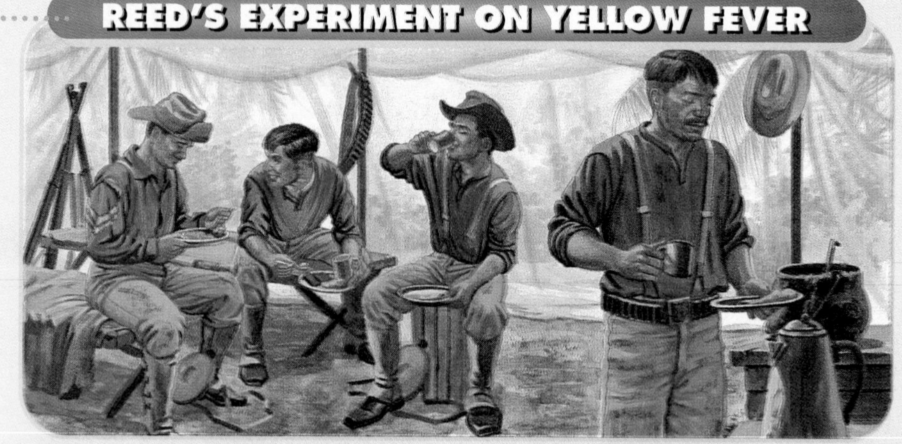

Figure 1–13
(a) In Reed's experiment, one group of volunteers lived in the filthy clothing and bedding of yellow fever patients, and they were kept isolated from mosquitoes. The second group lived in clean clothing and surroundings but were not protected from being bitten by mosquitoes. (b) Only those who were bitten by mosquitoes developed yellow fever. As a result, Reed concluded that mosquitoes transmitted this disease.

(b) THE RESULTS

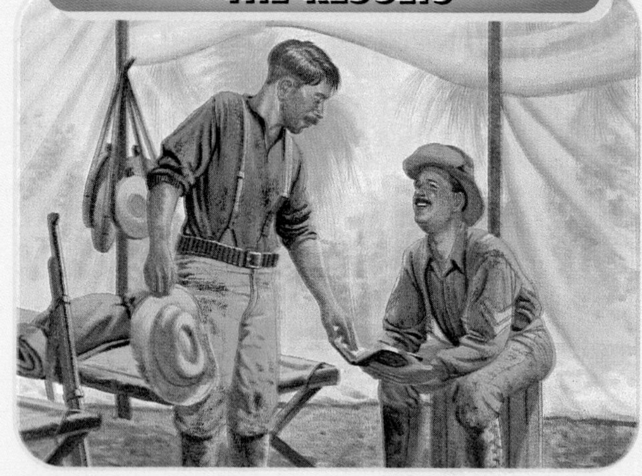

The Results

What happened? Not a single volunteer in the first group developed yellow fever. But many of the volunteers in the second group—including the three doctors—became sick with yellow fever. One of the doctors, Jesse Lazear, died from the disease.

The results of the experiment were very clear. Yellow fever was not spread by poor sanitation, person-to-person contact, or food. Instead, yellow fever was spread by mosquitoes. As a result, the United States government declared war on the mosquito. And within 90 days, the city of Havana, Cuba, was free of yellow fever.

Today, Jesse Lazear is remembered as a hero, and Findlay and Reed are recognized as the people who led the way to conquering a terrible disease. In addition, the work of these scientists had effects well beyond the fields of science and medicine.

Figure 1–14
(a) *Like Cuba, the country of Panama was once plagued by yellow fever. But when the disease was under control, the United States government decided to build the Panama Canal.* (b) *This photograph shows* Aëdes aegypti, *the mosquito that transmits yellow fever.*

For example, yellow fever had been a common disease in the Central American country of Panama. Once the disease was under control, people in the United States began seriously to consider building the Panama Canal.

Beyond Yellow Fever

As you know, the application of the scientific method in fighting disease did not stop with the conquest of yellow fever. In the decades since Reed's work, biology has changed from a science that only studies the world around us to one that makes changes in that world.

As a result, we live at one of the most remarkable times in human history. Biology will influence life in the twenty-first century—the century in which you will spend most of your life—to an extent never approached before. These are exciting times to be alive, and they are especially exciting times to study the science of biology.

INTEGRATING SOCIAL STUDIES

Why was controlling yellow fever important in building the Panama Canal? How did the Panama Canal affect the countries of Central America?

Section Review 1–3

1. **Explain** the role of testing in science.
2. **Describe** how Dr. Walter Reed demonstrated that mosquitoes transmitted yellow fever.
3. **MINI LAB** How might water and sunlight influence the growth of bread mold? **Design an experiment** to test your hypothesis.
4. **BRANCHING OUT ACTIVITY** Suppose that both you and your neighbor have bird feeders, but more birds eat from your neighbor's feeder. Why might the neighbor's feeder be a more popular place than yours? **Formulate a hypothesis** to answer this question, and **design an experiment** to test it.

CHAPTER 1

Laboratory Investigation

Looking Closely at Living Things

As you know, living things come in a variety of sizes, forms, shapes, and colors. Yet, all living things share certain characteristics. You can discover some of these characteristics by carefully observing a living thing.

Problem

What are the characteristics of life? **Observe** a living organism to discover possible answers.

Materials (per group)

a live organism
magnifying glass
metric ruler
laboratory balance
penlight or flashlight
white and dark paper

Procedure

1. **Obtain a live organism from your teacher. Be sure to follow any special precautions that your teacher gives you for handling the organism.**
2. **Observe the organism for 5 minutes, using your senses of sight, smell, and hearing. Record your observations in a data table similar to the one shown.**
3. **On a sheet of paper, sketch the organism. Indicate its color, shape, and size, as well as any other characteristics that you observe. Label any parts that you can identify.**
4. **Observe the organism through the magnifying glass. Sketch any interesting features that you see.**
5. **Use a metric ruler to measure the length, width, and height of the organism. In addition, measure any noticeable projections on the organism. Record all measurements.**
6. **If the organism can be placed on a balance, find and record the organism's mass.**

7. **Use a penlight or flashlight to shine a beam of light on the organism, and observe its response. Record your observations.**

8. **Use a sheet of dark paper to shade the organism from the light, and observe its response. Record your observations.**

Observations

Share your observations with your classmates, who studied different organisms. How are the organisms different? How are they similar?

Analysis and Conclusions

1. Which characteristics of life did you observe in the organism you studied?
2. For the characteristics of life that you did not observe in the organism, discuss whether or not you believe the organism has them. Use your observations as evidence for your answer.
3. Are you convinced that the organism you studied is alive? Explain your answer.
4. Among the organisms your class studied, was there a common color, shape, mass, or other physical characteristic? Discuss the significance of your answer.

DATA TABLE

Organism's name	
Color	
Shape	
Measurements	
Mass	
Response to light	

More to Explore

Design an experiment to show how your organism is affected by temperature or some other factor. Be sure your experiment does not harm your organism.

Study Guide

Summarizing Key Concepts

The key concepts in each section of this chapter are listed below to help you review the chapter content. Make sure you understand each concept and its relationship to other concepts and to the theme of this chapter.

1–1 The Characteristics of Life

- Science is a process of thinking and learning about the world around us. Biology is the science of life.
- Living things are made of cells, grow and develop, obtain and use energy, respond to their environment, and are able to reproduce.
- Biologists may ask questions at many different levels of organization—from the chemical level to the global level.

1–2 The Scientific Method and Biology

- The scientific method is a system of asking questions, developing explanations, and testing those explanations against the reality of the natural world.
- The steps of the scientific method include stating a problem, proposing a hypothesis for the problem, testing the hypothesis in an experiment, and analyzing the results and drawing conclusions.
- A scientific experiment uses one or more control groups and one variable. A variable is the factor that differs among the test groups. A control group provides a benchmark to measure the variable's effects.
- A scientific theory comes from a series of hypotheses that have been confirmed many times. Theories represent scientists' best explanations of the natural world.

1–3 The Scientific Method and Yellow Fever

- To obtain evidence for or against Findlay's hypothesis, Walter Reed realized he had to conduct a controlled experiment.
- From the results of their experiment, Reed and his associates demonstrated that yellow fever was transmitted by mosquitoes, not personal contact.

Reviewing Key Terms

Review the following vocabulary terms and their meaning. Then use each term in a complete sentence.

1–1 The Characteristics of Life

science
biology
organism
homeostasis

1–2 The Scientific Method and Biology

scientific method
hypothesis
experiment
control
variable
theory

Recalling Main Ideas

Choose the letter of the answer that best completes the statement or answers the question.

1. Biology is the study of
   - ✓**a.** living things.
   - **b.** growth.
   - **c.** matter.
   - **d.** reproduction.
2. Not every organism is able to
   - **a.** grow and develop.
   - **b.** obtain energy.
   - ✓**c.** move independently.
   - **d.** respond to its environment.
3. Biologists study life at
   - **a.** the chemical level only.
   - **b.** the cellular level only.
   - **c.** the global level only.
   - ✓**d.** many different levels.
4. The factor that differs among the test groups of an experiment is called a
   - **a.** theory.
   - ✓**b.** variable.
   - **c.** control.
   - **d.** hypothesis.
5. Artists and historians are not scientists because they do not rely on
   - **a.** observations.
   - **b.** careful study.
   - **c.** provable facts.
   - ✓**d.** the scientific method.
6. In Walter Reed's experiment on yellow fever, both test groups were subjected to
   - **a.** unsanitary living conditions.
   - **b.** mosquito bites.
   - **c.** contact with yellow fever patients.
   - ✓**d.** a strict set of living conditions.
7. A hypothesis can be described as a(an)
   - **a.** controlled experiment.
   - ✓**b.** possible explanation.
   - **c.** answerable question.
   - **d.** proven theory.

Putting It All Together

Using the information on pages xxx to xxxi, complete the following concept map.

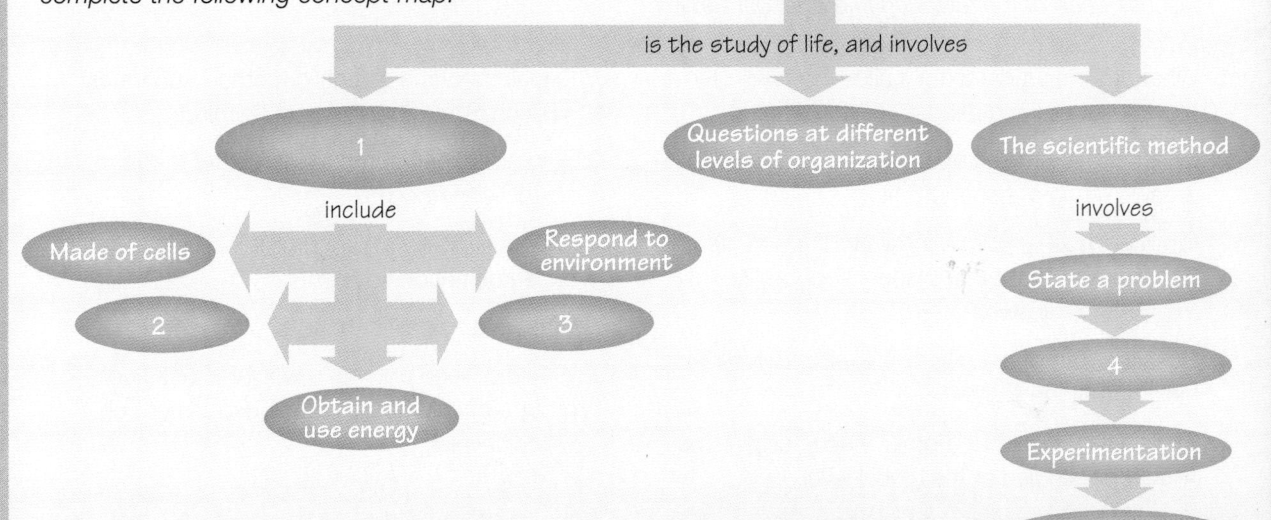

Assessment

Reviewing What You Learned

Answer each of the following in a complete sentence.

1. Give three examples of questions about living things.
2. What is biology?
3. List five major characteristics of life.
4. What is an organism?
5. From which sources do organisms obtain energy?
6. How do corals exhibit the characteristics of life?
7. Which field of biology is concerned with the relationships among organisms?
8. Describe the scientific method.
9. What is a hypothesis? Describe the role of the hypothesis in the scientific method.
10. Give an example of an observation a biologist might make.
11. What is a controlled experiment?

Expanding the Concepts

Discuss each of the following in a brief paragraph.

1. What are the fundamental units—or building blocks—of all living things?
2. Give three examples of a stimulus and how an organism might respond to it.
3. Earth is the only planet known to support life. Do you think that other planets support life? Explain your answer.
4. Describe four of the different levels at which biology may be studied.
5. Explain why corals and other marine organisms should be studied at the global level.
6. Explain how the scientific method separates science from other fields of learning.
7. Why are experiments fundamental to the scientific method?
8. Why is reliability an important factor in scientific experimentation?
9. What is a scientific theory?
10. Why did Reed and his associates use two groups of volunteers in their experiment on yellow fever?

Extending Your Thinking

Use the skills you have developed in this chapter to answer the following.

1. **Generalizing** What do you think is the size range of the organisms that live on Earth? What factors might limit this size range?
2. **Designing an experiment** Design an experiment to show which of four soil mixtures would be best for growing corn in a backyard garden.
3. **Formulating a hypothesis** In labrids, a type of small fish found along coral reefs, the length of the females and young males is typically 6 centimeters or less, while the length of the adult males is typically much greater. Formulate a hypothesis based on this evidence.
4. **Evaluating** A crystal of calcium carbonate can grow over time. Does this mean that the crystal is alive? Explain your answer.
5. **Applying** Suppose that a bed of unusual fossils has been discovered at a construction site. Which type of biologists should be invited to study the site? Explain your answer.
6. **Applying** People often develop and test hypotheses as part of their daily lives. What hypotheses have you recently developed or tested?
7. **Predicting** You have been asked to observe a maple tree over the course of one year. Based on your knowledge of the characteristics of life, predict some of the observations that you will make.

Applying Your Skills

Where's the Jelly?

Every scientific experiment needs a logical procedure—a series of steps that are clear to follow and that serve the experiment's purpose. But as you will discover, writing and following even the simplest procedures are not so easy!

1. Write a procedure for making a peanut butter and cracker sandwich from the materials your teacher provides.
2. Exchange procedures with one of your classmates.
3. Follow your classmate's procedure exactly as it is written.

• GOING FURTHER •

4. Evaluate your classmate's procedure. Share the evaluation with your classmate.
5. Did your classmate accurately interpret and follow the procedure you had written? Could you improve the procedure in any way?

CHAPTER 2

The Chemistry of Life

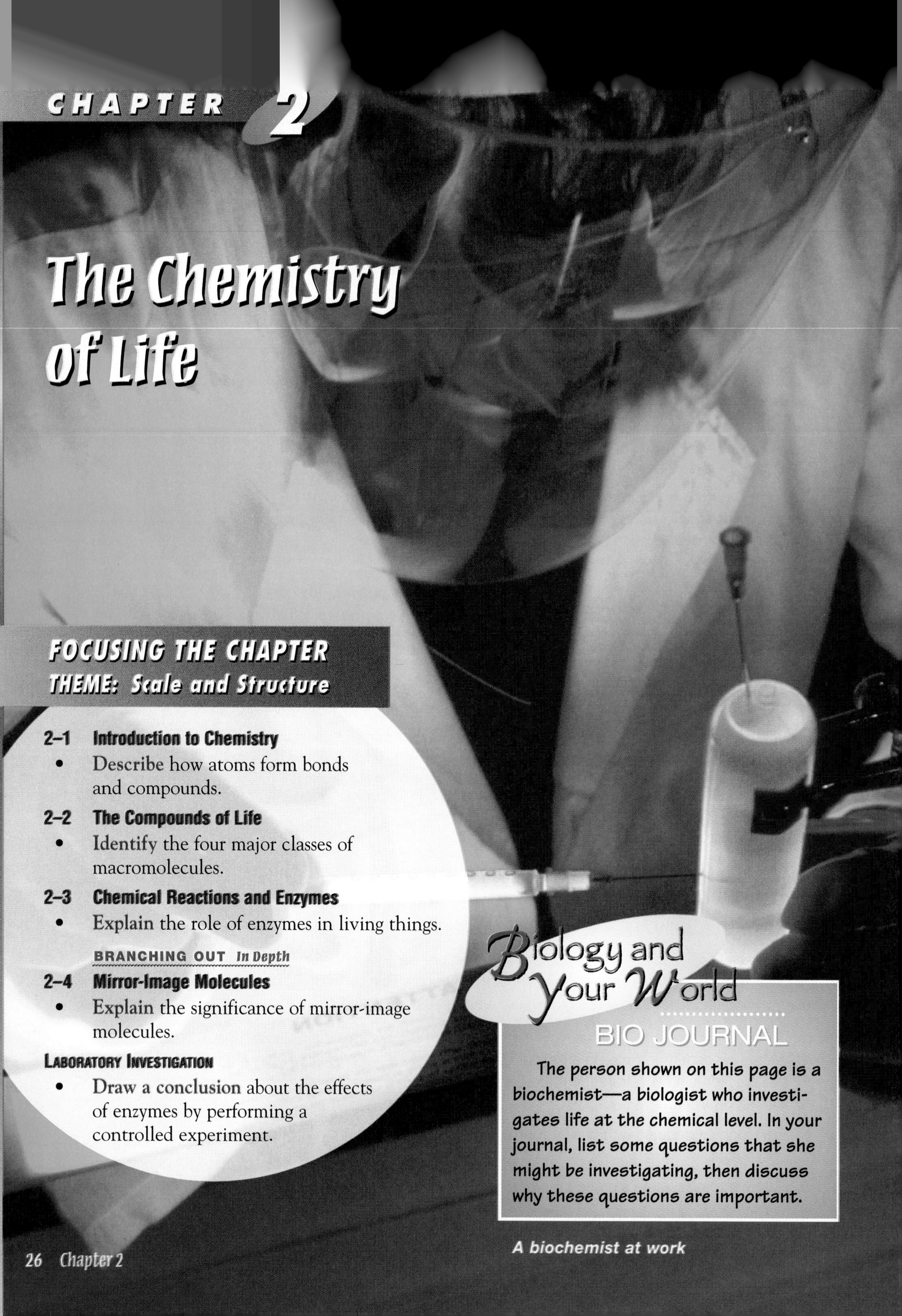

FOCUSING THE CHAPTER THEME: Scale and Structure

2–1 Introduction to Chemistry
- Describe how atoms form bonds and compounds.

2–2 The Compounds of Life
- Identify the four major classes of macromolecules.

2–3 Chemical Reactions and Enzymes
- Explain the role of enzymes in living things.

BRANCHING OUT *In Depth*

2–4 Mirror-Image Molecules
- Explain the significance of mirror-image molecules.

LABORATORY INVESTIGATION
- Draw a conclusion about the effects of enzymes by performing a controlled experiment.

Biology and Your World

BIO JOURNAL

The person shown on this page is a biochemist—a biologist who investigates life at the chemical level. In your journal, list some questions that she might be investigating, then discuss why these questions are important.

A biochemist at work

Introduction to Chemistry

GUIDE FOR READING

- **Explain** how atoms of elements differ.
- **Compare** the different bonds that atoms form.
- **Describe** the properties of the H^+ ion and the OH^- ion

WHY ARE LIVING THINGS SO different from nonliving things? One way to examine this question is at the chemical and molecular levels. As you will discover, the arrangement of chemical elements in living things is far more complex than anything found in the nonliving world.

Atoms and Compounds

A chemical element is a substance that cannot be broken down into any other substance. And the smallest unit of a chemical element is a particle called the **atom.**

Atoms are extremely small—so small, in fact, that 100 million atoms placed side by side would measure only about 1 centimeter, or about the width of your pinky finger. Yet despite their tiny size, atoms are made of subatomic particles, which are even smaller.

Parts of the Atom

In the center of every atom is a compact core called a nucleus. Although the nucleus takes up only a small fraction of an atom's volume, it accounts for almost all—99.9 percent—of an atom's mass.

The nucleus contains two types of subatomic particles—the **proton** and the **neutron.** A proton carries a positive charge while a neutron carries no charge, and both particles have approximately the same mass. Strong forces bind these two kinds of particles together in the nucleus.

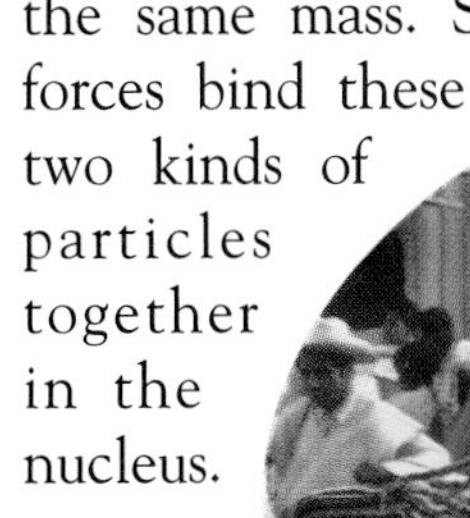

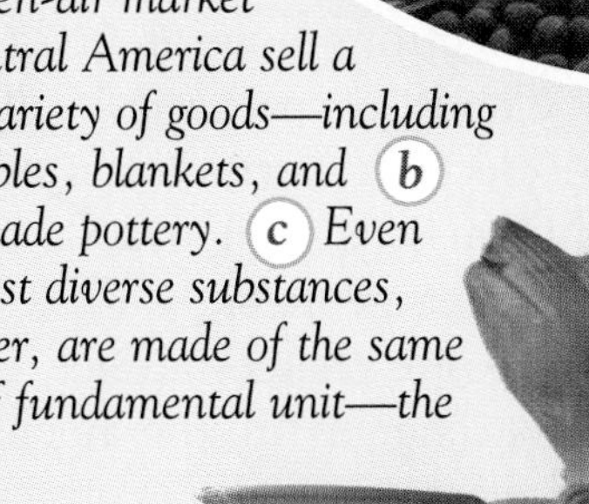

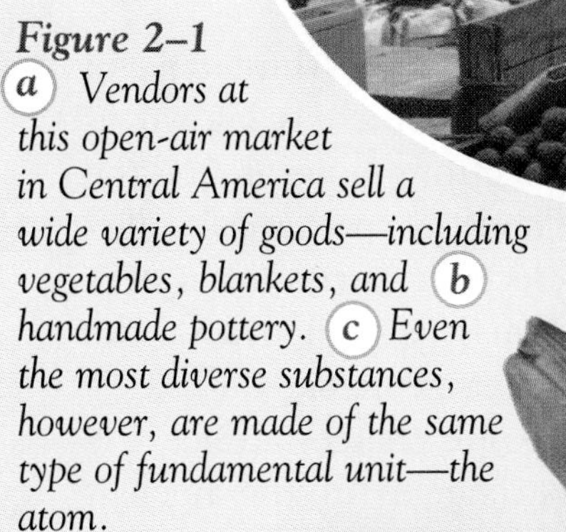

Figure 2–1 *(a) Vendors at this open-air market in Central America sell a wide variety of goods—including vegetables, blankets, and (b) handmade pottery. (c) Even the most diverse substances, however, are made of the same type of fundamental unit—the atom.*

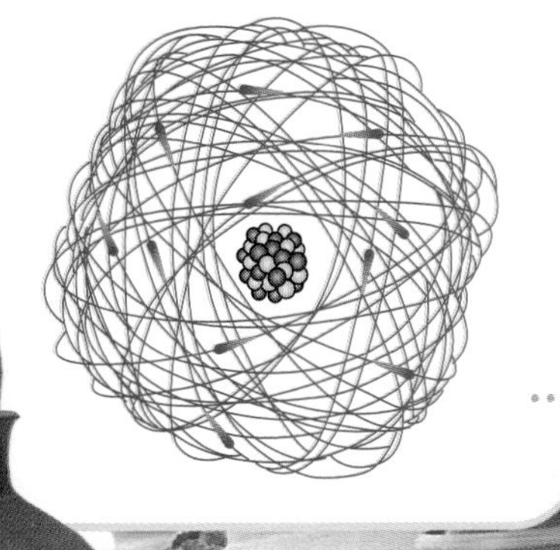

Figure 2–2
An atom's electrons are arranged in distinct energy levels. The electrons in helium, neon, and argon each fill the energy levels, which makes these atoms especially stable.

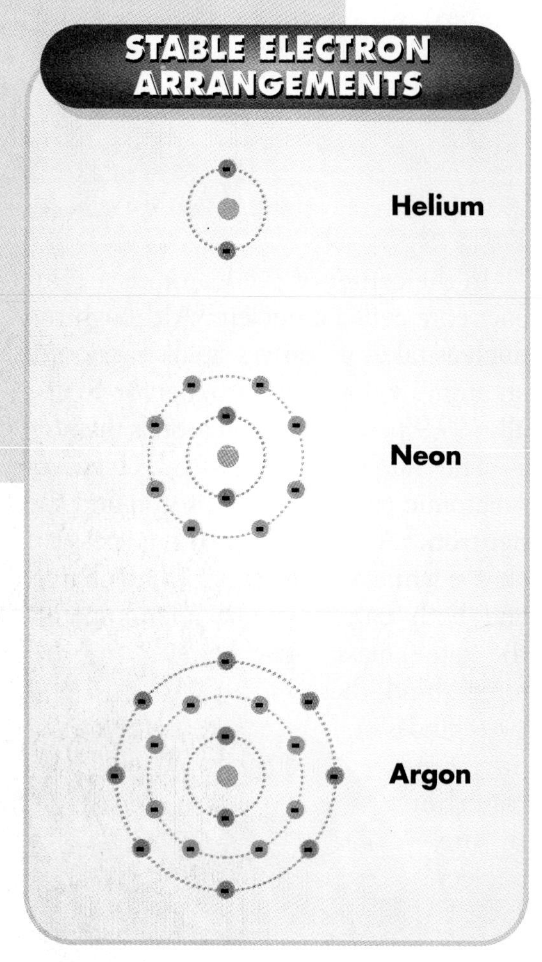

Atoms also contain a third type of subatomic particle, which is known as the **electron.** Electrons move around the nucleus in clouds of pathways called orbitals. An electron carries a negative charge, and its mass is about 1/2000 the mass of a proton or neutron.

Because opposite charges attract, the strong positive charges in the nucleus help to hold electrons in their orbitals. And because atoms have equal numbers of protons and electrons, they are electrically neutral, carrying no overall charge.

Why are atoms of one element, such as oxygen, different from the atoms of another element, such as gold? **The key difference between the atoms of different elements is the number of protons and electrons they contain.** An oxygen atom, for example, has 8 protons in its nucleus and 8 electrons surrounding the nucleus. A gold atom, by contrast, has 79 protons and 79 electrons.

Bonds and Compounds

There are 92 different chemical elements found in nature, and another 20 have been artificially produced in research laboratories. But if these 112 elements were the only substances found in nature, the world would be a very dull place.

Fortunately for us, the atoms of most elements link up with each other in different arrangements and combinations. The links between atoms are called chemical bonds. And a substance that is formed by the bonding of atoms in definite proportions is called a **chemical compound.** Water, salt, sugar, and ammonia are examples of common chemical compounds.

Why do atoms form bonds? The answer lies with the arrangement of their electrons. Among the atoms with fewer than 20 electrons, the most stable electron arrangements are found in helium (with 2 electrons), neon (with 10 electrons), and argon (with 18 electrons).

For the other small atoms—including atoms of hydrogen, carbon, nitrogen, and oxygen—electrons are shared or transferred, as though the atoms were trying to obtain the electron arrangement of helium, neon, or argon. As you will discover, shared and transferred electrons serve to bond atoms together.

Ionic Bonds

Consider the sodium atom and the chlorine atom shown in ***Figure 2–3.*** When these atoms approach each other, both become more stable when one electron transfers from the sodium atom to the chlorine atom. As a result, both atoms become **ions.** An ion is an atom that has gained or lost one or more of its electrons, thus acquring a net electrical charge.

Because opposite charges attract each other, the positively charged sodium ion (Na^+) attracts the negatively charged chloride ion (Cl^-). **The strong attraction between oppositely charged ions is called an ionic bond.** Strong **ionic bonds** link ions in simple compounds, like sodium chloride, and they help to hold together different parts of much larger chemical compounds.

Covalent Bonds

A **covalent bond** is another type of bond. **In a covalent bond, electrons are shared between two atoms.** For example, consider the two oxygen atoms in ***Figure 2–3.*** Individually, each oxygen atom is surrounded by 8 electrons—not an especially stable arrangement. But when each atom shares 2 electrons with the other atom, both acquire a stable arrangement of 10 electrons.

The more electrons in the covalent bond, the more strongly the two atoms are joined. A covalent bond can be formed from 2 electrons (called a single bond), 4 electrons (a double bond), or 6 electrons (a triple bond).

Because shared electrons help both atoms to be stable, a covalent bond usually does not break easily. Therefore, a group of atoms united by covalent bonds typically acts as a single unit, called a **molecule.** A molecule may contain as few as 2 atoms or it may contain atoms in the thousands or millions. DNA, the molecule of genetic inheritance, is a molecule made of millions of atoms. Sucrose, or ordinary table sugar, contains only 45 atoms. And one of the most important molecules in nature—the water molecule—contains only 3 atoms.

☑ **Checkpoint** How do ionic and covalent bonds differ?

Chemical Formulas

Most chemical compounds can be described by a kind of shorthand notation known as a chemical formula. A chemical formula indicates the elements that form the compound and the proportions in which they combine. Sodium chloride, for example, has the chemical formula NaCl. This formula indicates that sodium chloride contains one sodium ion (Na^+) for every chloride ion (Cl^-).

For compounds composed of molecules, a chemical formula also indicates the numbers of each atom in the molecule. For example, the formula for glucose, $C_6H_{12}O_6$, indicates that a molecule of glucose contains 6 carbon atoms, 12 hydrogen atoms, and 6 oxygen atoms.

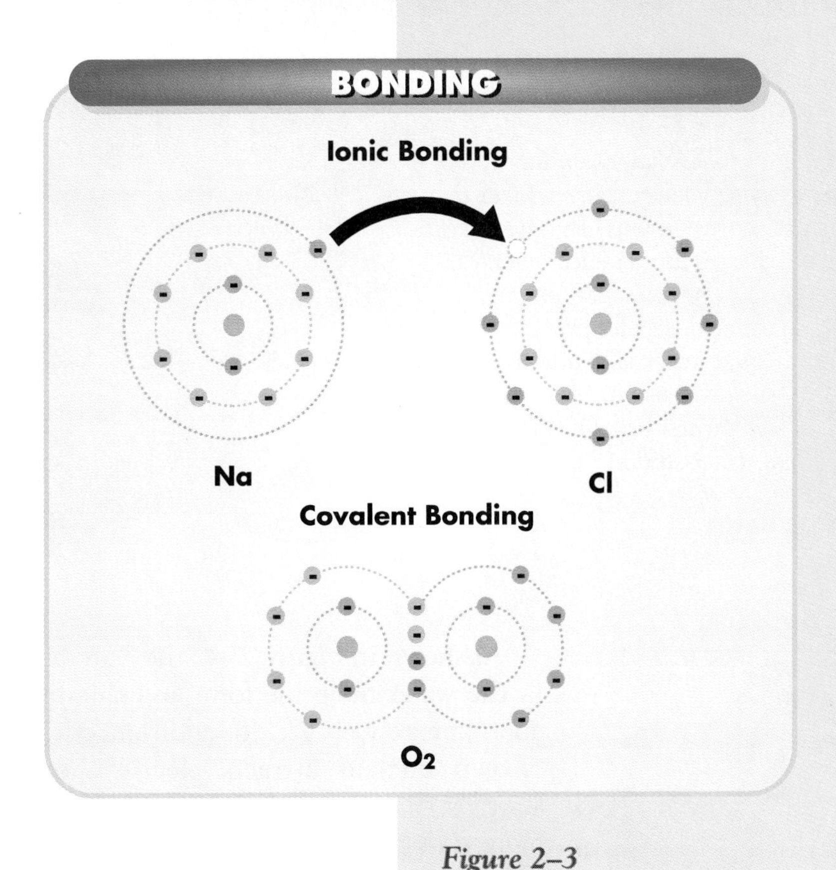

Figure 2–3
A sodium atom (Na) readily loses an electron to a chlorine atom (Cl), changing both atoms into ions and creating an ionic bond between them. In a molecule of oxygen (O_2), 2 oxygen atoms share 4 electrons, creating a covalent bond.

Water 

Liquid water is found inside every living cell—as well as outside most cells. Even the cells of organisms that live on dry land are bathed in a liquid that contains water.

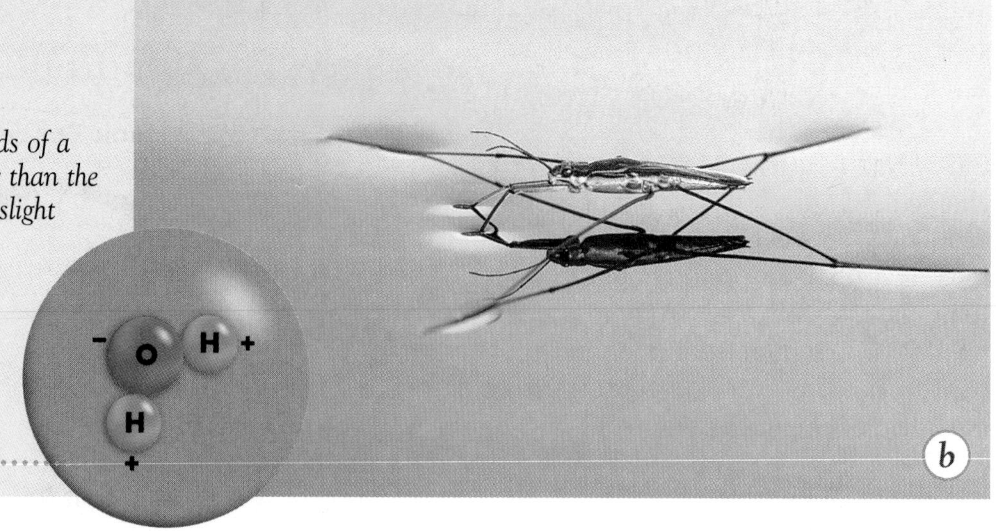

Figure 2–4
(a) *The charges on the ends of a water molecule are weaker than the charges on ions. But even slight charges attract each other.* (b) *The attraction among water molecules provides liquid water with a high surface tension, which is why water can support a water strider.*

The Water Molecule

As shown in ***Figure 2–4,*** the 3 atoms in the water molecule form an angle of approximately 105°. In addition, the oxygen nucleus attracts electrons so strongly that the electrons spend much more time near it than they do near the hydrogen nuclei. As a result, the oxygen atom acquires a slight negative charge and the 2 hydrogen atoms acquire slight positive charges.

Because of this charge separation, water is a polar molecule. A polar molecule has one end with a slight positive charge and another end with a slight negative charge. Molecules without positive and negative ends are called nonpolar.

Solutions

When water dissolves a substance, the liquid that results is known as a **solution.** Many compounds in cells are in solution, which is one reason why water is so vital to life.

As a general rule, water is good at dissolving polar and ionic compounds. ***Figure 2–5*** shows how water dissolves sodium chloride (NaCl).

H^+ and OH^- Ions

Another important property of water is its ability to dissociate into ions of its own. A water molecule will occasionally dissociate to produce hydrogen ions (H^+) and hydroxide ions (OH^-):

$$H_2O \rightarrow H^+ + OH^-$$

WATER DISSOLVING SODIUM CHLORIDE (NaCl)

Na^+

Cl^-

100 ml

BS 2021

Figure 2–5
(a) *Sodium chloride (NaCl) dissolves in water because the negative ends of water molecules cluster around the* Na^+ *ions, while the positive ends cluster around the* Cl^- *ions.* (b) *Water does not dissolve nonpolar compounds, such as cooking oil, which is why the cooking oil rests apart from the water in this separating funnel.*

Bio Frontier Connections

Acid Rain

Over the past 25 years, many trees in the American Northeast have come to look like those shown in the photograph. What is causing this damage? Some scientists suspect that the culprit is acid rain. Although rainwater is almost always acidic, some rainwater there has been found to be nearly as acidic as soda water.

Recent studies suggest that acid rain may wash away important substances from the soil, depriving large trees of nutrients they need to survive. Acid rain also kills fishes and other aquatic life. However, carefully controlled studies have shown that acid rain alone does not affect the growth of many types of trees—including pine, oak, and poplar.

The Causes of Acid Rain

What causes acid rain? One important cause is the burning of coal. Coal is composed principally of carbon, but it also contains small amounts of sulfur. When coal is burned, the sulfur reacts with oxygen to form sulfur dioxide (SO_2). Sulfur dioxide reacts with water in the atmosphere to form sulfuric acid (H_2SO_4), producing acid rain.

The Clean Air Act

To prevent acid rain, would it help to reduce the sulfur dioxide released into the atmosphere? In 1990, the United States government enacted a law to do just that. This law, called the Clean Air Act, required coal-burning power plants to reduce their sulfur dioxide emissions by 50 percent in 1995 and to cut them even more by the year 2000.

Has the Clean Air Act reduced acid rain, and will it do so in the years ahead? The question is not easy to answer. From 1972 to 1990, total SO_2 emissions decreased by about 30 percent, but the average acid content of rainwater in the eastern United States changed little. The Clean Air Act also forces reductions in particles released into the air. Many of these particles are alkaline, or basic, and thus help to lower the acidity of rainfall.

Acid rain destroyed these fir trees in North Carolina.

Changing Times for Coal Miners

Whether or not it has environmental effects, the Clean Air Act has had an economic impact. For many years, coal mines provided thousands of jobs in the eastern United States. But because the coal in much of this region is especially high in sulfur content, many mines are closing. Workers are without jobs and without the skills to get new jobs. However, some coal-production jobs have shifted to western regions of the United States, where the coal has less sulfur content.

Making the Connection

If a law to reduce acid rain led to a loss in your family's income, would you support the law? Should the government compensate or relocate the coal miners who lost their jobs? What do you think?

pH SCALE

Solution	pH
High H^+	0
	1
Stomach acid	
	2
Lemon juice	
Vinegar	3
	4
Banana	
Coffee	5
	6
Saliva	
Pure water	7
Blood	
	8
Borax	9
	10
Limewater	
	11
	12
Bleach	
	13
High OH^-	14

Figure 2–6 *The pH scale indicates the concentrations of H^+ ions and OH^- ions in a solution. The two ions are at equal concentrations at pH 7.*

The H^+ ion and the OH^- ion are two of the most reactive ions in nature. As shown in ***Figure 2–6,*** the higher the concentration of H^+ or OH^- ions, the more reactive the solution.

Acids and Bases

What is the source of extra H^+ ions or OH^- ions in a solution? Certain compounds dissociate in water to produce these ions. A compound that produces H^+ ions is called an **acid.** One acid that is produced in your own stomach is hydrochloric acid (HCl). Hydrochloric acid dissociates in water to produce H^+ ions and Cl^- ions:

$$HCl \rightarrow H^+ + Cl^-$$

A compound that produces OH^- ions is called a **base.** One example is sodium hydroxide (NaOH), which dissociates to produce Na^+ ions and OH^- ions:

$$NaOH \rightarrow Na^+ + OH^-$$

However, many bases do not themselves contain OH^- ions. Instead, they produce OH^- ions by accepting an H^+ ion from a water molecule. One base that reacts with water in this way is ammonia (NH_3):

$$NH_3 + H_2O \rightarrow NH_4^+ + OH^-$$

In living things, compounds that are related to ammonia are especially important bases.

Checkpoint What is an acid? A base?

pH

To indicate the strength of an acidic or a basic solution, scientists use the **pH scale.** As illustrated in ***Figure 2–6,*** the pH of a solution decreases with acidity. Therefore, a pH of 1 or 2 indicates a very acidic solution, and a pH of 12 or 13 indicates a very basic solution. Pure water contains equal amounts of H^+ ions and OH^- ions, and it has a pH of 7.

Biologists are often interested in the pH of solutions in living organisms. In your stomach, for example, the solution of hydrochloric acid (HCl) has a pH of approximately 2. This solution is far more acidic than human cells can tolerate. Normally, the stomach is lined with a material that protects its cells from the acid. But when that lining breaks down, the result is a painful condition called an ulcer.

Section Review 2–1

1. **Explain** how the atoms of elements differ.
2. **Compare** the different bonds that atoms form.
3. **Describe** the properties of the H^+ ion and the OH^- ion.
4. **Critical Thinking—Applying Concepts** Silica is a hard, glassy material that does not dissolve in water. Suppose silica crystals are accidentally mixed with sodium chloride. Describe how the mixture could be separated.

The Compounds of Life

SECTION 2–2

GUIDE FOR READING

- **Describe** the special properties of carbon.
- **Identify** the four major classes of macromolecules.

MINI LAB

- **Observe** how different foods react with Lugol's solution.

HOW DO THE COMPOUNDS OF life differ from the compounds of the nonliving world? The answer begins with the element carbon. Indeed, science-fiction writers often use the phrase "carbon-based life forms" to describe the living things of Earth. As you will discover, this description is quite accurate!

Carbon—A Special Element

As illustrated in **Figure 2–7,** a carbon atom has 6 electrons. In order to gain a stable arrangement of 10 electrons, a carbon atom typically forms 4 covalent bonds. And because the carbon atom is relatively small, the bonds it forms are significantly short and strong. As a result, carbon is able to form long, stable chains of atoms.

The bonds in the carbon chains can be single, double, or triple covalent bonds, and the chains can even close upon themselves to form rings and loops. In addition, a carbon chain can include or be attached to atoms of other elements, including nitrogen, oxygen, and phosphorus. **Although a few other elements form chains, none matches carbon in forming chains of different shapes, sizes, and complexity.**

With some exceptions, compounds that contain at least 2 carbon atoms are called **organic compounds.** All the other compounds are called **inorganic compounds.** As you will discover, living things make and use a tremendous number and

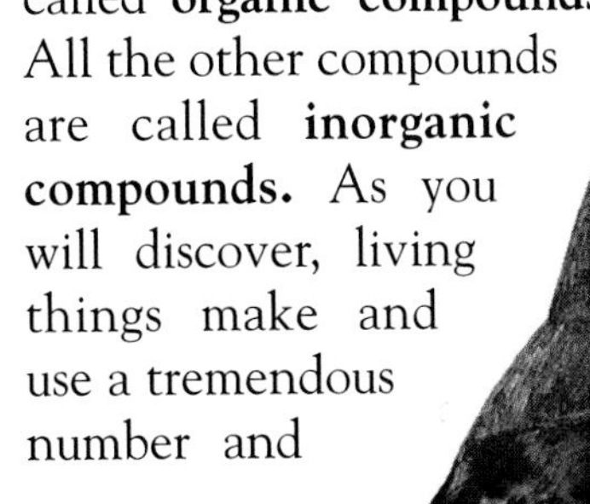

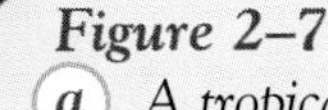

Figure 2–7 (a) A tropical rain forest is home to an incredible diversity of organisms—from leafy green palms and red heliconia flowers to (b) a colorful keel-billed toucan. (c) At the molecular level, life in the rain forest—and every place else on Earth—depends on carbon. Carbon is the principal element in all of the large molecules that organisms make and use.

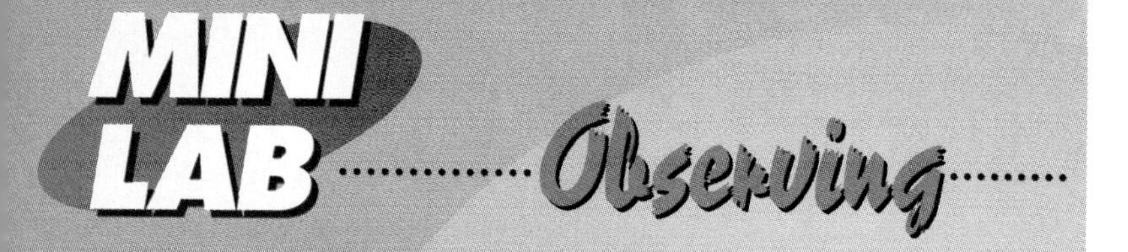

The Starch Test

PROBLEM *Which foods contain starch?* ***Observe*** *the results of adding Lugol's solution to different foods.*

PROCEDURE  

1. Place pieces of a soda cracker into a test tube.
2. Add 5 drops of Lugol's solution to the test tube. Lugol's solution is a test for starch. If the solution turns dark blue or black, starch is present. Record your observations.
3. Repeat the procedure using small amounts of peeled potato, white bread, oatmeal, and granulated sugar.

ANALYZE AND CONCLUDE

1. Of the foods you tested, which contain starch? Which do not contain starch?
2. Why might this procedure not indicate starch in a dark-colored food, such as a graham cracker?

variety of organic compounds. And they do so with a speed, accuracy, and efficiency that is far greater than any chemical process in the nonliving world.

Macromolecules

Most complex organic molecules are **polymers,** meaning "many units." Polymers are large molecules assembled from small, individual molecules. The small molecules are called **monomers,** meaning "single units". You can think of monomers as the individual letters and other characters of the English language. When these characters join together, they can form a nearly infinite variety of words, sentences, or paragraphs—the polymers. ●

INTEGRATING LANGUAGE ARTS

Why can a huge variety of English words and sentences be created from only 26 letters? Do all languages use letters to build words?

Many organic polymers are so large that they are known as **macromolecules,** meaning "giant molecules." **The four major classes of macromolecules are carbohydrates, lipids, proteins, and nucleic acids.** Let's take a closer look at each of these four classes.

Carbohydrates

Sugars and starches are members of the class of compounds called **carbohydrates.** Carbohydrates are made of carbon, hydrogen, and oxygen. They generally contain 2 hydrogen atoms for each oxygen atom, the same ratio found in water.

The smallest carbohydrates are the simple sugars. The simple sugars include galactose (found in milk), fructose (found in fruits), and glucose (found in the cells of every organism). Each of these simple sugars has the chemical formula $C_6H_{12}O_6$. However, they are different because each has a slightly different arrangement of its atoms.

Sugars are easy for cells both to make and to break down. They serve as a convenient way for cells to store chemical energy, and their breakdown provides cells with energy for all sorts of activities, including cell movement.

While organisms use simple sugars individually, they can also assemble them into polymers. For this reason, a simple sugar is also called a monosaccharide, meaning "single sugar." Two simple sugars joined together form a disaccharide. Ordinary table sugar, or sucrose, is one example of a disaccharide.

A polysaccharide consists of a large number of monosaccharides joined together. You can think of polysaccharides as warehouses for simple sugars. Plants store excess sugar in the form of a polysaccharide, called starch. Animals store excess sugar by making a slightly different polysaccharide, called glycogen.

☑ **Checkpoint** How do organisms use polysaccharides?

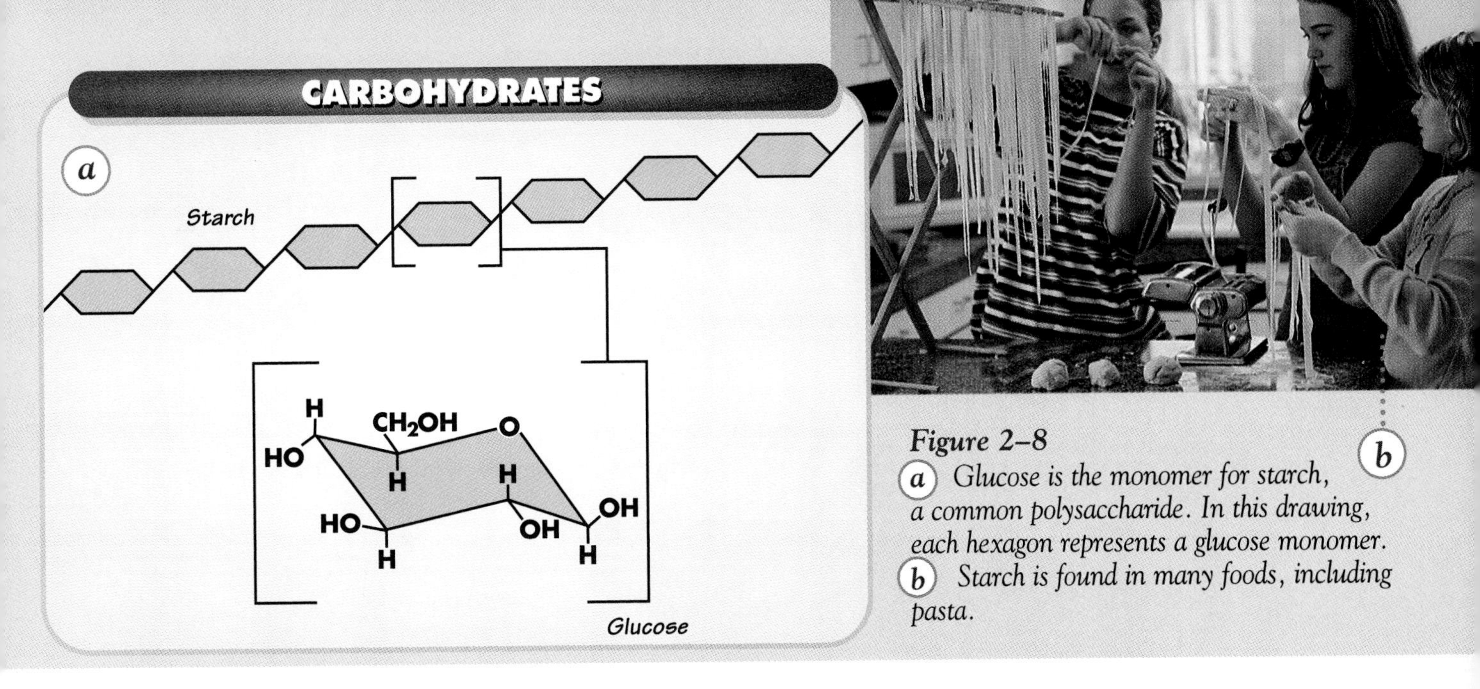

Figure 2–8
(a) *Glucose is the monomer for starch, a common polysaccharide. In this drawing, each hexagon represents a glucose monomer.* (b) *Starch is found in many foods, including pasta.*

Lipids

Another class of macromolecules made from carbon, oxygen, and hydrogen atoms is known as the **lipids.** Lipids are waxy, fatty, or oily compounds. Like carbohydrates, lipids can be used to store and release energy. But lipids have other uses as well.

Many lipids are formed by combining smaller compounds such as fatty acids and glycerol. Fatty acids are long chains of carbon and hydrogen acids that have an acidic carboxyl group (–COOH) attached to one end. Glycerol is a 3-carbon alcohol that contains three hydroxyl groups (–OH). Other lipids are based on a series of carbon rings, such as a compound called cholesterol.

☑ ***Checkpoint*** What are lipids?

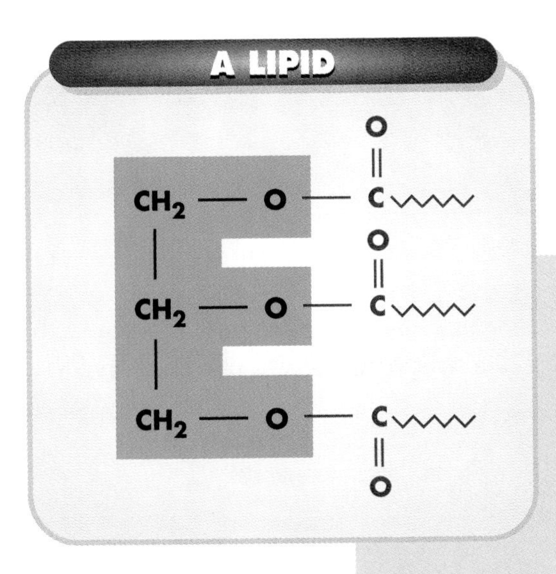

Proteins

An especially important class of organic compounds is the **proteins.** Proteins are polymers of molecules called **amino acids.** There are 20 common amino acids, but each one has a central carbon bonded to an amino group (–NH_2), a carboxyl group (–COOH), a hydrogen atom, and a fourth group abbreviated by the letter R. The R group is what makes the amino acids different from each other. In glycine, the R group is a hydrogen atom. In alanine, the R group is a methyl group (–CH_3).

Two amino acids can be joined together in a reaction between the amino group (–NH_2) of one amino acid and the carboxyl group (–COOH) of the other amino acid. The bond that forms from this reaction is called a peptide bond. For this reason, a chain of amino acids is often called a polypeptide. A complete protein contains one or more polypeptides, and it may contain a few other chemical groups.

Figure 2–9
A glycerol molecule can combine with three fatty acids to form a lipid. The fatty acid portion of the lipid contains long tails of carbon and hydrogen atoms, represented here as zigzag lines.

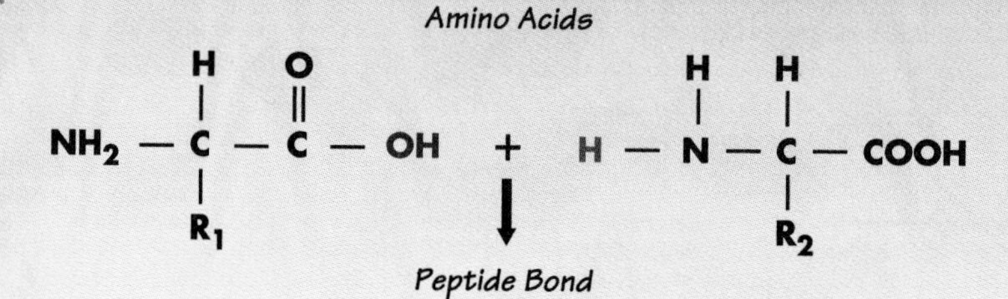

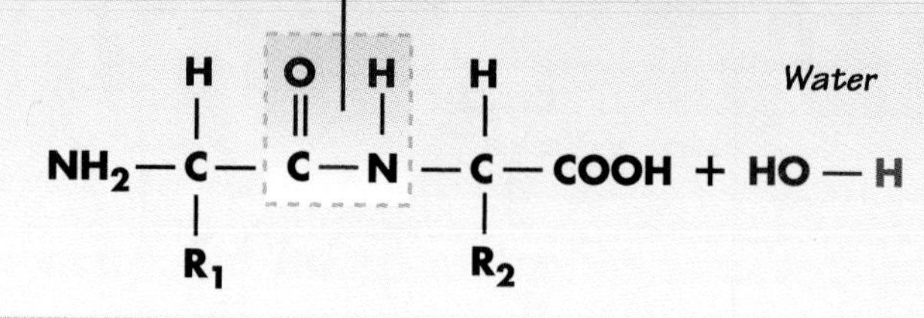

Figure 2–10
(a) The amino acids in proteins are joined by peptide bonds, which form between –COOH groups and –NH_2 groups. (b) Proteins are made linearly—one peptide bond after another. However, they bend and fold to form complex, three-dimensional structures.

Proteins are the principal components of a number of different structures, including feathers, skin, and muscles. They also help chemical reactions to proceed, pump small molecules in and out of cells, and they can even produce motion.

Checkpoint What are the monomers of proteins?

Figure 2–11
DNA is one type of nucleic acid. It consists of two strands twisted about each other, forming a double helix.

Nucleic Acids

The information-carrying molecules of the cell are the **nucleic acids.** The nucleic acids also are the molecules of inheritance—the molecules that parents pass on to their offspring.

There are two principal kinds of nucleic acids: DNA (deoxyribonucleic acid) and RNA (ribonucleic acid). As the beginnings of their names indicate, DNA contains the sugar deoxyribose and RNA contains the sugar ribose.

Nucleic acids are assembled from monomers called nucleotides. Each nucleotide contains three parts: a phosphate group, an organic compound known as a nitrogenous base, and a 5-carbon sugar—either ribose or deoxyribose. DNA and RNA each contain four different kinds of nucleotides. The arrangement of these nucleotides determines the information that the nucleic acid contains.

Section Review 2–2

1. **Describe** the special properties of carbon.
2. **Identify** the four major classes of macromolecules.
3. **Critical Thinking—Interpreting Data** Suppose that a cell lost its supply of simple sugars and its ability to make its own simple sugars. Which macromolecules could the cell no longer produce? Explain your answer.
4. **MINI LAB** When you added Lugol's solution to the soda cracker, what did you **observe?** What does your observation indicate about the soda cracker?

Chemical Reactions and Enzymes

GUIDE FOR READING

- **Describe** the function of enzymes.

INDIVIDUALLY, DRY CHOCOLATE cake mix, raw eggs, and vegetable oil are not especially tasty or good to eat. But if you were to mix them together, then bake the mixture in the oven, the result—a chocolate cake—would be delicious!

Of course, a living thing is not an oven, and the materials inside it are far more organized and specialized than the materials in cake batter. But in some respects, the processes involved in baking a cake are quite similar to processes that take place in living things.

Chemical Reactions

When you bake a cake, you are controlling a series of **chemical reactions.** A chemical reaction is a process that changes one set of substances into a new set of substances. In this example, the cake mix, eggs, and oil react with each other to produce a chocolate cake—a substance with far different properties from its individual ingredients.

Reactions in Living Things

Do chemical reactions take place in living organisms? Yes, absolutely! Every day, you take in the ingredients of food, water, and oxygen. You use these raw materials to produce energy, to build new cells, and to assemble important compounds. Your body uses a huge number of chemical reactions in each of these processes—and in many others.

$CO_2 + H_2O$

To better understand the chemical reactions that take place in living things, let's examine one important example. This example involves carbon dioxide (CO_2) and water (H_2O). Carbon dioxide is a waste product of cells. In humans and other animals, the blood transports carbon dioxide from the cells to the lungs, where it is eliminated from the body with every breath.

Figure 2–12
Flour, eggs, sugar, and other ingredients undergo a chemical reaction to become a new substance—a cake. Living things need to control the chemical reactions inside them, just as a chef needs to control the reactions in cake batter for the cake to come out just right.

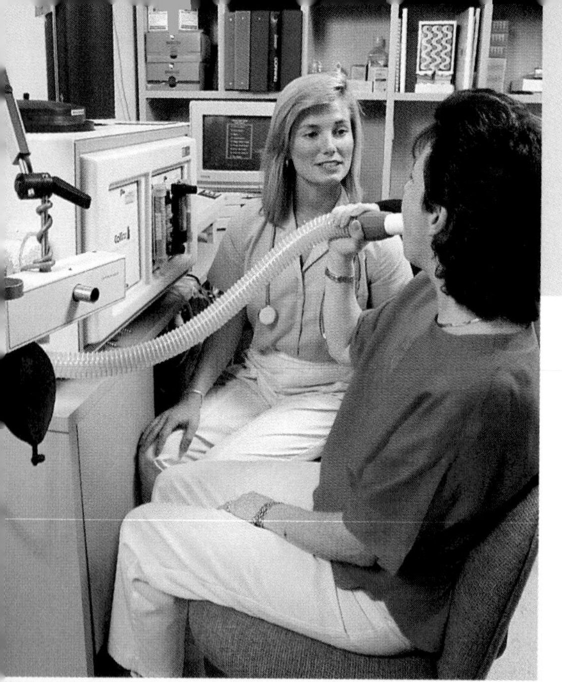

Figure 2–13
CAREER TRACK *A respiratory therapist works with people who have trouble breathing. The machine shown in this photograph analyzes the gases that a person exhales.*

When carbon dioxide enters the blood, it reacts with water to form a compound called carbonic acid (H_2CO_3). This reaction is described by the following equation:

$$CO_2 + H_2O \rightarrow H_2CO_3$$

This reaction is important because carbon dioxide dissolves only minimally in water, while carbonic acid dissolves in water to a much greater extent. By converting carbon dioxide to carbonic acid, the bloodstream is able to accept much more carbon dioxide than it could accept otherwise. Then, when the blood reaches the lungs, carbonic acid is rapidly converted back into carbon dioxide and water, and the carbon dioxide is exhaled from the lungs with every breath.

☑ **Checkpoint** Why is the reaction between carbon dioxide and water important?

Enzymes 

There is one problem with the reaction between carbon dioxide and water, however—it occurs very slowly. It occurs so slowly, in fact, that carbon dioxide would build up in the body faster than the bloodstream could take it away.

How does the body speed up this reaction? The answer is that it produces a molecule that serves as a **catalyst** for the reaction. A catalyst is a substance that speeds up a chemical reaction without itself being used up in the reaction.

In living things, the molecules that serve as catalysts are called **enzymes**—an extremely important group of organic molecules. **Enzymes catalyze almost every important chemical reaction in living things.** By speeding up chemical reactions, enzymes allow living things to carry out a dazzling collection of chemical feats—far more impressive than the reactions chemists create in their laboratories.

How Enzymes Work

Almost all enyzmes are proteins. The enzyme that catalyzes the reaction between carbon dioxide and water is a protein called carbonic anhydrase. In one second, a single molecule of carbonic anhydrase helps to form 600,000 molecules of carbonic acid. Carbonic anhydrase allows the reaction between carbon dioxide and water to occur 10 million times faster than it would occur otherwise.

How do enzymes perform such astounding feats? They do so by reducing the amount of energy it takes for reactions to occur.

Every chemical reaction involves making and breaking chemical bonds. Without an enzyme, the molecules of a reaction would need to collide with each other with enough energy to break existing bonds and form new ones. Enzymes, however, bind the reactions' components individually. This process requires much less energy and, therefore, allows the reaction to run faster.

Active Sites and Substrates

The components of the reaction that bind to the enzyme are called **substrates.** And the region of the enzyme where the substrates bind is called the **active site.**

Some enzymes bind their substrates at the active site in exactly the right orientation for the reaction to occur. Other

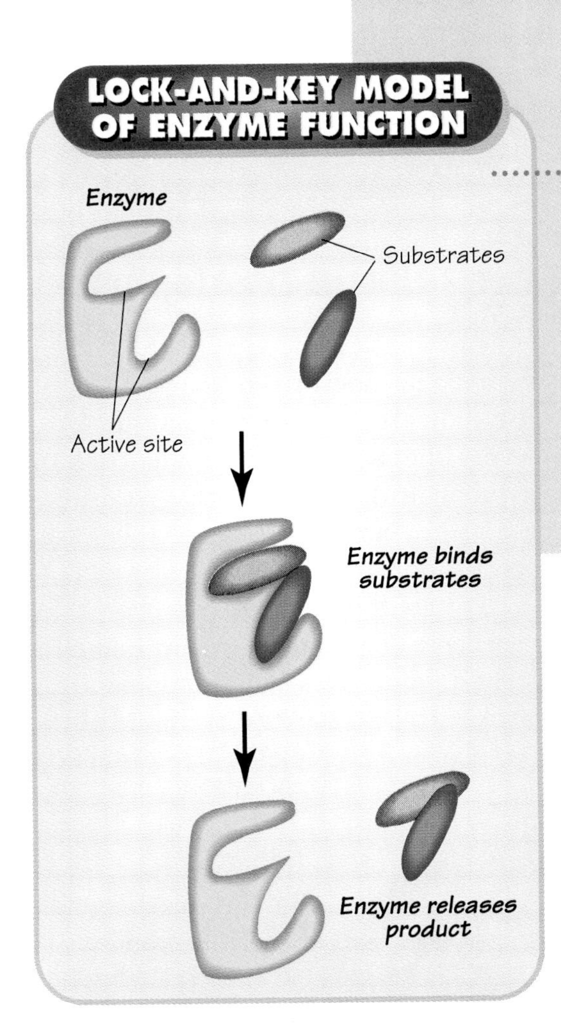

Figure 2–14 *(a) An enzyme binds its substrates at the active site, placing them in just the right position for a reaction to occur. (b) An enzyme and its substrates are often compared to a lock and key, because only substrates of the proper shape can fit in the enzyme's active site.*

enzymes bend or twist the substrates at the active site, straining or even breaking chemical bonds in order to start the reaction. When the reaction is complete, the reaction's products are released from the active site, the enzyme is free to bind new substrates, and the process can start all over again.

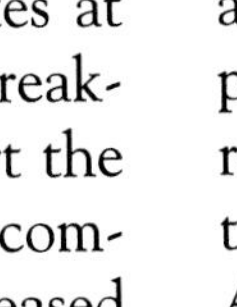

Interestingly, a single enzyme is able to catalyze one and only one chemical reaction. Why is this so? In general, the reason is because the shape of an active site precisely fits the shape of its substrates. In fact, scientists often compare an enzyme and its substrates to a lock and key, as shown in **Figure 2–14**.

☑ **Checkpoint** What is an active site? A substrate?

What Enzymes Do

Enzymes regulate chemical pathways, catalyze the synthesis and breakdown of important compounds, help a cell to store and release energy, and transfer information from one part of a cell to the next. As you will discover in your study of biology, enzymes are involved in digestion, respiration, reproduction, vision, movement, thought, and even in the making of other enzymes. The world of living things could not survive without them!

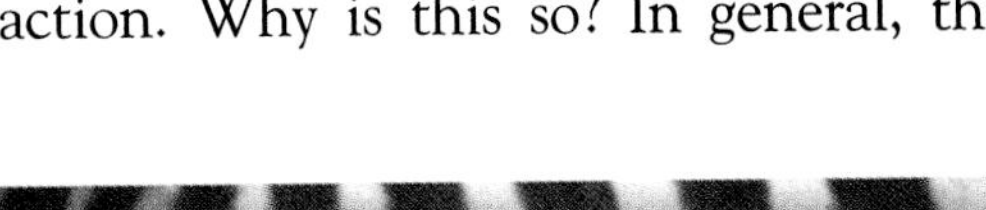

Section Review 2–3

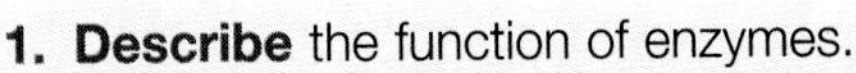

1. **Describe** the function of enzymes.
2. **Outline** the steps in which an enzyme catalyzes a reaction.
3. **Critical Thinking—Applying Concepts** Changing the temperature or pH changes an enzyme's shape. Describe how changing the temperature or pH would affect the function of carbonic anhydrase.

SECTION 2–4

BRANCHING OUT In Depth

Mirror-Image Molecules

Guide for Reading

- **Explain** how different molecules can have the same atoms and bonds.
- **Give examples** of the types of amino acids and sugars found in nature.

MINI LAB

- **Construct a model** of two mirror-image molecules.

PROTEINS HAVE A COMPLEX, three-dimensional structure, and this structure is very important to their function. The active site of an enzyme, for example, needs to have exactly the right shape to bind its substrates.

As scientists looked closely at the three-dimensional structures of amino acids, sugars, and other organic molecules, they discovered some very surprising facts about them. You will discover some of these surprises in this section.

Two Types of Tryptophan

Figure 2–15 shows the structure of tryptophan, one of the amino acids. It also shows the structure of another molecule, one that is almost identical to tryptophan. In fact, the two molecules have the same atoms and the same bonds, and they look almost identical.

How are they different? They are mirror images of each other! **Many complex organic molecules have nonidentical mirror images.** The two types of tryptophan are examples of **stereoisomers.** Stereoisomers have the same atoms and the same bonds, but the atoms are oriented differently in space.

Why are the two mirror images of tryptophan not the same? To answer this question, think about a pair of mirror-image objects in every-day life—such as a pair of gloves. A pair of gloves consists of a right-handed glove

(a)

TWO MIRROR-IMAGE MOLECULES

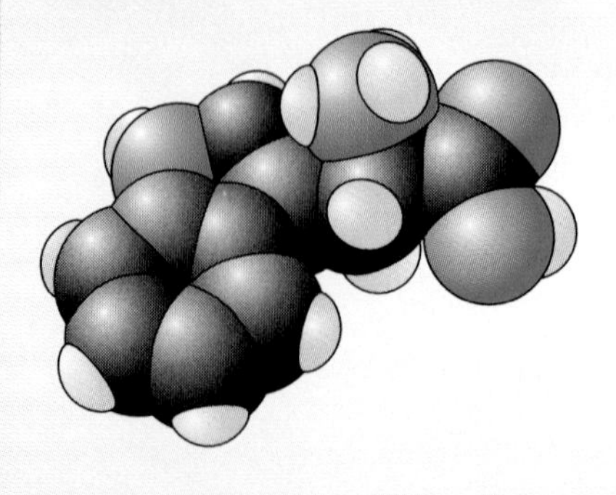

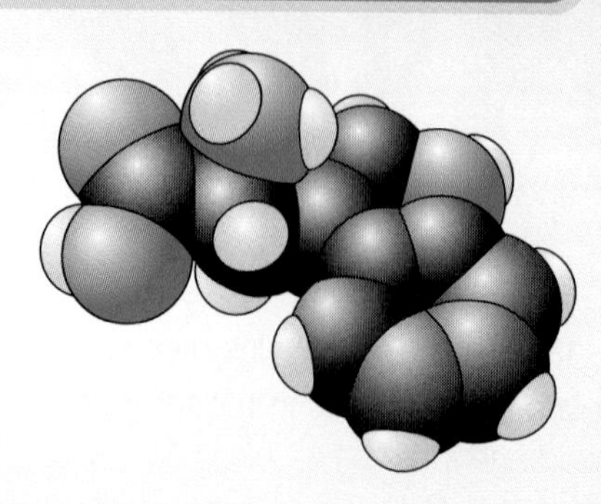

(b)

Figure 2–15 (a) *Like this fox—and many complex three-dimensional objects—some molecules have nonidentical mirror images.* (b) *These computer-generated models represent L-tryptophan and D-tryptophan, two mirror-image molecules.*

and a left-handed glove, and each glove is the mirror image of the other. Despite the fact that a left-handed glove is almost the same as a right-handed one, no matter how you turn or twist the left-handed glove, you cannot make it into a right-handed one.

The two types of tryptophan are related to each other in the same way. In fact, chemists describe them as left-handed and right-handed molecules. One type is called L-tryptophan. The letter L stands for *levo*, the Latin word for left-handed. The other type is called D-tryptophan. The letter D stands for *dextra*, the Latin word for right-handed.

Checkpoint Why is a pair of gloves like L-tryptophan and D-tryptophan?

A Mirror-Image Puzzle

Surprisingly, with very few exceptions, the amino acids found in living organisms are all of one type—the L type. Like a left-handed glove—one that fits only a left hand—the enzymes in living cells will fit only left-handed amino acids.

Sugars also exist in mirror-image forms. Cells, however, use and produce only right-handed sugars. The 6-carbon sugar D-glucose is one of the most common sugars in the cell. But although L-glucose can be made in the laboratory, it is not used in nature. It's almost as though living organisms decided many hundreds of millions of years ago to use only one type of amino acid and one type of sugar.

Mirror, Mirror

PROBLEM ***How would you construct a model** of two mirror-image molecules?*

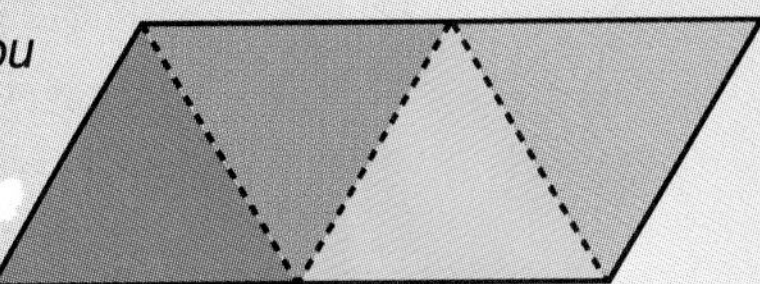

PROCEDURE

1. On a sheet of construction paper, draw the four colored triangles in the arrangement shown. The sides of the triangles should be the same length and be at least 10 cm.
2. Cut the paper along the outer edge of the arrangement.
3. Fold along the borders of the triangles to create a three-dimensional shape called a tetrahedron. Tape together the sides of the tetrahedron.
4. Construct a second tetrahedron by switching the positions of the blue and green triangles in the original arrangement and repeating steps 1 to 3.

ANALYZE AND CONCLUDE

1. Compare the two tetrahedrons you created. How are they similar? Different? Explain your answers.
2. How do the two tetrahedrons relate to the structures of D-tryptophan and L-tryptophan?

Section Review 2–4

1. **Explain** how two molecules can have the same atoms and bonds, yet still be different.
2. **Give examples** of the stereoisomers of amino acids and sugars that are found in living things.
3. **MINI LAB** How would you **construct a model** of mirror-image molecules?
4. **BRANCHING OUT ACTIVITY** **Write** a science-fiction story about a planet where the molecules of life are the mirror images of those on Earth. In your story, explore the problems a visitor from Earth might encounter on this planet.

CHAPTER 2

Laboratory Investigation

Three, Two, One . . . BLASTOFF!

The enzyme catalase speeds up the breakdown of hydrogen peroxide (H_2O_2) into water (H_2O) and oxygen gas (O_2). The reaction is described by the following equation:

$$2\ H_2O_2 \rightarrow 2\ H_2O + O_2$$

Problem

How does the concentration of an enzyme affect the rate of a reaction? Perform a controlled experiment to **draw a conclusion** about the function of enzymes.

Materials (per group)

potato extract solution
1% hydrogen peroxide solution
8 50-mL beakers
distilled water
filter paper disks
forceps
paper towels
glass-marking pencil

Solution	Potato Extract	Distilled Water
0% Potato extract	0 mL	20 mL
25% Potato extract	5 mL	15 mL
50% Potato extract	10 mL	10 mL
75% Potato extract	15 mL	5 mL
100% Potato extract	20 mL	0 mL

Procedure

1. **Catalase is found in potato extract. Using 5 of the 50-mL beakers, prepare the 5 solutions of potato extract that are described in the table shown. Label each beaker to indicate the percentage of potato extract in the solution.**
2. **Label each of the remaining 3 beakers H_2O_2. Pour 25 mL of the 1% hydrogen peroxide solution into each beaker.**
3. **Using the forceps, dip a filter paper disk into the beaker labeled 0% potato extract. Keep the disk in the solution for 4 seconds, then remove it.**
4. **Place the disk on a paper towel for 4 seconds to remove any excess liquid.**
5. **Using the forceps, transfer the filter paper disk to the bottom of one of the beakers labeled H_2O_2. The enzyme in the potato extract catalyzes the formation of bubbles of oxygen gas, which causes the disk to rise to the surface.**
6. **Release the filter paper disk. Have one person in your group measure how long it takes for the bubbles to carry the disk to the top of the beaker. Record the time in a data table similar to the one shown on the next page.**

7. Repeat step 6 two more times, using the other two beakers labeled H_2O_2.
8. Repeat steps 3 to 7 for each of the four remaining potato extract solutions.
9. Calculate the average rising time for each of the potato extract solutions. Record this information in your data table.

DATA TABLE

Beaker	Rising Time Trial 1	Rising Time Trial 2	Rising Time Trial 3	Rising Time Average
0% Potato extract				
25% Potato extract				
50% Potato extract				
75% Potato extract				
100% Potato extract				

Observations

Construct a graph that plots the concentration of potato extract (on the X axis) versus the average rising time (on the Y axis).

Analysis and Conclusions

1. Suppose you had dipped a filter paper disk in a 30% potato extract solution. Using the graph, predict how long it would take this disk to rise to the top of a beaker of H_2O_2.
2. How does the concentration of the enzyme affect the rate of the breakdown of hydrogen peroxide? Use the results of this experiment to justify your answer.

More to Explore

Design an experiment to show how the concentration of hydrogen peroxide affects the rate of its breakdown.

Study Guide

Summarizing Key Concepts

The key concepts in each section of this chapter are listed below to help you review the chapter content. Make sure you understand each concept and its relationship to other concepts and to the theme of this chapter.

2–1 Introduction to Chemistry

- The key difference among the atoms of different elements is their number of protons and electrons.
- The links between atoms are called chemical bonds. In ionic bonds, electrons are transferred. In covalent bonds, electrons are shared.
- Water is a polar molecule, which allows water to dissolve ionic substances such as NaCl.
- The pH of a water solution indicates its concentration of H^+ ions and OH^- ions. Acidic solutions have a low pH, and basic solutions have a high pH.

2–2 The Compounds of Life

- Carbon atoms form the backbone of every large molecule found in living organisms. Most carbon compounds are classified as organic compounds.
- A polymer is a large molecule made of repeated units called monomers. Macromolecules are large organic polymers.
- The four major classes of macromolecules are the carbohydrates, lipids, proteins, and nucleic acids.

2–3 Chemical Reactions and Enzymes

- Enzymes catalyze almost every important chemical reaction in living things. Almost all enzymes are proteins.
- Enzymes bind substrates, or the components of a reaction, at a region called the active site.

2–4 Mirror-Image Molecules

- Many complex organic molecules have nonidentical mirror images.
- Stereoisomers have the same atoms and the same bonds, but the atoms are oriented differently in space.

Reviewing Key Terms

Review the following vocabulary terms and their meaning. Then use each term in a complete sentence.

2–1 Introduction to Chemistry

atom
proton
neutron
electron
chemical compound
ion
ionic bond
covalent bond
molecule
solution
acid
base
pH scale

2–2 The Compounds of Life

organic compound
inorganic compound
polymer
monomer
macromolecule
carbohydrate
lipid
protein
amino acid
nucleic acid

2–3 Chemical Reactions and Enzymes

chemical reaction
catalyst
enzyme
substrate
active site

2–4 Mirror-Image Molecules

stereoisomer

Recalling Main Ideas

Choose the letter of the answer that best completes the statement or answers the question.

1. An atom's nucleus contains protons and

a. neutrons. **c.** ions.
b. electrons. **d.** elements.

2. Atoms of different elements always contain different numbers of

a. neutrons. **c.** ions.
b. protons. **d.** nuclei.

3. The three atoms in water form a(an)

a. straight line. **c.** angle of about 105°.
b. angle of 90°. **d.** ring.

4. Which pH indicates an acidic solution?

a. 10 **c.** 7
b. 8 **d.** 5

5. The backbone of large organic molecules is made mostly of

a. carbon. **c.** nitrogen.
b. hydrogen. **d.** oxygen.

6. To which class of macromolecules do sugars belong?

a. carbohydrates **c.** proteins
b. lipids **d.** nucleic acids

7. Which macromolecules are waxy or oily?

a. carbohydrates **c.** proteins
b. lipids **d.** nucleic acids

8. The role of enzymes is to

a. slow down reactions.
b. inhibit reactions.
c. speed up reactions.
d. store energy.

9. D-glucose and L-glucose are examples of

a. enzymes. **c.** polysaccharides.
b. stereoisomers. **d.** amino acids.

Putting It All Together

Using the information on pages xxx to xxxi, complete the following concept map.

THE COMPOUNDS OF LIFE

have backbones of

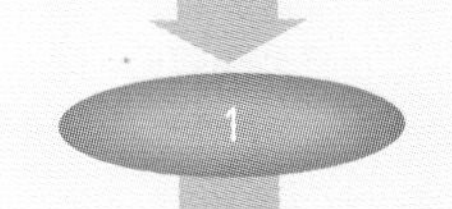

form large molecules called

2 include four main classes

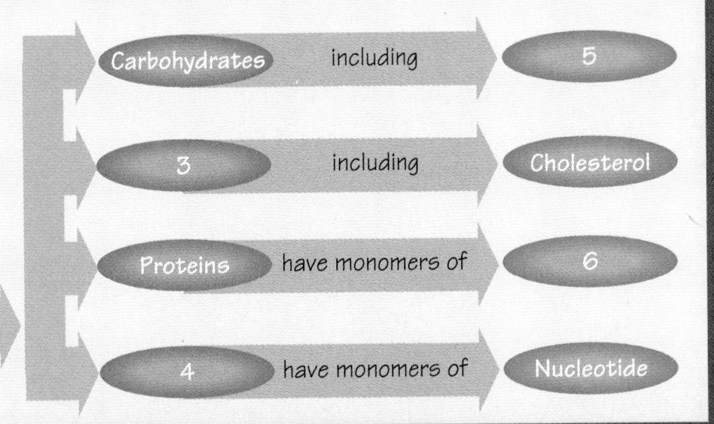

Assessment

Reviewing What You Learned

Answer each of the following in a complete sentence.

1. What is an atom?
2. Identify the three fundamental particles that make up an atom.
3. What does the formula H_2O indicate about water?
4. Why is water a polar molecule?
5. Describe the difference between an atom and an ion.
6. Define the terms acid and base.
7. Give examples of an acidic pH, a neutral pH, and a basic pH.
8. What is an organic molecule?
9. What is a polymer?
10. Identify two different types of lipids.
11. What are the monomers of a nucleic acid?
12. How is carbon dioxide eliminated from the human body?
13. What is the active site of an enzyme?
14. Give an example of a pair of mirror-image molecules.

Expanding the Concepts

Discuss each of the following in a brief paragraph.

1. Describe the way that bonds form between small atoms such as hydrogen, carbon, nitrogen, and oxygen.
2. Compare single, double, and triple covalent bonds.
3. Explain the difference between an ionic bond and a covalent bond.
4. Could a chemical formula, such as $C_5H_{10}O_5$, indicate more than one chemical compound? Explain your answer.
5. Why is water able to dissolve sodium chloride but not cooking oil?
6. Give examples of an important acid and an important base in living things.
7. Why is carbon able to form long chains of atoms?
8. Identify three ways in which organisms use carbohydrates.
9. Describe how amino acids are joined together in a protein.
10. How are the four classes of macromolecules similar? Different?
11. Compare enzymes with other catalysts.
12. Describe the lock-and-key model of enzyme function.
13. Discuss the differences between D- and L- amino acids.

Extending Your Thinking

Use the skills you have developed in this chapter to answer the following.

1. **Drawing conclusions** A nitrogen atom contains 7 electrons. Describe the bond between the 2 nitrogen atoms in a molecule of nitrogen gas (N_2).
2. **Predicting** Certain lipid molecules have regions that are polar and regions that are nonpolar. Predict how these lipid molecules would behave in water.
3. **Constructing formulas** At a pH of 7, the –COOH end of an amino acid loses an H^+ ion, while the $-NH_2$ end gains an H^+ ion. Draw the structure of this form of an amino acid and label any charged ends. Has the amino acid acquired a net charge?
4. **Analyzing** A student adds a small amount of the enzyme catalase to a solution of hydrogen peroxide (H_2O_2). The solution produces bubbles of oxygen gas for 10 minutes, then the bubbling stops. Next, the student adds more catalase to the solution. Will more bubbles of oxygen gas be produced? Explain your answer.
5. **Recognizing patterns** Look at the structure of a molecule of acetone. Is acetone found in distinct left- and right-hand versions? Explain.

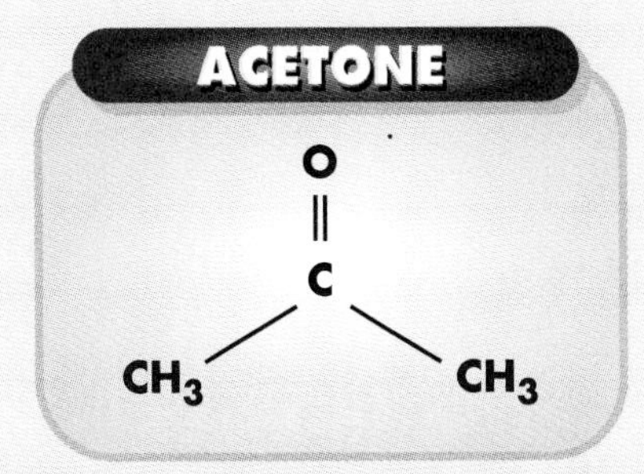

6. **Observing** When you look at a soda cracker with the unaided eye, can you observe the carbohydrates that compose it? Explain why or why not.
7. **Constructing a model** Construct a model of a carbon atom. Label the nucleus, protons, neutrons, and electrons in your model. In what ways is your model accurate? In what ways is it inaccurate?

Applying Your Skills

Read Those Labels!

Carbohydrates, proteins, and lipids are a part of every healthful diet. Just how much of these compounds are in the foods you eat? Perform this activity to find out.

1. Select ten packaged foods that are a part of your diet.
2. For each product, record the serving size and the total grams of carbohydrates, proteins, and lipids (or fats) in each serving. This information is printed on the product's packaging, typically in a table called Nutrition Facts.
3. Which foods provide the most carbohydrates, proteins, and lipids? Which provide the least?

• GOING FURTHER •

4. Bring to class the packaging containing the nutrition facts for your favorite snack. Then construct a table that lists the number of grams of carbohydrates, proteins, and lipids for each snack that you and your classmates brought. Which snack do you suspect is the most healthful? The least healthful? Explain your answers.

CHAPTER 3

Cell Structure and Function

FOCUSING THE CHAPTER THEME: *Scale and Structure*

3–1 Microscopes and Cells
- **Define** the cell theory.

3–2 Cell Boundaries
- **Describe** the cell membrane and cell wall.

3–3 Inside the Cell
- **Identify** the different organelles of the cell and their functions.

BRANCHING OUT *In Depth*

3–4 The Origin of the Eukaryotic Cell
- **Explain** the endosymbiont hypothesis.

LABORATORY INVESTIGATION
- **Compare** the characteristics of plant cells and animal cells.

Biology and Your World

BIO JOURNAL

The photograph on this page shows cells from a human intestine. In your journal, describe the structures inside the cells. In what ways are the cells similar? In what ways are they different?

Cells in human small intestine (magnification: 800X)

SECTION 3–1

Microscopes and Cells

Guide for Reading

- **Define** the cell theory.
- **Identify** the magnification powers of different types of microscopes.

MINI LAB

- **Observe** how the compound microscope changes an image.

THE TYPICAL CELL IS ONLY about 10 micrometers wide—far too small for the unaided eye to see. So how do biologists study cells? Although biologists study cells with a variety of different tools, arguably their most important tool is the microscope. A microscope is an instrument that produces magnified images of tiny structures. Without microscopes, we might never have discovered cells at all!

Early Microscopes

In the 1600s, Dutch businessman Anton van Leeuwenhoek discovered another purpose for the glass lenses he was grinding. By placing several magnifying lenses at the proper distances from each other, Leeuwenhoek created instruments that could produce magnified images of very, very small objects. These instruments were some of the world's first microscopes. One of Leeuwenhoek's microscopes is illustrated in **Figure 3–1.**

INTEGRATING PHYSICS

Why can lenses magnify an image?

Leeuwenhoek's microscopes are, in fact, examples of the most familiar type of microscope, the light microscope. In a light microscope, rays of light are bent through lenses to produce an enlarged image.

Discovery of Cells

Leeuwenhoek used his light microscopes to look at drops of pond water and other liquids. To his amazement, he discovered that many

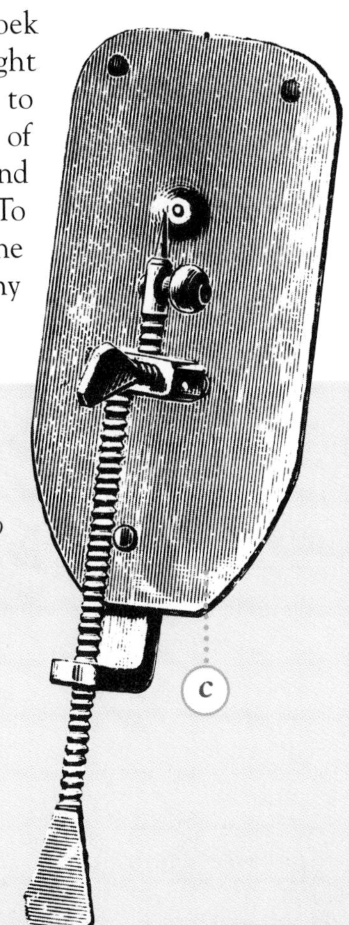

Figure 3–1
(a) *Anton van Leeuwenhoek was the first person to use a microscope to identify living things.* (b) *Robert Hooke used his microscope to study cork, and he made these drawings of his observations. Hooke was looking not at living cork cells but at the cell walls of dead cells.* (c) *Although Leeuwenhoek's microscope may look simple or primitive, it could magnify objects a few hundred times.*

Figure 3–2
The cell theory applies to all organisms—from ⓐ *a koala and eucalyptus tree in Australia to* ⓑ *a saguaro cactus in Arizona.*

of these liquids were filled with tiny living things. He called these tiny things *animalcules*, meaning "little animals."

At about the same time, Robert Hooke, an English physicist, used a microscope to observe flowers, insects, spider webs, and slices of cork. In 1665, Hooke published a book of drawings of his observations. He pointed out that the woody parts of plants contained tiny rectangular chambers, which he called **cells.** Hooke chose this name because the chambers reminded him of the tiny rooms in a monastery, which are also called cells.

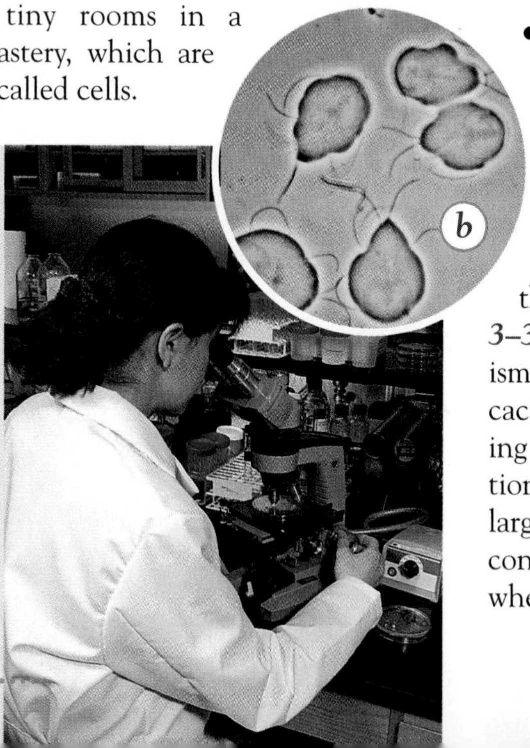

Figure 3–3
CAREER TRACK
ⓐ *Cytotechnologists typically study cells by preparing slides, then examining the slides under a compound microscope.* ⓑ *A compound microscope produced this image of* Giardia lamblia, *which cause an intestinal disease (magnification: 100X).*

Hooke believed that only plants were made of cells—an idea that scientists did not challenge for nearly 200 years. But in 1839, the German biologist Theodor Schwann found that some animal tissues closely resembled the cellular tissues of plants. As Schwann looked at animal tissues with better microscopes, he gradually came to the conclusion that animals too were made of cells.

The Cell Theory

In the meantime, Robert Brown, a Scottish biologist, had discovered an object near the center of many cells, a structure now called the nucleus. German biologist Matthias Schleiden expanded on Brown's work, suggesting that the cell's nucleus plays a role in cell reproduction. In 1855, German physician Rudolf Virchow further studied cell reproduction. Virchow proposed that animal and plant cells are produced only by the division of cells that already exist.

The discoveries and observations of these scientists make up what is now called the **cell theory.** The cell theory forms a basis for the way biologists study living things. **The cell theory states:**

- **All living things are composed of cells.**
- **Cells are the smallest working units of living things.**
- **All cells come from preexisting cells by cell division.**

The cell theory applies to all organisms. Some organisms, such as the *Giardia lamblia*, shown in **Figure 3–3,** contain only one cell. Other organisms, such as the koala and the saguaro cactus, contain millions of cells—all acting together to help the organism function as a single unit. Yet no matter how large the organism or how many cells it contains, each of its cells was produced when another cell divided in two. It is interesting to note that, in most cases, an organism begins its life as nothing more than a single cell.

Checkpoint What is the cell theory?

Modern Microscopes

It is not surprising that biologists learned more and more about the cell as better and more powerful microscopes were developed. Let's take a look at some of the different kinds of microscopes that biologists use today.

Compound Light Microscope

The most common and familiar type of microscope is the **compound light microscope**—or compound microscope, for short. The word compound indicates that the microscope contains more than one lens. You probably will use this type of microscope in your biology classroom or laboratory.

Most compound microscopes can magnify an image up to 1000 times. This makes the compound microscope useful for studying many kinds of cells and small organisms. In addition, these cells and organisms can sometimes be studied while they are still alive.

Electron Microscope

In the 1920s, German physicists discovered a way to use magnets to focus a beam of electrons, similar to the way a glass lens focuses a beam of light. They used this discovery to build a device with which you are already familiar—the television set. In addition, their work led to the development of the **electron microscope.** An electron microscope uses a beam of electrons instead of light to examine a sample.

An electron microscope can magnify images as much as 1000 times larger than a light microsope can magnify them. For this reason, the electron microscope can show much smaller structures than an ordinary light microscope can reveal.

Checkpoint What is an electron microscope?

Is Seeing Believing?

PROBLEM *What affects the image that you* ***observe*** *through a compound microscope?*

PROCEDURE 

1. Obtain a slide with a typewritten label. Place the slide on the stage of a compound microscope. Use the stage clips to hold the slide in place.
2. Use the low-power objective to bring the letters on the label into focus. Record the image that you see.
3. While looking through the eyepiece, slowly move the slide in different directions along the stage. Then record your observations.
4. Switch to the high-power objective and record the image that you see.

ANALYZE AND CONCLUDE

1. In what ways did the compound microscope alter the image of the letters?
2. How did moving the slide affect the image?
3. What are the advantages of using the high-power objective? What are the disadvantages?

Types of Electron Microscopes

There are two basic kinds of electron microscopes. A transmission electron microscope (TEM) shines a beam of electrons through a sample, then magnifies the image onto a fluorescent screen. A scanning electron microscope (SEM) uses a thin beam of electrons to scan a sample's surface. The SEM collects the electrons that bounce off the sample, then forms an image on a television screen.

INTEGRATING PHYSICS

How does a television screen change a beam of electrons into a picture?

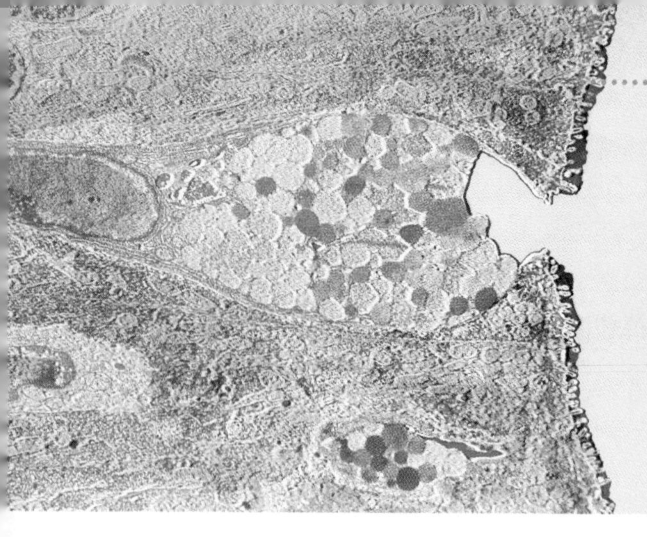

Figure 3–4
(a) A transmission electron microscope produced this photograph of cells in the human small intestine. A computer added color to the photograph—the actual cells do not have the colors shown here (magnification: 2400X). (b) A scanning electron microscope produced this false-color photograph of pollen from a daisy (magnification: 4200X).

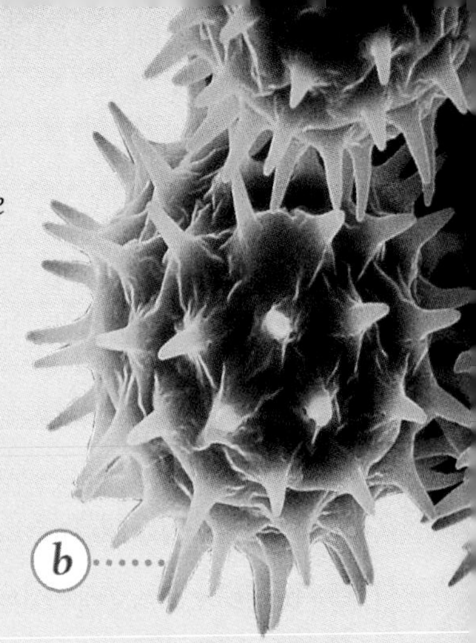

Although electron microscopes are powerful, they do have limitations. The sample must be kept in a vacuum, which is a space without air, because collisions between the electron beam and air particles would blur the image. As a result, electron microscopes can be used to study only nonliving specimens.

In addition, specimens for a TEM generally must be cut into very thin slices so that the electron beam can pass through them. And while the SEM produces realistic, often dramatic pictures of a sample's surface, it does not reveal the internal structure of a sample.

Scanning Probe Microscope

In the 1980s, researchers developed a new class of microscope that does not use lenses of any kind to produce images. Because these microscopes trace the surfaces of a sample with a tiny tip known as a probe, they are called **scanning probe microscopes.**

Scanning probe microscopes have revolutionized the study of surfaces. They have even produced pictures of individual atoms and molecules, as shown in ***Figure 3–5.*** Unlike electron microscopes, scanning probe microscopes do not require that specimens be placed in a vacuum. Researchers are eagerly searching for the best ways to apply these powerful new instruments in biology.

Figure 3–5
Scanning probe microscopes can provide pictures of very small objects—including atoms and molecules. This false-color photograph shows the surfaces of lipid molecules (magnification: 10,000,000X).

Section Review 3–1

1. **Define** the cell theory.
2. **Identify** the magnification powers of the compound light microscope and the electron microscope.
3. How do different kinds of microscopes **compare?**
4. **Critical Thinking—Sequencing** Describe the events that led to the development of the cell theory. If microscopes had not been invented, do you think the cell theory would have been developed? Explain.
5. **MINI LAB** Compare the image you **observe** through a compound microscope with the object on the stage of the microscope.

Cell Boundaries

SECTION 3–2

Guide for Reading

- **Discuss** the roles of the cell membrane and cell wall.
- **Describe** passive transport and active transport.

WHEN POWERFUL MICROSCOPES are used to look at cells, do all the cells look alike? The answer is no, they do not. In fact, two cells can differ as much from each other as can two completely different organisms.

However, even the most distinctive cells have some structures in common, as you will see as you read this chapter. Let's begin by looking at the structures that enclose and contain the cell.

Cell Membrane

Every cell has a **cell membrane** along its outer boundary. **The principal role of the cell membrane is to separate and protect the cell from its surroundings.** In this sense, the cell membrane is similar to the walls that surround and protect a house or the walls that divide the individual apartments in an apartment building.

However, a cell could not survive if its cell membrane were an ordinary wall or container. To stay alive, every cell must take in raw materials and eliminate waste products. This means that the cell membrane must allow certain substances to permeate it, or pass through it. For this reason, the cell membrane is described as selectively permeable.

Why does the cell membrane allow some substances to pass through but deny passage to other substances? To answer this question, let's first look at the cell membrane at the molecular level.

Figure 3–6
Cells in the same organism can be very different from each other. (a) This human nerve cell has long, spindly projections (magnification: 924X). (b) Human fat cells, on the other hand, are round and bloblike (magnification: 175X).

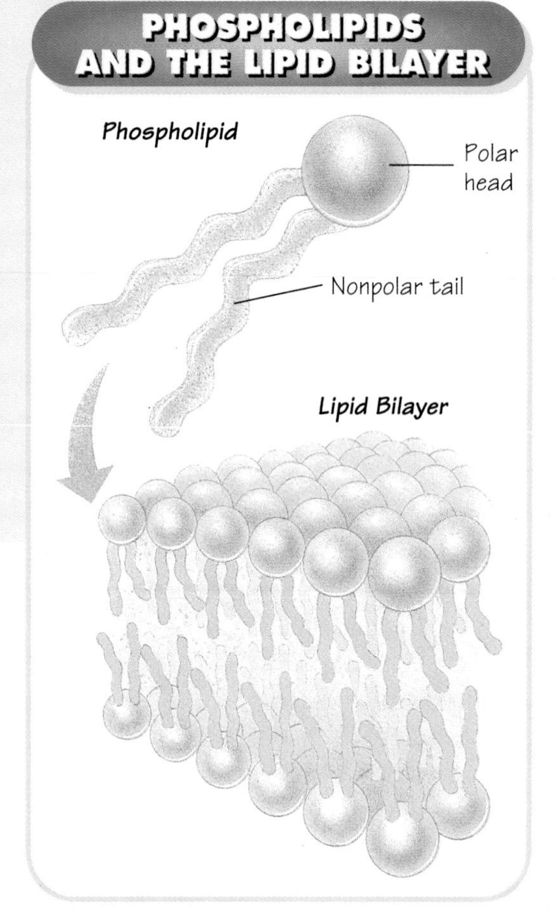

Figure 3–7
In water, a group of phospholipids forms a structure called a lipid bilayer. The polar heads are found on the outside of the lipid bilayer, and the nonpolar tails are found on the inside.

Lipid Bilayer

Cell membranes are built around a core of lipid molecules, including ones known as phospholipids. As shown in ***Figure 3–7,*** a phospholipid has a polar end called the head and a nonpolar end called the tail. Recall that polar substances tend to attract water and nonpolar substances tend to avoid water. As a result, a collection of phospholipids in water is typically found in a double-layered pattern.

This pattern is known as the **lipid bilayer.** The polar heads group together on the outside of the lipid bilayer because they are attracted to water. The nonpolar tails group together on the inside of the lipid bilayer because they avoid water. Because water is the main component of all cells, it is perhaps not surprising that all cell membranes contain a lipid bilayer.

The lipid bilayer provides cell membranes with a tough, flexible barrier that effectively protects the cell from many

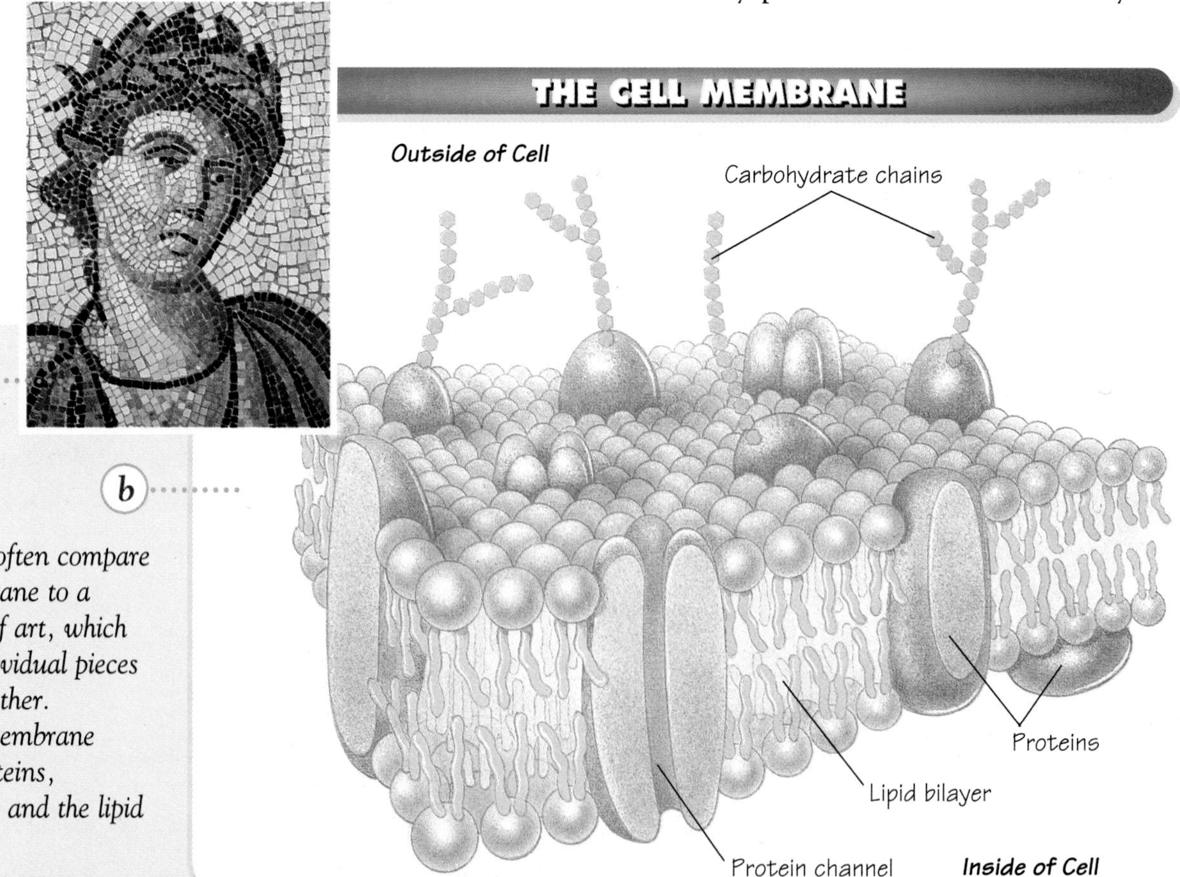

Figure 3–8
(a) Biologists often compare the cell membrane to a mosaic work of art, which is made of individual pieces assembled together.
(b) The cell membrane consists of proteins, carbohydrates, and the lipid bilayer.

Figure 3–9
Tough, thick cell walls surround the cells in this cross section of a pine stem. Cell walls are found in plants, algae, and some bacteria, but not in animals (magnification: 100X).

substances—although not all of them. In general, substances that dissolve in lipids can easily pass through the cell membrane, but lipid-insoluble substances cannot pass through. You will discover the consequences of this fact as you read on in this section.

☑ **Checkpoint** What is the lipid bilayer?

Other Cell Membrane Components

Lipids aren't the only molecules that make up the cell membrane, however. Most cell membranes have proteins embedded in the lipid bilayer, and many of these proteins have carbohydrates or other types of molecules attached to their outer surfaces.

What roles do these different molecules play? Some proteins form channels and pumps that help to move material across the cell membrane. Other proteins protect the membrane. And many of the carbohydrates act as chemical identification cards, allowing the organism to recognize which cells belong in the organism and which are foreign.

☑ **Checkpoint** What other molecules make up the cell membrane?

Cell Wall

In many organisms—including plants, algae, and bacteria—a **cell wall** is located outside the cell membrane. Animal cells, however, do not have cell walls. This is one of the important differences between plant cells and animal cells.

The cell wall helps to support and protect the cell. Unlike the walls of your house or apartment, however, most cell walls are very porous. They allow water, gases, and other substances to pass through easily.

Most cell walls are made of fibers of carbohydrate and protein. In the cell walls of plants, the principal carbohydrate is cellulose. Cellulose is a tough, flexible compound that gives plant cells much of their strength and rigidity. Cellulose is also the main component of both wood and paper. This means that you are looking at the cell walls of tree trunks as you read this page!

☑ **Checkpoint** What is the cell wall?

Passive Transport

How does the cell transport substances across the cell membrane? In a kind of transport known as **passive transport,** the substances literally transport themselves. **In passive transport, substances cross the cell membrane without the cell expending energy.**

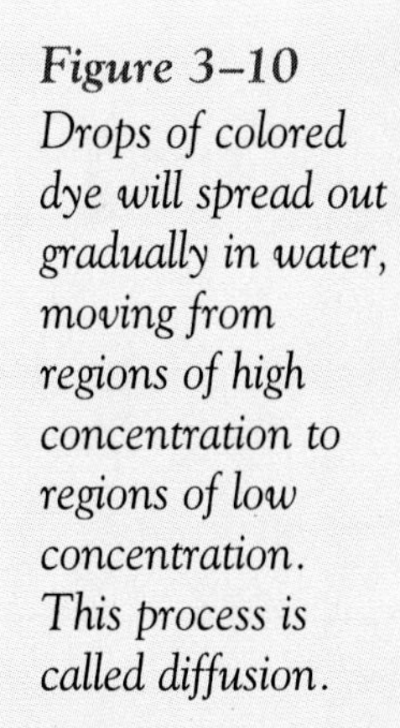

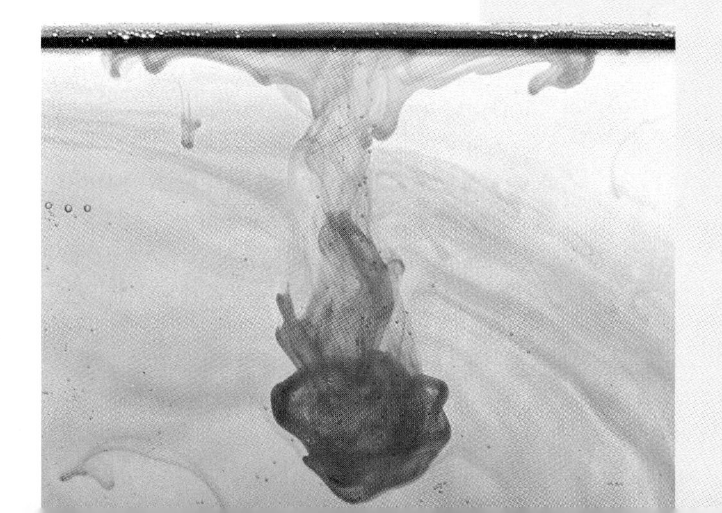

Figure 3–10
Drops of colored dye will spread out gradually in water, moving from regions of high concentration to regions of low concentration. This process is called diffusion.

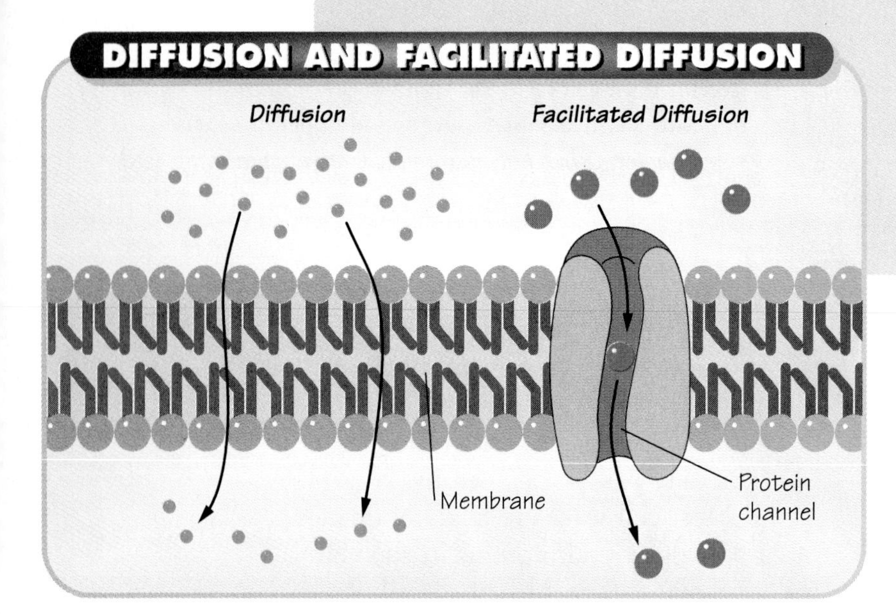

Figure 3–11
Some substances can diffuse directly across the lipid bilayer. Other substances can diffuse only through special protein channels, a process called facilitated diffusion.

Diffusion

Suppose you walk into a crowded room with four of your friends. If each of you is free to go your own way, before long you will spread out into the crowd and lose track of each other.

This process also happens in liquids—only at the molecular level. Because molecules in a liquid do not have fixed positions, a group of molecules that are close together in one instant will generally not be together for very long.

The random movement of molecules causes **diffusion.** Diffusion is the process by which substances spread through a liquid or gas. In diffusion, substances move from regions of high concentration to regions of low concentration. For example, when a drop of dye is added to water, the dye diffuses through the water until it becomes evenly distributed, as shown in ***Figure 3–10*** on page 55.

What happens if a membrane is placed in the solution? In one sense, nothing changes—diffusion still causes substances to move from regions of high concentration to regions of lower concentration. Substances such as alcohol, water, and small lipids easily cross the lipid bilayer, and so they diffuse directly across the cell membrane.

Checkpoint What is diffusion?

Facilitated Diffusion

Many membranes contain protein channels that allow some substances to pass through. The cell membranes of red blood cells, for example, contain a protein channel that allows free passage to one substance and one substance only—glucose. As a result, glucose can diffuse either into or out of the cell through this channel.

Because the diffusion of glucose is facilitated, or helped, by the protein, this process is called facilitated diffusion.

Figure 3–12
This membrane allows water but not sugar to pass through it. Thus, water moves into the sugar solution by the process of osmosis.

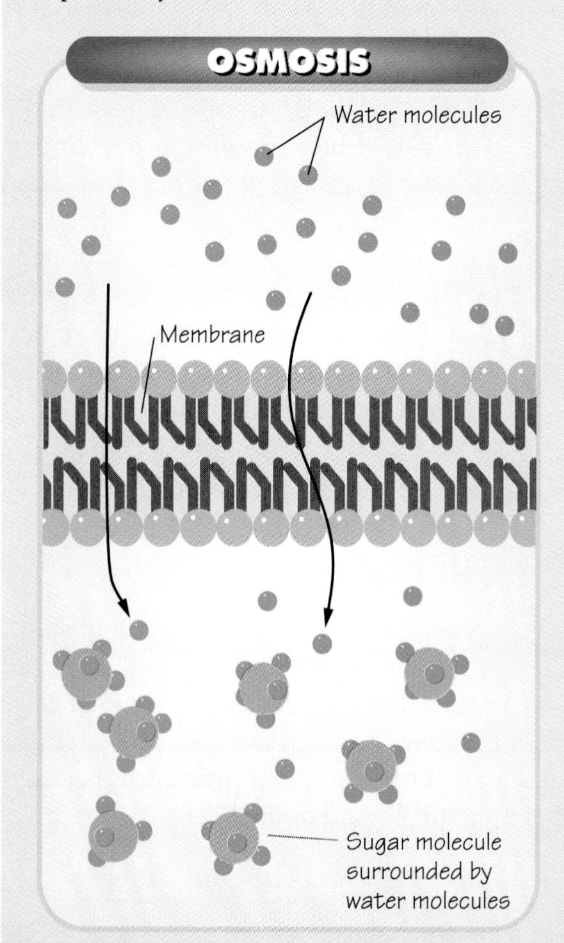

Figure 3–13
Red blood cells have different volumes in solutions of different concentrations. (a) Water moves into red blood cells by osmosis when the surrounding solution is hypotonic, or less concentrated than the cells. (b) Water moves out of red blood cells by osmosis when the surrounding solution is hypertonic, or more concentrated than the cells. (c) In an isotonic solution—a solution of equal concentration with the cells—there is no net movement of water across the membranes (magnification of each: 11,500X).

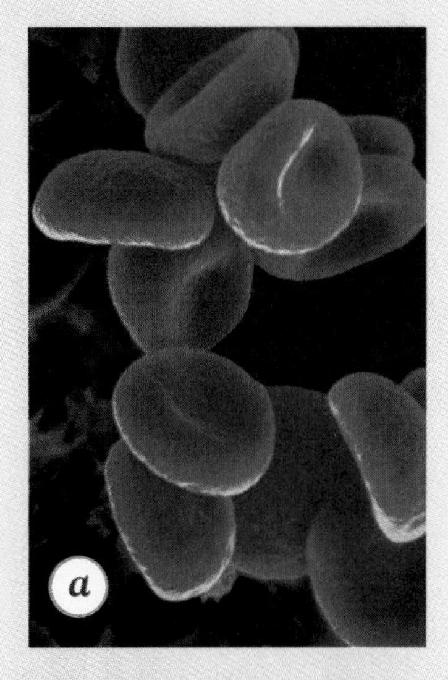

In facilitated diffusion, substances diffuse across the cell membrane through special channels in proteins. Researchers have discovered hundreds of different channel proteins—each specific for an ion, sugar, salt, or other substance.

Osmosis

As you have read earlier, only substances that dissolve in lipids can pass through the cell membrane. However, water is an important exception to this rule. Water is not lipid soluble, yet it passes through most biological membranes very easily. This fact has important consequences for the cell.

The diffusion of water through a selectively permeable membrane is called **osmosis.** ***Figure 3–12*** shows a membrane that, like the cell membrane, is permeable to water but is not permeable to sugar. On one side of the membrane is a sugar solution, and on the other side is pure water.

As the figure shows, there is a net movement of water into the sugar solution. This occurs because water, like any other substance, diffuses from regions of high concentration to regions of low concentration. In this example, the pure water has a higher concentration of water than the sugar solution does.

☑ ***Checkpoint*** What is osmosis?

Osmotic Pressure

When water moves by osmosis, it can produce powerful pressure. How powerful? Enough to destroy a cell.

The cytoplasm of a typical cell is filled with salts, sugars, and other dissolved substances. Therefore, if a dilute water solution should come in contact with a cell, water would rapidly enter the cell by osmosis. If uncontrolled, the cell would swell like a balloon and burst.

All cells must deal with osmotic pressure. In general, cells control this powerful force in one of three ways:

- **They use a cell wall.** As you have read, bacteria and plants are surrounded by strong, tough cell walls. This prevents the cell from expanding, thus counteracting the osmotic pressure.
- **They pump out the water.** Many single-celled organisms pump out water as quickly as it enters. These organisms rely on specialized structures such as the contractile vacuole.
- **They bathe cells in blood.** Most large animals prevent their cells from being in direct contact with dilute water—even if the animal drinks water or swims in it. But cells still need water in order to survive. Therefore, these animals bathe their cells in blood or bloodlike liquids. An animal's blood and its cells have nearly the same concentrations of dissolved substances.

Biology AND HEALTH Connections

Death by Osmosis

Have you ever had an earache? How about a strep throat or an infected cut or scrape? Each of these conditions—and many others—is typically caused by bacteria. Bacteria are single-celled prokaryotic organisms. There is an almost countless variety of bacteria, and they can be found almost everywhere on Earth. Many types of bacteria live their entire lives inside the body of a human or other animal yet cause no ill effects. Other bacteria, however, can easily harm an animal once they infect it. And all bacteria can cause serious damage if their numbers grow beyond tolerable limits.

Penicillin—The Bacteria Killer

Until the 1940s, doctors had only mildly effective ways to treat bacterial infections, let alone cure them. All that changed, however, with the discovery of powerful drugs known as antibiotics. The first of these drugs was penicillin. Today, penicillin is still widely used to help kill bacteria and fight infections.

Penicillin works by weakening the bacteria's cell walls. Bacteria form cell walls by binding together special proteins and carbohydrates. In many ways, penicillin looks just like one of these proteins. Thus, when penicillin is present, the bacteria will incorporate it into new cell walls. As a result, the fibers of the cell walls are linked together more loosely than is normal.

E. coli *bacteria can be killed with penicillin (magnification: 20,000X)*

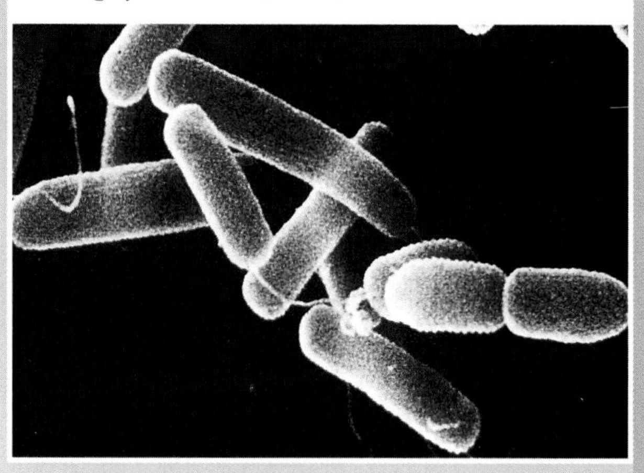

Exploding Bacteria!

Bacteria depend on their cell walls to control osmotic pressure. Because of osmosis, bacteria are constantly subjected to water pressure across their cell membranes, and only tough cell walls keep bacteria from bursting. Because the cell wall is looser and weaker when penicillin is present, the cell wall will stretch, crack, and finally break as water streams across the cell membrane. In a matter of seconds, so much water enters the cell that it literally explodes.

The picture may not be pretty, but it is exactly how penicillin kills bacteria. Death by osmosis!

Penicillium notatum—*the green mold that produces penicillin*

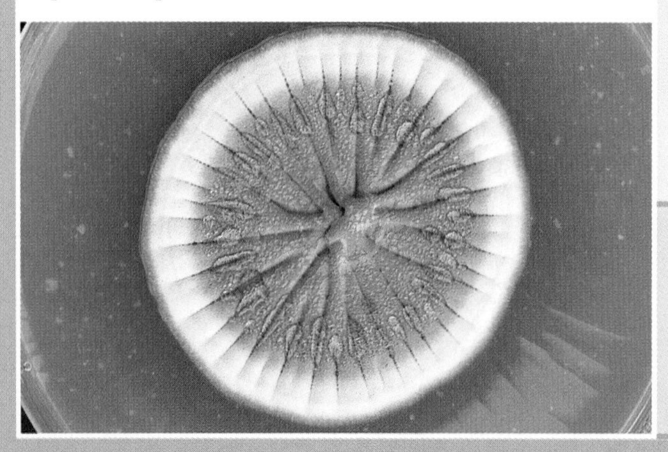

Making the Connection

Why does penicillin not kill human cells? Do you think a person should take penicillin on a regular basis? Why or why not?

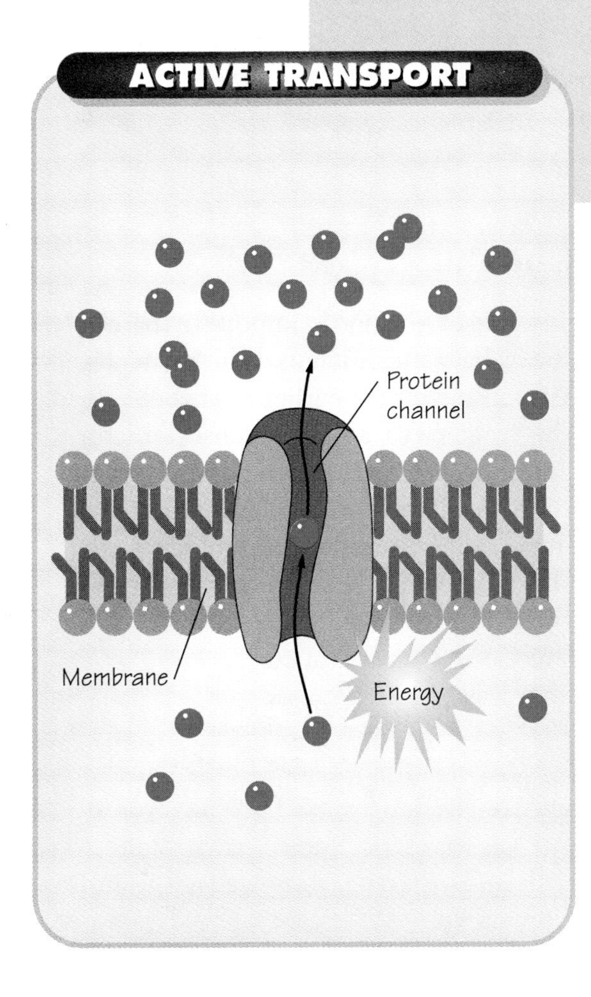

Figure 3–14
In active transport, a cell expends energy to move substances against a concentration difference.

Active Transport

As you have just read, passive transport moves substances from regions of high concentrations to regions of lower concentrations. Can a cell move substances in the opposite direction—from regions of low concentration to higher concentration? The answer is yes. In fact, cells constantly move certain substances against a concentration difference. However, they have to pay a price to do this. That price is energy.

Movement of a substance against a concentration difference is called active transport, and it always requires energy. The process of **active transport** is often compared to a pump. Most animal cells, for example, have membrane proteins that pump sodium ions out of the cell and pump potassium ions into it. Active transport is not limited to small ions or molecules, however. Many cells expend energy to transport large molecules, clumps of food, or even whole cells.

One way in which cells import larger materials is literally to turn part of the cell membrane inside out. This process is called endocytosis, and it is illustrated in ***Figure 3–15***. When the particles are very large, the process is sometimes called phagocytosis. In another process, called exocytosis, cells use energy to expel material.

Figure 3–15
In endocytosis, a cell wraps its membrane around a particle, then turns its membrane inside out to take in the particle. In this false-color photograph from a scanning electron microscope, a cell is taking in a small particle by endocytosis (magnification: 6000X).

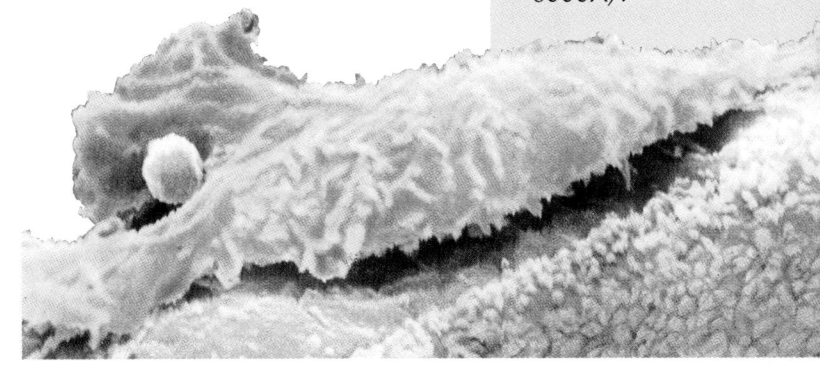

Section Review 3–2

1. **Discuss** the roles of the cell membrane and the cell wall.
2. **Describe** the processes of passive transport and active transport.
3. **Critical Thinking—Applying Concepts** "Water, water, everywhere; Nor any drop to drink." This quote from *The Rime of the Ancient Mariner* refers to sea water. If you drink sea water, your body actually loses water. Explain why this happens.

SECTION 3–3

Inside the Cell

Guide for Reading

- **Describe** the composition and function of the nucleus.
- **List** and **describe** the organelles of the cytoplasm.

MINI LAB

- **Interpret** the changes observed when a paramecium takes in food.

WHAT DOES A FACTORY NEED to operate efficiently? Certainly, it needs the proper machines and workers to assemble the factory's products, to package the products, and to ship them for delivery. The factory also needs a source of energy to power the machines, with mechanics on hand to fix the machines if they break. And the factory needs a main office, or control center, from which all the work can be regulated.

Like a factory, the cell meets its needs by using distinct parts. As you will discover, the parts of the cell function much like the different parts of the factory.

Nucleus

One of the most important parts of the cell is the **nucleus** (plural: nuclei). As you have read earlier in this chapter, the nucleus is a large, dense structure contained in the cells of many organisms. In fact, the nucleus is such an important organelle that biologists classify organisms into two categories—those that do have a nucleus and those that do not have a nucleus.

Prokaryotes and Eukaryotes

The organisms made of cells that contain nuclei are called the **eukaryotes**—*eu-* means "true" and *-karyon* means "nucleus." Eukaryotes range in size from tiny, single-celled organisms to organisms composed of millions of cells, including all large plants and animals. The organisms that do not contain nuclei are called **prokaryotes**—*pro-* means "before." Most prokaryotes are small, single-celled organisms. They include the bacteria and relatives of bacteria.

Figure 3–16
(a) *This photograph of plant cells reveals only some of the cells' distinct parts (magnification: 500X).* (b) *Like this factory, a cell uses different parts to accomplish different tasks.*

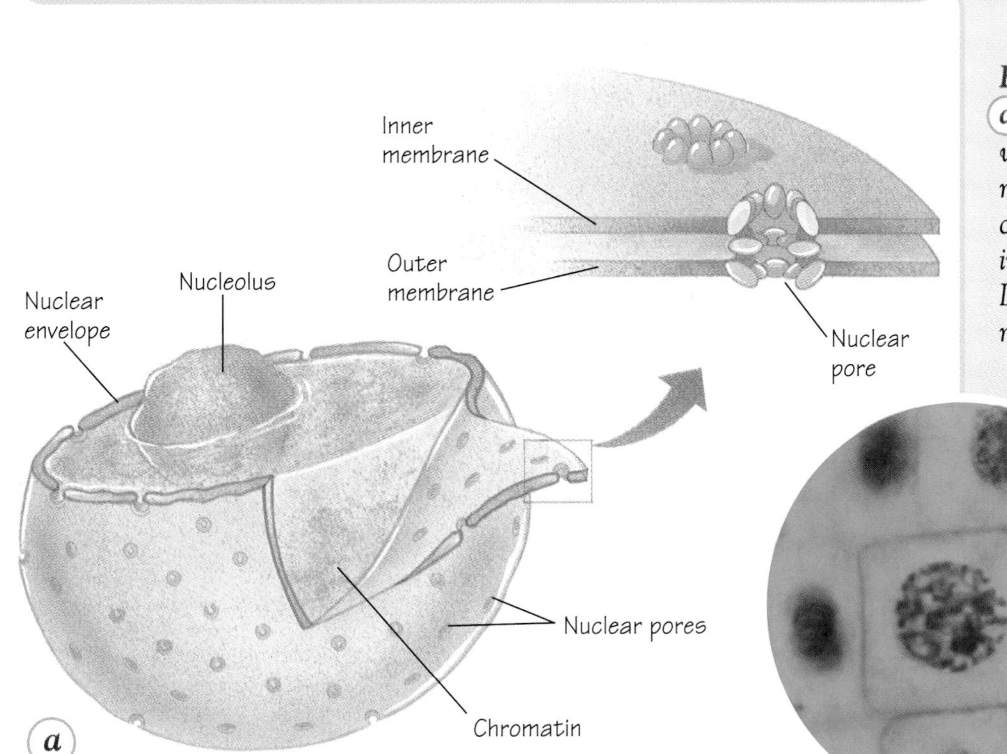

Figure 3–17 *(a) The nucleus stores DNA, which is spread throughout the nucleus in a material called chromatin. When needed, instructions are copied from DNA and sent through the nuclear pores to the rest of the cell. (b) In this photograph of onion cells, the nuclei are the round structures in the center of each cell. (magnification: 100X).*

Role of the Nucleus

Why is the nucleus so important? **The nucleus contains nearly all of a cell's DNA.** Recall that DNA is the molecule that contains coded instructions for making proteins and other important molecules. By storing DNA in the nucleus, the cell is better able to control and regulate its use.

The DNA molecules and proteins form a material called chromatin. Chromatin is usually spread throughout the nucleus. But when a cell divides, the chromatin condenses to form larger structures called **chromosomes.** As shown in ***Figure 3–17,*** chromosomes are large enough to be visible under the compound microscope.

If the nucleus were the control center of a factory, then the chromosomes would be the plans or blueprints for the factory. When necessary, instructions from these blueprints are sent out of the control center to the factory floor.

☑ **Checkpoint** What are chromosomes?

Structures in the Nucleus

Most nuclei also contain a small, dense region known as the nucleolus. For many years, the function of the nucleolus was a mystery. But biologists now know that the nucleolus is the site where ribosomes are assembled. You will learn more about ribosomes later in this section.

The nucleus is surrounded by two distinct membranes that together are called the nuclear envelope. The envelope contains many nuclear pores, or tiny holes that allow material to move into and out of the nucleus.

Cytoplasm 

The portion of the cell outside the nucleus is known as the **cytoplasm**—*cyto-* means "cell" and *-plasm* means "fluid." At one time, biologists thought that the cytoplasm was a simple fluid, as the name cytoplasm suggests. But as biologists studied cells with the compound microscope and other microscopes, they realized that the cytoplasm is quite complex.

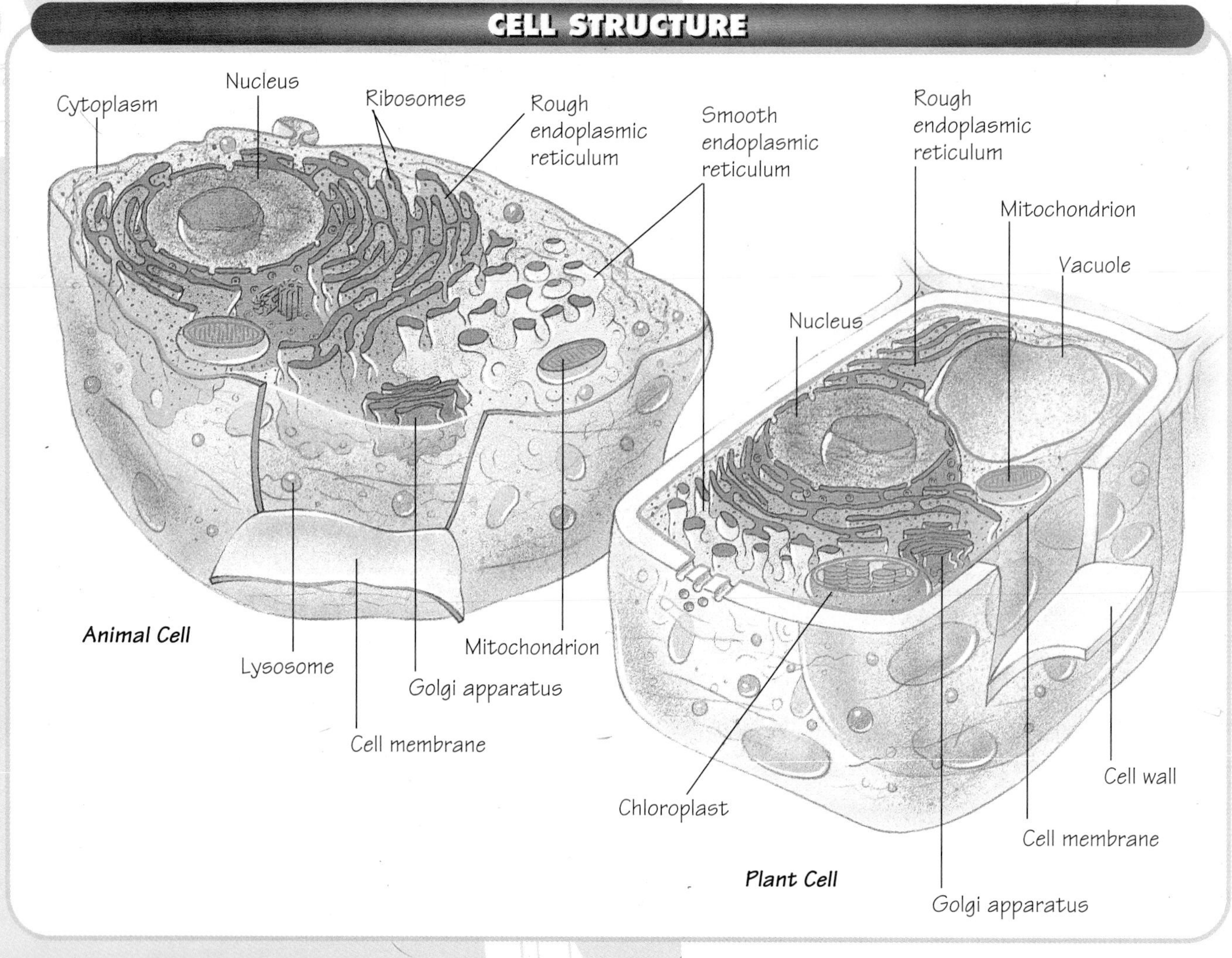

CELL STRUCTURE

Figure 3–18
Both plant and animal cells contain a variety of organelles. Although some organelles are specific to either plant or animal cells, others—such as the mitochondrion—are found in both types of cells.

Organelles

The cytoplasm contains many individual parts called **organelles.** The word organelle means "little organ." **An organelle is a small structure that performs a specialized function within a cell, just as a machine performs a specialized function in a factory.** Understanding the cell's organelles is the key to understanding the cell as a whole.

Figure 3–18 shows many of the organelles in a typical animal cell and those in a typical plant cell. Let's take a look at each of these organelles.

☑ ***Checkpoint*** What is an organelle?

Ribosomes

Most cells contain small structures called **ribosomes.** Ribosomes are tiny particles made of RNA and protein. A ribosome is only 25 nanometers wide, or about 0.25 percent of the width of a typical cell. Some cells have ribosomes in the thousands or tens of thousands.

What purpose do ribosomes serve? Ribosomes are the sites where proteins are assembled. Later on, you will learn more about ribosomes and their role in making proteins. For now, you can picture each ribosome as a small assembler that produces proteins—an assembler that gets its instructions from the nucleus.

☑ ***Checkpoint*** What are ribosomes?

Endoplasmic Reticulum and Golgi Apparatus

In many cases, the cell can use a protein immediately after a ribosome has

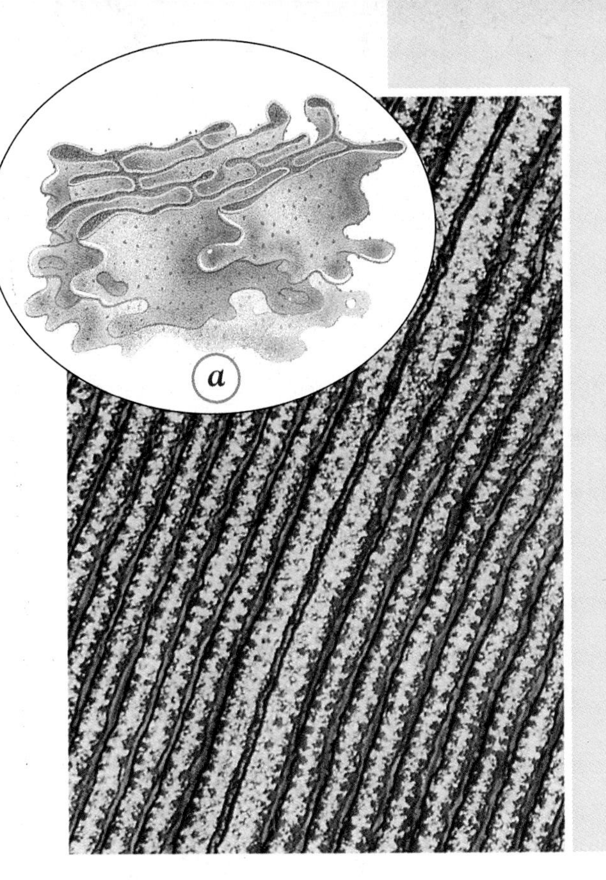

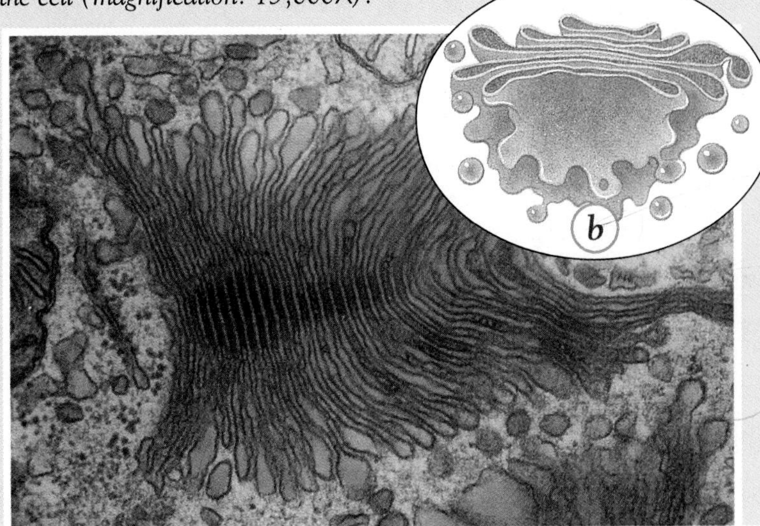

Figure 3–19
(a) The endoplasmic reticulum (ER) shown here is called rough ER because it is studded with ribosomes. Proteins are assembled at ribosomes, then travel through the ER network (magnification: 50,000X). (b) In the Golgi apparatus, proteins are modified, packaged, and shipped to their destinations in the cell (magnification: 13,000X).

assembled it. However, some proteins need further processing. Many proteins need to be delivered to specific locations within the cell, while others need to be packaged for export, or removal from the cell. In addition, some proteins need special components attached to them before they can be used.

Processing and transporting proteins and other macromolecules is the function of the **endoplasmic reticulum**—or ER, for short—and the **Golgi apparatus.** Both the endoplasmic reticulum and the Golgi apparatus are networks of membranes within the cell. The term endoplasmic reticulum means "network inside the cell." The Golgi apparatus is named after Camillo Golgi, the Italian scientist who first identified it.

As illustrated in ***Figure 3–19,*** ribosomes stud the surface of one form of ER, giving it a "rough" appearance. This type of ER is called rough ER. After a protein is assembled at a ribosome in the rough ER, it travels through the ER network into the Golgi apparatus, where special enzymes may attach carbohydrates or lipids. The modified protein then travels out of the Golgi apparatus to locations throughout the cell or outside the cell.

Together, the ER and the Golgi apparatus work like the packaging and shipping divisions of a factory. They modify and add components to proteins, then ship them to their final destinations.

☑ **Checkpoint** What is the function of the ER and the Golgi apparatus?

Lysosomes

Every well-organized factory needs a cleanup crew, and the cell is no exception. The workers of the cleanup crew of the cell are the **lysosomes.** Lysosomes are saclike membranes filled with chemicals and enzymes that

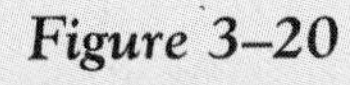

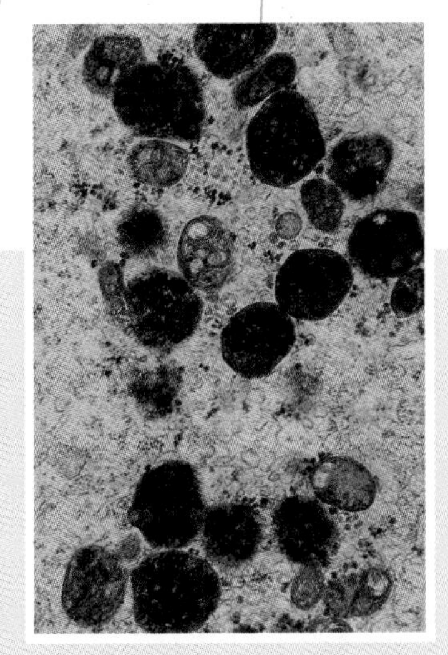

Figure 3–20
The round black structures in this photograph are lysosomes—the cell's cleanup crews. Lysosomes are used to break down nonfunctioning organelles or foreign materials in the cell (magnification: 95,000X).

A Colorful Paramecium

PROBLEM *How does a paramecium take in food? **Interpret** observations to answer this question.*

PROCEDURE  

1. Prepare a slide of a live paramecium. Focus the slide under the low-power objective of a microscope.
2. Obtain a small sample of a yeast suspension. The yeast suspension has been treated with an acid/base indicator that is red above pH 5 and blue below pH 3.
3. Use a toothpick to transfer a small drop of the yeast suspension to the edge of the slide. Observe the paramecium under the microscope for 5 minutes. Record your observations.

ANALYZE AND CONCLUDE

1. Describe how the paramecium reacted to the yeast.
2. What changes did you observe in the paramecium? Which organelles were involved in these changes?
3. Propose an explanation for any color changes that you observed.

can break down almost any substance within the cell.

In some cases, lysosomes will fuse with a damaged organelle and literally break it apart into basic chemical compounds. As a result, the lysosomes both clear away the damaged organelle and recycle its components for other uses in the cell.

Cytoskeleton

Every large factory building needs internal beams for support. On a smaller scale, a cell must cope with stresses and strains that are just as powerful as those on a building. So it is no surprise that the cell has its own support structures.

Eukaryotic cells contain a supporting framework called the **cytoskeleton**—meaning "cell skeleton." The cytoskeleton has many components, including structures called microtubules and microfilaments. Microtubules are hollow tubes of protein about 25 nanometers wide. Microfilaments are also made of protein but are only about 7 nanometers wide. Together, microtubules and microfilaments provide a tough, flexible framework of support for the cell.

A large factory also needs ways to move materials from one end of the factory floor to the other. In the cell, the cytoskeleton takes care of this job, too. Proteins can attach to the cytoskeleton and move organelles along it, just as a locomotive moves on a railroad track.

Cytoskeletal proteins are responsible for other types of cellular movement, too. For example, microtubules make up hairlike structures called cilia and flagella. Cilia and flagella are found on the surfaces of many cells. In some organisms, the beating of cilia or flagella propel the cell from one location to the next.

Vacuoles

Many cells store materials in saclike structures called **vacuoles.** In animal cells, vacuoles may store proteins, fats, or carbohydrates. In plant cells, a large, central vacuole often stores water and dissolved salts.

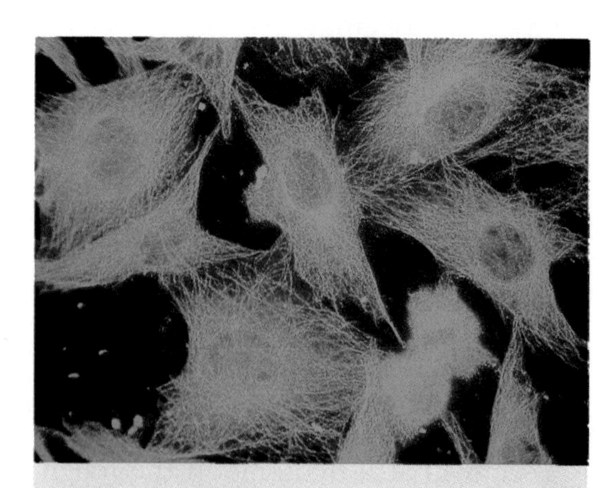

Figure 3–21
The cells in this photograph were specially treated to highlight their cytoskeleton frameworks (magnification: 275X).

Vacuoles also provide plant cells with support. The pressure inside the vacuoles allows a plant to grow quickly and to support heavy structures such as leaves and flowers.

Mitochondria and Chloroplasts

To complete our picture of the cell as a busy and productive factory, we need one last component—a supply of energy. Without energy, a cell could not produce proteins, move molecules through its cytoplasm, or perform any of the other activities necessary for life.

If the owners of a factory had to produce their own energy, they might choose to produce energy by burning a chemical fuel, such as coal or natural gas. Or they might choose to harvest energy from the sun. In either case, they would need to build a specialized power plant to convert the energy source into a usable form of energy, such as electricity.

Like a factory, the cell can produce its energy from either a chemical fuel or the sun. The organelle that produces energy from a chemical fuel is called the **mitochondrion** (plural: mitochondria). And the organelle that harvests the energy of sunlight is called the **chloroplast.** Mitochondria are found in all sorts of eukaryotic organisms, including most plants and animals. Chloroplasts, however, are found only in plants and certain types of algae.

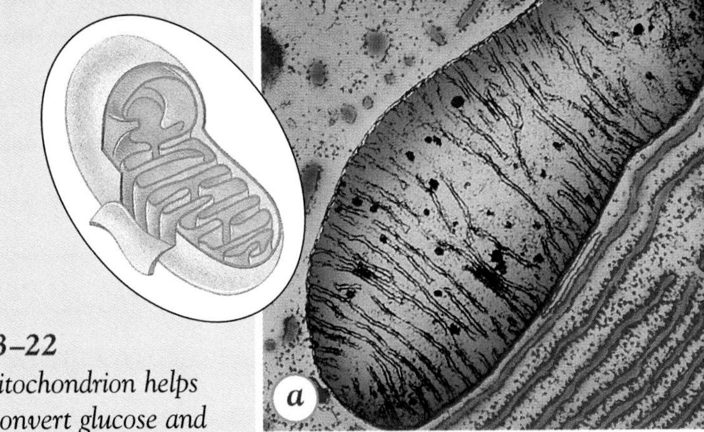

Figure 3–22
(a) A mitochondrion helps the cell convert glucose and oxygen into a usable form of energy (magnification: 85,000X). (b) Chloroplasts, which are found in plant cells but not in animal cells, convert sunlight into chemical energy (magnification: 1850X).

For the mitochondrion, the source of chemical fuel is not coal or natural gas but complex organic molecules, such as glucose or other sugars. And unlike power plants, neither a mitochondrion nor a chloroplast produces electricity. Instead, both organelles produce molecules that serve as small packets of chemical energy. In the next chapter, you will learn much more about these molecules and how they are produced.

☑ **Checkpoint** What is a chloroplast? A mitochondrion?

Section Review 3–3

1. **Describe** the composition and function of the nucleus.
2. **List** and **describe** the organelles of the cytoplasm.
3. **Critical Thinking—Drawing Conclusions** Red blood cells, which do not contain mitochondria, transport oxygen gas through the bloodstream. Based on this information, would you conclude that oxygen crosses red blood cell membranes by active transport or by passive transport? Explain.
4. **MINI LAB** A paramecium produces acid to help digest food. Use this fact to **interpret** the color changes you observed as the paramecium took in dye-stained food.

BRANCHING OUT **In Depth**

The Origin of the Eukaryotic Cell

GUIDE FOR READING

- **Define** the endosymbiont hypothesis.

EUKARYOTES HAVE NUCLEI AND prokaryotes do not have nuclei—but that is only the beginning of the differences between them. Eukaryotes also have mitochondria, chloroplasts, and a host of other organelles that prokaryotes do not have.

How did eukaryotes and prokaryotes come to be so different? No one can answer this question for sure. However, one scientist has proposed a very interesting hypothesis.

The Work of Lynn Margulis

Lynn Margulis, a scientist from the University of Massachusetts, focused her attention on two organelles—the mitochondrion and the chloroplast. As Margulis noted, both organelles have several unusual properties. First, they contain their own DNA. While other organelles rely solely on the DNA in the nucleus, mitochondria and chloroplasts use their own DNA to produce many important compounds.

Second, both the mitochondrion and the chloroplast are surrounded by two membranes. Most membrane-bound organelles—including the Golgi apparatus, the ER, and the lysosome—are surrounded by only one membrane.

Third, in many respects the mitochondrion and chloroplast reproduce separately from the rest of the cell. The cell can produce other organelles individually, but mitochondria seem to come only from other mitochondria, and chloroplasts seem to come only from other chloroplasts.

These facts raised several questions in Margulis's mind. Why do mitochondria and chloroplasts have their own DNA? Why do they have a second membrane, while other organelles have only one? And why do they reproduce separately from the rest of the cell?

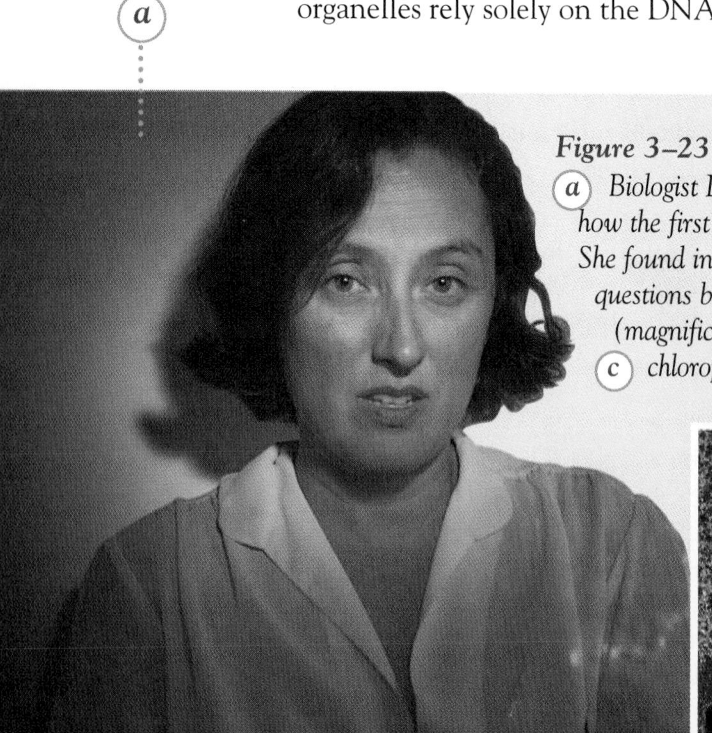

Figure 3–23
(a) Biologist Lynn Margulis was interested in how the first eukaryotes came into existence. She found intriguing answers to her questions by studying (b) mitochondria (magnification: 75,000X) and (c) chloroplasts (magnification: 6500X).

The Endosymbiont Hypothesis

Margulis proposed an interesting hypothesis that answered each of the questions she raised. **Margulis proposed that billions of years ago, eukaryotic cells arose as a combination of different prokaryotic cells.** She called this hypothesis the **endosymbiont hypothesis.** *Endo-* means "inside," and a symbiotic relationship is a close association between two organisms—sometimes of benefit to both organisms. The term endosymbiont highlights the idea that one organism might actually have been living inside the other.

Checkpoint What is the endosymbiont hypothesis?

Margulis's Model

According to Margulis, both mitochondria and chloroplasts had ancestors that were free-living organisms. These organisms formed endosymbiotic relationships with larger cells. Over time, the offspring of these organisms lost their independence, becoming organelles in larger cells.

This model answers each of Margulis's questions about chloroplasts and mitochondria. These organelles have their own DNA and reproduce separately because they were once independent organisms. Their inner membranes could be the remnant of the cell membrane of the free-living organism, and the outer membranes could be the cell's membrane surrounding the "foreign" cell.

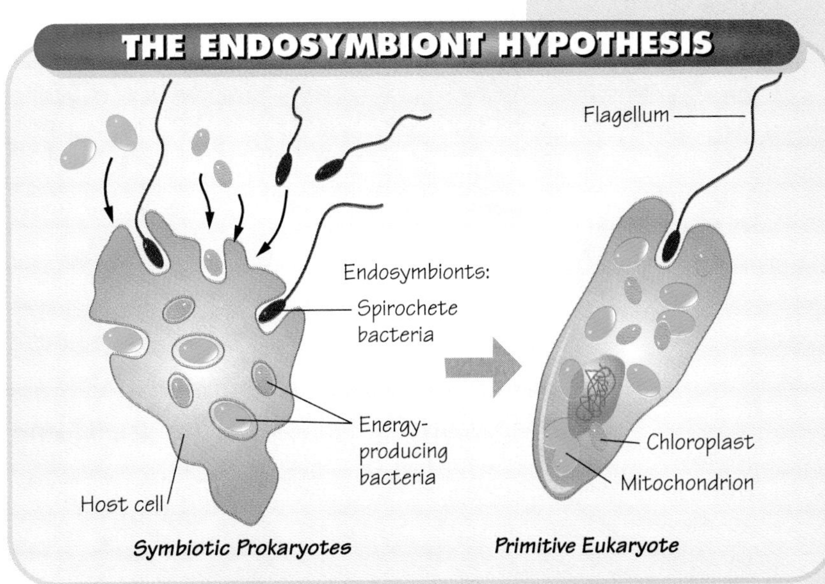

Figure 3–24 *According to the endosymbiont hypothesis, a collection of prokaryotic bacteria joined together billions of years ago. Eventually, they lost their independence and became a primitive eukaryote.*

Further Evidence

There is further evidence for the endosymbiont hypothesis. For example, detailed studies of the DNA molecules in chloroplasts show that they are much more like the DNA of prokaryotes than they are like the DNA found in the nucleus of plants. This result would be expected if chloroplasts were once independent prokaryotic organisms.

Researchers have also discovered that mitochondria and chloroplasts contain their own ribosomes and make many of their own proteins. These ribosomes, however, are smaller and chemically different from those found in the rest of the eukaryotic cell. Instead, they resemble the ribosomes found in prokaryotes.

Section Review 3–4

1. **Define** the endosymbiont hypothesis.
2. **Describe** the evidence in favor of the endosymbiont hypothesis.
3. **BRANCHING OUT ACTIVITY** Think of two organisms—either real or imaginary—that might form an endosymbiotic relationship. In a series of drawings, **describe** the organisms and the endosymbiotic relationship that you propose they form.

Laboratory Investigation

Inside Plant and Animal Cells

Ever since the first microscopes were invented, biologists have been studying the structure of all living cells. In this investigation, you will compare the structures of plant cells and animal cells.

Problem

How do the structures of plant cells and animal cells **compare?**

Materials (per group)

scalpel
tweezers
onion
medicine dropper
glass slide
coverslip
iodine solution
paper towel
prepared slide of human cheek cells
microscope

Procedure

1. **Using tweezers, peel the thin, transparent skin from the inner surface of an onion, as shown.**

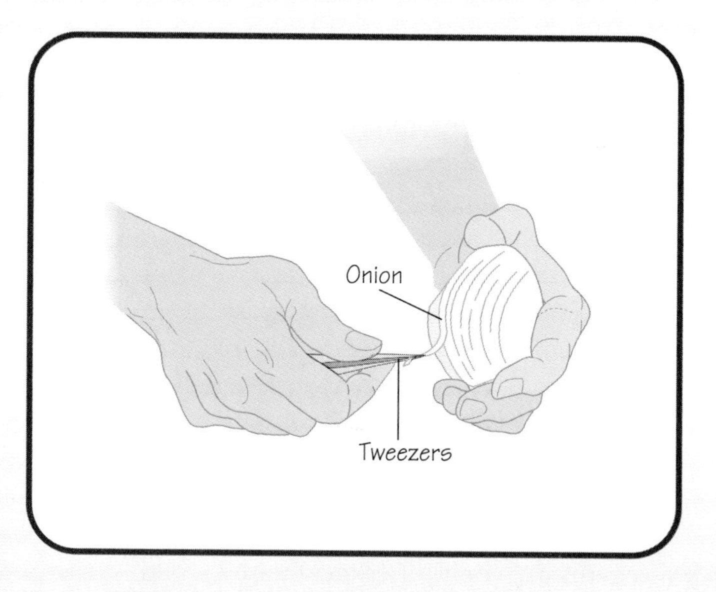

2. **With a scalpel, cut a small piece of the skin from the section that was removed from the onion. CAUTION: *Be careful when using a scalpel or any other sharp instrument.* Place the small piece of onion skin on a clean glass slide.**
3. **Add a drop of water to the piece of onion skin and cover with a coverslip.**

4. **Using the medicine dropper, place a drop of iodine solution at one end of the coverslip. Holding a piece of paper towel near the opposite edge, draw the iodine solution underneath the coverslip.**
5. **Examine the onion skin slide under the low-power objective of the microscope. Sketch and label what you observe.**
6. **Repeat step 5 using the high-power objective.**
7. **Repeat steps 5 and 6 using the prepared slide of the human cheek cells.**

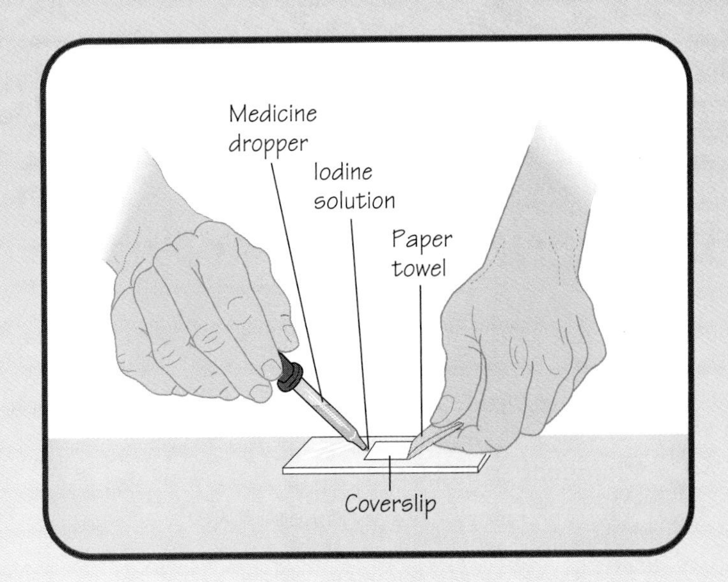

Observations

1. What is the shape of the onion skin cells? The shape of the human cheek cells?
2. Describe the general structures of the onion cells and the cheek cells.

Analysis and Conclusions

1. How are plants and animals similar in structure? How are they different?
2. What was the purpose of adding the iodine solution to the onion cells?
3. In onion plants, the liquid outside the cells is significantly less concentrated than the liquid inside the cells. This concentration difference creates high osmotic pressure. Describe how onion cells respond to osmotic pressure.
4. Of the cell structures presented in this chapter, which did you not see in either the cheek cell or the onion cell? Discuss the reasons why you did not see these structures.
5. Based on your observations, draw and label a generalized structure of a plant cell and an animal cell.

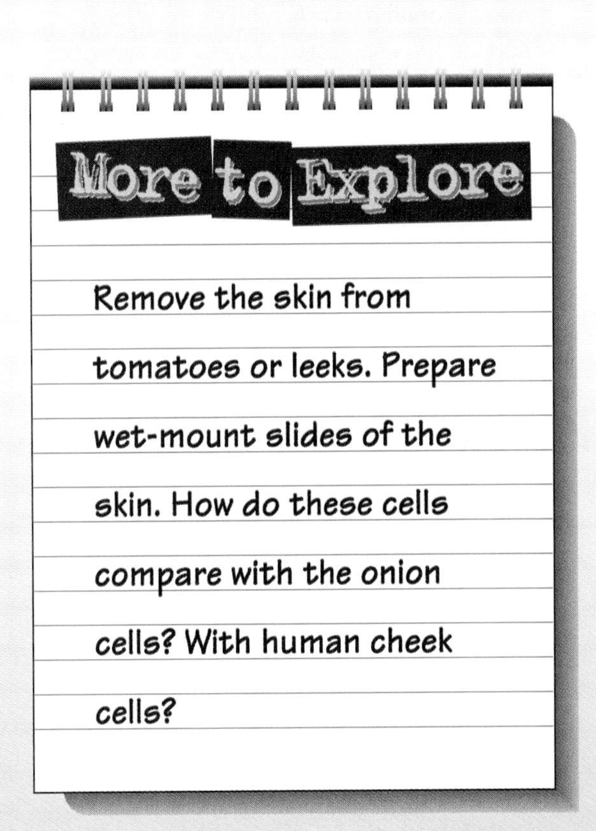

Study Guide

Summarizing Key Concepts

The key concepts in each section of this chapter are listed below to help you review the chapter content. Make sure you understand each concept and its relationship to other concepts and to the theme of this chapter.

3–1 Microscopes and Cells

- Anton van Leeuwenhoek used light microscopes to identify small living things. Robert Hooke was the first person to identify cells.
- The cell theory states that living things are made of cells, that cells are the smallest working units of living things, and that cells arise from the division of other cells.
- The compound light microscope is arguably the most useful tool in biology. Electron microscopes and scanning probe microscopes are especially powerful microscopes.

3–2 Cell Boundaries

- The cell membrane separates and protects the cell from its surroundings.
- The cell membrane contains a double layer of lipids, called the lipid bilayer. The cell membrane also contains proteins, carbohydrates, and other compounds.
- In passive transport, substances move from regions of high concentration to low concentration. In active transport, cells use energy to move substances against a concentration difference.

3–3 Inside the Cell

- The nucleus contains nearly all of a cell's DNA.
- An organelle is a small structure that performs a specialized function within a cell.
- The different organelles include the ribosomes, endoplasmic reticulum, Golgi apparatus, lysosomes, cytoskeleton, vacuoles, mitochondria, and chloroplasts.

3–4 The Origin of the Eukaryotic Cell

- Lynn Margulis proposed the endosymbiont hypothesis, which states that billions of years ago eukaryotic cells arose as a combination of different prokaryotic cells.

Reviewing Key Terms

Review the following vocabulary terms and their meaning. Then use each term in a complete sentence.

3–1 Microscopes and Cells

cell
cell theory
compound light microscope
electron microscope
scanning probe microscope

3–2 Cell Boundaries

cell membrane
lipid bilayer
cell wall
passive transport
diffusion
osmosis
active transport

3–3 Inside the Cell

nucleus
eukaryote
prokaryote
chromosome
cytoplasm
organelle
ribosome
endoplasmic reticulum
Golgi apparatus
lysosome
cytoskeleton
vacuole
mitochondrion
chloroplast

3–4 The Origin of the Eukaryotic Cell

endosymbiont hypothesis

CHAPTER 3

Recalling Main Ideas

Choose the letter of the answer that best completes the statement or answers the question.

1. The smallest working units of living things are
   a. nuclei. **c.** cells.
   b. organelles. **d.** atoms.

2. Organisms that have cells without nuclei are called
   a. eukaryotes. **c.** prokaryotes.
   b. animals. **d.** plants.

3. Which of the following could produce an image of individual atoms?
   a. transmission electron microscope
   b. scanning electron microscope
   c. scanning probe microscope
   d. Van Leeuwenhoek's microscope

4. A lipid bilayer forms the core of the
   a. cytoplasm. **c.** nucleus.
   b. cell membrane. **d.** ribosome.

5. Which structure is found in the cells of plants but not in animals?
   a. cell membrane **c.** mitochondrion
   b. cell wall **d.** lysosome

6. The nucleus stores almost all of a cell's
   a. ribosomes. **c.** proteins.
   b. DNA. **d.** amino acids.

7. Which organelle harvests energy from sunlight?
   a. lysosome **c.** chloroplast
   b. mitochondrion **d.** ribosome

8. The principal role of the endoplasmic reticulum and the Golgi apparatus is to
   a. store food. **c.** produce energy.
   b. move the cell. **d.** package proteins.

9. According to the endosymbiont hypothesis, early ancestors of mitochondria were
   a. free-living organisms.
   b. ribosomes.
   c. chloroplasts.
   d. nuclei.

Putting It All Together

Using the information on pages xxx to xxxi, complete the following concept map.

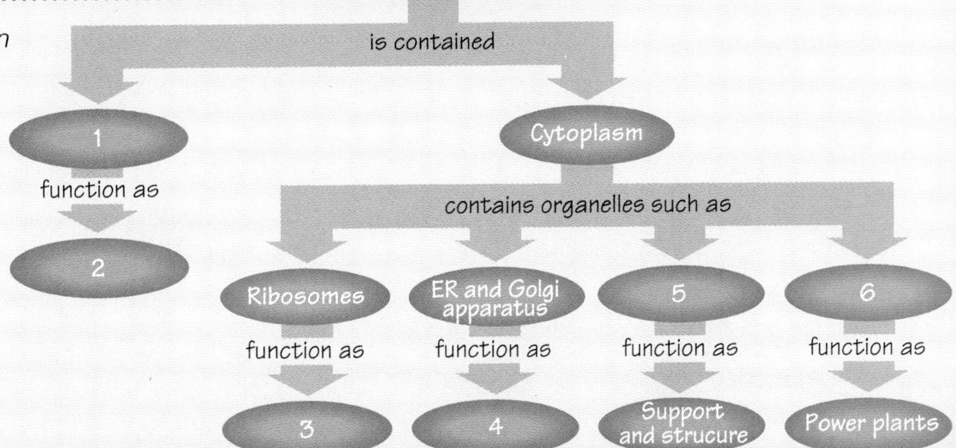

Assessment

Reviewing What You Learned

Answer each of the following in a complete sentence.

1. How did Anton van Leeuwenhoek contribute to the study of biology?
2. How powerful is the compound light microscope?
3. What is a TEM? An SEM?
4. What is the purpose of the cell membrane?
5. How are phospholipids arranged in a cell membrane?
6. Give an example of water moving by osmosis.
7. What is the role of the nucleus?
8. Which organelle acts as the cell's "cleanup crew"?
9. What is the function of the endoplasmic reticulum?
10. Which organelle stores food for the cell?
11. For what purpose do cells use microtubules and microfilaments?
12. Who first proposed the endosymbiont hypothesis?

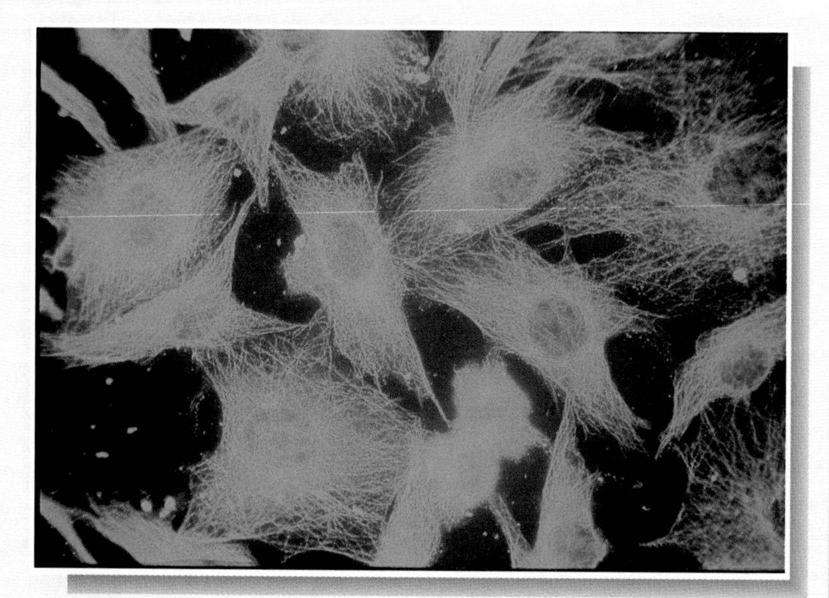

Expanding the Concepts

Discuss each of the following in a brief paragraph.

1. What is the cell theory?
2. Compare a light microscope with an electron microscope. What are the advantages and disadvantages of each?
3. Identify the polar and the nonpolar regions of a phospholipid. Why are these regions significant?
4. Compare passive transport with active transport.
5. What is osmotic pressure? How do cells respond to it?
6. Where are proteins assembled? What organelles process proteins after they are assembled?
7. Describe the role of lysosomes.
8. What two roles do vacuoles serve in plant cells?
9. Compare a mitochondrion with a chloroplast.
10. Describe the facts about mitochondria and chloroplasts that led to the development of the endosymbiont hypothesis.

Extending Your Thinking

Use the skills you have developed in this chapter to answer the following.

1. **Observing** While looking through a compound light microscope, you observe an air bubble below and to the left of the image of a cell. What is the actual spatial relationship between the cell and the air bubble? Explain.
2. **Applying concepts** A beaker contains two salt solutions divided by a membrane. The solution is higher on the left side of the beaker than on the right side. The membrane is permeable to water but not to salt. Which side of the beaker contains the more concentrated salt solution? Explain your answer.
3. **Designing an experiment** Design an experiment to determine the water concentration of a peeled potato. (*Hint:* When placed in a sugar solution, a peeled potato will either gain or lose water through osmosis.)
4. **Interpreting data** The photograph at right was produced by an electron microscope. What structures can you identify? Can you determine whether the photograph shows a plant cell or an animal cell? Explain.
5. **Evaluating evidence** Suppose that researchers discover a way to separate mitochondria from the rest of the cell, then develop the mitochondria into free-living organisms. Would this discovery strengthen, weaken, or have no effect on Margulis's endosymbiont hypothesis? Explain.

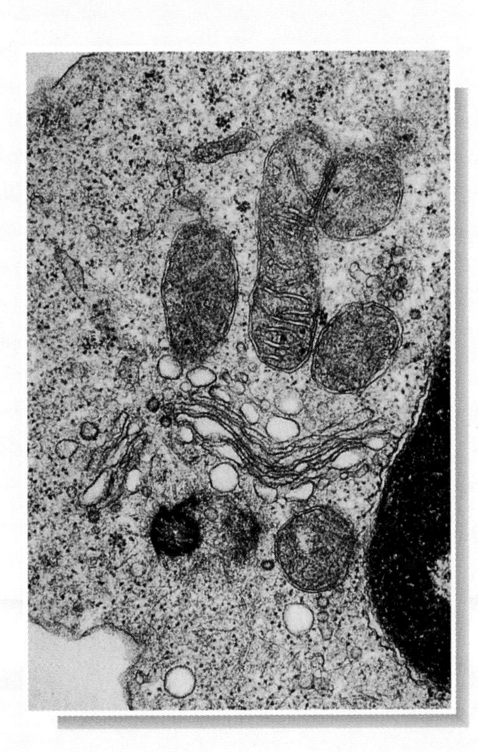

Applying Your Skills

A Model Membrane

A diagram in a textbook can provide a great deal of information, but scientists often use three-dimensional models to better understand complex structures.

1. Using ***Figure 3–8*** on page 54 as a guide, construct a model of the cell membrane. You may use Styrofoam™ balls, toothpicks, construction paper, or any other materials available in your classroom.
2. Label the components of your model.
3. In what ways is your model accurate? In what ways is it inaccurate?

• GOING FURTHER •

4. Research other models of the cell membrane. Use the results of your research to improve the model you constructed.
5. Using your model as a visual aid, present a report to your class on the structure and function of the cell membrane.

CHAPTER 4

Energy and the Cell

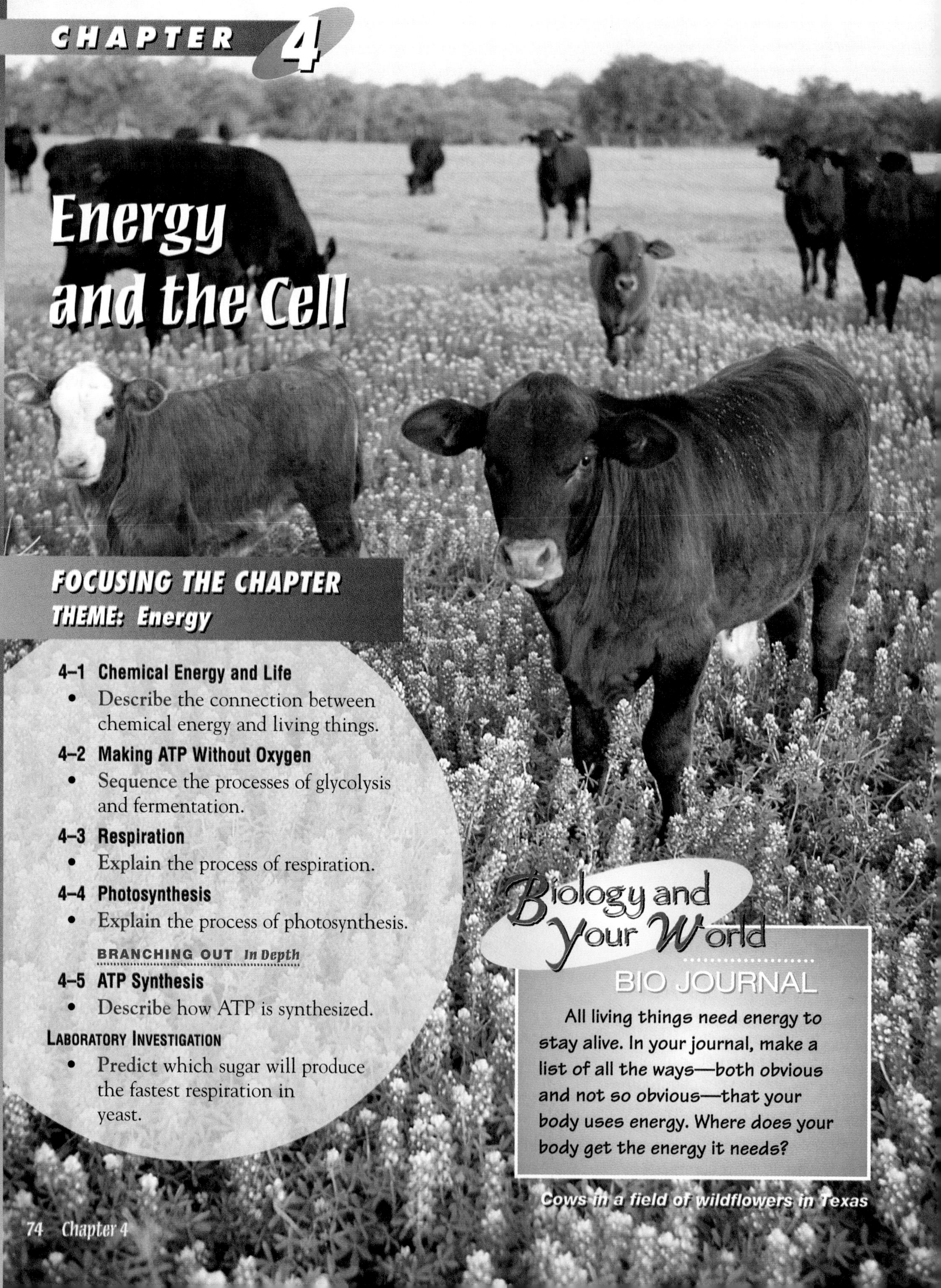

FOCUSING THE CHAPTER
THEME: Energy

4–1 Chemical Energy and Life
- Describe the connection between chemical energy and living things.

4–2 Making ATP Without Oxygen
- Sequence the processes of glycolysis and fermentation.

4–3 Respiration
- Explain the process of respiration.

4–4 Photosynthesis
- Explain the process of photosynthesis.

BRANCHING OUT *In Depth*

4–5 ATP Synthesis
- Describe how ATP is synthesized.

LABORATORY INVESTIGATION
- Predict which sugar will produce the fastest respiration in yeast.

Biology and Your World

BIO JOURNAL

All living things need energy to stay alive. In your journal, make a list of all the ways—both obvious and not so obvious—that your body uses energy. Where does your body get the energy it needs?

Cows in a field of wildflowers in Texas

Chemical Energy and Life

GUIDE FOR READING

- **Explain** why energy is so important to living things.
- **Describe** how energy is stored in ATP and released from ATP.

EVERYTHING THAT IS ALIVE needs energy. The need for energy is easy to see in a group of athletes sprinting for the finish line, in a cheetah chasing a herd of antelopes, or in a flock of geese flying southward. Energy also is required for less obvious tasks. A tree standing in a meadow is silently using enormous amounts of energy to draw water from deep inside the Earth. A sleeping fox draws on its energy reserves to rebuild and replace the cells and tissues that were lost during the day's hunting.

Cells use energy for virtually everything they do. Energy is required to build new proteins, to pump ions across the cell membrane, to copy genetic information, and even to move. Where do your cells get their energy? The simple answer is that energy comes from the food you eat. And the energy in that food ultimately came from the sun.

Cells and Energy

Why is energy so important to living things? **Without the ability to produce and use energy, living things would cease to exist.** Living cells need supplies of energy for thousands of activities at the same time. How do they get it? Imagine thousands of construction workers needing energy for their power drills and sanders. How could you supply energy to them? One way would be via a battery-recharging center that supplied the workers with batteries for their power tools. Each battery would carry a small amount of energy. When that energy was used up, the worker would snap in a new battery and send the spent battery back for recharging.

Figure 4–1
All organisms use energy. (a) Although it is not obvious, these flowers need a tremendous amount of energy to transport materials throughout the plants. (b) In order to fly, these Canada geese also need a great deal of energy.

ATP

Obviously, cells don't have batteries, but they do have a compound that works almost like a battery. This compound is called **ATP,** or adenosine triphosphate. As its name suggests, ATP has three phosphate groups. A similar compound, ADP (adenosine diphosphate), has two phosphates, and AMP (adenosine monophosphate) has just one. Try to think of AMP as an uncharged battery. **It takes a significant amount of energy to attach a phosphate to AMP to make ADP. This energy is stored in the phosphate bond, in much the same way that electricity is stored in a battery.**

A similar amount of energy is required to attach the third phosphate, converting ADP to ATP. Why is ATP so important? ATP is similar to a fully charged battery, ready to supply energy to do the work of the cell. How does ATP do this? **Throughout the cell, enzymes that require energy have binding sites for ATP and similar energy-carrying molecules. When one of these enzymes needs energy to complete a chemical reaction, one phosphate breaks off, converting ATP to ADP.** Breaking the phosphate bond releases just enough energy to pump an ion across a membrane, attach an amino acid to a growing protein, or flick a cilium a fraction of a micrometer. The molecule of ADP is then available to store energy by forming ATP again. This process continues over and over—storing energy and then releasing it as needed.

Checkpoint What makes up an ATP molecule?

How Do Cells Make ATP?

It might seem remarkable that so many different functions within the cell are energized by ATP, but it also makes good sense. The energy released by converting ATP to ADP enables the cell to power just about everything it does. To meet its potential energy needs in everything from growth to movement, all that a cell has to do is recharge its chemical batteries by attaching phosphates to AMP and ADP to make ATP.

Figure 4–2
A molecule of ATP consists of the amino acid adenine, a sugar called ribose, and three phosphate groups.

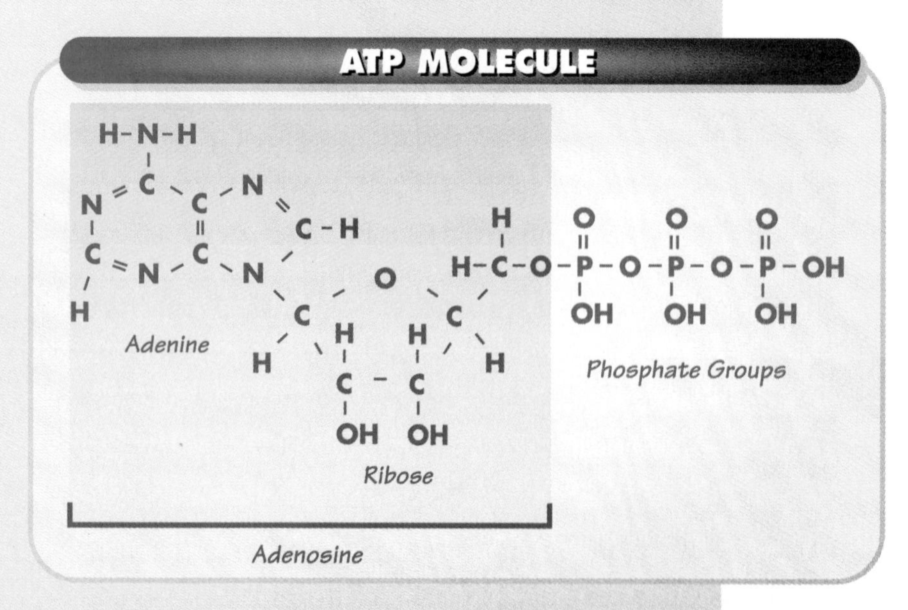

Releasing Energy From Food

Where do cells get the energy they need to replenish their supply of ATP? Animal cells get that energy from the food that is consumed. Most food molecules contain a great deal of energy. For example, a single glucose molecule contains more than 90 times the chemical energy released by splitting the third phosphate off ATP to make ADP.

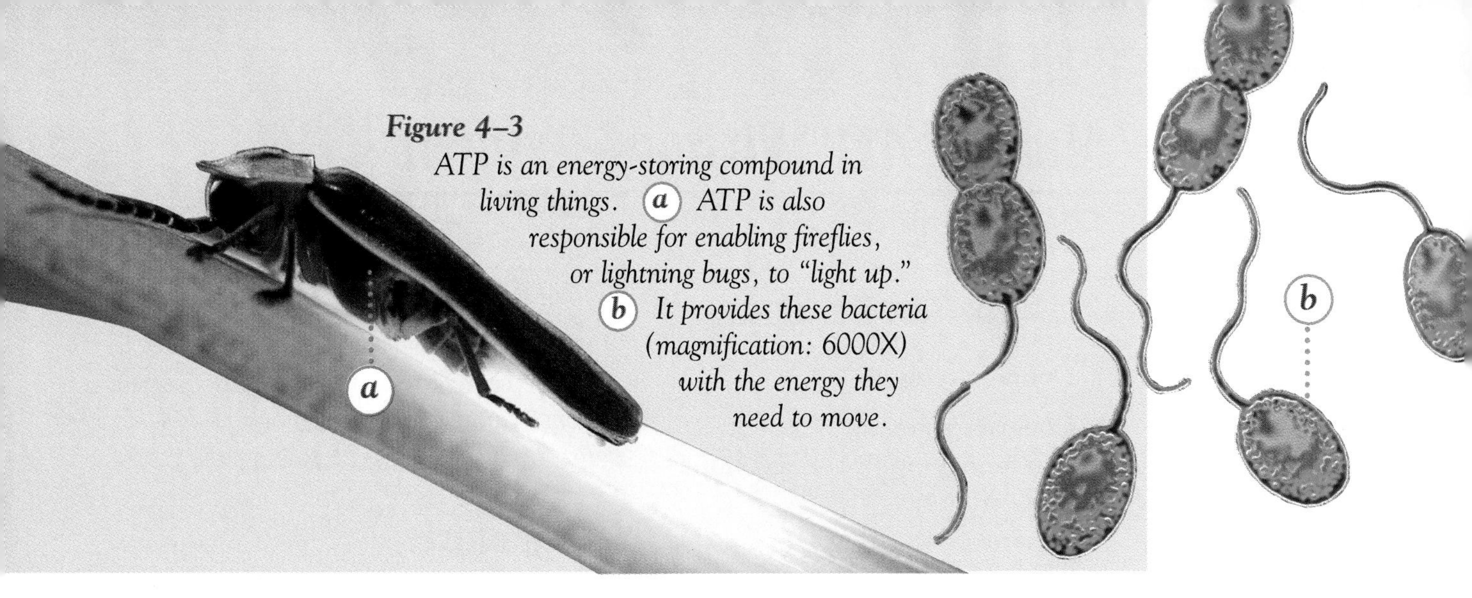

Figure 4–3
ATP is an energy-storing compound in living things. (a) *ATP is also responsible for enabling fireflies, or lightning bugs, to "light up."* (b) *It provides these bacteria (magnification: 6000X) with the energy they need to move.*

In the laboratory, the chemical energy in glucose can be released by burning it. This chemical reaction requires oxygen and produces carbon dioxide and water. In fact, the reaction of glucose with oxygen releases so much energy that it gives off heat and light—it burns. But a living cell has to control that reaction and release the energy a little bit at a time. This is where ATP comes in. It helps to trap the energy.

Checkpoint What kinds of energy are produced in the reaction of oxygen and glucose?

Chemical Energy

Before you consider how cells go about trapping the energy in glucose, it might be useful to think about where that energy comes from. It certainly does not come from the nuclei of the atoms involved in the reaction. The nuclei are unchanged by the reaction. The only possible source of all that energy are the electrons that surround the carbon, hydrogen, and oxygen nuclei—the ones that form the chemical bonds between those atoms.

Compounds such as glucose store energy in their chemical bonds. This energy is released when the bonds are broken and new bonds storing less energy are formed. In other words, all that heat and light energy comes simply from rearranging the electrons that form the chemical bonds between the atoms of glucose.

How can cells capture energy from the chemical bonds of glucose? Recall that the reaction of glucose and oxygen releases an enormous amount of energy. Cells, however, need to capture energy in small amounts. Therefore, the reactions in living things take place in small steps, releasing small amounts of energy at a time. This enables cells to make the ATP they need.

INTEGRATING CHEMISTRY

What is the chemical formula for this reaction? What are the reactants? The products?

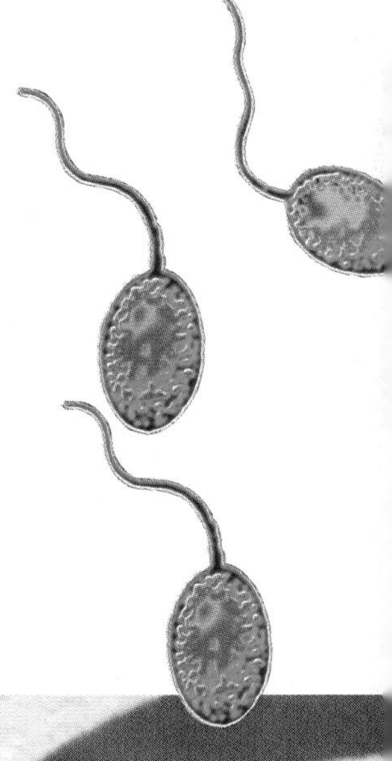

Section Review 4–1

1. **Explain** why energy is so important to living things.
2. **Describe** how energy is stored in ATP. How is it released?
3. **Critical Thinking—Drawing Conclusions** What is the relationship between energy and chemical bonds?

SECTION

4–2 Making ATP Without Oxygen

GUIDE FOR READING

- **Define** glycolysis.
- **Compare** the processes of lactic acid fermentation and alcoholic fermentation.

MINI LAB

- **Predict** the effect temperature has on the rate of fermentation.

TO RELEASE ENERGY FROM glucose a little bit at a time, the cell must take apart the glucose a little at a time. The cell does this by breaking the glucose down in a series of chemical reactions. These reactions take place in the cytoplasm of the cell, and they happen very quickly. It takes no more than a few milliseconds for most cells to produce thousands of molecules of ATP in this way. These ATP molecules provide quick energy for many cellular activities, including muscle contraction.

Each of these reactions is catalyzed by its own enzyme, and each is carefully controlled by the cell. Surprisingly, the first few reactions require energy—that is, the cell must use 2 molecules of ATP to begin the process.

Glycolysis

The series of reactions in which a molecule of glucose is broken down is called glycolysis. The word **glycolysis** means "sugar-breaking." The process of glycolysis, shown in ***Figure 4–5,*** begins when 2 molecules of ATP are used to convert the glucose molecule into a high-energy 6-carbon sugar with 2 phosphates. The 6-carbon sugar is broken down into two 3-carbon molecules (PGAL). The 2 PGAL molecules go through several more chemical reactions and produce pyruvic acid. These reactions produce 4 molecules of ATP for each molecule of glucose. This is a net gain of 2 molecules of ATP for each glucose molecule broken down.

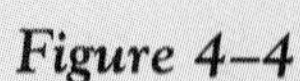

Figure 4–4
(a) Although the burning of wood is somewhat similar to the breakdown of glucose because energy is produced, the energy that is released from burning wood is much more dramatic. (b) In the absence of oxygen, the breakdown of glucose occurs by fermentation. The air bubbles in bread are a result of the fermentation of tiny yeast.

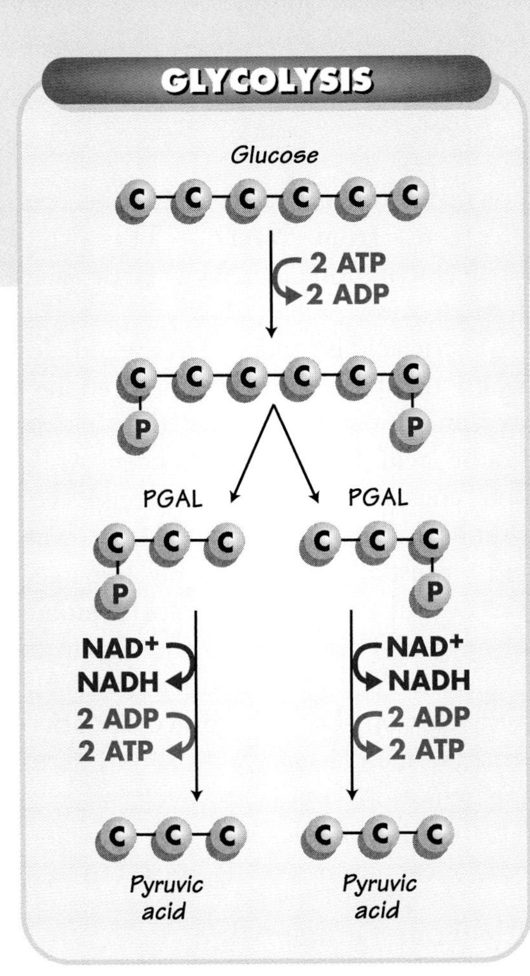

Figure 4–5
The complete process of glycolysis generates 4 molecules of ATP and 2 NADH molecules. NADH is formed when NAD^+ accepts a pair of high-energy electrons and a hydrogen ion (H^+).

In addition to the net gain of 2 molecules of ATP, glycolysis also produces 2 pairs of high-energy electrons. These electrons are passed to NAD^+, an electron carrier, to form NADH. In order for glycolysis to continue, there must be a constant supply of NAD^+.

The energy in these electrons can only be utilized when oxygen is available. However, if oxygen is not available, NADH cannot get rid of the high-energy electrons. Within a few seconds, all the cell's available NAD^+ molecules would be converted to NADH. Lacking a constant supply of NAD^+, ATP production would stop, and the cell would be in peril.

☑ **Checkpoint** What is the net gain of ATP in glycolysis?

Fermentation

Not surprisingly, the cell has a mechanism to get rid of the electrons if oxygen is not available. It passes the high-energy electrons from NAD^+ back to the same carbon atoms from which they came. What does this accomplish? It allows the cell to keep using its NAD^+ molecules over and over again, and that allows glycolysis to continue. It also has an interesting consequence—the cell accumulates the compounds that accept those electrons.

This process—the regeneration of NAD^+ to keep glycolysis running—is called **fermentation.** If oxygen is present, the cell can use the high-energy electrons in NAD^+ to make ATP, so fermentation is not necessary. Therefore, in nature, fermentation occurs only in cells and organisms that lack oxygen. There are two basic types of fermentation.

Lactic Acid Fermentation

In most animals, the pyruvic acid that accumulates due to glycolysis is

Figure 4–6
(a) During rapid exercise, such as running, muscle cells begin to produce lactic acid by lactic acid fermentation. This process provides these runners with the energy they need. (b) The process of alcoholic fermentation is responsible for turning the grapes in this vineyard into wine.

Temperature and Fermentation

PROBLEM *How can you **predict** the effect of temperature on the rate of fermentation?*

PROCEDURE  

1. Add enough yeast suspension to two test tubes so that it comes to within 3 cm of the top. Place a stopper with a rubber tube into each test tube. Put the test tubes aside for now.
2. Obtain 4 beakers from your teacher. Fill two of the beakers halfway with Bromthymol blue solution. **CAUTION:** *Bromthymol blue can stain your skin and clothing.*
3. Using a glass-marking pencil, label the remaining beakers A and B. Fill beakers A and B about two-thirds full with warm water. Using a thermometer and very warm water or ice, try to keep the temperature of the water in beaker A at 30°C and the water in beaker B at 20°C.
4. Place one test tube in beaker A and the other in beaker B. Put the free end of each rubber tube into a beaker of Bromthymol blue.
5. Record the time that it takes for each beaker of Bromthymol blue to change color.

ANALYZE AND CONCLUDE

1. Which beaker of Bromthymol blue turned color faster?
2. What effect did temperature have on the rate of fermentation?
3. If a third water bath with a temperature of 10°C was included in the experiment, would the rate of fermentation be faster or slower than that of beaker A or B? Explain your answer.

converted to lactic acid when it accepts electrons from NAD^+. This kind of fermentation is called lactic acid fermentation.

Lactic acid is produced in the muscles when the body cannot supply enough oxygen to muscles to meet their needs. **Lactic acid fermentation** occurs when you engage in vigorous exercise. For example, if you run, ride a bike, or swim fast for just a few seconds, the large muscles of your arms and legs quickly run out of oxygen. These muscles then begin to make the ATP they need by lactic acid fermentation. Lactic acid accumulation in these muscles causes a painful, burning sensation familiar to every athlete. This is why muscles may feel sore after only a few minutes of vigorous activity.

Alcoholic Fermentation

Another kind of fermentation occurs in yeasts. Yeasts are one-celled organisms that are used in baking or brewing. **In the absence of oxygen, pyruvic acid is broken down to produce alcohol and carbon dioxide instead of lactic acid. This process is known as alcoholic fermentation.**

Alcoholic fermentation causes bread dough to rise. When the yeast in the dough is starved for oxygen, it gives off tiny bubbles of carbon dioxide. The same carbon dioxide is the source of bubbles in beer and sparkling wines. To brewers, alcohol is a welcome byproduct of fermentation.

Section Review 4-2

1. **Define** glycolysis.
2. **Compare** lactic acid fermentation and alcoholic fermentation.
3. **Critical Thinking—Inferring** What would happen to wine if there was an air leak in the fermentation tank?
4. **MINI LAB** **Predict** the effect that temperature has on the rate of fermentation.

Respiration

Guide for Reading

- **Define** respiration.
- **Explain** the Krebs cycle.

AS USEFUL AS IT IS, GLYCOLYSIS releases only a small amount of the chemical energy that was originally stored in glucose. Most of that energy—about 90 percent—is still unused. To convert the rest of that chemical energy to ATP, the cell needs oxygen. Why does the cell need oxygen? In a sense, oxygen is the best electron acceptor. If a cell has access to oxygen, it can tap nearly all the energy available for its purposes.

The Process of Respiration

How does the cell use oxygen? You may recall that outside the cell, when glucose reacts with oxygen, it produces carbon dioxide and water and releases so much energy that it actually burns. The cell cannot afford to waste energy on a bonfire. Instead, it takes apart the pyruvic acid molecule formed in glycolysis, a little at a time. In the breakdown of pyruvic acid, the cell takes a few high-energy electrons with every step and uses their energy to make ATP. This process, called **respiration,** requires oxygen. **Respiration is the release of energy from the breakdown of food molecules in the presence of oxygen.**

In eukaryotic cells, respiration takes place inside the mitochondrion. You may recall that the mitochondrion is the cell organelle where energy is produced. The process of respiration begins when the pyruvic acid enters the mitochondrion.

The Krebs Cycle

The chemical bonds in pyruvic acid are broken apart in a series of reactions called the Krebs cycle, named for its discoverer, Dutch scientist Hans Krebs.

Figure 4–7
In order to release the maximum amount of energy from glucose, animals need oxygen. (a) All animals, including this grizzly bear, get the energy they need from the food—in this case, salmon—that they eat. (b) Dolphins must come to the surface of the water in order to get the oxygen they need.

Figure 4–8
In the presence of oxygen, pyruvic acid gives off carbon dioxide and attaches to coenzyme A to form acetyl CoA. Acetyl CoA enters the Krebs cycle and joins with a 4-carbon compound to form citric acid. Each turn of the cycle produces 2 molecules of carbon dioxide, 1 molecule of GTP, which, in terms of energy, is equivalent to ATP, 3 molecules of NADH, and 1 molecule of $FADH_2$. $FADH_2$ is formed when FAD accepts a pair of high-energy electrons and H^+.

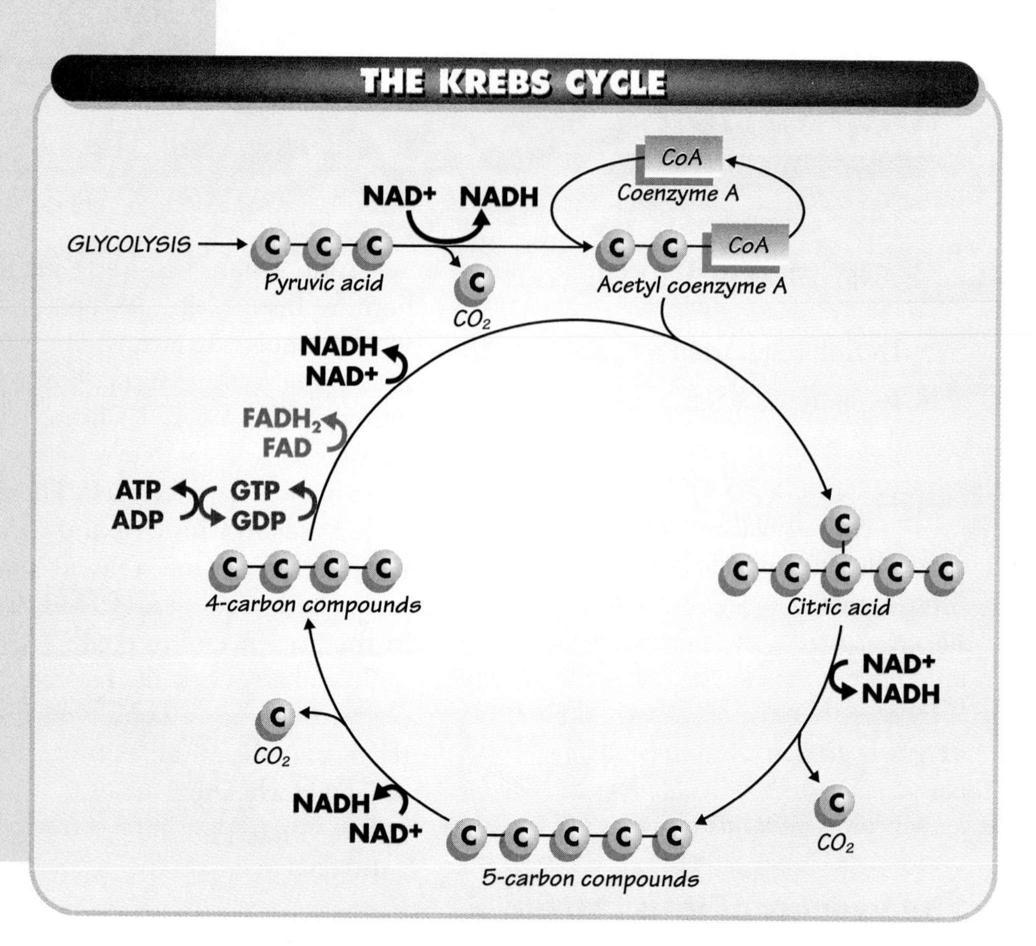

At first glance, the **Krebs cycle** may look complicated, but what actually happens is remarkably simple. Rather than break down pyruvic acid directly, the cell first breaks off 1 carbon atom and releases it in the form of carbon dioxide. Then it combines the remaining 2 carbons with a 4-carbon compound. The 2 carbons plus the 4 carbons make a 6-carbon compound. The 6-carbon compound formed in this way is called citric acid. For this reason, the Krebs cycle is sometimes called the citric acid cycle.

☑ ***Checkpoint*** How is citric acid formed?

The Breakdown of Citric Acid

Citric acid is gradually broken apart. First, one carbon is broken off, and then a second carbon breaks off. After a few more steps, the very same 4-carbon compound is produced that was used to start the cycle. The fact that it returns to its own starting point means that the Krebs cycle can go around and around, pulling apart the high-energy bonds that were left after glycolysis, and releasing carbon in the form of carbon dioxide.

In fact, the carbon dioxide that is released in the Krebs cycle is the source of all the carbon dioxide in your breath. Every time you exhale, you expel the carbon dioxide produced by the Krebs cycle in trillions of mitochondria throughout your body.

Electron Carriers

If the carbon atoms in glucose are exhaled in the form of carbon dioxide, what happens to all those high-energy electrons whose energy the cell hoped to trap? The electrons are passed to two electron carriers: NAD^+ and FAD. The two electron carriers can each accept a pair of high-energy electrons, hold them for a short time, and then pass them along to another compound. Between the two electron carriers, they

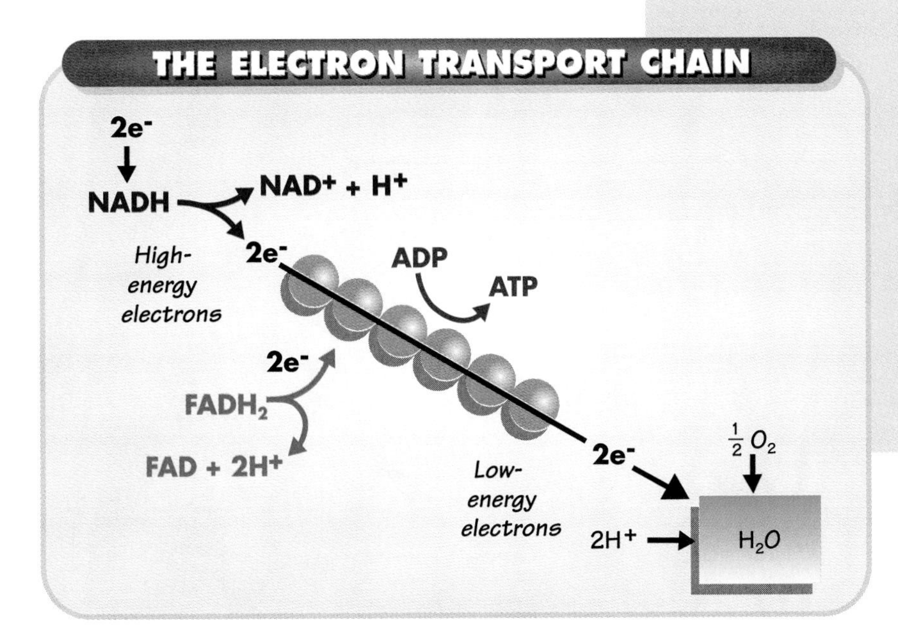

Figure 4–9
As electrons from the Krebs cycle move down the electron transport chain, their energy level is reduced. This energy is used to produce ATP. At the end of the electron transport chain, the energy-depleted electrons combine with oxygen and hydrogen ions to form water.

remove five pairs of high-energy electrons during each turn of the Krebs cycle. What does the cell do with the high-energy electrons? That's where oxygen comes in.

Checkpoint What do the electron carriers do?

The Electron Transport Chain

The electron carriers, NAD^+ and FAD, take the high-energy electrons directly to the inner membrane of the mitochondrion, where a series of special molecules is waiting for them. These molecules are known as the **electron transport chain,** and they receive the high-energy electrons from the carriers. The carriers then go back to the Krebs cycle for more electrons. Incidentally, the electrons that were passed to NAD^+ in glycolysis can also be brought into the mitochondrion and passed to the electron transport chain.

The high-energy electrons are then passed from one molecule in the electron transport chain to the next. With each transfer, the energy level of the electrons is gradually lowered. The energy of these electrons is used to produce ATP from ADP. Compared to the process of glycolysis, the amount of ATP produced in the mitochondrial electron transport chain is very large. Roughly 36 molecules of ATP are produced for each molecule of glucose, compared to just 2 molecules of ATP produced in glycolysis.

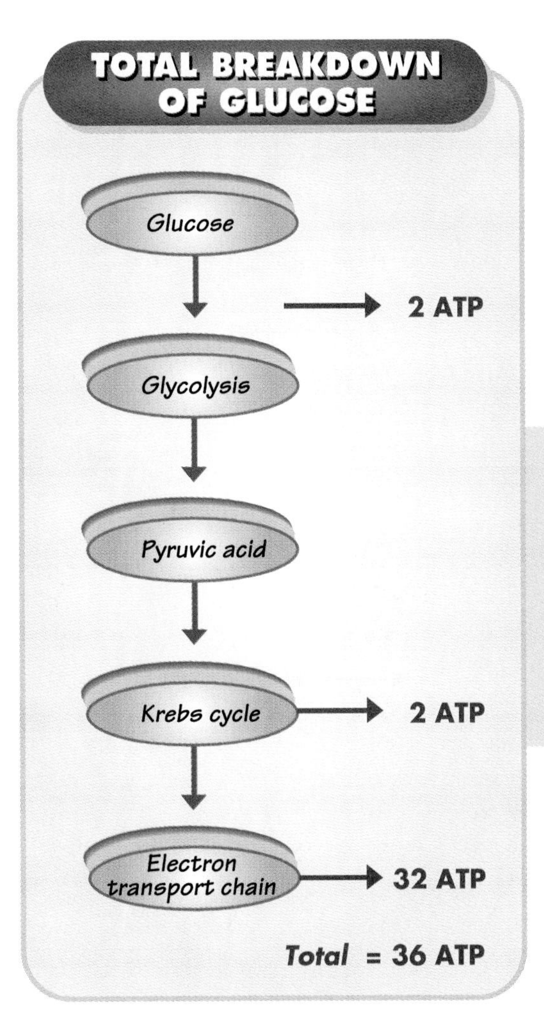

Figure 4–10
The complete breakdown of glucose in the presence of oxygen yields a total of 36 molecules of ATP.

Visualizing Respiration

Respiration is the release of energy from the breakdown of food molecules in the presence of oxygen. The process of respiration can be summarized in the following equation:

$$C_6H_{12}O_6 + 6O_2 \rightarrow 6CO_2 + 6H_2O + \text{Energy}$$

Glucose + Oxygen → Carbon dioxide + Water + Energy (ATP + heat)

reactants products

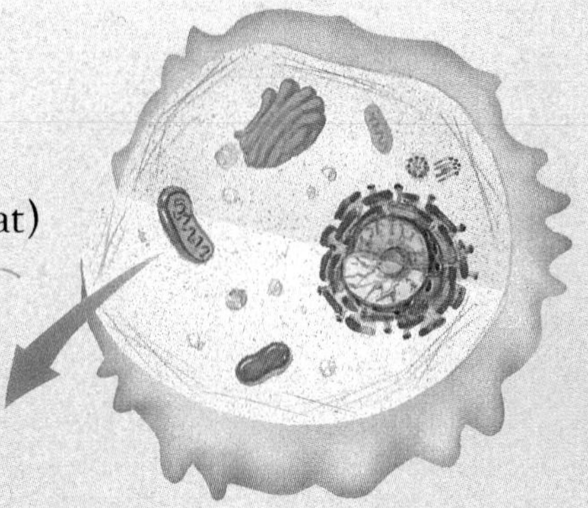

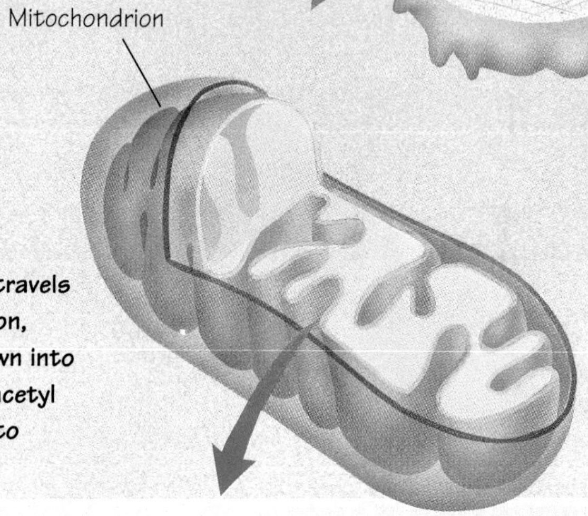

Mitochondrion

1 In the cytoplasm, glucose is broken down through glycolysis into 2 molecules of pyruvic acid.

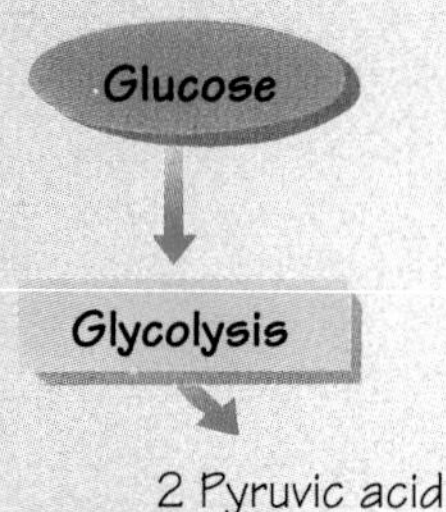

2 Pyruvic acid

2 The pyruvic acid travels into the mitochondrion, where it is broken down into CO_2 and a 2-carbon acetyl group that is bound to Coenzyme A (CoA).

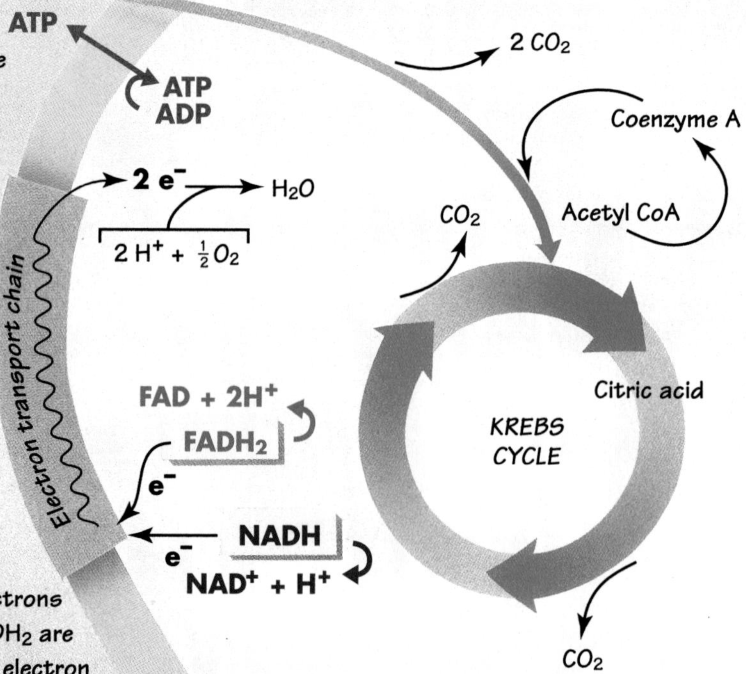

6 At the end of the chain, an enzyme combines the electrons from the electron transport chain, some of the hydrogen ions from fluid in the cell, and oxygen to form H_2O. Some of the energy from the hydrogen ions is used to form ATP.

ATP

ATP
ADP

2 CO_2

Coenzyme A

$2e^-$ → H_2O

$2H^+ + \frac{1}{2}O_2$

CO_2

Acetyl CoA

Electron transport chain

FAD + $2H^+$

$FADH_2$

e^-

Citric acid

KREBS CYCLE

NADH

e^-

$NAD^+ + H^+$

CO_2

3 Acetyl CoA then passes the 2 carbons of the acetyl group into the Krebs cycle, where they join with a 4-carbon compound to form a 6-carbon compound, citric acid.

5 High-energy electrons from NADH and $FADH_2$ are passed through the electron transport chain. As the electrons move down the chain, their energy level is decreased.

4 At two places in the cycle, CO_2 is released. At four places in the cycle, a pair of high-energy electrons is accepted by the electron carriers NAD^+ and FAD, forming NADH and $FADH_2$.

Figure 4–11 *The delicious-looking food on this table is broken down into simple forms that can be used by the body to produce energy.*

The Role of Oxygen and Breathing

At the very end of the chain, the electrons are passed to oxygen. Each oxygen atom accepts a pair of electrons and takes a pair of hydrogen ions (H^+) from the material inside the cell to form water, H_2O. Believe it or not, this is the reason that we need to breathe oxygen. The steady supply of oxygen that is taken in is needed throughout the body for just one reason—to accept the electrons at the end of the electron transport chain. Without oxygen, electron transport could not take place, the Krebs cycle would stop, and ATP would not be produced.

☑ **Checkpoint** What is the role of oxygen in the electron transport chain?

Energy and Food

Even though glucose has been used as the example to show how food energy is utilized to produce ATP, the very same pathways are used for other food compounds. For example, a complex carbohydrate such as starch is broken down into simple sugars, most of which can then be converted into glucose. Most lipids and many proteins can be broken down into molecules that either enter glycolysis or the Krebs cycle at one of several places. Like a furnace that can burn wood, coal, or oil, the cell can generate chemical energy in the form of ATP from just about any source.

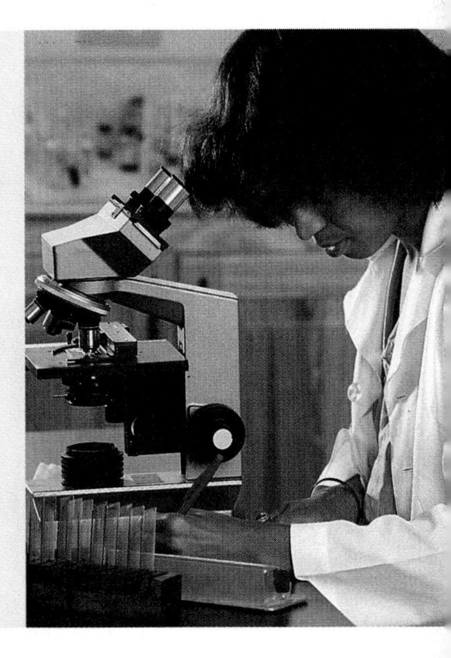

Figure 4–12
CAREER TRACK
In the laboratory, biological technicians help scientists by setting up equipment, performing experiments, and gathering information.

Section Review 4–3

1. **Define** respiration.
2. **Explain** the Krebs cycle.
3. **Critical Thinking—Analyzing** What is the relationship between breathing and the process of respiration?

SECTION 4–4

Photosynthesis

Guide for Reading

- **Define** photosynthesis.
- **Compare** the light-dependent reactions and light-independent reactions of photosynthesis.

MINI LAB

- **Design an experiment** to find out what happens to white light as it passes through a prism.

WHERE DOES THE ENERGY IN food come from? This depends on the source of the food, of course. The ultimate source of all food energy is the sun. Green plants and microscopic organisms are able to trap the energy of sunlight and use that energy to build the high-energy complex molecules that make up food. Whether an animal is a meat-eater or a plant-eater, the source of energy in its diet can be traced back to plants. Plants obtain the energy they need from the sun.

The Process of Photosynthesis 

Photosynthesis is the process by which green plants use the energy of sunlight to produce carbohydrates. In an overall sense, **photosynthesis** is the reverse of respiration. In respiration, the cell used the high-energy electrons found in glucose to provide energy to make ATP. The products of respiration—carbon dioxide and water—contain low-energy electrons. What does the cell have to do to make the process run in reverse? Quite simply, it must find a way to take those low-energy electrons and raise their energy levels to the point where their atoms can rearrange to form glucose. The cell must trap the sun's energy and pass it along to these electrons.

Sunlight

The sun bathes the Earth in a steady stream of sunlight. Sunlight provides

Figure 4–13 *Light in the form of sunlight is responsible for the process of photosynthesis in (a) green algae (magnification: 500X) and in (b) trees.*

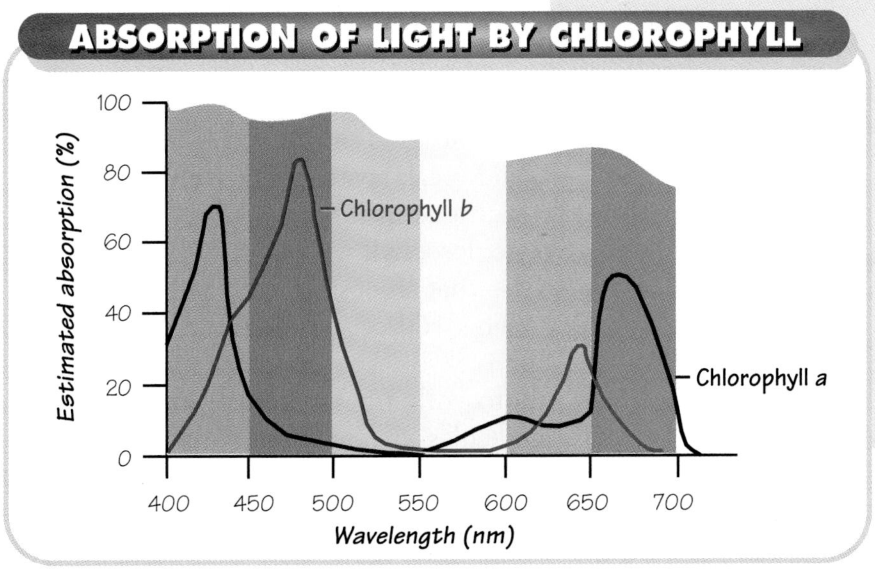

ABSORPTION OF LIGHT BY CHLOROPHYLL

Figure 4–14
Chlorophyll a *and chlorophyll* b *absorb slightly different wavelengths of light. The height of each line represents the amount of light of a particular color absorbed by each pigment. Neither form of chlorophyll absorbs light well in the middle of the spectrum, which is why plants are green.*

the energy to warm the Earth and to drive the process of photosynthesis. What your eyes perceive as white light from the sun is actually a mixture of different wavelengths of light. Many of these wavelengths are visible to your eyes and make up what is known as the visible spectrum. Your eyes see the different wavelengths of the visible spectrum as different colors.

Chlorophyll

Plants contain **pigments.** Pigments are colored substances that reflect or absorb light. These pigments help plants to gather the sun's energy. The principal pigment of green plants is **chlorophyll.** Chlorophyll absorbs light very well in the blue and in the red regions of the spectrum, as you can see in ***Figure 4–14.*** However, chlorophyll does not absorb light in the green region of the spectrum. This is what gives chlorophyll its green color and explains why plants appear green to the human eye.

Because light is a form of energy, a compound that absorbs light also absorbs the energy from the light. When chlorophyll absorbs light, much of that energy is transferred directly to electrons in the chlorophyll molecule. In other words, chlorophyll absorbs light energy and produces its own high-energy electrons.

☑ ***Checkpoint*** What is chlorophyll?

Light-Dependent Reactions

There are two stages of photosynthesis. The first stage consists of the **light-dependent reactions.** Light-dependent reactions get their name because they require the direct involvement of light.

Electron Transport

The chlorophyll in green plants is found inside photosynthetic membranes of the chloroplast in clusters called photosystems. Light absorption by a photosystem produces high-energy electrons. These electrons move to an electron transport chain in the membrane, and are passed through the chain from one electron carrier to the next. As electrons are passed along the chain, their energy level is reduced. Some of this energy is used to produce ATP.

At the end of the photosynthetic electron transport chain, additional light energy is absorbed, raising the energy level of the electrons again. These high-energy electrons are passed to an electron carrier, $NADP^+$, forming NADPH. Then the NADPH transfers the electrons—energy and all—to a chemical reaction elsewhere in the cell.

Replacing Electrons

The electrons that were removed from the chlorophyll need to be replaced, or in just a few milliseconds, chlorophyll

MINI LAB .... Experimenting ....

Breaking Out of Prism

PROBLEM *How is white light separated into the colors of the visible spectrum?* ***Design an experiment*** *to find out.*

PROCEDURE

1. Obtain a sheet of white paper, a flashlight (or a lamp), and a prism.
2. Formulate a hypothesis to explain what will happen to the light when it is passed through a prism.
3. Design an experiment to test your hypothesis.

ANALYZE AND CONCLUDE

1. What happened to the light as it passed through the prism?
2. What do you think the prism does to the wavelengths of light, which enables you to see the different colors?
3. What colors of the visible spectrum did you observe?

would lose so many electrons that it would break apart. Fortunately, the photosynthetic membrane has a way to supply chlorophyll with electrons to replace those that are lost.

The source of electrons is water. Enzymes on the inner surface of the photosynthetic membrane split water molecules in two. Four electrons are removed from 2 molecules of water, leaving 4 hydrogen ions and 2 oxygen atoms. The electrons are returned to the chlorophyll. This reaction is the source of nearly all the oxygen in the Earth's atmosphere.

The light-dependent reactions produce two important products—ATP and high-energy electrons carried by $NADP^+$. These compounds have an important role in the plant cell: They provide energy to make carbohydrates.

☑ ***Checkpoint*** What are the light-dependent reactions?

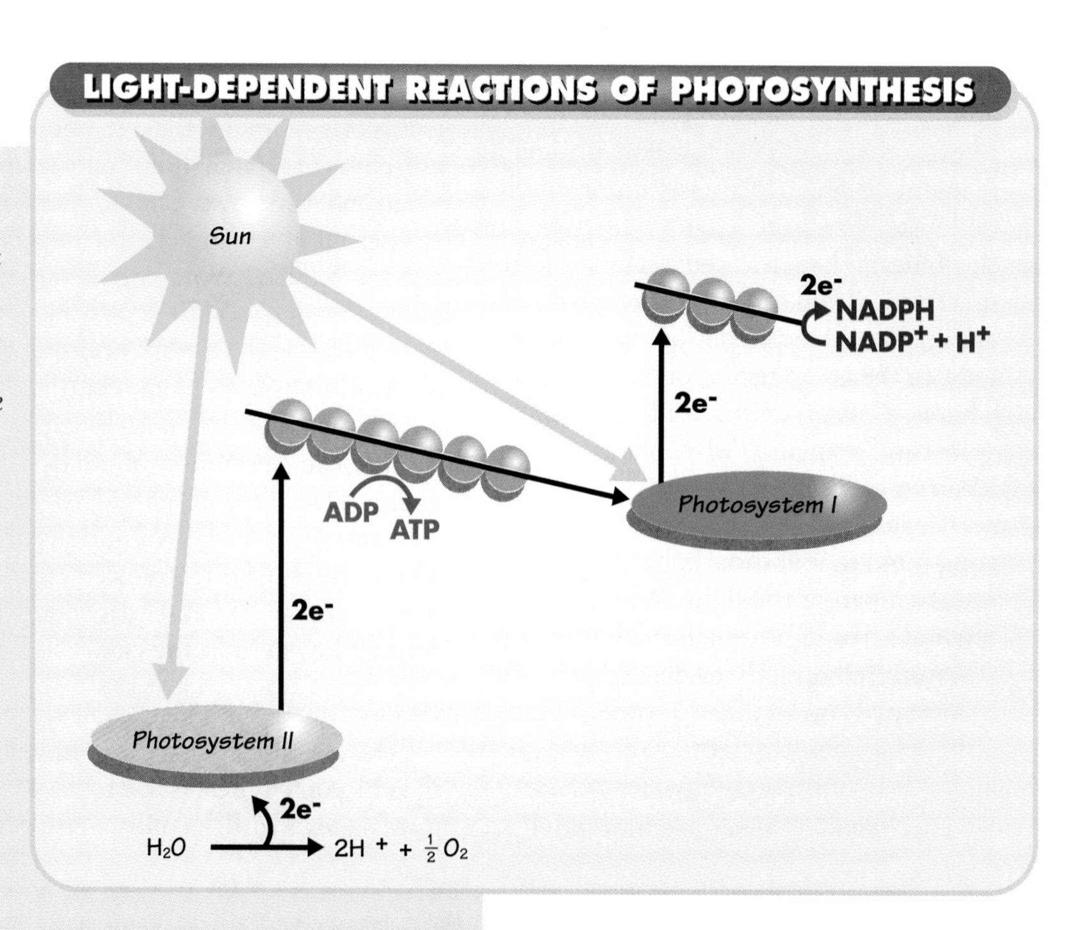

Figure 4–15
During the light-dependent reactions, light is absorbed by clusters within the photosynthetic membranes called photosystems. These photosystems (I and II) produce high-energy electrons that move to electron transport chains in the membrane. Here the electrons are passed from one electron carrier to the next, producing NADPH and ATP, both of which are needed for the second stage of photosynthesis.

Visualizing Photosynthesis

Photosynthesis is the process by which plants use the energy of sunlight to produce carbohydrates. Photosynthesis can be summarized in the following equation:

$$6CO_2 + 6H_2O + \text{Energy} \rightarrow C_6H_{12}O_6 + 6O_2$$

$$\text{Carbon dioxide} + \text{Water} + \text{Energy (sun)} \rightarrow \text{Glucose} + \text{Oxygen}$$

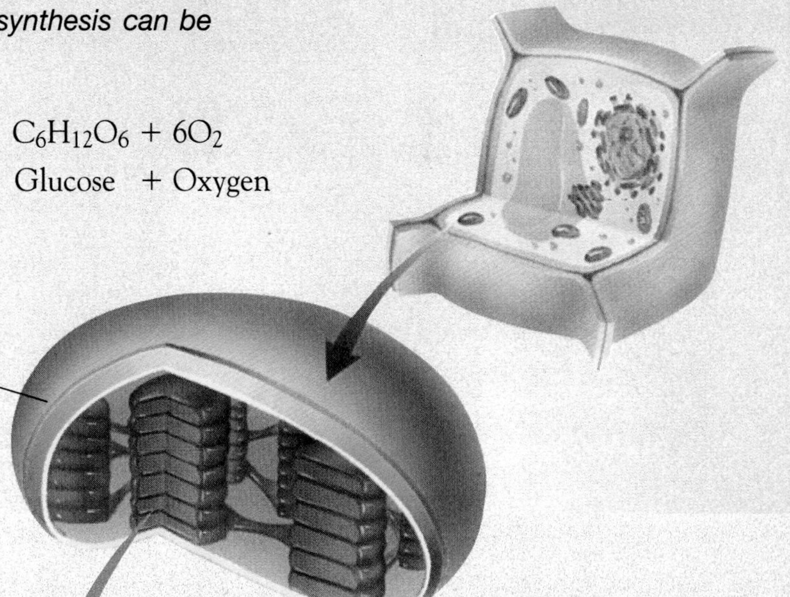

Chloroplast

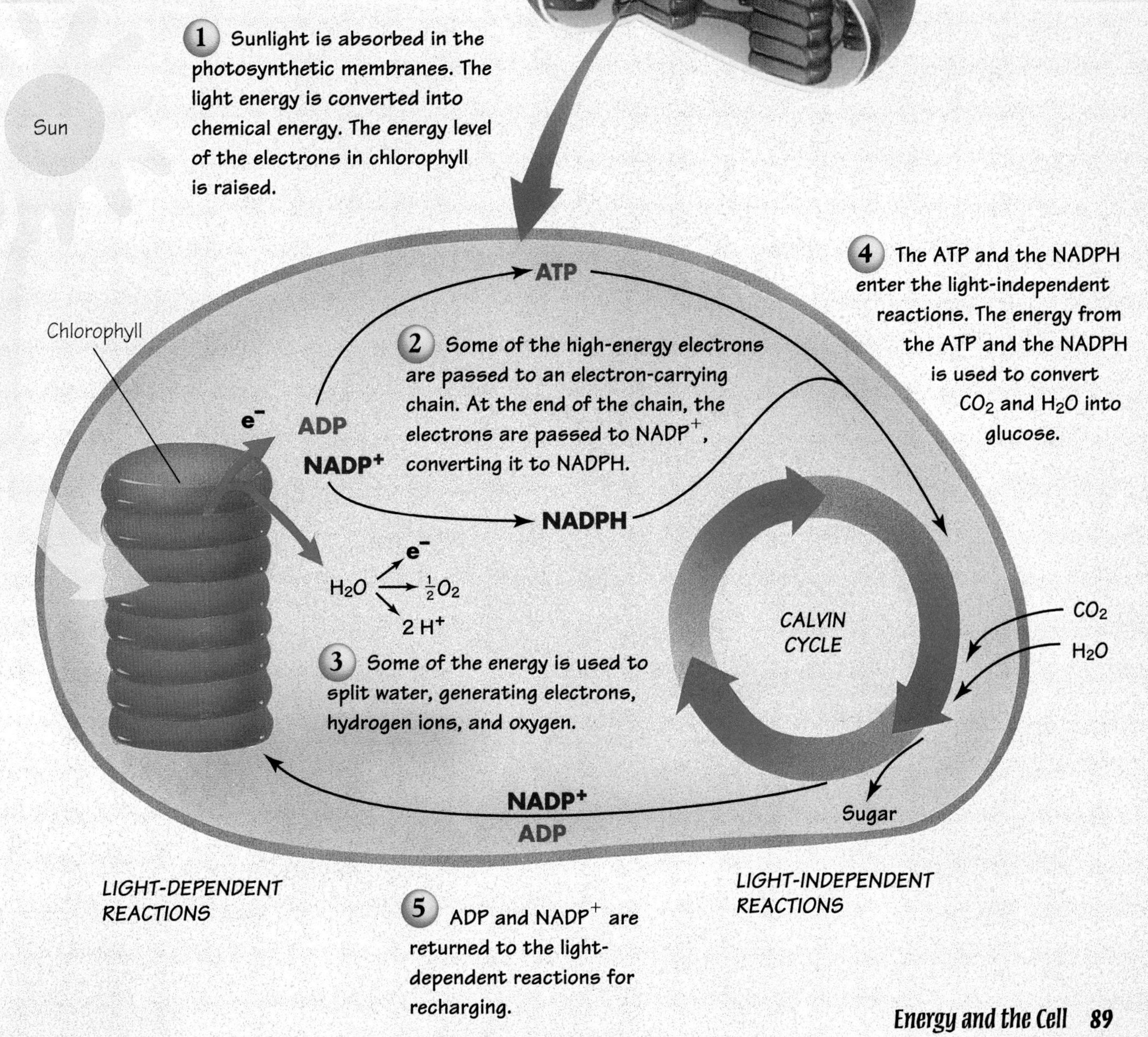

Sun

1 Sunlight is absorbed in the photosynthetic membranes. The light energy is converted into chemical energy. The energy level of the electrons in chlorophyll is raised.

Chlorophyll

e^-

ATP

ADP

$NADP^+$

2 Some of the high-energy electrons are passed to an electron-carrying chain. At the end of the chain, the electrons are passed to $NADP^+$, converting it to NADPH.

NADPH

$H_2O \rightarrow e^-$, $\frac{1}{2}O_2$, $2\,H^+$

3 Some of the energy is used to split water, generating electrons, hydrogen ions, and oxygen.

4 The ATP and the NADPH enter the light-independent reactions. The energy from the ATP and the NADPH is used to convert CO_2 and H_2O into glucose.

CALVIN CYCLE

CO_2

H_2O

Sugar

$NADP^+$

ADP

LIGHT-DEPENDENT REACTIONS

LIGHT-INDEPENDENT REACTIONS

5 ADP and $NADP^+$ are returned to the light-dependent reactions for recharging.

Problem Solving

DESIGNING AN EXPERIMENT

Variations in Light and Dark

As you enter the laboratory, your teacher tells you that she has a mystery for you to solve. You are to determine if photosynthesis can take place in the dark.

You are to formulate a hypothesis for the mystery and then test your hypothesis by designing an experiment. Your experiment will require the materials and clues your teacher has given you.

The materials provided to you include Bromthymol blue, a drinking straw, *Elodea* (an aquatic plant), and beakers. Your teacher gives you two clues: (1) Bromthymol blue will turn yellow in the presence of carbon dioxide but remains blue when oxygen is present, and (2) drawings of the experimental setup and the control setup.

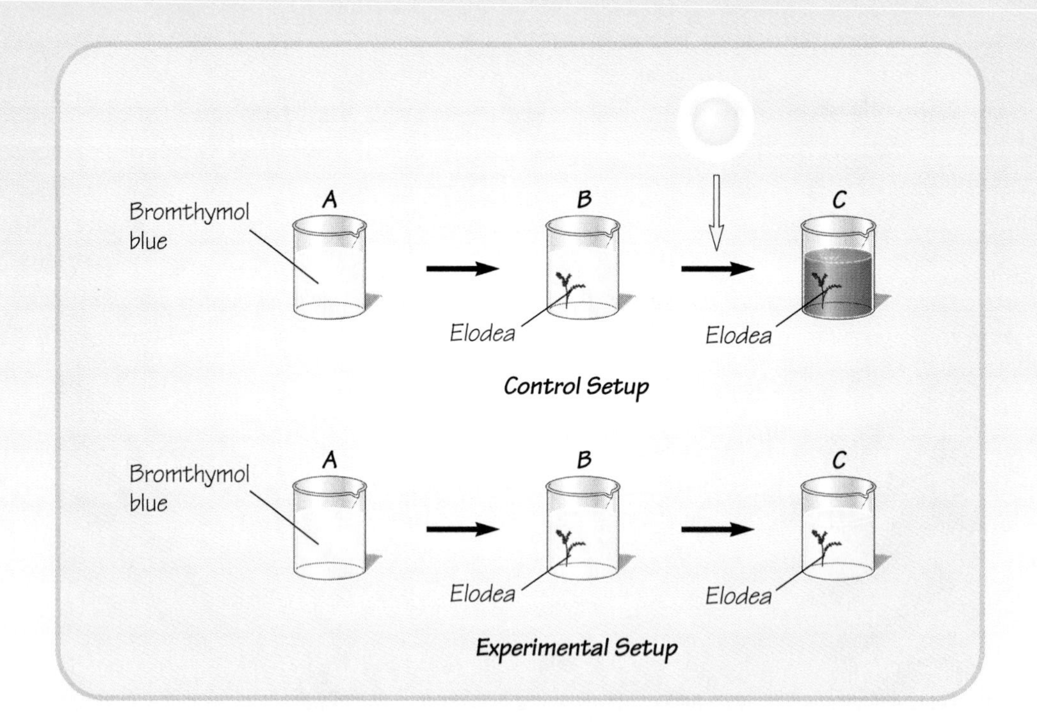

THINK ABOUT IT

1. Formulate a hypothesis for the mystery.
2. Outline your procedure.
3. What is the purpose of the control setup?
4. Based on the data provided and your knowledge of biology, does photosynthesis take place in the dark? Explain your answer.

Figure 4–16
During the Calvin cycle, carbon dioxide combines with a 5-carbon compound to form 2 molecules of PGA. Using the energy provided by ATP and NADPH produced by the light-dependent reactions, PGA can then be converted into PGAL, which, in turn, is used to make sugars.

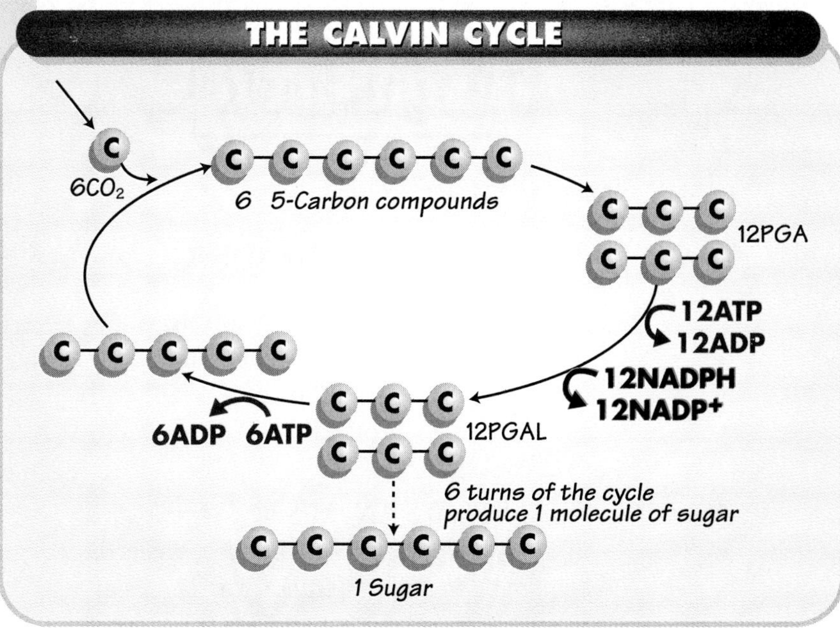

The Light-Independent Reactions

The second stage of photosynthesis consists of the **light-independent reactions.** This set of reactions gets its name because it does not directly involve light. **The light-independent reactions convert the energy from the ATP and NADPH into a form that can be stored indefinitely—sugars.**

The raw materials from which the cell builds sugars are carbon dioxide and water. Carbon dioxide from the atmosphere is combined with a 5-carbon sugar to produce two 3-carbon molecules called PGA. The 2 PGA molecules then enter a chemical pathway that uses the energy from ATP and NADPH. The pathway is called the **Calvin cycle,** after American scientist Melvin Calvin, who discovered it.

Each turn of the Calvin cycle uses 1 molecule of carbon dioxide and 2 atoms of hydrogen. After six turns, a 6-carbon sugar molecule is produced. As photosynthesis continues, the Calvin cycle works steadily, producing energy-rich sugars and removing carbon dioxide from the atmosphere. The plant is able to use the sugars produced in this way both to meet its energy needs and to build larger and more complex molecules, such as cellulose, that it needs for growth and development.

The products of the light-dependent reactions in photosynthesis are used to provide the energy to build energy-containing sugars from low-energy compounds. In this way, the two sets of reactions work together—the light-dependent reactions trap the energy of sunlight in chemical form and the light-independent reactions use that chemical energy to produce stable, high-energy sugars from carbon dioxide and water.

Section Review 4–4

1. **Define** photosynthesis.
2. **Compare** the light-dependent reactions and the light-independent reactions.
3. **Critical Thinking—Synthesizing** In the autumn, green plants stop producing chlorophyll. How does this explain why leaves turn color in the fall?
4. **MINI LAB** How would you **design an experiment** to find out what happens to white light when it passes through a prism?

SECTION 4–5 BRANCHING OUT In Depth

ATP Synthesis

Guide for Reading

- **Describe** the role that photosynthetic and mitochondrial membranes play in the production of ATP.

ATP PLAYS AN IMPORTANT ROLE *in every living cell. ATP is also a central compound in both respiration and photosynthesis. As you have read, ATP is produced from ADP in both mitochondria and chloroplasts. In both places, ATP is made by an electron transport chain that is associated with a membrane. Is this just a coincidence or is there some special connection between these membranes and ATP production?*

Membranes and ATP

In the early 1960s, British scientist Peter Mitchell thought that he could explain the connection between the membranes and ATP. Mitchell suggested that the changes that took place in the membrane during electron transport were essential in the production of ATP. In fact, Mitchell argued, the purpose of electron transport is to produce different electrical charges on each side of the membrane and then to use those differences to power the production of ATP.

ATP and Photosynthesis

Recall that during light-dependent reactions of photosynthesis, some of the high-energy electrons are passed to an electron-carrying chain. At the end of the chain, some of the energy is used to "split" water. When the water molecule is broken apart, the oxygen is released and the electrons are returned to chlorophyll. But where do the hydrogen ions go?

The hydrogen ions are released inside the photosynthetic membrane when the electrons are removed. This produces a high concentration of hydrogen ions (H^+) inside the membrane, giving it a positive charge. Outside the membrane, there is a low concentration of hydrogen ions, which gives it a negative charge.

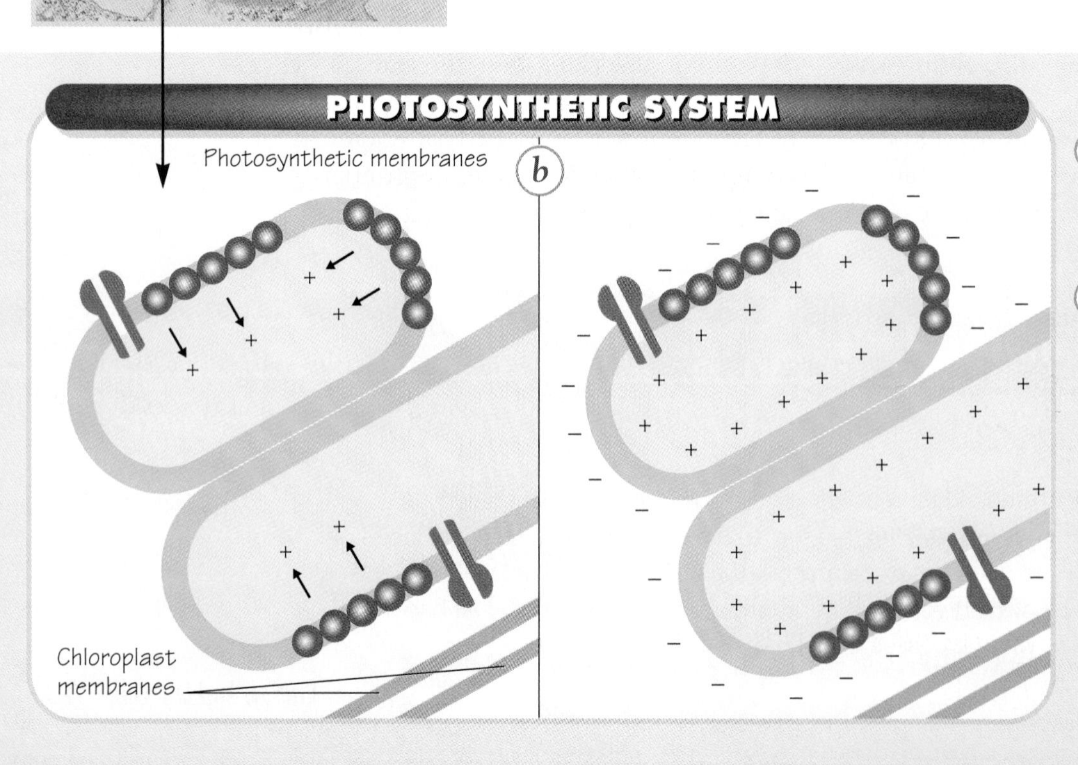

Figure 4–17
(a) The boxed area shows part of a chloroplast's photosynthetic membrane (magnification: 4500X). (b) Electron transport pumps H^+ ions across this membrane (left), resulting in a high H^+ ion concentration inside the membrane (right). In mitochondria a similar process takes place. However, a high H^+ ion concentration builds up in the space between the inner and outer membrane.

ATP and Respiration

During the process of respiration, something very similar takes place in the mitochondrion. The electron carriers, NADH and $FADH_2$, pass their electrons to the electron transport chain. As these electrons are passed down the chain, they provide the energy to pump hydrogen ions out of the membrane. Because there are more hydrogen ions outside the membrane, there is a more positive charge. Inside the membrane, where there are fewer hydrogen ions, there is a negative charge.

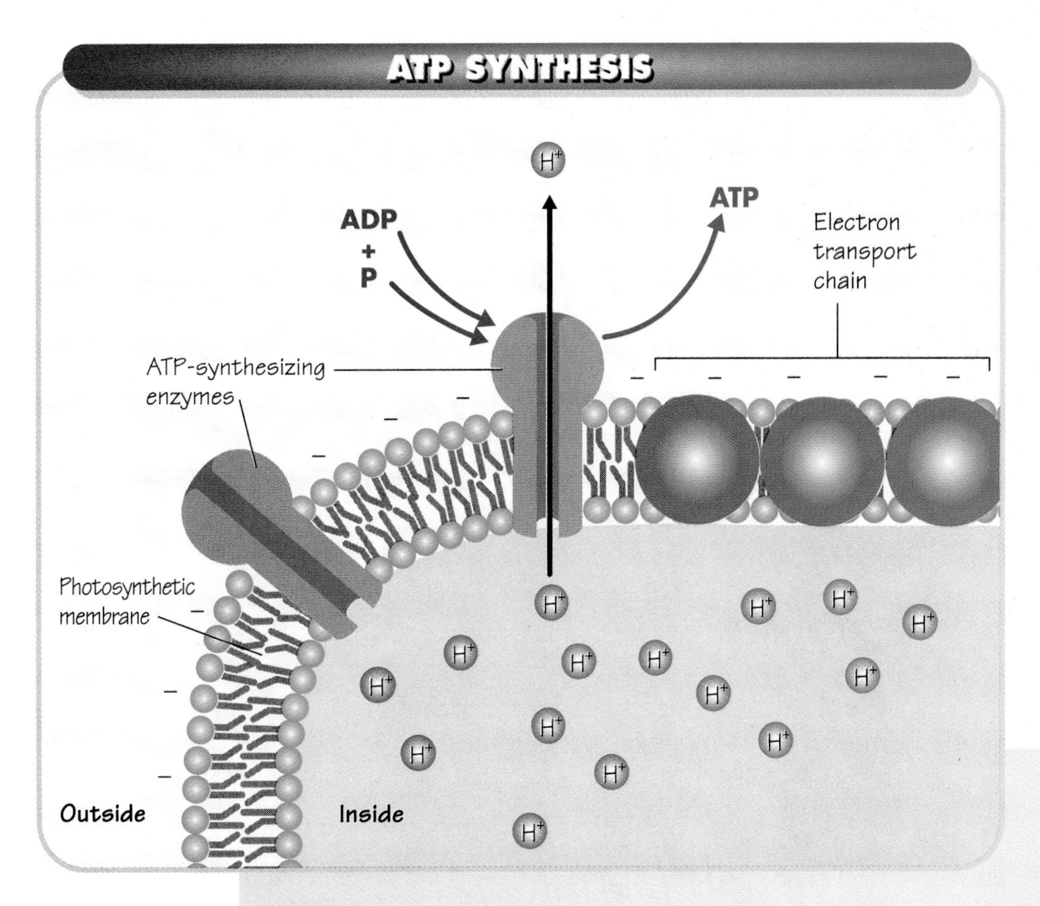

Figure 4–18
The force of the outward movement of H^+ ions that were pumped inside the photosynthetic membrane by electron transport provides the energy to produce ATP. A similar process produces ATP in mitochondria.

Chemiosmosis

The difference in electrical charges across the photosynthetic and mitochondrial membranes is a source of energy. This energy could be used to attach a phosphate to ADP to make ATP. ***Figure 4–18*** shows how this occurs. The photosynthetic and mitochondrial membranes are impermeable to hydrogen ions. However, there is an enzyme that seems to have a channel right through its center. The channel allows the hydrogen ions (H^+) to pass through it, drawn by the strong negative charges (from OH^- ions) on the other side of the membrane. This enzyme is called an ATP-synthesizing enzyme, meaning an "ATP-maker."

As the hydrogen ions pass through this channel, the energy from the movement is used to attach a phosphate to ADP, making ATP. This process of ATP formation in chloroplasts and mitochondria is called **chemiosmosis.**

Section Review 4–5

1. **Describe** the role of photosynthetic and mitochondrial membranes in the production of ATP.
2. **BRANCHING OUT ACTIVITY** Using reference books, find out more about Peter Mitchell. **Summarize** your findings in a brief report.

CHAPTER 4

Laboratory Investigation

Tiny Bubbles

Yeast are single-celled organisms that use sugar as a food source. In this investigation, you will observe the substances produced by yeast cells from the breakdown of food molecules.

Problem

Predict which kind of sugar produces the most rapid fermentation in yeast cells.

Materials (per group)

5 large test tubes
5 disposable plastic pipettes
weight to fit on pipette stem
thermometer
yeast solutions
- yeast-sucrose
- yeast-glucose
- yeast-lactose
- yeast-molasses
- yeast-water

glass-marking pencil
watch with second hand or timer

Procedure   

1. Label 5 test tubes from 1 to 5.
2. Fill the bulb section of a pipette with the yeast-sucrose solution. To fill the pipettes, pull up as much liquid as possible into the stem by squeezing the bulb and then slowly releasing it. Turn the pipette upside down and tap the pipette to move the liquid into the bulb. Keep the pipette upside down.
3. Attach a weight to the pipette stem just above the bulb and place the pipette (still upside down) into test tube 1, which is three-quarters

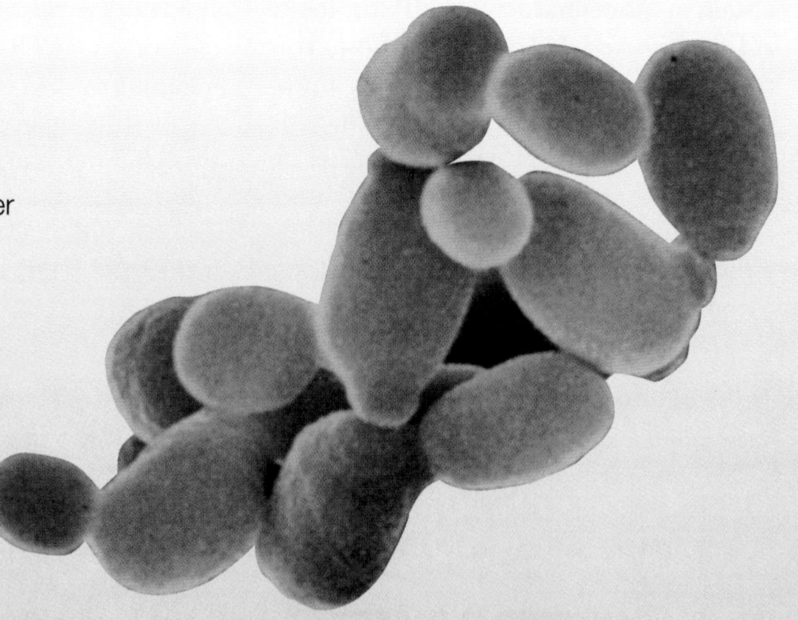

full with warm water (about 37°C). The pipette should be completely covered by the water and have about 2 to 3 cm of water above the tip.

4. You should observe tiny bubbles being released from the tip of the pipette. Count the number of bubbles released over a period of 10 minutes and record your results.

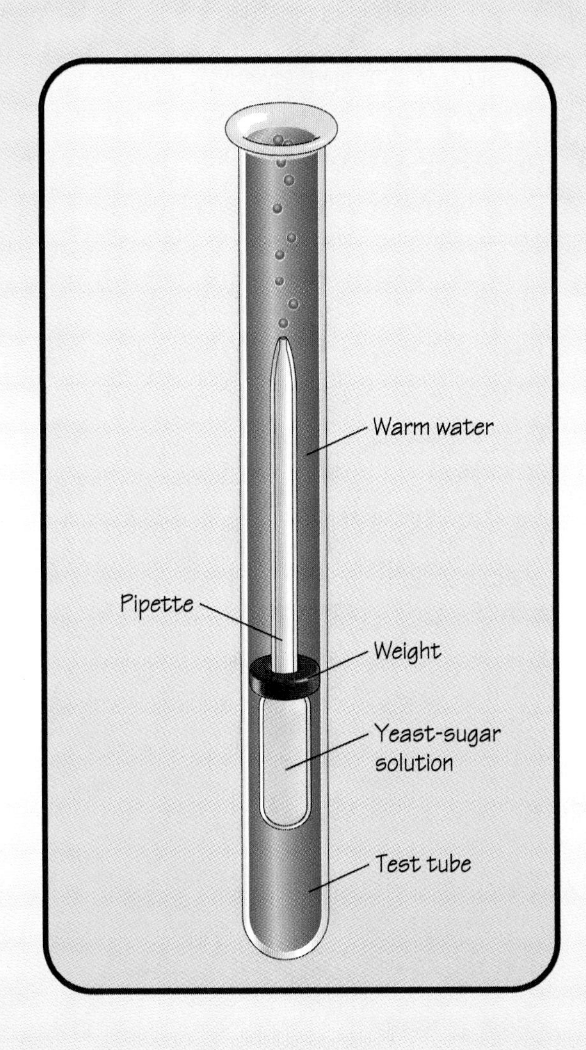

5. Repeat steps 2 to 4 for each of the remaining solutions. Record your observations for each.

6. Collect data from the class and calculate the average number of bubbles produced by each solution. Construct a graph of your data and the class average.

Observations

1. What differences do you note in the number of bubbles that you see?
2. How did your data compare with the class average?

Analysis and Conclusions

1. What was the purpose of the yeast-and-water solution?
2. What is the name of the gas that is inside the bubbles released from the solutions? Can you design a test to identify this gas?
3. Explain how the counting of the gas bubbles is a way of measuring fermentation.
4. Which sugar was the best food source for yeast? Give evidence to support your answer.
5. How does the molasses differ from the other sugars that you used?

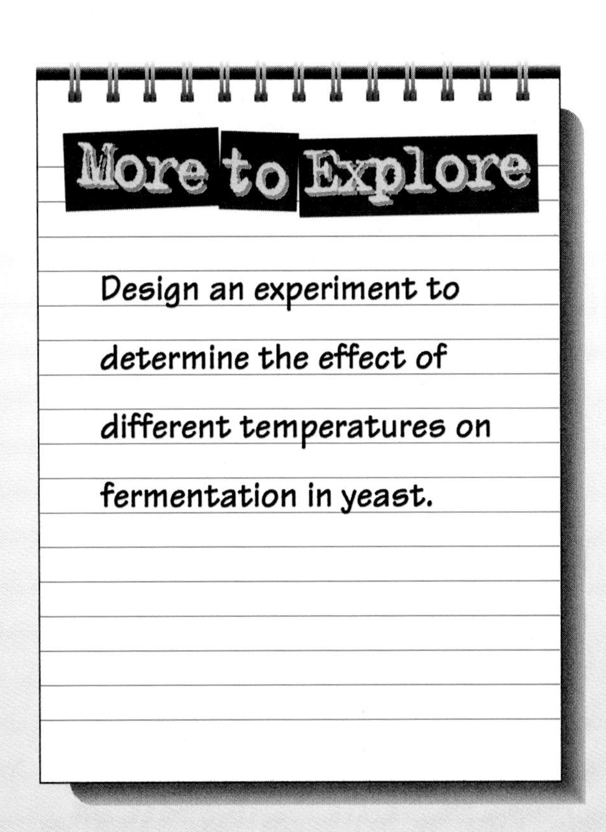

Study Guide

Summarizing Key Concepts

The key concepts in each section of this chapter are listed below to help you review the chapter content. Make sure you understand each concept and its relationship to other concepts and to the theme of this chapter.

4–1 Chemical Energy and Life

- Without the ability to produce and use energy, living things could not survive.
- Energy in the cell comes from ATP. The high-energy bonds in ATP are used for storing and releasing energy.

4–2 Making ATP Without Oxygen

- Glycolysis is the series of reactions in which a molecule of glucose is broken down.
- Fermentation is the process that allows glycolysis to continue without oxygen, producing ATP. There are two types of fermentation—lactic acid and alcoholic fermentation.

4–3 Respiration

- Respiration is the release of energy from the breakdown of food molecules in the presence of oxygen.
- The pyruvic acid from glycolysis is broken down further in the Krebs cycle.

4–4 Photosynthesis

- Photosynthesis is the process by which green plants use the energy of the sun to produce carbohydrates. There are two sets of reactions that take place—the light-dependent reactions and the light-independent reactions.

4–5 ATP Synthesis

- The difference in electrical charges across the photosynthetic and mitochondrial membranes is a source of energy. This energy could be used to attach a phosphate to ADP to make ATP.
- ATP is generated when hydrogen ions move from a high concentration to a low concentration. This is called chemiosmosis.

Reviewing Key Terms

Review the following vocabulary terms and their meaning. Then use each term in a complete sentence.

4–1 Chemical Energy and Life

ATP

4–2 Making ATP Without Oxygen

glycolysis
fermentation
lactic acid fermentation
alcoholic fermentation

4–3 Respiration

respiration
Krebs cycle
electron transport chain

4–4 Photosynthesis

photosynthesis
pigment
chlorophyll
light-dependent reaction
light-independent reaction
Calvin cycle

4–5 ATP Synthesis

chemiosmosis

Recalling Main Ideas

Choose the letter of the answer that best completes the statement or answers the question.

1. The cell's main energy-storing compound is

a. FADH. **c.** ATP.
b. NAD. **d.** AMP.

2. Pyruvic acid is a product of

a. respiration. **c.** fermentation.
b. photosynthesis. **d.** glycolysis.

3. During vigorous activity, large body muscles produce

a. alcohol. **c.** glucose.
b. lactic acid. **d.** starch.

4. During respiration, the final acceptor of electrons in the electron transport chain is

a. oxygen. **c.** water.
b. carbon dioxide. **d.** ATP.

5. Carbon is released from the Krebs cycle in the form of

a. glucose. **c.** citric acid.
b. carbon dioxide. **d.** water.

6. The pigment in green plants, where photosynthesis takes place, is

a. chlorophyll.
b. mitochondrion.
c. the ATP-synthesizing enzyme.
d. the photosynthetic membrane.

7. The Calvin cycle is part of

a. the light-dependent reactions.
b. respiration.
c. the light-independent reactions.
d. fermentation.

8. The movement of hydrogen ions across a membrane to generate energy to form ATP is called

a. chemiosmosis. **c.** respiration.
b. photosynthesis. **d.** fermentation.

Putting It All Together

Using the information on pages xxx to xxxi, complete the following concept map.

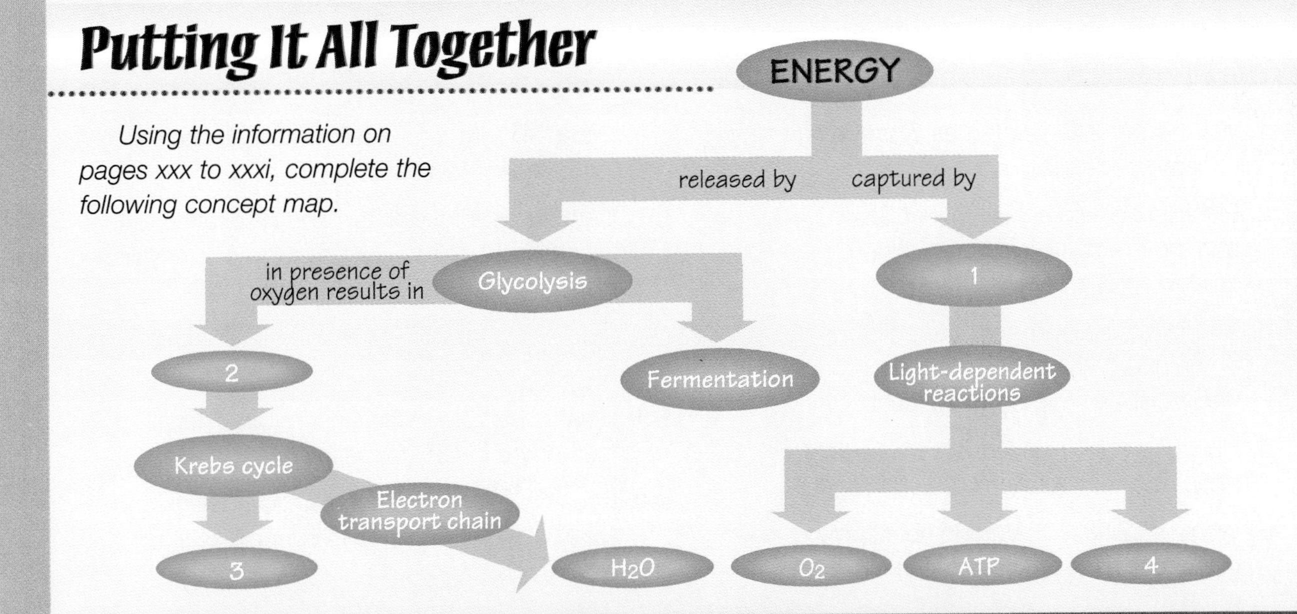

Reviewing What You Learned

Answer each of the following in a complete sentence.

1. What is the ultimate source of all energy used by life on our planet?
2. Describe the components of an ATP molecule.
3. Observe the equation for respiration:

   $C_6H_{12}O_6 + 6O_2 \rightarrow 6CO_2 + 6H_2O$

   What are the products? The reactants?
4. Where is energy stored in macromolecules?
5. What is PGAL, and where do you find it?
6. Where does glycolysis occur in a eukaryotic cell?
7. How many molecules of ATP are produced in the breakdown of glucose in the absence of oxygen? What is this process called?
8. What is the electron transport chain?
9. In essence, photosynthesis is the reverse of what biological process?
10. How are the functions of FAD, NAD^+, and $NADP^+$ similar?

Expanding the Concepts

Discuss each of the following in a brief paragraph.

1. What is the role of enzymes in providing energy for living systems?
2. Electron carriers are key molecules in glycolysis and the Krebs cycle. Explain their role in each process.
3. If oxygen was not present to accept electrons at the end of the electron transport chain, what would happen to the overall reaction?
4. How does alcoholic fermentation differ from lactic acid fermentation?
5. What portions of the visible spectrum do green plants use as their energy source?
6. Compare light-dependent reactions and light-independent reactions.
7. Compare photosynthesis and respiration to the processes that occur in an aluminum recycling plant.
8. Diagram the Calvin cycle and the production of a sugar.
9. What is the role of the inner mitochondrial membrane in ATP formation?
10. Does chemiosmosis occur in both plants and animals? Give evidence to support your answer.

Extending Your Thinking

Use the skills you have developed in this chapter to answer the following.

1. **Comparing** Photosynthesis and respiration have often been described as opposite reactions. List both similarities and differences of these two major reactions.
2. **Predicting** How might the buildup of carbon dioxide in the atmosphere affect the process of photosynthesis? Is this something we need to be concerned about? Explain your answer.
3. **Observing** Write the chemical equation for the reaction of glucose and oxygen. What physical characteristic of this reaction can you observe? From what reactant is the oxygen in the resultant product, water, derived?
4. **Designing an experiment** Design an experiment to determine the optimal light wavelength for a plant.
5. **Applying concepts** Isotope tracer technology is a major tool used in understanding living reactions. How might Calvin have used this technology to better understand the many reactions of photosynthesis?

Applying Your Skills

The Energizer

Everything alive requires energy. Energy for the cell is stored in the compound ATP and is then released as needed. Actually, ATP energizes the cell in much the same way that a battery provides energy for a flashlight. Just as a battery needs to be recharged to provide that energy, cells need to replenish their supply of ATP. ATP is like a fully charged battery, ready to supply energy to the cell for the cell to do its work.

1. Working in a group, obtain a flashlight bulb, some insulated wire, and several D-cell batteries.
2. Connect one battery to the wire and the bulb. Is there any evidence that there is energy in your electrical system?
3. Now add a second battery, end to end with the first battery, and then add the wire and bulb to your system. Did the light bulb become brighter or did it stay the same?

• GOING FURTHER •

4. If you add a third battery, end to end to the other batteries, will the bulb get even brighter?
5. Explain how the chemical energy in the batteries is similar to the energy stored in the bonds of ATP.

CHAPTER 5

Cell Division and Specialization

FOCUSING THE CHAPTER THEME: Systems and Interactions

5–1 Cell Growth and the Cell Cycle
- **Describe** cell growth and **explain** the cell cycle.

5–2 Cell Division
- **Summarize** the process of mitosis.

5–3 Cell Specialization and Organization
- **Sequence** the levels of organization in multicellular organisms.

BRANCHING OUT *In Depth*

5–4 Controlling the Cell Cycle
- **Explain** how the cell cycle is controlled.

LABORATORY INVESTIGATION
- **Observe** the various phases of mitosis in a plant cell.

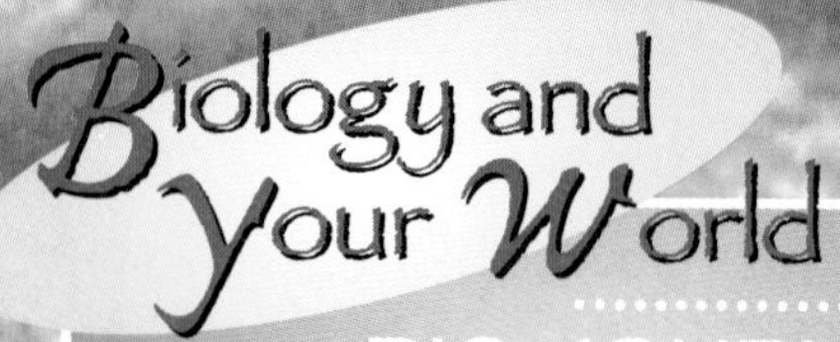

BIO JOURNAL

Think about an organization, such as a club or team, of which you're a member. In your journal, explain how your organization is similar to the organization in multicellular organisms.

Cell division in an animal cell

SECTION 5–1

Cell Growth and the Cell Cycle

Guide for Reading

- **Define** the cell cycle.
- **Describe** the four phases of the cell cycle.

MINI LAB

- **Analyze** the relationship between the size of a cell and diffusion.

A BEAR, LIKE ALL animals, is made up of cells. Bear cubs, like other newborn animals, are much smaller than adults. What happens to the cells of a bear cub as it grows into an adult? Does a living thing get larger because its cells increase in size? Or does it get larger because it produces more and more cells? In most cases, a living thing grows because it produces more and more cells. On the average, the cells of an adult bear are no larger than those of a bear cub—there are just a lot more of them.

Cell Growth

Is there a reason why organisms grow by producing more cells? Couldn't an organism grow just by allowing its cells to get larger and larger? Or are there limits that prevent cells from getting too big? In a way, there are limits as to how large a cell can become.

Suppose that a cell were to double its diameter, as shown in ***Figure 5–2*** on the next page. As you can see, the internal volume of the cell is now eight times as great, while the surface area of the cell is only four times as great. ● Clearly, the larger a cell gets, the more difficult it is to get things in and out of it. If cells grew too large, they would not be able to supply their own needs, and growth would come to a stop. This is one of the main reasons why cells do not grow much larger even if the organism itself does.

INTEGRATING MATH

Calculate the volume/surface-area ratio for a cube in which each side is 4 cm.

Checkpoint Why is a small cell more efficient than a large cell?

Figure 5–1
(a) Although the cells in these bear cubs are the same size as the cells in the adult bear, the adult bear has many more cells. The same is true for (b) this adult oak tree and (c) the oak sapling.

THE RATIO OF SURFACE AREA TO VOLUME IN CELLS

Cell Size	Volume (length x width x height)	Surface Area (number of surfaces x length x width)	Surface Area/ Volume Ratio
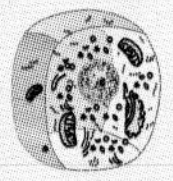	1 cm x 1 cm x 1 cm = 1 cm^3	6 x 1 cm x 1 cm = 6 cm^2	6 cm^2/1 cm^3 = 6/cm
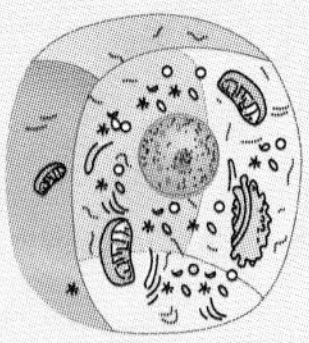	2 cm x 2 cm x 2 cm = 8 cm^3	6 x 2 cm x 2 cm = 24 cm^2	24 cm^2/8 cm^3 = 3/cm

Figure 5–2
As a cell doubles its diameter, its surface area increases by four, while its internal volume increases by eight.

The Cell Cycle

In most animals and plants, cells increase in size and then divide into two cells. These two new cells increase in size and then divide again. The process continues over and over. This regular series of events, called the **cell cycle,** occurs in eukaryotic cells. **The cell cycle is the period of time from the beginning of one cell division to the beginning of the next.** What happens to the cell during the cell cycle? Basically, the cell doubles its contents so that it is ready to divide into two completely independent cells.

The Cell Divides

Thousands of events take place during the cell cycle, but two of those events are so important that they are used as landmarks to define everything else. One of the events is **cell division.** Cell division is the process in which the cell divides into two independent cells, called daughter cells. In eukaryotic cells, this process is called **mitosis.** The period of time when mitosis takes place is the M phase of the cell cycle.

☑ ***Checkpoint*** What is cell division?

The Cell Copies Its Chromosomes

The other major event in the cell cycle is the copying of the chromosomes that contain the cell's genetic information. When the cell copies this information, it synthesizes, or makes, a duplicate set of DNA molecules. As a result, this part of the cell cycle is called the S phase.

Because most cells do not begin DNA synthesis right after cell division, there is a time gap between the end of one M phase and the beginning of an S phase. This time gap is called the G_1 phase. A similar gap, the G_2 phase, occurs between the end of the S phase and the beginning of the M phase.

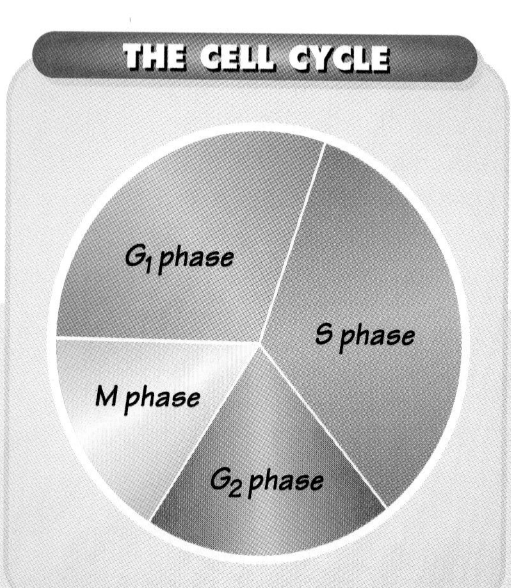

Figure 5–3
The cell cycle consists of four phases. During G_1, most of the cell's growth and activity take place. After the cell leaves G_1, it moves into the S phase, where chromosome replication takes place. After the S phase, the cell enters G_2, where it makes final preparations for cell division. Cell division takes place during the M phase.

Problem Solving

INTERPRETING DATA

To Be or Not to Be . . .

On the fictional planet Cellmion, laboratory technicians are able to manufacture living boxes that can be used to create more complex life forms. The Cellmions have mastered this technology with one exception—the surface-area-to-volume ratio. If there is too little surface area for the volume of the living box, the substances needed for life are unable to enter the boxes in sufficient quantities. In addition, waste products, which are unable to leave the box quickly enough, will poison the box. As the head laboratory technician, you have determined that the optimal surface-area-to-volume ratio is 3:1. If a living box has a smaller ratio—for example, 2:1 or 1:1—the box will die because it has too little surface area to support its volume. Boxes with a 3:1 ratio or greater will live.

As the technicians create the boxes, you must check their surface-area-to-volume ratios. Laboratory technicians who have made boxes that will die because of a low surface-area-to-volume ratio must be retrained. The chart provides you with the names of the technicians and the dimensions of the boxes they have made.

LIVING BOXES

Technician	*Box Length*	*Box Width*	*Box Height*
Androm	1 cm	1 cm	1 cm
Shama	3 cm	3 cm	3 cm
Marcha	2 cm	1 cm	1 cm
Berich	4 cm	2 cm	1 cm

• THINK ABOUT IT •

1. How do you calculate the surface area of the boxes? The volume?
2. How do you calculate the surface-area-to-volume ratio?
3. Which technicians' boxes will live?
4. Which of the technicians will need to be retrained?

How Big Is Too Big?

PROBLEM ***Analyze*** *the relationship between cell size and diffusion.*

PROCEDURE   

1. Using a plastic spoon, place three agar blocks of different sizes into a 250-mL beaker. Cover the blocks with 0.4% sodium hydroxide solution. **CAUTION:** *Do not touch the blocks or the solution. If the solution splashes on your skin, wash it off immediately.*
2. Calculate the surface-area-to-volume ratio for each block. Record your calculations.
3. After 10 minutes, use tongs to remove the agar blocks and gently blot them dry with paper towels.
4. Cut each block in half using a scalpel. **CAUTION:** *Be careful with sharp instruments.* You will see that a dark-pink color has diffused into each block.
5. Using a metric ruler, measure the distance the pink color has diffused in each block. Record your measurements.

ANALYZE AND CONCLUDE

1. Which block has the greatest proportion of pink color?
2. Compare your answer to the surface-area-to-volume ratios you computed. How does the ratio relate to diffusion?

The Length of the Cell Cycle

Not all cells move through the cell cycle at the same rate. In the human body, most muscle and nerve cells do not divide at all after they have developed. In contrast, the rapidly dividing cells of an embryo can complete the cell cycle in as little as 30 minutes. However, the average length of the cell cycle in a typical human adult cell is about 20 hours.

In general, when a cell stops growing, it stops in the G_1 phase. Although there are a few exceptions, when a cell leaves the G_1 phase and enters the S phase, it will continue through the rest of the cell cycle and enter cell division.

Controlling Cell Growth

One of the most striking aspects of cell behavior in multicellular organisms such as humans is how carefully cell division and cell growth are controlled. Cells in certain places of the body, including the brain and the heart, rarely divide—if they divide at all. In contrast, the cells of the skin and the digestive tract divide rapidly throughout life, replacing cells that have broken down due to daily wear and tear.

Why is the cell cycle in humans regulated so carefully? One important reason is that a mistake in regulating the cycle can be fatal. Cancer, a disease in which some of the body's cells grow uncontrollably, causes thousands of deaths every year. The various types of cancer seem to have one thing in common. To one degree or another, cancer cells have lost the normal ability to regulate the cell cycle.

Section Review 5-1

1. **Define** the cell cycle.
2. **Describe** each of the four phases of the cell cycle.
3. **Critical Thinking—Inferring** What could a cell biologist learn about the cell cycle by studying cancer cells?
4. **MINI LAB Analyze** the relationship between cell size and diffusion.

Cell Division

SECTION
5–2

GUIDE FOR READING

- **Describe** the four phases of mitosis.
- **Compare** cytokinesis in plant and animal cells.

IN MOST PROKARYOTIC CELLS, cell division is a simple matter of a single cell separating into two daughter cells. Most bacteria and other prokaryotes have a special mechanism that makes sure a cell copies its genetic information before cell division begins. Prokaryotes also ensure that each of the two daughter cells gets its own copy of that information before the cells completely separate.

In a sense, eukaryotic cells do much the same thing. But the more complex structure of eukaryotic cells and the fact that their nuclei contain many chromosomes explain why cell division in eukaryotes is more complex.

Interphase

There are a few eukaryotic cells in which chromosomes are visible all the time, but these are the exceptions. As a general rule, chromosomes are not clearly visible—under either the light microscope or the electron microscope—during the G_1, S, and G_2 phases of the cell cycle. These three phases are usually called **interphase** because they are the phases that occur in between cell divisions.

During interphase, the long strands of DNA and protein that make up the chromosomes unfold. The individual chromosomes are still there, but they are difficult to see. In fact, in interphase the chromosomes are most active, using the information they contain to direct cell growth and development.

☑ **Checkpoint** What is interphase?

Figure 5–4
Most cells undergo cell division. (a) These bacteria, which are magnified 30,000 times, are in the final stages of cell division. (b) The plant cells, on the other hand, are about to begin the process of cell division. The plant cells are magnified 1600 times.

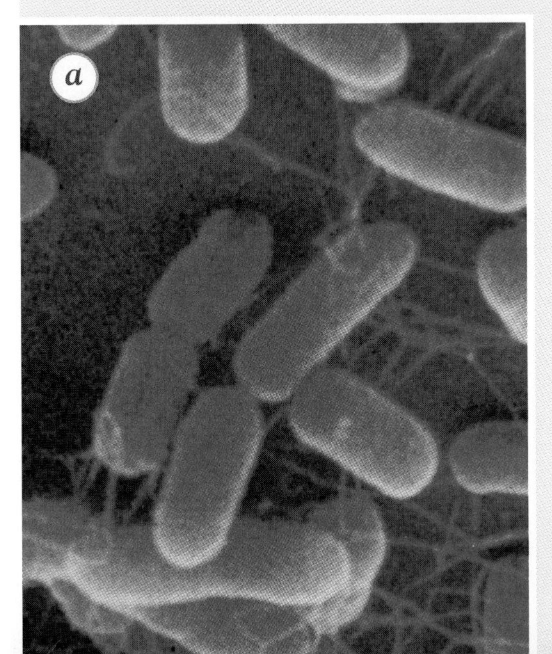

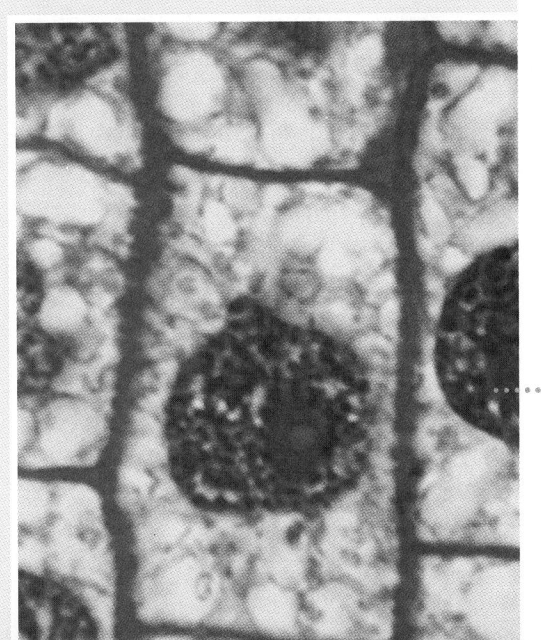

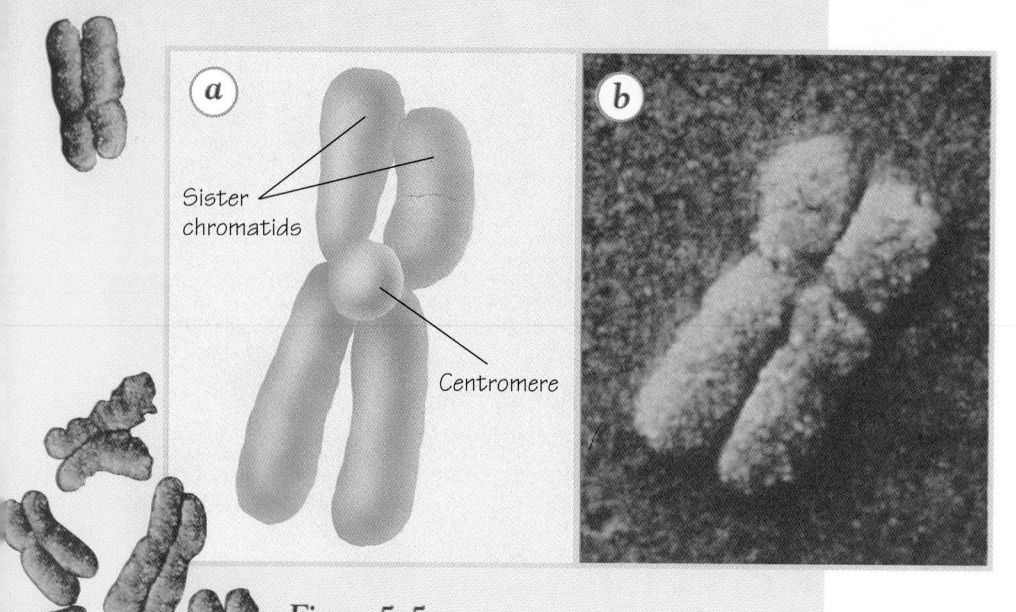

Figure 5–5 *(a) This diagram of a human chromosome shows that it consists of two chromatids attached at a centromere. (b) If you looked through a scanning electron microscope, this is what a human chromosome would look like (magnification: 40,000X).*

In animal cells, the **centrioles**—tiny structures that help to organize microtubules—separate from each other and move to opposite sides of the nucleus. Each pair of centrioles serves as a focal point for the growth of a cluster of microtubules that span the nucleus. This cluster of microtubules is called the **spindle.** Some of the microtubules of the spindle attach themselves directly to the chromosomes. Plant cells do not have centrioles, but they organize a nearly identical spindle in much the same way.

Checkpoint What are chromatids?

Mitosis 

After interphase, the cell is ready for mitosis, or cell division. **Mitosis is divided into four phases: prophase, metaphase, anaphase, and telophase.** As you read about the process of mitosis, refer to page 107.

Prophase

The first phase of mitosis, **prophase,** is generally the longest of the four phases, taking as much as 50 percent of the total time required to complete mitosis. Often, the first clue that prophase has started is the appearance of chromosomes.

The cell nucleus changes dramatically in prophase, as the chromosomes come together and form thick, threadlike structures that are visible under the light microscope. At this point, each chromosome consists of two identical strands called **chromatids.** Because the chromatids are identical copies of the same chromosome, they are often called sister chromatids. The chromatids are attached at an area called the **centromere.**

Metaphase

The second phase of mitosis is called **metaphase.** Metaphase is the shortest phase of mitosis and may last only a few minutes. During metaphase, the chromosomes complete their attachment to the spindle and line up across the center of the cell. When metaphase is complete, each chromosome is aligned so that one of its chromatids is closer to one pole of the spindle and the other chromatid is closer to the other pole.

Anaphase

At the end of metaphase, the centromeres that hold the sister chromatids together suddenly split. The chromatids move toward the two poles of the spindle. This sudden movement marks the beginning of **anaphase.** Anaphase is the phase of mitosis in which the duplicated chromosomes separate from each other.

In a matter of a few minutes, the chromosomes move apart and are concentrated into two groups—one at each pole of the spindle. Anaphase is complete when the movement of chromosomes stops.

Visualizing the Cell Cycle

In eukaryotic cells, the cell cycle has two major phases—interphase and mitosis.

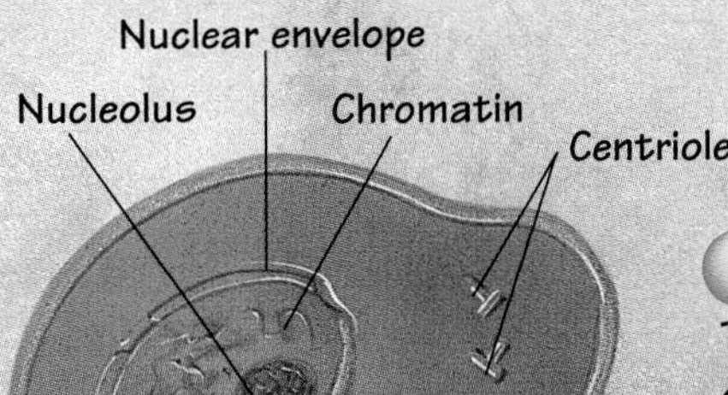

INTERPHASE

1 Interphase

The chromosomes have been copied and are spread out in the nucleus. The centrioles have also been copied.

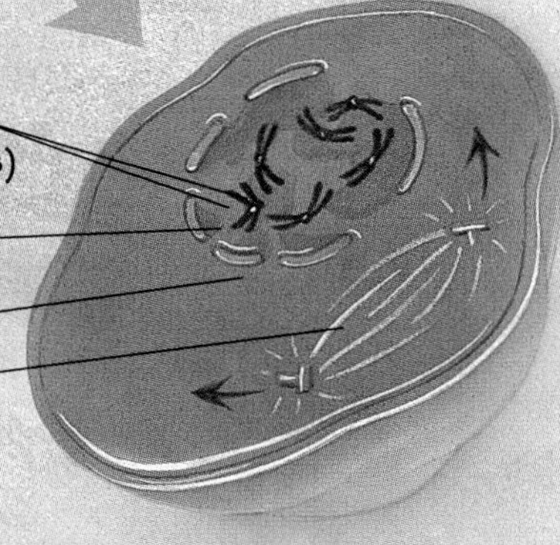

PROPHASE

2 Prophase

The chromosomes come together, and the nucleolus disappears. The centrioles begin to separate. The spindle begins to form. The nuclear envelope breaks down.

METAPHASE

3 Metaphase

The chromosomes attach to the spindle and line up in the middle of the cell.

ANAPHASE

Sister chromatids separate

4 Anaphase

The centromeres holding sister chromatids split. The microtubules shorten, pulling one chromatid from each chromosome toward each pole of the cell.

TELOPHASE

Nuclear envelope

5 Telophase

The chromosomes have separated to opposite sides of the cell, and the spindle breaks apart. The nuclear envelope begins to reform.

CYTOKINESIS

6 Cytokinesis

The cytoplasm and its contents divide into two individual daughter cells. Each daughter cell receives one nucleus and about half the cytoplasm. After telophase and cytokinesis, the two daughter cells return to interphase.

Figure 5–6
(a) *During cytokinesis in most animal cells, the cell membrane is pulled inward by a ring of filaments. This continues until the cytoplasm is pinched into two nearly equal parts, each containing its own nucleus and cytoplasmic organelles.*
(b) *The process of cytokinesis is almost complete in this human kidney cell (magnification: 1740X).*

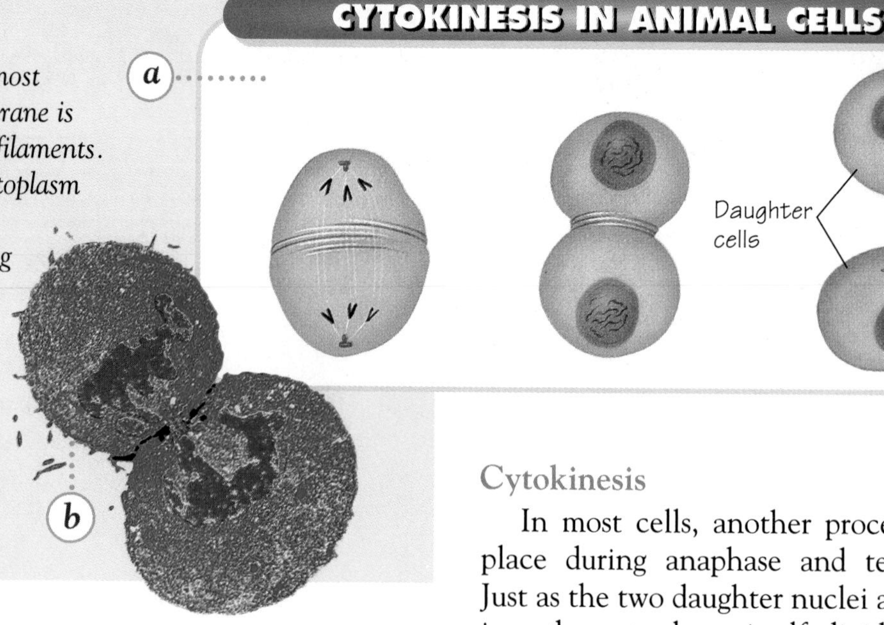

Telophase

The fourth phase of mitosis is called **telophase.** In this phase, the microtubules of the spindle begin to break apart. The chromosomes that are clustered at each of the two poles begin to spread out. A nuclear membrane forms around each cluster, and two distinct nuclei gradually form within the cell. These two daughter nuclei will each go to one of the cells produced by the division, so that each of these new cells contains a nucleus.

Cytokinesis

In most cells, another process takes place during anaphase and telophase. Just as the two daughter nuclei are forming, the cytoplasm itself divides. This process is called **cytokinesis.** After telophase and cytokinesis are complete, the process begins again.

Figure 5–7
(a) *In plant cells, a structure called the cell plate forms midway between the two nuclei. The cell plate grows gradually outward until a new cell wall appears between the daughter cells.*
(b) *In this plant cell, the cell plate has almost completely fused with the cell wall, creating two independent daughter cells (magnification: 1200X).*

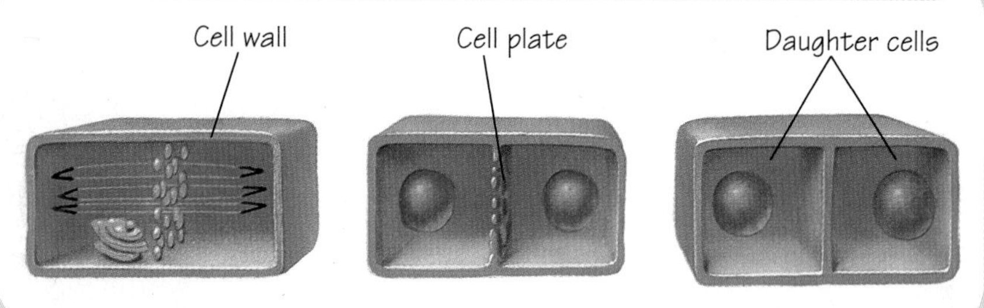

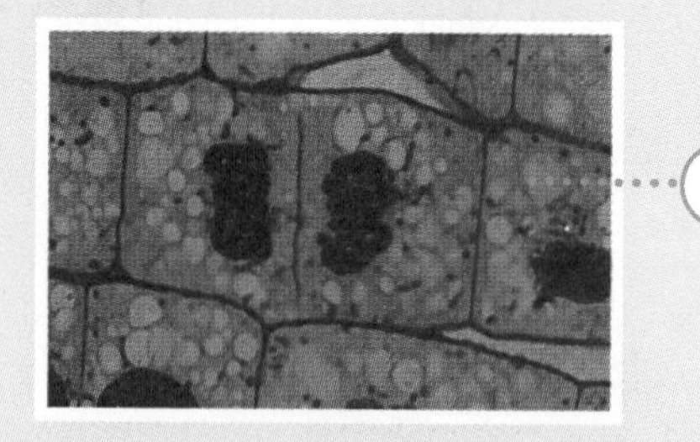

Section Review 5–2

1. **Describe** the four phases of mitosis.
2. **Compare** the process of cytokinesis in plant and animal cells.
3. **Critical Thinking—Sequencing** Describe the order of events in the process of mitosis.

Cell Specialization and Organization

Guide for Reading

- **Define** cell specialization.
- **List** the four levels of organization in a multicellular organism.

MINI LAB

- **Observe** the characteristics of blood.

EACH OF US BEGAN LIFE AS A single cell. By the time we were born, that single cell had gone through millions of cell divisions—proof of the importance of mitosis. A multicellular organism, however, is much more than a large group of cells, in the same way that a society is much more than just a large group of people. And just as a society is made up of individuals or groups of individuals who have certain specialties, each cell, or group of cells in a multicellular organism, has a different specialty. These different specialties allow the organism to continue functioning.

Specialized Cells

Cells in a multicellular organism tend to be specialized. **Cell specialization means that specific cells are uniquely suited to carry out specific functions.** Some cells may be specialized to move, while other cells may specialize in responding to the environment. Still other cells may make products that the rest of the organism needs.

What gives a cell the ability to do one job so much better than other cells? Let's take a close look at two important specialists from the hundreds of different types of cells in the human body.

Macrophages

In every large society there is a need for law and order, for protection. The trillions of cells in our bodies have a similar need, and one type of cell that provides this protection is the macrophage. The macrophages travel throughout the body in the bloodstream.

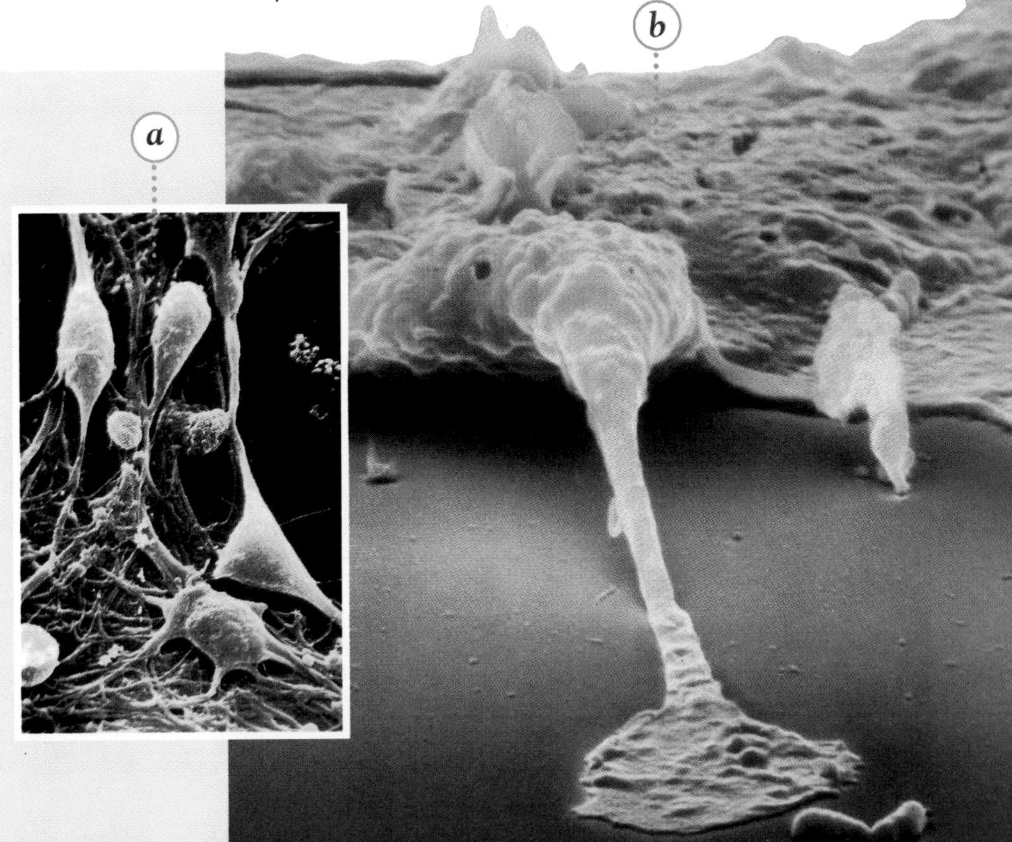

Figure 5–8 *The human body has hundreds of specialized cells. (a) This false-color scanning electron micrograph shows the intricate network of neurons that carry messages throughout the human body. (magnification: 235X). (b) Macrophages help protect the body from disease. In this photograph, a macrophage is about to engulf a bacterium (magnification: 4000X).*

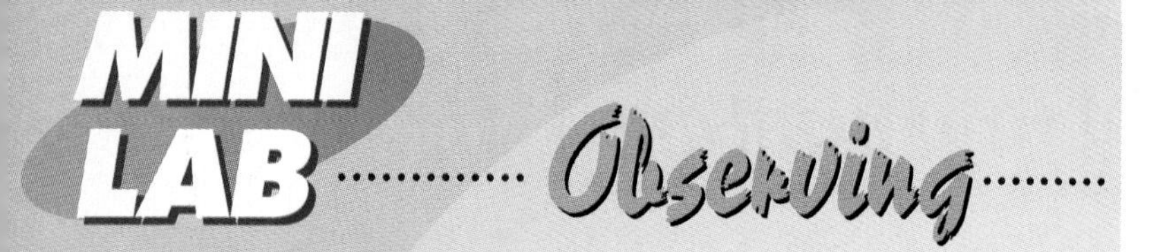

Red, White, and Blood

PROBLEM ***Observe*** *the characteristics of blood.*

PROCEDURE

1. Obtain a prepared slide of human blood.
2. Use the low-power objective to examine the slide under a microscope. Switch to high power.
3. Locate a field in which you can see both red and white blood cells. The nuclei of the white blood cells will appear purple due to the stain.

ANALYZE AND CONCLUDE

1. Why do you think is it necessary for white blood cells to have a nucleus? Why don't the red blood cells have a nucleus?
2. When an infection is present, the number of white blood cells increases. If your blood has a higher than normal percentage of white blood cells, what might a doctor conclude?
3. Should blood be considered a tissue? Explain why.

At the site of a wound or an infection, macrophages appear in great numbers and begin to disarm the "bad guys"—bacteria and other invading organisms. The cytoplasm of a macrophage is filled with specialized granules containing chemicals that attack the cell walls of bacteria. If these chemicals fail, the cell membrane of a macrophage may then surround the invader and completely engulf it. Within a few minutes the bacterium has been taken inside the macrophage, and in a few hours it is completely destroyed.

☑ ***Checkpoint*** Why are macrophages specialized cells?

Neurons

Communication is important in any large group of individuals, and it is just as important in a large group of cells. Without communication, one hand literally wouldn't know what the other hand was doing. How do cells in different parts of the body communicate? One of the most important ways is by means of the nervous system, a specialized network of cells that works almost like a network of telephone operators.

Some of these message-carrying cells, or neurons, are among the longest and thinnest cells in the body. Rapid movements of charged molecules across their cell membranes produce electrical impulses. These impulses carry messages from one end of the cell to the next, helping to relay information and control movements. These impulses also coordinate the activities of the most complicated society of cells in the entire world—the human brain.

Levels of Organization

Cell specialization is only part of the story of how a multicellular organism is put together. Many jobs are far too complex for a single cell to handle on its own. Groups of specialized cells may be necessary in such cases.

Tissues

A **tissue** is a group of similar cells that perform similar functions. The two specialized cells that you have just read about are, in fact, members of a tissue. The cells that protect the body from infection are one part of a tissue—blood. The cells that carry impulses from one neuron to the next are part of nerve tissue.

Organs

Although hundreds or even thousands of cells may make up a tissue, some tasks are too complicated to be carried out by just one type of tissue. In these cases, an **organ,** or a group of tissues that work together to perform a specific function, is needed. Each muscle in your body is an individual organ, and so are each of your eyes. These organs contain many different tissues that work together to carry out an essential task, such as movement or vision.

Organ Systems

In many cases, even a complex organ is not sufficient to carry out a series of specialized tasks. In these cases, an **organ system,** or a group of organs, performs several closely related functions. For example, the organs of the digestive system all work together to digest the food that you eat. There are eleven major organ systems in the human body, such as the muscular system, the skeletal system, and the nervous system.

The four levels of organization—cells, tissues, organs, and organ systems—are the same for nearly all multicellular organisms. The division of labor among the cells in these levels is one of the things that makes multicellular life possible. Specialized cells, such as nerve and muscle cells, are able to exist precisely because other cells are specialized to obtain the food and oxygen that these cells need. This overall specialization and interdependence is one of the remarkable characteristics of living things.

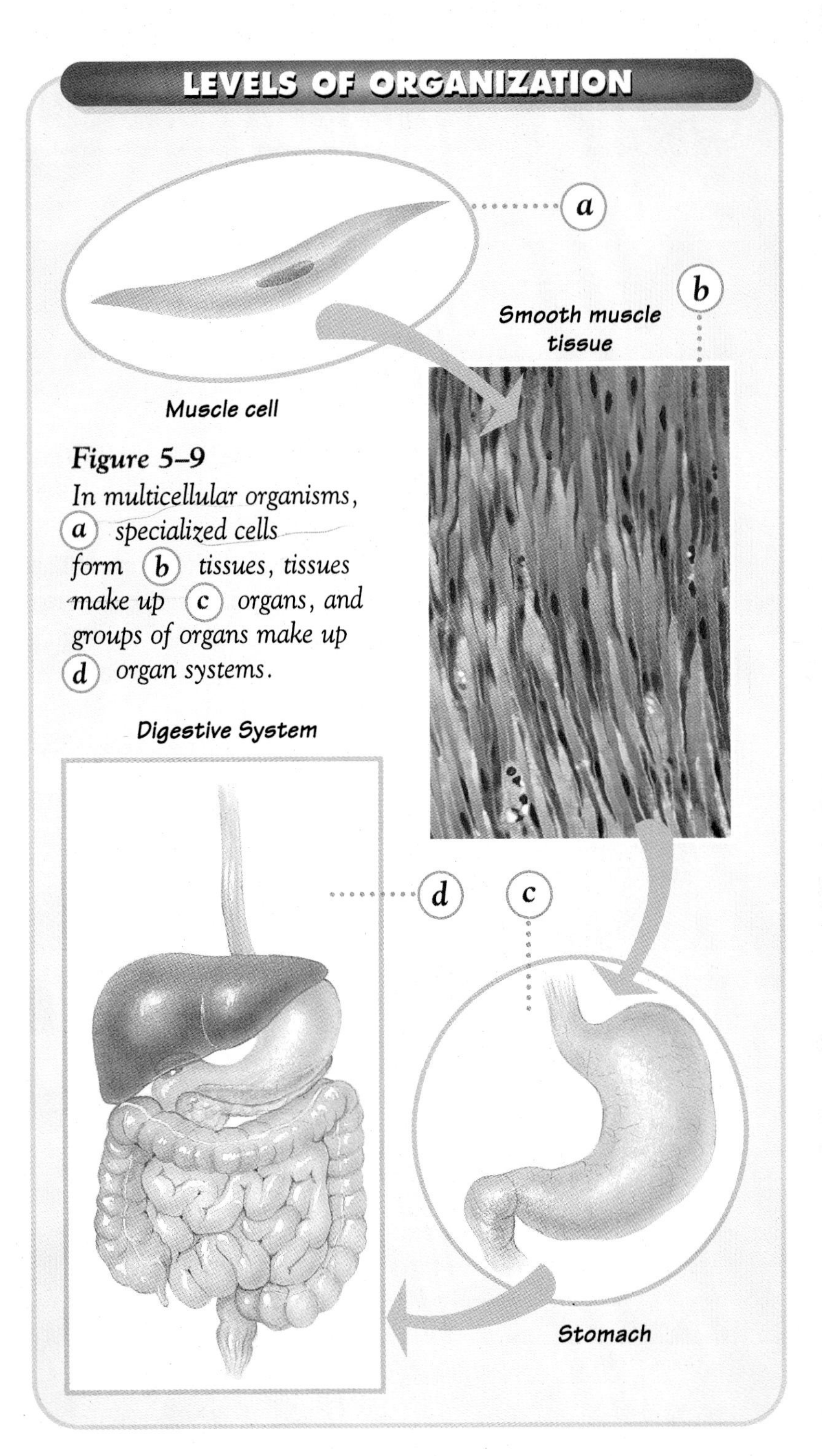

Figure 5–9
In multicellular organisms, (a) specialized cells form (b) tissues, tissues make up (c) organs, and groups of organs make up (d) organ systems.

Section Review 5-3

1. **Define** cell specialization.
2. **List** the four levels of organization in multicellular organisms.
3. **Critical Thinking—Applying Concepts** Why is blood classified as a tissue?
4. **MINI LAB** **Observe** the characteristics of blood.

SECTION 5–4

BRANCHING OUT **In Depth**

Controlling the Cell Cycle

GUIDE FOR READING

- **Describe** the role of cyclin in the cell cycle.

CELL GROWTH AND DIVISION in a large organism are carefully regulated. This means that something must control whether a cell is allowed to divide or whether it can enter the next phase of the cell cycle. For years biologists wondered what that something might be. At long last, it seems as though they have the answer.

Nearly 20 years ago, scientists were trying to find out whether there was a signal that caused the egg cell of a frog to begin dividing after it was fertilized by a sperm cell. They discovered that the dividing cell contained a protein that would cause a spindle to form if it was injected into a nondividing egg cell. To their surprise, they discovered that the amount of this protein in the cell rose and fell in timing with the cell cycle.

Cyclins and the Cell Cycle

Scientists called the protein that caused a spindle to form **cyclin,** because the amount of this protein changed in time with the cell cycle. It turned out that cyclin was not just found in frogs. And injections of cyclin could cause just about any cell to enter mitosis!

Cyclins regulate the timing of the cell cycle in all eukaryotic cells that have been studied—from tiny yeast cells to humans. It just so happens that there are many different cyclins, including those that regulate different phases of the cell cycle. For example, the cyclins that were first discovered in frogs are now called M-phase cyclins because they regulate the entry of the cell into mitosis.

As a cell goes through the cell cycle, cyclin is made at a fairly constant rate and gradually builds up inside the cell. When the cyclin

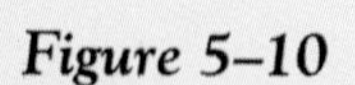

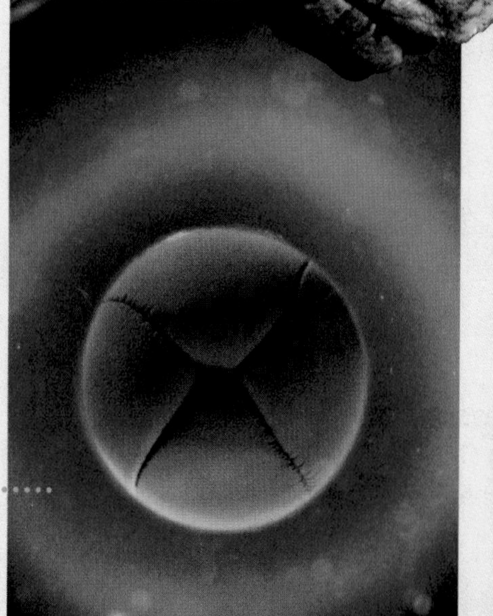

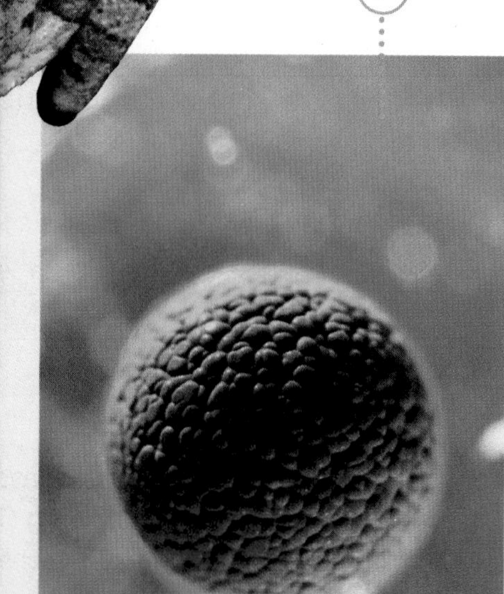

Figure 5–10
Cyclin, one of the proteins that can control the cell cycle, was first discovered in (a) a frog's cells. This protein causes (b) a fertilized frog egg to undergo cell division, forming a few cells (magnification: 20X) and (c) then forming many more cells (magnification: 20X).

Figure 5–11
Cyclin can control the cell cycle at each of three main points: late in the G_1 phase, early in the S phase, and near the border between the G_2 phase and mitosis.

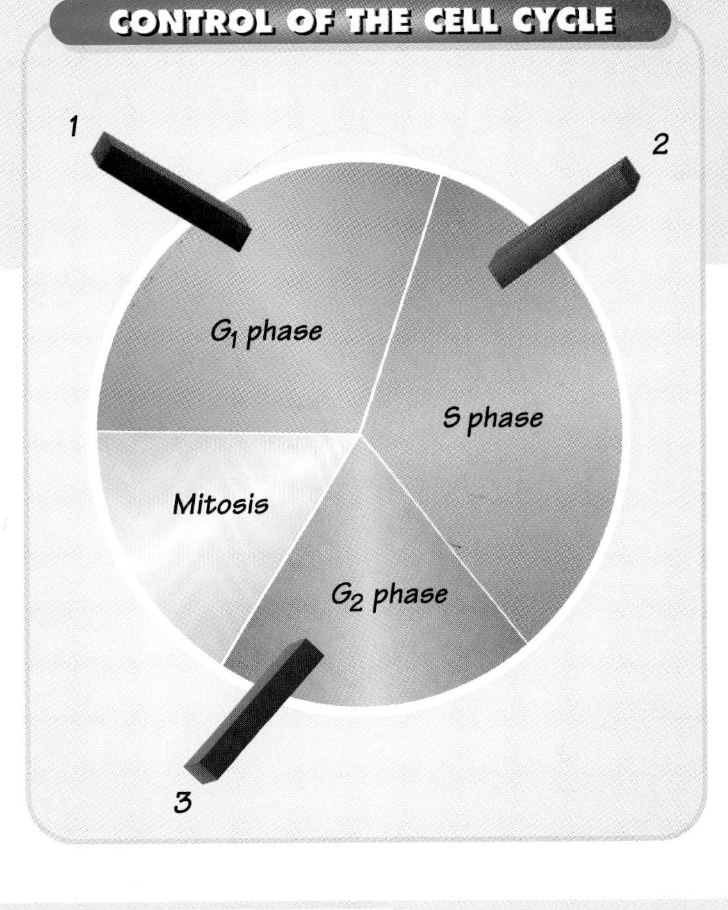

level reaches a critical point, it triggers the cell to enter mitosis. Then something remarkable happens. Once the cell has entered mitosis, it destroys nearly all its cyclin. This means that the cell must start to make cyclin again before it can enter mitosis a second time. The time it takes for a cell to make more cyclin determines how long it will take the cell to move through the cell cycle.

Checkpoint What is cyclin?

The Cell Cycle and Cancer

You may recall that the ability to control cell growth is vital to an organism's survival, and that the cell cycle is the point at which growth is actually controlled. Cancer cells do not respond to the usual signals that keep other cells from growing uncontrollably. Very often, the reason they don't respond to the signals is because of a defect in cell cycle regulation. Cyclin and the proteins that interact with it have turned out to be the master proteins in regulating the cell cycle.

Figure 5–12
CAREER TRACK
Histologic technicians, who prepare tissue samples for further study, generally work in laboratories and hospitals.

Section Review 5–4

1. **Describe** the role of cyclin in the cell cycle.
2. **BRANCHING OUT ACTIVITY** Cyclin is a protein that regulates the timing of the cell cycle. Taxol is a powerful drug derived from the Pacific yew (*Taxus bervifolia*), an evergreen that grows in the northwestern United States. Research the role of taxol in fighting uncontrolled cell growth. **Communicate** your findings in a brief written summary.

Laboratory Investigation

Mitosis

The tip of a root is a good place to look for cells that are in the process of dividing because roots grow at the tip. In this investigation, you will prepare slides from garlic root tips. If your preparation is a good one, you may be able to find all the phases of mitosis.

Problem

How many phases of mitosis can you **observe** in a plant cell?

Materials (per group)

- forceps
- garlic root tips
- 2 watch glasses or Petri dish covers
- 50% hydrochloric acid-50% ethyl alcohol solution (HCl-EtOH solution)
- Carnoy's fixative
- microscope slide
- scalpel
- medicine dropper
- aceto-orcein stain
- coverslip
- paper towel
- compound microscope

Procedure

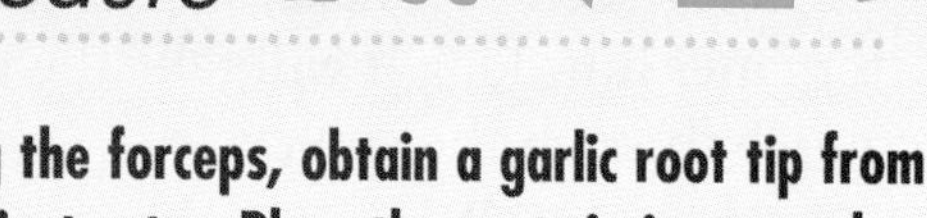

1. **Using the forceps, obtain a garlic root tip from your instructor. Place the root tip in a watch glass.**
2. **Carefully pour HCl-EtOH solution over the root tip so that the entire tip is immersed in the solution. CAUTION: *If the solution touches your skin, wash the area immediately with running water.* Soak the root tip in the solution for 5 to 6 minutes.**
3. **Using the forceps, transfer the root tip to the second watch glass. Pour Carnoy's fixative over the root tip. Leave the root tip in the fixative for 3 minutes.**
4. **Using the forceps, transfer the tip to a microscope slide.**

5. **Holding the root tip with the forceps, use the scalpel to cut off 2 cm from the root tip. CAUTION:** ***Be careful when using sharp instruments.*** **Discard the rest of the root tip.**
6. **Use the medicine dropper to add just enough aceto-orcein stain to cover the remaining tip. CAUTION:** ***Aceto-orcein will stain clothing.*** **Leave the root tip in the stain for 5 minutes.**

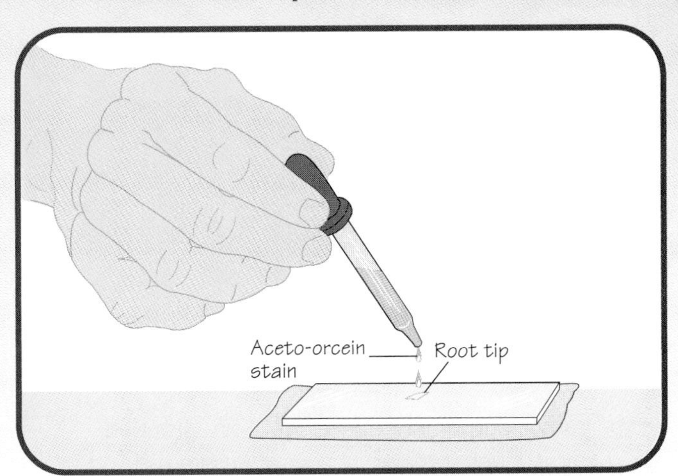

7. **Place a clean coverslip over the root tip. Place a folded paper towel on a flat surface and fold the paper towel over the slide. Gently press down on the paper towel with your thumb so the paper towel will absorb the excess liquid.**

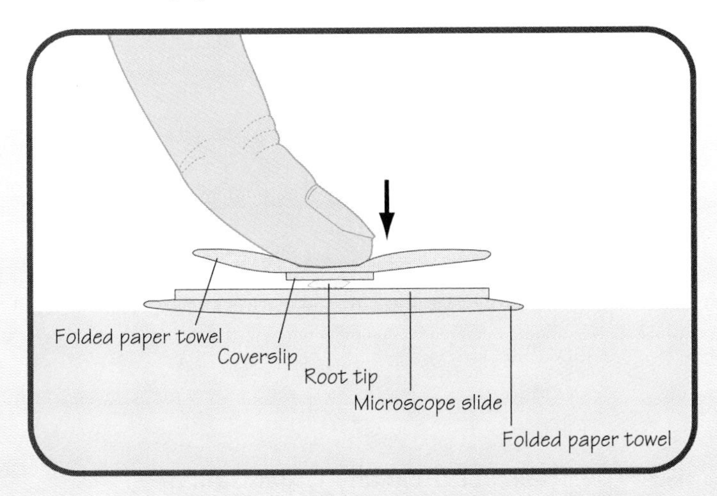

8. **Place the slide under low power on your microscope. Focus the microscope. Then scan the slide until you find some stained cells.**
9. **Switch the microscope to high power and locate as many phases of mitosis as you can.**

Observations

1. Sketch any cells that are in a phase of mitosis.
2. Label the sketch with the name of the appropriate phase.

Analysis and Conclusions

1. How do the cells that are in a phase of mitosis differ from cells that are not dividing?
2. What was the purpose of the hydrochloric acid-ethyl alcohol solution? (*Hint:* Consider how plant cells are held together.)
3. Did the aceto-orcein stain one part of the cell more than any other? If so, which one?
4. In which phase of mitosis were most of the cells you observed? Why do you think this was so?

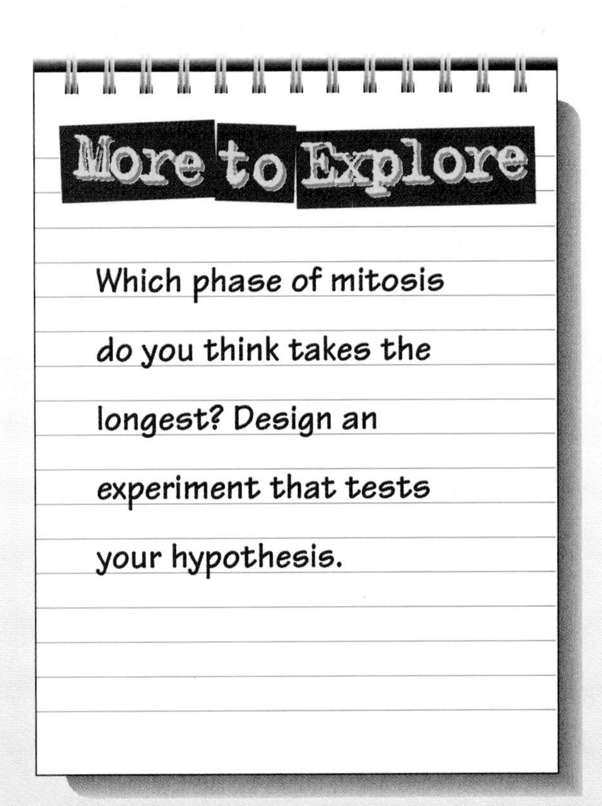

Study Guide

Summarizing Key Concepts

The key concepts in each section of this chapter are listed below to help you review the chapter content. Make sure you understand each concept and its relationship to other concepts and to the theme of this chapter.

5–1 Cell Growth and the Cell Cycle

- As a cell increases in size, its volume increases at a faster rate than its surface area.
- The cell cycle is the period of time from the beginning of one cell division to the beginning of the next.
- Cell division is the process in which the cell divides into two independent cells. In eukaryotic cells, this process is called mitosis.
- There are four stages in the cell cycle—G_1, S, G_2, and M—during which a cell duplicates its contents and divides.

5–2 Cell Division

- The four phases of mitosis are prophase, metaphase, anaphase, and telophase.
- The two identical parts of a chromosome, called chromatids, are attached to each other at a centromere.
- In most cells, the cytoplasm divides as well as the nucleus.

5–3 Cell Specialization and Organization

- Many cells are specialized, or uniquely suited, to carry out specific functions.
- In multicellular organisms, cells are grouped into tissues, organs, and organ systems.

5–4 Controlling the Cell Cycle

- Cyclins are proteins that regulate the timing of the cell cycle.

Reviewing Key Terms

Review the following vocabulary terms and their meaning. Then use each term in a complete sentence.

5–1 Cell Growth and the Cell Cycle

cell cycle
cell division
mitosis

5–2 Cell Division

interphase
prophase
chromatid
centromere
centriole
spindle
metaphase
anaphase
telophase
cytokinesis

5–3 Cell Specialization and Organization

tissue
organ
organ system

5–4 Controlling the Cell Cycle

cyclin

Recalling Main Ideas

Choose the letter of the answer that best completes the statement or answers the question.

1. When a cell doubles its diameter, its internal volume increases by

a. 2. **c.** 6.
b. 4. **d.** 8.

2. G_1, G_2, and S are often called

a. interphase. **c.** metaphase.
b. prophase. **d.** telophase.

3. Chromosomes first appear in the longest phase of mitosis, called

a. cytokinesis. **c.** anaphase.
b. metaphase. **d.** prophase.

4. Chromosomes line up across the middle of the cell during

a. prophase. **c.** anaphase.
b. metaphase. **d.** telophase.

5. Each chromosome contains two strands, called

a. centromeres. **c.** chromatids.
b. centrioles. **d.** microtubules.

6. Duplicated chromosomes separate from each other during

a. prophase. **c.** anaphase.
b. metaphase. **d.** telophase.

7. The cytoplasm divides during

a. cytokinesis. **c.** anaphase.
b. telophase. **d.** metaphase.

8. What regulates the timing of the cell cycle?

a. cyclin **c.** chromatids
b. centromeres **d.** centrioles

9. A group of tissues that work together form a

a. cell. **c.** muscle.
b. organ. **d.** organ system.

Putting It All Together

Using the information on pages xxx to xxxi, complete the following concept map.

CELLS
divide by
1
2
Mitosis
made up of 3 phases
divided into 4 phases
G_1
3
G_2
Prophase
4
Anaphase
5

Assessment

Reviewing What You Learned

Answer each of the following in a complete sentence.

1. Why is a large cell less efficient than a small cell?
2. What is the cell cycle?
3. What happens during the S phase of the cell cycle?
4. What is cancer?
5. What three phases make up interphase?
6. What are the phases of mitosis?
7. Describe the chemical and physical structure of a chromosome.
8. What function do centrioles have in a dividing animal cell?
9. What do macrophages do?
10. Describe multicellularity and the division of labor.
11. What are the four levels of organization in a multicellular organism?
12. What functions do cyclins have?

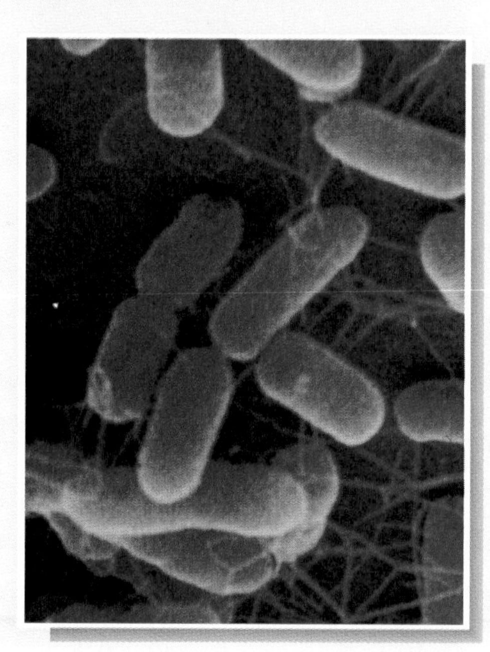

Expanding the Concepts

Discuss each of the following in a brief paragraph.

1. How might a cell improve its cell surface-area-to-volume ratio without changing its actual volume?
2. How much must the diffusion rate increase if a cell of one unit enlarges to three units, maintaining a constant level of material exchange?
3. If a normal body cell had 23 pairs of chromosomes, how many chromosomes would you expect to find during the G_2 phase? How many chromatids? Explain your answers.
4. Which phase of the cell cycle would be best to examine if you wanted to observe a particular organism's chromosomes? Explain why.
5. What differences can you **observe** in plant and animal cells during cytokinesis?
6. What are the advantages of specialized tissues in multicellular organisms?
7. What factors are involved in the regulation and determination of cell division?
8. **Analyze** the differences among chromosomes, chromatids, and chromatin.
9. Evolution tends to progress from simple to complex. How does this statement parallel the progression from cells to tissues to organs to organ systems in the development of living things?

Extending Your Thinking

Use the skills you have developed in this chapter to answer the following.

1. **Interpreting data** During an experiment in which you were calculating the mass of nuclear DNA, you came across a group of cells that contained twice the normal DNA. Suggest what is happening to these cells.
2. **Designing an experiment** Design an experiment in which you could estimate the relative length of each phase of mitosis.
3. **Inferring** Mitosis is often called duplication division. Discuss abnormalities in mitosis that might lead to variation within a species that reproduces asexually.
4. **Using the writing process** Discuss the role of both microtubules and microfilaments in plant and animal mitosis. Write a three- or four-paragraph summary.
5. **Applying concepts** Cancer is characterized by abnormal cell division. Using the information you have learned about cell division, suggest reasons why it is difficult to cure cancer.

Applying Your Skills

Only Skin Deep

Your body requires new cells to be made to replace worn-out or damaged cells. These new cells must be identical to the original cells. An example of this process occurs any time your skin becomes damaged from sunburn or worn out through normal, everyday washing or aging. How does your body make more of the same kind of skin cells?

1. Obtain 60 cm of yarn, 100 cm of string, and a pair of scissors.
2. Cut four pieces of yarn in 3-cm lengths, four pieces in 5-cm lengths, and four pieces in 7-cm lengths.
3. Cut two 25-cm lengths of string. Then cut the rest as needed.
4. Using the yarn and string, illustrate the processes of interphase and mitosis. Begin with six pieces of yarn—two of each length.
5. Explain to your teacher what each step is and what the yarn and string represent.

• GOING FURTHER •

6. Using glue, paste your yarn representation of one of the phases of the cell cycle onto a piece of poster board. Display your poster in the classroom.

Genetics

CHAPTERS

"(A)ny living cell carries with it the experiences of a billion years of experimentation by its ancestors."

— Max Delbruck

CAREER TRACK

As you explore the topics in this unit, you will discover many different types of careers associated with biology. Here are a few of these careers:

- Horticulturist
- Genetic Counselor
- Geneticist
- Genetic Engineering Research Assistant

Roses growing in a garden

CHAPTER 

Introduction to Genetics

FOCUSING THE CHAPTER THEME: Systems and Interactions

6–1 The Science of Inheritance
- **Identify** the basic principles of heredity.

6–2 Meiosis
- **Describe** how gametes are produced.

6–3 Analyzing Inheritance
- **Explain** how probability applies to genetics.

BRANCHING OUT *In Depth*

6–4 A Closer Look at Heredity
- **Describe** the different ways in which genes can be expressed.

LABORATORY INVESTIGATION
- **Interpret** data from a model of crossing-over.

Biology and Your World

BIO JOURNAL

The Irish setters shown on this page are members of the same family. In your journal, describe the ways in which the family members resemble one another. Why do you think that puppies resemble their parents in some ways and differ from them in other ways?

A female Irish setter and her puppies

The Science of Inheritance

Guide for Reading

- **Define** heredity and **identify** the units of heredity.
- **Explain** the principles of dominance, segregation, and independent assortment.

MINI LAB
- **Classify** dominant and recessive traits.

HAVE YOU EVER WONDERED why offspring resemble their parents? For example, why will puppies grow up to look like the adult dogs that produced them? Why will new maple trees grow from the seeds of adult maple trees?

If you have asked questions such as these, you are not alone. Since the beginning of recorded history, people have wondered why humans, animals, and plants grow up the way they do.

Parents and Offspring

Offspring resemble their parents because of their **heredity**—their biological inheritance. **An organism's heredity is the set of characteristics it receives from its parents.** Today, the study of heredity is known as **genetics.** As you will discover, genetics is one of the most important and useful branches of biology. By applying the principles of genetics, farmers are raising more useful animals and plants, physicians are preventing and treating many diseases, and scientists are discovering some of the secrets of life on our planet.

For thousands of years, people thought that the heredity of a living thing was merely a blend of the characteristics of its parents. After all, most animals look a little bit like the mother and a little bit like the father. In addition, if a big animal and a small animal produced offspring, the offspring would typically be medium-sized—a blend of the sizes of the two parents.

Figure 6–1
The same basic principles of genetics apply to a wide variety of organisms, including flowering plants, such as (a) this maple tree, and animals, such as (b) humans and (c) grizzly bears.

Figure 6–2
Although organisms resemble their parents in many ways, sometimes their characteristics are quite different. Deer with brown fur were the parents of this white deer, called an albino.

Is heredity really that simple? No, it isn't. For example, two deer with brown fur were the parents of the white deer shown in ***Figure* 6–2.** The white fur of this deer could hardly be described as a blend of the colors of both of its parents' fur.

☑ **Checkpoint** What is heredity?

Gregor Mendel

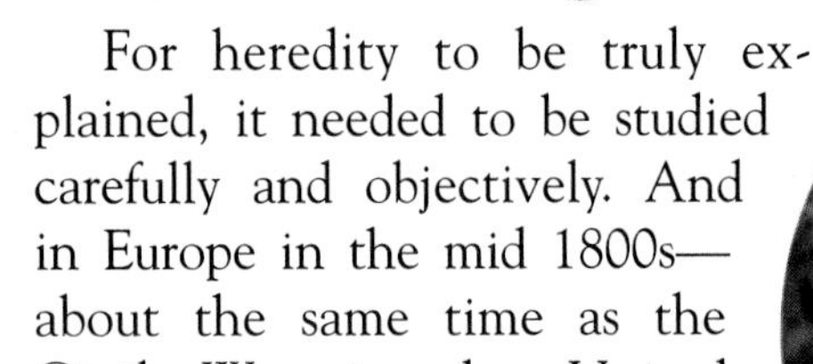

For heredity to be truly explained, it needed to be studied carefully and objectively. And in Europe in the mid 1800s—about the same time as the Civil War in the United States—one man did just that. His name was Gregor Mendel.

Gregor Mendel was born in 1822. He lived for most of his life in the town of Brno, which is now part of the Czech Republic. After becoming a priest, he spent several years studying science and mathematics at the University of Vienna. When Mendel left Vienna, he returned to the monastery in Brno, where he taught science in a local high school and supervised the monastery's garden. History tells us little about Mendel's work as a teacher, but we know a great deal about his work in the garden.

☑ **Checkpoint** What was Mendel's contribution to the study of biology?

Mendel's Work on True-Breeding Pea Plants

Before Mendel arrived at the monastery, the previous gardeners had developed different true-breeding stocks of pea plants. A true-breeding stock always passes its characteristics to the next generation. For example, one true-breeding stock of pea plants might always produce tall plants with green pods, while another stock might always produce short plants with yellow pods.

What would you do if you had different stocks of true-breeding pea plants? Would you mate them and observe their offspring? That is exactly what Mendel did. (a)

Figure 6–3
(a) *Gregor Mendel may have looked like an ordinary gardener, but his achievements were far from ordinary. His work provides the basis of our understanding of heredity.* (b) *Mendel performed his experiments on* Pisum sativum, *the garden pea plant.*

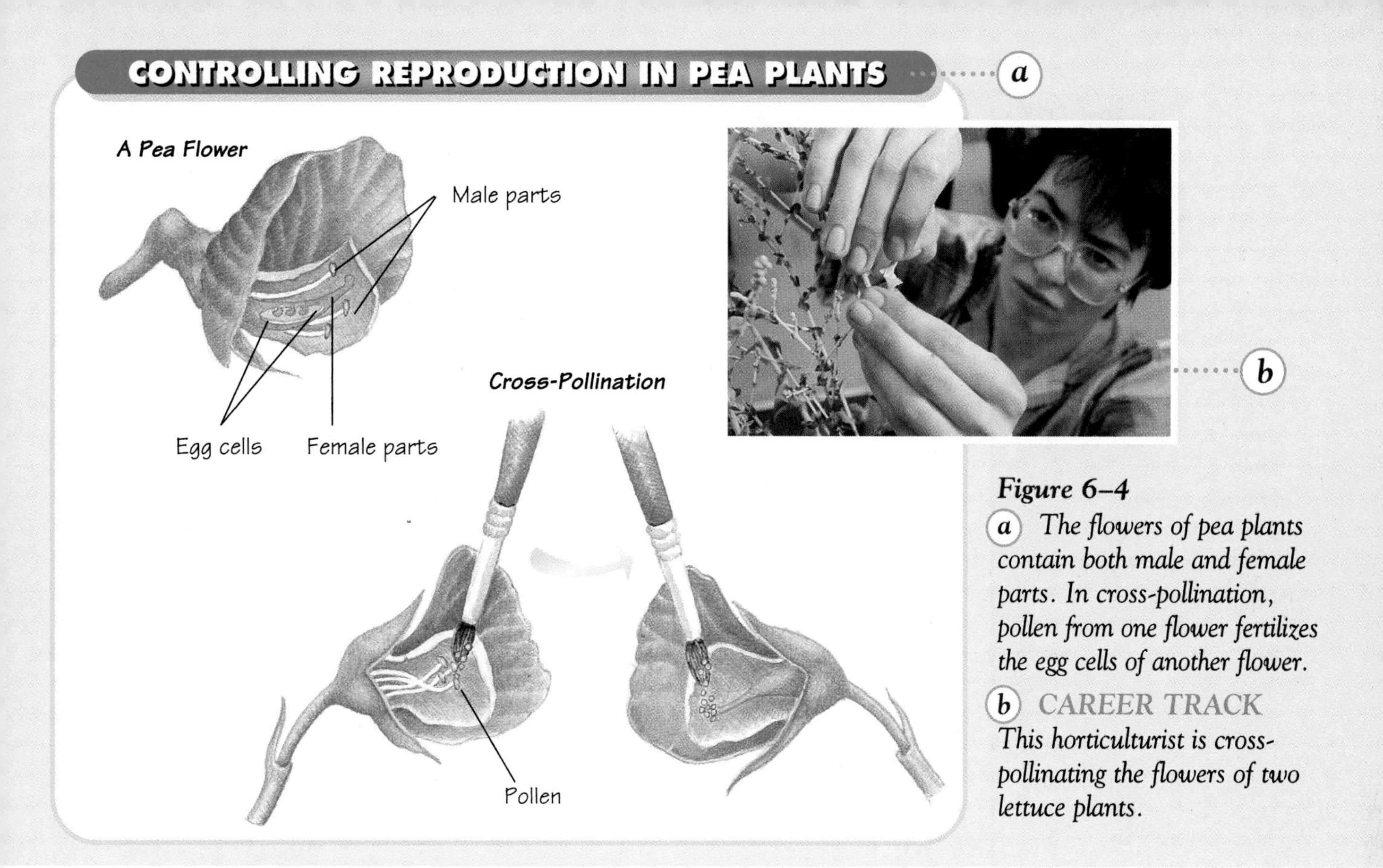

Figure 6–4
(a) The flowers of pea plants contain both male and female parts. In cross-pollination, pollen from one flower fertilizes the egg cells of another flower.
(b) CAREER TRACK This horticulturist is cross-pollinating the flowers of two lettuce plants.

Reproduction in Pea Plants

Like many plants, pea plants use parts of their flowers to reproduce. One part produces pollen—the male sex cells—and another part produces egg cells—the female sex cells. When pollen fertilizes an egg cell, a seed for a new plant is formed.

Pea plants normally reproduce by self-pollination, in which pollen fertilizes egg cells on the same flower. In effect, seeds that are produced from self-pollination have only one plant as a parent.

However, pea plants can also cross-pollinate. In cross-pollination, pollen from the flower on one plant fertilizes the egg cells of a flower on another plant. The seeds produced from cross-pollination have two plants as parents.

To perform his experiments, Mendel had to select the pea plants that mated with each other. Therefore, he needed to prevent flowers from self-pollinating and to control their cross-pollinating. How did Mendel accomplish this task? First, he cut away the male parts of a flower. Then, as shown in ***Figure 6–4,*** he dusted that flower with pollen from a second flower.

Checkpoint What is self-pollination? Cross-pollination?

Seven Traits

With this technique, Mendel could choose any two pea plants to cross-pollinate—or cross, for short. But how did he choose which type of pea plants to cross? Mendel had several different stocks of pea plants, and each was true-breeding for a variety of different characteristics.

To simplify his investigation, Mendel chose to study only seven **traits** in pea plants, illustrated in ***Figure 6–5*** on page 126. A trait is a characteristic that distinguishes one individual from another.

Looking back at Mendel's work, we realize that one of his most important decisions was to study just a small number of traits. Also, each of these traits has two contrasting forms. For example, seed shape is either round or wrinkled and pod color is either green or yellow.

MENDEL'S SEVEN F_1 CROSSES ON PEA PLANTS

	Seed Shape	Seed Color	Seed Coat Color	Pod Shape	Pod Color	Flower Position	Plant Height
P Generation	Round × Wrinkled	Yellow × Green	Gray × White	Smooth × Constricted	Green × Yellow	Axial × Terminal	Tall × Short
F_1 Generation	Round	Yellow	Gray	Smooth	Green	Axial	Tall

Figure 6–5 *Mendel performed F_1 crosses for each of the seven traits shown here. To Mendel's surprise, the traits of just one parent appeared in the F_1 generation.*

The F_1 Generation

Using true-breeding stocks for each of the seven traits, Mendel crossed pea plants that showed one form of a trait with pea plants that showed the other form, as illustrated in ***Figure 6–5.*** The offspring of these crosses are called **hybrids.** A hybrid is an offspring of parents with different characteristics.

Mendel called the hybrids the F_1 generation, and he called the cross that produced them an F_1 cross. The letter F stands for *filius*, which in Latin means "son." The true-breeding plants Mendel called the P generation. The letter P stands for the Latin word *parentis*, meaning "of the parent."

To Mendel's surprise, the traits of the parents did not blend in the F_1 generation. Instead, the traits of just one parent appeared in the offspring. The traits of the other parent seemed to have vanished!

☑ Checkpoint What are hybrids?

The F_2 Generation

If Mendel had stopped with the F_1 cross, he might not be remembered today. But he was curious about what had happened to the traits that seemed to have disappeared in the F_1 generation. So he decided to take the next logical step. He crossed the plants of the F_1 generation among themselves. This second cross he called the F_2 cross, and the plants that resulted he called the F_2 generation.

Incredibly enough, for each of the seven traits, the form that had vanished in the F_1 generation reappeared in the F_2 generation! Moreover, as illustrated in ***Figure 6–6,*** they reappeared in approximately one fourth of the plants in the F_2 generation.

Mendel knew that this pattern could not be a simple coincidence. Indeed, he realized that it signified something important about the nature of heredity.

☑ Checkpoint What is the F_1 generation? The F_2 generation?

Genes

Because the traits did not blend, Mendel reasoned that some indivisible unit must determine each of the traits he investigated. Mendel called this unit a *Merkmal,* which is the German word for "character." Today, the unit that determines traits is called a **gene.** Therefore, true-breeding tall pea plants contain genes for tallness, and true-breeding short pea plants contain genes for shortness.

☑ ***Checkpoint*** What is a gene?

Alleles

How many copies of each gene does a pea plant contain? Mendel concluded that for each of the seven traits he investigated, a pea plant must contain at least two genes—one from each parent.

Today, the different forms of a gene are called **alleles** (uh-LEELZ). For example, the gene that determines height in pea plants has two alleles. One allele produces a tall plant and another allele produces a short plant. While some genes have only two alleles, many genes have three, four, or even dozens of different alleles.

As illustrated in ***Figure 6–7*** on page 128, all the plants in the P generation contain two copies of an allele. Each tall plant contains two copies of the allele for tallness. And each short plant contains two copies of the allele for shortness.

When pea plants reproduce, each parent produces sex cells—either pollen or egg cells. Unlike other cells, the sex cells contain only one copy of each gene. Therefore, when a sex cell from a true-breeding tall plant unites with one from a true-breeding short plant, the seed it produces contains one allele for tallness and one allele for shortness.

Dominant and Recessive

As Mendel had observed, all the plants in the F_1 generation were just as tall as the tall plants in the P generation. But according to his model, the plants in the two generations contained different alleles. Why does a plant with two alleles for tallness grow to the same height as a plant with only one allele for tallness?

To answer this question, Mendel proposed a simple but very important idea. Mendel called the allele for tallness the **dominant** allele. The allele for shortness he called the **recessive** allele. In other words, in pea plants that have both alleles, only the dominant allele—the allele for tallness—is expressed.

Today, biologists represent a dominant allele with a capital letter and a recessive allele with a lowercase letter. Thus, for pea plants, a capital letter T represents the allele for tallness and a lowercase letter t represents the allele for shortness.

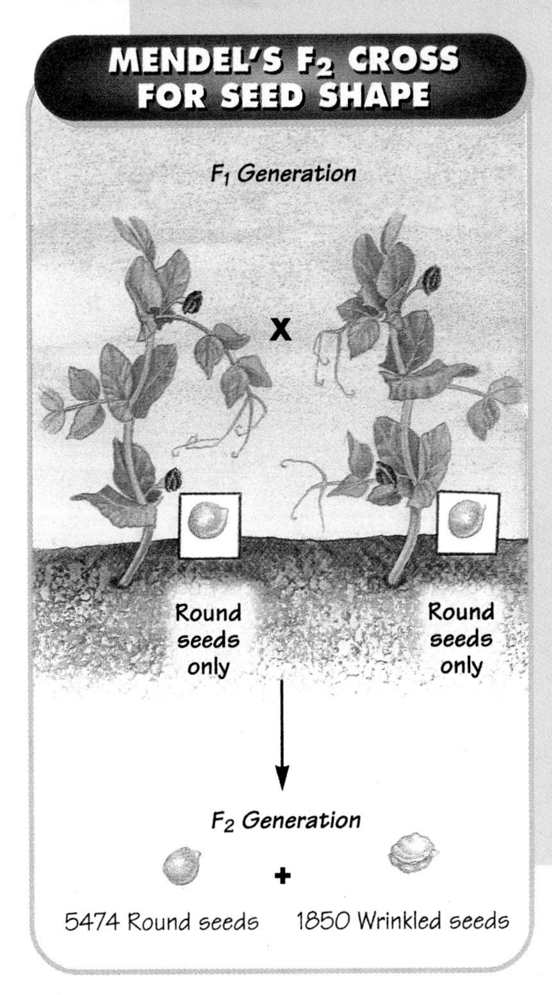

Figure 6–6 *In his crosses for seed shape, Mendel found that wrinkled seeds disappeared in the F_1 generation but reappeared in roughly one fourth of the plants of the F_2 generation. Mendel believed that this pattern was significant—and he was right!*

Phenotype and Genotype

An organism's **phenotype** is the form of a trait that it displays. Pea plants, for

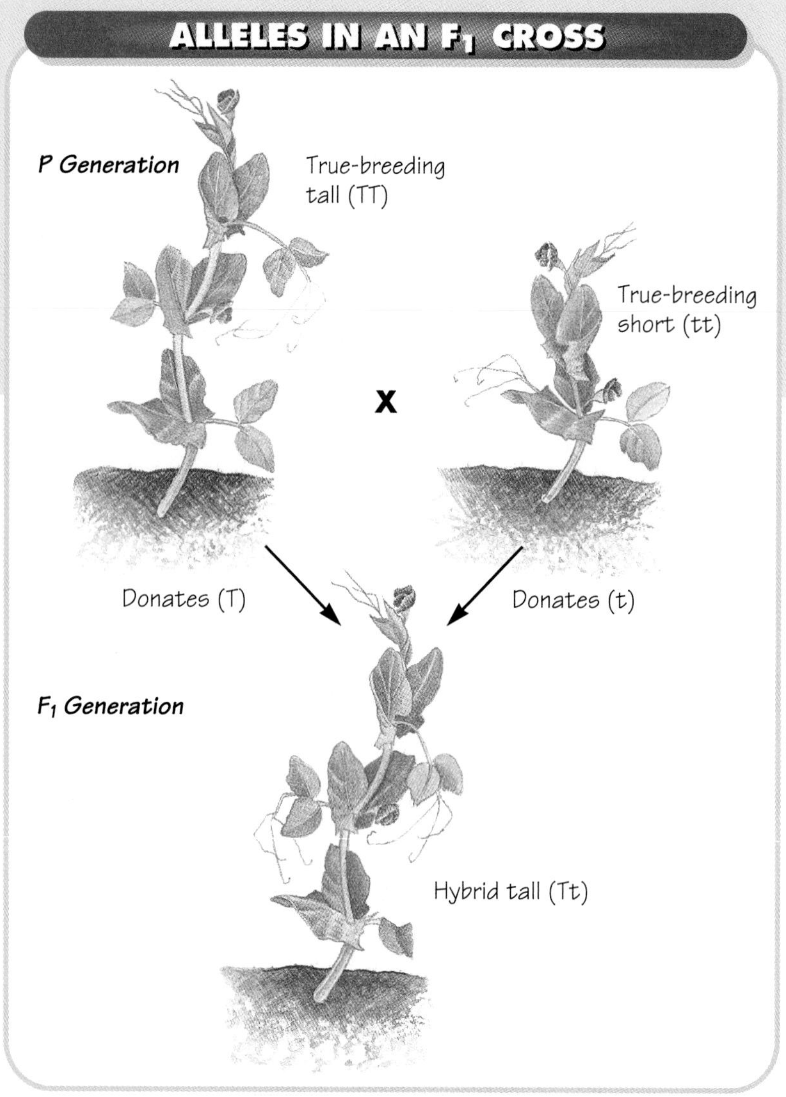

ALLELES IN AN F_1 CROSS

P Generation

True-breeding tall (TT)

True-breeding short (tt)

X

Donates (T)

Donates (t)

F_1 Generation

Hybrid tall (Tt)

Figure 6–7
Mendel used the idea of dominant and recessive alleles to explain the results of his experiments. He concluded that a pea plant contains two alleles for each of the seven traits that he studied. When the alleles differ, the dominant allele determines the form of the trait.

example, express either the phenotype for tallness or the phenotype for shortness. But Mendel had the insight to conclude that identical phenotypes could be produced by more than one **genotype**—an organism's genetic composition.

Today, organisms that have an identical pair of alleles for a trait—such as TT or tt—are said to be **homozygous** (hoh-moh-ZIGH-guhs) for the trait. The prefix *homo-* means "same," and the suffix *-zygous* means "joined together." Organisms that have a mixed pair of alleles—such as Tt—are said to be **heterozygous** (heht-er-oh-ZIGH-guhs) for the trait. The prefix *hetero-* means "different."

In his F_1 cross, Mendel crossed plants that were homozygous for tallness (TT) with plants that were homozygous for shortness (tt). All the offspring were heterozygous for height (Tt), yet were just as tall as the homozygous tall plants.

☑ **Checkpoint** What is the difference between genotype and phenotype?

Segregation

Mendel reasoned that when plants produced pollen and egg cells, the two copies of each gene would need to undergo **segregation**—a process that separates the two alleles of a gene. Remember that all the plants in the F_1 generation have the heterozygous genotype for height (Tt). Therefore, when a plant with the genotype Tt produces pollen or egg cells, half the sex cells will carry the allele for tallness (T) and the other half will carry the allele for shortness (t).

In an F_2 cross, half the pollen cells and half the egg cells carry the recessive allele (t), so exactly $1/2 \times 1/2$, or 1/4 of the offspring, should have two copies of the recessive allele (tt). Therefore, 1/4 of the offspring should be short plants (tt) and 3/4 of the offspring should be tall plants (TT or Tt). Indeed, these ratios are almost exactly what Mendel's F_2 cross produced.

Independent Assortment

Mendel's first experiments left one important question about genes unanswered: Does the inheritance of one gene affect the inheritance of another? To answer this question, he performed an experiment to follow two different genes as they passed from one generation to the next.

PUNNETT SQUARE FOR AN F_2 CROSS

Heterozygous tall (Tt)

	T	t
T	TT	Tt
t	Tt	tt

Heterozygous tall (Tt)

Homozygous short (tt)

Figure 6–8
This Punnett square illustrates Mendel's F_2 cross for plant height. The possible gametes from one parent (Tt) are written along the top of the square, and the possible gametes from the other parent (Tt) are written along the left. The boxes in the square show the four possible genotypes of the offspring.

From his earlier experiments, Mendel knew that the gene for seed shape had two forms—round (R) and wrinkled (r)—and that the allele for roundness was dominant. He also knew that the gene for seed color had two forms—yellow (Y) and green (y)—and that the allele for yellow color was dominant.

To follow both genes, Mendel first performed an F_1 cross between plants that were true-breeding for round yellow seeds (RRYY) and plants that were true-breeding for green wrinkled seeds (rryy). As he expected, every plant in the F_1 generation had round yellow seeds. Mendel reasoned that each of these plants received the dominant alleles (RY) from one parent and the recessive alleles (ry) from the other parent, providing them with the heterozygous genotype for both traits (RrYy).

Mendel then crossed the F_1 plants. When the F_1 plants produce pollen and egg cells, would the allele for round seeds (R) always stay with the allele for yellow seeds (Y)? Would the allele for wrinkled seeds (r) always stay with the allele for green seeds (y)? Or instead, would the alleles segregate independently, allowing for other combinations of alleles?

As shown in ***Figure 6–9,*** Mendel's second cross produced plants with distinct ratios of different combinations of

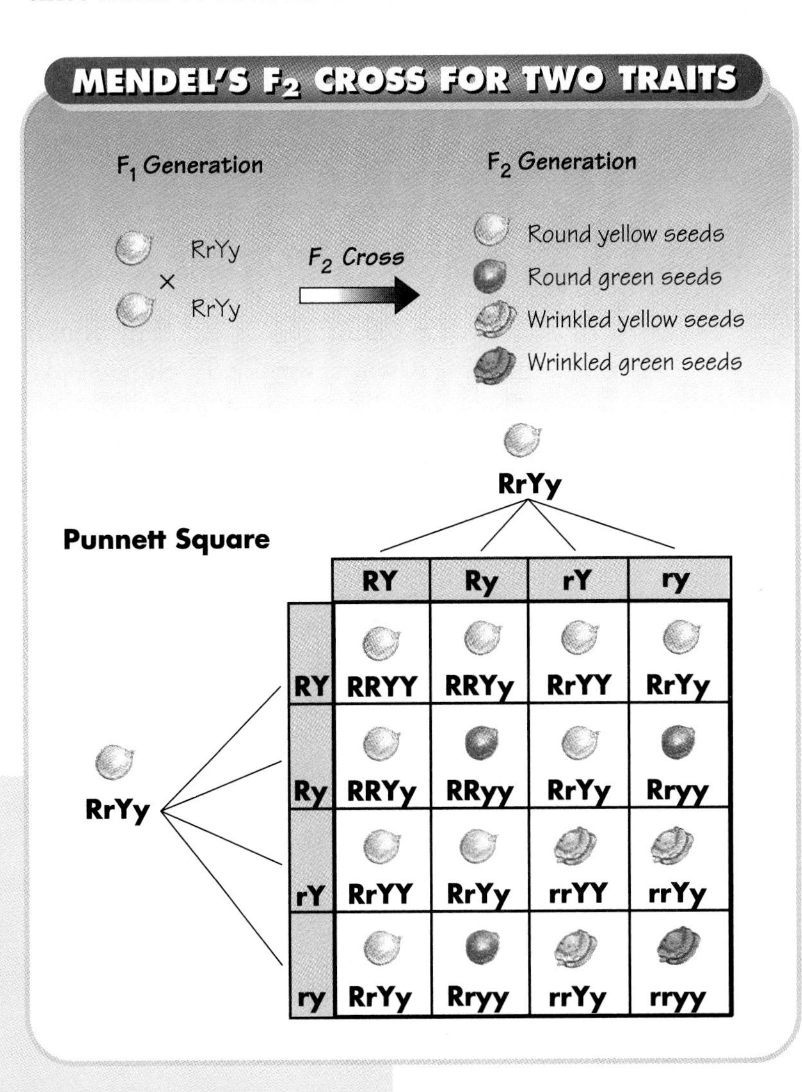

Figure 6–9
Mendel's F_2 cross for seed shape and seed color produced four seed phenotypes in a 9:3:3:1 ratio. This result showed that the genes for seed color and seed shape assort independently.

You and Your Genes

PROBLEM *How can you **classify** the genotypes and phenotypes for four traits that you display?*

PROCEDURE

1. Using the information in the table, classify your phenotype as dominant or recessive for each trait.

	Dominant	Recessive
Freckles	Present	Absent
Earlobes	Free	Attached
Hair	Curly	Straight
Fingers	Mid-digit hair	Hairless

2. Poll the class to find out how many students display each phenotype.

ANALYZE AND CONCLUDE

1. For which traits do you have the dominant phenotype? The recessive phenotype?
2. For which traits can you determine your genotype? Explain your answer.

seed shape and color—including round green seeds and wrinkled yellow seeds. This clearly meant that the alleles for seed shape segregated independently of those for seed color—a principle known as **independent assortment.** Put another way, genes that segregate independently—such as the genes for seed shape and seed color in pea plants—do not influence each other's inheritance.

☑ **Checkpoint** What is independent assortment?

Mendel's Principles

You might think that the publication of Mendel's work brought him fame and led to an immediate revolution in biology. But science doesn't always work that way.

In fact, Mendel's work went almost unnoticed for more than 30 years. Then, around the early 1900s, a number of scientists rediscovered Mendel's work and saw that its principles could be applied to animals as well as to plants.

Today, biologists can look back on Mendel's work and note four important principles that he developed:

- **Individual units, called genes, determine biological characteristics.**
- **For each gene, an organism receives one allele from each parent. The alleles separate from each other—a process called segregation—when reproductive cells are formed.**
- **If an organism inherits different alleles for the same trait, one allele may be dominant over the other.**
- **Some genes segregate independently.**

Section Review 6–1

1. **Define** heredity and **identify** the units of heredity.
2. **Explain** dominance, segregation, and independent assortment.
3. **Critical Thinking—Constructing Diagrams** A geneticist crosses a pea plant that is true-breeding for round green seeds (RRyy) with one that is true-breeding for wrinkled yellow seeds (rrYY). Construct a Punnett square to describe this cross. Compare it to ***Figure 6–9*** on page 129.
4. **MINI LAB** **Classify** yourself as expressing the dominant or recessive form for each of the four traits you investigated.

Meiosis

SECTION 6–2

Guide for Reading

- **Define** meiosis.

MINI LAB

- **Construct a model** of meiosis.

BECAUSE OUR KNOWLEDGE OF genetics has increased so much since Mendel's time, it's easy to overlook the fact that Mendel's genes were just hypothetical units. Mendel was not at all sure where genes might be located or how he would identify them. Fortunately, his descriptions of how genes behave were so specific that it was not long before biologists were certain they had found the location of genes in the cell.

Chromosomes

Because genes affect the entire organism, you might suspect that every cell has a copy of every gene. Where in the cell might you find genes? Remember that during mitosis—the process in which eukaryotic (yoo-kar-ee-AHT-ihk) cells (cells with a nucleus) produce daughter cells—structures called chromosomes appear. The chromosomes separate from each other when the cell divides, with one chromosome given to each daughter cell.

Could chromosomes actually contain genes? When chromosomes separate from each other during mitosis, are they in fact dividing genetic information between the two daughter cells? The idea that chromosomes contain genes is certainly an intriguing one—to say the least!

Forming Gametes

The best way to answer these questions is to watch what happens when an organism forms **gametes.** The term gamete is another name for a reproductive cell, such as a pollen cell or an egg cell.

According to Mendel's principles, an organism has two copies of each gene, but only one copy is passed on to an offspring. This means that when an organism forms gametes, the two copies of each gene need to separate precisely from each other.

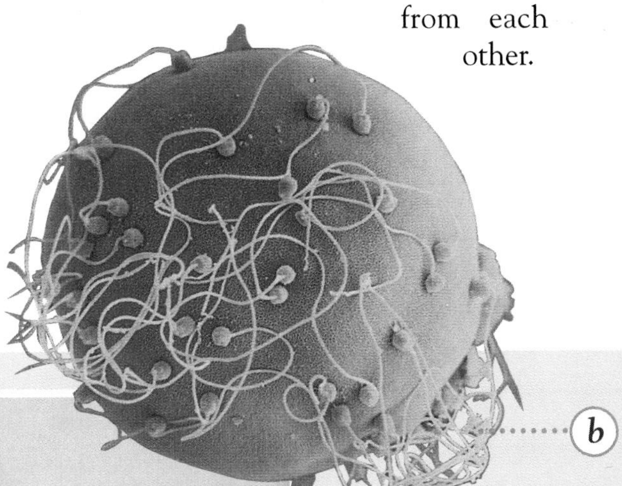

a

b

Figure 6–10
Both flowering plants and animals reproduce through gametes—specialized reproductive cells. (a) Each of these lilies contains a central pistil, which houses egg cells, and several stamens, which produce pollen. (b) This animal egg cell is surrounded by sperm cells. Typically, only one sperm fertilizes an egg (magnification: 480X).

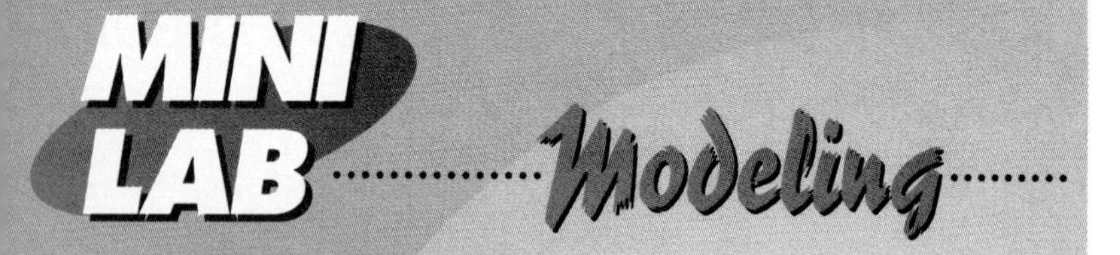

Toothpick Meiosis

PROBLEM *How can you **construct a model** of meiosis?*

PROCEDURE

1. Construct a model of a diploid cell that has 6 chromosomes. To model the chromosomes, use colored toothpicks, pipe cleaners, or other materials.
2. Move the chromosomes and redraw the cell boundaries to model the different stages of meiosis.
3. Repeat steps 1 and 2, this time aligning the chromosomes differently in Metaphase I.

ANALYZE AND CONCLUDE

1. What are the end products of meiosis?
2. Compare Metaphase I with Metaphase II.
3. Which step of meiosis determines the genetic composition of the gametes? Explain your answer.

Haploid and Diploid Cells

Gametes, in fact, contain exactly half the number of chromosomes found in other cells in the organism. For example, most human cells contain 46 chromosomes, but human gametes—sperm and egg cells—contain 23 chromosomes. The fusion of sperm and egg brings together chromosomes from both parents, just as Mendel's principles require.

The gametes, which contain just one set of chromosomes, are described as **haploid** cells. The number of chromosomes in a haploid cell is often represented by the letter n. In humans, $n = 23$.

Cells that contain a double set of chromosomes are described as **diploid.** The number of chromosomes in a diploid cell is represented by the term 2n. For humans, $2n = 46$.

☑ ***Checkpoint*** What is a haploid cell? A diploid cell?

The Phases of Meiosis

How does a diploid organism produce haploid gametes? It does so by **meiosis** (migh-OH-sihs)—a special process of cell division. **In meiosis, the number of chromosomes in a diploid cell is reduced by half, producing haploid gametes.**

At one time, meiosis was called reduction division. Reduction division is not a bad name for meiosis, because meiosis reduces the number of chromosomes in the cell by one half—from diploid (2n) to haploid (n).

Study the sequence of steps in meiosis, presented in the feature on the next page. At first, you may think that meiosis looks much like mitosis—the process by which eukaryotic cells divide. And unfortunately, the words meiosis and mitosis sound just enough alike to make it easy to confuse the two. But don't be fooled. There is something very different about meiosis—something that would have made Gregor Mendel smile.

☑ ***Checkpoint*** What is meiosis?

Meiosis and Genetic Diversity

In the first division of meiosis, the separation of each homologous chromosome pair is a random event. As a result, the haploid gametes could contain a great many different combinations of chromosomes.

Chromosome Combinations

How many chromosome combinations are possible? If an organism's cells have 6 chromosomes ($2n = 6$) that means that they have 3 pairs of homologous chromosomes. Because each pair can segregate in two different ways, there are 2^3 ($2 \times 2 \times 2$), or 8, possible combinations of chromosomes in the gametes.

Visualizing Meiosis

Meiosis takes place in two stages, called meiosis I and meiosis II. Let's take a look at the steps of each stage, using a cell that has 6 chromosomes (or 2n = 6) as an example.

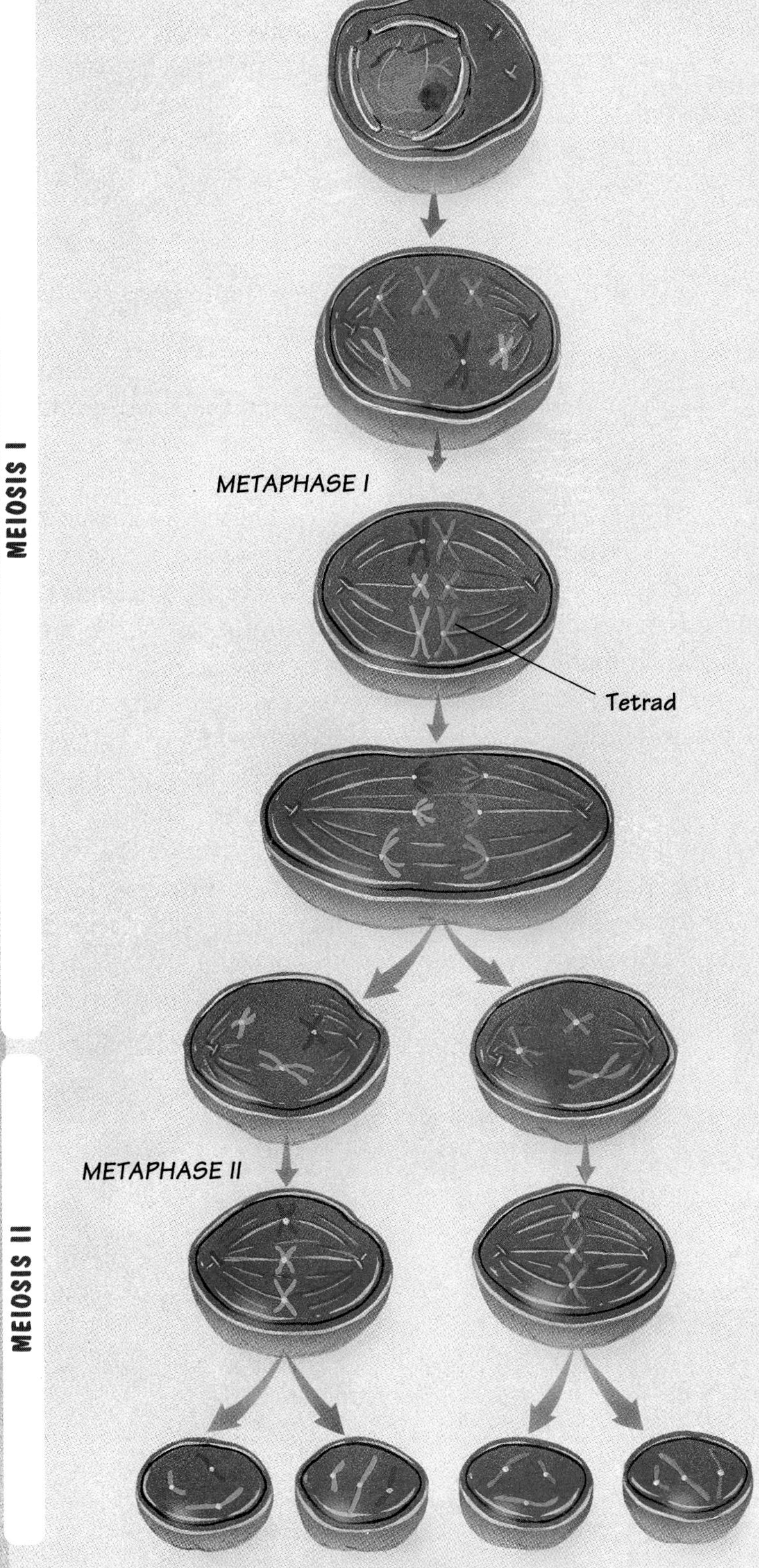

1. At the beginning of meiosis, each chromosome is replicated, forming duplicate chromatids joined at their centromeres.

2. Prophase I resembles prophase of mitosis. Chromosomes uncoil and the spindle apparatus forms.

3. At metaphase I, corresponding chromosomes, called homologous chromosomes, pair together and line up in the center of the cell. The paired chromosomes form structures called tetrads.

4. At anaphase I, the homologous chromosomes are pulled toward opposite sides of the cell.

5. The end products of meiosis I are two haploid (n) daughter cells. For the cells shown here, n = 3. In meiosis II, the two daughter cells undergo a second round of cell division.

6. In metaphase II, notice that chromosomes line up just as they line up in the metaphase stage of mitosis.

7. The end products of meiosis II are four haploid cells that may develop into gametes. In most male animals, all four cells develop into gametes. In females, only one cell does so.

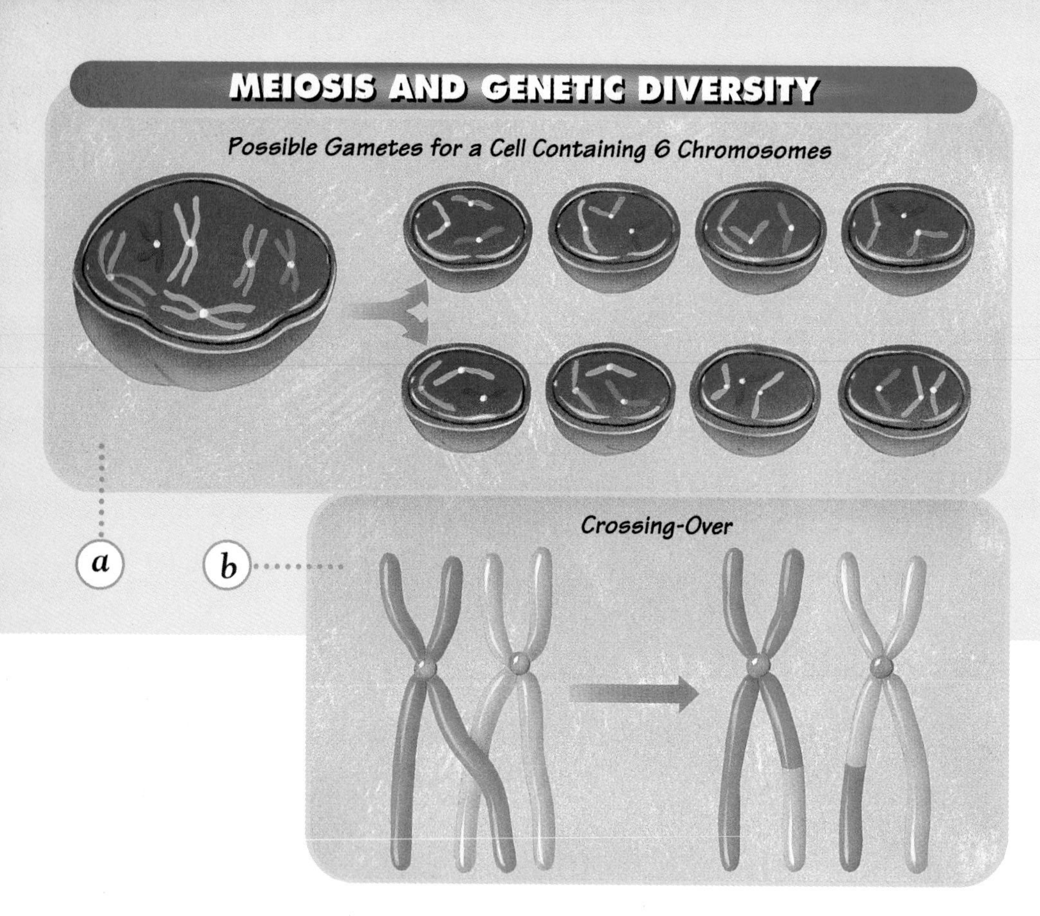

Figure 6–11
(a) *Through meiosis, a cell with 6 chromosomes can produce any of* 2^3*, or 8, different gametes. Human cells, which have 46 chromosomes, could produce any of* 2^{23} *different gametes!*
(b) *During meiosis, homologous chromosomes may exchange pieces of themselves. This process is called crossing-over, and it further increases the genetic diversity of gametes.*

Although eight combinations is not that many, think about the combinations in a human cell. Human cells have 23 pairs of chromosomes. Thus, the chromosomes can segregate in 2^{23} possible ways. That is more than 8 million possibilities!

Crossing-Over

You might think that 8 million ways to segregate 23 chromosome pairs is more than enough to shuffle the genetic deck. But that is just the beginning. Think about groups of genes that are found on the same chromosome. These genes often are inherited together, and therefore are said to be linked genes. But there is an important exception to gene linkage.

During meiosis, when homologous chromosomes are paired in a tetrad, a piece of one chromosome may change places with a piece of the other. This exchange between homologous chromosomes is called **crossing-over.** As shown in ***Figure 6–11,*** the effect of crossing-over is the switching of alleles from one chromosome to another.

When a crossing-over event occurs between two alleles on the same chromosome, the alleles could become located on different chromosomes. This means that crossing-over produces even more possible combinations of genetic material!

Section Review 6–2

1. **Define** meiosis.
2. **Sequence** and **describe** the process of meiosis.
3. **Explain** how crossing-over increases genetic diversity.
4. **Critical Thinking—Predicting** Suppose that a pair of homologous chromosomes fails to separate during the first round of meiotic division. How will this affect the gametes that are produced?
5. **MINI LAB** How can you **construct a model** of meiosis?

Analyzing Inheritance

GUIDE FOR READING

- **Relate** probability to genetics.

IF YOU KNOW THE GENOTYPES of both parents in a cross, can you determine the most likely genotypes of their offspring? The answer is yes. Indeed, one of the most useful aspects of genetics is that it has predictive value. Put another way, we can use genetics and mathematical principles to tell us the chances that certain events will happen.

Probability

Think about an ordinary event with an uncertain outcome, such as the flip of a coin. Flipping a coin has two possible outcomes—the coin may land either heads up or tails up. Each outcome has an equal probability, or likelihood, of occurring. Therefore, the probability of a single coin flip landing heads up is 1 out of 2, or 1/2. ●

If you flip a coin four times in a row, what is the probability that it will land heads up each time? Because each coin flip is an independent event, the probability of each coin landing heads up is 1/2. Therefore, the probability of flipping four heads in a row is

$$1/2 \times 1/2 \times 1/2 \times 1/2 = 1/16$$

As you can see, you have 1 chance in 16 of flipping heads four times in a row. The fact that we multiplied the individual probabilities together illustrates an important point—past outcomes do not affect future ones. Even if you flipped three heads in a row, the probability of the fourth coin landing heads up is still 1/2.

INTEGRATING MATHEMATICS

If you flipped a coin 50 times, how many times would you expect it to land heads up?

Probability and Genetics

Does probability apply to events in genetics? It definitely does. **Probability applies to genetics because the formation of gametes depends on random events.** Remember that a gamete is equally likely to contain one chromosome or its homologous chromosome,

Figure 6–12
Probability applies to outcomes in genetics, such as (a) whether a sheep has white wool or black wool and (b) whether a Persian cat has long hair or short hair.

Problem Solving

INTERPRETING DIAGRAMS

Cystic Fibrosis

Cystic fibrosis is an example of a genetic disease, or a disease that is inherited. A recessive allele causes cystic fibrosis, which means that a person who inherits only one copy of the allele will be unaffected. If two such people have a child, however, the child could inherit two copies of the allele. This genotype causes the disease.

Cystic fibrosis affects the glands that produce mucus, which is a slick, slimy liquid that coats the lining of the lungs, intestines, and other organs. In people with cystic fibrosis, the mucus is significantly thicker and stickier than normal. The abnormal mucus clogs the breathing and digestive passages, which typically leads to infections and malnutrition. Fifty years ago, a baby born with cystic fibrosis usually would die within a few years. Today, thanks to better diagnosis and treatment, many people with cystic fibrosis survive into adulthood and lead productive lives. Researchers continue to study this disease.

To learn more about the way cystic fibrosis is inherited, study the hypothetical cross presented in the Punnett square shown below. Identify the phenotype of each parent and potential offspring, then calculate the probability that a child of these parents would have cystic fibrosis.

PUNNETT SQUARE

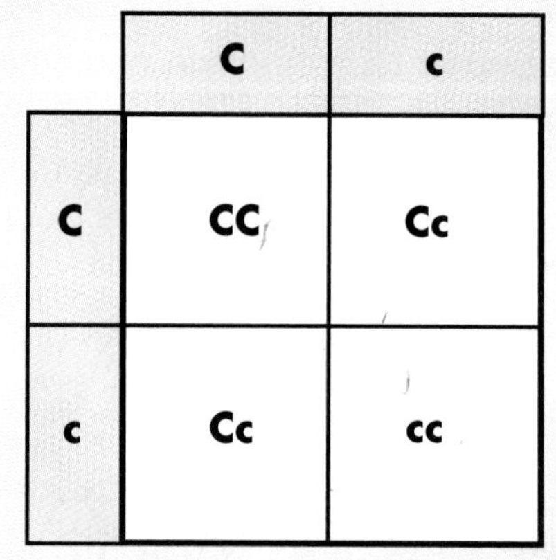

	C	c
C	CC	Cc
c	Cc	cc

Key: **C** normal allele
c disease allele

THINK ABOUT IT

1. A healthy man and a healthy woman have a child who has cystic fibrosis. Identify the genotypes for this trait for both parents and for their child.
2. This man and woman want another child. Calculate the probability that the second child would have cystic fibrosis.
3. Calculate the probability that a third child would have cystic fibrosis.
4. More serious genetic diseases are caused by recessive alleles than by dominant alleles. Explain why this is the case.

Figure 6–13
In this cross, 2 of the 4 possible offspring have white flowers. Therefore, the probability that the cross will produce a plant with white flowers is 2 out of 4, or 1 out of 2, or 1/2.

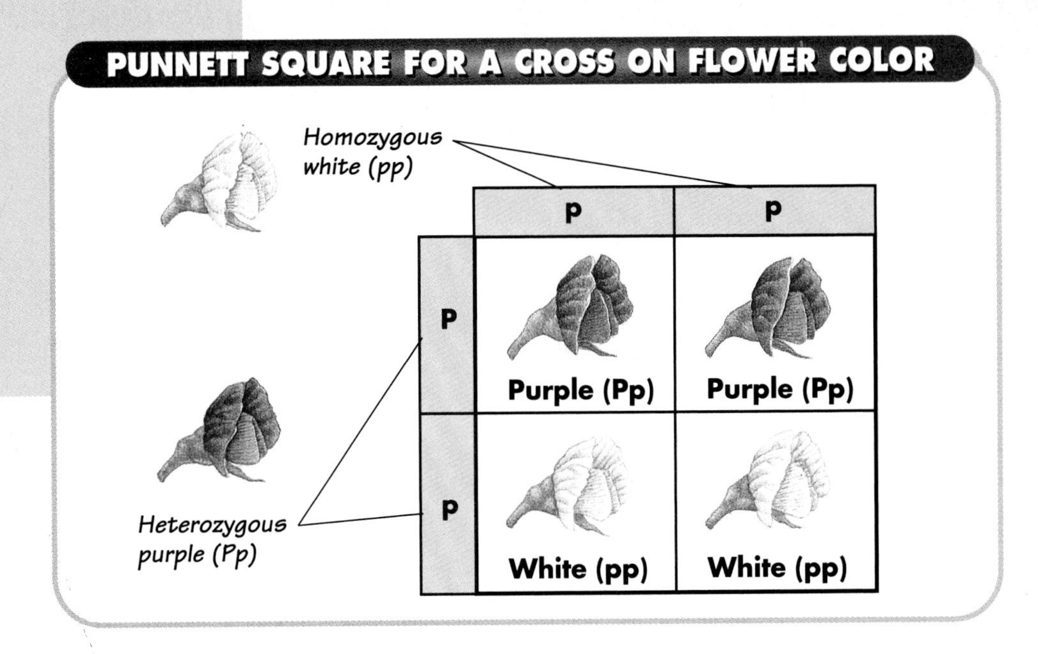

just as a coin is equally likely to land heads up or tails up. Therefore, if an organism has two different alleles for the same gene, the chances are 1 in 2, or 1/2, that one of its gametes carries one of the alleles.

Purple Flowers or White Flowers?

Let's take flower color in pea plants as an example. Suppose you cross a plant that is heterozygous for purple flowers (Pp) with one that is homozygous recessive for white flowers (pp). What is the probability that a single seed from this cross will produce a plant with white flowers?

As shown in ***Figure 6–13,*** the white flower produces only gametes that contain the allele for white flowers (p), while the purple flower produces two types of gametes—one with the allele for purple flowers (P) and one with the allele for white flowers (p). Therefore, half the seeds will be heterozygous for purple flowers (Pp) and half will be homozygous recessive for white flowers (pp). The probability that an offspring will produce white flowers is 1 out of 2, or 1/2.

Four Seeds

Suppose you plant four seeds from the cross just described. What is the probability of producing four plants with white flowers? Like flipping coins and getting four heads in a row, the probability is $1/2 \times 1/2 \times 1/2 \times 1/2 = 1/16$.

These numbers give an idea of what the term probability means. If you plant four seeds from this cross, you would expect half to produce white flowers and half to produce purple flowers. However, that doesn't mean that the seeds will always produce two plants with white flowers and two with purple ones. As you've just seen, there is a probability of 1 in 16 that all four will have white flowers.

Section Review 6–3

1. **Relate** probability to genetics.
2. **Calculate** the probability that a cross between individuals that are heterozygous for a trait (Aa × Aa) will produce a heterozygous offspring.
3. **Critical Thinking—Interpreting Data** In cats, the allele for short hair (S) is dominant over the allele for long hair (s). A cat with short hair is mated with a cat with long hair, producing five kittens with short hair. Can you identify the genotypes of both parents and their offspring? How certain can you be of your answers?

A Closer Look at Heredity

GUIDE FOR READING

- **Identify** the factors that cause genes to be expressed in different ways.

A COMPLEX ORGANISM MAY *have thousands or tens of thousands of genes. As you might suspect, some of those genes have patterns of inheritance that are a little more complicated than others. These complications make life interesting, and they explain why not every inherited trait follows exactly the same pattern.*

Incomplete Dominance and Codominance

Many genes have more than one allele or have alleles that are neither dominant nor recessive. Genes with these kinds of alleles give rise to a variety of different phenotypes.

Snapdragons, for example, have two alleles for flower color—red (R) and white (r). Snapdragons that are homozygous red (RR) produce red flowers, and those that are homozygous white (rr) produce white flowers. However, the heterozygous plants (Rr) are neither red nor white. They're pink!

Which allele is dominant? The answer is that neither allele is completely dominant. Although the presence of the red allele does produce red color in the heterozygous plant (Rr), the red color is not as intense as it is in the homozygous plant (RR). Therefore, the red allele displays **incomplete dominance** over the white allele. In incomplete dominance, the heterozygous phenotype is somewhere in between the two homozygous phenotypes.

Another case in which both alleles affect phenotype is known as **codominance.** Interesting examples of codominance are found in roan horses and erminette chickens, as illustrated in ***Figure 6–16.*** In codominance, both alleles of a gene are expressed.

Figure 6–14
Not all genes display the simple behavior that you have studied so far. (a) In rabbits, the gene that determines coat color has four alleles. (b) In snapdragons, neither of the alleles that control flower color is dominant.

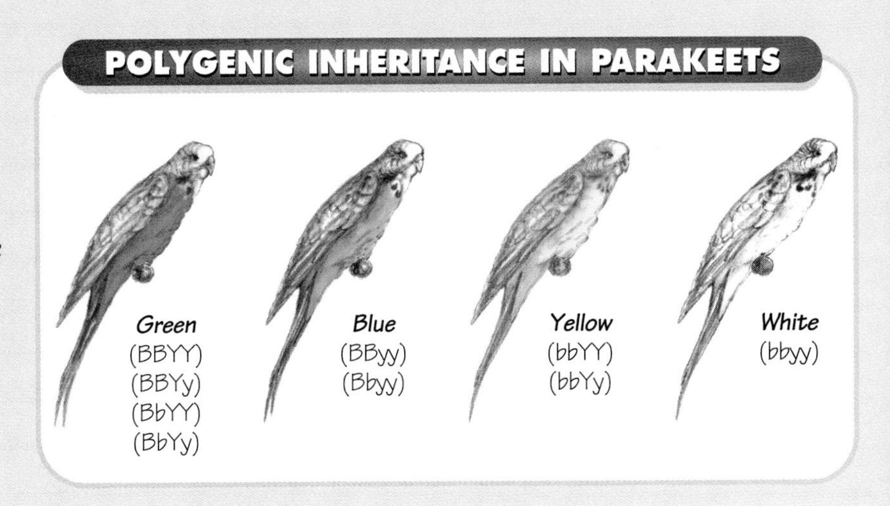

Figure 6–15
In parakeets, feather color is controlled by two genes—one that controls blue color and one that controls yellow color. Green parakeets have at least one dominant allele for each gene, whereas white parakeets have only the recessive alleles.

Multiple Alleles

As you learn the basic principles of genetics, it's easy to focus on genes that have just two alleles. But genes that have multiple alleles are very common in nature. For example, coat color in rabbits is determined by a single gene that has four well-known alleles. In humans, genes that have multiple alleles include the genes for blood group and eye color.

Polygenic Traits

Many inherited traits are controlled by more than one gene. This is particularly true of the genes that control body shape and form. For example, although your facial appearance is inherited, no single gene determines the exact shape of your mouth or the position of your ears. Traits that are controlled by more than one gene are said to be **polygenic traits.** Feather color in parakeets is one example of a polygenic trait, as shown in ***Figure 6–15.***

Because polygenic traits are controlled by more than one gene, their inheritance can be complicated. Polygenic traits often show a very wide range of phenotypes. For example, the range of skin colors in humans comes about partly because at least four different genes control this trait.

Figure 6–16
(a) In Erminette chickens, the gene for feather color has two codominant alleles—one for black feathers and one for white feathers. Thus, heterozygous chickens have both types of feathers.
(b) Roan horses have two colors in their coats—another result of codominant alleles.

1. **Identify** the factors that cause genes to be expressed in different ways.
2. **Describe** four different types of genetic expression.
3. **BRANCHING OUT ACTIVITY** **Organize** a scheme of inheritance for three different traits in a mythical animal or plant. Each trait should follow one of the patterns of inheritance discussed in this section.

Laboratory Investigation

Mapping a Chromosome

How do scientists determine the location of a gene on a chromosome? One method is to study an event called crossing-over—the exchange of genes between homologous chromosomes. As you will see, the farther apart that genes are located on a chromosome, the more often that crossing-over separates them.

Problem

How are chromosomes mapped? **Interpret** the results of an experiment to discover the answer.

Materials (per group)

- pipe cleaner
- metric ruler
- paper
- pencil or pen

Procedure

1. **With a metric ruler, draw a vertical line 15 cm long on a sheet of paper. This line represents a chromosome. The pipe cleaner represents its homologous chromosome.**
2. **Mark the bottom of the line with a small horizontal line. Measuring from the horizontal line, mark off points 1 cm, 3 cm, 6 cm, 10 cm, and 15 cm. Draw a small horizontal line at each point. The horizontal lines represent the locations of genes on the chromosome.**
3. **Label the horizontal lines A through F, starting from the bottom horizontal line.**
4. **Construct a data table similar to the one shown above Observations.**
5. **Place the sheet of paper 15 cm from the edge of the table. Standing at least 30 cm away from the table, toss the pipe cleaner so it crosses the chromosome—the vertical line on the sheet of paper.**
6. **Assume a crossing-over occurs at the point where the pipe cleaner touches the line. Determine which genes (B through F) have separated from gene A. For example, if the pipe cleaner falls between genes C and D, then genes D, E, and F have separated from gene A. If the pipe cleaner falls between genes D and E, then genes E and F have separated from gene A. Place a check mark in each appropriate box in your data table.**
7. **Repeat steps 5 and 6 to produce 50 crossing-overs.**
8. **Count the total number of check marks for each gene. Record each total in the second column of your data table.**

DATA TABLE

Gene	Times Separated From Gene A		Frequency of Separation From Gene A	Gene Location	
	Check	Total		Calculated	Actual
B					
C					
D					
E					
F					

Observations

1. To calculate each gene's frequency of separation from gene A, divide the total number of separations by 50. Record the frequency of separation in the third column of your data table.
2. To calculate the location of each gene, multiply its frequency of separation by 15. Record these values in the fourth column of your data table.

Analysis and Conclusions

1. Which genes separated most frequently from gene A? Which genes separated least frequently from gene A?
2. Describe the relationship between the locations of two genes on a chromosome and the frequency with which crossing-over separates them. Use the data generated to support your answer.
3. How accurately did this procedure determine the locations of the genes? Explain why the procedure did not determine the exact locations of the genes.

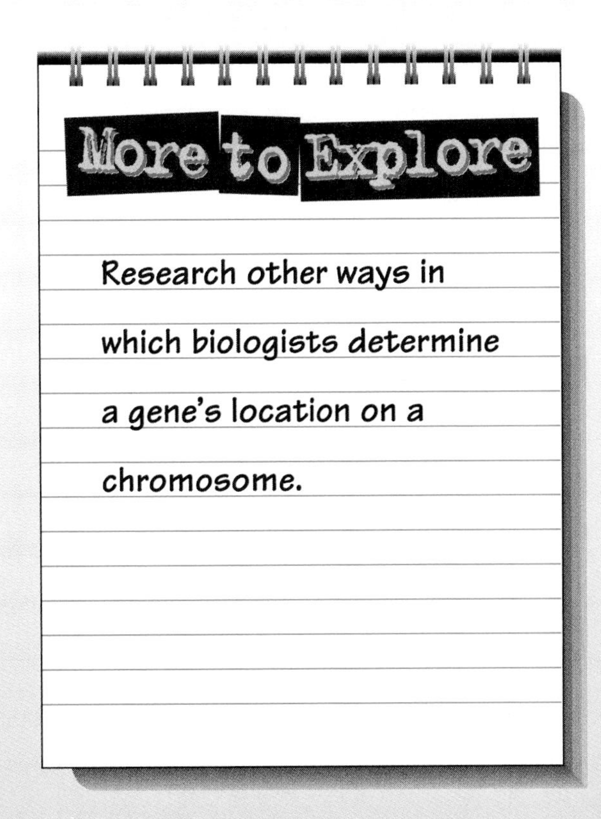

Study Guide

Summarizing Key Concepts

The key concepts in each section of this chapter are listed below to help you review the chapter content. Make sure you understand each concept and its relationship to other concepts and to the theme of this chapter.

6–1 The Science of Inheritance

- An organism's heredity is the set of characteristics it receives from its parents. Today, the study of heredity is known as genetics.
- As Mendel's experiments showed, traits are determined by units called genes. An organism contains two copies of every gene because it receives one copy from each parent.
- Genes have different forms, called alleles. When an organism receives two different alleles for the same trait, only the dominant allele is expressed.
- An organism's phenotype is the form of a trait it displays, and its genotype is its genetic composition. Different genotypes can produce the same phenotype.
- When sex cells are formed, the alleles undergo segregation, or separate from each other. Genes that are assorted independently do not influence each other's inheritance.

6–2 Meiosis

- Genes are located on chromosomes. Gametes contain a single set of chromosomes and are described as haploid cells. Cells that contain a double set of chromosomes are described as diploid.
- In meiosis, the number of chromosomes in a diploid cell is reduced by half, producing haploid gametes.

6–3 Analyzing Inheritance

- Probability applies to genetics because the formation of gametes depends on random events.

6–4 A Closer Look at Heredity

- Many genes have more than one allele, or have alleles that are neither dominant nor recessive. Genes with these kinds of alleles give rise to a variety of diffferent phenotypes.

Reviewing Key Terms

Review the following vocabulary terms and their meaning. Then use each term in a complete sentence.

6–1 The Science of Inheritance

heredity	recessive
genetics	phenotype
trait	genotype
hybrid	homozygous
gene	heterozygous
allele	segregation
dominant	independent assortment

6–2 Meiosis

gamete	meiosis
haploid	crossing-over
diploid	

6–4 A Closer Look at Heredity

incomplete dominance
codominance
polygenic trait

Recalling Main Ideas

Choose the letter of the answer that best completes the statement or answers the question.

1. Who was the first person to study heredity scientifically?
   a. Mendel **c.** Watson
   b. Punnett **d.** Calvin
2. Crossing different true-breeding stocks produces offspring called
   a. the P generation. **c.** cross-overs.
   b. the F_2 generation. **d.** hybrids.
3. The different forms of a gene are called
   a. traits. **c.** gametes.
   b. alleles. **d.** hybrids.
4. Which represents a heterozygous genotype for height in pea plants?
   a. (TT) **c.** (tt)
   b. (Tt) **d.** (T) or (t)
5. Gametes are described as
   a. haploid cells. **c.** triploid cells.
   b. diploid cells. **d.** adult cells.
6. During meiosis, homologous chromosomes pair together to form structures called
   a. chromatids. **c.** gametes.
   b. tetrads. **d.** centromeres.
7. What are the end products of meiosis?
   a. 2 diploid cells **c.** 2 haploid cells
   b. 4 diploid cells **d.** 4 haploid cells
8. Crossing-over is the exchange of genetic information between
   a. reproductive cells.
   b. diploid cells.
   c. any 2 chromosomes.
   d. homologous chromosomes.
9. The red, pink, and white colors of snapdragons are controlled by a gene that shows
   a. incomplete dominance.
   b. codominance.
   c. polygenic dominance.
   d. multiple alleles.

Putting It All Together

Using the information on pages xxx to xxxi, complete the following concept map.

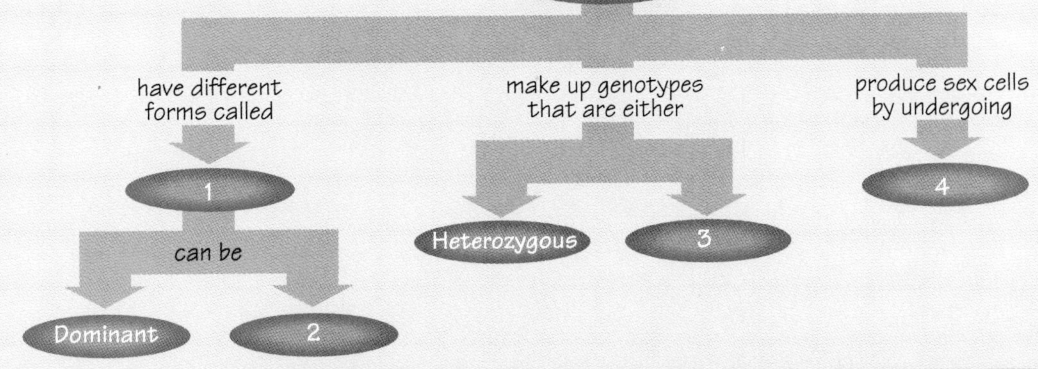

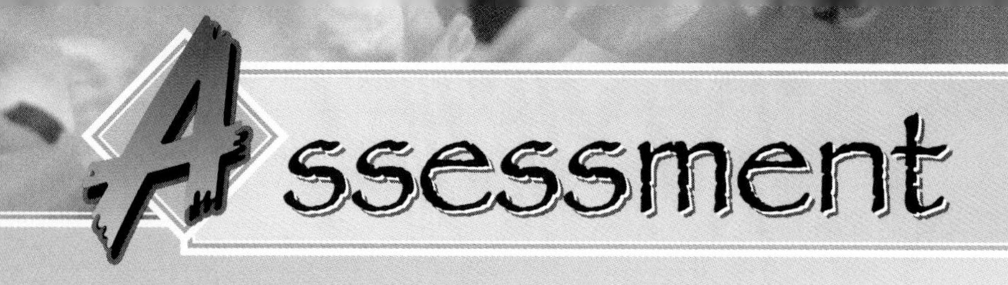

Reviewing What You Learned

Answer each of the following in a complete sentence.

1. What is an organism's heredity?
2. Describe the contributions that Gregor Mendel made to the study of biology.
3. Explain the difference between self-pollination and cross-pollination.
4. In Mendel's experiments, what was the P generation? The F_1 generation? The F_2 generation?
5. Why can organisms that have the same phenotype for a trait have different genotypes for the trait?
6. What are haploid cells? Diploid cells?
7. What are the end products of meiosis?
8. What is crossing-over?
9. Compare a homozygous genotype and a heterozygous genotype.
10. What is segregation?
11. Define polygenic traits.

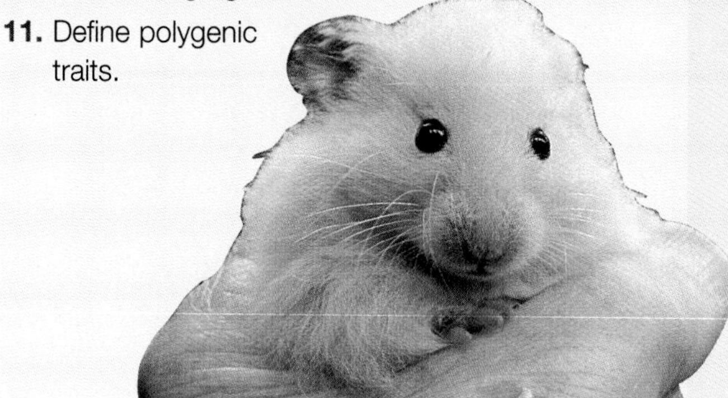

Expanding the Concepts

Discuss each of the following in a brief paragraph.

1. Why did Mendel use true-breeding stocks of pea plants in his experiments?
2. When Mendel cross-pollinated two pea plants, why did he cut off the male flowering parts of one pea plant?
3. In Mendel's F_1 crosses, explain why one form of a trait seemed to disappear in the F_1 generation and then reappeared in the F_2 generation.
4. In Mendel's F_2 cross on height in pea plants, 3/4 of the plants in the F_2 generation were tall and 1/4 of the plants were short. Discuss the significance of these ratios.
5. Do all genes have only two alleles, with one allele dominant over the other? Explain your answer.
6. Describe the principle of independent assortment. How did Mendel show that the genes for seed shape and seed color are assorted independently in pea plants?
7. What is a tetrad? Discuss why the formation of tetrads is important in meiosis.
8. If an organism has 5 chromosomes, how many different gametes could it produce? Explain your answer.
9. A geneticist crosses two organisms that are heterozygous for a trait (Aa $\times$ Aa). Construct a Punnett square for this cross. Then identify the genotype and phenotype of each of the possible offspring.
10. What is probability? How does it relate to genetics?
11. Compare incomplete dominance with codominance.
12. Why can multiple alleles provide many different phenotypes for a trait?

Extending Your Thinking

Use the skills you have developed in this chapter to answer the following.

1. **Relating** Describe how Mendel's principles of inheritance relate to the segregation of chromosomes in meiosis.
2. **Calculating** In fruit flies, the allele for a gray body (G) is dominant over the allele for a black body (g). In a cross between a black fruit fly (gg) and a heterozygous gray fruit fly (Gg), what is the probability that one offspring will have a gray body? That two offspring will have gray bodies?
3. **Interpreting data** In hamsters, the allele for black fur (B) is dominant over the allele for brown fur (b), and the allele for long hair (L) is dominant over the allele for short hair (l). A student performs an F_1 cross for these traits (BBLL × bbll), then performs an F_2 cross. Of the 102 offspring in the F_2 generation, 77 have long black hair and 25 have short brown hair. Explain the significance of this result.
4. **Classifying** In snapdragons, the allele for red flowers shows incomplete dominance over the allele for white flowers. If you know the color of a snapdragon flower, can you classify the genotype as homozygous or heterozygous? Explain your answer.
5. **Drawing conclusions** In humans, the allele for free earlobes (F) is dominant over the allele for attached earlobes (f). In one family, both parents and three of the children have free earlobes and one child has attached earlobes. From this information, can you conclude that each parent is heterozygous for ear shape? Explain your answer.
6. **Constructing a model** Construct a model of crossing-over. You may use straws, pipe cleaners, or toothpicks to represent the chromosomes. Why does crossing-over increase genetic diversity?

Applying Your Skills

Dog Breeding

In dogs, the allele for a spotted coat (S) is dominant over the allele for a solid coat (s). Suppose that you own a male dog with a spotted coat and a female dog with a solid coat and you allow the dogs to mate.

1. Can you calculate the probability that the first offspring will have a solid coat? Explain your answer.
2. Suppose that one parent of the female dog had a solid coat. Does this information change your answer to question 1? Explain.
3. Suppose that one parent of the male dog had a solid coat. Does this information change your answer to question 1? Explain.

• GOING FURTHER •

4. Other traits in dogs include hair length (short is dominant over long), hair texture (wiry is dominant over silky), and hair curliness (curly is dominant over straight). Is the dominant form of these traits necessarily more common than the recessive form? Explain why or why not.

CHAPTER 7

Human Inheritance

FOCUSING THE CHAPTER
THEME: Unity and Diversity

7–1 The Human Genetic System
- **Explain** why humans are not ideal organisms for the study of genetics.

7–2 Sex-Linked Inheritance
- **Describe** some of the genetic disorders carried on the sex chromosomes.

7–3 Human Genetic Disorders
- **Classify** some genetic disorders based on their cause.

BRANCHING OUT *In Depth*

7–4 Special Topics in Human Genetics
- **Describe** some of the recent advances in the field of genetics.

LABORATORY INVESTIGATION
- **Observe** the differences among fingerprints of different people.

Biology and Your World

BIO JOURNAL

People often resemble each other but don't look exactly alike, except for identical twins. Think about the people you know. In what ways are they similar? In what ways are they different? Answer these questions in your journal.

A group of teenagers

The Human Genetic System

Guide for Reading

- **Explain** how most human traits are inherited.
- **Define** multiple allele.

WHAT MAKES EACH PERSON unique and different? Obviously, every individual has a different set of experiences. Each person grows up with different surroundings and with a different viewpoint of the world. Added to these, however, are the biological differences that individuals inherit in their genes. These also help to make each person unique. Each one of us is born with a biological inheritance so large that it would take a library full of books to do it justice.

How much do we know about the library of human genetics? Today, it's fair to say that we have just begun to understand how large that library is and have begun to read some of its books. They are interesting, to be sure, but we have a lot of reading to do.

A Model System?

Are humans ideal organisms to use for the study of genetics? Unfortunately, they are not. If you wanted a perfect organism to study inheritance, you would surely pick something that was much simpler, that reproduced more quickly, that produced more offspring, and that took up less space in the lab! You might even make the same choice that geneticists made years ago—the common fruit fly, *Drosophila melanogaster*.

The fruit fly was first suggested as the perfect organism for genetic studies nearly 100 years ago by the American geneticist Thomas Hunt Morgan. Morgan realized that the fruit fly had several advantages for the study of genetics. First, it can produce a new generation in just a few weeks, making it possible to conduct experiments in a reasonable amount of time. Second, a fruit fly is small and thus easy to maintain in large numbers in the laboratory. Third, a fruit fly has a relatively simple genetic system, with just 8 chromosomes in a diploid cell.

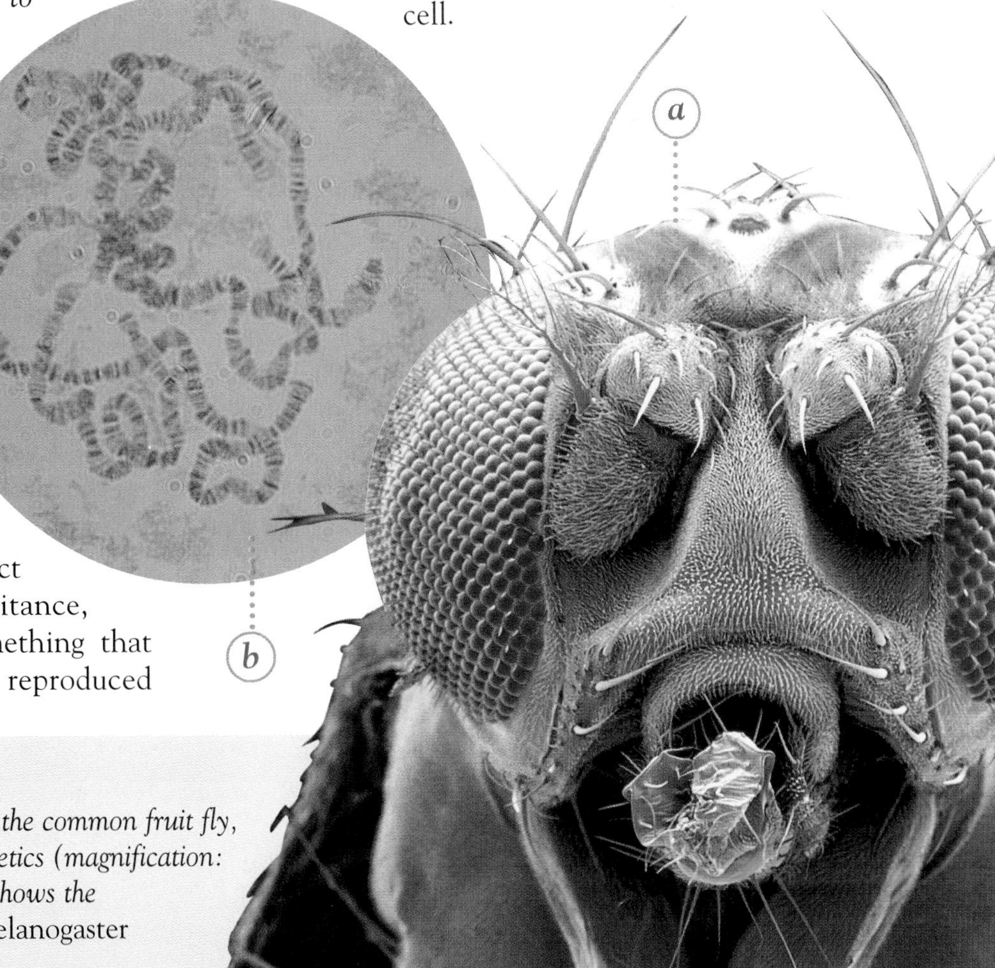

Figure 7–1
(a) Drosophila melanogaster, *the common fruit fly, is often used in the study of genetics (magnification: 140X).* (b) *This photograph shows the chromosomes of* Drosophila melanogaster *(magnification: 500X).*

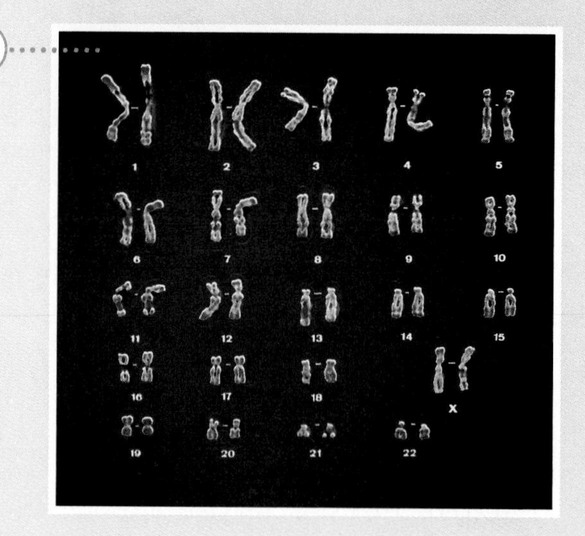

Figure 7–2
(a) In order to prepare a karyotype, a geneticist will literally cut out each individual chromosome from a photograph, match it to its corresponding homologous pair, and then arrange them in numerical order. (b) This photograph shows a complete karyotype of a female, with all the chromosomes arranged in numbered homologous pairs.

Biologists have used the fruit fly as a model system for the study of genetics for many years. These tiny animals have helped biologists learn how heredity functions in other organisms, including humans. In fact, many major discoveries in genetics were made in the fruit fly first and then were applied to the more complex genetic system—the human—later.

☑ **Checkpoint** Why is the fruit fly a model organism for the study of genetics?

The Human Cell

Diploid human cells have a total of 46 chromosomes—23 from each parent. As you may recall, chromosomes are easiest to see during mitosis. However, in order to see all 46 chromosomes at once, cell biologists have to use a special process. This process involves growing a sample of cells in colchicine, a poison that breaks down microtubules. This action prevents cells from completing mitosis. Before long, most of the cells are trapped in metaphase, and their chromosomes are fully condensed and easy to see and photograph. To analyze the chromosomes, cell biologists will literally cut out the photograph of each chromosome and then group them together, as shown in ***Figure 7–2.*** A picture of chromosomes put together this way is known as a **karyotype** (KAR-ee-uh-tighp).

A human karyotype reveals that each cell has 22 pairs of homologous (similar in structure) chromosomes that are numbered from 1 to 22 in order of decreasing size—chromosome 1 being the largest. The 22 pairs are called **autosomes,** or autosomal chromosomes. Each cell also has one pair of **sex chromosomes,** called the X and Y chromosomes. Female cells have two X chromosomes. Male cells have one X chromosome and one Y chromosome.

☑ **Checkpoint** What is a karyotype?

Human Reproductive Cells

The human reproductive cells—sperm and eggs—are produced by meiosis in the male and female reproductive systems. Sperm and eggs are both haploid—that is, they contain 23 chromosomes, only half the total number of chromosomes in human body cells. Therefore, they normally carry just one sex chromosome each. During fertilization, a sperm and an egg unite to form a zygote that contains 46 chromosomes.

Just like any other organism, human reproductive cells go through meiosis. Homologous chromosomes pair during the first meiotic division, and crossover events between human chromosomes produce genetic recombinations.

The sex chromosomes form tetrads with each other. This action ensures that each egg normally carries a single X chromosome. During meiosis in males, the X and Y chromosomes pair and then

SEX DETERMINATION

Male

		X	Y
Female	X	XX	XY
	X	XX	XY

Figure 7–3
Female cells contain only X chromosomes, and male cells contain both X and Y chromosomes. As a result, the male cells determine the sex of a baby. As you can see from this Punnett square, there is a 50-50 chance of having one sex or the other.

are separated in the first meiotic division. This means that half of a male's sperm cells carry an X chromosome and the other half carry a Y chromosome. Think about this fact as you look at ***Figure 7–3,*** and see if you can explain why males and females are born in nearly equal numbers.

☑ **Checkpoint** Is an egg haploid or diploid?

Pedigree Analysis

A human generation spans more than 20 years, making it almost impossible to carry out the kinds of experiments that are possible with other organisms, such as fruit flies. In order to study the inheritance of human traits, biologists have to rely on family histories and medical records to provide the information they need. One of the best ways to summarize this information is to construct a **pedigree.** A pedigree is a diagram that follows the inheritance of a single trait through several generations in a family.

In a pedigree, squares represent males and circles represent females. Vertical lines connect parents and their children, and horizontal lines connect male and female parents. In a family, the symbols for the children are placed from left to right in birth order, with the oldest child on the extreme left. If a pedigree illustrates an inherited recessive trait, the squares or circles representing males and females with this trait are shaded. If a person is heterozygous for the trait (or a hybrid), the square or circle is half shaded.

☑ **Checkpoint** What is a pedigree?

Genes and People

Because Mendel's principles of genetics also apply to humans, it is possible to use genetics to analyze human inheritance. **Many human traits are inherited by the action of genes that have dominant and recessive alleles.** Other traits are determined by genes that have more than two alleles.

Dominant and Recessive Alleles

The Rh blood group is one example of a trait that is determined by a single-gene, two-allele system—positive and negative. The positive allele (Rh^+) is dominant, so persons who have two positive alleles (Rh^+/Rh^+) or one positive (Rh^+) allele and one negative (Rh^-) allele are said to be Rh-positive. Those with two negative alleles (Rh^-/Rh^-) are Rh-negative.

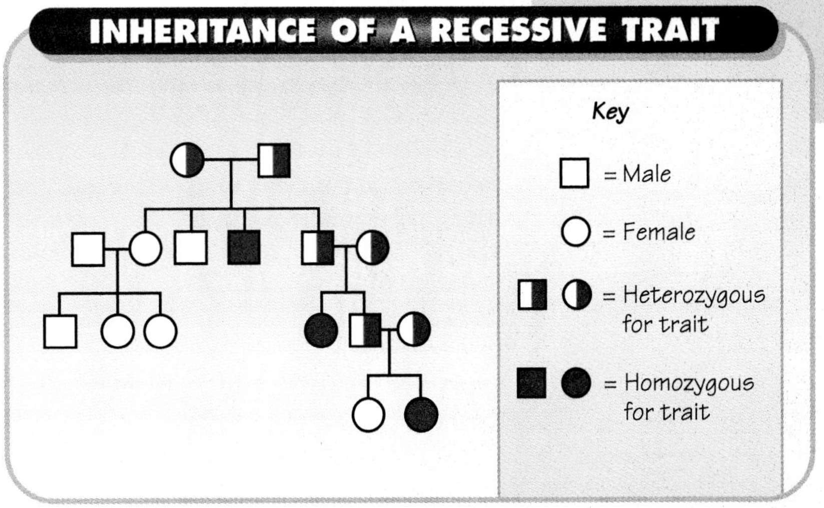

Figure 7–4
This pedigree traces the inheritance of a single recessive trait in a family.

INHERITANCE OF A RECESSIVE TRAIT

Key
□ = Male
○ = Female
◨ ◑ = Heterozygous for trait
■ ● = Homozygous for trait

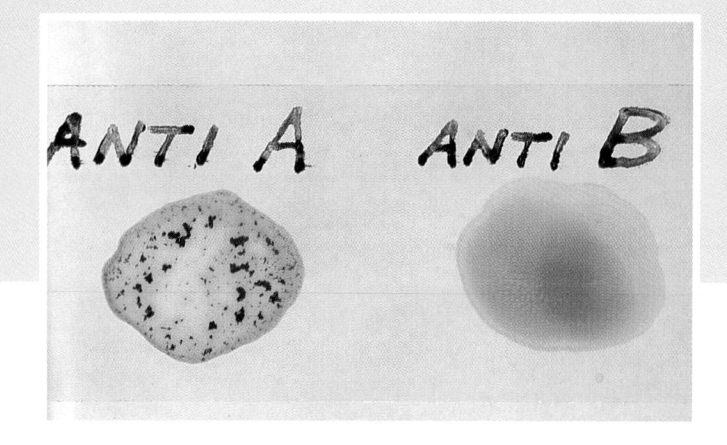

Figure 7–5
Blood groups are one example of a multiple allele. The blood sample on the left has formed clumps, indicating that two incompatible blood groups were mixed. The blood sample on the right has no clumps, indicating that compatible blood groups were mixed.

Multiple Alleles

A multiple allele is a type of gene that is determined by more than two alleles. The ABO blood group is an example of a **multiple allele.** There are three alleles for this gene—I^A, I^B, and i. To complicate matters further, the I^A and I^B alleles are codominant, meaning they are both dominant over the allele i. Alleles I^A and I^B are responsible for producing molecules called antigens, which can be recognized by the immune system on the surface of the red blood cell. Allele i does not produce any antigens.

INTEGRATING CAREERS

A phlebotomist is a person trained to withdraw blood. What kind of training does a phlebotomist need?

When medical workers describe blood groups, they usually mention both groups at the same time. For example, they may say that a patient has AB-negative blood, meaning that the person has both A and B antigens from the ABO gene and the negative allele from the Rh blood group.

Blood groups are very important, especially in medical procedures involving blood transfusions. ● Physicians must take care to be sure that the blood groups of the donor and the recipient are compatible. A transfusion of incompatible blood could cause a violent, or even fatal, reaction.

Figure 7–6
A person with blood group AB can safely receive a blood transfusion from any of the other blood groups. A person with blood group O, on the other hand, can receive a blood donation only from another individual with blood group O.

BLOOD GROUPS

Blood Group	Alleles	Antigen on Red Blood Cell	Safe Transfusions	
			To	From
A	I^AI^A or I^Ai	A	A, AB	A, O
B	I^BI^B or I^Bi	B	B, AB	B, O
AB	I^AI^B	A, B	AB	A, B, AB, O
O	ii	none	A, B, AB, O	O

Section Review 7–1

1. **Explain** how most human traits are inherited.
2. **Define** multiple allele.
3. **Critical Thinking—Drawing Conclusions** Why do you think that people with blood group O are called universal donors? (*Hint:* Look at **Figure 7–6**).

Sex-Linked Inheritance

GUIDE FOR READING

- **Describe** how sex is determined in fruit flies.
- **Explain** why the sex-linked disorders are more common in males than in females.

MINI LAB

- **Construct a model** showing the inheritance of sex.

GENES THAT ARE LOCATED ON the 22 pairs of autosomes are inherited according to the principles described by Mendel. Does this also apply to genes located on the sex chromosomes? Should you expect to find a special pattern of inheritance for genes located on the X or the Y chromosome? The answer to these questions is yes, although it may surprise you that the answers did not come from studies of human inheritance. The understanding that genes could be sex-linked came from work on the fruit fly.

Sex Determination

The diploid cells of a fruit fly have only 8 chromosomes, but like human cells, they contain 2 sex chromosomes. **Female fruit flies have two X chromosomes, and male fruit flies have one X chromosome and one Y chromosome.** Like humans, the segregation of sex chromosomes in fruit fly meiosis ensures that male and female fruit flies will be born in nearly equal numbers.

This means that in fruit flies and in humans, the determination of sex is chromosomal. In other words, whether an individual is male or female depends on the sex chromosomes that the individual inherits. This has some very interesting consequences for the **sex-linked genes,** or the genes located on the sex chromosomes.

Morgan's Experiments

One day in 1909, Thomas Hunt Morgan noticed something strange in one of the bottles of fruit flies that he had in his lab. Normally, the flies have deep-red eyes. But in this bottle, one of the male flies had white eyes, which

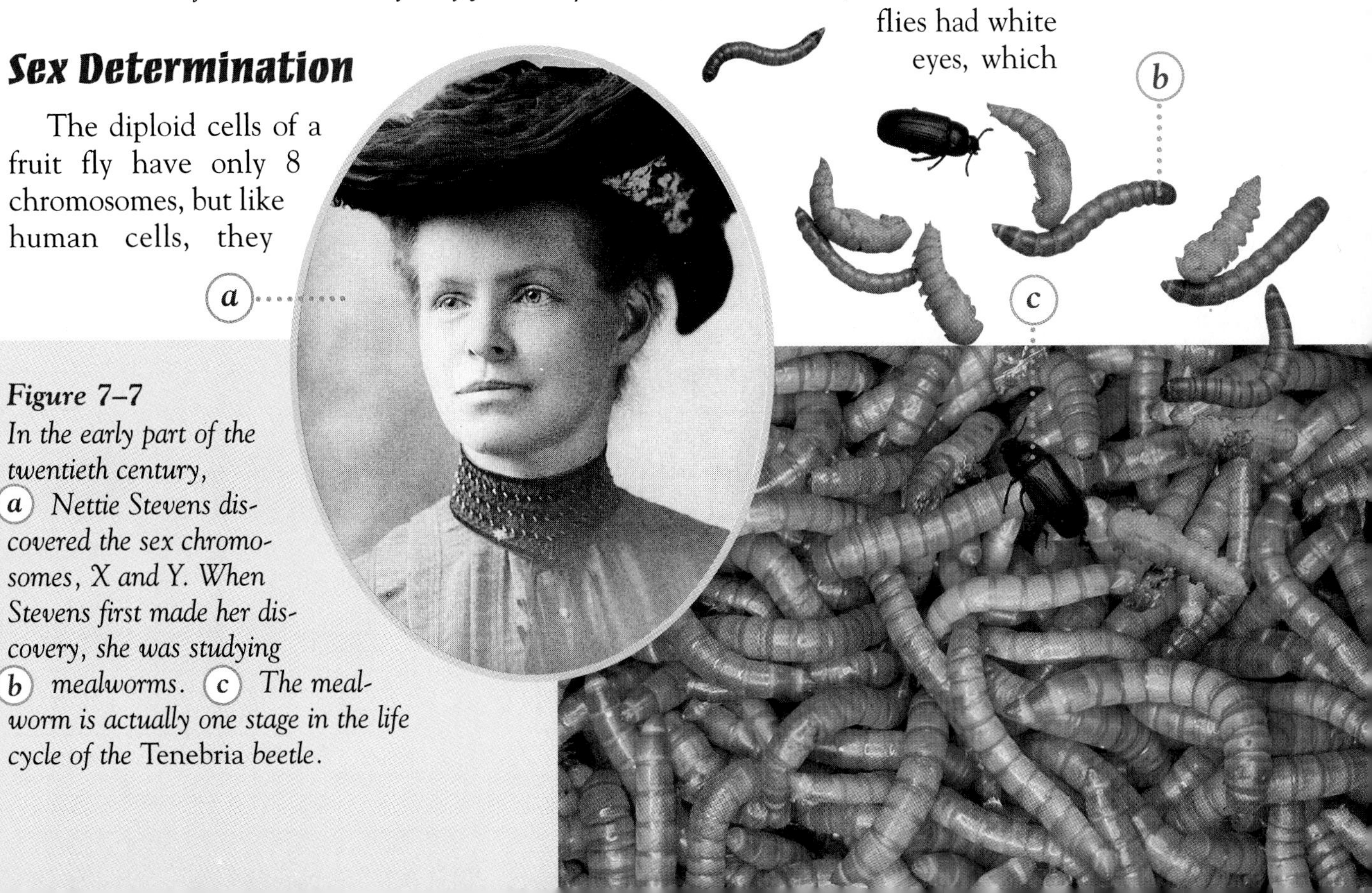

Figure 7–7
In the early part of the twentieth century, (a) Nettie Stevens discovered the sex chromosomes, X and Y. When Stevens first made her discovery, she was studying (b) mealworms. (c) The mealworm is actually one stage in the life cycle of the Tenebria beetle.

immediately caught Morgan's attention. He decided to find out how the white-eye trait was inherited.

Morgan crossed the white-eyed male with a female that had the normal red eye color. All the flies in the F_1 generation had red eyes. This meant that red eye color probably was a dominant allele, and therefore white was a recessive allele.

Morgan then took two flies from the F_1 generation and crossed them. Because each of these flies should be heterozygous for eye color—a white allele from the male and a red allele from the female—Morgan expected that 1/4 of the F_2 offspring would have white eyes and 3/4 would have red eyes. When he did this cross, this is exactly what happened. However, something very strange also occurred—all the white-eyed flies were male!

Morgan thought that these results had to be more than a coincidence. He quickly realized what was going on, and he made two assumptions—the allele for white eye color was recessive and the gene for eye color was located on the X chromosome. Suddenly, everything fell into place.

Genes on the Sex Chromosomes

Look at ***Figure 7–8.*** Because the original white-eyed fly was a male, only the females in the F_1 generation inherited its X chromosome with the white-eye allele. All the male flies inherited a Y chromosome from their white-eyed male parent and an X chromosome from their red-eyed female parent. When the F_1 generation of fruit flies were crossed, producing the F_2 generation, there was only one way to produce a white-eyed fly—half the egg cells produced by the F_1 females must carry an X chromosome with the white-eye allele. When one of those egg cells was fertilized by a sperm carrying a Y chromosome, a white-eyed male was produced.

Because male flies have just one X chromosome, each allele on that chromosome helps to determine the fly's phenotype. A female that inherits a recessive allele on one X chromosome may inherit a dominant allele on the other X chromosome. However, any allele that a male inherits on its X chromosome is expressed, whether or not it is recessive.

What sort of cross would be needed to produce a white-eyed female? Recall that

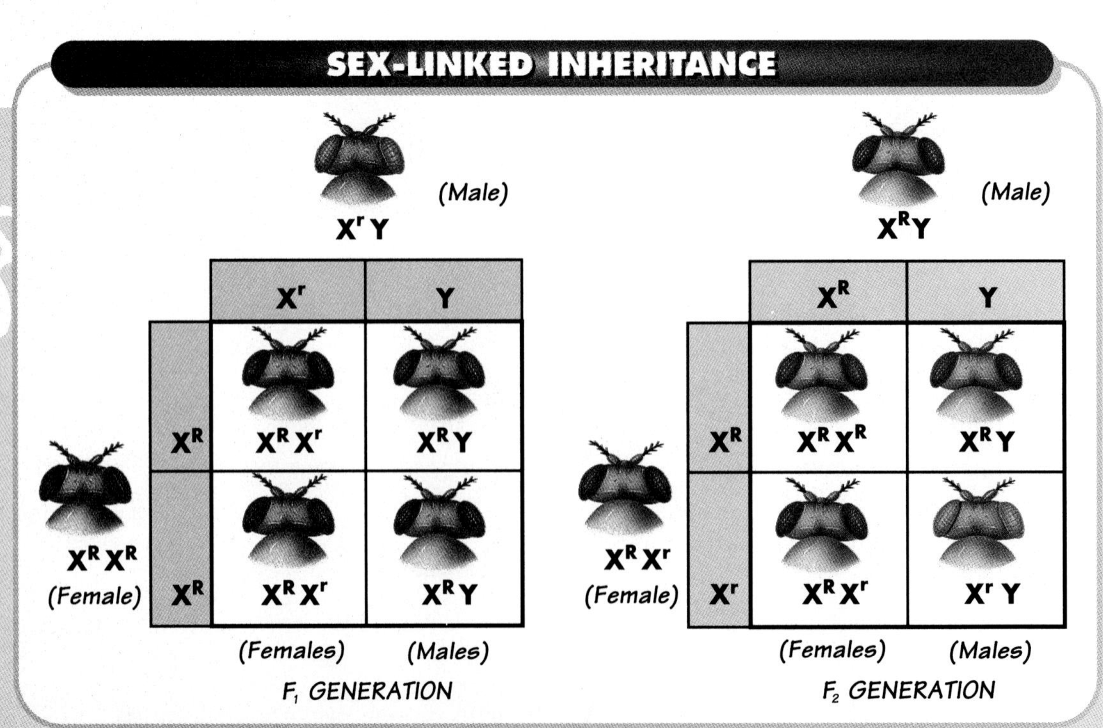

Figure 7–8
The Punnett squares illustrate Morgan's experiments of sex-linked inheritance in fruit flies. White-colored eyes are symbolized by X^r, and red-colored eyes are symbolized by X^R.

the white-eye allele is recessive. Therefore, in order to have white eyes, a female fly needs to inherit two X chromosomes with the recessive allele.

☑ **Checkpoint** Why is a recessive allele carried on the X chromosome always expressed in a male?

Sex-Linked Genes in Humans

Although the details of sex determination in humans are a bit different from fruit flies, the way in which genes on the human X chromosome are inherited is very much the same. In humans, sex-linked genes are almost always located on the large X chromosome. Although both males and females carry a copy of the X chromosome, only males carry the smaller Y chromosome that contains a few genes related to male sexual development. Therefore, the most important genes—those needed in both males and females—cannot be located on the Y chromosome.

How can you recognize sex-linked inheritance? **Because males have just one X chromosome, any X-chromosome-linked gene a male inherits, recessive or not, is expressed.** In addition, because males pass their X chromosomes along to their daughters, sex-linked genes will tend to move from fathers to their daughters. Then these genes may show up in the sons of those daughters.

Figure 7–9
This scanning electron micrograph shows the human sex chromosomes, X and Y. Notice how much larger the X chromosome is compared to the Y chromosome (magnification: 534X).

Human Beans

PROBLEM *How can you* ***construct a model*** *of the inheritance of sex?*

PROCEDURE

1. Obtain two jars and label one "female" and the other "male."
2. Place 10 red beans in the jar labeled "female."
3. Place 5 red beans and 5 white beans in the jar labeled "male." The red beans will represent X chromosomes, and the white beans will represent Y chromosomes.
4. With your eyes closed, pick a pair of beans, one from each jar. Record the color of each bean.
5. Repeat step 4 until all the beans have been picked.

ANALYZE AND CONCLUDE

1. Why was one bean selected from each jar?
2. How many male offspring were produced? How many female offspring?
3. What does this tell you about the chances of an individual offspring being born either male or female?

Colorblindness

One of the important genes carried on the X chromosome is responsible for normal color vision. The dominant allele produces normal color vision. The recessive allele causes colorblindness. As many as 10 percent of all males in the United States suffer from at least one form of colorblindness, which is an inability to see certain colors properly.

Like the white-eyed fruit flies, a colorblind human male carries a single X chromosome with a recessive allele for colorblindness. How did he inherit this X chromosome? A human male inherits

Figure 7–10
Queen Victoria of England was a carrier for the hemophilia gene. She passed the gene on to her offspring, who then passed it on to their offspring, spreading hemophilia through most of the royal families of Europe.

the X chromosome from his mother and the Y chromosome from his father.

Red-green colorblindness is about ten times more common among males than it is among females. Can you figure out why this is the case? If you think about the alleles that would have to be present on the X chromosomes of the parents of a colorblind female, you should be able to explain why only 1 female in 100 shows red-green colorblindness.

Checkpoint Is colorblindness caused by a dominant or a recessive allele?

INTEGRATING HEALTH

Why must blood be carefully screened for infectious diseases before being given to anyone?

Hemophilia

Two other important genes carried on the X chromosome help to control blood clotting. Individuals with the recessive allele for one of these genes are unable to produce one of the clotting factor proteins that normally help blood to clot. This condition is called hemophilia. For a person with hemophilia, even small cuts can present serious problems.

Hemophilia affects roughly 1 male in every 10,000. On the other hand, few females—less than 1 in 1 million—suffer from this disorder. Although no cure has been developed, it is possible to treat this disorder with injections of clotting factor proteins taken from the blood of healthy individuals. ●

Checkpoint What is hemophilia?

Duchenne Muscular Dystrophy

In the United States, 1 out of 3000 males between the ages of 3 and 6 develops Duchenne muscular dystrophy, a genetic disorder that causes a sudden weakness in muscles. The gene that is responsible for this disorder is also carried on the X chromosome. The recessive allele produces a defective protein that causes the muscles to weaken and break down, eventually causing death.

Researchers are trying to find a cure for this disorder. One possibility involves inserting a normal allele into the muscle cells of people with muscular dystrophy.

Section Review 7–2

1. **Describe** how sex is determined in fruit flies.
2. **Explain** why sex-linked disorders are more common in males than in females.
3. **Critical Thinking—Relating Concepts** If the gene for colorblindness were carried on the Y chromosome, what kind of pattern would you expect to see among males? Among females?
4. **MINI LAB** After **constructing your model** of the inheritance of sex, what can you conclude?

Human Genetic Disorders

Guide for Reading

- **Describe** why many genetic disorders are carried on autosomes.
- **Identify** some genetic disorders caused by nondisjunction.

MINI LAB
- **Interpret the data** from a pedigree of a recessive trait in a family.

SOMETIMES THE EXISTENCE OF a gene is discovered because one of its alleles produces an unusual trait—one that is very different from the rest of the population. Other genes are noticed because their alleles produce genetic disorders. From this information, you may get the impression that human genetics is nothing more than the study of everything that can go wrong. This is not the case, however. For every allele that causes a genetic disorder, there is also a normal, functional allele that works just fine in most people.

Autosomal Genetic Disorders

Most human genes are located on 1 of the 22 pairs of autosomes, rather than on the X and Y sex chromosomes. As you have read, some genetic disorders—colorblindness, hemophilia, and Duchenne muscular dystrophy—are carried on the sex chromosomes. **However, the majority of human genes, and therefore the majority of genetic disorders, are carried on the autosomes.** Albinism, cystic fibrosis, Tay-Sachs disease, sickle cell anemia, PKU, and Huntington disease are some examples of genetic disorders that are carried on autosomes.

Albinism

Albinism (AL-buh-nihz-uhm) is a genetic disorder caused by a recessive allele on chromosome 11. Individuals who have two copies of this allele are unable to produce melanin, the pigment responsible for most human skin color. People with albinism have no pigment in their hair or skin. In addition,

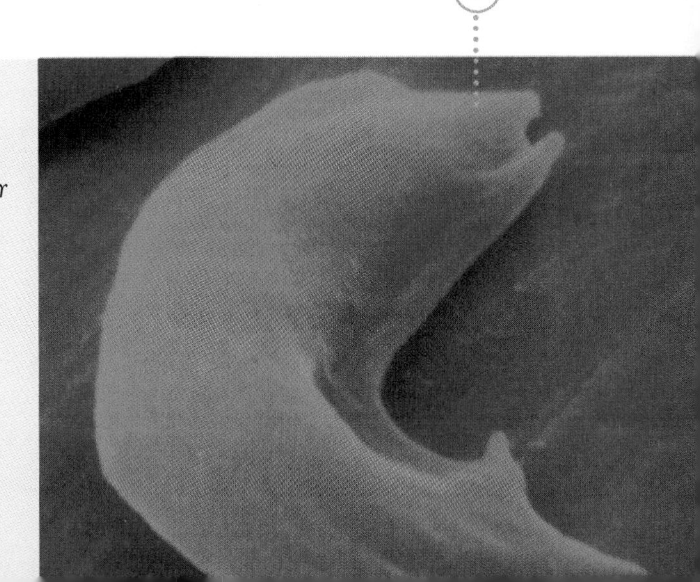

Figure 7–11
Sickle cell anemia is an example of a genetic disorder carried on the autosomes. (a) Normal red blood cells have a flattened disk shape (magnification: 4202X), whereas (b) red blood cells from a person who has sickle cell anemia have a crescent shape (magnification: 18,000X).

Figure 7–12
(a) This giraffe has inherited the allele for albinism. As a result, its coat is white rather than (b) spotted, as in the other giraffes.

they are sensitive to light and, therefore, must avoid excessive exposure to bright sunlight.

Cystic Fibrosis

In the United States, cystic fibrosis (SIHS-tihk figh-BROH-sihs) is the most common fatal genetic disease. It is found in people of European ancestry and affects approximately 1 child in 2500. Cystic fibrosis is caused by a recessive allele on chromosome 7.

Individuals with two copies of this allele make a defective cell membrane protein that interferes with the movement of chloride ions into and out of the cell. Soon chloride ions begin to build up inside the cells, causing water from the surrounding liquid to enter the cells. As a result, the surrounding liquid becomes thick and heavy, clogging the lungs and breathing passageways.

Tay-Sachs Disease

Like cystic fibrosis, Tay-Sachs (TAY SAKS) disease is a fatal genetic disorder caused by a recessive allele. Tay-Sachs disease is most common in Jewish families of Eastern European ancestry. Children who are born with Tay-Sachs disease suffer from a rapid breakdown of the nervous system beginning at age 2 or 3.

Sickle Cell Anemia

Sickle cell anemia is a blood disorder that is characterized by crescent, or sickle-shaped, red blood cells. Sickle cell anemia is caused by a recessive allele that produces an alternate form of hemoglobin—the red blood cell protein. Interestingly, sickle cell anemia is common in those parts of the world where malaria is also common. Malaria is an infectious disease that causes severe chills and fevers, and may even cause death.

Individuals with two copies of the sickle cell allele suffer from sickle cell anemia and have serious medical problems. People who are heterozygous for the sickle cell allele, however, are generally healthy. In addition, they have the important benefit of being resistant to malaria. In the United States, sickle cell anemia is most common among people of African ancestry whose families trace their ancestry to regions of Africa where malaria is common.

INTEGRATING SOCIAL STUDIES

In which parts of Africa is malaria common? Use reference books to find out.

Checkpoint What is sickle cell anemia?

PKU

Roughly 1 child in every 15,000 is born with phenylketonuria (fehn-uhl-keet-oh-noor-ee-uh), or PKU. PKU is another genetic disorder caused by a recessive allele. PKU can cause severe mental retardation. Fortunately, there is both a test and a treatment for PKU. In fact, most states require all newborn infants to be tested for PKU. If the infant has the disorder, a special diet can help prevent damage to the nervous system.

Huntington Disease

Not all genetic disorders are carried on recessive alleles. Huntington disease, for example, is a rare genetic disorder caused by a dominant allele located on chromosome 4. Most individuals with this disorder have no symptoms until their late 30s or 40s, when they begin to lose control over their muscles. Later, as the disease progresses, the nervous system begins to break down, and most patients die within 15 years after symptoms of the disorder first appear.

Because Huntington disease appears in middle age, most people who are at risk for the disorder have already had children by the time they find out that they carry the allele. Because the allele is dominant, a heterozygous person with Huntington disease has a 50-50 chance of passing the disorder along to one of his or her children.

☑ *Checkpoint* What kind of allele causes Huntington disease?

MINI LAB .... Interpreting ....

Climbing a Family Tree

PROBLEM *How can you **interpret the data** from a pedigree showing inheritance of a recessive trait in a family?*

PROCEDURE

1. Draw a pedigree for a family showing 2 parents and 4 children. Include the following in the pedigree:
   a. The 2 oldest children should be males and the 2 youngest children should be females.
   b. The second son has sickle cell anemia.
   c. The older son is married and has a daughter.
   d. The oldest daughter is married and has a normal daughter and a son who has sickle cell anemia.
2. Using this information and your knowledge about constructing pedigrees, fill in the pedigree. Record the genotypes for each member of the pedigree next to his or her symbol. Use N for a normal allele and n for the sickle cell allele.

ANALYZE AND CONCLUDE

1. Can you determine the genotype for the people whose symbols are not shaded in? Explain your answer.
2. If the youngest daughter married, what would be the probability of her having a child with sickle cell anemia? Explain your answer.

Chromosome Number Disorders

Every human life begins with a single cell, the zygote, which is formed by the fusion of a sperm and an egg. Normally, a human sperm and egg each contain 22 autosomes and 1 sex chromosome. As you have seen, when these sex cells are formed in meiosis, each chromosome pair separates during the first meiotic division.

Every now and then, however, something goes wrong and a chromosome pair fails to separate correctly. The most common error of this type in meiosis is called **nondisjunction,** which literally means "not coming apart." When nondisjunction occurs, abnormal numbers of chromosomes are produced in the sex cells. If such a sex cell produces a zygote, a genetic disorder results.

☑ *Checkpoint* What is nondisjunction?

Turner Syndrome

One type of nondisjunction in which sex chromosomes fail to separate in meiosis can result in Turner syndrome. During meiosis, either a sperm or an egg is produced without a sex chromosome. When such a cell fuses with a sex cell carrying a single X chromosome, the zygote will be XO. The O indicates that a sex chromosome is missing. The karyotype for people who have Turner syndrome is written as 45XO.

Only females can be afflicted with Turner syndrome. Because their sex organs do not fully develop, these females cannot have children. However, most people with Turner syndrome are able to lead otherwise full and healthy lives.

Klinefelter Syndrome

Nondisjunction can also produce males whose cells contain an extra chromosome. This abnormality, symbolized as 47 XXY, is called Klinefelter syndrome. Mental retardation is often associated with Klinefelter syndrome, although its extent varies from one person to the next. The extra X chromosome interferes with meiosis and prevents these individuals from reproducing.

Down Syndrome

Nondisjunction can occur in autosomes as well as in sex chromosomes. If the two copies of an autosome do not separate correctly in meiosis, an individual can be born with cells that contain three copies of a chromosome. This condition is called **trisomy.** The most common form of trisomy is Down syndrome. **In Down syndrome, there is an extra copy of chromosome 21.**

INTEGRATING HEALTH

Why is Down syndrome more common in infants of mothers over the age of 35? Use reference material to find out.

Down syndrome results in heart and circulatory problems, a weakened immune system, and mental retardation. The degree of retardation varies greatly. Although some people who have Down syndrome are severely retarded, others are able to function quite well in society.

Scientists are only beginning to learn why a little extra genetic information causes so many problems. Current theories suggest that an extra copy of so many genes upsets the balance by which genes are regulated, leading to the disorders associated with trisomy.

Chromosome Deletions and Translocations

As you may recall, crossing-over during meiosis results in genetic recombination. When you read about crossing-over, it may have occurred to you that if something went wrong with the process, chromosomes might break. Well, occasionally, this does happen.

Many genetic disorders result from pieces of chromosomes breaking off and getting lost in meiosis. These are called **chromosome deletions.** Other disorders arise when pieces of broken chromosomes become reattached to another chromosome. These are called **chromosome translocations.**

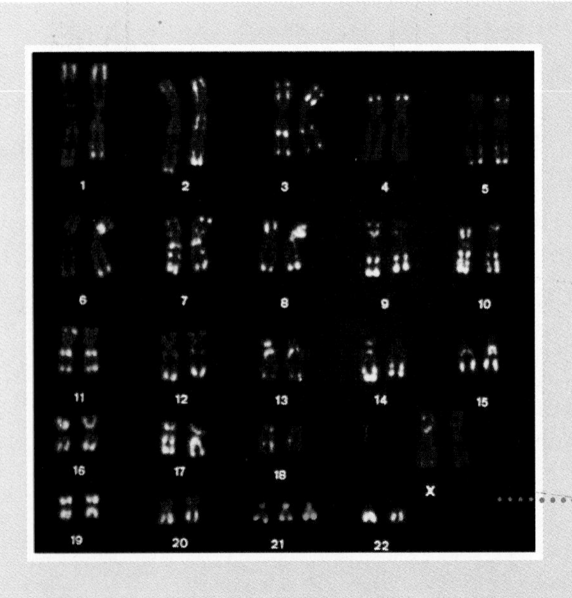

Figure 7–13
(a) Down syndrome is characterized by the presence of an extra copy of chromosome 21, as shown in this karyotype. (b) Many people with Down syndrome, including the television star Christopher Burke, live active lives.

CHROMOSOME DELETION

a b c d e f → a c d e f

Figure 7–14 (a) A chromosome deletion occurs when a piece of the chromosome is broken off. (b) In some cases, pieces of chromosomes become attached to another chromosome. This is called chromosome translocation.

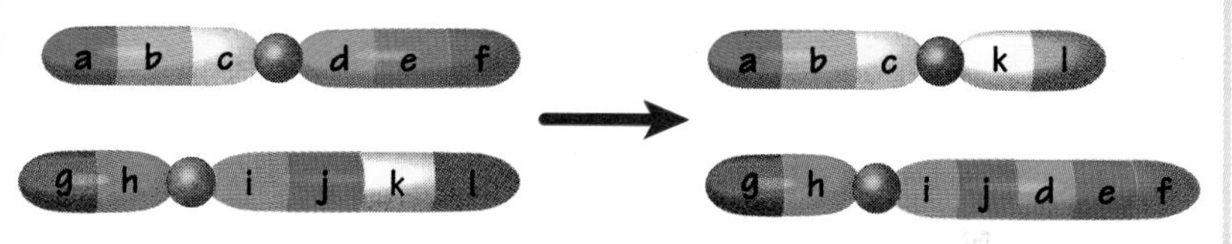

Prenatal Diagnosis

A rapidly expanding list of genetic disorders—including Down syndrome, Tay-Sachs disease, and Huntington disease—can now be detected before birth in the cells of a developing fetus. These cells can be grown in the laboratory for a few days and then analyzed by a variety of techniques to determine whether a baby is likely to be born with a genetic defect. The problem is how to collect such cells without harming either the mother or the developing fetus.

Currently, there are two ways to do this. One technique, **amniocentesis** (am-nee-oh-sehn-TEE-sihs), involves withdrawing a small amount of fluid from the sac surrounding the fetus. The fluid contains cells from the fetus which can be examined for abnormalities. Cells from a fetus can also be obtained by another technique, called **chorionic villus** (kor-ee-AHN-ihk VIHL-uhs) **sampling.** In this technique, tissue surrounding the fetus is removed and examined. Because the cells come directly from the fetus, chorionic villus sampling provides a quicker way to examine cells than does amniocentesis.

Both of these techniques have made it possible to detect a large number of genetic disorders, enabling at-risk parents to know before birth whether their child may suffer from a genetic disorder. As the understanding of genetics advances, careful testing may even make it possible to treat such disorders before birth to maximize the chances of delivering a healthy baby.

Section Review 7–3

1. **Describe** why many genetic disorders are carried on autosomes.
2. **Identify** some genetic disorders caused by nondisjunction.
3. **Critical Thinking—Synthesizing Information** Why do people with Huntington disease have a greater chance of passing this disorder on to their children than people who have Tay-Sachs disease or cystic fibrosis?
4. **MINI LAB Interpret the data** from a pedigree of a recessive trait in a family.

Special Topics in Human Genetics

Guide for Reading

- **Define** what a Barr body is and explain why it is not found in males.

"KNOW THYSELF" IS SOMETIMES given as a philosopher's first challenge to students. It's wonderful advice. How can we hope to know the world around us if we don't seek to understand ourselves first?

In one sense, knowing ourselves is precisely what we try to do when we study human genetics. Understanding our own inheritance helps us to understand more completely our place in the world of living things.

Can we look back on all that we have learned about human genetics and conclude, after so much scientific effort, that now we really do know ourselves? Not at all. In this section, we will briefly examine some questions about human genetics that remain unanswered and continue to puzzle biologists.

X-Chromosome Inactivation

Human cells normally contain two copies of each autosome. However, the X chromosome is different. Recall that females have two copies of the X chromosome, while males have just one. As you have read, the X chromosome contains important genes that control everything from color vision to blood clotting. If just a single X chromosome is enough to regulate these tasks in male cells, how does the cell adjust to the extra X chromosome in female cells?

Turning Off the X

The answer to this question was discovered by the British geneticist Mary Lyon. She noted that most cells from human females have a dense region in the nucleus, called a Barr body. **Lyon suggested that the Barr body was**

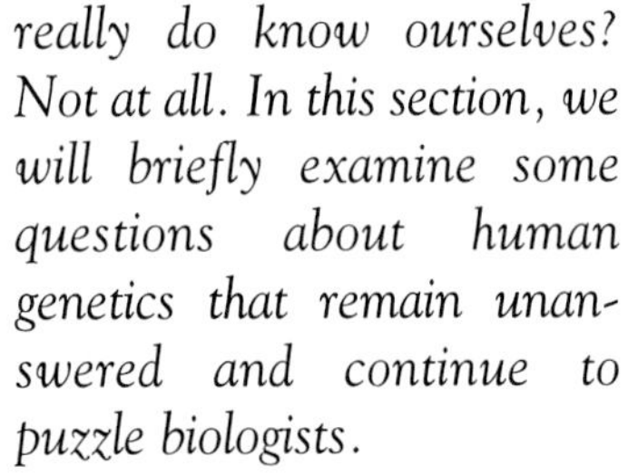

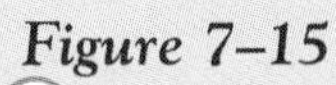

Figure 7–15
(a) This female cat and her litter of kittens illustrates the hypothesis that one X chromosome is inactivated in females. The female cats are calico with spots of orange and black, whereas the male cats have spots of only one color—orange or black. (b) The dark-purple sphere in the middle of this human female cell is a Barr body. Barr bodies were first discovered in cats.

Figure 7–16
CAREER TRACK
Genetic counselors often review family histories and provide advice to people who are considering having children.

actually a condensed turned-off X chromosome. The reason Barr bodies aren't found in male cells is that their single X chromosome is still active.

It happens that the same process also occurs in other mammals. In cats, for example, a gene controlling the color of coat spots is located on the X chromosome. As a cat embryo develops, a single X chromosome in many of its cells is randomly and permanently switched off. If one X chromosome has an allele for brown spots and the other X chromosome has an allele for black spots, the cat's white fur will have a mixture of brown and black spots.

Interestingly, male cats, which have just one X chromosome, can have spots of a single color only. In fact, this is an almost foolproof way to tell the sex of a spotted cat! If a cat with a spotted coat has three colors—white with spots of tan and black—you can be sure it's a female.

☑ **Checkpoint** What is a Barr body?

What Turns Off the X?

Careful studies of female human cells show that one of the X chromosomes has been inactivated in nearly every tissue in the body. The X chromosome that has been inactivated varies from tissue to tissue and sometimes from cell to cell. How does this happen? This is the big question, and it remains unanswered.

Recent research has shown that a handful of genes is still functioning in the X chromosome that has been turned off. Maybe these genes control the activity of the rest of the chromosome. If this is the case, biologists may learn how to turn off all the genes in a single chromosome at once. This and many other discoveries could be very helpful in the treatment of genetic disorders.

Gene Imprinting

Should it make any difference whether you inherit a gene from your mother or from your father? You may not think so. In fact, until a few years ago, any geneticist would have told you that once a sperm and an egg fuse, there is no way to tell whether a particular gene came from one parent or the other.

Now, however, it turns out that this may not be true. A few years ago, a team of researchers traced a genetic disorder called Angelman syndrome to a mutation of chromosome 15. Patients with this disorder are quite short and suffer from obesity. Meanwhile, another group of researchers studying Prader-Willi syndrome, another genetic disorder, traced it to chromosome 15 as well. Individuals suffering from Prader-Willi syndrome are usually hyperactive, very thin, and of normal height. When the two groups

Bio FRONTIER Connections

Who Owns Your Genes?

Arthur had just graduated from college and was looking for a job. He found an ad in the newspaper, applied for the job, and went on the interview. The woman interviewing Arthur told him that in order to be hired and covered under the health insurance plan, he had to agree to a complete physical examination and some genetic testing. Arthur agreed and had the tests done.

Test Results

The test results came back and showed that Arthur was a carrier of the gene for cystic fibrosis. Because Arthur only carried the gene, the company wasn't concerned by his test results. The company did mention to Arthur that if he married and had children, his wife and children would have to be tested as well.

Five years later, Arthur got married, and his wife was tested. The results of her test were disturbing. She, too, carried the gene for cystic fibrosis. Again, the company was not concerned because she too was just a carrier. However, if Arthur and his wife were to have a baby, the company wanted the baby to be tested prenatally because the child would have a one in four chance of having cystic fibrosis.

Future Testing

In the future, there will be tests for many, if not all, genetic diseases. The Human Genome Project is identifying and mapping all the human chromosomes. All the genes will be located in their position on each chromosome and identified. Once the genes have been identified, scientists will try to develop tests for identifying any problems. Only then can possible cures for these diseases be found.

DNA sequencing machine

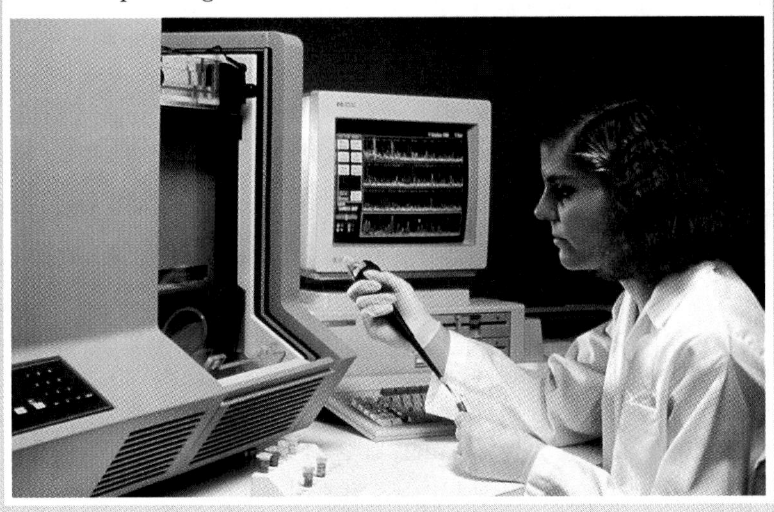

Scientists at universities, private companies, and the government are working together to map all our genes. Many genes have been placed in their proper locations, but there are at least 100,000 genes to map. It may take years, but when this project is over, human heredity will no longer be a mystery.

If employers and insurance companies gain access to genetic information, they might raise insurance rates or refuse to employ or insure people who carry certain genes. Today, medical records and all test results are private. But in the future, companies may make genetic testing a requirement for employment.

Making the Connection

If you were Arthur, would you allow the company to test your child prenatally? What if the company threatened to fire you if you didn't submit to the test? What are some of the pros and cons of genetic testing?

compared notes, they were shocked. Both disorders were caused by exactly the same gene!

Why are the disorders different even though they are caused by the same apparent defect? As it turns out, a person with Prader-Willi syndrome inherits the defective allele from the mother, whereas someone with Angelman syndrome inherits the defective allele from the father. Why should these findings make a difference? That's a good question. Some researchers suggest that certain genes are marked, or imprinted, with a genetic marker that is different in males and in females—which could account for the difference. It will be interesting to find out how widespread gene imprinting is and whether it plays an important role in human genetics.

Ethical Issues in Genetics

An important new set of ethical issues has been raised by the rapid advances in human genetics. Simply stated, these questions revolve around the proper ways to use genetic information. Does anyone own genetic information? When a new gene is found, do its discoverers have special commercial rights? Do insurance companies have the right to conduct genetics tests before they issue policies? Do individuals have a right to keep their genetic information to themselves?

Many people have already had to face these questions. For example, in 1989 a test for the Huntington disease allele was developed. For the first time, it was possible for people who might carry a copy of the allele to be tested before they developed symptoms. The test was a great relief to many people who wanted to know whether or not they could pass the disorder along to their children. However, because there is no way to treat Huntington disease, many people chose not to be tested. Quite simply, they did not wish to find out whether or not they carried this fatal allele. What is the right choice in such a situation? Do people have an obligation to be tested for such alleles? If the test is positive, what are their responsibilities to their children or potential children?

These questions don't have easy answers. But in a democracy, every citizen should be prepared to play a role in determining the rules that we will live by. Science can help to shed some light on these important issues, and you owe it to yourself to develop an understanding of them. However, science by itself cannot provide the moral values or judgments that will be needed to make these important choices and decisions. Answers to these ethical questions will not come from scientific experts, but from the wisdom and strength of people in a free society.

Section Review 7–4

1. **Define** what a Barr body is and explain why it is not found in males.
2. **BRANCHING OUT ACTIVITY** Find a recent newspaper or magazine article written about one of the genetic disorders in this chapter. **Summarize** the article in a few paragraphs.

CHAPTER 7

Laboratory Investigation

Only the Prints Can Tell

Fingerprints are often used to solve mysteries and crimes. Although there are similar patterns in fingerprints, no two fingerprints are alike. In this investigation, you will identify what makes fingerprints unique.

Problem

Observe how fingerprints can be used to identify a person.

Materials (per group)

ink pad
2 index cards
magnifying glass

Procedure

1. Obtain an ink pad, two index cards, and a magnifying glass.
2. On an index card like the one shown below, write your name, the words left hand, and the numbers 1 to 5.
3. On the second index card, include the same information except write the words right hand instead of left hand.

Your Name

Hand ________

1	2	3	4	5

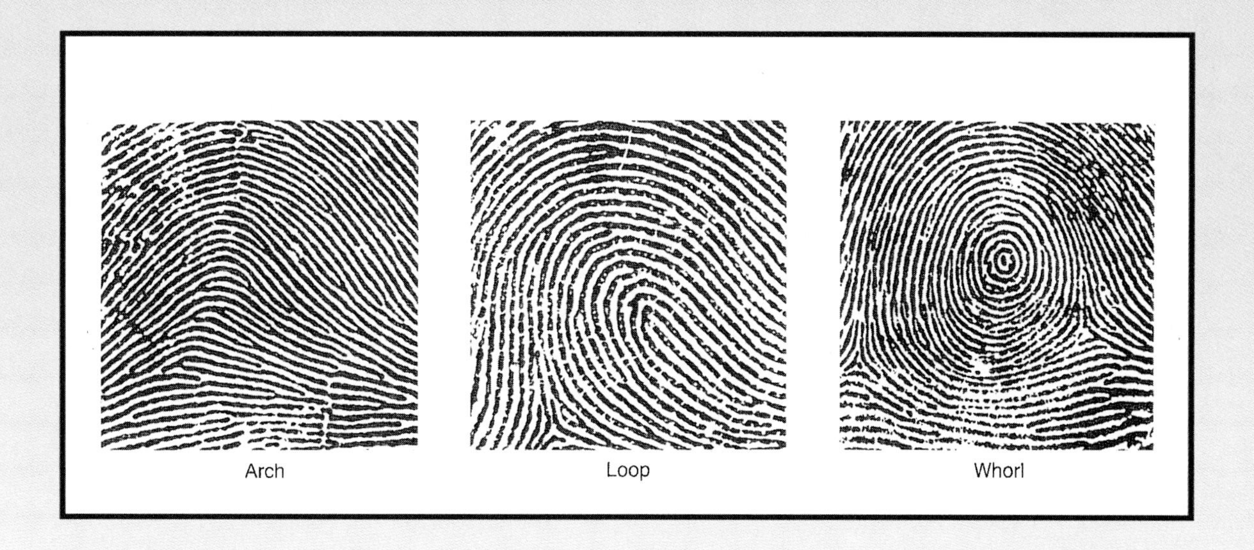

4. Carefully roll the tip of one finger from left to right on the ink pad.

5. Roll the inked fingertip on the index card in the spot numbered 1. Lift your finger straight up after making the print to avoid smudging the print.

6. Repeat steps 4 and 5 until all fingertips of both hands appear in order on the index card.

Observations

1. For each fingerprint, determine whether you have an arch, a loop, or a whorl, and record it under the prints.
2. Fingerprints are formed before birth. The genotype LL forms a whorl, Ll forms a loop, and ll forms an arch pattern. For each fingerprint, record your genotype for each fingertip.
3. For each fingerprint, draw an imaginary line from the center of your pattern to the triradius. The triradius is the triangular area formed by intersecting ridges. Count the number of ridges for each fingertip and record the number under your prints.

Analysis and Conclusions

1. Study each of your fingerprints and describe any patterns you notice.
2. Compare your fingerprints with those of others in the class and describe the results.
3. How can fingerprints be used to identify a person? Explain your answer.

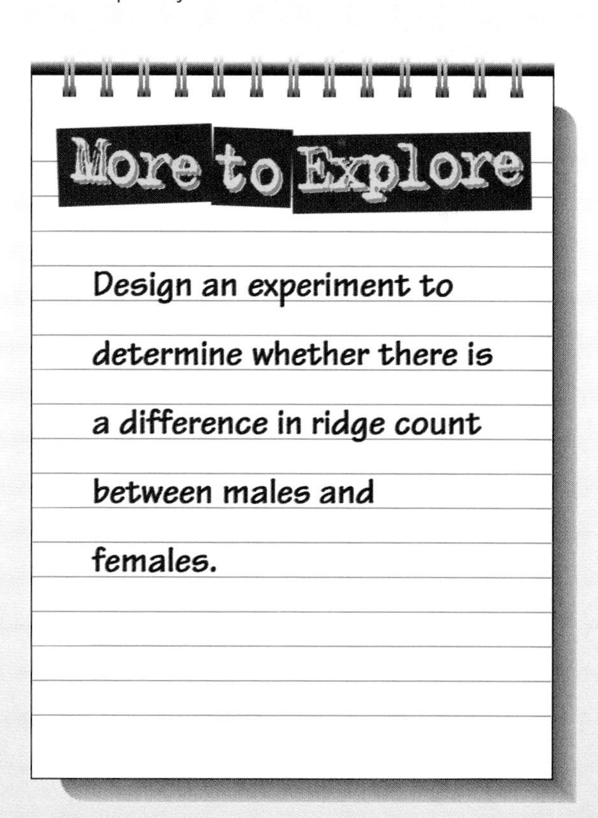

Study Guide

Summarizing Key Concepts

The key concepts in each section of this chapter are listed below to help you review the chapter content. Make sure you understand each concept and its relationship to other concepts and to the theme of this chapter.

7–1 The Human Genetic System

- Many human traits are inherited by the action of genes that have dominant and recessive alleles.
- Other traits are determined by more than two, or multiple, alleles.

7–2 Sex-Linked Inheritance

- The sex of an individual—male or female—depends on the sex chromosomes that the individual inherits.
- Because males have only one X chromosome, any X-chromosome-linked gene a male inherits, recessive or not, is expressed.
- Colorblindness, hemophilia, and Duchenne muscular dystrophy are examples of disorders carried on the sex chromosomes.

7–3 Human Genetic Disorders

- The majority of human genes, and therefore the majority of genetic disorders, are carried on the autosomes.
- Albinism, cystic fibrosis, Tay-Sachs disease, sickle cell anemia, PKU, and Huntington disease are examples of autosomal genetic disorders.
- When a chromosome pair fails to separate during meiosis, it is called nondisjunction.
- Many genetic disorders can be detected before birth by either amniocentesis or chorionic villus sampling.

7–4 Special Topics in Human Genetics

- New discoveries in human genetics may help in the treatment of genetic disorders.

Reviewing Key Terms

Review the following vocabulary terms and their meaning. Then use each term in a complete sentence.

7–1 The Human Genetic System

karyotype
autosome
sex chromosome
pedigree
multiple allele

7–2 Sex-Linked Inheritance

sex-linked gene

7–3 Human Genetic Disorders

nondisjunction
trisomy
chromosome deletion
chromosome translocation
amniocentesis
chorionic villus sampling

Recalling Main Ideas

Choose the letter of the answer that best completes the statement or answers the question.

1. How many chromosomes do human cells have?

   a. 46 **c.** 22
   b. 23 **d.** 2

2. An example of a multiple allele is

   a. the Rh blood group.
   b. trisomy.
   c. nondisjunction.
   d. the ABO blood group.

3. A genetic disorder in which the blood does not clot properly is

   a. PKU. **c.** hemophilia.
   b. Huntington disease. **d.** Tay-Sachs disease.

4. People who are resistant to malaria often are heterozygous for which genetic disorder?

   a. sickle cell anemia **c.** PKU
   b. Turner syndrome **d.** colorblindness

5. When cells contain three copies of a chromosome, it is called a

   a. chromosome deletion.
   b. trisomy.
   c. chromosome translocation.
   d. multiple allele.

6. Down syndrome results from a(an)

   a. extra chromosome.
   b. chromosome translocation.
   c. chromosome deletion.
   d. missing chromosome.

7. A Barr body is a(an)

   a. turned-off X chromosome.
   b. trisomy.
   c. turned-off Y chromosome.
   d. autosome.

Putting It All Together

Using the information on pages xxx to xxxi, complete the following concept map.

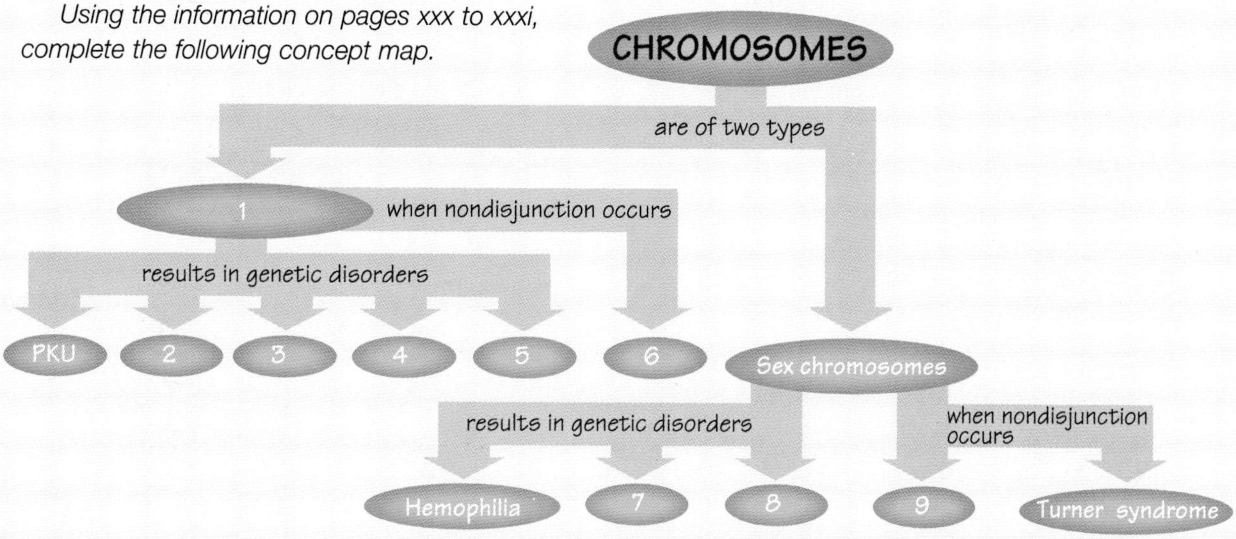

CHROMOSOMES
are of two types
1
when nondisjunction occurs
results in genetic disorders
PKU
2
3
4
5
6
Sex chromosomes
results in genetic disorders
when nondisjunction occurs
Hemophilia
7
8
9
Turner syndrome

Reviewing What You Learned

Answer each of the following in a complete sentence.

1. Why is *Drosophila melanogaster* an excellent organism for the study of genetics?
2. What are diploid cells? Where would you expect to find them in your body?
3. What can you observe in a normal human karyotype?
4. What is an autosome? What is a sex chromosome?
5. How do the following disorders—hemophilia, sickle cell anemia, and Klinefelter syndrome—differ from one another?
6. Give one example of a multiple allele.
7. Why are people who have sickle cell anemia resistant to malaria?
8. **Construct a model** of meiosis in which nondisjunction has occurred.
9. Do males have Barr bodies? Explain your answer.
10. What is gene imprinting?

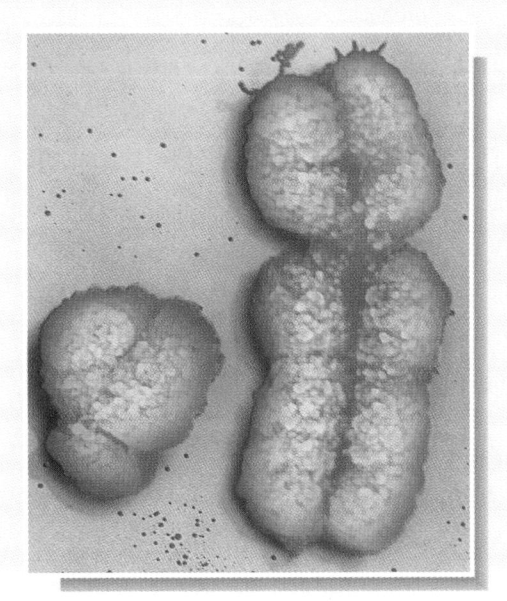

Expanding the Concepts

Discuss each of the following in a brief paragraph.

1. Why is the ratio of male to female offspring usually about 50-50?
2. Explain how blood types in humans are the results of multiple alleles.
3. Diagram a P_1 cross between a red-eyed male fruit fly and a white-eyed female fruit fly. If red eyes are a dominant sex-linked trait, what frequencies would you expect in the F_1 and F_2 generations?
4. Describe a cross that would clearly show the concept of sex-linked inheritance in humans.
5. How could you determine whether peppermint stripes in carnations was autosomal dominant or autosomal recessive?
6. Why is hemophilia more common in males than in females?
7. Compare the expected ratios of a sex-linked disease, such as colorblindness, with that of an autosomal disease, such as cystic fibrosis.
8. Describe three genetic disorders caused by nondisjunction.
9. If a translocation occurred between two different chromosomes in only one parent, what would be the possible genetic combinations of the offspring?
10. Explain in genetic terms why cats with three-colored coats are usually female.

Extending Your Thinking

Use the skills you have developed in this chapter to answer the following.

1. **Interpreting** Draw a three-generation pedigree of an imaginary family that addresses hair color and sex.
2. **Comparing** For the purpose of blood transfusions, why is O blood considered the universal donor, whereas AB is considered the universal acceptor?
3. **Making judgments** You are the director of a research team that has just developed a new strain of genetically engineered cattle with a new gene that produces excessive growth hormone. These animals reach maturity in three fourths the usual time and their meat is of exceptional quality. However, they have many bone deformations. Should these animals be made available to ranchers to help solve the food requirements of humans around the world? Support your position.
4. **Hypothesizing** Discuss how the trait for sickle cell anemia could develop and become prevalent in the gene pool.
5. **Evaluating** If you were a genetic counselor, what types of information would you seek in counseling a couple that had cystic fibrosis and Huntington disease in their respective family histories?
6. **Applying concepts** Some diseases are carried as dominant genes. Marfan's syndrome is a dominant allele that causes extreme height and a weakened aorta in both sexes. Explain how a pedigree for a family with Marfan's syndrome would differ from the pedigree for a recessive disease.

Applying Your Skills

Genetic Disorders

Cystic fibrosis is the most common fatal genetic disorder in the United States. People with cystic fibrosis have difficulty breathing due to an accumulation of thick mucus in the lungs. Researchers doing genetic research at the University of Michigan are hopeful that they will be able to isolate some of the DNA mutations that cause cystic fibrosis.

1. Working in a group, research the topic of cystic fibrosis. Find out what causes this disorder and why cystic fibrosis patients rarely live past the age of twenty.
2. People with cystic fibrosis have an excess amount of sodium chloride in their sweat, making it very salty. What effect does the sodium chloride in the lung cells have on the mucus?

• GOING FURTHER •

3. In your journal, describe your group's findings regarding cystic fibrosis.
4. What new research is being conducted that may help to find a cure for cystic fibrosis?

PTER 8

DNA and RNA

FOCUSING THE CHAPTER
THEME: Systems and Interactions

8–1 Discovering DNA
- **Explain** how scientists discovered the role of DNA.

8–2 DNA Structure and Replication
- **Describe** the structure of DNA and how DNA replicates.

8–3 RNA
- **Describe** transcription and translation.

BRANCHING OUT *In Depth*

8–4 Controlling Gene Expression
- **Discuss** how cells control gene expression.

LABORATORY INVESTIGATION
- **Observe** the DNA extracted from wheat germ.

Biology and Your World

BIO JOURNAL

Create a code in which one letter is substituted for another. Write the key for the code in your journal, then create a coded message. How might a molecule contain coded information?

Computer-generated model of transfer RNA (red) binding to an enzyme (blue)

Discovering DNA

SECTION 8–1

GUIDE FOR READING

- **Identify** the composition of genes.
- **Describe** the results of experiments by Griffith, Avery, and Hershey and Chase.

VISIT A LARGE PUBLIC LIBRARY and you will find novels, magazines, newspapers, and a host of books that contain information on almost any subject—including science, art, travel, history, and sports, to name just a few. In fact, a public library might best be described as a storehouse of information for the community.

Do cells have libraries? In a sense, yes. The library of a cell is its genetic information—information organized on chromosomes. If we compare chromosomes to the stacks of shelves in a library, then each gene would be a book on a shelf—a book that contains the information that is needed to produce a trait.

The Language of Genes

The information in books consists of marks on paper—marks that we recognize as letters, punctuation, or other symbols of language. But these marks are useless unless the reader knows the language in which the book was written. The word ostrich, for example, signifies a long-necked, flightless bird, but it does so only to someone who can read the English language.

How is genetic information stored in chromosomes? And by what code—or language—does the cell interpret this information? To answer these questions, biologists first had to learn which molecule contains genetic information. In other words, they had to find the molecule of heredity.

INTEGRATING LANGUAGE ARTS

Do all written languages use letters and words? Research this interesting topic.

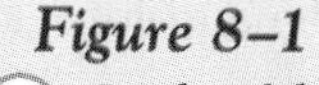

Figure 8–1
(a) In this false-color photograph, the blue structures are chromosomes (magnification: 2200X). (b) Like a stack of shelves in a library, a chromosome contains highly organized information. (c) You can compare a book on a stack of shelves to a gene on a chromosome.

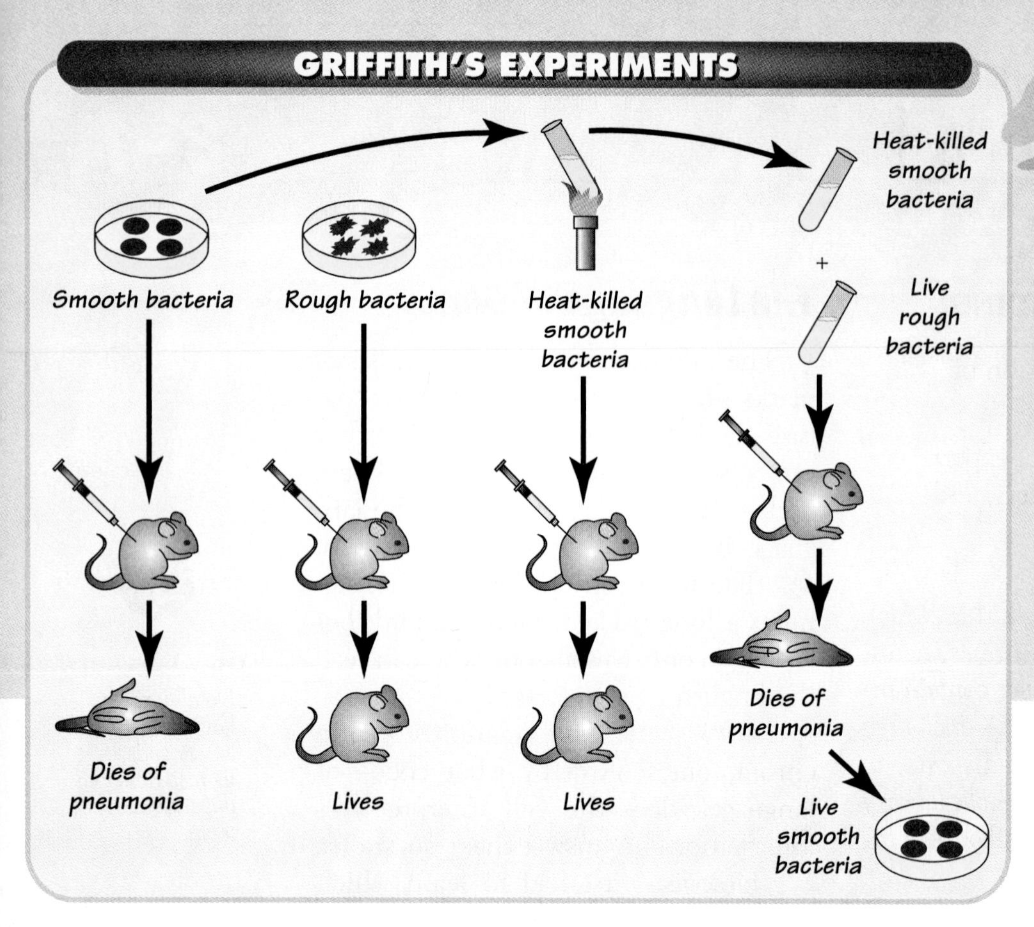

Figure 8–2
Griffith injected mice with the different samples of bacteria shown here. He found that heat-killed smooth bacteria and live rough bacteria were harmless individually but lethal in combination! Griffith concluded that some factor changed the live bacteria from rough to smooth—a process he called transformation.

Griffith and Transformation

One of the first important steps in finding the molecule of heredity was taken in 1928 by a British scientist named Frederick Griffith. Griffith was trying to find better ways to fight pneumonia—a serious and sometimes fatal infection of the lungs.

In his laboratory, Griffith had isolated two types of bacteria that cause pneumonia. He called these types "smooth" and "rough" because the smooth bacteria produced shiny, smooth colonies, whereas the rough bacteria produced dull, rough-edged colonies.

Griffith found that when samples of each type of bacteria were injected into laboratory mice, pneumonia developed only in the mice injected with the smooth type. The mice injected with the rough bacteria remained healthy.

Griffith's Experiments

In one experiment, Griffith heated a sample of smooth bacteria until all the bacteria were dead, then injected the sample into mice. The mice did not develop pneumonia, so Griffith concluded that smooth bacteria needed to be alive to cause pneumonia.

Griffith then mixed a sample of heat-killed smooth bacteria with a sample of live rough bacteria. By themselves, neither type of bacteria should have caused the mice to develop pneumonia. But when Griffith injected mice with the mixture of the two types of bacteria, something remarkable happened. The mice became sick with pneumonia, and many of them died.

Which type of bacteria caused the pneumonia in these mice? Griffith removed the lungs of the dead mice, recovered live bacteria from them, and grew colonies of the bacteria in laboratory culture dishes. The colonies that developed were the glistening, smooth type of the pneuomia-causing bacteria.

Griffith's Conclusions

Griffith immediately recognized what had happened. Some molecule or group of molecules had changed the harmless rough bacteria into

deadly smooth bacteria. Because one type of bacteria was transformed into another, Griffith called this process **transformation.**

☑ *Checkpoint* What is transformation?

Avery and DNA

In 1944, Canadian biologist Oswald Avery realized that studying the process of transformation might be the key to identifying the molecule of heredity. Avery and his co-workers assumed that in Griffith's experiments, genes had been transferred from the dead smooth bacteria to the live rough bacteria. They reasoned that if they found out which molecule was needed for transformation to occur, they might be able to identify the molecule that makes up genes.

In a series of experiments, Avery's team carefully treated heat-killed bacteria with enzymes that destroyed proteins, lipids, carbohydrates, and other molecules. None of these treatments affected transformation. But transformation was blocked when they destroyed a molecule called **DNA**—deoxyribonucleic acid.

From these experiments, Avery reached a simple conclusion about DNA and genes. Genes are made of DNA.

The Hershey-Chase Experiment

Although many biologists accepted Avery's conclusion, a few were still skeptical. In 1952, American scientists Alfred Hershey and Martha Chase carried out another experiment that further supported the conclusion that genes are made of DNA.

Viruses

Hershey and Chase were investigating viruses—tiny particles that can invade and replicate within host cells. A type of virus that infects bacteria is known as a bacteriophage, or phage, for short.

Hershey and Chase knew that a phage contains both protein and DNA. They reasoned that if they could determine which of these molecules enters a bacterium during an infection, they would know which molecule makes up the genes of the phage.

☑ *Checkpoint* What is a bacteriophage?

Radioactive Isotopes

How can protein and DNA molecules be followed, or traced, when a phage attacks a bacterium? One way to trace molecules is to take advantage of an interesting property of matter—that not all of an element's atoms are identical!

Elements have different isotopes, or atoms that have the same numbers of protons but different numbers of neutrons in their nuclei. For example, while the nuclei of the different hydrogen isotopes each contain 1 proton, they contain 0, 1, or even 2 neutrons.

Although the common isotopes of many elements are stable, some isotopes are unstable, which means that their nuclei release small particles or rays of energy. This release of particles or energy is called radioactivity.

Figure 8–3 *Are genes made of DNA or protein? Martha Chase and Alfred Hershey answered this question with a clever experiment.*

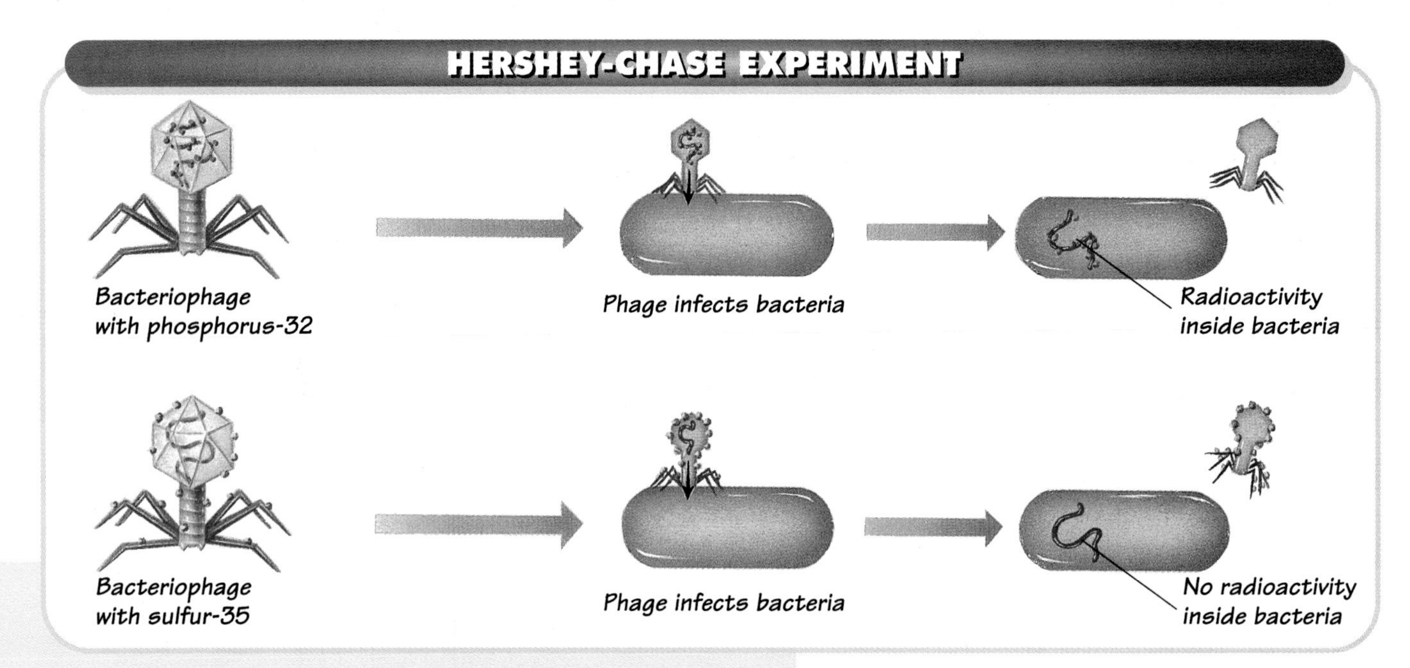

Figure 8–4
Hershey and Chase used radioactive isotopes to label the DNA and proteins of bacteriophages. Because only the phages' DNA entered the bacteria, they concluded that the phages' genes are made of DNA.

INTEGRATING HEALTH

Which radioactive isotopes are used in the diagnosis or treatment of certain illnesses? Write a report on your findings.

Radioactive isotopes become part of molecules just as stable isotopes do. And under the proper safety conditions, radioactivity can be readily observed in the laboratory. As a result, radioactive isotopes provide biologists with a useful way to label molecules, allowing them to be identified as they move from one place to the next. ●

Checkpoint What are isotopes?

Labeling DNA and Proteins

As shown in ***Figure 8–4,*** Hershey and Chase prepared two samples of the phage: one with phosphorus-32, a radioactive isotope of phosphorus, and one with sulfur-35, a radioactive isotope of sulfur. Phosphorus is part of DNA but not of proteins, so phosphorus-32 labeled the phages' DNA. And sulfur is part of proteins but not of DNA, so sulfur-35 labeled the phages' proteins.

Hershey and Chase allowed both samples of bacteriophages to infect bacteria, then analyzed the bacteria for radioactivity. Only one of the samples made the bacteria radioactive. This was the sample grown in phosphorus-32—the isotope that labeled DNA.

The Conclusions

Hershey and Chase concluded that the genetic material of a bacteriophage was DNA, not protein. In fact, their experiment convinced biologists around the world that genes were made of DNA.

Section Review 8–1

1. **Identify** the composition of genes.
2. **Describe** the results of experiments by Griffith, Avery, and Hershey and Chase.
3. **Critical Thinking—Predicting Results** A biology student repeats the Hershey-Chase experiment using a radioactive nitrogen isotope instead of phosphorus-32. Describe the results of this experiment. Would the results show that genes were made of DNA?

DNA Structure and Replication

SECTION 8–2

Guide for Reading

- **Identify** the structure of the DNA molecule.
- **Explain** how DNA replicates.

MINI LAB

- **Construct a model** of DNA.

YOU MIGHT BE THINKING THAT Hershey and Chase solved the great mystery of heredity. However, the really difficult questions had yet to be asked. If genes are made of DNA, then the information that genes contain must be coded in the DNA molecule. How can a molecule carry information? And how can that information be copied every time a cell divides? To begin answering these questions, let's take a look at the structure of the DNA molecule.

DNA NUCLEOTIDES

Nitrogenous base

Adenine

Deoxyribose

Guanine

Phosphate group

Cytosine

Thymine

DNA Structure 

DNA is a nucleic acid—one of the four major classes of large molecules called macromolecules. All nucleic acids are polymers of **nucleotides.** A nucleotide is made of three parts: a phosphate group, a nitrogenous (nitrogen-containing) base, and a 5-carbon sugar, which in DNA is deoxyribose (dee-ahks-ee-RIGH-bohs).

DNA is made of four different types of nucleotides. These nucleotides are adenine (AD-uh-neen), cytosine (SIGHT-oh-seen), guanine (GWAH-neen), and thymine (THIGH-meen). As shown in ***Figure 8–5,*** each of these nucleotides contains a different nitrogenous base.

Chargaff's Rules

In 1950, American biochemist Erwin Chargaff discovered a curious fact about the nucleotides in DNA. Chargaff studied the nucleotide composition of many different samples of DNA, and he found

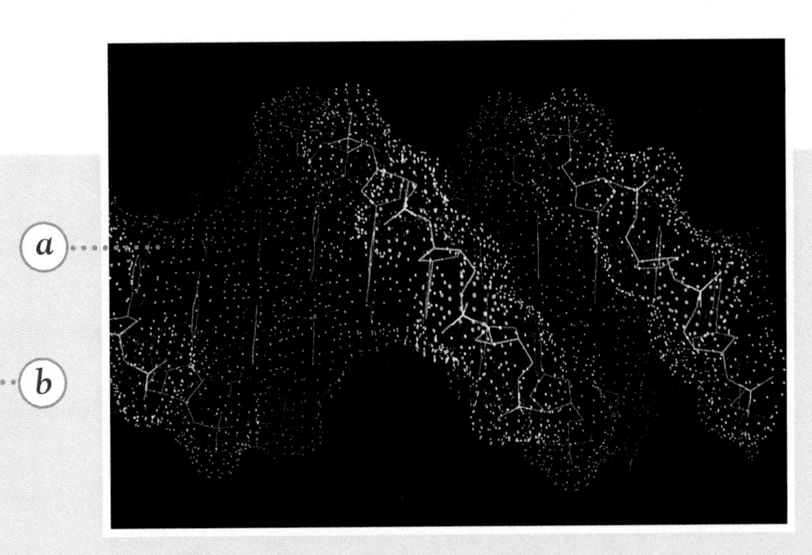

Figure 8–5
(a) A computer generated this image of the structure of DNA. (b) The backbone of DNA is a chain of nucleotides joined together by bonds between the phosphate group of one nucleotide and the deoxyribose sugar of another.

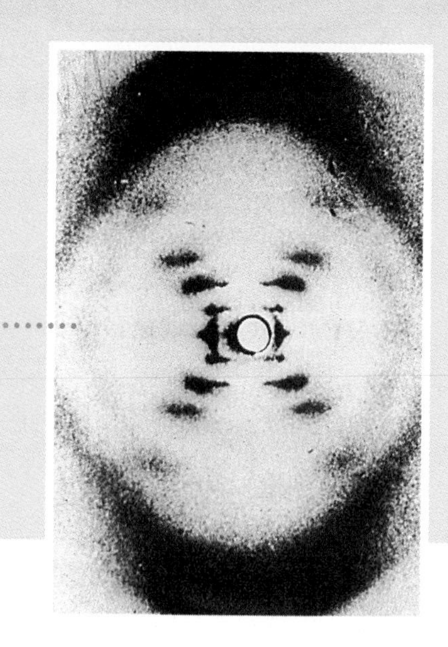

Figure 8–6
(a) *Rosalind Franklin used X-rays to study all sorts of substances, including viruses and coal.*
(b) *Franklin's X-ray image of DNA provided a vital clue to determining DNA's structure.*

that the amount of each type of nucleotide differed in each sample. However, the amounts also followed a distinct pattern. The amounts of adenine (A) and thymine (T) were almost always equal, as were the amounts of cytosine (C) and guanine (G).

This pattern has proven true for the DNA of almost every organism, and it is now known as Chargaff's rules. However, neither Chargaff nor any of his contemporaries had any idea why DNA should follow this pattern.

Rosalind Franklin

In 1951, the English scientist Rosalind Franklin was studying the DNA molecule with a technique called X-ray diffraction. In this technique, a powerful X-ray beam is aimed at a sample, then the scattering pattern of the X-rays is recorded on film. Working with Maurice Wilkins, another English scientist, Franklin produced better and better scattering patterns of the DNA molecule. In 1952, she produced the image shown in ***Figure 8–6.***

By itself, Franklin's X-ray pattern does not reveal the structure of DNA. But the pattern does contain important clues, similar to the clues that footprints provide at the scene of a crime.

In the winter of 1953, Franklin's work came to the attention of the American scientist James Watson. Watson and Francis Crick, a British scientist, had also been investigating the structure of DNA. When Watson saw Franklin's X-ray patterns for the first time, he realized they contained just the clues that he needed. Watson wrote: "The instant I saw them my mouth fell open and my heart began to race."

The Watson-Crick Model

Only a few weeks after seeing Franklin's X-ray work, Watson and Crick had solved the structure of DNA. They believed that Franklin's X-ray pattern suggested that DNA contains two strands, each twisted around the other. This twisted pattern, which is similar to the threads on a screw, is called a helix.

To see if DNA could have such a structure, Watson and Crick built models of two short strands of nucleotides and placed them next to each other, keeping the sugar-phosphate chains on the outside. They then wound the two strands around each other, producing a model with two helix-shaped strands, as shown in ***Figure 8–7.*** **The structure of the DNA molecule is a double helix.** The double-helix model of DNA accounts for most of the features seen in Franklin's X-ray diffraction pattern.

Bonds Between the Strands

However, Watson and Crick faced one big question: What held the two strands together? It took the scientists only a few days to find the answer.

As shown in ***Figure 8–7,*** the nitrogenous bases of each strand were arranged very close to each other along the center of the double helix. When Watson and Crick sketched the structures of the nitrogenous bases, they found that hydrogen bonds could form between adenine (A) and thymine (T), and also between cytosine (C) and guanine (G). These pairs of nitrogenous bases are called base pairs, and the hydrogen bonds between the nitrogenous bases in a base pair provide the force that holds the two strands together.

Suddenly, everything became clear to Watson and Crick. The sequence of nucleotides on one strand is matched perfectly to a complementary sequence on the other strand. In addition, they realized that base-pairing explained Chargaff's rules. Chargaff's rules state that DNA contains equal amounts of adenine and thymine, as well as equal amounts of cytosine and guanine. These are exactly the same pairs that form hydrogen bonds with each other. Therefore, for every adenine in a double-stranded DNA molecule, there had to be one thymine. And for every cytosine, there had to be one guanine.

☑ **Checkpoint** What is the shape of the DNA molecule?

Significance of the Double Helix

Solving the structure of DNA was one of the great scientific achievements of the century. In 1962, the Nobel prize—the highest award the international community can give for a scientific discovery—was given to Watson, Crick, and Wilkins. Rosalind Franklin

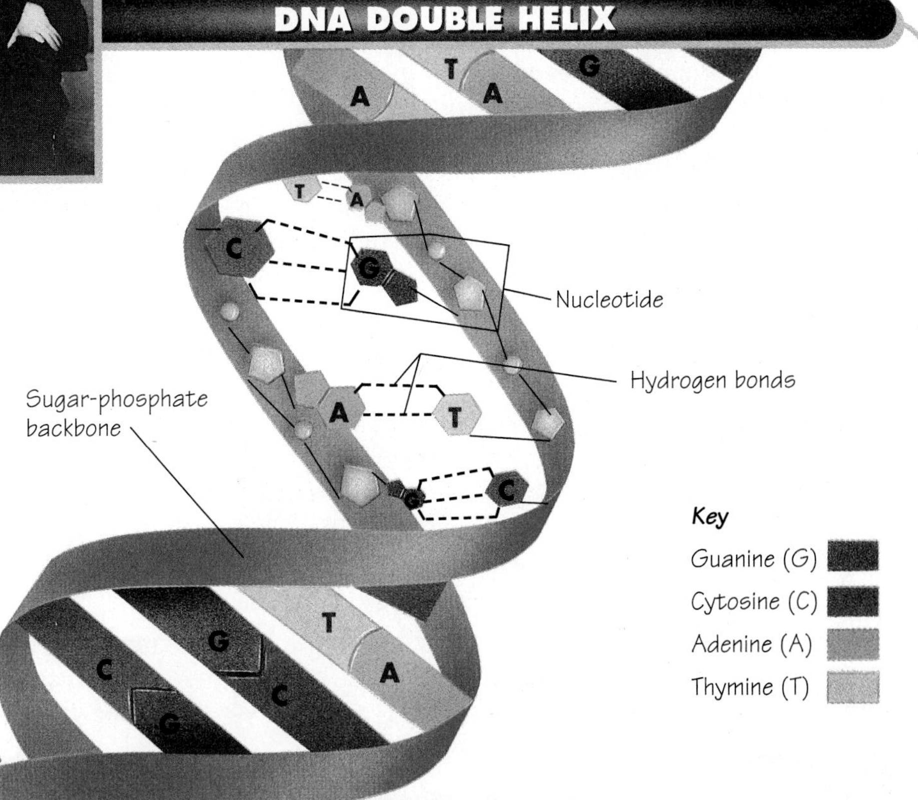

Figure 8–7
(a) *Building on Franklin's work, James Watson (left) and Francis Crick (right) determined the double-helix structure of DNA.*
(b) *Sugar and phosphate groups form each strand of the double helix, while hydrogen bonds hold the strands together. The hydrogen bonds form only between adenine and thymine or between cytosine and guanine.*

Bio FRONTIER Connections

DNA Fingerprinting

Every human has a different set of physical characteristics, which means that every human has unique DNA. Therefore, in theory, people can be conclusively identified from samples of their DNA—just as they can be identified from their fingerprints.

In fact, scientists have developed several procedures to identify a DNA sample, and these procedures are called DNA fingerprinting. Using DNA fingerprinting, the DNA from a drop of blood or a skin sample can be matched to the person who left the sample behind.

CAREER TRACK *This geneticist is preparing DNA for electrophoresis—one step in DNA fingerprinting.*

Techniques of DNA Fingerprinting

You might think that DNA fingerprinting involves identifying all the nucleotides in a DNA molecule. However, such a technique is not feasible—at least not with today's technology. Why is this so? A molecule of human DNA contains billions of nucleotides—far too many to analyze quickly.

How is DNA fingerprinting done today? In one technique, a DNA molecule is split apart at specific locations, creating fragments that vary in length from one person to the next. Proponents of this technique believe it to be sensitive enough to identify one person from 100 thousand to 100 million others.

The results of DNA fingerprinting are often presented as evidence in criminal trials.

Uses of DNA Fingerprinting

DNA fingerprinting and other types of DNA analysis are being used for a variety of interesting purposes. Geneticists are analyzing DNA to study hereditary diseases. Archaeologists are analyzing DNA to identify blood stains on ancient artifacts and in caves. And one group of historians is using DNA to trace the lineages of European royal families.

But arguably the most controversial applications of DNA fingerprinting have been in the criminal justice system. Although the police often use DNA evidence to identify suspects, juries have not always accepted this evidence when a case comes to trial. Observers have argued that many jury members do not trust DNA fingerprinting because it is too new or too difficult to understand.

The Debate Ahead

Although DNA fingerprinting impresses many experts, others argue that, at best, it is an imperfect means of identification. The strengths and weaknesses of DNA fingerprinting are sure to be debated and tested in the years ahead.

Making the Connection

If you were on a jury, would you accept the results of DNA fingerprinting as evidence? What questions about DNA fingerprinting would you want to ask?

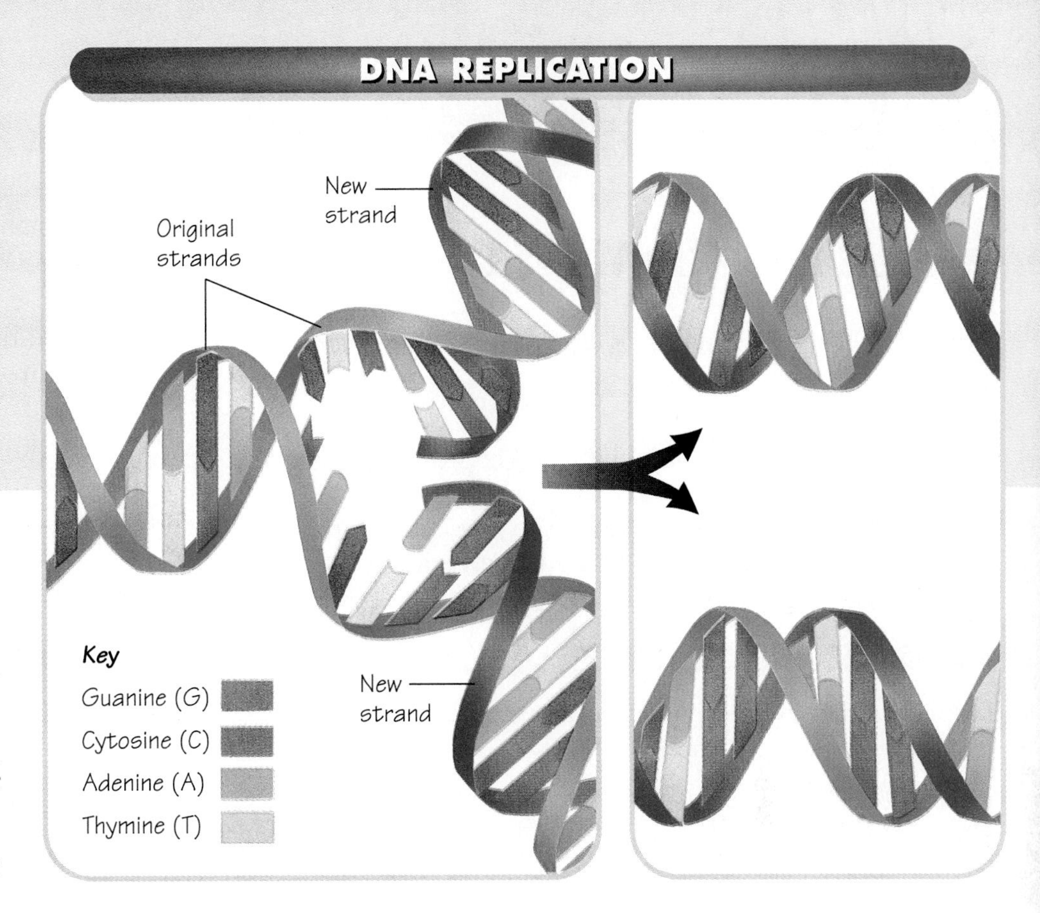

Figure 8–8
In DNA replication, each strand serves as a pattern for the synthesis of a complementary strand. The result is two double-stranded DNA molecules, each an exact copy of the original molecule.

undoubtedly would have shared in this prize, but she died in 1958, and Nobel prizes are given only to living scientists.

Watson and Crick published their model of DNA structure in a scientific paper that was just one page long. However, that brief paper contained a sentence that explained one more thing about DNA: how the molecule could copy itself. "It has not escaped our notice," wrote Watson and Crick, "that the specific pairing we have postulated immediately suggests a possible copying mechanism for the genetic material."

DNA Replication 

Before Watson and Crick developed the double-helix model, biologists had difficulty imagining how any molecule could be duplicated. However, when it was clear that DNA had two strands bound to each other by base-pairing, it was easy to see how DNA was copied. If the two strands of a DNA molecule are separated, the sequence of bases on each strand makes it possible to form an exact duplicate by following base-pairing rules.

DNA is copied in a process called **DNA replication.** DNA replication takes place during the synthesis phase (S-phase) of the cell cycle, and the replication is carefully controlled by a group of enzymes. **In DNA replication, enzymes separate the two strands, then synthesize new strands that base-pair to the original strands.** This process is illustrated in ***Figure 8–8.***

When DNA replication is complete, a cell has duplicated all its genetic information. The cell is then ready to begin cell division.

☑ **Checkpoint** How does DNA replicate?

DNA and Chromosomes

In eukaryotic cells, genetic information is contained in chromosomes located in the cell nucleus. Does this mean that chromosomes are made of DNA? Although chromosomes do contain DNA, their structure is significantly more complex. In fact, eukaryotic chromosomes actually contain more protein than DNA.

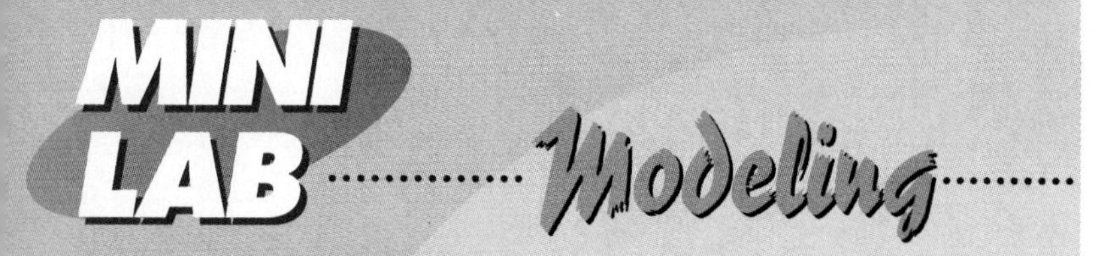

Build Your Own DNA

PROBLEM *How can you **construct a model** of DNA?*

PROCEDURE

1. Construct a three-dimensional model of DNA. The model should clearly show sugars, phosphate groups, and the four different nitrogenous bases. Use pipe cleaners, plastic straws, toothpicks, or any other materials available.
2. Create a key that identifies the different parts of your model.

ANALYZE AND CONCLUDE

1. In what ways does your model accurately represent DNA? In what ways is it inaccurate?
2. What are Chargaff's rules? How does your model follow these rules?
3. Watson and Crick believed that their model of DNA held clues to how DNA was copied. Do you see any such clues? Explain your answer.

Chromatin

Chromosomes are made of a material called **chromatin.** Chromatin consists of DNA and a number of different proteins. Some of the most important of these proteins are histones, a class of proteins that bind directly to DNA.

Nucleosomes

In 1973, the American scientists Don and Ada Olins and Christopher Woodcock discovered that histones form tiny particles that they called nucleosomes. What do nucleosomes do? They may help to fold and package DNA. Folding and packaging DNA is an important job, because the DNA in a single chromosome may be as much as 10,000 times as long as the chromosome itself! In addition, recent evidence suggests that nucleosomes play a role in regulating the way genes are transcribed into RNA.

Scientists have learned a great deal about the structure of DNA and the way DNA and proteins are organized in chromosomes. However, we still have a long way to go in understanding DNA.

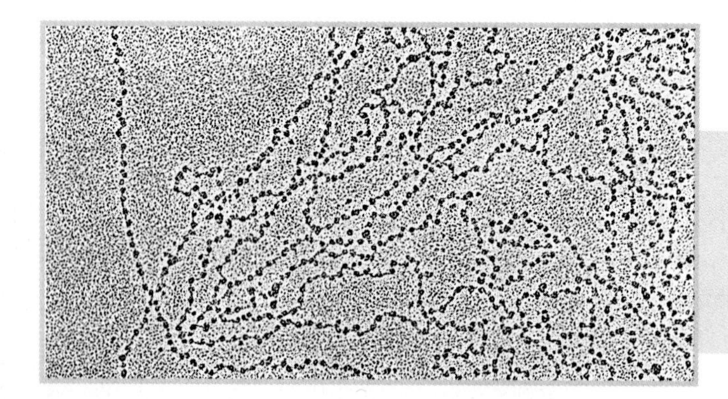

Figure 8–9
Nucleosomes look like beads on a string when viewed through an electron microscope (magnification: 100,000×).

Section Review 8–2

1. **Identify** the structure of the DNA molecule.
2. **Explain** how DNA replicates.
3. **Describe** the different components of chromatin.
4. **Critical Thinking—Interpreting Models** How does the double-helix model of DNA explain Chargaff's rules?
5. **MINI LAB** How can you **construct a model** of DNA?

Guide for Reading

- **Identify** the role of RNA.
- **Compare** RNA with DNA.
- **Describe** the processes of transcription and translation.

MINI LAB

- **Interpret** the genetic code.

THE DOUBLE-HELIX MODEL shows that a DNA molecule can be replicated, just as you might expect for a molecule that is found in every cell and that contains genetic information. But how is the genetic information decoded? And after it is decoded, what does it say?

As you will see, the sequence of nucleotides in DNA really does contain a code. And to put that code to use, the cell first makes a copy of the coded message.

RNA Structure

To decode the genetic information contained in DNA, the cell uses a molecule called **RNA**—ribonucleic acid. **RNA is the principal molecule that carries out the instructions coded in DNA.**

Like DNA, RNA is a nucleic acid—a macromolecule formed by nucleotides. However, RNA differs from DNA in three principal ways:

- **The sugar in RNA is ribose. In DNA, the sugar is deoxyribose.**
- **RNA is generally single-stranded. Recall that the structure of DNA is a double-stranded helix.**
- **RNA contains uracil** (YOOR-uh-sihl) **instead of thymine. Like thymine, uracil can form hydrogen bonds with adenine.**

✓ *Checkpoint* How does RNA differ from DNA?

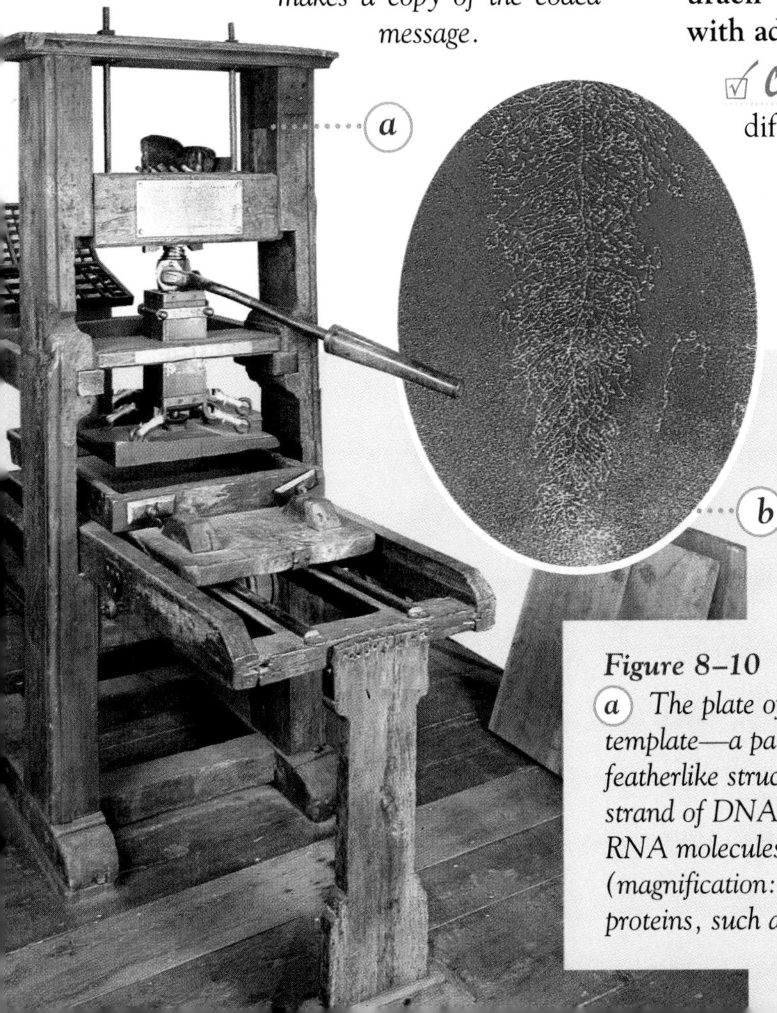

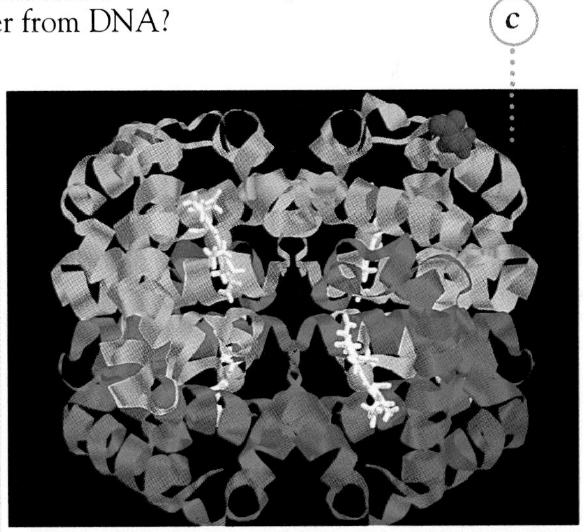

Figure 8–10
(a) The plate of this old-fashioned printing press is an example of a template—a pattern from which copies can be made. (b) In the featherlike structure shown in this photograph, the center line is a long strand of DNA. The DNA serves as a template for the synthesis of RNA molecules, which appear as branches off the center line (magnification: 6700X). (c) Together, DNA and RNA code for proteins, such as the hemoglobin protein modeled here.

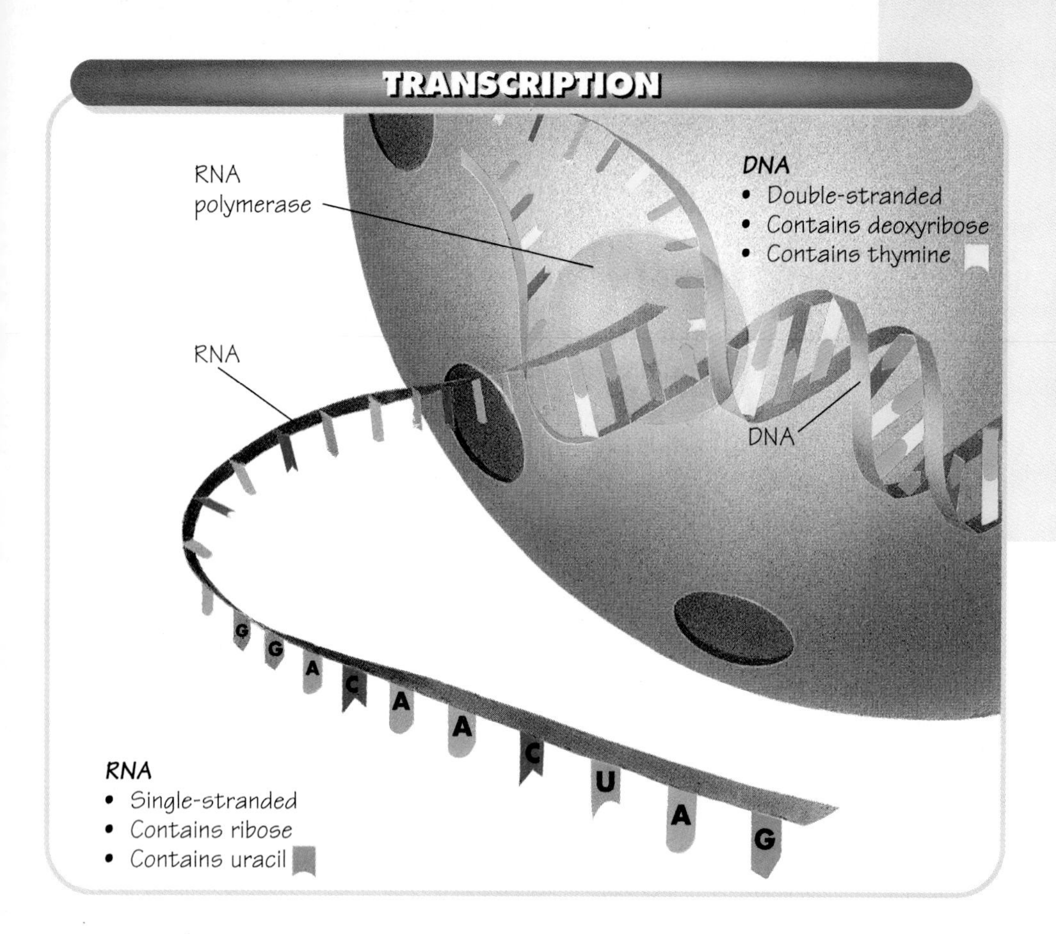

Figure 8–11
In transcription, DNA serves as a template to produce RNA. The RNA then enters the cytoplasm and is used to make proteins. While DNA and RNA are very similar, they differ in the ways listed in the diagram.

Transcription 

Transcription is the process by which RNA molecules are made. **In transcription, part of the nucleotide sequence of a DNA molecule is copied into RNA.** For this reason, DNA is said to act as a template for RNA. A template is a pattern, or guide, from which a copy can be made.

RNA Polymerase

Transcription is carried out by RNA polymerase, an enzyme that binds directly to a molecule of DNA. As shown in ***Figure 8–11,*** RNA polymerase produces a strand of RNA, one nucleotide at a time, by matching base pairs with the nucleotides in DNA. For example, a nucleotide sequence of AACT in DNA would be copied as UUGA in RNA. This careful base-pairing ensures that each RNA molecule carries a copy of the coded instructions in the DNA sequence.

Where on the DNA molecule does RNA polymerase begin transcription? And where does it stop? The answers lie in special "start" and "stop" sequences in DNA. These sequences determine exactly which parts of the DNA will be copied into RNA.

☑ **Checkpoint** What does RNA polymerase do?

RNA as a Message

By making RNA, the cell achieves at least two goals. First, a single sequence in DNA may be copied again and again into RNA, making hundreds or even thousands of identical copies. This is especially important for a cell that is growing rapidly and might need multiple copies of the instructions that DNA contains. Second, by making RNA, the cell is able to keep its DNA in reserve, controlling access to it and carefully regulating its use and replication.

Forms of RNA

What does RNA do? It does quite a few things, but in most cells the main forms of RNA are all involved in making proteins. There are three principal forms of RNA:

- **Messenger RNA** Most genes contain instructions for the assembly of amino acids into polypeptides, which are long chains of amino acids. The form of RNA that carries copies

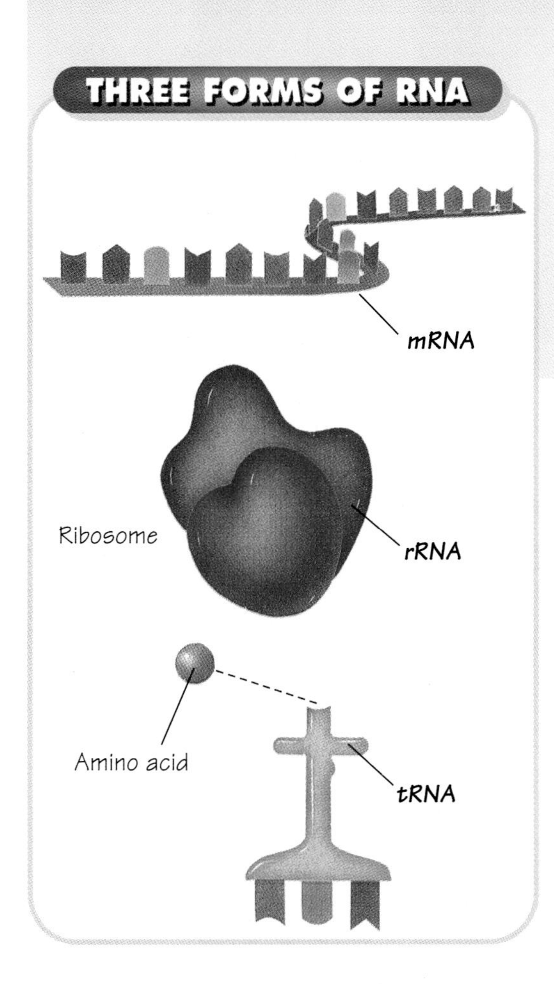

Figure 8–12
Cells use three main forms of RNA. The nucleotide sequence in messenger RNA (mRNA) determines the amino acid sequence of a polypeptide. Ribosomal RNA (rRNA) is an important component of ribosomes. And each type of transfer RNA (tRNA) binds to a specific amino acid.

of these instructions is known as **messenger RNA** (mRNA). As its name implies, mRNA serves as a messenger from DNA to the rest of the cell.

- **Ribosomal RNA** Proteins are assembled at particles called ribosomes. Ribosomes are made of several dozen proteins, as well as a form of RNA known as **ribosomal RNA** (rRNA).
- **Transfer RNA** When a protein is assembled, molecules of another form of RNA transfer one amino acid after another to the ribosome. This form of RNA is called **transfer RNA** (tRNA).

The mRNA, rRNA, and tRNA in a cell are each produced by transcription of their own genes in DNA. As you will see, each form of RNA plays a distinct role in the production of proteins.

Checkpoint What are the three main forms of RNA?

Genetic Code

Proteins are made from polypeptides, and polypeptides are made from 20 different amino acids. The order of the amino acids determines the polypeptide's properties.

What do the bases in DNA have to do with the sequence of amino acids in a polypeptide? The instructions coded in DNA specify the order in which the 20 different amino acids are put together. The language of the instructions in DNA and RNA is called the **genetic code.**

Only 4 Letters

As you have read, RNA contains 4 different nucleotides: A, U, C, and G. In effect, this means that the code is written in a language of only 4 letters.

How can a code that uses only 4 letters carry instructions for 20 different amino acids? The answer is that the nucleotides in mRNA are read in groups of 3. Put another way, each word of the genetic code is 3 letters long. Thus, in theory, the genetic code could signify 64 different words, because 4^3 equals 64.

INTEGRATING MATHEMATICS

How many different 3-letter words can 4 letters signify? How many different 2-letter words can 4 letters signify?

Codons

The group of 3 nucleotides in mRNA that specifies an amino acid is known as a **codon.** You can think of codons as the words of the genetic message. For example, consider the following sequence of nucleotides in mRNA:

AAACACGGU

This sequence is read as 3 codons:

AAA–CAC–GGU

And each codon represents a different amino acid:

AAA–CAC–GGU
Lysine–Histidine–Glycine

Figure 8–13 shows all 64 possible codons of the genetic code. Notice that more than one codon can specify the same amino acid. For example, 6 different codons specify the amino acid leucine, and 6 others specify the amino acid arginine.

In most molecules of mRNA, the AUG codon is the start signal, or initiator codon. AUG also codes for the amino acid methionine, which means that methionine is usually the first amino acid when a polypeptide chain is assembled.

Three codons—UAA, UAG, and UGA—serve as stop signals. Like a period at the end of a sentence, these codons signify the end of a genetic message—and the end of the polypeptide.

☑ ***Checkpoint*** What is a codon?

Translation

In eukaryotic cells, mRNA emerges from the nucleus and enters the cytoplasm. There, the codons in mRNA are translated into amino acids—and **translation** is exactly what biologists call this process. **In translation, nucleotides in mRNA are decoded into a sequence of amino acids in a polypeptide.**

You can think of mRNA as a plan or blueprint for a polypeptide. However, blueprints do not produce buildings all by themselves—builders are needed to follow them.

In cells, the polypeptide builders are the ribosomes—small particles made of rRNA and protein. And the molecules that actually bind the individual amino acids are molecules of tRNA. In effect, ribosomes "read" the codons in mRNA, allowing tRNA molecules to bring the proper amino acids to form a polypeptide chain. To better understand this process, study the illustration on the next page.

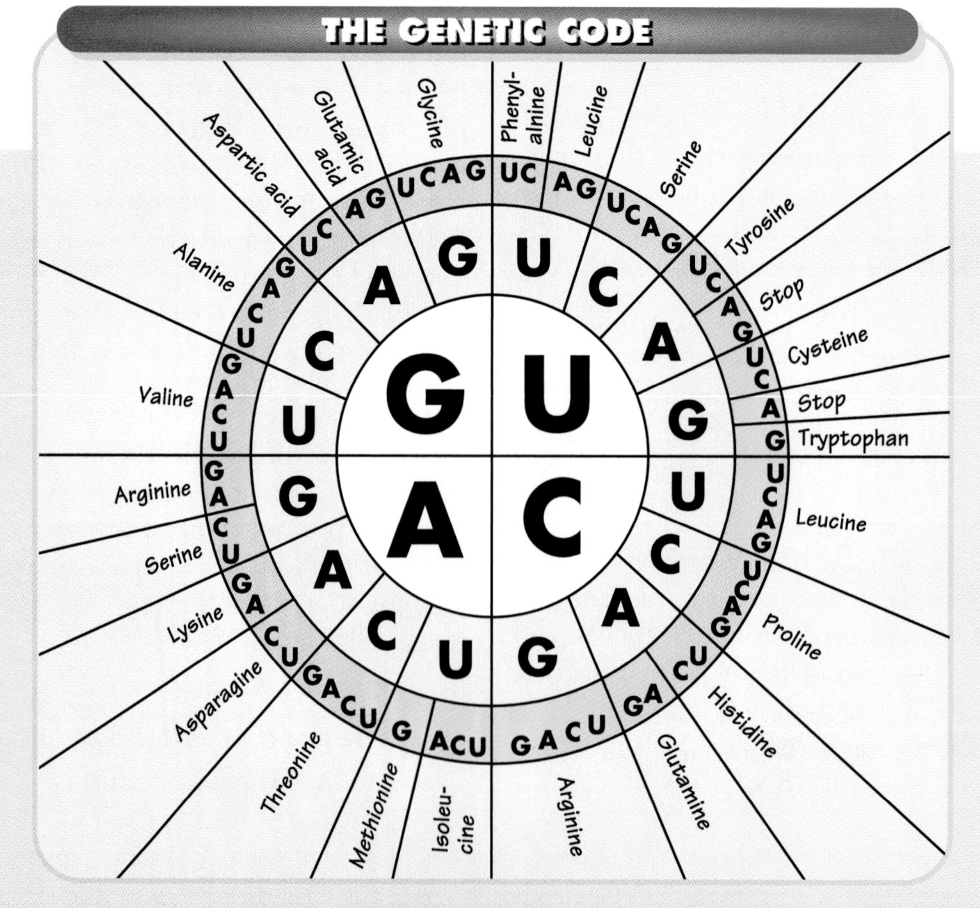

Figure 8–13
You can use this table to "decode" a codon by starting at the middle of the circle and moving outward. For example, the codon GAU codes for aspartic acid.

Visualizing Protein Synthesis

The different forms of RNA work together to change a coded message into a protein—a process called translation.

1 mRNA is transcribed in the nucleus, then enters the cytoplasm and attaches to a ribosome. Translation begins at AUG, the start codon.

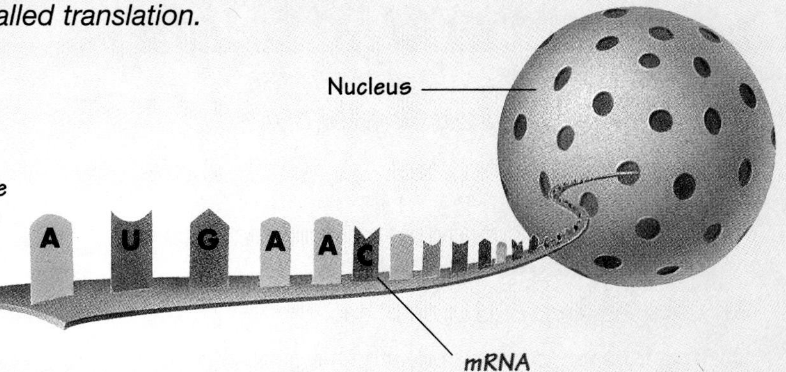

2 Each tRNA contains an anticodon—three nucleotides that base-pair to a specific codon in mRNA. The ribosome positions the start codon to attract its anticodon, which is part of the tRNA that binds methionine. The ribosome also binds the next codon and its tRNA.

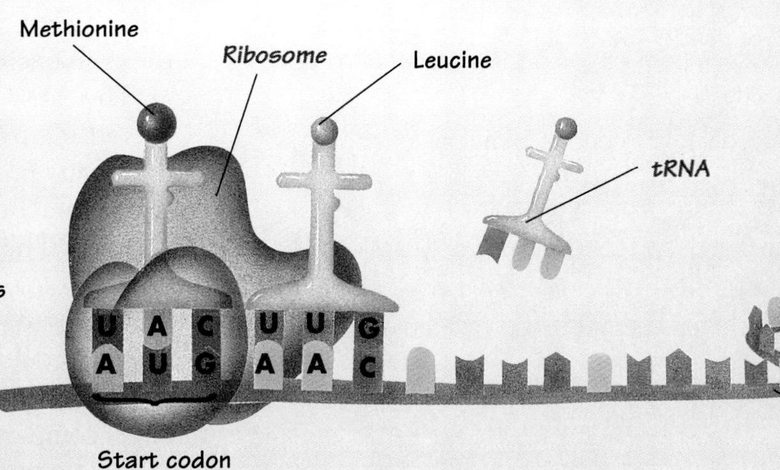

3 The ribosome joins the two amino acids and cuts the bond between methionine and its tRNA. The tRNA floats away from the ribosome, allowing it to bind another methionine.

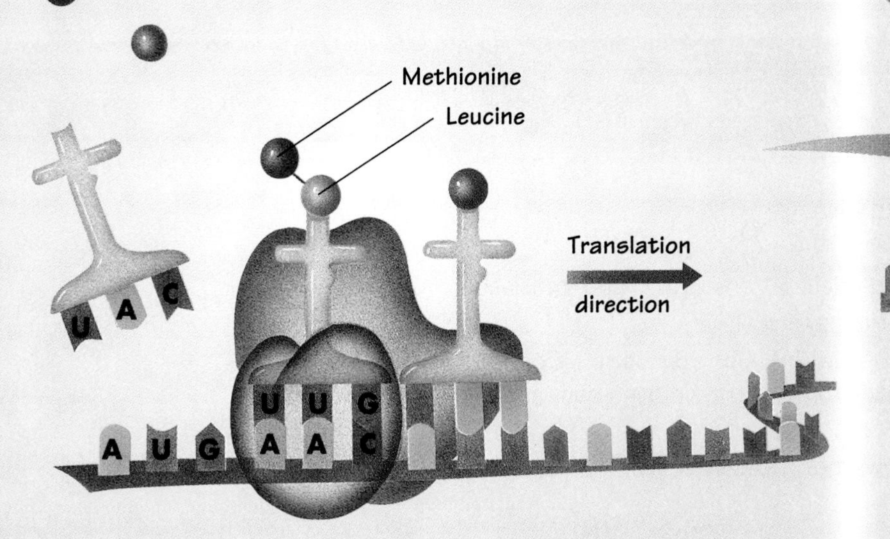

4 The ribosome advances along the mRNA, binding new tRNA molecules and amino acids.

5 The process continues until the ribosome reaches one of three stop codons. The result is a complete polypeptide.

Left to Right or Right to Left?

PROBLEM *How does a cell **interpret** DNA?*

PROCEDURE

1. Assume that a gene contains the following sequence of nucleotides: **GACAAGTCCACAATC**

   Write this sequence on a piece of paper.
2. Reading from left to right, transcribe the gene into a molecule of mRNA.
3. Reading the codons from left to right, translate the mRNA into a polypeptide.
4. Repeat step 3 reading the codons from right to left.

ANALYZE AND CONCLUDE

1. Why did steps 3 and 4 produce different polypeptides?
2. Do cells decode nucleotides in one direction only or can they decode them in either direction? Formulate a hypothesis to answer this question.

Genes and Proteins

As you have just read, the three principal forms of RNA work together to produce polypeptides. Through transcription, a section of DNA directly determines the sequence of nucleotides in a molecule of mRNA. And through translation, the nucleotides in mRNA are translated into the amino acids of a polypeptide. In this way, a section of DNA—a gene—directs the synthesis of a protein.

If you are wondering why proteins are so important, that's a good question! Recall that in your studies of genetics, you saw that genes control the color of flowers, the height of plants, the type of human blood, and even the sex of a newborn baby. But genes code for proteins—what do proteins have to do with these characteristics?

The answer is that proteins have everything to do with them! Remember that many proteins are enzymes, and enzymes catalyze and regulate chemical reactions. Thus, a gene could control flower color by coding for an enzyme that helps to produce a pigment. Another gene could determine blood type by coding for an enzyme that helps to produce complex molecules in red blood cells. The proteins from other genes regulate the rate and pattern of growth throughout an organism, controlling its size and shape.

In fact, proteins are the keys to almost everything that living cells do. By coding for proteins, DNA holds the key to life itself.

Section Review 8-3

1. **Identify** the role of RNA.
2. **Compare** RNA with DNA.
3. **Describe** the processes of transcription and translation.
4. **Explain** the importance of proteins.
5. **Critical Thinking—Making Comparisons** Compare the three different forms of RNA. Which forms could be used over and over again to produce different polypeptides?
6. **MINI LAB** How does a cell **interpret** the genetic code?

Controlling Gene Expression

GUIDE FOR READING

- **Explain** why cells control gene expression.
- **Describe** how mRNA is edited.

EVEN THE SIMPLEST LIVING organism contains thousands of genes. Are all those genes active all the time? Do cells copy every gene into mRNA?

The answer to both questions is no. Like the thousands of books in a library, not all the genes on a cell's chromosomes are read at the same time. In fact, only a small fraction of the cell's genes are actively copied at any given time.

Gene Expression

A cell that transcribed all its genes at the same time would be making lots of proteins that it did not need, thus wasting energy and raw materials. Fortunately for cells, they use their genes more efficiently. **Cells regulate gene transcription because they do not always need a gene's product.**

A gene that is transcribed into mRNA is said to be expressed, and a gene that is not being transcribed is said to be unexpressed. Put another way, an expressed gene is "turned on" and an unexpressed gene is "turned off."

How does a cell efficiently regulate gene transcription? How does a cell recognize when to turn genes on and when to turn them off? The first answers to these questions came from studies done on the genes of a bacterium called *Escherichia coli*.

Expressing Lac Genes

Escherichia coli—*E. coli*, for short—contains about 2000 genes. Three of these genes are called the *lac* genes. Each *lac* gene codes for a protein, and

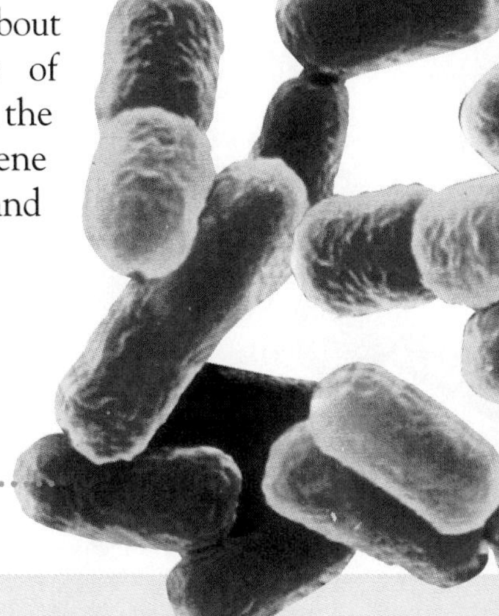

Figure 8–14
(a) *Dairy products contain lactose, a kind of sugar. To break down lactose, organisms need a special enzyme to separate the two rings of the lactose molecule.* (b) *E. coli, a type of bacteria, produces the enzyme that separates the two rings, but it does so only when lactose is present. Thus, E. coli must be able to "turn on" and "turn off" the gene for this enzyme (magnification: 3700X).*

each protein must be expressed for the cell to use lactose, a type of sugar.

When does it make sense for *E. coli* to express *lac* genes? Obviously, it makes sense only when lactose is actually present! In fact, because of the energy needed to make the three proteins, expressing the *lac* genes might even be a waste of precious resources. Therefore, an efficient bacterium would express the *lac* genes only when lactose is present. And that is exactly what the bacterium does!

The Promoter

To one side of the three *lac* genes is a special region of DNA called the **promoter.** The promoter is a binding site for RNA polymerase.

RNA polymerase is able to recognize the promoter, and there it attaches itself to the chromosome. It then moves along the DNA molecule until it finds the first gene. At that point it starts transcription, making a single mRNA molecule from the three genes that can be translated into three proteins.

The Repressor and Operator

To regulate the expression of the *lac* genes, the bacterium uses a protein called a *lac* **repressor.** A repressor is a DNA-binding protein that blocks a gene's transcription.

The *lac* repressor binds to another special region of DNA called the **operator.** When the *lac* repressor binds to the operator, RNA polymerase cannot move past the operator to reach the *lac* genes. Like a lock on the library door, the repressor blocks the expression of the *lac* genes.

☑ ***Checkpoint*** What is the *lac* repressor? The operator?

Unlocking the Padlock

Once the repressor is bound to the operator, how can the *lac* genes be expressed? In fact, when lactose is not present, the repressor remains bound to the operator and the genes' transcription is blocked. However, when lactose is present, something remarkable happens.

As shown in ***Figure 8–15,*** the repressor has a binding site for lactose. When lactose binds to the repressor,

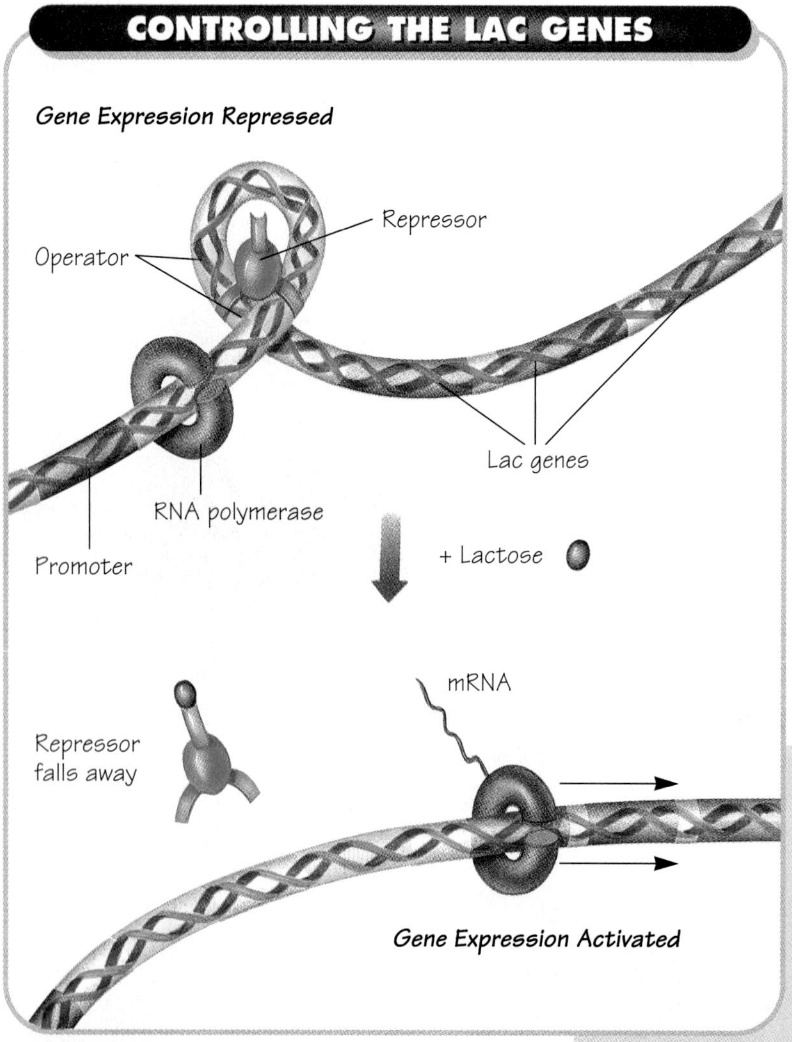

CONTROLLING THE LAC GENES

Figure 8–15
How does the repressor "turn off" the lac genes? Researchers have shown that it binds to the gene at two different places, thus tying the gene in a knot! Gene expression is repressed only when lactose is not present.

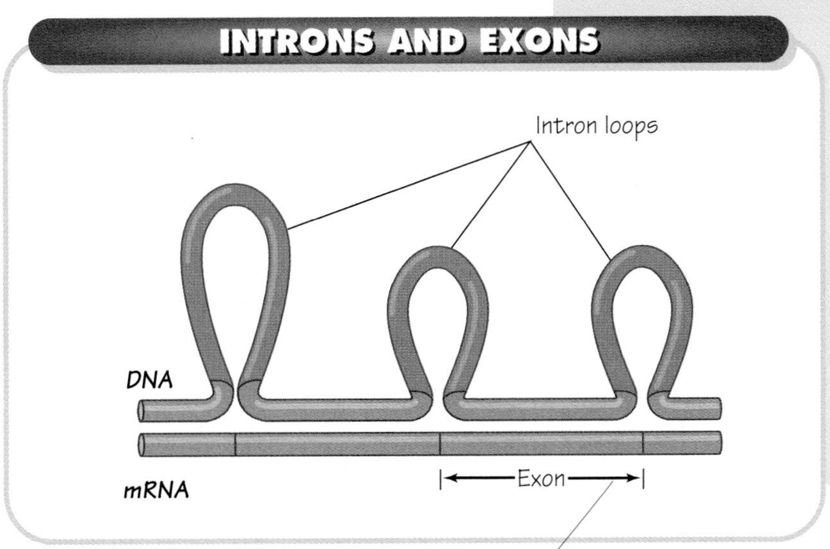

INTRONS AND EXONS

Intron loops

DNA

mRNA

Exon

Figure 8–16
After a molecule of mRNA is transcribed, sequences called introns may be removed. The remaining sequences, called exons, are then spliced together. As a result, the final version of mRNA matches to separate pieces of the DNA that coded for it, not to the entire gene.

the repressor changes shape and falls off the DNA. Suddenly, the lock is gone, and the *lac* genes are ready to be transcribed.

Not all genes work exactly as the *lac* genes do. In eukaryotic cells, for example, the control of gene expression can be especially complicated. However, many genes follow the general pattern of DNA-binding proteins—such as the *lac* repressor—that control where and when transcription happens.

☑ **Checkpoint** Why are the *lac* genes expressed when a lactose molecule is present?

Genes in Pieces

In 1976, scientists Philip Sharp and Susan Berget made a surprising discovery about gene expression in eukaryotes. They carefully compared the DNA sequence of a gene with the base sequence of mRNA for the very same gene. They expected the two sequences to match perfectly—but they were amazed to see something quite different!

As Sharp and Berget discovered with further research, mRNA often undergoes a process similar to editing, in which parts of the molecule are removed and discarded. The discarded parts are called **introns**—short for intervening sequences. **After the introns are removed, the remaining parts are spliced together to form the final version of mRNA.** The remaining parts are called **exons**—short for expressed sequences.

Why do genes make mRNA with introns and exons? Biologists still have no good answer for this question. Some mRNA molecules may be spliced in different ways, making it possible for a single gene to produce several different molecules of mRNA. In addition, introns and exons may play a role in evolution, making it possible for very small changes in DNA sequences to have dramatic effects in gene expression.

Section Review 8–4

1. **Explain** why cells control gene expression.
2. **Describe** how mRNA is edited.
3. **BRANCHING OUT ACTIVITY** **Construct a model** of the *lac* genes and the promoter, repressor, and operator. Use your model to explain how the expression of the *lac* genes is controlled.

Laboratory Investigation

Extracting DNA

DNA is stored in the nucleus of every eukaryotic cell. In this investigation, you will perform a DNA extraction—a procedure that removes the DNA from cells. In this investigation, you will extract the DNA from wheat germ, the fat-rich part of the seed of the wheat plant. A similar procedure, however, could extract the DNA from many different organisms.

Problem

How can you **observe** DNA?

Materials (per group)

200-mL plastic cup
distilled water
1.5 g raw wheat germ
hot plate
1000-mL beaker
thermometer
liquid dishwashing detergent
3 g meat tenderizer
baking soda solution
ice
10-mL graduated cylinder
10 mL alcohol solution
glass stirring rod
tongs

Procedure

1. **Mix 100 mL of distilled water and the wheat germ in the 200-mL plastic cup.**
2. **Place the plastic cup in a water bath, as shown in the diagram.**

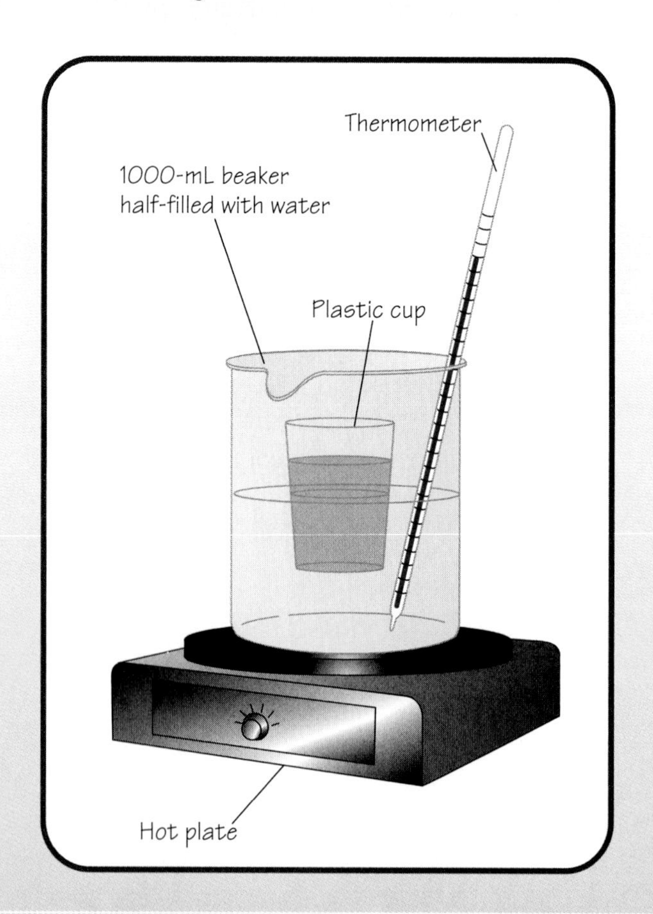

3. **Heat the water bath to 50°C. As you perform steps 4 through 6, adjust the hot plate as necessary to maintain the water temperature near 50°C. DNA will denature, or lose its structure, at temperatures above 60°C.**
4. **Add 5 mL of liquid dishwashing detergent to the plastic cup. Liquid dishwashing detergent breaks apart cell membranes.**
5. **Add the meat tenderizer to the plastic cup.**
6. **Add 10 mL of the baking soda solution to the plastic cup. Stir the mixture, then wait 10 minutes.**
7. **Turn off the hot plate. Using the tongs, transfer the plastic cup to a container of ice. Keep the plastic cup in the ice for 15 minutes.**
8. **Using the 10-mL graduated cylinder, slowly add the alcohol solution to the plastic cup. The alcohol solution should form a second layer on top of the wheat germ solution.**
9. **The DNA of the wheat germ precipitates at the interface of the two layers. Place the tip of the glass stirring rod at the interface and twist it to spool the DNA.**

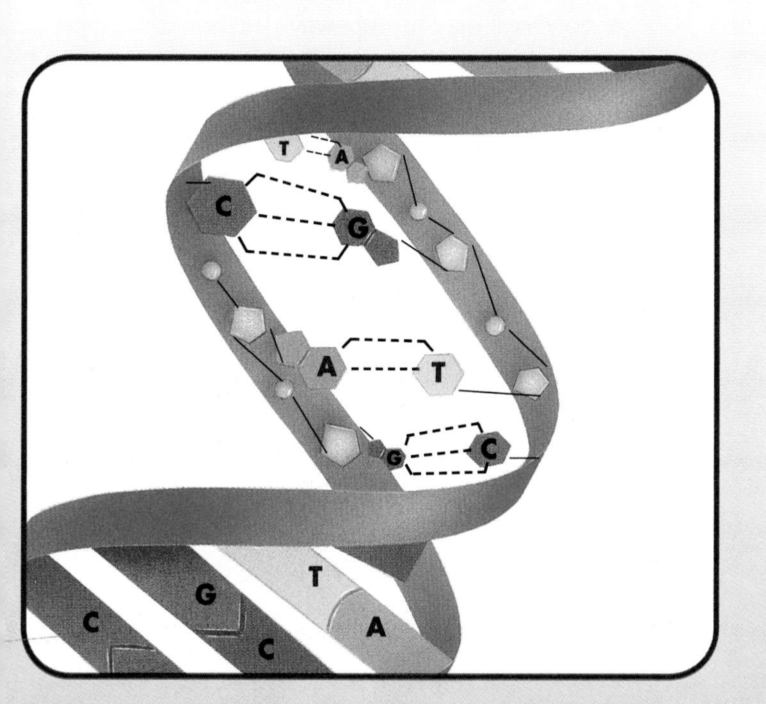

Observations

Describe the DNA you produced.

Analysis and Conclusions

1. What is the role of DNA in the cell?
2. Explain the purpose of adding liquid dish-washing detergent in step 4.
3. Can you directly observe the double-helix structure in the DNA sample you produced? Explain your answer.
4. Explain why a procedure similar to the one in this investigation could be used to extract DNA from the cells of other organisms.

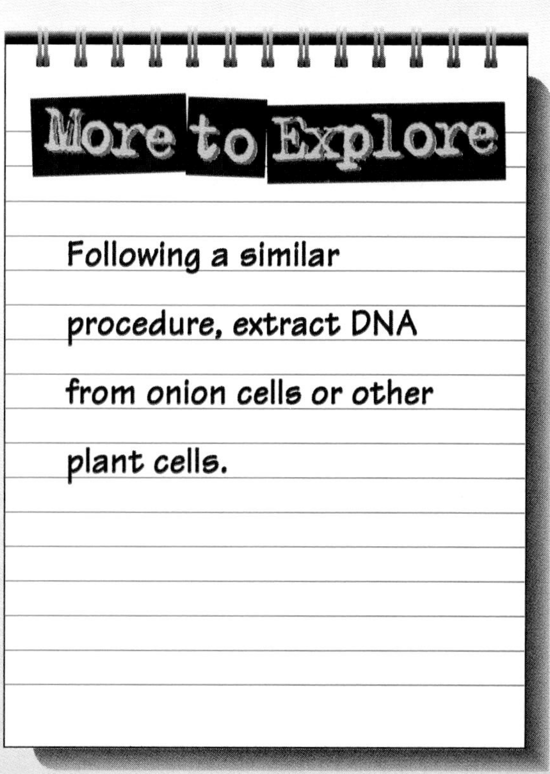

Summarizing Key Concepts

The key concepts in each section of this chapter are listed below to help you review the chapter content. Make sure you understand each concept and its relationship to other concepts and to the theme of this chapter.

8–1 Discovering DNA

- Frederick Griffith discovered transformation, a process that changes one type of bacteria into another type.
- DNA is the molecule of heredity—the molecule that forms genes. The role of DNA was demonstrated by the work of Oswald Avery and the Hershey-Chase experiment.

8–2 DNA Structure and Replication

- Watson and Crick determined that DNA has the structure of a double helix. Hydrogen bonds between base pairs link the two strands together.
- In DNA replication, enzymes separate the two strands, then synthesize new strands that base-pair to the original strands.
- Chromosomes are made of chromatin, which contains DNA and proteins. Biologists suspect that proteins help to fold and package DNA.

8–3 RNA

- RNA carries out the instructions coded in DNA. The three principal forms of RNA are mRNA, rRNA, and tRNA.
- In transcription, a nucleotide sequence in DNA is copied into RNA. Transcription is carried out by an enzyme called RNA polymerase.
- A codon is a group of three nucleotides in mRNA that codes for an amino acid. In translation, the codons in mRNA are translated into a polypeptide.

8–4 Controlling Gene Expression

- Cells regulate gene transcription because they do not always need the genes' products.
- After an mRNA molecule is transcribed, sections of it may be cut away. These sections are called introns. The remaining sections, called exons, are expressed.

Reviewing Key Terms

Review the following vocabulary terms and their meaning. Then use each term in a complete sentence.

8–1 Discovering DNA

transformation
DNA

8–2 DNA Structure and Replication

nucleotide
DNA replication
chromatin

8–3 RNA

RNA
transcription
messenger RNA
ribosomal RNA
transfer RNA
genetic code
codon
translation
anticodon

8–4 Controlling Gene Expression

promoter
repressor
operator
intron
exon

Recalling Main Ideas

Choose the letter of the answer that best completes the statement or answers the question.

1. What did Griffith name the process that changed rough bacteria to smooth bacteria?

a. transformation **c.** translation
b. transcription **d.** replication

2. Hershey and Chase showed that the genetic material of a bacteriophage is

a. protein. **c.** DNA.
b. sugar. **d.** RNA.

3. According to Chargaff's rules, the amount of adenine (A) in DNA equals the amount of

a. thymine (T).
b. cytosine (C).
c. guanine (G).
d. cytosine (C) and guanine (G).

4. In the Watson-Crick model of DNA, the two strands of the double helix are united by hydrogen bonds between

a. sugar molecules. **c.** nitrogenous bases.
b. phosphate groups. **d.** nitrogen atoms.

5. Which molecule(s) are found in chromosomes?

a. DNA only **c.** proteins only
b. RNA only **d.** DNA and proteins

6. RNA polymerase is the enzyme that carries out

a. transcription. **c.** transformation.
b. translation. **d.** DNA replication.

7. In translation, amino acids bond to molecules of

a. DNA. **c.** rRNA.
b. mRNA. **d.** tRNA.

8. The *lac* repressor falls off the operator in the presence of

a. lactose. **c.** rRNA.
b. RNA polymerase. **d.** tRNA.

9. The portions of mRNA that are removed and discarded are called

a. introns. **c.** nucleosomes.
b. exons. **d.** tRNA.

Putting It All Together

Using the information on pages xxx to xxxi, complete the following concept map.

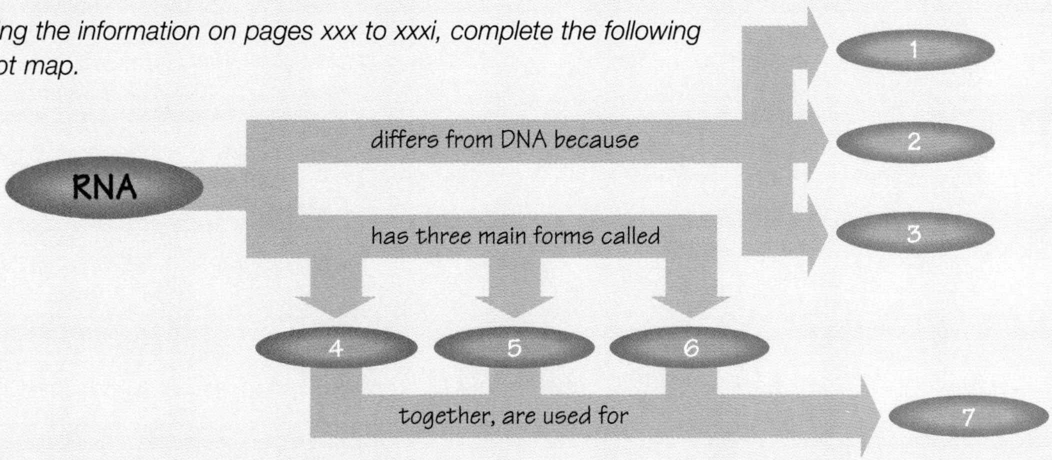

RNA
differs from DNA because
1
2
3
has three main forms called
4
5
6
together, are used for
7

Assessment

Reviewing What You Learned

Answer each of the following in a complete sentence.

1. What is bacterial transformation?
2. How do the isotopes of an element differ?
3. How can radioactive isotopes be used to label molecules?
4. In the Hershey-Chase experiment, which phage molecule entered the bacteria? How did Hershey and Chase identify this molecule?
5. How did Rosalind Franklin contribute to the understanding of the structure of DNA?
6. Describe the composition of a nucleotide and name the four different nucleotides in DNA.
7. Identify three ways in which RNA differs from DNA.
8. Identify the three main forms of RNA.
9. Where in the cell does translation take place?
10. What is a codon?
11. How does tRNA recognize a codon in mRNA?
12. When does *E. coli* express the *lac* genes?
13. What is an intron? An exon?

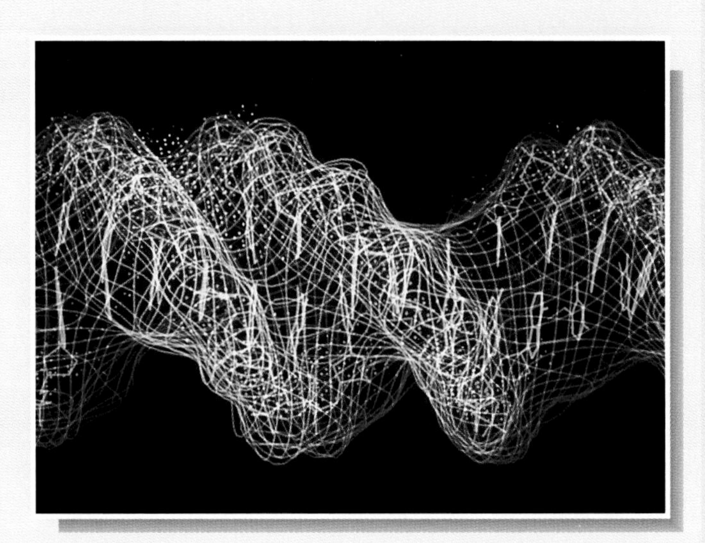

Expanding the Concepts

Discuss each of the following in a brief paragraph.

1. Describe Avery's experiments. Why did his results suggest that DNA was the molecule of transformation?
2. Explain how Hershey and Chase showed that DNA is the molecule that contains a bacteriophage's genetic information.
3. How did Watson and Crick develop the double-helix model of DNA? Describe this model.
4. Is DNA the only molecule that chromosomes contain? Explain your answer.
5. Compare transcription with DNA replication.
6. DNA contains four different nucleotides. How can just four nucleotides code for all the different proteins in a cell?
7. Describe the role of each form of RNA in protein synthesis.
8. Why are proteins important to an organism?
9. Explain how the promoter, repressor, and operator combine to regulate the expression of the *lac* genes in *E. coli.*
10. Describe possible reasons why cells produce mRNA with introns and exons.

Extending Your Thinking

Use the skills you have developed in this chapter to answer the following.

1. **Identifying assumptions** A biology student identifies the nucleotide sequence AAC in a strand of mRNA. In the genetic code, AAC codes for asparagine. Will asparagine necessarily appear in the polypeptide? Explain your answer.
2. **Interpreting data** A polypeptide contains the amino acid sequence of histidine–isoleucine–valine–glycine. Using the genetic code and the rules of base-pairing, identify a sequence of nucleotides in DNA that code for this amino acid sequence. Is more than one such sequence possible in DNA?
3. **Observing** Look at the X-ray diffraction pattern shown in ***Figure 8–6*** on page 176 and describe what you observe. What evidence can you see of the double-helix structure of DNA in this pattern?
4. **Constructing models** Using construction paper, pipe cleaners, straws, or other materials available in your classroom, construct models of mRNA, tRNA, and amino acids. Use your models to explain protein synthesis.
5. **Interpreting diagrams** The table below lists the percentages of the four nucleotides in the DNA of different organisms. What conclusions can be drawn from this table? How does the structure of DNA explain these conclusions?

NUCLEOTIDE PERCENTAGES

Source of DNA	A	T	G	C
Streptococcus	29.8	31.6	20.5	18.0
Yeast	31.3	32.9	18.7	17.1
Herring	27.8	27.5	22.2	22.6
Human	30.9	29.4	19.9	19.8

Applying Your Skills

DNA Replication

DNA replication relies on the principles of base-pairing—adenine pairs with thymine, and cytosine pairs with guanine.

1. Construct a model of the following double-stranded sequence of DNA:

   AAGCTTATGG
   | | | | | | | | | |
   TTCGAATACC

   Model the nucleotides with colored paper clips, straws, pipe cleaners, or any other materials available.
2. Separate the strands of your model. Using ***Figure 8–8*** on page 179 as a guide, construct models of the new strands that DNA replication would produce.
3. Describe the products of DNA replication.

• GOING FURTHER •

4. If DNA were single-stranded instead of double-stranded, could it be replicated by the procedure you just modeled? Explain your answer.
5. Using your model as a visual aid, present a report to your class on DNA replication.

CHAPTER 9

Genetic Engineering

FOCUSING THE CHAPTER THEME: Systems and Interactions

9–1 Breeding New Organisms
- **Describe** how selective breeding and mutations can produce new organisms.

9–2 Manipulating DNA
- **Explain** how recombinant DNA is produced.

9–3 Engineering New Organisms
- **Describe** the achievements of genetic engineering.

BRANCHING OUT *In Action*

9–4 The New Human Genetics
- **Identify** RFLPs and discuss the Human Genome Project.

LABORATORY INVESTIGATION
- **Design an experiment** to determine the effects of radiation on seeds.

Biology and Your World

BIO JOURNAL

The researcher in this photograph is performing gel electrophoresis—one step toward identifying the nucleotide sequence in a strand of DNA. In your journal, explain why this information might be useful.

A laboratory technician performing gel electrophoresis on DNA

Breeding New Organisms

SECTION 9–1

GUIDE FOR READING

- **Describe** the process of selective breeding.
- **Define** a mutation.

MINI LAB

- **Construct a model** to show the effects of a point mutation.

WHEN THE EUROPEANS FIRST settled the beautiful, rugged land that is now the northwestern United States, they noticed that the Palouse and Nez Percé owned an exceptional breed of spotted horses. The horses were strong, durable, and gentle, as well as unusually colorful and attractive. The English-speaking settlers named these animals palouse horses. Over time, the name palouse was slurred first to palousey and, eventually, to appaloosa—the name by which the breed is known today.

Breeding the Appaloosa

Spaniards introduced horses to North America in the early 1500s. Over a span of 200 years, both the Palouse and Nez Percé had acquired horses and learned how to train and breed them. ● From Spanish horses of mixed color and temperament, the two tribes produced the appaloosa—the first recognized breed of horses in North America and still a favorite among horse owners today.

How did the tribes accomplish this feat? First, they realized that the Spanish horses showed a great deal of variability. Some of the horses were fast, and others were slow. Many horses bore plain colors, but a few had bright, attractive colors.

Then, as people of many cultures have done, the Palouse and Nez Percé set about to breed the kind of animal they wanted. Over the years, colorful horses must have been mated with horses that were very easy to handle. Naturally, the goal was to produce offspring that would have the best characteristics of both parents.

INTEGRATING SOCIAL STUDIES

What was the relationship between the Spanish explorers and the Native Americans they encountered?

Figure 9–1
(a) The Palouse and Nez Percé well understood the principles of selective breeding, as demonstrated by how quickly they bred the appaloosa horse. (b) Appaloosa horses are prized for their beauty, strength, and gentleness.

Figure 9–2
Perhaps the greatest selective breeder of all time was Luther Burbank, shown here holding a tuber that he developed. Burbank developed more than 800 varieties of plants, including many important crops.

Selective Breeding

The breeding of the appaloosa is an example of **selective breeding**—one of the oldest ways in which humans have produced organisms that best suit their purposes. **In selective breeding, individual organisms with desired characteristics are chosen to produce a new generation.** Selective breeding has produced nearly all domestic animals—including cats, dogs, and horses—as well as most crop plants.

☑ ***Checkpoint*** How was the appaloosa produced?

Hybridization

A cross between dissimilar individuals is known as **hybridization.** Hybridization is one of the breeder's most important tools because it provides a way to combine the best characteristics of two organisms. For example, hybridization could be used to combine the disease resistance of one plant with the food-producing capacity of another.

Inbreeding

When a breed of organisms has been established, the breeder must decide how to maintain it. Sometimes this is done by **inbreeding**—the continued breeding of closely related individuals. The many breeds of dogs—from beagles to poodles to schnauzers—are each maintained by inbreeding.

Inbreeding helps to ensure that the characteristics that make each breed special will be preserved. However, inbreeding also carries with it a special set of problems. Because most members of a breed are genetically similar, crossing individuals of the same breed may bring together two recessive alleles for a genetic defect. Excessive inbreeding has caused several problems in many dog breeds, including blindness and joint deformities in both shepherds and retrievers.

☑ ***Checkpoint*** What is inbreeding?

Mutations

Selective breeding is a powerful tool for producing new organisms, but it depends entirely on the variation that already exists in a species. Where does that variation come from? And is there a way to increase the variation in a species? The answers to both questions involve a molecule that you have already studied—DNA.

DNA codes for an organism's genetic characteristics, and it is found in every cell of the organism. As you have read earlier, a cell copies its DNA before it divides. But although the copying procedure usually works perfectly, every now and then it produces a mistake. These mistakes are called **mutations**—from a Latin word meaning "change." **A mutation is an inheritable change in genetic information.** Mutations may occur in any cell and in any gene.

Chromosomal Mutations

Changes involving the number or structure of a cell's chromosomes are called **chromosomal mutations.** As shown in ***Figure 9–3,*** chromosomal mutations may move a gene's location on a chromosome or change the number of copies of a gene.

Some chromosomal mutations take place during meiosis. When a whole set of chromosomes fails to separate during meiosis, gametes with extra sets of chromosomes can be produced. These gametes may produce triploid (3n) or even tetraploid (4n) organisms. An organism with 3 or more sets of chromosomes in its cells is said to be polyploid.

In animals, polyploidy is usually fatal. However, for unknown reasons, plants tolerate polyploidy much better than animals do. In fact, it is a healthy condition in many plants—often making them larger and stronger than diploid (2n) plants. Crops such as bananas, wheat, and citrus fruits are each produced from polyploid plants.

Gene Mutations

While a change that affects an entire chromosome is called a chromosomal mutation, a change that affects only an individual gene is known as a **gene mutation.** A gene mutation may involve only one nucleotide or many nucleotides. But no matter how large or how small the change, a gene mutation can have dramatic consequences for an organism.

A gene mutation that involves a single nucleotide—such as the substitution of one nucleotide for

Figure 9–3 *Chromosomal mutations include the four types shown here, as well as mutations that involve whole sets of chromosomes. Gene mutations affect only a single gene. Notice that a nucleotide substitution produces fewer changes than a nucleotide deletion.*

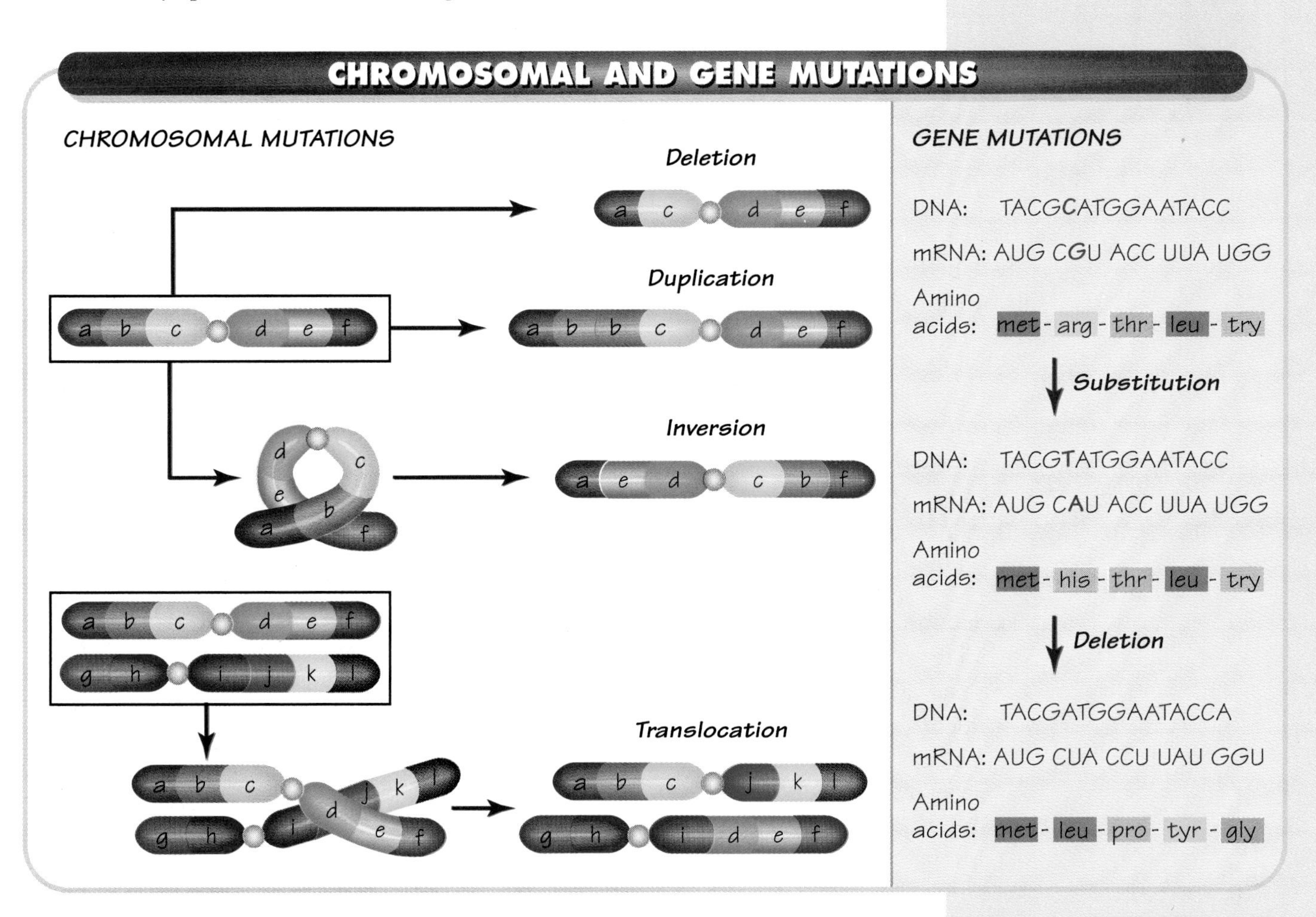

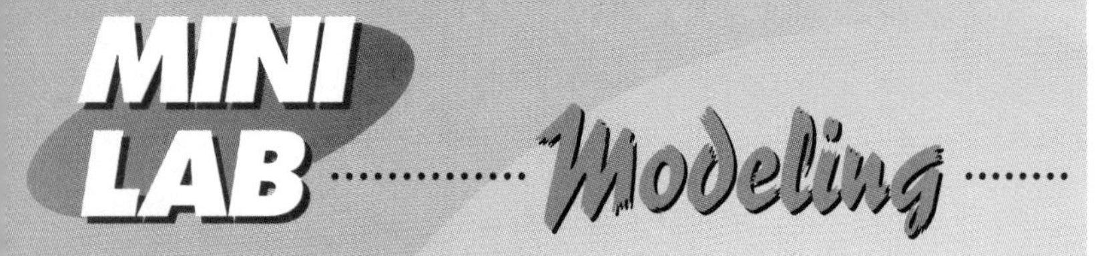

Mind Your Mutations!

PROBLEM *How can you **construct a model** of the effects of a point mutation?*

PROCEDURE

TACTGTCTCACCATT

1. Assume that the letters above represent the nucleotide sequence of a gene. Using the rules of base-pairing and the genetic code, determine the polypeptide for which this gene codes.
2. Design a procedure to show how the polypeptide is affected by a random point mutation in the gene.
3. Repeat the procedure at least five times. Include one substitution, one insertion, and one deletion mutation.

ANALYZE AND CONCLUDE

1. Which mutation changed the polypeptide the most? The least? Discuss the significance of your answers.
2. Can a gene mutation not affect a polypeptide? Explain your answer.

another—is called a **point mutation.** As shown in ***Figure 9–3*** on page 199, the substitution of a nucleotide can change one of the amino acids for which the gene codes.

However, much bigger changes can result from another kind of point mutation—the insertion or deletion of a nucleotide. Why is this so? Recall that the genetic code is read in groups of three nucleotides known as codons. When a nucleotide is inserted or deleted, the grouping of the nucleotides is shifted. This type of mutation is known as a **frameshift mutation**—so called because it shifts the "reading frame" of the genetic message. Frameshift mutations can cause tremendous changes to a polypeptide.

☑ ***Checkpoint*** What is a point mutation? A frameshift mutation?

Mutations and Breeding

Most mutations are harmful. However, mutations occasionally produce very desirable characteristics. For this reason, breeders sometimes try to increase the rates at which mutations occur, thus increasing the chances of producing a beneficial one. X-rays, ultraviolet light, and certain chemicals can each increase an organism's mutation rate.

Another way to increase the chances of obtaining beneficial mutations is to work with large numbers of organisms. Bacteria can be grown in the millions, allowing researchers to study even the rarest mutations in bacterial genes.

Section Review 9–1

1. **Describe** the process of selective breeding.
2. **Define** a mutation.
3. **Explain** why breeders may increase an organism's mutation rate.
4. **Critical Thinking—Comparing** Compare a mutation that deletes a sequence of three nucleotides from a strand of DNA with a mutation that deletes only one nucleotide.
5. **MINI LAB** How can you **construct a model** to show the effects of a point mutation?

Manipulating DNA

GUIDE FOR READING

- **Explain** how biologists edit a DNA molecule.
- **Define** cell transformation.

MINI LAB

- **Construct a model** of restriction enzymes.

SELECTIVE BREEDING IS A SIMPLE way to influence which genes are passed on to new individuals. But what would the world be like if farmers and ranchers didn't have to wait for generations of selective breeding to produce the combinations of genes they seek? What if genes could be cut from one organism, then spliced into another? What if scientists could read the genetic message, then rewrite it any way they chose?

In many ways, we do not need to imagine any of these things as science fiction. It is possible to do every one of them right now.

The Tools of Editing

When a movie or television show is put together, film editors usually combine pieces from several videotapes in order to assemble different scenes for their proper impact. What tools do film editors need to accomplish this task? First, they need a way to view the videotape—to translate the encoded information into pictures or sounds. Second, they need a way to cut out and separate different sections of the videotape. And third, they need a way to splice the different sections together.

In a sense, a chromosome is like a reel of videotape—both contain long stretches of information in a coded form. And not surprisingly, editing a piece of DNA requires tools that work in a similar way to the tools of film editing.

Are tools actually available for editing DNA? Yes, they are! **Biologists have tools to cut, separate, and read DNA sequences and to splice together these sequences in almost any order.** Let's take a look at each of the tools that biologists use to edit DNA.

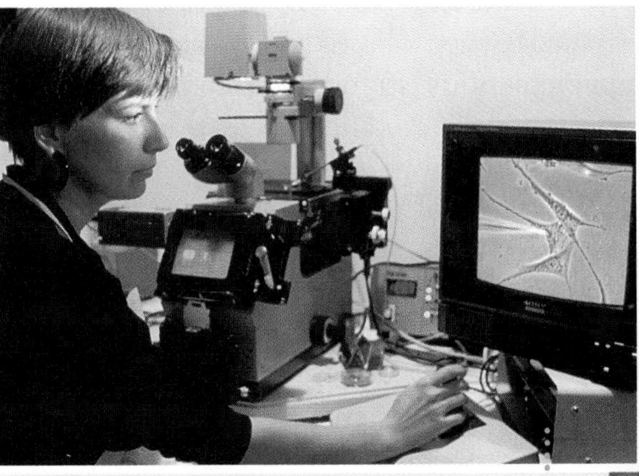

Figure 9–4
(a) *With the aid of a microscope and a television monitor, a tiny needle can be used to inject DNA into an animal cell.*
(b) CAREER TRACK *This genetic engineering research assistant is interpreting the results of a DNA sequencing procedure.*

So Many Restrictions!

PROBLEM *How can you **construct a model** of the action of restriction enzymes?*

PROCEDURE

1. Write a 100-character sequence of the letters A, C, G, and T in a random order.
2. Copy the sequence onto each of three different strips of paper. Use different-colored pencils or paper to identify each copy.
3. Cut the strips to model the action of the three restriction enzymes shown in ***Figure 9–5.*** Use a different strip for each enzyme.

ANALYZE AND CONCLUDE

1. Which restriction enzyme produced the most pieces? The fewest pieces?
2. Why do biologists use more than one restriction enzyme to cut DNA?

Tool #1—Cutting DNA

DNA can be cut at specific places by proteins known as **restriction enzymes.** More than 100 restriction enzymes are known, and each one cuts DNA at a specific sequence of nucleotides. As shown in ***Figure 9–5,*** restriction enzymes are amazingly precise. Like a key that fits only one lock, a restriction enzyme will cut a DNA sequence only if it matches the sequence perfectly.

Restriction enzymes make it possible to cut enormous DNA molecules into smaller, precisely sized fragments. Biologists can then work on pieces of DNA that contain a few hundred nucleotides, instead of many millions.

☑ ***Checkpoint*** What is a restriction enzyme?

Tool #2—Separating DNA

What good is a mixture of DNA fragments of various sizes? It's not much good at all—unless it is possible to separate them. Fortunately, DNA fragments can be separated by a technique known as electrophoresis (ee-lehk-troh-fuh-REE-sihs). In electrophoresis, DNA fragments separate as they move through a special gel—a watery solid with a consistency similar to dessert gelatin.

First, DNA fragments are placed at one end of the gel. Then the gel is placed in an electric field. Because DNA fragments carry negative charges, they move toward the positively charged electrode. This separates the fragments because the smaller fragments slip through the gel faster than the larger fragments.

Tool #3—Reading DNA

More than 20 years ago, researchers developed a way to read the sequence of small, single-stranded pieces of DNA—typically fewer than 200 nucleotides. First, the pieces are placed in test tubes with the enzyme DNA polymerase. The enzyme is then allowed to make a new complementary strand, occasionally using chemically modified nucleotides that halt the assembly of the new strand at certain places. The new strands are then separated from each other by electrophoresis, producing a pattern of bands such as the one shown in ***Figure 9–4*** on page 201. This pattern reveals the base sequence of the original strand.

DNA sequencing has become a routine laboratory procedure. It is so routine, in fact, that automated machines now carry out the sequencing reactions, and computers analyze the sequencing gels.

Tool #4—Splicing DNA

To a film editor, the term "splicing" means joining two sections of film by taping their ends together. Can

Figure 9–5
Restriction enzymes cut DNA only at specific sequences. In this diagram, the labeled nucleotides indicate how the restriction enzymes EcoR*1*, Bam*1*, *and* Hae3 *each cut DNA.*

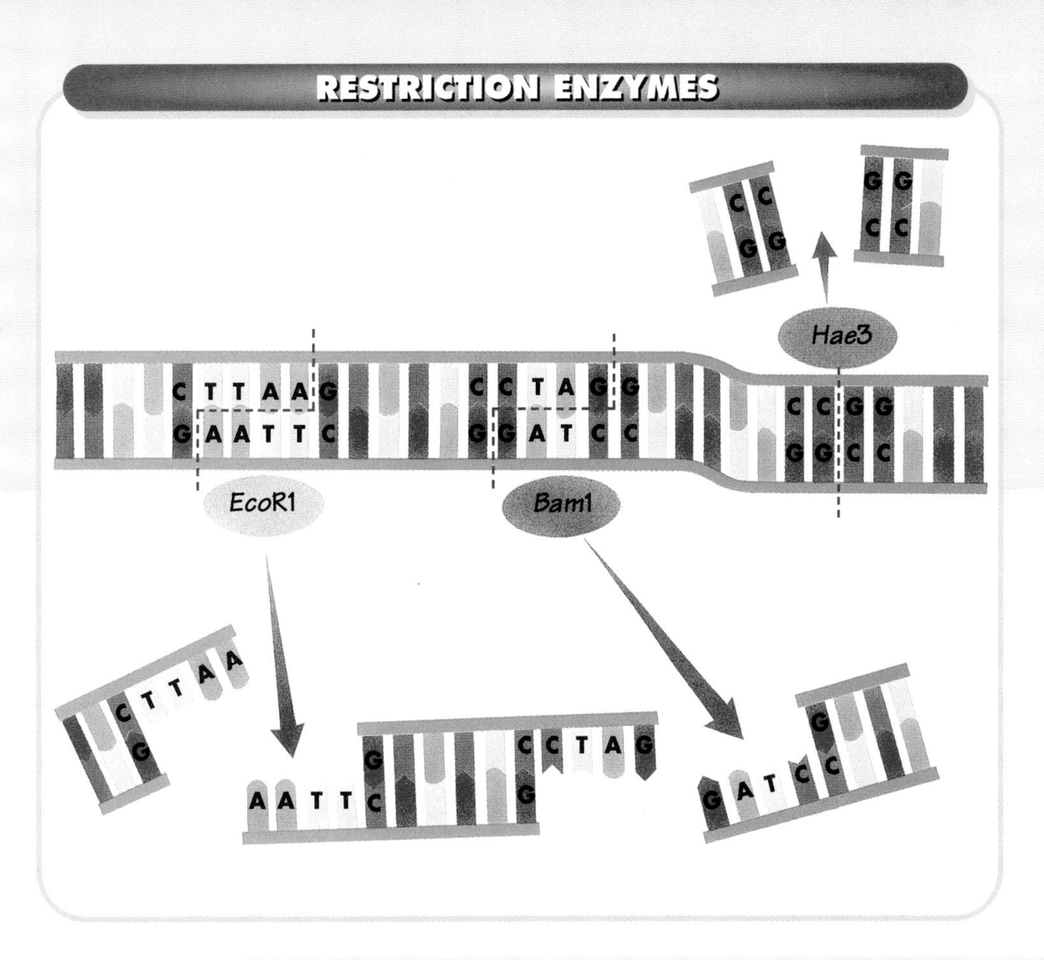

sections of DNA also be spliced? The answer is yes, they can!

When some restriction enzymes cut DNA, they leave a short, single-stranded region on each side of the cut, and these regions act like the sticky side of a piece of tape. So by mixing two DNA fragments cut with the same enzyme, the single-stranded regions hold the ends of the fragments together. Other enzymes can then be used to permanently join the fragments.

The joined pieces of DNA act like a single DNA molecule. This kind of DNA is known as **recombinant DNA** because it is made by combining DNA from two different sources.

Why would anyone want to join together two pieces of DNA? Doing so makes it possible to design DNA

Figure 9–6
(a) Plasmids are circular pieces of DNA found in bacteria (magnification: 160,000X). (b) Plasmids can be isolated, cut, and combined with DNA from other organisms. Any DNA made from two or more sources is called recombinant DNA.

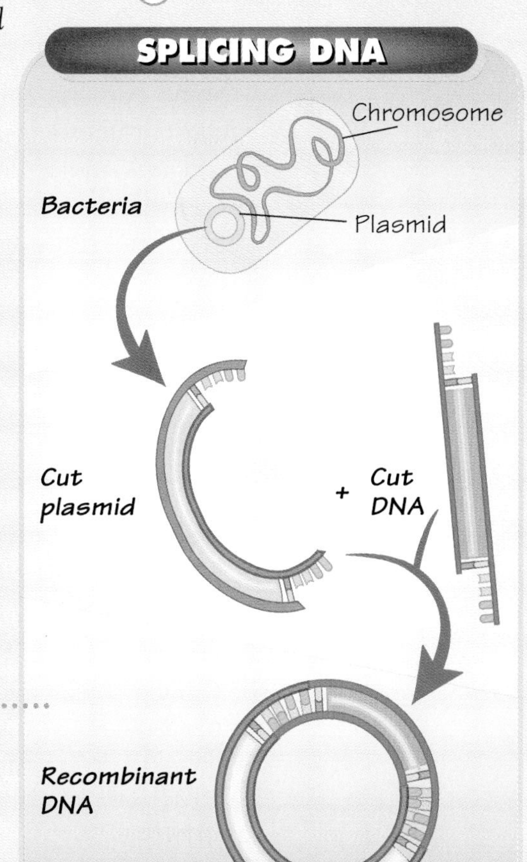

Figure 9–7
This tree tumor, called a crown gall, is caused by bacteria that insert plasmids into the tree's cells. In laboratories, researchers are using a similar process to transform a variety of plant cells.

molecules for all sorts of special purposes. Two genes could be fused into one, the control region of one gene could be placed next to another, and several genes from different sources could be assembled into the same DNA molecule.

Checkpoint What is recombinant DNA?

Cell Transformation

After a recombinant DNA molecule is assembled, the next step is **cell transformation**—putting the recombinant DNA inside a live cell. **In cell transformation, new genes are inserted into a cell, thus changing the cell's genetic makeup.**

Recombinant DNA can be used to transform cells in many ways. Let's take a look at some of them.

Transforming Bacteria

Some bacteria, in addition to their regular chromosomes, contain small circular DNA molecules called **plasmids.** As you will see, plasmids can be used as agents of cell transformation.

First, pieces of DNA are joined to isolated plasmid molecules. These recombinant plasmids are mixed with a culture of bacteria that do not contain plasmids. Under the right conditions, a few of the bacteria in the culture will take up the plasmids.

The transformed bacteria can be isolated, then used to grow millions of transformed bacteria. Each of the bacteria inherits both the plasmid and the recombinant DNA.

Checkpoint What are plasmids?

Transforming Eukaryotes

Biologists routinely transform bacteria, which are prokaryotes. Eukaryotes are more complex than prokaryotes, however, and it is significantly harder to make their cells accept new DNA molecules. But the task is not impossible. Yeasts, which are single-celled eukaryotes, have plasmids of their own. Yeast plasmids are now used to transform yeast cells in a way that is very similar to bacterial transformation.

In addition, biologists have developed ways to transform animal and plant cells, which do not have plasmids. One way is to use a tiny needle to inject new DNA into a cell, as shown in ***Figure 9–4*** on page 201. Sometimes the cell breaks down and destroys the new pieces of DNA, but at other times the cell takes them into the nucleus and inserts them into chromosomes. The result is a transformed cell.

Section Review 9–2

1. **Explain** how biologists edit a DNA molecule.
2. **Define** cell transformation.
3. **Critical Thinking—Analyzing** Almost every organism has DNA that is made of the same four nucleotides and translated by the same genetic code. Explain why this fact is significant in cell transformation.
4. **MINI LAB** How can you **construct a model** of the effects of restriction enzymes?

Engineering New Organisms

GUIDE FOR READING

- **Describe** the achievements of genetic engineering.

BY MANIPULATING DNA, ARE biologists now able to create a griffin—a mythical creature that is half eagle and half lion? For many reasons, the answer is no. However, biologists have been able to produce some fascinating organisms that may be just as remarkable as any mythical one.

Manipulating Genes

The cutting and splicing of genes and DNA from different sources is sometimes called **genetic engineering**—one of the newest branches of the biological sciences. **Through genetic engineering, researchers are able to insert new genes into almost any organism.**

Organisms that have been transformed with genes from other organisms are said to be **transgenic.** As you have read, plasmids can be used to produce transgenic bacteria and transgenic yeasts. And if a complete organism is grown from a transformed animal or plant cell, the result is a transgenic animal or a transgenic plant.

What good will come from genetic engineering? How will it affect your life? These are important questions, and we will present some answers in this section.

Genetically Engineered Proteins

If a gene is properly inserted into a yeast or a bacterium, the yeast or bacterium is "tricked" into producing large amounts of the protein for which the gene codes. For example, transgenic bacteria are now producing a human protein called insulin. People with a disease called diabetes do not make enough insulin of their own, and the insulin produced by the bacteria is both inexpensive and effective.

INTEGRATING HEALTH

What are the different forms of diabetes? How are they treated?

In fact, transgenic cells are used to make proteins for treating cancer, heart attacks, and other diseases and disorders. Researchers are also trying to create transgenic bacteria to produce vaccines.

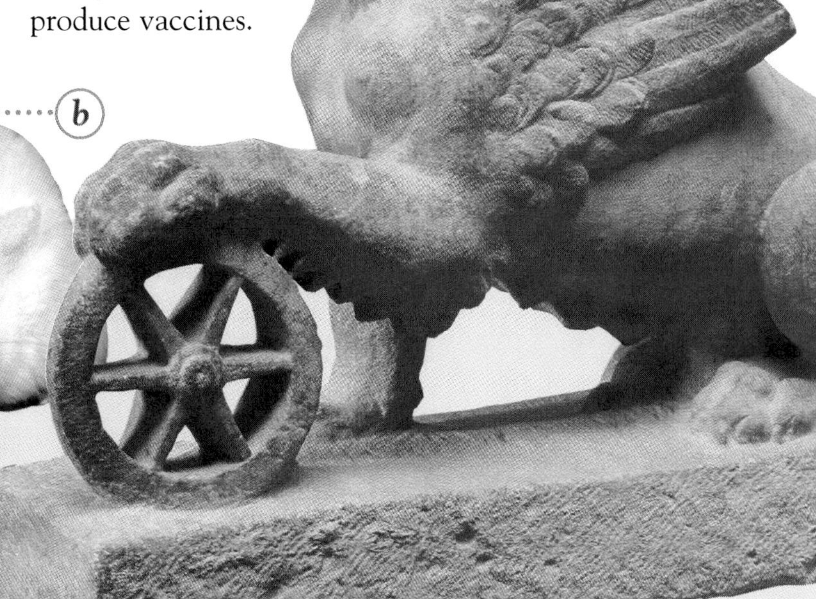

Figure 9–8
(a) This griffin is a combination of an eagle and a lion. While genetic engineering cannot produce creatures from mythology, it has created (b) a giant mouse, shown to the right of a mouse of normal size.

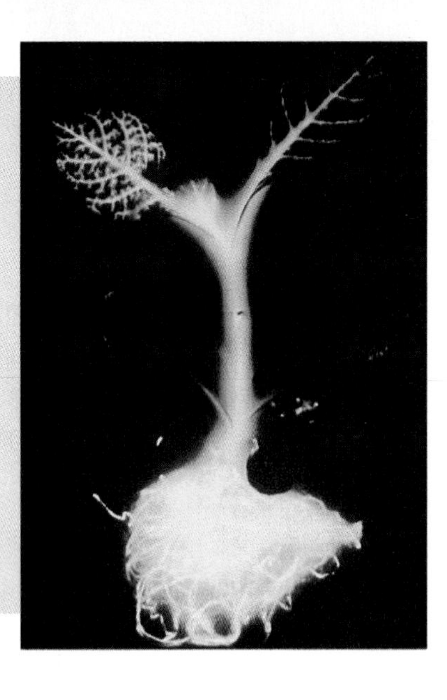

Figure 9–9
By transferring a gene from a firefly to a tobacco plant, researchers created this transgenic tobacco plant that glows in the dark! The gene codes for the enzyme luciferase, the catalyst for a reaction that produces light.

Transgenic Plants and Animals

Researchers have found ways to stimulate transformed plant cells to grow into transgenic plants. Their results include a tomato that stays fresher longer than other tomatoes, a strain of soybeans that is highly resistant to weed-killing chemicals, and the unusual tobacco plant shown in ***Figure 9–9.*** Currently, work is underway to produce transgenic plants that resist insect pests, make their own fertilizer, or produce more food than ordinary plants do.

To produce a transgenic animal, researchers first transform an animal's egg cell, then allow the egg to be fertilized and to mature. In the future, this procedure may be used to create all sorts of useful animals. Transgenic cattle, for example, could be created with a gene that increases milk production.

Checkpoint How can a transgenic animal be produced?

Human Gene Therapy 

In the laboratory, human cells can be transformed just like other animal cells. However, transforming cells in a living human has not proved to be as easy a task.

For example, consider the latest efforts in treating cystic fibrosis (CF)—a serious genetic disease that is inherited as an autosomal recessive trait. In one experimental treatment, researchers transformed viruses with healthy alleles for the CF gene. Then they sprayed the viruses into the air passageways of CF patients. Sure enough, the viruses carried the healthy allele into the patients' cells, and their symptoms improved! Unfortunately, when the transformed cells were replaced by other cells, the symptoms of cystic fibrosis reappeared.

The experiments in gene therapy raise many important questions. Should researchers carry out experiments that permanently change human DNA? Should we use genetic engineering only to cure certain disorders? Or should we also try to make people taller, stronger, or more disease resistant?

Genetic engineering will affect everyone, and you should expect to play a role in forming public policy about it. We urge you to learn as much as you can about genetic engineering, then to use both your knowledge and your ethical and moral training to address the important questions that will be raised.

Section Review 9–3

1. **Describe** the achievements of genetic engineering.
2. What are some of the ethical questions that genetic engineering raises? **Discuss** these questions.
3. **Critical Thinking—Inferring** To produce a transgenic animal, scientists typically transform an unfertilized egg cell. Why is the cell transformation done at such an early stage of an animal's life?

BRANCHING OUT **In Action**

The New Human Genetics

SECTION **9–4**

GUIDE FOR READING

- **Define** RFLPs and **explain** why biologists study them.
- **Describe** the purpose of the Human Genome Project.

A NEW KIND OF GENETICS IS taking shape. In the old genetics, researchers could only observe the effects of genes. They recorded plant size, flower color, blood type—anything that provided clues about the genes that were hidden and invisible within the cell.

The new genetics, however, is not restricted to indirect clues. Rather than looking for the effects of genes, biologists now can look at the genes themselves. As you will see, the new genetics is already affecting everyday life, and it will become even more important in the years ahead.

"RIHF-lihps"

How can a specific nucleotide sequence be isolated from a huge DNA molecule? One way is to cut the DNA into small pieces and separate the pieces on a gel. Then, a specific sequence can be located with a probe—often, a chemically treated sequence of nucleotides. The probe binds to complementary sequences in the DNA pieces, revealing the sequences as dark bands.

The bands identified in this procedure are sometimes called **RFLPs** (RIHF-lihps), which stands for Restriction Fragment Length Polymorphism. Like a gene, a RFLP is inherited, and different individuals have different sets of RFLPs in their DNA. **By using probes to identify RFLPs, biologists can identify and classify an individual's DNA.** Let's take a look at some applications of this technique.

☑ *Checkpoint* What is a RFLP?

Probing for Genetic Diseases

Consider people at risk for Huntington disease—a serious genetic disease caused by an autosomal dominant allele. Huntington disease slowly damages the nervous system, eventually leading to death. However, the disease's symptoms do not appear until late middle age.

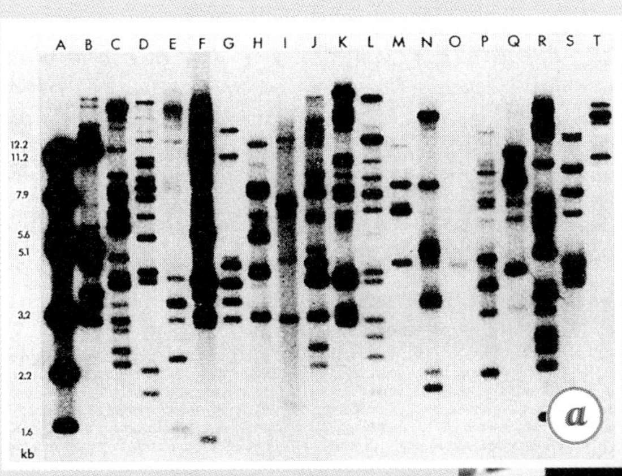

Figure 9–10
Researchers probe for RFLPs to distinguish the DNA of individuals of the same species. (a) *This photograph shows how one probe bound to the DNA of 20 different* Xanthomonas *bacteria, labeled A through T. The dark bands indicate the RFLPs to which the probe attached.* (b) *Each of these young women has a different set of RFLPs in her DNA.*

Biology AND YOU .... Connections

Patenting Life

When inventors produce a new machine or a new way of doing something, they can protect their inventions with a patent. A patent provides the exclusive rights to build and sell an invention. Thomas Edison, for example, held a patent on the phonograph for many years.

Should it be possible to patent living things? At the beginning of the twentieth century, it was not. Therefore, when a plant breeder produced a new high-yielding strain of a crop plant, anyone who acquired a few seeds could produce that plant. As a result, breeders had little financial incentive to develop better plants.

Patent laws now protect the developers of improved crop plants such as this high-yielding cotton.

Changes to the Patent Laws

In 1930, however, the United States government changed the patent laws so that some plants and other organisms could be protected. This law benefited many plant breeders, as well as the descendants of Luther Burbank, arguably the greatest plant breeder of all time. Burbank's estate was granted 17 patents on the varieties of plants he had produced.

Today, patents are granted not only for plants bred by ordinary means, but also for those created by genetic engineering. Without patents, genetic researchers might not have developed these plants—many of which are potentially very beneficial.

For example, one company has engineered a strain of soybeans that is resistant to weed-killing chemicals. Other companies have developed strains of cotton plants and potato plants that actually produce a bug-killing compound. And one laboratory has produced plants that absorb mercury and other harmful compounds. These plants may prove useful in clearing pollutants from contaminated soil.

Patent Laws Today

Under the patent laws of 1930, any genetically engineered organism can be patented. Although these laws are more than half a century old, they are helping to protect the rights of genetic researchers and are spurring the growth of whole new industries.

However, no one in 1930 could have foreseen what is happening in biology today. Indeed, biotechnology companies are testing the limits of the old laws every day. Some companies have even filed for patents on parts of the human DNA sequence!

The Future

What can you expect in the future? As biologists continue to develop the techniques of genetic engineering, laws will be needed to deal specifically with the kinds of organisms that researchers will be creating. We hope that you—the decision makers of the future—will find the best ways to prosper from this new technology.

Making the Connection

Does your state have laws to regulate genetic engineering? What advantages and disadvantages do disease-resistant crops offer?

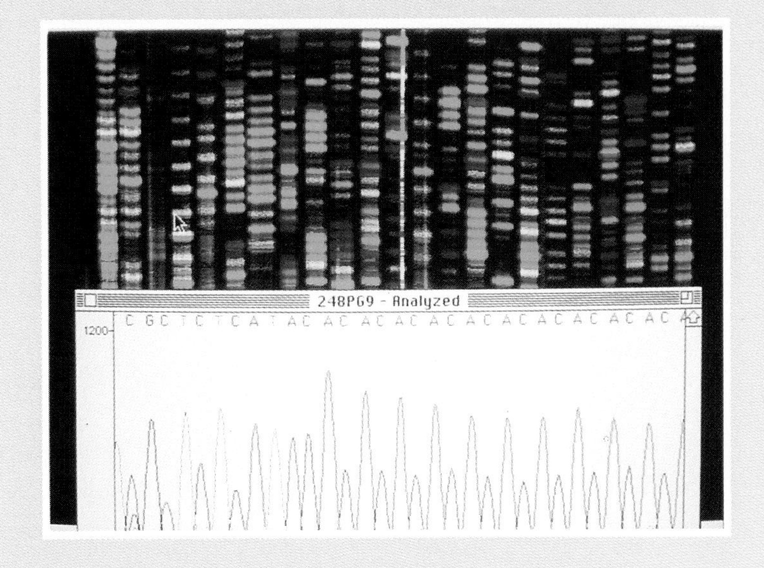

Figure 9–11
Computers are routinely used to determine the sequence of short pieces of DNA. However, even with the aid of computers and other tools, the Human Genome Project is an extraordinarily complex undertaking.

Before genetic testing, the children of people who had Huntington disease had to wait many years before learning whether they, too, would suffer from this condition. Today, however, researchers have identified RFLPs that are inherited with the allele that causes Huntington disease. Thus, by probing DNA for these RFLPs, geneticists can determine whether a person carries this allele.

Similar tests have been developed for other genetic disorders, including cystic fibrosis, sickle cell anemia, and Tay-Sachs disease. These tests are particularly useful for couples deciding whether they should have children.

DNA Fingerprinting

Some probes detect RFLPs that are highly variable from one individual to another. Researchers quickly realized that identifying these RFLPs could make it possible to distinguish the DNA of one individual from another. Today, using RFLPs to identify individuals is popularly known as **DNA fingerprinting.**

DNA fingerprinting is used for all sorts of purposes, including criminal investigations. DNA fingerprinting can identify suspects of serious crimes—as well as clear people of crimes that they did not commit.

The Human Genome Project

A human cell contains 46 chromosomes, with 3 billion pairs of DNA nucleotides. Such a tremendous amount of information certainly seems overwhelming. **However, a number of years ago, scientists throughout the world decided to identify the complete nucleotide sequence in human DNA. This project is now known as the Human Genome Project.** Its goal is huge, but one that many scientists feel is worth achieving.

What good will come from the Human Genome Project? Information from the project has already identified a host of genes associated with genetic disorders. It has provided clues to the genetic basis of cancer and heart disease, and it may eventually give important clues to that most interesting of all questions—what is there about our genes that truly makes us who we are?

Section Review 9–4

1. **Define** the term RFLP. **Explain** why biologists study RFLPs in DNA.
2. **Describe** the purpose of the Human Genome Project.
3. **BRANCHING OUT ACTIVITY** Research the progress of the Human Genome Project. **Discuss** the findings of your research with your class.

Laboratory Investigation

DESIGNING AN EXPERIMENT

The Effects of Radiation on Seeds

Mutations occur naturally in all organisms. However, an organism's mutation rate increases when it is exposed to certain chemicals or types of radiation. In this investigation, you will design an experiment to show the effects of X-ray exposure on seeds.

Problem

Do irradiated seeds grow differently from non-irradiated seeds? **Design an experiment** to answer this question.

Suggested Materials

irradiated seeds and nonirradiated seeds from the same organism
2 Petri dishes
paper towels
pots with soil
glass-marking pencil

Suggested Procedure

1. Design an experiment to determine whether irradiated seeds grow differently from non-irradiated seeds. Include two test groups in your experiment, with one test group as a control. Make sure that the test groups are treated identically.

2. To grow the seeds, follow any special directions on the seeds' packages. Otherwise, place the seeds in Petri dishes and cover the seeds with water, as shown in the diagram.

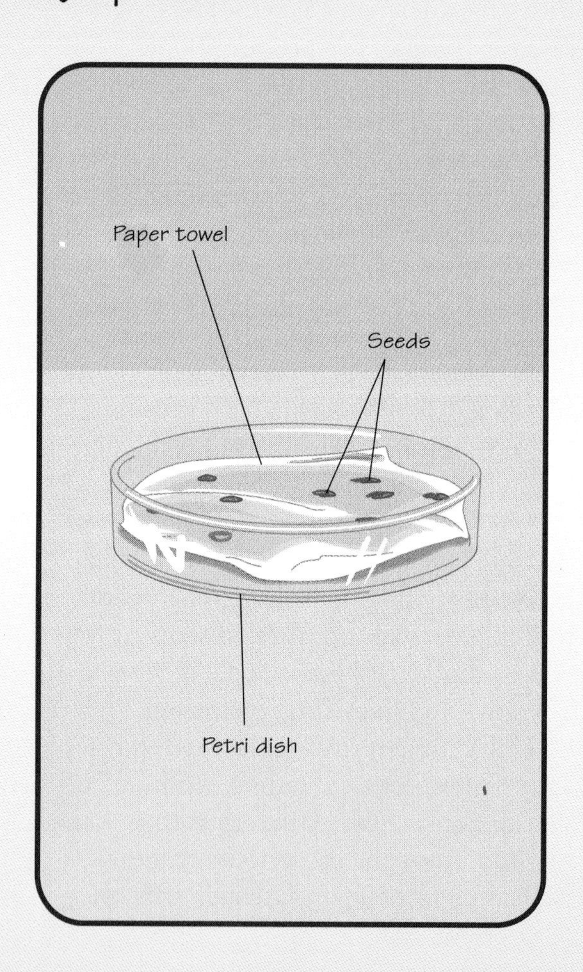

3. Use the glass-marking pencil to label each Petri dish. Every Petri dish that your class uses in this experiment should have a unique label. Store the Petri dishes in a location designated by your teacher.

4. Observe the growth of the seeds over the course of 2 weeks. Record your observations in a data table similar to the one shown. Add water to the Petri dishes as necessary to keep the seeds moist.
5. After 2 weeks, carefully transfer five young plants in each test group from the Petri dishes to pots containing soil.
6. Label the pots and place them in sunlight or under a fluorescent lamp. Water the plants regularly. Observe their growth for the next few weeks. Record your observations in a data table similar to the one shown.

DATA TABLE

Day	Nonirradiated Seeds	Irradiated Seeds
1		
2		
3		
4		

Observations

1. Describe the roots produced by the seeds. In what ways are they similar? Different?
2. Describe any leaves that the young plants produced. In what ways are the leaves similar in the two groups? In what ways are they different?
3. Note any other similarities or differences among the plants that the seeds produced.

Analysis and Conclusions

1. Compare how the irradiated seeds and the nonirradiated seeds grew into plants.
2. Did the irradiated seeds contain mutant DNA? How certain can you be of your answer? Explain.
3. How could you best determine whether a seed contains mutant DNA?
4. Formulate a hypothesis to explain the results of your experiment. Could a different hypothesis also explain the results? Discuss this possibility.

Study Guide

Summarizing Key Concepts

The key concepts in each section of this chapter are listed below to help you review the chapter content. Make sure you understand each concept and its relationship to other concepts and to the theme of this chapter.

9–1 Breeding New Organisms

- In selective breeding, individual organisms with desired characteristics are chosen to produce a new generation.
- A mutation is an inheritable change in genetic information. A chromosomal mutation involves a change in the number or structure of chromosomes. A gene mutation affects only an individual gene.
- Breeders may stimulate mutation rates to increase the chance of finding a beneficial mutation.

9–2 Manipulating DNA

- Biologists have tools to cut, separate, and identify DNA sequences, as well as to splice together these sequences in almost any order.
- In cell transformation, new genes are inserted into a cell, thus changing the cell's genetic makeup. Small DNA molecules called plasmids can be used to transform bacteria and other organisms.

9–3 Engineering New Organisms

- Through genetic engineering, researchers are able to insert new genes into almost any organism.
- An organism with new genes is said to be transgenic. Scientists have produced many transgenic organisms, and they are working to put new genes in human cells to cure genetic disorders.

9–4 The New Human Genetics

- By using probes to identify RFLPs, biologists can identify and classify an individual's DNA.
- The goal of the Human Genome Project is to identify the complete nucleotide sequence of human DNA.

Reviewing Key Terms

Review the following vocabulary terms and their meaning. Then use each term in a complete sentence.

9–1 Breeding New Organisms

selective breeding
hybridization
inbreeding
mutation
chromosomal mutation
gene mutation
point mutation
frameshift mutation

9–2 Manipulating DNA

restriction enzyme
recombinant DNA
cell transformation
plasmid

9–3 Engineering New Organisms

genetic engineering
transgenic

9–4 The New Human Genetics

RFLP
DNA fingerprinting

Recalling Main Ideas

Choose the letter of the answer that best completes the statement or answers the question.

1. Selective breeding involves

a. inbreeding only.
b. hybridization only.
c. inbreeding and hybridization.
d. mutations only.

2. Polyploidy is an example of a

a. chromosomal mutation.
b. gene mutation.
c. point mutation.
d. frameshift mutation.

3. The substitution of a thymine (T) for an adenine (A) in a gene is an example of a

a. chromosomal mutation.
b. point mutation.
c. frameshift mutation.
d. RNA mutation.

4. To cut DNA at specific sequences, biologists use

a. restriction enzymes. **c.** DNA polymerase.
b. electrophoresis. **d.** recombinant DNA.

5. What is the name of DNA that is made from two different sources?

a. mutant DNA **c.** RFLPs
b. plasmid DNA **d.** recombinant DNA

6. Inserting new genes into a cell is a process called

a. DNA replication. **c.** cell transformation.
b. RNA transcription. **d.** gene therapy.

7. Transgenic bacteria that produce insulin were transformed with

a. insulin.
b. the virus that produces insulin.
c. the gene that produces insulin.
d. the mRNA that is translated into insulin.

8. An individual's DNA can be classified by identifying special sequences called

a. RNA. **c.** RFLPs.
b. restriction enzymes. **d.** polymerases.

Putting It All Together

Using the information on pages xxx to xxxi, complete the following concept map.

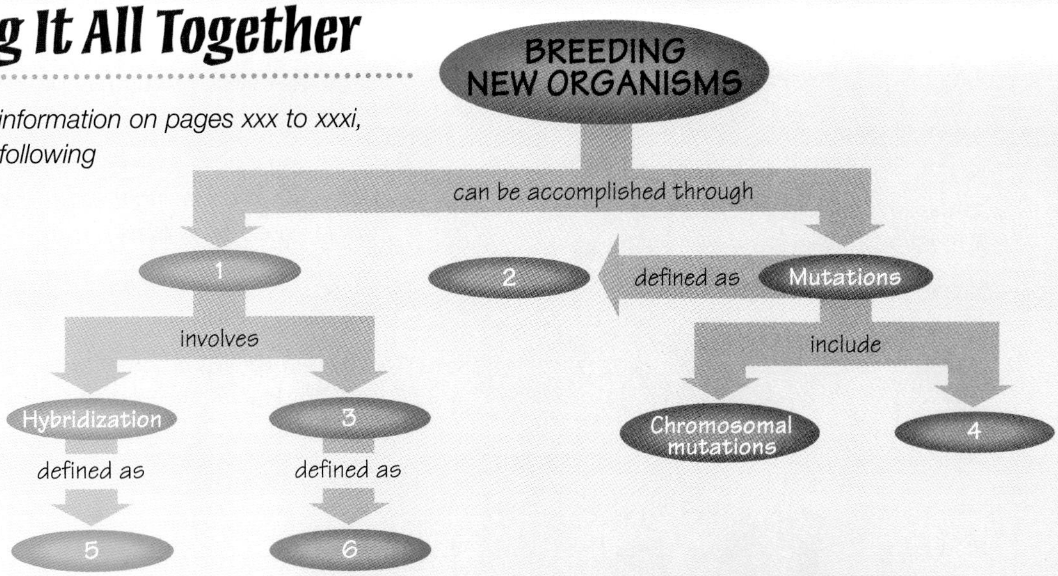

BREEDING NEW ORGANISMS
can be accomplished through
1
2
defined as
Mutations
involves
include
Hybridization
3
Chromosomal mutations
4
defined as
defined as
5
6

Assessment

Reviewing What You Learned

Answer each of the following in a complete sentence.

1. What is the purpose of selective breeding?
2. Compare hybridization and inbreeding.
3. What are mutations? Give an example of a chromosomal mutation and a gene mutation.
4. Describe a frameshift mutation.
5. For what purpose do biologists use restriction enzymes?
6. Why can two DNA fragments be spliced together easily?
7. What is recombinant DNA?
8. Define cell transformation.
9. Of bacterial cells, plant cells, and animal cells, which are easiest to transform?
10. Describe a plasmid.
11. Give three examples of transgenic organisms and their uses.
12. What is a RFLP? Why do biologists probe DNA for RFLPs?
13. Describe the Human Genome Project.

Expanding the Concepts

Discuss each of the following in a brief paragraph.

1. Why do breeders use both hybridization and inbreeding to maintain a breed?
2. Explain how some mutations can be silent, causing no changes in a gene's function.
3. Explain why a frameshift mutation can completely change the way a gene is interpreted.
4. Describe the different tools used to edit DNA.
5. How does electrophoresis separate DNA fragments of different sizes?
6. Discuss the role of plasmids in transforming bacterial cells.
7. Discuss two different ways in which biologists transform eukaryotic cells.
8. How can bacteria be "tricked" into producing large quantities of a human protein, such as insulin?
9. Why is replacing genes in human cells more difficult than replacing them in bacteria or other single-celled organisms?
10. Describe how radioactive probes are used to identify RFLPs in DNA.
11. Why can RFLPs be used to test a person's DNA for Huntington disease and other genetic diseases?
12. Explain the principles of DNA fingerprinting.

Extending Your Thinking

Use the skills you have developed in this chapter to answer the following.

1. **Analyzing arguments** Your friend proposes that by using the techniques of genetic engineering, biologists should be able to produce an organism with any combination of characteristics, such as the body of a frog and the wings of a bat. Do you agree or disagree with this proposal? Explain.
2. **Relating ideas** Compare the tools used in editing DNA with the tools used in editing videotape.
3. **Interpreting data** A certain gene codes for a protein made of 195 amino acids. After the gene undergoes a point mutation, it codes for a protein made of only 101 amino acids. Explain how this is possible.
4. **Predicting** Predict three ways in which transgenic organisms will be used in the future.
5. **Constructing a model** Using construction paper, pipe cleaners, straws, or other materials available in your classroom, construct a model of the electrophoresis of DNA fragments. Use your model to explain this process to your class.
6. **Designing an experiment** In a grassy field, you discover a patch of dandelions that are each roughly 2 centimeters longer than the other dandelions in the field. Design an experiment to determine whether this difference is caused by a mutation.

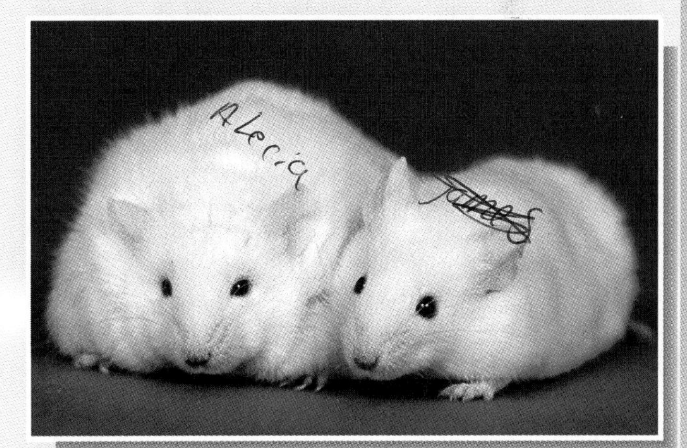

Applying Your Skills

The Debate About DNA

Should researchers combine DNA from any organisms of their choice? Should they try to transform human cells to cure genetic diseases? Should the government keep DNA records of its citizens? These questions—and questions like them—are sure to be debated in the years ahead, just as you will debate them now.

1. Select one of the above questions or formulate a similar question. Organize class members into two teams to debate the question. Assign each team one position to argue.
2. Research evidence that supports your team's position. Good sources of evidence include encyclopedias, magazines, textbooks, and scientific journals.
3. Conduct the debate, with your teacher as moderator.

• GOING FURTHER •

4. Create a policy statement that you feel offers a reasonable answer to the issue you debated.
5. Present your policy statement in a letter to your state legislator. Ask the legislator for his or her opinions on how genetic engineering and genetic testing should be regulated.

UNIT 3

Evolution

CHAPTERS

"I have called this principle, by which each slight variation, if useful, is preserved, by the term Natural Selection."

— Charles Darwin

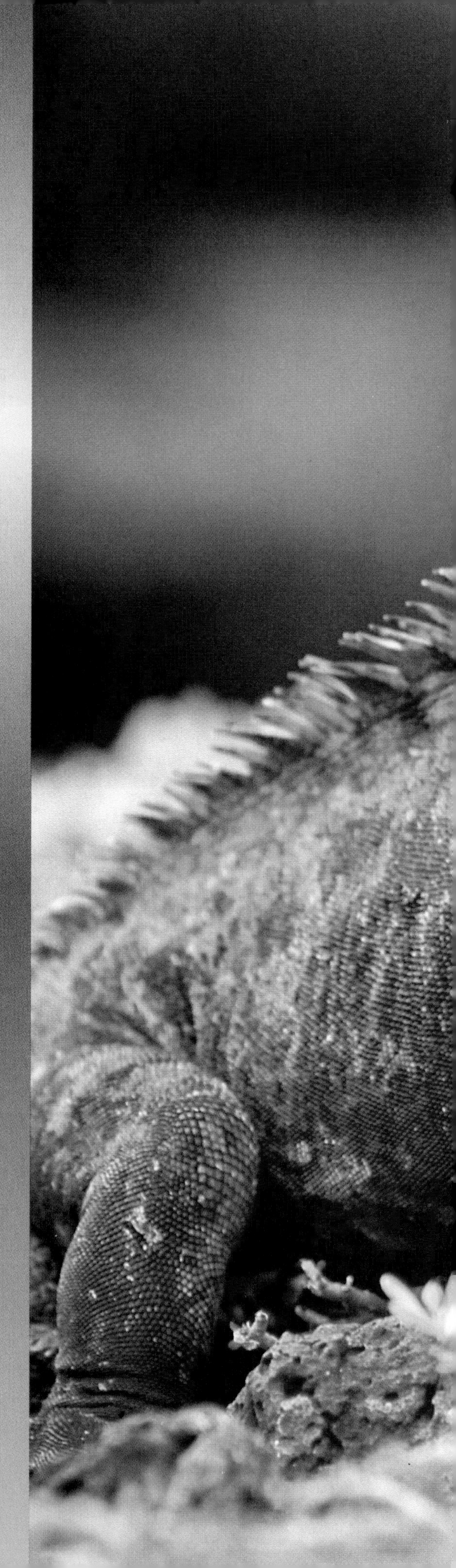

CAREER TRACK

As you explore the topics in this unit, you will discover many different types of careers associated with biology. Here are a few of these careers:

- Zoologist
- Genetic Engineer
- Taxonomist

A marine iguana in the Galapagos Islands

CHAPTER 10

Natural Selection

FOCUSING THE CHAPTER THEME: Evolution

Biology and Your World

BIO JOURNAL

Have you heard the saying "Variety is the spice of life"? What does it mean to you? In your journal, write a short essay about how the differences among people—their interests, abilities, and talents—can enrich more than divide a community.

Atlantic puffins on a sea cliff in Scotland

A Riddle: Life's Diversity and Connections

SECTION 10–1

GUIDE FOR READING

- **Define** the term evolution.
- **Identify** the observations of living things that puzzled Darwin on his voyage.

THE LIVING WORLD PRESENTS us with an extraordinary puzzle. Humans share the Earth with millions of other species—from bacteria to blue whales. What's more, among the vast majority of multicellular species, no two individuals are exactly alike. Yet all forms of life are regulated by similar messages of molecular instructions written in the living code of DNA. What scientific process can account for the incredible diversity on one hand and the unity of life on the other?

A Daring Voyage

The answer to this question lies in a process known as **evolution**—a collection of facts, observations, and hypotheses about the history of life. **Evolution means a change over a period of time.** And a systematic study of evolution began on one of the most important adventures in all of human history.

In late December 1831, the research ship HMS *Beagle* set sail from England for a five-year cruise around the world. There was nothing out of the ordinary about the

Figure 10–1 *(a) A group of flamingoes in Florida, (b) plant life in a cypress swamp in South Carolina, and (c) the bacteria that cause food poisoning—*Staphylococcus aureus *(magnification: 50,000X)—are examples of the variety of life forms that inhabit the Earth.*

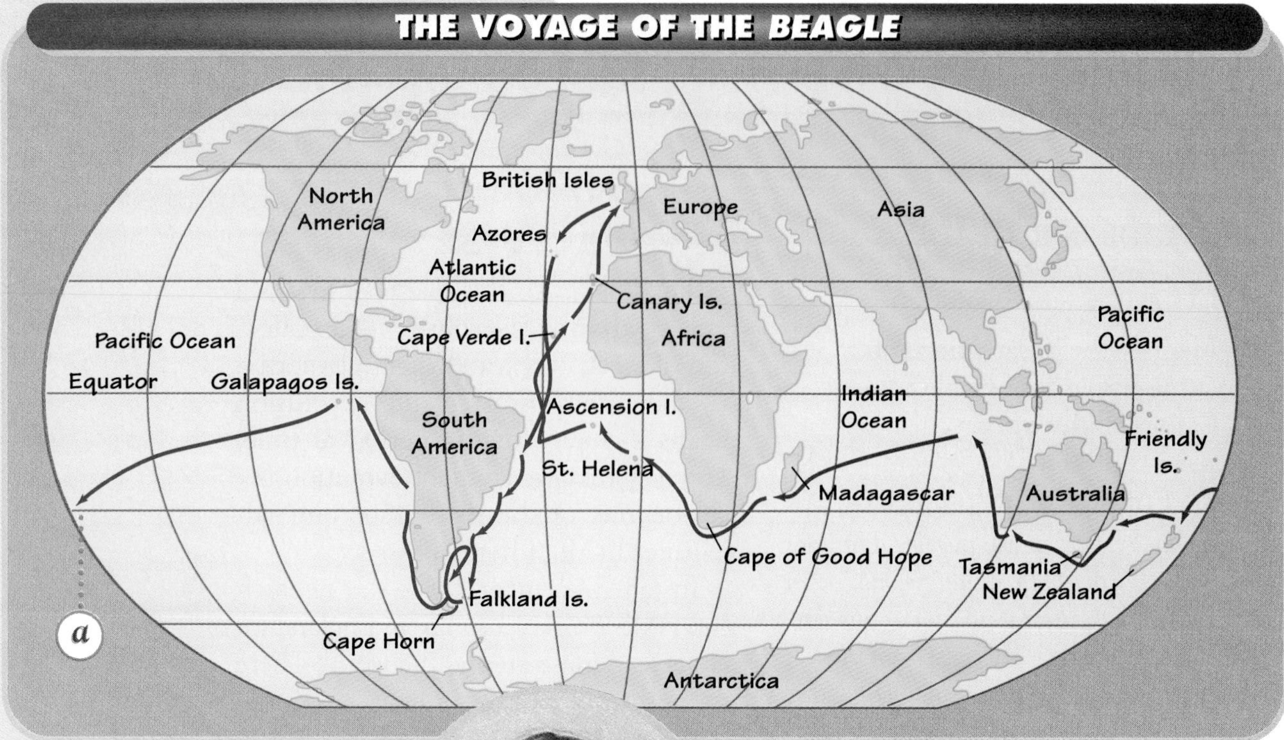

Figure 10–2
(a) The Beagle's voyage around the world lasted for a little over four years. Although the stop at the Galapagos Islands was for just one month, (b) Charles Darwin made the wealth of observations that led to his revolutionary hypothesis about the history of life on Earth.

Beagle, its captain, or its crew. But on this trip, the *Beagle* carried an educated, self-enlightened 22-year-old ship's naturalist named Charles Robert Darwin. No one—not even Darwin himself—knew that this voyage would begin a revolution in scientific thought that would radically change the way scientists look at life.

Darwin loved natural history. During the voyage of the *Beagle*, Darwin's powers of observation helped him to notice things that others had missed. At sea, Darwin studied specimens, read scientific books, and gave a great deal of thought to what he had seen. This was the key to his great contribution. Other explorers made long lists of little discoveries. But to Darwin, every individual find seemed to be another piece in a great puzzle that kept him involved for decades.

Pieces of the Puzzle

Wherever the *Beagle* went, Darwin collected plant and animal specimens. Time and again, he was thrilled by extraordinary sights that few people had imagined. The things that excited Darwin so long ago are just as astonishing today.

Living Diversity

As Darwin traveled, he was amazed by the tremendous variety of animals and plants that inhabited the Earth. Just one day's work in a Brazilian rain forest produced a collection of more than 68 different species of beetles! **Darwin called the variety of living things the diversity of life, or living diversity.** He wondered how this diversity came to be. Today, scientists know that living diversity is far greater than Darwin imagined. They estimate that Earth is home to between 5 and 50 million different species. Where did they come from?

Figure 10–3
(a) *The South American armadillo bears a striking resemblance to* (b) *the glyptodon, an extinct animal known only from its ancient fossilized remains.*

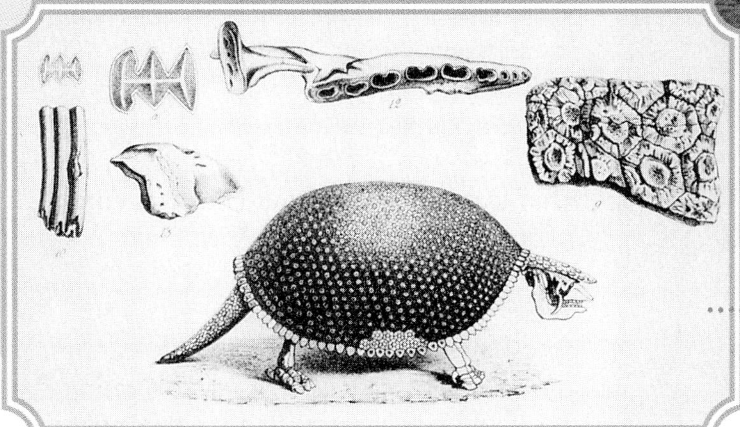

Fossil Diversity

In many places, Darwin collected bones and other traces of ancient organisms, called **fossils.** Some of these were rather ordinary, but many looked as strange as dragons or other imaginary beasts. **Since Darwin's time, researchers have uncovered the fossilized remains of many more unusual creatures.** In fact, scientists estimate that 99.9 percent of all species that have ever lived are now extinct. If that is correct, hundreds of millions of species have appeared, lived for a time, and vanished. Where did these marchers in the parade of life come from? And why did so many disappear?

Adaptation

As Darwin studied plants and animals, he noticed that every species seemed remarkably well suited to the life it leads. Each species had a combination of physical characteristics and behaviors that helped it catch food, withstand harsh conditions, or reproduce. These characteristics and behaviors, called **adaptations,** are found at every level of biology—from cell chemistry to animal behavior. **Biologists use the word adaptation to describe physical and behavioral traits that enable organisms to survive.**

Fitness

Darwin realized that for a species to survive, its members must do more than just stay alive. They must also reproduce. **Darwin used the word fitness to describe the ability of an individual to survive and reproduce.** Evolutionary biologists today define **fitness** in a more precise way. Fitness now is based on an organism's ability to successfully pass on its genes to its offspring.

☑ **Checkpoint** What is meant by the term adaptation? By the term fitness?

Figure 10–4
Although the hawk might not have much success with this prey, its sharp talons and hooked beak are adaptations that are essential for its survival.

Figure 10–5
The Galapagos Islands are the tips of undersea volcanoes that reach from the ocean floor to just above sea level.

The Galapagos Islands

The *Beagle* visited so many different places that the captain's log filled several books. But of all the *Beagle's* ports of call, Darwin was most influenced by his visit to a group of islands located 1000 kilometers west of South America. All these islands are small—some mere specks on the map. Several of them have climates that are quite different from the rest. The smallest islands at the lowest elevation above sea level are hot, dry, and nearly barren. Yet a strange assortment of plants and animals live there, including giant tortoises, marine iguanas, and some odd birds. These islands in the Pacific Ocean are the Galapagos Islands, and their inhabitants provided Darwin with his greatest inspiration.

Peculiar Creatures

Darwin was fascinated by the Galapagos wildlife, especially land tortoises and marine iguanas. He also saw several types of small brown birds looking for seeds. Although he collected several, he didn't find them unusual or important. As he examined these specimens, however, he noted that the birds showed a variety of beak shapes. He thought some of them were wrens, others were warblers, and still others were blackbirds. But that was all he noticed—at least at first.

There was another zoological tidbit that Darwin noted but did not think important at the time. The Vice-Governor of the islands told Darwin that giant tortoises varied in predictable ways from one island to another. Furthermore, the Vice-Governor boasted to Darwin that he could tell which island a particular tortoise came from by looking at its shell. Darwin later admitted in his notes that "I did not for some time pay sufficient attention to this statement."

The Journey Home

While heading home, Darwin had plenty of time—between bouts of seasickness—to look at his Galapagos specimens and think more about them. While examining several mockingbirds, he noticed that individuals collected from the islands of Floreana and Santiago looked different from one another. They also looked different from individuals collected on other islands. Darwin remembered what the Vice-Governor had told him about the tortoises. Darwin began to wonder whether animals living on different islands had once been members of the same species and had changed after they were isolated from one another. Was that possible? How could it be? If it were true, it would turn his whole view of the natural world upside down!

Exciting Time, Exciting Thoughts 

Darwin made his voyage during one of the most exciting periods in the history of Western science. Many explorers were traveling the world, expanding the horizons of knowledge. Great thinkers in several fields of science had begun to challenge established views about the natural world. To understand just how radical Darwin's thoughts would soon become, you must understand a few things about the world in which he lived.

The vast majority of Europeans in Darwin's day believed that the Earth and all forms of life were divine creations, produced a few thousand years ago over a span of one week. Since that original creation, both the Earth and its living species were thought to have remained fixed and unchanged. By the time Darwin set sail on the *Beagle*, there were numerous discoveries of evidence—fossils of extinct animals, for example—that this traditional view could not explain. Some scientists adjusted their beliefs to include not one period of creation but several creations—each preceded by a catastrophe, such as the Flood described in the Book of Genesis.

At first, Darwin accepted these beliefs. But as he traveled, much of what he observed did not fit neatly into this view of unchanging life. The more he saw, the more he wondered whether all species really did remain the same. Could living things change over time? If Darwin had lived a century earlier, he might have done little more than think about such things, as others had before him. But as Darwin wondered, he also read the latest scientific books and discussed his findings with other scientists interested in natural history. Slowly, his thinking began to follow a dramatically different path.

Figure 10–6
(a) No matter how different the limbs of many animals look, their bones are incredibly similar to one another. (b) The earliest stages of these three embryos look so similar that it is difficult to tell them apart. As these embryos grow, their homologous structures develop in very different ways, producing infant animals that can easily be recognized.

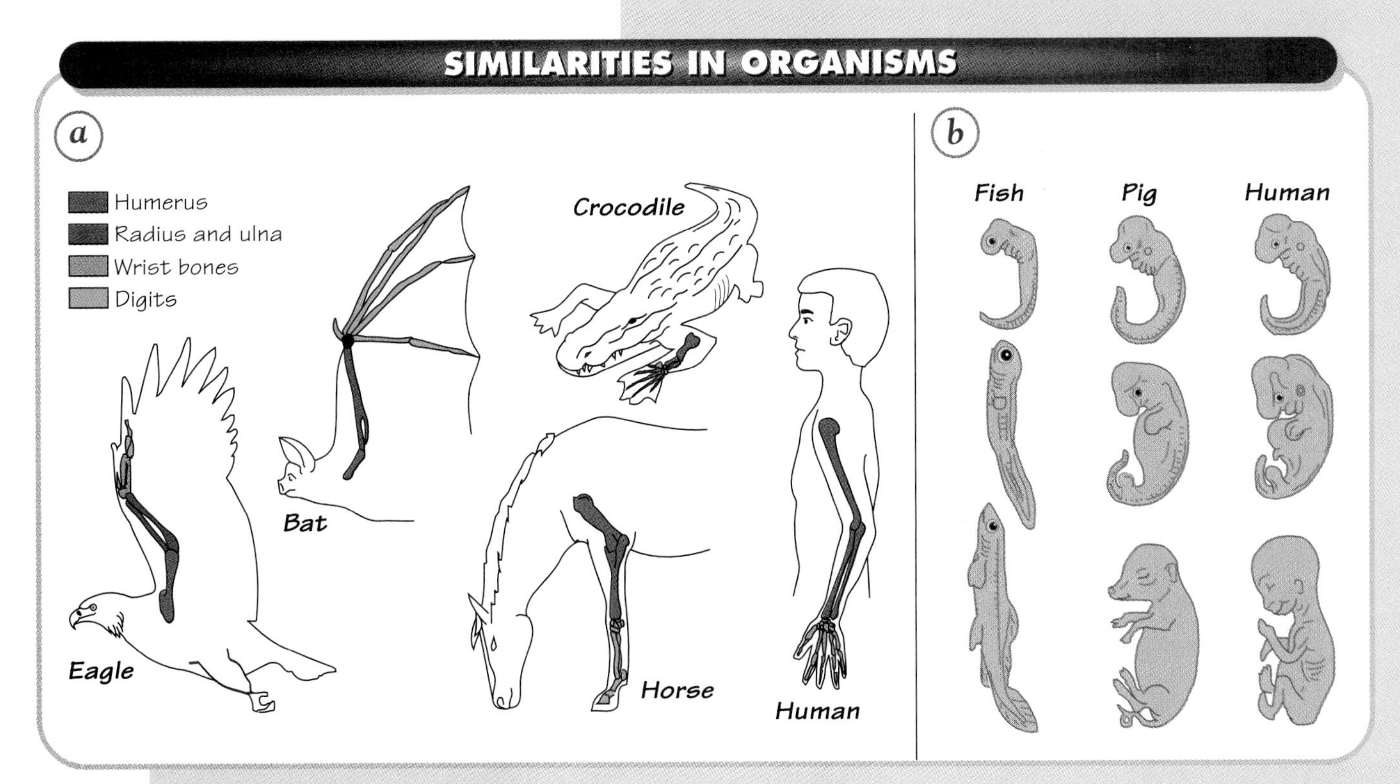

Figure 10–7
These flightless cormorants are found on the islands off the western coast of South America.

Anatomists also found structures that seemed to have little or no obvious purpose in the organism. These structures are known as **vestigial organs.** As more of these structures were found, they presented real puzzles to scientists, who were unable to explain their presence.

Checkpoint What are homologous structures?

Living Questions

During Darwin's lifetime, several interesting observations were made by anatomists who studied the structure of adult animals and by biologists who studied developing embryos. Among their findings were many phenomena that needed to be explained.

When looking at adult animals with four limbs, anatomists saw striking similarities among the bones of very different types of arms and legs. Structures such as those that are color-coded in ***Figure 10–6*** on page 223 are called **homologous structures.** Homologous structures develop from similar tissues in the early developmental stages of the organism. As you can also see, studies of developing embryos revealed even more surprising similarities.

INTEGRATING EARTH SCIENCE

What evidence did scientists uncover to support the hypothesis that Earth was constantly changing?

An Ancient, Changing Earth

While on board the *Beagle*, Darwin read the recent works of Charles Lyell, a geologist who agreed with a revolutionary proposal that had been made nearly a century earlier. Lyell put forth some convincing arguments that Earth had to be older than people thought. He argued that the modern world must have been shaped by the same geological forces that can be seen in action today. Thus, if river valleys were dug by a process as slow as erosion, Earth had to be older than a few thousand years.

Lyell then extended his thinking beyond geology to all of science. He proposed that scientists must explain the past in terms of events and processes that they could observe themselves.

Figure 10–8
Charles Lyell's hypothesis about the age of the Earth and the forces that shape it was based on observations such as (a) volcanic eruptions and (b) erosion.

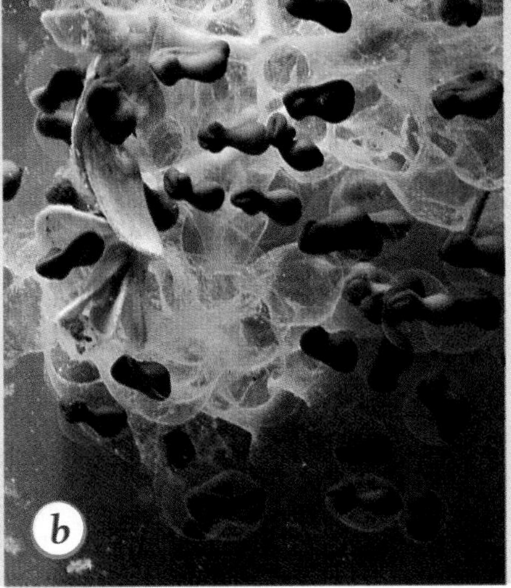

Figure 10–9
(a) *Plants such as the Australian* Eugenia corniflora, *laden with fruit, and* (b) *animals such as the spadefoot toad, whose young are shown here, illustrate the huge reproductive capacity of many organisms.*

In other words, Lyell argued that science could not call upon miracles and supernatural events to explain observed phenomena. For if that were the case, a scientific approach could never enable scientists to explain the past or make reliable predictions about the future.

Lyell's arguments made sense to Darwin after he witnessed a few geological events. In South America, Darwin observed a volcanic eruption—from a safe distance. Later, he wrote about a violent earthquake that lifted a stretch of rocky coastline more than 3 meters into the air.

With Lyell's ideas fresh in his mind, Darwin was able to understand how the fossils of marine animals, which he had found in rock layers several meters above sea level, got there. Geological phenomena such as volcanic eruptions and earthquakes could change the face of the Earth. And if Earth itself could change over time, why not life on Earth as well?

The Problem of Reproduction

Another influence on Darwin's thinking was the English economist Thomas Malthus. In 1788, Malthus observed that babies were being born faster than people were dying. If this growth continued, he realized, the world would be overrun with humans. And if that happened, Malthus suggested that war, famine, and disease would limit human populations.

When Darwin read Malthus's work, he realized that this reasoning applied even more to plants and animals, because they produce many more offspring than humans do. A mature maple tree can produce thousands of seeds in a single summer. One oyster can produce millions of eggs each year. If all the offspring of any species survived for several generations, they would overrun the world.

Obviously, this doesn't happen. Continents aren't covered with maple trees, and oceans aren't filled with oysters. The majority of the offspring of a species die, and only a few that survive succeed in reproducing. But what causes the death of those individuals? And what factor or factors determine which survive and reproduce and which do not? These questions will be answered in the next section.

Section Review 10–1

1. **Explain** what is meant by the term evolution.
2. **Identify** the observations of living things that puzzled Darwin on his voyage.
3. **Critical Thinking—Drawing Conclusions** What evidence do homologous structures and vestigial organs offer about the patterns of change in living things?

SECTION

10–2 Darwin's Solution

Guide for Reading

- **Describe** artificial selection.
- **Explain** evolution by natural selection.

MINI LAB

- **Interpret** the variation in height of students in your class.

AFTER DARWIN RETURNED TO England, the scientific community was soon buzzing with excitement. Darwin had turned over many specimens, including the Galapagos birds, to specialists at the Zoological Society of London. He soon learned that the mockingbirds he had collected were actually three separate species. Even more exciting was the news that these three species live only on the Galapagos Islands. And there were more surprises to come.

Figure 10–10
(a) The blue-footed booby, (b) the sally light-foot crab, and (c) the giant tortoise are some examples of the wildlife that inhabits the Galapagos Islands.

The Findings

To Darwin's astonishment, the small seed-eating birds he had thought were wrens, warblers, and blackbirds were actually all finches. And they too, like the mockingbirds, live only on the Galapagos Islands. The same was true of the tortoises, marine iguanas, and all the plants that Darwin had collected on the islands. For all these new species, the story was the same—each island species looked a great deal like a species on the mainland of South America. Yet the island species were distinctly different—both from the mainland species and from each other.

Not only was Darwin stunned by these discoveries, he was also disturbed by them. Years later, he wrote, "It was evident that such facts as these . . . could be explained on the supposition that species gradually became modified, and the subject haunted me." He began to record his thoughts and observations on the process that would later be called evolution.

Figure 10–11
Taken from Darwin's The Origin of Species by Means of Natural Selection, *these sketches clearly illustrate the difference in beak shape and size found among the finches of the Galapagos Islands.*

Although Darwin returned to England in 1836, it wasn't until 1844 that he first wrote down most of his ideas on how evolution might work. Then in 1859, nearly 30 years after he began his voyage on the *Beagle*, Darwin published his explanations in a book entitled *The Origin of Species by Means of Natural Selection*. Darwin published his work at that time only because someone else was about to be given credit for the same ideas. That someone was Alfred Russel Wallace, who had been studying natural history in Malaysia. Wallace sent Darwin a short essay in which he summarized exactly what Darwin had been thinking about for almost 25 years.

In *The Origin of Species*, Darwin accomplished two important tasks. First, he assembled an enormous quantity of evidence supporting the idea that life has changed, or evolved, over time. Second, he proposed a scientific hypothesis to explain how and why evolution occurs.

Darwin's book caused an immediate sensation. The first printing sold out the day it was released! Some people questioned or rejected Darwin's message. Others found his arguments brilliant. Let's look at what Darwin actually proposed in his book.

☑ **Checkpoint** What were Darwin's findings concerning the birds of the Galapagos Islands?

Of Farmers and Pigeons

Darwin began his discussion of evolution not with his discoveries on the Galapagos Islands but with observations made closer to home—observations of farm animals and agricultural crops in England. He knew that over the years, farmers had greatly altered and improved domesticated plants and animals. How did they do it?

No Two Alike

Plant and animal breeders told Darwin that no two individuals among their livestock and crop plants were exactly alike. Some were larger or smaller, heavier or lighter than others of their kind. Some cows and sheep gave more milk than others. Some plants produced larger fruit than others of their kind. Often this variation was inheritable, which means that it could be passed on to the next generation. Farmers told Darwin that although they could not cause this variation, they did know how to use it when they found it.

Artificial Selection

What the farmers did was simple. Looking over their stock, they decided which animals and plants had the characteristics they wanted to use for breeding. The farmers called this process picking, or selection. Generation after

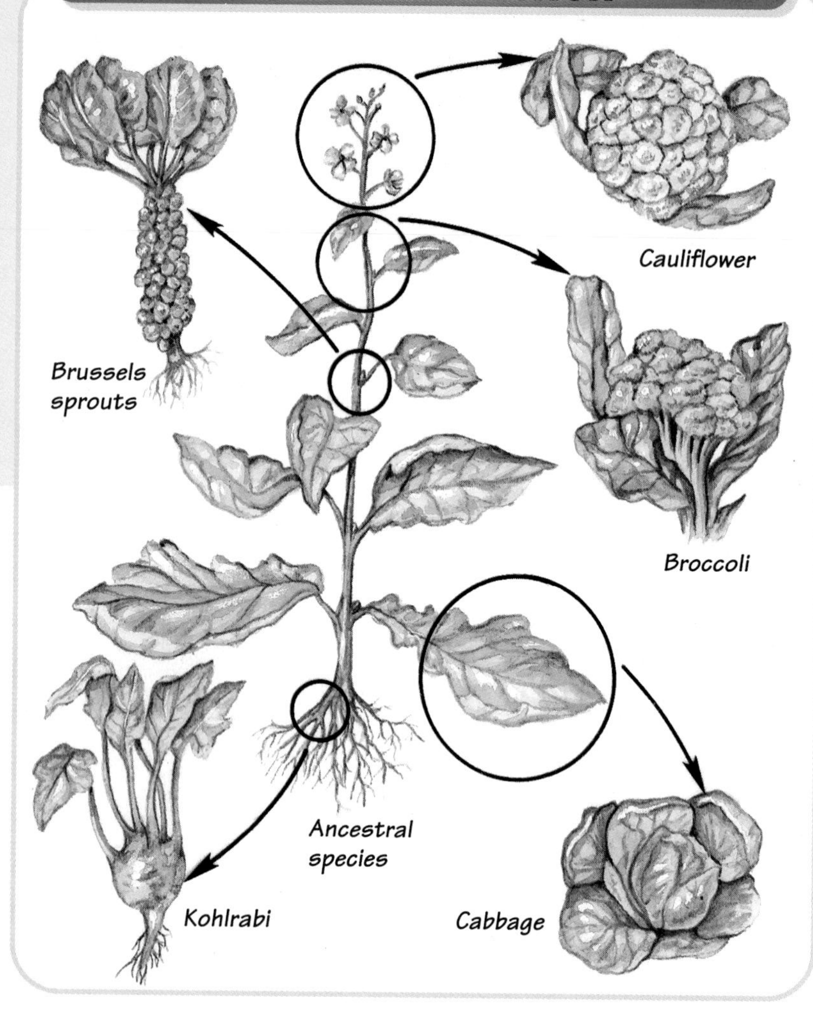

Figure 10–12
The vegetables shown here are the descendants of a single species. They have been produced from generations of artificial selection for leaves (kale), buds (Brussels sprouts and cabbage), flowers and stem (broccoli), stem (kohlrabi), and flower clusters (cauliflower).

generation, the farmers would select only the largest hogs, the fastest horses, or the cows that gave the most milk. **As Darwin put it, nature provided the variation and humans allowed only selected organisms to produce offspring. He called this process artificial selection.** Using **artificial selection** over many years, breeders had produced a wide range of plants and animals that looked very different from their ancestors.

☑ **Checkpoint** What is artificial selection?

Selection in Nature

Darwin was convinced that a process such as artificial selection occurred in nature. But how did this process work? His explanation of the process is where Darwin made his greatest contribution.

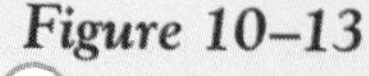

Figure 10–13
ⓐ *Despite their similar appearance, the ladybugs shown here differ in many subtle ways. Natural variation results in differing levels of fitness in members of a species.*
ⓑ CAREER TRACK *Zoologist Dr. Louise Emmons, who studies tree shrews in Danum Valley, Sabah Borneo, is interested in all aspects of the life of these animals—including adaptations that increase their fitness.*

Variation in Nature

Darwin pointed out that in nature, as in domesticated plants and animals, inheritable variation was common. Today this seems obvious because we can see variation in almost any plant or animal species. In Darwin's time, however, this was a revolutionary statement because living things, as part of creation, were thought to be perfect. Variations were occasional defects and nothing more. Of course, no one knew just yet why variation occurred or even how inheritance worked. Although Gregor Mendel had done work in genetics, his work was not published until 1866, and it was not widely recognized until much later.

Struggle for Existence

Darwin recognized a process in nature that could operate in much the same way as artificial selection did on farms and in fields. It was in this thinking that Darwin made his sharpest break with the past. Instead of calling upon a mysterious force to take the place of the farmer, Darwin searched for a scientific hypothesis to explain selection in nature. Remembering the work of Malthus, Darwin realized that high birth rates and a shortage of life's basic needs forced organisms into a constant struggle for existence. The fastest predators could catch the most prey. Only the best camouflaged, or best protected, prey survived the hunt.

Natural Selection

Here is the heart of Darwin's insight. Because each individual differs from all other members of its species, each has slightly different advantages and disadvantages in the struggle for existence. During that struggle, individuals that are well-suited to their environment will survive and reproduce more often than individuals not well-suited. Because individuals that survive and reproduce are the best-suited, or fittest, this process is called survival of the fittest.

Darwin proposed that generation after generation, the struggle for existence selects the fittest individuals to survive in nature. Darwin called this process **natural selection.** Natural selection explains how over time, species become better suited to their environment as they respond to various selection pressures. Thus, natural selection explains the riddles of adaptation and fitness.

☑ *Checkpoint* How do artificial selection and natural selection differ?

The Tall and Short of It

PROBLEM *How does height vary in a small human population?* ***Interpret data*** *to answer this question.*

PROCEDURE

1. Working with a partner, use a meterstick to measure your partner's height to the nearest centimeter. Record this information.
2. Reverse roles with your partner and repeat step 1.
3. Obtain the heights of all class members. Then construct a data table that contains 3-cm ranges from the smallest height recorded to the tallest height recorded. For example, if you had a 152- to 155-cm range, a person who is 154 cm tall would fit into this range.
4. Record the number of students who fall within each height range.
5. Plot the number of students within a height range as a function of height.

ANALYZE AND CONCLUDE

1. What is the range of heights and the average height?
2. Why do you think there is a variation in height?
3. How can you interpret the shape of the graph?
4. Are there any selective advantages to being very tall or very short?

Problem Solving

INTERPRETING GRAPHS

What's So Different About Pill Bugs and Cattails?

John and Sasha's teacher asked them if all living things of the same species were exactly the same. "Of course not," they said. So their teacher asked them to pick a species and then choose a characteristic of that species to study. She asked them to collect data as evidence to share with their class.

John went to a wooded area and lifted a rotting log from the forest floor. Tiny pill bugs dashed to and fro. He collected about 50 pill bugs, placed them in a jar, and brought them back to his classroom. There he fed and observed them until he was sure they were fully grown. Then he measured their body length.

Sasha walked to a pond and noticed that many cattails were growing near the edge of the water. Her teacher told her that the cattails were fully grown at that time of year. Sasha measured their long brown combs.

John's and Sasha's data are shown in the data table. Plot each data set to see whether they share any common features.

Pill Bug Measurements

Body length (in mm)	4	5	6	7	8	9	10	11	12
Number of pill bugs	3	4	5	8	10	9	6	4	1

Cattail Comb Measurements

Length of comb (in cm)	21	22	23	24	25	26	27	28	29	30	31	32	33
Number of combs	1	1	2	3	5	7	8	6	4	3	2	2	1

• THINK ABOUT IT •

1. **Compare** the pill bug and cattail graphs. **Describe** the pattern common to both graphs.
2. What is the significance of this pattern with respect to the process of natural selection?
3. A new predator that eats pill bugs has emigrated to the forest. This predator's vision allows it to see only medium-sized and large-sized pill bugs. If pill bugs do not reproduce until they are fully grown, what changes might take place (over many generations) in the population of pill bugs? **Predict** what would happen if the pill bugs were able to reproduce when they were only 4 mm in length.

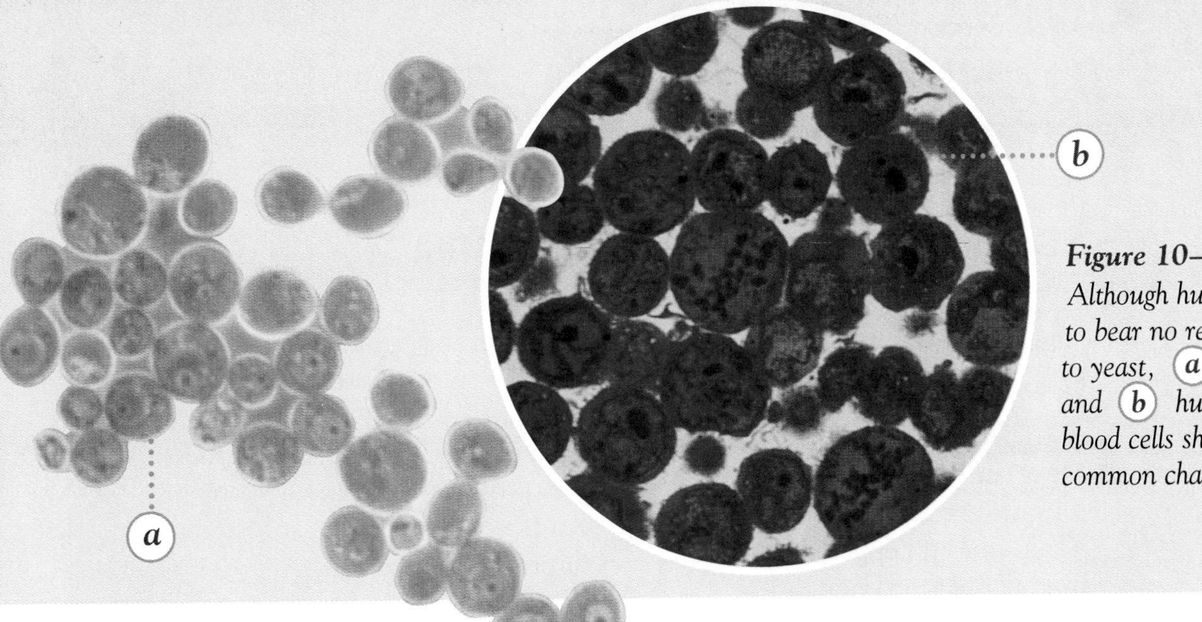

Figure 10–14 Although humans appear to bear no resemblance to yeast, (a) yeast cells and (b) human white blood cells share many common characteristics.

Common Descent

Finally, Darwin argued that living species were not spontaneously produced as we know them. Instead, modern organisms have been produced through a long, slow process of evolutionary change. Darwin further argued that just as each living organism is the descendant of its parents and grandparents, each living species has descended from other species over time.

Look back in time, and you will find ancestors that were common to humans and monkeys. Look farther back, and you will find ancestors that were common to humans, alligators, and fishes. And if you could look even farther back, you would find that all living things share common ancestors. Therefore, this principle is called **common descent.**

The Test of Time

Today, more than a century after Darwin, the overwhelming scientific evidence has caused virtually all scientists to agree that life has evolved. Evolutionary change is undeniable from a scientific perspective. Explaining how and why evolution occurs, however, is much more difficult. Evolution by natural selection has, over time, grown into an enormous collection of carefully reasoned and tested hypotheses.

As you will learn in the next chapter, a new generation of experiments and observations is supplying even more evidence that evolution has occurred in the past and is continuing, day by day, in the natural world around us. These experiments and observations reinforce the power and beauty of Darwin's insight.

Section Review 10–2

1. **Describe** artificial selection. Explain why it is practiced.
2. **Explain** evolution by natural selection.
3. **Critical Thinking—Analyzing and Evaluating** What are the key elements of Darwin's explanation of evolution? Can they be verified experimentally?
4. **MINI LAB** How does height vary in a small human population? **Interpret data** to answer this question.

SECTION 10–3 BRANCHING OUT In Depth

Darwin's Revolution

Guide for Reading

- **Discuss** the importance of evolution in the study of biology.

MINI LAB

- **Make an inference** about selection pressure and adaptation in peppered moths from population data.

EVOLUTION IS CONTROVERSIAL enough in certain circles that some people wonder why biologists insist on teaching it. The answer is simple. Evolution is the most powerful general statement ever made about living things.

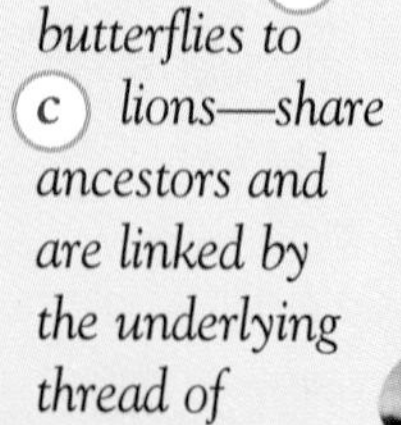

Figure 10–15 *Despite obvious differences, all living things—from (a) giant weevil beetles to (b) morpho butterflies to (c) lions—share ancestors and are linked by the underlying thread of evolution.*

Evolution

Evolution provides a single common language that enables biologists in all fields to share information with one another. Why is it that researchers can learn about human DNA by studying the genes of yeasts? Because humans and yeasts share common ancestors—and a surprising number of genes! Why is it so important to preserve endangered species? Because each living species has been produced by millions of years of evolution and may contain irreplaceable knowledge about how life works. **Quite simply, evolution provides a unifying principle that underlies all of biology—from the micro level of molecular genetics to the macro level of global ecology.**

The body of powerful and complex scientific thought that has built on the concept of evolution is constantly changing. Some of Darwin's original ideas have proved to be right on target. Others have been revised. But revision does not mean that evolutionary change itself is in question or that evolutionary concepts are a collection of vague guesses.

By comparison, consider that physicists still do not understand how gravity works, despite the fact that no one doubts

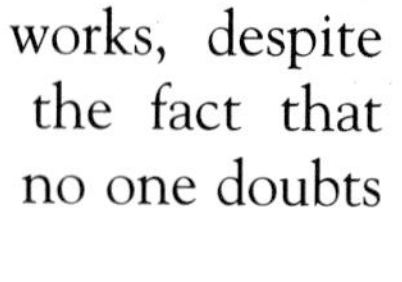

the existence of this powerful force. There is no doubt that if you jump up into the air, you will end up on the ground below. It makes no difference whether you understand or even believe in gravity. What goes up must come down. Just as definitely, life on Earth has evolved and is continuing to evolve all around us all the time.

MINI LAB ... Inferring

Now You See Me, Now You Don't

PROBLEM *What can you **infer** about selection pressure and adaptation in peppered moths from population data?*

PROCEDURE

Graph the data presented in the data table.

ANALYZE AND CONCLUDE

1. What has happened to the population of light-colored moths over the course of the study? Dark-colored moths?
2. What do you think caused these changes?
3. What is the adaptation in this species? What is the selection pressure in this study?

Year	Light-Colored Moths	Dark-Colored Moths
1	537	112
2	484	198
3	392	210
4	246	281
5	225	357
6	193	412
7	147	503
8	84	594
9	56	638
10	38	673

Life in Constant Motion

One of the most powerful predictions concerning evolution is that under certain circumstances life may be constantly changing. If enough variation exists in a species and if environmental conditions change, the species will evolve. One of the first examples to be documented of evolution in progress was a fascinating color change in peppered moths in England around 1760.

The idea of changing species also helps explain why disease-causing bacteria become resistant to every antibiotic that is used against them and why crop-damaging insects become resistant to pesticides. ● Concepts about evolution also set the stage for understanding that variety is indeed the spice of life! For most organisms, inheritable variation is essential to species survival.

INTEGRATING BIOLOGY AND SOCIETY

How can we slow the rate of development of antibiotic-resistant bacteria and pesticide-resistant insects?

Section Review 10-3

1. **Discuss** the importance of evolution in the study of biology.
2. **Examine** evolution in action.
3. **MINI LAB** What can you **infer** from population data about the selection pressure and adaptation in peppered moths?
4. **BRANCHING OUT ACTIVITY** Find out how the overuse of antibiotics can cause bacteria to become drug resistant. **Write** a short report on your findings.

Laboratory Investigation

Simulating Natural Selection

Natural selection can be observed in nature. However, studying selection in nature takes a very long time. In this laboratory investigation, you will play a game that will simulate natural selection.

Problem

How can you **construct a model** for the process of natural selection?

Materials (per group)

paper or cloth to cover lab table

10 paper squares of each of the following 10 colors: brown, green, yellow, gray, black, white, blue, purple, violet, and red

additional paper squares as needed

Procedure

1. Working in groups of 4, spread your cloth "habitat" on your lab table and obtain 10 paper squares of each of the 10 colors.
2. Have 1 person of your group act as the gamekeeper and the other 3 as the predators.
3. While the predators have their backs turned to the habitat, have the gamekeeper distribute the square pieces of colored paper, which represent the prey, on the table. The pieces must be randomly placed and must not overlap or hide other pieces.
4. When the gamekeeper says "Go!" the predators should turn around, face the habitat, and quickly pick up the first piece of paper they see, then turn their backs again.
5. Repeat step 4 until there are only 10 prey remaining in the habitat.
6. Count the number of "survivors" of each color. Record these numbers.
7. Allow each "survivor" to reproduce by adding 9 more pieces of the same color.
8. Repeat steps 2 to 7 until each person in the group has had a turn being the gamekeeper.

Observations

1. Construct a graph that shows the results of the simulation. The graph should include the original populations and the survivors at the end of each round.
2. How long did it take for any color to become extinct?
3. How long did it take for any color to become dominant (more than 50 percent of total)?

Analysis and Conclusions

1. What represents the selection pressure in this game?
2. What is the adaptation mechanism of the prey?
3. Does one color eventually become dominant? Why does this happen?

More to Explore

Design an experiment in which survival depends on escaping predators that must "feel" for their prey because the predators are blind. What kinds of variations in the prey will be significant for natural selection?

Study Guide

Summarizing Key Concepts

The key concepts in each section of this chapter are listed below to help you review the chapter content. Make sure you understand each concept and its relationship to other concepts and to the theme of this chapter.

10–1 A Riddle: Life's Diversity and Connections

- The variety of living things is called living diversity.
- Fossil diversity is the variety of the fossilized remains of creatures that lived long ago.
- Adaptations are physical and behavioral traits that enable organisms to survive.
- Fitness is the ability of an individual to survive and reproduce.

10–2 Darwin's Solution

- Artificial selection is a process used by breeders to produce a wide range of plants and animals that are functionally different from their ancestors.
- The struggle for existence and the survival of the fittest serve to select organisms for survival in nature. This is called natural selection.
- The principle of common descent states that each living species has descended from other species over time.

10–3 Darwin's Revolution

- Evolution provides a unifying principle that underlies all of biology—from the micro level of molecular genetics to the macro level of global ecology.
- A color change evolved by peppered moths in response to a change in their environment is a modern example of evolution at work.

Reviewing Key Terms

Review the following vocabulary terms and their meaning. Then use each term in a complete sentence.

10–1 A Riddle: Life's Diversity and Connections

evolution
fossil
adaptation
fitness
homologous structure
vestigial organ

10–2 Darwin's Solution

artificial selection
natural selection
common descent

Recalling Main Ideas

Choose the letter of the answer that best completes the statement or answers the question.

1. The collection of facts, observations, and hypotheses about the history of life is called

a. survival of the fittest.
b. adaptation.
c. evolutionary theory.
d. artificial selection.

2. Charles Darwin was unaware of the works of

a. Malthus.
b. Lyell.
c. Mendel.
d. Wallace.

3. Which of the following puzzled Darwin the least?

a. diversity in living things
b. adaptation of organisms to their environment
c. fossils of strange creatures
d. similarities among organisms

4. During Darwin's time, many people believed that the

a. Earth constantly changed.
b. Earth and its organisms did not change.
c. Earth was hundreds of millions of years old.
d. living things on Earth changed gradually.

5. Vestigial structures in organisms

a. look the same.
b. perform the same function.
c. serve little or no apparent function.
d. develop from the same embryonic tissues.

6. Farmers improve domestic plants and animals by

a. providing proper nutrients.
b. selective breeding.
c. common descent.
d. survival of the fittest.

7. Darwin proposed that natural selection occurs because of

a. natural variation.
b. the struggle for existence.
c. the survival of the fittest.
d. all of the above.

8. The principle that states that all living things share common ancestors is called

a. common descent.
b. natural selection.
c. artificial selection.
d. adaptive radiation.

Putting It All Together

Using the information on pages xxx to xxxi, complete the following concept map.

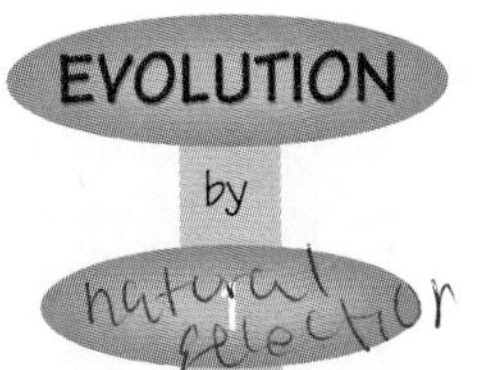

is a result of an organism's

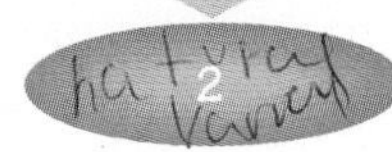

Survival of the fittest

Assessment

Reviewing What You Learned

Answer each of the following in a complete sentence.

1. What is evolution?
2. Of all the species that have lived on Earth, what percentage are now extinct?
3. State Darwin's theory of natural selection.
4. Give examples of two or more homologous structures.
5. What are vestigial organs?
6. What is meant by natural history?
7. In the context of the survival of a species, what did Darwin mean by the term fitness?
8. What is the principle of common descent?
9. Who was Alfred Russel Wallace and what role did he play in the publication of Darwin's findings?
10. Give an example of an inheritable variation in humans.

Expanding the Concepts

Discuss each of the following in a brief paragraph.

1. What is the role of adaptation in a species' survival?
2. How did the organisms that Darwin observed on the Galapagos Islands help him to formulate his theory of natural selection?
3. What impact did the works of Charles Lyell and Thomas Malthus have on Darwin's thinking?
4. How can you **interpret** the similarities among species that have been geographically isolated for thousands of years?
5. Compare artificial selection and natural selection.
6. Why is the evolutionary theory of natural selection considered one of the ten most important concepts in biology?
7. Why is the peppered moth example often termed "Darwinism in Action"?
8. Why do the birds of the Galapagos Islands demonstrate a great diversity of beak shapes and sizes?
9. If all living things are controlled by DNA, how does evolution account for the great diversity of life?
10. Generally, only a few of the many offspring produced in the reproductive cycle of an organism survive. What happens to the other offspring? Relate your answer to the theory of natural selection.

Extending Your Thinking

Use the skills you have developed in this chapter to answer the following.

1. **Sequencing** Develop a time line of important events for Darwin, beginning with the voyage of the *Beagle* and ending with his publication of *The Origin of Species by Means of Natural Selection.*
2. **Inferring** Explain why the study of life on islands usually demonstrates great living diversity and rapidly evolving species.
3. **Evaluating** Discuss the concept that adaptation on the organism's part is not an active process but a passive one. Use specific examples in your discussion.
4. **Constructing a model** Construct a model to simulate natural selection.
5. **Evaluating** Discuss how sudden changes in the environment—such as the release of radioactivity in the Chernobyl disaster—set evolution in motion.

Applying Your Skills

A Survival Scenario

Adaptations are inherited physical and behavioral traits that enable an organism to survive in its environment. What if an organism's environment were to suddenly change or be disrupted? Would the organism be able to survive?

1. Select an adaptive trait of an animal or plant and write a scenario describing how the trait might have evolved.
2. Write a second scenario reflecting the work of Malthus or Wallace.

• GOING FURTHER •

3. Have other students read the scenarios and identify the theory upon which each one is based.

CHAPTER 11

The Mechanisms of Evolution

FOCUSING THE CHAPTER THEME: *Patterns of Change*

Biology and Your World

BIO JOURNAL

Look at old photograph albums showing a family—siblings, parents, uncles, aunts, and grandparents—at various times in their lives. In your journal, write about the similarities you observed and share your thoughts on why you think the similarities range from significant to none.

Variation in poison dart frogs, Panama

Darwin "Meets" DNA

SECTION 11–1

Guide for Reading

- **Identify** the source of inheritable variation in organisms.
- **Define** species, fitness, and adaptation in genetic terms.

MINI LAB

- **Relate** common vegetables to the parts of the ancestral species from which they were artificially selected.

NEARLY A CENTURY after Darwin proposed his theory of natural selection to explain evolution, the theory itself went through an incredible period of growth and change. Biologists interested in evolution combined Darwin's insights with new developments in the study of inheritance. Indeed, scientists could now explain how evolution worked far better than Darwin ever could.

How Does Variation Arise?

Darwin knew nothing about how inheritable traits were passed from one generation to the next. He also had no idea where the inheritable variation that his theory of natural selection depended on might come from. When geneticists rediscovered Mendel's work during the early part of this century, evolutionary biologists realized that they had stumbled onto a gold mine.

When biologists recognized that genes in an organism's cells are carriers of traits, they realized that genes are also the source of inheritable variation. Because inheritable variation provides the raw material for natural selection, genes quickly became the center of attention in hypotheses and experiments aimed at understanding the mechanisms

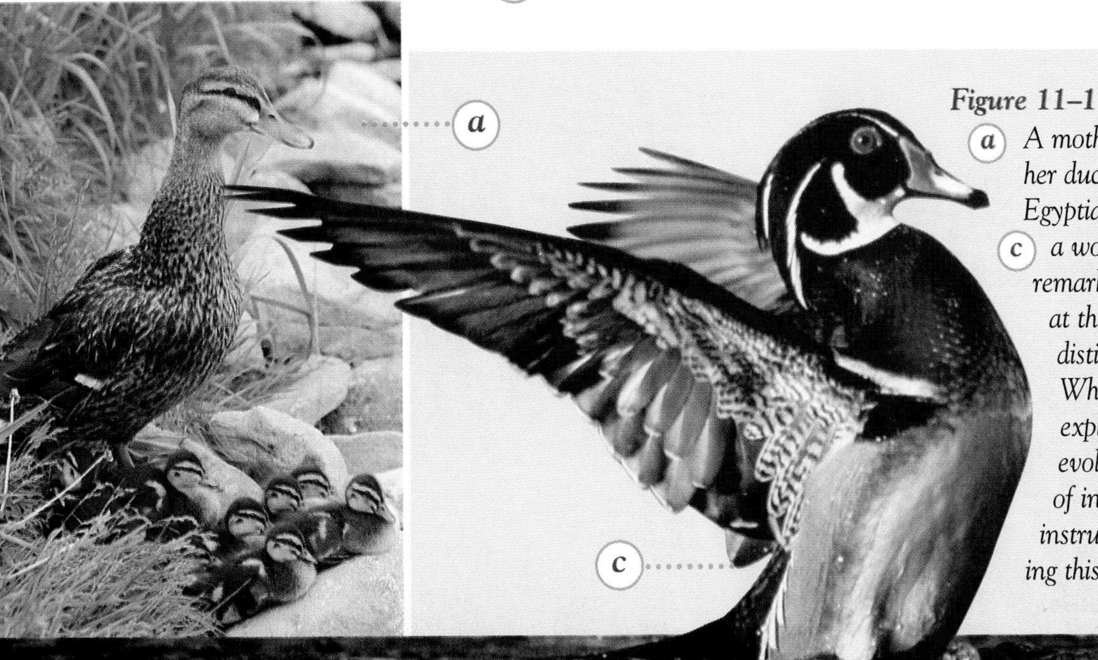

Figure 11–1 (a) A mother duck and her ducklings, (b) an Egyptian goose, and (c) a wood duck are remarkably similar and, at the same time, are distinctly different. What mechanism can explain how they evolved? The science of inheritance was instrumental in answering this question.

of evolution. The result was a new view of the principles of evolution—one based on genes. **Scientists now know that inheritable variation comes primarily from two kinds of changes in an organism's genetic material—mutations and gene shuffling.**

Mutations

Sometimes changes in the structure of an organism's DNA cause changes in the information carried in one or more of its genes. Such changes in DNA are called **mutations.** The rate at which mutations occur varies a great deal from one group of organisms to another. Bacteria, for instance, have fairly high mutation rates, whereas humans have fairly low mutation rates. Mutations are one source of variation in nature, and they can have powerful effects—both positive and negative—on organisms.

Gene Shuffling

Mutations are not the only source of inheritable variation. You do not look exactly like either of your biological parents, even though they provided you with all your genes. You probably look even less like any brothers or sisters you may have (except an identical twin). Yet mutant genes are not the likely source of differences among you. Instead, the differences are probably due to the extensive shuffling of genes that occurs when egg and sperm are produced. When the same genes are combined in different ways, they often interact differently and produce different results. Thus, sexual reproduction is an important source of variation in nature.

☑ **Checkpoint** What are the sources of variation in nature?

Figure 11–2
Studying heredity in (a) the evening primrose, Hugo de Vries, a Dutch botanist, was the first to recognize the nature of mutations. Mutations—random changes in an organism's DNA—have given rise to (b) the dramatic variation in the color of the tigers. Variation, although less obvious in (c) these emperor penguins, is very common in populations of organisms, and it arises in large part due to the shuffling of genes in sexual reproduction.

How Common Is Genetic Variation?

Darwin realized that inheritable variation is common in nature, but he had no way to measure how common it is. Today, biologists know that a significant percentage of genes in natural populations have at least two different forms, called alleles (uh-LEELZ). In insects, as many as 15 percent of all genes have more than one allele. In fishes, reptiles, and mammals, about 5 percent of all genes have more than one allele. This level of genetic difference between organisms is a very important source of raw material that can be influenced by natural selection.

Inheritable Variation

There are many kinds of inheritable variation. Some you can see with the unaided eye, such as the flower colors of Mendel's pea plants. Others are invisible because they involve hidden biochemical processes, such as enzyme action and protein synthesis. In addition, some traits are controlled by a single gene, and others are controlled by several genes. This makes a big difference in the way genetic variation is expressed in individual organisms.

Of course, natural selection never "sees" individual genes. In fact, natural selection doesn't even act directly on an organism's genotype—its complete genetic makeup. "Survival of the fittest" refers to whether or not an individual organism manages to survive and reproduce. For this reason, natural selection can affect only variations in phenotype—the characteristics produced by the interaction of an organism's genes and its environment. Not surprisingly, natural selection can have different effects on different kinds of traits.

Traits Controlled by Single Genes

If a trait is controlled by a single gene with two alleles, there are three possible genotypes. Depending on the trait involved and the way the gene operates, these genotypes may produce either two or three phenotypes, which in turn may have very different fitness. A distribution for three phenotypes of a single-gene trait is shown in **Figure 11–3.**

Traits Controlled by Several Genes

Many important traits, however, are polygenic, which means that they are

Figure 11–3
The graphs illustrate the distribution of phenotypes that would be expected for a trait if one, two, or many genes contributed to the trait. The last graph is a bell-shaped curve characteristic of a normal distribution.

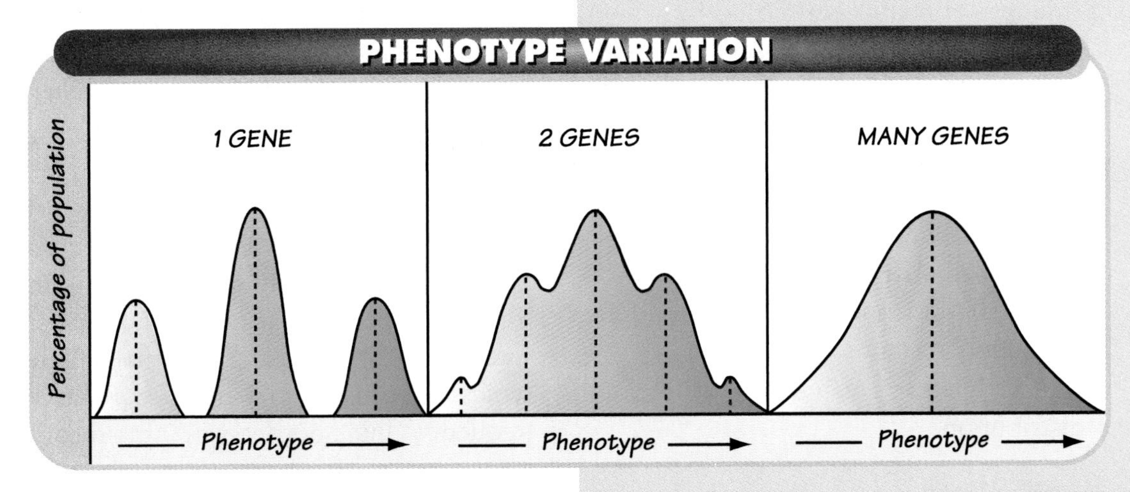

It's All Relative to Broccoli

PROBLEM *How can you **relate** common vegetables to the parts of the ancestral plant from which they were artificially selected?*

PROCEDURE

1. Working in a group, examine each vegetable and discuss its main features.
2. Decide on the feature of the ancestral *Brassica* plant—belonging to the cabbage and turnip group—that you think was artificially selected to breed this vegetable. Record this as your observation.
3. Organize your observations under these categories: selection for leaves, stem, flowers, roots, and buds.

ANALYZE AND CONCLUDE

1. All of the *Brassica* vegetables you observed have 18 chromosomes per cell, yet they are considered separate species. Why?
2. Do you think that hybrids (such as hybrid broccoli-Brussels sprouts) can be produced from two of the species?

controlled by two or more genes. Often, each of these genes has two or more alleles. As a result, there are a large number of possible genotypes and an even larger number of possible phenotypes for such traits. Height in humans is a good example of a trait controlled by several genes. If you were to measure and graph the height in a group of people, you would find that people aren't just tall, medium, or short. Instead, there is a wide range of heights. In fact, the graph would be shaped like a bell. This characteristic bell-shaped curve—called a normal distribution—is observed quite commonly in nature.

INTEGRATING MATH

What are the characteristics of a normal distribution? Discuss examples of traits that exhibit such a distribution.

Checkpoint What is a normal distribution?

Genes, Fitness, and Adaptation

Today, an understanding of genetics makes it possible to define fitness and adaptation in a more meaningful and measurable way. Each time an organism reproduces, it passes copies of its alleles on to its offspring. **An organism's evolutionary fitness can be defined as its success in passing genes to the next generation. An adaptation can be described as any genetically controlled trait that increases an individual's ability to pass along copies of its genes.**

Genes Also Define Species

In Darwin's time, biologists defined a species as a group of organisms that looked alike. **Today, a species is defined as a group of similar-looking organisms that can breed with one another and produce fertile offspring.** Members of two different species cannot interbreed. This is known as **reproductive isolation.**

Why is this genetic definition important? When members of a species interbreed, they share genes with one another. Researchers therefore say that individuals within a species share a group of alleles called a **gene pool.** Because of that common gene pool, a genetic change that occurs in one member of a species can spread through the population, as ***Figure 11–4*** illustrates. On the other hand, because members of different species do not share genes, each species evolves as a separate unit.

Swimming in the Gene Pool

The gene pool of each species contains a certain number of alleles for each trait. In a species that is not under pressure from natural selection, each allele typically occurs at a particular relative frequency. Put another way, this means

that each allele occurs a certain number of times compared with other alleles for the same gene.

Sexual reproduction by itself does not change the relative frequency of alleles in a population. Why? Think of the allele combinations produced by sexual reproduction as being similar to the different hands that can be dealt from a deck of playing cards. Shuffling and reshuffling the cards can produce an enormous number of different hands. But shuffling will not change the relative numbers of aces, kings, or jokers in the deck. The only way that this can be done is if hands containing certain cards are discarded. As you will soon discover, something similar to this happens when a species evolves.

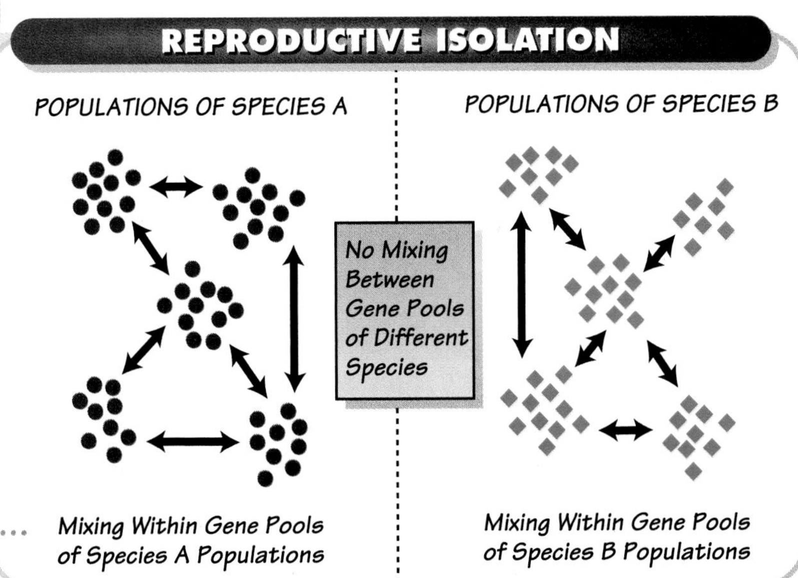

Figure 11–4
(a) The parent bird with five nestlings is more fit than its neighbor with only two nestlings. Fitness is defined as an organism's success in passing genes to the next generation. (b) Species A and B are reproductively isolated. Within each of the two separate gene pools of species A and species B, mixing occurs. However, no mixing occurs between the gene pools of the two species.

Section Review 11-1

1. **Identify** the source of inheritable variation in organisms.
2. **Define** species, fitness, and adaptation in genetic terms.
3. **Critical Thinking—Hypothesizing** How does variation depend on the size of the gene pool?
4. **MINI LAB** How do common vegetables **relate** to the parts of the ancestral species that were used in artificial selection?

SECTION 11–2

Evolution as Genetic Change

GUIDE FOR READING

- **Define** evolution in genetic terms.
- **Define** genetic drift.
- **Explain** how speciation can occur.

MINI LAB

- **Construct a model** of genetic drift.

WHAT DOES IT MEAN TO SAY that a species evolves? Of course, individual organisms of a species do not evolve. But a population of organisms, slowly, over many generations, does undergo changes due to inheritable variations and the forces of natural selection. Does this mean that evolution occurs in all species at all times and at the same rate? And is natural selection the only means by which evolutionary change takes place?

Evolution—Genetically Defined

Today, evolutionary biologists view evolution as a natural consequence of the nature of genes and DNA and the interactions between organisms and their environment. **In modern genetic terms, evolution can be defined as any change in the relative frequencies of alleles in the gene pool of a species.** You may recall that sexual reproduction alone will not cause such a change. What then can cause the allele frequencies in a gene pool to change? Let's examine some possibilities, or mechanisms, beginning with Darwin's choice—natural selection.

Figure 11–5
If you look closely at any species of organisms, such as (a) blue heelers (Australian cattle dogs), (b) coconut palms, or (c) dahlias, you will notice natural variation. Natural variation, which can be visible or invisible, reflects the frequency of different alleles for a given trait.

Natural Selection on Single-Gene Traits

As mentioned earlier, some traits important to evolutionary fitness are controlled by a single gene. What happens if a mutation occurs in one copy of that gene carried by a single individual? The answer depends on several factors, including chance. For starters, let's look at the role natural selection can play.

Assume that the mutation is not lethal. Or at least, assume that a single copy of this new allele doesn't kill an individual that carries it before that individual can reproduce.

Harmful Mutations

If a mutation decreases the ability of some member of a species to survive and reproduce, fewer copies of the allele will be passed on to future generations. Members of the same species that do not carry the allele will survive and reproduce. Thus, the allele will tend to become even less common in the population.

Helpful Mutations

Now suppose that the new allele increases the evolutionary fitness of individuals that carry a single copy. Over time, carriers of this allele would produce more offspring than the individuals that do not have the allele. As a result, over a few generations, the relative frequency of this allele would increase in the population. And if the allele has no negative effects, it may sooner or later be found in nearly all members of the species.

Natural Selection on Polygenic Traits

The situation with polygenic traits is both more interesting and more complicated. As you have read, the action of alleles on traits such as height produces a wide range of phenotypes. In most cases, the graphs of the variation of such traits display a bell-shaped curve, or normal distribution. The fitness of individuals located near one another on the curve will be similar. However, fitness can vary a great deal from one end of the curve to the other. And if fitness varies, natural selection can act. When it does act, natural selection can affect these distributions in any of three ways: It can stabilize the distribution, it can shift the distribution in one direction or the other, or it can disrupt the distribution.

Stabilizing Selection

Stabilizing selection occurs when organisms near the center of the curve are more fit than organisms at either end. This situation keeps the center of the curve at its current position and may or may not decrease the amount of variation above and below the average. Evolution as defined in this chapter is either minor or absent. See ***Figure 11–6*** on page 248.

The mass of human infants at birth seems to be under the influence of stabilizing selection. Infants that are born much smaller or larger than average are less healthy and have a greater chance of developing serious complications that may lead to death. Obviously, the fitness of those individuals is low. Data show that the average mass of humans at birth coincides quite well with the mass at which mortality is lowest.

Directional Selection

Directional selection occurs when individuals at one end of the curve have higher fitness than individuals in the middle or at the other end. This situation will cause the entire curve to move as the trait changes. This means both a change in the allele frequencies in the gene pool and that evolution occurs.

Among Darwin's seed-eating finches, the size of a bird's beak determines the size of the hard-shelled seeds it can crack open to eat. Larger beaks enable birds to feed on larger seeds. If the number of available small- and medium-sized seeds decreases, birds that are able to open larger seeds will have more access to food and therefore have higher fitness. The average beak size in this population of birds would be expected to increase as the species evolves.

Disruptive Selection

Disruptive selection occurs when individuals at both the upper and lower ends of the curve have higher fitness than individuals near the middle. In such situations, natural selection acts most strongly against individuals of an intermediate type. If the pressure of natural selection is strong enough and lasts long enough, this situation can cause the single curve to split into two curves, as illustrated in ***Figure 11–6,*** causing evolution to occur.

Sometimes an insect species that predators find tasty gains protection because it resembles another insect species that contains poisonous or distasteful chemicals. If the tasty species inhabits a very large geographical area, it may be under pressure from disruptive selection to imitate two different poisonous species in different places. In this situation, insects that resemble neither of the poisonous species have lower fitness than those who resemble either of the poisonous species. As a result, the population may be split into different forms as it evolves.

☑ **Checkpoint** How do stabilizing selection and disruptive selection differ?

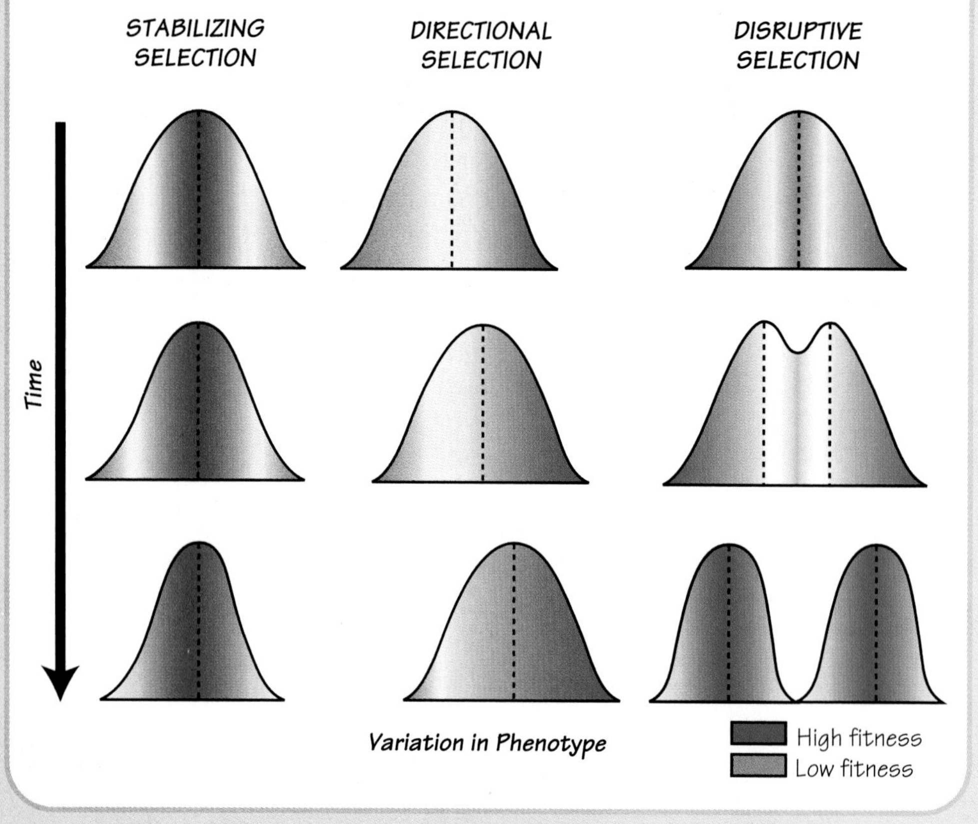

Figure 11–6
The effect of natural selection on a normal distribution of phenotype depends on how fitness varies with phenotype. Stabilizing selection occurs when the fitness of individuals is high near the center and low toward the ends of the distribution. Directional selection occurs when fitness is high at one end and low at the other end of the distribution. Disruptive selection occurs when fitness is high at both ends but low around the center of the distribution.

Bio FRONTIER Connections

Altering the Human Gene Pool

Cathy Sullivan couldn't believe it. A California team of doctors was working on preventing cystic fibrosis, a condition with which Cathy's 10-year-old daughter, Melissa, was born. Mel, as her parents called her, took various experimental drugs, had her back pounded three times a day to loosen the mucus that filled her lungs, and yet couldn't do all the things a normal 10-year-old can do.

Because cystic fibrosis is a genetic disorder caused by a recessive gene, both Cathy and her husband, Al, were carriers of the gene. But they were not sick. Melissa's sickness was an unfortunate consequence of nature's random actions.

Modern Treatment

Cystic fibrosis, or CF, is a condition whose treatment has come a long way. Today, there are drugs to help reduce the risk of lung diseases, thus enabling children with CF to live to adulthood. There is even a gene therapy for the condition, still experimental, whereby the non-CF gene is introduced into the patient's body in order to replace the CF gene.

Is Prevention Better Than the Cure?

Scientists have found a way to test an egg or a sperm for the gene for CF. In this procedure, eggs are removed from the mother and examined for a smaller cell that contains a set of chromosomes, called the polar body, that is discarded just before fertilization occurs. An egg will discard a polar body so that it will have half the number of chromosomes needed for fertilization. If a woman is a carrier, her polar body may or may not have the gene for CF. If the polar body does not have the gene, the egg does. By the same token, if the polar body has the gene for CF, the egg does not. Consequently, when this egg is fertilized by a sperm that does not have the CF gene, a child without CF will be produced.

CAREER TRACK *The genetic engineer is isolating gene-sized fragments of DNA in order to conduct recombinant DNA research, which has important applications in medicine, industry, and agriculture.*

The procedure seems to offer the perfect solution—at least in theory. But many people find the procedure, called preimplantation testing, questionable in principle. Why? Because although the notion of manipulating human genes to achieve desirable outcomes is intriguing to some, it is frightening to others. No one actually knows what the altering of human genes may mean to our future.

Making the Connection

How will altering human genes affect human evolution? Should we forge ahead with such medical practices before we learn more about their long-term effects? What do you think?

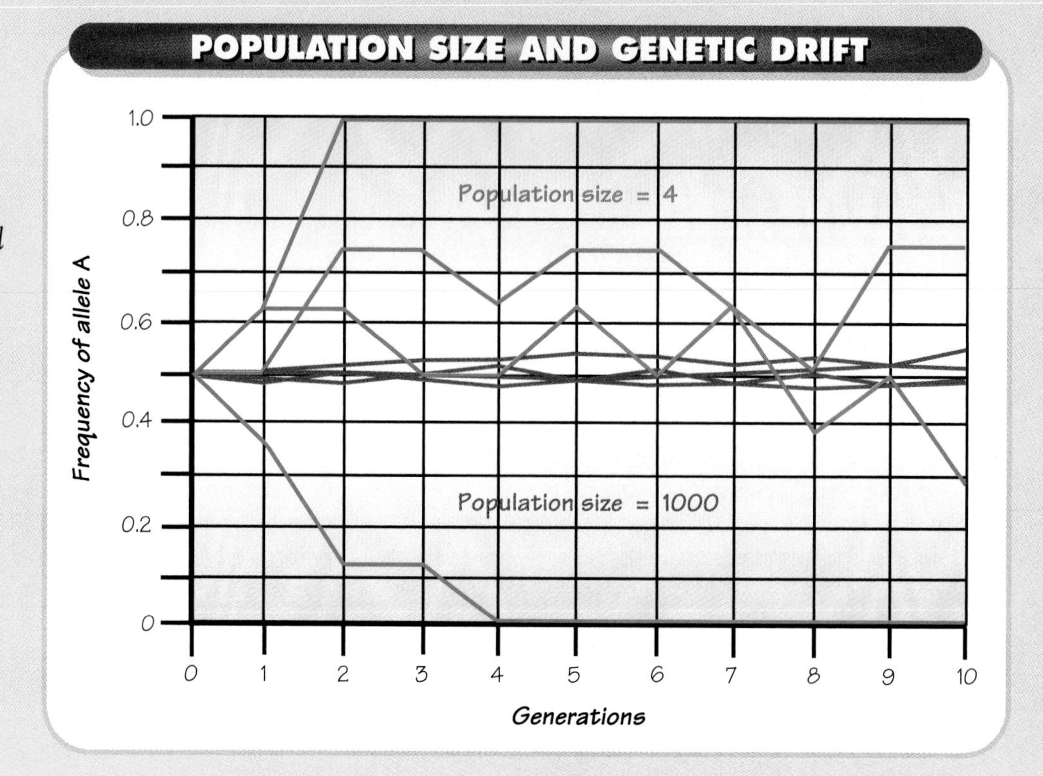

Figure 11–7
These computer simulations illustrate the effect of population size on genetic drift. The green lines represent an initial population of 4 individuals, and the red lines represent an initial population of 1000. In both cases, the initial population was composed of half A and half a alleles. After 10 generations, notice that in two simulations for population size 4, one allele became extinct by chance. All four simulations for a population size of 1000 show that the allele frequencies remained fairly constant.

The Role of Chance

Over the century and a half since Darwin, experiments and observations have supported natural selection in a remarkable number of cases. However, evolutionary biologists today also recognize the existence of other mechanisms of evolutionary change.

Recently, researchers realized that organisms can sometimes evolve whether or not natural selection is operating. **Studies have shown that an allele can become more or less common in a population simply by chance. Such a random change in allele frequency is called genetic drift.** How can this happen?

In some cases, individuals that carry a particular allele may leave more descendants than other individuals—not because they are more fit but just because of chance. Look at ***Figure 11–7*** to see how **genetic drift** is more likely to occur in small populations than in larger ones.

Often a small group of organisms will colonize a new habitat. These organisms may carry alleles in different relative frequencies from the larger population from which they came. Thus, the offspring they produce will be genetically different from the original population simply by chance.

☑ ***Checkpoint*** What is genetic drift?

The Birth of New Species

So far you have seen several ways in which changes may occur within a species. But how do these changes lead to **speciation**—the formation of new species? Remember that, by definition, members of a species can interbreed with one another but not with members of other species. This is called reproductive isolation. **Therefore, for speciation to take place, a population must evolve enough genetic changes—by mechanisms such as natural selection—so that breeding cannot occur between the emerging, genetically different groups.**

Speciation in animals and most plants usually occurs when two groups of individuals are physically separated from one another by a geographical barrier.

Visualizing Evolution in Galapagos Finches

From a small ancestral group of birds, the 13 species of birds known as Darwin's finches evolved. The following steps outline the probable mechanism of their evolution.

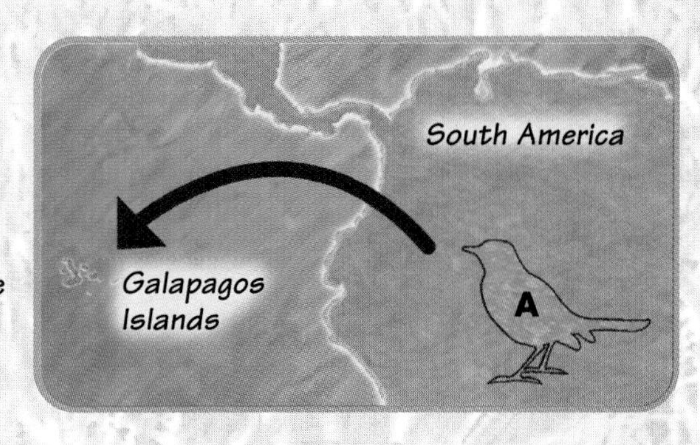

1 Arrival of the founders

A few finches, A, find their way to one of the islands. There they survive and reproduce.

2 Separation of populations

A few of the species A travel to another island. These two populations no longer share a common gene pool.

3 Changes in the gene pool

Over time, the isolated populations begin to evolve into two species, A and B, in ways determined by the conditions on their island home.

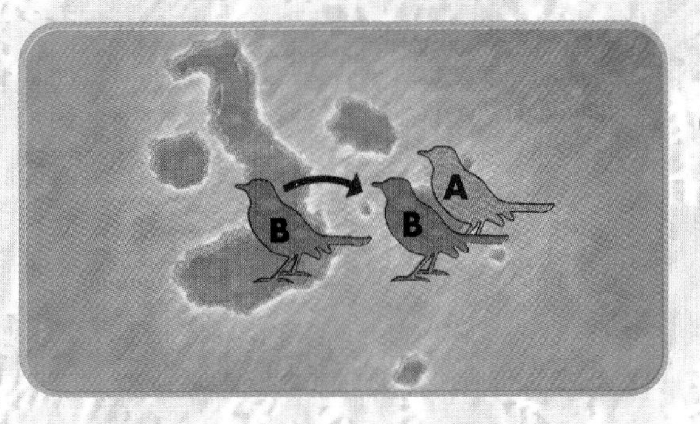

4 Reproductive isolation

Some of the birds, B, travel back to the original island, home to the founder species, A. Because the finches prefer to mate with birds that display similar phenotypes, A and B probably will not interbreed. The two gene pools are now separate, and the birds belong to different species.

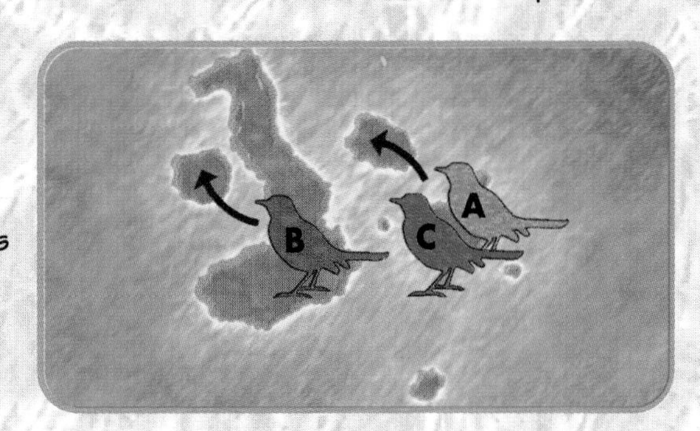

5 Sharing the same island

As the two birds, A and B, continue to share a single island, they compete with each other for food in times of drought. Directional selection leads to further changes in the species, causing B to evolve into C.

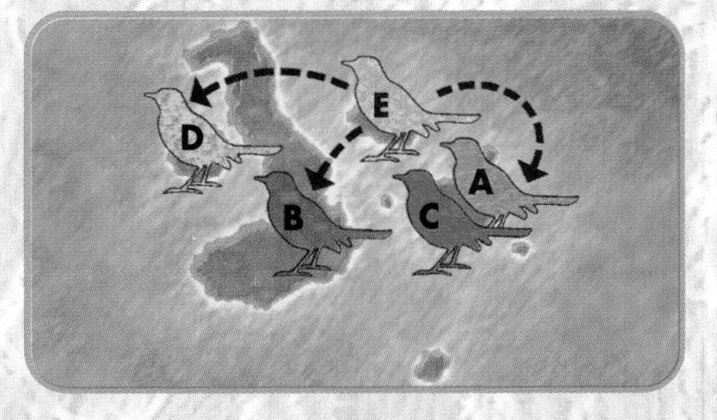

6 Repeating the process

If this process is repeated on the different islands, it could easily produce the 13 living species and the several extinct forms known as Darwin's finches on the Galapagos Islands.

Catch the Drift

PROBLEM *How can you **construct a model** of genetic drift?*

PROCEDURE

1. Without looking, scoop out a handful of beans from the container provided by your teacher.
2. Place the beans on your table. Count and record the number of beans of each type.
3. Repeat steps 1 and 2 two more times and record your observations in a data table.

ANALYZE AND CONCLUDE

1. If the container has 10 types of beans and 10 beans of each type, what is the percentage of each type of bean in the container?
2. Calculate the percentage of each type of bean in each handful of beans. How do these percentages compare with the percentages of the beans in the container? With each other?
3. Are the small samples (handfuls) representative of the larger sample (container)? Explain your results.

The illustration on page 251 shows how speciation might have occurred among the finches of the Galapagos Islands.

☑ *Checkpoint* When does speciation usually occur?

The Pace of Evolution

Evolutionary biologists agree that evolution occurs as the result of forces such as natural selection and genetic drift. But they differ on the relative importance of these forces and on the pace at which evolutionary change occurs.

Darwin may have overestimated the connection between geology and biology. Like Lyell, who argued that geological change was slow and steady, Darwin believed that evolutionary change also had to work the same way—very slowly. The view that evolutionary change occurs slowly and steadily over long periods of time is known as **gradualism.**

As researchers came to know the fossil record in greater detail, they discovered that evolutionary change did not always seem to proceed at the same rate. Sometimes species seemed to remain unchanged for very long periods of time. Now and again, there seemed to be periods of evolution that proceeded relatively quickly. This pattern is known as **punctuated equilibrium,** because it involves long periods of stability that are interrupted by episodes of rapid change.

Gradualism and punctuated equilibrium represent two extreme views of evolution. Most evolutionary biologists hold the view that evolution follows both patterns—in varying degrees at various times in evolutionary history.

Section Review 11-2

1. **Define** evolution in genetic terms.
2. **Define** genetic drift and **explain** how it can occur.
3. **Explain** how speciation can occur.
4. **Critical Thinking—Relating** How does natural selection affect the relative frequencies of alleles in a gene pool for each of the three distributions representing variation in a polygenic trait?
5. **MINI LAB** How can you **construct a model** of genetic drift?

How Species Change

Guide for Reading

- **Explain** how natural selection can lead to speciation in Darwin's finches.

DAPHNE MAJOR IS A TINY island—even for the Galapagos. The tip of this dead volcano is so small that you could walk around its crater rim in less than 20 minutes. It is a lonely, desolate place. Black lava rock bakes in the sun. There is no fresh water. There are no palm trees. But there are flocks of Darwin's finches. Thanks to these birds, researchers are studying processes that Darwin thought no one ever could. They are seeing natural selection in action. Day by day, as they watch, some birds survive and some die. Some reproduce and others don't. Species evolve—right before their eyes!

Darwin's Finches

On the desolate Galapagos Islands, Peter and Rosemary Grant have spent more than 20 years of their married life watching evolution in action. Before you can appreciate what they have been doing and why it is important, you need to learn a little more about the remarkable birds they study.

Researchers now know that the 13 species known today as Darwin's finches all evolved from a common ancestor. Darwin thought some of them were wrens, others were warblers, and still others were blackbirds because their beaks looked very different. Then as other researchers observed these birds, some interesting facts came to light. Each species has body structures—including beaks—and behaviors that enable it to exist in a very different way from its neighbors.

Some of these adaptations are difficult to believe. Some species eat different kinds of seeds, while others eat various types of insects. Some use their beaks as tools and others use cactus spines as tools. These species pick up a spine, shape it

Figure 11–8 The beaks of these Galapagos finches differ greatly as a result of the slow process of natural selection. Shown here are (a) Darwin's small tree finch, (b) large-billed ground finch, (c) Darwin's medium ground finch, and (d) cactus finch.

with their beaks, and use it to pull hidden grubs (larvae of some insects) out of dead branches. One species survives by eating leaves, a feat accomplished by only one other bird species in the world. Another species pulls bark from twigs and branches to expose the nutritious living tissue beneath. There is even a vampire finch that sits on the backs of larger birds, pecks them until they bleed, and then drinks their blood!

The variety of beak sizes and shapes that confused Darwin has evolved in ways that help the birds accomplish these varied tasks. If you think of beaks as tools, these birds began with a common set of pliers and evolved a whole tool kit. But how did these beaks diversify this way? Could these 13 species really have evolved from a single founding species?

A Galapagos Study

To show that this is a reasonable hypothesis—that the 13 species could have evolved from a single species—Peter and Rosemary Grant needed to do two things. First, they had to show that there is enough variation within each finch species so that natural selection can operate. Second, they had to document natural selection in action—something that had never been done before.

☑ **Checkpoint** What was the objective of the Grants' study?

Studying the Flock

The first task was to catch and identify as many birds as possible. That is why the Grants chose Daphne Major. The island is large enough to support fairly big finch populations. Yet it is small enough to allow the Grants to catch and identify nearly every bird. And once the birds were caught and tagged, the Grants could return, year after year, to keep a record of the individuals that lived and those that died, the ones that succeeded in breeding and those that did not. Today, the Grants not only know the age of all the birds they watch, they also know the history of each of these birds for three generations.

Looking for Variation

As the Grants catch each bird, they record the colors of their beaks and feathers, the body mass, and the lengths of

Figure 11–9
Notice the striking similarity between the beaks of some finches and these tools designed by human technology.

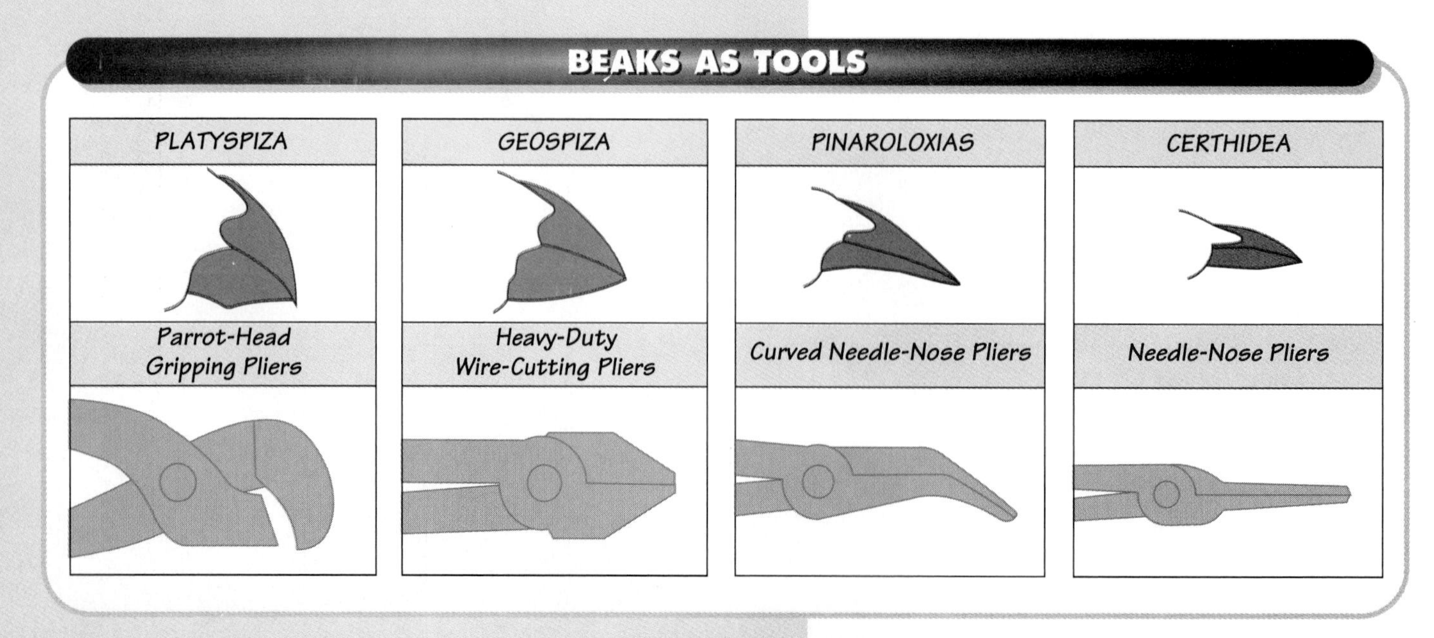

their wings, legs, and beaks. What they have found is fascinating. Many traits produce just the sort of normal distribution that is characteristic of polygenic traits such as human height. And individuals regularly vary by as much as 10 percent from the average measurements for their species—plenty of variation.

Looking for Natural Selection

According to Darwin, natural selection is constantly inspecting "... the slightest variations; rejecting those that are bad, preserving and adding up all that are good; silently ... working, whenever and wherever opportunity offers. ..." Yet when Darwin visited the Galapagos Islands, he did not see variation and selection in action. In fact, he often saw several species eating the same food. How could that be?

Actually, there is a simple explanation. Darwin visited these islands during the rainy season, when plant food is plentiful. At such times, many finches eat whatever they fancy. But it doesn't rain throughout the entire year. During the dry season, the food situation is dramatically different. Some types of food disappear. Other foods are scarce. Then different beaks come in handy.

As food becomes scarce, birds become feeding specialists—each selecting the type of food its beak is best suited to handle. The birds with big, heavy beaks go after big, thick seeds that no other species can crack open. Nevertheless, every year during the dry season, many birds die. Many young birds hatched during the wet season do not survive. Some older birds die, too. The situation is sometimes even more severe. During a long drought that lasted from 1976 through 1977, the Grants observed a population of 1200 birds dwindle to 180! Individual birds they had identified and measured survived and reproduced or died.

☑ *Checkpoint* Why did Darwin not observe natural selection in action?

What Selection Does

The results of the Grants' study were fascinating. When a large-beaked species encountered food scarcity, the birds that survived were those that had the largest beaks. Individuals with small beaks—falling at the other end of the bell curve—did not survive. So the average beak size in the surviving population increased. The change was so large and happened so quickly that a colleague made a startling calculation. It might take only between 12 and 20 droughts to change one species of finch into another! **From this study, you can imagine how several species of finches can, over time, evolve from an ancestral finch species—by the action of the environment on individual organisms with varying fitness.**

Section Review 11–3

1. **Explain** how natural selection can lead to speciation in Darwin's finches.
2. **Explain** the importance of variation to natural selection.
3. **BRANCHING OUT ACTIVITY** **Identify** the selection that occurred in the large-beaked bird species studied by the Grants as either stabilizing, directional, or disruptive selection. Construct a graph to illustrate the appropriate selection process.

CHAPTER 11

Laboratory Investigation

Mutations in Bacteria

The environment of bacteria that live in the human body undergoes a significant change when antibiotics, such as penicillin or streptomycin, are used. Although most bacteria are killed by an antibiotic, some may survive because they are genetically resistant to the drug. These resistant forms can be thought of as mutants.

Problem

How do antibiotics affect the growth of bacteria? **Formulate a hypothesis** to answer this question.

Materials (per group)

culture of bacteria
sterile swabs
sterile nutrient agar plate
glass-marking pencil
antibiotic filter paper discs
forceps
transparent tape
metric ruler

Procedure   

1. **CAUTION:** ***Before and after working with bacteria, you must wash your hands and working surfaces with soap and water and disinfectant.*** **Carefully turn the sterile nutrient agar plate over and place it on your table. Do not open the plate.**
2. **With a glass-marking pencil, draw two lines at right angles to each other so that the plate is divided into four equal areas, or quadrants. Label the quadrants from 1 to 4.**
3. **Remove the covering from the sterile swab. Carefully dip the swab in the culture of bacteria. CAUTION:** ***Be very careful when working with bacterial cultures.***
4. **Remove the cover of a nutrient agar plate and rub the bacteria-containing swab over the entire surface of the agar. Follow the directions of your teacher as to how to dispose of the swab.**
5. **With clean forceps, place the antibiotic discs on the nutrient agar, making certain that each disc is placed in the center of a numbered quadrant.**

6. Cover the plate and tape the plate closed.
7. Using the glass-marking pencil, write your initials on the bottom of the plate.
8. Place the plate upside down in the area designated by your teacher.
9. After 24 hours, observe the growth of bacteria around each disc. Record your observations.
10. Using a metric ruler, measure in millimeters the diameter of the zone of no bacterial growth, or zone of inhibition, that surrounds each of the antibiotic discs. Construct a data table and record the diameters of the zones of inhibition.
11. Carefully observe the zone of inhibition. Do you see a hazy secondary zone of bacterial growth? Record your observations in your data table.

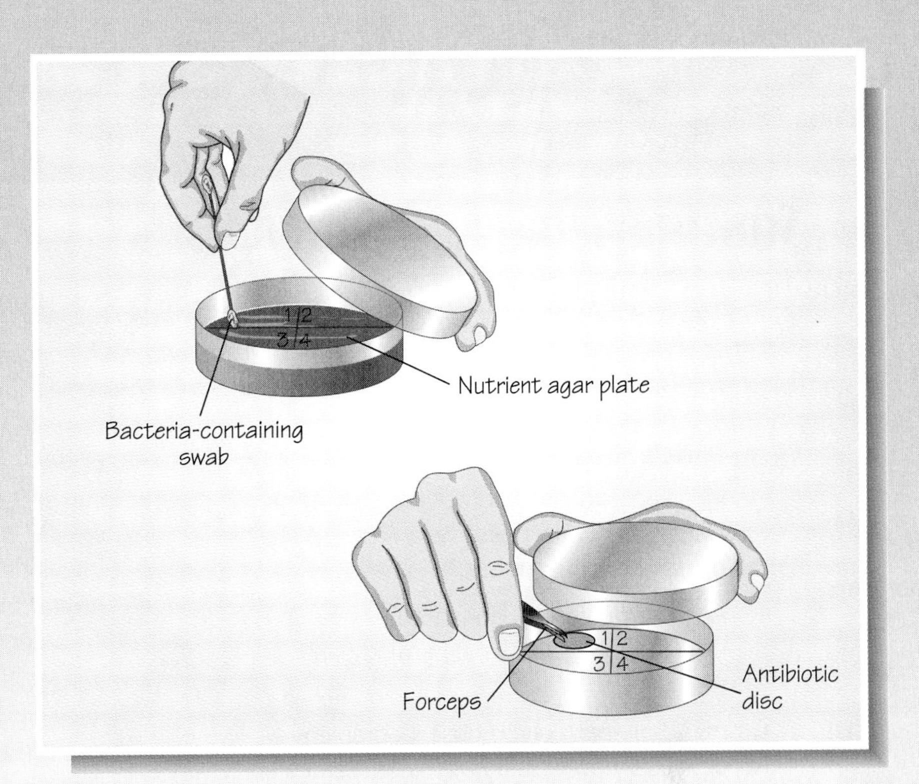

Observations

How is the growth of bacteria affected by the antibiotic discs?

Analysis and Conclusions

1. What factor represents the selection pressure in this experiment?
2. Many of the infectious bacteria—such as *Staphylococcus*—have evolved a resistance to antibiotics. How has this occurred? What is the selection pressure? The adaptation?
3. Do the results of your experiment support your hypothesis?

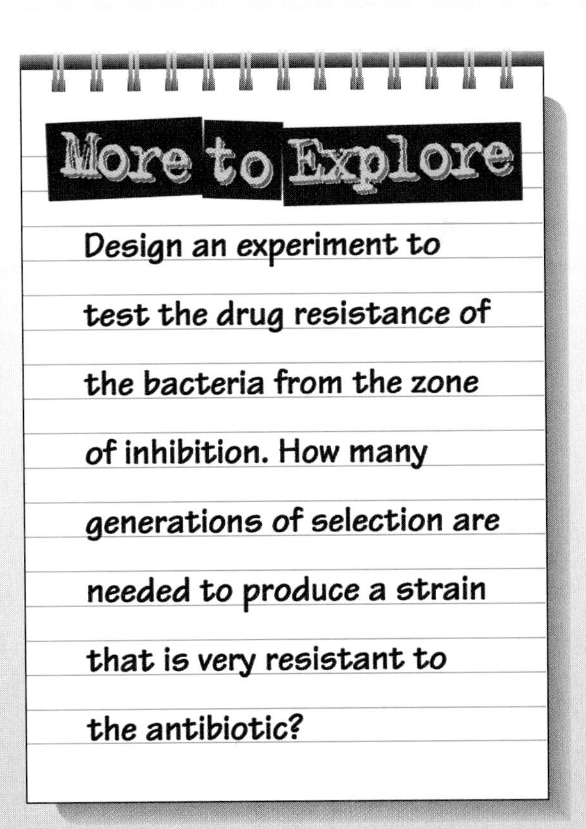

More to Explore

Design an experiment to test the drug resistance of the bacteria from the zone of inhibition. How many generations of selection are needed to produce a strain that is very resistant to the antibiotic?

Study Guide

Summarizing Key Concepts

The key concepts in each section of this chapter are listed below to help you review the chapter content. Make sure you understand each concept and its relationship to other concepts and to the theme of this chapter.

11–1 Darwin "Meets" DNA

- Inheritable variation comes primarily from two kinds of changes in genetic material—mutations and gene shuffling.
- An organism's evolutionary fitness can be defined as its success in passing genes to the next generation.
- An adaptation can be described as any genetically controlled trait that increases an individual's ability to pass along copies of its genes.
- A species is defined as a group of similar-looking organisms that can breed with one another and produce fertile offspring. The members of two different species cannot interbreed and are therefore said to be reproductively isolated from one another.

11–2 Evolution as Genetic Change

- In genetic terms, evolution is defined as any change in the relative frequencies of alleles in the gene pool of a species.
- Natural selection alters the allele frequencies in a gene pool by selecting helpful mutations and by selecting variations that represent the highest fitness.
- A random change in allele frequency is called genetic drift.
- For speciation to take place, a population must evolve enough genetic changes so that breeding cannot occur between the emerging genetically different groups.
- Gradualism and punctuated equilibrium are two views of the pattern of evolution.

11–3 How Species Change

- Organisms display a variation in phenotypes. The variation in phenotypes often corresponds to a variation in fitness. Natural selection chooses the fittest individuals to survive and reproduce, slowly causing the distribution of phenotypes to change. When the change is large enough to prevent interbreeding between the populations, a distinct species has evolved.

Reviewing Key Terms

Review the following vocabulary terms and their meaning. Then use each term in a complete sentence.

11–1 Darwin "Meets" DNA

mutation
reproductive isolation
gene pool

11–2 Evolution as Genetic Change

genetic drift
speciation
gradualism
punctuated equilibrium

Recalling Main Ideas

Choose the letter of the answer that best completes the statement or answers the question.

1. Which is a source of inheritable variation in organisms?
   - **a.** climate
   - **b.** struggle for existence
   - **c.** gene shuffling
   - **d.** punctuated equilibrium

2. A normal distribution of phenotypes is observed for traits controlled by
   - **a.** many genes.
   - **b.** 2 genes.
   - **c.** 1 gene.
   - **d.** 0 genes.

3. In genetic terms, evolution is
   - **a.** the diversity of living things.
   - **b.** a change in the relative frequencies of alleles in the gene pool of the species.
   - **c.** natural selection acting on inheritable variation.
   - **d.** a combination of gradual and rapid change in living things.

4. A helpful mutation causes the corresponding allele's frequency to
   - **a.** increase.
   - **b.** decrease.
   - **c.** remain unchanged.
   - **d.** double over 10 generations.

5. Natural selection acting on a normal distribution of phenotypes in which the fittest individuals correspond to the center of the graph is called
   - **a.** disruptive selection.
   - **b.** directional selection
   - **c.** stabilizing selection.
   - **d.** artificial selection.

6. A random change in the frequency of an allele in a gene pool is called
   - **a.** gradualism.
   - **b.** gene shuffling.
   - **c.** speciation.
   - **d.** genetic drift.

7. Speciation in nature usually occurs because
   - **a.** populations become separated by a geographical barrier.
   - **b.** only the fittest organisms survive.
   - **c.** of artificial selection.
   - **d.** of genetic drift.

8. Two populations that occupy the same habitat but do not share a gene pool
   - **a.** belong to the same species.
   - **b.** are reproductively isolated.
   - **c.** are geographically isolated.
   - **d.** undergo disruptive selection.

Putting It All Together

Using the information on pages xxx to xxxi, complete the following concept map.

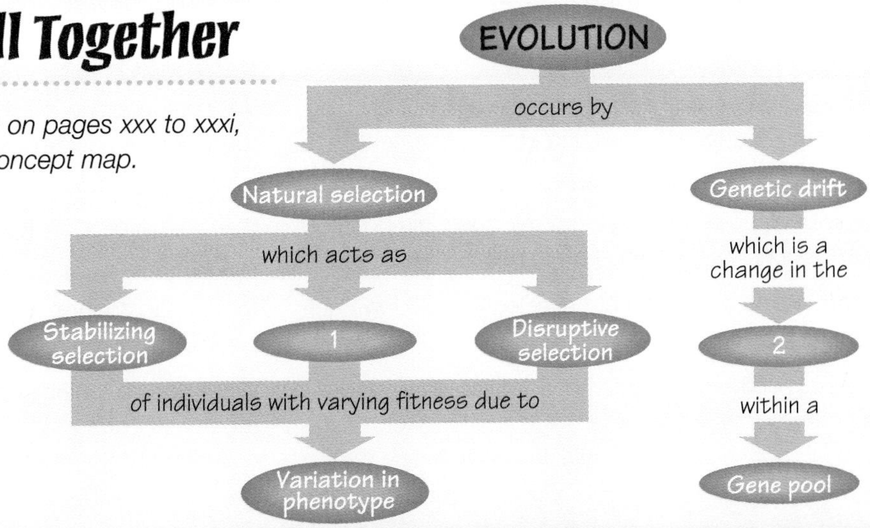

Assessment

Reviewing What You Learned

Answer each of the following in a complete sentence.

1. What area of biology has made possible a better understanding of evolution?
2. What is a gene? An allele?
3. What is a mutation?
4. How common is genetic variation in nature?
5. What is meant by a polygenic trait?
6. How are the genetic definitions of fitness and adaptation different from Darwin's?
7. What is meant by reproductive isolation? Define the term species in this context.
8. Explain the concept of a gene pool.
9. What happens to the frequency of an allele as a result of a harmful mutation? A helpful mutation?
10. Identify three ways in which natural selection affects the distribution of variation in polygenic traits.
11. What is genetic drift? In what kinds of situations is it more likely to occur?
12. What conditions give rise to speciation?
13. What is gradualism? Punctuated equilibrium?
14. Describe the results of the study of birds on Daphne Major, one of the Galapagos Islands.

Expanding the Concepts

Discuss each of the following in a brief paragraph.

1. Compare the variations produced by single-gene and polygenic traits.
2. Natural selection acts on phenotype and not on genotype. Explain the meaning of this statement.
3. Sexual reproduction shuffles genes within a gene pool, causing variation. Why does this not cause a change in relative allele frequencies?
4. Describe the action of natural selection on polygenic traits.
5. In which of the three types of selection (stabilizing, directional, or disruptive) is evolution likely to be minor or absent? Explain your answer.
6. How might genetic changes lead to the formation of new species?
7. Why are isolating mechanisms important to the formation of new species?
8. Explain why an organism's phenotype is usually the driving force for its evolution.
9. Why is the role of chance as the mechanism for genetic change more significant in smaller populations?
10. What does a biologist mean when stating that a particular organism is better adapted to its environment than another of that species?

Extending Your Thinking

Use the skills you have developed in this chapter to answer the following.

1. **Hypothesizing** In a small group of islands, it was noted that tortoise shells were of two major types—those with high neck arches and those with low neck arches. Formulate a hypothesis to explain the evolution of this difference. Then discuss the role of environmental and genetic factors that might have had selective value.
2. **Relating** Harmful gene mutations are normally harmful only in the homozygous recessive condition, in which the individual has two copies of the recessive allele. Discuss why the removal of the harmful gene from the population is almost impossible.
3. **Evaluating** "The complete control of detrimental diseases is highly unlikely." Discuss the validity of this statement.
4. **Constructing a model** Construct a model to simulate genetic drift due to a small population separating from a large population and establishing a new, distinct gene pool. Will evolution occur? Explain your reasoning.
5. **Hypothesizing** Orchids and insects are often quoted as examples of coevolution—the evolution of two or more closely interacting species that serve as the selection force for one another. Formulate a hypothesis to explain how species might coevolve.

Applying Your Skills

Natural Selection in Action

A mouse population exhibits variation in fur color. Mice that have more white fur are easier prey for hawks, and those with darker fur are better camouflaged in their environment. The allele for dark fur is dominant to the allele for white fur.

1. Over a number of years, there was an increase in hawk population. What will this do to the allele frequencies for fur color in mice? Write a paragraph to answer the question.
2. Draw a graph to help explain your answer.

• GOING FURTHER •

3. Over many generations of mice in this environment, what adaptation will allow the mice to survive and reproduce? Will this affect the hawk population? Explain your answer.

CHAPTER 12

The Origins of Biodiversity

FOCUSING THE CHAPTER THEME: Unity and Diversity

12–1 The Unity of Life

- **Discuss** adaptive radiation and convergent evolution as origins of biodiversity.
- **Examine** genetic evidence for the unity of life.

12–2 Biodiversity

- **Define** biodiversity and **explain** its importance.

BRANCHING OUT *In Depth*

12–3 DNA: A Storehouse of History

- **Explore** how changes in DNA can be used to mark time.

LABORATORY INVESTIGATION

- **Predict** the relatedness of several species from their amino acid sequences.

Biology and Your World

BIO JOURNAL

How do these butterflies differ from one another? How are they the same? Write the answers to these questions in your journal.

A display of diversity among butterflies

The Unity of Life

SECTION 12–1

GUIDE FOR READING

- **Compare** adaptive radiation and convergent evolution.
- **Examine** molecular evidence for the unity of life.

MINI LAB

- **Classify** various parts of organisms by structural or functional similarity.

THE VOLCANIC CRATER OF Solfatarra near Naples, Italy, isn't the sort of place you would expect to find much life. Superheated steam bubbles out of boiling mud that is laced with sulfuric acid. Yet these bubbling mud pots are home to bacteria that are as different from humans as any form of life could be. Unlike humans, the bacteria are poisoned by oxygen. Yet they thrive in water that would boil humans alive, and they eat sulfur. Nevertheless, some of their genes are astonishingly similar to genes in your cells! The story behind this remarkable genetic similarity is nothing less than the story of life itself.

Evolving Differences

Ever since Darwin, biologists have known that all organisms on Earth are related. But relationships among some organisms are easier to see than others. A chimpanzee looks a lot like its siblings, a bit less like a gorilla, and even less like a dog, a bird, or a dolphin. Yet if you were to look closely at all these animals, you would see similarities in body structures. One such similarity is that they all have backbones. Their arms, legs, wings, and flippers are formed from similar bones that have been modified in different ways.

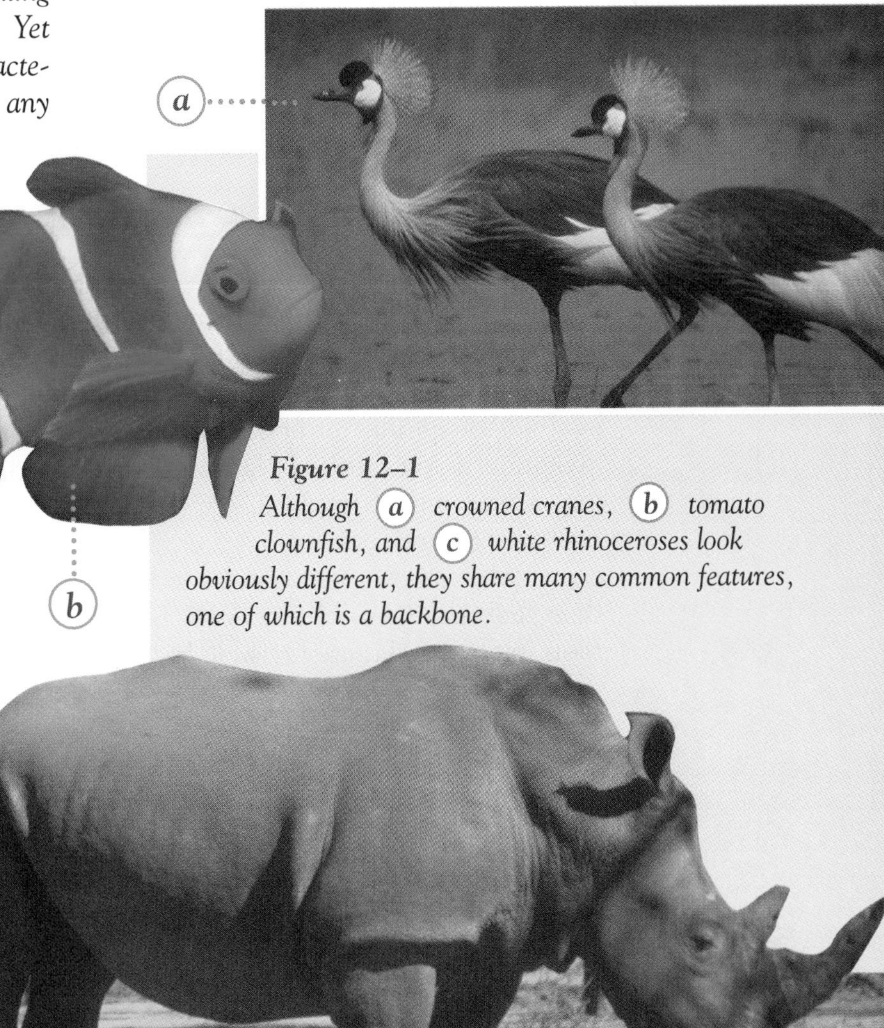

Figure 12–1
Although (a) crowned cranes, (b) tomato clownfish, and (c) white rhinoceroses look obviously different, they share many common features, one of which is a backbone.

ADAPTIVE RADIATION

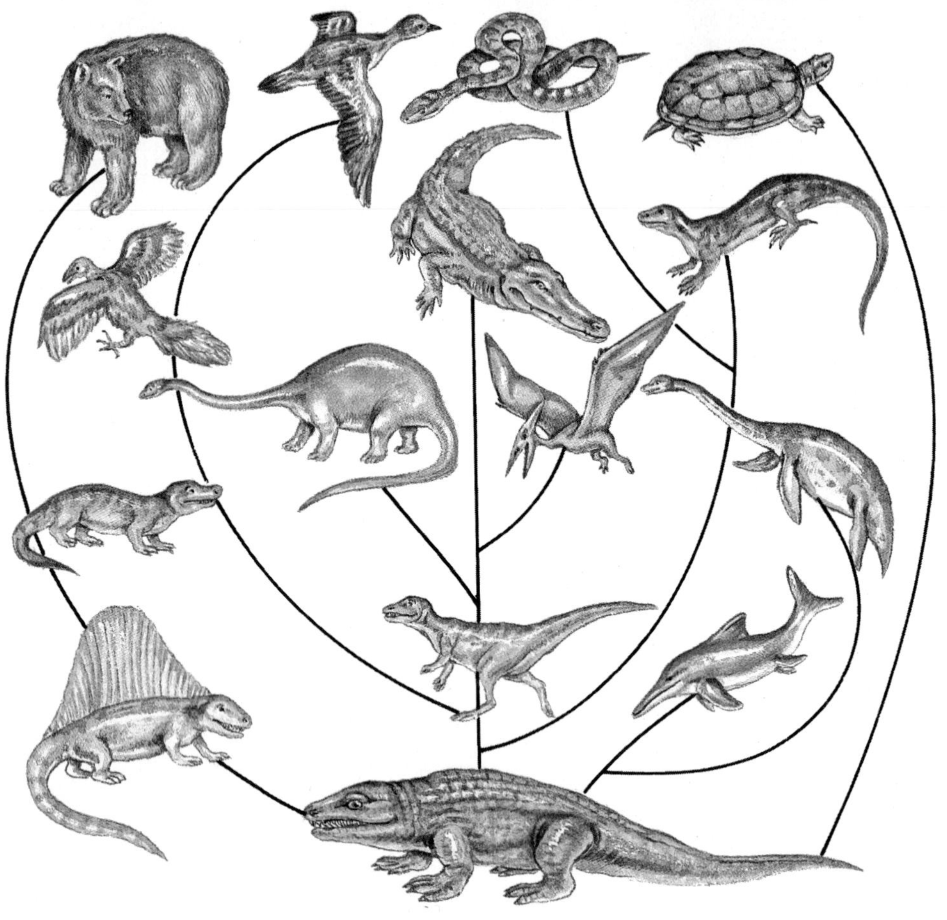

Figure 12–2
The illustration shows the formation of several different species from an adaptive radiation of ancient reptiles. Dinosaurs were some of the most spectacular products of this adaptive radiation.

You may recall that these homologous structures are formed from similar tissues.

How and why do such structures evolve? You can better understand their evolution if you think of plant and animal adaptations as biological tool kits. Each species and each major group of organisms has its own version of a tool kit that performs essential functions. Some tools in the tool kit are invisible because they involve biochemical processes such as respiration and protein synthesis. More visible parts—such as lungs, claws, or wings—are used for breathing, catching food, and moving around.

Recall that natural selection works by selecting inheritable variations in plant and animal characteristics. It is therefore easy for natural selection to change the shape or size of a particular tool in ways that change its function. You have seen how this process can increase beak size in Galapagos finches. Natural selection can work the same way for almost any trait that is under genetic control. But it is much more difficult (and therefore much less common) for natural selection to produce a kit with an entirely new set of tools. Simply put, evolution can modify existing tool kits much more easily than it can generate entirely new ones.

Adaptive Radiation

Every now and then, however, a new tool kit does evolve. Fossil records show several instances of diversity occurring in a newly evolved species in a relatively short period of time. This process is known as **adaptive radiation.** Adaptive radiation also occurs when an organism or a group of organisms colonizes a new area where other species that compete for life's necessities are lacking. In both these situations, selection and adaptation lead to the formation of a new species. **When a newly evolved species or a group of organisms in a new area evolve—sometimes somewhat quickly—into different species that live in different ways, this pattern of evolution is known as adaptive radiation.**

Figure 12–3
Although the wings of the (a) bald eagle, the (b) fishing bat, and the (c) dragonfly serve the same function, they have evolved from different parts of the animals' tool kits. The wings of these animals are an example of analogous structures.

Darwin's finches are an example of the type of adaptive radiation that often happens on isolated island groups. In this case, more than a dozen species evolved from a single founding species that colonized the islands from the mainland of South America.

☑ **Checkpoint** What is the process of adaptive radiation?

Convergent Evolution

Often, adaptive radiations in different groups of organisms produce species that are adapted to similar feeding habits or ways of moving from place to place. **Convergent evolution** is the name given to this process. **In convergent evolution, unrelated species may independently evolve superficial similarities because of adaptations to similar environments.**

For example, woodpeckers get their food by drilling into tree bark to uncover grubs and other insects. But several other animals—such as the honeycreeper, the striped opossum, and the aye-aye—also obtain their food in this way. There are also several groups of animals that fly. Bird and bat wings are similar but not identical. Close observation reveals that their front limb bones support flight structures in different ways. Insect wings are completely different. Structures such as these—which are similar in appearance and function but are developed from anatomically different parts—are called **analogous structures.** Analogous structures are the hallmark of convergent evolution.

☑ **Checkpoint** What is convergent evolution?

Problems With Distant Relations

Homologous structures in adult organisms and fossils often enable biologists to piece together evolutionary history and determine how various species are related. But this approach has its limits. Adult body parts cannot offer much help in deciding how humans might be related to animals such as insects or snails.

Looking at structures in the early stages of embryo development can help in some cases. But similarities in embryo development cannot help with single-celled organisms because they have neither embryos nor body parts that correspond to those of humans. Interestingly, the most exciting clues to evolutionary history and relationships among these organisms have come from the work of researchers who weren't even studying these topics!

MINI LAB .... Classifying ....

Structure or Function?

PROBLEM *How can you **classify** various parts of organisms by structural or functional similarity?*

PROCEDURE

1. Examine the photographs of the organisms provided by your teacher.
2. Identify a part in each organism that could be similar in structure or function to a part in another organism. Record this information.

ANALYZE AND CONCLUDE

1. What are the terms used to indicate these structures?
2. Classify the organisms into two groups: those with parts exhibiting structural similarities and those with parts exhibiting functional similarities.
3. What patterns of evolution are indicated by these similarities?

The Molecular Unity of Life

Recently, molecular biologists developed techniques to read information coded in DNA letter by letter. When that information first began to accumulate, most researchers used it to answer questions about the workings of genes, genetic diseases, or cell chemistry. These applications continue today, of course. But as more and more DNA data were gathered, evolutionary biologists became fascinated with DNA, and molecular biologists were astonished to find themselves interested in evolution. Thus began an entirely new age of exploration in evolution.

INTEGRATING CHEMISTRY

What is the chemical structure and function of proteins, amino acids, and DNA?

Reading DNA

What information did molecular biology provide? **As information on DNA sequences accumulated, biologists realized that many genes were shared by a wide range of organisms.** Researchers knew that amino acids—the basic building blocks of proteins—were common to all forms of life. But new data have shown that whole stretches of DNA and, therefore, entire sequences of amino acids in many proteins were practically identical in nearly every organism studied!

Of course, as you can see in ***Figure 12–4,*** the more closely two organisms are related, the more closely their genes resemble each other. And organisms that differ as much as bacteria differ from humans contain many different genes. Bacteria that live in hot springs and mud pots, for example, carry genes that help them to survive high temperatures. Humans lack those genes, but they do have genes that carry instructions on how to walk upright on two legs.

☑ ***Checkpoint*** What did researchers discover by studying DNA?

Identical Tool Kits

Interestingly, certain genes in bacteria that live in hot springs are almost identical to genes and proteins found in every other living organism—from yeast to fruit flies to humans. One common feature of all living things, for example, is the ribosome—the part of life's molecular tool kit that serves as a protein factory. When molecular biologists compared DNA sequences that build bacterial ribosomes with genes that direct the assembly of human ribosomes, they found them to be astonishingly similar. Certainly, these genes and proteins are homologous.

Now think about that for a minute. First, you should know that some hot-springs' bacteria are more similar to the very first forms of life on Earth than any other living organisms. Yet the genes that carry instructions to build their ribosomes are incredibly similar to ours! Both molecular biologists and evolutionary biologists find this fact amazing.

Figure 12–4
Cytochrome-c is a protein molecule that organisms need for cellular respiration. Each number in this illustration represents the number of amino acid substitutions in the cytochrome-c of various organisms, measured from the previous branch point.

CYTOCHROME-C FAMILY TREE

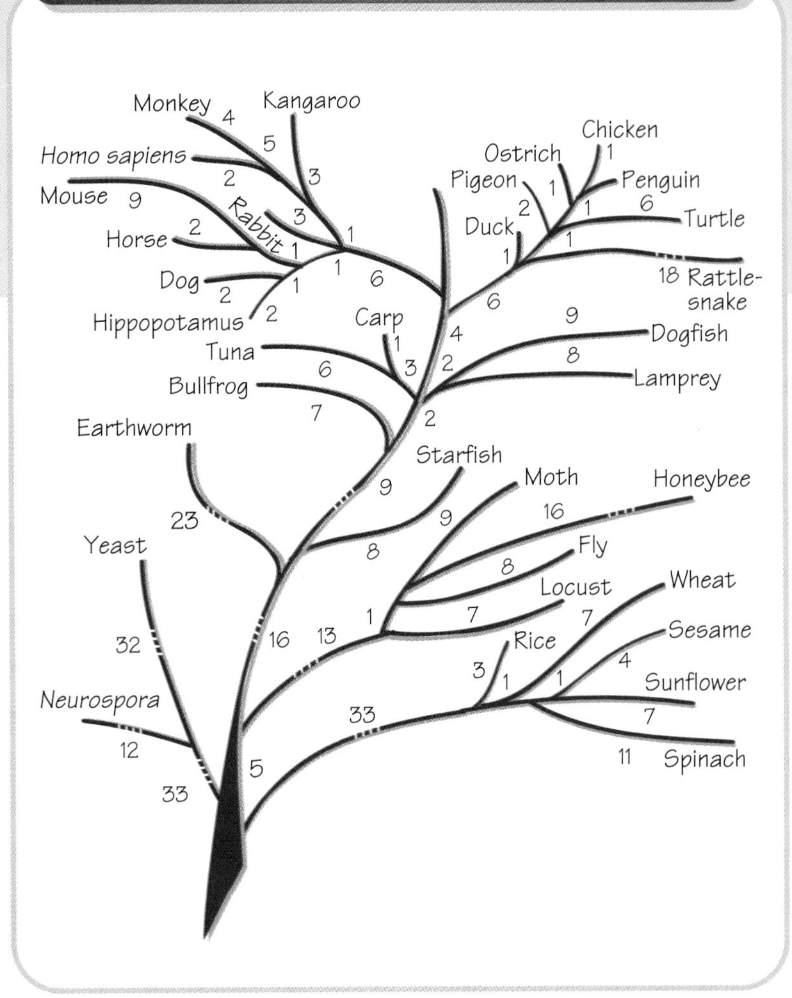

Muscle Proteins—In Yeast!

The molecular unity of life doesn't end there. Researchers were surprised to find a yeast gene that codes for a protein called myosin. Why were they surprised? Because most biologists know that myosin is found in the muscle cells of humans and other multicellular animals. In our bodies, long fibers of myosin interact with other proteins to cause muscles to contract, which enables us to move. But yeasts don't have muscles. And they don't move.

Or do they? Well, even in yeasts, certain cellular components move around within the cell. Myosin in yeasts interacts with other cell proteins to make that movement possible. Molecular biologists have found similarities among many other genes in a wide range of organisms, including plants. How can scientists explain these observations?

The original form of myosin made it possible for parts of cells to move. Since then, as life diversified, evolution of the original myosin genes produced the forms that help your body to move. Time and time again, diversity evolved by adding up changes in original tools, rather than by inventing new tools from scratch. Over time, the bits and pieces of different tool kits were mixed and modified to produce the great variety of living organisms on Earth.

Section Review 12-1

1. How do adaptive radiation and convergent evolution **compare?**
2. **Examine** molecular evidence for the unity of life.
3. **Critical Thinking—Relating** How can adaptive radiation explain the unity as well as the diversity of life?
4. **MINI LAB** How can you **classify** structures of organisms by structural and functional similarity?

SECTION 12–2

Biodiversity

Guide for Reading

- **Define** biodiversity.
- **Explain** the importance of genetic diversity.

IN THE WORDS OF EDWARD O. Wilson, Professor of Science at Harvard University, "*. . . imagine yourself on a journey upward from the center of the Earth, taken at the pace of a leisurely walk. For the first twelve weeks you travel through furnace-hot rock and magma devoid of life. Three minutes to the surface—five hundred meters to go—you encounter the first organisms: bacteria feeding on nutrients that have filtered into the deep water-bearing strata. You breach the surface and for ten seconds glimpse a dazzling burst of life—tens of thousands of species of microorganisms, plants, and animals within horizontal line of sight. Half a minute later almost all are gone. Two hours later only the faintest traces remain. . . .*"

The thin layer teeming with life at the Earth's surface is called the biosphere. And the diversity of life in the biosphere is a subject of utmost importance to humans.

What Is Biodiversity?

During the billions of years since life first appeared on Earth, genes and the organisms they produce have been constantly changing. Yet the similarities in the genes and DNA of all organisms testify that life got going only once, then diversified by evolving, combining, and shuffling changes in the molecular makeup of living organisms. One adaptive radiation after another has generated this **biodiversity**—the variety of living organisms. **Biodiversity is the variety of organisms, the genetic information they contain, and the biological communities in which they live.**

Figure 12–5 (a) *Mushrooms,* (b) *Volvox, a type of green alga (magnification: 15X), and* (c) *an arctic fox are a meager representation of Earth's staggering biodiversity.*

Biodiversity is an important part of the living world as we know it. Biodiversity is also an enormous and invaluable treasure.

☑ **Checkpoint** What is biodiversity?

Importance of Biodiversity

Although all forms of biodiversity are interconnected, researchers often talk about biodiversity on three levels: **ecosystem diversity, species diversity,** and **genetic diversity.** Biodiversity is important in sustaining many species, including the human species *Homo sapiens*. Unfortunately, human activity around the world seriously threatens diversity at each of these levels.

Ecosystem Diversity

Ecosystems are communities of organisms and their environments. Ecosystem diversity includes the variety of habitats, living communities, and ecological processes in the living world. The diversity of ecosystems on Earth is remarkable because organisms have adapted to nearly every part of our planet. And everywhere there is life, there are complex combinations of organisms.

Species Diversity

When most people talk about the diversity of life, they are probably referring to species diversity. Species diversity refers to the enormous variety of living organisms on Earth, and it is the easiest kind of diversity to see and relate to. About 1.4 million species have been identified and named so far, but different estimates suggest that there may be anywhere from 4 to 50 million more species awaiting discovery. The number and variety of species in an ecosystem can profoundly influence that system's stability, productivity, and value to humans. Sometimes the presence or absence of a single species can completely change the nature of life in an area.

Figure 12–6
Elephants feeding on and trampling through trees in the Moremi Wildlife Reserve in Botswana play a vital role in shaping the ecosystem. When elephants are eliminated from an area by actions such as poaching, trees grow and change what was once grassland into woodland.

Genetic Diversity

Genetic diversity refers to the sum total of all the different forms of genetic information carried by all organisms living on Earth today. Within each species, genetic diversity refers to the total variety of all alleles, which are the different forms of all the genes present in the gene pool of that species.

Sometimes genetic diversity in a species is found between groups of organisms living in different places. In India, farmers have traditionally cultivated thousands of different varieties of rice with different alleles and allele frequencies in their gene pools. In South America, the same is true of potatoes. Other times, the genetic makeup of individual organisms within a population may vary a great deal.

☑ **Checkpoint** What are the three levels of biodiversity?

Problem Solving

INTERPRETING DATA

Measuring Biodiversity

As the culmination of the year's activities, a teacher wanted her biology class to develop a display that showed the biodiversity on the high school campus. She issued this challenge to her students: Develop a way of measuring and displaying campus biodiversity.

First, the entire class had a discussion regarding this task and possible methodologies. After developing a method of measuring the number and distribution of different organisms, the class was divided into working groups to sample the various ecosystems surrounding the school. Then the groups worked together to compare the diversity among ecosystems and to display their work. If you were in this class, how would you suggest that the class accomplish each of these tasks?

NUMBER OF CURRENTLY KNOWN LIVING SPECIES

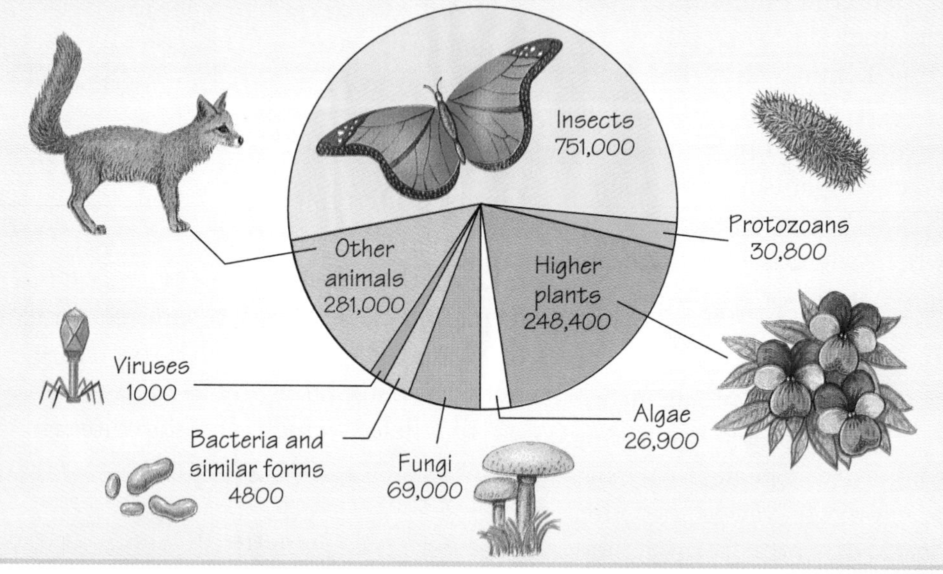

Insects 751,000
Other animals 281,000
Higher plants 248,400
Protozoans 30,800
Algae 26,900
Fungi 69,000
Bacteria and similar forms 4800
Viruses 1000

THINK ABOUT IT

1. Should each group's method of measuring biodiversity and the distribution of organisms be the same in order to make comparisons among ecosystems? Explain your answer.
2. If more than one class was involved in this project, would measurements of biodiversity and the distribution of organisms need to be done at the same time of year for valid comparisons to be made among ecosystems? Explain your answer.
3. How would you organize and present the data of the various groups so that the main findings, comparisons, and conclusions could be easily seen in the class display?

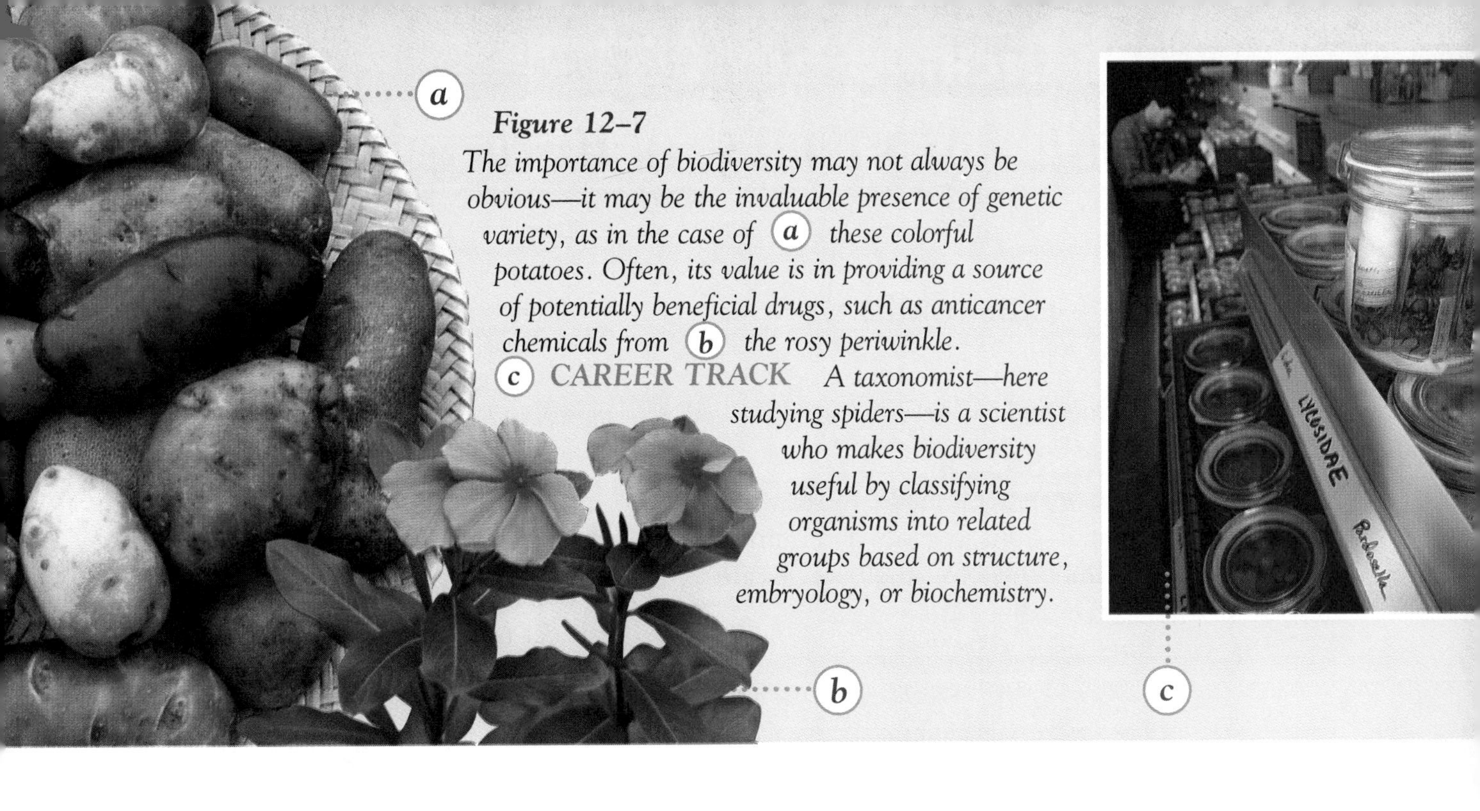

Figure 12–7
The importance of biodiversity may not always be obvious—it may be the invaluable presence of genetic variety, as in the case of (a) these colorful potatoes. Often, its value is in providing a source of potentially beneficial drugs, such as anticancer chemicals from (b) the rosy periwinkle. (c) CAREER TRACK *A taxonomist—here studying spiders—is a scientist who makes biodiversity useful by classifying organisms into related groups based on structure, embryology, or biochemistry.*

Importance of Genetic Diversity 

In a sense, genetic diversity provides the foundation on which the rest of biodiversity is built. **Genetic diversity gives rise to inheritable variation, which, as you have learned, provides the raw material for evolution.** Without inheritable variation on which natural selection can operate, the ability of most species to adapt to changing environments is limited. Because human activity is altering both local and global environments faster than ever, this adaptability may be essential for many species to survive.

What's more, if the genetic diversity within a species decreases too much, an individual organism's strength, ability to combat disease, and ability to reproduce may be seriously threatened. Thus, decreased genetic diversity can threaten a species with extinction.

An important point to remember is that the variety of genes carried by all living species is the result of millions of years of random mutation, natural selection, and genetic drift. An enormous number of diverse and potentially useful characteristics in different plants and animals have evolved as a result of these changes—a priceless and irreplaceable genetic resource for plant and animal breeders and genetic engineers. ●

INTEGRATING BIOLOGY AND SOCIETY

In what ways is genetic diversity important to society?

Thomas Eisner captures the potential of genetic diversity in this metaphor: "A biological species is not merely a hardbound volume of the library of nature. It is also a loose-leaf book, whose individual pages, the genes, might be available for selective transfer and modification of other species."

Section Review 12-2

1. **Define** biodiversity.
2. **Explain** the importance of genetic diversity.
3. **Critical Thinking—Hypothesizing** Formulate a hypothesis to explain why small populations of species are more likely to be threatened with extinction.

DNA: A Storehouse of History

GUIDE FOR READING

- **Explain** the concept of molecular clocks.

MINI LAB

- **Construct a model** to show how mutations in genes determine the degree to which species are related.

WHAT DO A QUARTZ CRYSTAL clock and a pendulum clock have in common? Any instrument used to mark time requires a periodically repeating process. In a pendulum clock, that process is the periodically swinging pendulum. In a quartz clock, that process is the periodic vibrations of a quartz crystal under certain conditions. What do pendulum clocks and quartz crystal clocks have to do with DNA and history?

Believe it or not, researchers have reason to think that changes in DNA that occur with some regularity may be a process they can use to mark time in their study of evolution. They hope that in the DNA sequences of various organisms they will find a record of the very deepest and most ancient links between all living organisms.

Molecular Clocks

Scientists hope that changes in DNA can act as **molecular clocks**—biological timekeepers that record how long ago living organisms shared a common ancestor. **The idea of molecular clocks comes from the notion that simple mutations in DNA should occur at a somewhat constant rate.** If the mutations are neither helpful nor harmful, they might accumulate at a constant rate over time. And if that's the case, it should be possible to look at homologous DNA sequences in different organisms to determine how long ago those genes branched off from a common ancestor.

In practice, this idea is complicated for several reasons. First, different positions in DNA sequences accumulate mutations faster than others. Second, although no one knows why, different branches on the evolutionary tree accumulate mutations at different rates. And third, some genes more than others are under stronger pressure from

Figure 12–8
The connection between (a) the clock in Grand Central Station in New York and (b) a DNA model is simply that changes in the DNA of organisms may give scientists a way to mark time in Earth's living history.

natural selection not to change. Thus, there is not just one molecular clock in the genome—a complete set of chromosomes of an organism—but many, all "ticking" at different rates.

☑ **Checkpoint** What changes in DNA serve to mark the passage of time?

Slow Clocks, Fast Clocks

Now here's the interesting point: As long as researchers can get some clues on the speed with which clocks in different parts of the genome "tick," differences in speed can be very useful for timing different kinds of evolutionary events.

To understand why, think of a conventional clock with an hour hand, a minute hand, and a second hand. If you want to time a very brief event—say, a 100-meter dash—you must pay close attention to the second hand because it's the only hand that ticks fast enough to tell the difference between one runner and another. However, if you want to time a marathon that lasts more than two hours, you need not only the minute hand but the hour hand as well. Why? Because the second hand and minute hand both move too quickly to time the entire event by themselves. Only the hour hand moves slowly enough to be in a unique position at both the beginning and the end of the race.

The same is true of molecular clocks. To time recent evolutionary events, researchers will use a clock that ticks fairly quickly. But to estimate how long ago humans and bacteria shared a common ancestor, they must use clocks that tick very slowly, because they are trying to time something that happened long ago.

Molecular Clocks

PROBLEM *How can mutations in genes help determine the degree to which species are related? **Construct a model** to answer this question.*

PROCEDURE

Answer the questions on the basis of the following hypothetical information: Gene A is composed of a DNA sequence that mutates at the rate of one mutation every 4 hours, gene B is composed of a DNA sequence that mutates at the rate of one mutation every 10 minutes, and gene C is composed of a DNA sequence that mutates at the rate of one mutation every 10 seconds.

ANALYZE AND CONCLUDE

1. After 24 hours, how many mutations would each gene have accumulated?
2. Which of the three genes in this model would be the best one to determine how long ago modern humans shared a common ancestor with a paramecium and a dog? Explain your answer.
3. Would there be any limitations to this model? Explain your answer.

Section Review 12–3

1. **Explain** the concept of molecular clocks.
2. **MINI LAB** How do mutations in genes determine the degree to which species are related? **Construct a model** to answer this question.
3. **BRANCHING OUT ACTIVITY** Make a list of various timekeeping methods that are used today. **Identify** the periodically repeating processes that serve to measure time.

Laboratory Investigation

Who's Related to Whom?

Natural selection and evolution would predict that species which diverged from one another relatively recently in the history of life on Earth will share more genetic similarities than species that diverged from one another earlier. Because proteins are programmed by genes, a comparison of the amino acid sequences of their proteins would indicate the relatedness among species.

Problem

How can you **predict** the degree to which species are related from the amino acid sequences in their proteins?

Materials

Data table of amino acid sequences of the protein cytochrome-c

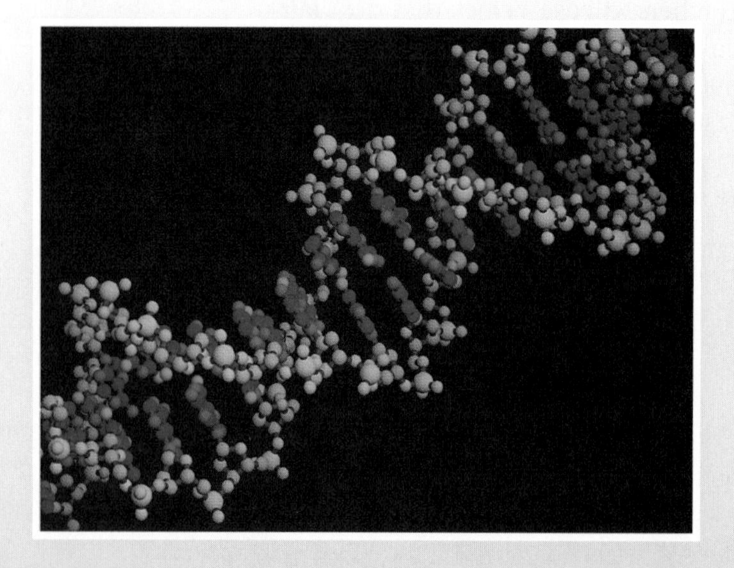

Procedure

1. Use the amino acid sequences to calculate the percentage difference between a fruit fly and a sunflower. To calculate the percentage difference between two species, determine the number of amino acids that differ between the two species, then divide that number by the total number of amino acids in the sequence. (The sequences provided have 60 amino acids.)
2. Make an approximate prediction regarding the percentage difference of the cytochrome-c amino acid sequences of a human and a chimpanzee. Record this information in a data table.
3. To test your prediction, repeat step 1 for a human and a chimpanzee.
4. Repeat steps 1 to 3 using the amino acid sequences of a horse and a donkey.
5. Repeat steps 1 to 3 using the amino acid sequences of a chicken and a turkey.
6. Repeat steps 1 to 3 using the amino acid sequences of birds, rattlesnakes, and mammals.
7. Repeat steps 1 to 3 using an animal and the sunflower (plant) or yeast (fungus).

AMINO ACID SEQUENCES IN CYTOCHROME-C

Human	GDVEKGKKIFIMKCSQCHTVEKGGKHKTGPNLHGLFGRKTGQAPGYSYTAANKNKGIIWG
Chimpanzee	GDVEKGKKIFIMKCSQCHTVEKGGKHKTGPNLHGLFGRKTGQAPGYSYTAANKNKGIIWG
Rhesus monkey	GDVEKGKKIFIMKCSQCHTVEKGGKHKTGPNLHGLFGRKTGQAPGYSYTAANKNKGITWG
Horse	GDVEKGKKIFVQKCAQCHTVEKGGKHKTGPNLHGLFGRKTGQAPGFTYTDANKNKGITWK
Donkey	GDVEKGKKIFVQKCAQCHTVEKGGKHKTGPNLHGLFGRKTGQAPGFSYTDANKNKGITWK
Chicken	GDIEKGKKIFVQKCSQCHTVEKGGKHKTGPNLHGLFGRKTGQAEGFSYTDANKNKGITWG
Turkey	GDIEKGKKIFVQKCSQCHTVEKGGKHKTGPNLHGLFGRKTGQAEGFSYTDANKNKGITWG
Rattlesnake	GDVEKGKKIFTMKCSQCHTVEKGGKHKTGPNLHGLFGRKTGQAVGYSYTAANKNKGITWG
Fruit fly	GDVEKGKKLFVQRCAQCHTVEAGGKHKVGPNLHGLIGRKTGQAAGFAYTNANKAKGITWQ
Baker's yeast	GSAKKGATLFKTRCELCHTVEKGGPHKVGPNLHGIFGRHSGQAQGYSYTDANIKNVLTWD
Sunflower	GDPTTGAKIFKTKCAQCHTVEKGAGHKQGPNLNGLFGRQSGTTAGYSYSAANKNMAVIWE

The letters represent the amino acids as shown:
G = glycine, A = alanine, V = valine, L = leucine, I = isoleucine, M = methionine, F = phenylalanine, W = tryptophan, P = proline, S = serine, T = threonine, C = cysteine, Y = tyrosine, N = asparagine, Q = glutamine, D = aspartic acid, E = glutamic acid, K = lysine, R = arginine, H = histidine

Observations

1. How different are humans and chimps in the first 60 amino acids of the cytochrome-c sequence?
2. What pattern of relatedness did you observe from the percentage differences among various species?

Analysis and Conclusions

1. Which species share recent common ancestors? Distant common ancestors?
2. Would you be able to draw a "family tree" of the organisms named, showing when they diverged from one another? Would such a family tree be very accurate?
3. Some positions on the amino acid sequence are the same for all cytochrome-c molecules shown. Why might this be so?
4. As a rule, what general conclusion can you draw regarding how closely related species are and how their cytochrome-c amino acid sequences compare?

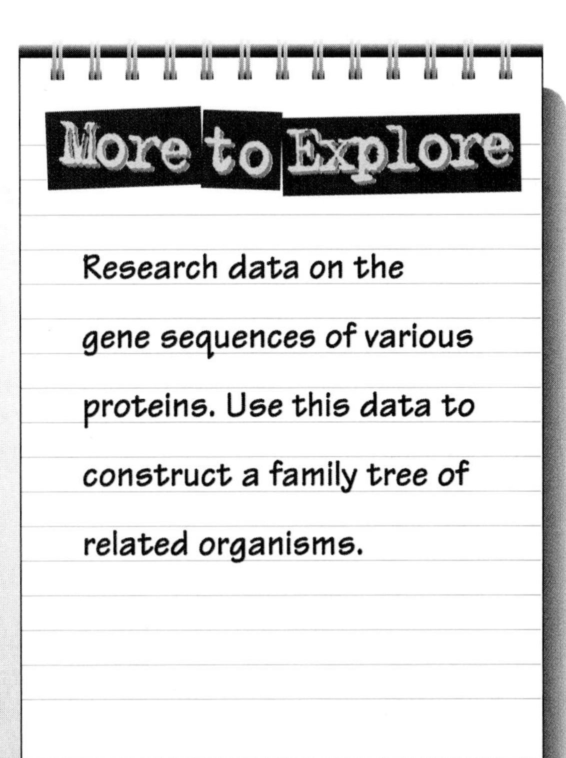

Study Guide

Summarizing Key Concepts

The key concepts in each section of this chapter are listed below to help you review the chapter content. Make sure you understand each concept and its relationship to other concepts and to the theme of this chapter.

12–1 The Unity of Life

- When a newly evolved species or a group of organisms in a new area evolve—sometimes somewhat quickly—into different species that live in different ways, the pattern of evolution is known as adaptive radiation.
- Unrelated species independently evolve superficial similarities because of adaptations to similar environments by a process known as convergent evolution.
- Analogous structures are similar in appearance and function but are developed from anatomically different parts.
- Information gathered on DNA sequences reveals that many genes are shared by a wide range of organisms. New data have shown that whole stretches of DNA and, therefore, entire sequences of amino acids in many proteins are practically identical in nearly every organism studied.
- A comparison of the amino acid sequences in the proteins of organisms can indicate the degree to which species are related.

12–2 Biodiversity

- Biodiversity is the variety of all the organisms, the genetic information they contain, and the biological communities in which they live.
- Ecosystem diversity consists of the variety of habitats, living communities, and ecological processes in the living world.
- Species diversity refers to the enormous variety of living organisms on Earth.
- Genetic diversity refers to the sum total of all the different forms of genetic information carried by all organisms living on Earth.
- Decreased genetic diversity can threaten a species with extinction.

12–3 DNA: A Storehouse of History

- Changes in DNA can act as molecular clocks—biological timekeepers that record how long ago living organisms shared a common ancestor.
- The idea of molecular clocks comes from the notion that simple mutations in DNA should occur at a somewhat constant rate.

Reviewing Key Terms

Review the following vocabulary terms and their meaning. Then use each term in a complete sentence.

12–1 The Unity of Life
adaptive radiation
convergent evolution
analogous structure

12–2 Biodiversity
biodiversity
ecosystem diversity
species diversity
genetic diversity

12–3 DNA: A Storehouse of History
molecular clock

Recalling Main Ideas

Choose the letter of the answer that best completes the statement or answers the question.

1. A new species or a species in a new area giving rise to many species, sometimes in a relatively short period of time, is called

a. convergent evolution. **c.** stabilizing selection.
b. adaptation. **d.** adaptive radiation.

2. Analogous structures are a sign of

a. directional selection. **c.** divergent evolution.
b. convergent evolution. **d.** adaptive radiation.

3. The more closely related two organisms are, the

a. greater the similarity in their genes.
b. less the similarity in their genes.
c. greater the occurrence of analogous structures.
d. greater the differences in their DNA sequences.

4. Structures that are similar in appearance and function but differ anatomically are said to be

a. analogous. **c.** vestigial.
b. homologous. **d.** convergent.

5. Which of the following is not a level of biodiversity?

a. genetic diversity **c.** species diversity
b. ecosystem diversity **d.** habitat diversity

6. Species diversity refers to the variety of

a. living organisms.
b. Earth's ecosystems.
c. alleles in the gene pool of a species.
d. species already extinct.

7. Genetic diversity refers to

a. the variety of alleles in the species.
b. phenotype variation.
c. genotype variation.
d. normal distribution.

8. Changes in DNA can mark time because

a. amino acid sequences in DNA change.
b. mutations occur at fairly constant rates.
c. mutations do not cause amino acid sequences to change.
d. all organisms have DNA.

Putting It All Together

Using the information on pages xxx to xxxi, complete the following concept map.

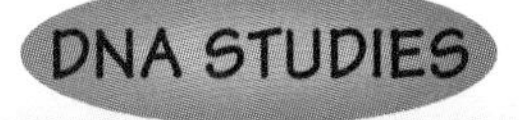

offer support to the concept of

Common descent

of the variety of living things on Earth called

the levels of biodiversity are

2

Species

3

Assessment

Reviewing What You Learned

Answer each of the following in a complete sentence.

1. Characterize adaptive radiation.
2. How do convergent evolution and adaptive radiation differ?
3. Give two examples of homologous and analogous structures. What criterion did you use to **classify** structures as analogous or homologous?
4. What are the applications of DNA studies?
5. How are the genes of bacteria and humans similar? How are they different?
6. How can common descent be considered evidence for evolution?
7. What are the levels of biodiversity? Which level is the foundation for the other two?
8. How might DNA act as a molecular clock?
9. What is a genome?
10. Why might some genes mutate at a faster rate than others?

Expanding the Concepts

Discuss each of the following in a brief paragraph.

1. Why is the study of DNA important to evolutionary biologists?
2. Can genes be homologous? Explain your answer.
3. If some human genes are similar in code and function to certain yeast genes, is it reasonable to assume that humans and yeasts share a common ancestor? Explain your answer.
4. How might gene modification be a key to understanding evolution?
5. Give evidence to support this statement: "Evolution can modify existing tool kits much more easily than it can generate entirely new ones."
6. How has human activity threatened each level of biodiversity?
7. Give evidence to support this statement: "The greater the genetic diversity, the greater the survival potential of a species."
8. Only a small portion of the DNA within a human cell is actually functional. What might be the purpose of the excess?
9. How might the number and variety of species actually influence an ecosystem's stability?
10. How can a molecular clock be used to determine the length of shared ancestry?

Extending Your Thinking

Use the skills you have developed in this chapter to answer the following.

1. **Hypothesizing** Ribosomes are remarkably similar throughout the six kingdoms. Formulate a hypothesis to explain this observation. Suggest a way to support your hypothesis.
2. **Analyzing** Obtaining and using energy is the primary function of all living things. Would you expect to find energy flow and control in genes or gene products from diverse organisms to be similar?
3. **Inferring** Homeoboxes are a series of repeatable DNA sequences in organisms. Find out how mutations to individual homeoboxes produce complex organisms such as an earthworm or a fruit fly.
4. **Constructing a model** Construct a model to simulate mutations in DNA as timekeepers.
5. **Evaluating** An enormous amount of money and time is being spent to collect, study, and use DNA for various purposes—such as to treat diseases and alter crop characteristics. Discuss the pros and cons of the collection, storage, and use of diverse genetic material.

Applying Your Skills

Observing Diversity in Action

Ecosystems usually contain an amazing diversity of species. Species diversity is the number of species living within an ecosystem.

1. Select an ecosystem in your area—a forest, field, pond, lake, grassland, or desert.
2. Within that ecosystem, identify as many species as possible.

• GOING FURTHER •

3. What might happen to the ecosystem if a new species with no natural predators was introduced?

UNIT 4

Ecology

CHAPTERS

"In the end, we will conserve only what we love, we will love only what we understand, we will understand only what we are taught."

— Baba Dioum

Rain forest in Olympic National Park in Washington State

CHAPTER 13

Energy and Nutrients

FOCUSING THE CHAPTER

THEME: Energy

13–1 Ecology: Studying Nature's Houses
- Define ecology and explain its importance.

13–2 Energy: Essential for Life's Processes
- Observe how energy flows through the biosphere.

13–3 Nutrients: Building Blocks of Living Tissue
- Describe how nutrients are cycled through the biosphere.

13–4 Food Webs: Who Eats Whom?
- Describe a food web and identify its trophic levels.

BRANCHING OUT *In Depth*

13–5 The Carbon Cycle: A Closer Look
- Explain how the carbon cycle is affected by human activity.

LABORATORY INVESTIGATION
- Analyze energy flow through a food web.

Biology and Your World

BIO JOURNAL

As living things, plants and animals share certain basic requirements to survive and grow. In your journal, list the needs that you share with those of a plant, an aphid, and a ladybug.

Adult ladybug feeding on aphids

Ecology: Studying Nature's Houses

GUIDE FOR READING

- **Define** ecology.
- **Explain** the importance of studying ecology.

THE STORIES SEEM TO POP UP in newspapers and on television everywhere nearly every day. "Mississippi Floods Hit the Midwest!" "American Songbirds Vanish . . . Why?" "New England Fisheries Disappear!" "As Ozone Holes Get Larger, Skin Cancer Risk Rises!" "Coral Reefs Die in the Florida Keys!"

At first glance, these headlines seem to deal with very different events. Do they have anything in common? And why are they mentioned here? Read on to find out the answers to these questions.

Ecological Research

All the events mentioned in these headlines have to do with **ecology**—its issues and concerns. **Ecology is the scientific study of interactions between different kinds of living things and between living things and the environments in which they live.** Ecology is named after the Greek word *oikos*, meaning "house," because ecology is the study of our house—the planet Earth—with all its living and nonliving parts.

Ecologists, the scientists who study ecology, examine different things in different ways, depending on the questions they ask. Ecologists ask questions about phenomena and organisms that range in size and scope from a single cell to the entire living planet.

Figure 13–1 (a) *This baby gorilla from Central Africa,* (b) *underwater coral reefs near Australia, and* (c) *forests in North Carolina represent examples of living things threatened by various changes in their environment.*

To study different parts of the biosphere, or the living world, ecologists use a wide range of techniques. Some use genetic fingerprints to identify single-celled organisms that inhabit the mud of coastal marshes. Others use radio transmitters to track migrating wildlife across continents. Still others use data gathered from satellites to monitor the temperature of the oceans. Today, ecological research makes use of three basic methods, summarized in ***Figure 13–2.*** ●

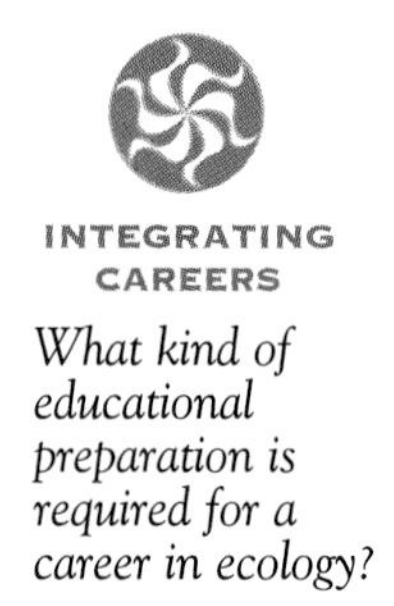

INTEGRATING CAREERS

What kind of educational preparation is required for a career in ecology?

Why Study Ecology?

Scientists are becoming increasingly aware of the important role that ecology plays in our lives. **Today, ecological research provides us with information that is necessary to understand and resolve many of the environmental and ecological issues that confront us.** In order to ask and answer the appropriate questions, gain an appreciation of the short-term and long-term effects of your actions, and make wise choices on controversial issues, you need an understanding of ecology.

Some Examples

To start your thinking, here are a few questions for you to consider:

- What has happened to the populations of certain food fishes in parts of the North Atlantic? How can families continue to earn a living from the sea? To answer these questions,

Figure 13–2
The table presents three research methods of ecology.

RESEARCH METHODS OF ECOLOGY

	Observations	*Experiments*	*Models*
Method	Careful observations of the natural world form the foundations of ecological science.	Hypotheses based on prior observations are tested by intelligently designed and carefully performed experiments. The effect of different variables is studied independently.	Models useful in studying ecological phenomena—such as the flow of energy and nutrients and the way in which organisms reproduce—usually consist of mathematical formulas and equations that are manipulated by computer. The best models are those that are driven by data gathered by observation, experimentation, or both.
Advantages	Observations often lead to valuable understanding of situations and phenomena. Making observations is usually the first step in asking important questions.	Experiments allow control of important variables and the chance to manipulate each variable separately.	Models allow tests of hypotheses over long periods of time. Models also allow simple manipulation of variables that may be difficult or impossible to vary in nature.
Disadvantages	Although observations often suggest connections between various events in nature, usually observations alone cannot prove those links.	When trying to study large, complex collections of organisms, it is possible that laboratory experiments differ from the real world in ways that experimenters do not realize.	Models are only as good as the data and thought that go into them. Models based on faulty hypotheses or incomplete data make faulty predictions.

Biology AND YOU .... Connections

The Mississippi River Floods—How Can Wetlands Help?

The summer of 1993 was a nightmare for the Midwest. Heavy rains swelled the Mississippi River system to five times normal widths. Levees burst. Millions of hectares of land were flooded, destroying homes, farms, and businesses.

The Role of Wetlands

As anyone who lives near the Mississippi knows, the river has always been prone to flooding. Nutrient-rich sediment spread by flood waters over thousands of years has made the region fertile for growing crops. But in the old days, the river's banks and tributaries were surrounded by habitats called wetlands. Wetlands—swamps, marshes, seasonal creeks, and wet-footed woods—act like giant sponges. During heavy rains, wetlands soak up enormous quantities of water and then release it slowly.

Over the last century, development projects drained and filled wetlands for farming and housing. Without natural "safety valves" in the water cycle, the river became more prone to rising suddenly during rainy spells. To contain it, many miles of dikes and levees were built. This simply moved the flood zone downstream, making it worse.

Hope for the Future

Applied ecological research is offering hope that future floods might not be as devastating. Researchers are finding out that restoring former wetlands to their original condition can be fast and inexpensive. They calculate that if only about 3 percent of the land in the upper Mississippi watershed is returned to its former wetland condition, disastrous floods could be avoided. Do you think we will put our understanding of ecology to use and test the scientists' hypothesis?

Making the Connection

What are the advantages of building levees and dikes? What are the benefits of preserving wetlands?

RESTORING WETLANDS

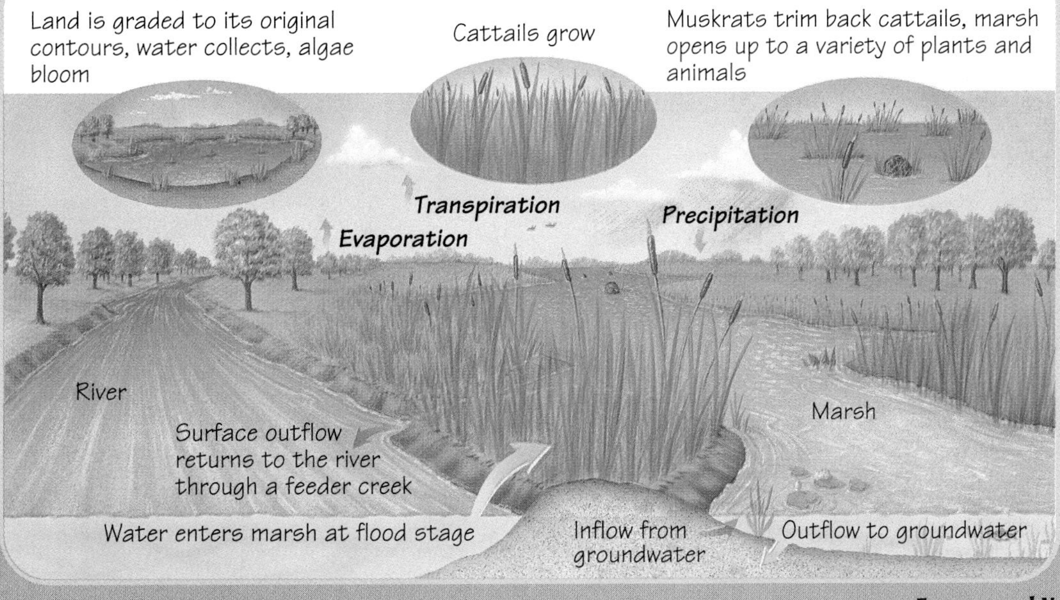

Land is graded to its original contours, water collects, algae bloom

Cattails grow

Muskrats trim back cattails, marsh opens up to a variety of plants and animals

Transpiration

Evaporation

Precipitation

River

Marsh

Surface outflow returns to the river through a feeder creek

Water enters marsh at flood stage

Inflow from groundwater

Outflow to groundwater

Figure 13–3
(a) Earth is home to an amazing variety of living and nonliving things that are closely connected.

CAREER TRACK
(b) Ecologists study the biosphere to understand these connections so that humans can use Earth's resources without causing long-term harm.

you must come to understand what fishes need in order to grow and reproduce in abundance.

- Is the Earth getting warmer? If so, how will this affect the way we live? To grasp this important and complicated question, you must piece together various kinds of information that you will learn while studying ecology. Even then, the answer may not be clear.

Checkpoint Why is it important for you to have an understanding of ecology?

Web of Interdependence

As a result of the work of dedicated ecologists, people around the world have begun to understand two fundamental ecological truths. First, our planet is home not only to humans but to many other forms of life as well. Second, the health of human society depends on the well-being of much of that life. From microscopic organisms to tall trees and from swamps to coral reefs, living things affect each other in many ways. To take care of our planet and ourselves properly, we must understand how and why organisms affect one another and their environment as they do.

At the core of every organism's interaction with its environment and with other organisms is the organism's need for energy and nutrients—energy to power life's processes and raw materials to build and maintain living tissue. As you study this chapter, you will discover that the processes of acquiring energy and nutrients join all life together in an intricate web of interdependence and coexistence. Therefore, the study of ecology must begin with the study of how different organisms obtain and use the energy and nutrients on which all life depends.

Section Review 13–1

1. **Define** ecology. **Explain** why it is important to study ecology.
2. **Describe** three methods that ecologists use in their studies. What are the advantages and disadvantages of each?
3. **Critical Thinking—Relating** Give two examples of the kinds of questions ecologists ask. Explain why they are important.

SECTION 13–2

Energy: Essential for Life's Processes

Guide for Reading

- **Explain** how energy flows through the biosphere.
- **Describe** how energy flows through the different trophic levels in the biosphere.

MINI LAB

- **Classify** the living things you see in your neighborhood.

ALL OF LIFE'S PROCESSES require energy. Although both animals and plants require energy, only plants and certain bacteria can collect energy from their environment and harness it to do biological work. An animal can store energy in its body in the form of complex chemicals such as carbohydrates, fats, or proteins, but the only way an animal can obtain energy is to eat other organisms. To understand how energy moves through the biosphere, you must begin where most energy enters living systems—in green plants.

Energy From the Sun

Green plants can do something that no animals can do. They can harness the energy from sunlight by a process called photosynthesis. Ultimately, the plants assemble simple substances into the building blocks of living tissue, such as carbohydrates, fats, and proteins.

Primary Producers

Plants are **primary producers** because they produce living tissue from nonliving sources, such as water, carbon dioxide, and energy. Primary producers are also called **autotrophs,** meaning "self-feeding." While plants get their energy from the sun, the chemoautotrophs, a group of bacteria, harvest energy from certain chemicals.

Consumers

Animals, as well as most bacteria, cannot capture energy from the sun as plants do. Yet animals need energy for their life processes. Animals also cannot make all the building blocks of living tissue from simple nonliving chemical substances

Figure 13–4 (a) *Swans,* (b) *seals, and even* (c) *sunflowers must obtain energy to live.*

Figure 13–5
(a) Energy from the sun is captured by green plants and other primary producers. (b) Herbivores such as sheep and (c) carnivores such as leopards are consumers and get their energy by eating other organisms.

available to them in the environment. As a result, animals must eat other organisms to obtain their energy and nutrients. For this reason, animals are called **consumers.** Animals are also called **heterotrophs,** because they must feed on other organisms to obtain energy.

Heterotrophs

There are many kinds of heterotrophs in the natural world. Herbivores obtain energy by eating autotrophs that have manufactured and stored proteins, carbohydrates, and other high-energy substances. Carnivores obtain energy by eating other animals. Omnivores are animals that eat both plants and animals. Parasites are organisms that live in or on other organisms and obtain their

Figure 13–6
Organisms depend on one another for food. Ecologists use the concept of trophic levels to study this interdependence.

nutrients from their living host. Although parasites "eat their victims alive," they do not necessarily kill them—at least not right away. Still other organisms, called **decomposers,** feed on the dead bodies of animals and plants or on their waste products.

☑ *Checkpoint* What is the difference between autotrophs and heterotrophs?

Energy Flow Through the Biosphere

Because energy cannot be recycled, or used again in the biosphere, it can be thought of as a flow—a one-way flow. **Arriving as sunlight, energy flows through the tissues of primary producers to the tissues of consumers and then to the tissues of decomposers.** In a sense, you can think of energy in the biosphere as a stream that flows steadily downhill, powering a series of water wheels (primary producers, consumers, and decomposers) along the way.

Trophic Levels

Each step in this series of organisms eating other organisms makes up a **trophic level,** or a feeding level. Theoretically, there is no limit to the number of trophic levels in the biosphere. **However, the greater the number of trophic levels between consumers and primary producers, the smaller the amount of energy that is available to the consumers compared to the energy originally captured by the primary producers.** This is because every time one organism eats another, much of the energy obtained is used up rather than stored. Why does this happen? Because all life's processes—from growth and reproduction to the powering of simple daily activities—require energy. In addition, most animals cannot use 100 percent of the energy potentially contained in the foods they eat.

As a result of these energy losses, only about 10 percent of the energy at one trophic level can be used by the consumers at the next trophic level. Thus, only 10 percent of the energy contained in plants ends up stored in the tissues of herbivores, and only 10 percent of the energy in herbivores can be stored in the tissues of carnivores. Further, only 10 percent of the energy in carnivores—that is, 10 percent of 10 percent of 10 percent, or 1 part in 1000 of the original amount—is available to carnivores that eat other carnivores!

☑ *Checkpoint* How can you explain the decrease in the energy that is available at successive trophic levels?

Life in Your Neighborhood

PROBLEM *How can you **classify** the organisms in your area?*

PROCEDURE

1. Choose an area around your school or around your home. Make a list of the different kinds of organisms—such as plants, insects, animals, and people—that you see.
2. Classify each organism as either a primary producer or a consumer.
3. Further classify each consumer as a carnivore or herbivore.

ANALYZE AND CONCLUDE

1. What organisms did you classify as primary producers? Consumers?
2. Did you observe any herbivores? What were they?
3. Did you observe any carnivores? If so, what were they?
4. What do you think would happen if all the primary producers in the area died out?

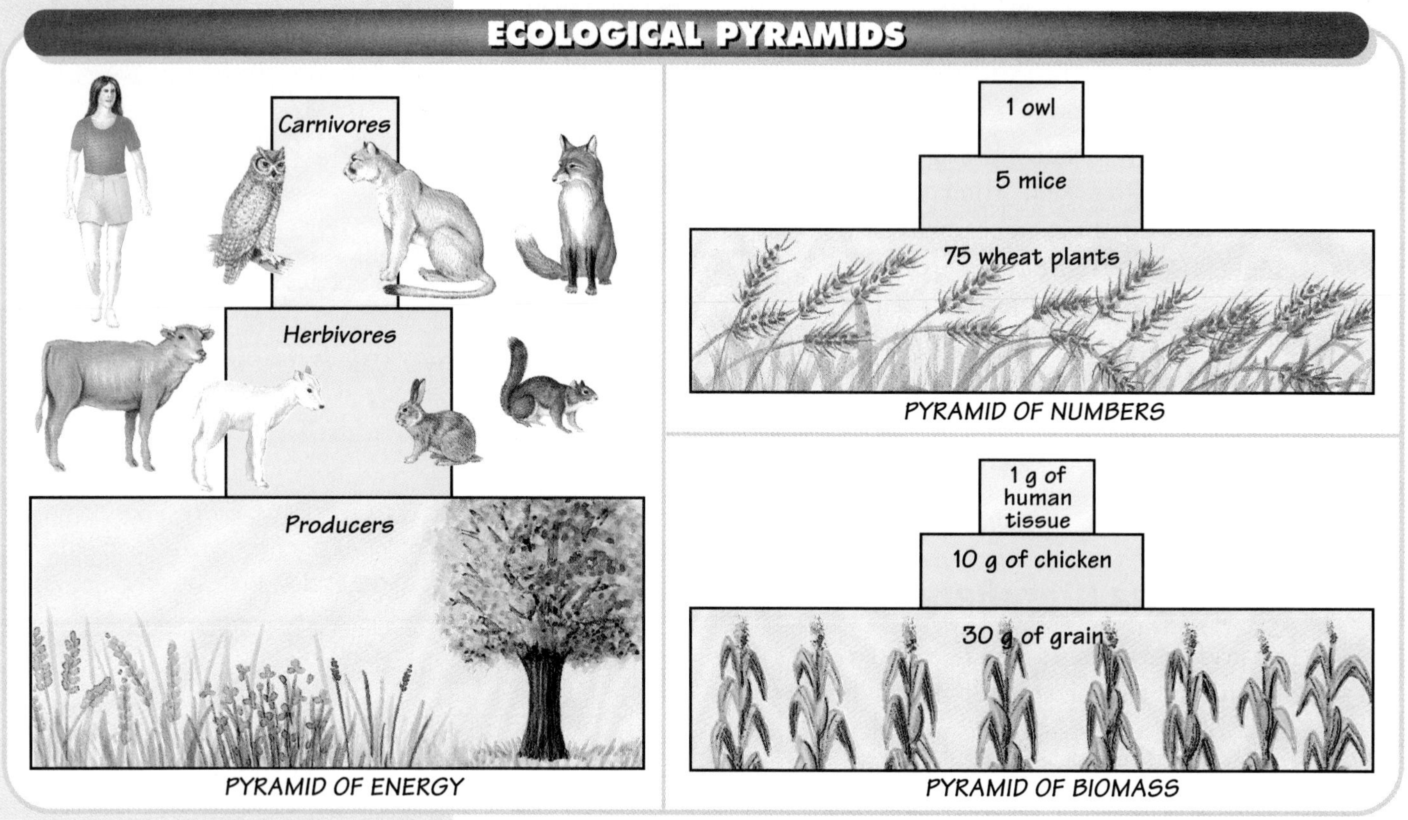

ECOLOGICAL PYRAMIDS

Carnivores
Herbivores
Producers
PYRAMID OF ENERGY

1 owl
5 mice
75 wheat plants
PYRAMID OF NUMBERS

1 g of human tissue
10 g of chicken
30 g of grain
PYRAMID OF BIOMASS

Figure 13–7
Ecological pyramids show the decreasing amounts of energy, living tissue, or number of organisms at successive trophic levels.

Ecological Pyramids

Figure 13–7 shows the ecologically important result of the energy flow from plants to the carnivores on the highest trophic level. The less energy available at any trophic level, the less living tissue that trophic level can have. Thus, the energy chain from primary producers to herbivores to carnivores creates what are called **ecological pyramids.** Three types of ecological pyramids are shown in ***Figure 13–7.***

In summary, only primary producers such as green plants can make the sun's energy available to the rest of Earth's living things. But energy is not all that organisms need. Organisms also need materials for growth and maintenance. And the foods they eat give them both—energy and nutrients. In the following section, you will learn more about nutrients, how organisms obtain them, and how they cycle through the biosphere.

Section Review 13-2

1. **Explain** how energy flows through the biosphere.
2. **Describe** how energy flows through the different trophic levels in the biosphere.
3. **Compare** the loss of energy at each trophic level. **Relate** this to ecological pyramids.
4. **Critical Thinking—Hypothesizing** What is the significance of the pattern of energy loss for animals on higher trophic levels? Following this line of reasoning, why might it be advantageous for a large animal such as a whale to feed on plankton, or tiny marine primary producers?
5. **MINI LAB** **Classify** the organisms you observed in your neighborhood into trophic levels based on who eats whom. What might you call such a sequence?

SECTION 13–3

Nutrients: Building Blocks of Living Tissue

Guide for Reading

- **Describe** how nutrients are recycled in the biosphere.
- **Define** nutrient limitation.

MINI LAB

- **Formulate a hypothesis** to explain the effect of fertilizer on plant growth.

AS YOU HAVE JUST READ, *organisms need more than energy in order to survive. Organisms also need nutrients—the chemical building blocks of life. Nutrients are the substances that organisms use to build living tissues and to grow.*

What Are Nutrients?

Autotrophs can manufacture substances such as carbohydrates, fats, and proteins from simple chemical nutrients that they can readily obtain from their environment. Green plants, for example, take up water, carbon dioxide, nitrogen, phosphorus, and potassium from their environment in substantial quantities. They require other substances, such as iron and magnesium, in smaller, or trace, amounts. Making use of the sun's energy, primary producers assemble these relatively simple nutrients into complex substances such as carbohydrates, proteins, and fats for their growth and maintenance.

Heterotrophs, on the other hand, cannot manufacture all the complex substances that they need from simple ingredients such as water and carbon dioxide. Therefore, animals must eat other organisms in order to obtain the nutrients—such as carbohydrates, fats, and proteins—they need. For the most part, carbohydrates

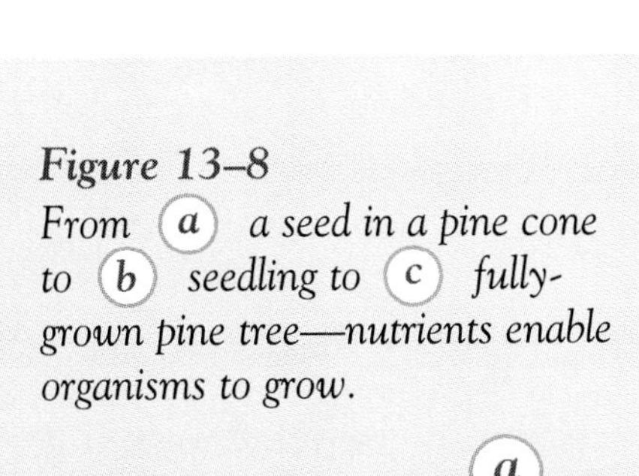

Figure 13–8
From (a) a seed in a pine cone to (b) seedling to (c) fully-grown pine tree—nutrients enable organisms to grow.

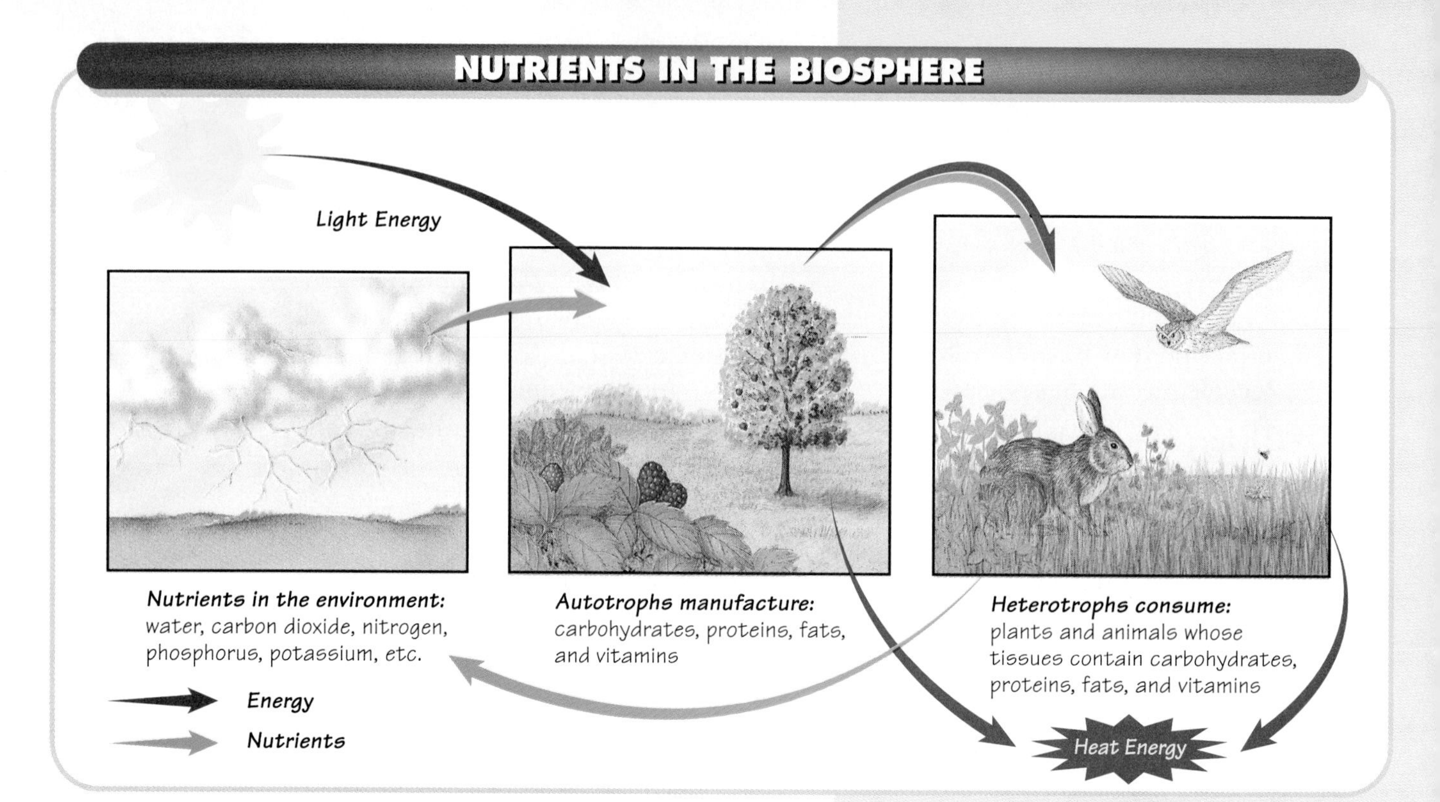

Figure 13–9
The arrows show the movement of nutrients and energy through the biosphere. Notice that nutrients are recycled, whereas energy is not.

and fats provide immediate energy or are stored in the animal's body, providing energy when it is needed. Proteins are broken down into amino acids, which can be reassembled into the particular protein the organism needs. Animals also need several vitamins manufactured by plants and many of the same trace elements that plants need.

Earlier in this chapter, you read that organisms in each trophic level eat members of the level beneath them to obtain both energy and nutrients. Put another way, energy and nutrients move together from one trophic level to the next in the form of plant or animal tissue. However, energy and nutrients move through the biosphere in very different ways.

You may recall that energy is always arriving from the sun and is continually being used by organisms to perform work. Strictly speaking, energy is not destroyed as organisms use it—it is converted into a different form. Ultimately, the energy ends up as the heat that is released into the biosphere. However, living things cannot harness and then reuse this heat energy. So although energy is conserved, it cannot be recycled. As a result, the energy that all living things need must be continually captured from its original source—the sun.

☑ **Checkpoint** What are nutrients?

Nutrient Cycles 

Nutrients, unlike energy, do not arrive on Earth from outer space. Like energy, however, nutrients are not destroyed as a result of their use by Earth's organisms. Nutrients either become part of living tissue or they are eliminated from the organism as waste products.

Nutrients that are available in fixed quantities on Earth are passed from one organism to another and from one part of the biosphere to another through the

pathways of nutrient cycles. These **nutrient cycles** pass the same nutrients over and over again through the different parts of the biosphere. As you may expect, understanding nutrient cycles is essential to understanding the biosphere.

Three important nutrient cycles are illustrated below and on the following two pages. As you examine these cycles, keep in mind that cycles are closed loops in which something flows continuously without a beginning or an end. Often, cycles consist of several processes that are going on at the same time.

Checkpoint How do nutrient cycles differ from energy flows?

Nutrient Limitation

Although nutrient cycles are ultimately global processes, over the short term they may be limited to smaller parts of the biosphere, called ecosystems. Ecosystems might appear to be self-contained and to function independently of one another. In fact, sometimes just the opposite is true. As nutrients move through ecosystems, they often have some interesting consequences.

An ecosystem's productivity is a measure of the rate at which energy is captured by its autotrophs. This productivity may be limited by a single nutrient

Visualizing Nutrient Cycles

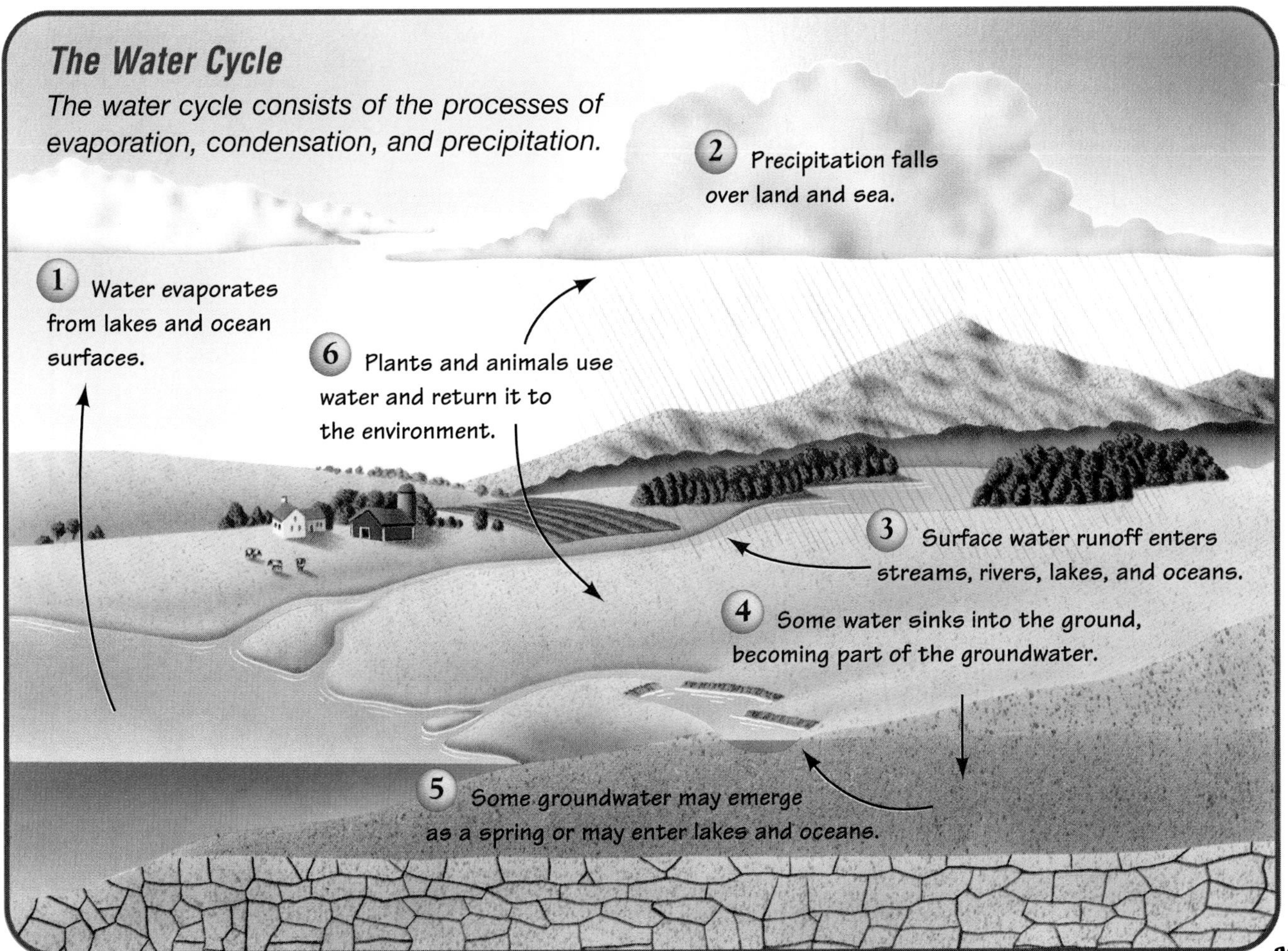

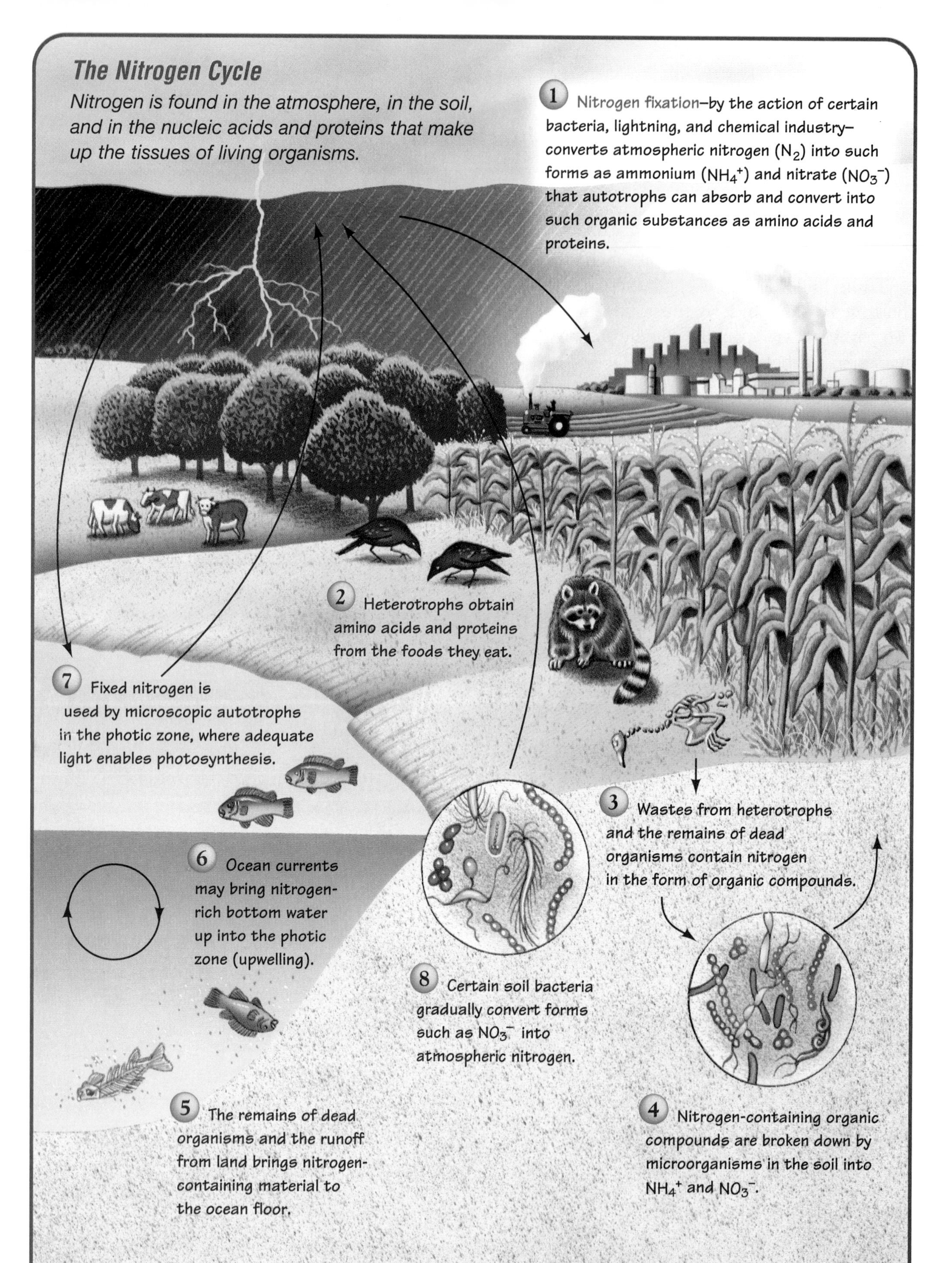
The Nitrogen Cycle
Nitrogen is found in the atmosphere, in the soil, and in the nucleic acids and proteins that make up the tissues of living organisms.
1 Nitrogen fixation—by the action of certain bacteria, lightning, and chemical industry—converts atmospheric nitrogen (N_2) into such forms as ammonium (NH_4^+) and nitrate (NO_3^-) that autotrophs can absorb and convert into such organic substances as amino acids and proteins.
2 Heterotrophs obtain amino acids and proteins from the foods they eat.
3 Wastes from heterotrophs and the remains of dead organisms contain nitrogen in the form of organic compounds.
4 Nitrogen-containing organic compounds are broken down by microorganisms in the soil into NH_4^+ and NO_3^-.
5 The remains of dead organisms and the runoff from land brings nitrogen-containing material to the ocean floor.
6 Ocean currents may bring nitrogen-rich bottom water up into the photic zone (upwelling).
7 Fixed nitrogen is used by microscopic autotrophs in the photic zone, where adequate light enables photosynthesis.
8 Certain soil bacteria gradually convert forms such as NO_3^- into atmospheric nitrogen.

The Carbon Cycle
Carbon is found on the Earth in three large reservoirs—in the atmosphere as CO_2 gas, in the oceans as dissolved CO_2, and underground as coal, petroleum, and calcium carbonate rock.
1 Autotrophs take up CO_2 gas from the atmosphere as a raw material for photosynthesis.
2 Organisms release CO_2 gas to the atmosphere as a byproduct of respiration.
3 Fossil fuels are the result of the decomposition of animals and plants buried for millions of years.
4 CO_2 moves back and forth between the ocean and the atmosphere.
5 Dissolved CO_2 is taken up by marine autotrophs. CO_2 is returned to sea water by the respiration of marine organisms.
6 Carbonate rock is formed at the bottom of the ocean as shells of dead organisms settle to the ocean floor.
7 CO_2 enters the atmosphere when terrestrial plants, fossil fuels, or their products are burned.
8 CO_2 enters the atmosphere when carbonate rocks are vaporized inside volcanoes.

Fertilizers—Nutrients Unlimited?

PROBLEM *How does lawn fertilizer affect the balance in an ecosystem?* ***Formulate a hypothesis.***

PROCEDURE   

1. Fill two wide-mouthed jars with pond water or water from an aquarium.
2. Add a few strands of *Elodea* or other living aquatic plants to each jar.
3. With a marking pencil, label one jar A and the other jar B.
4. Add one teaspoon of fertilizer to jar B. **CAUTION:** *Always wear protective gloves when working with fertilizer.*
5. Place the jars in a sunny location.
6. Observe the jars each day for 2 to 3 weeks. Record your observations.

ANALYZE AND CONCLUDE

1. Which jar was the control? Which jar contained the variable?
2. Describe any changes that occurred in the jars. What may have caused these changes?
3. Based on your results, what do you think happens when large amounts of fertilizers are washed into ponds and other bodies of water?

that is either in short supply or moves through the ecosystem very slowly. This phenomenon is called nutrient limitation. Farmers—well aware of this phenomenon—use fertilizers to boost productivity. Fertilizers usually contain varying amounts of three important plant nutrients—nitrogen, phosphorus, and potassium. These nutrients help plants grow larger and more quickly than they would in unfertilized soil.

Most lakes and streams carry moderate amounts of nitrogen and potassium but little phosphorus. When the detergent industry developed phosphate detergents, what do you suppose happened? As waste water containing these detergents made its way into lakes and streams, the added phosphates caused the algae in those bodies of water to grow at alarming rates. Soon algae covered the surfaces of ponds, lakes, and even some rivers. As a result, underwater plants died because sunlight could not penetrate the layers of algae. Bacteria began growing on the dying plants, using up much of the oxygen. As oxygen levels in the water decreased, fishes and other animals suffocated and died. Luckily, when investigators understood what was causing this problem, a solution was easily found. Detergent manufacturers simply developed nonphosphate detergents.

Section Review 13-3

1. **Describe** how nutrients are recycled in the biosphere.
2. **Sequence** the path of a drop of water through the water cycle.
3. **Define** nutrient limitation.
4. **Critical Thinking—Predicting** Based on the carbon cycle, what do you think might happen if vast areas of forest are cleared by burning?
5. **MINI LAB Formulate a hypothesis** to explain the effect of fertilizers on productivity. Can you think of any reasons why fertilizers should be used with caution?

Food Webs: Who Eats Whom?

SECTION 13–4

GUIDE FOR READING

- **Describe** a food web.

HOW DOES ENERGY FLOW AND how do nutrients cycle in a particular ecosystem? As you may already know, animals and plants don't flow or cycle through different ecosystems. Instead, they eat, grow, reproduce, eliminate their wastes, and die within the same ecosystem. In fact, many organisms have developed unique methods of obtaining food, growing, and reproducing. The intriguing question here is how these various methods interact to cause the energy flows and the nutrient cycles of Earth's biosphere. Read on to find out how this happens.

Food Chains and Webs

In the biosphere, all organisms are linked together into complex networks based, more or less, on who eats whom. At first these relationships were thought of as simple, straight-line **food chains.** A food chain is a sequence of organisms related to one another as predator and prey. In other words, the old "big fish eats little fish" story. But as you might have guessed, things in nature rarely happen this simply.

Almost every place you look in nature, you can find more than one type of primary producer. Often, you can find dozens. As you look closer, you will see that most plant-eating animals feed on at least two different kinds of plants and often more. Carnivores usually eat at least two different kinds of herbivores—and sometimes even each other. Scavengers eat leftovers from other animals. Bacteria and fungi decompose dead tissue and release essential nutrients in different forms.

Figure 13–10
(a) The squirrel feeds on acorns produced by (b) an oak tree. The squirrel in turn may become food for another organism.

Figure 13–11
This diagram shows a simplified forest food web. Notice that the organisms belong to at least three trophic levels.

Detritus (dee-TRIGHT-uhs) feeders, such as earthworms, are animals that digest a combination of bacteria, bodily wastes, and bits of decaying organisms. And different kinds of parasites draw their food from many different types of hosts.

A **food web** is the best way to illustrate how organisms feed on one another. **Food webs show the complex feeding relationships that result from interconnecting food chains.** As *Figure 13–11* shows, food webs have many crisscrossing strands.

In spite of the complexity of typical food webs, it is helpful to remember that they can all be described in terms of three categories of organisms—primary producers, consumers, and decomposers. Note that consumers as well as decomposers are classified as heterotrophs because they obtain their energy by feeding on other organisms.

☑ **Checkpoint** What is the difference between a food chain and a food web?

Food Web Diversity

Although the general rules of energy flow and food web interactions hold true in all ecosystems, some variations arise due to interactions that are unique to an ecosystem. Temperate woodlands in the United States and in Germany differ very little. Although they are home to different types of animals and plants, from an ecological standpoint they are almost identical.

Other ecosystems, however, can differ from one another significantly. In the terrestrial ecosystem shown in ***Figure 13–11,*** the predominant pathway of energy and nutrient cycling is from primary producers to herbivores to carnivores. Now let's examine two very different kinds of food webs.

Coastal Salt Marsh Food Web

Think of a typical Atlantic Coast salt marsh. At first, this ecosystem may appear much like any terrestrial grassland. However, few of the animals eat grasses directly. Instead, energy and nutrients take a different route, which supports a host of aquatic animals. Although some of the primary producers (plants) are eaten by herbivores (grasshoppers and snails), most of the primary producers die, decompose, and are converted into detritus. Each detritus particle acts as a tiny ecosystem because it consists of a piece of dead plant or animal matter coated with microscopic organisms of decay. As the detritus is eaten by detritus feeders that are eaten by carnivores, energy and nutrients are transferred from one organism to another. Such food webs are common in many coastal marine ecosystems.

☑ *Checkpoint* Identify the main energy and nutrient pathway in a coastal marsh ecosystem.

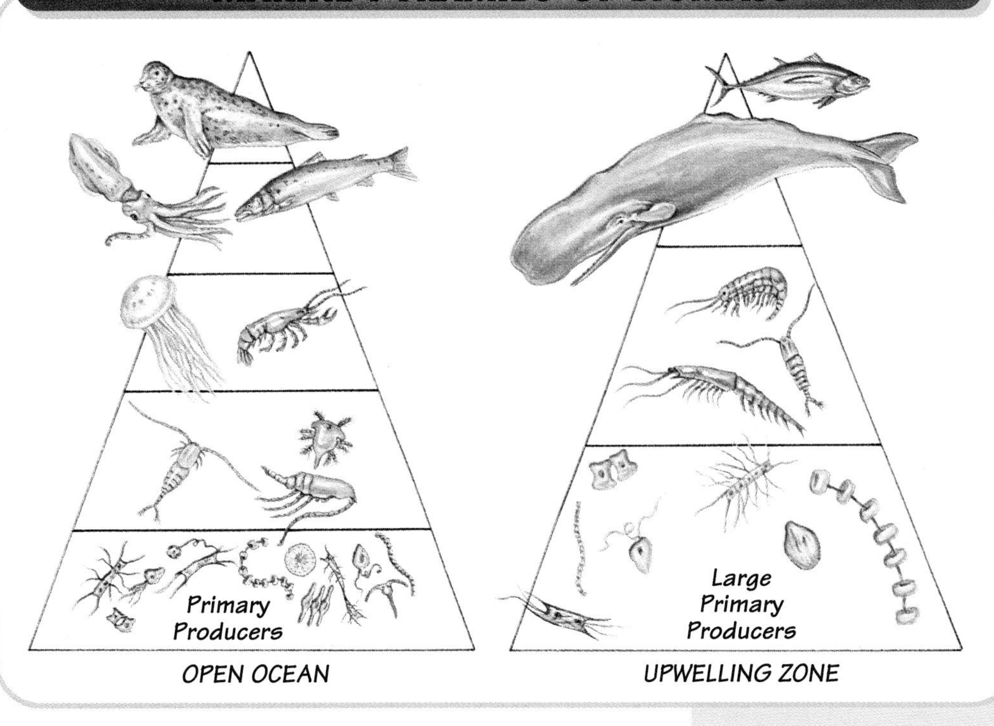

Figure 13–12 *Nitrogen-rich ocean water in the upwelling zone supports larger primary producers. As a result, fewer trophic levels separate large consumers, such as tuna and seal, from the first trophic level.*

Marine Food Web

Consider the "big fish eats little fish" scenario as an example of a food chain. How many trophic levels do you see in the first part of ***Figure 13–12***—between the primary producers and carnivores such as salmon and tuna? With so many different primary producers and consumers, this simple chain becomes an extremely tangled web. But that's not all, as the second half of ***Figure 13–12*** shows.

In parts of the open sea where nitrogen is in short supply, only tiny primary producers can grow. These primary producers are so small that only small consumers can eat them. But in areas where ocean currents bring up nitrogen-rich water from the depths to the surface, greater amounts of nitrogen can support larger primary producers. Notice the smaller number of trophic levels between the primary producers and the largest consumers in the second pyramid shown in ***Figure 13–12***.

Section Review 13–4

1. **Compare** a food web and a food chain.
2. **Explain** how the food web in a typical salt marsh differs from one in a terrestrial grassland.
3. **Critical Thinking—Relating** Why do upwelling areas support more fish life than most parts of the open sea?

SECTION 13–5

BRANCHING OUT In Depth

The Carbon Cycle: A Closer Look

GUIDE FOR READING

- **Explain** how the global carbon cycle is affected by human society.

ABOUT FORTY YEARS AGO, TWO scientists named R. Revelle and H. Seuss published an article describing an unplanned planetwide experiment. Human activity had begun "returning to the atmosphere and oceans the concentrated organic carbon stored in sedimentary rocks over hundreds of millions of years." What does that statement mean? Why is it important? And how has our understanding of this issue changed over the years? Read on.

The Greenhouse Effect

The article by Revelle and Seuss drew attention to a phenomenon called the greenhouse effect. If you have experienced the warmth inside a glass greenhouse on a sunny day even when the outside air temperature was much lower, you have a good idea of what the greenhouse effect is. What causes the air inside the greenhouse to be warmer than the air outside? The glass allows the sun's energy to enter the greenhouse, where it is transformed into heat energy. The glass, however, does not allow the heat energy to leave the greenhouse as easily. The energy thus trapped causes the temperature inside the greenhouse to rise. What does the greenhouse effect have to do with the carbon cycle?

Greenhouse gases, such as carbon dioxide, trap energy in the atmosphere

Figure 13–13
Human activities such as (a) the combustion of fossil fuels, (b) the burning of forests, and (c) deforestation contribute to the increase of the levels of greenhouse gases in Earth's atmosphere.

just as the glass in a greenhouse traps heat energy inside the greenhouse. Atmospheric carbon dioxide allows the sun's energy to reach the Earth's surface, where it is absorbed and ultimately converted to heat energy. Later, the heat energy radiates back into outer space. But carbon dioxide and other gases in the atmosphere absorb some of this heat energy, forming a "heat trap" around the Earth.

THE GREENHOUSE EFFECT

Figure 13–14 *The Earth's atmosphere acts something like the glass in a greenhouse, allowing energy in the form of sunlight to enter, but absorbing and holding some of that energy once it is converted to heat. This atmospheric effect is largely due to the presence of carbon dioxide gas.*

If Earth's temperature and climate are to remain stable, the planet must lose energy to space at the same rate at which energy arrives. If heat leaves faster, the planet will cool. (Without the presence of greenhouse gases, Earth would be 30°C cooler than it is today!) If heat leaves more slowly, extra energy in the form of heat will build up.

With this background information, you are now ready to examine some data to help you determine how the carbon cycle—and therefore the carbon dioxide in the atmosphere—has changed over the course of the last 100 years.

☑ **Checkpoint** How does the Earth's atmosphere act like a greenhouse?

Major Carbon Pathways

If you refer back to pages 294 and 295, you will notice that there are four different kinds of pathways that are simultaneously occurring in the carbon cycle:

1. Biological pathways: photosynthesis, respiration, and death and decay of plants and animals.

2.  Geochemical pathways: release of carbon dioxide to the atmosphere by volcanic activity and weathering of rocks, and the carbon dioxide exchange between the ocean and the atmosphere. ●

3. Biogeochemical pathways: burial and conversion of carbon from once-living organisms into fossil fuels (coal and petroleum).

4. Human-initiated pathways: mining and burning of fossil fuels, and the cutting and burning of forests.

By studying these processes over many years, scientists are attempting to find out how much carbon travels along the various pathways and the rate at which it travels. They also want to know how much carbon is stored in the atmosphere, in the oceans, on land, in carbonate rocks, and in living tissues. Measurable data such as these are necessary in order to predict the effects that human activities have on the carbon cycle.

Some researchers believe that most carbon travels through geochemical pathways. If this is true, the clearing of forests and the burning of coal and oil will not affect the global scheme of things much. Other scientists, however, think that a significant amount of carbon travels through biological and biogeochemical pathways. If this is true, then human activities that influence

INTEGRATING EARTH SCIENCE

How does CO_2 cycle through the atmosphere, the oceans, and the rocks in the Earth's crust?

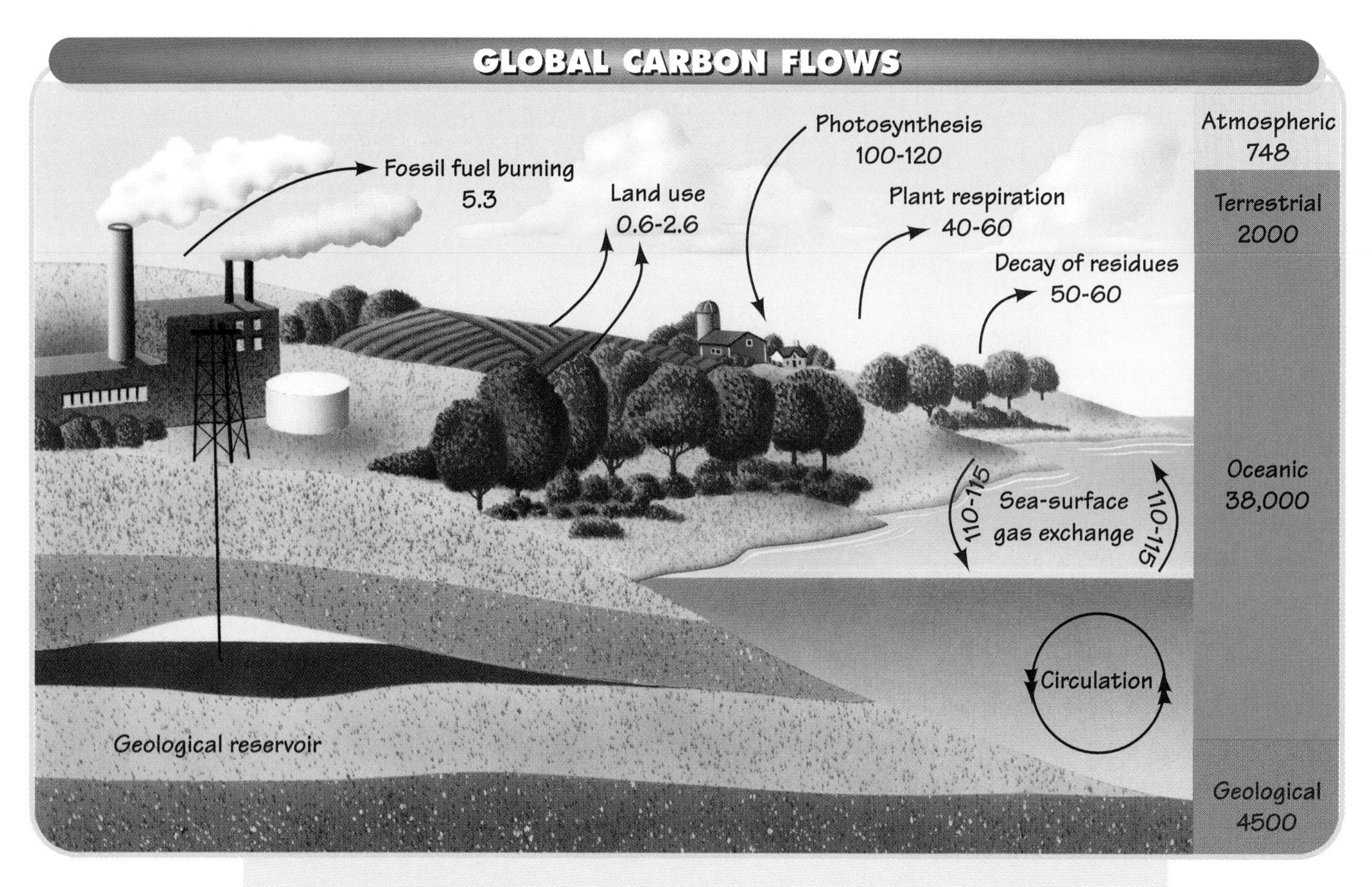

Figure 13–15
Carbon flows between the three major carbon reservoirs—oceans, atmosphere, and deposits in the Earth—are shown in units of gigatons per year. Although the amount of carbon entering the atmosphere as a result of human activity is relatively small, its cumulative effect in the future could be significant.

these pathways could have significant effects on how much carbon ends up in various places—specifically, as carbon dioxide in the atmosphere.

☑ **Checkpoint** What are the ways in which human activities affect the carbon cycle?

Experimental Data

What data have scientists gathered so far? If you look at ***Figure 13–15***, you will see information about rates of carbon flow along different pathways. ***Figure 13–15*** also shows the relative amounts of carbon stored in different parts of the biosphere. What do these data show? Notice that the amount of carbon returned to the atmosphere by the burning of fossil fuels and human land use is small in comparison with the other pathways. Also note that most of the carbon is stored in the oceans, very little is in the atmosphere, and a small but significant amount is in terrestrial ecosystems and fossil fuels.

Although many questions still remain—particularly about the future—certain things are becoming more clear. Notice that the amount of carbon that is cycled in and out of the ocean is roughly equal to the amount that is cycled in and out of terrestrial environments. Therefore, any process that significantly affects the biological pathways for

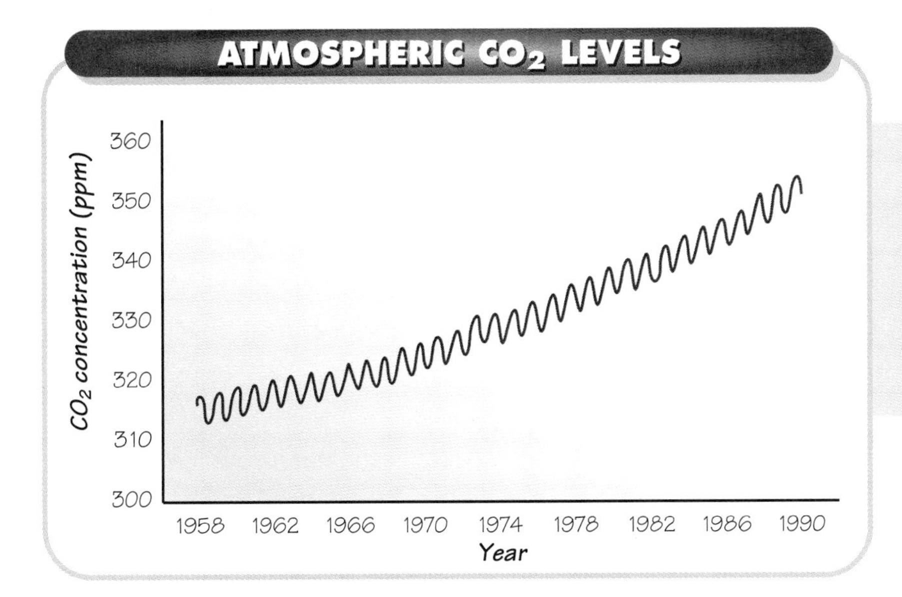

Figure 13–16 The graph shows atmospheric carbon dioxide concentrations measured over the past 30 years.

carbon on land, such as major deforestation, could affect the global flow of carbon. Also note that human activities—which did not amount to much until a century ago—now account for a small percentage of the carbon flow each year. Interestingly, this flow is one way, without a reverse human process that takes it back. Although one year's worth of such activity may not make much of a difference on a global scale, its effect could be significant and damaging if the activity continues year after year.

Atmospheric CO_2 Levels

Let's look at a final bit of data. ***Figure 13–16*** shows a graph of atmospheric carbon dioxide concentrations in parts per million measured at Mauna Loa, Hawaii, from 1958 through 1990. This graph clearly shows that the average concentration of carbon dioxide is rising slowly but steadily over time. **This gradual increase in atmospheric carbon dioxide levels is probably due to human activity. And as the insulating carbon dioxide blanket gets thicker, it could have a warming effect on Earth.**

As your knowledge of ecology grows, you will learn that your environment is not static. You will learn that the biological activity of organisms can affect even environments as large as Earth's atmosphere. As living organisms, humans too have a part in this activity. We affect our environment and are affected by it in ways that we are just now beginning to understand.

Section Review 13–5

1. **Explain** the greenhouse effect.
2. **Describe** how the global carbon cycle is affected by humans.
3. **BRANCHING OUT ACTIVITY** **Formulate a hypothesis** to explain the periodic fluctuation in atmospheric carbon dioxide levels. Suggest ways to test this hypothesis.

CHAPTER 13

Laboratory Investigation

Food or Feeders?

Within a community, organisms interact in many ways. Tracing the flow of energy within a community can help you to understand how the organisms interact.

Problem

How does the energy flow through a community affect the community's complexity and stability? **Analyze** the feeding relationships among organisms from two terrestrial communities.

Materials (per group)

2 large sheets of unlined paper
plain notebook paper
pen or pencil
several colored markers
tape or glue

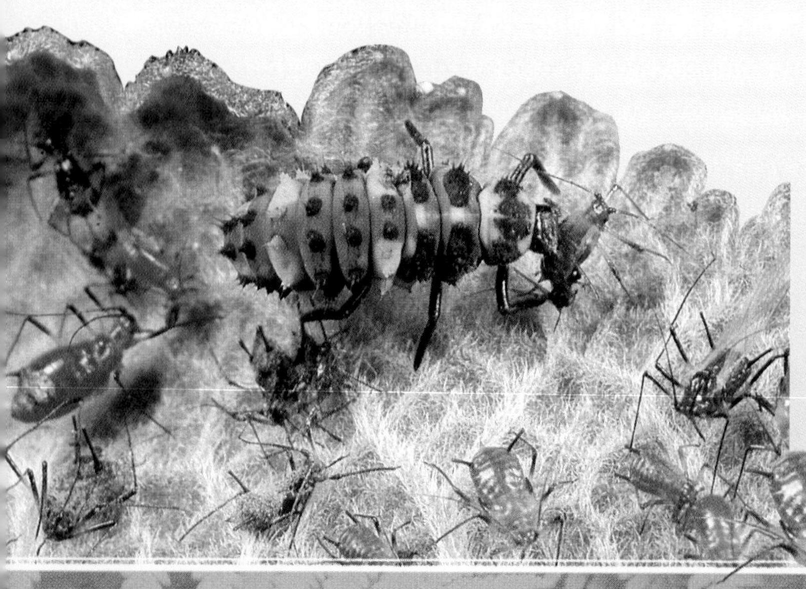

Procedure

1. On the next page, you will find a chart that contains two lists of organisms. One list includes organisms from a hickory/oak forest community in a temperate deciduous forest. The other is from a cultivated cornfield community.
2. Carefully read the lists and identify as many feeding relationships as you can. In many cases, one species may be linked to several others—either as food or as feeder.
3. Write the names of the organisms from the first community on a sheet of notebook paper. Cut the names out and arrange them on one of the large sheets of paper. Note: *Do not attach them to the sheet yet.*
4. Discuss how the organisms are to be arranged. Keep in mind that the names have to be connected to one another to indicate feeding links.
5. When you have decided on the arrangement, attach the names of the organisms to the large sheet with tape or glue.

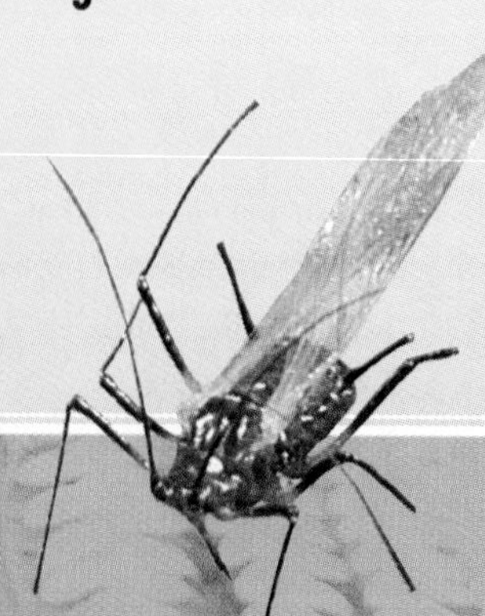

ORGANISMS IN DECIDUOUS FOREST AND CORNFIELD COMMUNITIES

	Hickory/Oak Forest Community	Cultivated Cornfield Community
Plant Species	White oak, black oak, tulip tree, white pine, birch, big tooth aspen, dogwood, sassafrass, viburnum	Corn
Animal Species	Invertebrates, such as a beetle, ant, sow bug, earthworm, snail, termite, moth, centipede, and spider; birds, such as cardinal, warbler, chickadee, woodpecker, flycatcher, and owl; other animals, such as raccoon, squirrel, chipmunk, black bear, opossum, wood mouse, vole, deer, and black racer (snake)	Raccoon, corn snake, woodchuck, field mouse, deer; invertebrates, such as corn borer, grasshopper, cricket, earthworm, butterfly, moth, fly; birds, such as sparrow, meadowlark, crow, hawk
Fungi and Bacteria	Various fungi and bacteria	Various fungi and bacteria

6. **Use the markers to construct food chains by drawing arrows from the food source to the feeder. Use different-colored markers to indicate different food chains. Make your food web as complex as possible.**

7. **Repeat steps 3 through 6 for the second list of organisms.**

Observations

1. Compare your webs to those created by other groups.
2. Which community—the hickory/oak forest or the cornfield—seems to be more complex?

Analysis and Conclusions

1. Are the food webs for the hickory/oak forest community the same for different groups? How are they different?
2. Are the food webs for the cornfield community the same for different groups? How are they different?
3. Which community has the greater variety of primary producers? The greater number of trophic levels?
4. Suppose a parasite destroyed most of the oak trees in the forest community and the corn plants in the cultivated field. How would each community be affected?

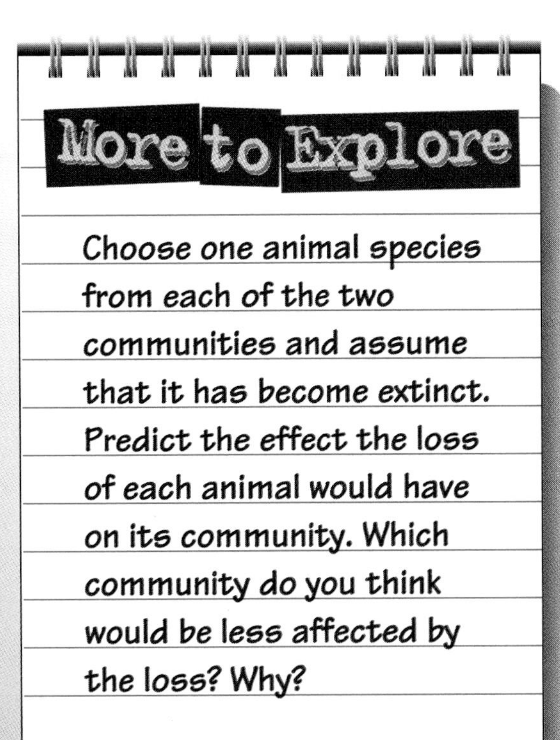

More to Explore

Choose one animal species from each of the two communities and assume that it has become extinct. Predict the effect the loss of each animal would have on its community. Which community do you think would be less affected by the loss? Why?

Study Guide

Summarizing Key Concepts

The key concepts in each section of this chapter are listed below to help you review the chapter content. Make sure you understand each concept and its relationship to other concepts and to the theme of this chapter.

13–1 Ecology: Studying Nature's Houses

- Ecology is the scientific study of interactions between different types of organisms and between organisms and their environments.
- Ecologists typically use one or more of three different research approaches—observations, experiments, and models—when studying the biosphere.

13–2 Energy: Essential for Life's Processes

- Primary producers (autotrophs) are able to make their own food. Consumers and decomposers (heterotrophs) must obtain energy from other organisms.
- Energy moves through the biosphere in a one-way flow. Arriving as sunlight, energy flows through the tissues of primary producers to the tissues of consumers and then to the tissues of decomposers.
- Only a small fraction of the energy at any trophic level can be used by the organisms in the next higher trophic level. This gives rise to ecological pyramids.

13–3 Nutrients: Building Blocks of Living Tissue

- Nutrients that are available in fixed quantities on Earth are passed from one organism to another and from one part of the biosphere to another through nutrient cycles.
- Three major nutrient cycles are the water cycle, the nitrogen cycle, and the carbon cycle.
- Productivity in an ecosystem may be limited by a single nutrient. This is called nutrient limitation.

13–4 Food Webs: Who Eats Whom?

- Organisms in an ecosystem are linked together by food webs.
- A food web shows the complex feeding relationships that result from interconnected food chains.

13–5 The Carbon Cycle: A Closer Look

- The gradual increase in atmospheric carbon dioxide levels is probably due to human activities. This increase could have a warming effect on Earth.

Reviewing Key Terms

Review the following vocabulary terms and their meaning. Then use each term in a complete sentence.

13–1 Ecology: Studying Nature's Houses

ecology

13–2 Energy: Essential for Life's Processes

primary producer
autotroph
consumer
heterotroph
decomposer
trophic level
ecological pyramid

13–3 Nutrients: Building Blocks of Living Tissue

nutrient cycle

13–4 Food Webs: Who Eats Whom?

food chain
food web

Recalling Main Ideas

Choose the letter of the answer that best completes the statement or answers the question.

1. Green plants are also called

a. autotrophs. **c.** herbivores.
b. heterotrophs. **d.** carnivores.

2. Animals that eat only other animals are called

a. herbivores. **c.** carnivores.
b. omnivores. **d.** decomposers.

3. In a food chain, herbivores are known as

a. consumers. **c.** producers.
b. decomposers. **d.** carnivores.

4. The process of converting atmospheric nitrogen into a form producers can use is called

a. nitrogen fixation. **c.** denitrification.
b. respiration. **d.** photosynthesis.

5. A hawk eats a snake that has eaten a mouse that has eaten some vegetation. This is an example of a

a. food web. **c.** pyramid of numbers.
b. food chain. **d.** pyramid of biomass.

6. In a series of trophic levels, the organisms farthest from the producers usually

a. constitute the least biomass.
b. are herbivores.
c. are autotrophs.
d. constitute the greatest biomass.

7. Most carbon on Earth is stored

a. in fossil fuels. **c.** in the oceans.
b. in the atmosphere. **d.** on land.

8. Measurements of carbon dioxide concentrations in the air over past decades have shown a

a. leveling off.
b. slow but steady increase.
c. dramatic increase.
d. slow but steady decline.

Putting It All Together

Using the information on pages xxx to xxxi, complete the following concept map.

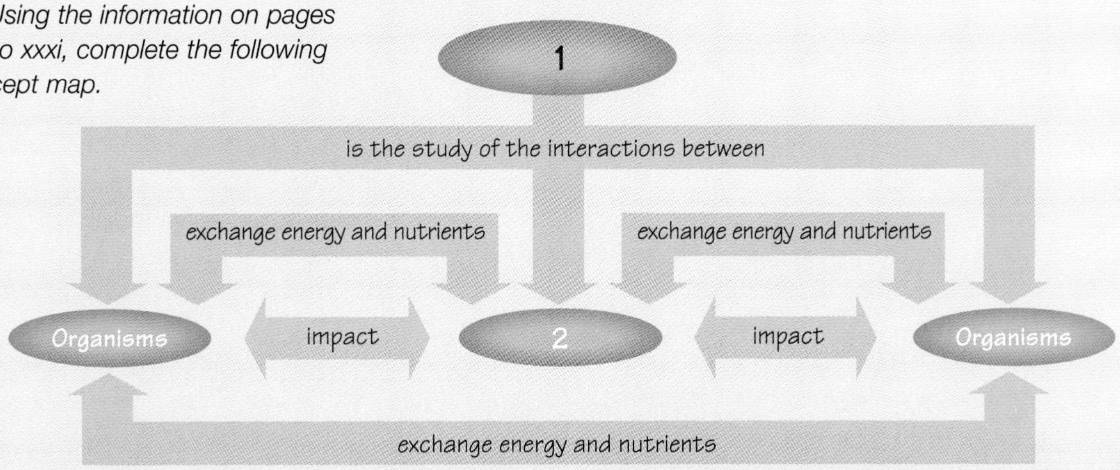

1
is the study of the interactions between
exchange energy and nutrients
exchange energy and nutrients
Organisms
impact
2
impact
Organisms
exchange energy and nutrients

Assessment

Reviewing What You Learned

Answer each of the following in a complete sentence.

1. Give a brief definition of ecology.
2. What are the research methods used by ecologists?
3. Neither herbivores nor carnivores are autotrophs. Is this statement true? Explain your answer.

4. What type of organism would be considered a primary producer?
5. What happens to the energy lost at each trophic level?
6. What is meant by a nutrient cycle?
7. Name the three states of water observed in the water cycle.
8. What is the primary role of a decomposer?
9. What are the common features of all food webs?
10. What are the different forms of nitrogen in the nitrogen cycle?
11. Why did the removal of phosphates in household detergents improve the quality of our nation's waterways?
12. Where is most of the carbon on Earth located?
13. What is the difference between a geochemical and a biogeochemical pathway?
14. What is the greenhouse effect?

Expanding the Concepts

Discuss each of the following in a brief paragraph.

1. Explain the basis for **classifying** organisms as autotrophs and heterotrophs.
2. Trace the flow of energy from its origin in the foods you ate for lunch today.
3. If you are located at the fourth trophic level in a food chain, how would you compare with organisms at the second trophic level?
4. Compare a food chain and a food web.
5. Why is nutrient limitation important to understanding the productivity of an ecosystem?
6. Why is it possible that a carbon atom in your ear lobe might once have been part of a dinosaur?
7. Why do most overpopulated countries base their diets on plants or plant products?
8. Discuss the difference between nutrients and energy.
9. Without nitrogen-fixing bacteria, what would be the future of life on this planet?

Extending Your Thinking

Use the skills you have developed in this chapter to answer the following.

1. **Predicting** In which areas of the world would you expect to find a large fishing industry and why?
2. **Relating** Recent Mississippi floods have been much worse than those in previous years. How can an understanding of ecology help to prevent similar occurrences in the future?
3. **Analyzing** Why must people be extremely careful when using insecticides to control unwanted pests in the environment?
4. **Relating** Burning tropical rain forests to provide space for farming has had many serious consequences on the environment. Discuss two of these consequences.
5. **Hypothesizing** Filling in salt marshes for new home sites has caused a noticeable decline in fishing industries hundreds of kilometers away. Discuss the factors that are associated with this scenario.

Applying Your Skills

The Food Web in Your Community

Food webs exist in every ecosystem. What does the food web in your surrounding ecosystem look like?

1. Form three groups to collect information on the following organisms living around your school. List five primary producers, three herbivores, two omnivores, one carnivore, two detritus feeders, and two decomposers.
2. Combine your information with the information from the other groups.
3. Combine the information from all three groups to make a poster of a food web.

• GOING FURTHER •

4. Write a paragraph explaining how this food web would be different from one in your backyard.

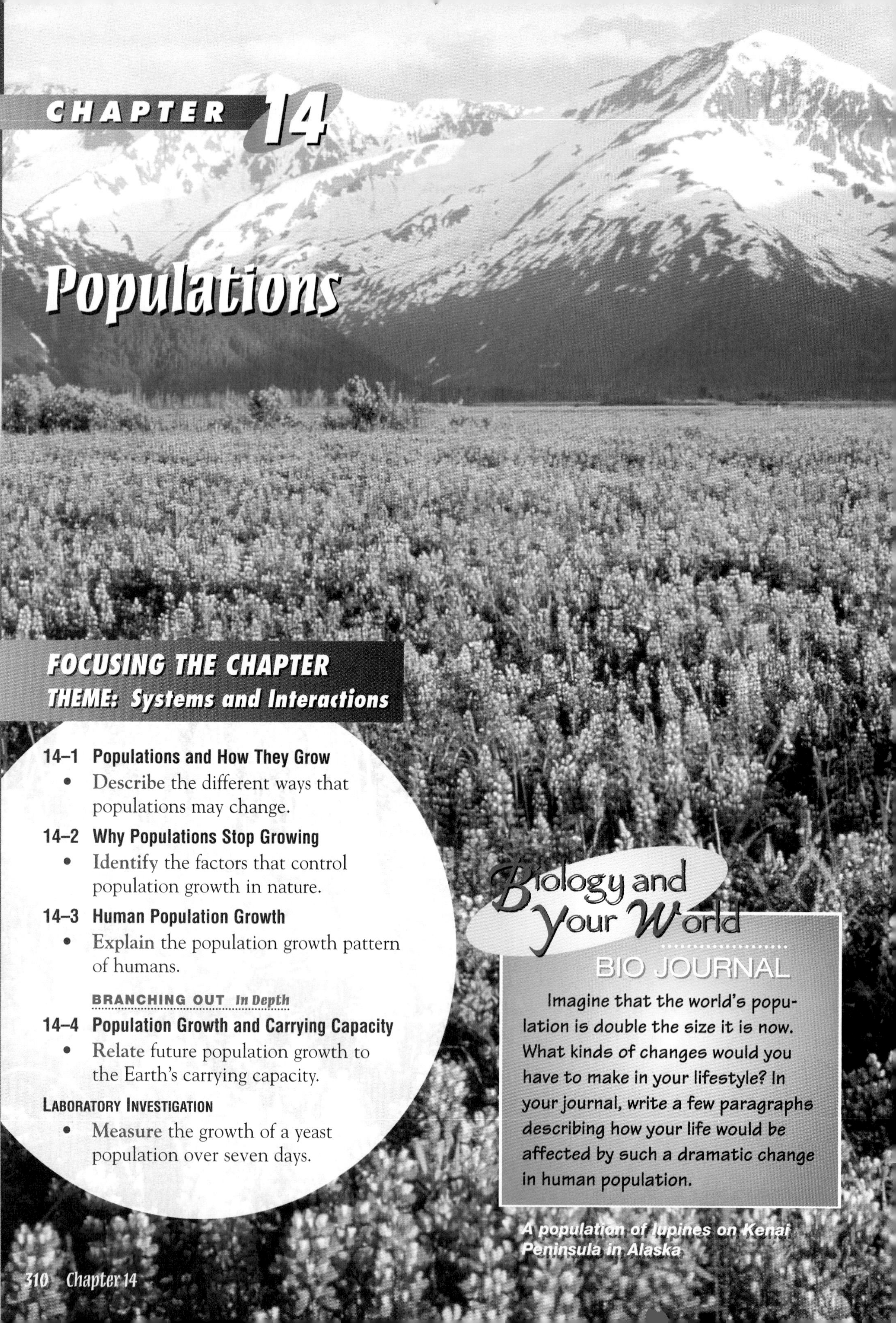

CHAPTER 14

Populations

FOCUSING THE CHAPTER THEME: *Systems and Interactions*

14–1 Populations and How They Grow
- Describe the different ways that populations may change.

14–2 Why Populations Stop Growing
- Identify the factors that control population growth in nature.

14–3 Human Population Growth
- Explain the population growth pattern of humans.

BRANCHING OUT *In Depth*

14–4 Population Growth and Carrying Capacity
- Relate future population growth to the Earth's carrying capacity.

LABORATORY INVESTIGATION
- Measure the growth of a yeast population over seven days.

Biology and Your World

BIO JOURNAL

Imagine that the world's population is double the size it is now. What kinds of changes would you have to make in your lifestyle? In your journal, write a few paragraphs describing how your life would be affected by such a dramatic change in human population.

A population of lupines on Kenai Peninsula in Alaska

Populations and How They Grow

Guide for Reading

- **Define** population.
- **Compare** population growth under ideal conditions and actual conditions.

MINI LAB

- **Construct a model** of human population growth.

THROUGHOUT FLORIDA, THE waterways are being strangled by hydrilla. In rivers, the tangled stems of this plant prevent boats from passing. In lakes, the hydrilla grows so thick that birds can walk across the water's surface. About 40 years ago, hydrilla plants made their way into a canal. The offspring of those plants now cover more than 45 square kilometers.

Meanwhile, in the Northeast, families who make their living from the sea are in trouble. For more than 300 years, their ancestors harvested cod, haddock, and flounder from the rich fishing ground called Georges Bank. In the last 20 years, the present-day families have noticed a drastic decrease in their fishing catch. Many of these families are afraid they'll lose the only jobs they have ever known.

Changing Populations

At first glance, these two stories seem to be very different. One is about plants growing out of control and the other is about disappearing fishes. Yet both situations involve changes in the size of what ecologists call a **population.** What is a population? **A population is a group of organisms of a single species that live in a given area.** A species is a group of organisms that reproduce fertile offspring.

INTEGRATING CAREERS

What are some of the different jobs open to an ecologist?

In nature, populations often stay about the same size from year to year. Sometimes, however, a particular population will grow very rapidly, like the hydrilla in Florida. Other times, populations shrink quickly, like the fish on Georges Bank. What accounts for these different situations?

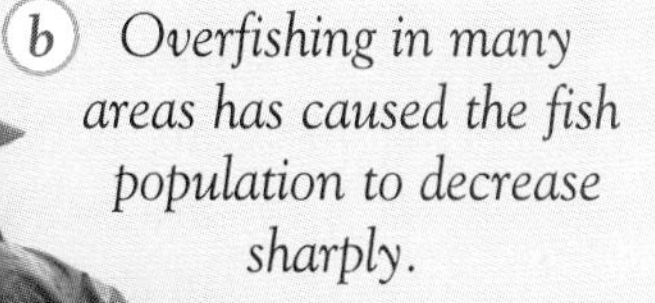

Figure 14–1
These photographs show two different stories of population growth. (a) Hydrilla plants, which were brought to this country from Sri Lanka because they grew so well in aquariums, were accidentally tossed into a canal in Florida. This action caused the hydrilla population to grow rapidly. (b) Overfishing in many areas has caused the fish population to decrease sharply.

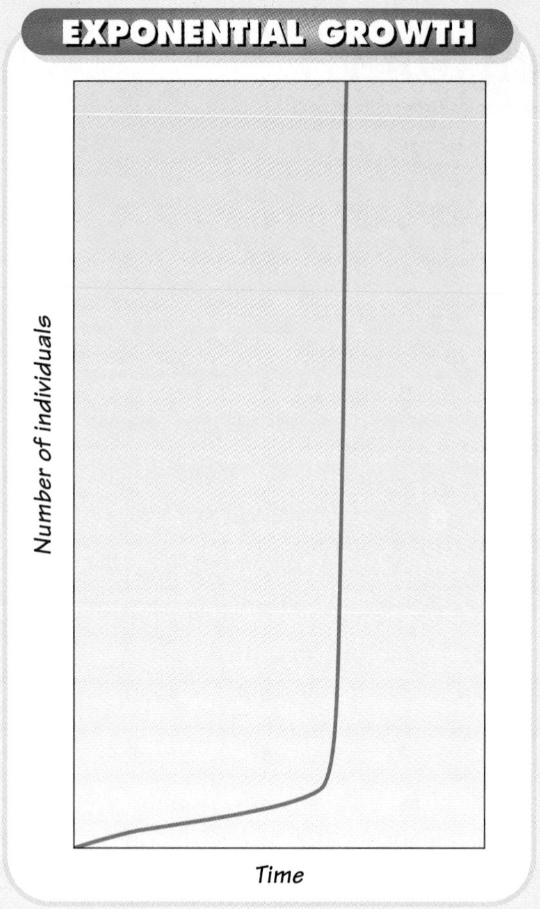

EXPONENTIAL GROWTH

Number of individuals

Time

Figure 14–2
This J-shaped graph shows the exponential growth of bacteria. If ideal conditions continue, the population of bacteria will continue to grow rapidly.

Growth Rate

Simply put, a population will change size depending on how many organisms are added to it and how many organisms are removed from it. This change in population size is called **growth rate.** Growth rate can be positive, negative, or zero.

You may recall that producers, consumers, predators, and prey are linked by complex food webs. So it should not surprise you to learn that an organism's population growth rate depends in large part on the other organisms with which it interacts.

☑ ***Checkpoint*** What is growth rate?

Why Populations Grow

If you provide a population of any species with ideal conditions, it will increase, or grow. That's because all healthy organisms—from bacteria to humans—reproduce faster than their death rate. Of course, that is only to be expected. If a species' characteristics did not include the ability to produce offspring faster than its members die, that species would quickly become extinct!

In general, a population will grow if more organisms are born in a given period of time than die during the same period. An ecologist would say that such a population's birth rate is higher than its death rate. Because birth rates usually are higher than death rates for healthy organisms, populations tend to grow unless something stops them.

☑ ***Checkpoint*** How does birth rate affect population size?

A Baby Boom

If a population lives with ideal conditions—such as an adequate food supply, protection from predators, and shelter—something interesting happens. First, the existing organisms reproduce. Soon, their offspring reproduce. And then, their offspring's offspring reproduce. **As long as ideal conditions continue, the larger a population gets, the faster it grows.** This type of growth is called **exponential growth.**

Let's examine a typical instance of exponential growth. Suppose a single bacterium divided every half hour. After the first half hour, there would be two cells. Half an hour later, there would be four cells. And after another half hour, there would be eight cells. Can you grasp what is happening? Look at ***Figure 14–2*** to see the number of bacteria that could result from one cell if nothing were to stop it. If this growth continued, bacteria would cover the planet!

☑ ***Checkpoint*** What is exponential growth?

Growth With Limits

Bacteria do not cover the planet; therefore, we conclude that exponential growth does not continue in natural populations for long. Something eventually stops it or at least slows it down. The graph in ***Figure 14–3*** shows a different type of growth curve. If you were to introduce a few organisms into a new environment, their numbers would begin to grow slowly. Soon the population would enter an exponential growth phase. **But because exponential growth does not continue for long, population growth would begin to slow down.**

Zero Population Growth

As you will see, there are several reasons why growth slows down. In the meantime, what do you think would happen to a population if its birth rate and death rate were the same? Population growth would stop. In other words, the population would enter a state called **zero population growth.** Zero population growth does not mean that the number of individuals in the population is zero. Rather, it means that the size of the population stays the same

But Not a Drop to Drink

PROBLEM *How many people can the Earth hold? What will happen if the birth rate keeps increasing? Will the Earth overflow with people? How can you **construct a model** of population growth?*

PROCEDURE

1. With a partner, label three large bowls Birth, Earth, and Death. Fill the Birth and Earth bowls three-quarters full with water. Leave the Death bowl empty.
2. Using a one-cup measuring cup, begin adding water from the Birth bowl to the Earth bowl. At the same time, have your partner use a quarter-cup measuring cup to remove water from the Earth bowl and put it into the Death bowl.
3. Continue this process until the Earth bowl is about to overflow.

ANALYZE AND CONCLUDE

1. Does this activity model positive or negative population growth?
2. What does the change in water level in the Earth bowl symbolize? How does it relate to population growth?
3. **Design an experiment** to model zero population growth. What modifications would have to be made to the experiment?

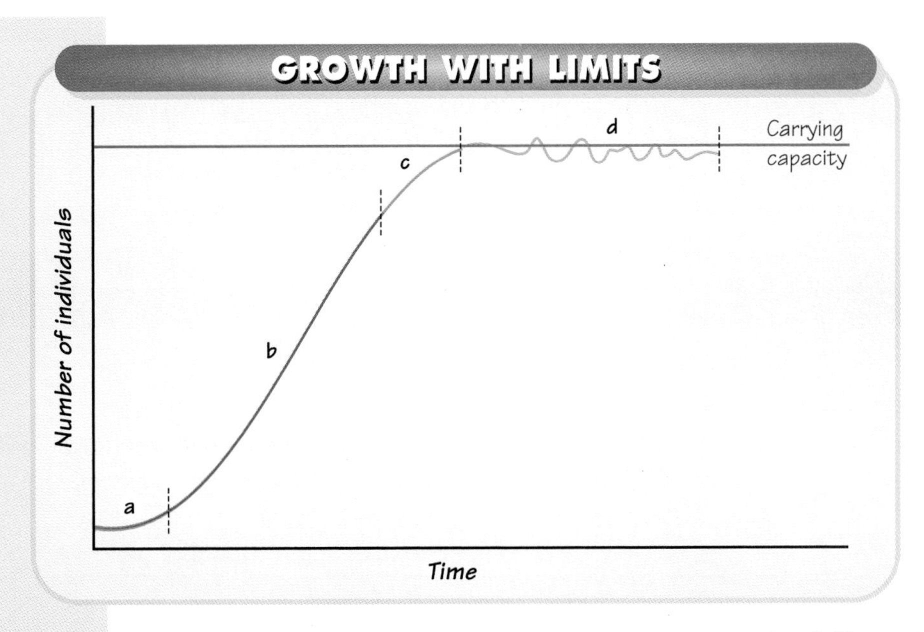

Figure 14–3
*In a growth-with-limits curve, the number of organisms begins to grow slowly, as shown in segment **a** of the graph. This slow growth is followed by an exponential growth phase, shown in segment **b**. As shown in segment **c**, eventually the population growth slows down, and at segment **d**, it levels off at carrying capacity.*

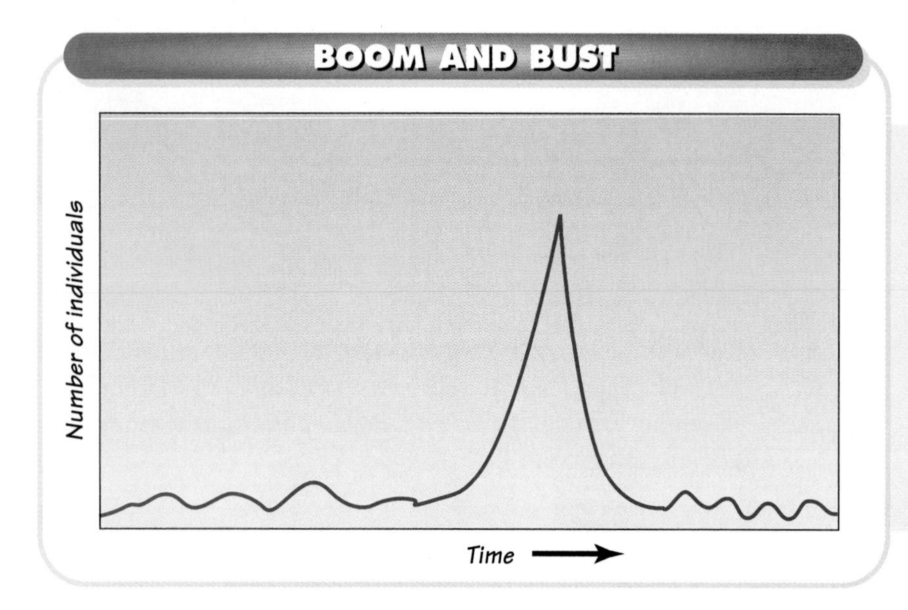

Figure 14–4
In a boom-and-bust growth curve, the population remains steady at a low rate before growing rapidly. When the population reaches its peak, the organisms die off dramatically. The population remains steady again until the next peak.

because its growth rate is zero. This situation is called the steady state.

☑ **Checkpoint** What is zero population growth?

Carrying Capacity

If you were to draw a horizontal line through the middle of the steady-state region, the line would tell you how large this stable population is. This line represents the largest number of individuals that can survive over long periods of time in a given environment. Scientists call this the **carrying capacity** of a particular environment for a particular species. When a population reaches the carrying capacity of its environment, a variety of factors act to stabilize it at that size.

If the population gets larger than the carrying capacity, either its birth rate falls or its death rate rises. If the population falls below the carrying capacity, its birth rate rises, its death rate drops, or both occur. In most natural populations, the steady state is not absolutely steady. It rises and falls somewhat from one year to the next.

☑ **Checkpoint** What is meant by carrying capacity?

Boom and Bust

Although many species in nature increase until they reach carrying capacity and then level off, it would be wrong to think that all species do. Some species grow exponentially until they reach a peak population size (the boom) and then crash dramatically (the bust). After the crash, the population may build right up again or may stay low for some time.

Section Review 14–1

1. **Define** population.
2. **Compare** the features of exponential growth and growth with limits.
3. **Critical Thinking—Inferring** What might cause the carrying capacity of a population to change?
4. **MINI LAB** After **constructing a model** of population growth, what can you conclude?

SECTION 14–2

Why Populations Stop Growing

GUIDE FOR READING

- **List** four density-dependent limiting factors.
- **Compare** density-dependent limiting factors and density-independent limiting factors.

WHAT KINDS OF FACTORS IN AN environment actually control population growth in nature? Population growth may be limited by several factors—some that depend on the size and density of the population and others that do not. Population density is the number of organisms in a given area. Acting separately or together, the population-limiting factors keep natural populations somewhere between extinction and covering the entire planet.

Density-Dependent Limiting Factors

Some population-limiting factors operate more strongly on large, dense populations than on small, less-crowded ones. These factors are called **density-dependent limiting factors.** Species whose populations are controlled by density-dependent limiting factors tend to have fairly stable populations. **Competition, predation, parasitism, and crowding are examples of density-dependent limiting factors.**

Competition

When populations become crowded, individual plants or animals may compete with one another for food, water, space, sunlight, or other things essential to life. Some individuals may obtain enough of what they need to survive and reproduce. Others may obtain enough to live but not enough to enable them to raise offspring. Still others may starve to death or die from lack of shelter.

Figure 14–5
Population growth may be limited by several factors. (a) An example of a density-dependent limiting factor—predation—is shown here between a marine iguana and a Galapagos hawk. (b) Populations such as orange trees, on the other hand, are controlled by density-independent limiting factors, such as snow and ice.

Competition can thus lower birth rates, increase death rates, or do both.

How can competition among members of a species be considered a density-dependent limiting factor? The more individuals there are in an area, the sooner they will use up the available resources. The fewer individuals there are, the less competition they have for the resources and the longer they will last. Competition for limited resources is often one of the most important factors in determining the carrying capacity of an environment for a particular species.

☑ **Checkpoint** What is the relationship between competition and carrying capacity?

Predation

You may recall that energy in the biosphere flows through producers and consumers. Most species serve as food for some other species. In nature, predators and the organisms they prey upon usually coexist over long periods of time. Over time, predator and prey become accustomed to each other's strengths and weaknesses.

Prey develop some remarkable defenses against predators. Predators in turn develop their own defenses, or counterdefenses. Strong jaws and sharp teeth, powerful digestive enzymes, or keen eyesight are examples of counterdefenses that have developed.

The presence of defenses and counterdefenses does not mean, however, that the number of predators and prey will always reach a balance. As you can see in ***Figure 14–6,*** populations of predators and prey almost always change in size over time.

☑ **Checkpoint** If an entire wolf population is killed off, what may happen to the deer population on which it preys?

Parasitism

Parasites are similar to predators in many ways. A **parasite** is an organism that takes nourishment from its host. Parasites live at the expense of their hosts, weakening them and causing disease and, in some cases, death.

You may wonder why parasitism acts as a density-dependent control on population size. Parasites work most effectively when hosts are present in large numbers. Why is this so? Parasites are often host-specific. This means that they grow best in members of a single species. Crowding helps parasites travel from one suitable host to another.

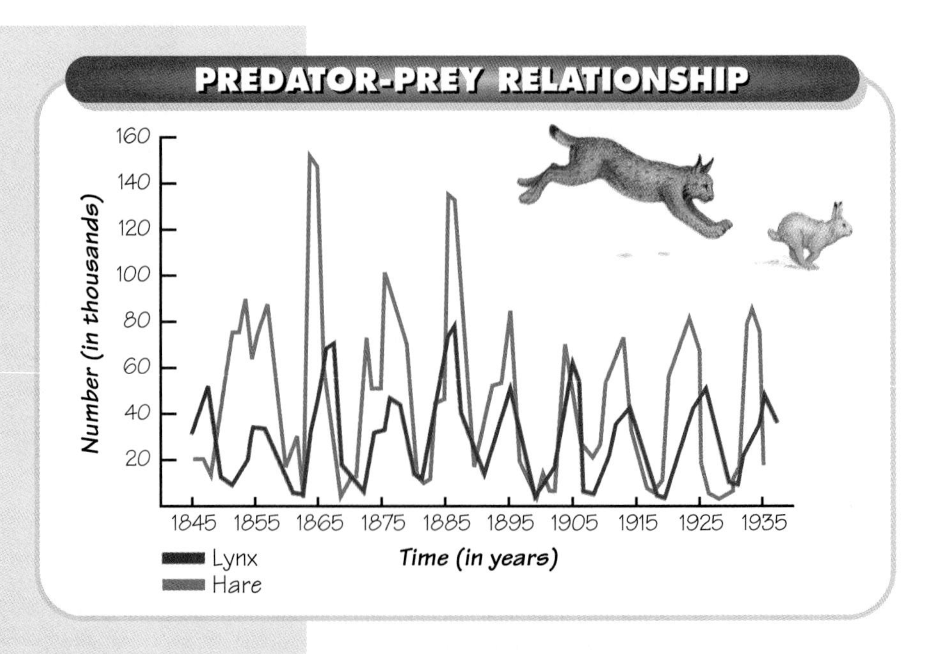

Figure 14–6
The relationship between populations of lynx (predators) and snowshoe hares (prey) changes over many years. As the number of hares increases, the number of lynx increases as well. The lynx eat more hares than the number being born, and the hare population decreases. As a result, the lynx begin to starve, causing their population to drop. With a decreased lynx population, the hares begin to recover and the cycle repeats itself.

Problem Solving

INTERPRETING DIAGRAMS

Flowering Plants

The sophomore class needs to raise money for their year-end dance. One Biology class decided to have a plant sale. The class reasoned that if they planted the seeds themselves and then transplanted the flowering plants into decorative flowerpots, they could make a lot of money.

Based on the prices for the seeds, fertilizer, potting soil, flats, and flowerpots, they determined that in order to break even, they would need to grow 25 to 29 plants per flat. If they grew 30 or more plants per flat, the class would make a profit.

Two different groups planted seeds. The students in Group A decided to try to grow 30 plants per flat. They carefully counted out enough seeds and evenly spaced the seeds in the potting soil. Those in Group B decided to try to grow more than 30 plants. They planted enough seeds for 50 plants in the same amount of space. The two groups' flats were placed next to each other on a windowsill and given the same amount of water and fertilizer each week. The results of each group's plant flats are shown below.

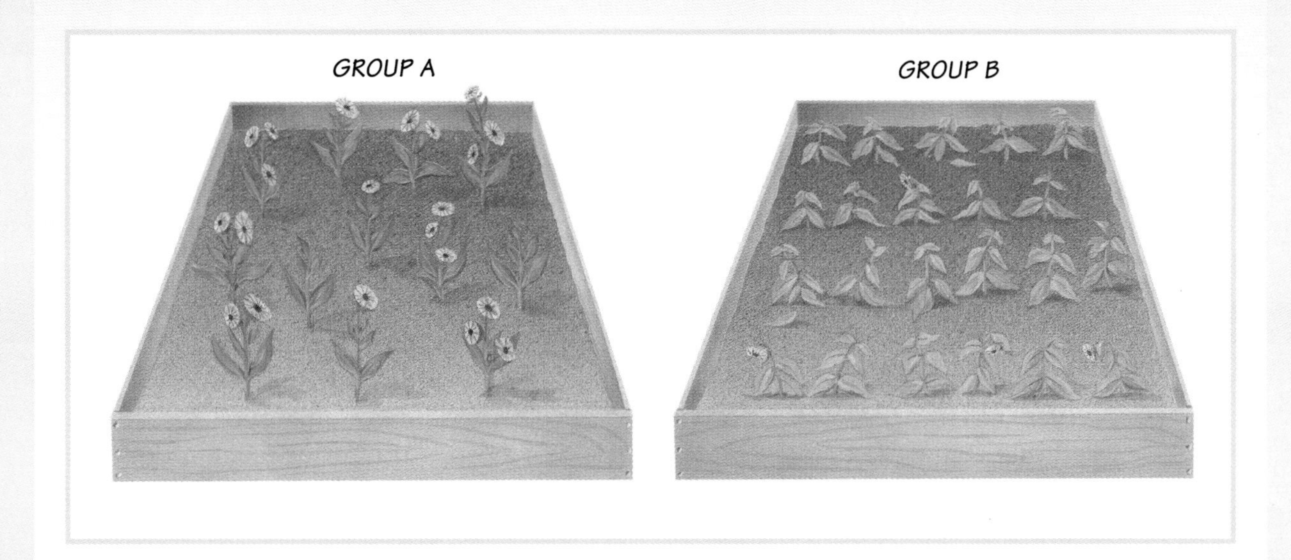

THINK ABOUT IT

1. What density-dependent factors might have affected the plant growth? Were these factors equally strong on both groups' plants?
2. What density-independent factors might have affected the plant growth? Were these factors equally strong on both groups' plants?
3. **Design an experiment** to determine the maximum number of plants per flat that would flower and grow if the flat measured 1' × 2' × 3".

Figure 14–7 *Stress caused by crowding is an example of a density-dependent limiting factor.*

Note that few parasites kill their hosts—at least, not right away. If a parasite killed its host too quickly, the parasite would have to find another host or it too would die. That's why under most circumstances, it is to a parasite's advantage not to be too deadly.

☑ **Checkpoint** Why is parasitism a density-dependent limiting factor?

Crowding and Stress

Most animals, including humans, have a built-in behavioral need for a certain amount of space. Both the males and the females of a species may need room to hunt for food. They may need a certain amount of space for nesting, or they may need a territory of a certain size. In such cases, the number of suitable territories regulates population size in a density-dependent manner.

Some organisms fight among themselves if they become overcrowded. Fighting can cause high levels of stress, which disturbs the finely tuned system of hormones that coordinate body functions. Often, the immune system may become weakened. Hormonal changes from stress can also upset animals' behavior so that they neglect, kill, or even eat their own offspring. All these factors limit population growth.

☑ **Checkpoint** How does crowding act as a density-dependent limiting factor?

Density-Independent Limiting Factors

Not all populations are controlled by density-dependent limiting factors. **Species that have boom-and-bust growth curves may be affected by factors that kill organisms regardless of how large the population is.** Because the density of the population does not matter in such cases, these factors are called **density-independent limiting factors.**

Weather is probably the most important density-independent limiting factor. An entire insect population can be destroyed by a rainstorm. They may also be harmed by unusually hot or cold weather, by disturbances such as fires, or by droughts, floods, or hurricanes. Human activities—such as a toxic waste spill, the spraying of pesticides, or clear-cutting a forest—may also act as density-independent limiting factors.

Section Review 14–2

1. **List** four density-dependent limiting factors.
2. **Compare** density-dependent limiting factors and density-independent limiting factors.
3. **Critical Thinking—Relating Concepts** Give an example of a density-independent limiting factor that has impacted a human population. Explain how it has impacted the population.

Human Population Growth

GUIDE FOR READING

- **Describe** how the human population is growing exponentially.
- **List** the three stages of demographic transition.

LIKE THOSE OF OTHER organisms, human populations tend to increase with time in the same ways and for the same reasons. What, if anything, will stop humans from covering the planet? How and why do human populations stop growing? Read on to find the answers to these questions.

Growth Increases

For most of human existence, human populations grew slowly. Why was this so? Population-limiting factors kept populations at a low level. Life was harsh. Food was difficult to find. Predators and parasites were everywhere. Death rates, therefore, for humans were quite high. Until recently, only half the children survived to adulthood. Families had many children, just to make sure that some of them would survive.

Human Population Grows Quickly

About 300 years ago, the world's human population started growing more rapidly. Over several hundred years, agricultural and industrial revolutions made human life easier. People developed ways to control some population-limiting factors. Thanks to modern agriculture, more nutritious foods became available. Recently, doctors learned to cure or prevent diseases that once killed large numbers of people. Better health care and nutrition dramatically reduced the number of infants who died. People started living healthier, longer lives.

INTEGRATING HEALTH

How do the leading causes of death today differ from those in 1900?

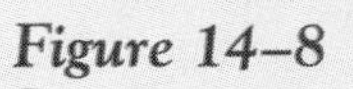

Figure 14–8
(a) *This artist's representation of the population of North and South America illustrates how the world's population continues to grow at an astonishing rate. In the United States, population growth is slow. However, some areas are very crowded,* (b) *as shown in this photograph of New York City.*

Figure 14–9
CAREER TRACK
As human population growth impacts animals and other wildlife, care must be taken to prevent organisms from becoming extinct. Here a wildlife biologist tags a frog-eating bat in a Panama rain forest. This enables wildlife biologists to monitor the bat population.

In other words, several factors combined to lower human death rates. More children than ever before survived to marry and have children of their own. At the same time, birth rates in most places continued to be as high as they had ever been. Because the birth rate was higher than the death rate, the human population grew. **Today, the world's human population is still growing exponentially.** At present, 180 people are born every minute. This means that there are 92 million more humans each year.

Checkpoint Why has the human death rate decreased?

Controlling Human Populations

Human population size cannot increase exponentially forever for the same reason that populations of bacteria don't do so. That much is obvious. Exactly why and how will human growth slow down? The answer to this question is far less obvious.

You have seen that the human population grows for the same reasons that

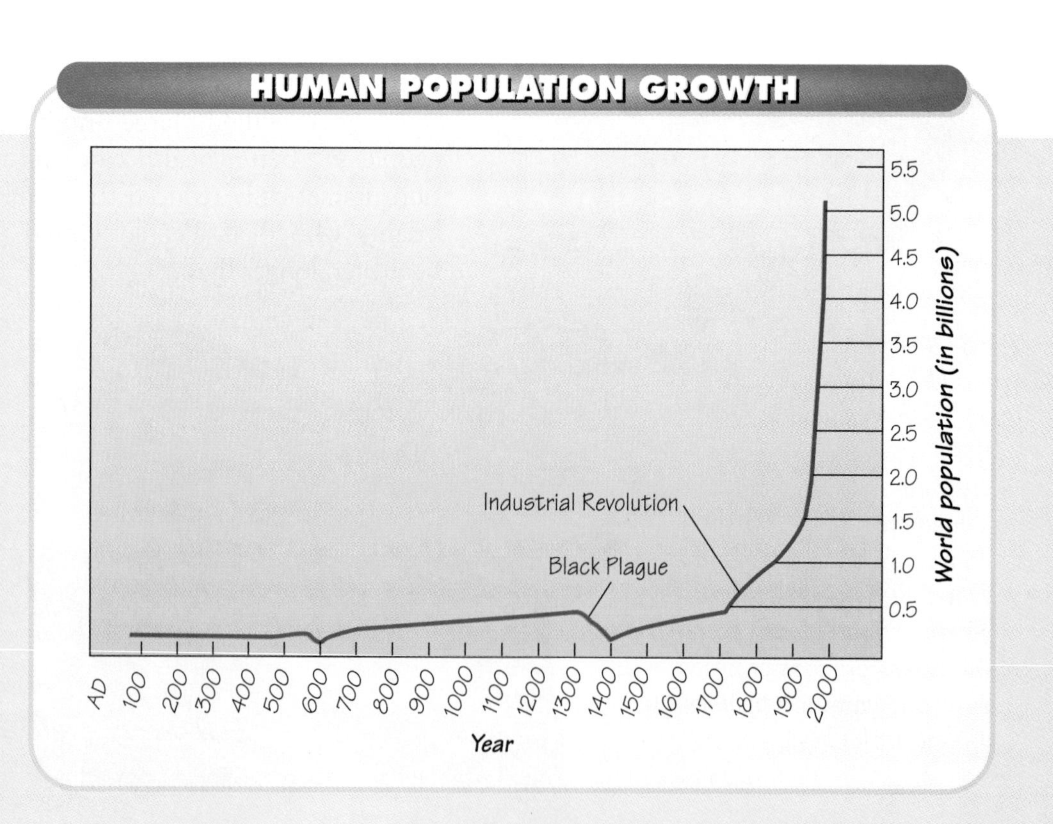

Figure 14–10
The global human population reached 1 billion in 1800. In a little more than 100 years, the population doubled to 2 billion, then doubled again to 4 billion less than 50 years later. And 13 years later, the population reached 5 billion. Notice the growth decrease in the 1300s due to the Black Plague and the beginning of the rapid increase in the 1700s due to the Industrial Revolution.

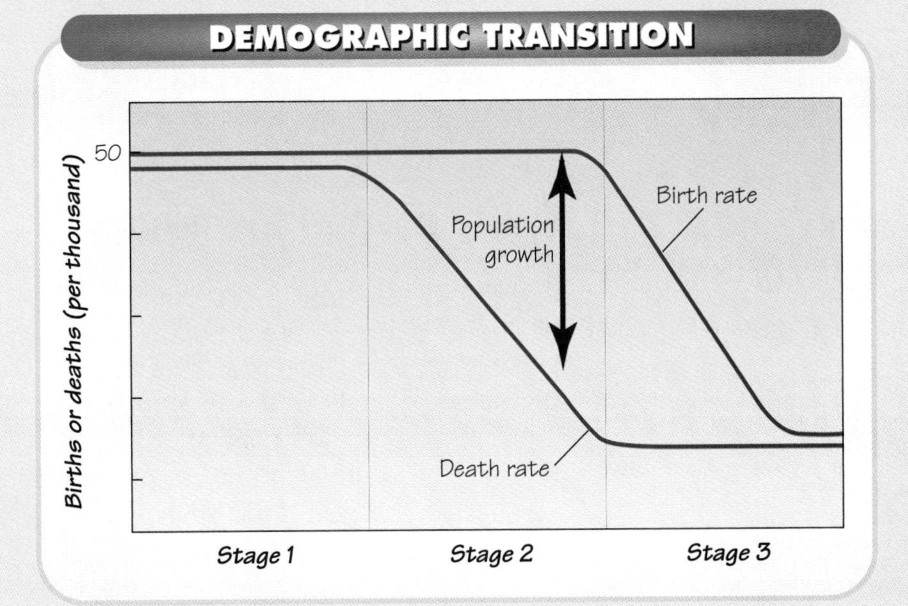

Figure 14–11 *Before the Industrial Revolution in the eighteenth century, populations in North America and Europe were in Stage 1 of the demographic transition. After industrialization, death rates decreased due to improved medical and scientific advances. Birth rates remained high, and Stage 2 was reached. As the economy began to improve and education became more widespread, birth rates began to decrease to match the death rate. North America and Europe are now in Stage 3.*

animal populations grow. The human population would stop growing if birth rates fell, death rates rose, or both occurred. But do human birth and death rates change for the same reasons as they do in other animal populations? To answer this question, let's look at some situations in which human population growth has indeed slowed or stopped.

Growth Slows Down

Over the last century, human population growth has slowed down dramatically in some countries. What factors have caused this change? Biologists say that these populations passed through the **demographic transition.** The demographic transition is a change in growth rate resulting from changes in birth rate.

The demographic transition consists of three stages. During the first stage, there is a high birth rate and a high death rate. Families have many children to make sure some will survive. Because both the birth rate and death rate are high, population growth is slow during this stage.

During the second stage, improvements are made in living conditions. Food production increases, there are advances in medicine, and sanitation is improved. As a result, more children live to adulthood and the death rate decreases. Because the birth rate remains high, the population grows rapidly.

In the third stage, the birth rate decreases for a variety of reasons. Because more children are surviving, families begin to have fewer children. The birth rate and death rate reach a balance at a lower level. Population growth slows down and may stabilize.

Section Review 14–3

1. **Describe** how the human population is growing exponentially.
2. **List** the three stages of demographic transition.
3. **Critical Thinking—Inferring** In what stage of the demographic transition is the United States?

SECTION 14–4 **BRANCHING OUT** In Depth

Population Growth and Carrying Capacity

Guide for Reading

- **Explain** why the world's population continues to grow.
- **Predict** the kind of human population growth there will be in the future.

MINI LAB

- **Compare** the total number of offspring over five generations in one-, two-, and three-children families.

FUTURE HUMAN POPULATION growth is an important issue. To many ecologists, the size of the human population is the single most important factor in determining the overall health of the Earth. Why? Because human activities have profound effects on both local and global environments. And population size affects all these activities.

World Population Growth

Despite fairly stable populations in most western countries and Japan, the global population is still growing exponentially. The last serious estimate in 1990 put the world's population at 5.3 billion. The United Nations has predicted that by the year 2100 that number may reach between 6 billion and 19 billion. Why will this number continue to grow? **The world's population continues to grow because most people live in countries that have not yet completed the demographic transition.**

The United States, Canada, Japan, and the countries of Europe have gone through all three stages of the demographic transition. In these countries, there is slow population growth. The population of the United States, for example, is still growing, but it is growing much more slowly than in the past. In some European countries, populations are even declining slightly.

Countries in Asia, Africa, and South America have not yet completed the demographic transition. These areas are home to approximately 80 percent of the world's population. As you can see in ***Figure 14–13,*** these areas are still undergoing rapid growth. This rapid

Figure 14–12
(a) Countries that have rapid growth patterns will continue to add large numbers of people to the world population in the future. (b) Overcrowding caused by a growing population is an issue in many parts of the world, as shown by this crowded bus in Vietnam.

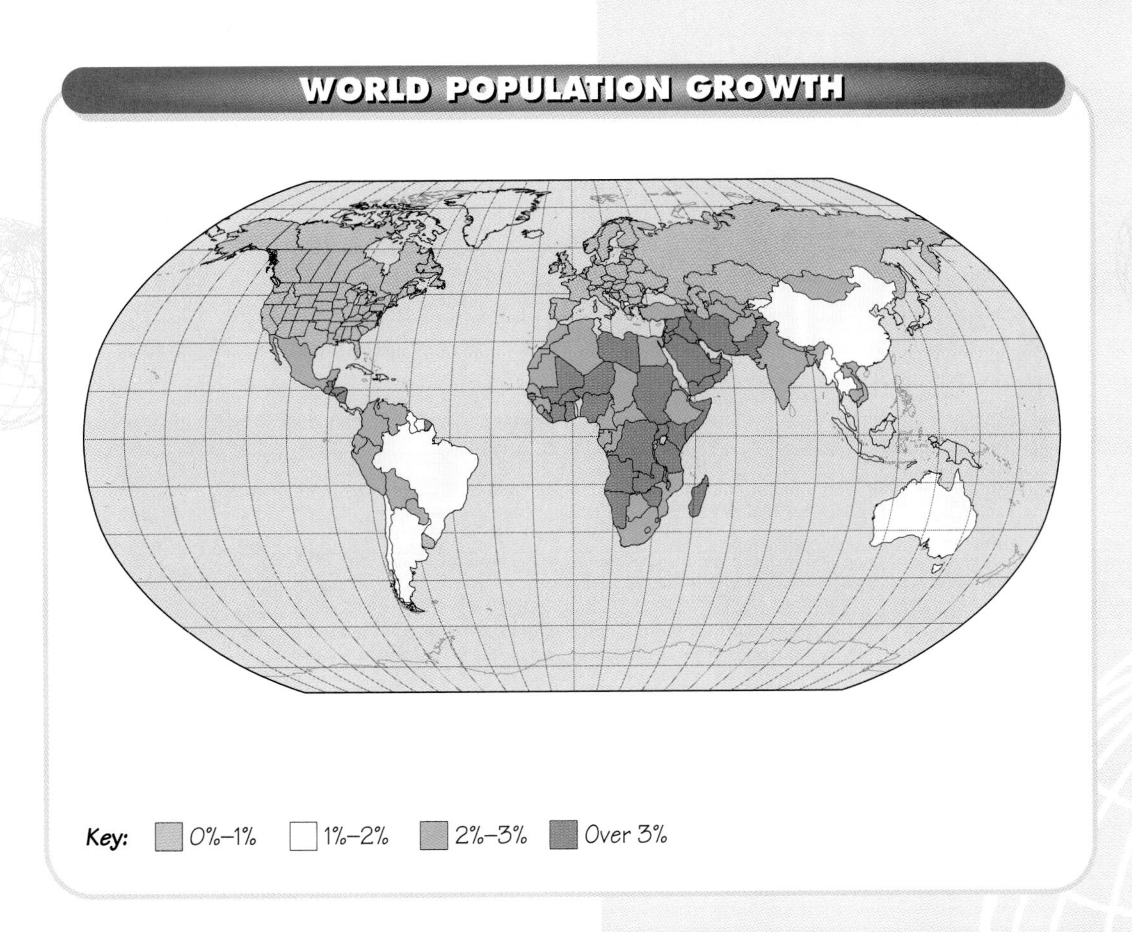

Figure 14–13
This map shows population growth rates in different parts of the world. Notice the high growth rate in Asia, Africa, and South America.

growth, coupled with a declining death rate, has intensified economic problems.

Future Population Growth

Future population growth in various countries depends, in part, on how many people of different ages are living in that country today. On the next page, *Figure 14–14* shows what are called the **age-structure diagrams** of human populations in three different countries. Demographers, or people who study population growth, use age-structure diagrams to make predictions about future growth.

Rapid Growth

In a country with rapid growth, such as Mexico, the majority of the population is under the age of 15. Older groups represent a much smaller percentage of the population. Although the birth rate in Mexico has decreased, there is still great potential for growth. How can you tell? The largest percentage of people has not yet reached their childbearing years. As this group reaches reproductive age, Mexico can expect to grow even more.

Slow Growth

In a country such as Sweden, there is slow population growth. There are almost equal numbers of people in each age group. The growth rate in Sweden is almost zero.

In the United States, too, there are almost equal numbers of people in each age group. This pattern predicts a slow, steady growth for the near future. Do you see the large number of people in the age

INTEGRATING CAREERS

What kind of degree must demographers have? What kinds of courses should they take?

Figure 14–14
On an age structure diagram, each bar represents the percentage of the population of individuals within a 5-year age group. The percentage of males in that age group is found to the left of the center line, and the percentage of females is to the right. In a country with slow growth, such as Sweden, there are almost equal numbers of people in each age category. In a country with rapid growth, such as Mexico, the larger percentage of people is under the age of 15. The United States has fairly stable growth.

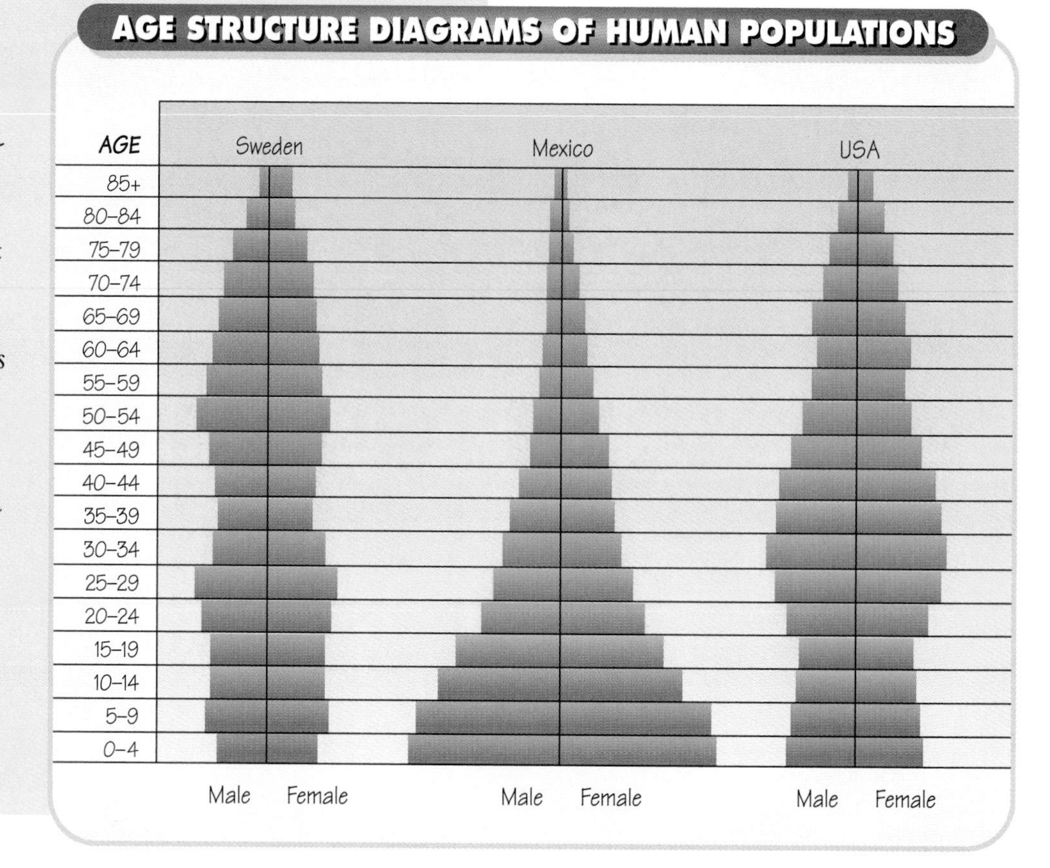

groups born just after World War II? These people, called the baby boomers, belong to the largest segment of the population.

Baby boomers have hit their childbearing years and yet population growth is still slow. How can that be? Due to many social changes, many women are delaying childbirth and having fewer children.

Earth's Carrying Capacity

Is world population growth going to be a problem in the future? Opinions differ about how long population growth will continue, whether it is a problem, and how important it is to slow it down. Ecologists argue that if population growth doesn't slow down, there could be serious and lasting damage to both the local and global ecology. Others disagree. Economists think that science,

Figure 14–15
The city of Nairobi in Kenya provides a backdrop to this topi and impala buck, illustrating how human growth is beginning to have an impact on other natural populations.

technology, and changes in society will help to control the human impact on the environment. The most important question, however, is what the Earth's carrying capacity for humans might be. And that is an incredibly difficult question to answer.

As a group, ecologists feel strongly that human population should not be viewed in isolation. Instead, human activity should be looked at as it affects the Earth's ability to provide the essentials of life, such as food, water, air, land, shelter, and minerals. That is easily said. But just how much human, plant, and animal life can a particular area support? How much of the biosphere can be covered by homes, farms, and highways without interfering with global life-support systems?

Different experts, using different approaches, have suggested that the Earth can hold anywhere between 5 billion and 20 billion people. It is interesting to note that the actual human population has now reached the lower part of that range. Within your lifetime, the number of people on Earth will double at least once.

How long will this growth continue? What does it mean for the future? Finding answers to these and many other questions is vital to the future of humans on Earth.

MINI LAB .... Comparing ....

A Baby Boom?

PROBLEM *How would the number of offspring in one-, two-, and three-children families over five generations* ***compare*** *if each offspring bears the same number of offspring as its parents had?*

PROCEDURE

1. Calculate the number of offspring that could be born in each generation for a one-child family over five generations. Record your calculations.
2. Repeat Step 1 for a two-child family and then a three-child family.

ANALYZE AND CONCLUDE

1. In the third generation, how many total offspring are there in the two-child family? In the three-child family?
2. In the fifth generation, how many total offspring are there in the one-child family? In the three-child family?
3. In a four-child family, how many offspring could there be in the fifth generation?

Section Review 14–4

1. **Explain** why the world's population is still growing rapidly.
2. **Predict** the kind of human population growth there will be in the future.
3. **MINI LAB** **Compare** the total number of offspring in one-, two-, and three-children families over five generations. Assume that each child will have the same number of offspring as its parents had.
4. **BRANCHING OUT ACTIVITY** In the age-structure diagram of the United States, there is a decrease in the percentage of people in the age category before the baby boom. **Formulate a hypothesis** to explain this decrease. Use reference material to find out what happened during that time that could account for a low birth rate.

CHAPTER 14

Laboratory Investigation

The Rise and Fall of Yeast

The population of yeast varies over time. In this investigation, you will monitor the population of a yeast culture over seven days. Will the yeast population grow and thrive? Perform this experiment to find out.

Problem

How can you **measure** the population of a yeast culture over a period of seven days?

Materials (per group)

4 25-mL beakers
glass-marking pencil
40 mL molasses solution
4 mL yeast solution
2 pipettes
microscope slide
coverslip
microscope
graph paper

Procedure 

1. Use the glass-marking pencil to label each beaker with the following information: group name and beaker number (1 through 4).
2. Add 10 mL of the molasses solution to each of the beakers.
3. Stir the yeast solution, then add 10 drops to each of the beakers.
4. Use a clean pipette to transfer 1 drop of solution from beaker 1 to a clean microscope slide. Cover with a coverslip.
5. Use the low-power objective to examine the slide under a microscope. Switch to high power. Use the fine adjustment to locate some yeast cells.
6. Count the number of yeast cells in the field of view. Each bud counts as a single yeast cell. Record your number in a data table similar to the one shown.
7. Repeat step 6 three more times. Calculate an average of the four counts. This is the average population of yeast in the culture.

DATA TABLE

Populations	Day 1	Day 3	Day 5	Day 7
	Beaker 1	Beaker 2	Beaker 3	Beaker 4
View 1				
View 2				
View 3				
View 4				
Average				

8. Store beakers 2, 3, and 4 in a dark, warm area where they can remain undisturbed for seven days.

9. On the third day, repeat steps 4 through 7 using beaker 2.

10. On the fifth day, repeat steps 4 through 7 using beaker 3.

11. On the seventh day, repeat steps 4 through 7 using beaker 4.

Observations

1. Use the graph paper to construct a graph representing the population growth of the yeast culture by plotting time horizontally and average population vertically.
2. Using your graph, identify the period of time when the population of yeast increased. Did the yeast population ever decrease?
3. Summarize the population density of the yeast over the seven-day period.
4. Predict what might happen to the yeast population after another two days.

Analysis and Conclusions

1. Identify and label the stages of growth on your graph.
2. What kind of growth does the yeast population follow?
3. Why does the population density of yeast in this culture change over time?

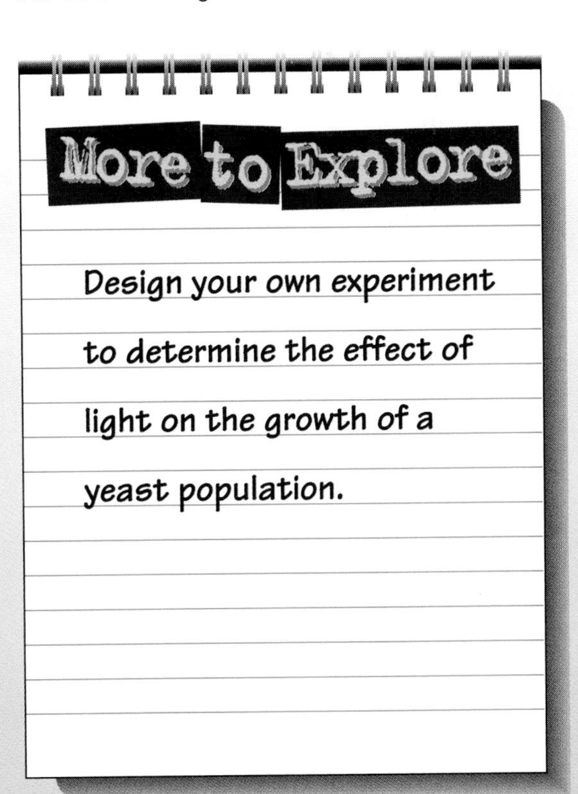

Study Guide

Summarizing Key Concepts

The key concepts in each section of this chapter are listed below to help you review the chapter content. Make sure you understand each concept and its relationship to other concepts and to the theme of the chapter.

14–1 Populations and How They Grow

- A population is a group of organisms of a single species that live in a given area.
- Given ideal conditions, a population will grow rapidly and without limits. This kind of growth is called exponential growth.
- In nature, most populations will grow exponentially for a short time and then level off at a steady state. The steady state represents the carrying capacity of the population.

14–2 Why Populations Stop Growing

- Population growth may be controlled by limiting factors—some that depend on population size and density and others that do not.
- Population-limiting factors that act more strongly on large, dense populations are called density-dependent limiting factors.
- Limiting factors that act on a population regardless of its size and density are called density-independent limiting factors.

14–3 Human Population Growth

- Human population has been steadily increasing since the beginning of time. The human population is now growing exponentially.
- Humans, unlike other organisms, have the ability to alter the carrying capacity of the environment in which they live.
- The demographic transition is a change in growth rate from a high birth rate and death rate to a low birth rate and death rate.

14–4 Population Growth and Carrying Capacity

- Most people in the world live in countries that have not yet completed the demographic transition. Therefore, the population continues to grow.
- An age-structure diagram is a graphic illustration of the distribution of males and females in a country according to age.

Reviewing Key Terms

Review the following vocabulary terms and their meaning. Then use each term in a complete sentence.

14–1 Populations and How They Grow

population
growth rate
exponential growth
zero population growth
carrying capacity

14–2 Why Populations Stop Growing

density-dependent limiting factor
parasite
density-independent limiting factor

14–3 Human Population Growth

demographic transition

14–4 Population Growth and Carrying Capacity

age-structure diagram

Recalling Main Ideas

Choose the letter of the answer that best completes the statement or that answers the question.

1. A population that has a death rate greater than its birth rate is said to be

a. increasing.
b. decreasing.
c. growing exponentially.
d. staying the same.

2. A growth curve characterized by a population that starts growing slowly and then increases rapidly before leveling off is a(an)

a. boom-and-bust curve.
b. exponential curve.
c. growth-with-limits curve.
d. age-structure curve.

3. In most situations, predators and their prey

a. destroy each other.
b. coexist over a long time.
c. cannot coexist.
d. compete for the same limited resources.

4. About 500 years ago, the human population

a. began to grow slowly.
b. began to grow exponentially.
c. reached carrying capacity.
d. reached zero population growth.

5. A group of organisms of a single species that live in a given area is called a

a. community.
b. niche.
c. food web.
d. population.

6. Density-dependent limiting factors act most strongly on which populations?

a. dense
b. small
c. large
d. scattered

7. The world's human population is currently in which phase?

a. zero population growth
b. slow population growth
c. rapid population growth
d. carrying capacity

8. In the second stage of the demographic transition,

a. birth rate is high and death rate is high.
b. birth rate is high and death rate is low.
c. birth rate is low and death rate is high.
d. birth rate is low and death rate is low.

Putting It All Together

Using the information on pages xxx–xxxi, complete the following concept map.

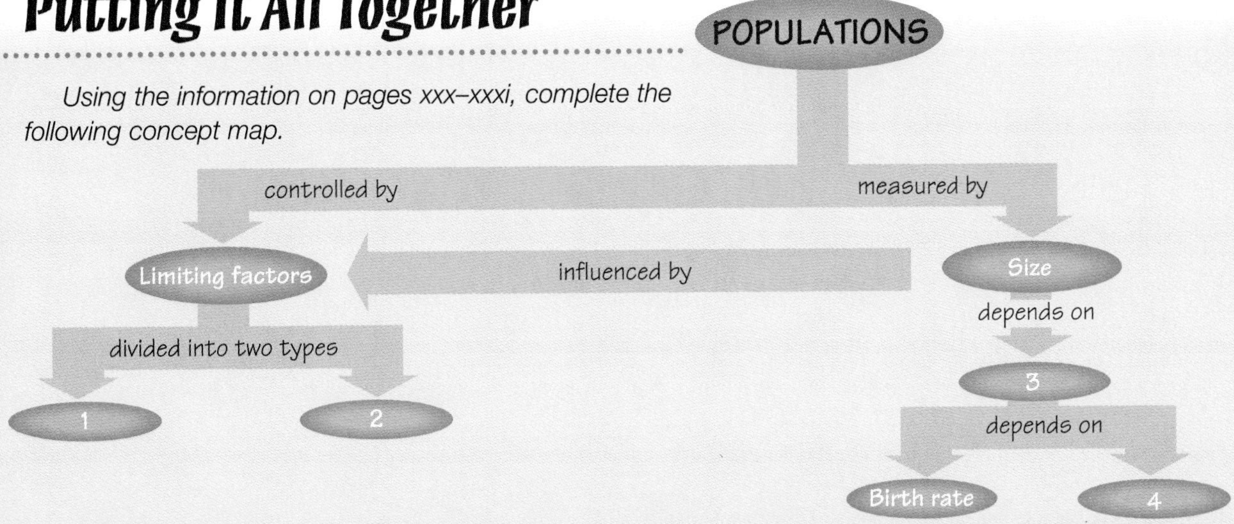

POPULATIONS
controlled by
measured by
Limiting factors
influenced by
Size
divided into two types
depends on
1
2
3
depends on
Birth rate
4

Reviewing What You Learned

Answer each of the following in a complete sentence.

1. Define and identify a population in your community.
2. What major factors influence population growth rates?
3. What is meant by the carrying capacity of an environment?
4. What happens when a population gets larger than its carrying capacity?
5. What are five examples of density-dependent limiting factors?
6. How does competition help to maintain a stable population?
7. What is a parasite?
8. **Compare** the density-dependent and density-independent factors that influence a population.
9. Why is weather an important density-independent limiting factor for certain organisms under certain conditions?
10. What is the demographic transition?
11. What is an age-structure diagram?
12. What kind of work does a demographer do?
13. Who are the baby boomers?

Expanding the Concepts

Discuss each of the following in a brief paragraph.

1. How does the removal of an organism in a food web affect the other organisms with which it interacts?
2. What factors influence exponential growth in a typical deer population in a natural forest?
3. Choose one of the density-dependent limiting factors and explain how it applies to human populations.
4. Parasites generally do not kill their host quickly. Why?
5. How can hormonal changes in an organism limit population growth?
6. **Construct a model** that illustrates the effect of one density-independent limiting factor on a plant population.
7. What have been the effects of improvements in living conditions on the birth and death rates of humans?
8. Why is it essential to understand human growth rates for the future of this planet?
9. Why can we not view the human population in isolation?

Extending Your Thinking

Use the skills you have developed in this chapter to answer the following.

1. **Drawing conclusions** Modern medicine has had a pronounced effect on both density-dependent and density-independent factors for humans. Explain how.
2. **Making judgments** What generally controls birth and death rates in humans? Are these factors similar to those that influence a deer population in a natural environment? Support your answer.
3. **Using the writing process** If you were the leader of a research group assigned to control a nonnative animal pest, what would be the focus of your initial research and why?
4. **Making predictions** You have observed the population growth curves of two different species of *Paramecia* in isolation. In identical environments, species *A* had a doubling time of 24 hours and species *B* had a doubling time of 18 hours. When placed together in a limited environment, what would you predict would happen to their growth rates?
5. **Drawing conclusions** In many isolated environments, such as the islands of Hawaii, most plant populations grew without predation from grazing animals. What would be the consequences of the introduction of sheep?

Applying Your Skills

What Goes Up Always Comes Down—Or Does It?

How do births and deaths in a population affect the population size? On a yearly basis, wildlife biologists collect information on certain animals that live in the wild. This information is used to help determine regulations to help control the size of the population through various management practices. You are going to collect some fictional data, graph your results, analyze the information, and then suggest future guidelines for this population.

1. Roll two dice. Each roll of the two dice is equal to one year's birth rate when multiplied by 1000. Record the birth rate. Roll the dice a total of ten times, recording the birth rate each time.
2. Repeat step 1. This time, however, the number represents the death rate.
3. Calculate the growth rate for each of the 10 years. Record this information.
4. Construct a graph that shows the growth rate over a 10-year period. The population size began at 100,000. (This is the carrying capacity.)
5. Graph the growth rate for 10 years.

• GOING FURTHER •

6. Analyze your data. Based on your population size after 10 years, write a paragraph on the future management of the population.

CHAPTER 15

Our Living Planet

FOCUSING THE CHAPTER
THEME: Systems and Interactions

15–1 Climate and Life
- **Explain** the significance of climate to environments and ecosystems.

15–2 Environments and Life
- **Compare** ecological succession and climax.

15–3 Earth's Biomes
- **Describe** the characteristics of Earth's major biomes.

BRANCHING OUT *In Depth*

15–4 Climate: A Complex Story
- **Explain** how Earth's climate patterns arise and **discuss** global climate change predictions.

LABORATORY INVESTIGATION
- **Design an experiment** to explore the impact of the greenhouse effect on global warming.

Biology and Your World

BIO JOURNAL

What does the term climate change mean to you? Talk to people from other age groups about climate change, then record their ideas in your journal. Do you have reason to think that the global climate may be changing?

Effect of strong winds on the growth of an evergreen tree

Climate and Life

GUIDE FOR READING

- **Explain** the importance of climate to life.
- **Define** environment and ecosystem.

HAVE YOU EVER WONDERED *why banana trees cannot grow in Anchorage, Alaska, or why tulips and blueberries do not grow in Hilo, Hawaii? Thanks to tropical sunshine, plenty of rain, and frost-free temperatures, bananas do, of course, grow well on the island of Hawaii. Tulips and blueberries, on the other hand, grow well in Alaska, where they find the cold winters they need. Why do different parts of the world support different kinds of plants and animals?*

Climate

Different parts of the globe have different **climates.** Climate is the temperature range, the average annual precipitation (rain or snow), humidity, and the amount of sunshine that a region typically experiences. The climate of an area is a powerful factor in determining the types of living organisms that the area can support. **Populations of plants and animals grow, remain stable, or vanish, depending in part on climate conditions such as temperature and rainfall. These conditions also affect the structure of food webs and the flow of nutrients.**

a

b

c

Figure 15–1 (a) *Conifers,* (b) *bananas, and* (c) *wheat naturally grow in parts of the world where climate conditions suit them best.*

What gives rise to a region's climate? The sun's energy, as it interacts with Earth's air, water, and land, causes global climate patterns. You will see that these global climate patterns, in turn, shape all life on Earth.

Earth's Thermostat

Earth is the only planet in the solar system that has temperatures that are acceptable to life as we know it, day in and day out, throughout the year. This is because the Earth's atmosphere serves as a natural thermostat. Carbon dioxide, water vapor, and a few other gases in the atmosphere allow solar energy to reach the Earth's surface, where it is absorbed and later converted into heat. These gases, however, do not allow the heat energy to leave the Earth quite as readily, keeping the heat energy trapped inside for a period of time. This natural function of the atmosphere is called the **greenhouse effect.**

☑ *Checkpoint* What is the greenhouse effect?

Earth's Climate Zones

Interactions between solar energy and the atmosphere are responsible for much more than the Earth's temperature. Solar energy is what powers global winds and ocean currents, which give rise to the variety of climate zones on the Earth.

For instance, tropical regions are much warmer than temperate regions, which are warmer than polar regions. Rainfall patterns arise in large part due to the interaction of prevailing winds, ocean currents, and landmasses.

How Climate Varies

Climate and its effects are often quite complicated. Locations that are hundreds of kilometers apart can have similar climates. For example, gardeners know that some plants that can grow in Mississippi and Louisiana can also be grown along the coastline of Oregon and Washington. And certain areas of central Alabama and Georgia can occasionally get as cold in winter as some of the islands off the coast of Alaska.

On the other hand, some locations that are quite close to one another can have dramatically different climates. In Hawaii, orchid and banana plants can grow just kilometers from the snow-covered Mauna Loa volcano. Parts of California's northern coast have lush redwood forest growth, yet areas a short distance inland are desertlike.

Climate also can vary on a much smaller scale. If you look closely at tree trunks in a North

Figure 15–2
This diagram shows examples of different microclimates within a temperate forest.

ENVIRONMENTAL FACTORS

Environment

Biotic Factors

Abiotic Factors

Figure 15–3 *The biotic factors of an environment include all the organisms that live there. The abiotic factors are the pond and soil as well as factors such as humidity and sunlight, which are not easily shown in a drawing.*

American forest, you will find moss growing on their northern sides. And if you walk through a city in the Northern Hemisphere in early spring, you will see that trees growing in the sun next to south-facing buildings have leaves, but similar trees growing across the street won't put out leaves until several days later. Conditions such as these that vary over small distances are referred to as the **microclimate** for that location.

☑ ***Checkpoint*** What is microclimate?

Environments and Ecosystems

An environment is a combination of physical and biological factors that influence life. Physical environmental factors, called **abiotic factors,** are the area's climate, the type of soil and its acidity, and the availability of nutrients. Biological environmental factors, called **biotic factors,** include all the living things with which an organism might interact.

When you look carefully at an organism in its **environment,** it is often difficult to separate the biotic and the abiotic factors. This is because organisms in nature affect each other's environments. A tree growing in a forest shades the ground beneath it. By dropping leaves that decay, the tree contributes to the amount of moisture-holding material in the soil. In this case, are shade and moisture in the soil abiotic factors or biotic factors? They are abiotic factors that arise due to the biotic factors in the environment.

Ecosystem

If you ask an ecologist where a particular organism lives, he or she might say "on a Caribbean coral reef," "in a temperate beech-maple forest," or "in an Amazon rain forest." These answers can be thought of as the biological "street address" of an organism. Like your street address, a biological address tells you more than just where an organism lives.

Figure 15–4 *Too much or too little of an abiotic factor such as heat, light, and humidity can be difficult for an organism to tolerate, as shown in this graph.*

ZONES OF TOLERANCE

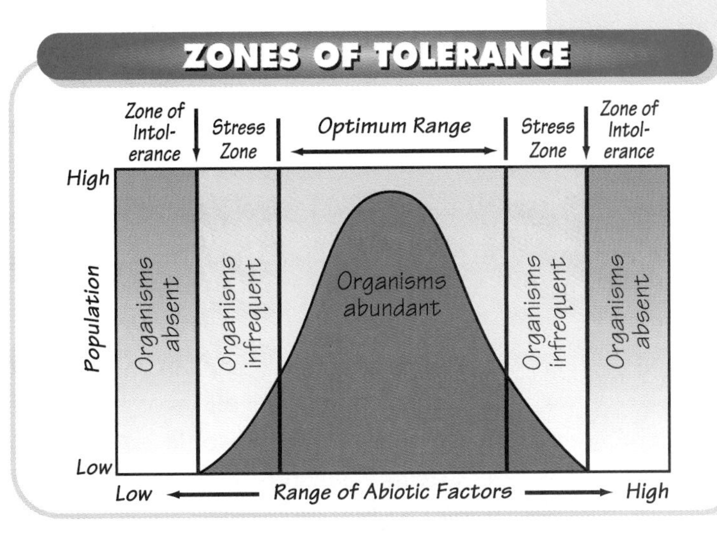

Figure 15–5 (a) *The scarlet macaw lives in the tropical rain forest of Central America, whereas* (b) *the black-tail prairie dog lives in the western grasslands of the United States.*

It tells you the type of climate the organism is accustomed to and the kinds of neighbors it is likely to have. **In the natural world, a combination of biotic and abiotic factors is called an ecosystem.** An **ecosystem** is a collection of organisms—producers, consumers, and decomposers—interacting with each other and with their physical environment.

☑ **Checkpoint** What is an ecosystem?

Niche

An address by itself does not tell you a great deal about a person—for example, what the person does for a living or what her favorite foods are. In a similar way, knowing an organism's ecosystem in itself does not tell you everything about the organism. A description of an organism's **niche,** however, does. A niche is the full range of physical and biological conditions in which the organisms in a species can live and the way in which the organisms use those conditions.

The biotic factors of the niche identify other organisms that a species interacts with in any way. They include the plants that an animal rests on or makes a home in, the prey it eats, the predators it may encounter, and so on. Besides a description of what it eats, an animal's niche also includes information about when it eats and where and how it finds its food.

Habitat

Many fishes that live on coral reefs eat plankton, or tiny marine animals, that swim or drift in the ocean current. You might be tempted to say that all these fishes have the same niche. But some species feed only during the daytime, while others feed only at night, making for distinctly different niches. What the fishes do share, however, is the same **habitat.** Habitat simply indicates the type of surroundings in which a species lives and thrives—defined in terms of the plant community and the abiotic factors. Organisms that share the same habitat do not necessarily compete with one another if they have different niches.

Section Review 15-1

1. **Define** climate and **explain** its significance to living things.
2. **Define** environment and ecosystem.
3. **Compare** biotic and abiotic environmental factors.
4. **Critical Thinking—Predicting** What do you think will happen if two species with similar niches move into the same habitat?

Environments and Life

SECTION 15–2

GUIDE FOR READING

- **Explain** ecological succession.
- **Define** climax community.

MINI LAB

- **Observe** ecological succession and a climax community.

EVER SINCE LIFE APPEARED ON this planet, organisms have been gradually changing the environments in which they live. Some organisms have had minor effects on ecosystems, and others have had profound effects on the entire planet.

You have seen how an environment's biotic and abiotic factors jointly determine and shape all life within it. Yet that is only part of what happens in the biosphere.

Life Affects Environments

Consider a forest or a grassland. Each of these is associated with a particular set of plants, shrubs, and trees that are most common to that ecosystem. Once these plants are established, they become part of the environment for all other plants and animals living in the area. For example, they offer food, nesting sites, and protection from the weather for the organisms that live there.

A remarkable example of organisms affecting the environment began nearly three billion years ago, when bacteria appeared that could photosynthesize in much the same way that plants do today. The bacteria released a dangerous toxic waste product that had never before been present in the atmosphere. This toxic waste product was oxygen. Of course, new forms of life eventually evolved aerobic respiration, which puts oxygen to good use. But that first global "pollution" changed the entire course of life on Earth.

Figure 15–6 (a) *The Whitsunday Islands are part of* (b) *the Great Barrier Reef, Australia. The Great Barrier Reef and the organisms it shelters, such as* (c) *the fish shown here, are an example of the interdependence between living things and their environment.*

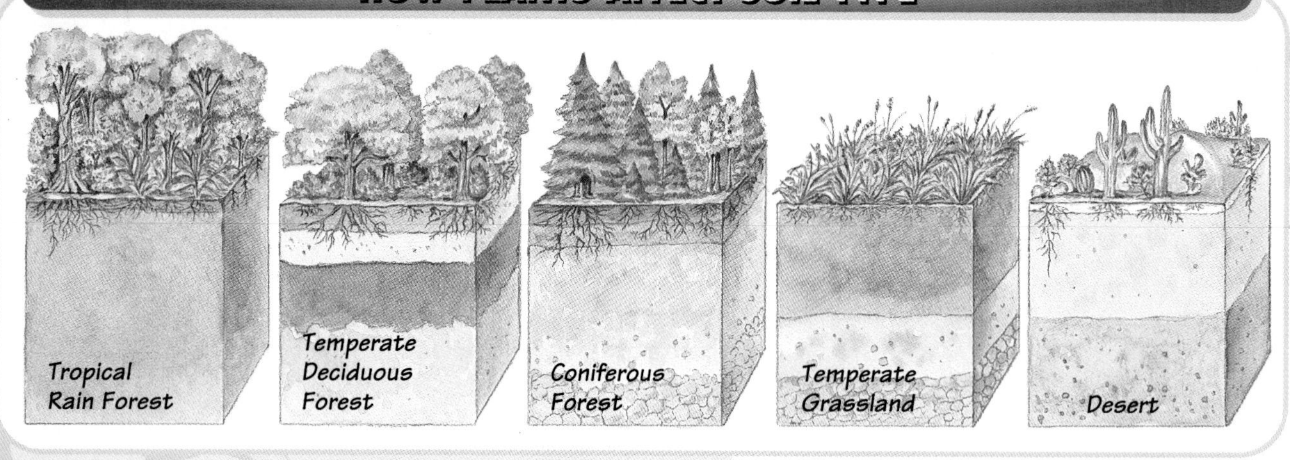

Figure 15–7
Plants affect the soil upon which they grow. Along with different patterns of rainfall, plants produce different types of soils in various ecosystems.

As you can see, the biotic and the abiotic factors of an environment continually interact with and affect each other. These interactions as well as their capacity to change ecosystems usually operate slowly, over a very long time scale. As you read ahead, you will discover that ecosystems respond to change by undergoing more changes!

Changes in Ecosystems

On the time scale of a human life, most ecosystems seem stable. But because organisms alter their surroundings, many ecosystems are constantly changing. As an ecological system changes, older inhabitants gradually die out and new plants and animals move in, causing further changes in the ecosystem.

Ecological Succession

Rakata Island in Indonesia was created by a violent volcanic eruption that destroyed the larger island of Krakatau (Krakatoa) in 1883. At first, only the hardiest of organisms—mosses, fungi, and a few tough grasses—were able to survive on the newly cooled lava rock. Over time, these organisms caused the

Figure 15–8
This series of diagrams illustrates how ecological succession gradually changes a pond into dry land covered with a forest.

rock to break down, producing a thin layer of soil. Further changes took place in the soil, enabling other types of plants to survive and grow. As the plant community grew and changed the soil, trees appeared. Soon animals that flew, swam, or drifted in from the nearby islands of Java and Sumatra made a home on Rakata. In less than 100 years, the new volcanic island was transformed into a tropical rain forest. This process, known as **ecological succession,** often occurs in natural environments for physical as well as biological reasons. **Ecological succession is the process by which an existing ecosystem is gradually and progressively replaced by another ecosystem.**

Ecological succession can also occur when human activities disturb an area. If a cleared field is left abandoned, grasses, wildflowers, and field animals move in. Over time, the seeds of bushes and small trees sprout. As these taller plants mature, birds and small mammals move into the area. Over many years—if soil and climate conditions are favorable—the field may become a forest.

✓ ***Checkpoint*** What is the process of ecological succession?

Successful Succession?

PROBLEM *How can you **observe** ecological succession and a climax community?*

PROCEDURE  

1. Obtain a clean jar with a cover and place a handful of dried plant material into the jar.
2. Fill the jar with boiled pond water or sterile spring water. Determine the initial pH of the water with pH paper.
3. Cover the jar and place it in an area that receives indirect light.
4. Examine the jar every day for the next few days. Test and record the pH each day.
5. When the jar appears cloudy, prepare microscope slides of water from various levels of the jar. Use a pipette to collect the samples.
6. View the slides under the low-power objective of a microscope and record your observations.

ANALYZE AND CONCLUDE

1. Why did you use boiled or sterile water?
2. Where did the organisms you saw come from?
3. Did the pH of the water change?
4. Was ecological succession occurring? Give evidence to support your answer.
5. Did your community reach a stable, or climax, condition?

Climax Community

Ecological succession proceeds until a relatively stable state is reached in the interaction between organisms and their environment. **The relatively stable collection of plants and animals that results when an ecosystem reaches such a state is called a climax community.**

To say that a **climax community** is stable does not mean that it never changes. On the contrary, fires, floods, or winds may destroy large areas of climax communities on land. Similarly, tsunamis, which are mistakenly called tidal waves, and hurricanes can destroy marine climax communities. When a small area within a larger ecosystem disappears because of any of these disasters, ecological succession begins and continues along toward a climax. It is interesting to note that most climax communities are not uniform over large areas. Rather, they are similar to a patchwork quilt and consist of different areas in different stages of ecological succession.

Figure 15–9
(a) The temperate forest in Nova Scotia, Canada, and (b) the lowland rain forest in Borneo, Indonesia, are examples of climax communities. (c) A climax community damaged by a fire—such as this one in Yellowstone National Park—begins to recover by the gradual process of ecological succession.

Don't think of natural disasters such as fire and drought as necessarily bad or harmful to a community. Some climax communities and the species that live in them are dependent on such catastrophes. The seeds of some plants—such as jack pines—will not sprout unless exposed to the heat of a fire. Other plants need the heat of a fire to make certain nutrients in the soil available to them.

Consider one final observation about ecological succession. Why does it occur slowly? Remember that ecological succession occurs because living things modify their environment. With the exception of humans, organisms usually modify their environment a little at a time, so ecosystem change is gradual. But what about situations in which human impact causes a more substantial change in the environment? Might this lead to an ecological upset or an ecological collapse and not ecological succession? These are questions that ecologists are trying to answer.

Section Review 15–2

1. **Explain** what is meant by ecological succession.
2. **Define** climax community.
3. **List** some ways in which climax communities can be disturbed.
4. **Critical Thinking—Sequencing** Arrange the following organisms according to their appearance in a process of ecological succession on a rock: mosses, shrubs, weeds and grasses, birch and pine, lichens.
5. **MINI LAB** How were you able to **observe** ecological succession and a climax community?

Earth's Biomes

SECTION 15–3

GUIDE FOR READING

- **Define** biome.
- **List** and **describe** some terrestrial and aquatic biomes.

MINI LAB

- **Predict** how leaves control the amount of water lost to the environment.

NOW YOU ARE READY TO answer the question posed in the first section of this chapter: Why don't bananas grow in Alaska and why can't blueberries grow in Hawaii? Each of these plants is adapted to and therefore thrives in a particular ecosystem—bananas are unique to a hot and wet tropical ecosystem, whereas blueberries are unique to a cooler, less wet ecosystem. Thus, bananas also grow well in Central America and Indonesia, where there are tropical rain forests.

What Is a Biome?

Earth's diverse living organisms inhabit a wide range of ecosystems. The type of ecosystem in a particular part of the world depends primarily on the climate conditions of the region. For the sake of convenience, biologists often refer to the world's major ecosystems by the name of their most common climax communities. **Ecosystems identified by their climax communities are called biomes.**

Figure 15–10
This graph shows the correlation between the climate of a region and the climax community, or biome, that it supports.

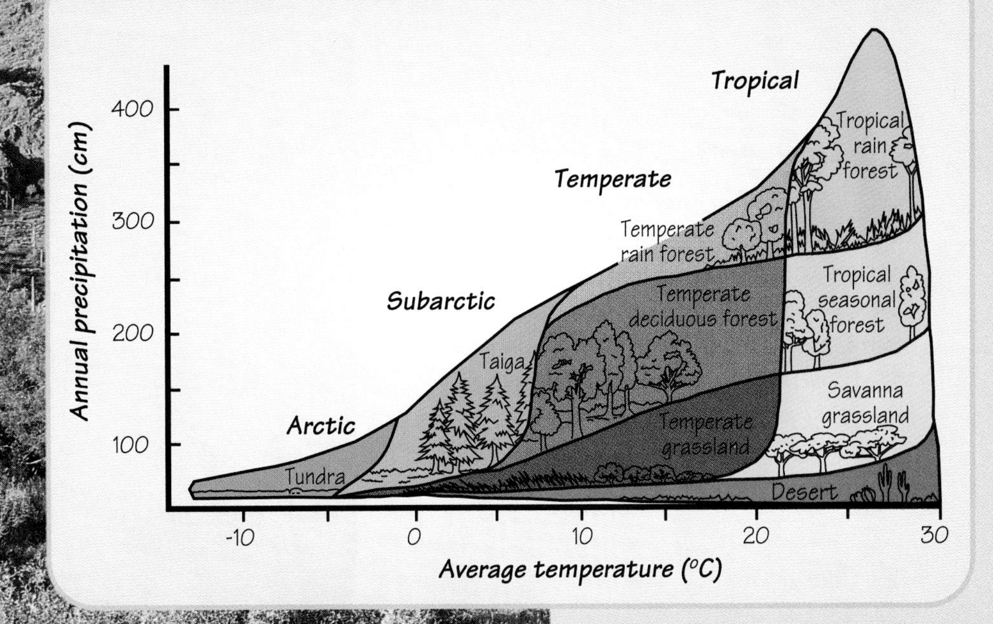

Visualizing Terrestrial Biomes

Biomes are the Earth's major types of ecosystems. Six terrestrial biomes are presented, along with a list of their main characteristics.

1 Tundra

ABIOTIC FACTORS: temperature range −40°C to 10°C, annual precipitation less than 25 cm, windy, permafrost (frozen ground)

BIOTIC FACTORS: vegetation—nearly treeless, mainly grasses, sedges, low flowering herbs and lichens; animals—arctic hare, lemming, arctic fox, musk ox, rock ptarmigan, snowy owl

6 Taiga

ABIOTIC FACTORS: temperature range −30°C to 20°C, annual precipitation 50 to 125 cm, soil thaws completely during summer

BIOTIC FACTORS: vegetation—coniferous trees, ferns, mosses, fungi (mushrooms); animals—snowshoe hare, timber wolf, shrew, lynx, weasel, black bear, woodchuck, woodpecker, chickadee, crossbill

5 Temperate Deciduous Forest

ABIOTIC FACTORS: temperature range −10°C to 25°C, annual precipitation 75 to 125 cm

BIOTIC FACTORS: vegetation—sugar maple, beech, yellow birch, pine, oak, shrubs, flowering plants, mosses, and ferns; animals—white-tailed deer, cottontail rabbit, gray squirrel, beaver, raccoon, opossum, woodpecker

2 Tropical Rain Forest

ABIOTIC FACTORS: temperature range 20°C to 30°C, annual precipitation greater than 200 cm

BIOTIC FACTORS: vegetation—broad-leafed evergreen trees, ferns, tangled lianas; animals—monkey, colorful birds, flying squirrel, tapir, anteater, ocelot, jaguar, agouti, and armadillo

3 Desert

ABIOTIC FACTORS: average annual temperature is 10°C in cool deserts to 20°C in hot deserts, annual precipitation less than 25 cm

BIOTIC FACTORS: vegetation—brush, cacti, small plants; animals—road runner, jack rabbit, kit fox, lizard, scorpion

4 Grassland

ABIOTIC FACTORS: temperature range −10°C to 25°C, but daily fluctuations more extreme than deciduous forest, annual precipitation 25 to 75 cm

BIOTIC FACTORS: vegetation—various grasses, small plants, mosses, and lichens; animals—large grazing herbivores such as bison and antelope in North America, zebra, wildebeest, elephant, and giraffe in Africa

Visualizing Aquatic Biomes

Biomes are the Earth's major types of ecosystems. Four aquatic biomes are presented along with a list of their main characteristics.

1 Open Water

ABIOTIC FACTORS: temperature range slight, little seasonal variation, light and nutrient availability slight to moderate

BIOTIC FACTORS: phytoplankton, fishes, dolphins, and whales

2 Fresh Water

ABIOTIC FACTORS: temperature range moderate (temperate freshwater) to slight (tropical freshwater), light and nutrient availability usually good

BIOTIC FACTORS: vegetation—algae, mosses, lichens; animals—insects, fishes, amphibians, and often reptiles and mammals

3 Estuaries

ABIOTIC FACTORS: temperature range extreme, annual precipitation highly seasonal, good light and nutrient availability

BIOTIC FACTORS: vegetation—algae, mosses, lichens, abundance of aquatic plants; animals—insects, shrimps, crabs, fishes, amphibians, birds

4 Rocky Intertidal

ABIOTIC FACTORS: exposure to air and sunlight alternating with being submerged by ocean water

BIOTIC FACTORS: algae, barnacles, snails, sea urchins, starfish, mussels

Terrestrial and Aquatic Biomes

The relationship between the major terrestrial **biomes** and two of the most important climate conditions—namely, temperature and rainfall—is graphically shown in ***Figure 15–10*** on page 341. The tundra biome on Earth is found above the Arctic Circle. The taiga biome is found in a wide band just below the Arctic Circle. Tropical rain forests and seasonal forests abound in the Earth's equatorial land areas. Temperate deciduous forests thrive throughout much of Europe and the eastern part of the United States. The desert biome occurs in large parts of Africa and the southwestern United States and in small parts of South America, Asia, and Australia. Savanna grasslands cover much of South America, Africa, and Australia, while temperate grasslands are found in the central United States, western Canada, and in a wide belt across southern Asia.

Aquatic biomes are classified as freshwater or marine ecosystems. It may surprise you to know that only 3 percent of the water on Earth is fresh water! And 98 percent of the fresh water (98 percent of 3 percent) is frozen in icecaps. Freshwater biomes are important not only because they provide us with water for drinking, they also provide us with water for irrigating crops and habitats for food fishes.

Do Leaves Have Waxy Skin?

PROBLEM *Can you **predict** how leaves control the amount of water loss to the environment?*

PROCEDURE

1. Obtain three paper towels and dampen them with water so they are wet but not dripping.
2. Cut two pieces of waxed paper the same size as a towel and sandwich one wet towel between the two pieces of waxed paper. Fasten the corners with paper clips and place this flat on a tray.
3. Roll up the two remaining towels. Place one roll on the tray uncovered and the other covered with a length of waxed paper. Secure the edges with paper clips.
4. Place the tray in direct sunlight. Predict which paper towel setup will lose the least amount of water.

ANALYZE AND CONCLUDE

1. How did the towels differ after 24 hours?
2. Which of the setups conserved water best?
3. Was your prediction correct? Explain your answer.
4. Explain how the leaves of desert plants conserve water.

Major marine biomes include estuaries (a zone in which a river joins the sea), rocky intertidal zones, and open water. The characteristics of aquatic biomes are presented on the previous page.

Section Review 15–3

1. **Define** biome.
2. **List** and **describe** some terrestrial and aquatic biomes.
3. **Critical Thinking—Relating** Use examples to illustrate how plants and animals within a given biome are suited to that biome.
4. **MINI LAB** How can you **predict** how leaves control the amount of water lost to the environment?

Climate: A Complex Story

Guide for Reading

- **Distinguish** between climate change and global warming.
- **Explain** why climate changes are difficult to predict.

IN 1995, A SEVERE SUMMER drought parched the northeastern part of the United States, Europe, and Australia. Torrential rains flooded the Mississippi basin and the west coasts of North and South America. In central England, the month of August was 3.4°C warmer than average, breaking all records since record keeping began in 1659. And the hurricane season in the Atlantic was one of the most severe in 125 years of record keeping.

Were these events unrelated? Or was there a common cause behind them? Are human actions changing global climate? If so, how will those changes affect the biosphere and human life?

What Causes Climate?

Ecologists are hard at work trying to find the answers to the questions just posed. In order to understand the issues and controversies that surround the subject of global climate, you must first learn what causes climate. This will help you understand how climate changes.

Global climate patterns are caused by the action of winds and ocean currents. Winds and ocean currents are in turn powered by solar energy that makes its way to Earth. How does this happen?

Winds

Because the Earth's surface is curved, different parts of the surface receive different amounts of solar energy. Near the equator, solar energy is more concentrated than it is at the North and South poles. As a result, the surface of the Earth is warmer at the equator than it is at the poles. See **Figure 15–12.**

Figure 15–11 (a) *Desertification, increased storm activity such as* (b) *tornadoes,* (c) *floods, and global warming are some of the climate changes scientists have predicted as possible consequences of phenomena such as the enhanced greenhouse effect.*

Perhaps you already know that warm air rises and cool air sinks. Therefore, the warmer air near the equator rises, while the cooler polar air sinks to the ground. Cool air then moves toward the equator, where it warms up and rises again. As shown in ***Figure 15–12***, this great solar heat pump sets up three large circuits of rising and falling air on each side of the equator.

As the Earth rotates on its axis beneath the atmosphere, the air flows move east or west relative to the ground. Especially important are surface air flows, called the trade winds. Trade winds have been the power behind sailing vessels for hundreds of years.

Checkpoint What causes wind?

INTEGRATING PHYSICS

Why does warm air rise and cool air sink?

Figure 15–12
Winds and ocean currents are the results of uneven solar heating of Earth's air and water. The interaction of these air and water flows with Earth's landmasses produces the rainfall patterns of the world.

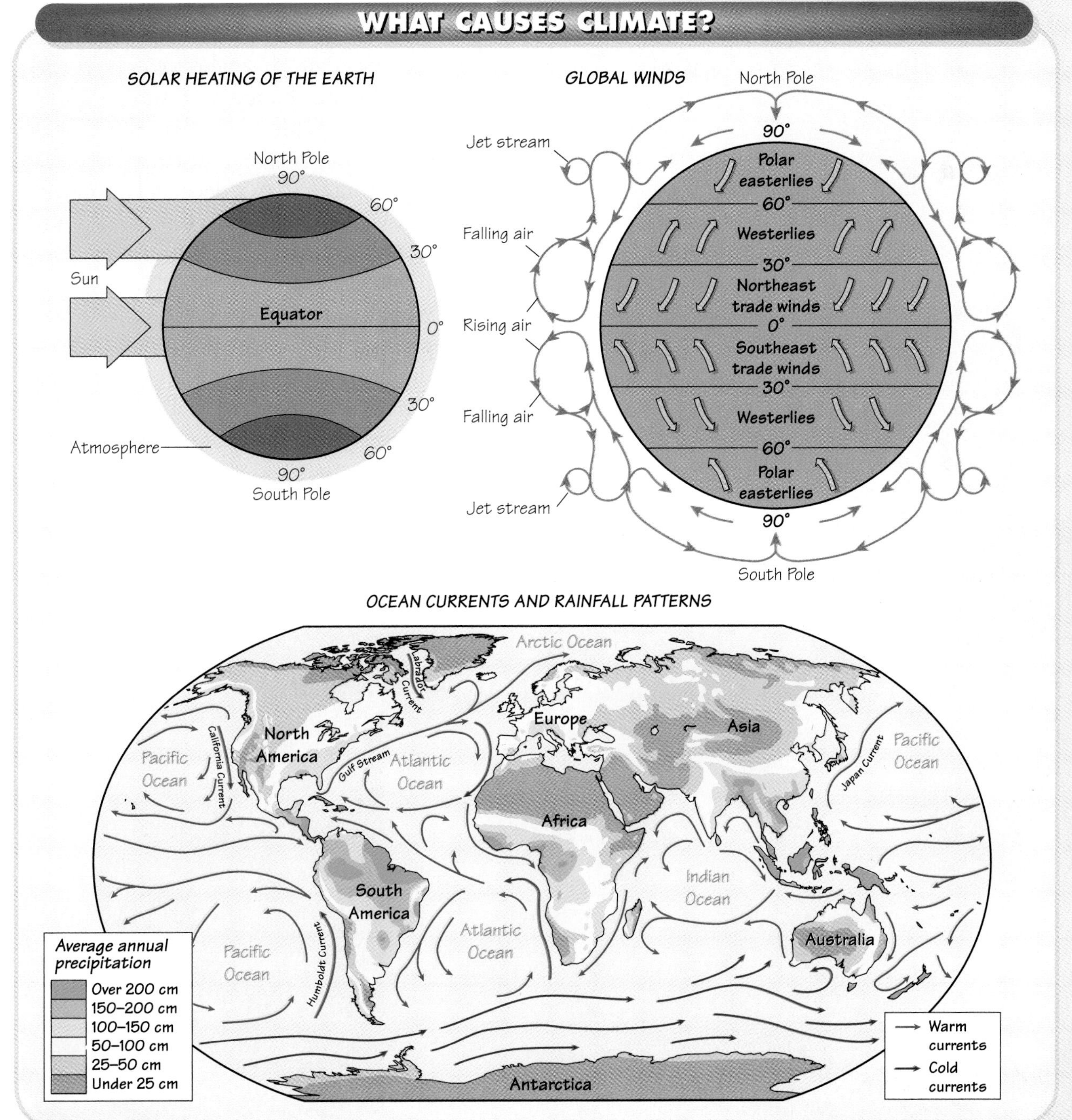

Ocean Currents

Like the atmosphere, oceans experience more solar heating at the equator than at the poles. As a result, warmer water at the ocean surface moves from the equator toward the cooler water at the poles. On the other hand, cold water near the poles sinks to the bottom of the ocean and travels along the ocean floor toward the equator. Surface water is also pushed around by the action of winds. Of course, continents get in the way and affect the path that water can take. The result of all these flows is the worldwide ocean current pattern, as shown in ***Figure 15–12*** on page 347.

Earth's Geography

As you have read earlier, Earth's climate is caused by the interaction of air and water currents with Earth's landmasses. As warm air moves over warm water, it picks up moisture in the form of water vapor. If the warm, moist air rises and later cools, the water vapor condenses and falls to the Earth as fog, rain, snow, or sleet. This often happens when warm, moist air meets high mountain ranges—such as the Sierra Madre and Rocky Mountains in North America, the Andes in South America, and the Himalayas in Asia.

☑ **Checkpoint** How do mountain ranges create rainfall patterns?

The Climate Controversy

Chances are that when you hear the words climate change, you think of global warming. This is because the global warming issue has been discussed and debated for more than a decade. Part of the controversy stems from a misunderstanding of basic terms. The enhanced greenhouse effect, climate change, and global warming are often used interchangeably, as if they were three different names for the same phenomenon. Let's examine each of these individually, separating facts from predictions. And remember that predictions are only as good as the data they are based upon.

Figure 15–13
(a) Winds force moist air up and over a mountain range, causing (b) large amounts of rain to fall on the windward side. Once over the top, the air becomes dry and picks up moisture from its surroundings, often giving rise to (c) a desert on the leeward side.

Enhanced Greenhouse Effect

Earlier in this chapter, you learned that the greenhouse effect is a natural phenomenon occurring in Earth's atmosphere. This effect has maintained a nearly stable temperature range on Earth for millions of years. However, when scientists discuss the **enhanced greenhouse effect,** they are referring to the fact that human activities are adding greenhouse gases to the atmosphere. There is little scientific debate that this is happening. Because humans are adding carbon dioxide and other greenhouse gases to the atmosphere, Earth will retain more heat from the sun than it has in the past.

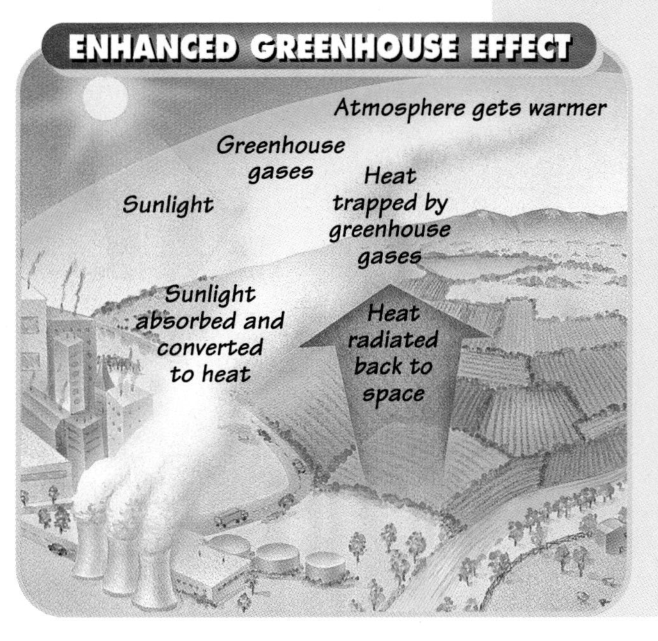

Figure 15–14 *Various human activities, such as the burning of fossil fuels, add to the greenhouse gases released to the atmosphere. As a result, the Earth will retain more heat than it has in the past.*

Researchers predict that concentrations of greenhouse gases in the atmosphere will double during the early part of the twenty-first century. You may want to take a look at ***Figure 13–16*** on page 303, which shows the atmospheric carbon dioxide levels for the past 35 years. This doubling of greenhouse gases will reduce the rate at which Earth loses energy to space by about 2 percent. If the rate at which energy is returned to space decreases, energy will build up in the atmosphere. You may not think that a 2-percent change is anything to worry about. But given the amount of energy the Earth receives and ultimately returns to space, the 2-percent reduction is equivalent to adding the energy content of 3 million metric tons of oil to the atmosphere every minute!

Most experts agree that energy will be added to the biosphere and the biosphere will adjust to it. Just how the biosphere will adapt is the question. As you can imagine, the worldwide system of atmospheric and oceanic currents carrying heat around the globe is complicated. The global carbon cycle that links various carbon reservoirs is also complex. Exactly what will the accumulation of energy do to these cycles and systems? This is where scientific questions and debates arise.

☑ **Checkpoint** What is the enhanced greenhouse effect?

Global Warming

Many scientists believe that a significant amount of the extra energy will remain in the biosphere in the form of heat, causing Earth's average temperature to rise. You may have heard of **global warming** scenarios, predicting the melting of polar icecaps, the rising of sea levels, and the flooding of coastal cities. **When the enhanced greenhouse effect (a fact) is used to predict a significant rise in Earth's average temperature (a possibility), the prediction is called global warming.**

What evidence do scientists have to support this prediction? All over the world, temperature records, similar to those shown in ***Figure 15–15*** on page 351, offer evidence that Earth has been warming over the last century.

Predicting how global temperatures will change in the future is not easy. Climate researchers rely on computer models of the oceans and atmosphere, trying to take into account as many relevant factors as possible, such as volcanic and

Biology AND EARTH SCIENCE Connections

El Niño

Usually the tropical region of the Pacific Ocean is subject to strong winds that blow from east to west. Sometimes, however, the winds are weaker than normal, allowing the warmer waters of the western Pacific to move toward the eastern end of the ocean and South America. This change of wind flow and water current is an event commonly called El Niño, meaning "the male child" in Spanish.

The red areas in the equatorial regions of the Pacific Ocean indicate the presence of the warmer waters of El Niño.

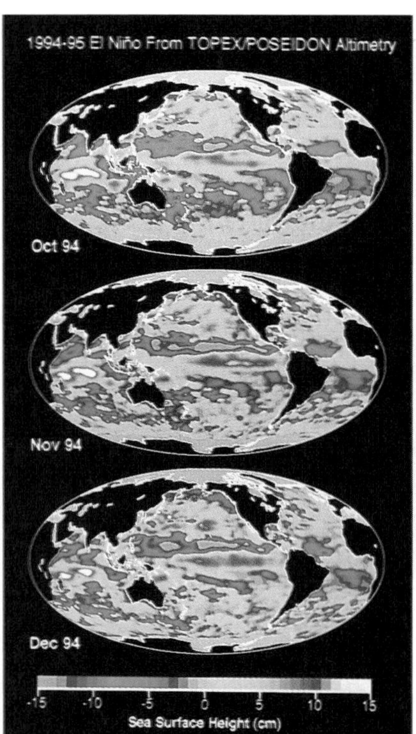

CAREER TRACK Oceanographers gather and interpret data such as ocean water temperatures to understand the importance of climate change.

Effects of El Niño

What changes can El Niño cause? It is often associated with increased thunderstorms and other climate changes in western South America and beyond. Moreover, off the coast of Peru, El Niño reduces a phenomenon called upwelling—the rising of cold, nutrient-filled water toward the surface. With fewer nutrients, the higher levels of water support smaller populations of fishes and other marine life.

In fact, the effects of an El Niño event are felt all over the world. Usually an El Niño event coincides with drought in Australia, floods in western North and South America, and a mix of heavy rain and drought in Africa.

Climate Change?

An El Niño event may last from a few weeks to several months. Until recently, the events occurred approximately once every 3 to 7 years. To the concern of scientists around the world, however, the frequency and severity of El Niño events seem to be increasing. Some researchers wonder whether this is a coincidence or one of the first signs of climate change. More data are needed before questions about such phenomena can be answered for certain.

Making the Connection

How might the enhanced greenhouse effect impact on El Niño occurrences in the future? Do you think the El Niño effects are short term or long term? Give evidence to support your answer.

Figure 15–15
This graph shows the annual surface temperature of the Earth from the years 1856 to 1991. Notice the warming trend during the last two decades.

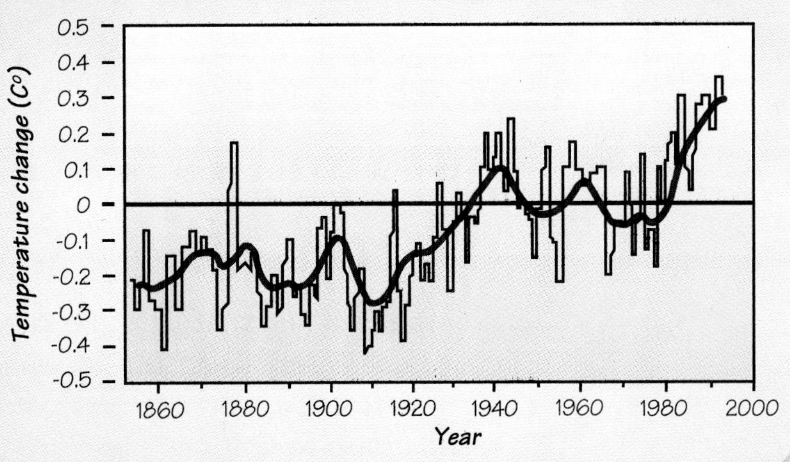

other geologic activities. The most recent models suggest that the average global surface temperature will increase by about 1 to 2°C by the year 2050.

Other scientists question whether the Earth's temperature will rise by a significant amount. But most scientists involved in this research agree that the Earth will undergo some change in climate. **Climate change includes any significant change in patterns of temperature, rainfall, humidity, storm activity, and cloud formation that occurs in any part of the world.** As you see, temperature is only one factor in the larger picture of global climate.

Climate Change: Is It a Problem?

Global warming is not the only possible climate change. Acting like giant conveyor belts, winds and ocean currents carry heat from warmer places to cooler ones. Many scientists are concerned that adding energy to the biosphere—a result of the enhanced greenhouse effect—will influence the air and water flows in ways that they do not yet understand.

They point out that such changes may not be gradual and predictable. Several climate models also suggest that these changes may begin long before there is any clear indication that the Earth is warming up. Why should we be concerned about climate change? The following example may provide the beginnings of an answer.

Recent observations of marine environments raise troubling questions for ocean ecosystems. Since 1950, surface currents off the coast of California have been warming up. This warming trend has caused populations of zooplankton—tiny marine organisms that are part of marine food chains—to decline. Some of these populations have decreased as much as 80 percent over the last 40 years! Not surprisingly, the populations of animals on higher trophic levels have also been declining. Is this trend permanent? Will it continue? No one knows for sure. But can we afford to ignore it?

Section Review 15–4

1. **Distinguish** between climate change and global warming.
2. **Explain** why climate changes are difficult to predict.
3. **Describe** the greenhouse effect. Why has it increased during the past decades?
4. **BRANCHING OUT ACTIVITY** Working in a group, **predict** some of the effects that rising temperatures could have on the Earth. Compare your predictions with those of your classmates.

CHAPTER 15

Laboratory Investigation

DESIGNING AN EXPERIMENT

Greenhouse in a Bottle

Scientists often find that using models of ecosystems is a convenient and useful way to collect information and make predictions. In this investigation, you will construct model ecosystems in plastic soda bottles to test the effects of a warming environment.

Problem

Design an experiment to observe the impact of the greenhouse effect on global warming.

Suggested Materials

2 outdoor thermometers
2 2-liter plastic bottles
dry potting soil
clear plastic wrap
rubber band
100-watt bulb on a ring stand
scissors
tape
graph paper
cardboard
one of the following:
- water
- wet and dry soil
- ice cubes
- sod or other plants

Suggested Procedure

1. **Remove all labels from the 2-liter bottles. Cut the tops from each bottle just where they begin to narrow.**
2. **Put 350 g of potting soil into one bottle. Tape an outdoor thermometer to the inner wall of the bottle, making sure the thermometer does not touch the soil and the calibrations face outward. Tape a piece of cardboard directly over the bulb of the thermometer.**
3. **Cover the opening of the bottle with plastic wrap and secure it with a rubber band.**
4. **Record the temperature in the bottle.**
5. **Hang the light bulb from a ring stand, then position the bottle on one side of the light.**
6. **Turn the light on for 15 minutes. Record the temperature in the bottle every 3 minutes. If necessary, extend the time until the temperature increases noticeably in the bottle.**
7. **Using a setup similar to the one you used in steps 1 to 6, design an experiment to test one of the following hypotheses:**

- **Air over water heats up at a different rate than that over land.**
- **Air over ice-covered surfaces heats up at a different rate from that over land that is not covered with ice.**
- **The presence of plants affects the rate of warming.**

8. Be sure to include a control in your experiment. After you write out the procedure, have your teacher check it.

9. Write out your hypothesis and predict the results of your experiment.

10. Carry out your planned experiment and record your results.

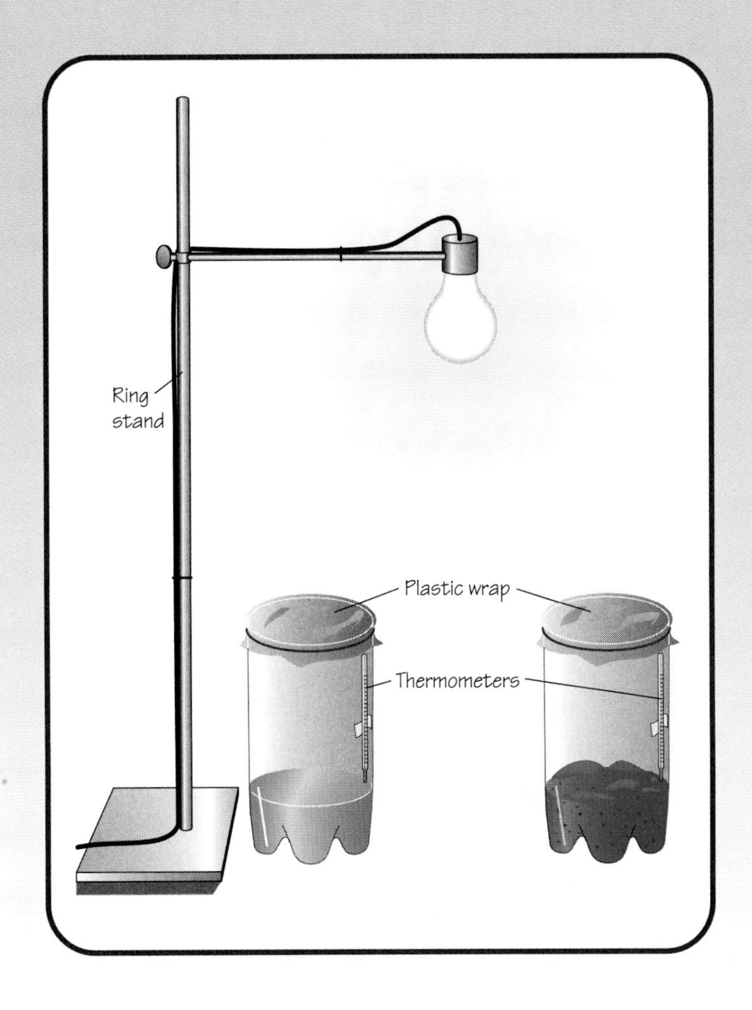

Observations

1. Make line graphs of the data from your experiments.
2. Compare the changes in temperature in each set of bottles over time.
3. What was the control in the experiment? What was the variable?

Analysis and Conclusions

1. How would your results have been affected if you had used two pieces of plastic wrap to cover the bottle?
2. Identify the parts of Earth that each part of the experimental setup represents.
3. In what real-life situations have you observed the greenhouse effect?
4. How did your prediction compare with the data you collected? Was your hypothesis supported?
5. Compare your results with those of your classmates. Was there a difference in the amount of warming in different environments?

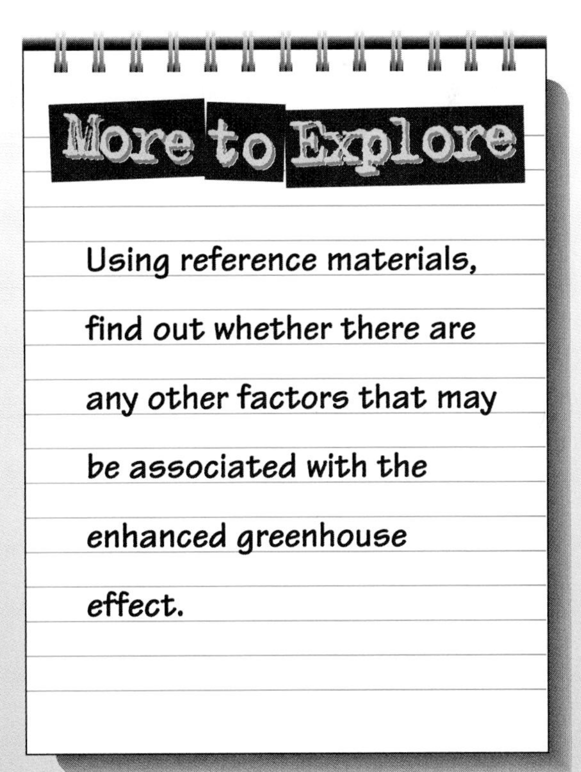

More to Explore

Using reference materials, find out whether there are any other factors that may be associated with the enhanced greenhouse effect.

Study Guide

Summarizing Key Concepts

The key concepts in each section of this chapter are listed below to help you review the chapter content. Make sure you understand each concept and its relationship to other concepts and to the theme of this chapter.

15–1 Climate and Life

- A region's climate is the temperature range, average annual precipitation, humidity, and amount of sunshine that the region has.
- Populations of plants and animals grow, remain stable, or vanish depending in part on climate conditions such as temperature and rainfall. These conditions also affect the structure of food webs and the flow of nutrients.
- Climate is one of the physical (abiotic) environmental factors that influence life. Biological (biotic) environmental factors are all the living things with which an organism might interact.
- An ecosystem is a particular combination of a physical environment and the organisms that live in that environment.

15–2 Environments and Life

- Ecological succession is the process by which an existing ecosystem is gradually and progressively replaced by another ecosystem.
- A fairly stable collection of organisms in an ecosystem is called a climax community.

15–3 Earth's Biomes

- Ecosystems identified by their climax communities are called biomes.

15–4 Climate: A Complex Story

- Winds and ocean currents are caused by differential solar heating of the Earth's surface. Winds and ocean currents, as they interact with Earth's geography, give rise to climate.
- Global warming is a prediction based on the enhanced greenhouse effect, or the increased heat in the biosphere due to an increase in levels of greenhouse gases.
- Climate change includes any significant change in patterns of temperature, rainfall, humidity, and storm activity that occurs in any part of the world.

Reviewing Key Terms

Review the following vocabulary terms and their meaning. Then use each term in a complete sentence.

15–1 Climate and Life

climate
greenhouse effect
microclimate
abiotic factor
biotic factor
environment
ecosystem
niche
habitat

15–2 Environments and Life

ecological succession
climax community

15–3 Earth's Biomes

biome

15–4 Climate: A Complex Story

enhanced greenhouse effect
global warming

Recalling Main Ideas

Choose the letter of the answer that best completes the statement or answers the question.

1. A region's climate does not depend on its
   - **a.** annual rainfall.
   - **b.** population.
   - **c.** average temperature.
   - **d.** distance from the equator.

2. The greenhouse effect occurs in the Earth's
   - **a.** atmosphere.
   - **b.** oceans.
   - **c.** ozone layer.
   - **d.** plant life.

3. Earth's climate patterns arise due to
   - **a.** the water cycle.
   - **b.** global warming.
   - **c.** Earth's biomes.
   - **d.** winds and ocean currents.

4. The process of slow change by which a forest recovers after a fire is called
   - **a.** ecological succession.
   - **b.** microclimate.
   - **c.** climate change.
   - **d.** climax.

5. Which of the following biomes receives the least annual precipitation?
   - **a.** rain forest
   - **b.** grasslands
   - **c.** taiga
   - **d.** desert

6. An organism's niche is a description of
   - **a.** its physical environment.
   - **b.** its biological environment.
   - **c.** the range of physical and biological conditions in which it can live.
   - **d.** its habitat.

7. Winds and ocean currents are caused by the
   - **a.** differences in solar heating of Earth's land and ocean surfaces.
   - **b.** unevenness of Earth's landmasses.
   - **c.** tilt of the Earth's axis.
   - **d.** cutting of tropical rain forests.

8. The enhanced greenhouse effect is due in large part to
   - **a.** global warming.
   - **b.** climate change.
   - **c.** human activity that adds greenhouse gases to the atmosphere.
   - **d.** microclimate change.

Putting It All Together

Using the information on pages xxx to xxxi, complete the following concept map.

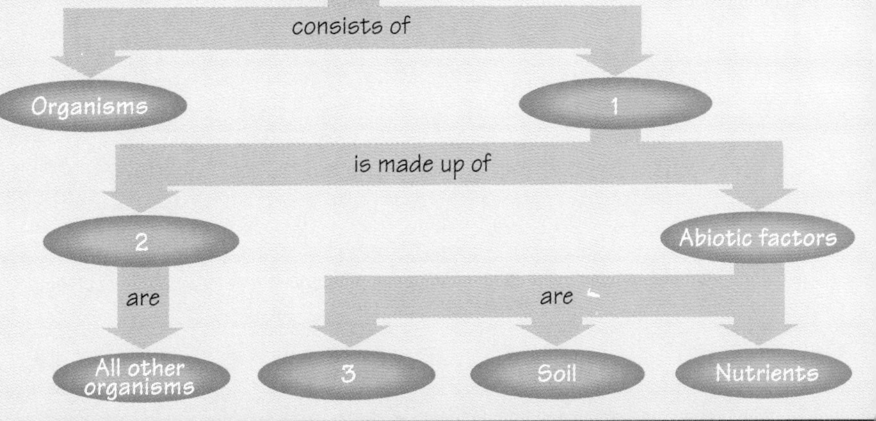

Reviewing What You Learned

Answer each of the following in a complete sentence.

1. Define climate and briefly describe the climate you **observe** in your area.
2. Give examples from your neighborhood of locations with different microclimates.
3. Contrast biotic and abiotic factors of an environment.
4. Characterize any four ecosystems, one of which represents the area in which you live.
5. What abiotic factors are associated with an organism's niche?
6. How does an organism's niche differ from its habitat?
7. Define ecological succession. Why does it occur?
8. Why might an oak forest be considered a climax community?
9. Identify and characterize three major biomes.
10. Distinguish between the enhanced greenhouse effect, global warming, and climate change.

Expanding the Concepts

Discuss each of the following in a brief paragraph.

1. Explain the significance of Earth's curvature to climate.
2. Why might the microclimate under a layer of leaves at the base of a tree be much different from the characteristic climate of the forest?
3. Can two organisms occupy the same niche? Support your answer.
4. Do organisms influence or change their habitat? Explain your answer.
5. Describe an area in your community in which human activity has altered natural succession.
6. How might one consider natural disasters as part of nature's order?
7. How do mountain ranges affect the climate on either side of the range?
8. What is meant by the "Earth's heat pump," and what is its connection to Earth's climate?
9. Discuss the role of carbon dioxide in the atmosphere and the predictions based on its rising concentrations.

Extending Your Thinking

Use the skills you have developed in this chapter to answer the following.

1. **Hypothesizing** Explain why the ocean currents in the Northern Hemisphere travel in a clockwise direction whereas those in the Southern Hemisphere travel in a counterclockwise direction.
2. **Relating** Why are the biomes at subsequent elevations of a mountain range similar to those you would encounter while traveling from the equator to the North Pole?
3. **Sequencing** The lava flows of the seventeenth century greatly devastated areas of the island of Hawaii. Trace the probable steps in ecological succession to the climax community that now exists.
4. **Evaluating** Why is global warming called prediction and not fact? Justify your answer.
5. **Predicting** Predict what might happen if a new species were to arrive in a habitat and begin to occupy the same niche as one of the species in that habitat.

Applying Your Skills

Designing a City of the Future

Through this activity you will explore how air pollution in a city can be minimized by the arrangement of living areas, work areas, and landscaping.

1. Design appropriate symbols for the following elements: single-family home, office building, apartments and townhouses, industrial plant, parks and forests, playground and ballfield, school, gas station, restaurant, highway, shopping mall, bus and subway routes, and walking, jogging, and bike trails.
2. Working in groups, imagine that you are urban planners and design a city. Decide on the placement of the elements in order to minimize air pollution. Make sufficient copies of the elements and mount them on cardboard. Then cut and paste these elements, following your design.
3. Present your design to other groups, giving the rationale for your layout.

• GOING FURTHER •

4. Invite an urban planner to your class to discuss and evaluate the layouts of all the groups.

CHAPTER 16

Humans in the Biosphere

FOCUSING THE CHAPTER THEME: Patterns of Change

16–1 This Island Earth
- **Examine** the capacity of human actions to change environments.

16–2 Agriculture
- **List** several strategies for sustainable agriculture.

16–3 Resources at Risk
- **Name** resources at risk and **identify** ways to protect the environment.

BRANCHING OUT *In Depth*

16–4 Conserving Biodiversity
- **Examine** the role of conservation biology in preserving wildlife habitats.

LABORATORY INVESTIGATION
- **Construct a model** compost column and **predict** the suitability of various materials for composting.

Biology and Your World

BIO JOURNAL

Imagine that it is the year 2020. In what ways might the world look different than it does now? How much of that change do you think may be caused by human activity? In your journal, write your predictions along with your reasons.

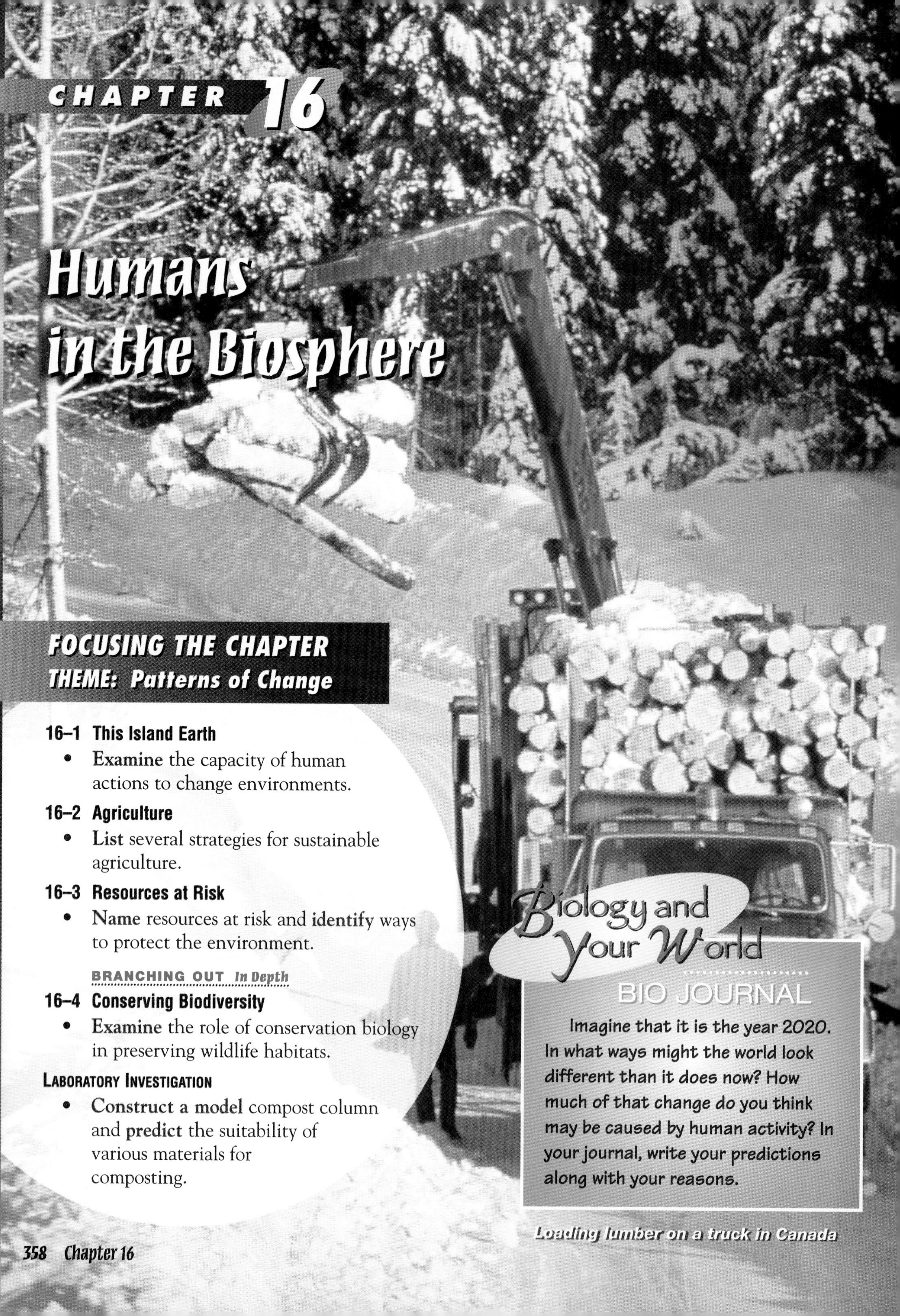

Loading lumber on a truck in Canada

This Island Earth

SECTION 16–1

Guide for Reading

- **Explain** how human actions alter environments.
- **Define** sustainability.

MINI LAB

- **Calculate** the concentration of DDT at several aquatic trophic levels.

CENTURIES AGO, THE FIRST humans to settle the islands we now call Hawaii learned to live on tiny specks of land. They had limited amounts of fresh water and land on which to live and grow crops. But more importantly, the culture and customs of these people reflected their awareness of the limitations. Their society was completely self-supporting, although the population was greater than it is today.

In a far different environment, the Anasazi, which in Navajo means "ancient people," built great cities in the canyons of the American Southwest. As their population grew, however, the Anasazi required more from the desert than the ecosystem could supply. When a 30-year drought struck, the Anasazi were forced to abandon their canyon cities—never to return.

Human Activity and the Environment

In the past, human cities were islands of humanity in the sea of the biosphere. The impact of individual cultures was limited to their immediate environment. But exponential human population growth has changed all that.

Figure 16–1 *(a) Despite limited resources such as fresh water, ancient Hawaiian cultures thrived. The Anasazi civilization built (b) elaborate cities and created (c) beautiful objects but did not survive a long period of drought. Perhaps these cultures offer important lessons as we look to the future.*

Figure 16–2
Although this eaglet managed to hatch from its egg, its sibling will not be as lucky. High concentrations of DDT in the tissues of its mother at the time of egg production results in thin-shelled eggs that may never hatch.

Today, roughly half of all land on Earth not covered with ice and snow has been used and altered in some way by human actions. **According to a recent study, human actions now use almost as much energy and transport almost as much material as all other species on Earth—plants and animals combined. Humans have become the greatest source of change in the biosphere.**

Yet the human species is still a part of the biosphere and depends on global food webs, nutrient cycles, and energy flows. You can think of the biosphere as an island, and although its resources and space are abundant, they are not limitless. What's more, because our understanding of the biosphere's life-support systems is still incomplete, we should be cautious about actions we take. The following example illustrates how the physical and biological systems in the biosphere sometimes respond to human actions in ways that threaten human health and well-being.

Poisons in the Food Web

Decades ago, researchers discovered an insect-killing chemical, called DDT. At first, DDT seemed to be the perfect weapon for insect control. Once sprayed in an area, it remained active for a long time and killed many different types of insects. By killing mosquitoes, DDT helped to reduce the spread of deadly diseases, such as malaria. DDT was found to be effective against body lice and common household insects. It was also sprayed over large areas of farmland to control agricultural pests.

For a while, everything appeared to be fine. But as the insecticide became part of the water runoff from farmland, it entered nearby rivers and streams.

BIOLOGICAL MAGNIFICATION

	DDT concentration (parts per million)
Fish-eating birds	20.00
Small fish	2.0
Zooplankton	0.20
Algae	0.04
Water	0.000003

Figure 16–3
Studies show that DDT concentration was magnified almost 7 million times as it passed from primary producers, such as algae, to the highest level of consumers, such as fish-eating birds.

First, the fishes in those streams began to die. Soon, fish-eating birds—such as eagles, pelicans, and ospreys—began to lay eggs that failed to hatch. Then, to their surprise, scientists discovered DDT in human body fat! Traces of DDT were even found in the flesh of penguins as far from civilization as the South Pole! What had happened?

After careful study, scientists discovered two characteristics that made this insecticide hazardous in the biosphere. First, DDT is **nonbiodegradable,** meaning that it cannot be broken down by the life processes of living things. This characteristic makes it much more potentially dangerous than **biodegradable** substances, which are broken down in the environment. Second, when an organism picks up DDT from its environment, it does not eliminate it from its body. In combination, these two characteristics led to a surprising chain of events.

Biological Magnification

Although DDT may be present in lakes and streams in very low concentrations, it can be picked up and stored by aquatic primary producers, such as algae. When herbivores eat the algae, they collect and store the DDT. The more they eat, the more DDT they store. In fact, in herbivores, the levels of DDT are ten times greater than they are in plants. When carnivores eat herbivores, DDT is concentrated even more. This process, called **biological magnification,** continues throughout the food web. Through the process of biological magnification, substances such as toxic metals and chemicals accumulate over time and are passed up the trophic levels at higher and higher concentrations.

Luckily, the trace amounts of DDT found in humans were discovered before any people were harmed. But the situation was far different for other species, such as fishes and birds. The American bald eagle, for example, was threatened with extinction due largely to the harmful effects of DDT on its eggs.

Today, the use of DDT is strictly controlled. Similar controls also apply to many other pesticides and potentially toxic substances that can be released into the environment.

☑ **Checkpoint** What is biological magnification?

When a Little Can Mean a Lot

PROBLEM *How would you **calculate** the amount of DDT that accumulates in organisms of an aquatic food chain?*

PROCEDURE

1. Obtain a bowl of red and white beans. The beans represent two different types of algae, or primary producers.
2. Label 10 small plastic bags "zooplankton," and number the bags 1 through 10. Zooplankton are tiny organisms that feed upon algae.
3. Place any combination of the red and white beans—up to a total of 15—into each plastic bag. Record the number of each type of bean in each bag.
4. Label a large plastic bag "killfish." Place the 10 zooplankton bags into the large plastic bag.

ANALYZE AND CONCLUDE

1. If a red bean represents 5 ppm of DDT and a white bean represents 10 ppm of DDT, calculate the ppms in each of the zooplankton.
2. Calculate the ppms of DDT the killfish could accumulate in a day if the killfish consumes an average of 10 zooplankton per hour.
3. If a flounder (a larger fish) consumes 10 killfish per day, how much DDT could the flounder accumulate in 30 days?
4. Predict how much DDT a human could accumulate in a year by eating a flounder each month.

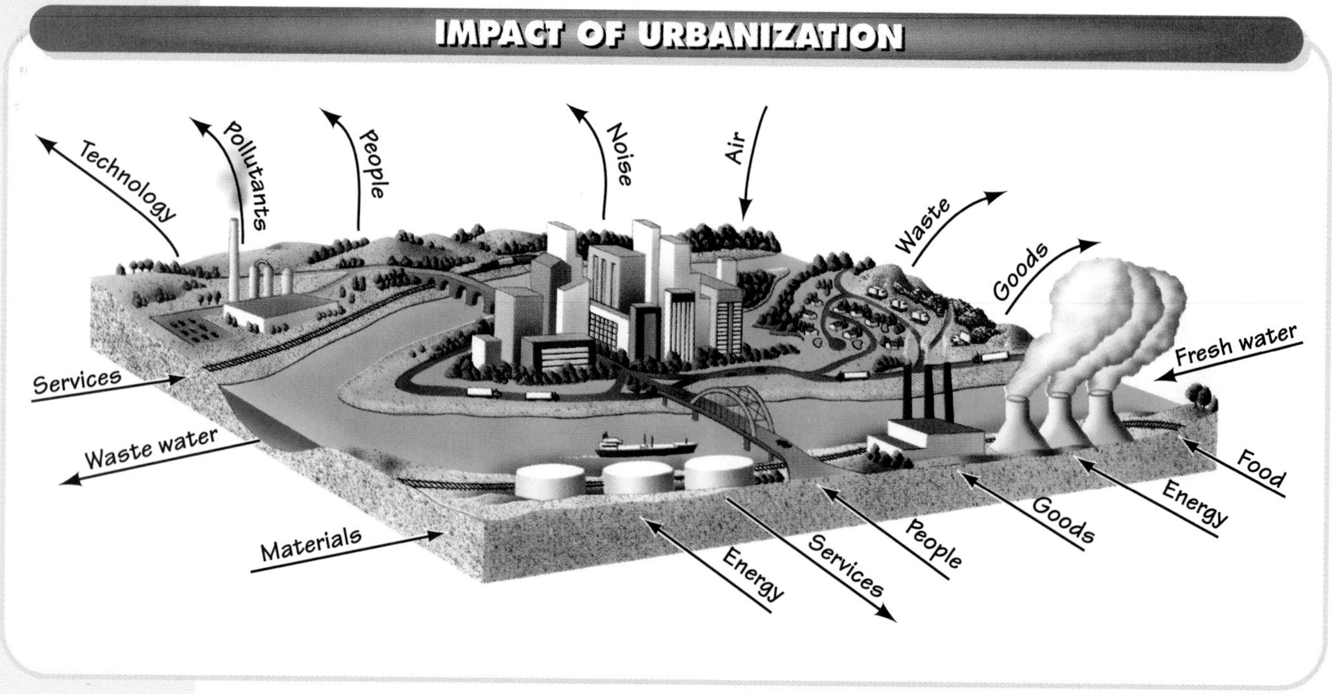

IMPACT OF URBANIZATION

Figure 16–4 Cities rely heavily on exchanging large quantities of materials and energy with the environment.

Charting a Course

The DDT disaster is one of several examples that has taught humanity a valuable lesson—we do not yet know enough about the workings of the biosphere to predict all the consequences of our actions. Given these limitations, can an understanding of ecology help us shape a positive future for ourselves and the biosphere? The answer is yes. And it centers on **sustainability,** a fundamental and important concept of ecology.

Human ways of living that are based on the principles of ecology will be ecologically sustainable. Ways of living that pursue economic growth while ignoring these principles are less likely to be ecologically sustainable. **Sustainable practices are in harmony with the biosphere and do not deteriorate its living and nonliving parts.** Simply put, we must realize that we are a part of an interdependent world and must learn to live within the limits of nature's systems in order for them to support us.

As you read the next sections of this chapter, you will have an opportunity to examine several human activities along with their consequences. This will help you to understand the role and the responsibility of humans in the biosphere.

Section Review 16-1

1. **Explain** how human actions are potentially more powerful than the actions of Earth's other organisms.
2. **Define** sustainability.
3. **Critical Thinking—Evaluating** Biological magnification is frequently quoted as an example to show the unexpected consequences of some human actions. Discuss the use of the word unexpected.
4. **MINI LAB** How can you **calculate** the concentration of DDT in organisms related by a food chain?

Agriculture

SECTION 16–2

GUIDE FOR READING

- **Describe** the key practices of the green revolution and **relate** them to their environmental impact.
- **Identify** several strategies for sustainable agriculture.

MINI LAB

- **Design an experiment** to see how planting winter rye grass in a summer-cultivated cornfield helps prevent soil erosion.

THE ORIGIN OF AGRICULTURE was among the most important events in human history. Why? Agriculture supplies humans with one of their most basic needs—a dependable supply of food. Without the concentrated and predictable food supply that farming provides, humans could not gather into cities in such overwhelming numbers. Today, even if we stopped driving cars and recycled everything we use, producing food for humanity would still have an enormous impact on the biosphere.

The Green Revolution

The Earth would not be able to support as many people as it does today if it were not for a dramatic improvement in the way humans grow food. During the 1950s, governments and researchers looked at the rapidly growing world population with alarm. How could farmers possibly grow enough food to prevent mass starvation? A global effort to increase food production resulted in what is called the **green revolution.** The green revolution led to a substantial increase in crop yields as a result of a few key practices.

Figure 16–5 The changing face of agriculture over the centuries is captured in images of sowing and plowing in (a) ancient Egypt and present-day (b) Walla Walla, Washington, and (c) Bangkok, Thailand.

Figure 16–6 (a) *Soaring crop yields were a result of several green revolution efforts, such as* (b) *large-scale irrigation.*

- Clear and plow large fields and plant a single highly productive crop. This strategy, called **monoculture,** makes sowing, tending, and harvesting more efficient.
- Use machinery powered by fossil fuels.
- Increase the use of irrigation.
- Boost crop production with chemical fertilizers.
- Control plant pests with chemical pesticides.

These practices produced remarkable results worldwide. In 20 years, Mexican farmers increased their production of wheat tenfold. The world's most densely populated countries—China and India—were able to produce most of the food they needed for the first time in years. The green revolution, probably more than any other single effort, helped to prevent global food shortages.

INTEGRATING TECHNOLOGY AND SOCIETY

How do the green revolution practices make use of technology to improve conditions for human society?

Checkpoint How did the green revolution produce high crop yields?

Environmental Impact

In many places, green revolution techniques continue to be enormously successful. But constant use of these techniques can cause problems. With what you know now about ecology, you should be able to understand why and how those problems have arisen. **The green revolution practices can cause a number of pressing environmental problems—such as pesticides in the environment, loss of soil fertility, and the dwindling of water resources.** Some of these problems are outlined in the pages that follow.

Pests and Diseases

To an insect that eats the leaves of corn plants, much of the farmland in the United States looks like a huge dinner table set with endless fields of identical tasty corn plants! Once a population of corn-eating insects gets established in such an area, it can grow unchecked because it is surrounded by an almost unlimited supply of food. The same is true for bacterial and fungal plant diseases.

The first response to the problem of pests and diseases was the use of chemical pesticides—a key green revolution practice. Pesticides were relatively cheap and seemed to solve the problem easily and cleanly. So they were applied in large quantities to field after field, year after year.

Bio FRONTIER Connections

Good Guys in the Badlands

Joe R. Weatherly isn't someone most people would think of as an environmentalist. He is a cattle rancher in Wheeler, Texas. Like more than 90 percent of American cattle ranches, his company—Heritage Beef Cattle Company—is a family operation. And in order to be able to pass the business on to future generations, the Weatherly family wants to keep the land healthy. So they use as many sustainable agriculture techniques as they can. For their work, the Weatherly family won the 1995 Environmental Stewardship Award from the National Cattlemen's Association. Since 1991, this award has recognized ranchers who practice environmentally sound strategies.

The Problem

Unlike the grazing habits of antelope, elk, deer, and bighorn sheep, beef cattle do not graze for a while in one area and then move on. Instead, left to themselves, beef cattle overgraze range land and destroy the streams that run through it. As a result, many areas of grassland became scrub brush unsuitable for grazing, and water supplies were in trouble.

The Solution

Soon ranchers began working with ecologists to see if they shared common ideas. Together they agreed that a solution must combine good science, workable conservation strategies, and economic sense. Because fire is an important part of prairie ecosystems, both groups worked out plans to allow some areas to burn in controlled ways. They also developed strategies to move cattle from place to place so they would not overgraze any single area. In addition, ranchers learned how to protect streams to guard water supplies.

Thanks to this combined effort, most range land today is either stable or in the process of improving. What's more, many ranchers are now working with environmental groups to protect endangered species that are found on their property.

Cattle grazing on properly managed rangeland

Making the Connection

What are other uses of land that can cause lasting harm to our resources or the environment? How can people begin to adopt sustainable ways of using their land?

MINI LAB ......Experimenting....

Holding Your Own

PROBLEM *How can planting winter rye grass in a summer-cultivated cornfield help to prevent soil erosion?* ***Design an experiment*** *to answer the question.*

SUGGESTED PROCEDURE

1. Formulate a hypothesis about the effect that planting rye grass has on soil.
2. Use plastic cups, potting soil, rye grass seed, radish seeds, and water to design and conduct an experiment to test your hypothesis.
3. Allow yourself two weeks to make observations and draw conclusions.

ANALYZE AND CONCLUDE

1. Did your experiment support your hypothesis? Explain your answer.
2. Why do you think farmers would plant an annual rye grass during the winter?

What's Wrong With Pesticides?

Over time, however, problems began to surface. The first to be noticed was biological magnification. But other problems were brewing as well. As farmers continued to use a pesticide, insects became resistant to it. At first, higher doses of pesticide worked. Then farmers switched to more powerful poisons. But sooner or later, the insects become immune to nearly any pesticide. Meanwhile, as more pesticides were used, streams, lakes, and underground water supplies were contaminated. The total environmental cost of pesticide use in this country is estimated at $8 billion each year. Still, insect damage continues to rise.

And there are other problems. Not only do pesticides kill the target pests, they also kill insects that are their natural predators. Because of biological magnification, pesticides often affect these beneficial insects on a higher trophic level more adversely than they affect the destructive pests at the lower trophic level. This means that as soon as pesticide use is stopped, destructive pests multiply uncontrollably because their natural predators are gone. Researchers estimate that the destruction of natural predators by pesticides costs farmers more than $500 million each year.

The Good Earth

Farmland in the United States is productive largely because vast areas of the country are covered with excellent topsoil. The Midwest was once a prairie ecosystem covered by rich soil. Tough, deep roots of long-lived grasses held the soil in place against the actions of wind and rain. When the prairie was converted to farmland, these grasses were removed. Today, single-season crops are planted and harvested, leaving large fields exposed to wind and rain between crops. As a result, soil is eroded faster than it can be replaced.

Just how rapid is soil loss? A typical field on the high plains of the Midwest loses an average of about 4 million kilograms of topsoil per square kilometer every year. In parts of the tropics, the situation is much worse. Poor farmers often must plant crops on steep hillsides. Some of these fields lose up to 150 million kilograms of topsoil per square kilometer each year. An astonishing 30 percent of the world's cropland has been abandoned over the last 40 years because erosion has stripped it of useful soil. Some researchers estimate the cost of soil erosion in the United States to be as high as $27 billion every year.

☑ ***Checkpoint*** How does cultivation of single-season crops lead to soil loss?

Figure 16–7
Prairie grass ecosystems—which once protected the rich topsoil from erosion by rain and wind—have been converted to farmland, leaving fields vulnerable to soil loss throughout the winter months.

Water, Water Everywhere?

Agriculture across much of the world depends on large-scale watering, called irrigation. More than 70 percent of all water consumed in the United States is used for agriculture. In some states, such as California, river water is transported by pipes and aqueducts over long distances to farmlands. Farmlands in Texas, Oklahoma, Kansas, Colorado, South Dakota, New Mexico, and Nebraska get their water from a huge underground deposit called the Ogalalla aquifer.

In many places, however, water is becoming scarce. Water in the Ogalalla aquifer, for example, is often described as "fossil water" because it has collected over millions of years and is not replaced each year by rainfall. In fact, so much water in this aquifer is being pumped out and used for agriculture that the entire aquifer is expected to run dry within 20 to 40 years!

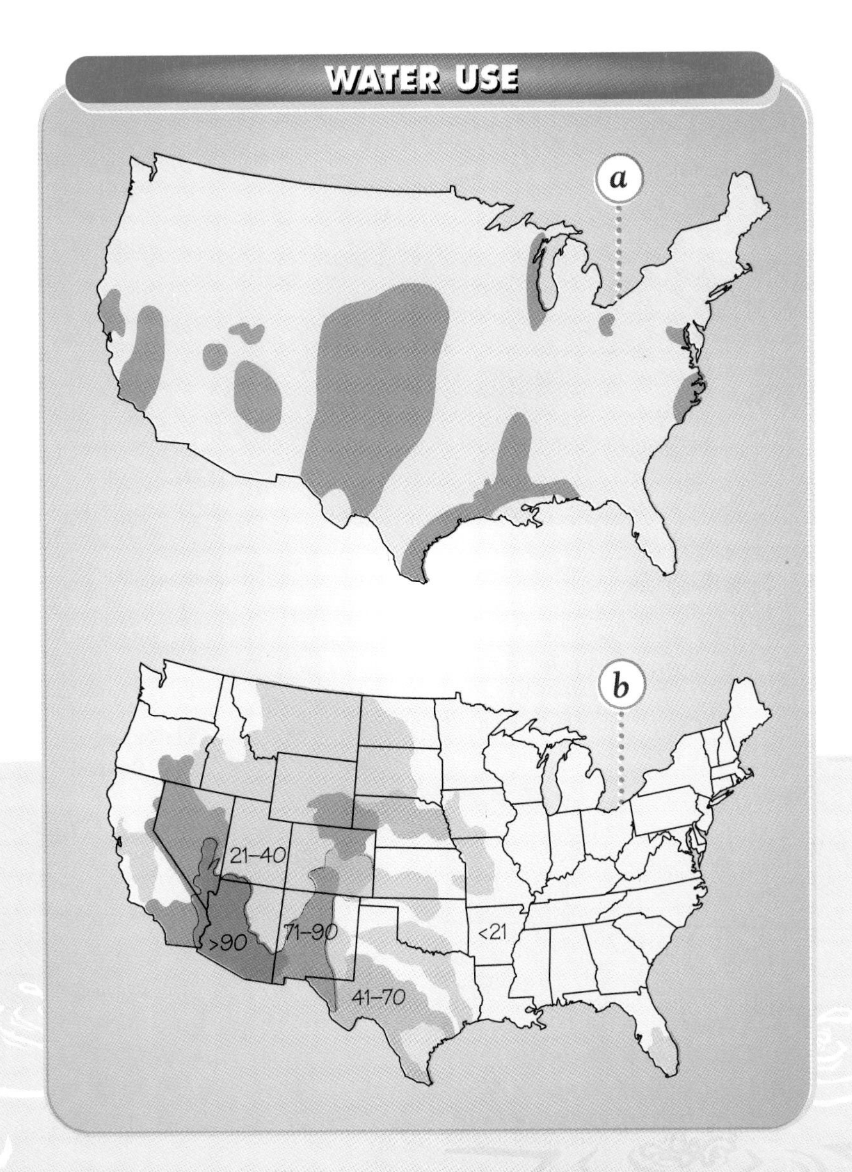

Figure 16–8
(a) This map shows areas of groundwater depletion, while (b) this map identifies areas where a high percentage of surface water is used.

Visualizing Agriculture in the Future

Agriculture in the future must use practices that are compatible with the principles of sustainability. Planting alternating rows (strips) of crops, cultivating in contoured strips at right angles to the slope of the terrain, and rotating crops from year to year are practices that conserve soil. Delivering water, drop by drop, directly to the roots of plants that need it conserves water. Crop rotation and the use of predators and parasites to control destructive insects are some of the practices that jointly constitute what is known as Integrated Pest Management. Decreasing meat consumption and increasing the direct consumption of grains is a more efficient use of existing supplies.

3 Contour Plowing

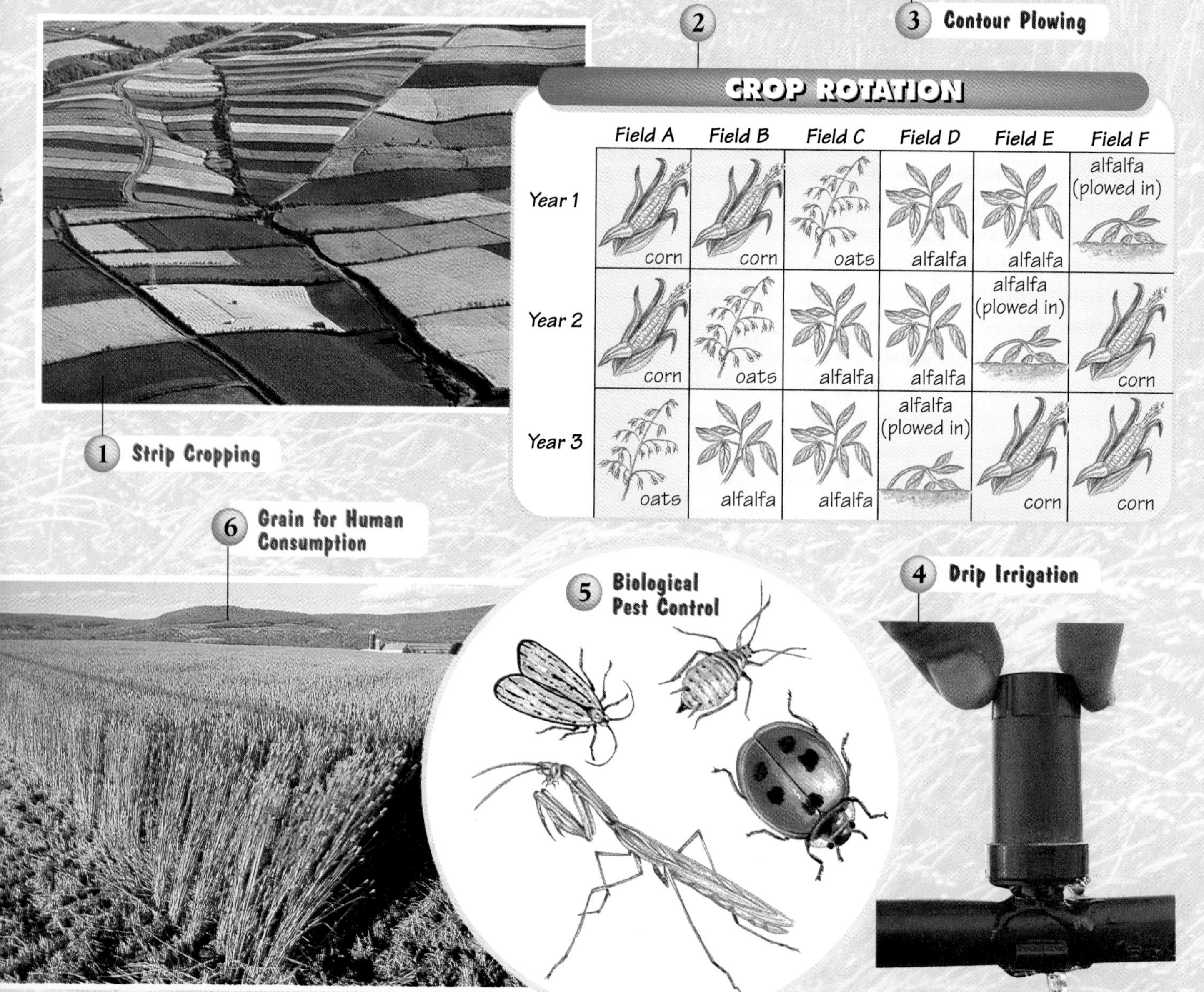

CROP ROTATION

	Field A	Field B	Field C	Field D	Field E	Field F
Year 1	corn	corn	oats	alfalfa	alfalfa	alfalfa (plowed in)
Year 2	corn	oats	alfalfa	alfalfa	alfalfa (plowed in)	corn
Year 3	oats	alfalfa	alfalfa	alfalfa (plowed in)	corn	corn

States such as California, Arizona, and Nevada are also running short of water—river water. Interestingly, most researchers estimate that no more than 30 percent of a river's average flow should be taken out for use each year if water shortages are to be avoided. Take a look at ***Figure 16–8*** on page 367 to see if this is being followed.

Agriculture in the Future

Perhaps you are beginning to see the challenge facing agriculture in the future. The goal of sustainable agriculture is to provide an adequate supply of food to feed a growing global human population while, at the same time, protecting the environment. This sounds good, but what does it mean?

Most ecologists would define a sustainable agricultural system as having the following characteristics:

- **Stability** The system must be able to operate without causing long-term harm to the soil, water, and climate on which it depends.
- **Flexibility** The system must be flexible enough to survive environmental stresses—such as droughts, floods, and extreme heat or cold.
- **Appropriate Technology** The system's needs for equipment and energy must be suitable for local environments and compatible with the abilities and training of the people using them.
- **Efficiency** The system should consume as little energy and material as possible. There should be a shift away from a reliance on fossil fuel energy sources toward renewable energy sources, such as solar energy.

How are these principles translated into practice? Visualizing Agriculture in the Future on the previous page presents a few specific strategies, or practices, that address the environmental problems you have read about earlier in this chapter. In addition, such simple practices as planting crops appropriate to local rainfall and covering soil with mulch—dead organic matter such as leaves and grass—to further prevent soil loss will go a long way in conserving precious resources and protecting the environment.

With proper management based on accurate information, many agricultural systems could be made much more sustainable than they are today. Will you be one of the people who helps humanity achieve that goal?

Section Review 16–2

1. **Describe** the key practices of the green revolution and **relate** them to their environmental impact.
2. **Identify** several strategies for sustainable agriculture.
3. **Critical Thinking—Sequencing** Sequence the use of the following integrated pest management (IPM) strategies so that they cause the least harm to the environment: low-impact nontoxic sprays, biodegradable poisons, crop rotation, biological control.
4. **MINI LAB** What type of plants might be most effective in reducing the erosion of topsoil from vacant fields in winter? **Design an experiment** to answer the question.

SECTION
16–3 Resources at Risk

GUIDE FOR READING

- **Identify** the major forms of pollution and their sources.
- **Define** biodiversity and **explain** its importance.

THE ALASKAN OIL PIPELINE cutting through untouched wilderness is symbolic of the dilemma human society faces today. Although we appreciate and love the beauty of natural ecosystems, we also manipulate and often exploit natural resources for our needs and wants. Can humans use our planet's resources without causing long-term harm to our environment or perhaps even to the biosphere?

Environmental Pollution

Although agriculture is the largest single human activity, many other things that people do also greatly affect the biosphere. In the past, cities and industries could grow rapidly and cheaply because people were not concerned about discarding waste products into the environment. When pollution problems became serious locally, solutions seemed simple. Too much liquid waste? Dump it into the ocean. Too much smoke? Build a higher smokestack so the wind will carry it away. This was not because people were careless about the environment—it was because no one really understood the long-term effects of these actions or what was at stake.

We now know that our waste products are polluting many parts of the continents we live on and the oceans around us. **A variety of human activities are affecting natural resources—such as clean air and water—that serve as the lifeblood of the biosphere.**

Some forms of pollution are caused by activities that are as basic to our lives as eating. We drive cars, heat homes, use electric lights, and keep food in refrigerators. We shop in malls filled with an amazing variety of products that go far

Figure 16–9
The Trans-Alaska oil pipeline symbolizes human reliance on resources that are nonrenewable and pollute the biosphere. The burning of fossil fuels such as coal and oil have largely contributed to today's highly industrialized society.

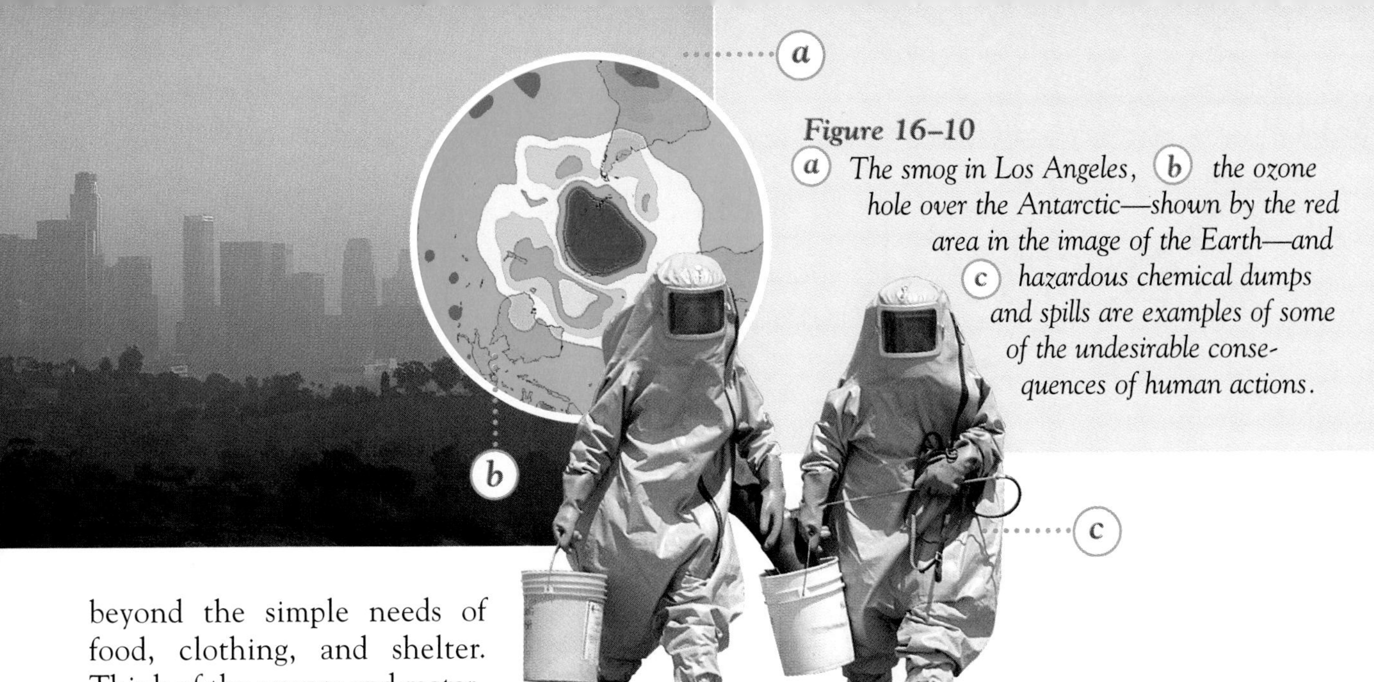

Figure 16–10
(a) The smog in Los Angeles, (b) the ozone hole over the Antarctic—shown by the red area in the image of the Earth—and (c) hazardous chemical dumps and spills are examples of some of the undesirable consequences of human actions.

beyond the simple needs of food, clothing, and shelter. Think of the energy and material requirements of the manufacturing industries and the waste products they generate—waste products that often end up polluting our air, water, and land resources.

One of the biggest sources of environmental pollution is the burning of fossil fuels—coal, oil, and natural gas. Nearly 90 percent of the energy used in the world today—for transportation, generating electricity, and heating buildings, for example—comes from fossil fuels.

Smog

Smog is due primarily to automobile exhausts and factory smokestacks. It threatens the health of the elderly and of people with asthma or other respiratory conditions.

Acid Rain

Many combustion processes, such as the burning of coal in power plants, release acidic gases into the atmosphere. Here gases combine with water vapor to form strong acids that fall to Earth as acid rain. Acid rain is harmful to plants as well as to animals.

Holes in the Ozone Layer

Ozone, a gas that is a pollutant at ground level, is a helpful barrier in the upper atmosphere. It shields the Earth from excessive, potentially harmful ultraviolet radiation. Pollutants such as CFCs (chlorofluorocarbons)

Figure 16–11
Precipitation that is 10 to 1000 times more acidic than normal is common in the northeastern United States. This diagram shows some of the effects of acid rain.

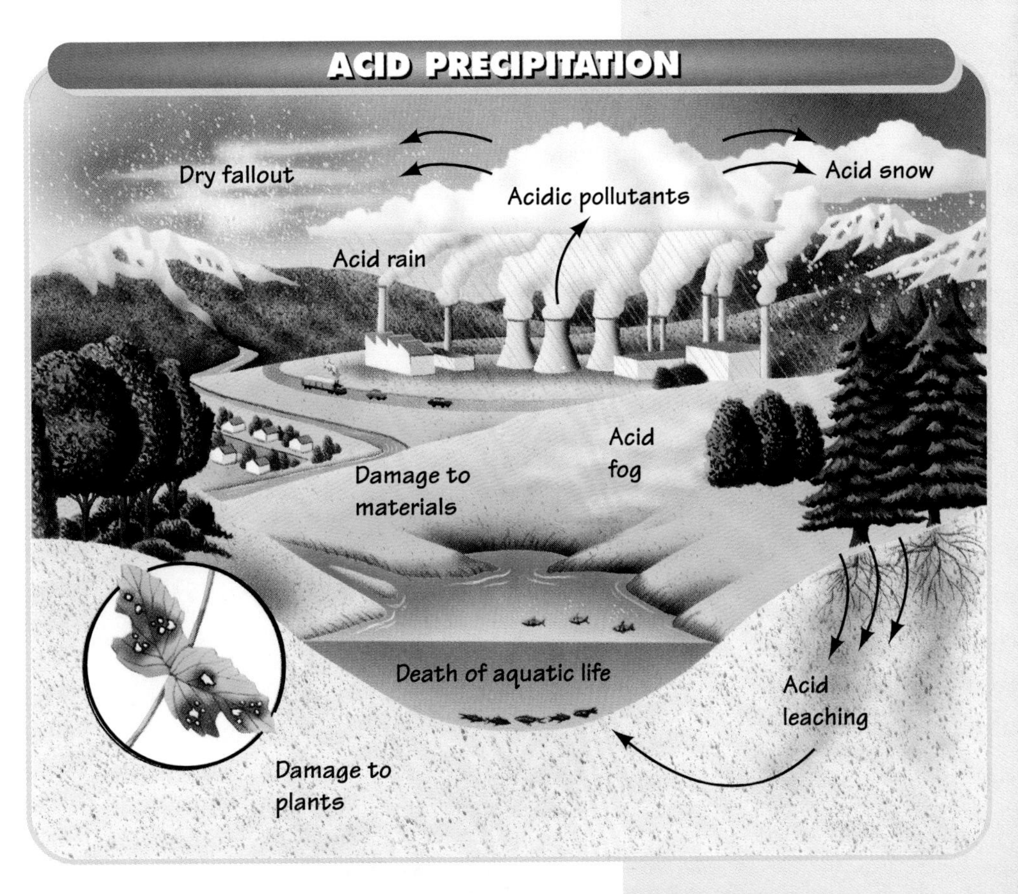

AIR POLLUTANT LEVELS

Pollutant level (% of 1977 level): 0, 10, 20, 30, 40, 50, 60, 70, 80, 90, 100

Year: 1975, 1980, 1985, 1990

Ozone, Nitrogen dioxide, Particulates, Carbon monoxide, Sulfur dioxide, Lead

Figure 16–12
The graph shows the levels of major air pollutants in selected areas of the United States. The reason for the remarkable drop in the level of lead is due to the use of unleaded gasoline.

rise up into the upper atmosphere and destroy the ozone. International agreements on controlling CFC use are expected to halt the release of these gases by 1998.

Lead

The levels of this toxic metal in the environment rose dangerously high largely due to the use of leaded gasoline in automobile engines. Adding lead to gasoline offered an inexpensive way of enhancing the performance of ordinary gasoline. Today, as you can see in **Figure 16–12,** the levels of lead in the environment are dropping as a result of the switch to lead-free gasoline. ●

INTEGRATING HEALTH

How is lead harmful to human health?

Chemical Wastes and Oil Spills

Improperly discarded hazardous chemicals pollute streams, rivers, oceans, and underground water supplies. The chemicals are harmful to aquatic plants and animals and pose a threat to humans as they are consumed and pass up the food chain through the process of biological magnification. The Environmental Protection Agency now regulates toxic waste disposal and promotes sustainable strategies such as "reduce, reuse, recycle, and discard safely."

Oil can spill into the ocean if tankers or barges run aground, hit another ship, or are damaged by storms. Oil spilled at sea is difficult to recover—even in calm weather. When oil is spilled during stormy weather, very little can be done. Oil slicks and tar balls are deadly to marine animals that swallow them or are coated with the oil. When oil blows into estuaries, animals and plants are smothered or poisoned quickly.

Figure 16–13
Sustainable energy use must rely on renewable and less-polluting energy sources, such as solar, hydroelectric, geothermal, and biomass. Solar energy can be used to generate electricity (a) in a solar home and (b) in an experimental solar power station.

Sustainable Strategies

How can we stop pollution without changing our lifestyles? It's a difficult question to answer. Just as there are strategies for sustainable agriculture, are there sustainable strategies for dealing with pollution? The answer is yes.

One significant step we can take to protect our environment is to shift from nonrenewable fossil fuel energy sources to renewable, nonpolluting energy sources such as solar and hydroelectric power. ***Figure 16–13*** offers a glimpse of the types of changes we must make if we are to protect our resources in the future.

Checkpoint What are some sources of pollution in the environment?

Biodiversity

You may be surprised to learn that one of Earth's greatest natural resources is **biodiversity,** or biological diversity. **Biodiversity is the genetically-based variety of living organisms in the biosphere.** The processes that bring forth new species as well as those that make species extinct have been occurring for hundreds of millions of years. The net result of these processes is the biodiversity you see all around you.

Value of Biodiversity

Biodiversity is considered a natural resource, and it is valuable to society in such diverse endeavors as agriculture, medicine, and recreation. Let's see how.

Many challenges facing agriculture may be solved with the help of wild plants. Most crop plants have surviving wild relatives around the world. Often, these wild plants contain valuable genes that can lend disease resistance, pest resistance, drought resistance, or other useful traits. Similarly, forestry, animal breeding, and aquaculture benefit from a diverse gene pool to draw upon as sources of desirable traits.

Modern medicine relies on a variety of plants as a source of a majority of commonly prescribed drugs. Painkillers, antibiotics, heart drugs, anticancer drugs, coagulants, anticoagulants, enzymes, hormones, and antidepressants are some of the drugs that have come from Earth's huge diversity of plant species.

Do you enjoy observing wild plants and animals in their natural habitats? Camping, hiking, birdwatching, visiting wild animal parks, zoos, and aquariums

Figure 16–14
Conserving biodiversity is important for many reasons. (a) *Appreciating the beauty of wildlife, such as this resplendent quetzal from Costa Rica, and* (b) *camping in unspoiled, serene surroundings are examples of popular recreational activities.* (c) CAREER TRACK *Botanists in Sri Lanka study local plant life in order to explore possible medicinal uses.*

Figure 16–15
The value of genetic diversity extends beyond visible differences such as the height of these corn plants. A greater genetic diversity may offer benefits such as resistance to disease and drought.

or going on a whale watch are a source of joy to many of us.

In addition to the benefits just mentioned, the systematic study of biodiversity provides human society with an invaluable understanding of the workings of our living planet. Such knowledge is essential if we are to use our resources responsibly.

☑ **Checkpoint** What is biodiversity and its value to humans?

Biodiversity Threatened

Various types of human activity are causing a decrease in Earth's biodiversity. Scientists think of biodiversity as the variety at different levels of organization.

Ecosystem diversity is the variety of living ecosystems in the world, and it is threatened by the development of land for housing, industry, and agriculture. In the United States, only 5 percent of the old-growth forests, or forests that existed when the first colonists arrived, remain today. Wetlands are disappearing at an alarming rate. In the tropical regions of the world, rain forests are vanishing even more quickly.

The number and variety of different life forms, or **species diversity,** is also declining rapidly across the world. As habitats are destroyed, species living in those habitats are placed at risk. A species in danger of becoming extinct is called an endangered species. Today, many species around the world fall into this category.

Scientists also extend the concept of biodiversity to **genetic diversity**—the variety of different forms of genes present in a population. Such diversity is essential to the survival of most species. Genetic diversity in the biosphere is at risk for several reasons. When species are reduced to small populations, the populations lose genetic diversity. This tends to make individual organisms weaker and more prone to birth defects. It can also make breeding animals less fertile and less able to fight off parasites and disease. Populations that stay too small for too long thus become more vulnerable to extinction.

All forms of biodiversity are threatened by climate change today as never before in the history of life on Earth. Why? In the past when Earth's climate warmed or cooled or when patterns of rainfall shifted, living things could move around. They could extend their ranges to the north or the south. But today, most natural environments have been reduced to small patches surrounded by cities, suburbs, and farms. If climate changes now, many organisms will have no place to go.

Section Review 16–3

1. **Identify** major forms of pollution and their sources.
2. **Define** biodiversity and **explain** why it is important.
3. **Critical Thinking—Evaluating** What problem is symbolized by the Alaskan oil pipeline winding through pristine natural ecosystems? Do you think the advantages outweigh the disadvantages?

Conserving Biodiversity

GUIDE FOR READING

- **Explain** the relationship among habitat size, population size, and species diversity.
- **Examine** the role of conservation biology in land resource management.

WHEN YOU THINK OF AN ISLAND, you probably imagine a small plot of land in an ocean of water. But a lake, too, can be thought of as an island—an island of water surrounded by land. New York's Central Park is an island of trees and grass in a sea of concrete and asphalt. Even a patch of forest can be a biological island if it is surrounded by plowed fields, housing developments, or shopping malls. The concept of a biological island is important to conservation efforts aimed at preventing the loss of a species.

Island Size

To a biologist, an island is any patch of habitat surrounded by a different habitat—meaning that organisms living on the island stay confined to the island and do not move into the surrounding habitat. Studies of islands show that there are fewer species on small islands than there are on larger ones. Data indicate that the number of species on islands doubles for every tenfold increase in island size. Why does species diversity relate to island size this way? There are several reasons.

Habitat Diversity

The larger an island is, the more likely it is to have a variety of habitats. The greater the variety of habitats on the island, the greater the variety of species able to live there.

Habitat Size

The larger an island is, the larger its habitats can be. Larger habitats can support larger populations. And large populations are more likely to survive over the long term than small populations. Why? Recall that density-independent limiting factors such as unpredictable changes in environmental conditions can cause sudden increases in death rates or decreases in birth rates. The larger a

Figure 16–16
Examples of biological islands include (a) land surrounded by water and (b) New York's Central Park.

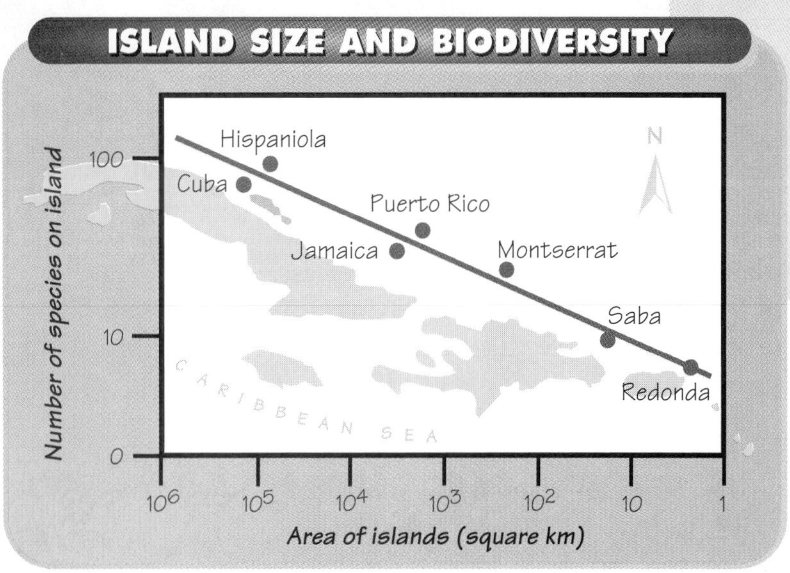

ISLAND SIZE AND BIODIVERSITY

Figure 16–17
The graph compares the diversity of reptiles and amphibians on seven Caribbean islands of various sizes. It shows that a 90-percent reduction in island size correlates with a 50-percent reduction in the number of species.

population is, the more likely it is to survive these changes. On the other hand, the smaller a population is, the more likely it is to become extinct when stressed.

Edge Effects

An island's size has an interesting and important effect on the relationship between the island's area and its perimeter, or the length of its edge. Large islands of forest and jungle offer much more space in the interior relative to space near their edges. Small forest or jungle islands, on the other hand, have much more edge for each unit of their area.

Why is this distinction important? Here's an example to illustrate the idea. Songbirds that summer in North America thrive when they build their nests inside large areas of temperate forest. They are not as successful near the edge of forest patches for an interesting reason. Animals such as opossums, raccoons, blue jays, and cowbirds thrive in and around forest edges. In fact, these species prefer to live near neighborhoods where they can rummage through garbage cans for food. When a large patch of forest is divided by roads, shopping malls, and houses, local populations of these animals increase in size. How does this affect songbird populations? Jays, raccoons, and opossums prey on songbird eggs and hatchlings. Cowbirds are parasites that lay eggs in the nests of songbirds. Cowbird hatchlings either throw songbird hatchlings out of the nest or steal food brought back by songbird parents so the songbird chicks starve.

So when a large forest is cut up into smaller forest islands, the forest edge increases at the cost of interior forest area. Populations of predators and parasites increase and destroy so many songbird eggs and hatchlings that the size of the songbird population falls dramatically.

The lesson from several similar studies is clear. **Larger habitats support larger populations of organisms, which in turn promotes greater species diversity.** How will this understanding guide human society in deciding the best way to use land resources? Only time will tell.

☑ ***Checkpoint*** Why do larger islands support a larger number of species?

Conservation Biology

From the United States to Indonesia, what were once giant ecosystems are now being reduced to island habitats in a sea of human environments. Most people would like to see precious habitats preserved. Most people also want to save endangered plant and animal species. But as populations grow, it will be impossible to leave entire forests, jungles, or wetlands untouched, and compromises will have to be made.

Conservation biology is a new discipline that has emerged due to a growing understanding of the importance of habitats and biodiversity. The primary goal of this discipline is **conservation,** or the managing of natural resources

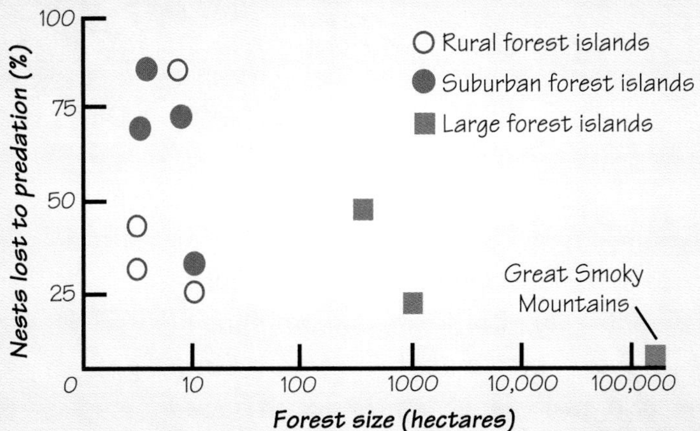

Figure 16–18
Recent research on island and habitat size is indicating that small forest fragments are a threat to many bird species. (a) This graph shows that nest predation is significantly higher in rural and suburban forest fragments than in large forests. (b) A yellow warbler, victimized by cowbird parasitism, is caring for a cowbird hatchling at the cost of two of its own smaller hatchlings.

in a way that maintains biodiversity. Knowledge and information from areas—such as genetics, ecology, and resource management—are integrated and used to achieve conservation goals.

How Big Is Big Enough?

Conservation management requires detailed information about the relationship between populations and habitat size. To save spotted owls in the Northwest, researchers must find out how large the fragments of old-growth forest must be for the birds to thrive. Is 1 hectare enough? Do they need 10? Or 100?

Habitat Conservation Research

One study designed to gather such data has been going on in Brazil since the 1970s, set up by the World Wildlife Fund. Brazilian landowners who wanted to develop jungle land were required by law to set aside half their land untouched. The landowners had to set aside the untouched land in chunks of different sizes! The jungle islands range in size from 1 hectare to nearly 1000 hectares.

Data gathered have revealed many fascinating interactions among populations in jungle islands. More experiments like this will provide the information we need to manage our world into the next century.

Section Review 16–4

1. **Explain** the relationship among habitat size, population size, and species diversity.
2. **Examine** the role of conservation biology in land resource management.
3. **List** a few examples of human behavior that may result in the loss of wildlife habitats.
4. **BRANCHING OUT ACTIVITY** Obtain a detailed map of your county. Of the following categories, **identify** the ones found in your county and **estimate** their percentage (by area): highly urban, suburban residential, rural/agricultural, and natural preserve.

CHAPTER 16

Laboratory Investigation

Turn Your Garbage Around

With the growing waste disposal problem, there is a great deal of interest in composting as a valuable way of reducing the amount of waste to be disposed and converting it into a useful product. Composting is a process by which biodegradable matter is decomposed by microorganisms such as bacteria and fungi. As a result, essential nutrients are recycled and make the soil more fertile.

Problem

Construct a model compost column and **classify** various materials according to their suitability for composting.

Materials (per group)

2 2-L plastic bottles
markers
knife
scissors
porous cloth
transparent tape
balance
rubber bands
pieces of cloth
materials for composting
water
slides
microscope

Procedure

1. Prepare a list of materials to be composted—both biodegradable and nonbiodegradable.
2. Show your list to your teacher for approval.
3. Predict which material will decompose the fastest, the next fastest, and so on.
4. Record your predictions in a data table.
5. Remove the labels from the bottles. Using the knife, make the first cut and then use scissors to cut the bottles as indicated in the diagram. **CAUTION:** *Be careful when using a knife.* Make windows in the bottle tops as shown, taping strips of porous cloth over them to allow air to circulate.
6. Cover the mouth of the longer bottle top with a piece of porous cloth. Secure this with a rubber band. In this way, liquid can percolate through the bottom of the compost column.
7. Invert the bottle top into the bottle base and tape them together securely with transparent tape.
8. Determine and record the mass of the materials to be composted using a balance. Use enough material to fill approximately half of the inverted bottle section of the column.

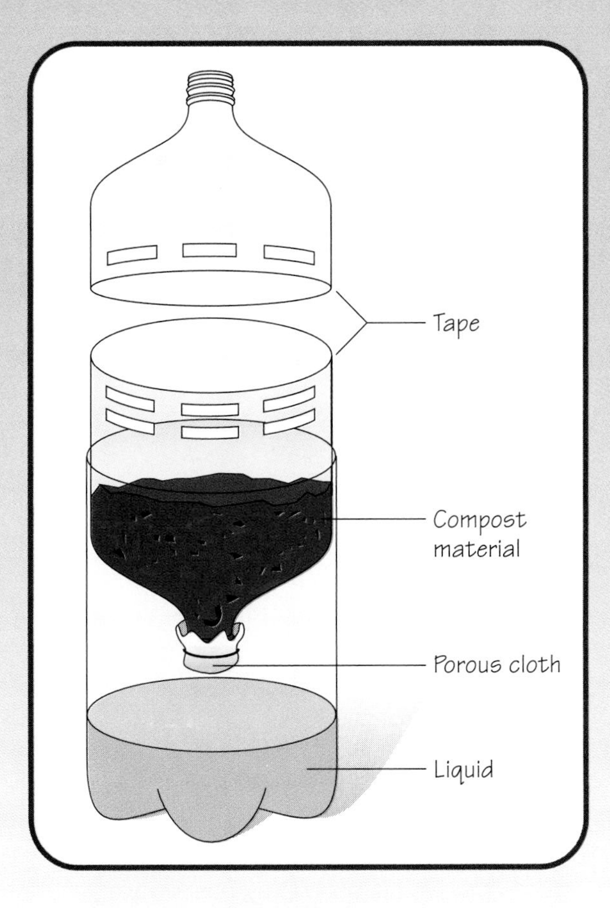

9. **Add approximately 250 mL of water to the material in the column. Then tape the remaining bottle top to the top of the column.**
10. **Place the column in an area that will provide the best environment for decomposition.**
11. **Examine the compost column every week for at least 4 weeks, and observe and record any changes.**
12. **During each observation, carefully loosen the tape from the bottom of the column to remove any liquid that has accumulated in the bottom section of the column.**
13. **Examine several drops of the liquid under the low-power objective of a microscope. Record your observations.**
14. **After 4 weeks, spread out a plastic trash bag on your table. Carefully empty the contents of the column onto the plastic bag.**
15. **Wearing rubber gloves, examine the compost and record your observations.**
16. **Record the mass of the compost material.**

Observations

1. Describe the changes you observed in your compost column.
2. Which material appears to decompose the fastest? The slowest?
3. What changes did you observe in the water samples from week to week?
4. How did the mass of the compost material compare with its initial mass?

Analysis and Conclusions

1. How do your results compare with your predictions? How can you explain any differences?
2. Calculate the percentage change in the mass of the compost material. How can you account for this difference?
3. Did the location of the compost column affect the results of the experiment in any way? You may consult with your classmates.
4. What can you conclude regarding the benefits of composting as one option for waste disposal?

Study Guide

Summarizing Key Concepts

The key concepts in each section of this chapter are listed below to help you review the chapter content. Make sure you understand each concept and its relationship to other concepts and to the theme of this chapter.

16–1 This Island Earth

- Human activities are the most important source of change in the biosphere.
- Nonbiodegradable substances accumulate in organisms and are passed up the trophic levels at higher and higher concentrations by the process of biological magnification.
- Humans must adopt ways of living that are ecologically sustainable—which means human actions must not cause deterioration of the environment.

16–2 Agriculture

- The green revolution is based on such key practices as monoculture (planting large fields with a highly productive crop), the use of chemical fertilizers and pesticides, machinery, and irrigation.
- Green revolution practices have harmful impacts on the environment, such as biological magnification of toxic substances, loss of soil, and the depletion and pollution of water resources.
- Sustainable practices—such as biological control of pests, crop rotation, contour strip farming, drip irrigation, and the cultivation of crops suitable to a region's rainfall—protect the environment.

16–3 Resources at Risk

- Environmental pollution occurs in large part because of the combustion of fossil fuels and the wastes generated by numerous manufacturing industries.
- Biodiversity is the genetically-based variety of living organisms in the biosphere.

16–4 Conserving Biodiversity

- Larger habitats support larger populations of organisms, which in turn allows greater species diversity.
- Conservation biology aims to maintain biodiversity by applying what we know about genetics, ecology, and resource management.

Reviewing Key Terms

Review the following vocabulary terms and their meaning. Then use each term in a complete sentence.

16–1 This Island Earth

nonbiodegradable
biodegradable
biological magnification
sustainability

16–2 Agriculture

green revolution
monoculture

16–3 Resources at Risk

biodiversity
ecosystem diversity
species diversity
genetic diversity

16–4 Conserving Biodiversity

conservation

Recalling Main Ideas

Choose the letter of the answer that best completes the statement or answers the question.

1. A substance that is broken down by living things in the environment is said to be

a. nonbiodegradable.
b. biodegradable.
c. sustainable.
d. biologically magnified.

2. Which of the following is not a strategy central to the original green revolution?

a. use of fertilizers
b. monoculture
c. irrigation
d. crop rotation

3. A major cause of soil erosion is

a. monoculture.
b. inadequate irrigation.
c. the use of chemical fertilizers.
d. biochemical pest control.

4. Which is not a strategy for sustainable agriculture?

a. crop rotation
b. growing crops suited to local rainfall
c. biological pest control
d. monoculture

5. Which of the following pollutants poses a substantially lower threat today than it did 10 years ago?

a. smog
b. acid rain
c. lead
d. oil spills

6. The value of biodiversity to agriculture is mainly through

a. genetic diversity.
b. species diversity.
c. ecosystem diversity.
d. habitat diversity.

7. The main goal of conservation biology is to

a. manage resources to conserve biodiversity.
b. combine the fields of genetics, ecology, and resource management.
c. save the spotted owl.
d. do conservation research.

8. Studies show that species are most likely to be threatened in

a. large rural forest fragments.
b. small urban forest fragments.
c. large urban forest fragments.
d. small rural forest fragments.

Putting It All Together

Using the information on pages xxx to xxxi, complete the following concept map.

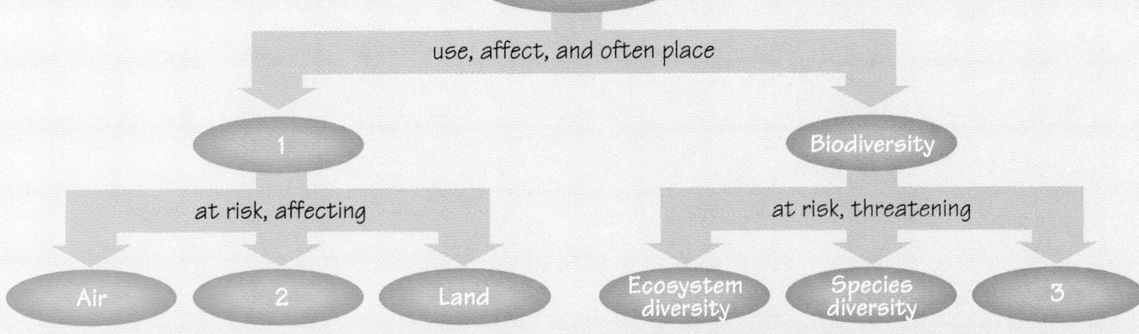

Assessment

Reviewing What You Learned

Answer each of the following in a complete sentence.

1. What has been the greatest source of change in the biosphere?
2. When is a substance considered to be biodegradable?
3. Give an example of biological magnification.
4. What was the green revolution and where did it occur?
5. What is monoculture? What are its advantages?
6. What are some examples of environmental pollution?
7. Define biodiversity.
8. What is a biological island? Give two examples.
9. Speaking biologically, how do smaller islands differ from larger islands?
10. What is meant by conservation?

Expanding the Concepts

Discuss each of the following in a brief paragraph.

1. Give an example of an ecologically unsustainable practice. Predict its consequences.
2. Compare the benefits and the costs of modern agriculture.
3. How do sustainable agricultural practices differ from key green revolution practices?
4. Why is it impossible to eliminate a crop pest completely?
5. Identify and discuss three benefits of biodiversity.
6. Describe five actions your community might take to reduce the pollution in your area.
7. Why do larger habitats have more biodiversity than smaller ones?
8. Distinguish between ecosystem, species, and genetic diversity.
9. Identify three resources currently at risk and indicate what might be done to alleviate the problem.

Extending Your Thinking

Use the skills you have developed in this chapter to answer the following.

1. **Hypothesizing** A monoculture of cotton was planted in the 1980s in many southern states. A new disease invaded the cotton plants, almost completely destroying them. Why did this occur?
2. **Evaluating** Water is often referred to as our most important resource. Do you agree? Why or why not?
3. **Calculating** A nonbiodegradable toxic substance is concentrated ten times at each trophic level. What will be its concentration in organisms at the fifth trophic level if the primary producers of the first trophic level store the substance at concentrations of 40 ppm?
4. **Experimenting** Can covering soil with mulch near the base of plants help reduce soil erosion? Design an experiment to answer the question.
5. **Evaluating** Some industries, such as agriculture, have established gene banks to maintain the genetic diversity important to that industry. Discuss the pros and cons of this practice.

Applying Your Skills

A Letter to the Editor

Humans are affecting the biosphere in many ways. In order to help maintain sustainability, everyone must become aware of the issues involved and be able to communicate information effectively and to express opinions based on facts.

1. Investigate a local problem that may affect the future ecology of your area.
2. Gather relevant facts and information from different sources to help you arrive at possible solutions.

• GOING FURTHER •

3. Write a letter about the problem to the editor of your local newspaper, stating your concern and offering one or more possible solutions.

UNIT 5

Life on Earth: An Overview

CHAPTERS

". . . the astonishing thing about the Earth, catching the breath, is that it is alive."

— Lewis Thomas

CAREER TRACK

As you explore the topics in this unit, you will discover many different types of careers associated with biology. Here are a few of these careers:

- Paleontologist
- Nursery Operation Technician
- Ocean Technician
- Animal Physiologist
- Anthropologist

A bee pollinating a poppy plant

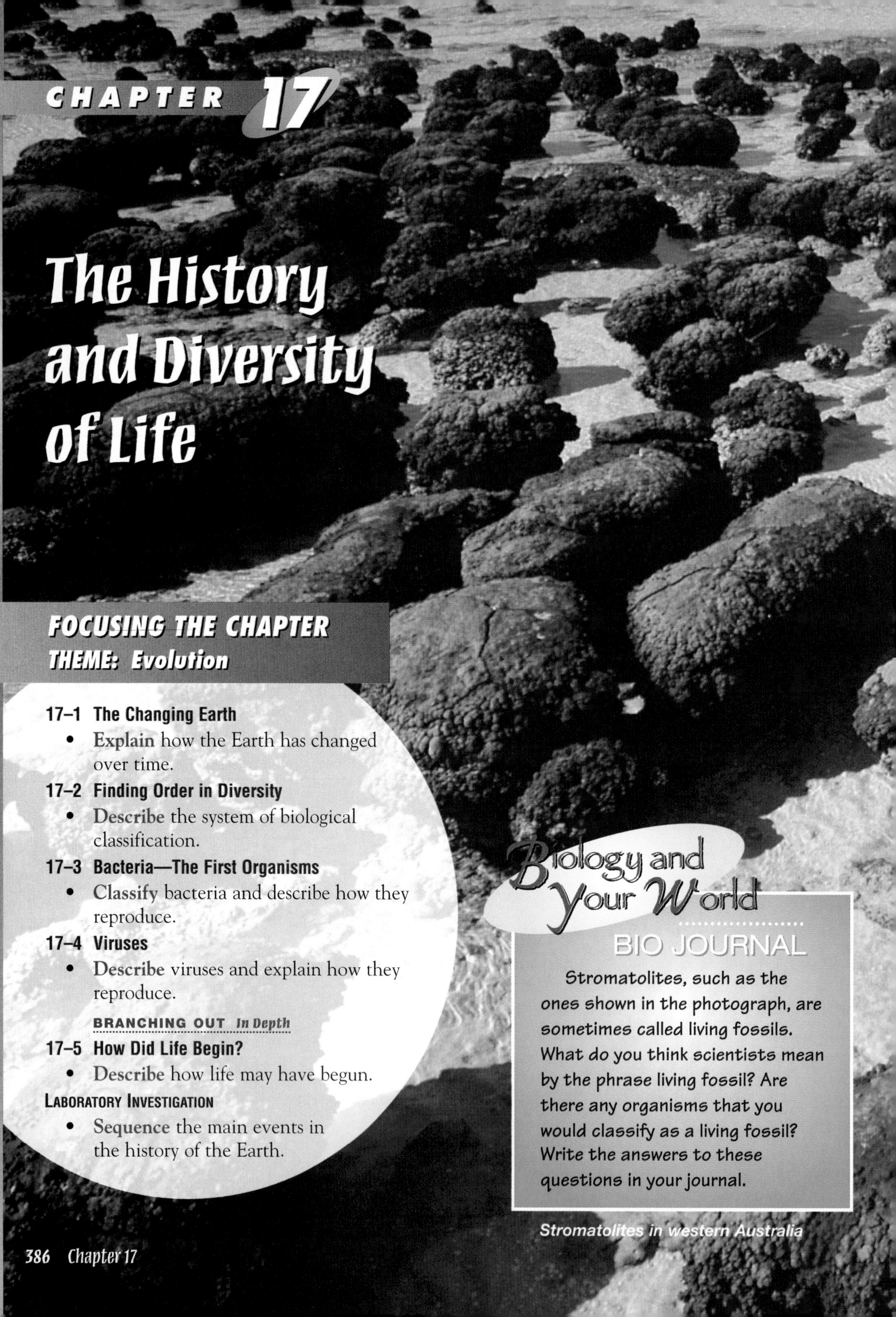

CHAPTER 17

The History and Diversity of Life

FOCUSING THE CHAPTER
THEME: Evolution

17–1 The Changing Earth
- **Explain** how the Earth has changed over time.

17–2 Finding Order in Diversity
- **Describe** the system of biological classification.

17–3 Bacteria—The First Organisms
- **Classify** bacteria and describe how they reproduce.

17–4 Viruses
- **Describe** viruses and explain how they reproduce.

BRANCHING OUT *In Depth*

17–5 How Did Life Begin?
- **Describe** how life may have begun.

LABORATORY INVESTIGATION
- **Sequence** the main events in the history of the Earth.

Biology and Your World

BIO JOURNAL

Stromatolites, such as the ones shown in the photograph, are sometimes called living fossils. What do you think scientists mean by the phrase living fossil? Are there any organisms that you would classify as a living fossil? Write the answers to these questions in your journal.

Stromatolites in western Australia

The Changing Earth

Guide for Reading

- **Describe** the importance of the fossil record.
- **Explain** the importance of the geologic time scale.

MINI LAB

- **Relate** an element's half-life to the amount of the original element that remains.

THE STORY OF LIFE ON EARTH is both fascinating and exciting. It is filled with mysteries and life-and-death struggles. The cast contains species as amazing as any monsters from science fiction. And Earth's changes over millions of years set the stage for this exciting drama.

Some of these changes were caused by sudden events, such as volcanic eruptions, collisions with meteors, or earthquakes. Other changes happened more slowly. Yet all these events—both sudden and gradual—are important to the history of life, because every change in our planet affects the daily lives of every man, woman, and child.

The Changing Planet

Many scientists hypothesize that our planet Earth was formed as pieces of cosmic debris became attracted to one another over hundreds of millions of years. As the Earth developed, it was struck by large objects, one of which may have been as large as the planet Mars. This catastrophic event produced incredible amounts of heat—enough, in fact, to melt the entire globe.

Once the Earth melted, the elements rearranged themselves according to their density. The most dense elements formed the Earth's core, while less dense elements formed the outer coverings of the Earth. Eventually, these elements cooled to form a solid crust. Still lighter elements, such as hydrogen and nitrogen, escaped from Earth's surface to form its first atmosphere.

Figure 17–1
(a) *Although this view of the planet Earth from space doesn't show it, Earth has had a long and interesting history.* (b) *Many scientists now believe that an asteroid collided with Earth—similar to what is shown in this artist's representation—causing mass extinctions of some forms of life.*

Had you been there at that time, you never would have recognized the Earth—it was so different from today's Earth. When the crust cooled enough for water to stay in liquid form, oceans covered most of the surface. The Earth's first atmosphere lacked oxygen and was composed mainly of carbon dioxide, hydrogen sulfide, and methane. A few deep breaths would have killed you!

Figure 17–2
The history of life has been recorded in many ways. (a) *Sometimes entire organisms, such as these ants, were trapped in plant sap that has solidified into amber.* (b) *Sometimes only footprints of an organism, such as these of a dinosaur, have been left behind.*

Fossils and Their Stories

What evidence do scientists have that helps them know when the Earth was formed? And how can they tell when organisms that are now extinct once lived? **To learn about life's past and to fully understand its present, researchers read the history preserved in the fossil record.** Unfortunately, the oldest rocks in this record are not very useful to scientists. Over time, most of these rocks have usually been subjected to enough heat and pressure to destroy any traces of life they may once have contained.

Layers of Rock

However, rocks formed about 3 billion years ago are a different story. These rocks often contain **fossils,** which are the remains or traces of ancient life.

Figure 17–3
Layers of sedimentary rock sometimes contain fossils that represent billions of years of Earth's history. (a) *When a fish dies, for example, it sinks to the bottom of a body of water. Over time, sediments pile up on top of the fish's remains until it is covered. The soft parts of the fish disintegrate and only its hardest parts are left. Millions of years later, changing circumstances bring the fish's remains, or fossil, to the surface, where erosion exposes more of it.* (b) *The photograph shows the layers of sedimentary rock that make up the walls of the Grand Canyon.*

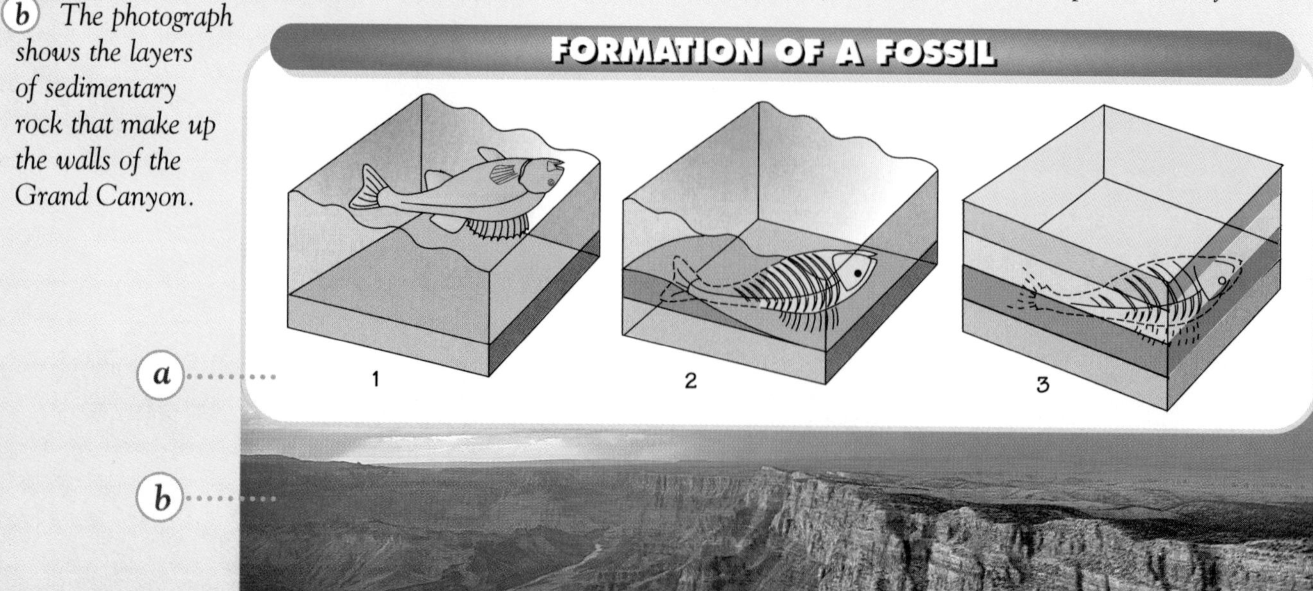

Fossils are most commonly found in **sedimentary rocks** that form when silt, sand, or clay builds up in the bottom of a river, lake, or ocean. ● As these sediments pile up, pressure on lower layers turns them into rocks such as sandstone or limestone. Over time, layers of sedimentary rocks can be laid down like layers in a cake. Each rock layer is identified with a time period in history, which is often named after a place where the rocks were found.

☑ **Checkpoint** What are fossils?

Putting the Past in Order

Because each fossil species appears at some point in time and disappears at another, scientists can recognize distinct groups of fossils in specific rock layers. Often, the assortment of fossils in one rock layer differs from fossils in other rock layers. By matching rock layers and fossils, geologists can identify rocks that were formed at the same time. They can also arrange the rock layers in chronological order—from the deepest, oldest layers to more recent layers closer to Earth's surface. Scientists can then compare the ages of the fossils in one layer with the ages of fossils found in other layers. This is called relative dating.

☑ **Checkpoint** What is relative dating?

Figure 17–4
Fossils are usually found in sedimentary rocks. Because younger sedimentary rocks normally lie on top of older sedimentary rocks, scientists can determine the course of changes in living things on the Earth. Scientists often compare similar fossils found in rock layers in different areas.

Radioactive Dating

Relative dating cannot give the actual age of rocks and fossils, so how are scientists able to calculate the age of the Earth? About 100 years ago, scientists discovered that certain elements are radioactive—that is, they break down from an unstable form into a more stable form over time. In addition, geologists realized that these radioactive elements could provide a series of clocks by which they could measure the age of rocks.

INTEGRATING EARTH SCIENCE

Igneous rocks are rocks formed from hot lava or magma. Why do you think fossils are usually not found in igneous rock?

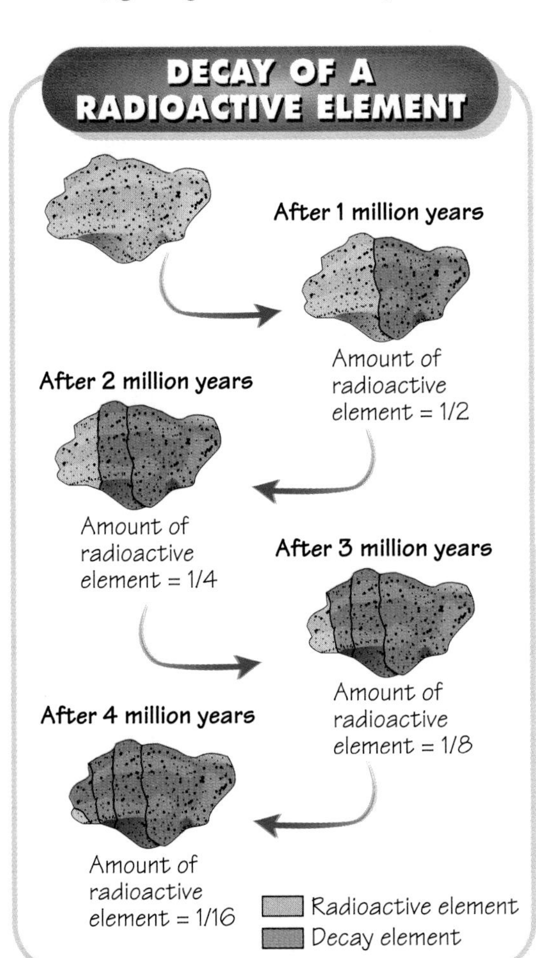

Figure 17–5
Radioactive elements break down, or decay, into stable elements at a steady rate, measured in a unit called a half-life. This sample element, for example, has a half-life of 1 million years. After 1 million years, 1/2 of the sample is radioactive and 1/2 is stable. After 4 million years, only 1/16 of the sample element remains radioactive.

When Can Half a Clock Tell Time?

PROBLEM *How does half-life* ***relate*** *to the amount of original element that remains?*

PROCEDURE

1. Place a round piece of filter paper in front of you. Label the center of the paper C-14 (carbon-14). At the top of the paper (the 12 o'clock position), write 0 years. At the 6 o'clock position, write 5770 years (the half-life of C-14). And at the 9 o'clock position, write 11,540 years.
2. Fold the filter paper in half down the middle, from top to bottom.
3. Repeat steps 1 and 2 using U-238 (half-life 4.5 billion years) and K-40 (half-life 1.3 billion years).

ANALYZE AND CONCLUDE

1. If the unfolded filter represents 12 g, what mass remains after folding the paper in half? After folding the paper in half again?
2. How many years does folding the paper in half represent? What is this time period called?

INTEGRATING MATH

If the half-life of an element is 10 years, how many years will it take until only 1/8 of the element remains?

How might these clocks work? Each radioactive element decays at a constant rate. This rate is measured by a unit called **half-life.** The half-life of an element is the amount of time needed for half the original element to decay into its more stable form.

The half-lives of radioactive elements vary enormously. Carbon-14, for example, has a half-life of 5770 years. At the end of 5770 years, one half of a given amount of carbon-14 decays to nitrogen-14. Carbon-14 is present in the atmosphere, so organisms are constantly taking it in. When the organism dies, it stops taking in carbon-14. By comparing the amounts of carbon-14 and nitrogen-14 in the organism, scientists can estimate its age. This kind of dating is called absolute dating because it allows scientists to calculate the actual age of the organism.

Carbon-14 is often used to date organic material, such as charcoal, that is less than 50,000 years old. Another radioactive element, Potassium-40, has a half-life of 1.3 billion years. It is used to date fossils that are at least 1 million years old. Because the half-life of uranium-238 is 4.5 billion years, it has been used to date the oldest rocks on Earth, roughly 4 billion years old.

☑ **Checkpoint** What is meant by the half-life of an element?

Figure 17–6
The geologic time scale is a record of the history of life. Many of the time periods get their names from the areas in which the rocks and fossils were found. For example, the Jurassic Period was named after the Jura Mountains of France and Switzerland, where rocks from that time period were discovered.

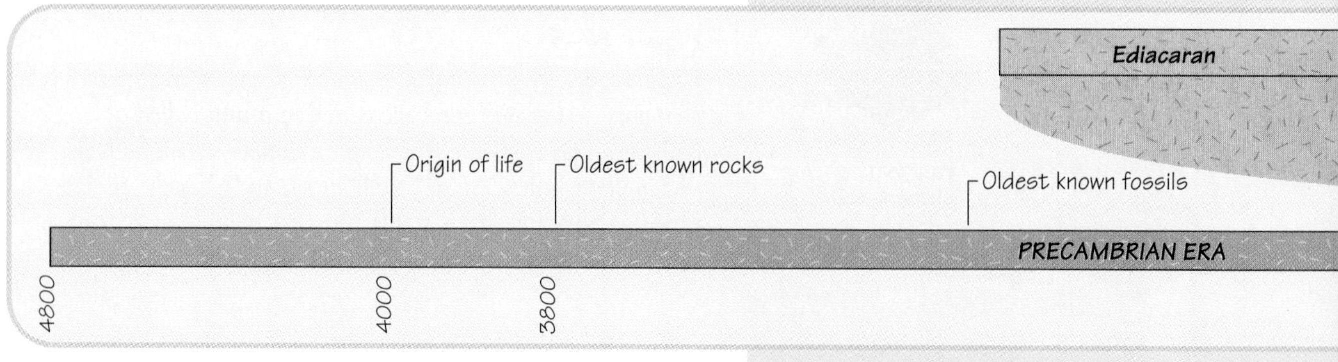

The Geologic Time Scale

Scientists have put together a history of life from almost 4 billion years ago to the present. Four billion years is an incredibly long time. We don't usually talk about time periods longer than a century. But to geologists and paleontologists, a thousand years is like the blink of an eye. **To make Earth's long history more manageable and to show how organisms have changed over time, scientists developed the geologic time scale.** Each unit of the **geologic time scale**—eons, eras, periods, and epochs—is based on information contained in rocks studied around the world.

Figure 17–7
CAREER TRACK
If fossils of ancient organisms interest you, you may enjoy a career in paleontology, which is a branch of geology.

The fossil record also shows that the history of life has not been one of steady progress. You will see that during some periods, life didn't change much at all. During other periods, major changes took place that gave rise to many new species.

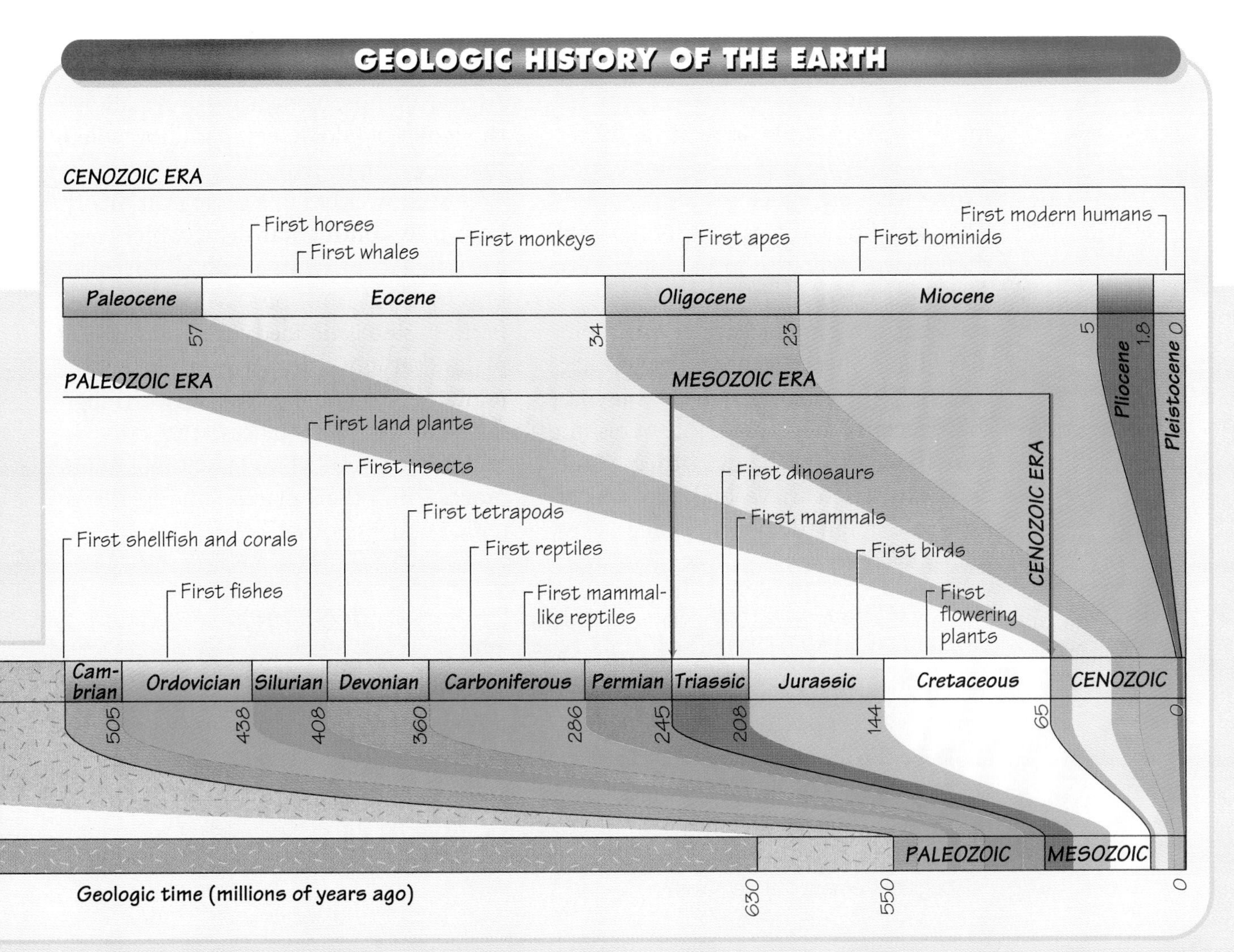

Figure 17–8
Mountains such as these, beautifully reflected in a lake, are formed when continental plates collide.

Over time, huge numbers of species have disappeared over relatively short periods of time. At least five times during the history of life, as many as 50 percent of all living species became extinct during what are called mass extinctions. The best-known mass extinction—although it wasn't the most disastrous—ended the era of dinosaurs.

What causes mass extinctions? Some may have been caused by climate change. Others may have been caused by the effects of huge meteorites crashing into the Earth. Whatever their cause, mass extinctions have had powerful effects in shaping the history of life.

Wandering Continents

There was still a puzzle to be solved before scientists could fully understand the history of the Earth. Nearly 150 years ago, geologists looking at maps of the globe noticed that major landmasses resembled pieces of a jigsaw puzzle. They saw that the east coast of South America could comfortably fit against the west coast of southern Africa. In the same way, North America and Greenland could fit neatly into Europe and northern Africa.

As time went on, more curious information was revealed. For example, rocks and fossils of a certain age from South America closely resemble those found in Africa. If those continents were always thousands of kilometers apart, how could that have happened?

The answer lies in a geological theory known as **continental drift,** which suggests that continents move slowly over hundreds of millions of years. The Earth's crust is divided into eight major plates that move. In places where continental plates collide, one of two things happens. In some places, the crust is folded and wrinkled into giant mountain ranges. In other places, one plate is forced under another.

Section Review 17–1

1. **Describe** the importance of the fossil record.
2. **Explain** the importance of the geologic time scale.
3. **Critical Thinking—Comparing** Compare relative dating and absolute dating.
4. **MINI LAB** How can you **relate** an element's half-life to the amount of the original element that remains?

Finding Order in Diversity

SECTION 17–2

GUIDE FOR READING

- **Explain** the importance of classification systems.
- **Classify** living things into six kingdoms.

MINI LAB

- **Classify** common objects into groups.

HUNDREDS OF MILLIONS OF species have come and gone on Earth during the history of life. This is quite a lot to think about. In order to even begin to speak about this staggering diversity, we need to develop a classification system to help us identify specific living things. In this way, we can give names to individual organisms and see how they relate to other organisms.

Why Do We Classify?

A classification system identifies objects and gathers them into groups whose members are similar to one another. Whether you are aware of it or not, you use classification systems all the time. You might, for example, classify people as teachers, students, or auto mechanics if you are speaking about something that involves the work they do.

This simple way of naming things is similar to scientific classification in two ways. First, it is based on a logical method of naming things. Second, all classification systems group objects into categories based on some feature they have in common. For example, the objects might look alike, function in similar ways, or be related to one another—if they are living things. This combination of

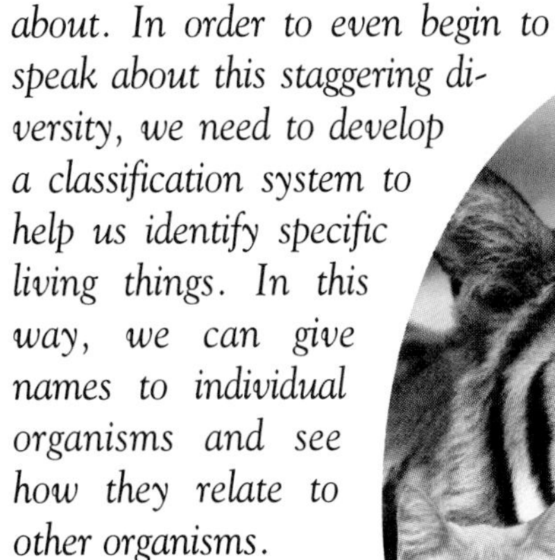

Figure 17–9
These three organisms are all members of the cat family because they have some common characteristics. Yet they differ in several ways. For this reason, (a) the common house cat is given the scientific name Felis domesticus, *(b) the tiger is called* Panthera tigris, *and (c) the black leopard is called* Panthera pardus.

widely understood names and groupings allows us to communicate information about objects easily and efficiently among ourselves.

Biological Classification

Obviously, any useful system of classification requires universal rules. In biology, those rules begin with the name assigned to each species. First, scientists assign a single universally accepted name to each species. Then, they group species into larger categories that have biological significance. The science of naming organisms and assigning them to these groups is called **taxonomy.**

INTEGRATING CAREERS

What kinds of work do you think botanists do?

The classification system used today was established by the Swedish botanist Carolus Linnaeus in the eighteenth century. ● Linnaeus gave each organism a two-part name in Latin, the language of science at the time. The house cat, for example, is called *Felis domesticus*. The first part of the name, *Felis*, refers to the **genus** (plural: genera) to which this cat belongs. A genus refers to the small group of organisms that are quite similar to one another, though different in certain respects.

The second part of the cat's name, *domesticus*, refers only to one particular species. Often, this part of the name is a Latin description of some important characteristic of the organism. *Felis domesticus* literally means domesticated cat. In this textbook, as in other scientific books, the genus and species name is printed in italics. In addition, the genus name is capitalized.

☑ **Checkpoint** What is taxonomy?

Species and Genus

The smallest group in the biological classification system is the **species,** which is a population of organisms that share similar characteristics and that interbreed in nature. If two species share many features but are different biological units, they are classified as different species within the same genus. For example, although the bobcat is a catlike animal, it is placed in a different species from the house cat and is called *Felis rufus*.

☑ **Checkpoint** What is a species?

Other Groupings

The genus *Felis* and three other genera of cats are grouped into larger units

Figure 17–10
This illustration shows the classification of a house cat. Notice that as you go from species to kingdom, the house cat has less in common with members of each group.

called **families.** All genera of catlike animals belong to the family Felidae.

Several families of similar organisms make up the next larger group—an **order.** Members of the cat, dog, bear, and raccoon families are placed in the order Carnivora.

Orders are grouped into **classes.** All members of the order Carnivora are placed in the class Mammalia, along with other animals that are warmblooded, have body hair, and produce milk for their young. Several classes are placed in a **phylum** (FIGH-luhm), which has a large number of very different organisms. These organisms share some important characteristics. For example, mammals are grouped with the birds, reptiles, amphibians, and several classes of fishes into the phylum Chordata. Lastly, all phyla belong to one of six large groups called **kingdoms.** Animals make up the kingdom Animalia.

☑ **Checkpoint** To which class do humans belong?

A Taxing Situation

PROBLEM *What rules do you use to **classify** common objects?*

PROCEDURE

1. As a group, think of a characteristic that will divide the following materials into two groups: plain straight pins, straight pins with colored plastic tops, safety pins of different sizes, buttons, a zipper, steel nails of various sizes, steel screws of various sizes, brass nails of various sizes, brass screws of various sizes, galvanized nails of various sizes, and galvanized screws of various sizes.
2. Divide each of the two groups into two smaller groups.
3. Further divide each smaller group into subgroups.
4. Continue this process until there is only one object in each group.

ANALYZE AND CONCLUDE

1. What characteristics do all these objects have in common? In what ways are they different?
2. How many groups does your classification system have?
3. On what basis did you choose to place each object?

Figure 17–11
Except for viruses, all organisms can be classified into one of the six kingdoms of life.

MAJOR CHARACTERISTICS OF THE SIX KINGDOMS

Kingdom	Archaebacteria	Eubacteria	Protista	Fungi	Plantae	Animalia
Cell Type	Prokaryotes	Prokaryotes	Eukaryotes	Eukaryotes	Eukaryotes	Eukaryotes
Cell Structures	Have cell walls that lack peptidoglycan	Have cell walls made up of peptidoglycan	Have a nucleus, mitochondria, some have chloroplasts	Have a nucleus, mitochondria, but no chloroplasts/ cell wall of chitin	Have a nucleus, mitochondria, chloroplasts/cell wall of cellulose	Have a nucleus, mitochondria, but no chloroplasts; no cell wall
Body Form	Unicellular	Unicellular	Mostly unicellular, some multicellular	Some unicellular, most multicellular	Multicellular	Multicellular
Nutrition	Autotrophic or heterotrophic	Autotrophic or heterotrophic	Autotrophic or heterotrophic	Heterotrophic (absorption)	Autotrophic	Heterotrophic
Examples	Methanogens, halophiles	Rhizobium	Amebae, paramecia	Yeasts, molds, mushrooms	Mosses, ferns, flowering plants, seaweeds	Sponges, worms, snails, insects, mammals

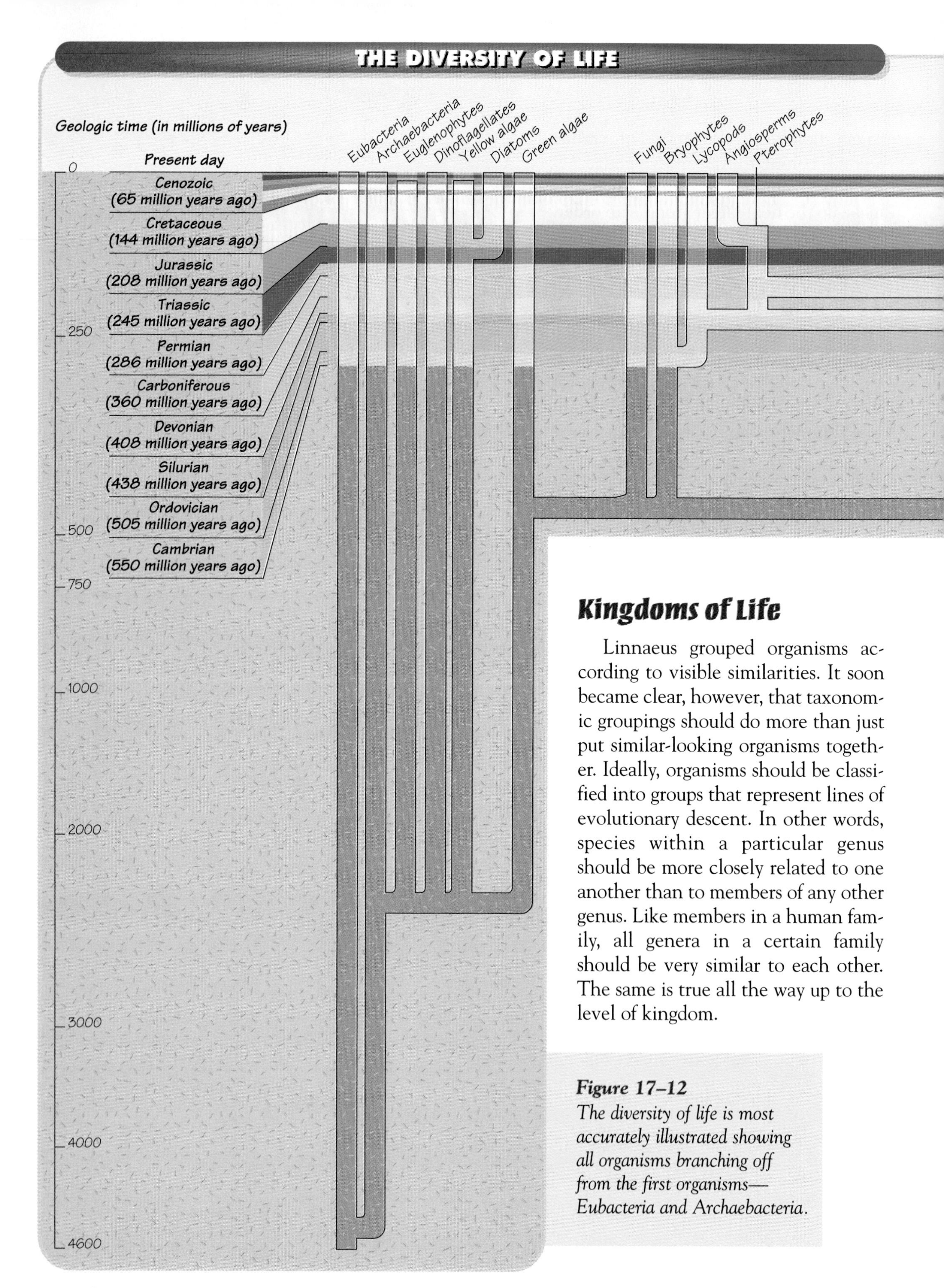

Kingdoms of Life

Linnaeus grouped organisms according to visible similarities. It soon became clear, however, that taxonomic groupings should do more than just put similar-looking organisms together. Ideally, organisms should be classified into groups that represent lines of evolutionary descent. In other words, species within a particular genus should be more closely related to one another than to members of any other genus. Like members in a human family, all genera in a certain family should be very similar to each other. The same is true all the way up to the level of kingdom.

Figure 17–12
The diversity of life is most accurately illustrated showing all organisms branching off from the first organisms—Eubacteria and Archaebacteria.

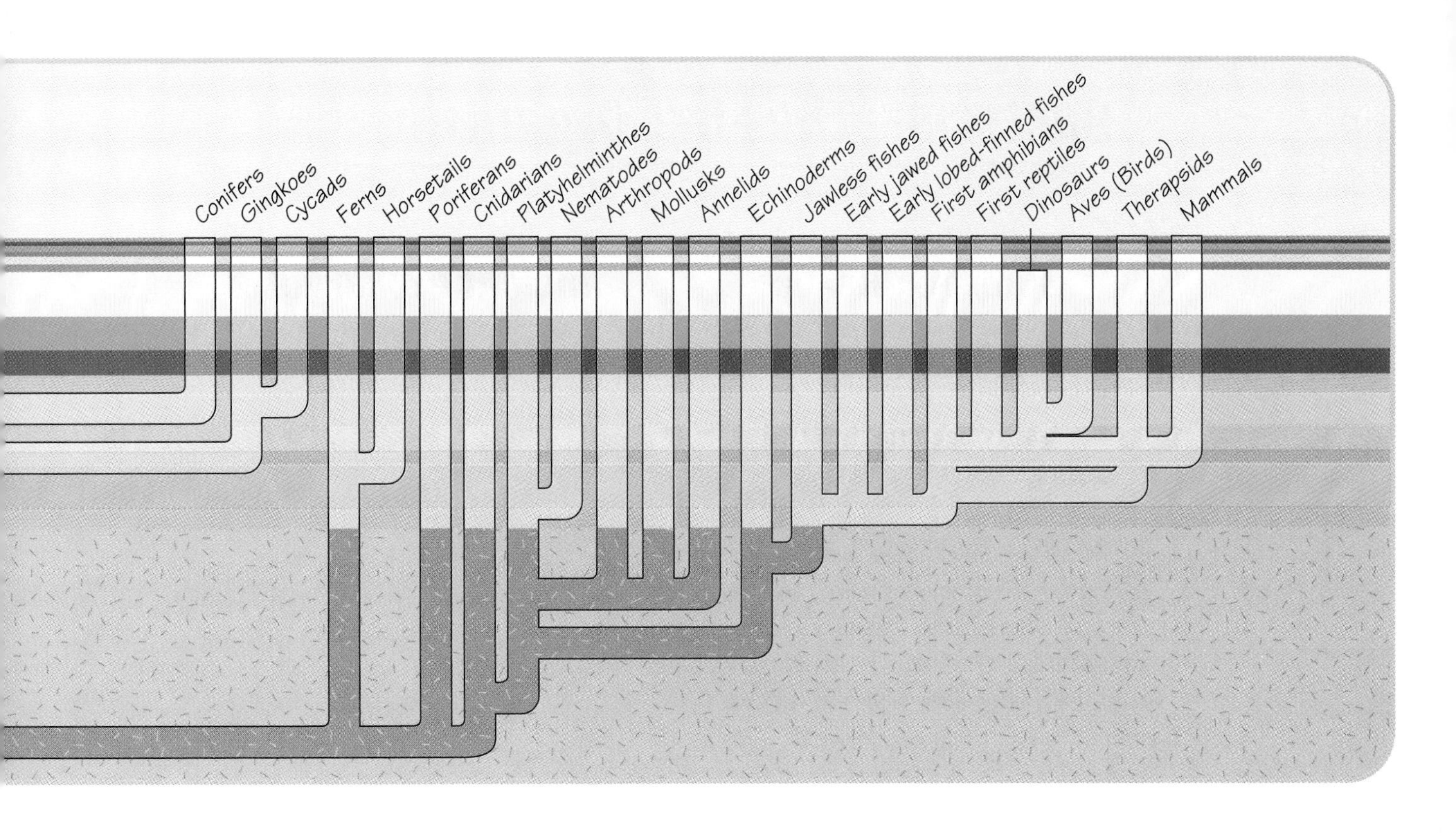

The Six Kingdoms

For decades, biologists recognized five kingdoms of life—Monera, Protista, Fungi, Plantae, and Animalia. In recent years, however, molecular techniques such as reading DNA and sequencing proteins were applied to evolutionary history and relationships. Biologists now realize that the two major groups of bacteria are so different that they deserve to be placed in two separate kingdoms—Eubacteria (yoo-bak-TEER-ee-uh) and Archaebacteria (ahr-kee-bak-TEER-ee-uh). **Today, most researchers agree that there should be at least six kingdoms—Eubacteria, Archaebacteria, Protista, Fungi, Plantae, and Animalia.** This textbook will use a six-kingdom classification system.

Diagramming Life's Diversity

What is the best way to represent life's diversity in pictures? A good way to start is to draw the evolutionary relationships among organisms to the best of our knowledge. This kind of diagram, shown in ***Figure 17–12,*** shows both that living organisms had common ancestors and that other descendants of those organisms are still living and evolving today.

Section Review 17–2

1. **Explain** the importance of classification systems.
2. Name the six kingdoms into which living things are **classified.**
3. **Critical Thinking—Drawing Conclusions** Why is taxonomy a continuing science?
4. **MINI LAB** What rules do you use to **classify** common objects?

SECTION
17–3 Bacteria—The First Organisms

GUIDE FOR READING

- **Describe** the structure of bacteria.
- **Classify** bacteria into two kingdoms.

AS AMAZING AS IT MAY SOUND, life on Earth has existed nearly as long as the Earth itself. Fossils of single-celled organisms that look surprisingly similar to modern organisms have been found in rocks that are more than 3.5 billion years old. This is remarkable, because it is believed that until at least 4 billion years ago, volcanic eruptions and collisions with comets and asteroids made any form of life impossible. How then did life begin? What characteristics did the first life forms have? Although several hypotheses have tried to explain how life may have arisen, we may never know the answers.

Ancient Prokaryotes

Nearly 3.5 billion years ago, advanced single-celled organisms were common on the Earth. Microscopic fossils—or microfossils—of that age looked remarkably similar to certain living photosynthetic prokaryotes. Prokaryotes are unicellular organisms that do not have a nucleus. Scientists hypothesize that the first life forms must have evolved without oxygen because Earth's first atmosphere didn't contain that highly reactive gas.

Over time, some cells evolved that had the ability to use the sun's energy to produce food. These photosynthetic prokaryotes were common in the shallow seas of the Precambrian Period. By 2.2 to 2 billion years ago, these single-celled organisms were producing vast quantities of oxygen.

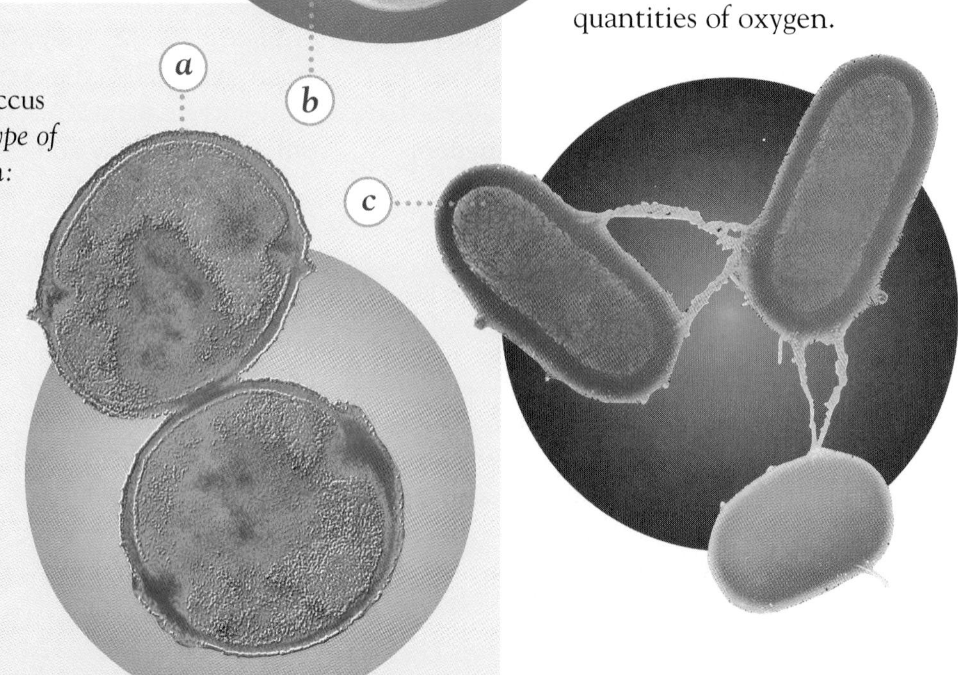

Figure 17–13
(a) One type of bacteria, Streptococcus pneumoniae is responsible for one type of pneumonia in humans (magnification: 30,000X). (b) Cyanobacteria are photosynthetic bacteria that some scientists believe may have been the predecessor of chloroplasts (magnification: 600X). (c) Bacteria reproduce by either binary fission or conjugation. In this photograph, the bacteria E. coli, which live in the intestines of humans, are undergoing conjugation (magnification: 11,250X).

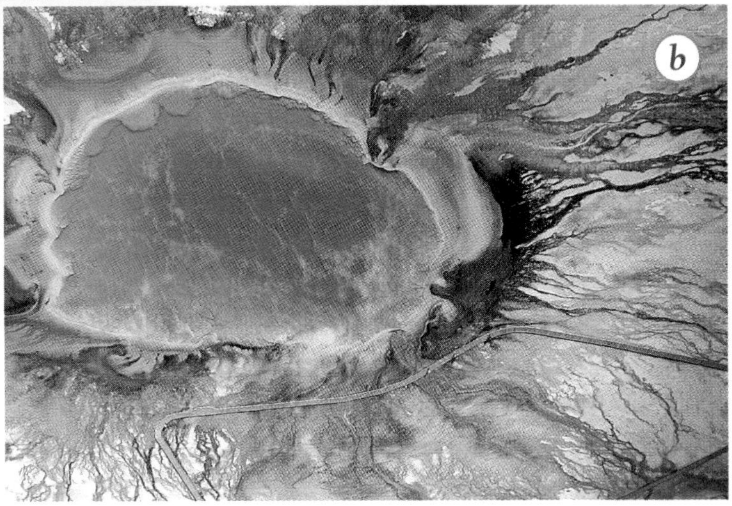

Figure 17–14
Bacteria can live in a variety of environments. (a) Some bacteria prefer cold environments and live on icebergs. (b) Other bacteria prefer the hot temperatures of these sulfur pools in Yellowstone National Park in Wyoming.

The oxygen gas started to accumulate in the atmosphere. As a result, the oxygen concentration in the atmosphere rose. Over several hundred million years, the oxygen concentration in the atmosphere continued to rise until it reached its present-day level.

This event brought about a worldwide crisis for organisms that had evolved in the absence of oxygen. The high level of free oxygen in the atmosphere drove some early life forms to extinction. Others were forced into deep, airless hiding places. And still others evolved new ways to use the oxygen to generate useful energy for the cell. The stage was set for the evolution of modern life.

☑ **Checkpoint** How old are the earliest fossils?

Living Prokaryotes

If you look at the geologic time scale, you will see that prokaryotes were the only form of life for nearly half of Earth's history. So it shouldn't surprise you to learn that their descendants—modern **bacteria**—have evolved incredibly diverse lifestyles.

You may not think about them much, but bacteria are everywhere. They live on the tops of mountains and the bottom of the ocean floor, on the surfaces of melting glaciers, and in very hot water in volcanic hot springs. As a group, bacteria are incredibly adaptable. Many are nearly indestructible. And they play a greater role in your daily life than you probably imagine. The number of bacteria living in just one part of your body—your intestines—at this very moment is greater than the total number of human beings who have ever lived on the Earth!

What Are Bacteria?

All bacteria are prokaryotes. **Bacteria have a cell membrane that is surrounded by a tough, protective cell wall. Their genetic material is contained on a single strand of circular DNA that is not surrounded by a nuclear envelope.** Bacteria also lack mitochondria, chloroplasts, and other cellular structures enclosed in membranes.

The classification and evolutionary relationships of bacteria have been completely changed in recent years. Why has this occurred? One reason is that researchers have examined the similarities and differences in ribosomal RNA, which serves as a "slow" molecular clock for timing evolutionary events. The data they gathered divide the major groups of living bacteria into two kingdoms. **The two kingdoms of bacteria are Eubacteria and Archaebacteria.** Members of these kingdoms evolved from a single, extinct common ancestor.

Checkpoint Are bacteria prokaryotes or eukaryotes?

Eubacteria

Members of the kingdom Eubacteria, or "true" bacteria, have thick, rigid cell walls made up of the carbohydrate peptidoglycan (pehp-tih-doh-GLIGH-kan). The cell wall surrounds a typical cell membrane. Eubacteria are extremely diverse in their ecology and metabolism. For example, some species of Eubacteria live in soil, while others infect larger organisms and produce disease. And still other species of Eubacteria are photosynthetic and make their own food using light energy.

INTEGRATING HEALTH

What are the different kinds of pneumonia? How are they treated?

Archaebacteria

Members of the kingdom Archaebacteria, or "ancient" bacteria, live in the most extreme environments imaginable—volcanic hot springs, brine pools, and black organic mud that lacks oxygen. Many can survive only in the absence of oxygen. Their cell wall lacks peptidoglycan, and their cell membrane contains certain lipids that are not found in any other organism.

How Do Bacteria Reproduce?

Bacteria have fairly simple reproductive lives. Most of the time, bacteria reproduce by **binary fission**—the division of the cell into two daughter cells, along with the duplication of their single chromosome. There are, however, some exceptions to this rule.

Certain bacteria, such as the common intestinal bacteria *Escherichia coli* (*E. coli*), sometimes undergo a process called **conjugation,** which is a simple form of sexual reproduction. During conjugation, part of the genetic information from one cell is transferred to another cell.

Certain other bacteria, such as *Streptococcus pneumoniae*, reproduce only by binary fission but have a habit of picking up bits and pieces of DNA from other bacteria. This process, which is called **transformation,** offers an unusual opportunity for a normally asexual organism to pick up new traits. Because *S. pneumoniae* can cause pneumonia in humans, particularly infants and toddlers, this ability can have serious consequences for human health. ●

Section Review 17-3

1. **Describe** the structure of bacteria.
2. What are the two kingdoms into which bacteria are **classified**?
3. **Critical Thinking—Hypothesizing** How have bacteria survived so successfully?

Viruses

Guide for Reading

- **Describe** the structure of a virus.
- **Explain** how a virus reproduces.

THEY ARE INCREDIBLY SMALL and impossibly simple. They are not alive because they are not cells and are not composed of cells. Medical science has yet to devise foolproof ways to stop them once they have infected a cell.

No one knows where they came from. Because they cannot survive without invading living host cells, they cannot have appeared before the first true life forms. Yet because they infect virtually every known organism, including bacteria, it is likely that they evolved fairly soon after the appearance of the first prokaryotes.

What Is a Virus?

What are these mysterious objects? They are viruses. **A virus is a tiny particle—made up of genetic material and protein—that can invade and replicate within a living host cell.**

Viruses are so fundamentally different from living things that they are not even called organisms. In a sense, you can think of viruses as pure information, similar to a computer program stored on a disk with no hardware of its own. By itself, a virus can do absolutely nothing. It doesn't respire, eat, excrete, or reproduce. It simply holds onto its information and waits to infect a suitable host.

Each virus particle contains a core of either DNA or RNA, surrounded by a protein coat. The outer surface of the virus may also be covered with an envelope composed of lipids and proteins. Often, these compounds help the virus bind to the host cell membrane and "trick" the host cell into taking the virus inside, where it can do its harm.

Checkpoint What is a virus?

Figure 17–15
Viruses come in all shapes and sizes, and they cause a variety of diseases in both animals and plants. (a) *The influenza virus causes the flu (magnification 100,000X).* (b) *The stripes on these Parrot tulips are caused by a virus. If the virus were not present, the tulips would be a solid color.*

Biology AND YOU .... Connections

Viral Evolution and the Flu

Most viruses are similar to organisms in that they constantly evolve in response to changes in their environment. The environment for viruses that infect humans includes our bodies, our domestic animals, and the medicines we take.

Some viruses, such as those that cause measles and smallpox, do not evolve quickly. Because of this slow evolution, vaccines were developed to help eliminate the viruses. A vaccine is an injection into the body of a weakened or mild form of the virus. Vaccines give the body the ability to recognize and attack an invading virus more quickly. Because viruses of this type don't change much, they always look the same to the body's immune system, so that a single vaccination can protect against infection for years.

Influenza viruses (magnification: 37,700X)

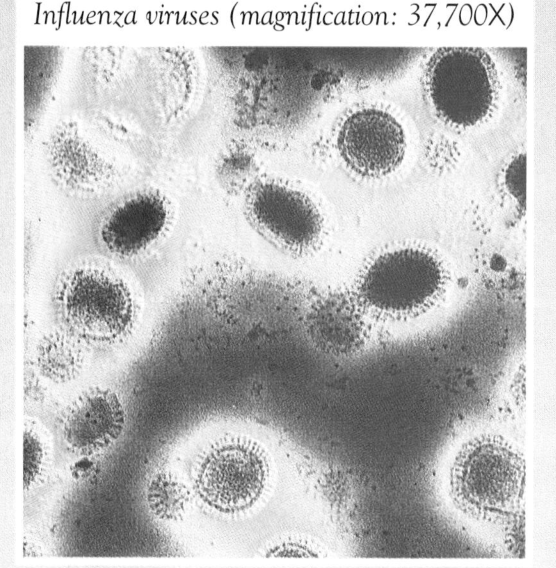

Flu Virus Adaptations

But some viruses—like those that cause influenza, the flu—have two survival adaptations that keep them in circulation. First, flu viruses mutate rapidly, constantly changing their appearance. So the strains of flu viruses that are present this year have evolved from last year, and they no longer look the same to the body. As a result, the body does not recognize the new viruses and so does not attack them, enabling them to infect the body.

More Than One Host

The second adaptation is that certain viruses have the ability to live in two or more different animal hosts. One strain of the virus, for example, may live in ducks and pigs, while another strain of the same virus may live in pigs and humans. If both these two strains infect the same cells in the same pig at the same time, it can spell trouble. Can you imagine why? Flu viruses happen to carry their genetic information on eight pieces of DNA. If two strains of the virus infect the same cell, some of those genes can get mixed up. If that happens, the result may be a new strain of the virus that can infect humans. This new strain carries proteins on its surface from the duck virus, which humans have never encountered before. And these new strains of viruses often cause the most serious diseases in the most people.

Making the Connection

Why are new flu vaccines developed every year? Why do you think flu vaccinations are recommended for health-care workers, the elderly, people with circulatory and respiratory problems, and people with immune system disorders?

Figure 17–16
The white blood cell in this electron micrograph is infected with the human immunodeficiency virus (HIV), the virus responsible for causing AIDS. The virus (shown here as green spheres) will continue to multiply and will eventually destroy the cell (magnification: 16,000X).

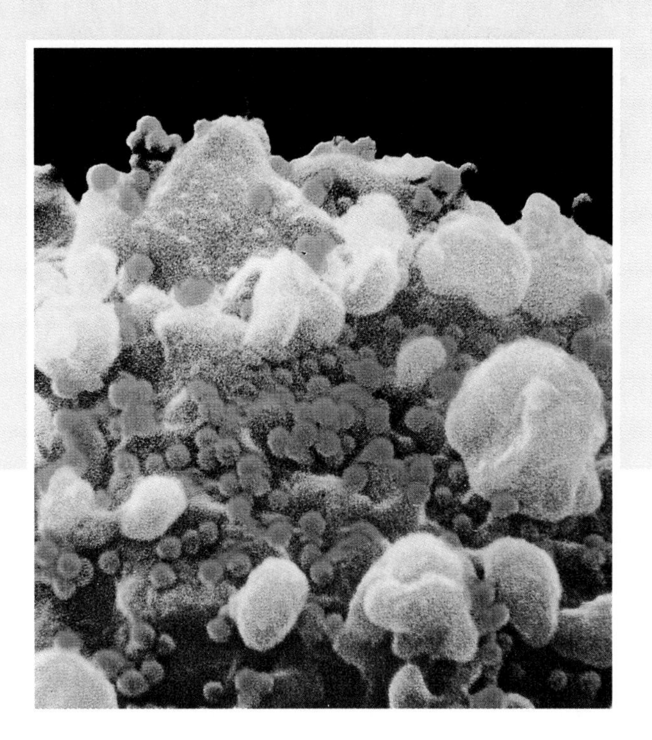

Types of Viruses

Just as living cells can have a variety of shapes and forms, so too can viruses. According to the nature of their genetic material and the way they infect cells, viruses can be divided into three groups—DNA viruses, RNA viruses, and retroviruses. Viruses infect virtually every known organism—from bacteria to plants to humans.

How Do Viruses Reproduce?

Because viruses are not alive, they can reproduce only when they invade, or infect, a living host cell. Viruses have three basic types of life cycles that occur within host cells.

In the lytic cycle, the virus injects its genetic material into a host and immediately takes over the cell's metabolic machinery. The host's DNA is used to make viral DNA, and the host cell manufactures new viral genes and proteins until it bursts, releasing many new viruses.

In the lysogenic cycle, viral genes do not go into action immediately. Instead, the viral DNA attaches to the circular DNA of a bacterium host. They may remain there for quite some time. When this occurs, viral genes are copied each time the infected cell divides.

The third type, retroviruses—which contain RNA—first make DNA copies of their RNA genes and then insert them into the host cell's chromosomes. There they may remain dormant for varying lengths of time before becoming active, directing the production of new viruses and causing the death of the host cell. Retroviruses were discovered only about 20 years ago. One example of a retrovirus is the Human Immunodeficiency Virus (HIV), which causes AIDS.

Section Review 17–4

1. **Describe** the structure of a virus.
2. **Explain** how a virus reproduces.
3. **Critical Thinking—Making Judgments** Why are viruses not classified into any of the kingdoms of life?

How Did Life Begin?

GUIDE FOR READING

- **Describe** the chemical conditions on the early Earth.

ONLY 200 TO 300 MILLION YEARS after the Earth had cooled enough to carry water in its liquid form, prokaryotes similar to modern bacteria were growing everywhere. For a long time, scientists believed that this span of time was much too short for something as complicated as a living cell to have evolved from random molecules. But in recent years, new data have shown that this process—though still astonishing—is much more probable than anyone had thought.

Building Blocks

Living cells are composed of four major classes of complex organic molecules: lipids, carbohydrates, proteins, and nucleic acids. Atoms do not put themselves together into these complex organic molecules on modern Earth. There are many reasons why, but two are easy to understand. First, the presence of oxygen in the atmosphere makes it unlikely that any organic molecules could accumulate without being broken down, or oxidized. Second, no sooner would organic molecules begin to accumulate than something—bacteria or some other form of life—would probably come along and gobble them up!

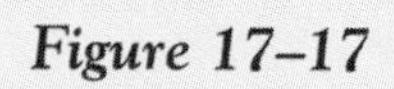

Figure 17–17
The conditions that were present on Earth when living organisms first appeared were very different from the conditions on Earth today. (a) Volcanic eruptions similar to the eruption of Pu'u'O'O in Hawaii and (b) electrical storms were common occurrences—and may have provided the energy that sparked life-generating chemical reactions on Earth.

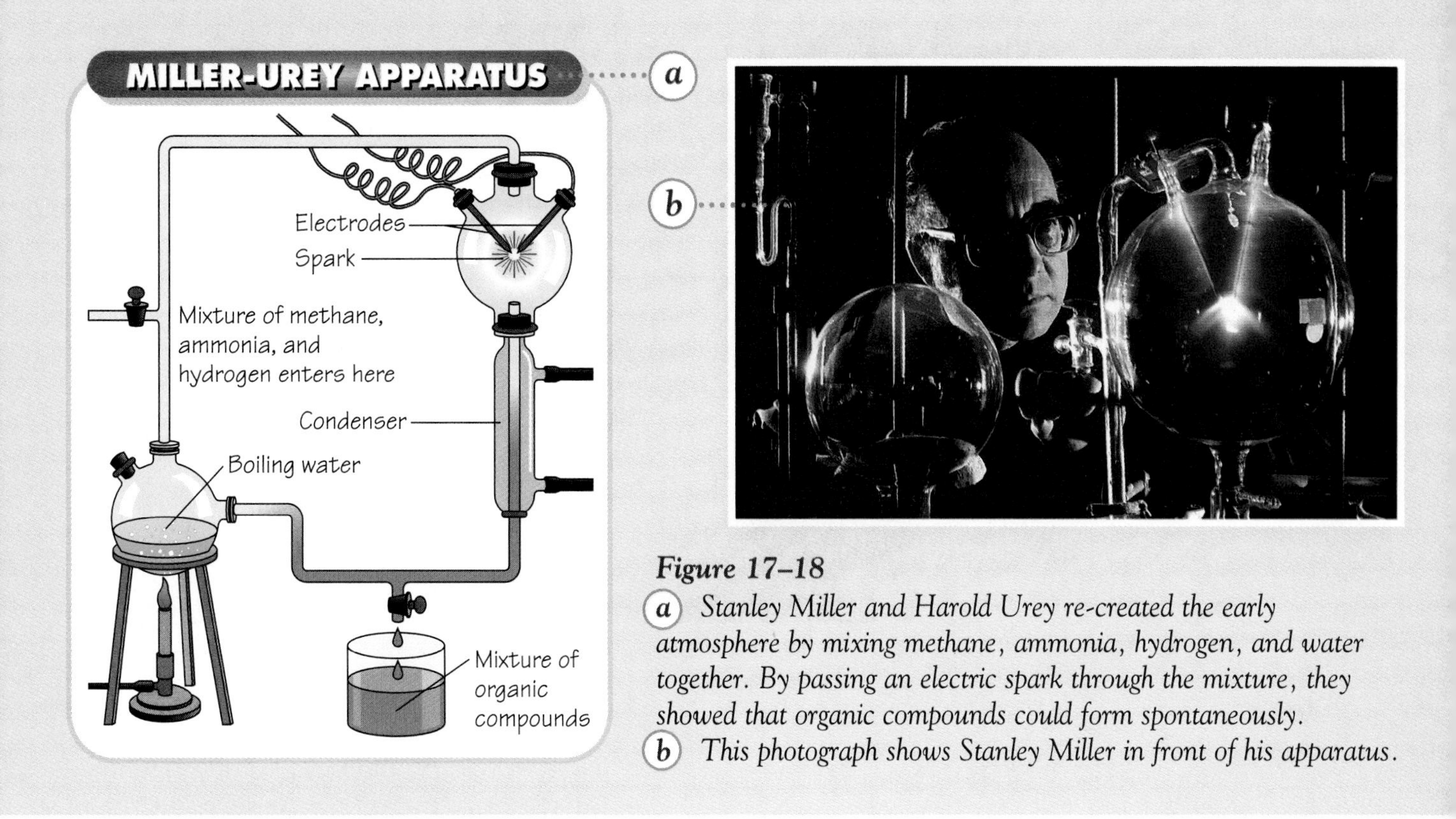

Figure 17–18
(a) *Stanley Miller and Harold Urey re-created the early atmosphere by mixing methane, ammonia, hydrogen, and water together. By passing an electric spark through the mixture, they showed that organic compounds could form spontaneously.*
(b) *This photograph shows Stanley Miller in front of his apparatus.*

On Early Earth

As you have read, the early Earth was a very different place from today's Earth. **Scientists hypothesize that Earth's early atmosphere was probably composed of hydrogen cyanide, carbon dioxide, carbon monoxide, nitrogen, hydrogen sulfide, and water vapor.** Could something have happened unexpectedly to cause organic molecules to form and accumulate in ways that they never could today?

In the early 1950s, American scientists Stanley Miller and Harold Urey tried to re-create the conditions that occurred on early Earth in a laboratory to see what might have occurred. They filled a flask with hydrogen, methane, ammonia, and water. They carefully kept oxygen out of the mixture and sterilized the equipment to make certain that no microorganisms could contaminate the results. Then they passed repeated electric sparks through the mixture to simulate the lightning storms that were probably a constant occurrence on the early Earth.

The First Organic Molecules

The results of this experiment were spectacular and exceeded Miller and Urey's wildest dreams. Over just a few days, their apparatus accumulated significant quantities of several amino acids—the building blocks of proteins that are produced only by living organisms. Since that time, similar experiments have produced a wide variety of surprisingly complex compounds. Not only does this list include sugars, but also the nitrogen bases—purines and pyrimidines that form the backbones of DNA and RNA!

Some years later, a sizable meteorite that crashed in Murchison, Australia, was carefully split open and analyzed. Among the compounds discovered were the same amino acids that had been produced in Miller and Urey's apparatus! Because these compounds were almost certainly not produced by living organisms, they supported Miller and Urey's experimental results.

Checkpoint What were the results of Miller and Urey's experiment?

INTEGRATING ASTRONOMY

What is a meteorite?

Figure 17–19
Proteinoid microspheres, shown in this scanning electron micrograph (magnification: 3000X), are hollow structures that have some of the characteristics of living systems. Under the right conditions, proteinoid microspheres can absorb material, grow, and even divide.

The Greatest Leap

Of course, a bubbling stew of organic molecules is a long way from a living cell. The leap from this point to the simplest forms of life we know is still the greatest gap in hypotheses about life's origins.

Studies have shown that large organic molecules can, under some conditions, organize themselves into tiny bubbles called proteinoid microspheres. Proteinoid microspheres aren't cells, but they do have some characteristics of living systems. They are surrounded by a two-layered membrane made of lipids, and they can grow in size by taking in more organic matter. Several hypotheses suggest different ways that these structures might have slowly acquired the characteristics of living cells.

The largest stumbling block in bridging the gap between nonliving and living still remains. All living cells are controlled by information stored in DNA, which is transcribed into RNA and then made into protein. This is a very complicated system, and each of these three molecules requires the other two—either to put it together or to help it work. DNA, for example, carries information but cannot put that information to use or even copy itself without the help of RNA and protein.

☑ **Checkpoint** What are proteinoid microspheres?

A Role for RNA?

How could this multi-step biochemical machinery ever have gotten started in the first place? The answer may have been found in the way that many scientific discoveries happen—by researchers who weren't even concerned with this question! Over the last 20 years, molecular biologists realized that they had not completely understood all the things RNA can do.

In protein synthesis, for example, one type of RNA—called messenger RNA—is usually described as a cross between a cellular copying machine and a messenger. Able to do nothing on its own, it simply copies the genetic information from DNA and brings it to the ribosomes—the cell structures in which proteins are made. What's more, in modern cells both messenger RNA and another type of RNA—transfer RNA—are produced only on instructions from DNA.

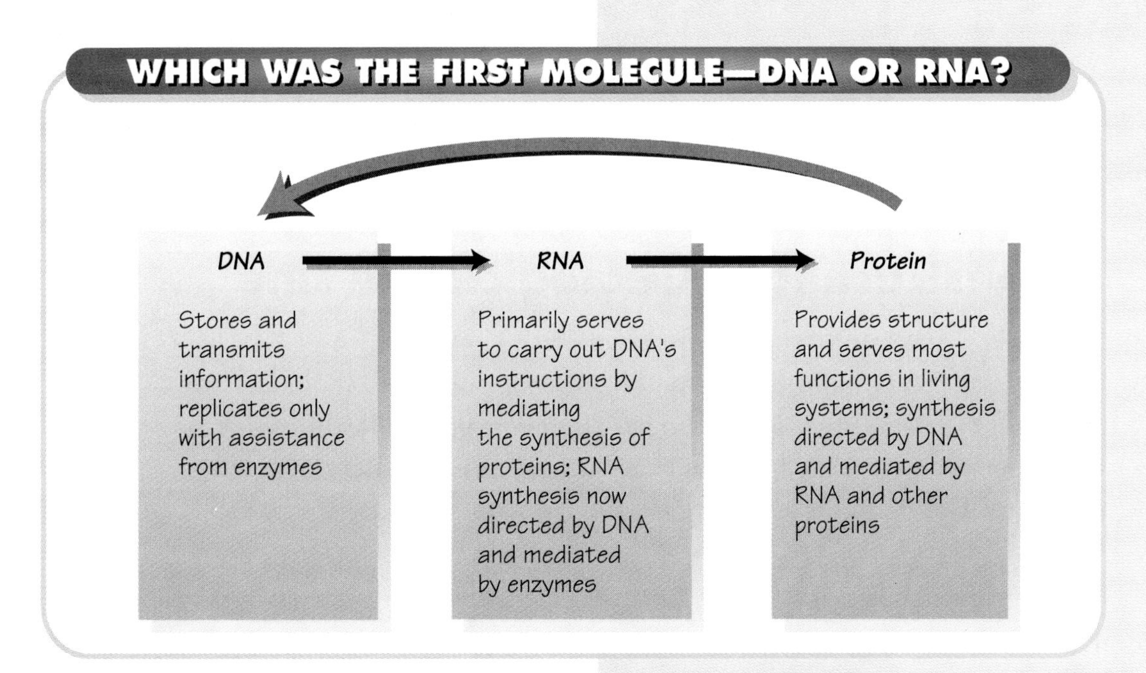

Figure 17–20
Some scientists hypothesize that RNA evolved first, followed by DNA and proteins.

More Than a Messenger

Yet under the right circumstances, RNA sequences can do some surprising things. Some sequences help DNA copy itself. Others process messenger RNA after it copies the information for DNA. Still others catalyze chemical reactions.

Recent evidence suggests that RNA molecules can grow and duplicate themselves entirely on their own! In a new series of experiments that duplicated probable conditions found on the early Earth, several scientists showed that small sequences of RNA could have formed on their own. Given enough time, the scientists think that these short RNA chains could have grown longer and may ultimately have taken on a life of their own.

As a result of these experiments, some scientsts now hypothesize that RNA, rather than DNA, may have been the first information-storage molecule. Such a molecule could also direct the formation of the first proteins.

At this point, several steps could have led to the modern system of DNA-directed protein synthesis. It may be possible—as many researchers now believe—that the modern DNA world evolved from an earlier RNA world. Perhaps future experiments will gather more data to confirm or refute this fascinating hypothesis.

Section Review 17–5

1. **Describe** the chemical conditions on the early Earth.
2. **Compare** proteinoid microspheres and living cells.
3. **BRANCHING OUT ACTIVITY** Look up the words biogenesis and abiogenesis. **Compare** the meanings of the two words.

Laboratory Investigation

Constructing a Geologic Time Line

In order to develop a true perspective of the vast spans of Earth's history during which different life forms existed, it helps to construct a time line. In this investigation, you will construct a time line representing the Earth's history and identify the main events of the beginnings of life on Earth.

Problem

How can you **sequence** some of the main events in the history of life on Earth?

Materials (per group)

meterstick
5 m adding machine paper
colored markers
photographs and artwork of different life forms
index cards
transparent tape

Procedure

1. **For your geologic time line, use a scale in which 1 mm = 1 million years, or 1 m = 1 billion years.**
2. **Mark the adding machine paper strip at appropriate intervals, such as every 25 million years, from the beginning of the Earth to the present.**

3. **Using the table provided, identify and label the major geologic eras, periods, and epochs on the geologic time scale according to their relative lengths.**
4. **Decorate the geologic time scale using the paper strip and the art and photographs of the life forms. Write brief descriptions of the life forms on the index cards. Art, photographs, and descriptions should be affixed to the times during which the first fossils of a particular group were formed.**

Observations

1. How long did it take for the first life forms to leave fossils?
2. In which period did the first multicellular organisms begin to leave fossils?
3. When in Earth's history did photosynthetic organisms first appear?
4. During which periods did the dinosaurs rule the Earth?

Analysis and Conclusions

1. Where on the time line do most life forms exist?
2. What percentage of your time line accounts for the Precambrian Era?
3. Assume that humans have been around for about 250,000 years. What percentage of your time line accounts for the history of modern humans?

MAJOR EVENTS IN THE HISTORY OF LIFE

Event	Millions of Years Ago
First apes	23 – 34
First whales	34 – 57
First horses	34 – 57
First monkeys	34 – 57
First flowering plants	65 – 144
First birds	144 – 208
First dinosaurs	208 – 245
First mammals	208 – 245
First reptiles	286 – 360
First insects	360 – 408
First land plants	408 – 438
First fishes	438 – 505

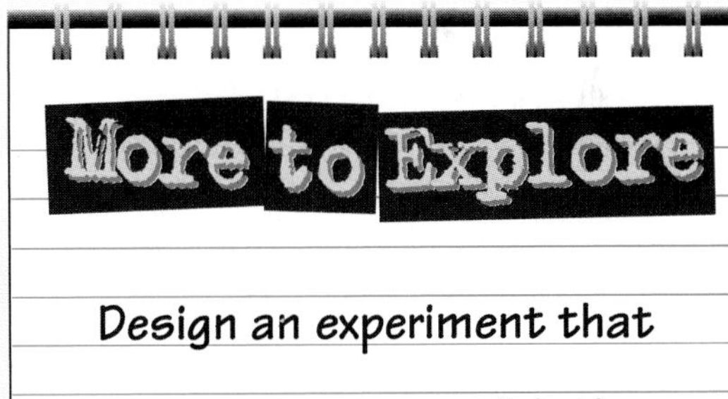

Design an experiment that equates one year with the mass of one penny. What would be the mass, in grams, of all the pennies that represent the age of the Earth? In kilograms?

Study Guide

Summarizing Key Concepts

The key concepts in each section of this chapter are listed below to help you review the chapter content. Make sure you understand each concept and its relationship to other concepts and to the theme of this chapter.

17–1 The Changing Earth

- To learn about life's past and to fully understand its present, researchers read the history preserved in the fossil record.
- Fossils are the remains or traces of ancient life, commonly found in sedimentary rocks.
- The age of fossils can be determined by two methods—relative dating and absolute dating.
- To make Earth's long history more manageable, scientists developed the geologic time scale.
- The theory of continental drift provides further evidence of evolution from matching rocks and fossils.

17–2 Finding Order in Diversity

- A classification system identifies objects and gathers them into groups.
- Taxonomy is the science of naming organisms and classifying them into groups.
- Living things are classified into six kingdoms: Eubacteria, Archaebacteria, Protista, Fungi, Plantae, and Animalia.

17–3 Bacteria—The First Organisms

- Bacteria have a cell membrane that is surrounded by a tough, protective cell wall. Their genetic material is contained on a single strand of circular DNA that is not surrounded by a nuclear envelope.
- There are two major kingdoms of bacteria—Eubacteria and Archaebacteria.
- Bacteria reproduce by binary fission, and some by conjugation or transformation.

17–4 Viruses

- A virus is a tiny particle—made up of genetic material and protein—that can invade and replicate within a living host cell.
- Because viruses are not alive, they can reproduce only when they invade, or infect, a living host cell.

17–5 How Did Life Begin?

- Scientists hypothesize that Earth's early atmosphere was composed of hydrogen cyanide, carbon dioxide, carbon monoxide, nitrogen, hydrogen sulfide, and water.

Reviewing Key Terms

Review the following vocabulary terms and their meaning. Then use each term in a complete sentence.

17–1 The Changing Earth

fossil
geologic time scale
sedimentary rock
continental drift
half-life

17–2 Finding Order in Diversity

taxonomy
species
genus
family
order
phylum
class
kingdom

17–3 Bacteria—The First Organisms

bacteria
conjugation
binary fission
transformation

17–4 Viruses

virus

Recalling Main Ideas

Choose the letter of the answer that best completes the statement or answers the question.

1. The Earth's first atmosphere did not contain

a. nitrogen. **c.** oxygen.
b. hydrogen sulfide. **d.** carbon dioxide.

2. Which is the most helpful element for determining the age of a 10,000-year-old fossil?

a. potassium-40 **c.** uranium-238
b. carbon-14 **d.** nitrogen-14

3. The system of taxonomy used today was developed by

a. Miller. **c.** Urey.
b. Linnaeus. **d.** Aristotle.

4. Related phyla are grouped into a

a. family. **c.** species.
b. genus. **d.** kingdom.

5. The first organisms on Earth were thought to be

a. viruses. **c.** prokaryotes.
b. mitochondria. **d.** eukaryotes.

6. All bacteria reproduce by

a. transformation. **c.** conjugation.
b. the lytic cycle. **d.** binary fission.

7. A virus particle contains a core of

a. DNA. **c.** DNA or RNA.
b. RNA. **d.** protein.

8. Viruses can reproduce by

a. the lytic cycle. **c.** conjugation.
b. transformation. **d.** binary fission.

Putting It All Together

Using the information on pages xxx to xxxi, complete the following concept map.

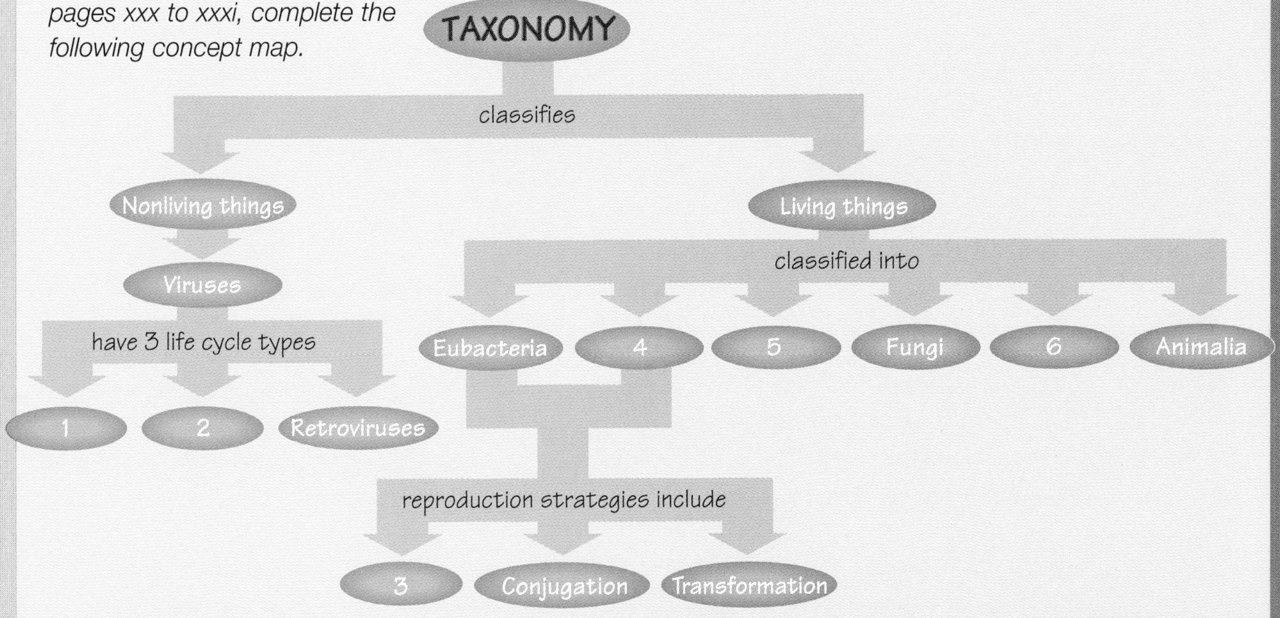

TAXONOMY
classifies
Nonliving things
Viruses
have 3 life cycle types
1
2
Retroviruses
Living things
classified into
Eubacteria
4
5
Fungi
6
Animalia
reproduction strategies include
3
Conjugation
Transformation

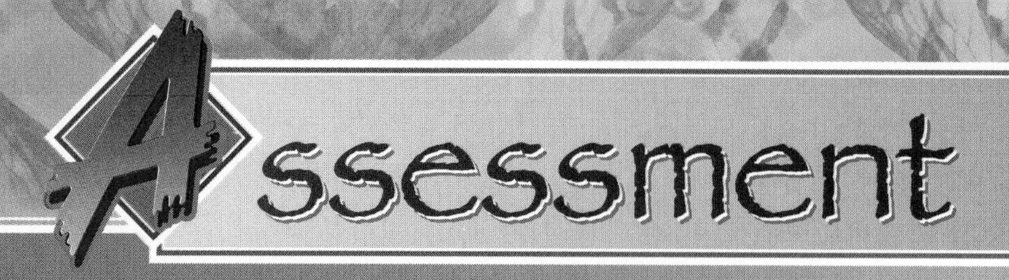

Reviewing What You Learned

Answer each of the following in a complete sentence.

1. Are fossils that are found in the deepest layers of sedimentary rocks older or younger than fossils found closer to the surface?
2. Approximately how old are the oldest known rocks that contain fossils?
3. What is the geologic time scale?
4. During which geological periods did dinosaurs rule the Earth?
5. How did Linnaeus **classify** organisms?
6. What are the six kingdoms of life?
7. Which important gas was missing from the ancient or early atmosphere?
8. What are prokaryotes?
9. What is a retrovirus?
10. What are proteinoid microspheres?

Expanding the Concepts

Discuss each of the following in a brief paragraph.

1. If the half-life of an element is 5 years, **relate** how long it will take until only 1/16 of the original unstable element remains.
2. If you were trying to find out the exact age of a fossil, which type of dating would you use—relative or absolute? Explain your answer.
3. Why are sedimentary rocks so important in studying the history of the Earth?
4. Two groups of organisms are in different genera, but they are included in the same family. What does this tell you about the two groups?
5. In a library, books are classified into different categories. What are two other classification systems used in everyday life?
6. Reproduction in prokaryotes can be asexual or sexual. Give an example of each type.
7. Why are bacteria often called the most successful organisms on the Earth?
8. Why are viruses unable to reproduce on their own?
9. Two of the three types of life cycles found in viruses are lytic and lysogenic. Describe these cycles and indicate which cycle is potentially more dangerous to humankind.
10. How did Miller and Urey's work explain how life might have developed on this Earth?

Extending Your Thinking

Use the skills you have developed in this chapter to answer the following.

1. **Comparing** Develop a chart that compares the similarities and differences of eubacteria and archaebacteria.
2. **Analyzing** Could viruses have been the first life form on Earth? Develop an argument that supports your position.
3. **Hypothesizing** Beginning with the gases of the primitive atmosphere, sequence the steps that might have led to the first living organism. Remember that chemical evolution would have had to precede biological evolution.
4. **Analyzing data** A scientist finds a new organism but is unsure to which kingdom it belongs. The organism's characteristics include: It is unicellular, has a cell wall made up of peptidoglycan, has a circular DNA and ribosomes, but it lacks a nuclear envelope. To which kingdom does the organism probably belong?

Applying Your Skills

Building a Model of a Bacteriophage

Viruses have a variety of shapes and forms. One type of virus is a bacteriophage. A bacteriophage is a virus that infects a bacterium. For this activity, you will build a model of a bacteriophage.

1. Look at the photograph of the bacteriophage.
2. Make a three-dimensional model of a bacteriophage using the following materials: machine screw, 2 acorn nuts that fit the screw, pliers, 3 pieces of floral wire, and a pipe cleaner.
3. Diagram the bacteriophage on a sheet of paper and label its parts.

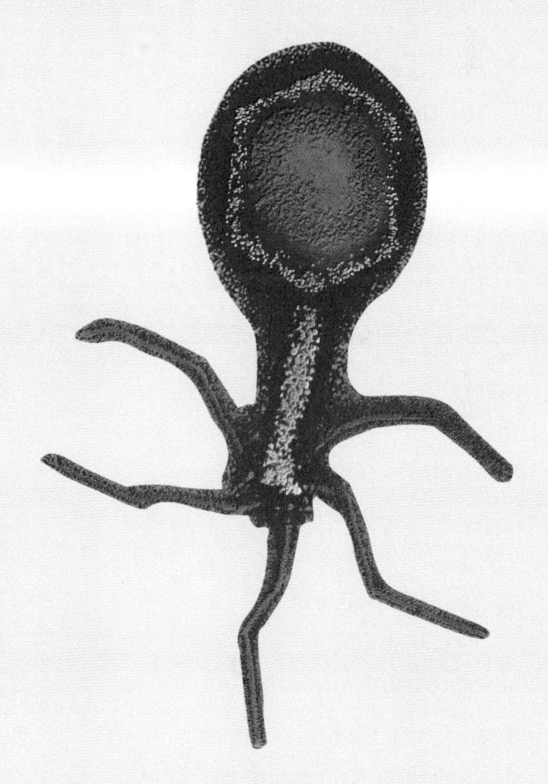

• GOING FURTHER •

4. What part of your model represents the outer protein coat?
5. Where is the RNA or DNA located in your model virus?
6. Compare your model with models of other members of your group and other members of your class. How are the models alike? How are they different?
7. In your journal, write one or two paragraphs describing your model. State how your model is similar to and different from the virus in the photo. How could you improve your model?

CHAPTER 18

Protists, Fungi, and Plants

FOCUSING THE CHAPTER
THEME: Scale and Structure

18–1 Protists
- **Classify** protists and **describe** their major characteristics.

18–2 Fungi
- **Describe** the characteristics of fungi.

18–3 Multicellular Plants
- **Identify** the major characteristics of plants.

BRANCHING OUT *In Depth*

18–4 The Coevolution of Plants and Animals
- **Describe** an example of the coevolution of plants and animals.

LABORATORY INVESTIGATION
- **Predict** the role of vascular tissue in plants.

Biology and Your World

BIO JOURNAL

How have these trees adapted to their surroundings? Answer this question in your journal.

Redwoods, firs, and rhododendrons in California

Protists

Guide for Reading

- **Describe** the characteristics of protists.
- **Compare** algae and protozoans.
- **Explain** how protists reproduce.

SOME SWARM AT THE OCEAN'S surface. Others dig through mud in ponds and rivers. Some act like animals, feeding on smaller organisms. Some act like plants, producing complex organic compounds from sunlight, water, and inorganic nutrients. Others decompose dead organic matter. Many are deadly parasites. As the oldest eukaryotes, they have played vital roles in the history of life for more than a billion years.

What Is a Protist?

These eukaryotes, better known as **protists,** are a fascinating yet puzzling group of organisms. What is a protist? Surprisingly, there is no simple answer to that question because the organisms classified into the Protist kingdom are such a diverse group. **All protists are eukaryotes, but these organisms do not share a set of unique characteristics.** While most protists are unicellular, quite a few are multicellular. Many are microscopic, but the single-celled *Caulerpa* can grow up to a meter in length. And some multicellular species grow 70 meters long! Most protists need oxygen to survive, but a few are poisoned by it. Some protists are autotrophic, and others are heterotrophic. Some reproduce sexually, some asexually, and still others alternate between the two. The differences go on and on!

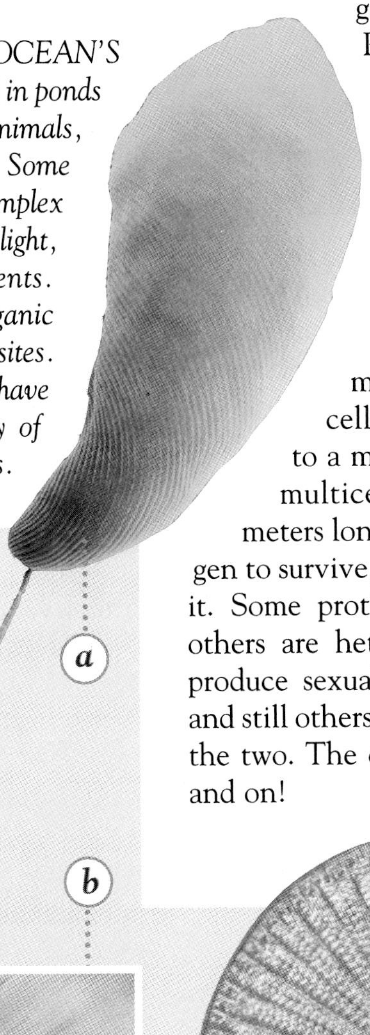

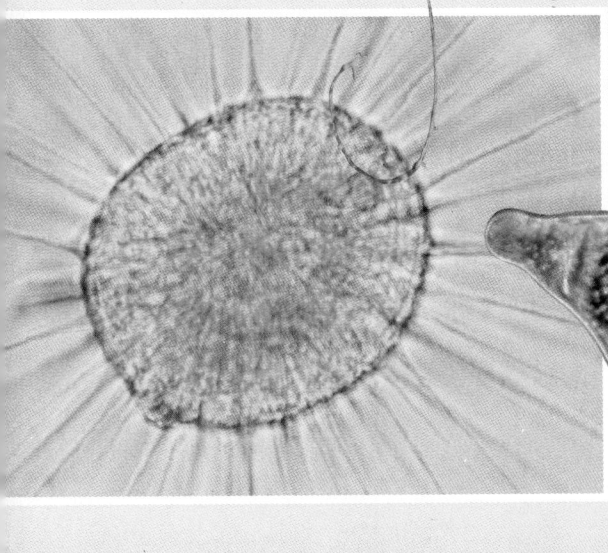

Figure 18–1
As a group, protists do not share any unique characteristics. These photographs of (a) a euglena (magnification: 2000X), (b) a heliozoan (magnification: 31X), and (c) diatoms (magnification: 63X) show just how diverse protists really are.

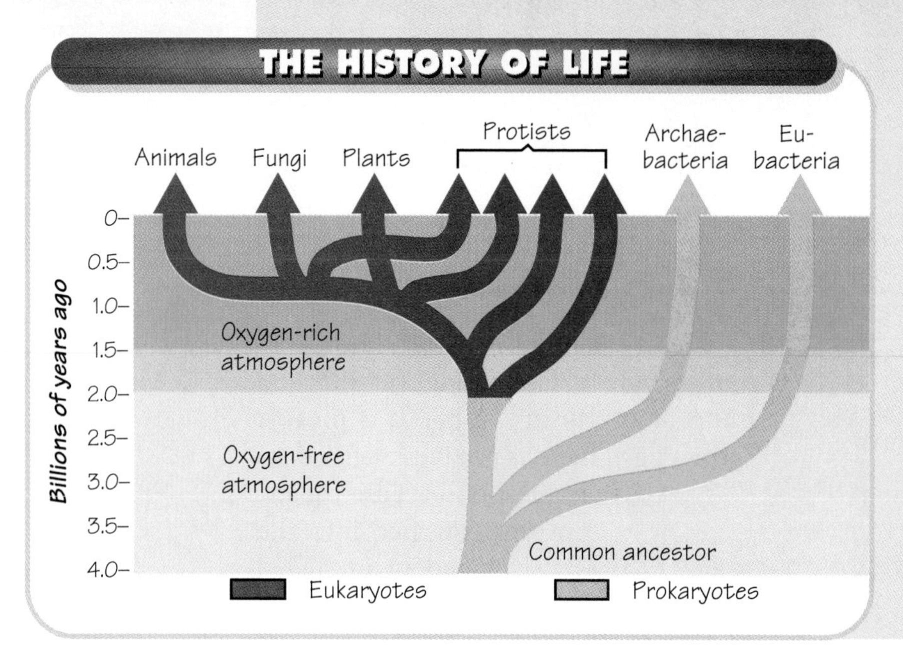

Figure 18–2
This modern evolutionary tree illustrates the history and relationships of the six kingdoms of life. The base of the tree represents the common ancestor of all eukaryotic organisms. Notice that once oxygen became available, organisms quickly evolved.

Evolution of Protists

The best way to approach the study of protists is to follow the evolutionary history of life. You may recall that the first living cells were prokaryotes, which resembled living bacteria. One of those ancient prokaryotes became the ancestor of all eukaryotes.

The First Eukaryotes

Sometime between 2 and 1.8 billion years ago, a great leap occurred in the evolution of life. As ***Figure 18–3*** illustrates, one form of prokaryotic cell grew larger and evolved internal cell membranes. Some of those membranes formed a nuclear envelope that surrounded its DNA. Other membranes became involved in producing and transporting proteins inside the cell. The result was the ancestor of all eukaryotes.

Then, something remarkable happened. One or more prokaryotic organisms entered this ancestral eukaryotic cell. But these invading organisms didn't infect their host, as parasites would have done. And the host didn't digest them, as it would have digested prey. Instead, the smaller prokaryotes began living inside the larger eukaryotic cell. Over time, a remarkable relationship called **symbiosis** (sihm-bigh-OH-sihs) evolved. This particular form of symbiosis, in which both the host and guest benefit, is called mutualism.

The smaller cells had the ability to use oxygen to generate the energy-rich molecule ATP. Over time, the smaller prokaryotes evolved together with their new hosts and became mitochondria—the cellular powerhouses of most protists and all multicellular organisms.

Other groups of prokaryotes moved into several different groups of evolving eukaryotic cells. These prokaryotes had the ability to perform photosynthesis. The prokaryotes within evolving eukaryotic cells further evolved into chloroplasts, and the partnerships gave rise to all photosynthetic eukaryotes, including algae and plants.

☑ ***Checkpoint*** What is symbiosis?

Evolution of Sexual Reproduction

About 300 million years after single-celled eukaryotes first appeared, the group started evolving very quickly. Why did this happen? Eukaryotes evolved the ability to reproduce sexually.

Prokaryotes, which reproduce by binary fission, simply copy their single chromosome and send one daughter chromosome to each daughter cell. This fast and effective means of reproduction

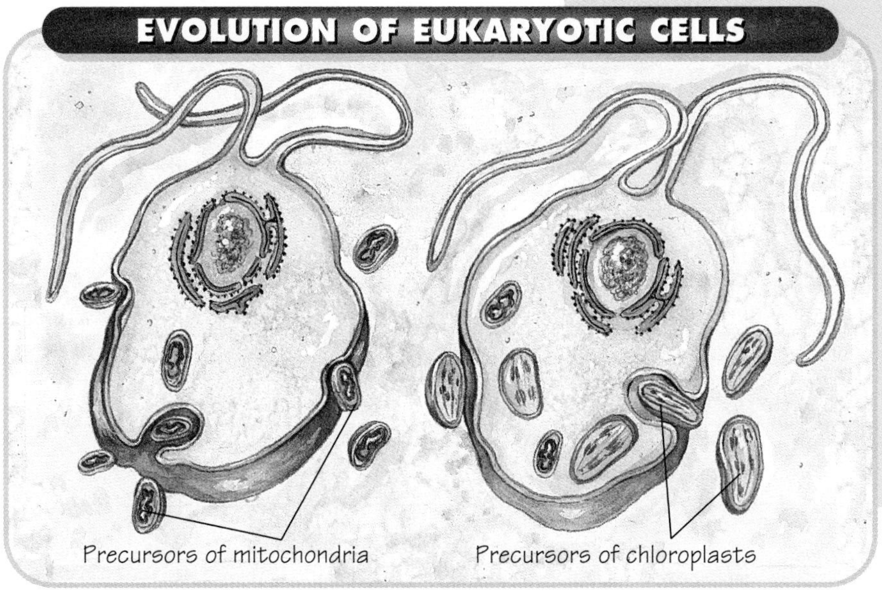

EVOLUTION OF EUKARYOTIC CELLS

Precursors of mitochondria

Precursors of chloroplasts

Figure 18–3
A widely accepted hypothesis states that the evolution of modern eukaryotic cells involved the formation of symbiotic relationships between two different kinds of ancient organisms. According to this hypothesis, mitochondria and chloroplasts are the descendants of ancient prokaryotes that took up residence inside the ancestors of modern eukaryotic cells.

has served bacteria well for billions of years. However, daughter cells produced in this way are exact replicas of their parents—unless a mutation occurs.

Eukaryotes that reproduce sexually, on the other hand, produce a lot of genetic variation among their offspring. You may recall that a great deal of chromosome shuffling occurs during independent assortment in meiosis. In addition, crossing-over can result in the exchange of pieces of DNA between members of a chromosome pair. This process can rearrange genes into different combinations. In this way, the next generation is formed by the fusion of gametes carrying genes from two different parents.

Because all these processes increase genetic variation, sexual reproduction provides more raw material for natural selection to operate. For that reason, soon after sexual reproduction evolved, eukaryotes experienced an extraordinary adaptive radiation—which is one species giving rise to many new species in a short period of time. Between 1.25 and 1 billion years ago, a great burst of evolutionary change produced the rest of living protist groups, as well as the ancestors of plants, animals, and fungi.

☑ **Checkpoint** Why did the eukaryotes suddenly begin to evolve quickly?

Living Protists

Ancestors of modern protists separated from one another a very long time ago. For that reason, living protists have been evolving alongside one another for about a billion years. That's an awfully long time, and many things have happened in each surviving group. Several groups have become photosynthetic, and others have become parasites. Still others have evolved varied ways of surviving.

Because differences among living protists are so extreme, most scientists agree that they should be divided into many different kingdoms. As a compromise, this chapter recognizes the diversity among protists but discusses the organisms in two groups—based on whether they are autotrophic or heterotrophic.

Autotrophic Protists

Algae are photosynthetic autotrophs that look and act like plants in certain ways. According to current estimates, there are about 30,000 species of algae in the oceans and in fresh water. Today, between 30 and 40 percent of all photosynthesis that takes place on Earth is performed by these protists. Major groups of algae, which vary greatly, are shown in **Figure 18–4** on page 418. Some are nearly as small as bacteria, while

CLASSIFICATION OF PROTISTS

Phylum	Examples	Nutrition
Euglenophyta (Euglenoids)	*Euglena*	Photosynthetic autotrophs
Pyrrophyta (Dinoflagellates)	*Gonyaulax* (causes red tide)	Photosynthetic autotrophs
Chrysophyta (Golden algae) (Diatoms)	*Orchromonas* *Navicula*	Photosynthetic autotrophs
Chlorophyta (Green algae)	*Chlamydomonas, Ulva*	Photosynthetic autotrophs
Rhodophyta (Red algae)	*Polysiphonia*	Photosynthetic autotrophs
Phaeophyta (Brown fungi)	*Fucus*	Photosynthetic autotrophs
Zoomastigina (Zooflagellates)	*Trypanosoma* (causes African sleeping sickness)	Heterotroph
Sarcodina	*Ameba*	Heterotroph
Sporozoa	*Plasmodium* (causes malaria)	Heterotrophic parasites
Ciliophora	*Paramecium*	Heterotroph
Myxomycota (Plasmodial slime molds)	*Physarum*	Heterotroph
Acrasiomycota (Cellular slime molds)	*Dictyostelium*	Heterotroph

Figure 18–4
Protists are often classified according to the way they obtain nutrients. Protists that make their own food by photosynthesis are called autotrophs, whereas those that get their food from another source are called heterotrophs.

others—such as giant kelp—form dense forests that may reach 70 meters.

Three groups of algae—Euglenophyta (yoo-glee-nuh-FIGHT-uh), Pyrrophyta (pigh-roh-FIGHT-uh), and Chrysophyta (krihs-uh-FIGHT-uh) are mostly single-celled. Chlorophyta (klor-oh-FIGHT-ah), Rhodophyta (roh-duh-FIGHT-uh), and Phaeophyta (fee-ah-FIGHT-uh) are mainly multicellular. Among multicellular species, most of the cells are nearly identical.

Checkpoint What are algae?

Heterotrophic Protists

Heterotrophic protists, some examples of which are shown in ***Figure 18–6***, have been commonly known as protozoans. Most protozoans are quite active and can usually be seen with a compound light microscope. **All heterotrophic protists spend most of their lives as unicellular organisms. Some species, such as slime molds, gather into groups at some time during their life cycle.**

Protozoans have evolved many different feeding strategies, and living species may act as predators, decomposers, or

Figure 18–5
Algae, the autotrophic protists, are found in moist and sunny environments where they can carry out photosynthesis. (a) *Notice the single chloroplast in the center of the Spirogyra, a strandlike green algae (magnification: 200X).* (b) *In most red algae, a red pigment masks the other pigments, such as chlorophyll, giving the algae its characteristic red color.*

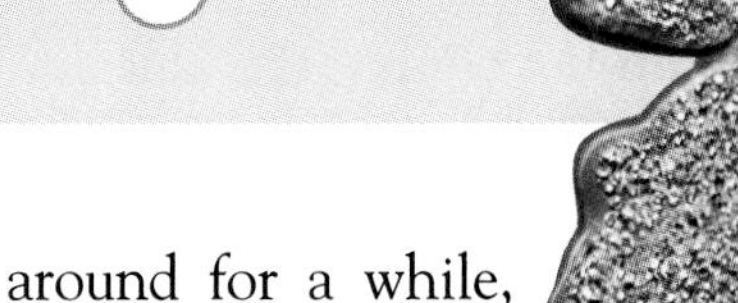

Figure 18–6
Some representative examples of heterotrophic protists include (a) Myxophyta, *a slime mold, and* (b) Amoeba proteus, *a sarcodine found in fresh water (magnification: 160X).*

parasites. Sporozoa may look like simple organisms, but they are complex parasites that cause serious diseases in humans and other animals. One species of sporozoa, the genus *Plasmodium*, causes malaria, a disease that kills more humans each year than any other disease.

☑ ***Checkpoint*** What are protozoans?

How Do Protists Reproduce?

Life cycles among protists are enormously varied. Because these organisms have been evolving for a billion years, it isn't surprising that they have evolved more life-cycle variations than the fungi, plant, and animal kingdoms combined.

One group, Euglenophyta, branched off from the others before sexual reproduction evolved. *Euglena* and its relatives reproduce only asexually. All other groups reproduce sexually at least occasionally.

Cellular slime molds spend most of their lives as individual, free-moving amebalike cells. But under certain circumstances, they gather together and form a sluglike mass. This mass moves around for a while, then forms a fruiting body that releases **spores** that develop into new amebalike cells on their own, without needing to join with another cell.

Algae show several variations in a life cycle that alternates between a sexual stage and an asexual stage. The green algae *Ulva* shows a pattern of reproduction that also occurs in green plants. This species illustrates a process known as **alternation of generations.** In alternation of generations, diploid (2n) and haploid (n) cells switch back and forth. Diploid cells have the normal number of chromosomes for that species. Haploid cells have half the normal number of chromosomes. The diploid generation is called the sporophyte because it undergoes meiosis to produce haploid spores. These spores grow into haploid male and female cells called gametophytes (gah-MEET-uh-fights). Gametophytes produce gametes—eggs and sperm. When an egg and sperm fuse, they produce a diploid cell called a zygote (ZIGH-goht), which grows into a sporophyte again.

Section Review 18–1

1. How would you **describe** the characteristics of protists?
2. **Compare** algae and protozoans.
3. **Explain** how protists reproduce.
4. **Critical Thinking—Comparing** Describe the differences between the sporophyte and gametophyte generations in *Ulva*.

SECTION

18–2 Fungi

GUIDE FOR READING

- **Classify** fungi into four phyla.
- **Explain** how fungi reproduce.

MINI LAB

- **Predict** the conditions that are needed for the growth of mold.

TO MOST PEOPLE, THE WORD fungus means death and decay. Molds spoil food, mildew ruins books and clothes, and mushrooms sprout from rotting logs. Although some fungi are harmful, causing serious diseases in humans and plants, other fungi are beneficial. Fungi help trees and many other seed plants extract nutrients from the soil. Without them, many of the great evergreen forests of North America would wither away and die. Other fungi are used to make bread and many cheeses. And the first antibiotics, as well as many other drugs, have been processed from fungi.

What Is a Fungus?

Fungi are heterotrophic eukaryotes that have cell walls. Some fungi—such as yeast—are single-celled and microscopic. Most fungi, however, are multicellular. Some fungi are parasites that live on plants or animals, causing them harm. Other fungi obtain food by digesting the dead remains of other organisms. And still other fungi live in a mutual relationship with other organisms—ranging from bacteria to animals.

Although many fungi look like plants, several characteristics show that they are more closely related to animals. Similar to plants, fungi have cell walls. But the cell walls of fungi do not contain cellulose. Instead, they are reinforced with **chitin** (KIGH-tihn), a complex carbohydrate that is also found in insect skeletons. Also, many fungi can store energy in the form of glycogen—something that animals do all the time, but that plants cannot.

☑ **Checkpoint** What is chitin?

(a)

Figure 18–7 *Fungi are a diverse group of organisms.* (a) *These bracket fungi are growing on a dead tree. Other types of fungi, such as these* (b) *fly agaric mushrooms and* (c) *Amanita fulva mushrooms, are poisonous and if eaten could be fatal.*

(b)

(c)

Living Fungi

Not much is known about the evolution of fungi, but researchers agree that they have been around since at least the Silurian Period (about 430 million years ago). It seems likely that fungi began evolving on land probably when the first plants appeared.

All fungi, as you can see in ***Figure 18–8*** have the same basic structures. The most basic structure in a fungus is a threadlike filament called a **hypha** (HIGH-fuh; plural: hyphae). Oddly enough, the cells in the hyphae may have a single nucleus, two nuclei, or many nuclei. In most common types of fungi, hyphae grow down into whatever the fungus is feeding on, forming a cottony mass called a **mycelium** (migh-SEE-lee-uhm). Most of the mycelium grows into and through the food source, secreting enzymes that break it down into compounds that can be absorbed into the fungus's cells. When the conditions are right, parts of a mycelium may become organized into the fruiting bodies you would recognize as a mushroom. **The living fungi are divided into four phyla according to their life cycles—Zygomycota** (zigh-goh-migh-KOHT-uh), **Ascomycota** (as-kuh-migh-KOHT-uh), **Basidiomycota** (buh-sihd-ee-uh-migh-KOHT-uh), **and Deuteromycota** (doo-ter-uh-migh-KOHT-uh).

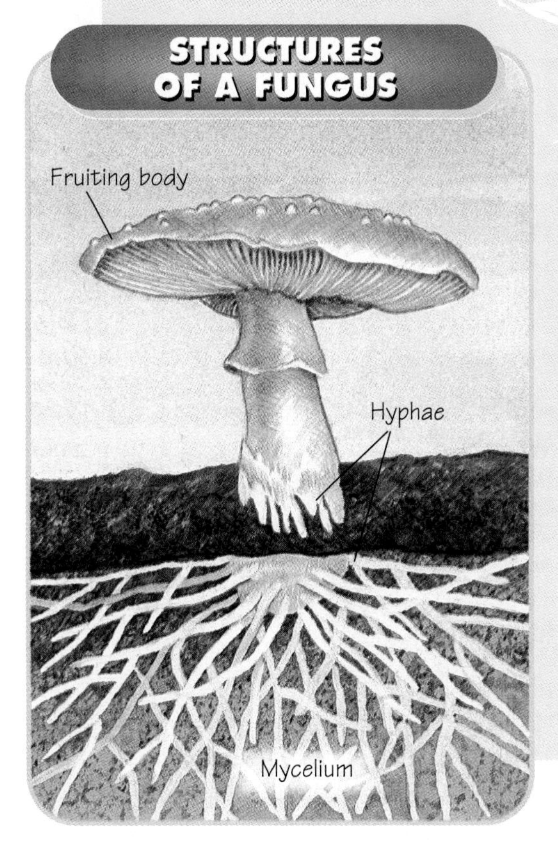

Figure 18–8 *Like most fungi, mushrooms have the same basic structures—a fruiting body, hyphae, and mycelium.*

☑ ***Checkpoint*** What is a mycelium?

Zygomycetes

Fungi that belong to the phylum Zygomycota are called zygomycetes. These fungi are often called bread molds because this group includes the black mold that is sometimes found on bread.

CLASSIFICATION OF FUNGI

Phylum	*Examples*
Zygomycota (Common mold)	*Rhizopus* (Black bread mold)
Ascomycota (Sac fungi)	Yeasts, morels, truffles, *Neurospora* (Red bread mold)
Basidiomycota (Club fungi)	Mushrooms, puffballs, bracket fungi, rusts, jelly fungi, toadstools
Deuteromycota (Imperfect fungi)	*Penicillium*, *Aspergillus*, ringworm and athlete's foot fungus, black spot on roses fungus, tomato blight fungus, cucumber scab fungus

Figure 18–9 *Fungi are classified according to how they reproduce. Some of the most common fungi are shown in their respective phyla.*

Problem Solving

INTERPRETING INFORMATION

The Fruit Detective

It is summer, and you are faced with a dilemma. Whenever you leave tomatoes, blueberries, or strawberries on the kitchen counter, they spoil within a few days—often before you can eat and enjoy them. Although oranges eventually spoil, they stay fresh somewhat longer than the others.

Your friend was visiting one day, and you explained the problem to her. Your friend needed a science project for her summer school class and decided to use your spoiling fruit problem as her project. She knew that the decomposition of food—its spoiling—is caused by the growth of microorganisms. She was determined to find out which type of microorganism was growing on the fruit. She gathered up the spoiled fruit and headed for the school lab.

To find out which type of organism was spoiling the fruit, your friend observed the spoiled areas of the fruit with a magnifying lens. She observed black and blue-green fuzzy growths on the fruit. She then made some wet-mount slides of the visible growth from the fruit and observed them under a microscope. The photographs illustrate what she observed.

THINK ABOUT IT

1. **Compare** the characteristics of the organisms shown in the photograph with those organisms described in the chapter. Which type of organism is spoiling the fruit? How do you know?
2. What conditions favoring the growth of these organisms might be present in the summertime?
3. What could you do to keep the fruit fresh longer? **Explain** your answer.

Ascomycetes

The sac fungi, or the fungi that belong to the phylum Ascomycota, include as many as 30,000 recognized species of mildews, molds, and yeasts. This is the largest phylum in the kingdom Fungi. Other members of this phylum are edible—the rare and expensive delicacies known as morels and truffles.

Basidiomycetes

Members of the phylum Basidiomycota include the fungi whose fruiting bodies form mushrooms. Although many mushrooms are good to eat, some can be poisonous. Unfortunately, the difference between the delicious ones and the deadly ones may be difficult to detect. Therefore, never gather your own mushrooms for dinner! Other members of this phylum include puffballs, bracket fungi, rusts, and smuts.

Deuteromycetes

The phylum Deuteromycota is sometimes called *fungi imperfecti*, or "imperfect" fungi. These fungi are called

Fuzzy Food

PROBLEM *How can you* ***predict*** *the conditions that are needed to grow mold?*

PROCEDURE

1. Line the bottoms of three plastic containers with damp paper towels.
2. Place a slice of bread into each of the three containers.
3. Place the containers in three different places—such as a refrigerator, a windowsill, and in a desk drawer—where they will remain undisturbed for four days.
4. Based on where you put the containers, predict in which containers mold will grow.
5. On the fifth day, use a hand lens to observe the changes in each container. Draw a diagram of what you observe.

ANALYZE AND CONCLUDE

1. In which, if any, containers did mold grow?
2. Were your predictions correct?
3. What conditions were needed for mold to grow?

Figure 18–10
Some representative fungi from the different phyla include (a) Rhizopus stolonifer, *the black bread mold, a member of the phylum Zygomycota;* (b) *black truffles, a rare and expensive delicacy, a member of the phylum Ascomycota; and* (c) *honey mushrooms, a member of the phylum Basidiomycota.*

Figure 18–11
One type of asexual reproduction in fungi involves the production of spores. In puffballs, spores can be dispersed by the slightest touch—in this case, falling raindrops. Once released, these lightweight spores can be carried great distances by the wind.

imperfect because researchers have never been able to identify their sexual stages. One member of this phylum is *Penicillium*, the source of the antibiotic penicillin. Other members of this phylum produce plant and animal diseases, such as tomato blight and athlete's foot, respectively.

Lichens

Lichens—familiar to anyone who hikes or walks in the woods—are the colorful scalelike patches on tree trunks or rocks. But to many people's surprise, lichens are not a single organism. They are a symbiotic partnership between a fungus and a photosynthetic organism, such as a cyanobacterium or a green alga. The alga carries out photosynthesis, providing the fungus with a source of organic substances. The fungus, in turn, provides the alga with water and minerals that it has collected from the surfaces on which it grows. Because both partners benefit from this relationship, lichens are able to survive in some of the world's harshest environments, where neither partner could survive alone.

☑ **Checkpoint** What are lichens?

How Do Fungi Reproduce?

Life cycles of most fungi are extremely complicated and vary from phylum to phylum. **Fungi can reproduce both sexually and asexually.** Asexual reproduction occurs either by the production of spores or by the fragmentation of the hyphae. The spores produced in many fungi are small and lightweight. When they are released, the spores are carried by the wind. If they land in suitable places, they germinate and grow into hyphae again.

Sexual reproduction involves two mating types. When the hyphae of opposite mating types meet, some of their filaments fuse together. The nuclei from the two mating types fuse and immediately undergo the process of meiosis to produce tiny spores.

Section Review 18–2

1. **Classify** fungi into four phyla.
2. **Explain** how fungi reproduce.
3. **Critical Thinking—Applying Concepts** Why do you think fungi were once classified as plants?
4. **MINI LAB** **Predict** the conditions that are needed to grow mold.

Multicellular Plants

SECTION 18–3

Guide for Reading

- **Classify** plants into two major groups.
- **Describe** the structures that enabled plants to live on land.
- **Explain** how plants reproduce.

THE PLANT KINGDOM INCLUDES some of the most beautiful and important organisms on Earth. Because members of the plant kingdom are the only primary producers that live entirely on land, the food they provide makes terrestrial life possible for humans and other members of the animal kingdom. Plants are not only used for food, they are used for clothing, shelter, medicines, and even gifts. Researchers are learning that plants have powerful effects in shaping global environments. And because plants have a long evolutionary history, they have developed hundreds of relationships with members of other kingdoms. Life as we know it couldn't exist without them.

What Is a Plant?

Plants vary remarkably in shape, size, and habitat. Some mosses and orchids are so small that an adult plant could fit on the tip of a pencil eraser! Other plants—such as the giant redwoods—are the tallest organisms that have ever lived. Some plants live in the freezing Arctic, and others inhabit Earth's hottest areas.

What characteristics tie all these different organisms into a single kingdom? All plants are multicellular, photosynthetic eukaryotes whose cells are enclosed and supported by cell walls made of cellulose. The vast majority of plants reproduce sexually and have life cycles that involve alternation of generations. In addition, many plants can reproduce asexually.

☑ **Checkpoint** What do all plants have in common?

Figure 18–12 *Ferns were among the earliest plants found on Earth. They range in size from (a) tree ferns—the largest of the ferns—to (b) sword ferns, common in many gardens.*

Figure 18–13
(a) The first forests were dominated by ferns and other plants without seeds. (b) Some of the earliest vascular plants did not have leaves and were very primitive.

As with many other forms of life, plants evolved in water. Water provides an environment that supports the plant body and surrounds the plant with nutrients. During their evolutionary history, plants had to adapt to a series of changes in order to survive in a dry, nonsupporting environment.

Evolution of Plants 

The first ancestors of living plants started leaving traces in the fossil record during the late Ordovician Period, about 450 million years ago. At that time, continental drift and changes in global climate set the stage for the evolution of modern organisms. At the beginning of the Silurian Period, about 430 million years ago, a global cold spell ended and continental drift began moving large landmasses from the South Pole closer to the equator. Together, these changes created many warm, swampy habitats that served as ideal places for the evolution of terrestrial life.

Challenges of Life on Land

Plants are most closely related to the green algae. Researchers know this because algae and plants share many molecular similarities and cell structures. The photosynthetic process of plants also resembles that of green algae.

Environment

The first challenge in adapting to life on land was the dry environment. All cells need a constant supply of water. In order for plants to survive out of water, they needed to develop a way to prevent water loss.

Structures

Another adaptation plants had to make was the development of rigid structures to support the plant on land. This support enabled plants to hold their photosynthetic leaves up to the sun.

Plants also needed to develop structures that would anchor them in place and absorb water and nutrients from the soil. Once the water and nutrients were absorbed, they needed to be transported to the leaves, where photosynthesis takes place. To meet this need, plants had to develop an internal transport system.

Reproduction

Finally, plants also had to evolve new methods of reproduction. The life cycles

Figure 18–14
Bryophytes were the earliest plants. (a) Hornworts, (b) mosses, and (c) liverworts are three groups of modern bryophytes.

of algae require water in which gametes can swim or young plants can float. Before plants could survive in any but the wettest habitats on land, they had to evolve new ways of transporting sperm to eggs that did not require water. All these requirements were met as plants became fully adapted to life on land.

The Evolution of Two Groups

Early on, the ancestors of plants branched into two groups. The first group was the bryophytes (BRIGH-oh-fights), **which resemble algae in many ways.** Bryophytes evolved only partial solutions to the environmental challenges of living on land. **The second group, the tracheophytes** (TRAY-kee-oh-fights), **developed specialized tissue as a way of adapting to these challenges.**

Bryophytes

The bryophyte group includes mosses, liverworts, and hornworts. Bryophytes are well-equipped to survive in wet places, where they may grow in great numbers. Bryophytes evolved rapidly during the Devonian and Carboniferous periods (about 286 to 408 million years ago), when many land environments were wetter and more humid than they are today.

Modern bryophytes never grow more than a few centimeters tall. Some species, such as *Sphagnum*, or peat moss, grow almost completely submerged in water. Others grow well on wet tree trunks, rocks, and soil. Bryophytes can grow and reproduce only when wet, so they can survive only in places where, for at least part of the year, plenty of water is available. Some moss species can survive dry periods, but in order to do so they must stop growing.

☑ **Checkpoint** What are bryophytes?

Tracheophytes

Although bryophytes live on land, they still depend upon water to survive. The tracheophytes were the first "true" land plants because they evolved ways of decreasing this dependence on the constant presence of standing water and humid conditions. Tracheophytes have **vascular tissues** that transport water and nutrients throughout the plant. As a result, tracheophytes can grow much larger and can live in a wider range of habitats than bryophytes can.

Ferns

Ferns and several extinct relatives were the first tracheophytes. Ferns have vascular tissue in addition to roots, strong

Figure 18–15
Gymnosperms, such as this sitka spruce, were the first plants that did not require standing water in order to reproduce.

stems, and leaves. Ferns and club mosses grew into dense forests that covered huge areas during the Devonian and Carboniferous periods. Some of them grew as much as 30 to 40 meters tall! The partially decomposed and compressed remains of these dense forests formed the great coal deposits that gave the Carboniferous Period its name.

INTEGRATING EARTH SCIENCE

How is coal formed from ferns and other plants? Research the topic and write a brief summary.

Today, most ferns that live in areas with changing seasons die back to the ground each year. In tropical rain forests, some ferns grow into tall, graceful trees. Some ferns can survive long droughts as adult plants. Yet because all ferns have life cycles that require wet conditions, they are rare or absent in many dry terrestrial habitats.

☑ **Checkpoint** What is vascular tissue?

Gymnosperms

The next major adaptation to terrestrial life was the evolution of a life cycle that freed plants from the need for standing water in order to reproduce. The first seed-bearing tracheophytes that evolved were the **gymnosperms,** which include cycads, ginkgoes, and conifers. Gymnosperms carry their seeds exposed to the air, usually in a cone-shaped structure. A **seed** is a reproductive structure that includes a developing plant and a food reserve enclosed in a resistant outer covering.

Many gymnosperms appeared in the fossil record about 360 to 408 million years ago, during the Devonian Period, when mosses and ferns were prevalent. Millions of years later, continental drift and climate change caused many terrestrial environments to become much drier. That gave gymnosperms an advantage over earlier plants, and by the time dinosaurs walked the Earth, forests were filled with gymnosperms. The Petrified Forest in Arizona contains the fossilized trunks of ancient conifers that were not very different from their modern descendants.

Today, the most common gymnosperms are the conifers, known for their distinctive cones and needlelike leaves. The cones produce and carry seeds. Although conifers are often called evergreens, their leaves don't last forever. The leaves—which normally remain on the plants for 2 to 14 years—exhibit several adaptations that help these plants survive the cold, dark winters and short growing seasons of the northern regions where many conifers live.

☑ **Checkpoint** What are gymnosperms?

Angiosperms

Angiosperms are the plants whose beautiful flowers are enjoyed as gifts and decorations in all cultures throughout the world. But more importantly, angiosperms are food sources for humans and other animals. These fully terrestrial plants are latecomers to the history of life. Their first fossils didn't appear in the fossil record until about 65 to 144 million years ago, during the Cretaceous Period. But once they appeared, angiosperms rapidly evolved and spread over the Earth.

DIFFERENCES AMONG MONOCOTS AND DICOTS

Structure	Monocots	Dicots
Seeds	One cotyledon	Two cotyledons
Flowers	Flower parts in threes or multiples of three	Flower parts in fours or fives or multiples of four or five
Leaves	Veins in leaves are parallel to each other	Veins in leaves form a branching network
Stems	Vascular bundles scattered throughout the stem	Vascular bundles arranged in a ring around the stem

Figure 18–16 *The distinguishing traits of monocots and dicots are illustrated in this chart. Notice the differences among the seeds, flowers, leaves, and stems.*

The key to their success involved the evolution of flowers and seeds. Both adaptations enabled angiosperms to reproduce and grow to maturity more quickly than gymnosperms. Once dinosaurs, insects, and mammals appeared and began eating plants, this rapid reproduction may have been a selective advantage.

Angiosperms are divided into two main groups—the **monocots** and **dicots.** Within the seeds of angiosperms are structures that contain food for the developing plant. These structures are called cotyledons (kaht-uh-LEED-'nz). Monocots—such as rice, wheat, corn, lilies, orchids, tulips, and palms—have one cotyledon. Dicots—such as tomatoes, roses, maples, and the daisylike plants, including sunflowers—have two cotyledons. Both groups evolved an astonishing variety of sizes, leaf shapes, and flower patterns, as shown in ***Figure 18–16.***

☑ **Checkpoint** What are angiosperms?

Adapting to Land

Some of the structures that have allowed plants to thrive on land include roots, leaves, vascular tissue, and stems. Let's take a look at these structures and their functions.

Roots

Roots anchor plants in place in the soil and keep them from being blown away by strong winds. They also absorb water and dissolved inorganic nutrients from the soil. Roots can perform both these tasks because they branch as they grow into the soil, making a dense network.

Leaves

Leaves provide the surface area over which the plant can capture sunlight for photosynthesis. The broader and flatter a leaf is, the more sunlight it can capture. But there's a tough functional tradeoff here. The greater the leaf area, the more tissue surface that is exposed to dry air, and the more water the plant

Figure 18–17 *Angiosperms, which are the flowering plants, are the most complex and advanced tracheophytes. Angiosperms are characterized by their flowers, as shown by this cherry tree in full bloom.*

Figure 18–18 *All multicellular green plants, including this trillium, have the same basic structures—leaves, stems, and roots.*

can lose by evaporation. For that reason, the leaves of most land plants are covered with a waxy, waterproof covering called a cuticle.

On the other hand, a totally waterproof barrier presents another problem. Leaves must exchange oxygen and carbon dioxide with the air around them in order to carry out photosynthesis and respiration. A barrier that is waterproof usually doesn't let gases in and out either. That's why cuticles are dotted with tiny openings called **stomata** (singular: stoma) that can open and close as needed to allow for gas exchange and to prevent the loss of too much water.

☑ **Checkpoint** What is a cuticle?

Vascular Tissues

As plants evolved from the size of mosses to large sunflowers and maple trees, they faced a serious challenge. They had to transport water efficiently from roots to leaves, while moving the products of photosynthesis from leaves to other parts of the plant. The solution was the evolution of two types of specialized vascular tissue, called **xylem** (ZIGH-luhm) and **phloem** (FLOH-ehm).

Xylem carries water and dissolved inorganic nutrients from roots to branches and leaves. Cellulose-reinforced cell walls of xylem tissue are thick and strong, so they are also a major source of strength in woody plants such as trees.

Phloem carries the products of photosynthesis and certain other substances from one part of a plant to another. Depending on the time of day, season, and condition of the plant, phloem may transport these substances upward or downward.

☑ **Checkpoint** What are two types of vascular tissue?

Stems

Stems hold leaves up to the sun and position leaf surfaces to capture as much light as possible. Stems also conduct water, nutrients, products of photosynthesis, and other materials through the plant by means of their vascular tissues.

How Do Plants Reproduce?

Many changes in plants over time have been adaptations to drier habitats, and life cycles are no exception. **All plant life cycles involve alternation of generations between sporophyte and gametophyte.** But in different plants, the relationship between gametophyte and sporophyte changes. In mosses, the gametophyte is the longest part of the cycle. In flowering plants, the sporophyte part of the cycle is much longer.

Gametes and Spores

What you recognize as moss is the gametophyte generation of the moss life cycle. These gametophytes are either male or female and produce either sperm or eggs. For reproduction to be successful, the plants must be close enough to water so that the sperm can swim to fertilize an egg. The zygote produced by fertilization grows into the sporophyte, a slender stalk with a spore capsule on the end. Inside this capsule, meiosis produces spores. The spores are dispersed and produce gametophyte plants.

Ferns show a slightly different pattern. The gametophytes produce sperm and eggs. Water must be present so the sperm can swim to the eggs for fertilization. But in this case, the zygote grows into an independent sporophyte that is much larger than the gametophyte and can live on its own for years. This sporophyte is what you recognize as a fern. The fern sporophyte produces spores, usually in clusters on the undersides of its leaves.

Figure 18–19
CAREER TRACK
Nursery operation technicians work in greenhouses, garden centers, or botanical gardens. They are responsible for the care and maintenance of plants. In this photograph, a nursery operation technician is growing rare wildflowers.

Pollen and Seeds

Seed plants have gone still further in freeing themselves from the need for water. The plants you see are all members of the dominant sporophyte generation of seed plant life cycles. Where are the gametophytes? They have been reduced to small clusters of a few cells that grow inside structures called cones in gymnosperms and flowers in angiosperms.

The entire male gametophyte is contained in a tiny structure called a pollen grain. A pollen grain produces sperm that do not have to swim because the pollen is carried to the eggs by wind, insects, beetles, birds, or bats. This process is called **pollination.** Once the pollen lands on the female part, a long tube containing the sperm begins to grow down inside the flower until it reaches the egg.

After fertilization, the egg grows into a tiny plant called an embryo, which grows for a short time before becoming dormant inside the seed. The seed itself is supplied with food and a tough outer covering. Protected this way, the seeds of many flowering plants can survive heat, drought, and bitter cold—sometimes for years—before germinating when conditions are right. This adaptability allows angiosperms to grow in many habitats where mosses and ferns cannot.

Section Review 18–3

1. **Classify** plants into two major groups.
2. **Describe** the structures that enabled plants to live on land.
3. **Explain** how plants reproduce.
4. **Critical Thinking—Applying Concepts** Why are tracheophytes able to grow larger than bryophytes?

The Coevolution of Plants and Animals

GUIDE FOR READING

- **Define** coevolution.

MINI LAB

- **Design an experiment** to see how maple seeds are adapted for dispersal.

NO QUESTION ABOUT IT—spring is one of the most wonderful times of the year. From Florida to Washington and from California to Maine, the sight and fragrance of flowers in bloom are sources of delight. But flowers are only part of the story of springtime. Sit quietly for a few minutes near an apple tree in bloom or by a lawn where yellow daisies have poked their way into the sunlight. You'll notice that a lot of work is going on. Bees are drifting among the trees, and hummingbirds may be winging their way from flower to flower. What are they doing? And why are they attracted to the flowers?

Pollination

The answers to these questions may explain why angiosperms have been so successful. In order to carry out two of the most important steps in their life cycle—pollination and seed dispersal—many angiosperms rely on the help of animals.

Why is pollination so important to plants? It is usually best for a species if sexual reproduction occurs between different individuals. Insects and other pollinators help plants reproduce sexually with other plants of the same species. Pollination, therefore, helps to maintain genetic diversity—the variety of different genes present in a population.

Pollination by Wind

Gymnosperms rely on the wind to carry pollen from male to female flowers. Pollen grains are small and light, and they can be scattered for many kilometers in a strong breeze. Gymnosperms

Figure 18–20
Many plants rely on animals for pollination. (a) Birds, such as this green-crowned brilliant hummingbird, pick up pollen as they remove nectar. (b) Insects, such as honeybees, are also important pollinators. Here, a honeybee already covered in pollen is collecting even more!

must produce and release enormous amounts of pollen to make sure that this hit-or-miss strategy works. You may have seen entire lakes and ponds covered with bright-yellow dust in the spring and early summer. This dust is actually pollen from pines and other conifers.

Pollination by Insects

Many angiosperms, on the other hand, are pollinated by insects that visit one flower after another. Bees, for example, will visit flowers that produce sweet secretions called nectar. As a bee feeds on the nectar, pollen is dusted onto its body. When the bee leaves one flower, it will carry that pollen directly to another nectar-containing flower, transferring the pollen as it moves. How does the bee find flowers with nectar? That's where the scent and color of the flower come in. They serve as signals that help bees find a particular kind of flower.

The spread of pollen from one plant to another by an animal is called **vector pollination.** Vector pollination is much more efficient than pollination by the wind because the insect or other animal takes the pollen directly from one flower to the next. This relationship, called **coevolution,** is beneficial to both organisms. **Coevolution is the evolution of structures and behaviors in two different organisms in response to changes in each organism over time.** Natural selection has favored flower structures and markings that are bright and colorful, making it easier for the bees to locate those flowers. Flowers that are pollinated by night-flying moths are often white and have a strong fragrance that attracts insects that hunt using their sense of smell.

☑ **Checkpoint** What is coevolution?

Pollination by Birds and Mammals

Bees and other insects are not the only animals involved in pollination. Many birds, especially hummingbirds, sip liquid nectar from deep within flowers. As they visit one flower after another, they spread pollen from one to another. Other birds that pollinate flowers include the honeycreepers of Hawaii and the brush-tongued parrots of Australia.

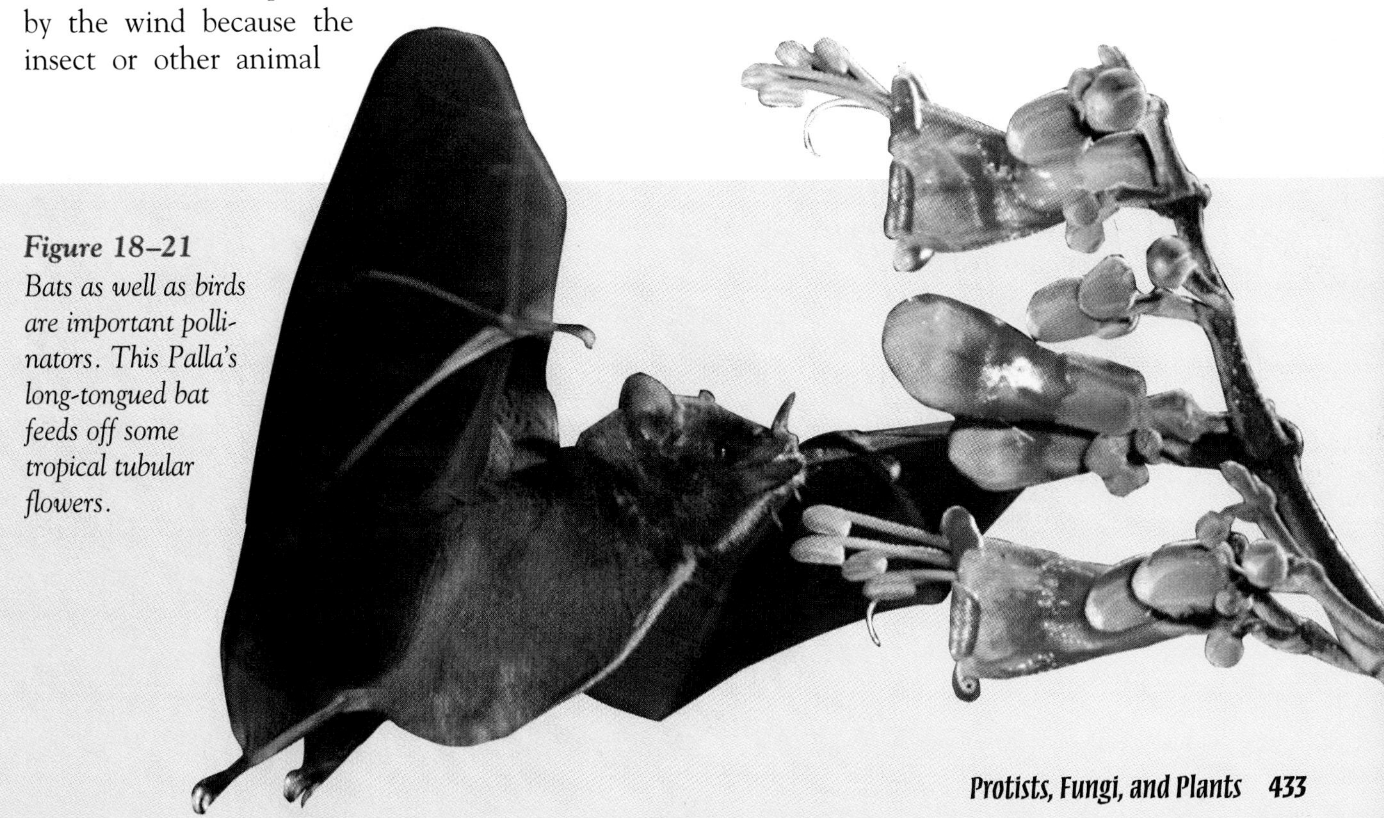

Figure 18–21
Bats as well as birds are important pollinators. This Palla's long-tongued bat feeds off some tropical tubular flowers.

MINI LAB ····· Experimenting ·····

Traveling Seeds

PROBLEM *How is a maple fruit adapted for seed dispersal? **Design an experiment** to find out.*

PROCEDURE

1. Obtain three to four maple fruits.
2. Carefully observe a seed and note its characteristics.
3. Formulate a hypothesis to explain how the fruit is adapted for seed dispersal.
4. Design an experiment to test your hypothesis.

ANALYZE AND CONCLUDE

1. By which method of seed dispersal do you think maple fruits are dispersed?
2. Which characteristics of the maple fruit help in the dispersal of the seed? How do you know?

A few flowers are pollinated by small pollen-eating bats. The poor vision of bats requires that these flowers be strongly scented to attract the animals, which come out only at night. Flower-feeding bats may lack teeth and have long brushlike tongues that quickly gather nectar and pollen. Bananas, as well as the saguaro cactus of Arizona, are pollinated by bats.

Checkpoint What are four ways in which pollen is spread from flower to flower?

Seed Dispersal

Pollination is only one part of the reproductive battle for plants. Once the seeds are fertilized and ready to sprout, it is a great advantage for them to travel some distance from the parent plant before germinating. What is the advantage of seeds traveling? If all the seeds landed at the foot of their parent plants, they would all be competing with their relatives for water, nutrients, and sunlight! Seed dispersal makes it possible for a species to move into new environments where conditions may be more favorable.

Dispersal by Wind

Many angiosperm seeds are dispersed by the wind. Some of these seeds are encased in a winglike structure that helps them to glide or spin as they fall to the Earth. Others, such as milkweed and dandelions, are lightweight and have feathery attachments that allow them to float and drift on the slightest breeze.

Figure 18–22
The fruits of this red maple have winglike structures that help in the dispersal of the fruits by wind.

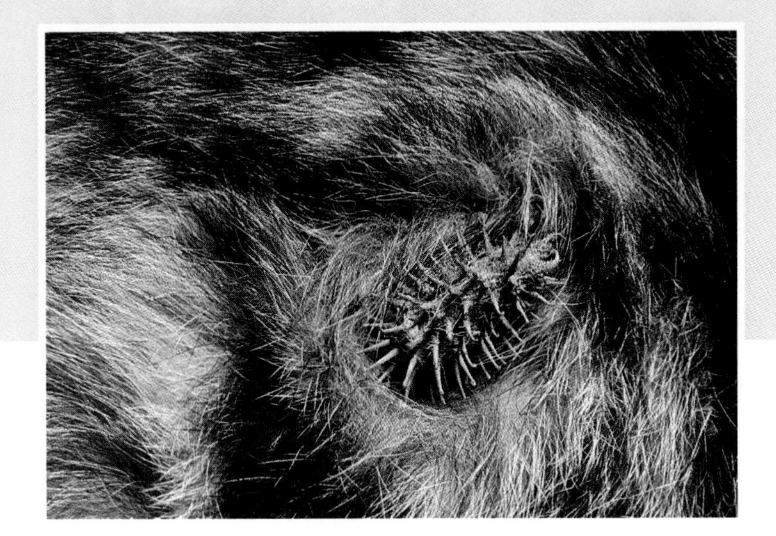

Figure 18–23
Some plants rely on animals for seed dispersal. Notice the small hooks on this cocklebur, which keep it securely attached to a cat's fur.

Dispersal by Animals

Many angiosperms have coevolved with animals in ways that help to spread their seeds. For example, some fruits have tiny hooks or barbs that enable them to catch a ride on the fur of a passing animal. By the time the seeds are finally removed, the animal may have carried them for many kilometers, giving them a free ride to a new location.

Other plants have evolved a very different strategy. Have you ever wondered why plants produce such brightly colored, sweet-tasting nutritious fruits? It is because those fruits attract a wide range of animals—from birds to monkeys—that eat them. The seeds inside these fruits are usually very strong and tough. For that reason, some of them survive being chewed and passed through the animals' digestive tract. As the animal walks or flies or grazes, the seeds may be carried great distances. They may then be deposited far from the parent plant, right in the middle of a handy batch of natural fertilizer!

Figure 18–24
To spread their seeds, plants have enlisted the help of animals. The bright-red color of these berries, along with their sweet taste, will attract animals, such as this cedar waxwing, that eat the berries and then deposit the seeds some distance from the parent plant.

Section Review 18–4

1. **Define** coevolution.
2. **MINI LAB** How would you **design an experiment** to find out how maple seeds are adapted for seed dispersal?
3. **BRANCHING OUT ACTIVITY** Farmers sometimes rely on wind pollination for their corn crops. Find out how farmers plant their crops to take full advantage of wind pollination. **Summarize** your findings in a brief report.

Laboratory Investigation

Vascular Plant Tissues

The vascular system is the system of tubes that carries water and minerals throughout a plant. The vascular system is made up of two types of tissues—xylem and phloem. In this investigation, you will discover the role of the xylem and phloem.

Problem

Can you **predict** the role of a vascular system in a plant?

Materials (per group)

- celery stalk
- blue food coloring
- metric ruler
- clear plastic container
- scalpel
- microscope

DATA TABLE

Time	*Height (in millimeters)*
5 minutes	
10 minutes	
15 minutes	
20 minutes	
24 hours	

Procedure

1. Fill the plastic container halfway with water. Add several drops of blue food coloring to make the solution a deep shade of blue.
2. Using the scalpel, cut off about 1 cm of the base of the celery stalk. **CAUTION:** *Be careful when using a scalpel—it is very sharp.*

3. Place the celery into the food-coloring solution, being sure the freshly cut base is in the solution.
4. Construct a data table similar to the one shown.

5. **After 5 minutes, measure the height in millimeters of the food coloring in the celery. Record your measurement.**
6. **Measure the height of the food coloring in the celery after 10, 15, and 20 minutes. Record each of your measurements.**
7. **What do you predict will happen to the food coloring in the celery after 24 hours? Write your prediction on a sheet of paper.**
8. **After 24 hours, measure the height of the food coloring in the celery.**
9. **Using the scalpel, cut off a thin slice about 2 to 3 mm from the base of the celery stalk. CAUTION:** ***Be careful when using a scalpel—it is very sharp.***
10. **Observe the cross section of the celery under a microscope. Identify the xylem and make an illustration of what you see.**

Observations

1. What happens to the food coloring in the celery stalk?
2. What tissue is responsible for transporting the food coloring?
3. Does the vascular system of the celery stalk extend throughout its entire length? How do you know?
4. Does the movement of the liquid occur quickly or slowly? Support your answer using your data.

Analysis and Conclusions

1. What was your prediction? Was it correct?
2. What adaptation to life on land does this activity illustrate?
3. Why is a plant's vascular system important?

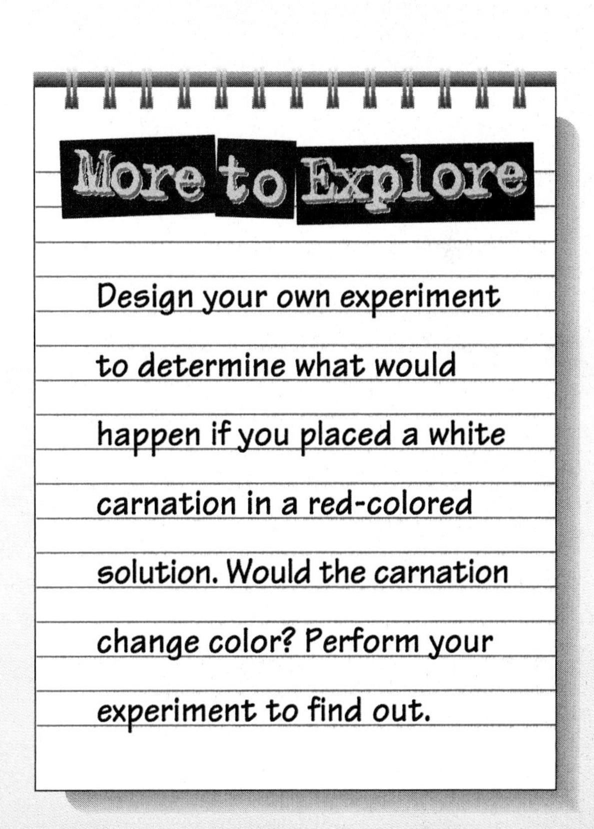

Study Guide

Summarizing Key Concepts

The key concepts in each section of this chapter are listed below to help you review the chapter content. Make sure you understand each concept and its relationship to other concepts and to the theme of this chapter.

18–1 Protists

- Protists are eukaryotic organisms that do not share a unique set of characteristics.
- Algae are photosynthetic autotrophs that look and act like plants in certain ways.
- All protozoans spend some of their lives as unicellular organisms. Some species, however, gather into groups at some time during their life cycles.
- Life cycles among protists are enormously varied.

18–2 Fungi

- Living fungi are divided into four phyla according to their life cycles—Zygomycota, Ascomycota, Basidiomycota, and Deuteromycota.
- Fungi can reproduce both sexually and asexually.

18–3 Multicellular Plants

- During their evolutionary history, plants had to adapt to a series of changes in order to survive in a dry, nonsupporting environment.
- Early on, ancestors of plants branched into two groups—bryophytes and tracheophytes.
- All plant life cycles involve alternation of generations between sporophyte and gametophyte.

18–4 The Coevolution of Plants and Animals

- Coevolution is the evolution of structures and behaviors in two different organisms in response to changes in each organism over time.

Reviewing Key Terms

Review the following vocabulary terms and their meaning. Then use each term in a complete sentence.

18–1 Protists

protist
spore
symbiosis
alternation of generations

18–2 Fungi

chitin
mycelium
hypha

18–3 Multicellular Plants

vascular tissue
angiosperm
gymnosperm
monocot
seed
dicot
root
phloem
leaf
stem
stoma
pollination
xylem

18–4 The Coevolution of Plants and Animals

vector pollination
coevolution

Recalling Main Ideas

Choose the letter of the answer that best completes the statement or answers the question.

1. To which kingdom do algae belong?

a. protist **c.** plant
b. fungi **d.** animal

2. The process of alternation of generations first appeared in

a. algae. **c.** plants.
b. protozoans. **d.** fungi.

3. Yeast is an example of a

a. protist. **c.** fungus.
b. plant. **d.** bryophyte.

4. In which structure can cells have a single nucleus, two nuclei, or many nuclei?

a. mycelium **c.** hypha
b. chitin **d.** phloem

5. An example of vascular tissue is

a. xylem. **c.** stoma.
b. cuticle. **d.** hypha.

6. Which of the following groups of plants lack vascular tissue?

a. bryophytes **c.** tracheophytes
b. monocots **d.** dicots

7. The waxy, waterproof covering of plants is a

a. stoma. **c.** seed.
b. cuticle. **d.** mycelium.

8. The evolution of structures and behaviors in different organisms in response to evolutionary changes in the other is called

a. symbiosis. **c.** coevolution.
b. pollination. **d.** vector pollination.

Putting It All Together

Using the information on pages xxx to xxxi, complete the following concept map.

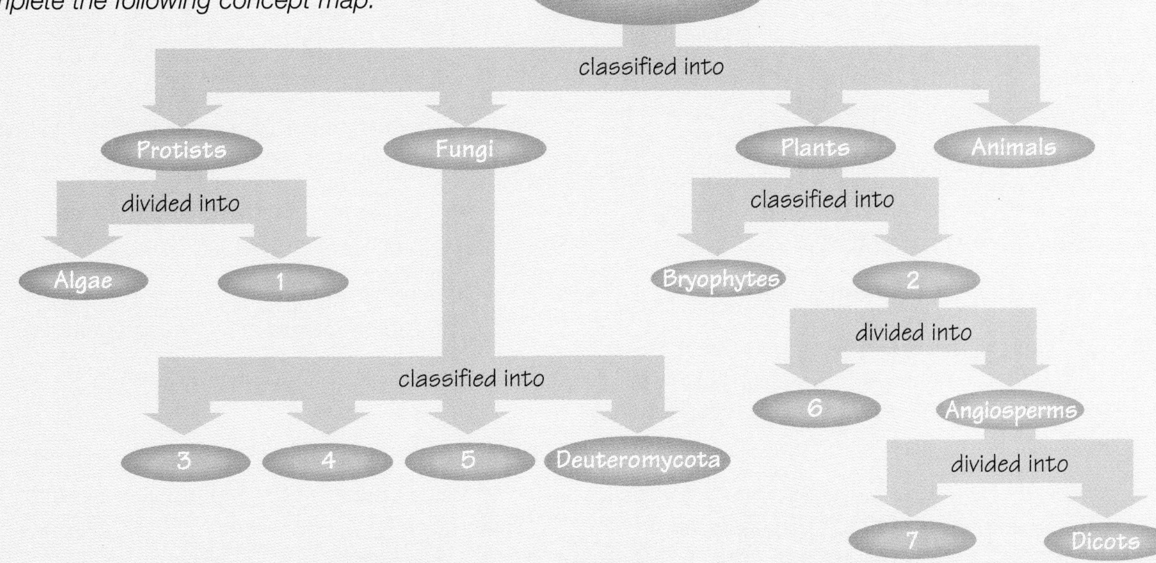

Assessment

Reviewing What You Learned

Answer each of the following in a complete sentence.

1. What is one characteristic that all protists share?
2. Define symbiosis.
3. What is adaptive radiation?
4. Which group of protists exhibits alternation of generations?
5. What substance makes up the cell walls of fungi?
6. What are hyphae?
7. List the four phyla of fungi.
8. To which phylum do mushrooms belong? Bread mold?
9. What characteristic do all species of the phylum Deuteromycota share?
10. What characteristics do all plants share?
11. Why do mosses need water?
12. Why have the bryophytes been limited in their evolutionary pathway?
13. What is vascular tissue?
14. What advantages do ferns have over mosses?
15. Compare a monocot and a dicot.

Expanding the Concepts

Discuss each of the following in a brief paragraph.

1. What effect did the development of sexual reproduction have on the evolution of organisms?
2. Compare algae and protozoans.
3. What characteristics make euglenophytes different from other protists?
4. How do spores differ from gametes?
5. Alternation of generations is a characteristic of the green algae known as *Ulva,* as well as of most mosses and ferns. Describe this unusual life cycle.
6. Fungi are characterized by their life cycles. List some examples of the major types of fungi.
7. How are fungi similar to protists? To plants?
8. What early evidence indicates that plants are probably most closely related to green algae?
9. Discuss four adaptations that plants had to make in order to survive on land.
10. Discuss how the development of seeds helped to extend the evolutionary advantages of plants.
11. What advantages do angiosperms have over gymnosperms?
12. List the four structures that allowed plants to thrive on land. Give the function of each structure.

Extending Your Thinking

Use the skills you have developed in this chapter to answer the following.

1. **Using the writing process** There are many exceptional "partnerships in nature." Write an essay that describes how the relationship between animals and plants is important to the overall existence of life on this planet.
2. **Comparing** Describe the similarities and differences between reproduction and development in bryophytes and angiosperms.
3. **Predicting** Why would you expect to find more fungi growing in a forest than in a field?
4. **Predicting** A conifer releases thousands of seeds. Each of these seeds has the potential to grow into a new conifer. Why do you think an organism such as a conifer produces so many seeds?
5. **Designing an experiment** You notice that the plants in the principal's office are dying even though they are being watered and are getting enough sunlight. Upon closer examination, you discover lots of dust on their leaves. Design an experiment to find out and explain what is happening to the plants.

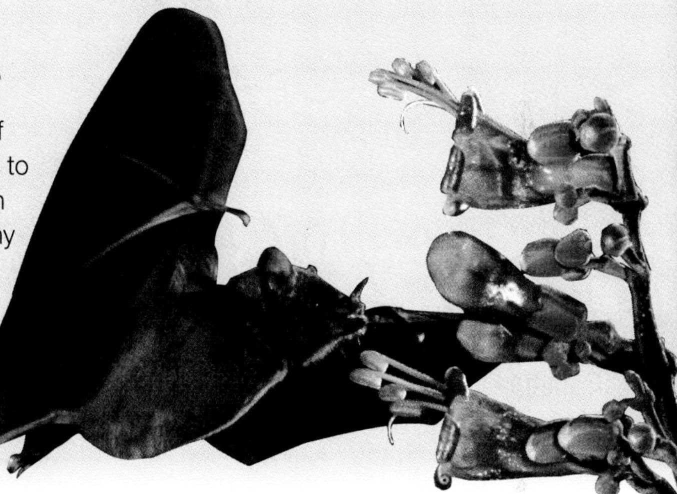

Applying Your Skills

Living Together

Use your knowledge of symbiotic relationships to create a new organism. Your new organism does not have to be real, but the two organisms that compose it should already exist. Your new organism will live in one of the following environments: on the top of a mountain above the tree line; in the deep trenches of the Pacific Ocean; in the back of a cave where no light can penetrate; or in an area that reaches −10°C in the winter and 40°C in the summer.

1. Design a new organism that is composed of two organisms—each from a different kingdom. Draw your new organism and label its parts.
2. Describe the adaptations this organism has made in order to survive in its environment.
3. What is the symbiotic relationship between the two organisms?

• GOING FURTHER •

4. What would happen to your new organism if its environment suddenly changed? What additional adaptations would your organism have to make?

CHAPTER 19

Animals: Invertebrates

FOCUSING THE CHAPTER

THEME: Unity and Diversity

19–1 Evolution of Multicellular Animals

- **Describe** some of the adaptations animals made to life on land.

19–2 A Survey of Living Invertebrates

- **Describe** the different phyla of living invertebrates.

19–3 Form and Function in Invertebrates

- **Compare** the body systems of the living invertebrates.

BRANCHING OUT *In Depth*

19–4 Specialized Reproductive Cycles

- **Describe** some of the specialized reproductive cycles of invertebrates.

LABORATORY INVESTIGATION

- **Compare** the nervous systems of a hydra and a planarian.

Biology and Your World

BIO JOURNAL

Does the organism in this photograph look like a plant or an animal? In your journal, make a list of the characteristics of both plants and animals. When you have completed the chapter, look at your lists and see if you can identify the organism as a plant or an animal.

Coral in the Solomon Islands

SECTION 19–1

Evolution of Multicellular Animals

GUIDE FOR READING

- **Explain** the relationship between evolution and animal complexity.
- **Describe** the characteristics of animals.

FOR MORE THAN 80 PERCENT of Earth's history, our planet was inhabited by bacteria, single-celled eukaryotes, and algae. Then, about 550 million years ago, at the beginning of the Cambrian Period, multicellular animals appeared, it seemed, quite suddenly. An incredible assortment of complex animals were produced by a great burst of evolutionary change called, appropriately, the Cambrian explosion. How and why did so many different kinds of animals appear so quickly? Could they have come from even earlier multicellular animals? If so, what were they like? To find out the answers to these questions, read on.

The Earliest Known Animals

When fossils from the Cambrian Period were first discovered, they posed a major puzzle. Scientists could not find any evidence of older, simpler multicellular animal life from which these animals might have evolved. Yet the odd creatures of this era were too complex to have evolved directly from single-celled eukaryotes. Who had their ancestors been, and what happened to them?

The End of an Era

Between 650 and 600 million years ago, the Earth saw many changes. Landmasses that had once been joined together broke up and moved apart. A series of ice ages chilled the globe. The entire Earth was gripped in a deep freeze.

Figure 19–1
(a) The Burgess Shale is an area in the Rocky Mountains of British Columbia, Canada, that has large deposits of shale. (b) The Burgess Shale, which contains fossils of trilobites, is formed from small particles of mud and clay and can often be broken into flat pieces, as shown here. (c) This close-up of a Burgess Shale trilobite provides a better view as to what these animals looked like when they were a dominant group during the Cambrian Period.

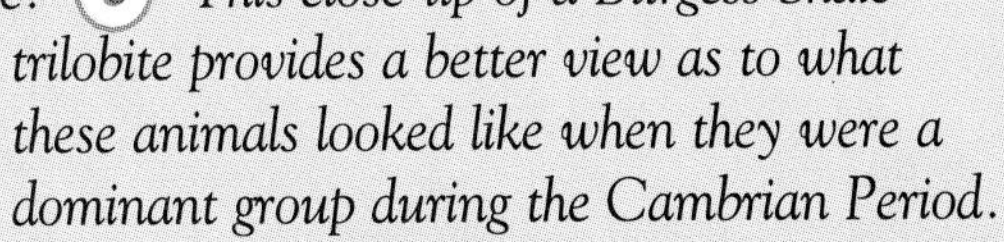

Scientists have discovered that at about this time, fossils of early photosynthetic organisms became rare. They also noticed that fossils of free-swimming single-celled eukaryotes disappeared too. What had happened? In addition to the cold, something—or several things—might have appeared and eaten the earlier forms of life. But who might those hungry new animals have been?

New Information From Australia

The first clear answers came from discoveries made in the Ediacaran Hills of Australia. There, fossils that dated from about 580 to 560 million years ago revealed a strange collection of animals. Some researchers think that these animals belong to such living groups as jellyfishes and worms. Others feel that these fossils belong to other animal groups that quickly became extinct. In either case, these fossils prove that simple multicellular animals had begun to evolve before the Cambrian Period. These findings were so important that paleontologists added what they call the Ediacaran Period to the end of the Precambrian Era.

Another discovery from the same period is even more intriguing. Researchers looked closely at rocks from the Ediacaran Hills and concluded that many other animals must have lived at that time—even though their bodies were never preserved. Studies have revealed a whole series of burrows and tracks in the mud that could have been made by several types of soft-bodied organisms. This discovery provides further evidence that multicellular animals evolved long before the Cambrian explosion, even though few fossils have survived.

Figure 19–2
This fossil of a polychaete worm was found in the Ediacaran Hills in Australia. It lived during the Precambrian Era and is estimated to be 600 million years old.

The Cambrian Explosion

By the beginning of the Cambrian Period, the number of different kinds of animals found in the fossil record was increasing with astonishing speed. These animals were unbelievably diverse in size and shape. The ancestors of almost all major living animal groups appeared in the fossil record for the first time. But there were also many animals that do not resemble animals living today.

There are a few reasons why the number of organisms found in the fossil record was increasing. One reason was that new animal groups were evolving quickly. Another reason was that more and more groups of these animals were evolving larger bodies, skeletons, and hard body parts, such as shells.

Hard skeletons and other hard body parts were among the most important evolutionary trends of this time. Skeletons gave animals the support they needed for arms, legs, wings, and flippers. Shells and other hard body parts offered protection from the elements and from predators.

By this time, these animals had also evolved specialized cells. Specialized cells are different groups of cells that evolved over time into different functions and shapes in ways that enable them to perform certain tasks more efficiently than generalized cells could.

As the animals evolved, they became increasingly more complex. For example, specialized cells organized into tissues, which are groups of similar cells that work together. Groups of tissues assembled to form organs—collections of tissues that work together to perform a specific function. And groups of organs that perform related functions evolved into organ systems.

Checkpoint What are specialized cells?

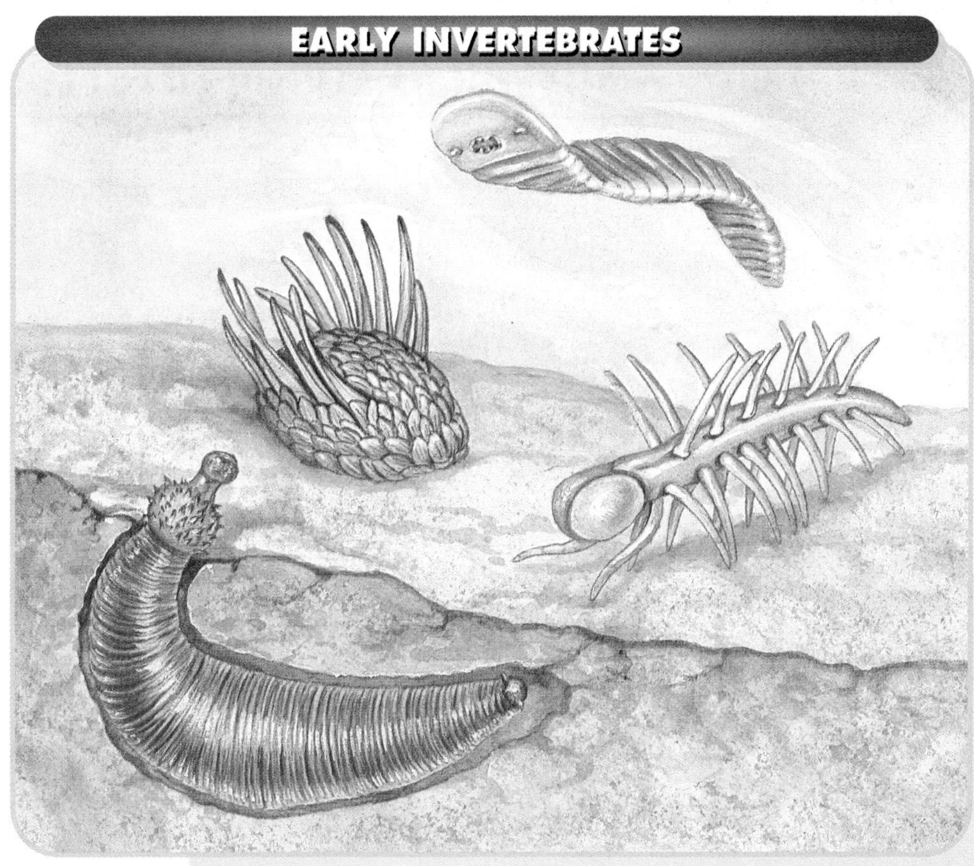

Figure 19–3
During the Cambrian explosion, the number of organisms increased dramatically. This artist's drawing shows some of the invertebrates from this time. In a clockwise direction from the top, these animals are Odontogriphus, Hallucinogenia, Wiwaxia, *and* Ottoia.

Body Plans

Cambrian fossils are fascinating because they show different ways in which tissues and organs can be assembled to produce animals. What does that mean? **To survive, all animals must perform the same functions: body support and movement, feeding and digestion, respiration, excretion, internal transport, and response to the environment.** But if all animals do the same things, why don't we all look alike? Because each major group of animals evolved its own ways of performing these functions.

Over evolutionary time, each phylum combined a particular type of breathing device, a certain type of body support system, and its own variations on other bodily functions. The result is a unique body plan for each phylum. Living phyla represent one set of body plans that work successfully. But Cambrian animals tried out a wide range of body parts and combinations that are no longer found in living species. Thus, animals of the Cambrian Period enabled researchers to study strategies for multicellular life as they were first being evolved.

Section Review 19–1

1. **Explain** the relationship between evolution and animal complexity.
2. **Describe** the characteristics of animals.
3. **Critical Thinking—Inferring** Why are fossils of animals with hard body parts found more frequently than those of soft-bodied animals?

SECTION 19–2

A Survey of Living Invertebrates

GUIDE FOR READING

- **Identify** three important trends in invertebrate evolution.
- **Classify** living invertebrates into different phyla.

MINI LAB

- **Observe** a planarian and identify its head and tail.

ABOUT 97 PERCENT OF ALL THE animal species on Earth belong to one category of animals. They range in size from microscopic dust mites to sponges large enough for humans to hide in. They can be found in the air and at the bottom of the sea. What is this group of animals?

These animals are invertebrates—animals without vertebrae. Invertebrates don't have the sort of bony backbone, or vertebral column, that other animals, including humans, have. The name invertebrate defines a lot of very different animals in an odd way—by describing a characteristic that they don't have, rather than one that they all share.

Trends in Invertebrate Evolution

Because there are so many invertebrate phyla that evolved over millions of years, it is helpful to point out important trends that occurred in their evolutionary history. **The common ancestors of nearly all multicellular animals had evolved two distinct cell layers separated by a jellylike middle layer.** The outer cell layer, or **ectoderm,** develops into skin and other body coverings and also gives rise to the nervous system. The inner cell layer, or **endoderm,** grows into the tissues and organs of the digestive tract.

Jellyfishes are some of the few living animals that retain a primitive jellylike layer between the ectoderm and the endoderm. Most other animals have a third tissue layer. This middle layer,

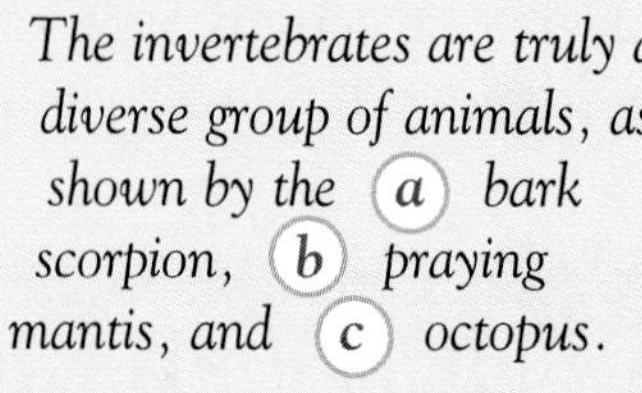

Figure 19–4 *The invertebrates are truly a diverse group of animals, as shown by the (a) bark scorpion, (b) praying mantis, and (c) octopus.*

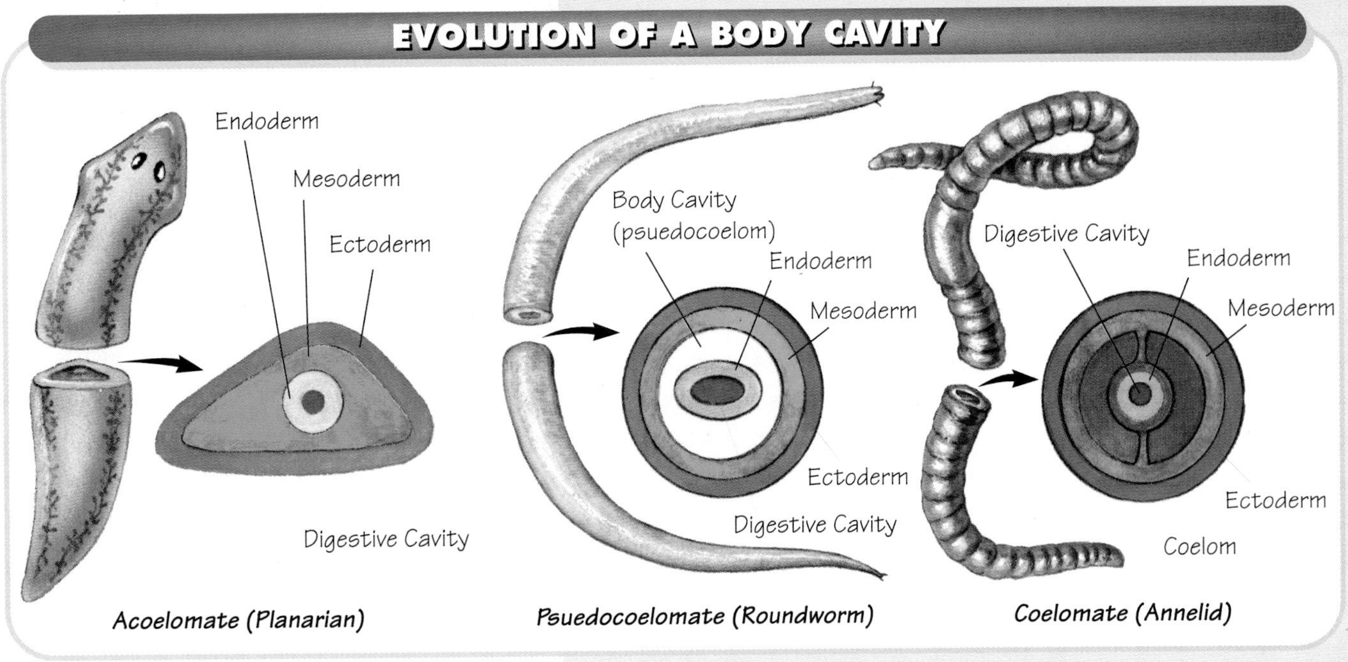

EVOLUTION OF A BODY CAVITY

Endoderm
Mesoderm
Ectoderm
Digestive Cavity
Acoelomate (Planarian)

Body Cavity (psuedocoelom)
Endoderm
Mesoderm
Ectoderm
Digestive Cavity
Psuedocoelomate (Roundworm)

Digestive Cavity
Endoderm
Mesoderm
Ectoderm
Coelom
Coelomate (Annelid)

Figure 19–5
The evolution of a coelom was an important trend in the evolution of invertebrates. Some invertebrates have no coelom and are therefore referred to as acoelomates. Others have a body cavity between the mesoderm and the endoderm that resembles a coelom and are called pseudocoelomates. Still others, called coelomates, have a true coelom that houses the digestive tract and other internal organs. Mollusks were the first phylum to have a true coelom.

or **mesoderm,** develops into the body's skeleton and muscles.

A second important evolutionary trend in invertebrates is the existence of a mesoderm-lined cavity. This body cavity is called a **coelom** (SEE-lohm). Why should that be important? A body cavity provides an open space inside the body within which organs can grow and function without being squeezed or twisted by body movements. The fluid in some coeloms also plays a role in carrying food, wastes, or dissolved gases from one part of the body to another. Animals that have a body cavity are called coelomates (SEE-loh-mayts).

A third important evolutionary trend in invertebrates is the evolution of a body plan that is built up from several body compartments. These body compartments are called **segments.** The presence of segments allows an animal to increase in body size with a minimum of new genetic material because certain structures are repeated in each segment. In addition, as animals become more complex, different segments become specialized for specific functions.

What is a coelom? 1

Figure 19–6
An arthropod, such as this lobster, has a body plan that is built up from several compartments.

The Movement Onto Land

Animals that moved from sea to land faced adaptive challenges similar to those that faced plants making the same evolutionary journey. To live on land, animals, like plants, had to evolve ways to perform essential survival tasks without losing too much water in the process.

It is important to remember, however, that land animals are not necessarily any more advanced than their aquatic cousins. Some successful animal groups live their entire lives in the water. Other groups live entirely on land, and still others live in both environments.

Figure 19–7
Sponges, such as this pink vase sponge, come in a variety of shapes, sizes, and colors. They are the simplest of all invertebrates.

Sponges

Sponges—the most primitive of all living multicellular animals—make up the phylum Porifera (por-IHF-er-ah). Sponge cells are relatively independent, and they live together for mutual benefit. Most sponges live almost entirely in the sea, although a few species are found in fresh water.

Cnidarians

Jellyfishes, corals, sea anemones, and hydras belong to the phylum Cnidaria (nigh-DAIR-ee-ah). Cnidarians are mostly marine, although a few species, such as

Figure 19–8
As invertebrates evolved, they moved from having no symmetry to having radial symmetry and finally to having bilateral symmetry. Animals that show bilateral symmetry have a definite head end and tail end, as well as a front and a back.

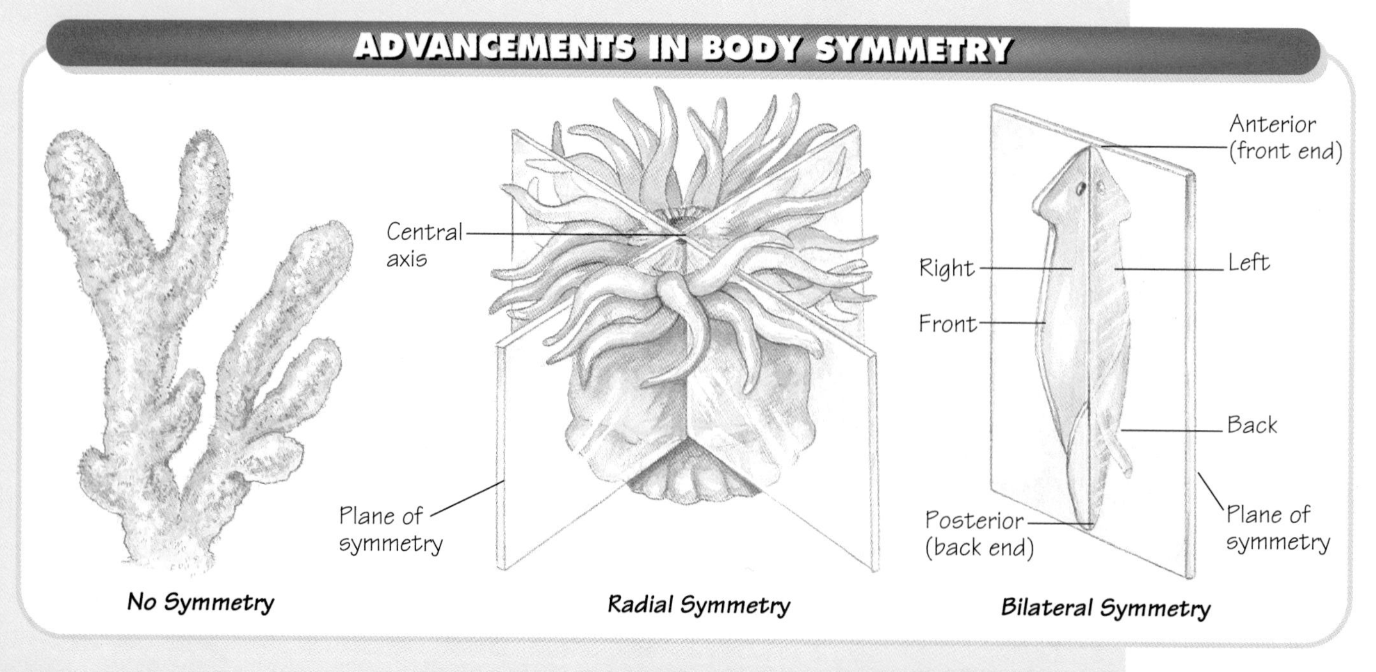

Figure 19–9
Cnidarians, including this coral, are the first invertebrates to show radial symmetry.

the hydras, live in fresh water. Cnidarians have **radial symmetry.** Animals with radial symmetry have body parts that repeat around an imaginary line drawn through the center of their body, as shown in ***Figure 19–9.*** Cnidarians are some of the most colorful and beautiful of all invertebrates.

☑ **Checkpoint** What is radial symmetry? 1

MINI LAB .... Observing ......

Which End Is Up?

PROBLEM *What can you **observe** about the shape and structure of a planarian to help identify the head end from the tail end?*

PROCEDURE 

1. Observe a planarian and then sketch it on a sheet of paper. Label one end anterior and the other end posterior.
2. Using an arrow, indicate the direction of movement of the planarian through the environment.

ANALYZE AND CONCLUDE

1. What things about the shape and structure of the planarian helped you label the anterior and the posterior ends?
2. As an organism moves through the environment, what are the advantages of having sense organs at the anterior end?

Platyhelminthes

Flatworms make up the phylum Platyhelminthes (pla-tee-hehl-MIHN-theez). They are the simplest animals that show **bilateral symmetry.** Animals with bilateral symmetry have left and right sides that are mirror images if you were to draw an imaginary line through their center. They also have specialized front and back ends as well as upper and lower sides. Most flatworms also show **cephalization** (sehf-uh-lih-ZAY-shun), which means they have a front end that is developed enough to merit being called a head.

☑ **Checkpoint** What is bilateral symmetry? 2

Nematodes

Roundworms, or nematodes, make up the phylum Nematoda (nehm-uh-TOHD-uh). Roundworms were among the first animals to evolve a tubelike digestive system, with a mouth at one end and an anus at the other. Some roundworms are microscopic, while others grow to more than a meter long.

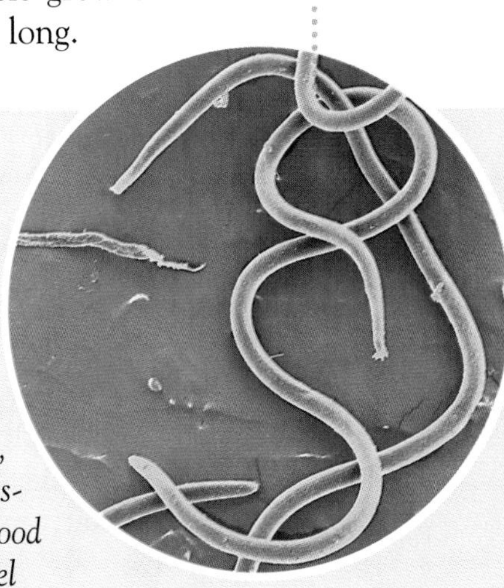

Figure 19–10
(a) This flatworm, a platyhelminth, shows bilateral symmetry—symmetric left and right sides. (b) In roundworms, or nematodes, the digestive system is very efficient because food can enter the mouth and travel through the digestive tract to the anus (magnification: 57X).

Figure 19–11
Annelids were the first phylum of animals to show segmentation. If you look carefully at this clamworm, you can see its body segments.

Mollusks

Members of the phylum Mollusca—which includes clams, snails, and squids—have managed to colonize almost every habitat on Earth. Many live in the sea, some in fresh water, and others on land. Some mollusks, such as snails, protect their body with a shell. Others, such as sea slugs, avoid predation by literally leaving a bad taste in the predator's mouth. Clams, mussels, oysters, and scallops have two shells connected by a flexible hinge. When they are in danger, they clamp their shells closed. Octopuses and squids can emit clouds of dark-colored ink to confuse predators, allowing the squid or octopus to escape.

Annelids

Segmented worms, or annelids, belong to the phylum Annelida (an-uh-LIHD-uh). This phylum includes a wide range of fascinating animals—from common earthworms to exotic marine worms with feathery rainbow-colored gills. Earthworms and their relatives recycle enormous amounts of decaying organic matter in the soil.

Echinoderms

INTEGRATING LANGUAGE ARTS

What does the word echinoderm mean? Use a dictionary to look up the word parts echino- *and* -dermis.

Starfishes, sea urchins, and sea lilies all belong to the phylum Echinodermata (ee-kigh-noh-DER-muh-tuh). Like some other invertebrates, echinoderms live their lives entirely in the water. They can be recognized by their spiny skin and by their five-part radial symmetry. Some echinoderms, such as starfishes, are found on nearly every rocky coast around the world. Others, such as the beautiful sea lilies, are found mostly on coral reefs and in the deep sea.

Arthropods

The jointed-leg animals—arthropods—that make up the phylum Arthropoda (AHR-throh-pahd-uh) are by far the largest and most diverse of all the animal phyla. Some experts estimate that there may be as many as 10 million species of arthropods! Their most important characteristics are an external skeleton and jointed legs. The three largest groups of arthropods are the chelicerates (kuh-LIHS-er-ayts), the crustaceans (kruhs-TAY-shuhnz), and the insects.

Spiders and scorpions are part of an ancient group named after their mouthparts—their chelicerae (kuh-LIHS-er-ee), hence, the subphylum Chelicerata. Most chelicerates are carnivorous, and a few, such as the black widow spider, have a

Figure 19–12
As a group, echinoderms have evolved a great variety of forms. However, all echinoderms, such as this sea cucumber, must live in water.

Figure 19–13
The three largest groups of arthropods are the chelicerates, the crustaceans, and the insects. (a) The sally light-foot crab is a crustacean, while (b) the leaf-cutter ants are one of the more than 900,000 types of insects.

poisonous bite that can be dangerous to a human. But these animals attack only by accident or when threatened.

Lobsters, shrimps, and crabs make up the class Crustacea. Crustaceans live primarily in the water. They range in size from tiny water fleas to Alaskan king crabs, whose legs can stretch more than two meters across.

At least half of all living animal species belong to the class Insecta. Insects are so diverse in form and habits that it is difficult to pick one that truly represents the entire group.

Invertebrate Chordates

Invertebrate chordates are peculiar animals that show a link between terrestrial animals with backbones and ancient ancestors in the sea. These animals are members of our own phylum, Chordata (kor-DAYT-uh). Chordates have an endoskeleton with a stiff rod, called a notochord, to which muscles are attached. They are found in marine environments throughout the world.

Figure 19–14
Sea squirts are tunicates, which are examples of invertebrate chordates. Sea squirts get their name from the fact that many species will shoot water at their attackers when touched.

Section Review 19–2

1. **Identify** three important trends in invertebrate evolution.
2. **Classify** the living invertebrates into the different phyla.
3. **Critical Thinking—Recognizing Relationships** What type of symmetry do humans have?
4. **MINI LAB** What can you **observe** about a planarian that helps to identify its head and tail?

SECTION

19–3 Form and Function in Invertebrates

GUIDE FOR READING

- **List** the three different kinds of skeletons.
- **Compare** open and closed circulatory systems.

MINI LAB

- **Design an experiment** to observe the movements of an earthworm.

THE VARIOUS INVERTEBRATE groups have evolved an astonishing variety of shapes and body structures that perform the essential functions of life. You can think of these different collections of body parts as "tool kits" for survival. Over time, each phylum has evolved its own variation of body parts that serve its own functions. The result is a body plan unique to each phylum. Some body systems are simple, and others are quite complex. Some are efficient, while others appear to be inefficient.

Support and Movement

Almost all multicellular animals use some kind of musclelike tissue to move around and perform other functions. All muscles and musclelike tissues work by contracting, or getting shorter. This is the only way that muscle tissue can generate force to perform useful work.

When it comes to helping animals move around, muscles by themselves don't operate very well. That's why most multicellular animals that move rapidly combine a muscle system with some sort of skeleton that provides firm support. **The three types of skeletons found in the animal kingdom are hydrostatic skeletons, exoskeletons, and endoskeletons.**

Hydrostatic Skeletons

In some animals, the muscles surround and are supported by a water-filled body cavity. This type of skeleton is called a **hydrostatic** (high-droh-STAT-ihk) **skeleton.** Hydrostatic skeletons do not contain hard structures for muscles to pull against. Instead, when the muscles contract, they push against the water in the

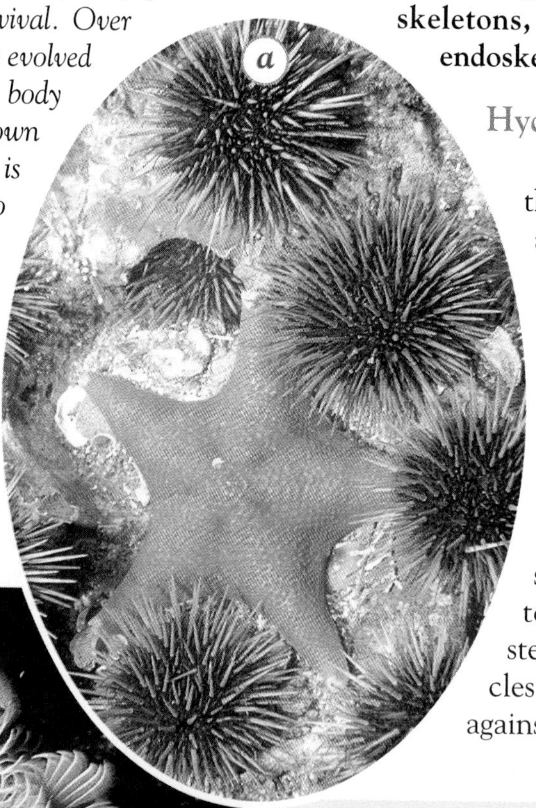

Figure 19–15
Invertebrates have evolved adaptations for survival. (a) Although starfishes move slowly through the water, they are carnivorous predators. (b) Feather duster worms, on the other hand, are annelid worms that use their feathery structures for feeding and respiration.

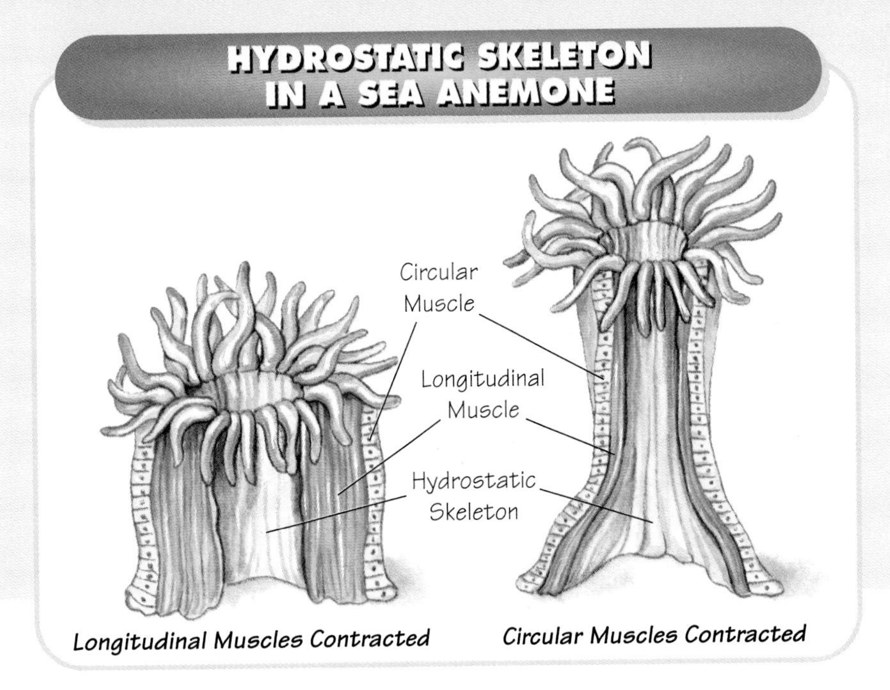

Figure 19–16
A hydrostatic skeleton is basically a fluid-filled tube with soft walls that contain two layers of muscle. Circular muscle forms a band around the circumference of the tube. Longitudinal muscles run lengthwise to the tube. When the longitudinal muscles contract, the sea anemone becomes short and fat. When the circular muscles contract, it becomes long and thin.

body cavity, as shown in ***Figure 19–16.*** Obviously, hydrostatic skeletons do not allow an animal to move very fast.

☑ **Checkpoint** What is a hydrostatic skeleton?

Exoskeletons

An external skeleton is called an **exoskeleton.** Arthropods are the best examples of organisms with exoskeletons. In arthropods, the exoskeleton is made of a tough carbohydrate called **chitin** (KIGH-tihn). Muscles are attached to the inside of the exoskeleton. All exoskeletons are thin and flexible at the joints, allowing the skeleton to bend and flex.

The arthropod exoskeleton is adaptable and successful, but it has two major drawbacks. One drawback is that in order to grow, the animal must shed the skeleton and grow a new one! The other drawback is that an exoskeleton is heavy for the amount of support it provides. For this reason, animals with exoskeletons cannot grow very large.

Endoskeletons

A skeleton that is located within the body is called an **endoskeleton.** Sponges and some echinoderms have endoskeletons, but the best examples are found among the vertebrates.

☑ **Checkpoint** What is an endoskeleton?

Feeding and Digestion

Because all animals, including humans, are heterotrophs, we all must eat in order to survive. As animals become more complex, their digestive systems become more specialized. Invertebrates have evolved many different ways of digesting their food.

Digestion Inside Cells

Sponges, the simplest multicellular animals, filter tiny particles of food from the water. Food is digested inside these cells, and the resulting nutrients are distributed among other cells within the sponge. Because this type of digestion takes place inside cells, it is known as **intracellular digestion.**

In many cnidarians and flatworms, larger food particles are taken into a **gastrovascular cavity.** A gastrovascular cavity is a digestive sac with only a single opening to the outside. Food enters through this opening, and indigestible solid wastes leave through the same opening. The food particles are broken down into smaller and smaller particles. Ultimately, the smallest particles are taken up by cells lining the cavity, and digestion is completed within the cells.

☑ **Checkpoint** What is a gastrovascular cavity?

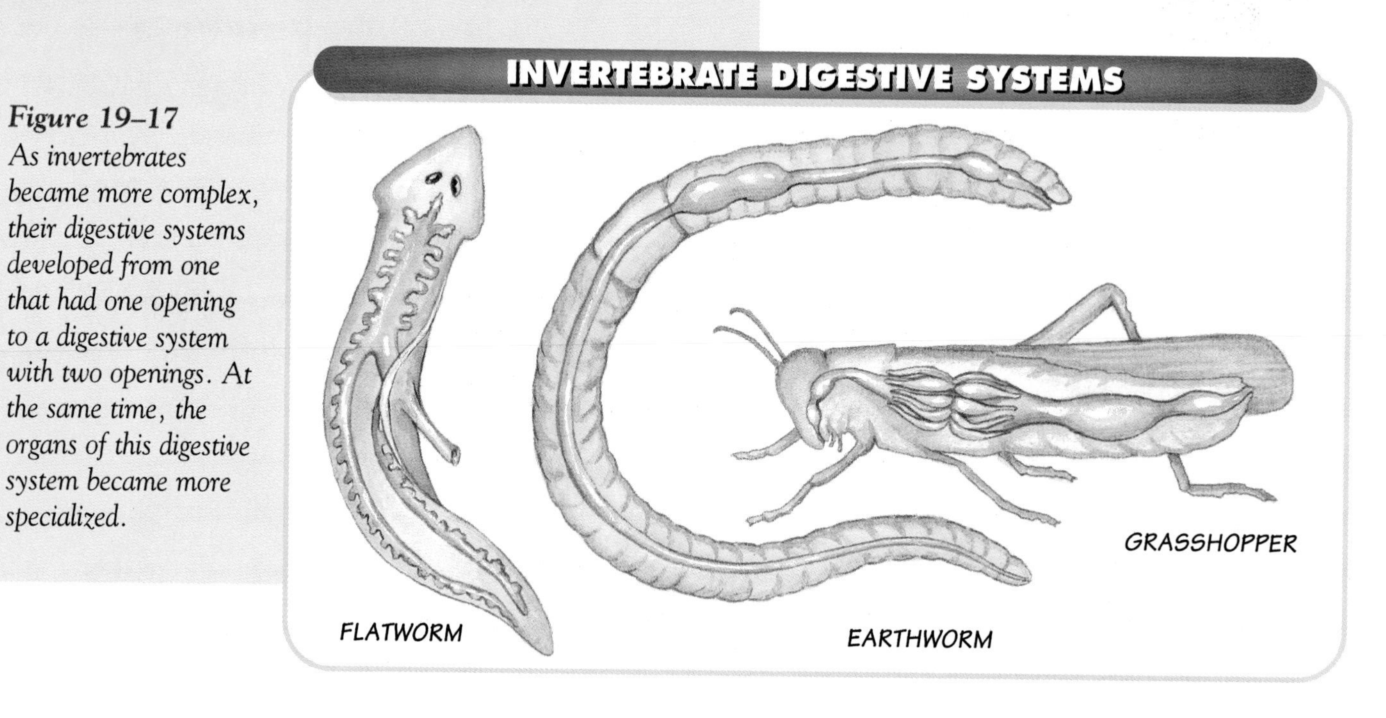

Figure 19–17
As invertebrates became more complex, their digestive systems developed from one that had one opening to a digestive system with two openings. At the same time, the organs of this digestive system became more specialized.

Digestion Outside Cells

More highly evolved digestive systems—such as those in annelids, mollusks, arthropods, and chordates—have two openings. Food enters through a mouth and leaves through an anus. Because the digestive tract in such animals forms a tube inside the body wall, this digestive tract is known as a "tube-within-a-tube" body plan.

Almost all the food—except for certain fats—is digested within the digestive cavity. Because this digestion takes place outside the cells, it is called **extracellular digestion.** The final products of digestion are absorbed into blood vessels that line the gastrovascular cavity.

Internal Transport

Living cells of multicellular animals require a constant supply of oxygen and nutrients, as well as a way to eliminate carbon dioxide and poisonous wastes. Small, thin soft-bodied animals may provide for these needs by simple diffusion through the body surface. But more complex multicellular animals have evolved some sort of circulatory system—a collection of one or more pumps and tubes that move important substances around within the body. The pumping part of such systems is called the heart. The fluid being pumped around the body is called blood, and the tubes that carry it are called blood vessels. In invertebrates,

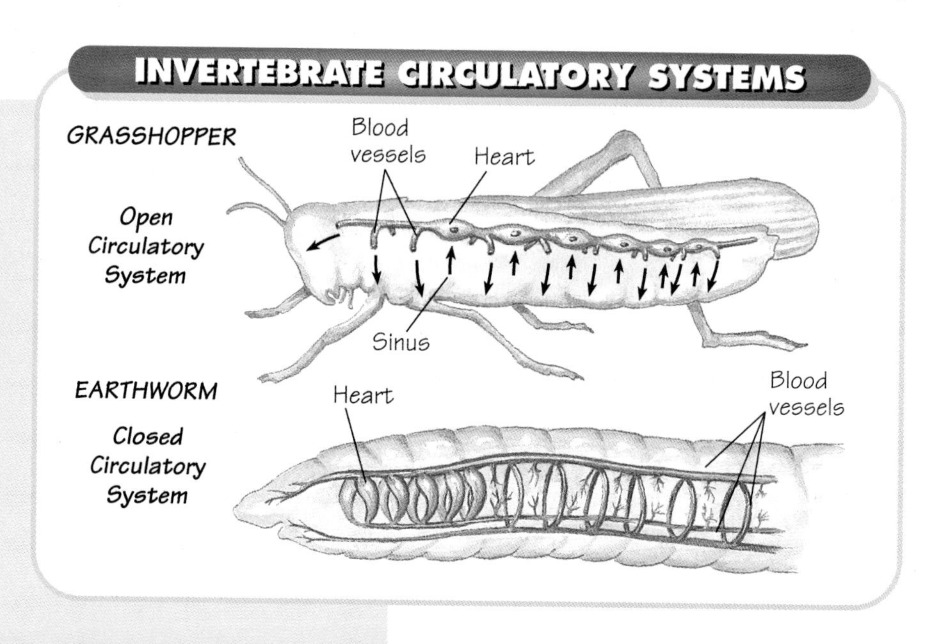

Figure 19–18
In an open circulatory system, the blood leaves blood vessels and comes in direct contact with the surrounding tissues. Here, it collects in the body sinuses before making its way back to the heart. In a closed circulatory system, the blood is completely contained within a network of blood vessels.

there are two main types of circulatory systems—open circulatory systems and closed circulatory systems.

Open Circulatory System

Open circulatory systems are systems in which blood from the heart is not entirely contained in blood vessels. Instead, the blood from the heart is pumped through a series of vessels that carry the blood and release it directly onto body tissues. The blood flows through tissues and collects in openings called sinuses. These sinuses connect with a larger sinus around the heart, which takes in the blood once again.

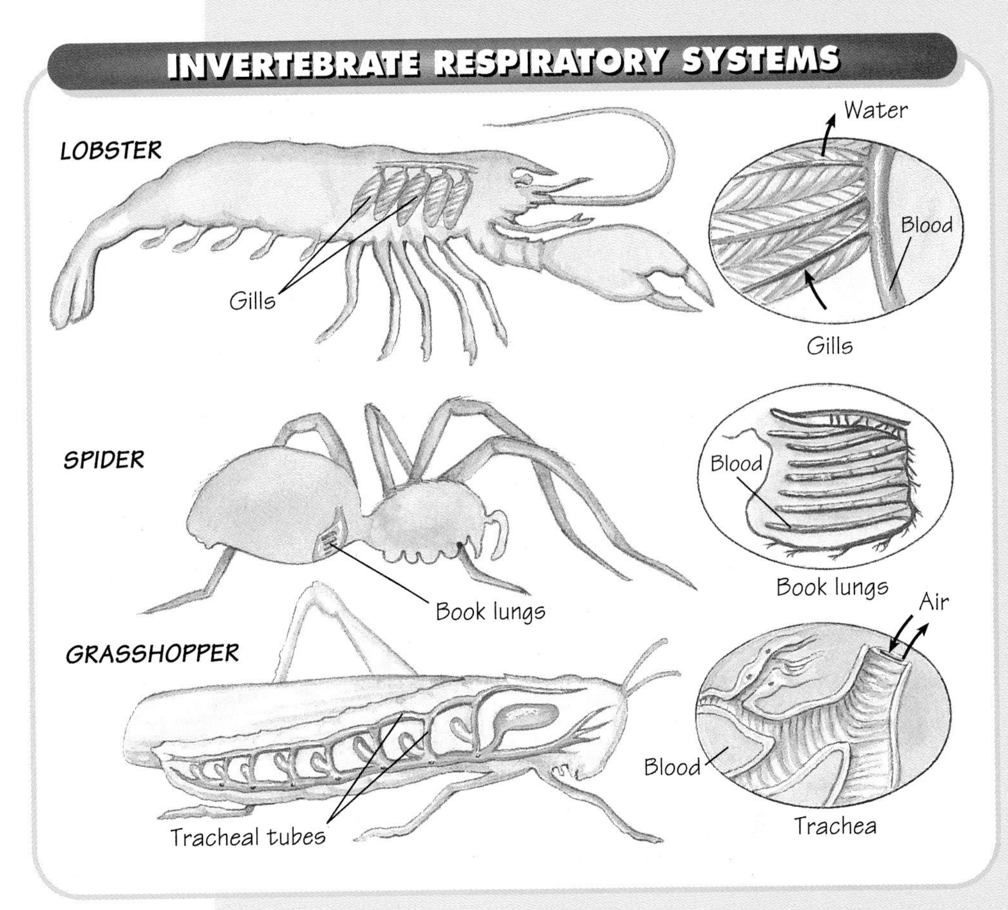

Figure 19–19
The respiratory structures of invertebrates vary. Water-dwelling invertebrates such as lobsters have gills, while some land-dwelling invertebrates such as spiders have book lungs that resemble internal gills. Grasshoppers, which are insects, have tracheal tubes.

Closed Circulatory System

Closed circulatory systems keep blood contained in a system of closed vessels that pass through various parts of the body and return the blood back to the heart. In such systems, blood does not come into direct contact with tissues. Instead, oxygen, carbon dioxide, and other substances diffuse in and out of the blood through very thin walls of tiny blood vessels. This system provides a more rapid and efficient control of blood flow through the body.

Respiration

Small soft-bodied animals may be able to exchange oxygen and carbon dioxide with their environment simply by diffusion through their body surfaces. But once animals get large enough to require a circulatory system, they often also require a specialized organ for respiration.

Regardless of where an animal lives, its respiratory system must deal with two facts of life. First, the respiratory system must have a large surface area that is in contact with the environment, so that gas exchange by diffusion is adequate to support the demands of the organism. Second, all respiratory organs must keep their gas exchange surfaces wet, because diffusion can take place only across moist membranes. If an animal's respiratory surface dries out, gas exchange would stop and the animal would suffocate.

Animals that live in water have no problem with the requirements you have just read about because their environment keeps respiratory surfaces wet.

MINI LAB ... Experimenting ....

To Catch a Worm

PROBLEM *How does an earthworm move through its environment?* ***Design an experiment*** *to answer this question.*

SUGGESTED PROCEDURE 

1. Formulate a hypothesis that you want to test.
2. Obtain an earthworm and a lab tray from your teacher.
3. Observe the movements of the earthworm.
4. Carefully touch the earthworm and observe its reactions.

ANALYZE AND CONCLUDE

1. What was your hypothesis?
2. How did the results of the experiment support your hypothesis?
3. Based on your observations, what type of skeleton do you think an earthworm has?

Many of the worms and cnidarians simply respire through body surfaces. Others—such as annelid worms, mollusks, and crustaceans—have respiratory organs called gills. Gills are feathery structures that expose a large surface area to the environment. Gills are rich in blood vessels that bring blood close to the surface for gas exchange.

Terrestrial invertebrates have evolved several different types of organs for breathing air. Spiders and their relatives have structures called book lungs, which are sheetlike layers of thin tissue that contain blood vessels. Insects have a series of air-filled tubes called trachea that bring air to each body cell. Trachea branch extensively and reach deep into most of their body tissues.

Excretion

For most animals, excretion—the process of eliminating toxic waste products—is closely tied with maintaining proper water balance within the body. Cells that break down amino acids produce a highly toxic water-soluble waste product—ammonia. The ammonia is carried through the body dissolved in blood and body fluids. As a result, getting rid of ammonia usually also involves getting rid of water.

INVERTEBRATE EXCRETORY SYSTEMS

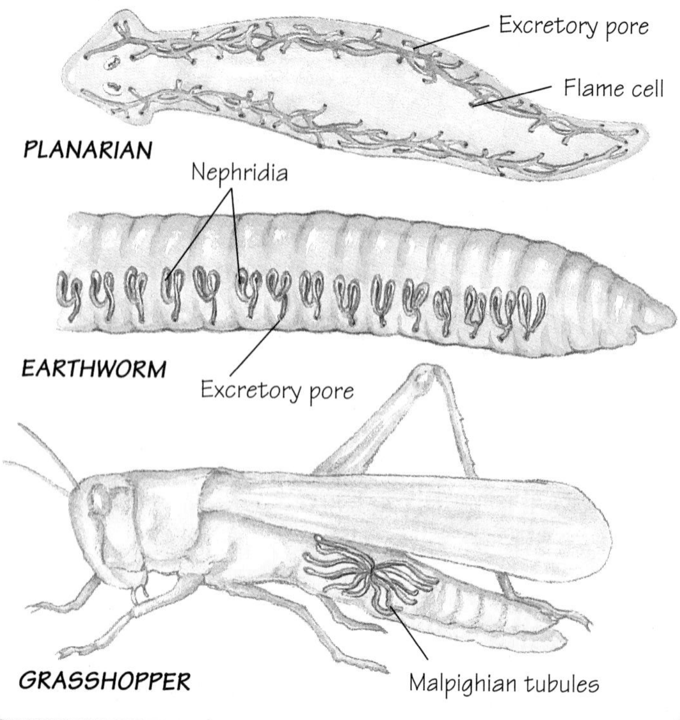

Figure 19–20
Excretory systems help invertebrates control the amount of water in their tissues. Some flatworms, such as planarians, rely on flame cells to remove excess water and ammonia from the body. Other invertebrates, such as earthworms, have nephridia that remove the wastes in the form of urea. In grasshoppers, Malpighian tubules remove the wastes in the form of uric acid.

Bio FRONTIER Connections

Starfishes or Coral Reefs?

One of the most beautiful coral reefs in the world—the Great Barrier Reef—is located off the coast of Australia. This reef stretches for kilometers, and sea life of all kinds live around it and in it. Colorful fishes, sea anemones, snails, and eels make it one of the most interesting areas on Earth.

Coral is a cnidarian that feeds off microscopic animals and plants that float by in the water. The coral has tentacles that catch and push food into its mouth. Because coral is a simple soft-bodied creature, it needs special protection. The animal itself secretes the hard outer skeleton, which is made of calcium carbonate. Millions of coral animals create a coral reef.

A Coral's Enemies

But the protection is not perfect, because the coral has enemies. One of these enemies is the crown of thorns starfish.

Oceanographers who study reefs began to notice a change in the Great Barrier Reef in the 1980s. Areas of the reef were beginning to die. When the damage to the Great Barrier Reef was first noticed, scientists also noticed that the crown of thorns starfish was increasing in numbers. The crown of thorns is a brown starfish with poisonous spines, which can cause serious injury to humans. It is known to eat up to 50 square centimeters of coral each year. A number of outbreaks of this starfish have occurred in the past—one in the 1960s and another in the 1980s. Scientists who are seeing an increase in the population of crown of thorns are worried. The starfishes are a threat not only to the Great Barrier Reef, but to reefs throughout the South Pacific and elsewhere.

Crown of thorns starfish devouring coral

Outbreaks of Starfish

Scientists have studied past outbreaks of this starfish by taking core samples of the coral. When the scientists drilled into the coral, they retrieved samples of coral that were hundreds of years old. This is because new coral always forms on top of the old coral. The cores showed fossils of crown of thorns at many levels, some as old as 7000 years!

Some scientists who have studied the problem think that humans are responsible for the outbreak of this starfish. The natural enemies of this starfish are being killed off by overfishing and, therefore, more starfishes are around to eat the coral. Other scientists say that this isn't the case because reefs in other areas of the Pacific show an increase in starfish numbers as well.

Making the Connection

In the 1980s, Australian divers went onto the reef with poison needles and killed the starfishes by injecting them with disease-causing organisms. Do you think this was a good idea? Why or why not? Tourists also are harming the reef by using boats and leaving garbage. Should tourists be banned from the reef or from other natural sites?

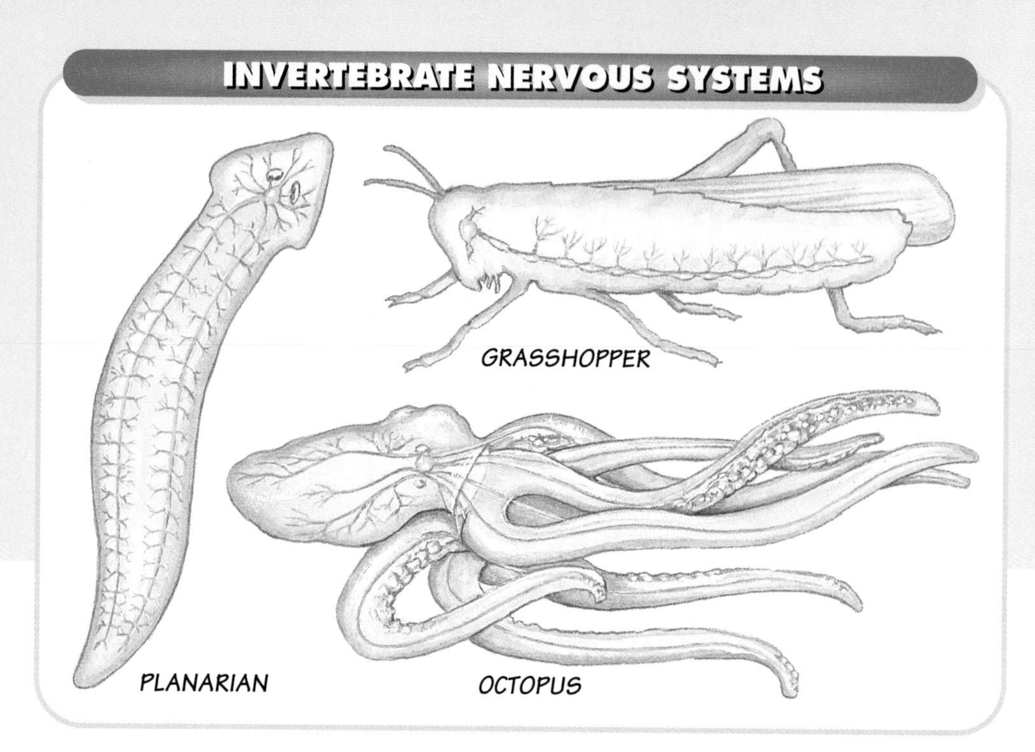

Figure 19–21
A nervous system allows an animal to respond to its environment. Notice how the nervous system of invertebrates contains a greater concentration of nerve cells in the head region as they become more complex.

Excretion in Water

Many marine invertebrates—cnidarians, for example—are thin-bodied and surrounded by water. Ammonia leaves the body of these animals by diffusion through body surfaces or through thin gill membranes.

Some freshwater flatworms have excretory systems containing **flame cells** that remove water and some water-soluble wastes such as ammonia at the same time. The flame cells form a network that empties water and wastes through tiny openings in the animal's skin. Many of the wastes, however, still leave the body through diffusion.

Other groups—such as annelids, mollusks, and invertebrate chordates—have evolved structures called **nephridia** (neh-FRIHD-ee-uh; singular: nephridium). Nephridia take the bodily fluids, remove the wastes, and return water and other important solutes to the body. Waste products are eliminated from the body in the form of urine.

☑ **Checkpoint** What are flame cells?

Excretion on Land

Terrestrial invertebrates have a different problem. They must get rid of body wastes while conserving water. Some groups convert ammonia into urea, which is soluble in water and is much less toxic than ammonia. Urea can be concentrated and stored in a smaller amount of water than an equivalent amount of ammonia. This waste-containing liquid, called urine, is collected and expelled from the body. Other groups, such as insects and some spiders and scorpions, convert wastes into uric acid, which is removed from the body by structures called **Malpighian** (mal-PIHG-ee-uhn) **tubules.** Much of the uric acid precipitates out of solution, forming a paste that is added to the insects' other solid wastes. Thus, their excretion process loses little water.

☑ **Checkpoint** What are Malpighian tubules?

Response

All animals that respond rapidly to their environment do so because of a collection of organs called the nervous system. Amazingly, the basic operations of individual nerve cells are virtually identical—from the most primitive to the most complex animals. The pattern in which nerve cells are arranged, however, changes dramatically from phylum to phylum.

Centralization

The most primitive invertebrates, including such cnidarians as hydras, have a loose, netlike arrangement of nerve cells. This nerve net spreads throughout their body. Other cnidarians, however, show the beginnings of **centralization,** in which the nerve cells are more concentrated and form nerve cords or nerve rings around the mouth. Nerve cords or nerve rings more efficiently send messages from one body part to another.

Cephalization

As animals began traveling in a head-first direction, nerve cells and sensory cells that gather information from the environment began to concentrate in the head. In primitive flatworms, this does not amount to much more than a few clumps of nerve cells called ganglia (GAN-glee-uh) in the head region. But in insects and some mollusks, ganglia grow larger and become organized into structures that can be called brains. From these brains, collections of nerve cells called nerve cords carry information to and from other parts of the body.

Along with the development of brains, invertebrates began to evolve more and more specialized sensory cells. Flatworms, for example, have simple eyespots that detect the presence or absence of light. Insects, on the other hand, have well-developed eyes that detect complex patterns and colors.

Figure 19–22
CAREER TRACK
Ocean technicians study invertebrates as well as vertebrates that live in the sea. They record data and chart the movements of many animals, studying their behaviors and habitats.

Reproduction

Although nearly every form of reproduction is found among the invertebrates, they are all capable of sexual reproduction. Sexual reproduction helps create and maintain genetic diversity. For that reason, life cycles that involve sexual reproduction are believed to improve a species' ability to survive and cope with environmental change. Many invertebrates, including most of those that reproduce asexually at times, engage in sexual reproduction that involves males that produce sperm and females that produce eggs.

Section Review 19–3

1. **List** the three kinds of skeletons found in invertebrates.
2. **Compare** open and closed circulatory systems.
3. **Critical Thinking—Sequencing** Explain how the nervous system has evolved from cnidarians through invertebrate chordates.
4. **MINI LAB** What can you observe about an earthworm's movements? **Design an experiment** to answer this question.

SECTION 19–4 BRANCHING OUT In Depth

Specialized Reproductive Cycles

GUIDE FOR READING

- **Describe** an advantage of hermaphrodites.
- **Compare** external and internal fertilization.

THE ABILITY TO REPRODUCE IS essential to the survival of any species. As a result, life cycles of invertebrates are important—both to the animals and to humans who may be interested in either breeding them or getting rid of them. Interestingly, there is not a single general trend in reproductive style that stretches across the invertebrates. Particular styles of reproduction vary in various phyla, but even the most highly evolved invertebrates use a wide range of strategies for perpetuating their kind.

Asexual Reproduction

Some animals—such as sponges, certain cnidarians, and some echinoderms—can regrow lost body parts. This is a simple form of asexual reproduction, known as regeneration. A planarian, for example, can be cut into several pieces, each of which will grow into a complete worm.

Other invertebrates, such as sponges and cnidarians, also reproduce asexually in a process called budding. Budding is a form of reproduction in which a parent organism grows a bud that becomes an identical individual. It then becomes larger and usually breaks free.

Hermaphrodites

In many species of mollusks, annelids, and echinoderms, individuals have both male and female reproductive structures. These individuals are **hermaphrodites** (her-MAF-ruh-dights). A hermaphrodite has both male and female reproductive organs and produces both sperm and eggs. Usually, though, two individuals

Figure 19–23
Invertebrates show a variety of reproductive techniques. (a) These nudibranchs are hermaphrodites and exchange sperm during mating. (b) Hydras reproduce asexually by budding. Notice the large bud forming on the right side of the hydra.

mate by exchanging sperm, rather than by self-fertilization. If you think about it, you can see how this strategy has advantages in small populations or when individuals are widely scattered. **If any two sexually mature hermaphrodites meet, they can mate with one another, producing two groups of offspring.** In species with separate sexes, individuals must be of different sexes to mate, and they produce only one group of offspring per mating.

☑ *Checkpoint* How do hermaphrodites reproduce?

Figure 19–24 *Notice the brown eggs being released as this* Acorpora *coral is spawning. Spawning is an example of external fertilization.*

External and Internal Fertilization

External fertilization and **internal fertilization** are two basic ways in which sperm cells and egg cells are brought together in sexual reproduction. **External fertilization occurs when eggs and sperm meet outside the organism's body.** This process is called spawning. Generally, the eggs give off a chemical attractant that causes sperm to swim toward them and fertilize them in water. Fertilized eggs typically develop into larvae (singular: larva) that move and drift about in the current. A larva is the immature stage of an organism and is unlike the adult form in appearance. This style of reproduction is simple, but it has some disadvantages. For one thing, it is not very efficient because many eggs and sperm never meet. Therefore, animals that reproduce this way release large quantities of eggs and sperm when they spawn. In addition, this sort of fertilization can work only underwater or in very wet places.

The other option, of course, is internal fertilization. **During internal fertilization, the eggs and sperm meet inside the body of the egg-producing individual.** In more complex invertebrates, this meeting is arranged by mating, in which the male deposits sperm inside the female's reproductive tract. Fertilization takes place within the body of the female, which may then either care for the eggs in some way or deposit them on a rock or other convenient place.

Section Review 19–4

1. **Describe** one advantage of hermaphrodites.
2. **Compare** internal and external fertilization.
3. **BRANCHING OUT ACTIVITY** **Construct a chart** in which you compare the body systems of all living invertebrate phyla, including the method of reproduction. Add artwork or photographs to illustrate the different phyla. Display your chart in the classroom.

CHAPTER 19

Laboratory Investigation

Comparing Nervous Systems

As the animal phyla move from simple to complex, the phyla develop more specialized features. In this investigation, you will compare the nervous systems of a cnidarian (hydra) and a platyhelminthe (planarian) to determine which phylum has a more advanced nervous system.

Problem

How can you **compare** the response to stimuli in a hydra and a planarian?

Materials (per group)

culture of live hydras
culture of live planarians
Petri dish or small clear container
dissecting probe or straight pin
microscope
medicine dropper

Procedure   

1. Use a medicine dropper to transfer one hydra to the center of the Petri dish. The hydras can be seen if you hold the culture up to the light.
2. Place the Petri dish on the microscope stage. Use the low-power objective of the microscope to focus on the hydra and observe its movements for 5 minutes.
3. After some time the hydra will stretch and lengthen. With the dissecting probe, gently touch the extended tentacle. Allow the hydra to recover from the stimulus, then touch the dissecting probe to another part of the organism. Record your observations.

4. **Place the hydra back in the culture and rinse out the Petri dish.**
5. **Use the medicine dropper to transfer a planarian from the culture to the center of the Petri dish.**
6. **Use the medicine dropper to give the planarian more water (one dropper full should be enough).**
7. **Place the Petri dish on the microscope stage and use the low-power objective of the microscope to focus on the planarian.**
8. **After some time—when the planarian is extended and moving smoothly across the dish—gently touch the head area with the dissecting probe. Allow the planarian to recover from the stimulus, then touch different parts of its body. Observe how the planarian reacts to each probe. Record your observations.**

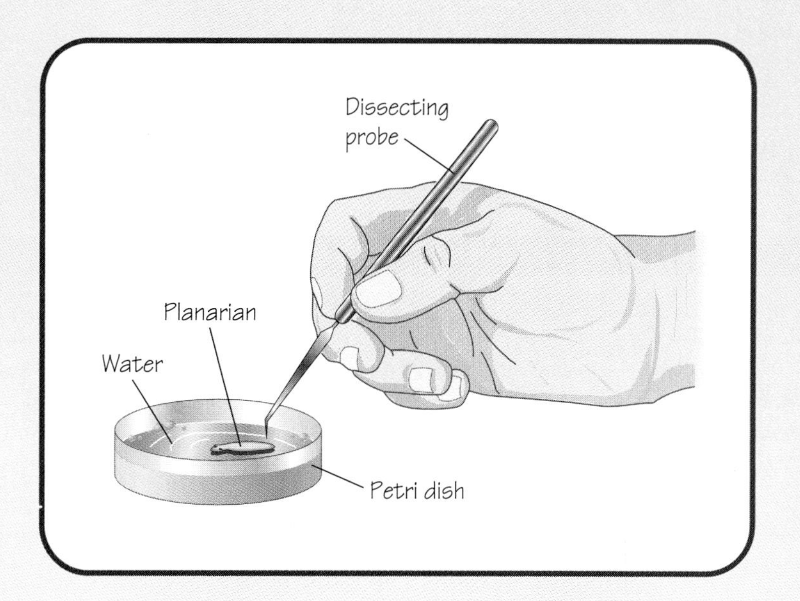

Observations

1. How did the hydra react to each probe? Indicate the body part touched and the response.
2. How did the planarian react to each probe? Indicate the body part touched and the response.
3. Did the planarian react differently when you touched its head and its tail?

Analysis and Conclusions

1. Did the animals respond differently from each other when probed? Describe each animal's reaction to being touched by the stimulus.
2. Which of the two animals do you think has a more advanced nervous system?
3. Which of the two animals you observed showed cephalization? Give examples to prove your point.

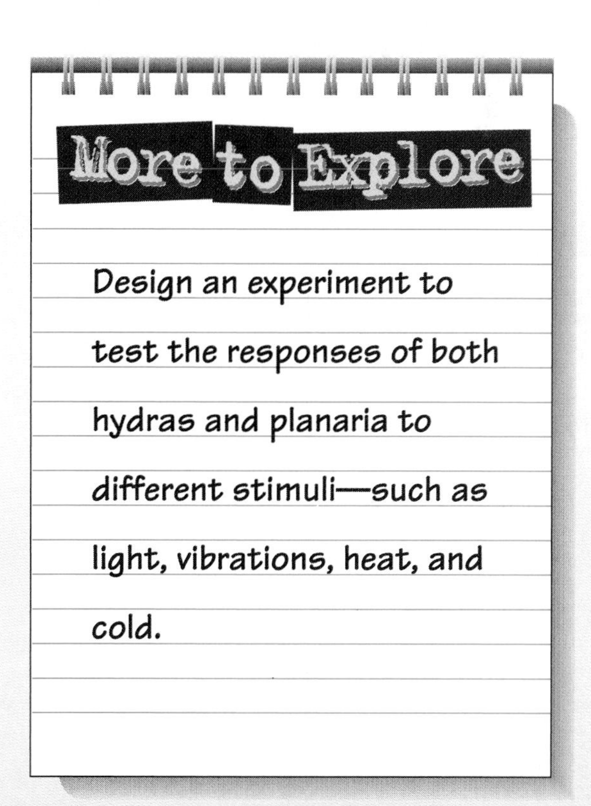

Study Guide

Summarizing Key Concepts

The key concepts in each section of this chapter are listed below to help you review the chapter content. Make sure you understand each concept and its relationship to other concepts and to the theme of this chapter.

19–1 Evolution of Multicellular Animals

- As animals evolved, they became increasingly more complex.
- All animals must perform the same functions: body support and movement, feeding and digestion, respiration, excretion, internal transport, and response to the environment.

19–2 A Survey of Living Invertebrates

- The common ancestors of all multicellular animals had already evolved two distinct cell layers separated by a jellylike middle layer.
- Another important evolutionary trend is the existence of a coelom, or mesoderm-lined cavity.
- The evolution of body plans that are built up from several body compartments is another important evolutionary trend.

19–3 Form and Function in Invertebrates

- The three main types of skeletons found in the animal kingdom are hydrostatic skeletons, exoskeletons, and endoskeletons.
- In some invertebrates, digestion takes place inside the cells. In others, digestion takes place outside the cells.
- Open circulatory systems are systems in which blood from the heart is not contained in blood vessels. Closed circulatory systems keep blood contained in a system of closed vessels that pass through various parts of the body and return blood back to the heart.

19–4 Specialized Reproductive Cycles

- Hermaphrodites are organisms that have both male and female reproductive organs and produce both eggs and sperm.
- External fertilization occurs when eggs and sperm meet outside the organism's body.
- During internal fertilization, the eggs and sperm meet inside the body of the egg-producing individual.

Reviewing Key Terms

Review the following vocabulary terms and their meaning. Then use each term in a complete sentence.

19–2 A Survey of Living Invertebrates

ectoderm
endoderm
mesoderm
coelom
segment
radial symmetry
bilateral symmetry
cephalization

19–3 Form and Function in Invertebrates

hydrostatic skeleton
exoskeleton
chitin
endoskeleton
intracellular digestion
gastrovascular cavity
extracellular digestion
flame cell
nephridium
Malpighian tubule
centralization

19–4 Specialized Reproductive Cycles

hermaphrodite
external fertilization
internal fertilization

Recalling Main Ideas

Choose the letter of the answer that best completes the statement or answers the question.

1. Which is not a characteristic of animals?

a. eukaryotic
b. autotrophic
c. heterotrophic
d. multicellular

2. Tissues that develop from the endoderm form the

a. skeletal system.
b. nervous system.
c. skin.
d. digestive system.

3. The mesoderm-lined body cavity found in some invertebrates is called a

a. segment.
b. nephridium.
c. coelom.
d. mesoderm.

4. The first animals to show bilateral symmetry were the

a. flatworms.
b. annelids.
c. sponges.
d. mollusks.

5. The largest animal phylum contains the

a. echinoderms.
b. nematodes.
c. arthropods.
d. platyhelminthes.

6. An external skeleton is a(an)

a. hydrostatic skeleton.
b. exoskeleton.
c. endoskeleton.
d. coelom.

7. Excretory structures in flatworms are known as

a. flame cells.
b. Malpighian tubules.
c. nephridia.
d. trachea.

8. Which animals have both male and female reproductive organs?

a. pseudocoelomates
b. acoelomates
c. hermaphrodites
d. coelomates

Putting It All Together

Using the information on pages xxx to xxxi, complete the following concept map.

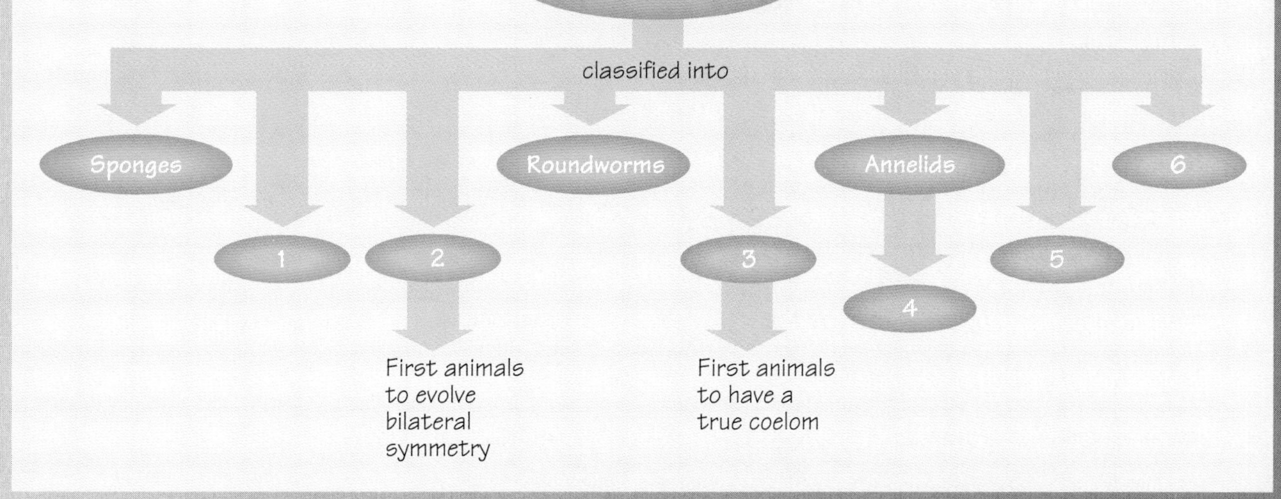

Reviewing What You Learned

Answer each of the following in a complete sentence.

1. When did the earliest known animals appear on Earth?
2. What is the Cambrian explosion?
3. What characteristics do all animals share?
4. What is an ectoderm? An endoderm? A mesoderm?
5. In which phylum did a true coelom first appear?
6. What type of organisms make up the phylum Porifera?
7. What are some examples of echinoderms?
8. Name the three largest groups of arthropods.
9. What characteristic does the phylum Chordata have that no other invertebrate phyla have?
10. What can you **observe** in an animal with radial symmetry? Bilateral symmetry?
11. How do an open circulatory system and a closed circulatory system compare?
12. Define cephalization.
13. Distinguish between intracellular and extracellular digestion.
14. What are nephridia? Flame cells? Malpighian tubules?
15. Compare internal fertilization and external fertilization.

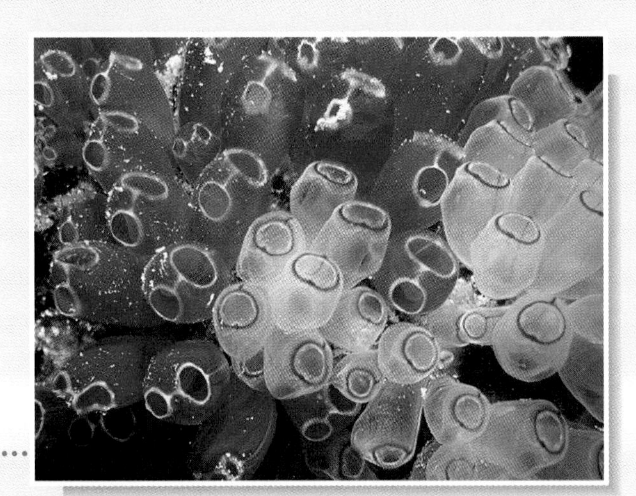

Expanding the Concepts

Discuss each of the following in a brief paragraph.

1. How might the development of specialized tissues help in the rapid explosion of animal forms?
2. What problems did animals face as they moved onto land?
3. Why was the development of the mesoderm so important?
4. A major structural division of the animal kingdom is the acoelomates, psuedocoelomates, and coelomates. Why is the evolution of coelomates so important?
5. Why is segmentation an important development in the evolution of animals?
6. How might the development of jointed appendages give rise to the largest and most diverse group of animals, the arthropods?
7. How would you **design an experiment** to illustrate how a hydrostatic skeleton helps an organism to move?
8. Describe the three types of invertebrate skeletons.
9. A complete digestive system (mouth to anus) allows better processing of food materials for extracellular digestion. Explain why this is true.
10. Describe the evolution of the respiratory system from gills to trachea.

Extending Your Thinking

Use the skills you have developed in this chapter to answer the following.

1. **Hypothesizing** How can crabs, which have a rigid nonliving exoskeleton, grow larger while being restricted by this exoskeleton? Endocrinologists have demonstrated that two hormones may be involved—one that increases the production of urine and one that decreases the production of urine. Develop a hypothesis to explain how they might work.
2. **Using the writing process** Write a one-page summary of the various adaptations that have occurred in the respiratory system of organisms that migrated from life in water to life on land. Start with single cells and moist membranes.
3. **Analyzing** Maintaining a proper water balance for organisms is crucial, no matter where they live. Combining the problem of water balance with toxic products produced by an organism causes an extremely dangerous situation to develop. How have organisms been able to solve this difficult situation, especially as they move to terrestrial environments?
4. **Hypothesizing** The fossil record indicates extreme diversity of the invertebrate form. Today, only a few invertebrates, such as the common cockroach, have survived the test of time, while other organisms, such as the trilobites, have disappeared. Develop a hypothesis to explain this situation.
5. **Problem solving** One group of people demonstrated that almost 85 percent of all blindness throughout the world is preventable. Much of the blindness is caused by various invertebrates (especially certain worms), protists, or the lack of certain vitamins. How would you set out to improve this situation, keeping in mind the necessity of maintaining the balance of nature?

Applying Your Skills

Worm or Beetle?

Is a mealworm really a worm? A mealworm may look like a worm, but it is really the larval stage of the Tenebria *beetle.*

1. Working in a group, observe three or four mealworms for five minutes, using a magnifying glass.
2. Identify as many physical characteristics of the mealworm as you can, then record these in your journal.
3. Draw a picture of your mealworm, identifying the various body parts.
4. Gently touch the mealworm with a pencil eraser. Record its response.
5. Place an object in front of the mealworm. How does the mealworm react? Does it go around, over, or under the object?

• GOING FURTHER •

6. From the chapter, select at least six invertebrate organisms and describe their method of locomotion. Does each have "feet"?
7. What are the common characteristics of the invertebrates you selected? Differences?

CHAPTER 20

Animals: Vertebrates

FOCUSING THE CHAPTER THEME: Unity and Diversity

20–1 Evolution of Vertebrates
- **Describe** some of the evolutionary advances in vertebrate history.

20–2 A Survey of Living Vertebrates
- **Describe** the different phyla of living vertebrates.

20–3 Form and Function in Vertebrates
- **Compare** the body systems of living vertebrates.

BRANCHING OUT *In Depth*

20–4 Reproductive Adaptations to Life on Land
- **Describe** some of the reproductive adaptations to life on land.

LABORATORY INVESTIGATION
- **Design an experiment** to compare a single-loop circulatory system and a double-loop circulatory system.

Biology and Your World

BIO JOURNAL

Judging from the look of the humpback whale in this photograph, you may think that it is a fish. However, a whale is a mammal, just as you are. In your journal, make a list of the characteristics that these two groups of vertebrates have in common. After you have read the chapter, revise your list. Include photographs and/or drawings.

Humpback whale breaching in Alaska

Evolution of Vertebrates

SECTION 20–1

Guide for Reading

- **Classify** vertebrates as a subphylum of chordates.
- **Identify** some of the trends in vertebrate evolution.

MORE THAN 99 PERCENT OF all chordates are vertebrates. Their skeletons and other hard body structures have left behind an excellent fossil record. As a result, we know a great deal about their evolutionary history. As you climb the vertebrate family tree from past to present, you will learn about the important evolutionary steps our vertebrate ancestors have made.

The Vertebrate Family Tree

Our family tree is rooted in the far distant past. The variety of fossilized organisms preserved in the Burgess Shale during the Cambrian Period included a peculiar creature that was different from all the others. This creature, *Pikaia,* was first thought to be a kind of worm. But researchers who took a closer look decided that it was actually the first known member of the phylum Chordata.

Why did scientists choose to place *Pikaia* in the chordate phylum, rather than among the worms it seems to resemble? The presence of a **notochord,** a flexible supporting structure along its back, provides the reason. A notochord is a characteristic unique to chordates.

☑ Checkpoint What is a notochord?

What Is a Vertebrate?

Humans, fishes, amphibians, birds, and reptiles are classified differently from *Pikaia* and the other invertebrate chordates by being placed in the chordate subphylum Vertebrata. Why are all these organisms in a separate subphylum? Because unlike their invertebrate chordate counterparts, vertebrates possess a notochord only during their early stages of development. As the vertebrate develops, the notochord is then replaced by a stronger supporting structure called

Figure 20–1 The great diversity of the vertebrates is shown by (a) the emerald boa, (b) the regal angelfish, and (c) the poison arrow frog.

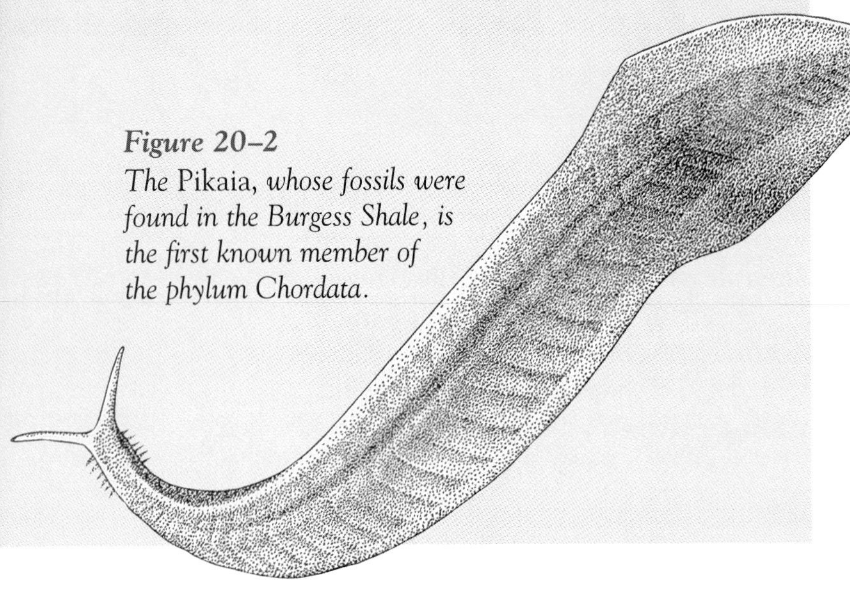

Figure 20–2
The Pikaia, whose fossils were found in the Burgess Shale, is the first known member of the phylum Chordata.

a backbone, or vertebral column. The backbone is made up of individual segments of bone called **vertebrae** (singular: vertebra). The backbone is the central part of the vertebrate endoskeleton because it provides a place for the bones of the skull, arms, and legs to attach. It also provides a means of attachment for muscles and protects the nerve cord.

Checkpoint What are vertebrae? 1

Fishes—Early Vertebrates

Vertebrates evolved for a long time in the sea, so it shouldn't surprise you to learn that our common ancestors are found among the fishes. In fact, several of the most important evolutionary stages in vertebrate evolution occurred among ancient fishes.

The earliest and most primitive fossil vertebrates are often called jawless fishes. These animals didn't lack mouths, of course. All of us have to eat! But their mouths weren't very helpful, because without real jaws they could have no teeth and couldn't bite. They could only suck in water containing food particles, like ocean-going vacuum cleaners.

The first notable trend in vertebrate evolution was the development of true bony jaws. These support structures made it possible for muscles to create a hard bite and provided a place for teeth

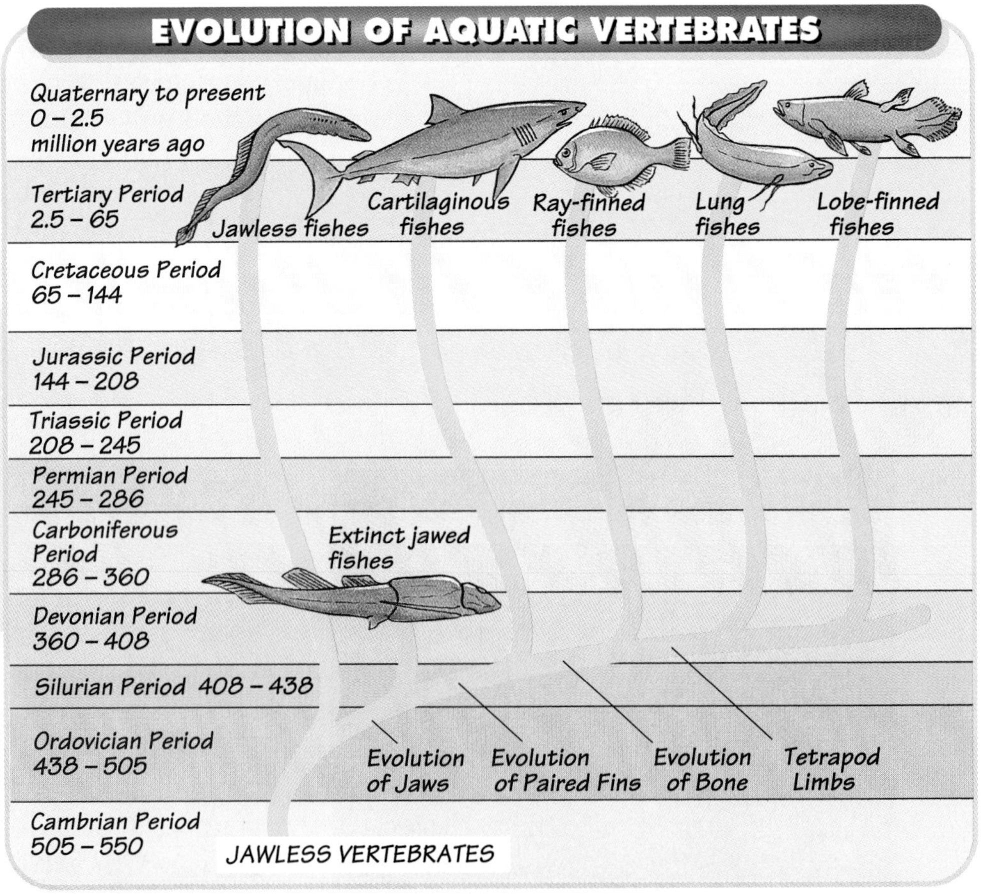

EVOLUTION OF AQUATIC VERTEBRATES

Quaternary to present
0 – 2.5
million years ago
Tertiary Period
2.5 – 65
Cretaceous Period
65 – 144
Jurassic Period
144 – 208
Triassic Period
208 – 245
Permian Period
245 – 286
Carboniferous Period
286 – 360
Devonian Period
360 – 408
Silurian Period 408 – 438
Ordovician Period
438 – 505
Cambrian Period
505 – 550
Jawless fishes
Cartilaginous fishes
Ray-finned fishes
Lung fishes
Lobe-finned fishes
Extinct jawed fishes
Evolution of Jaws
Evolution of Paired Fins
Evolution of Bone
Tetrapod Limbs
JAWLESS VERTEBRATES

Figure 20–3
This evolutionary tree of aquatic vertebrates shows how fishes evolved over time. The emergence of tetrapod limbs led to the evolution of land vertebrates.

to attach. Jaws thus transformed a simple opening into an adaptable, useful feeding tool.

The next evolutionary trend was the development of paired pectoral and pelvic limb girdles. Limb girdles connect limbs with the backbone in a way that enables muscles to work efficiently. The pectoral girdle connects the backbone to the front limbs (or arms), and the pelvic girdle connects the backbone with the rear limbs (or legs). Fishes didn't have arms or legs, but limb girdles enabled them to evolve complex and useful fins.

One group of vertebrates evolved a skeleton made of a strong, resilient material called **cartilage.** Other vertebrates evolved skeletons made of true bone.

One group of vertebrates, the lobe-finned fishes, evolved fins that were different from those of other fishes in an important way. **The bones in the fleshy fins of the lobe-finned fishes evolved into bones that support arms, legs, wings, and flippers in all higher vertebrates.** Because most vertebrates have a body plan that includes four limbs, we are all called **tetrapods,** which means "four-legged."

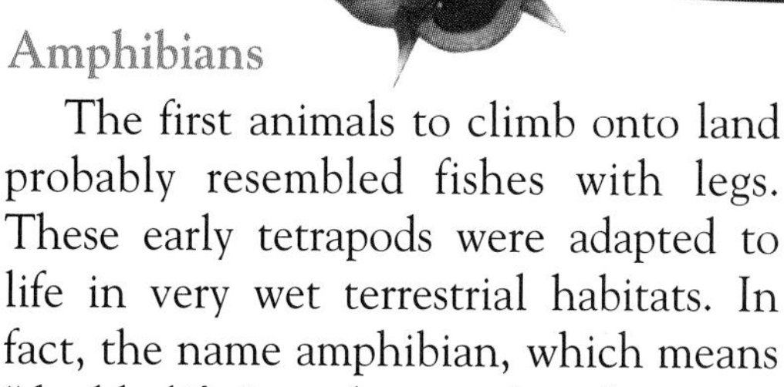

Figure 20–4
This Pacific hagfish is an example of a modern jawless fish. Jawless fishes are the ancestors of modern vertebrates.

Amphibians

The first animals to climb onto land probably resembled fishes with legs. These early tetrapods were adapted to life in very wet terrestrial habitats. In fact, the name amphibian, which means "double life," emphasizes that these animals live their lives both in water and on land.

As amphibians moved from water to land, they evolved two adaptations. The first was the tetrapod limbs. The second was a set of lungs and breathing tubes that enabled them to breathe air.

The descendants of amphibians evolved many adaptations that served them well in wet environments. Amphibians were very common in the warm, swampy fern forests of the Carboniferous Period, about 260 million years ago. These animals gave rise to the ancestors of living amphibians and the ancestors of vertebrates who live completely on land.

Challenges of Life on Land

Animals able to live on land didn't appear overnight. Adapting to terrestrial life involved more than just evolving legs and clambering out of the water! Vertebrates colonizing terrestrial habitats faced the same challenges that had to be overcome by terrestrial plants and invertebrates. They had to breathe air, protect themselves from drying out, support themselves against the pull of gravity, and reproduce without the help of standing water.

Figure 20–5
The coelacanth is the surviving species of the lobe-finned fishes—the ancestors of all tetrapods.

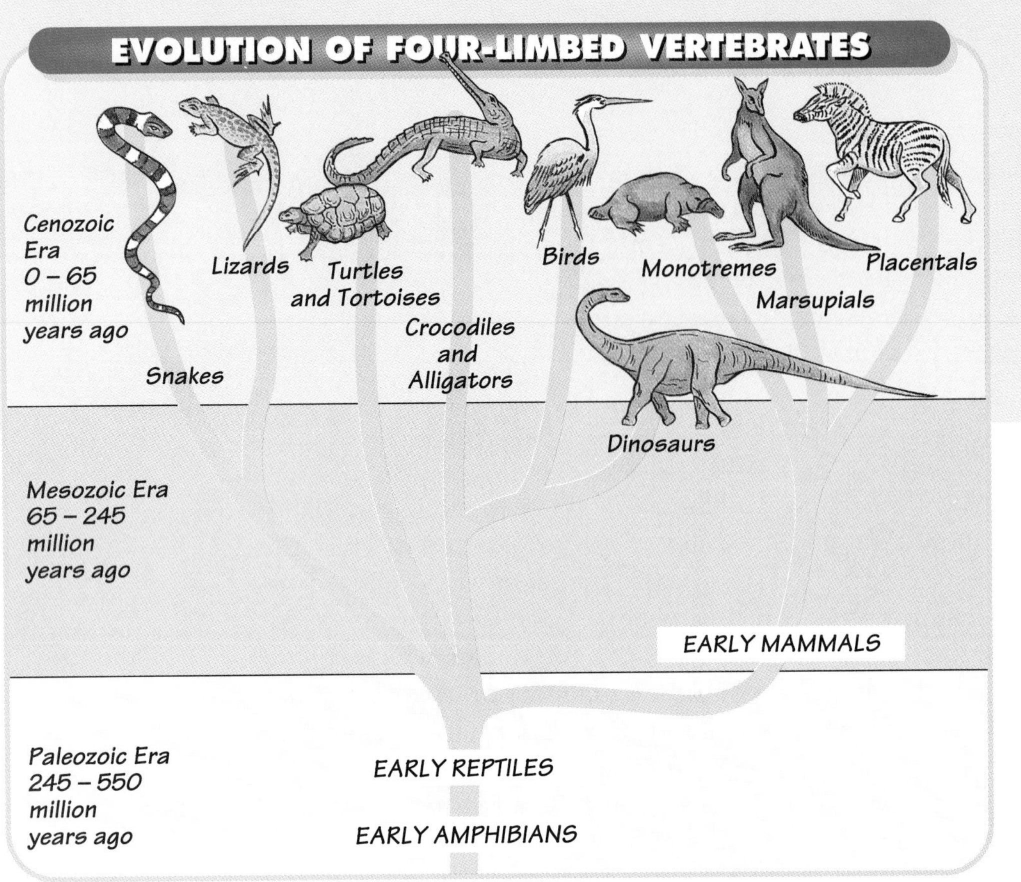

Figure 20–6
All modern reptiles, birds, and mammals evolved from amphibians, early reptiles, and early mammals. This evolutionary tree shows how all the vertebrates are related to one another.

As a group, amphibians need to keep their bodies moist and must prevent the rapid loss of water through their skin. In addition, amphibians must return to water to reproduce.

The First Reptiles

To colonize permanently dry habitats, animals needed a way to reproduce that didn't require water. The first animals to evolve this adaptation were the reptiles.

Unlike amphibians, reptiles reproduce by internal fertilization. Their fertilized eggs don't need water to develop because they are surrounded by a protective shell and several layers of membranes that protect the embryo.

As the Carboniferous Period drew to a close, terrestrial habitats changed. As the Permian Period began, the climate became drier. Many lakes and swamps disappeared. Under these drier conditions, reptiles evolved and diversified rapidly. By the end of the Permian Period, a great variety of reptiles had become widespread.

The Great Permian Extinction

The Permian Period ended about 245 million years ago in the most devastating mass extinction in the entire history of life on Earth. Almost 95 percent of all species of marine animals disappeared. This change in life on Earth was so extraordinary that paleontologists have declared this time not just the end of the Permian Period, but the end of the entire Paleozoic Era—the era of ancient life.

☑ **Checkpoint** Why did the end of the Permian Period also mark the end of the Paleozoic Era?

Figure 20–7
Amphibians were the first vertebrates to develop tetrapod limbs and lungs. Early amphibians from the Carboniferous Period may have resembled this artist's drawing.

The Mesozoic Era

The end of the Paleozoic Era cleared the stage for the next era in the history of life. The Mesozoic, or middle life, Era—the age of the ruling reptiles—could now begin.

Dinosaurs—The Ruling Reptiles

Few animals have captured human imagination as much as the dinosaurs and the other great reptiles of the Mesozoic Era have. Two separate groups of large aquatic reptiles prowled the seas. Ancestors of modern turtles, crocodiles, lizards, and snakes populated many terrestrial habitats. And dinosaurs, the great and terrible lizards, were everywhere.

Until about 30 years ago, paleontologists thought of dinosaurs as giant versions of living snakes and lizards. ● Recent research has revealed how interesting many of these animals actually were. Some dinosaurs traveled in great herds. Others lived in small family groups, caring for their eggs and young in carefully constructed nests. New and interesting information about dinosaurs is discovered every day!

Birds

If you, like many people, wish you could have seen a live dinosaur, you might have to go no farther than your window or the nearest pet store. Research has shown that modern birds evolved from a group of small flesh-eating dinosaurs! The most important piece in this puzzle is shown in **Figure 20–8.** *Archaeopteryx*, a 150-million-year-old fossil, is classified by researchers as the first known bird. If you remove the fossilized feathers, the bones of this animal could pass for those of a small running dinosaur.

The First Mammals

Our mammalian ancestors evolved over a long period of time. The first to appear over 300 million years ago were the odd mammallike reptiles. These animals and their descendants left behind many fossils with characteristics that changed slowly from those of reptiles to those of primitive mammals. That's one reason it is difficult to say exactly when true mammals appeared.

INTEGRATING CAREERS

What are some of the career options open to paleontologists? What kinds of courses must they take?

The Great Cretaceous Extinction

At the end of the Cretaceous (krih-TAY-shuhs) Period, 65 million years ago, a mass extinction ended the reign of the dinosaurs. Despite a great deal of research, geologists and paleontologists still do not agree on what caused this mass extinction. Many believe that a huge asteroid smashed into what is now Mexico's Yucatan Peninsula. ● That impact would have blasted a large dust cloud into the atmosphere, blocking the sunlight and lowering temperatures around the world. Other researchers propose a more gradual global cooling

INTEGRATING SOCIAL STUDIES

Use a map of Mexico to find the Yucatan Peninsula. In which part of the country is it located?

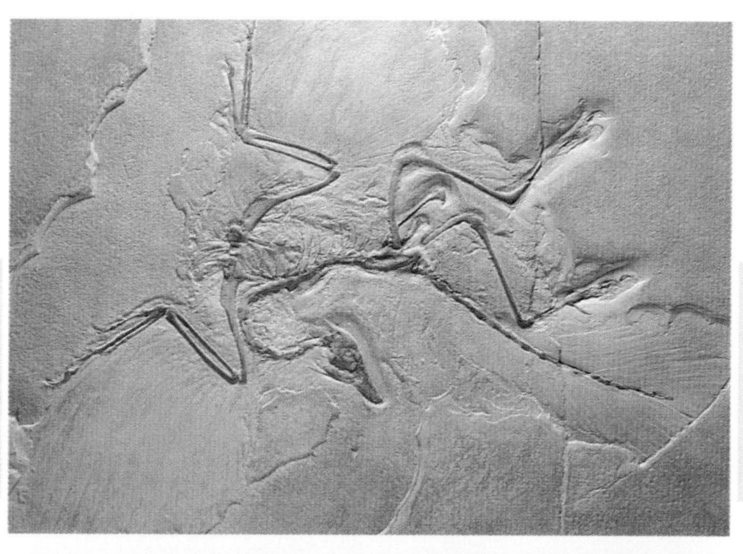

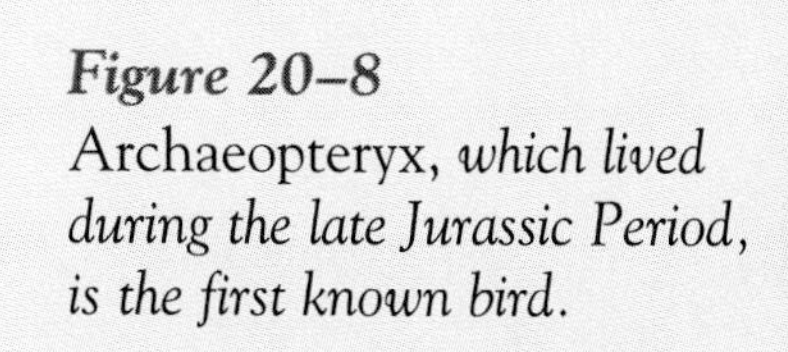

Figure 20–8
Archaeopteryx, *which lived during the late Jurassic Period, is the first known bird.*

caused by increased volcanic activity. Many volcanic eruptions could have created the same sort of dust cloud.

Whatever the cause, dinosaurs and the other ruling reptiles were wiped out, leaving a clean slate for the evolution of animals that were larger than insects and birds.

The Age of Mammals

The Earth that the mammals inherited was a changing patchwork of environments. During the late Mesozoic Era, the continents began to drift apart. First, two great landmasses separated into a northern continent and a southern continent. These landmasses isolated groups of ancient mammals that began to evolve along different paths. By the end of the Cretaceous Period, most modern continents, except for Antarctica and Australia, had separated. This caused some mammals to flourish in the southern continents, while other mammals evolved in the northern continents.

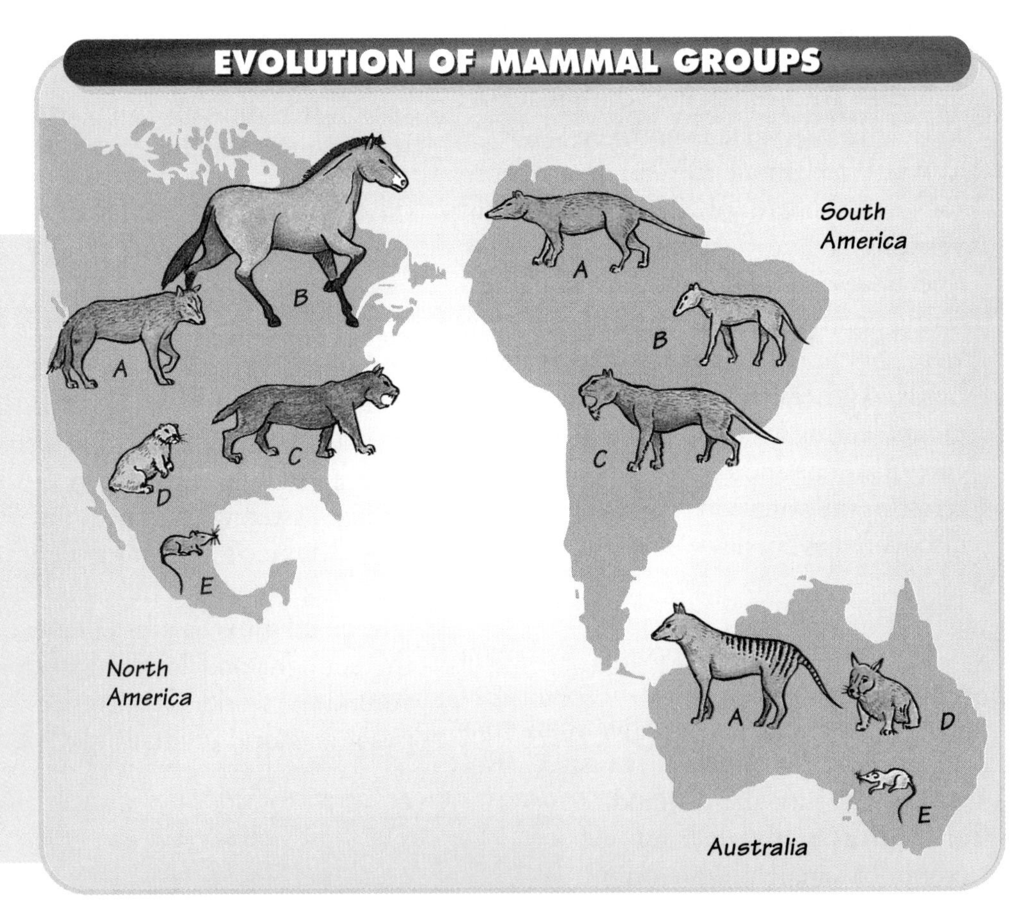

Figure 20–9
As the continents moved apart early in the Cenozoic Era, ancestors of different mammal groups were isolated from one another. In adapting to similar environments, some of these unrelated animals evolved similar appearances and behaviors. Species labeled with the same letter have similar adaptations, although they are unrelated to each other.

Section Review 20–1

1. **Classify** vertebrates as a subphylum of chordates.
2. **Identify** some of the trends in vertebrate evolution.
3. **Critical Thinking—Making Comparisons** In what ways are the vertebrate adaptations to life on land similar to plant adaptations to life on land?

A Survey of Living Vertebrates

Guide for Reading

- **Describe** the characteristics of living vertebrates.

MINI LAB

- **Design an experiment** to find out what factors affect the breathing rate of a fish.

THEY RANGE IN SIZE FROM TINY fishes a few centimeters long to blue whales, the largest animals ever to live on Earth. They walk, crawl, run, hop, slither, burrow, swim, and fly. They soar over mountains, roam ocean canyons, ride on Arctic ice floes, plod across trackless deserts, and call through the night in tropical rain forests. They are not the most numerous animals, and they are by no means the most diverse. They are the organisms most of us think of when we think of animals. They—or, more accurately, we—are the vertebrates.

Classification of Vertebrates

As you may recall, all vertebrates have a vertebral column, or backbone. What other characteristics do the vertebrates have in common? **Besides a backbone, most vertebrates have two sets of appendages (arms and legs), a closed circulatory system with a ventral heart, and either gills or lungs for breathing.**

Jawless Fishes

Lampreys and hagfishes are the only living vertebrates without jaws. They are not living fossils but specialized parasites and scavengers. Lampreys attach themselves to fishes or other aquatic animals and rasp at their flesh to feed on their blood and body fluids. Hagfishes feed on dead and decaying carcasses.

Cartilaginous Fishes

Sharks, skates, and rays make up the class of cartilaginous fishes called Chondrichthyes

Figure 20–10 *Vertebrates are classified into many different classes based on their characteristics. (a) The giant panda is an example of a mammal, (b) the swan is a bird, and (c) the sea turtle belongs to a class of reptiles.*

Figure 20–11
Fishes and amphibians are two diverse classes of vertebrates. (a) The orange sea horse is one of many bony fishes. (b) As an adult, the yellow-eyed salamander, a member of the order Urodela, lives in moist woods.

(cahn-DRIHK-theez), whose skeletons are made up of cartilage. Some, such as the great white shark, are fearsome predators. Despite their bad reputation, only a few species attack humans—and then only rarely. Many other species eat small fishes, mollusks, or plankton.

Bony Fishes

This enormous and diverse class of fishes, the Osteichthyes (ahs-tee-IHK-theez), contains more than half of all living vertebrate species. They inhabit nearly every aquatic habitat imaginable—from tropical streams to the sea beneath Arctic ice. There are two subclasses of bony fishes—ray-finned fishes and lobe-finned fishes.

The ray-finned fishes consist of more than 20,000 species and include nearly every fish. Ray-finned fishes have a well-developed system of bones, and many have specialized jaws and teeth that enable them to eat a variety of foods.

Although few species of lobe-finned fishes are alive today, they are of great evolutionary importance because they are the ancestors of all tetrapods.

Amphibians

The class Amphibia, or amphibians, have never fully recovered from the great Permian extinction. Only about 2500 species survive, and many of these species have been endangered or threatened. Amphibians lay their eggs in water, where they spend at least part of their life. As adults, amphibians live on land and breathe through lungs. Most have moist skin and cannot tolerate long periods of dryness.

Newts and salamanders, members of the order Urodela, are amphibians with tails. They lay their eggs in water, where they hatch into young that resemble the adult form. The larvae have gills that disappear and are replaced by lungs as they develop into adults.

Frogs and toads make up the order Anura. They do not have tails as adults. Their eggs hatch in water into larvae called tadpoles, which look completely

Figure 20–12
Reptiles are classified into three orders. (a) The Jackson's chameleon is a member of the order Squamata, which includes lizards and snakes. (b) This alligator, found in the Florida Everglades, is a member of the class Crocodilia.

different from the adults. Tadpoles live in water until they develop into adult frogs.

☑ *Checkpoint* What are the two orders of amphibians?

Reptiles

Reptiles, members of the class Reptilia, were the first fully terrestrial vertebrates. They reproduce by internal fertilization and produce leathery-shelled eggs that prevent water loss. They also have dry, scaly skin that retains water inside the body. Reptiles breathe air by means of developed lungs.

Tortoises and turtles belong to the order Chelonia (kuh-LOH-nee-uh). These animals, which have a very long fossil record, are encased in a protective shell of bony plates. Some live in fresh water, others in the sea, and still others in habitats as dry as deserts.

Crocodiles, alligators, and caimans make up the order Crocodilia. This order also has a long fossil record. These animals are now mostly in tropical and subtropical regions along fresh- or saltwater coastlines, swamps, and marshes.

The order of lizards and snakes—Squamata (skwah-MAH-tuh)—is the most abundant group of living reptiles. There are more than 2500 species that live in habitats ranging from oceans to deserts. Lizards and snakes look quite different but are actually closely related.

☑ *Checkpoint* What are the three orders of reptiles?

Birds

The class Aves, or birds, are defined by a unique feature—feathers. Feathers are used not only for flight, but also to conserve body heat. This is a large and diverse class, containing 8600 species in 27 different orders.

MINI LAB .... Experimenting ....

A Fishy Story

PROBLEM *What factors affect the breathing rate of a fish? **Design an experiment** to answer the question.*

SUGGESTED PROCEDURE  

1. Fill a large beaker about halfway with water from an aquarium. Use a net to move one fish from the aquarium to the beaker. Let the fish adjust to its new environment for at least 5 minutes.
2. Observe how the fish moves. Look at its fins and general body shape.
3. Using a clock with a second hand, count how many times the fish opens its mouth per minute. Observe any movement of the gill covers (operculum) during the same time period. Record your data. Return the fish to the aquarium.
4. Using a similar procedure, design an experiment to determine what effect the presence of another fish has on the first fish's breathing rate.
5. Write out your hypothesis and, with your teacher's approval, carry out your experiment.

ANALYZE AND CONCLUDE

1. How many times did the fish open its mouth per minute?
2. Did you observe any coordinated movement of the gill covers with the opening and closing of the mouth?
3. What effect did the presence of another fish have on the first fish? Give evidence to support your answer.

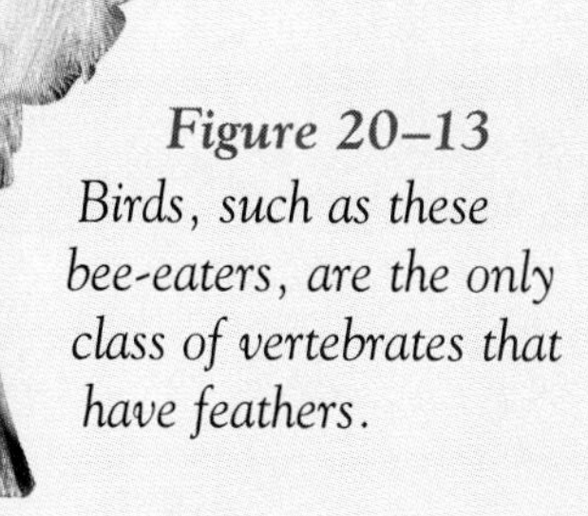

Figure 20–13 *Birds, such as these bee-eaters, are the only class of vertebrates that have feathers.*

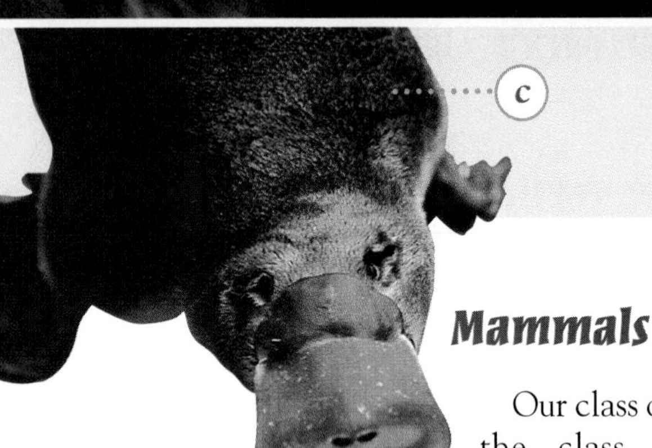

Figure 20–14
Mammals are divided into three groups based on the way their young develop. (a) The orangutans are placental mammals, (b) the gray kangaroo with its joey is an example of a marsupial, and (c) the duckbilled platypus is an example of a monotreme.

Mammals

Our class of vertebrates—the class Mammalia—is named for the presence of **mammary glands,** which enable females to nourish their young with milk. Mammals are also characterized by the presence of hair, whose primary function is to help retain body heat. There are three main groups of living mammals: monotremes, marsupials, and placental mammals.

Monotremes, or egg-laying mammals, are very rare. The duckbill platypus and two species of spiny anteaters are the only monotremes in existence today. Monotremes show a curious mix of features. Like reptiles, they lay leathery eggs. But like all mammals, monotremes have body hair and, once their eggs hatch, they nourish their young with milk.

Marsupials, like most other mammals, bear their young alive. Marsupials are born incredibly early, sometimes only eight days after fertilization, while they are little more than embryos. The embryos crawl to a pouch called the marsupium (mahr-SOO-pee-uhm), where they receive nourishment from their mother and complete their development. Such living marsupials as koalas and kangaroos live in Australia, but a few, such as the opossum, live in North, South, and Central America.

The largest group of living mammals—placental mammals—get their name from the **placenta,** an organ that connects the mother with her developing embryo. Nutrients, oxygen, wastes, and carbon dioxide are exchanged through the placenta. There are about 4500 living mammals that are divided into 16 different orders.

Section Review 20–2

1. **Classify** the living vertebrates into different classes and orders.
2. **Critical Thinking—Summarizing** Which classes of vertebrates rely on outer coverings to conserve body heat?
3. **MINI LAB** How could you **design an experiment** to determine what factors affect the breathing rate of a fish?

Form and Function in Vertebrates

SECTION 20–3

GUIDE FOR READING

- **Describe** a single-loop circulatory system and a double-loop circulatory system.
- **Compare** the two techniques of body temperature control in vertebrates.

MINI LAB

- **Design an experiment** to find out how animals maintain their internal body temperature.

VERTEBRATES HAVE EVOLVED a series of adaptations to terrestrial life and to a wide range of habitats. Some adaptations involve hard body parts that can be traced in the fossil record. Others involve soft tissue structures and functions that can be compared only among living animals. Moving from the oldest fishes up to primates, most body systems became more complex as vertebrates adapted to different environments.

Support and Movement

Vertebrates have an endoskeleton with bones that are surrounded by muscles and skin. The central element of that skeleton is the vertebral column, or backbone. This series of connected vertebrae, along with muscles and ligaments that attach to them, help support body mass and make it possible to control body movements.

As vertebrates adapted to terrestrial life, the position of both pairs of limbs changed. Early amphibians had limbs that stuck out almost horizontally from the body. These animals walked by bending from side to side, much as fishes swim. But as animals grew larger and heavier, horizontal limbs couldn't offer enough support. It isn't surprising, then, that many reptiles, including dinosaurs, evolved limbs that grew vertically. Vertical limbs could support mass more efficiently. Many mammals also have vertical limbs, allowing them to stand erect with their legs straight under them.

Figure 20–15
Vertebrates have developed body systems and features to help them survive. (a) Because camels are able to conserve water, they can survive in the desert without water for fairly long periods of time. (b) Most mammals, including these elephants, stay with their young and protect them until they are able to live on their own. (c) A lizard's specialized mouthparts and digestive system enable it to eat a crunchy grasshopper.

Feeding and Digestion

As part of the adaptive radiations of vertebrates, different species evolved ways of life that depend on nearly every kind of food in nature. But each potential food poses its own set of challenges. Meat is easy to digest, but it must first be caught and cut up for swallowing. Plant food is often tough and must be pulverized and shredded before being swallowed. Plant parts such as leaves and stems are also full of hard-to-digest cellulose.

Every group of vertebrates has evolved adaptations that enable them to capture and digest different foods. Some of these adaptations occur in the teeth of mammals. After food is swallowed, the way it is handled also depends on its chemical makeup. Meat is easily digested and can be passed quickly through a short gastrovascular cavity. So carnivores have short digestive tracts that secrete enzymes that break down meat proteins.

Because leaves and other plant parts are the toughest to handle, herbivores such as cows spend a great deal of time chewing and rechewing their food. They also have stomachs with colonies of bacteria that help to digest cellulose.

Respiration

Exchanging oxygen and carbon dioxide with the environment is a universal requirement of living things. Vertebrates have evolved many strategies that help them adapt to different environments.

Fishes and Amphibians

Fully aquatic vertebrates such as fishes rely almost entirely on gills for respiration. Amphibians, however, show a transitional pattern between water breathing and air breathing. Tadpoles have gills, whereas adult amphibians have lungs that are not well developed. Amphibians also do not have muscles that can inflate and deflate lungs. Most adults, therefore, depend on the exchange of gases through their thin, wet skin.

Reptiles

Reptiles—fully adapted to dry environments—could not breathe through their dry skin, so their lungs became more efficient. The inside of the reptilian lung is divided into many small chambers. These chambers greatly increase the amount of surface area available for gas exchange. In addition, reptiles evolved muscles that help inflate and deflate lungs to pump air in and out.

Figure 20–16
CAREER TRACK
Animal physiologists study animals and their habits. In some parts of the world, rhinoceroses are killed for their horns. In this photograph, two animal physiologists prepare to remove the rhinoceros's horn in order to protect it from being hunted.

EVOLUTION OF THE RESPIRATORY SYSTEM

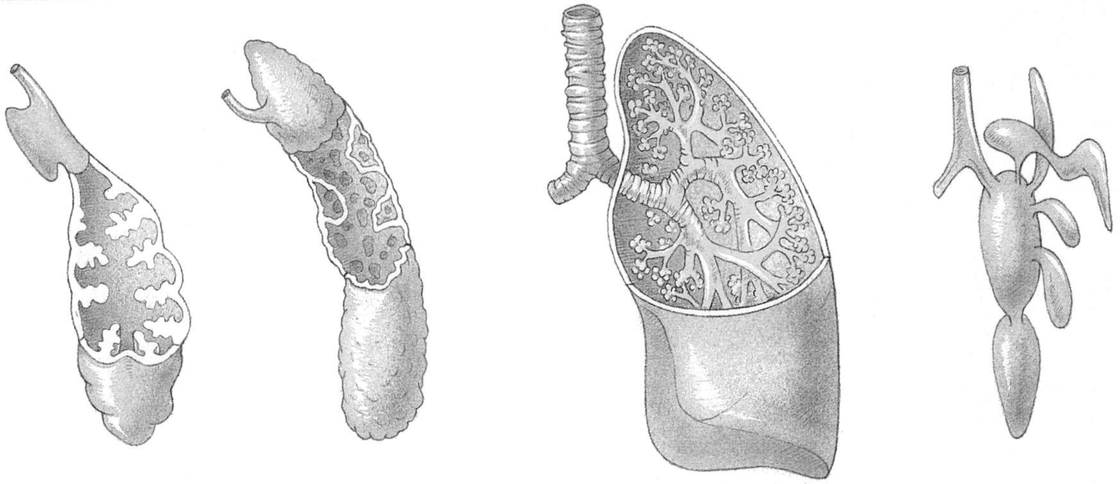

AMPHIBIAN

REPTILE

MAMMAL

BIRD

Figure 20–17
As vertebrates evolved, their respiratory structures became more specialized. Notice that the branching of the air tubes increases as you move from amphibians to mammals. Birds have the most advanced respiratory structures, which provide a continuous flow of fresh air into their lungs.

Mammals

Mammals require a higher rate of gas exchange because they have a higher metabolic rate—the rate at which they use food and oxygen. As a result, mammalian lungs have evolved in ways that have made them even more efficient. Air tubes called **bronchi** (BRAHN-kigh; singular: bronchus) enter the lungs and then branch extensively. They end in bubblelike structures called **alveoli** (al-VEE-uh-ligh; singular: alveolus) that are richly supplied with tiny capillaries. This allows for efficient gas exchange. However, the structure of the lungs is somewhat inefficient. Because air must move in and out of the same passageways, there is always some stale air that remains in the lungs after each breath.

Birds

Birds have the highest requirements for lung efficiency of any vertebrate group. Why? Bird flight is an intense long-term exercise that requires a large, steady oxygen supply. As a result, bird lungs have branched bronchi and alveoli similar to mammalian lungs but with an important difference. Air passageways in bird lungs connect to large air sacs in certain bones. Air is pumped through the lungs in a one-way flow. As a result, the gas-exchange surfaces constantly come in contact with fresh, oxygenated air.

Internal Transport

Circulatory systems in vertebrates have come a long way from the simple structures in ray-finned fishes. Typical fishes have a single-loop circulatory system. **In a single-loop circulatory system, blood is pumped from the heart to the gills and flows from gills to the rest of the body tissues before returning to the heart.** The heart itself consists of two chambers—an atrium that receives blood and a ventricle that pumps it out again.

Vertebrates with lungs for respiration have a double-loop circulatory system. **In a double-loop circulatory system, the first loop carries blood between the heart and the lungs. The second loop carries blood between the heart and the rest of the body.**

Problem Solving

INTERPRETING DIAGRAMS

My, What Big Teeth You Have!

You can tell a lot about an animal by its teeth. The back teeth, or premolars and molars, are jagged in carnivores and flat in herbivores, which allows the animal to grind and shred its food. The canines, or eye teeth, are extremely long and pointed. That's how the animal stabs its prey and holds on to it. The front teeth, the incisors, are sharp and wedge-shaped, like a chisel blade, to allow the animal to bite and cut its food.

In the drawings below are three vertebrate skulls. One of the known skulls belongs to a carnivore, a wolf. Another skull belongs to a herbivore, a deer.

Based on the information presented here and the drawings, determine whether the unknown skull belongs to a carnivore, a herbivore, or an omnivore. Then answer the questions below.

VERTEBRATE SKULLS

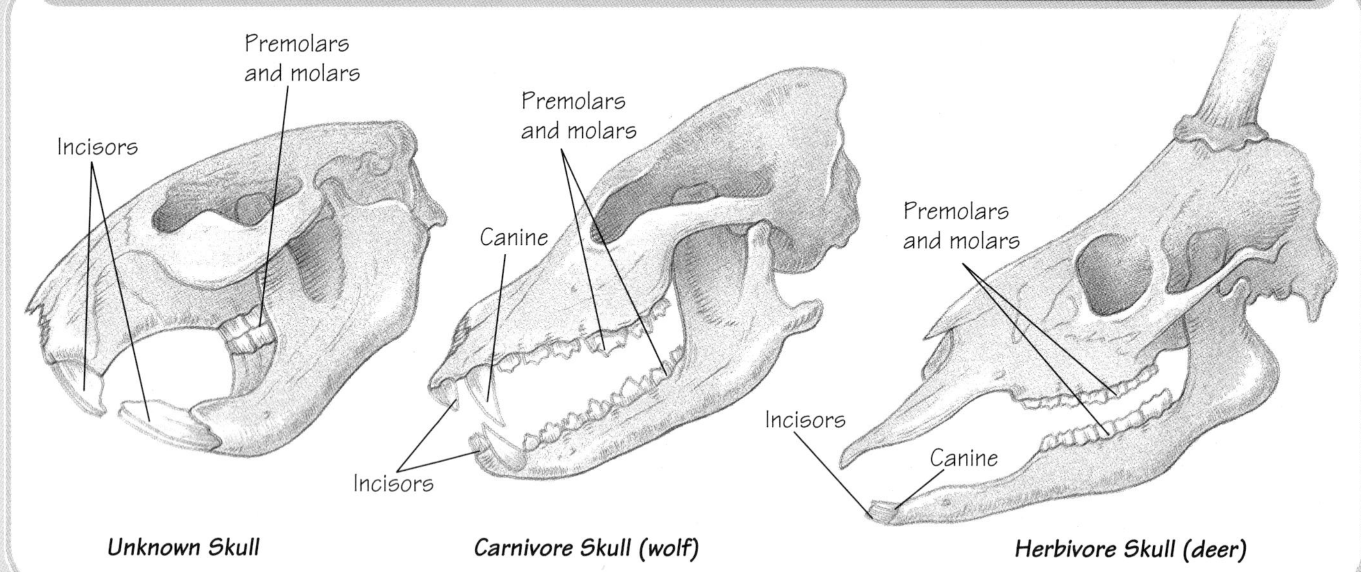

Unknown Skull *Carnivore Skull (wolf)* *Herbivore Skull (deer)*

THINK ABOUT IT

1. In the unknown skull, is the shape of the molars flat or jagged? Based on their shape, **predict** the function of the teeth.
2. What is the size of the incisors in the unknown skull? Based on their shape, **predict** the function of these teeth.
3. **Compare** the positions of the teeth in the jaws of the wolf and the deer in both form and function.
4. **Identify** the unknown skull as that of a herbivore, a carnivore, or an omnivore. Give reasons to support your answer.

Figure 20–18
The vertebrate circulatory system becomes more complex as you move from fishes to mammals. Fishes have a single-loop circulatory system with a two-chambered heart. After the appearance of lungs, vertebrates evolved a double-loop circulatory system. Amphibians and most reptiles have a three-chambered heart, including two atria and one ventricle. Mammals and birds have the most advanced circulatory systems. Their hearts have four chambers—two atria and two ventricles—eliminating the possibility of mixing oxygen-rich blood with oxygen-poor blood.

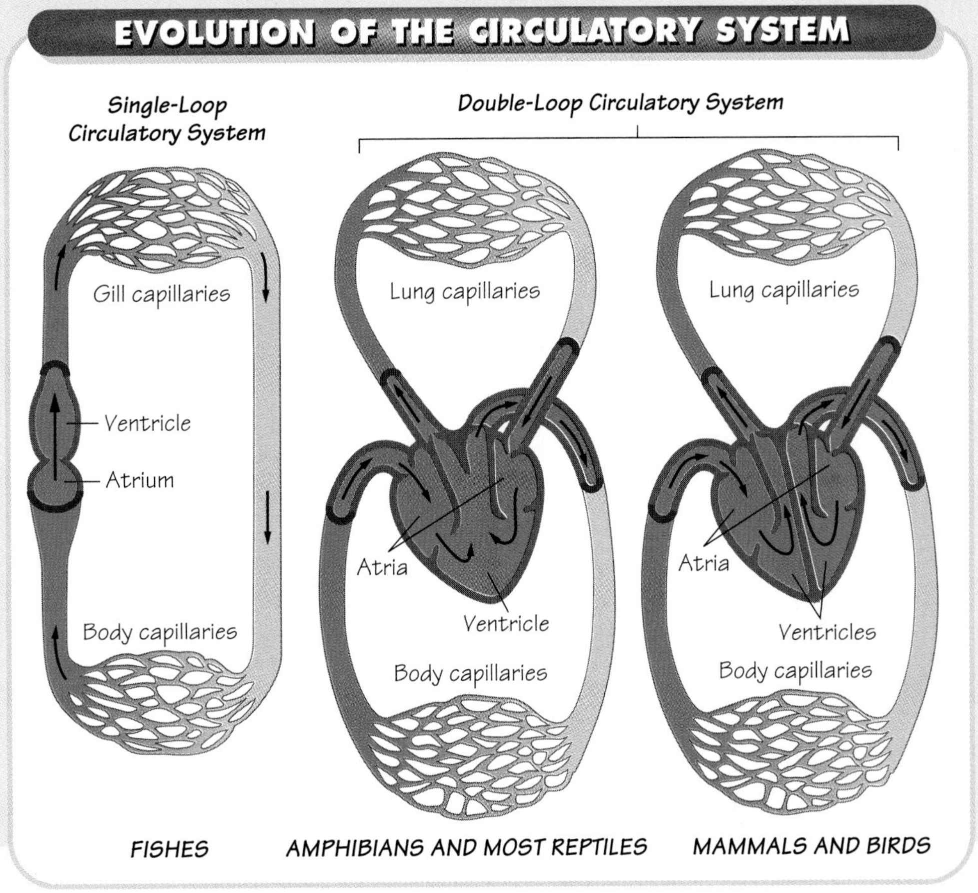

Temperature Control

The ability to control body temperature is an enormous asset. Many chemical reactions, including those important to living things, operate differently at various temperatures. Vertebrates use a variety of techniques to control temperature. All include a source of heat for the body, a method of conserving that heat, and a method of eliminating excess heat when necessary.

Most fishes, amphibians, and reptiles rely on interactions with their environment to help them control body temperature. These animals may bask in the sun to warm up or find shelter in the shade or in underground burrows to cool down. Such animals are known as **ectotherms,** which means "heat from the outside." Although they are often called coldblooded, the term is not accurate.

Mammals and birds generate heat in body tissues through chemical reactions in the body. In addition, mammals and birds have layers of fat and either fur or feathers to keep that heat from leaving the body. When these animals are cold, they shiver to cause muscles to generate more heat. When they are hot, they either pant, as dogs do, or sweat, as humans do, to cool the skin. These animals, often called warmblooded, are more properly called **endotherms,** which means "heat from within."

Excretion

You may recall that the need to eliminate nitrogen-containing wastes is tied in with the need to maintain water balance within the body. As with invertebrates, vertebrate wastes are produced in the form of ammonia. Aquatic vertebrates often get rid of ammonia through diffusion. Fishes lose a great deal across gill membranes, while amphibians lose some through their skin. But these animals also have the beginnings of a **kidney,** the organ used to excrete

How to Keep Warm

PROBLEM *How do fur and feathers help animals maintain their internal body temperature? **Design an experiment** to answer the question.*

PROCEDURE

1. Formulate a hypothesis that you want to test.
2. Using different-sized containers, a thermometer, water, timer, feathers, and pieces of fur, design an experiment that will allow for the collection of data (changes in temperature). Construct a table for your data.
3. Include a control and prepare a list of numbered directions.
4. After your teacher checks your proposed experiment, carry it out.

ANALYZE AND CONCLUDE

1. Does your data support or reject your hypothesis? Explain your answer.
2. What can you conclude about the insulating properties of feathers? Fur?
3. Is insulation of any benefit during the summer? Explain your answer.

nitrogen and other nonsolid wastes. Kidneys became more complex as vertebrates adapted to drier and drier habitats. The mammalian kidney is the most complex.

Like many invertebrates, some vertebrates convert ammonia into less-toxic compounds to make it easier to concentrate and eliminate. Mammals, some adult amphibians, and some cartilaginous fishes convert ammonia to urea, a less-toxic substance. The urea is excreted from the body as urine. Birds and reptiles convert nitrogenous wastes into a semisolid paste—uric acid. Uric acid is also less toxic than ammonia and requires less water to flush it out of the body.

Response

All vertebrates—from jawless fishes to humans—show a great deal of cephalization, the concentration of nerves and sense organs in the head. Even in simple vertebrates, the bundle of nerves and neural connections in the head is large enough to be called a **brain.** From the brain, a long, thick collection of nerves called the **spinal cord** runs down the

Figure 20–19
Because amphibians and reptiles, such as (a) the common collared lizard sunning itself on a rock, rely on the environment to maintain their body temperature, they are called ectotherms. Birds and mammals, such as (b) these sled dogs cuddled up in straw to help keep their bodies warm, can maintain their body temperature internally and are called endotherms.

EVOLUTION OF THE BRAIN

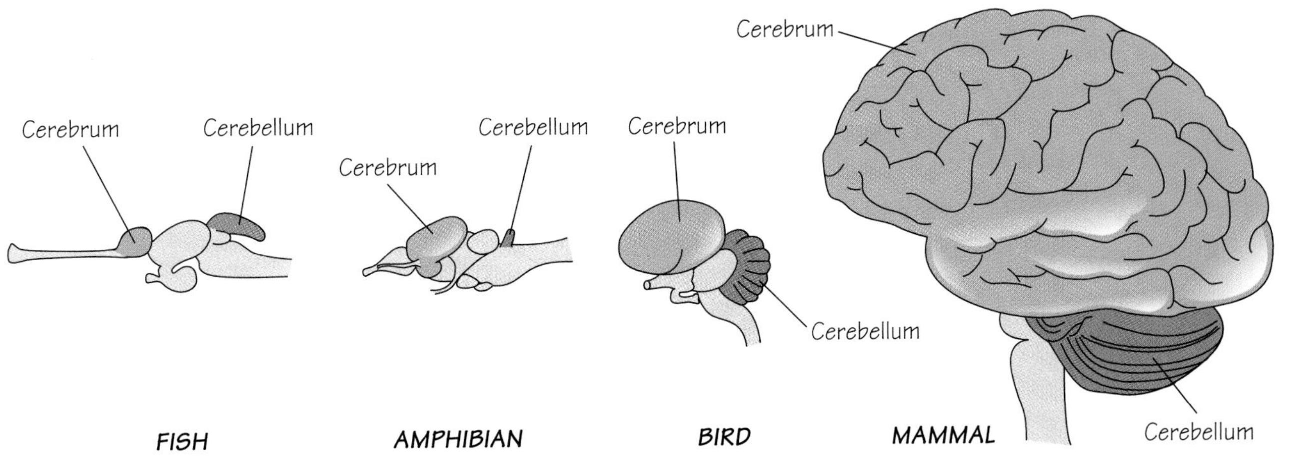

Figure 20–20
Although all vertebrate brains have a cerebrum and a cerebellum, the sizes of each varies. In mammals, the cerebrum, the part of the brain responsible for processing and interpreting information, is the most prominent.

back, protected by the vertebral column. At each joint between the vertebrae, pairs of nerves run in and out of the spinal cord to connect to muscles, organs, and sensory receptors in the skin and throughout the body.

Accompanying their larger brains, vertebrates developed increasingly complex behaviors, and different parts of the brain grew proportionately larger in different groups. The cerebrum, where most thinking takes place, grew steadily from fishes to humans. The cerebellum, the part of the brain largely responsible for coordinating balance and movement, is best developed in birds and mammals.

Reproduction

Almost all vertebrates reproduce sexually. There are, however, a few species of lizards, fishes, and amphibians that develop from unfertilized eggs. In some vertebrates, such as codfish and frogs, fertilization is external. In others—reptiles, birds, mammals, cartilaginous fishes, and certain amphibians—fertilization occurs inside the body of the female. As you move through the vertebrate classes from fishes to mammals, there is a trend from external fertilization to internal fertilization, with some exceptions, of course.

Section Review 20–3

1. **Describe** a single-loop circulatory system and a double-loop circulatory system.
2. **Compare** the two techniques of body temperature control in vertebrates.
3. **Critical Thinking—Drawing Conclusions** Are humans carnivores, herbivores, or omnivores? Give evidence to support your answer.
4. **MINI LAB** How would you **design an experiment** to find out how animals maintain their body temperature?

20–4 Reproductive Adaptations to Life on Land

GUIDE FOR READING

- Discuss the importance of the amniotic egg.

FISHES AND FROGS RELY ON water for reproduction. However, similar to land plants, the more advanced and successful groups had to evolve reproductive strategies that did not depend on water for either fertilization or egg development. That's why one of the most important adaptations in all of terrestrial vertebrate evolution was the amniotic egg.

Development of the Amniotic Egg

Most fishes and nearly all living amphibians reproduce by external fertilization. When adults mate, males deposit sperm and females deposit eggs in water, where external fertilization takes place. The eggs must develop and hatch in water, where they produce larvae that live entirely in water. In amphibians, larvae feed and grow in the water until they develop into adults. These eggs are relatively simple and easy for females to produce, so the females often lay hundreds, or even thousands, at a time.

Beginning with reptiles, however, reproduction followed a different pattern. Reptiles reproduce by internal fertilization, in which males deposit sperm inside the body of the female. Fertilized eggs develop for a time inside the female's body, where the eggs are provided with a food

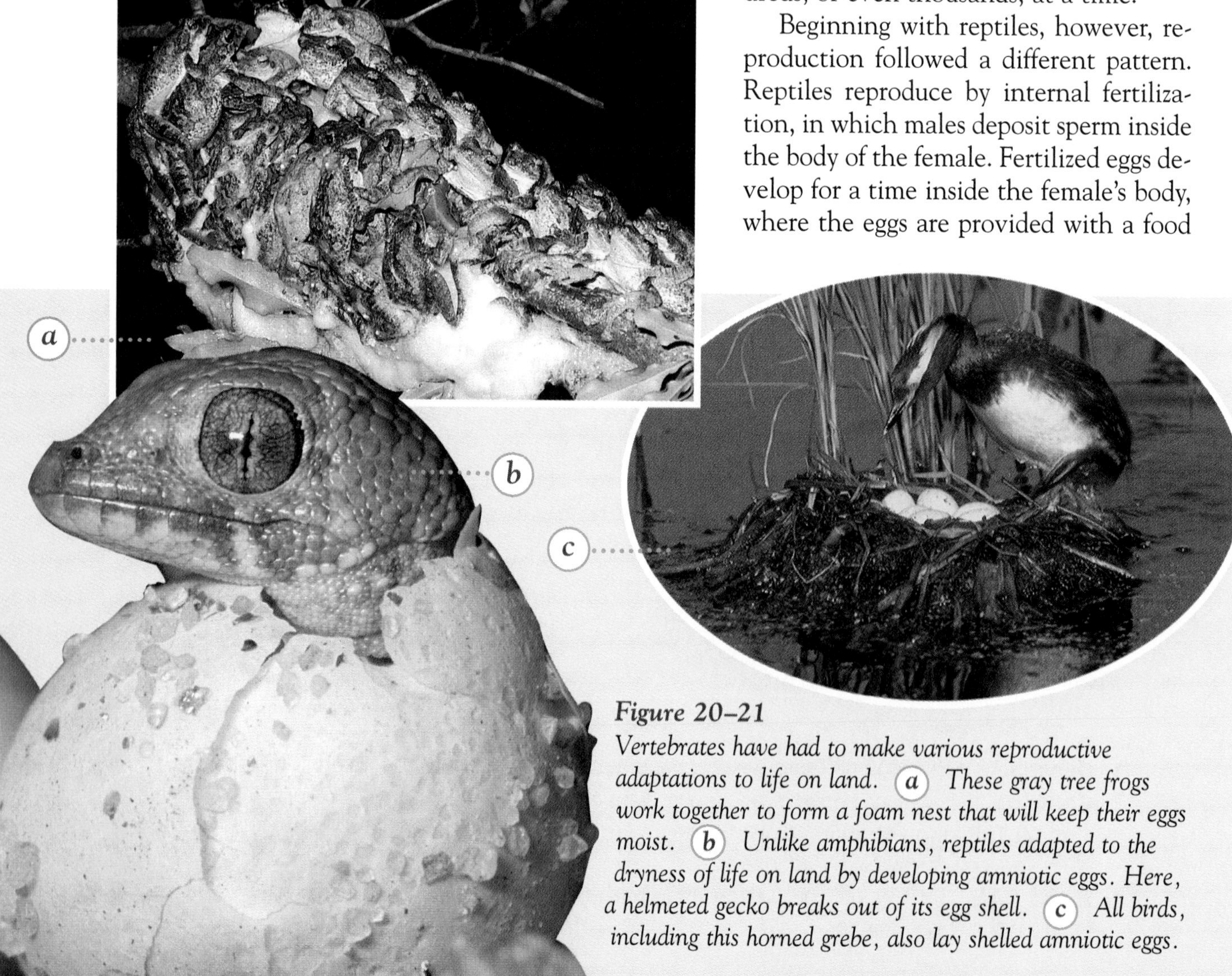

Figure 20–21
Vertebrates have had to make various reproductive adaptations to life on land. (a) These gray tree frogs work together to form a foam nest that will keep their eggs moist. (b) Unlike amphibians, reptiles adapted to the dryness of life on land by developing amniotic eggs. Here, a helmeted gecko breaks out of its egg shell. (c) All birds, including this horned grebe, also lay shelled amniotic eggs.

AMNIOTIC EGG

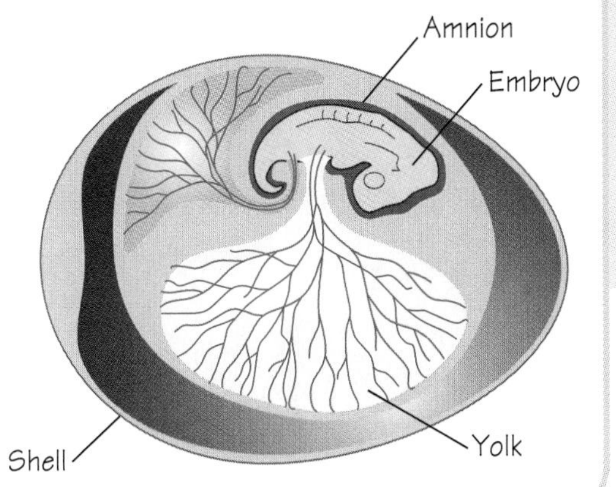

Figure 20–22
The development of the amniotic egg was an important trend in vertebrate evolution. In addition to a shell, amnion, and other membranes, the amniotic egg contains yolk, which is rich in nutrients. The yolk is used by the developing embryo until it is ready to hatch.

supply—the yolk—and are wrapped in several membranes. One of those membranes is the **amnion,** hence the name **amniotic** (am-nee-AHT-ihk) **egg.** The internal membranes bathe the developing embryo in liquid and receive and store its wastes. The entire structure is encased by a shell that allows the exchange of oxygen and carbon dioxide but keeps water inside. Eggs constructed in this way do not need to develop in water. As a rule, fewer of these eggs are laid, and either males or females or both take care of them until they hatch.

Amniotic eggs were an important adaptation to the survival of land animals. This adaptation was passed on from the early reptiles to modern reptiles, birds, and mammals.

☑ **Checkpoint** What is an amniotic egg?

Methods of Reproduction

As important as the amniotic egg has been in vertebrate evolution, it was not the only strategy evolved by vertebrates nor was it the final one. Aquatic and terrestrial vertebrates have also evolved various styles of handling both aquatic and amniotic eggs.

The simplest way to handle eggs, in the water or out, is the one you would usually think of—to lay them. In this strategy, eggs complete their development and hatch outside the female's body. Such animals are described as being **oviparous** (oh-VIHP-uh-ruhs).

Some sharks and bony fishes produce eggs but retain those eggs inside the female until the eggs hatch. In most cases, the embryos receive food stored in the yolk sac. Thus, they receive no further nutrients directly from the mother during development. These animals are described as being **ovoviviparous** (oh-voh-vigh-VIHP-uh-ruhs).

Still other animals, including all mammals except monotremes, retain developing embryos inside the body of the female for long periods. Generally, such eggs do not contain a yolk, or they may contain a small amount of yolk that is used up early in development. Thus, the females must provide additional nutrients during embryonic growth and development. Such animals are **viviparous** (vigh-VIHP-ah-ruhs).

Section Review 20–4

1. **Discuss** the importance of the amniotic egg.
2. **BRANCHING OUT ACTIVITY** **Construct a chart** in which you compare the body systems of all the different classes of vertebrates. Add artwork or photographs to illustrate the different classes. Display your chart in the classroom.

Laboratory Investigation

DESIGNING AN EXPERIMENT

Vertebrate Circulatory Systems

The heart is a part of the transport system that pumps blood throughout the body. In this investigation, you will compare a two-chambered, single-loop circulatory system in fishes to a four-chambered, double-loop circulatory system in birds and mammals.

Problem

How do a double-loop circulatory system and a single-loop circulatory system compare? **Design an experiment** to answer the question.

Suggested Materials

4 8-oz plastic bottles
1 sheet each of red and blue construction paper
red and blue yarn
scissors
scalpel
clear mailing tape

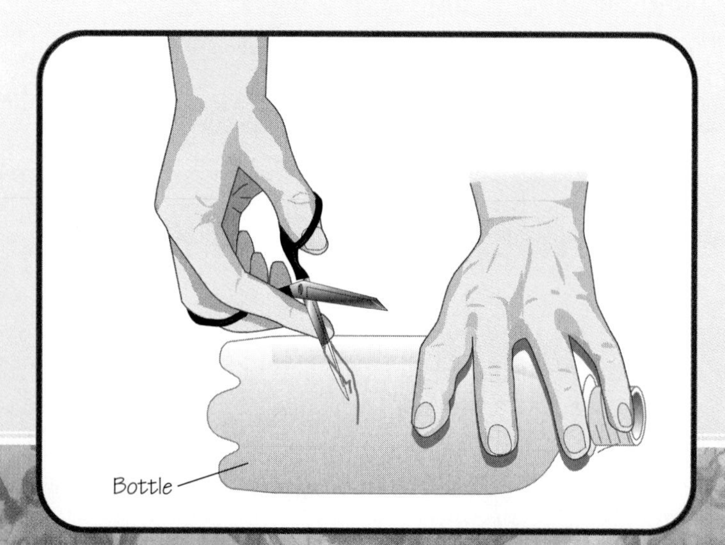

Suggested Procedure 

1. Remove the labels from two of the plastic bottles.
2. Using a scalpel, pierce a hole in one of the bottles where it tapers to form the bottom of the bottle. Place the point of a scissors in the hole and cut around the bottle until the bottom of the bottle is removed. **CAUTION:** *Be careful when using sharp instruments.*
3. Cut a sheet of blue construction paper just large enough to fit inside the uncut bottle. You will have to roll the paper tightly so that it can be inserted into the bottle. Remove the cap and insert the rolled paper into the bottle. This bottle represents the atrium filled with oxygen-poor blood.
4. Insert the remainder of the blue construction paper into the cut bottle. This represents the ventricle filled with oxygen-poor blood.
5. Insert the bottle representing the atrium into the bottle representing the ventricle. They should fit snugly together. Tape them together if necessary.

6. Remove the bottle cap from the "ventricle" and insert a 15- to 20-cm length of blue yarn into the neck of the bottle. Secure one end of the yarn by placing the bottle cap back on. This piece of yarn represents the blood vessel leading to the gills.
7. Unravel the free end of the blue yarn.
8. Unravel both ends of a 15- to 20-cm length of red yarn.
9. Tie the unraveled free ends of the blue yarn to one of the unraveled free ends of the red yarn to represent a capillary network in the gills.
10. Using another 15- to 20-cm length of blue yarn, unravel one end. Tie these unraveled ends to the remaining unraveled free ends of the red yarn. This will represent a capillary network of the body cells.
11. Tape the remaining free end of the blue yarn to the "atrium." This completes the two-chambered, single-loop model.
12. Using a procedure similar to the one given in steps 1 to 11, determine how you would construct a model of a four-chambered, double-loop circulatory system.

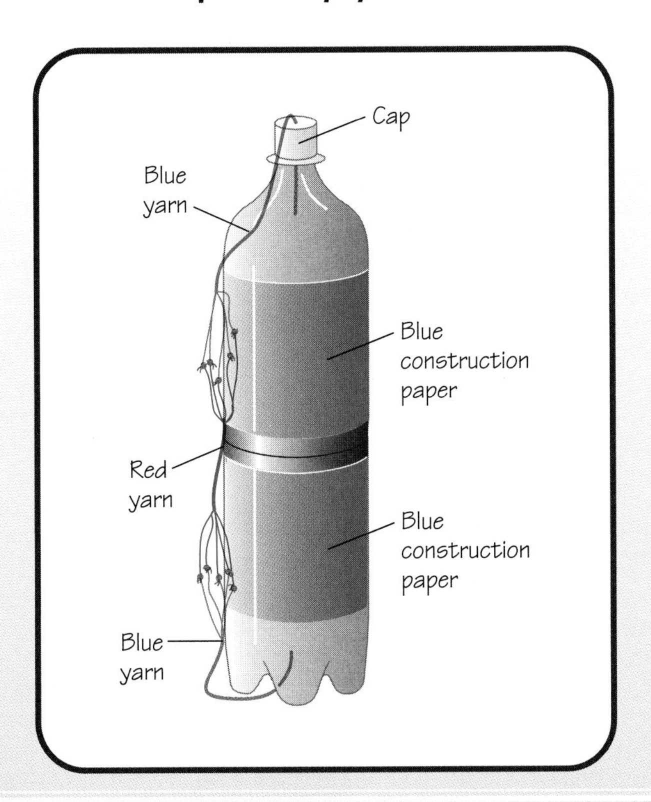

Observations

1. What modifications did you have to make to your single-loop system in order to construct your double-loop system?
2. In the double-loop system, what does the red construction paper symbolize?

Analysis and Conclusions

1. In the single-loop system, where did the blood go after leaving the ventricle? Where did the blood go after leaving the ventricle in the double-loop system?
2. In the single-loop system, where did the blood go after leaving the atrium? Where did the blood go after leaving the atrium in the double-loop system?
3. Why were the red and blue yarns unraveled and joined?
4. Compare the similarities and differences between the two systems you constructed.

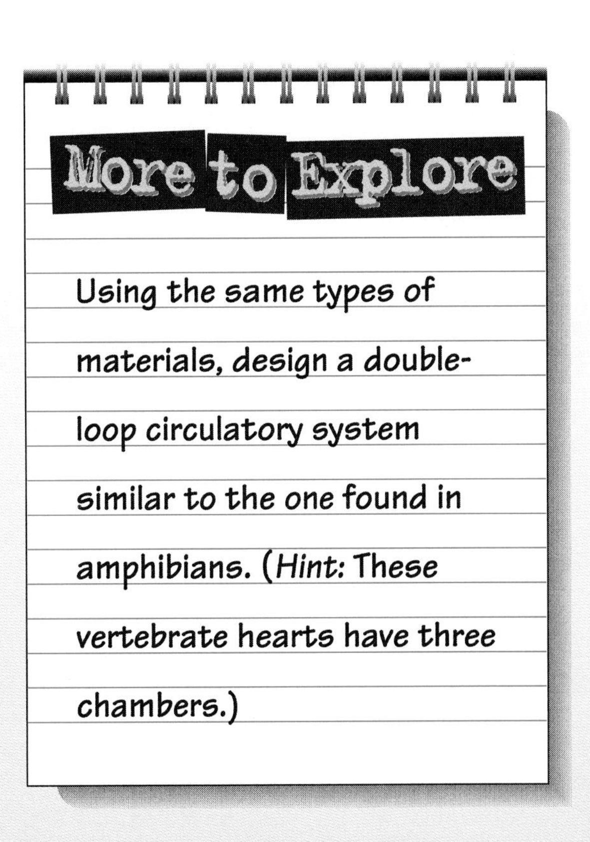

Study Guide

Summarizing Key Concepts

The key concepts in each section of this chapter are listed below to help you review the chapter content. Make sure you understand each concept and its relationship to other concepts and to the theme of this chapter.

20–1 Evolution of Vertebrates

- Mammals, fishes, amphibians, birds, and reptiles are classified in the chordate subphylum Vertebrata.
- The notable trends in vertebrate evolution were the development of true bony jaws, the development of paired pectoral and pelvic limb girdles, and the development of bones.

20–2 A Survey of Living Vertebrates

- Besides a backbone, most vertebrates have two sets of appendages, a closed circulatory system with a ventral heart, and either gills or lungs for breathing.

20–3 Form and Function in Vertebrates

- Primitive vertebrates had limbs that stuck out almost horizontally from the body. Reptiles evolved limbs that grew vertically and could support body mass more efficiently.
- Vertebrates have evolved a variety of jaw structures suited to the foods they eat.
- Fishes have a single-loop circulatory system. Amphibians, reptiles, birds, and mammals have a double-loop circulatory system.
- Most fishes, amphibians, and reptiles are ectotherms. Mammals and birds are endotherms.
- Most fishes and aquatic vertebrates excrete nitrogenous wastes in the form of ammonia. Mammals and some cartilaginous fishes excrete urea. Birds and reptiles excrete uric acid.

20–4 Reproductive Adaptations to Life on Land

- Amniotic eggs were an important adaptation to the survival of land animals.
- Vertebrates are either oviparous, ovoviviparous, or viviparous.

Reviewing Key Terms

Review the following vocabulary terms and their meaning. Then use each term in a complete sentence.

20–1 Evolution of Vertebrates

notochord
vertebra
cartilage
tetrapod

20–2 A Survey of Living Vertebrates

mammary gland
placenta

20–3 Form and Function in Vertebrates

bronchus
alveolus
ectotherm
endotherm
kidney
brain
spinal cord

20–4 Reproductive Adaptations to Life on Land

amnion
amniotic egg
oviparous
ovoviviparous
viviparous

Recalling Main Ideas

Choose the letter of the answer that best completes the statement or answers the question.

1. The first vertebrates that did not need water for reproduction were the
   a. reptiles. **c.** birds.
   b. mammals. **d.** amphibians.

2. Which was not a trend in vertebrate evolution?
   a. development of bony jaws
   b. development of gills
   c. development of paired limb girdles
   d. beginnings of arms and legs

3. The largest group of fishes is the
   a. jawless fishes. **c.** cartilaginous fishes.
   b. bony fishes. **d.** lobe-finned fishes.

4. Which animal does not belong with the others?
   a. newt **c.** toad
   b. salamander **d.** tortoise

5. Which is not a group of mammals?
   a. monotremes **c.** placentals
   b. lampreys **d.** marsupials

6. Which vertebrate group has the most advanced respiratory system?
   a. amphibians **c.** reptiles
   b. birds **d.** mammals

7. Ectotherms obtain the heat they need from
   a. the environment. **c.** food.
   b. the body. **d.** chemical reactions.

8. The cerebellum is best developed in
   a. fishes and amphibians.
   b. amphibians and reptiles.
   c. birds and mammals.
   d. reptiles and mammals.

9. An animal whose eggs develop and hatch outside the female's body is called
   a. oviparous. **c.** ovoviviparous.
   b. viviparous. **d.** amniotic.

Putting It All Together

Using the information on pages xxx to xxxi, complete the following concept map.

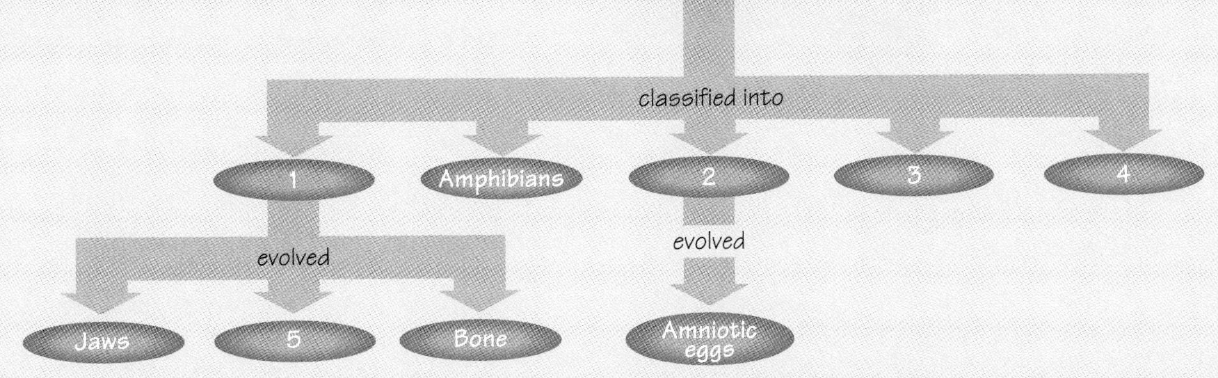

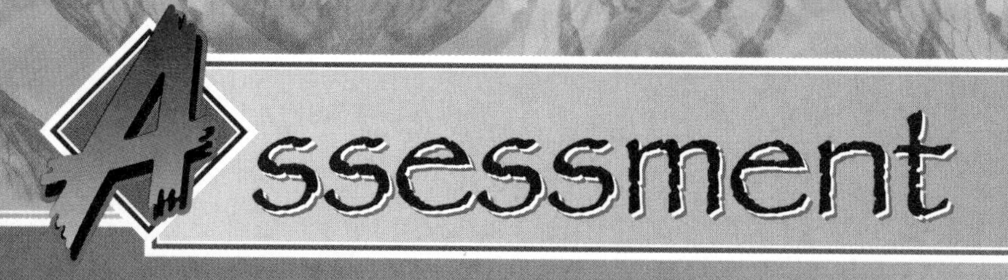

Assessment

Reviewing What You Learned

Answer each of the following in a complete sentence.

1. What is a notochord?
2. How are chordates different from vertebrates?
3. What is cartilage?
4. What is a tetrapod?
5. How can you distinguish a reptile from an amphibian?
6. What is another name for the Mesozoic Era? Why did it get this name?
7. Discuss the changes that led to the rapid and diverse adaptive radiation of the mammals.
8. List the orders of amphibians. Give an example of each.
9. Name three orders of reptiles.
10. How are mammals different from other vertebrates?
11. Name the three groups of living mammals.
12. What is a placenta?
13. What is the amnion?

Expanding the Concepts

Discuss each of the following in a brief paragraph.

1. What was the importance of the evolution of jaws?
2. Explain the importance of the evolution of limb girdles.
3. How are amphibians more advanced than fishes?
4. In what ways are reptiles more advanced than amphibians?
5. Scientists believe that birds evolved from reptiles. How are birds different from reptiles? How are they similar?
6. Discuss how changes in global geography affected the distribution of mammals.
7. **Design an experiment** in which you compare the respiratory system of an amphibian with that of a mammal.
8. Describe the types of skeletal changes that were required for movement on land.
9. **Design an experiment** in which you compare a single-loop circulatory system and a double-loop circulatory system.
10. Describe the differences between endotherms and ectotherms.
11. Discuss the role of the amniotic egg in the evolution of reproduction on land.
12. Compare the three methods of reproduction among vertebrates.

Extending Your Thinking

Use the skills you have developed in this chapter to answer the following.

1. **Analyzing** The kidney and its ability to maintain the internal water balance of vertebrates can be directly related to the organism's habitat. Discuss the structure and function of the excretory system, primarily the kidney, as vertebrates moved from water to land.
2. **Analyzing** At the end of the Pleistocene Era, there was a mass extinction of large animals, such as the mammoth, the mastodon, and the giant sloth in North America. There are two theories concerning this mass extinction. One theory states that many of the animals could not make the broad environmental adjustments they needed in order to survive. The second theory states that as the population of humans grew and moved into the many new habitats, these animals were killed off by humans. Do you agree with either of these theories, a combination, or some other explanation? Defend your answer.
3. **Using the writing process** The Age of Reptiles lasted approximately 150 million years. Its demise was considered the end of the Cretaceous Period. Compared with the reptiles, humans have existed on Earth for a very short time. Do you think humans will exist on Earth for at least as long as reptiles? Write a one-page response to defend your answer.
4. **Drawing conclusions** Groups of animals that have evolved later are often referred to as advanced. Does this mean that they are better animals? Explain your answer.

Applying Your Skills

Vertebrate Census

Scientists need to keep accurate data of both the kinds and numbers of animals that currently exist on Earth. Vertebrates live all over the world. If you observe and record carefully, you may be surprised at the diversity of vertebrates in your own area.

1. For a two-week time period, keep a list of all the vertebrates you observe in your area.
2. Identify the vertebrates by both their common name and their scientific name.
3. Record the number of each organism you see every day during this time period.

• GOING FURTHER •

4. What do you notice about the vertebrates in your area? How are they similar? Different?
5. If you lived in a different part of the country, would you expect your list to be different? If so, how?

CHAPTER 21

Human History

FOCUSING THE CHAPTER
THEME: Evolution

21–1 The Origins of *Homo sapiens*
- Discuss the evidence and various hypotheses to explain how humans evolved.
- Describe the effects of Native American isolation on the history of the Americas.

BRANCHING OUT *In Depth*

21–2 Coevolution of Humans and Parasites
- Explain how antibiotic-resistant bacteria can develop.

LABORATORY INVESTIGATION
- Construct a model to simulate the spread of an infectious disease.

Biology and Your World

BIO JOURNAL

The Incas thrived in inhospitable environments such as the desert coasts and the high mountains of South America. In your journal, describe how an Incan community living at an elevation of 3000 meters, such as the one shown here, might go about meeting its fundamental needs of food, clothing, and shelter all year round.

Incan ruins high in the Andes Mountains at Machu Picchu, Peru

The Origins of Homo sapiens

Guide for Reading

- **List** the characteristics that primates share.
- **Discuss** the origins of *Homo sapiens*.
- **Explain** why people in the Americas lacked resistance to infectious diseases that were common in Europe and Asia.

IN 1924, RAYMOND DART, A South African anatomist, was given a piece of rock from a limestone quarry in Taung, South Africa, that had been loosened in an explosion. After weeks of carefully chipping away at the surrounding limestone, Dart saw a skull—later identified as a child's skull—emerge. It had humanlike features in the shape and relative size of the brain and teeth. Although Dart reported his find in a British journal, it did not receive much attention from the scientific community for more than twenty years.

Since then, other fossils have been uncovered that support Dart's finding. Today, the species to which this "Taung child" belonged is considered to be one of the closest relatives of modern humans. But let's begin the story of the evolution of humans with the primates, an order within the class Mammalia.

The Primates

When Carolus Linnaeus imposed order on life's diversity, he gave special attention to the group of mammals that included humans. He named our order Primata, which means "the first" in Latin. Just what are we "first" in? Linnaeus emphasized the primates' intelligence. Yet some living primates and many fossil species could safely be described as below average in intelligence for mammals. When the first primates appeared, in fact, there was little to distinguish them from other mammals besides an increased ability to coordinate the function of the eyes and the front limbs to perform certain tasks.

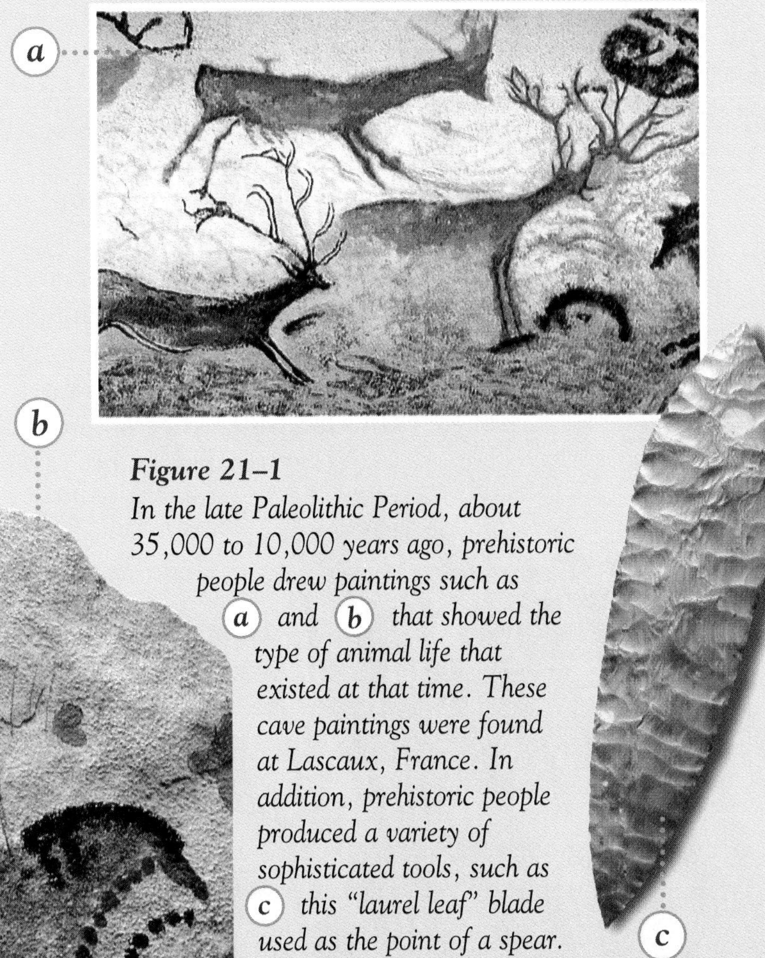

Figure 21–1
In the late Paleolithic Period, about 35,000 to 10,000 years ago, prehistoric people drew paintings such as (a) and (b) that showed the type of animal life that existed at that time. These cave paintings were found at Lascaux, France. In addition, prehistoric people produced a variety of sophisticated tools, such as (c) this "laurel leaf" blade used as the point of a spear.

Figure 21–2
(a) The earliest primates probably resembled animals such as this modern tree shrew, whose five-digited paws can be seen grasping food. (b) The tarsier, a tiny prosimian living entirely in trees, has enlarged skin pads on its hands and feet that help it to grasp and leap from branch to branch. Notice the tarsier's large eyes, which enable it to see more clearly at night.

(a) (b)

As primates evolved, some achieved the highest intelligence of any animals on Earth. **As a group, primates evolved several distinctive characteristics—a flatter face, flexible fingers, and a well-developed cerebrum.** Primates typically have a flat face, so both eyes face forward with widely overlapping fields of view. This feature gives primates excellent **binocular vision.** That means we can see a clear three-dimensional image of the world.

Primates also have flexible fingers that can grasp, hold, and manipulate objects. That ability is combined in many species with arms that join to the body at a flexible shoulder joint. Because primate arms can swing in broad circles around the shoulder, monkeys and apes can run along branches and swing from tree to tree. Many primates also have a large and well-developed cerebrum that makes complex behaviors possible.

The Primate Family Tree

Living primates include a wide variety of animals. Members of two groups, collectively called **prosimians** (proh-SIHM-ee-uhnz), don't look much like what most of us think of as monkeys or apes. Those that do are called higher primates, and they belong to the group called **anthropoids** (AN-thruh-poydz).

Early in their history, ancestors of living anthropoids were separated by continental drift into two main groups. One group, the New World monkeys, evolved in the Americas. These animals have long, prehensile tails that can coil up to grasp branches, like a fifth hand, when

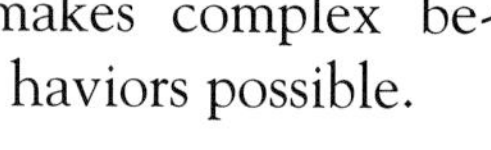

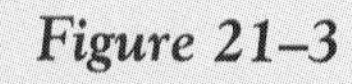

Figure 21–3
Significant trends in primate evolution are the increased care of offspring, as seen in (a) ring-tailed lemurs, classified as prosimians, and social interaction, as seen in (b) orangutans, classified as hominoids.

(a) (b)

Figure 21–4
Although there is still a debate among scientists as to which group of prosimians is the probable ancestor of anthropoids, this primate family tree shows the tentative sequence of evolutionary steps that led to the evolution of hominids.

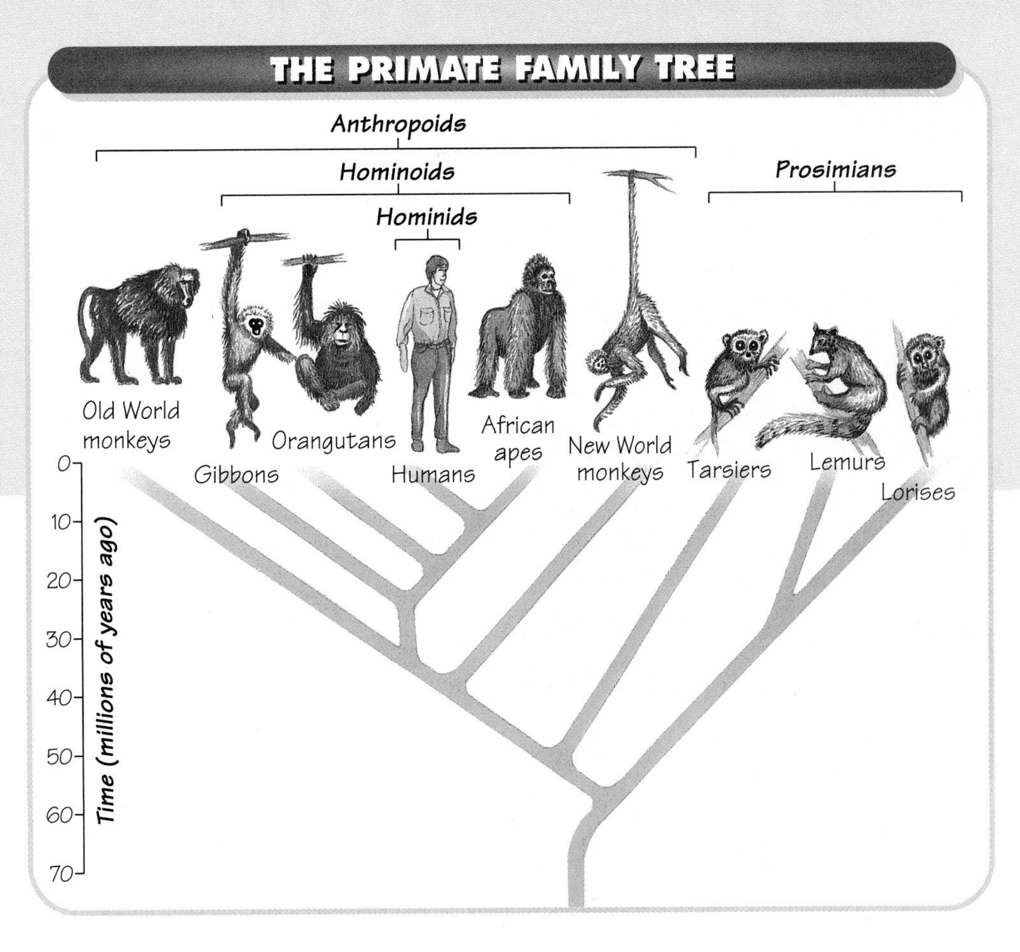

swinging through trees. The second group, which evolved into the Old World monkeys and great apes, arose in Africa, the Middle East, and Asia. Old World monkeys, such as baboons and macaques (muh-KAHKS), often lack tails, but when they do have tails, they are not prehensile. From this second group evolved the **hominoids,** which include gorillas, chimpanzees, and *Homo sapiens*. As ***Figure 21–4*** shows, the hominoid line gave rise to a small group known as the **hominids.** Although the early hominids were not yet human, they did show several evolutionary trends that set them apart from other hominoids.

The history described thus far is supported by an enormous volume of scientific data of several types that virtually all scientists interpret in the same way. The fossil record clearly documents links among many living and extinct primates. Studies in molecular biology shed additional light on relationships between humans and other living great apes.

☑ **Checkpoint** What are the two main groups within the primate family?

Hominid Origins

When we look closely at the branch of the primate family tree that ultimately produced humans, things get more interesting, more complicated, and more controversial for several reasons. First, fossil evidence suggests that the ancient hominid line evolved with incredible speed. Changes in some of our ancestors' features—teeth, skull, and legs—are seen in the fossil record. But uniquely human features—intelligence, memory, speech, and language—leave no fossil remains.

In addition to paleontologists, anthropologists also have research interests in human origins. They bring to the hunt a different set of tools, techniques, and strategies for interpreting data. As a result, the study of human origins is a changing and controversial field. Still, we know much more about our past today than we did twenty years ago, and new pieces of the puzzle fall into place every day.

Australopithecines

Most researchers now agree that the first hominids belonged to the genus *Australopithecus* (aw-stray-loh-PIHTH-uh-kuhs), meaning "southern ape." This

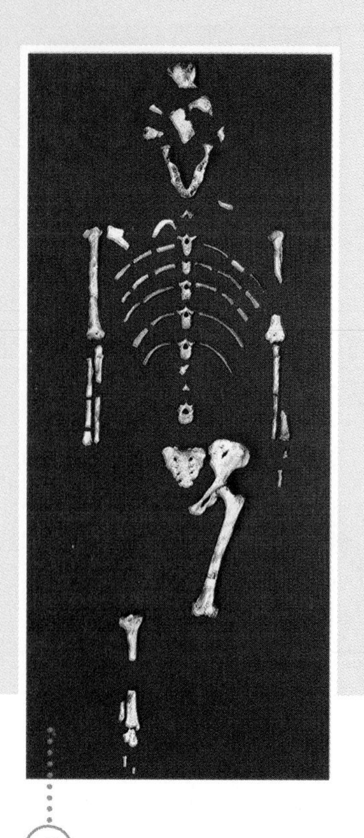

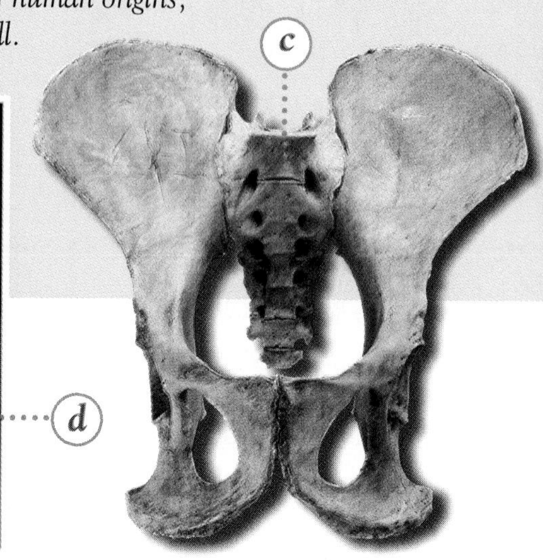

Figure 21–5
(a) In 1974, a research team led by Donald Johansen discovered Lucy—the name given to this skeleton—along with a collection of other fossils in Hadar, Ethiopia. Lucy is considered to be the most complete and oldest australopithecine fossil found to date. By comparing (b) the reconstructed pelvis of the hominid Lucy with (c) that of a chimpanzee, you can see the structural difference that permits hominids to walk upright. (d) CAREER TRACK *As part of her study of human origins, this anthropologist closely examines a primitive skull.*

was the genus into which Raymond Dart placed the fossil of the child's skull uncovered in South Africa. Since then, a large number of australopithecine fossils have been found. In particular, a fairly complete *Australopithecus* skeleton was discovered by American anthropologist Donald Johansen in 1974, shown in ***Figure 21–5.***

Based on the fossil evidence gathered so far, scientists estimate that various species of *Australopithecus* lived from nearly 4 million years ago to about 1.4 million years ago. It also seems clear that our own genus, *Homo*, emerged in Africa about 1.8 million years ago. But beyond these basics, questions greatly outnumber answers, as illustrated in the figure on page 499. It now seems as though at least nine hominid species—of the genera *Australopithecus*, *Paranthropus*, and *Homo*—have come and gone over the last 4 million years.

The story line becomes only slightly clearer with the emergence of *Homo erectus* in Africa. It seems that at least 1 million years ago, groups of this remarkable species left Africa and traveled as far away as what are now China and Indonesia. Various populations used stone tools, hunted animals for food, and lived in groups. In some areas, they sought shelter in caves. Although remains of *H. erectus* have been found in many places in the Eastern Hemisphere, the species ultimately disappeared—either evolving into another species or replaced by a new competitor.

By this time, Europe and other northern regions were being chilled by an ice age. The first human ancestors to adapt to these difficult conditions were the rather intelligent hominids known as Neanderthals. Now usually classified as *Homo neanderthalensis*, they lived between 200 and 35 thousand years ago in the Middle East, western Asia, and Europe. Gradually, they were replaced by the first true modern humans, *Homo sapiens*, about 30,000 years ago.

Visualizing Homo sapiens Origins

Until a few decades ago, humans were thought to have evolved as the result of a single lineage from Australopithecus *through* Homo erectus *to* Homo sapiens. *As more fossil evidence became available, more hominid species dating back to similar time frames were uncovered, rendering the earlier notion unacceptable. Scientists have proposed various hypotheses to account for the origins of modern humans, three of which are presented here. Notice that all the hypotheses leave the origins of* Homo erectus *uncertain.*

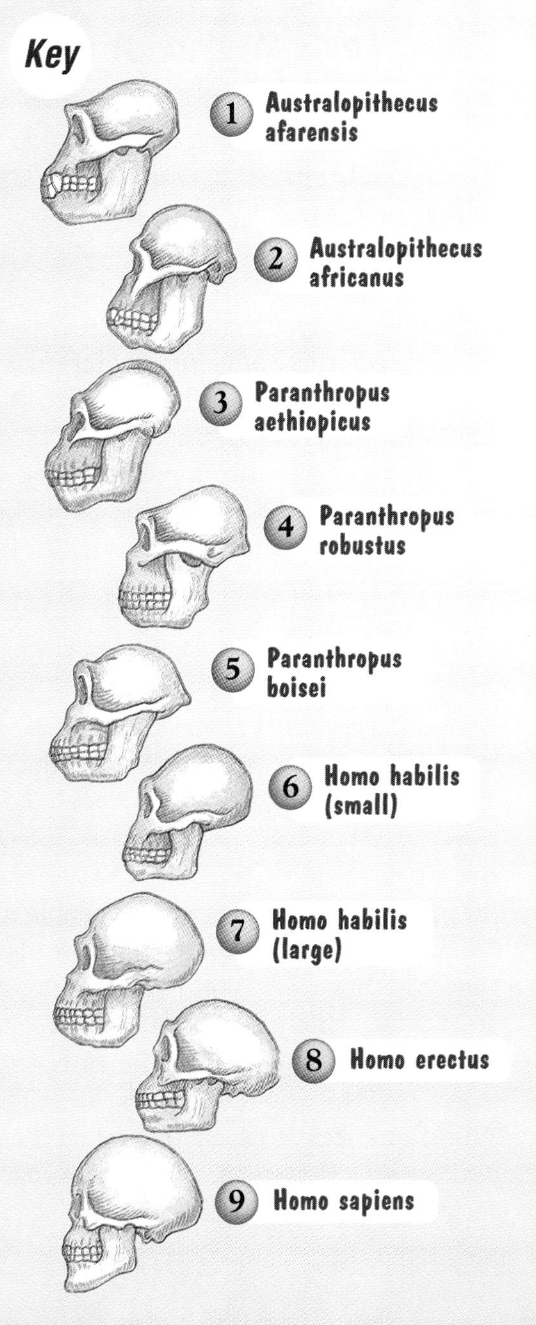

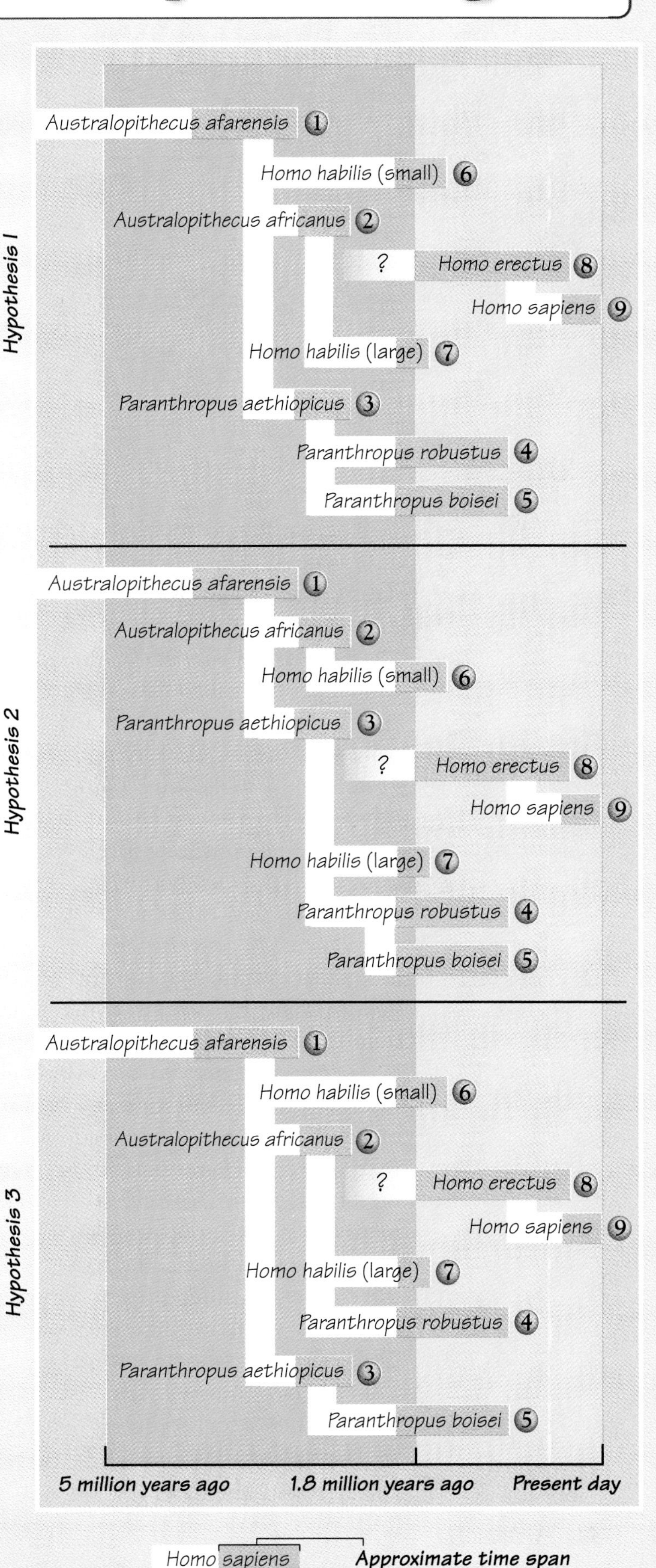

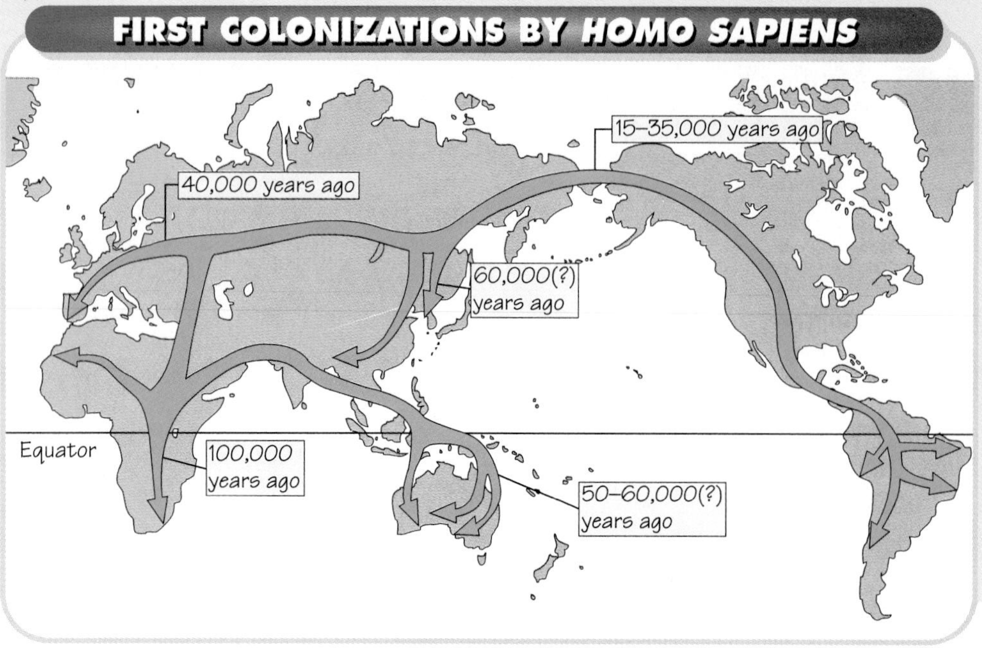

FIRST COLONIZATIONS BY *HOMO SAPIENS*

40,000 years ago
15–35,000 years ago
60,000(?) years ago
Equator
100,000 years ago
50–60,000(?) years ago

Figure 21–6
Based on fossil and genetic evidence gathered so far, scientists estimate that Homo sapiens *evolved in Africa 100,000 years ago and, from there, spread throughout the world over thousands of years. The map provides approximate time frames for the first colonization of different parts of the world by* Homo sapiens.

Homo sapiens

The earliest fossils that anthropologists agree should be called *Homo sapiens*, dating back about 100,000 years, were discovered in the caves of Skhul and Qafzeh in Israel. About 40,000 years ago, similar people, called Cro-Magnons—after the site in France where their fossils were first found—were living in Europe. From bones, tools, cave paintings, and other evidence across Europe, it is clear that Cro-Magnons were intelligent.

It is still not clear just when, how, and where they and other modern humans first evolved or how they made their way around the world. Some anthropologists believe that *H. sapiens* evolved in a number of places from scattered populations of *H. erectus*. **To most biologists interested in this question, a wide range of fossil and genetic evidence indicates that all living *Homo sapiens* originated in Africa.** From that ancestral stock, a relatively small group wandered out of Africa through the Middle East. As they traveled into different environments, they evolved slightly different characteristics—both at random and in response to local environments. The result is the treasure that is human diversity today.

☑ **Checkpoint** Where did the first *Homo sapiens* originate?

Prehistoric Migrations

The entire history of hominid evolution occupies a few moments of geologic time—barely an instant in life's 4.5-billion-year saga. On this time scale, *Homo sapiens* appeared only a few seconds ago. Precisely because history is so brief compared with prehistory, it is easy to suppose that ancient evolutionary events are irrelevant to human life. But that idea is simply not true. Many of the most important events in recorded history have been and are being shaped by the history of our species and its evolutionary interactions with other organisms.

Homo sapiens in the New World

Sometime around 30,000 years ago, wandering groups of *Homo sapiens* crossed from what is now Siberia to North America. These first Native Americans colonized this hemisphere from Canada to the tip of South America. Once here, they lived in isolation from the people in Africa, Europe, and Asia. They were also isolated from Old World monkeys and great apes, which lived near humans in other parts of the world. The only primates in this hemisphere were New World monkeys, whose ancestors had been separated from ours millions of years earlier by continental drift.

Figure 21–7
This painting shows a scene in the town of Marseilles, France, during a plague epidemic a few centuries ago. Epidemics of infectious diseases took countless human lives in the days prior to the discovery of antibiotics.

Humans and Parasites

Meanwhile, in the Eastern Hemisphere, parasites that had evolved for millions of years alongside other hominids and great apes were never far from *Homo sapiens*. Many of these parasites coevolved with animals closely related to our species, so they needed only minor evolutionary changes to infect humans. Certain other parasites could move back and forth between domestic animals and their human keepers. Over the centuries, as civilizations in Africa, Europe, and Asia spread and came into contact with each other, **epidemics**—the sudden and rapid spread of disease—caused by these parasites broke out and swept along trade routes.

Once epidemics started, bacteria and viruses thrived in crowded, unsanitary cities. The list of illnesses is long and frightening. Plague, smallpox, malaria, yellow fever, typhoid, measles, and cholera were among the worst. During the Middle Ages, a single epidemic could kill half the inhabitants of a European city. What does this mean in terms of natural selection? There was intense pressure on Europeans and Africans to evolve resistance to these diseases. In various ways, these humans "learned" to live with their parasites.

The story was quite different for the pre-Columbian Aztec and Mayan civilizations, in present-day Mexico and Central America, and the Incas, who inhabited the western coast of South America. Although these cultures had large cities and trading networks, they were free from diseases such as plague and smallpox. Why? **Infectious diseases common in Europe and Asia were unknown to Native Americans because they evolved in isolation from the humans in the Eastern Hemisphere.** In other words, humans in the Eastern Hemisphere and the Western Hemisphere followed different evolutionary pathways, which played a significant role in shaping the course of history.

INTEGRATING HISTORY

When and where did the Mayan and Incan civilizations flourish? What were some of their unique accomplishments?

Section Review 21–1

1. **List** characteristics that primates share.
2. **Discuss** the origins of *Homo sapiens.*
3. **Explain** why people in the Americas lacked resistance to infectious diseases common in Europe and Asia.
4. **Critical Thinking—Evaluating** Why is it difficult to determine accurately the exact sequence of the evolution of *Homo sapiens?*

SECTION 21–2

BRANCHING OUT In Depth

Coevolution of Humans and Parasites

GUIDE FOR READING

- **Describe** how bacteria become antibiotic resistant.

MINI LAB

- **Design an experiment** to determine the effect of location on the number of microorganisms present.

IN 1519, SPANISH CONQUISTADOR Hernando Cortes arrived in Mexico with about 600 soldiers, some guns, and a few dozen horses. The great Aztec empire he encountered was home to between 25 and 30 million people. Yet the tiny band of Spaniards managed to topple the enormous and powerful Aztec empire. Francisco Pizarro had similar results in subduing the Incas in Peru. How was this possible?

Besides the old Native American legends and the alliances between the Spaniards and the discontented people who hated their Aztec overlords, there was disease. Evolutionary history and disease-causing parasites were very important in determining the outcome of the struggle. What's more, the ongoing coevolutionary relationship between humans and parasites continues to affect all our lives today.

Shaping History

Although the Spaniards didn't realize it, they landed in the Americas carrying invisible allies far more powerful than guns and horses. They were the disease-causing viruses and bacteria to which Europeans were largely immune, but against which Native Americans had no natural defenses. As soon as the Spaniards landed, these parasites spread among the local people. By the night of a battle in which Aztec warriors drove the Spaniards out of their capital, a smallpox epidemic was killing thousands of Aztec civilians.

Epidemics of diseases such as smallpox, measles, and plague continued to spread with unimaginably devastating results. Within decades, nearly 90 percent of the Native American population in Mexico had sickened and died! What's more, epidemics spread from Mexico to Guatemala to Peru, home of the Incas, where they killed just as

Figure 21–8
The Spanish conquest of the Aztecs and the Incas in the Americas in the sixteenth century was due not just to the (a) shields, (b) weapons, and (c) horses of the conquistadors, but also to the invisible allies they brought in the form of disease-carrying microorganisms.

many people—including the Incan leader and his son. By the time Francisco Pizarro arrived in Peru in 1532, the Incan empire was falling apart because of disease. Thus, had it not been for the different prehistories of Europe and the Americas, history might have taken a different turn.

MINI LAB ... Experimenting ...

A Breath of Fresh Air

PROBLEM *What locations in your neighborhood would have the greatest number of microorganisms? **Design an experiment** to answer this question.*

SUGGESTED PROCEDURE 

1. Using sterile flasks of nutrient broth—a liquid growth medium used by scientists to culture microorganisms—design an experiment that will allow you to compare the presence of microorganisms in the air at various locations. Choose a variety of locations, such as enclosed, crowded areas and open fields. Include a control in your procedure.
2. Predict which location will have the greatest number and which will have the least number of microorganisms.
3. Have your teacher approve your experimental procedure.

ANALYZE AND CONCLUDE

1. How did the growth of microorganisms vary with location?
2. How did the results compare with your predictions?
3. Did your control work effectively? If not, why not?

Impact on Medicine 

Recently, the ongoing battle between humans and disease has taken a new and disturbing turn, as disease-causing bacteria have evolved in response to changes that humans have made in their environment. One of the most important of those changes is the widespread use of **antibiotics**—compounds that kill bacteria without harming the body cells of humans or other animals.

When antibiotics were first introduced, they did such a superb job of controlling bacterial diseases that they earned the name "wonder drugs." Within a few decades, antibiotics were in widespread use. In hospitals, patients are bombarded with antibiotics to prevent infections. On farms where animals are raised in crowded conditions, antibiotics are added to food to prevent illness and the growth of harmful bacteria. And in the general population, people use antibiotics to treat even minor infections. All this antibiotic use, however, makes sense only in the short term.

In the long term, can you see what this means from the perspective of bacteria? **As antibiotics become a permanent part of their environment, bacteria begin to evolve antibiotic resistance in response to this new and powerful form of natural selection.** What is the result? Today, patients are being infected by "super bugs"—strains of bacteria immune to the effects of five or more antibiotics. These **drug-resistant bacteria** are appearing everywhere, and they present a growing threat to human health that may someday surpass AIDS.

INTEGRATING BIOLOGY AND SOCIETY

Who discovered the first antibiotic? How did that discovery change human society?

Antibiotic Resistance

When prescribed antibiotics are used properly, the drugs kill enough bacteria to help the body's natural defenses destroy the rest. If the drug therapy and the body's defenses together destroy all the bacteria, there's no problem.

But a few individual bacteria in nearly every population carry mutant genes that make them resistant to one antibiotic or another. If antibiotics are used improperly, enough of the antibiotic-resistant bacteria may survive to multiply and begin a new strain in which most individuals will be antibiotic resistant.

Problem Solving

INTERPRETING DATA

Comparing Antibiotic Resistance

Your friend is assisting the head laboratory technician at the state microbiology laboratory, who is working on a problem occurring in area hospitals. A few types of bacteria are becoming resistant to commonly used antibiotics. Two hospitals have sent her cultures to conduct antibiotic-sensitivity tests to determine which bacteria are resistant to the antibiotics.

After adding agar, a solid nutrient material, to two Petri dishes labeled A and B, the technician swabs the surface of the agar in dish A with the bacterial culture from hospital A, then swabs dish B with the culture from hospital B. On each dish she places six discs, each containing a different antibiotic. She covers and incubates the dishes at 35°C for 18 hours. The zones of inhibition—regions of no bacterial growth that surround each disc—are illustrated below. Also listed are the diameters of the zones of inhibition that correspond to bacterial resistance or susceptibility to the antibiotic.

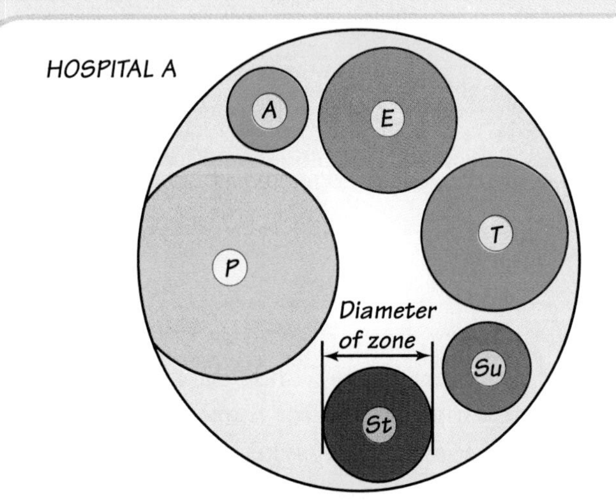

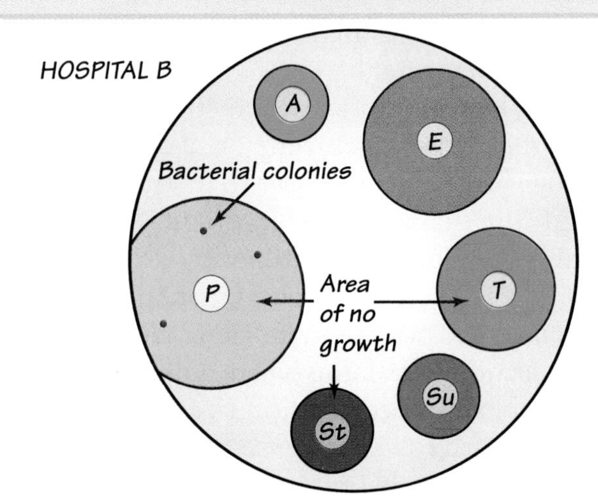

Antibiotic	R	I	S
Ampicillin (A)	≤ 11	12–13	≥ 14
Penicillin G (P)	≤ 20	21–28	≥ 29
Sulfonamides (Su)	≤ 12	13–16	≥ 17

Antibiotic	R	I	S
Erythromycin (E)	≤ 13	14–17	≥ 18
Streptomycin (St)	≤ 11	12–14	≥ 15
Tetracycline (T)	≤ 14	15–18	≥ 19

Key (All numbers represent diameter in millimeters.)

R Resistant, unlikely that the bacterium will be killed by the antibiotic

I Intermediate, uncertain whether the bacterium will be killed by the antibiotic

S Susceptible, likely that the bacterium will be killed by the antibiotic

THINK ABOUT IT

1. Determine the antibiotics to which bacteria from hospital A are susceptible. Determine the antibiotics to which bacteria from hospital B are susceptible.
2. Which hospital might be in danger of developing an antibiotic-resistant strain of bacteria? **Explain** your answer.

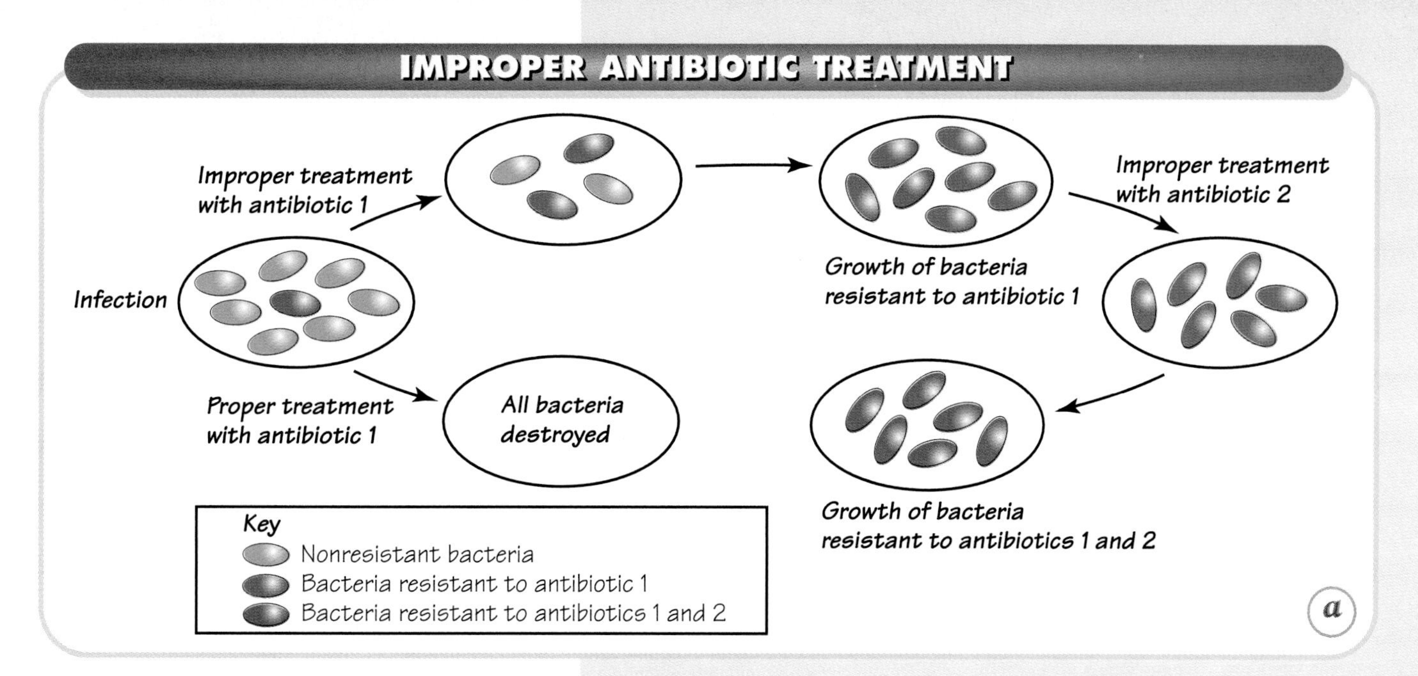

For a time, this new strain may still be susceptible to other antibiotics. But if those drugs are used improperly, new strains of bacteria can emerge that will be resistant to several drugs at once.

☑ **Checkpoint** What are drug-resistant bacteria?

Figure 21–9 *(a) This illustration shows how the improper use of antibiotics can give rise to the evolution of drug-resistant bacteria. (b) By developing new antibiotics, a pharmacologist, shown here inspecting a culture of disease-causing bacteria, helps to prevent this potentially serious situation.*

Medicine Versus Natural Selection

Problems with drug-resistant bacteria have not yet caused a crisis, but a half dozen common disease-causing bacteria have already evolved antibiotic resistance. Nearly one third of the strains of *Salmonella*—bacteria that cause severe and sometimes fatal diarrhea—are resistant to several antibiotics. Tuberculosis, which was under control in the United States for decades, is increasing because several strains of tuberculosis-causing bacteria are resistant to every antibiotic currently in use.

What can be done? Physicians and researchers are racing to battle the power of evolutionary change in the microbial world. New drugs are being developed. And new treatment strategies—aimed at slowing down evolution of drug-resistant bacteria—are being used to help patients.

Section Review 21–2

1. **Describe** how bacteria become resistant to antibiotics.
2. **MINI LAB** What effect does a specific location have on the number of microorganisms present? **Design an experiment** to answer the question.
3. **BRANCHING OUT ACTIVITY** Interview some health care professionals to find out the bacterial diseases that are difficult to treat with antibiotics because of the evolution of antibiotic-resistant bacteria. **Explain** how this has arisen.

Laboratory Investigation

The Spread of Disease

The way in which a disease spreads through a population demands the careful collection and analysis of data. When an outbreak of a serious infectious disease occurs, scientists must track down the disease and determine its origin. In this investigation, you will simulate the spread of an infectious disease and determine the original carrier of the disease.

Problem

How can you simulate the spread of a disease in a community? **Construct a model** for the simulation.

Materials (per group)

large test tube of stock solution
clean test tube
large pipette
medicine dropper

Procedure 

1. Select one stock solution from a numbered set of stock solutions provided by your teacher. Record the number in your data table.
2. Carefully fill the pipette with the stock solution and transfer it to the clean test tube.
3. At your teacher's signal, begin circulating among your classmates until the teacher tells you to stop. Using the medicine dropper, exchange a dropperful of your solution with the person closest to you. Make the exchange by putting a dropperful of the solution from your clean test tube into the clean test tube of the contact. You should also receive a dropperful of the solution from your contact's test tube. Record the name of that person as Contact 1 in a data table similar to Data Table 1.
4. Repeat step 3 and record the name of this person as Contact 2.
5. Repeat step 3 and record the name of this person as Contact 3.
6. Your teacher will now add several drops of an indicator to your test tube to determine whether you have been infected.
7. After performing the indicator test for the presence of infection for all the students in the class, your teacher will record the names and contacts of the infected individuals. Record this information in a data table similar to Data Table 2.

DATA TABLE 1

Your Stock Number	Contact 1	Contact 2	Contact 3

DATA TABLE 2

Infected Person	Contact 1	Contact 2	Contact 3

Observations

1. How many individuals were infected by the end of the simulation? How many were not infected?
2. How many infected individuals were there at the end of the first round of contacts?

Analysis and Conclusions

1. Using the class data, eliminate the names of those who were not infected. From this, try to find the original source of the infection by examining the remaining sequence of contacts.
2. Were you able to identify correctly the original carrier of the disease? If not, specify what information or test is required to identify the original source.
3. Make a diagram of the transmission route.
4. Suppose you came into contact with as many people as possible during a specified period of time. What effect would this have on the outcome of this simulation?

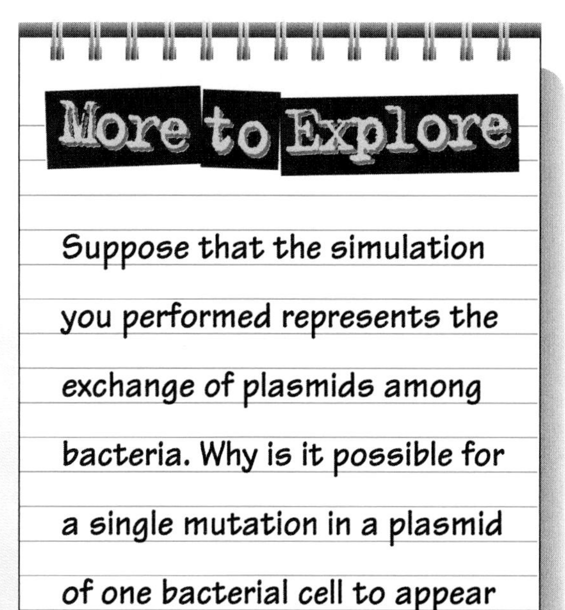

More to Explore

Suppose that the simulation you performed represents the exchange of plasmids among bacteria. Why is it possible for a single mutation in a plasmid of one bacterial cell to appear rapidly in other bacteria?

Study Guide

Summarizing Key Concepts

The key concepts in each section of this chapter are listed below to help you review the chapter content. Make sure you understand each concept and its relationship to other concepts and to the theme of this chapter.

21–1 The Origins of *Homo sapiens*

- Primates share a combination of distinctive characteristics, such as a flatter face, flexible fingers, and a well-developed cerebrum.
- The sequence of the evolution of humans within the order Primata is primates, anthropoids, hominoids, hominids, then humans.
- Researchers agree that the first hominids belong to the genus *Australopithecus.* It has been estimated that various species of this genus lived from nearly 4 million years ago to about 1.4 million years ago.
- Scientists propose different hypotheses to explain the process of evolution of *Homo sapiens* from the *Australopithecus* species.
- After studying a wide range of fossil and genetic evidence, many biologists believe that all living *Homo sapiens* originated in Africa.
- Native Americans, having arrived 30,000 years ago from Siberia, remained isolated from the people in Africa, Europe, and Asia.

21–2 Coevolution of Humans and Parasites

- When Europeans arrived in the Americas in the sixteenth century, they brought diseases to which they had evolved a substantial degree of immunity. Because Native Americans had not evolved any immunity to these diseases, a large percentage of the Native American population perished from epidemics of European diseases.
- Antibiotics—drugs that kill disease-causing bacteria—are becoming a permanent part of the environments in which bacteria live. So bacteria are under powerful pressure from natural selection to evolve drug resistance.
- The consequences of the evolution of drug-resistant bacteria are potentially serious.

Reviewing Key Terms

Review the following vocabulary terms and their meaning. Then use each term in a complete sentence.

21–1 The Origins of *Homo sapiens*

binocular vision
prosimian
anthropoid
hominoid
hominid
epidemic

21–2 Coevolution of Humans and Parasites

antibiotic
drug-resistant bacterium

Recalling Main Ideas

Choose the letter of the answer that best completes the statement or answers the question.

1. Which animals have long, prehensile tails?

a. prosimians **c.** Old World monkeys
b. hominoids **d.** New World monkeys

2. Which are not anthropoids?

a. apes **c.** humans
b. lemurs **d.** orangutans

3. The first hominids are believed to belong to the genus

a. *Australopithecus.* **c.** *Paranthropus.*
b. *Homo.* **d.** *Dryopithecus.*

4. Which of the following species was probably the first to evolve?

a. *Homo erectus* **c.** *Homo habilis*
b. *Homo sapiens* **d.** *Homo neanderthalensis*

5. Cro-Magnons were intelligent people who lived

a. more than 4 million years ago.
b. 1.8 million years ago.
c. more than 100,000 years ago.
d. less than 100,000 years ago.

6. Infectious diseases killed large numbers of Native Americans because

a. the Native Americans were immune to European diseases.
b. Europeans were immune to Native American diseases.
c. the Native Americans had no immunity to European diseases.
d. Europeans had no immunity to Native American diseases.

7. Antibiotics are drugs that

a. destroy microorganisms.
b. kill bacteria.
c. destroy viruses.
d. prevent infection.

8. Drug-resistant bacteria are

a. immune to all antibiotics.
b. susceptible to all antibiotics.
c. present only in the United States.
d. immune to one or more antibiotics.

Putting It All Together

Using the information on pages xxx to xxxi, complete the following concept map.

are believed to have evolved in

about 100,000 years ago from early

Assessment

Reviewing What You Learned

Answer each of the following in a complete sentence.

1. List the characteristics that primates share.
2. Which genus marks the emergence of the hominids? How did its name arise?
3. What did Raymond Dart find?
4. What are prosimians? Anthropoids?
5. Compare Old World monkeys and New World monkeys.
6. Based on fossil evidence, when and where did the Neanderthals live?
7. Who were the Cro-Magnons? Describe some of their characteristics.
8. What factors contributed to the fall of the Incan empire?
9. How did the early separation of Eastern Hemisphere and Western Hemisphere *H. sapiens* lead to serious consequences when they came together?
10. What is an antibiotic?
11. How could a "super bug" develop in a hospital that is designed to be disease free?
12. What is an epidemic?

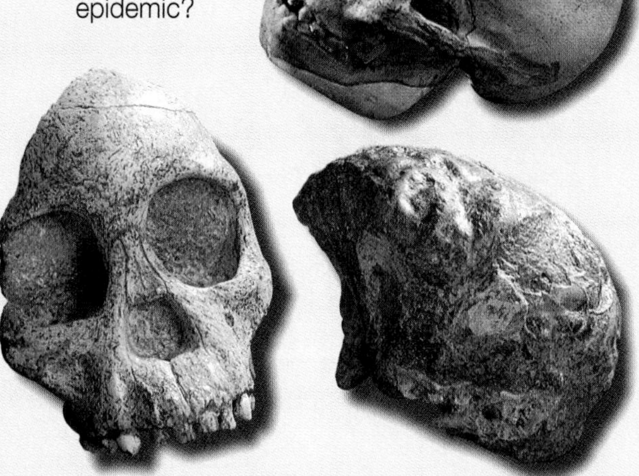

Expanding the Concepts

Discuss each of the following in a brief paragraph.

1. Why is the study of human origins a constantly changing and controversial field?
2. Describe the evidence that has led to the current view of hominid evolution.
3. Compare the hypotheses outlining the sequence of evolutionary steps leading up to *Homo sapiens.*
4. What evidence supports the theory that Cro-Magnons were an intelligent people?
5. What role can epidemics play in influencing the process of evolution?
6. How can you explain the fact that epidemics generally broke out along major trade routes?
7. Why is the evolution of drug-resistant strains of bacteria potentially very serious?
8. When the doctor prescribes an antibiotic medication for a bacterial infection, you often begin to feel better even before you have taken all the medication. Although you feel better, why is it important for you to take all the medication?

Extending Your Thinking

Use the skills you have developed in this chapter to answer the following.

1. **Hypothesizing** In ancient Hawaii, the native bird population thrived. These birds evolved in isolation and had few predators or diseases. After whalers began frequenting these islands, the native bird population from the seashore up to an elevation of 600 meters above sea level began to die. Formulate a hypothesis to explain this catastrophe.
2. **Experimenting** How does the number of microorganisms vary with location—such as city, farm, country, mountain, or seashore? Design an experiment to answer this question.
3. **Communicating** Write an essay to defend this statement: Humans have increased the evolution rate of many domestic and disease-causing organisms.
4. **Evaluating** Do you think there is a need for a worldwide organization to track diseases that affect plants, animals, and humans? Suppose this organization costs billions of dollars each year to maintain. Do you still think it is needed? Give reasons for your answer.
5. **Evaluating** Do you think that the extensive use of antibiotics in domesticated food animals is a problem for humans? Explain your answer.

Applying Your Skills

Antibiotics: The Wonder Drugs?

Before the 1940s, there were few known treatments for bacterial diseases. Today, penicillin is an antibiotic commonly used against certain strains of bacteria.

1. Research how penicillin was discovered. Write a brief essay on the role penicillin has played in human society since it was first discovered.
2. In a small group, discuss what might happen if all the bacteria in the world were destroyed.

• GOING FURTHER •

3. Tuberculosis (TB), thought by health officials to have been eradicated, is on the rise in the United States. As a group, conduct research to find out more about the cause of this occurrence.

UNIT 6

From Bacteria to Plants

CHAPTERS

"On any possible, reasonable, or fair criterion, bacteria are—and always have been—the dominant forms of life on Earth."

— Stephen Jay Gould

CAREER TRACK

As you explore the topics in this unit, you will discover many different types of careers associated with biology. Here are a few of these careers:

- Virologist
- Microbiologist Technician
- Phycologist
- Agricultural Technician
- Plant Propagator

A giant kelp forest off Santa Barbara Island in California

CHAPTER 22

Bacteria and Viruses

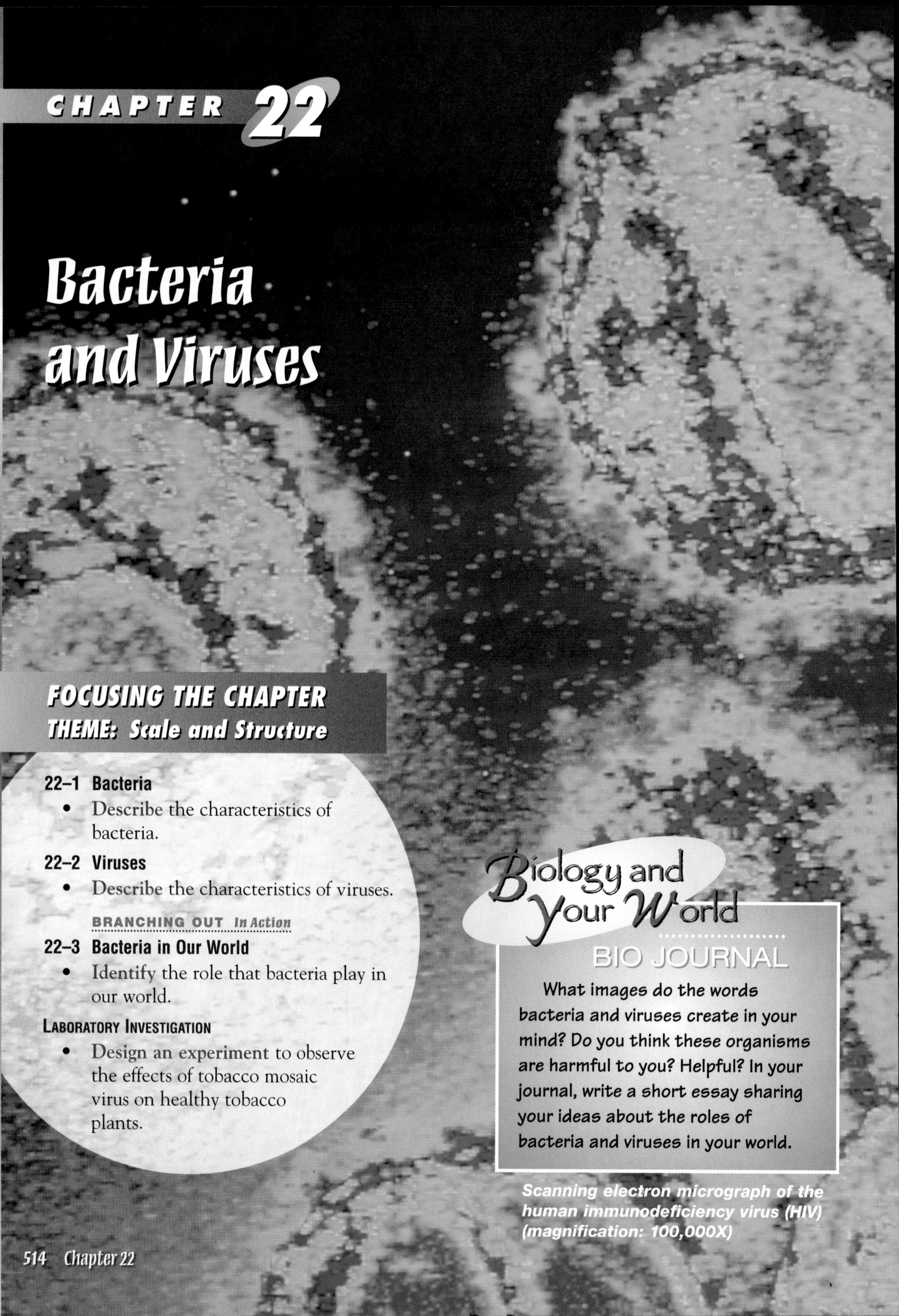

FOCUSING THE CHAPTER THEME: *Scale and Structure*

22–1 Bacteria
- Describe the characteristics of bacteria.

22–2 Viruses
- Describe the characteristics of viruses.

BRANCHING OUT *In Action*

22–3 Bacteria in Our World
- Identify the role that bacteria play in our world.

LABORATORY INVESTIGATION
- Design an experiment to observe the effects of tobacco mosaic virus on healthy tobacco plants.

Biology and Your World

BIO JOURNAL

What images do the words bacteria and viruses create in your mind? Do you think these organisms are harmful to you? Helpful? In your journal, write a short essay sharing your ideas about the roles of bacteria and viruses in your world.

Scanning electron micrograph of the human immunodeficiency virus (HIV) (magnification: 100,000X)

Bacteria

SECTION 22–1

GUIDE FOR READING

- **Describe** the structure of bacterial cells.
- **Classify** bacteria.
- **Explain** how bacteria reproduce and grow.

MINI LAB

- **Observe** some characteristics of bacterial cells through the microscope.

IF YOU WERE ASKED TO POINT out a single living thing from the world around you, what organism would come to mind first? Would it be a tree outside your school? A bird flying overhead? A squirrel gathering food? Any of these would be good choices, of course, but it is likely that you would pick something large enough for everyone to see. It's only natural to think first of the organisms that you see every day. However, there is another world around us that we generally don't think about or see, a world that contains the most abundant and most successful forms of life on the Earth—bacteria.

Bacterial Structure

Bacteria are found everywhere—from mountaintops to sulfur springs to the depths of great oceans. Bacteria are so small that there may be millions of them in a pinch of soil. Billions of bacteria even live inside your own body, residents of your digestive system.

Bacteria get their name from a Greek word meaning "little stick." As **Figure 22–2** on page 516 shows, the name fits. Bacteria are **prokaryotes**—cells without nuclei. Most bacteria are much smaller than eukaryotic cells—cells with nuclei. The length of a typical bacterial cell ranges from 1 to 10 micrometers. The length of a eukaryotic cell ranges from 10 to 100 micrometers.

INTEGRATING MATH

How many bacterial cells of typical length can line up across the width of a coin that is 1 cm in diameter?

Bacterial cells have a relatively simple structure. **Genetic information in bacterial cells is found on a single circular DNA molecule—or chromosome—in the cytoplasm. The cytoplasm is surrounded by a cell membrane made up of lipids and proteins.**

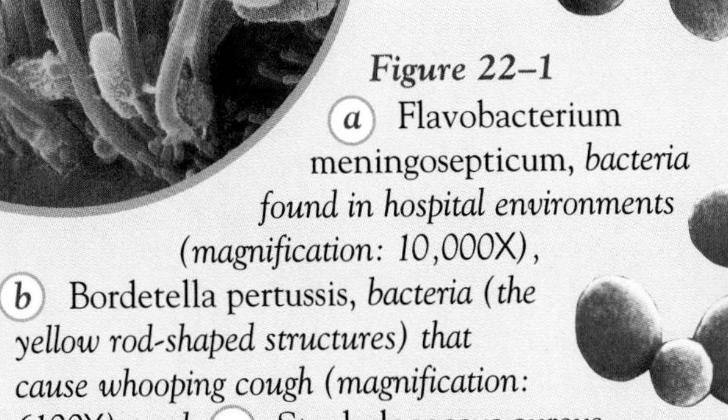

Figure 22–1 (a) Flavobacterium meningosepticum, *bacteria found in hospital environments (magnification: 10,000X),* (b) Bordetella pertussis, *bacteria (the yellow rod-shaped structures) that cause whooping cough (magnification: 6100X), and* (c) Staphylococcus aureus, *bacteria that cause many infections in humans (magnification: 34,000X), are a few examples of the numerous kinds of bacteria known to exist.*

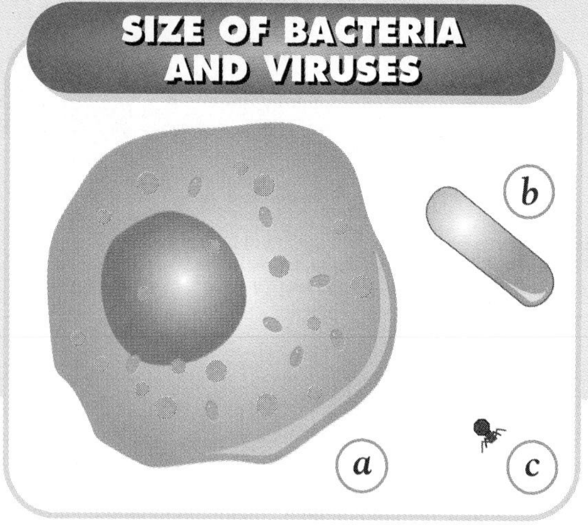

SIZE OF BACTERIA AND VIRUSES

Figure 22–2
This diagram illustrates the relative size of a typical (a) *eukaryotic cell,* (b) *bacterium, and* (c) *virus.*

The Bacterial Cell Wall

The cell membrane of a bacterium is surrounded by a cell wall composed of complex carbohydrates. The cell wall protects the bacterium from injury and helps to regulate the movements of molecules into and out of the cell.

Because the cell wall is so important, bacteria are often identified according to the type of cell wall they have. Biologists determine the type of cell wall by using a technique known as Gram staining, named after its inventor, Hans Christian Gram, a Danish physician. Bacteria that take up Gram's purple stain are known as gram-positive. Those that do not take up the stain are said to be gram-negative. What do these categories mean? Gram-negative bacteria have an extra lipid-containing layer around their cell walls. This layer not only keeps out the stain, it enables the gram-negative bacteria to resist many drugs used to fight bacterial infections. For this reason, an infection with gram-negative bacteria is usually very serious.

☑ **Checkpoint** Explain the function of the bacterial cell wall.

Cell Shape and Movement

Bacteria are often described according to their shape. Rod-shaped bacteria are called **bacilli** (buh-SIHL-igh; singular: bacillus). Spherical-shaped bacteria are known as **cocci** (KAHK-sigh; singular: coccus). And spiral-shaped bacteria are called **spirilla** (spigh-RIHL-uh; singular: spirillum).

Bacteria may also be described by how they move. Only a very small number of bacteria have no means of movement. Most can move, and some move very quickly. Some can change their shape to wriggle forward—almost like snakes—and others glide slowly along the surfaces of solid objects. Many bacteria have whiplike flagella that propel them through liquids with amazing speed.

Classifying Bacteria

Biologists have struggled for years to find the best way to classify different types of bacteria. It hasn't been easy. Many bacteria that look similar are actually very different in other respects, and there is no

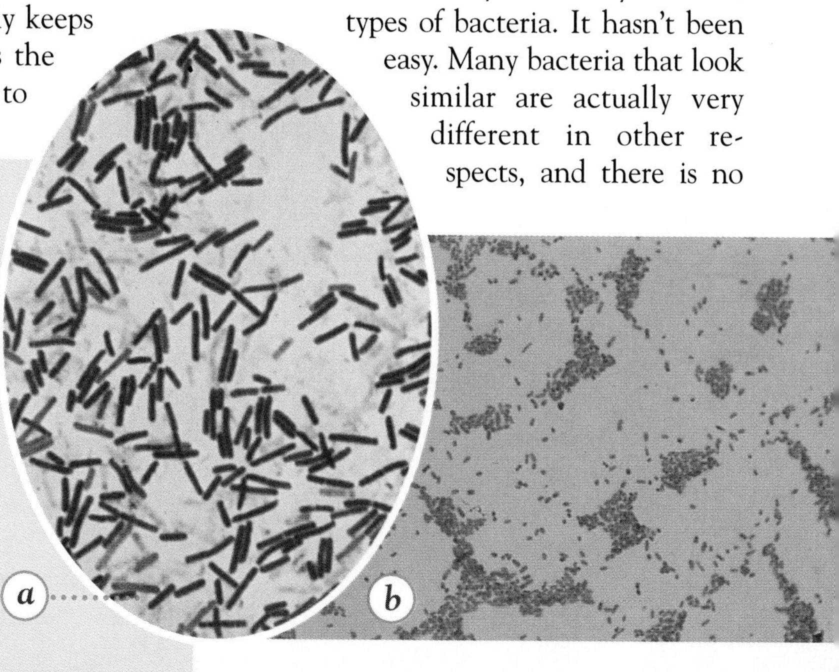

Figure 22–3
Because (a) Clostridium botulinum *take up the purple stain in a Gram-staining test, they are known to be gram-positive (magnification: 600X).* (b) Haemophilus influenzae, *on the other hand, do not and are considered gram-negative (magnification: 625X). Gram-negative bacteria have an extra lipid layer around their cell wall that keeps the purple stain out, along with other foreign substances, such as antibacterial drugs.*

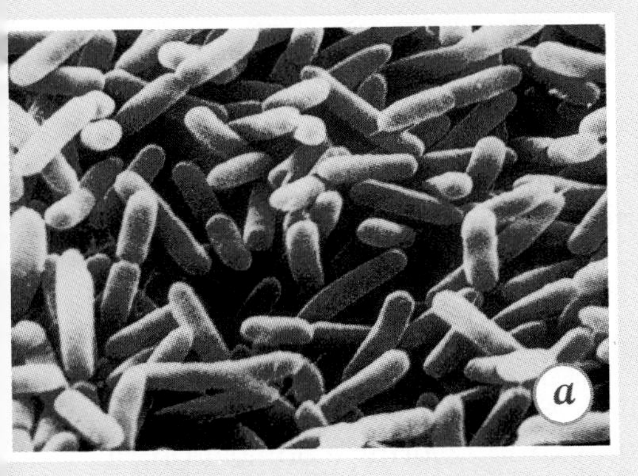

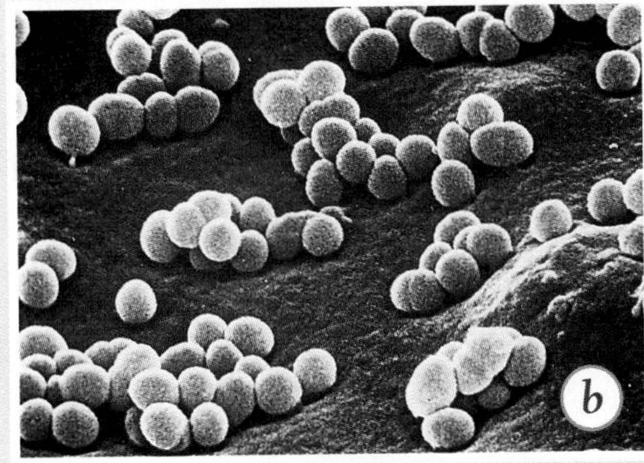

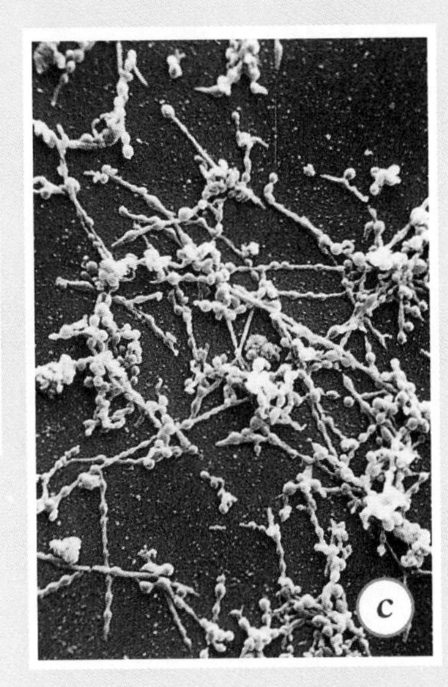

Figure 22–4
Bacteria can be grouped according to shape. (a) *Rod-shaped bacteria are called bacilli (magnification: 20,000X),* (b) *spherical-shaped bacteria are called cocci (magnification: 40,000X), and* (c) *spiral-shaped bacteria are called spirilla (magnification: 45,000X).*

simple way to decide which characteristics are most important. In recent years, however, it has become possible to compare the DNA sequences of bacteria, and scientists have been able to determine in this way which groups of bacteria are most closely related.

This has led to a startling discovery—one group of bacteria differs almost as much from another group of bacteria as prokaryotic cells do from eukaryotic cells! **As a result, the bacteria are now divided into two separate kingdoms—the Archaebacteria and the Eubacteria.**

Archaebacteria

The Archaebacteria are sometimes called "ancient" bacteria because they resemble the first known prokaryotes. The prefix *archae-* comes from a Greek word meaning "ancient." Although they resemble other bacteria, the Archaebacteria display striking differences in their cell membranes, biochemical pathways, and ribosomes. Archaebacteria include organisms that live in very harsh environments. For example, one group lives in oxygen-poor environments, such as thick mud and the digestive tracts of animals. These Archaebacteria are called methanogens because they produce methane gas.

Eubacteria

Eubacteria, or the "true" bacteria, are the largest and most diverse group of bacteria. They are literally found everywhere on the planet, and they practice just about every possible means of "making a biological living."

Many Eubacteria are photosynthetic, meaning that they capture and use the energy from sunlight. Other Eubacteria capture chemical energy from their surroundings—including even inorganic molecules such as hydrogen sulfide, sulfur, and iron—and use it for their needs.

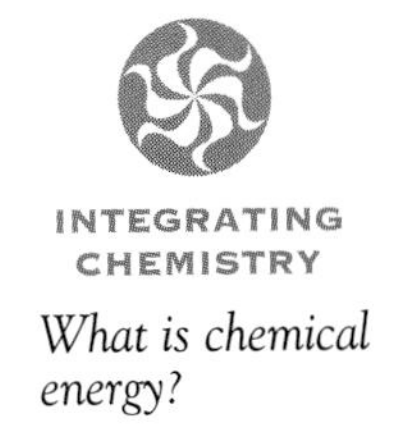

INTEGRATING CHEMISTRY

What is chemical energy?

Checkpoint How do Archaebacteria and Eubacteria differ?

Figure 22–5
Some Archaebacteria thrive in very hot environments, such as the Chromatic Spring of the Upper Geyser Basin, Yellowstone National Park, which owes its brilliant colors to various species of the bacteria.

Not Exactly Bean Soup

PROBLEM *How can you **observe** some characteristics of bacterial cells through the microscope?*

PROCEDURE  

1. Obtain a small amount of bacterial culture prepared by your teacher.
2. Place a drop of the culture on a microscope slide and observe under high power using a low light setting.
3. Look for bacterial cells of different shapes and for bacteria that are moving.
4. Record your observations.

ANALYZE AND CONCLUDE

1. What is the appearance of the bacterial culture? Does it have an odor? How can you account for the odor and appearance?
2. What cell shapes did you observe?
3. Are any of the cells moving?

Bacterial Reproduction and Growth

Under the proper conditions, bacteria can grow at astonishing rates. Some can completely reproduce themselves in as little as 20 minutes. This means that in the course of a single day, one cell can go through 72 generations—or 3 generations every hour. A single cell reproducing that quickly for a full day would give rise to more than 4 billion cells!

INTEGRATING MATH

Bacterial growth is an exponential growth. Give an example of another exponential growth process.

Binary Fission

When a bacterium has grown to the point where it has roughly doubled in size, it replicates its chromosome and divides in half, producing two identical daughter cells. This is known as **binary fission,** a type of reproduction. **Because binary fission does not involve the exchange of genetic information, it is an asexual form of reproduction.**

Conjugation

A few bacteria are able to reproduce by **conjugation.** During conjugation, a bridge of protein forms between two bacteria, allowing some DNA to be transferred from one cell to the other. When conjugation is over, the connection between the cells is broken. As a result of conjugation, some bacterial cells receive new genetic information and have a different set of genes than before. **Biologists consider conjugation to be a form of sexual reproduction because it results in new combinations of genes.** These new combinations may be useful to the bacteria. They help ensure that even if the environment changes, at least a few bacteria will have the combinations of genes needed to survive.

Spore Formation

"When the going gets tough, the tough get going." That may be the motto of spore-forming bacteria. When faced with difficult conditions, many bacteria form tough structures called **spores.** One type of spore, an **endospore,** is formed inside a bacterial cell. The endospore develops a thick wall that encloses part of the cytoplasm and a DNA molecule.

Endospores are resistant to heat, drying, radiation, and even to chemical disinfectants. **The endospores of some bacteria can survive for years, and when conditions are right, they can become activated and begin to grow.** Endospores enable bacteria to survive harsh conditions that might otherwise kill them. They make it difficult to completely eliminate bacteria.

Checkpoint Which is an asexual form of reproduction—binary fission or conjugation?

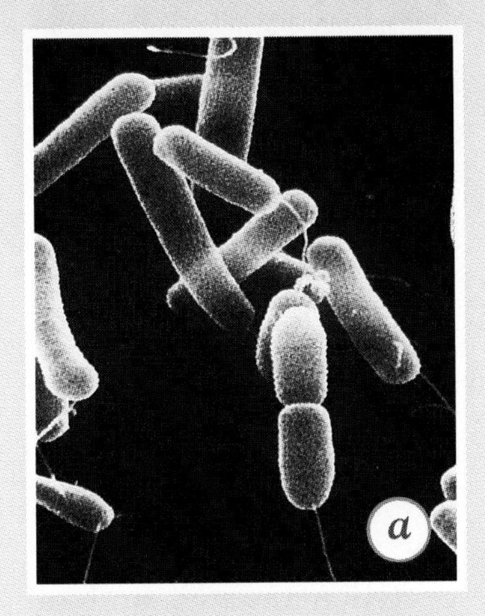

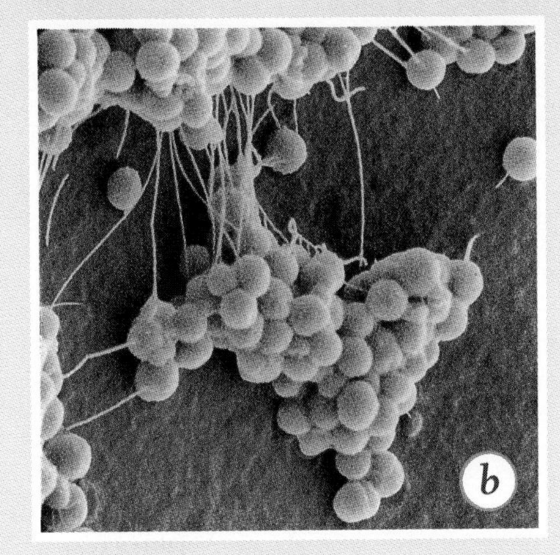

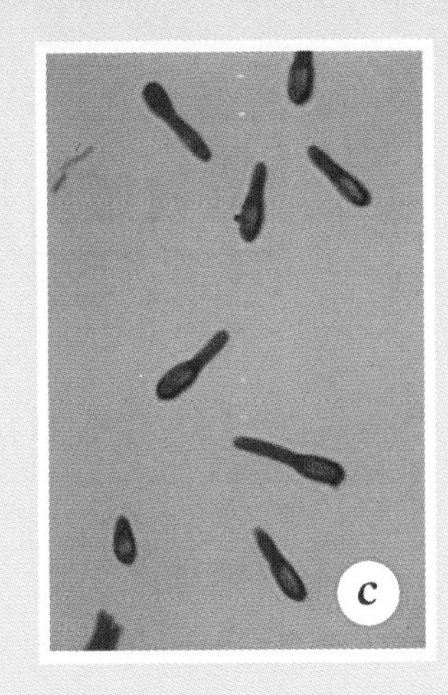

Figure 22–6
The scanning electron micrographs show (a) binary fission in Escherichia coli *(magnification: 20,000X), (b) conjugation in* Staphylococcus epidemidis *(magnification: 3145X), and (c) endospores in* Clostridium tetani *(magnification: 1000X).*

Bacteria and Disease

The French chemist and bacteriologist Louis Pasteur was the first person to show convincingly that certain bacteria cause disease. Pasteur established what has come to be known as the germ theory of disease when he showed that bacteria were responsible for a number of animal and plant diseases.

You may have encountered one of the most widespread disease-causing bacteria, *Streptococcus pyrogenes*, responsible for the severe sore throat called strep throat. Other diseases caused by bacteria include diphtheria, tuberculosis, typhoid fever, tetanus, syphilis, gonorrhea, cholera, Lyme disease, and bubonic plague.

Bacteria cause these diseases in one of two general ways. They may attack the cells and tissues of the body directly, breaking them down and using their materials for nourishment. Or they may release toxins (poisons) that travel throughout the body, interfering with the normal functions of the body.

For example, *Salmonella* is a bacterium that grows in foods such as meat, poultry, and eggs. If these foods are not properly cooked, *Salmonella* may get to the dinner table before you do, releasing poisons into the food. The symptoms of food poisoning range from a mild stomach upset to vomiting and diarrhea.

One of the best ways to fight bacterial diseases is by the use of vaccines, which help prevent an infection from getting started. If an infection does develop, drugs that can destroy bacteria—known as **antibiotics**—are powerful weapons that can treat most bacterial diseases.

INTEGRATING HEALTH

How do physicians treat bacterial diseases? Are there any reasons to be cautious with the use of such treatment?

Section Review 22-1

1. **Describe** the structure of bacterial cells.
2. **Classify** bacteria.
3. **Explain** how bacteria reproduce and grow.
4. **Critical Thinking—Relating** What is the relationship between the bacterial cell wall and bacterial resistance to antibiotics?
5. **MINI LAB** What characteristics of bacterial cells did you **observe** through a microscope?

SECTION

22–2 Viruses

GUIDE FOR READING

- **Describe** the structure of a virus.
- **Explain** how viruses infect cells.

THE YEAR IS 1892. DIMITRI Ivanovsky, a 28-year-old biologist in Russia, has been asked by the Czarist government to help solve a major problem. In the early 1800s, tens of thousands of hectares of land in the Ukraine were planted with tobacco, a crop imported from America. By the time Ivanovsky had graduated from college, farmers throughout this region of Russia depended upon the crop for their existence. But something was going wrong. Every year, more and more of the crop was being lost to a strange disease. Ivanovsky's investigation into the cause of the tobacco disease is a fascinating story. It tells about the discovery of another interesting segment in the spectrum of life's diversity—the viruses.

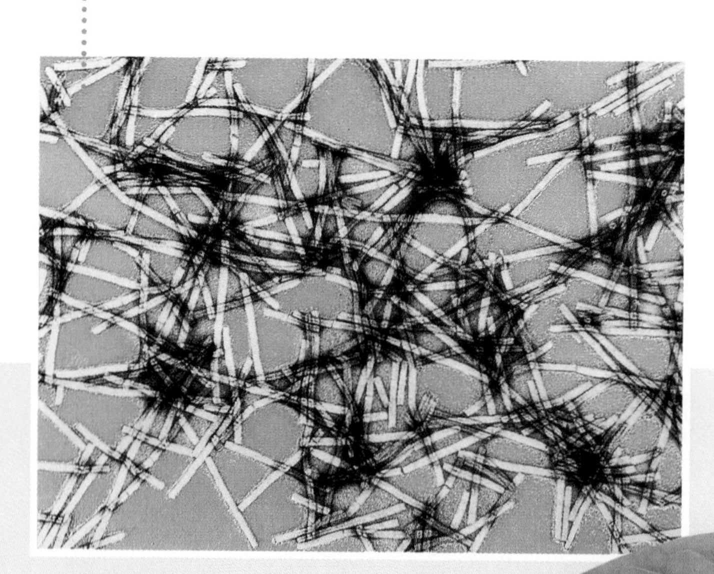

Figure 22–7
(a) *Tobacco mosaic virus (magnification: 34,000X), which was discovered by Dimitri Ivanovsky, causes* (b) *the leaves of tobacco plants to develop a pattern of spots.*

The Discovery of Viruses

Ivanovsky observed once-healthy tobacco leaves develop yellow spots. He watched the entire plant wither and die. Ivanovsky suspected a disease, and he performed a simple experiment to test his idea. He crushed some leaves from a diseased plant and collected the juice. Then he placed a few drops of juice on the leaves of a healthy plant. In a few days, yellow spots appeared on the leaves precisely where the drops had been placed. Ivanovsky had proven that tobacco mosaic disease—as it is now known—could be passed from one plant to the next.

Ivanovsky thought that a bacterium might be causing the disease, so he examined the juice under a microscope. To his surprise, there were no bacteria in the juice. In fact, there were no cells in the juice at all. The young scientist concluded correctly that the infected plants must contain an invisible infectious particle.

What Is a Virus?

A few years later, Dutch scientist Martinus Beijerinck confirmed Ivanovsky's experiments. He showed that not only were these particles invisible under a microscope, they were so small they could pass through the pores of the finest filters.

Bio FRONTIER Connections

Eliminating Viruses

At one time, smallpox was a dreaded disease that affected people all over the world, killing about 40 percent of its victims. Those who didn't succumb to the disease were seriously disfigured. Smallpox begins with fever and vomiting, followed by a skin rash that quickly spreads over the entire body. The rash forms itchy blisters, which result in unsightly scars when the victim scratches them.

Today, thanks to an intensive, worldwide vaccination campaign by the World Health Organization (WHO), smallpox has been eliminated as a disease. Vaccination for smallpox has been unnecessary since 1977. In 1980, WHO was able to announce that smallpox had been eradicated.

The Virus Lives On

Although smallpox is no longer an immediate threat to human health, the virus that causes it—called variola—still exists. Two samples of smallpox virus are preserved for research purposes in special Level 4 containment laboratories. One of these Level 4 labs is located at the Centers for Disease Control and Prevention in Atlanta, Georgia. The other is in the Russian State Research Center of Virology and Biotechnology in Koltsovo. Level 4 labs, which contain the most dangerous viruses, are floodproof and fireproof. As an added precaution, the viruses are kept in airtight spaces between floors to prevent their escape.

By 1994, scientists studying the virus had completely identified the virus's genetic code. Little reason remained to continue stockpiling the virus. So in 1996, the 190 member nations of WHO decided to destroy all remaining samples of smallpox virus on June 30, 1999. But even after the virus is destroyed, WHO plans to keep some smallpox vaccine on hand . . . just in case.

CAREER TRACK *This virologist is working in a Level 4 containment laboratory at the Centers for Disease Control in Atlanta, Georgia. The steel laboratory is equipped with air locks and decontamination alarms to minimize the chance of highly infectious organisms from escaping.*

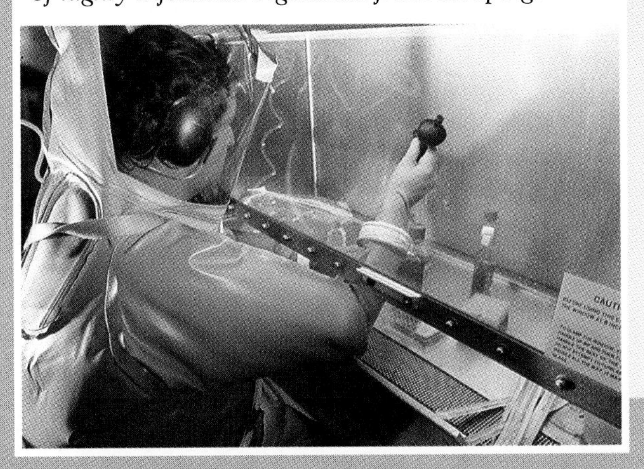

The Future of Laboratory Viruses

Smallpox isn't the only dangerous virus being studied in containment labs. Ebola virus, which is named after a small river in Zaire where it first appeared, is one of the deadliest viruses on Earth. It is also being studied, as is HIV, the virus that causes AIDS. But scientists differ as to what the future of these and other laboratory viruses should be.

Some scientists think that all deadly viruses should be destroyed. Others fear that in an attempt to destroy them, some viruses might escape from their confinement. Still others think that the scientific value of these viruses, especially in the event of future outbreaks, outweighs the risks of keeping them. As you see, scientists often hold differing—even strongly opposing—viewpoints.

Making the Connection

Should samples of viruses—such as the smallpox virus—that are only kept in containment labs be destroyed? Why or why not? What information could scientists gain from studying these viruses?

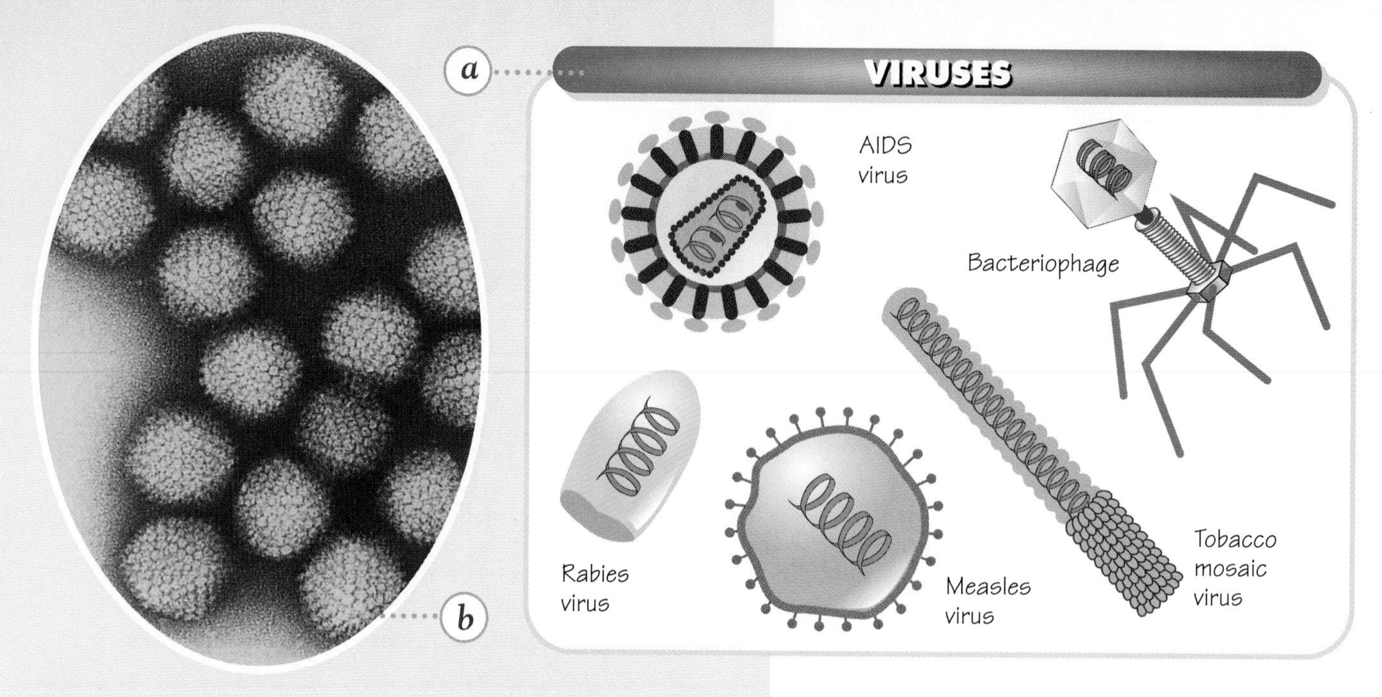

Figure 22–8
(a) The shapes of viruses vary greatly and are determined by their protein coats. (b) The electron micrograph shows Adenovirus*—one of many viruses that cause the common cold in humans (magnification: 160,000X).*

Knowing that the Latin word for poison was *virus*, Beijerinck wrote that the tobacco mosaic disease was caused by a filterable virus. Today, we refer to these particles as viruses.

A **virus** is a nonliving particle that contains DNA or RNA and that can infect a living cell. Viruses are much smaller than cells, as shown in ***Figure 22–2*** on page 516. Are viruses just smaller forms of living cells? Not at all. **Viruses consist only of genetic information—in the form of DNA or RNA—surrounded by a protein coat.** Although some viruses also contain other materials, such as lipid membranes and enzymes, the viruses cannot live outside a living cell.

Checkpoint What is a virus?

How Viruses Infect Cells

However lifeless it may seem outside a living cell, once it makes contact with a cell that it can infect, a virus goes to work. **A virus infects a bacterial cell by attaching to the membrane of the cell. It then penetrates or fuses with the cell's membrane, releasing its own genetic material into the cell's cytoplasm.**

As the illustration on the next page shows, once inside the host cell, viruses express their genes in a way that enables them to use the host to reproduce. The host cell makes copies of the virus that can ultimately infect other cells!

Some viruses take over and destroy the host cell. They produce enzymes that cut the host cell's DNA into pieces, grab control of its ribosomes to make viral proteins, and then use up all the cell's resources to make more viruses. This process is known as a **lytic infection.**

Other viruses are more subtle. In a **lysogenic infection,** viruses insert their genetic material right into the DNA of a host. The viral genes then go about making mRNA, like the genes of the host cell. However, these viral mRNAs will direct the synthesis of viral proteins, gradually converting the cell into a living factory for making new viruses.

An example of a lysogenic infection is the lambda virus, a **bacteriophage**—a virus that infects bacteria (the name means "bacteria eater"). When lambda infects a host cell, the virus inserts its double-stranded DNA molecule into the bacterium. In most cases, this DNA is then inserted into the host cell DNA,

Visualizing a Viral Infection

Two ways in which a virus infects a bacterial cell are lytic infection and lysogenic infection.

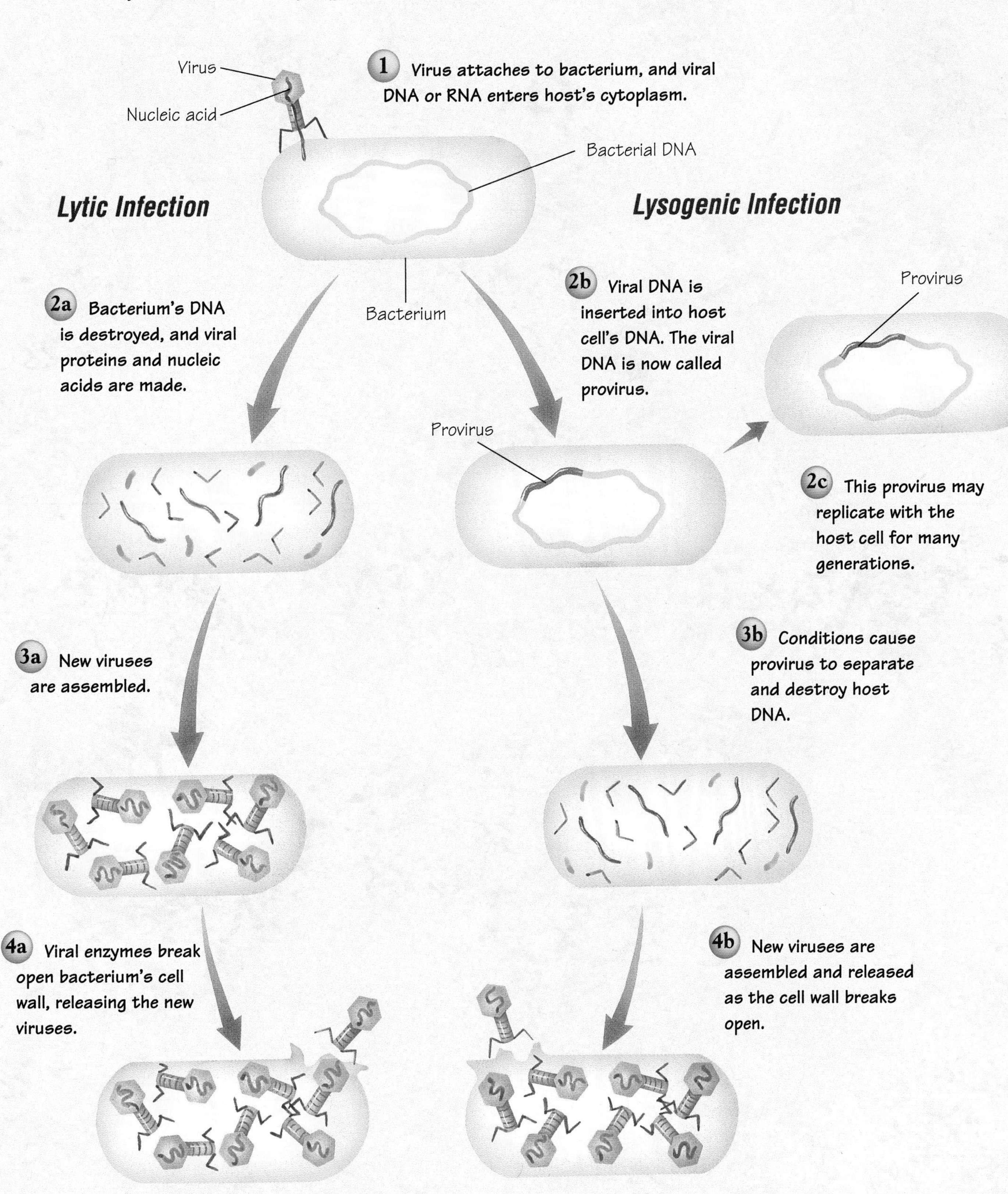

Figure 22–9
Polio virus, one of the smallest viruses, is only 28 nanometers in diameter. Research in the area of three-dimensional structures of large molecules produced this computer-generated image of the protein coat of type I polio virus. Such studies aim to develop antiviral drugs.

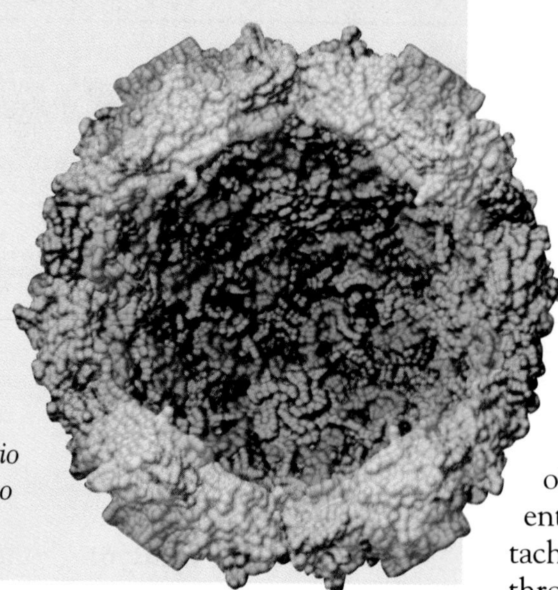

characteristic of a lysogenic infection. The viral genes—now part of the bacterium's own chromosome—are known as a **provirus.** The provirus may remain part of the cell's own DNA for many generations, passed on from one cell to the next.

The provirus state is not permanent. Any one of a number of factors, including radiation or chemicals that damage DNA, may cause lambda to activate genes that suddenly produce viral coat proteins. Lambda DNA then emerges from the host chromosome, produces copies of viral DNA, assembles scores of new viruses, and destroys the host cell.

☑ **Checkpoint** What is a provirus?

Viruses and Disease

All viruses are parasites. A parasite depends entirely upon another living organism for its existence in a way that harms that organism. Because viruses infect living cells, they often harm those cells, producing sickness and disease.

Polio Virus

Polio virus, an example of one of the smaller animal viruses, contains a single RNA strand, roughly 7500 nitrogen bases in length. Polio virus RNA is enclosed in a protein coat that has a highly regular structure resembling the surface of a soccer ball. When a polio virus enters the human body, it usually attaches to the surface of a cell in the throat or nose. Sometimes the virus travels throughout the body and attacks cells in the nervous system. When this happens, the nerve cells that control muscles are destroyed, producing paralysis and sometimes even impairing the ability to breathe.

Polio was once a serious infectious disease that killed or paralyzed thousands of people in the United States each year. Fortunately, highly effective vaccines now given to young children have almost completely eliminated this disease.

Retroviruses

The **retroviruses** are a class of viruses that, like polio virus, contain RNA as their genetic material. Retroviruses received their name from the fact that their genetic information is copied backwards—from RNA into DNA, instead of from DNA to RNA. The prefix *retro-* means backward. Retroviruses are responsible for some types of cancer in humans and animals. One type of retrovirus produces the disease known as AIDS.

Section Review 22–2

1. **Describe** the structure of a virus.
2. **Explain** how viruses infect cells.
3. **Critical Thinking—Comparing** Identify the main differences between lytic and lysogenic viral infections.

Bacteria in Our World

GUIDE FOR READING

- **Describe** the role of bacteria in nature.

MINI LAB

- **Formulate a hypothesis** about the effect on bacteria of washing hands with soap and water.

BACTERIA HAVE BEEN ON OUR planet longer than any other form of life. Fossils indicate that bacteria first appeared more than 3 billion years ago. It should not be surprising, therefore, that during those 3 billion years bacteria have found ways to live in just about every corner of the Earth. They have carved out a niche for themselves in habitats as varied as the barren tundra, the dark ocean depths, and thick boiling mud.

Bacteria in Nature

Careful estimates of the numbers of bacteria found in the soil, the ocean, and deep within the Earth suggest that the mass of living bacteria is greater than that of all other organisms combined! Given their tremendous numbers, what roles do bacteria play in the natural world? **In nature, bacteria do a little of everything—from photosynthesis to nutrient recycling.**

Decomposers and Recyclers

From the moment a plant or animal dies, thousands of bacteria attack and digest the dead tissue, breaking it down for nourishment and energy. Gradually the body tissues of the organism fall apart and crumble into the soil, as these armies of bacteria do their work. The raw materials are returned to the Earth, and the cycle of life can begin again. Although the bacteria are not the only organisms involved in this process, they are one of the most significant.

☑ ***Checkpoint*** How are nutrients recycled by bacteria?

Figure 22–10
Throughout the biosphere—in (a) animals and (b) plants, as well as (c) in the rushing waters of this mountain stream, often completely unnoticed—bacteria are present in huge numbers.

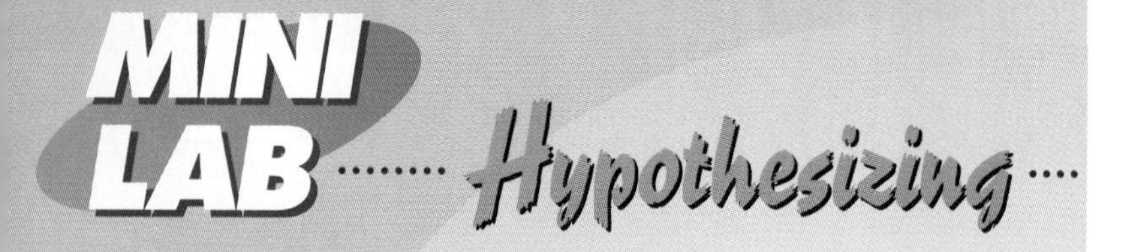

"Did You Wash Your Hands?"

PROBLEM *What effect does washing hands with soap have on bacteria?* ***Formulate a hypothesis.***

PROCEDURE

1. Obtain a Petri dish containing nutrient agar.
2. With a marking pencil, draw a line across the bottom of the Petri dish to divide it into two halves, marked A and B.
3. Open the dish and gently draw a finger across the agar on side A of the dividing mark. Quickly close the dish.
4. Wash your hands with soap and water, then dry them. Open the dish and draw a washed finger across the agar on side B of the dividing mark.
5. Tape the dish closed and give it to your teacher to be incubated. Do you think there will be a difference in the number of microorganisms in sides A and B? Record your hypothesis.
6. When you get your dish back, examine it for any changes and record your observations.

ANALYZE AND CONCLUDE

1. Did the experiment support your hypothesis?
2. Is the amount of bacterial growth on the two sides of the dish different? If so, explain why.
3. Is it advisable to wash your hands before preparing or eating food?

Symbiotic Relationships

You might not like to dwell on it, but your intestines are inhabited by large numbers of bacteria. Safely tucked away inside your digestive system, these bacteria are provided with a warm, safe place, plenty of nourishment, and even free transportation. What do you get in return? These bacteria help you to digest your food. They even make a number of vitamins that you cannot make yourself. This kind of relationship, one in which both organisms benefit, is known as **mutualism.** Most animals have mutualistic relationships with bacteria in their digestive tracts.

Natural Fertilizers

Nearly all organisms require a supply of nitrogen in order to survive. Although the Earth's atmosphere contains plenty of nitrogen gas, plants and other organisms cannot use nitrogen gas directly. Fortunately, many bacteria can carry out a process called nitrogen fixation, in which they take nitrogen gas and convert it to a chemical form that plants can use, producing a natural fertilizer.

☑ **Checkpoint** What is mutualism?

Bacteria and Humans

People have used bacteria, often without knowing it, for thousands of years. As you might expect, we have exploited the very things that bacteria do best—breaking down material for nourishment.

Food Processing

Bacteria are used to prepare a wide variety of foods. Some cheeses, sour cream, buttermilk, and yogurt are made by allowing certain types of bacteria to grow in milk products. Other forms of bacteria are used to make pickles, sauerkraut, and vinegar.

Sewage Treatment

Sewage water contains human wastes, discarded foods, and even chemical wastes. Because bacteria can break down each of these materials naturally, tons of bacteria are added to sewage at water-treatment plants. As they grow, these bacteria break down compounds in the wastes into simpler materials. Carefully used, bacterial sewage treatment produces purified water, nitrogen, and carbon dioxide, leaving a sludge that can be used as crop fertilizer.

Figure 22–11
(a) The Duffin Creek Water Pollution Control Plant in Picarin, Ontario, Canada, relies on bacteria to break down the organic as well as the chemical wastes in wastewater. (b) Most yogurt- and cheese-making processes are dependent on the action of bacteria. In this photograph, a worker examines the curds that will be processed to make Gouda cheese.

Bacteria in Mining

Believe it or not, bacteria play a major role in the mining industry. Most of the copper ore in the United States is a relatively low grade, which means that it contains only a small amount of copper. Extracting that copper would be a difficult job without *Thiobacillus ferrooxidans*. This bacterium grows naturally in copper-containing ore, producing sulfuric acid, which reacts with copper to form copper sulfate. Copper sulfate is water soluble, so miners simply allow this bacterium to grow in the ore, wash the copper sulfate out as it forms, and later retrieve the copper from the copper sulfate solution.

Checkpoint How are bacteria used by humans?

Bacteria and Health

Although most bacteria are not harmful, the fact that some of them can cause disease is reason enough to try to control the growth of bacteria in certain situations. Sterilization—the killing of bacteria with heat or chemicals—is the simplest way to control bacteria. Most bacteria cannot survive high temperatures, so hospital instruments are sterilized with high-pressure steam. Of course, a hospital room cannot be dropped into boiling water, but it can still be sterilized by the use of disinfectants—chemicals that kill bacteria.

The control of dangerous bacteria is of particular importance while handling and processing food. The growth of bacteria can be slowed down by low temperatures. This is how refrigerators make food last longer. Food can be sterilized by cooking it at high temperatures, and many foods—particularly meats—should not be eaten without thorough cooking to kill harmful bacteria.

INTEGRATING CHEMISTRY

What chemical changes does the copper in copper ore undergo before it is produced as copper metal?

Section Review 22–3

1. **Describe** the role of bacteria in nature.
2. **MINI LAB** What effect does washing hands with soap and water have on bacteria? **Formulate a hypothesis.**
3. **BRANCHING OUT ACTIVITY** Conduct library research to find different methods used to preserve foods. In a brief report, **describe** one of these methods and **explain** how it works.

CHAPTER 22

Laboratory Investigation

DESIGNING AN EXPERIMENT

Observing the Effects of the Tobacco Mosaic Virus

The tobacco mosaic virus (TMV) studied by Dimitri Ivanovsky still commonly infects field-grown tobacco today. Even though the virus cannot be seen, the symptoms caused by the virus can easily be observed. In this investigation, you will make an extract of TMV and then place it on the leaves of some healthy tobacco plants. The presence of spots on the healthy plants indicates that they have been infected with TMV.

Problem

What effect does the tobacco mosaic virus have on healthy tobacco plants? **Design an experiment** to answer this question.

Suggested Materials

young tobacco plants in pots with soil
tobacco sample
spatula
carborundum powder
dibasic potassium phosphate solution
small beakers
mortar and pestle
cotton swabs
marking pencil
small cards

Suggested Procedure

1. **Formulate a hypothesis that you want to test.**
2. **Divide the dibasic potassium phosphate solution between two beakers. CAUTION: *Be careful when using this solution.* Dip a cotton swab into one of the beakers to moisten it.**
3. **With a spatula, sprinkle the swab with a small amount of carborundum powder. Gently stroke some of the leaves of one plant with the swab. The carborundum is an abrasive and will make microscopic scratches on the surface of the leaf. Try not to rub too hard or you will destroy the leaf entirely. Put this plant aside for now.**
4. **Place the tobacco sample into a mortar, slowly add the dibasic potassium phosphate solution from the other beaker to the sample, and grind the mixture with the pestle. Allow the solid particles to settle. The extract contains the virus. Dampen a cotton swab with a small amount of the liquid in the mortar. Repeat step 3 using another tobacco plant.**

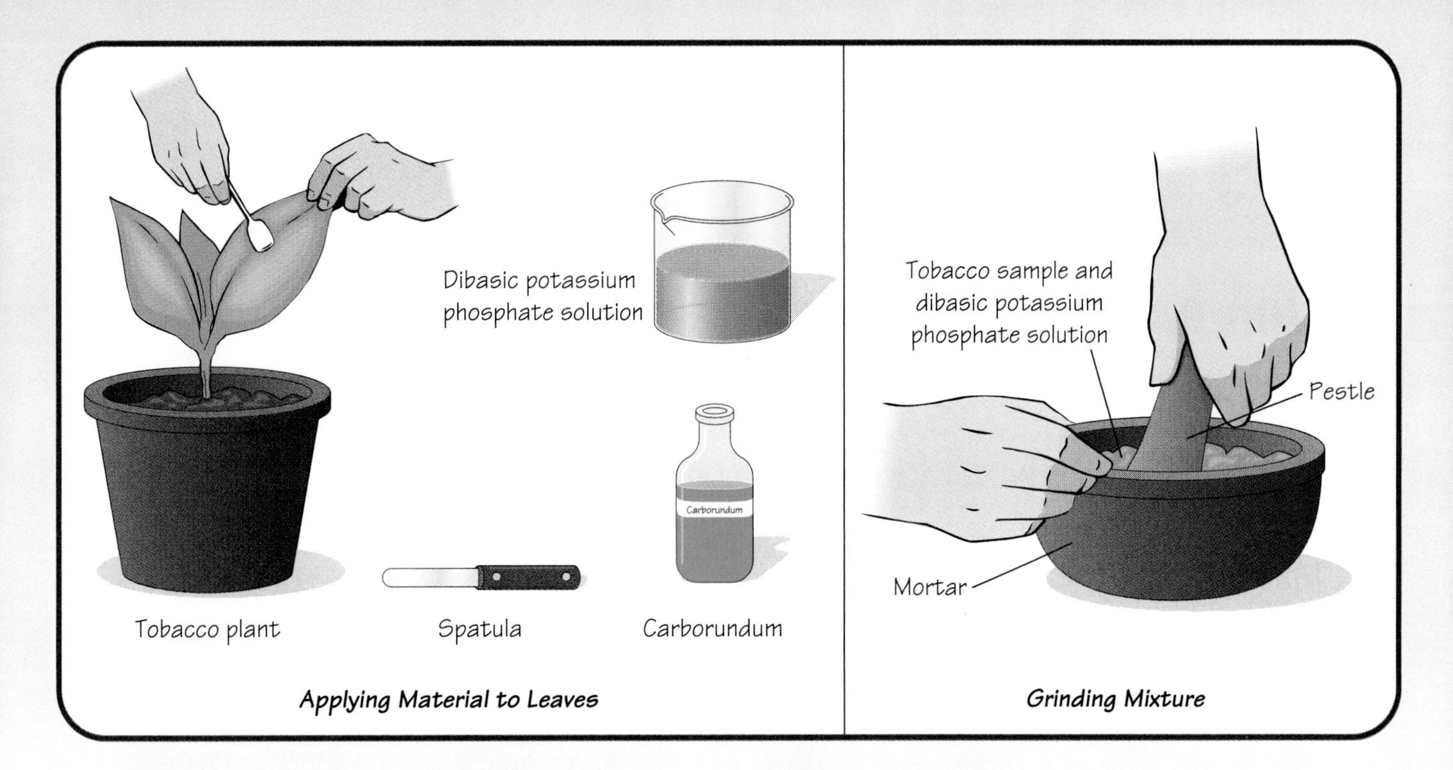

5. Plan to observe the plants for a two-week period. Make sure that the plants get sufficient water and light during that time. You may want to indicate on a small card which plant was swabbed with the virus extract and which one was not. Discard the cotton swabs, as directed by your teacher, and wash all equipment used with soap and water. Be sure to wash your hands with soap and water too.

Observations

1. What changes did your group see in each plant in the first 5 days? Over the next 5 days?
2. Did your group notice whether any changes were confined to one spot, or did they spread throughout the entire plant?

Analysis and Conclusions

1. What was your group's hypothesis for this investigation?
2. Did the result of the experiment support your hypothesis? Can you think of other hypotheses that the results also support?
3. What was the purpose of the plant that was not swabbed with the virus extract? Why was it prepared first?
4. What is the relationship between the treatment of the leaves with carborundum powder and what happens when you scrape a knee or an elbow in a fall?
5. Many plant growers do not allow any tobacco products near plants in their greenhouses. What reason can you give for this?

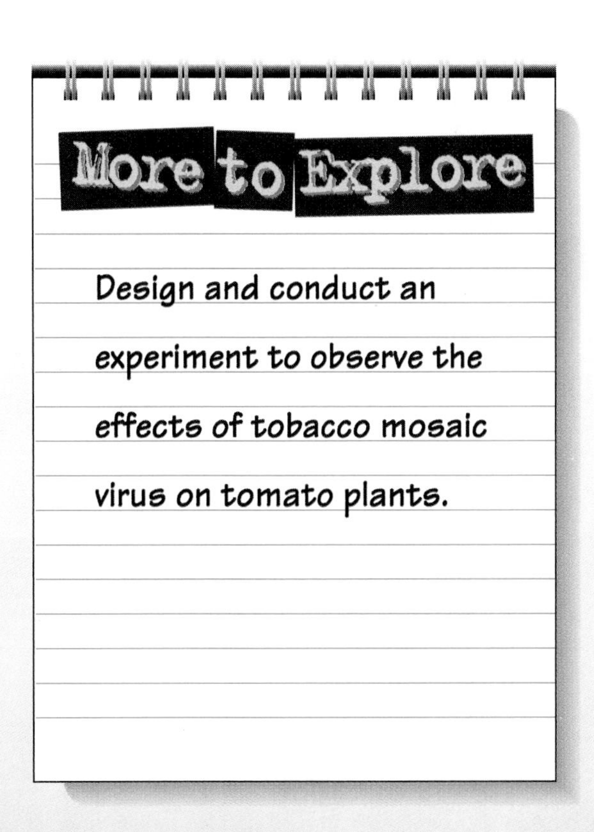

Study Guide

Summarizing Key Concepts

The key concepts in each section of this chapter are listed below to help you review the chapter content. Make sure you understand each concept and its relationship to other concepts and to the theme of this chapter.

22–1 Bacteria

- Bacteria are prokaryotes, which are cells without nuclei.
- Genetic information in bacterial cells is found on a single circular DNA molecule, or chromosome, in the cytoplasm. The cytoplasm is surrounded by a cell membrane composed of lipids and proteins.
- The basic shapes of bacteria are rodlike, spherical, and spiral.
- Bacteria are divided into two phyla—the Archaebacteria and the Eubacteria.
- Reproduction in bacteria occurs by binary fission and conjugation.

22–2 Viruses

- A virus is a nonliving particle that contains DNA or RNA and that can infect a living cell.
- Viruses contain genetic information—in the form of DNA or RNA—surrounded by a protein coat.
- A virus infects a cell by first attaching itself to the cell's membrane. It then penetrates or fuses with that membrane, releasing its own genetic material into the cytoplasm. Inside the cell, the virus expresses its genes in ways that produce new copies of the virus particle. These ultimately infect other cells.

22–3 Bacteria in Our World

- In nature, bacteria fix nitrogen, decompose and recycle dead matter, and help animals to digest food.
- Humans use bacteria to prepare foods, process sewage, and mine copper.

Reviewing Key Terms

Review the following vocabulary terms and their meaning. Then use each term in a complete sentence.

22–1 Bacteria

prokaryote
bacillus
coccus
spirillum
binary fission
conjugation
spore
endospore
antibiotic

22–2 Viruses

virus
lytic infection
lysogenic infection
bacteriophage
provirus
retrovirus

22–3 Bacteria in Our World

mutualism

Recalling Main Ideas

Choose the letter of the answer that best completes the statement or answers the question.

1. Bacteria are cells without nuclei, or

a. eukaryotes. **c.** prokaryotes.
b. chromosomes. **d.** endospores.

2. Where are Eubacteria found?

a. in very harsh environments
b. in bodies of water only
c. on land only
d. almost everywhere on the planet

3. In binary fission, a

a. bacterium replicates its chromosome and divides in half.
b. bacterium forms a thick wall that encloses part of its cytoplasm and a DNA molecule.
c. bridge of protein forms from one bacterium to another.
d. bacterium captures chemical energy from its surroundings.

4. Drugs that destroy bacteria are called

a. vitamins. **c.** antibiotics.
b. bacilli. **d.** vaccines.

5. A virus is a

a. living particle that contains DNA or RNA and that can infect a living cell.
b. nonliving particle that contains only DNA and that can infect a living cell.
c. nonliving particle that contains DNA or RNA and that can infect a living cell.
d. living particle that contains only RNA and that can infect a living cell.

6. When a virus infects a cell, what does it release into the cell's cytoplasm?

a. its own poisonous cells
b. its own genetic material
c. bacteria
d. a provirus

7. Which is not an example of the benefit of bacteria to humans?

a. Bacteria help humans to fight disease.
b. Bacteria help humans to digest food.
c. Bacteria help humans to make vitamins.
d. Bacteria prevent food poisoning.

Putting It All Together

Using the information on pages xxx to xxxi, complete the following concept map.

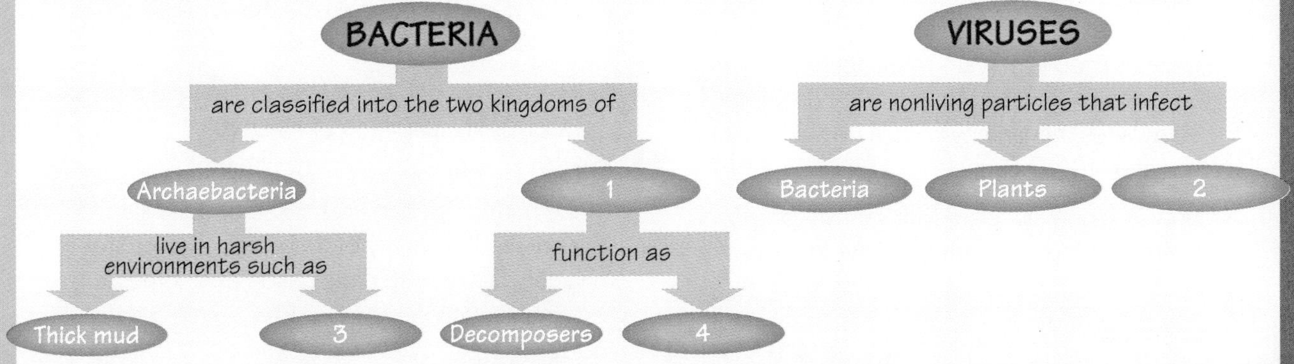

BACTERIA
are classified into the two kingdoms of
Archaebacteria
1
live in harsh environments such as
Thick mud
3
function as
Decomposers
4

VIRUSES
are nonliving particles that infect
Bacteria
Plants
2

Reviewing What You Learned

Answer each of the following in a complete sentence.

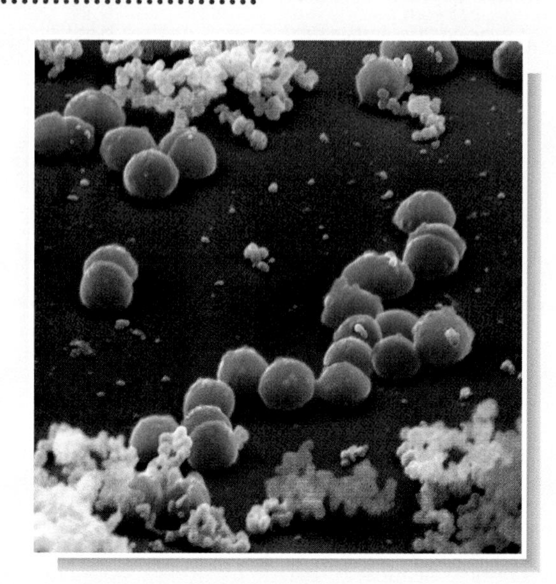

1. Bacteria are considered prokaryotes. How do they differ from other organisms?
2. What distinguishes a gram-positive bacterium from a gram-negative bacterium?
3. Identification of bacteria often begins with their shape. What are the commonly **observed** shapes of bacteria?
4. How do Archaebacteria differ from Eubacteria?
5. How can endospores enable bacteria to withstand long periods of unfavorable conditions?
6. What is the germ theory of disease?
7. What are some human diseases caused by bacteria?
8. What is a virus?
9. What is a provirus?
10. How do lytic and lysogenic infections differ?
11. What is a retrovirus?
12. Identify some roles bacteria play in the natural world.
13. How do humans use bacteria?

Expanding the Concepts

Discuss each of the following in a brief paragraph.

1. Why are the gram-negative bacteria potentially more difficult to treat than the gram-positive bacteria?
2. Some bacteria, under proper conditions, can reproduce very quickly, even as fast as 20 minutes per reproductive cycle. If you were infected with a single bacterium and it began to reproduce at a rate of once every 30 minutes, how many bacteria would exist in you after $10\frac{1}{2}$ hours?
3. How did the work of Dimitri Ivanovsky demonstrate the scientific method of investigation?
4. Why are viruses not considered alive by most scientists?
5. Describe what happens when a virus enters another cell, such as a bacterial cell.
6. Give evidence to support this statement: "All viruses are parasites."
7. How does the polio virus enter and use the human body?
8. The roles of bacteria on Earth are astounding. Discuss at least three of these roles.
9. Discuss how binary fission in bacteria increases the number of bacterial cells but does little to increase the variety of these cells. Compare binary fission with conjugation as another form of reproduction.

Extending Your Thinking

Use the skills you have developed in this chapter to answer the following.

1. **Designing an experiment** Bacterial conjugation is a form of sexual reproduction employed by many bacteria. How might conjugation be used to determine the sequence of DNA for a particular strain of bacteria?
2. **Evaluating** Recently, publications and motion pictures have presented frightening scenarios in which highly infectious bacteria or viruses are spread through the biosphere. The potential disaster they imply is disturbing. Do you believe this potential truly exists? Defend your answer.
3. **Evaluating** The use of genetic engineering has intensified the evolutionary potential of organisms. Many of these transformed organisms have been developed to benefit humans in various areas. One such organism is a bacterium that has the ability to digest crude oil. These bacteria help deal with oil spills, particularly in the cleanup of coastal zones. Do you consider the manipulation of organisms to benefit humans or the environment a proper and responsible use of scientific knowledge?
4. **Using the writing process** Bacteriological warfare is potentially so detrimental that it was outlawed during World War I. Numerous meetings between nations were convened to write up a set of rules so that such methods would never be used again. But today all major nations conduct research and develop new organisms to be used "if necessary." Write an essay expressing your views on this subject, discussing actions that must be taken to safeguard our planet from this potential danger.
5. **Hypothesizing** Do you think humans will or can completely control all infectious organisms? Formulate a hypothesis to answer this question.

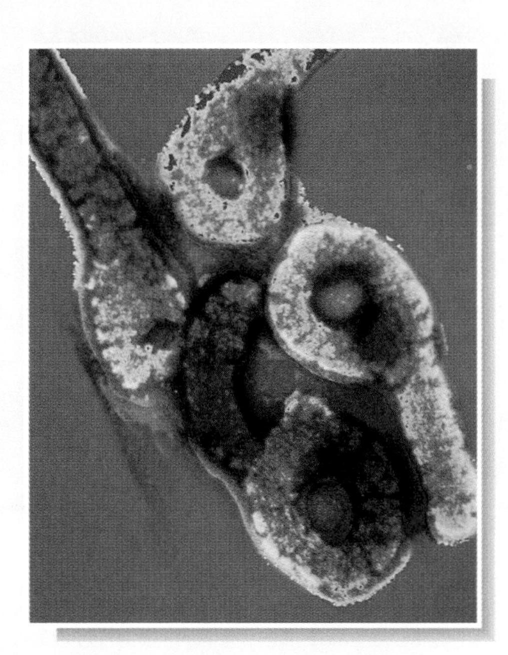

Applying Your Skills

Modeling Viruses and Bacteria

Even though bacteria and viruses are microscopic, both play a very important role in our everyday lives. To better understand how something functions, scientists often make a model of the object or process of interest.

1. Working with a partner, using simple, readily available materials, make a model of a bacterium and a virus, keeping their relative sizes in mind.

• GOING FURTHER •

2. Include important parts of viruses and bacteria and explain the functions of these parts.

CHAPTER 23

Protists and Fungi

FOCUSING THE CHAPTER THEME: Unity and Diversity

23–1 Protists
- **Classify** the protists.

23–2 Fungi
- **Identify** the four phyla of fungi.

BRANCHING OUT *In Depth*

23–3 Protist and Fungal Diseases
- **Describe** different diseases that protists and fungi cause.

LABORATORY INVESTIGATION
- **Design an experiment** to observe the structure of a paramecium and its response to different stimuli.

Biology and Your World

BIO JOURNAL

In your journal, describe the organisms shown in the photographs on this page and the next. Where have you encountered similar organisms?

Cup fungus on a forest floor in Peru

Protists

Guide for Reading

- **Classify** the protists.
- **Describe** the animallike, plantlike, and funguslike protists.

MINI LAB

- **Design an experiment** to observe protists.

A NIGHT OF HEAVY RAIN HAS soaked the thick mat of dead leaves and branches on the forest floor. As mist rises in the morning sunlight, something seems to have come alive on this carpet of debris. A thin yellow coating has appeared on a few of the branches. In a few days, it will grow and concentrate into a thick living mass.

The organisms that make up this yellow mass serve an important role in the forest—they break down dead plant matter, recycling it into usable components. Let's explore these organisms more closely.

Kingdom Protista 

The organisms shown in ***Figure 23–1*** are members of the kingdom Protista. Traditionally, Protista is considered to be the first eukaryotic kingdom. Indeed, the organisms in this kingdom are known as protists—meaning "the very first" in Greek. Most protists are single-celled organisms, although quite a few consist of hundreds or thousands of cells.

The kingdom Protista may contain as many as 200,000 species—ranging from microscopic organisms to those large enough to be seen with the unaided eye. Although some protists live their lives in ways you never notice, others produce the oxygen you breathe, and some cause especially deadly diseases. Altogether, protists affect us all in ways that cannot be ignored.

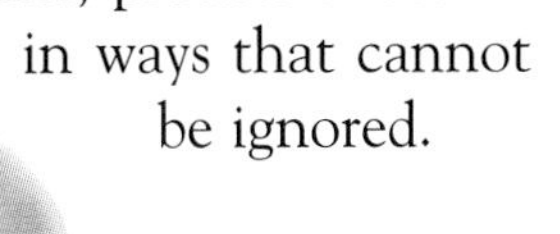

Figure 23–1 *The protists include some especially interesting and beautiful organisms—and some bothersome ones as well.* (a) *Colonies of* Synura *can contaminate water supplies, giving the water a fishy, oily taste (magnification 400X).* (b) *Because of its trumpetlike body, this protist was named* Stentor, *after a loud-voiced hero of the Trojan War (magnification: 125X).* (c) *A diatom is surrounded by an intricate, glasslike shell (magnification: 4600X).*

Single-Celled Eukaryotes

Biologists have argued for years over the best way to classify single-celled eukaryotes, and this issue may never be settled. Why the disagreement? Many of these organisms have more in common with one or two of the multicellular kingdoms—fungi, plants, and animals—than they do with each other. However, they do not develop the specialized multicellular structures that are found in each of the other three eukaryotic kingdoms. These facts make many single-celled eukaryotes extremely difficult to classify.

Classifying Protists

Currently, the kingdom Protista is defined in terms of what it is not! **Any eukaryote that is not classified as a fungus, a plant, or an animal is classified as a protist.** As you will see, the protists are an extremely diverse group—they solve the problems associated with being alive in remarkably different ways.

Biologists often describe a protist by first noting to which of the three multicellular kingdoms it is most similar. This is exactly the strategy that we will follow.

☑ ***Checkpoint*** What is a protist?

Animallike Protists

The animallike protists are also called protozoans, a name that means "first animals." There are four phyla of animallike protists—Zoomastigina (zoh-oh-mas-tuh-JIGH-nuh), Sarcodina (sar-koh-DIGH-nuh), Ciliophora (sihl-ee-AHF-uh-ruh), and Sporozoa (spor-oh-ZOH-uh). As you will see, protists are categorized into these phyla based on the way they move.

Flagellates

Organisms in the phylum Zoomastigina move through water using whiplike structures called **flagella** (singular: flagellum.) For this reason, the zoomastiginans are sometimes called flagellates.

Flagellates absorb food directly through their cell membranes. Some flagellates live in ponds and streams, where they live off dead and decaying organic matter. Others, however, live within the bodies of living animals.

Termites, for example, harbor large numbers of different flagellates in their digestive system. These flagellates help to digest the cellulose fibers in wood—something very useful to the termite. This relationship is an example of mutualism, a relationship between two organisms that benefits both organisms.

Other flagellates are parasites—organisms that harm animals in which they live. A parasitic flagellate

Figure 23–2
(a) *A termite is able to digest wood because of* Trichonympha, *a flagellate that lives inside the termite's digestive tract.* (b) *This* Trichonympha *is breaking down some wood particles, shown as colored streaks in the wide end of its body (magnification: 133X).*

Figure 23–3
(a) An ameba is a shape-shifting, bloblike protist. (b) Amebas take in a food particle by extending a pseudopod around it (magnification: 31X).

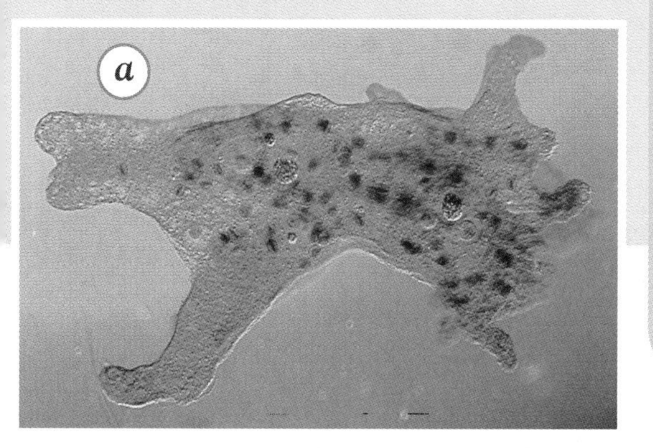

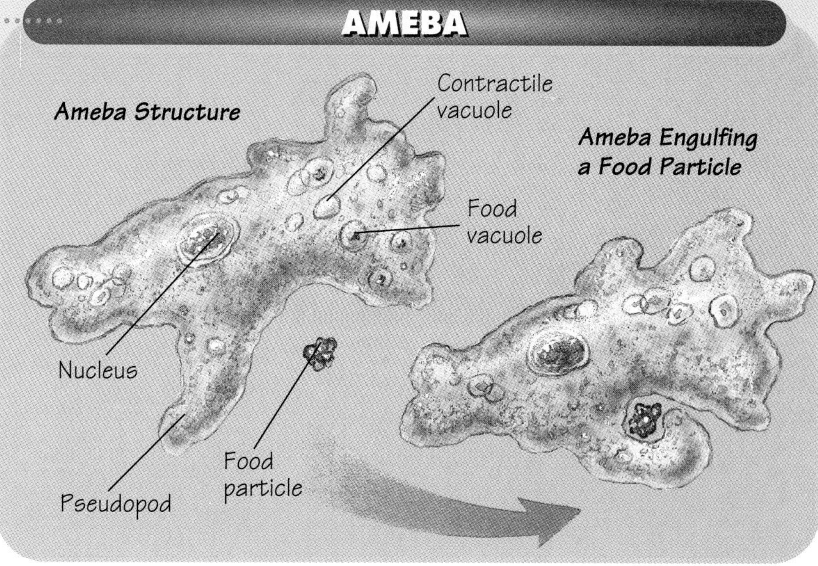

called a trypanosome lives in the human bloodstream and causes African sleeping sickness.

Flagellates reproduce asexually by cell division. However, many have a sexual phase to their life cycle.

Checkpoint What is a flagellate?

Sarcodines

Protists in the phylum Sarcodina move by pushing out temporary projections of cytoplasm. These projections are called **pseudopods** (SOO-doh-pahdz), meaning "false feet."

The best-known sarcodines are the amebas. Amebas have neither cilia, nor flagella, nor cell walls, and they even lack a definite cell shape. In fact, the word ameba comes from a Greek word that means "change."

An ameba moves by extending its pseudopod, as shown in ***Figure 23–3.*** Cytoplasm streams into the pseudopod, and the rest of the cell follows. Amebas also use pseudopods to capture and take in food particles.

Although there are thousands of different species of amebas, they are greatly outnumbered by another class of sarcodines called the Foraminifera (fuh-ram-uh-NIHF-er-uh). Foraminifers are so common that they have produced huge deposits of microscopic shells on the warmer regions of the ocean floor.

The sarcodines also include a stunning group of organisms known as the heliozoans—meaning "sun animals." Heliozoans produce thin shells of silica (SiO_2), the substance used to form glass.

Checkpoint What is a sarcodine?

Figure 23–4
(a) Foraminifers and other sarcodines produce thin shells of calcium carbonate, which they extract from sea water (magnification: 63X). (b) A heliozoan projects microtubules that support thin spikes of cytoplasm, making it look like a microscopic pin cushion (magnification: 310X).

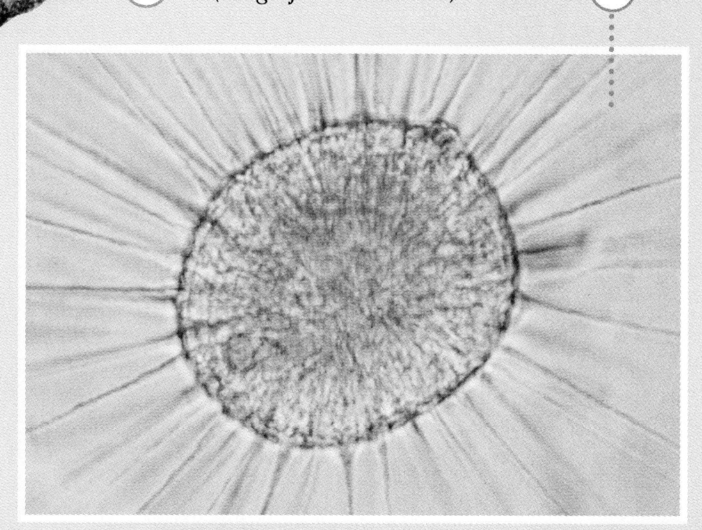

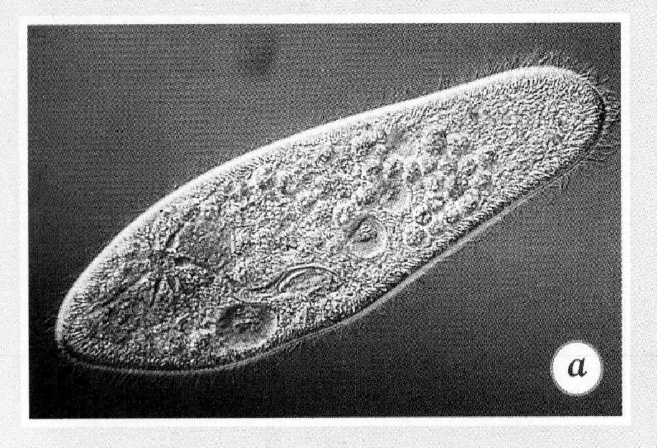

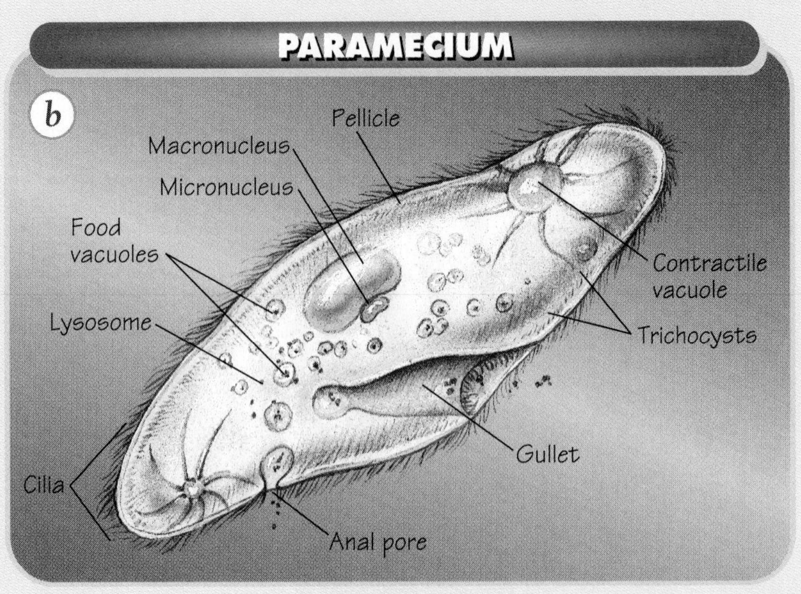

Figure 23–5
(a) A paramecium may contain only one cell, but that cell is extraordinarily complex and organized (magnification: 100X). (b) Each structure performs a specialized task.

Sporozoans

The members of the phylum Sporozoa do not have structures specialized for movement. All sporozoans are parasites, and they reproduce by forming small, single-celled structures called **spores**—the structures that give the phylum its name. Through spores, sporozoans can pass from one host to the next.

Sporozoans can infect fishes, worms, insects, birds, and humans. In fact, a sporozoan causes malaria—one of the most serious infectious diseases on Earth.

Sporozoans often have complex life cycles involving more than one host. The sporozoan that causes malaria, for example, lives part of its life in a mosquito.

☑ **Checkpoint** How do sporozoans reproduce?

Ciliates

The members of the phylum Ciliophora are the protists with **cilia**—short, hairlike projections that are similar to flagella. For this reason, these protists are often called ciliates. The coordinated beating of cilia pulls the ciliate quickly through the water, just as the rowing of hundreds of oars pulled boats in ancient times.

One especially complex and interesting ciliate is the paramecium, an organism common in freshwater ponds. As shown in ***Figure 23–5,*** a paramecium contains a variety of different structures that perform different tasks. For example, the **contractile vacuole** expels the water that diffuses into the paramecium by osmosis.

A paramecium has a **gullet** for taking in food particles that are processed into food vacuoles. Waste products are discharged when a used-up food vacuole fuses with the cell membrane at a region known as the **anal pore.** And embedded in the cell membrane are tiny bottle-shaped structures called **trichocysts.** Trichocysts are discharged when the cell is damaged or shocked, and they produce barbed projections that can damage predators.

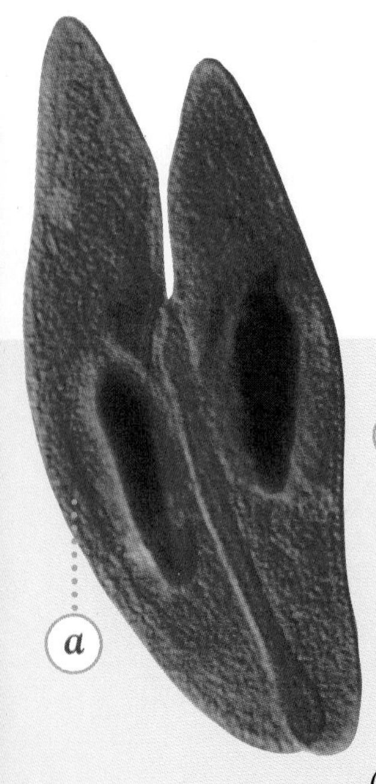

Figure 23–6
(a) Two paramecia exchange genetic information in a process called conjugation. New combinations of genes might help a paramecium to adapt to changes in its environment (magnification: 150X). (b) A paramecium reproduces by binary fission, in which it literally divides in two (magnification: 80X).

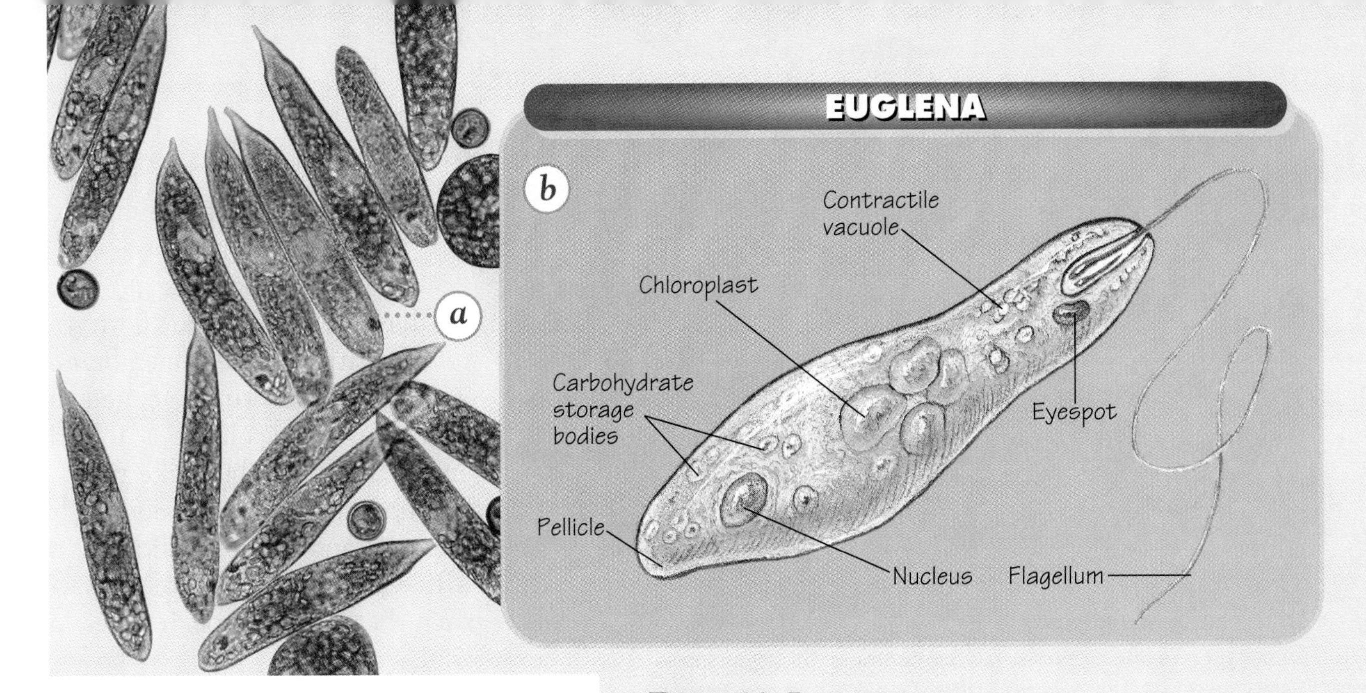

Figure 23–7
Euglenas are plantlike protists. (a) They get their green color from pigments in their chloroplasts—the structures where photosynthesis occurs (magnification: 200X). (b) In other respects, euglenas are similar to animallike protists.

One of the most remarkable features of ciliates is that they have two kinds of nuclei—a small **micronucleus** and a much larger **macronucleus.** Why does a ciliate need two nuclei? Evidence indicates that it uses the macronucleus as a "working library" of genetic information—a site for keeping multiple copies of the genes that it uses every day. The micronucleus, by contrast, contains backup copies of all the cell's genes.

Using the paramecium as an example, let's see how the two nuclei function in the ciliate's life cycle. Most of the time, a paramecium reproduces by binary fission. First, both the macronucleus and micronucleus are replicated. Then the cell splits in two, leaving each daughter cell with a single micronucleus and a single macronucleus.

Paramecia reproduce differently under stress, however, such as at extremes of temperature or when food is limited. At these times, they engage in a form of sexual reproduction known as **conjugation.** In conjugation, two paramecia exchange genetic information from their micronuclei. The result is two paramecia that are genetically identical but different from the way they were before the process began. Conjugation may provide the genetic diversity that paramecia need to meet environmental challenges.

☑ **Checkpoint** What is conjugation?

Plantlike Protists

The plantlike protists contain the green pigment chlorophyll and are capable of photosynthesis. Although most of these protists can move from place to place, they are described as plantlike because they are capable of photosynthesis. We will discuss three phyla of plantlike protists—Euglenophyta (yoo-glee-nuh-FIGHT-uh), Pyrrophyta (pigh-roh-FIGHT-uh), and Chrysophyta (krihs-uh-FIGHT-uh).

The plantlike protists are also examples of **algae,** a general term that describes all single-celled photosynthetic organisms. Most biologists include three other phyla of algae—red algae, brown algae, and green algae—among the plantlike protists. However, these algae are so similar to green plants that we will present them in another chapter.

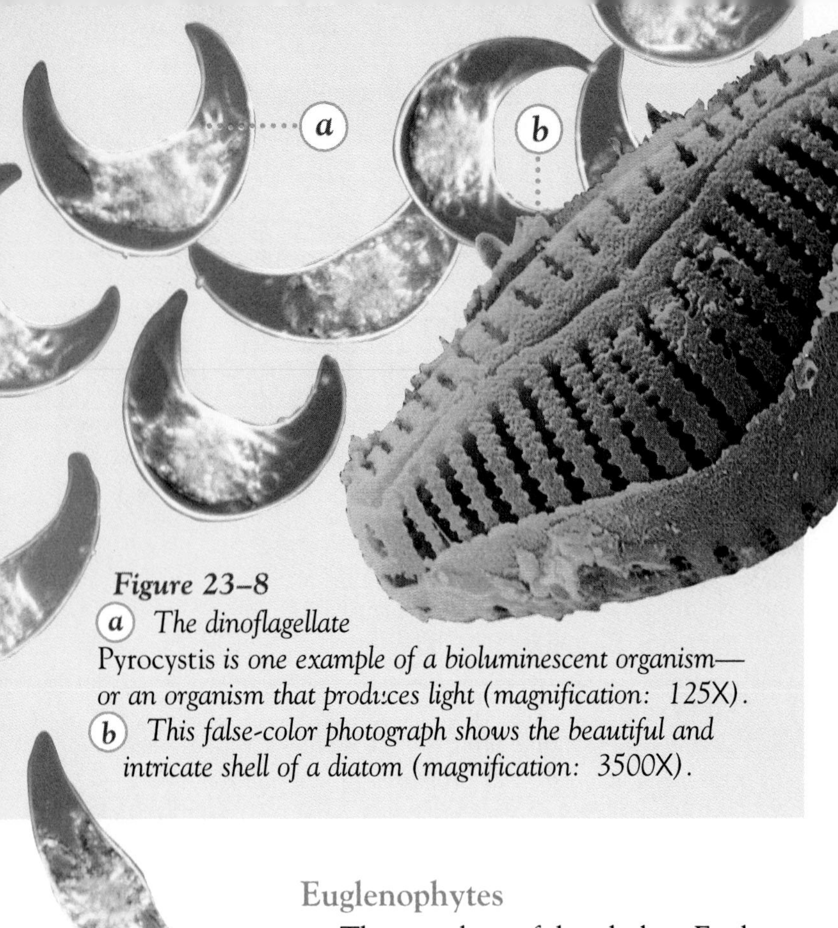

Figure 23–8
(a) *The dinoflagellate* Pyrocystis *is one example of a bioluminescent organism—or an organism that produces light (magnification: 125X).* (b) *This false-color photograph shows the beautiful and intricate shell of a diatom (magnification: 3500X).*

Euglenophytes

The members of the phylum Euglenophyta are remarkably similar to the flagellates. In fact, the only real difference between some species is the presence or absence of chloroplasts.

The phylum gets its name from the genus *Euglena*. A typical euglena is about 50 micrometers in length and has two flagella, as shown in ***Figure 23–7*** on page 539. The longer of the two flagella spins in a way that pulls the cell rapidly through water, making euglenas excellent swimmers.

Most euglenas are found in ponds and lakes, although a few saltwater species have been discovered. Euglenas reproduce asexually through binary fission.

Euglenas contain 10 to 20 oval-shaped chloroplasts that carry out photosynthesis. They also have a light-sensing structure called an eyespot—a cluster of red pigment near the flagella. With the aid of the eyespot, euglenas typically move rapidly to the brightest parts of their habitat. When light is not available, they can absorb food from the water around them, in much the same way that the animallike flagellates do.

While euglenas do not have cell walls, they do have an intricate cell membrane that is sometimes called a pellicle. The pellicle is folded into a series of ribbonlike ridges, each supported by microtubules. It also is tough and flexible, allowing a euglena to squirm and crawl along surfaces when there is not enough water to let it swim.

☑ **Checkpoint** What is the eyespot? What is the pellicle?

Dinoflagellates

Members of the phylum Pyrrophyta are often called the dinoflagellates (digh-noh-FLAJ-uh-lihts). Like the euglenophytes, most dinoflagellates are photosynthetic, although a few species seem to have lost their chloroplasts and live as heterotrophs. Most dinoflagellates are found in oceans, but a few live in lakes and ponds.

When agitated by a sudden movement in the water, many dinoflagellates undergo a chemical reaction that produces light. In the ocean on a dark night, this light is a remarkable, eerie sight. The light also gives the phylum its name—Pyrrophyta means "fire plants."

Chrysophytes

The chrysophytes include species commonly known as yellow-green algae, golden algae, and diatoms. The chloroplasts of these organisms contain bright-yellow pigments that give the phylum its name—Chrysophyta means "golden plants."

Like plants, chrysophytes contain cell walls. However, the chrysophytes' cell walls contain the carbohydrate pectin rather than cellulose.

The most abundant of the chrysophytes are the diatoms. In fact, diatoms are among the most abundant organisms on the Earth. You can see evidence of this fact on regions of the ocean floor that are littered with huge deposits of diatom remains.

Diatoms produce thin, delicate shells made of silica—the main ingredient in glass. The typical diatom shell contains two parts that fit together snugly, a bit like the two halves of a Petri dish. Some of these shells are etched with fine lines and markings that give diatoms a jewellike brilliance. If you look at a few diatom samples under a microscope, you may come to the same conclusion that we have—diatoms are among the most beautiful organisms on Earth!

☑ *Checkpoint* What is a diatom?

MINI LAB ····· Experimenting ·····

Protists in the Grass

PROBLEM *Under which conditions do protists grow best?* ***Design an experiment*** *to find out.*

SUGGESTED PROCEDURE 

1. Place a handful of grass in each of two clean glass jars. Add water until each jar is 3/4 full. Seal the jars tightly with lids.
2. Design an experiment to see whether protist growth in the jars is affected by temperature, motion, or sunlight.
3. With your teacher's permission, perform your experiment. After 3 days, open the jars in a well-ventilated area and gently stir their contents. Study drops of water from each jar under the compound microscope.

ANALYZE AND CONCLUDE

1. Did you observe protists in the water from either jar? If so, describe their appearance.
2. What conclusions can you draw from your experiment? Explain your answer.

Funguslike Protists

Some of the most interesting protists lack chlorophyll and absorb food through their cell walls. Because these characteristics also apply to fungi, these protists are known as the funguslike protists. The funguslike protists include the cellular slime molds, acellular slime molds, and water molds.

A few funguslike protists were once classified as fungi. However, fungi have cell walls made of chitin, which these organisms do not. And like other protists, funguslike protists contain centrioles, which true fungi lack.

Cellular Slime Molds

Members of the phylum Acrasiomycota (uh-kras-ee-oh-migh-KOH-tuh) are commonly called cellular slime molds. These organisms begin their life cycles as individual amebalike cells. In fact, they look so much like amebas that only an expert can tell them apart! As the slime mold grows, however, the resemblance ends very quickly.

When an individual cell begins to run out of moisture or food, it sends out chemical signals that attract other cells of the same species. Within a few days, thousands of cells will aggregate into a large sluglike mass, as shown in ***Figure 23–9*** on page 542.

This mass migrates for several centimeters, then stops and forms a **fruiting body**—a reproductive structure that produces spores. When the spores are scattered, each can give rise to a single amebalike cell that starts the cycle all over again.

How do individual, free-living cells signal each other first to aggregate, then to form a specialized structure like the

Figure 23–9
(a) *To reproduce, slime molds produce fruiting bodies that contain haploid (n) spores.* (b) *By growing on a dead log, this acellular slime mold—* Physarum polycephalum*—is recycling the log into materials other organisms can use.* (c) Saprolegnia, *a water mold, commonly grows on dead insects and fishes.*

fruiting body? That's a question that biologists are still asking! For decades, biologists around the world have been investigating these and other questions about the cellular slime molds. The answers may hold clues to how multicellular organisms are organized.

Acellular Slime Molds

Members of the phylum Myxomycota (mihks-uh-migh-KOH-tuh) are called the acellular slime molds. Just like the cellular slime molds, the acellular slime molds begin their life cycles as amebalike cells that grow into large masses. In the acellular slime molds, however, the mass is actually a single cell with thousands of nuclei.

Acellular slime molds stream across leaves and twigs, gobbling up bacteria and other organic matter along the way. Eventually, they form fruiting bodies that form haploid (n) spores. When the spores contact moist soil, they germinate to produce flagellated cells. These cells fuse to produce diploid (2n) amebalike cells. In this manner, the cycle starts all over again.

Checkpoint What is an acellular slime mold?

Water Molds

Have you ever seen a thin whitish fuzz growing on a dead fish or other marine animal? If so, you have seen a water mold—a member of the phylum Oomycota (oh-uh-migh-KOHT-uh). Like fungi, water molds thrive on dead or decaying matter. However, they have cell walls made of cellulose and produce spores that swim rapidly—two characteristics that fungi do not share.

Most water molds live in water, but a few made their way onto land. These include parasites of such crops as grapes, avocados, and potatoes.

Section Review 23–1

1. How do biologists **classify** protists?
2. **Describe** the animallike, plantlike, and funguslike protists.
3. **Critical Thinking—Analyzing** Suppose that you have identified an amebalike organism under the compound microscope. What further information would help you to classify it?
4. **MINI LAB** How can you **design an experiment** to observe protists?

Fungi

SECTION 23–2

GUIDE FOR READING

- **Describe** how fungi take in food.
- **Identify** the four phyla of fungi.

MINI LAB

- **Identify** the reproductive structures in bread mold.

WHEN YOU HEAR THE WORD fungus, what comes to mind? Chances are, it isn't anything pretty. To put it mildly, fungi have a bad reputation. Some spoil our food, others rot our lumber or destroy trees, while still others can make us sick. The fungi have a well-deserved reputation as pests and destroyers.

Do fungi have any value at all? In fact, the fungi fill a very important role in nature—they are the world's champions of decomposition! Without fungi breaking down organic matter, the raw materials that go into each season of life would be lost forever—and the Earth would be a barren place.

Kingdom Fungi 

The kingdom Fungi (FUHN-jigh) includes a wide variety of organisms with different structures and different life cycles. However, all fungi (singular: fungus) have certain features in common.

Fungi are heterotrophs, and they obtain food by extracellular digestion and absorption. This means that they produce enzymes that break down food particles outside their bodies, then absorb the digested molecules. This process separates fungi from animals and plants, which obtain their energy in other ways.

Except for yeasts, which are unicellular, the cells of a fungus are organized into filaments called **hyphae** (HIGH-fee; singular: hypha). And the hyphae form a tangled mass called a **mycelium** (migh-SEE-lee-uhm). In most fungi, individual cells are separated by perforated cross walls. The holes in these cross walls join the cytoplasm of adjacent cells, and this allows nutrients to move quickly through the hyphae.

Figure 23–10 Fungi have evolved a variety of relationships with other organisms. (a) Sulfur shelf fungi grow off tree trunks. (b) Fungi called mycorrhizae grow on the roots of many plants. The plants receive water, minerals, and nutrients that the mycorrhizae absorb from the soil (magnification: 50X). (c) These lichens are a result of a mutualistic partnership between fungi and algae. The fungi get food from the algae and in return provide the algae with water, minerals, and protection.

All fungi can reproduce asexually. They do so when cells or hyphae break off and begin to grow on their own. Some fungi also produce spores that can scatter a great distance away.

In addition, many fungi have a sexual phase to their life cycles. Typically, these fungi have hyphae of two mating types, called + (plus) and − (minus). When a + hypha fuses with a − hypha, the two nuclei come together in one cell. The two nuclei fuse to form a diploid (2n) nucleus, which eventually undergoes meiosis to produce haploid (n) spores.

Classifying Fungi

There are four phyla of fungi—Zygomycota (zigh-goh-migh-KOHT-uh), **Ascomycota** (as-koh-migh-KOHT-uh), **Basidiomycota** (buh-sihd-ee-oh-migh-KOHT-uh), **and Deuteromycota** (doo-ter-oh-migh-KOHT-uh). Let's take a close look at each of these phyla.

Zygomycetes

The zygomycetes—members of the phylum Zygomycota—are named for a thick-walled diploid spore called a zygospore. A zygospore is small, light enough to drift in the wind, and amazingly resistant to damage. Zygomycotes also produce spores asexually, which can help them to grow rapidly.

One common zygomycete is *Rhizopus stolonifer.* When a zygospore of this species settles on a piece of bread, the diploid nucleus undergoes meiosis. Haploid hyphae break through the spore wall and quickly grow through the porous bread, producing enzymes that break down the bread into absorbable nutrients.

Ascomycetes

The phylum Ascomycota is named for the ascus, a tough sac that contains the spores produced by sexual reproduction. These spores are called ascospores, and

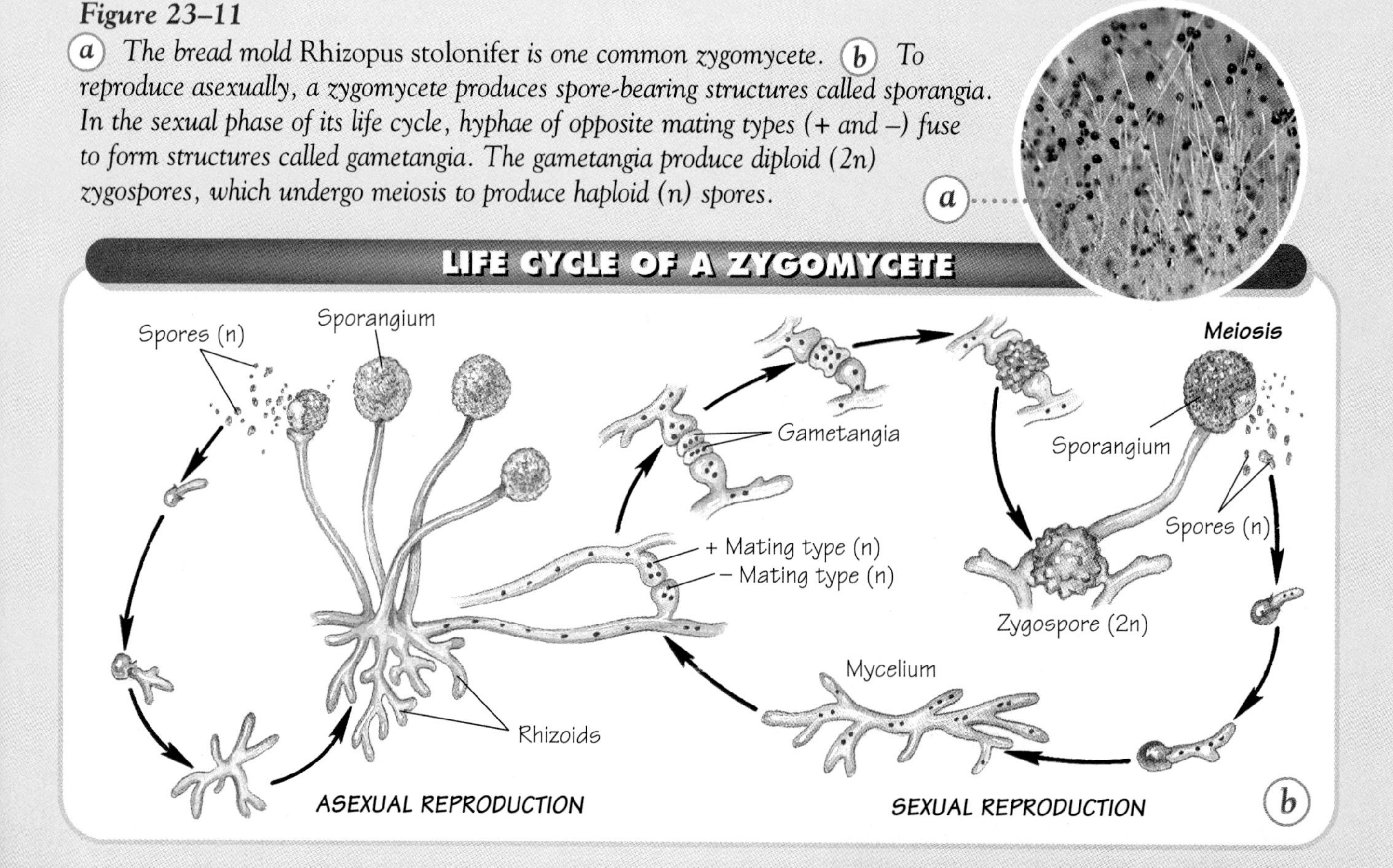

Figure 23–11
(a) *The bread mold* Rhizopus stolonifer *is one common zygomycete.* (b) *To reproduce asexually, a zygomycete produces spore-bearing structures called sporangia. In the sexual phase of its life cycle, hyphae of opposite mating types (+ and –) fuse to form structures called gametangia. The gametangia produce diploid (2n) zygospores, which undergo meiosis to produce haploid (n) spores.*

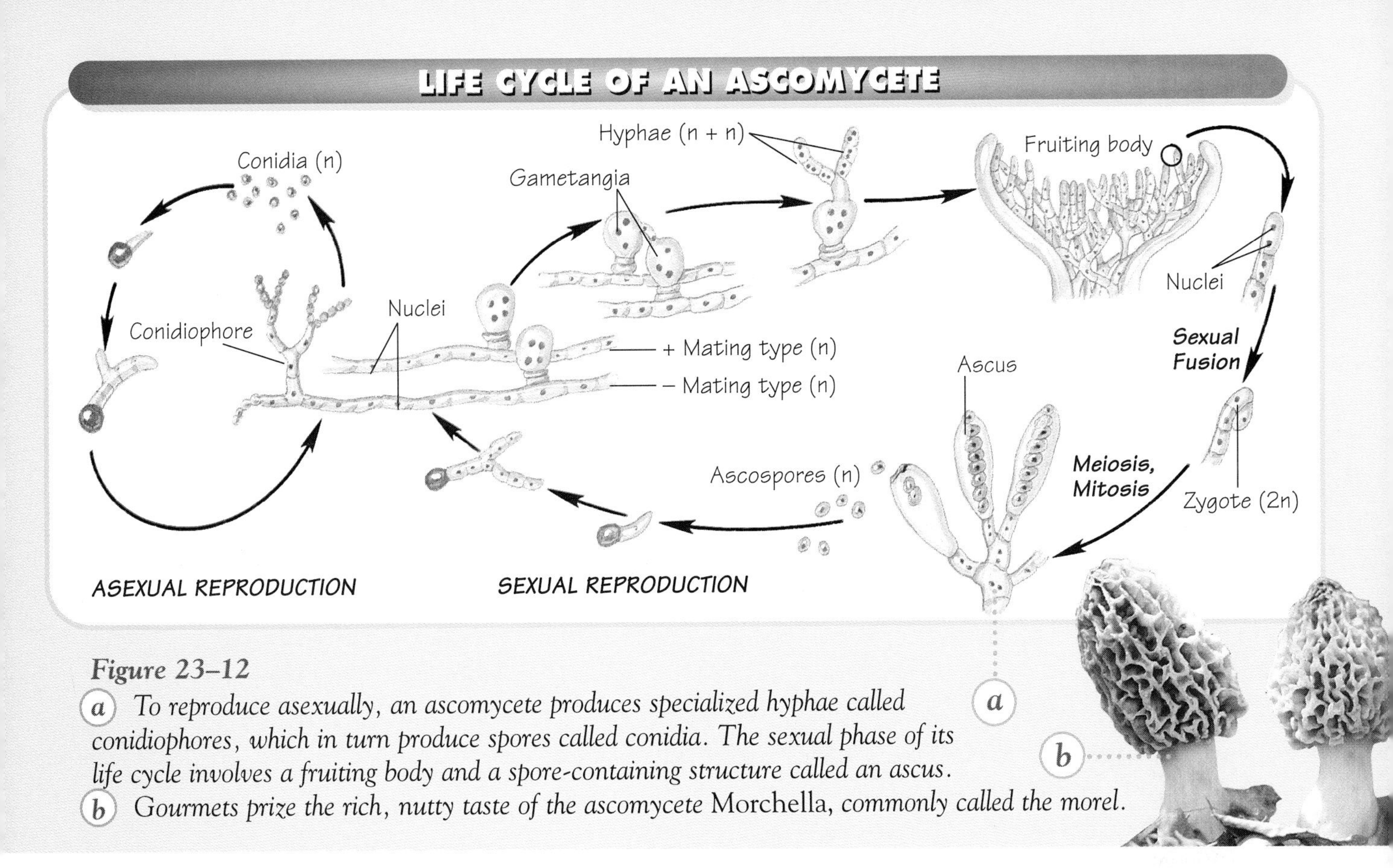

Figure 23–12
(a) *To reproduce asexually, an ascomycete produces specialized hyphae called conidiophores, which in turn produce spores called conidia. The sexual phase of its life cycle involves a fruiting body and a spore-containing structure called an ascus.* (b) *Gourmets prize the rich, nutty taste of the ascomycete* Morchella, *commonly called the morel.*

they are produced by hyphae of opposite mating types. First, haploid nuclei from different hyphae fuse together, eventually forming a diploid nucleus. Then, the diploid nucleus quickly enters meiosis and gives rise to 4 new haploid nuclei. In most ascomycetes, one or two rounds of mitosis follow meiosis, producing either 8 or 16 haploid ascospores.

Like the slime molds, ascomycetes form spores in a structure called a fruiting body. In many ascomycetes, such as common bakers' yeast, the fruiting body is small and insignificant. However, it is quite spectacular in other organisms, such as the morel shown in ***Figure 23–12.***

Basidiomycetes

Members of the phylum Basidiomycota are sometimes called the "club fungi," and they include nearly all the organisms commonly called mushrooms.

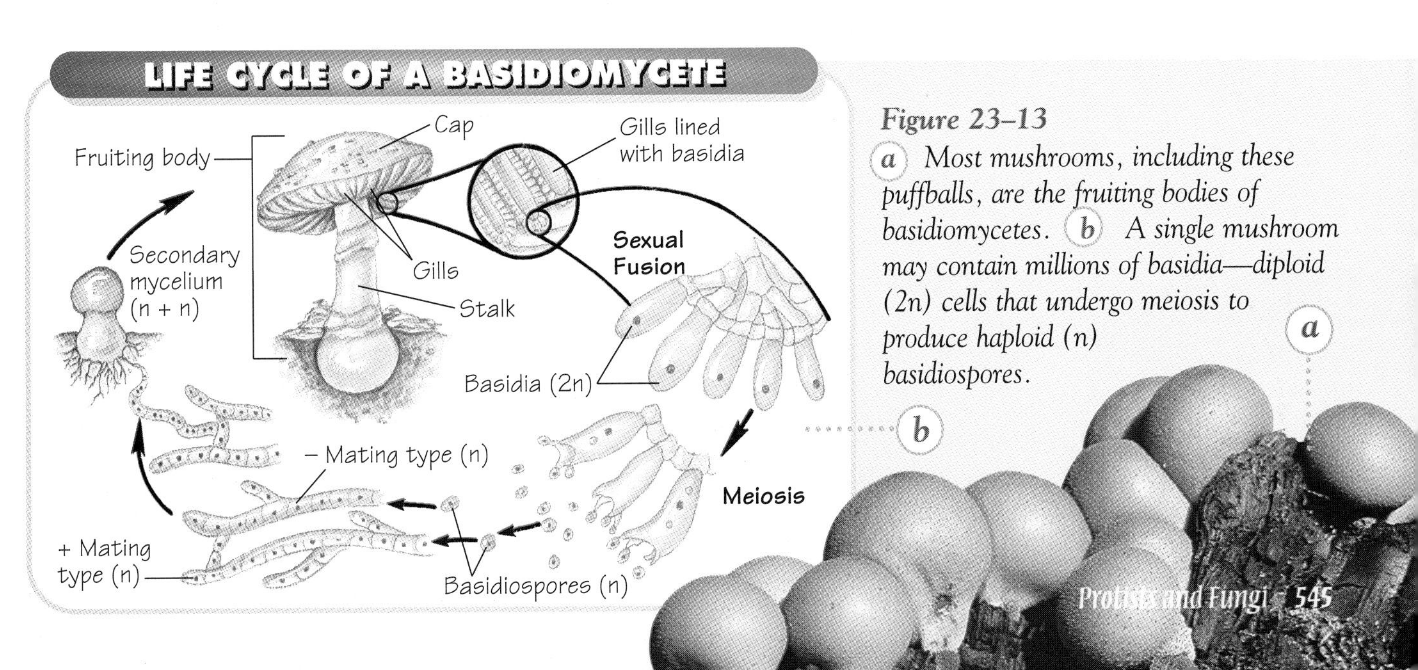

Figure 23–13
(a) *Most mushrooms, including these puffballs, are the fruiting bodies of basidiomycetes.* (b) *A single mushroom may contain millions of basidia—diploid (2n) cells that undergo meiosis to produce haploid (n) basidiospores.*

Sealed With Tape

PROBLEM *How can you **identify** reproductive structures in bread mold?*

PROCEDURE 

1. Touch the sticky side of a 2-cm piece of cellophane tape to the black "fuzzy" area of a bread mold culture.
2. Gently stick the tape to a glass slide. Observe the slide under the compound microscope. Draw a diagram of your observations.

ANALYZE AND CONCLUDE

1. Describe the reproductive structures of bread mold. What structures do they contain?
2. Why does a single bread mold culture produce a large number of reproductive structures?

The phylum also includes puffballs, toadstools, and the bracket fungi that appear on the trunks of trees.

The above-ground structure of a basidiomycete is actually just its fruiting body. Beneath the soil is a mycelium that has slowly digested its way through organic matter to reach the surface.

Just like the ascomycetes, the basidiomycetes have hyphae of opposite mating types. In basidiomycetes, however, the + and − hyphae donate nuclei to produce a structure called a secondary mycelium. A secondary mycelium can grow beneath the soil for many years. Some reach enormous sizes, often many meters in diameter. When conditions are right, the secondary mycelium produces a fruiting body—a mushroom—that forces its way upward through the soil.

Within the mushroom, haploid nuclei fuse and undergo meiosis, producing clusters of haploid spores called basidiospores. These spores appear in "gills" on the underside of the mushroom. A single mushroom can produce as many as a million spores—each with the potential of producing a new organism.

☑ ***Checkpoint*** What are some examples of basidiomycetes?

Deuteromycetes

Members of the phylum Deuteromycota are commonly called the imperfect fungi because they are believed to reproduce only through asexual spores. The deuteromycetes include a number of well-known organisms. Both athlete's foot and ringworm are produced by deuteromycetes, as is the mold from which we harvest the antibiotic penicillin.

Figure 23–14
This false-color electron micrograph shows the fruiting body of the deuteromycete Penicillium *(magnification: 11,500X).*

Section Review 23–2

1. **Describe** how fungi take in food.
2. **Identify** the four phyla of fungi.
3. **Critical Thinking—Comparing** How do the four different phyla of fungi compare?
4. **MINI LAB** How can you **identify** reproductive structures in bread mold?

Protist and Fungal Diseases

GUIDE FOR READING

- **Describe** diseases caused by protists and fungi.

MAKE NO MISTAKE ABOUT IT: Our world would grind to a halt without protists and fungi. These organisms fill all sorts of vital roles—from producing oxygen to decomposing dead organic matter. However, while many protists and fungi are some of nature's most useful organisms, others are some of its most damaging . . . and deadly!

Protist Diseases

Protists are so diverse that you should expect at least a few of them to have evolved into disease-causing parasites. For better or for worse, that is exactly the case.

African Sleeping Sickness

In central Africa, flagellates of the genus *Trypanosoma* cause a disease called African sleeping sickness. These protists are passed from person to person through the bite of the tsetse fly.

Trypanosomes destroy blood cells, producing fever, chills, and rashes. Even worse, however, is the damage they can cause to the nervous system. Affected individuals may lose consciousness and lapse into a deep, sometimes fatal sleep—the sleep that gives the disease its name.

Checkpoint What is African sleeping sickness?

Intestinal Diseases

Several protists infect the human digestive system, gaining entry through contaminated water. These protists include *Entamoeba* and *Giardia,* both of which attack the wall of the intestine. An *Entamoeba* infection produces a disease called amebic dysentery, and a *Giardia* infection causes a disease called giardiasis. Both infections cause diarrhea and severe abdominal pain.

Amebic dysentery is most common in areas with poor sanitation, but even crystal-clear mountain streams may be contaminated with *Giardia*. This organism produces tough microscopic cysts that can be killed only by thoroughly boiling the water that contains them.

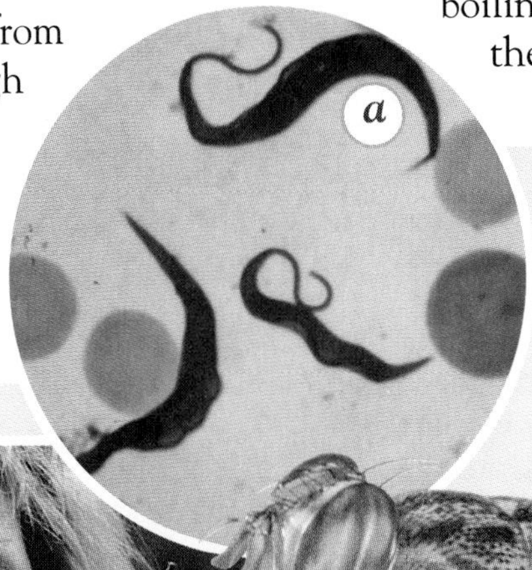

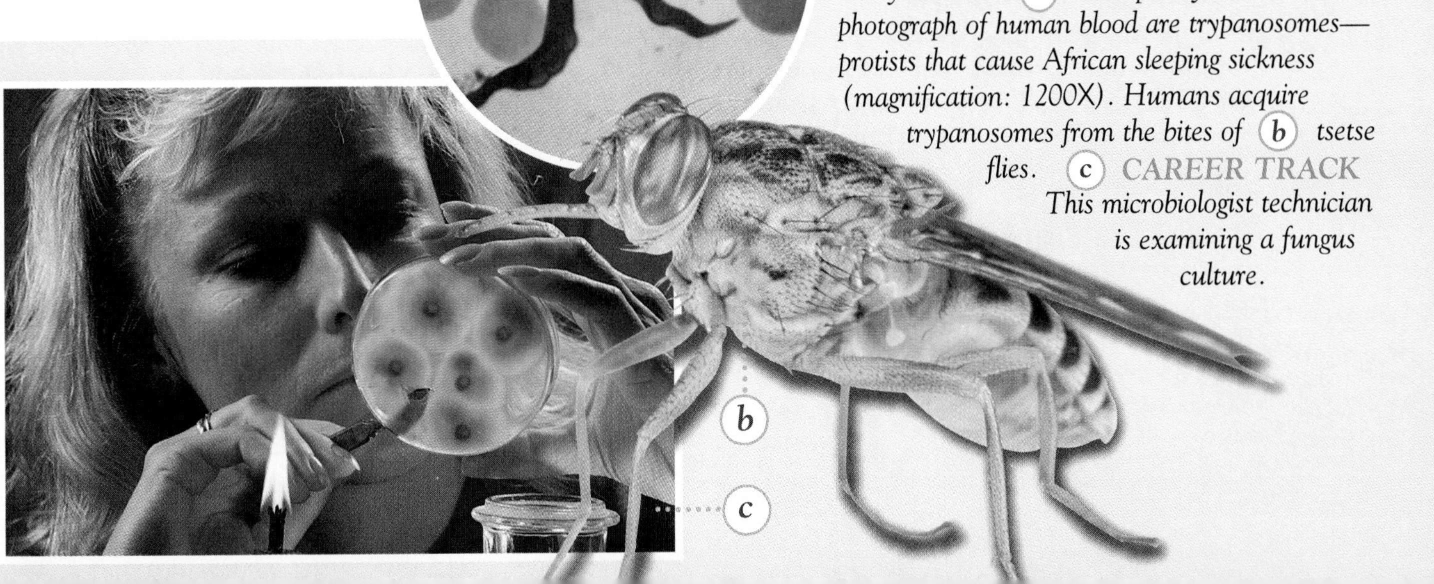

Figure 23–15 *Disease-causing protists and fungi are found in many habitats. (a) The spindly cells in this photograph of human blood are trypanosomes—protists that cause African sleeping sickness (magnification: 1200X). Humans acquire trypanosomes from the bites of (b) tsetse flies. (c) CAREER TRACK This microbiologist technician is examining a fungus culture.*

Biology AND YOU ... Connections

Potatoes, Trout, and People

Have protists and fungi evolved into parasites of plants and animals? Not surprisingly, the answer is yes. In fact, these organisms have devastated many plant and animal species over the years—and their attacks continue today.

The Great Potato Famine

By 1845, the potato was the major food crop of Ireland. Unfortunately, the summer that year was unusually wet and cool—ideal conditions for the water mold *Phytophthora infestans.* The threadlike hyphae of this mold invade the roots, stems, and leaves of the potato plant, eventually killing it entirely.

Phytophthora destroyed almost 60 percent of Ireland's potato crop in 1845, and the damage was even worse the following year. Over a million people died from the Great Potato Famine, as the event has become known. It also led to the emigration of at least 2 million people from Ireland to the United States.

Phytophthora infestans *destroying a potato plant*

Damage to Other Crops

Although the Great Potato Famine was especially devastating, it is hardly the only example of protists or fungi destroying crops. In 1935, for example, wheat rust destroyed more than 25 percent of the grain harvest in the United States. And even today, millions of dollars are lost each year to protists and fungi that attack fruits, grains, vegetables, and other crops.

Whirling Trout Disease

Fans of both fishing and dining value the rainbow trout, a colorful fish native to the northwestern United States. In the last few years, however, many rainbow trout have been seen swimming in circles and chasing their own tail—and dying soon thereafter. This strange behavior was named whirling trout disease.

As biologists discovered, the disease is caused by *Myxobolus cerebralis,* an animallike protist that infects both the rainbow trout and the worms they eat. In trout, the protist attacks cartilage and the nervous system.

Hope for the rainbow trout may lie with another fish, the brown trout, which lives in the same streams as rainbow trout but is resistant to *Myxobolus.* The brown trout is native to Europe, where *Myxobolus* is common. It is thus believed to have evolved into a resistant strain. Wildlife biologists hope that the rainbow trout can do the same.

Rainbow trout are susceptible to infections from an animallike protist, Myxobolus cerebralis.

Making the Connection

How did the Great Potato Famine in Ireland affect life in the United States? Aside from potatoes and rainbow trout, what other plants and animals are affected by protist parasites?

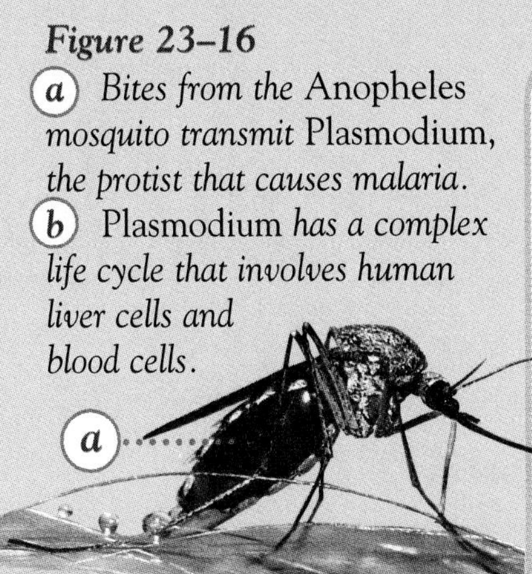

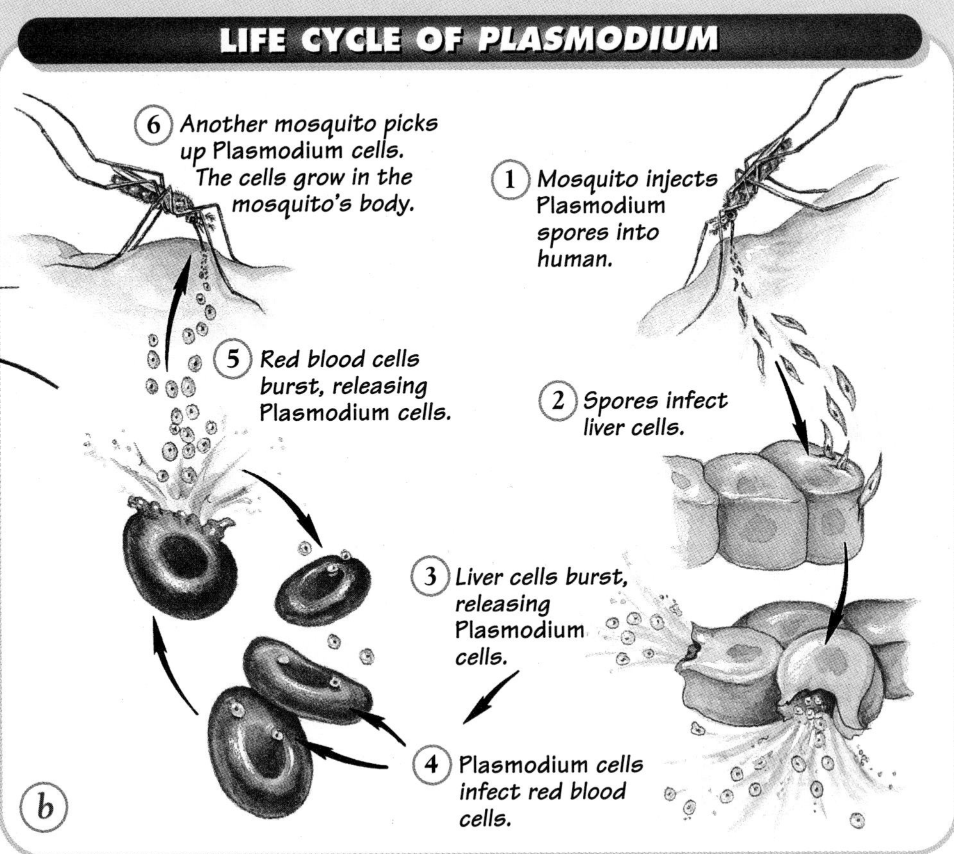

Figure 23–16 (a) *Bites from the* Anopheles *mosquito transmit* Plasmodium, *the protist that causes malaria.* (b) Plasmodium *has a complex life cycle that involves human liver cells and blood cells.*

Malaria

By far the most serious disease that protists cause is malaria. More than 200 million people suffer from malaria, and as many as 2 million people die from it each year.

Malaria is caused by *Plasmodium,* a sporozoan that is transmitted to humans through the bite of the *Anopheles* mosquito. *Plasmodium* enters the bloodstream through the mosquito's saliva, then infects liver cells and red blood cells. In a 48- to 72-hour cycle, infected cells burst open, dumping waste products and toxins into the bloodstream. This produces fever and chills—the characteristic symptoms of malaria.

Medical scientists have developed a number of vaccines that seem to block the growth of the parasite. Unfortunately, these vaccines are only partly effective. For the immediate future, the best means of controlling malaria is controlling the mosquitoes that carry it.

Fungal Diseases

Although fungi cause a few serious diseases, many common fungal diseases are relatively harmless and easy to treat. A deuteromycete of the genus *Tinea*, for example, attacks human skin but does not invade the body any farther. **A *Tinea* infection of the scalp causes a disease called ringworm, and a *Tinea* infection between the toes causes athlete's foot.** In both diseases, *Tinea* spores produce hyphae that penetrate the skin's outer layers, causing a red, inflamed sore. Fortunately, *Tinea* infections can be cured by applying antifungal medicines to the infected area.

INTEGRATING HEALTH

Where is malaria most common? How can it be treated?

Section Review 23–3

1. **Describe** diseases caused by protists and fungi.
2. **Explain** how malaria is transmitted.
3. **BRANCHING OUT ACTIVITY** **Research** other diseases that protists or fungi cause. Are any of these diseases common where you live?

CHAPTER 23

Laboratory Investigation

DESIGNING AN EXPERIMENT

Observing a Paramecium

A paramecium may be a single-celled organism, but it is hardly simple or primitive. In fact, it reacts to stimuli—changes in its environment—in a variety of different ways. In this investigation, you will observe a paramecium under the microscope and see how it responds to different stimuli.

Problem

How does a paramecium respond to different stimuli? **Design an experiment** to answer this question.

Suggested Materials

paramecium culture
medicine dropper
compound microscope
microscope slide with depression
coverslip
methyl cellulose
0.5-percent salt solution
penlight or flashlight
dilute acetic acid
colored yeast culture

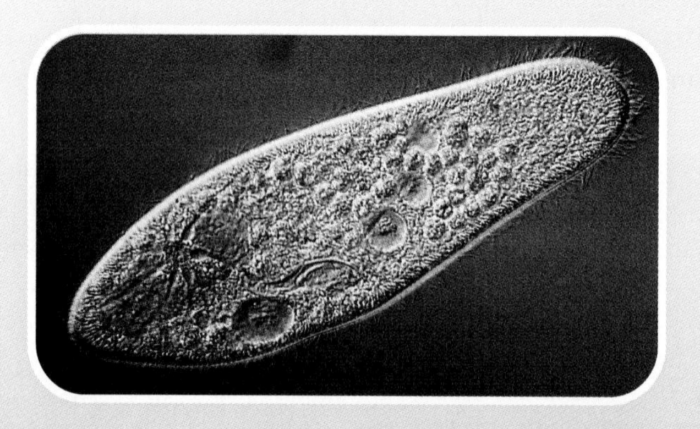

Suggested Procedure

1. Use the medicine dropper to transfer a drop of the paramecium culture to the microscope slide.

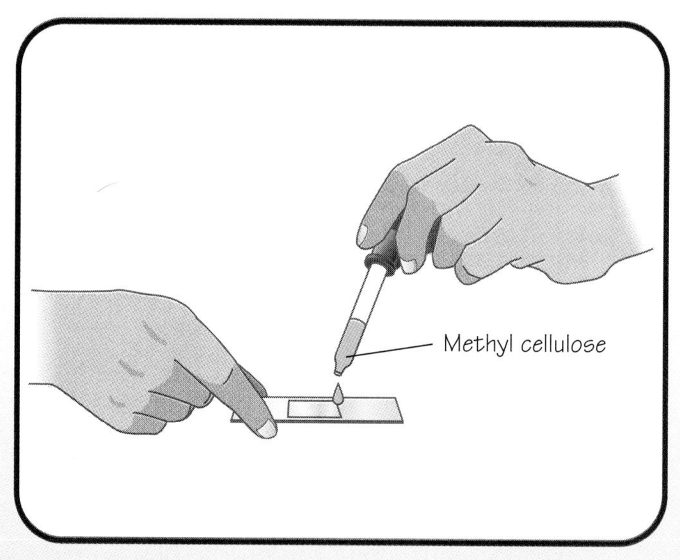

2. Add a drop of methyl cellulose to the slide. Methyl celluose slows down a paramecium's movement.

3. **Place a coverslip on the slide. Use the low-power objective to locate a paramecium on the slide, then switch to the high-power objective to observe its structure. Draw a detailed diagram of the paramecium.**
4. **While observing the paramecium under the high-power objective, count the number of times the contractile vacuole fills and empties in 30 seconds.**
5. **Place a drop of 0.5-percent salt solution at one edge of the coverslip. Count the number of times the contractile vacuole fills and empties in 30 seconds.**
6. **Flush out the salt solution by adding several drops of water at one edge of the coverslip.**
7. **Using a similar procedure, design an experiment to determine how a paramecium responds to the following stimuli:**
   - **light**
   - **dilute acid solution**
   - **yeast**
8. **With your teacher's approval, perform the experiment you designed. CAUTION:** ***Be careful when using an acid. It may burn your skin.***

Observations

1. In your drawing of the paramecium, label any structures that you can identify.
2. Describe any changes in the activity of the contractile vacuole when the 0.5-percent salt solution was added.
3. Describe the paramecium's response to the dilute acid solution.
4. How did the paramecium respond to light?
5. How did the paramecium respond to the yeast culture?

Analysis and Conclusions

1. Why is a paramecium classified as an animal-like protist and not a plantlike protist or a funguslike protist?
2. How does a paramecium move from place to place?
3. Why does a paramecium need a contractile vacuole? Use the data collected in this investigation to support your answer.
4. How does a paramecium respond to light? Why might such a response be useful?
5. Describe the paramecium's activity in the presence of the yeast culture. What purpose does this activity serve?

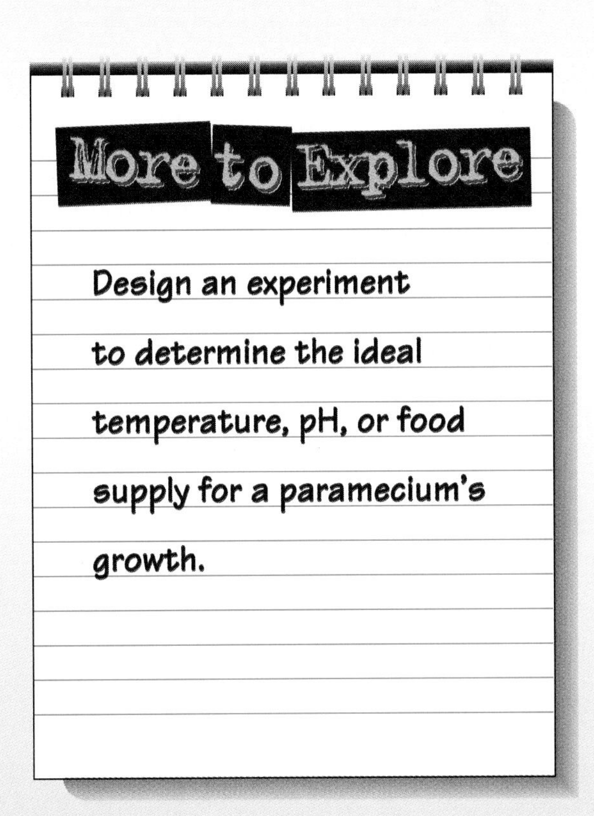

Summarizing Key Concepts

The key concepts in each section of this chapter are listed below to help you review the chapter content. Make sure you understand each concept and its relationship to other concepts and to the theme of this chapter.

23–1 Protists

- A eukaryote that is not a fungus, plant, or animal is classified as a protist.
- Animallike protists are classified into phyla by the way they move. Flagellates propel their bodies with whiplike flagella. Sarcodines move with "false feet" called pseudopods. Ciliates move through the beating of hairlike cilia. Sporozoans are parasites that have no special structures for movement.
- Plantlike protists are capable of photosynthesis. They include the euglenophytes, which resemble flagellates; dinoflagellates, many of which can produce light; and chrysophytes, which include the diatoms.
- Funguslike protists lack chlorophyll and absorb food through their cell walls. They include the slime molds and water molds.

23–2 Fungi

- Fungi are heterotrophs, and they obtain food by extracellular digestion and absorption.
- The cells of fungi are organized into filaments called hyphae. Hyphae form a tangled mass called a mycelium.
- The four phyla of fungi are Zygomycota, Ascomycota, Basidiomycota, and Deuteromycota. The structure commonly called a mushroom is typically the fruiting body of a basidiomycete.

23–3 Protist and Fungal Diseases

- Protists cause African sleeping sickness and intestinal diseases. Malaria is caused by *Plasmodium,* a sporozoan that is transmitted to humans through the *Anopheles* mosquito.
- A species of the genus *Tinea,* a deuteromycete, causes ringworm and athlete's foot.

Reviewing Key Terms

Review the following vocabulary terms and their meaning. Then use each term in a complete sentence.

23–1 Protists

flagellum
pseudopod
spore
cilium
contractile vacuole
gullet
anal pore
trichocyst
micronucleus
macronucleus
conjugation
alga
fruiting body

23–2 Fungi

hypha
mycelium
ascus

Recalling Main Ideas

Choose the letter of the answer that best completes the statement or answers the question.

1. Animallike protists are classified into phyla by the way they
   a. reproduce. **b.** move. **c.** take in food. **d.** produce shells.

2. A flagellum functions as a
   a. whiplike motor. **b.** spore. **c.** glasslike shell. **d.** nucleus.

3. An ameba moves by extending its
   a. cilia. **b.** flagella. **c.** pseudopod. **d.** nucleus.

4. A paramecium expels water through its
   a. contractile vacuoles. **b.** gullet. **c.** trichocysts. **d** nucleus.

5. Which of these phyla contain only parasites?
   a. flagellates **b.** sarcodines **c.** ciliates **d.** sporozoans

6. Unlike a paramecium, a euglena contains
   a. chloroplasts. **b.** mitochondria. **c.** cilia. **d.** flagella.

7. Which is a funguslike protist?
   a. water mold **b.** sporozoan **c.** heliozoan **d.** club fungus

8. In many fungi, spores are produced in structures called
   a. hyphae. **b.** mycelia. **c.** fruiting bodies. **d.** nuclei.

9. Mushrooms, including toadstools and bracket fungi, are examples of
   a. water molds. **b.** ascomycetes. **c.** basidiomycetes. **d.** deuteromycetes.

10. The *Anopheles* mosquito transmits the protist that causes
    a. African sleeping sickness.
    b. amebic dysentery.
    c. malaria.
    d. ringworm.

Putting It All Together

Using the information on pages xxx to xxxi, complete the following concept map.

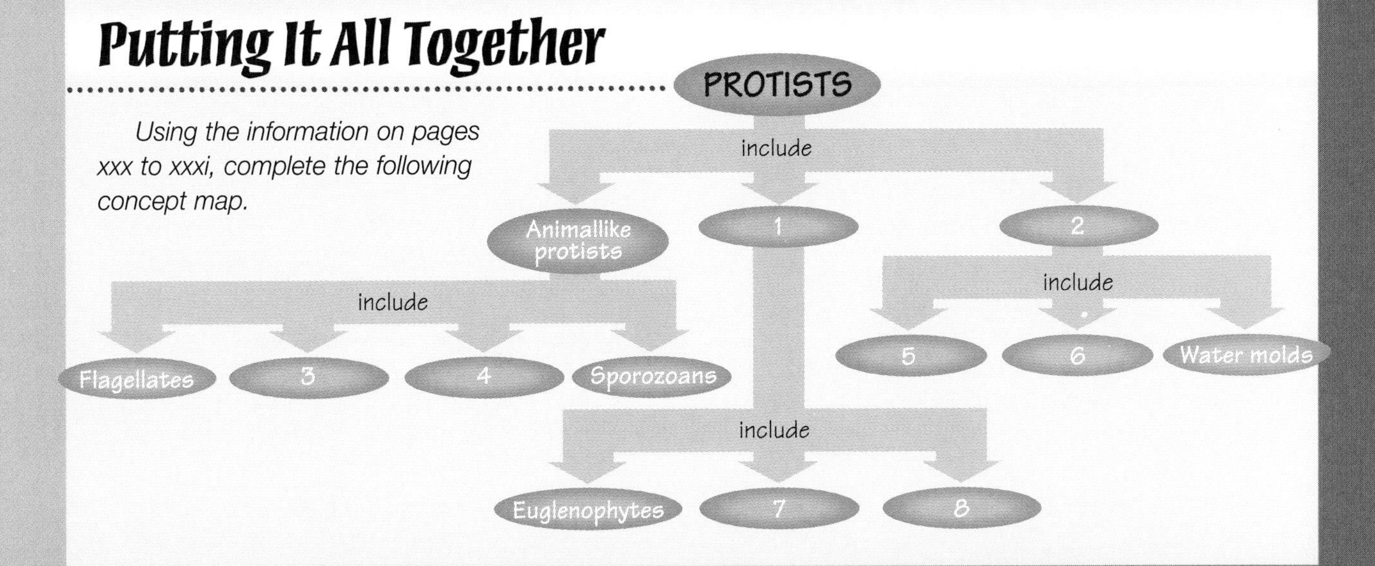

Assessment

Reviewing What You Learned

Answer each of the following in a complete sentence.

1. What are the four major phyla of animallike protists?
2. How do amebas take in food?
3. What type of movement categorizes the ciliates?
4. Describe the role of conjugation in paramecia.
5. What are algae?
6. Identify three phyla of plantlike protists.
7. What are the chrysophytes?
8. How does a paramecium use its trichocysts?
9. Identify three phyla of funguslike protists.
10. Describe the main role of fungi in nature.
11. How do nutrients pass through the cells of a fungus?
12. What is an ascus?
13. What is a secondary mycelium?
14. Identify three diseases that protists or fungi can cause.

Expanding the Concepts

Discuss each of the following in a brief paragraph.

1. Why is it difficult to classify protists?
2. Compare the four phyla of animallike protists.
3. Give an example of a mutualistic relationship between a protist and an animal.
4. Describe the role of the micronucleus and macronucleus in a paramecium.
5. Compare a euglena and a paramecium.
6. Describe the external structure of a diatom.
7. Compare a cellular slime mold to an acellular slime mold.
8. Describe the process of sexual reproduction in fungi.
9. In basidiomycetes, what is the role of the structure commonly called a mushroom?
10. Discuss the life cycle of *Plasmodium,* the protist that causes malaria.
11. Compare ringworm and athlete's foot.

Extending Your Thinking

Use the skills you have developed in this chapter to answer the following.

1. **Interpreting diagrams** The diagram shown below illustrates the six kingdoms of organisms. In what ways does the diagram accurately represent protists?

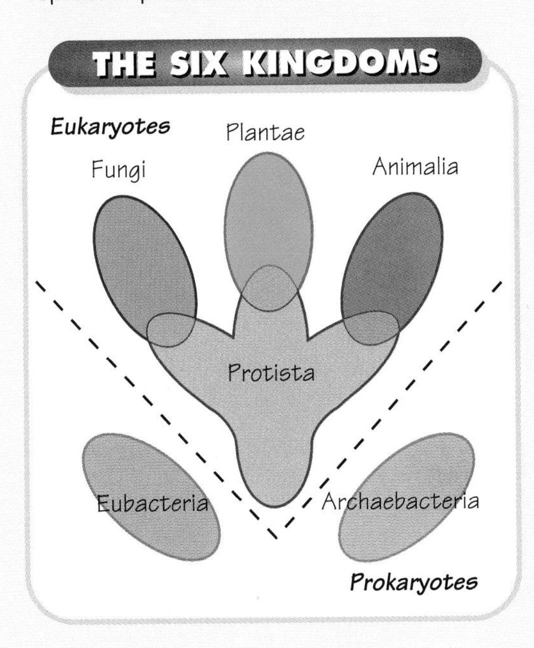

2. **Identifying** What structures can you identify in the mushroom shown on the opposite page? To what phylum does this organism belong? Explain your answer.
3. **Giving an example** Many protists and fungi can reproduce sexually or asexually. Give a specific example of each type of reproduction and describe the advantages it offers the protist or fungus.
4. **Designing an experiment** An orchid grower thought that fungi and protists in the soil could damage an orchid. So he sterilized the soil by heating it to 100°C. To his alarm, the orchid withered and eventually died. Develop a hypothesis to explain this result. Then, design an experiment to test your hypothesis.
5. **Using the writing process** Write a short story or play about a world that does not have protists or fungi. Your story should predict the problems this world would face and suggest some possible solutions.

Applying Your Skills

From the "Mouths" of Protists and Fungi

Some protists and fungi have relatively simple life cycles, while the lives of others are far more complex. If protists and fungi could talk about their lives, what would they say?

1. Your teacher will assign you a specific protist or fungi. Research the life cycle of this organism. Identify its methods of reproduction, movement, and energy production, as well as its role in nature.
2. Write a short story, poem, or play in which the narrator or main character is your assigned organism. Your work should include all the relevant information that you researched.

• GOING FURTHER •

3. Read your work to your classmates. You may use props or other visual aids to enhance your presentation.
4. Work with a classmate to present a comparison of two different protists or fungi.

CHAPTER 24

Multicellular Algae, Mosses, and Ferns

FOCUSING THE CHAPTER THEME: Unity and Diversity

24–1 Multicellular Algae
- Classify algae.
- Describe reproduction in green algae.

24–2 Bryophytes
- List some of the adaptations bryophytes need to live on land.
- Identify the main stages in the life cycle of bryophytes.

24–3 Seedless Vascular Plants
- Discuss the significance of vascular tissue.
- Describe the life cycle of a fern.

BRANCHING OUT *In Action*

24–4 Algae in Our World
- Discuss the role of algae in nature.
- List ways in which humans use algae.

LABORATORY INVESTIGATION
- Compare the characteristics of different types of green algae.

Biology and Your World

BIO JOURNAL

Have you ever looked carefully at the plant life in or near a pond, lake, or rocky seashore? How do you think these plants are suited to their aquatic environment? Based on your observations, make a list in your journal of the ways in which aquatic plants and terrestrial plants differ.

Tree ferns in Border Ranges National Park, South Queensland, Australia

Multicellular Algae

SECTION 24–1

Guide for Reading

- **Compare** the characteristics of multicellular algae and plants.
- **Explain** what is meant by alternation of generations.

AT HIGH TIDE ON A WINTER'S day, the Maine coastline seems like a cold and barren place. Icy water crashes against gray rock, and there is hardly a hint of anything alive. Just six hours later, at low tide, the scene is different. As the pull of the moon's gravity drops the waterline, it reveals a damp forest of green and brown clinging to the rocks. Even as waves toss them violently, these hardy underwater seaweeds hang on, bringing life to the cold salt water and providing food and shelter for scores of animals. These seaweeds are dramatic reminders of the way in which plant life has found its way into every corner of the planet—and the way in which plants continue to transform the world in which we live.

Algae—Protists or Plants?

The seaweeds you have just read about are **algae**—photosynthetic aquatic organisms that are classified as protists. Although most algae are unicellular, some are multicellular, such as the seaweeds that are seen on rocky seashores at low tide. We have already discussed three of the algae phyla—*Euglenophyta*, *Pyrrophyta*, and *Chrysophyta*—along with other protists. Remember that the algae in these phyla are unicellular. Because many of the algae in the remaining phyla, like plants, are multicellular, they have been included along with plants in this chapter. These algae also have reproductive cycles that are very similar to those in plants. Therefore, even though the multicellular algae are protists, you might think of them as "honorary" plants.

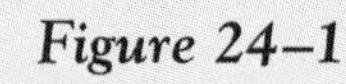

Figure 24–1
Multicellular algae are photosynthetic aquatic organisms and are classified as protists. Some examples include (a) Acetabularia crenulata, *a green alga,* (b) Corallina officinalis, *a red alga, and* (c) *a fluorescent red alga.*

Figure 24–2
Most brown and red algae are saltwater organisms. Many of the brown algae, such as (a) bull kelp, form stalks with leaflike blades and hollow bladders that fill with gas to help the alga stay afloat. Red algae are especially abundant in the warm waters of the tropics, and many, such as (b) the red coralline algae, are involved in the building of coral reefs.

Red Algae

Red algae get their name from phycoerythrin (figh-koh-uh-RIHTH-rihn), a bright-red pigment used to supplement chlorophyll in photosynthesis. They belong to the phylum Rhodophyta (roh-dah-FIGHT-uh), which means "red plants" in Greek. Despite this name, many red algae are green, purple, or reddish-black, depending on the particular combination of additional pigments they contain. These extra pigments make it possible for many red algae to live at great depths, where little sunlight penetrates.

INTEGRATING PHYSICS

How does the color of algae depend upon the pigments present in the algae?

Brown Algae

The brown algae contain carotenoids, or yellow-orange pigments, called xanthophylls (ZAN-thuh-fihlz) and fucoxanthins (fyoo-koh-ZAN-thihnz), that help them gather sunlight for photosynthesis. When these pigments are combined with the green pigment chlorophyll, the result is a dingy-brown color that explains the name of their phylum, Phaeophyta (fee-oh-FIGHT-uh), which means "dusky plants" in Greek.

All brown algae are multicellular, and some of them are so large you might have trouble believing they are protists. The brown algae known as giant kelps can be as long as 100 meters. One of the most common forms is *Fucus*, which is found almost everywhere on the eastern coast of the United States and is sometimes known as rockweed for the way in which it attaches itself to rocks.

Checkpoint What is the source of the color of brown algae?

Green Algae

While most brown and red algae are found in salt water, most green algae live in fresh water. Green algae are remarkably similar to green plants and are placed in the phylum Chlorophyta (klawr-uh-FIGHT-uh), which means "green plants" in Greek. **Like plants, green algae have cell walls made of cellulose, the green pigments chlorophyll *a* and *b*, and they store food in the form of starch.** Although most green algae are aquatic, a few of them are found on land in wet soil or growing on moist tree trunks.

Many green algae, such as *Chlamydomonas* (klam-uh-DAH-muh-nuhs), live all or most of their lives as single cells. Some, however, form multicellular structures. *Spirogyra* (spigh-roh-JIGH-ruh) consists of long filaments that produce tangles of greenish fuzz in lakes and ponds. And *Ulva* (UHL-vuh), found along the shoreline, has large bright-green leaflike sheets that help to explain its common name—sea lettuce.

Problem Solving

INTERPRETING GRAPHS

Marveling at Multicellular Algae

One of your friends recently visited a public aquarium for a behind-the-scenes tour. She had the opportunity to see several of the hundreds of species of multicellular algae housed in the aquarium. A scientist who worked there told her that green plants evolved long ago from green algae. He gave your friend a chart and asked her to compare the characteristics of multicellular algae with those of plants.

Your friend was also given a graph that showed the action spectrum for photosynthesis and the absorption spectra of some pigments—namely, chlorophyll *a* and *b* and the carotenoids. The absorption spectra showed the wavelengths of light absorbed by each pigment. The action spectrum showed the wavelength of light at which photosynthesis takes place.

If green algae and green plants both contain chlorophyll *a*, why doesn't the graph show the peak for chlorophyll *a* in the (green) 560-nanometer range? The brown algae contain chlorophyll *c*, which is not shown on the graph. Is it reasonable to think that chlorophyll *c* might absorb light in the green range? Why or why not?

MULTICELLULAR ALGAE AND GREEN PLANTS

	Photosynthetic Pigments	*Stored Product From Photosynthesis*	*Major Component of Cell Wall*
Brown Algae	Chlorophyll *a* Chlorophyll *c* Carotenoids	Lamarin Mannitol	Cellulose
Red Algae	Chlorophyll *a* Chlorophyll *d* Carotenoids	Floridean starch Mannitol	Cellulose
Green Algae	Chlorophyll *a* Chlorophyll *b* Carotenoids	Starch	Polysaccharides Cellulose in many
Green Plants	Chlorophyll *a* Chlorophyll *b* Carotenoids	Starch	Cellulose

ABSORPTION SPECTRA

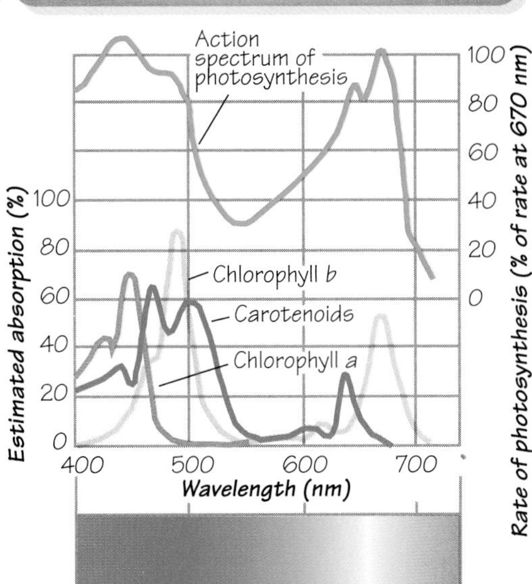

THINK ABOUT IT

1. What can you **conclude** about the evolution of green plants from a comparison of the characteristics of the three types of multicellular algae and green plants?
2. How might differences in the types of pigments present in the green, red, and brown algae account for the differences in their color? **Explain** the significance of these differences.
3. Using information from this chapter, **list** at least one other similarity between green plants and multicellular algae.

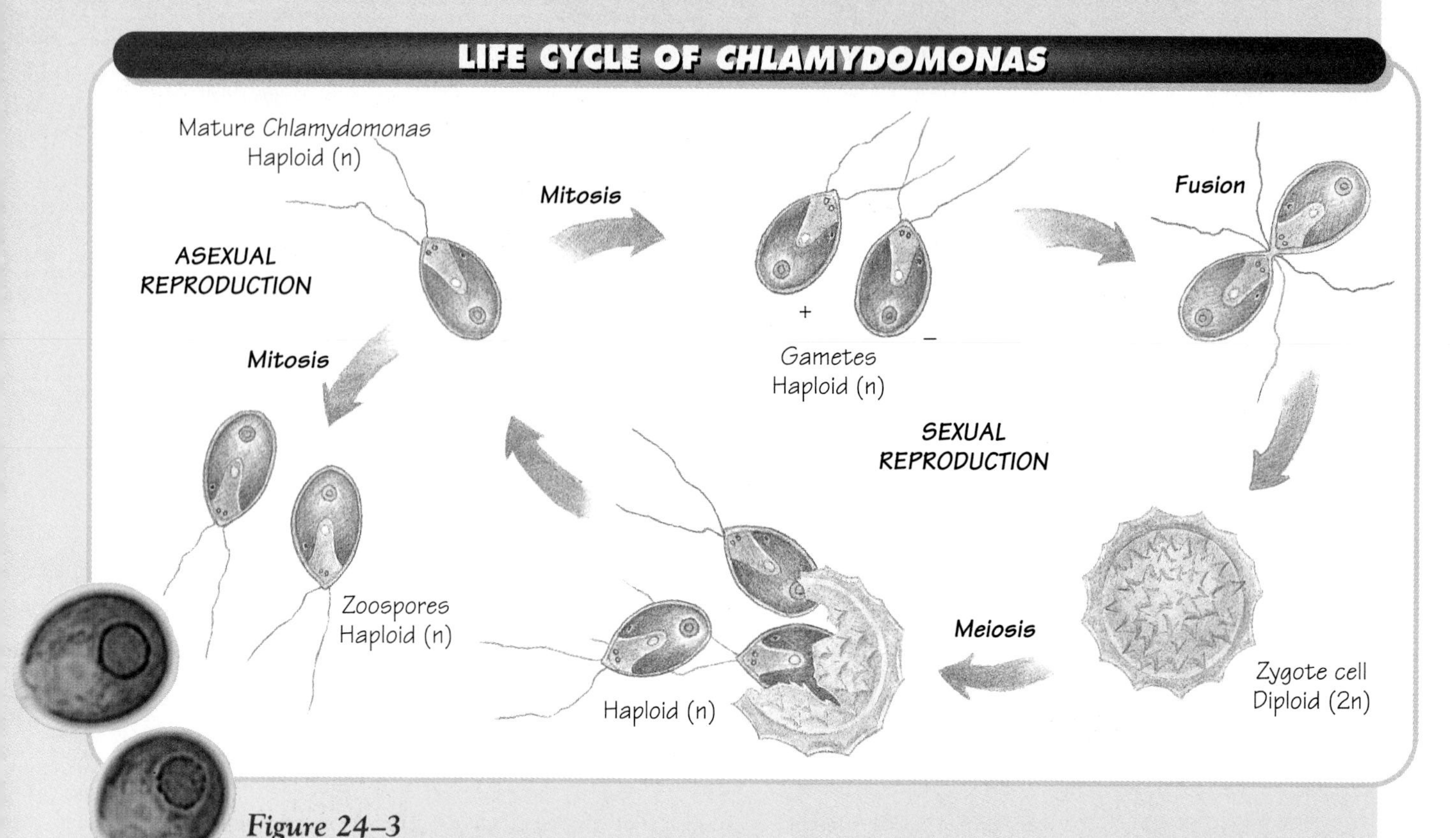

Figure 24–3
The life cycle of Chlamydomonas, *a unicellular green alga, involves asexual as well as sexual reproduction. In asexual reproduction, haploid (n) cells divide by mitosis to produce the next generation of haploid cells. In sexual reproduction, haploid (n) gametes are produced that fuse to form a diploid (2n) zygote. The zygote produces four haploid cells by meiosis.*

Reproduction in Algae

Most green algae have life cycles that involve both sexual and asexual stages. Their reproductive patterns are remarkably similar to plants, and therefore we will look at them in detail. Let's start with *Chlamydomonas*.

Reproduction in *Chlamydomonas*

Chlamydomonas cells have a single set of chromosomes, which is to say they are haploid (n). Under ordinary conditions, *Chlamydomonas* undergoes asexual reproduction, as illustrated in ***Figure 24–3.*** Notice that mature cells divide by mitosis to produce haploid cells, which are called **zoospores.** A zoospore (ZOH-oh-spor) is a reproductive cell that can produce a new individual simply by cell division.

If conditions are unfavorable, especially when the cells are starved for nitrogen, *Chlamydomonas* enters a sexual phase of its life cycle. The cells divide to produce haploid (n) **gametes** of two different mating types. A gamete is a reproductive cell that can produce a new individual only after fusing with a gamete of the opposing mating type.

The zygote grows a thick, protective wall that enables it to survive extremely harsh conditions—something it has to do when a pond dries up or freezes over. When favorable conditions return, the zygote goes through meiosis and produces four haploid cells that break out of the protective wall and swim away. These cells begin the life cycle all over again.

☑ **Checkpoint** What is a zoospore? A gamete?

Reproduction in *Ulva*

In some ways, the life cycle of *Ulva* is similar to that of *Chlamydomonas*, but

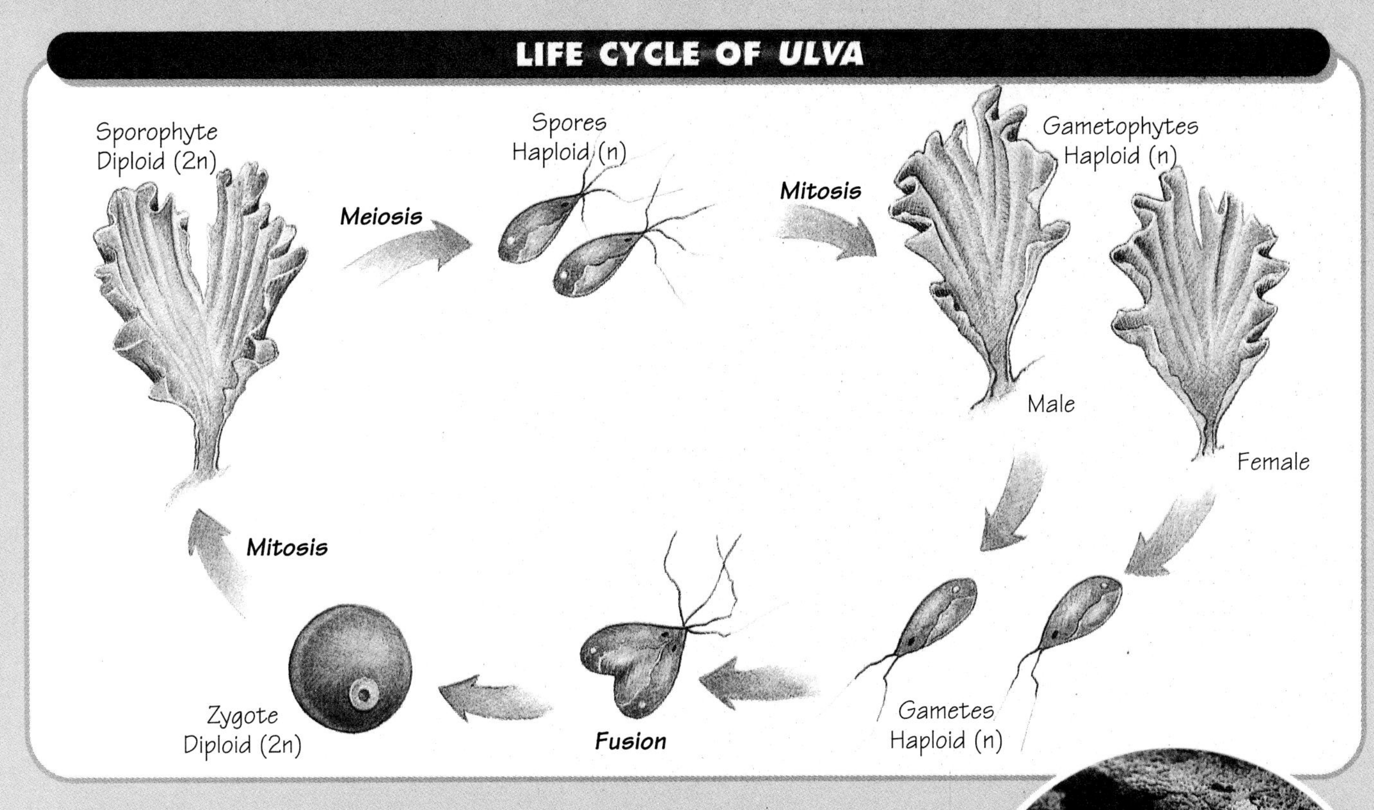

Figure 24–4
Ulva, *a multicellular green alga, has a life cycle that displays the pattern of alternation of generations. Sporophyte plants are diploid (2n) and produce haploid (n) spores by meiosis. The spores germinate to produce male and female gametophytes, which produce haploid (n) gametes. Fertilization of the gametes forms a dipoid (2n) zygote, which gives rise to the diploid sporophyte.*

there are some important differences. As you can see in ***Figure 24–4,*** the diploid and haploid phases of *Ulva*'s life cycle are both multicellular. Even though the "plants" of these two stages look almost identical, they produce completely different kinds of reproductive cells.

The life cycle of *Ulva* contains the first hint of a pattern that you will see over and over in plants. Notice that the cycle contains two separate generations—the **sporophyte** (spore plant) and the **gametophyte** (gamete plant). **Because the organism always alternates between sporophyte and gametophyte phases, this pattern of reproduction is called the alternation of generations.** Nearly all plants follow the pattern of **alternation of generations** in their life cycles.

Section Review 24–1

1. **Compare** the characteristics of multicellular algae and plants.
2. **Explain** what is meant by alternation of generations.
3. **Critical Thinking—Comparing** How is reproduction in *Chlamydomonas* similar to that in *Ulva?* How is it different?

SECTION

24–2 Bryophytes

Guide for Reading

- **Identify** some of the adaptations bryophytes needed to live on land.
- **Describe** the life cycle of a moss.

MINI LAB

- **Compare** the male and female structures of a moss plant.

LIVING ON LAND IS A TOUGH business. To thrive on land, a plant at the very least must be able to resist drying out by drawing enough water from the soil when water is not directly available. In addition, because the buoyant effect of water is absent, a land plant must develop structures to support itself against gravity. Although a few algae do manage to survive on land, their thin walls and lack of specialized tissues make it difficult for them to thrive in dry environments.

Adaptations of the Bryophytes

How have plants responded to the challenges posed by land? The first land plants were most likely close relatives of the green algae that could live out of water at least part of the time. About 400 million years ago, two important groups of multicellular plants evolved from these pioneering algae—bryophytes and tracheophytes.

Bryophytes, meaning "moss plants," are only partially adapted to life away from water. **Bryophytes have waxy coverings that protect them against water loss. In addition, special structures enclose their reproductive cells to prevent them from becoming dry.** Bryophytes display a distinct alternation of generations. And in most species, the sporophyte develops as an embryo inside the female gametophyte, which also helps to protect it from drying out.

Figure 24–5
Bryophytes are marginally adapted to life on land and can be considered to be on the boundary between aquatic and terrestrial plants. Bryophytes include (a) *the bug-on-a-stick moss,* Boxbaumia aphylla, *which gets its name from the shape of the stalk and the capsule of its sporophyte,* (b) *hornworts, such as* Ceratophyllum demersum, *and* (c) *liverworts, such as* Lunularia.

To a limited extent, bryophytes are able to gather water from moist soil and pass it from cell to cell. These plants are anchored to the soil by **rhizoids** (RIGH-zoidz), thin filaments that absorb water and nutrients from the soil. However, bryophytes do not have special fluid-conducting tissue, which is one of the reasons why they have remained small. In fact, the tallest known bryophytes are only about 20 centimeters tall!

Checkpoint What are rhizoids?

Mosses

The most widespread bryophytes are the mosses. They are found throughout the world in damp locations, and for some animals they prove to be an important source of food. **Like all bryophytes, mosses have a life cycle that involves alternation of generations between a haploid gametophyte and a diploid sporophyte.**

When a moss spore lands on wet soil, it germinates and grows into a tangle of thin filaments known as a **protonema** (proht-oh-NEE-muh). As the protonema gets larger, its filaments become more organized. Eventually, it grows upward, producing a leafy moss plant. These green moss plants are the gametophyte stage of the moss life cycle.

Gamete Formation and Fertilization

Mosses produce gametes—sperm and eggs—in special structures at the tips of the gametophyte plants. Some species produce both sperm and eggs on the same plant, whereas in other species the two sexes are on separate plants.

Although the gametes are protected against drying out, there is only one way for the flagellated sperm to get to the egg cells—that is, by swimming to them. This means that mosses cannot reproduce unless they are soaked with water. If there is enough water around for the sperm to swim to the eggs, sperm and egg fuse to form a diploid zygote that grows into a sporophyte. In mosses, the sporophyte grows right out of the body of the gametophyte and is dependent on the gametophyte for food and moisture.

Spore Formation

As the sporophyte matures, it develops a capsule at the top of a stalk. In the capsule, cells from the sporophyte undergo meiosis to produce haploid spores. Upon maturing, the capsule breaks open and the spores are scattered by the wind.

Checkpoint What are the stages in the life cycle of a moss?

Male or Female?

PROBLEM *How do the male and female structures of a moss plant compare?*

PROCEDURE 

1. Obtain a set of moss samples from your teacher.
2. To observe the samples under a microscope, prepare a wet-mount slide of each sample by adding a drop of water and then covering with a coverslip.
3. Observe under the low-power objective of a microscope and make a drawing of your observations.
4. Identify each moss sample as male or female by using ***Figure 24–6*** on page 564.

ANALYZE AND CONCLUDE

1. What is the function of each of the structures you observed?
2. What is the relationship between the shape of the structure and its function?
3. From which structure will the new sporophyte grow after fertilization?
4. Which stage in the moss's life cycle do these plants represent?

LIFE CYCLE OF A MOSS

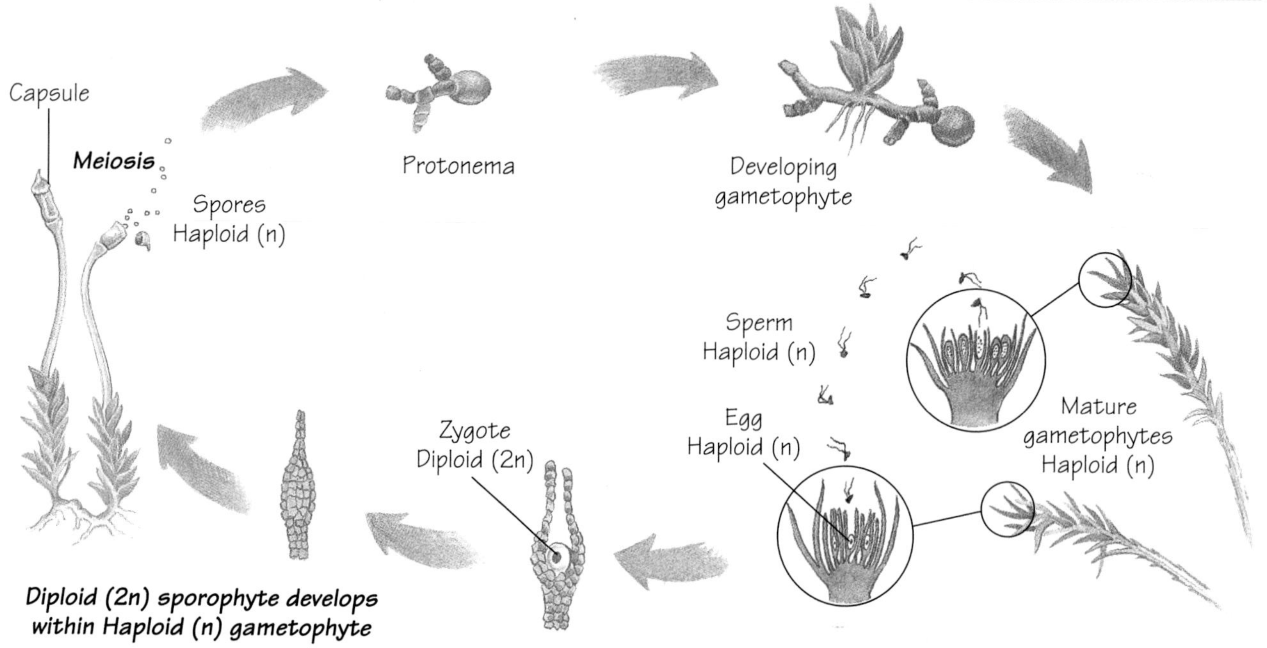

Capsule
Meiosis
Spores
Haploid (n)
Protonema
Developing gametophyte
Sperm
Haploid (n)
Egg
Haploid (n)
Mature gametophytes
Haploid (n)
Zygote
Diploid (2n)
Diploid (2n) sporophyte develops within Haploid (n) gametophyte

Figure 24–6
Reproduction in mosses involves alternating generations of diploid (2n) sporophytes and haploid (n) gametophytes. The leafy moss plants are the gametophytes that produce the sperm and the eggs. Upon fertilization, which requires standing water, the diploid (2n) sporophyte emerges within the gametophyte plant. The mature sporophyte develops a capsule in which haploid (n) spores are produced by meiosis. The spores then germinate and grow into gametophytes.

Hornworts and Liverworts

Hornworts and liverworts, which are also bryophytes, are similar to the mosses in many ways. The name liverwort comes from the way in which the lobes of liverwort gametophytes resemble the lobes of a liver.

Hornworts and liverworts are particularly dependent upon water and are generally found in places that remain damp year-round. They are covered with a waxy cuticle to protect against water loss, they have rhizoids that attach them to the soil, and they show an alternation of generations in their life cycles.

Section Review 24–2

1. **Identify** some of the adaptations bryophytes needed to live on land.
2. **Describe** the life cycle of a moss.
3. **Critical Thinking—Hypothesizing** From which algal group—brown, red, or green—do you think bryophytes evolved? Give evidence to support your hypothesis.
4. **MINI LAB** How do the male and female structures of a moss plant **compare**?

Seedless Vascular Plants

GUIDE FOR READING

- **Explain** the main characteristic of vascular plants.
- **Describe** the life cycle of a fern.

MINI LAB
- **Calculate** the number of spores produced by a mature fern frond.

FOSSILS INDICATE THAT THE very first land plants appeared roughly 450 million years ago. For a while, these plants were quite small, not unlike today's bryophytes. Vast land areas were still unoccupied because they were too dry for the survival of the bryophytes. Then, about 420 million years ago, something happened, and a sudden explosion of plant diversity took place. Much larger plants appeared—some of which were the size of small trees. These new plants were not only bigger, they were also able to invade land that was too dry to support bryophytes or algae.

Origins of Vascular Plants

How did the new plants manage to survive on much drier land? Fossils show that the plants had an evolutionary novelty, a type of cell that was specialized to conduct water. These cells, which are called **tracheids** (TRAY-kee-ihdz), are hollow, have thick cell walls to resist pressure, and are connected end to end like a series of drinking straws. Plants with tracheids, called **vascular plants,** can draw water from the ground and transport that water through the tracheids to reach the higher parts of the plants.

Vascular plants, or tracheophytes—which means "plants with tracheids"—developed vascular systems that draw water from deep in the ground and carry it to great heights. The development of a vascular system made it possible

Figure 24–7
Plants with vascular tissues are able to draw water from land. Among these are (a) *club mosses, such as* Lycopodium, (b) *horsetails, such as* Equisetum arvense, *used in colonial times as scouring pads to clean pots and pans, and* (c) *brachen ferns.* (d) *These structures, called fiddleheads, are unfurling fern fronds, which are eaten as a delicacy at this tender stage.*

MINI LAB .... Calculating ......

Spores in a Sorus

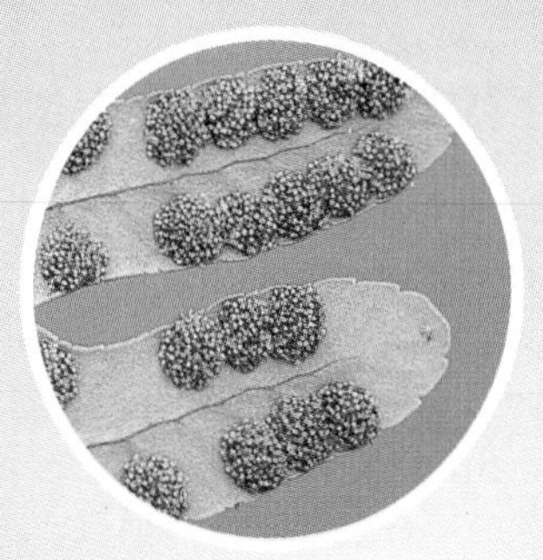

PROBLEM *How can you calculate the number of spores produced by a mature fern frond?*

PROCEDURE

1. Obtain a fern frond. Draw and label the frond, identifying the sori.
2. Using a dissecting needle, gently scrape some sori from a frond onto a microscope slide. **Caution:** *Be careful using a dissecting needle.* Add a drop of water and cover with a coverslip.
3. Observe the slide under the low-power objective of a microscope, then make a drawing of what you see.
4. Calculate the number of spores in a single sorus.

ANALYZE AND CONCLUDE

1. How many leaflets are on your fern frond?
2. What is the average number of sori on each leaflet? From your estimated number of spores, calculate the number of spores produced by a mature fern frond.
3. Why aren't there as many fern plants growing on Earth as the number of fern spores produced?

for these plants to develop true roots, stems, and leaves. From forests to grasslands to deserts, these plants now dominate Earth's landscapes. Because they are such a diverse lot, the tracheophytes are further classified into several divisions. In this chapter, we will consider only those tracheophytes that do not form seeds—the seedless vascular plants.

☑ *Checkpoint* What are vascular plants?

Club Mosses and Horsetails

Not long after the first vascular plants appeared on Earth, two groups rapidly diversified and gave rise to an astonishing variety of species. Some, including the club mosses, were placed in the Division Lycophyta (ligh-koh-FIGHT-uh), and others, such as the horsetails, were placed in the Division Sphenophyta (sfee-noh-FIGHT-uh). Plants from these two groups formed the Earth's first great forests, and for more than 100 million years were the largest of the land plants.

Many of these species grew to heights of up to 40 meters. That's as tall as a 12-story building! Today these plants are represented by just a few species, most of which grow in damp, well-shaded places in the forest. These plants have large, independent sporophytes. It's usually the sporophytes that we notice, because the gametophytes are much smaller and grow along the forest floor.

☑ *Checkpoint* How are club mosses and horsetails classified?

Ferns

Ferns are found throughout the world and are some of the most popular houseplants among plant enthusiasts. The Division Pterophyta (ter-oh-FIGHT-uh), in which ferns are placed, includes about 12,000 species, most of which grow in warmer regions. Ferns have an excellent vascular system, well-developed underground stems called **rhizomes,** and large leaves known as **fronds.**

☑ *Checkpoint* What are rhizomes? Fronds?

Sporophyte

The large, visible fern plant that you are probably familiar with is the sporophyte in the fern's life cycle. **Like the life cycle of other plants, the life cycle of the fern involves alternation of generations.** Fern sporophytes are diploid and produce haploid spores by meiosis. Spores form on the undersides of

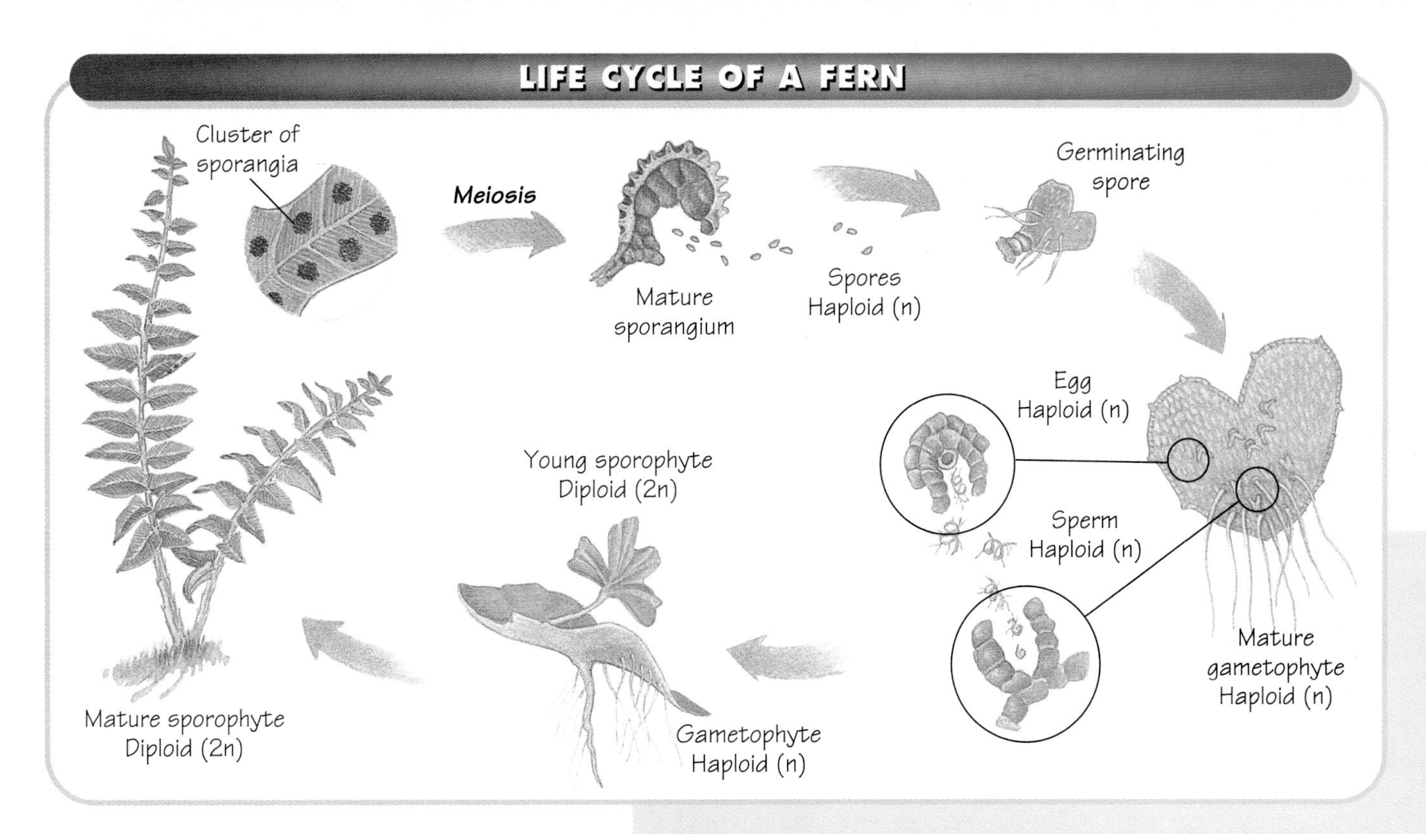

Figure 24–8
The fern life cycle displays the characteristic pattern of alternation of generations of diploid (2n) sporophytes and haploid (n) gametophytes. The gametophytes are tiny and grow from germinating spores on the forest floor. Fertilization of the haploid (n) gametes requires watery conditions and results in the formation of a diploid (2n) zygote. The zygote develops into the mature sporophyte, the large fern plant with fronds and sori.

the fronds in little chambers known as **sori** (singular: sorus). When the spores mature, these chambers burst open, releasing the tiny spores that may be scattered by wind or water. When a spore settles on rich, damp soil, it germinates and begins to grow into a tiny plant.

Gametophyte

The plants that develop from spores are independent haploid gametophytes. Fern gametophytes are small heart-shaped plants just one cell thick. Most are no more than a centimeter or two across. On the underside of the gametophyte, sperm and eggs develop in tiny reproductive organs. When the ground is moist enough, the flagellated sperm cells emerge from the underside of the plant and swim into the open chambers containing eggs. The fertilized zygote that results from the fusion of sperm and egg is a diploid cell that develops into an embryo. That embryo grows right out of the gametophyte and becomes the large, independent sporophyte.

Section Review 24–3

1. **Explain** the main characteristics of vascular plants.
2. **Describe** the life cycle of a fern.
3. **Critical Thinking—Hypothesizing** What advantage might there be for a plant's life cycle to undergo alternation of generations?
4. **MINI LAB** How can you **calculate** the number of spores produced by a mature fern frond?

SECTION 24–4

BRANCHING OUT **In Action**

Algae in Our World

GUIDE FOR READING

- **Explain** the role of algae in nature.
- **List** ways in which algae benefit humans.

AS YOU HAVE SEEN, THE multicellular algae, bryophytes, and seedless vascular plants are three extremely diverse groups of organisms. Each has a long fossil history, and each has helped to shape the patterns of life on Earth for many millions of years. What roles do these organisms play in the natural world, and how do humans use them? In this section, we will examine the role of algae in our world.

Shaping the Environment

All the photosynthetic organisms have had major roles in shaping Earth's environment. They are at the base of the food chain in local habitats throughout the biosphere. Algae and photosynthetic prokaryotes, for example, are food for most of the life in the ocean, and without them animal life underwater would not be possible. **The algae produce an enormous share of atmospheric oxygen. They also provide the oxygen dissolved in fresh and salt water that makes aquatic animal life possible.**

Algae, along with mosses, are among the few organisms that can grow on rocks and gravel. Their actions help to break down minerals and mix them with organic matter to form soil, making it possible for other plants to follow them.

☑ *Checkpoint* How do photosynthetic algae contribute to the biosphere?

Benefit to Humans

Many algae are used by humans around the world. **Algae are used in the processing of foods, chemicals, and even wastewater.**

Figure 24–9
Algae are widely used by a variety of organisms. (a) This otter wraps itself in strands of brown kelp to keep from being swept away by waves. Many people enjoy eating foods such as (b) ice cream and (c) sushi, which are made using types of red and brown algae.

Sources of Food

Have you ever eaten algae? Whether or not you knew it, the answer is almost certainly yes. Large, thin sheets of the red alga *Porphyra* are dried and used as a food wrapping, called *nori* in Japanese. Small pieces of fish or vegetables are placed on a bed of rice and wrapped in *nori* to produce the dish called sushi.

You've never eaten sushi? Well, chances are you've had ice cream, marshmallows, or a candy bar. Nearly all brands of these foods contain *algin*, a carbohydrate produced by brown algae and used as a thickening agent. Carrageenan (kar-uh-GEEN-uhn) is a carbohydrate produced by red algae and is used to stabilize pie fillings, puddings, and even toothpaste. Many algae are also rich sources of vitamin C and iron, and they are used to enrich other foods.

Figure 24–10
The chemical industry has found many uses for algae and their products, such as in (a) *the manufacture of paints.*

(b) CAREER TRACK *The diverse uses of algae by humans is the result of the research performed by phycologists. Phycologists study the structures and functions of algae.*

Uses in Industry

The range of chemical compounds produced by algae makes them especially useful to industry. Chemicals from algae are used to make plastics, waxes, transistors, deodorants, paints, lubricants, and even artificial wood. Algae have important uses in scientific laboratories. The compound agar-agar, extracted from certain red algae, is used to make the nutrient mixtures in which biologists grow bacteria and other organisms.

Wastewater Treatment

Wastewater is usually high in nutrients. These nutrients can upset the balance of life in freshwater streams and lakes if released into these waters. Many wastewater treatment plants use large, shallow holding ponds stocked with algae that absorb the nutrients. The treated water, with a smaller concentration of nutrients, can be safely released into the environment—thanks to the algae.

Section Review 24–4

1. **Explain** the role of algae in nature.
2. **List** several uses of algae that benefit humans.
3. **BRANCHING OUT ACTIVITY** Visit a pet shop that sells fishes and aquariums for household setups. Find out how an aquarium is designed to manage the growth of algae. In a brief report, **describe** what you learned.

Laboratory Investigation

Diversity of Green Algae

Green algae are members of the phylum Chlorophyta. All organisms within this phylum contain chlorophyll and are able to perform photosynthesis. In this investigation, you will observe several samples of green algae and compare their characteristics.

Problem

How do the different multicellular forms of green algae **compare?**

Materials (per group)

- living cultures of *Chlamydomonas, Spirogyra, Volvox,* and *Ulva*
- microscope slides
- medicine dropper
- coverslips
- dissecting needle
- microscope

Procedure 

1. **Using a medicine dropper, place one drop of the *Chlamydomonas* culture on a microscope slide. Cover with a coverslip as shown in the diagram. Observe under the low-power objective of a microscope. Draw a diagram of what you see and label the cell wall, nucleus, chloroplast, and cytoplasm.**
2. **Obtain a strand of *Spirogyra* and place it on a slide. Separate the strand into a 2-cm segment using a dissecting needle. CAUTION: *Be careful when using a dissecting needle.* Add a drop of water if necessary and cover with a coverslip. Focus the image under the low-power objective and then switch to the high-power objective. Observe the shape and draw a diagram of at least two cells of the filamentous alga. Label the cell wall, nucleus, chloroplast, and cytoplasm.**
3. **Place one drop of the *Volvox* culture on a microscope slide and cover with a coverslip. Again, observe the organism under the low-power objective and then switch to the high-power objective. Observe the structure of the organism and observe any movement that occurs. Draw a diagram of *Volvox* and label a chloroplast and flagellum.**

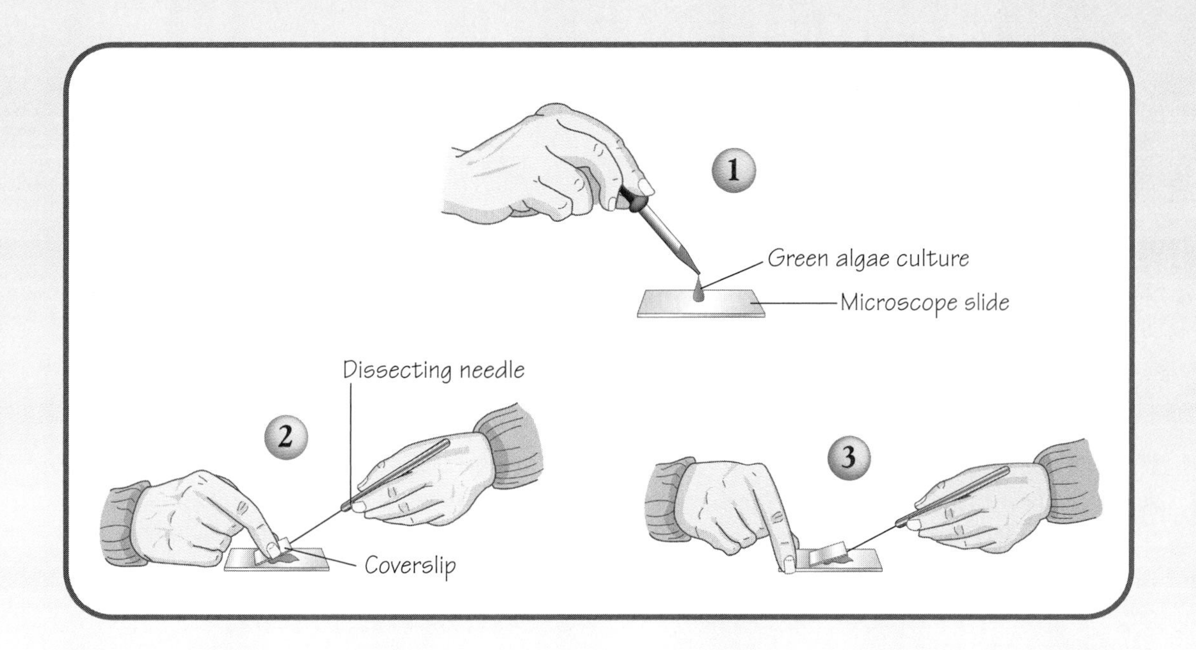

4. Separate a small piece of the *Ulva* from the specimen sample. Add a drop of water and then cover with a coverslip. As before, focus the image under the low-power objective of a microscope and then switch to high power. Determine the number of cells in the field of view. Draw a diagram of at least three cells, then label the nucleus, cell wall, chloroplast, and cytoplasm.

Observations

1. Describe the structure of each of the green algae you observed.
2. What features or characteristics do the green algae share?
3. Describe the motion of the *Volvox* you observed under the microscope.

Analysis and Conclusions

1. Compare the characteristics of the green algae you observed. Discuss their shape, structure, and motility. Construct a table to illustrate your answer.
2. Compare the arrangement of the chloroplasts among the specimens you observed. Why are they arranged in this way?
3. Why is there so much diversity within such a small classification of organisms? Explain your answer.

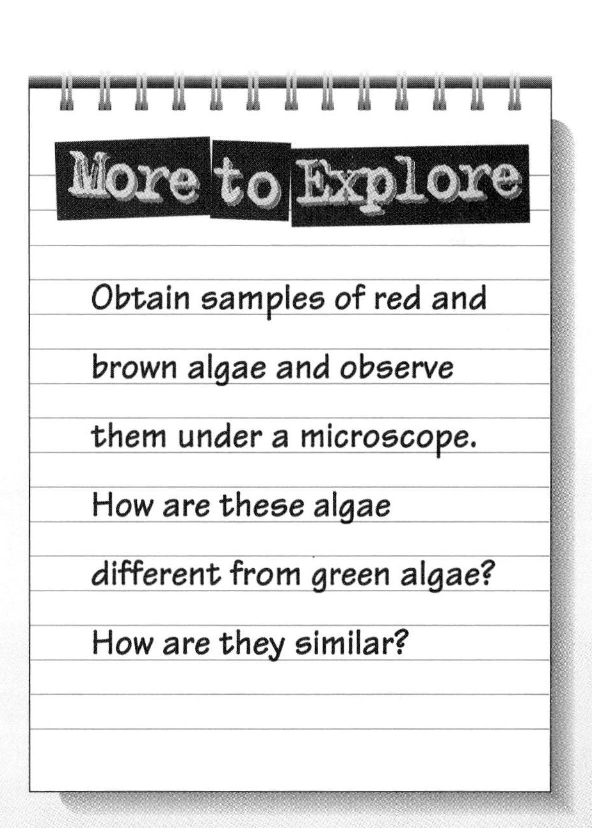

Study Guide

Summarizing Key Concepts

The key concepts in each section of this chapter are listed below to help you review the chapter content. Make sure you understand each concept and its relationship to other concepts and to the theme of this chapter.

24–1 Multicellular Algae

- Algae are photosynthetic aquatic organisms that are classified as protists.
- Brown and red algae are multicellular marine organisms. Most green algae are unicellular freshwater organisms.
- Reproductive patterns in green algae are similar to those in plants. Haploid (n) *Chlamydomonas* cells divide by mitosis to produce haploid zoospores. Under unfavorable conditions, *Chlamydomonas* cells divide to produce gametes that fuse to form a diploid (2n) zygote. The zygote later undergoes meiosis to produce haploid cells.
- The multicellular green alga *Ulva* alternates between generations of diploid (2n) sporophytes and haploid (n) gametophytes.

24–2 Bryophytes

- Bryophytes—such as mosses, hornworts, and liverworts—have waxy coverings, which protect them against water loss, and special structures that enclose their reproductive cells to keep them from becoming dry.
- Like all bryophytes, mosses have a life cycle that involves alternation of generations between a haploid (n) gametophyte and a diploid (2n) sporophyte.

24–3 Seedless Vascular Plants

- Tracheophytes, such as club mosses, horsetails, and ferns, are vascular plants—plants with fluid-conducting tissue.
- The life cycle of a fern alternates between the generations of sporophyte and gametophyte.

24–4 Algae in Our World

- Algae produce an enormous amount of atmospheric oxygen and much of the oxygen dissolved in fresh and salt water that makes aquatic life possible. Algae and photosynthetic prokaryotes are food for most of the life in the ocean.
- Many algae are used by humans around the world. Algae are used in the processing of foods, chemicals, and wastewater.

Reviewing Key Terms

Review the following vocabulary terms and their meaning. Then use each term in a complete sentence.

24–1 Multicellular Algae

alga
zoospore
gamete
sporophyte
gametophyte
alternation of generations

24–2 Bryophytes

rhizoid
protonema

24–3 Seedless Vascular Plants

tracheid
vascular tissue
rhizome
frond

Recalling Main Ideas

Choose the letter of the answer that best completes the statement or answers the question.

1. Why are multicellular algae considered plants?

a. They live on land.
b. They are protists.
c. Their reproductive cycles are similar to those of plants.
d. They are multicellular organisms.

2. The algae that live at great ocean depths are

a. red algae. **c.** unicellular algae.
b. brown algae. **d.** multicelluar algae.

3. Which is not a green alga?

a. *Ulva* **c.** *Volvox*
b. *Fucus* **d.** *Spirogyra*

4. What enables bryophytes to absorb water and nutrients from the soil?

a. vascular tissue **c.** ribosomes
b. tracheids **d.** rhizoids

5. Moss sperm and eggs form a zygote that grows into a

a. capsule. **c.** gametophyte.
b. zoospore. **d.** sporophyte.

6. In asexual reproduction, *Chlamydomonas* cells produce

a. haploid gametes by mitosis.
b. haploid gametes by meiosis.
c. diploid zoospores by meiosis.
d. haploid zoospores by mitosis.

7. In reproduction in *Ulva,* diploid plants produce

a. haploid spores by meiosis.
b. haploid spores by mitosis.
c. diploid spores by meiosis.
d. diploid spores by mitosis.

8. Large ferns used as decorative houseplants represent which stage in the fern's life cycle?

a. gametophyte **c.** zygote
b. sporophyte **d.** zoospore

9. Which functions do algae perform in nature?

a. provide food for sea animals
b. produce atmospheric oxygen
c. provide oxygen to fresh and salt water
d. all of the above

Putting It All Together

Using the information on pages xxx to xxxi, complete the following concept map.

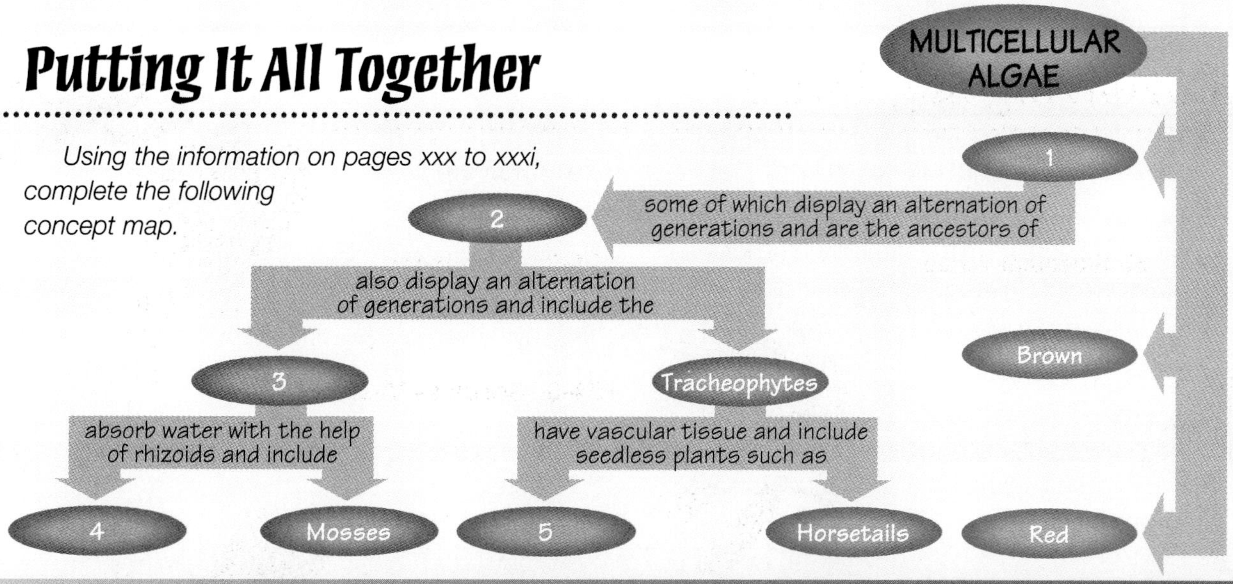

MULTICELLULAR ALGAE
1
some of which display an alternation of generations and are the ancestors of
2
also display an alternation of generations and include the
3
Tracheophytes
absorb water with the help of rhizoids and include
4
Mosses
have vascular tissue and include seedless plants such as
5
Horsetails
Brown
Red

Assessment

Reviewing What You Learned

Answer each of the following in a complete sentence.

1. What are algae? How are they classified?
2. List ways in which algae are similar to plants. How are they different from plants?
3. Describe the characteristics of brown algae.
4. Describe the characteristics of red algae.
5. How are green algae different from red and brown algae?
6. What is a gamete? A zoospore?
7. Define alternation of generations.
8. How were the bryophytes able to live on land?
9. Sequence the steps in the life cycle of a moss.
10. What is a tracheid?
11. What is a sorus?
12. Why can ferns survive in more extreme conditions than can mosses?
13. How do humans use algae?

Expanding the Concepts

Discuss each of the following in a brief paragraph.

1. **Compare** the reproductive cycle of *Ulva* with that of *Chlamydomonas.*
2. What were some of the major adaptations plants evolved in order to live on land?
3. Compared to the size of most land plants, mosses are very small. Formulate a hypothesis to explain this fact.
4. Draw and label a diagram of the life cycle of a moss.
5. Approximately 420 million years ago, plants began to spread to all areas of the Earth. What might explain this tremendous increase in land plants?
6. Which group of algae are the most likely ancestors of land plants? Give reasons to support your answer.
7. How does alternation of generations differ in ferns and in mosses?
8. How does the development of a spore differ from that of a zygote?
9. Why must mosses—and even ferns—have a wet environment at least part of the year in order to reproduce?
10. Discuss the significance of the role of algae in nature.

Extending Your Thinking

Use the skills you have developed in this chapter to answer the following.

1. **Synthesizing** Examine the life cycles of *Ulva,* mosses, and ferns. Draw a generalized, labeled diagram that represents the main elements shared by the cycles. What would be the title of this diagram?
2. **Calculating** Choose a flowering plant in or around your home. Select a segment of that plant and count the number of blossoms on the selected segment. From this number, estimate the total number of flower blossoms on the entire plant.
3. **Relating** In the early stages of space exploration, scientists considered sending single-celled algae into space with astronauts to provide many of the astronauts' needs. Identify what these needs are and predict how the algae would meet them.
4. **Inferring** Some people who tire of taking care of their pet fishes get rid of them by releasing them into lakes and streams. Many of these fishes die because they are placed in environments with which they are incompatible. Hawaii, for example, has lost a number of native fish species because of the release of algae-eating fishes into native streams. Explain why the native species might have died.
5. **Predicting** Green algae have fewer accessory pigments—pigments other than chlorophyll that absorb light energy and pass it to chlorophyll—than do red or brown algae. Based on this information, would you expect to find green algae in shallow or deep water? Where would you expect to find red algae? Brown algae?

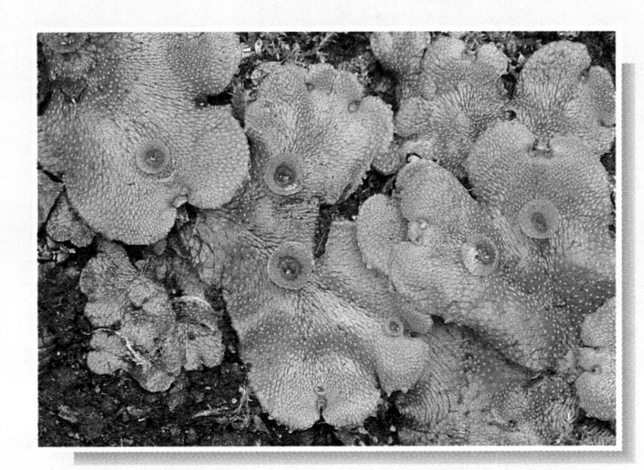

Applying Your Skills

Have You Eaten Your Algae Today?

Algae are used in the manufacture and processing of several foods. In this activity, you will read food labels and ingredient lists in order to identify the use of algae.

1. For one week, keep a record of the food products you eat that contain algae. Be sure to look at the labels of foods such as cereals, milk, and ice cream.
2. At the end of the week, prepare a list of the forms of algae or algae extracts that were used. Which type of algae is used most often in the food products you researched?
3. Share your list with other students and prepare a combined list for your class.

• GOING FURTHER •

4. Write an essay about the role of algae in your life.

CHAPTER 25

Plants With Seeds

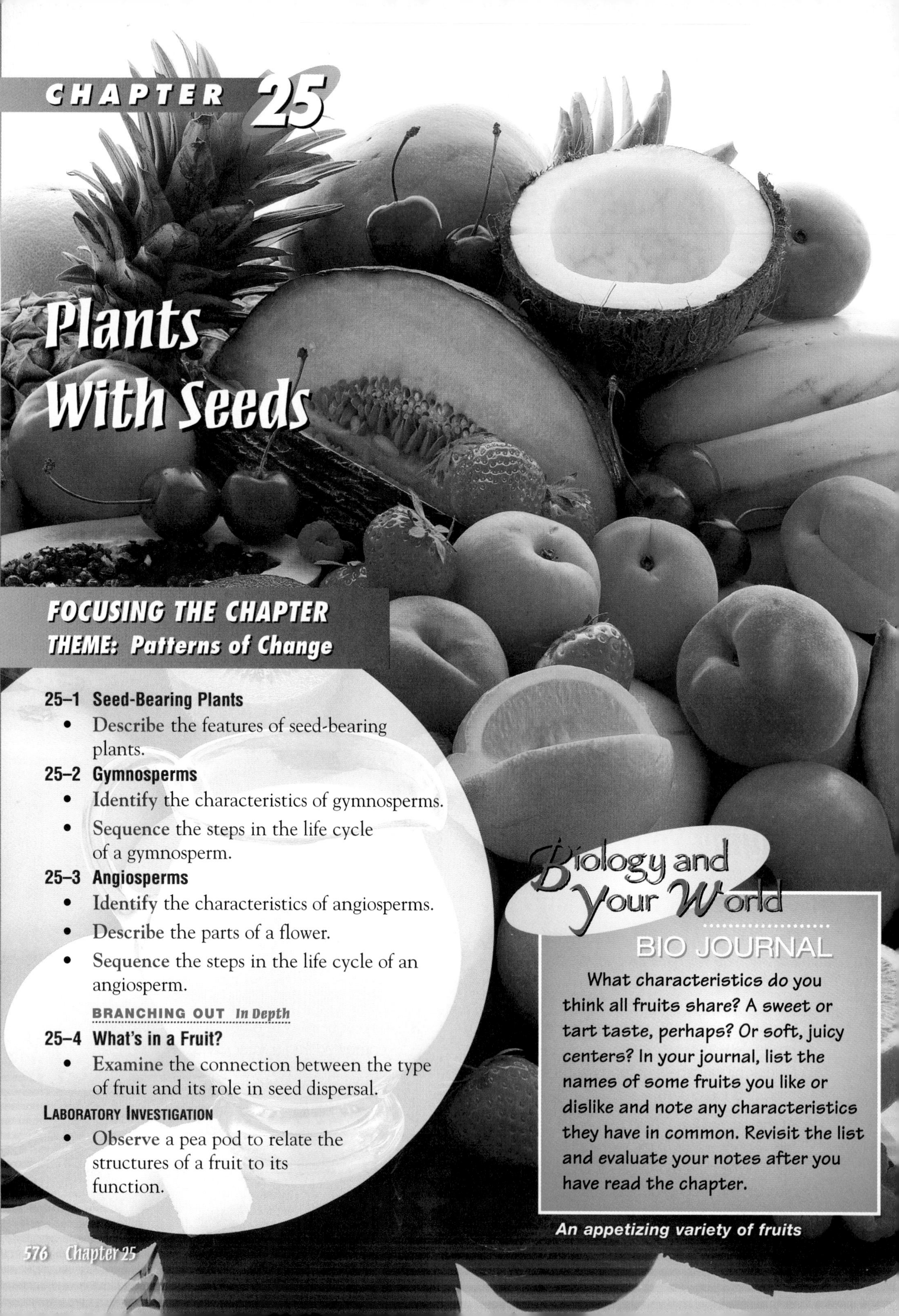

FOCUSING THE CHAPTER THEME: Patterns of Change

25–1 Seed-Bearing Plants
- **Describe** the features of seed-bearing plants.

25–2 Gymnosperms
- **Identify** the characteristics of gymnosperms.
- **Sequence** the steps in the life cycle of a gymnosperm.

25–3 Angiosperms
- **Identify** the characteristics of angiosperms.
- **Describe** the parts of a flower.
- **Sequence** the steps in the life cycle of an angiosperm.

BRANCHING OUT *In Depth*

25–4 What's in a Fruit?
- **Examine** the connection between the type of fruit and its role in seed dispersal.

LABORATORY INVESTIGATION
- **Observe** a pea pod to relate the structures of a fruit to its function.

Biology and Your World

BIO JOURNAL

What characteristics do you think all fruits share? A sweet or tart taste, perhaps? Or soft, juicy centers? In your journal, list the names of some fruits you like or dislike and note any characteristics they have in common. Revisit the list and evaluate your notes after you have read the chapter.

An appetizing variety of fruits

Seed-Bearing Plants

SECTION 25–1

GUIDE FOR READING

- **Define** seed.
- **Explain** why seed-bearing plants are able to live everywhere on Earth.

MINI LAB

- **Observe** the structures of a seed.

DURING THE CARBONIFEROUS Period, most of the Earth was warm and humid. Great forests of seedless plants covered the damp landscape. Then, about 290 million years ago, the Earth's climate changed. The next 50 million years, known as the Permian Period, were marked by the spread of glaciers and widespread drought. In these dry climates, plants that depended on standing water found it more and more difficult to reproduce. By contrast, any plant that was less dependent on water was immediately favored by natural selection.

These conditions affected animals as well as plants. It was during this period that the reptiles, which lay eggs with shells that can survive on dry land, first began to dominate the water-dependent amphibians. Not surprisingly, at the same time, the seed-bearing plants began to out-compete their seedless neighbors.

Seeds

As you know, algae, mosses, and ferns are still present on the Earth, but these plants form only a tiny fraction of the world's great forests and grasslands. The plants that grow everywhere today and the plants that you know from lawns and gardens and fields and farms are quite different from mosses and ferns. These familiar plants have one very important feature in common with each other that makes them different from the plants we examined in the last chapter—their ability to form **seeds.**

The single most important group of plants on land are the seed-bearing plants. These plants all belong to the subphylum Spermopsida (sper-MOP-sih-duh). Why is the ability to form seeds so important? The best way to answer that question is to see exactly how a seed is put together.

Figure 25–1 *Plants such as (a) pine, (b) water lilies, and (c) scarlet plume ensure their future generations by the production of seeds.*

MINI LAB .... Observing ....

Peanut With a Purpose

PROBLEM *How can you **observe** the various structures of a seed?*

PROCEDURE

1. Obtain a peanut in its shell. Describe the shell covering.
2. Carefully break open the shell and remove the contents.
3. Select one of the peanuts (the seed) and remove its papery red covering. Then gently separate the two halves of the peanut.
4. Examine the two halves with a hand lens or dissecting microscope. Sketch and label what you see.

ANALYZE AND CONCLUDE

1. What does the entire peanut, including the shell, represent? Where did you find the embryo?
2. What are the names of the two parts found inside the red covering? What is their function? How does their function relate to the importance of peanuts to humans?

A seed is a reproductive package that contains a plant embryo and a supply of stored food inside a protective coating. Seeds are resistant to drying, and most seeds can survive even the most extreme conditions of drought and heat for several years. By producing seeds, a plant seals its next generation into protective packages that can endure the worst that nature may subject them to and then spring to life when conditions are just right.

How are seeds different from the spores that are produced by other land plants? Most seeds are tougher and more resistant to drying than are spores. Seeds contain a fully formed plant embryo, rather than the single cell that is usually found in most spores. Seeds contain stored food—often enormous amounts of it—which spores lack. And many seeds have special tissues or structures that aid in their distribution, making it possible for seeds to find new places to grow far from their parent plants.

The very first seed-bearing plants appeared right alongside the great seedless vascular plants of the Carboniferous Period. As they diversified over millions of years, the seed-bearing plants gradually took over one part of the Earth after another. What has made these plants so successful? It is their independence from water! **The seed-bearing plants, unlike all other plants, do not require standing water for reproduction. This means that they can grow just about anywhere and reproduce at times of the year that are much too dry for ferns or mosses to reproduce.**

☑ *Checkpoint* What is a seed?

Reproduction

Seed-bearing plants, like other plants, display alternation of generations. In ferns, as you may recall, the most visible stage of the cycle is the sporophyte, a large diploid (2n) plant. The fern gametophyte is a tiny independent haploid (n) plant that produces male and female gametes. These gametes require standing water for fertilization, and that is why ferns, despite their vascular tissues, are still dependent upon damp conditions.

The common large plants you see around you are the diploid (2n) sporophytes of seed-bearing plants. Where are their gametophytes? The answer may surprise you. Their gametophytes live inside the sporophyte!

Cones and Flowers

The tiny gametophytes of the seed-bearing plants do not have an independent life of their own. Instead, they live inside special parts of the sporophyte

Figure 25–2
Reproduction in seed plants is characterized by the alternation of generations. The mature plant, a diploid (2n) sporophyte, produces reproductive structures known as cones and flowers. The haploid (n) gametes are produced within these structures, and upon fertilization, give rise to the diploid (2n) zygote. The zygote develops into the seed, which is capable of growing into a sporophyte.

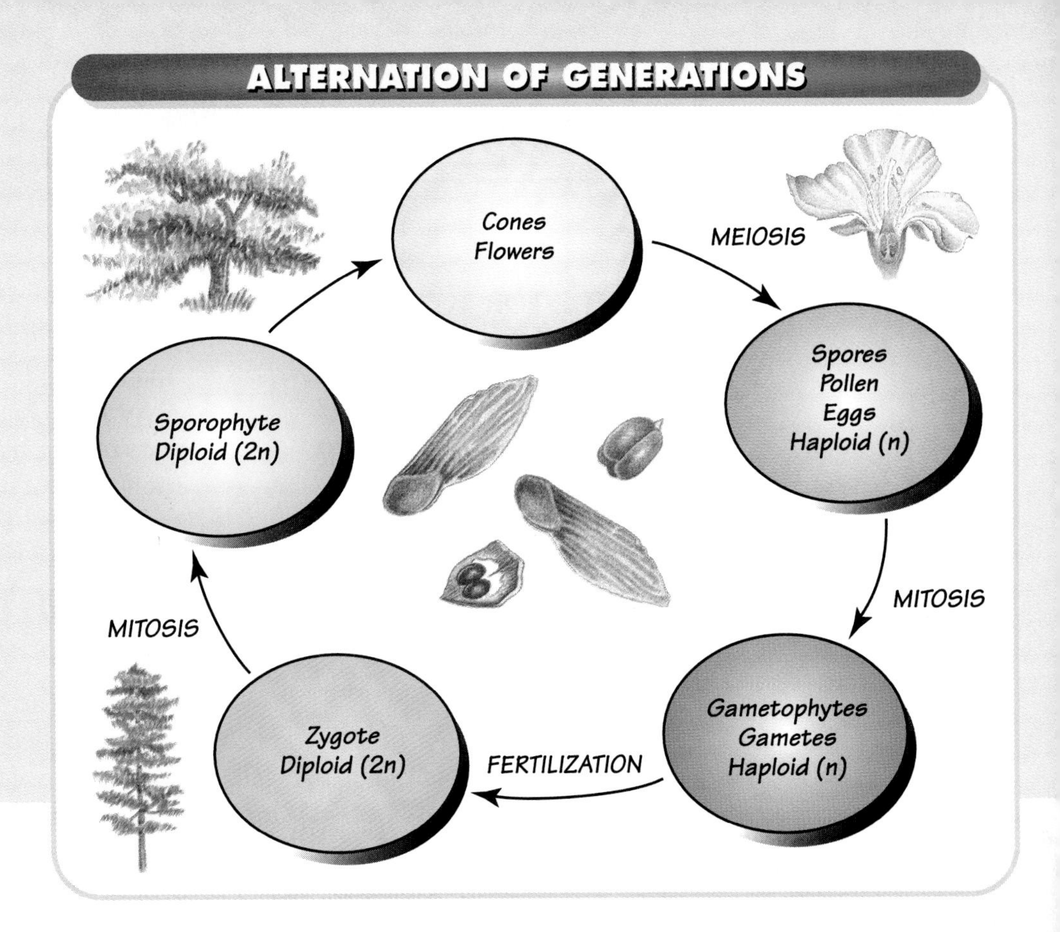

plant—parts known as cones and flowers. Pine cones and flowers are home to the gametophyte generation.

Spores

In plants that you have read about so far, the sporophytes produce haploid spores that grow to produce the gametophyte generation. Does this happen in the seed-bearing plants? Yes, but the spores produced by seed plants come in two very different forms—a tiny spore, called **pollen,** and a much larger spore, known as the **egg.**

As you will see, each of these spores grows into a tiny gametophyte that is contained within the much larger body of the sporophyte. Although we will examine the details of this process a bit later, one point is worth emphasizing right now. The fact that the sporophyte contains the gametophyte means that seed plants do not need standing water for reproduction. Millions of years of evolution have produced a kind of plant that, in effect, creates its own little microenvironment for its gametophyte generation. If seed-bearing plants have one unique ability that allowed them to spread to every corner of the planet, this is it—the ability to reproduce without water.

Section Review 25–1

1. **Define** seed.
2. **Explain** why seed-bearing plants are able to live everywhere on Earth.
3. **Critical Thinking—Comparing** How do spores and seeds differ?
4. **MINI LAB** What structures of a seed can you **observe** by examining a peanut?

SECTION

25–2 Gymnosperms

GUIDE FOR READING

- **Describe** the characteristics of gymnosperms.
- **Outline** the stages in the life cycle of a gymnosperm.

MINI LAB

- **Relate** the structures of male and female cones to their function.

DID YOU KNOW THAT THE world's tallest, largest, and oldest trees all belong to a group of seed-bearing plants called conifers? In California, a coastal redwood has been measured at 114 meters in height. The General Sherman sequoia in Sequoia National Park, also in California, is 11 meters in diameter and has an estimated mass of 2,040 metric tons. And a living bristlecone pine of the western Great Basin has been estimated to be 4900 years old! In this section, you will not only examine the conifers, but also three other groups of seed-bearing plants.

Divisions of Gymnosperms

Today, there are five divisions of living seed-bearing plants. **Four of the divisions, collectively known as gymnosperms, consist of plants that bear their seeds on the surfaces of reproductive structures called scales.** The **scales** are usually grouped into clusters familiarly known as cones. Because their seeds are exposed on the surfaces of the scales, these plants are called **gymnosperms,** which means "naked seeds."

Seed Ferns

The first seed-bearing plants were known as seed ferns. These seed-bearing plants had leaves that were similar to those of ferns. The seed ferns are known only because of the fossils they formed. As far as we know, not a single seed fern exists today.

Figure 25–3
(a) *A closeup of a ginkgo tree shows the "naked" seeds that are characteristic of gymnosperms.* (b) *The Ginkgo tree, the only living member of the Division Ginkgophyta, is often planted in cities to provide shade and to beautify the area.* (c) *Although it may look like a palm tree,* Cycas revoluta *is a member of the Division Cycadophyta.*

Figure 25–4
(a) Welwitschia mirabilis, *one of the rare group of plants classified as gnetophytes, grows in the Namib Desert in southern Africa. The plant shown here is a female plant with mature seed cones.* (b) *Conifers, such as this bristlecone pine, are some of the oldest living organisms. Forests of conifers provide homes for other plants and countless animals—from insects to birds.*

Cycads

Cycads, which are members of the Division Cycadophyta (sigh-KAD-oh-fight-uh) are palmlike plants that first appeared more than 225 million years ago. Cycads were especially common in the forests that were home to the great dinosaurs of the Triassic Period. Cycads are not palm trees, but they are often confused with them. The confusion is easy to understand, especially because one of the most common cycads is called the sago "palm." These beautiful plants are still common in tropical regions of Mexico, Australia, the West Indies, and Florida.

Ginkgoes

Remarkable trees that make up the Division Ginkgophyta (GIHNK-goh-fight-uh) are called ginkgoes. These plants were common in fossils that are dated to be more than 200 million years old. European scientists thought that these organisms were extinct until they discovered that the Chinese have carefully cultivated them as ornamental trees for centuries. The only living species in this class, commonly known as the ginkgo tree, *Ginkgo biloba,* is so similar to its fossil relatives that it may be the oldest living species of seed-bearing plants.

Gnetophytes

Gnetophytes—members of the Division Gnetophyta (NEE-toh-fight-uh)—are rare plants with many characteristics that remind scientists of flowering plants. In fact, some recent evidence suggests that the first plants with true flowers may have evolved from a gnetophyte.

Conifers

Conifers, which make up the Division Coniferophyta (koh-NIHF-er-oh-fight-uh), are sometimes called evergreens. The members of this class include some of the most important of all land plants—spruce, pine, cedar, redwood, fir, and yew trees. Forests of conifers cover vast areas of North America, China, Europe, and Australia.

The leaves of conifers are long and thin—you probably know them as needles. These thin leaves are coated with a tough, waxy substance. Conifer leaves are well adapted to prevent water loss, and most will remain on the plant for years. This, of course, is where the name evergreen comes from, although a few conifers—notably the bald cypress—lose their leaves each year.

Checkpoint What are the four living divisions of gymnosperms?

Visualizing the Life Cycle of a Gymnosperm

The pine tree, a gymnosperm, bears cones that produce gametophytes—pollen and eggs. Fertilization leads to the formation of a seed that germinates and grows into a new sporophyte.

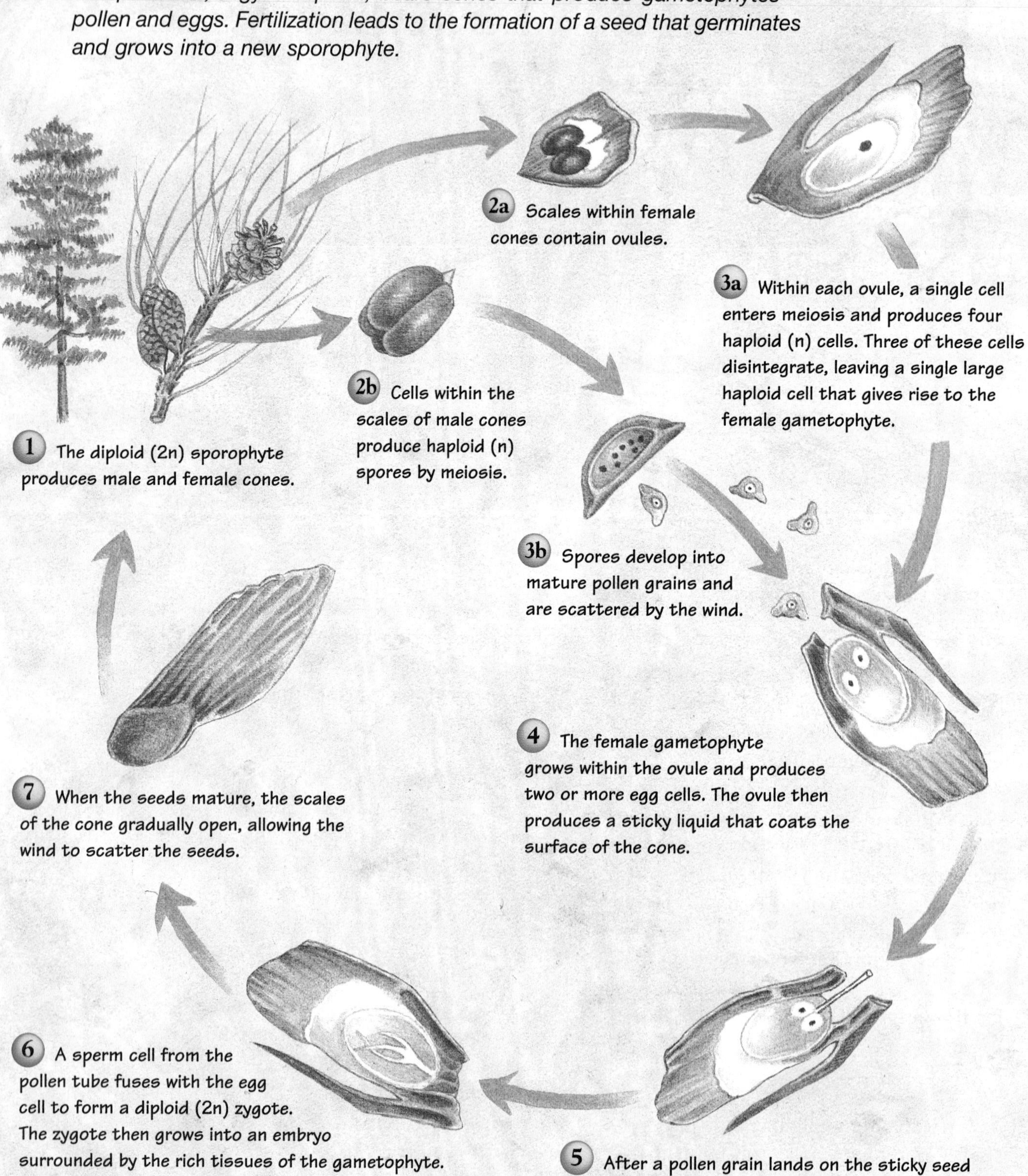

Reproduction in Conifers

As you might expect, the life cycle of a gymnosperm displays alternation of generations—generations of sporophytes and gametophytes that develop within the sporophytes. Most gymnosperms have two kinds of cones. The **male cones** produce pollen, and the **female cones** produce eggs. Because the female cones eventually contain mature seeds, they are sometimes called **seed cones.** Refer to the illustration showing the life cycle of a gymnosperm on page 582.

In a typical gymnosperm, such as a pine tree, the process of fertilization and seed formation may take as long as a year. Female seed cones that appear in one growing season will be fertilized the same year, but they will not be ready to release their seeds until the following season.

Now think for a moment about the way in which the conifer seed is put together. Botanists like to point out that the seed is a product of and contains tissues from three alternating generations! The embryo itself is a diploid sporophyte. It is surrounded by the food-storing tissues of the haploid gametophyte that produced the egg cell. And the whole structure is enclosed in a seed coat formed from the sporophyte generation that produced the seed.

Relating

Cones and More Cones

PROBLEM *How do the various structures found on male and female pine cones* ***relate*** *to their function?*

PROCEDURE 

1. Obtain a pine tree branch with pine cones.
2. Examine a pollen cone. Dust some of the pollen grains on a microscope slide, prepare a wet-mount slide, and observe through the low-power objective of a microscope. Sketch a pollen grain.
3. Look at a seed cone. Observe the scales and note their arrangement. Gently shake the cone. Observe what happens.
4. Remove one of the scales and examine its base. Even if the seeds have been shed, an impression of the seed still remains.
5. Examine a seed cone that has been soaked in water.

ANALYZE AND CONCLUDE

1. How is the structure of the pollen grain related to its function?
2. How is the structure of a seed related to its function?
3. What function do the scales of a seed cone serve? How do the scales on a soaked cone compare with a dry one?

Section Review 25–2

1. **Describe** the main characteristics of gymnosperms.
2. **Outline** the stages in the life cycle of a gymnosperm.
3. **Critical Thinking—Inferring** Why do the male cones of conifers, such as pine, produce large quantities of pollen?
4. **MINI LAB** How do the structures of a male and a female cone **relate** to their functions?

SECTION 25–3

Angiosperms

Guide for Reading

- **Describe** the main characteristics of angiosperms.
- **Identify** the parts of a flower.
- **Outline** the stages in the life cycle of an angiosperm.

MINI LAB

- **Design an experiment** to show how the structures of a flower are related to its function.

THE DOMINANT FORM OF plant life on land today is a class of seed plants known as angiosperms, or flowering plants. They are believed to have evolved from gymnosperms that began a mutually beneficial interaction with animal species such as insects. As the insects benefited by feeding on protein-rich pollen, the plants benefited by gaining a more reliable means of pollination and seed dispersal. In this section, you will read about adaptations in structure and function that led to the overwhelming success of flowering plants.

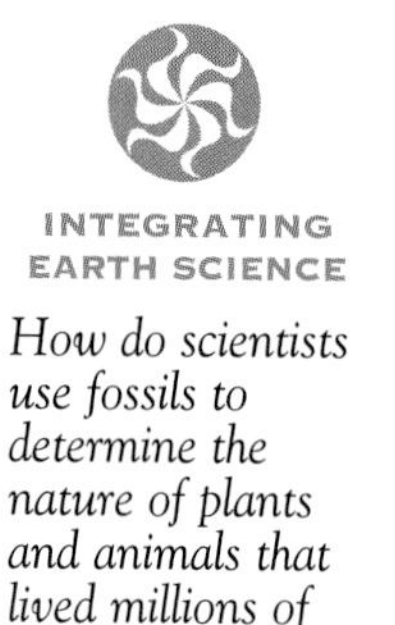

INTEGRATING EARTH SCIENCE

How do scientists use fossils to determine the nature of plants and animals that lived millions of years ago?

From Gymnosperms to Angiosperms

How are **angiosperms,** members of the Division Anthophyta different from gymnosperms? **Unlike the exposed seeds of gymnosperms, angiosperms produce seeds encased in a protective tissue of the sporophyte known as the ovary.** The combination of seed and **ovary** is known as a **fruit.** The angiosperms have one other distinguishing feature—a specialized reproductive structure known as a **flower.** Hence, the name "flowering plants."

Angiosperms are so common on Earth that you might think they have been around since land plants first appeared. That is not the case, however. They are actually the youngest of all the major groups of plants. The oldest fossils of flowering plants appeared in the

Figure 25–5
Flowering plants such as (a) the blue passion in the rain forest of Central America and (b) the magnificent Gul-Mohur tree in India owe their success to a mutually beneficial relationship they have evolved with insects and other animals for pollination and seed dispersal. An example of such a relationship is (c) the goldenrod plant and the ambush bug sitting on it.

fossil record about 130 million years ago, a time known as the Cretaceous Period. ● The fossil record indicates that these new flowering plants slowly evolved as they adapted to one environment after another. Then, about 65 million years ago, the angiosperms assumed the dominant position they hold today, making up more than 90 percent of land plants around the world.

What was the secret that enabled flowering plants to succeed? There is little doubt that it was their method of reproduction. Because of their independence from water, gymnosperms were able to displace seedless plants. But gymnosperms reproduce very slowly and must rely on air currents to spread their pollen and seeds. As you will soon discover, angiosperms developed a means of reproduction that is faster and leaves much less to chance.

☑ *Checkpoint* What are angiosperms?

The Flower

What structure made the angiosperms' means of reproduction faster and more predictable? It was the flower, of course! Flowers are the reproductive organs of angiosperms. Flowers come in many sizes and types. A typical flower produces both male and female gametophytes. However, there are also many exceptions to this rule. Corn, for example, has separate male and female flowers on the same plant. Some plants, such as willows, even produce male and female flowers on separate plants.

As you read about a typical flower, keep in mind that many of the plants you see around you will not be typical. **Flowers are formed from four types of specialized leaves—sepals, petals, stamens, and carpels.**

MINI LAB .... Experimenting ....

Flower Power

PROBLEM *How are flower structure and function related?* ***Design an experiment*** *to answer the question.*

SUGGESTED PROCEDURE  

1. Using a flower, forceps, razor blade, and hand lens, design an experiment to locate and examine the parts of a flower.
2. Obtain your teacher's approval of the procedure and carry out the experiment.

ANALYZE AND CONCLUDE

1. What types of specialized leaves did the flower have? How are these structures related to what they do?
2. Are the anthers located above or below the stigma? Why do you think this is so?
3. Would this flower most likely be self- or cross-pollinated? Explain your answer.
4. Based on the number of each type of specialized part, is the flower a monocot or a dicot?

Figure 25–6
A corn plant produces separate male and female flowers on the same plant. The tassel is the part of the flower that produces male gametophytes, while the silk is the part of the flower that contains the female gametophyte.

Sepals and Petals

The **sepals** in many flowers are truly leaflike, retaining their green color during the life of the flower. They enclose and protect the developing flower bud and open as the flower blooms.

The **petals** of most flowers are brightly colored, and they sometimes serve to attract insects and other animals to the flower. In a sense, the petals are a flower's "advertisement" to the animal world that it is ready for the business of reproduction. Because they do not produce reproductive cells themselves, the sepals and petals are sometimes called sterile leaves.

Stamens and Carpels

The remaining specialized leaves of flowers are fertile leaves located inside the petals. The male leaves, which produce pollen, are known as **stamens.** Stamens usually are thin **filaments** that emerge from the flower and contain sacs known as **anthers,** where pollen is developed and released. Most flowers have several stamens.

The female leaves of the flower are known as **carpels.** A single flower may have one or many carpels. Each carpel has a broad base containing an ovary. The ovary contains one or more **ovules,** where female gametophytes are produced. The diameter of the carpel narrows into a stalk called the **style,** which ends in a sticky tip known as the **stigma.** This is where pollen grains frequently land. Some flowers have several carpels fused together to form a single reproductive structure called a **pistil.**

Reproduction in Angiosperms

Reproduction in angiosperms also follows the characteristic pattern of alternation of generations of sporophytes and gametophytes. The production of gametophytes, fertilization of eggs, and development of seeds take place within the structures of flowers. Refer to the illustration showing the life cycle of an angiosperm on the next page.

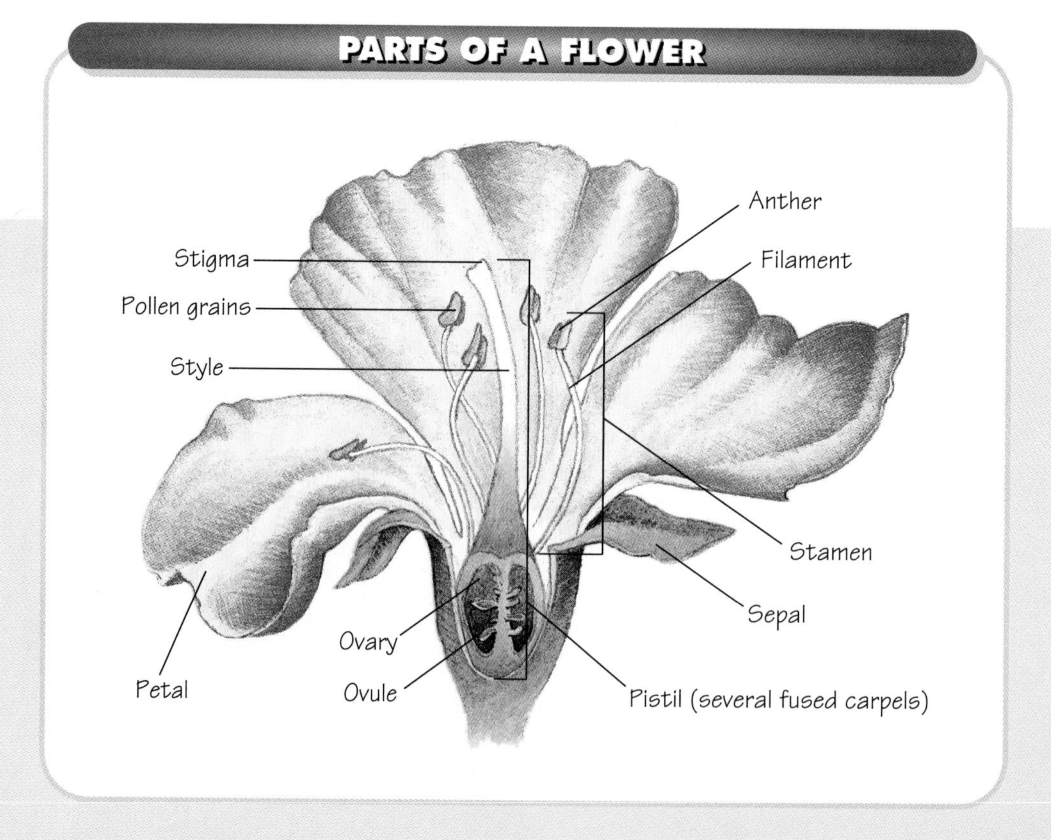

Figure 25–7
This flower is a complete flower, which means that it contains all four floral parts—sepals, petals, stamens, and carpels.

Visualizing the Life Cycle of an Angiosperm

A mature flowering plant is a diploid sporophyte that produces haploid gametophytes inside the flowers. Fertilization leads to the formation of a seed that germinates and grows into a sporophyte.

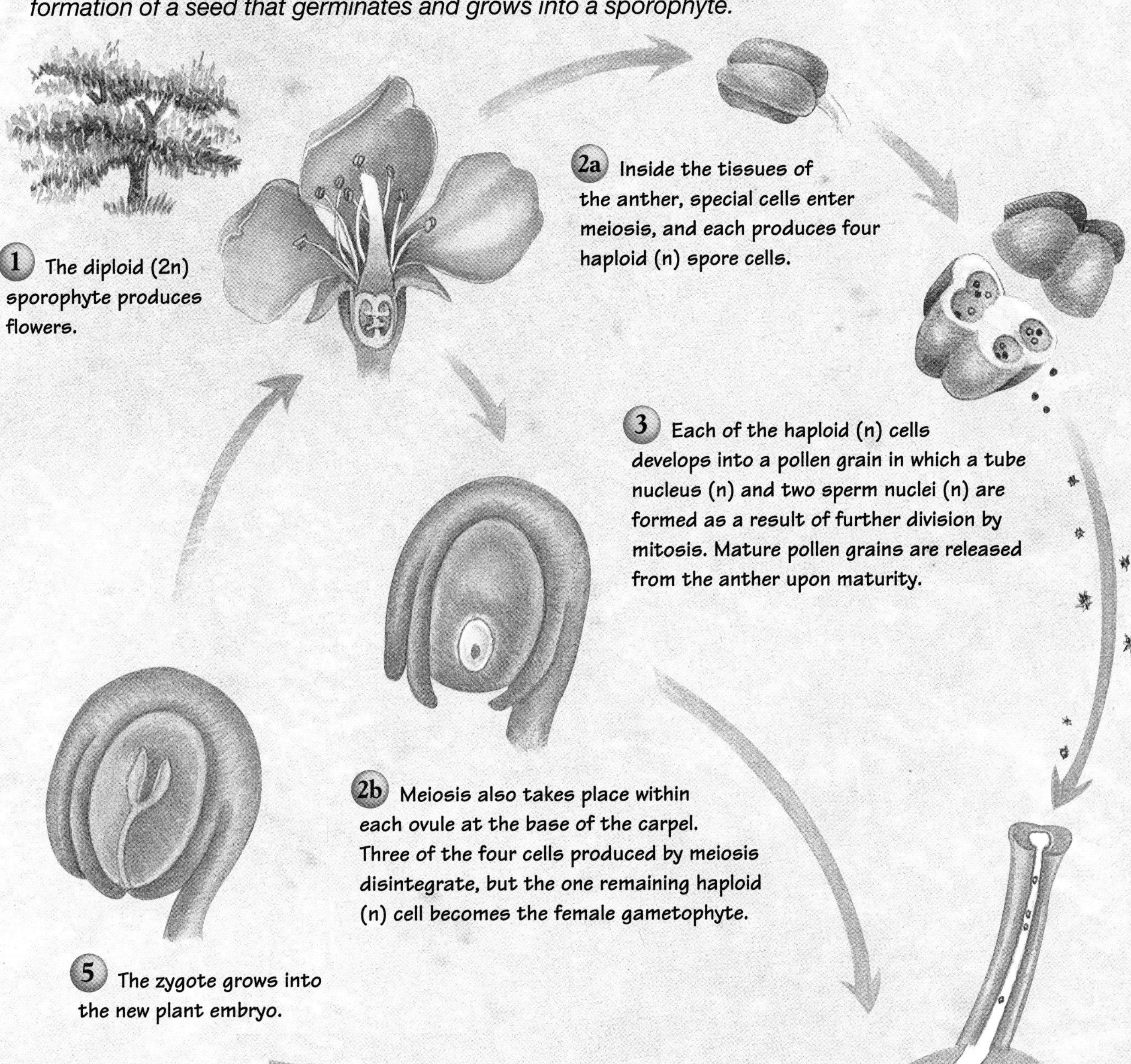

Figure 25–8
(a) Annatto, (b) peppers, and (c) thornless blackberries are examples of fruits—the mature, ripened ovaries of angiosperms within which angiosperm seeds are formed and nourished until they are dispersed.

An important feature distinguishes the life cycle of an angiosperm from the life cycle of a gymnosperm. In angiosperms, a second fertilization event takes place, in which another sperm nucleus fuses with two of the haploid nuclei in the embryo sac to produce a triploid (3n) cell. This cell grows and divides quickly to form a food-rich tissue known as an **endosperm,** which surrounds the embryo. The endosperm serves to nourish the growing plant after the seed sprouts, until the new plant is self-sufficient. Because two fertilization events take place inside each embryo sac, the whole process is called **double fertilization.**

Double fertilization may be a prime reason why angiosperms have been so successful. As you may remember, gymnosperms also build up a reserve of stored food for the seed. However, in gymnosperms this reserve is produced by the gametophyte before fertilization takes place. As a result, if an ovule is not fertilized, all those resources are wasted. In angiosperms, the resources are never wasted. If an ovule fails to be fertilized, the endosperm does not form either, and food is not prepared for a zygote that never forms.

☑ **Checkpoint** What is an endosperm?

Seed Formation

As you have just seen, fertilization causes the formation of a diploid (2n) plant embryo and a layer of triploid (3n) endosperm tissue. Fertilization sets in motion a number of other events as well. The wall of the ovule toughens and forms a seed coat that protects the developing seed. In many plants, the petals and stamens fall away, leaving the carpel, which contains the developing seeds.

The Fruit

As nutrients pour into the flower, many of them are taken up by the growing endosperm tissue inside the seed. However, in most plants, nutrients also flow into the wall of the ovary, which surrounds the seeds. Gradually, the wall of the ovary thickens and joins with other parts of the flower stem to form a fruit. This is the characteristic for which angiosperms are named—the way in which they form seeds enclosed inside the walls of the ovary.

Fruits and Vegetables

It's important to realize that a fruit in plant reproduction simply refers to an ovary and the seeds it contains. By this

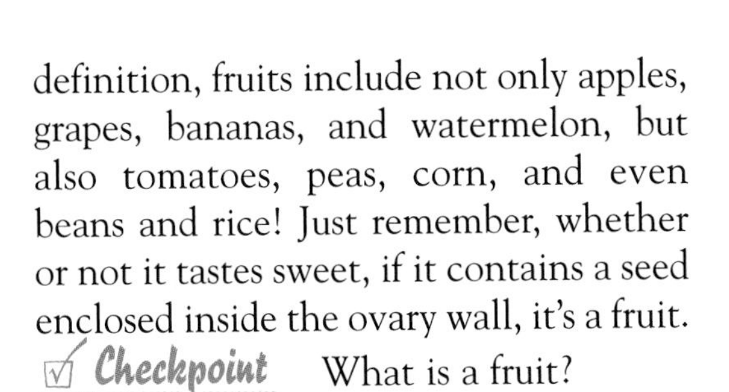

Figure 25–9
One way in which monocots and dicots differ is in the arrangement of vascular tissue. (a) *A cross section of a corn stem, a monocot, reveals a scattered distribution of vascular bundles (magnification: 7X).* (b) *In the stem of a sunflower, a dicot, vascular bundles are arranged in a circle (magnification: 8X).*

definition, fruits include not only apples, grapes, bananas, and watermelon, but also tomatoes, peas, corn, and even beans and rice! Just remember, whether or not it tastes sweet, if it contains a seed enclosed inside the ovary wall, it's a fruit.

☑ **Checkpoint** What is a fruit?

Speed of Reproduction

Another reason for the success of angiosperms is the speed of their reproductive cycle. The small number of cells in the gametophyte stages of angiosperm reproduction means that most flowering plants can reproduce much more quickly than most gymnosperms can.

Monocots and Dicots

The diverse plants that make up the angiosperms are further classified into two groups based on the number of **cotyledons** (kaht-uh-LEED-'nz)—the large seed leaves that contain stored food to nourish the plant embryos of seeds. Some plants have one cotyledon and are known as monocotyledons, or **monocots** for short. Monocots constitute the Class Monocotyledonae, which includes grasses, irises, and cattails. Other plants have two cotyledons and are known as dicotyledons, or **dicots.** The dicots constitute the Class Dicotyledonae, which includes roses, clover, tomatoes, oaks, and daisies. The flowering trees—such as maple, oak, elm, apple, and dogwood—are all dicots.

Although the primary distinction between monocots and dicots is the number of seed leaves, it is not the only one. Veins in monocot leaves usually lie parallel to each other, whereas veins in dicot leaves usually form a branching network. While monocot flowers usually have floral parts in multiples of three, dicot flowers usually show multiples of four or five.

Section Review 25–3

1. **Describe** the main characteristics of angiosperms.
2. **Identify** the parts of a flower.
3. **Outline** the stages in the life cycle of an angiosperm.
4. **Critical Thinking—Relating** What adaptations have led to the dominance of angiosperms on Earth?
5. **MINI LAB** How would you **design an experiment** to determine the relationship between flower structures and their function?

What's in a Fruit?

Guide for Reading

- **Explain** the role that fruits play in the survival strategy of plants.

WHAT'S IN A FRUIT? HAVING just read the previous section, the answer might seem simple. Fruits are mature ovaries that contain seeds. And within each seed is an embryo along with enough stored food to support that embryo's early growth. The fruits that a plant produces are home to the plant's next generation, its chances to be part of the future. This means that a fruit is more than just a reproductive structure—it is a key to the survival strategy of a whole species.

Perhaps you have marveled at the amazing variety of fruits found in nature. This section offers an opportunity for you to make the connection between the different types of fruit and the dispersal strategy they support.

Types of Fruits

As you have read, the double fertilization that takes place in flowering plants produces two distinct cells. One is the diploid zygote that forms the embryo. The other is the triploid endosperm. The embryo and endosperm are surrounded by a protective seed coat.

In flowering plants, all of this occurs inside the ovary, which contained the female gametophyte before fertilization took place. When a seed is formed, the ovary doesn't just disappear. Not only are the seeds of flowering plants enclosed within the ovary, but in many plants the ovary wall undergoes dramatic changes as the seeds develop. For many plants, these changes in the ovary are the key to survival in a difficult world.

The simplest fruits are those that consist of a single seed enclosed by a single ovary wall. Grains, such as wheat and corn, fit this description. In most grains, the wall of the ovary is so thin that it actually fuses to the seed coat. Each kernel of corn, therefore, is not just a seed but an ovary as well.

Figure 25–10
Fruits of angiosperms display an amazing variety of sizes, shapes, and forms. (a) Eat a grape, and you may crunch on a few seeds—that is, unless you happen to be eating a "seedless" grape. (b) Open up a fruit from the tropical papaya and observe (c) hundreds of tiny black seeds.

In acorns and chestnuts, the ovary wall hardens and forms a protective shell around the seed. Fruits such as these are called **nuts.** Instead of a tough casing, the ovary walls of peaches and cherries are soft and fleshy. The flesh of these fruits—which are known as **drupes**—encloses a single tough, stony seed.

There's no rule that says an ovary must contain just one seed. Many plants produce dozens or even hundreds of seeds in a single ovary. This is the case with **berries,** in which the soft ovary wall usually encloses many seeds. Examples of berries include the grape and, believe it or not, the tomato.

Legumes produce seeds within a pod that splits open on two sides. Peas and beans are legumes. Do you remember the last time you ate peanuts from the shell? Peanut seeds too are produced in a pod that splits open along a two-sided seam. And that means that peanuts are not nuts at all, but legumes!

Some of the fruits we prize most highly are more complex than they seem. Apples, for example, do contain seeds surrounded by an ovary wall. But most of the fruit is actually formed by the stem surrounding the ovary. If you slice an apple or pear open crosswise, the boundary between the ovary and the surrounding tissue will be apparent.

Checkpoint What are some of the different types of fruits found in nature?

Fruits and Seed Dispersal

Why should fruits come in such a bewildering variety of shapes, sizes, and styles? Why should some be sweet and edible and others be little more than a protective sac around the seed? We can find the answers by watching what happens to the fruit after it is produced by the plant.

Some of the simplest fruits, such as those of grasses and grains, are formed in species where seeds simply fall to the ground at the base of the parent plant. Most of these species are **annuals,** meaning that they live for just one growing season. Therefore, if seeds are dropped to the ground, they have a good chance of finding good growing conditions next year, when the generation that produced them has died.

This is not the case with **perennials**—the plants that live for many growing seasons. These plants need to disperse their seeds to give the new seedlings a chance to grow far enough away from their parent plant so they do not compete for sunlight, water, and nutrients. This is where fruits come in. **Because plants cannot move, they have to depend on other forces to scatter their seeds about. Fruits have evolved many ways to ensure that the seeds within them get the best possible chance to sprout under suitable conditions.**

Figure 25–11
(a) *A pineapple, which is formed as a result of the fusing of the ovaries of a cluster of flowers, is called a multiple fruit.*
(b) *A tomato is classified as a berry because it contains an ovary that encloses many seeds.* (c) CAREER TRACK *This agricultural technician monitors wheat, grown with salt water irrigation in Ashalim, Negev Desert, Israel. Wheat is a simple fruit that has a single seed enclosed by a single ovary wall.*

Problem Solving

INTERPRETING GRAPHS

Plants and Rodents: Friends or Foes?

While looking for ideas for a science fair project, you come across a study of the relationship between the Japanese wood mouse and the Asian skunk cabbage. According to this study, the seeds of the cabbage cannot be dispersed by water or wind due to the terrestrial location of the cabbage and the mass of its seeds. The researchers hypothesized that seed-dispersing animals bring the seeds to new locations, away from already formed colonies of skunk cabbage.

The researchers found that Japanese wood mice eat the skunk cabbage seeds and store them for the winter. The eaten seeds that are not well chewed give rise to seedlings after leaving the animal's digestive tract. Other seeds are dropped during the eating process. Many of the seeds stored for the winter never are recovered and survive to produce seedlings also.

The graphs illustrate the distance the mice dispersed the seeds and the quantity of seeds they dispersed.

DISTANCE OF SEED DISPERSAL BY MICE

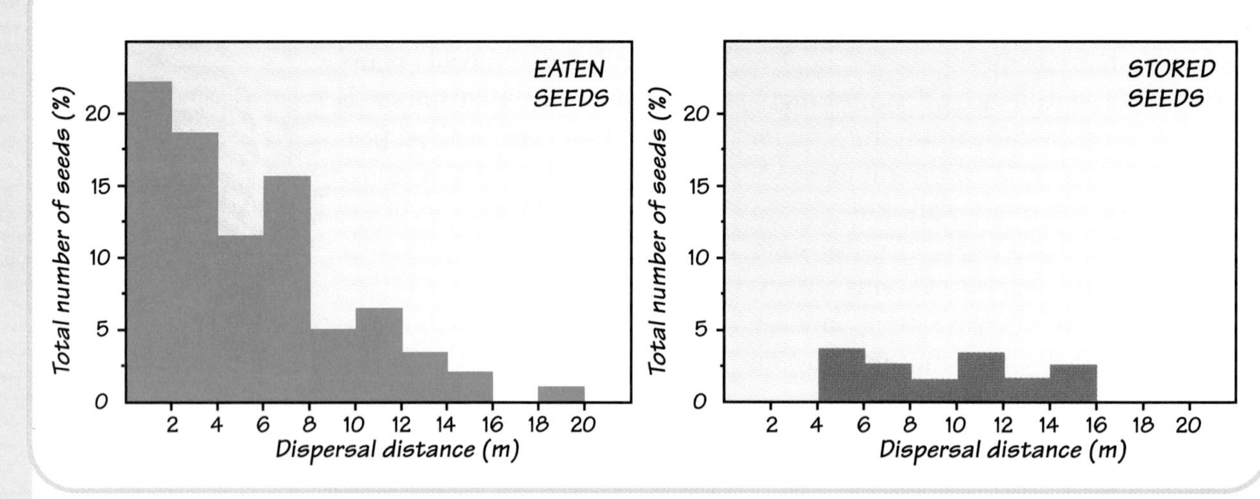

THINK ABOUT IT

1. A total of 80 seeds were dispersed by the mice, 68 of which were eaten and 12 of which were stored. What number of seeds does each bar on each graph represent?
2. **Calculate** the average distance that eaten seeds and stored seeds were dispersed.
3. On the average, do the mice transport the seeds a longer distance for storage or for immediate consumption?
4. **Explain** how seed dispersal is important to the survival of the Asian wood cabbage. Do you think Japanese wood mice are effective dispersers of Asian skunk cabbage seeds?
5. If you performed similar research for your science fair project, what combinations of animals and plants might you study?

Dispersal by Wind

Many seeds are released directly into the air, depending on the wind to help distribute them. Maple and ash trees have winged fruits that carry their seeds many meters from the parent plant. The fruit of the dandelion has a parasol of tiny filaments, allowing it to drift in the wind over great distances.

Figure 25–12

Fruits and seeds often include special features that help in seed dispersal. Milkweed fruits burst open when mature, releasing seeds that are dispersed by wind—aided substantially by attachments of tufts of silk.

A few plants produce fruits that provide their seeds with jet propulsion to help them find a spot in which to grow. The dwarf mistletoe, a parasitic plant that grows on the branches of pine trees, produces sticky seeds enclosed within a fluid-filled chamber. As the fruit matures, the fluid pressure builds until it's so strong it blows away the end of the fruit, pushing out the seed at speeds as great as 100 kilometers an hour! This live artillery enables the plant to spread quickly.

Dispersal by Water

Plants that live on or near the water often rely on this means of seed transportation. Some of their fruits contain air pockets to keep the seeds afloat until they have germinated. The most spectacular example of a waterborne fruit is the coconut palm. The fruits of this plant are packed with corklike tissue and air spaces, enabling it to float for long periods. Coconuts sprout quickly when washed ashore. They are one of the most successful seaside plants in the world.

Dispersal by Animals

Many flowering plants use animals to distribute their seeds. Some fruits have "bribes" that entice the animals to help disperse them—the edible fleshy parts of fruits like apples, grapes, and blackberries. When an animal eats the fruit, it enjoys the tasty ovary wall. But the tough seeds inside the fruit may pass through the animal's digestive system unharmed. As the animal walks or flies or grazes, the seeds may be carried great distances. Finally, they leave the digestive system and are deposited in a new location, where they may sprout and grow.

As animals ourselves, we may think that plants are at our mercy. But when an animal pulls off a branch, chomps on a flower, or gobbles up a fruit, the plant is also using the animal—to find a home for its offspring.

Section Review 25–4

1. **Explain** the role that fruits play in the survival strategy of plants.
2. **BRANCHING OUT ACTIVITY** Visit a farmers' market or the produce section of your local supermarket. **Identify** the produce items that can be classified as fruits and **formulate a hypothesis** to explain how they may be dispersed.

Laboratory Investigation

Structure of a Pea Pod

After pollination occurs, the ovary of a fruit matures into a fruit containing seeds. A pea pod meets this description, even though it usually is not thought of as a fruit. Peas and their pods usually are consumed before they mature and dry, so people rarely see their natural development. Like all fruits, the pea pod develops from a flower, and parts of the flower can still be identified after pollination. In this investigation, you will observe the features of the pea pod that are flower remnants and study how seeds form.

Problem

What are the parts of a fruit that are important to its reproduction? **Observe** a pea pod to help you answer this question.

Materials (per group)

pea pod (or string bean or lima bean pod)
soaked pea
hand lens
scalpel or razor blade
dissecting needle

Procedure

1. **With a hand lens, examine the external appearance of the pea pod. Record your observations.**
2. **Find the stalk that attaches the pod to the plant. Locate the sepals, which are the remaining parts from the base of the flower. Record the number of sepals.**

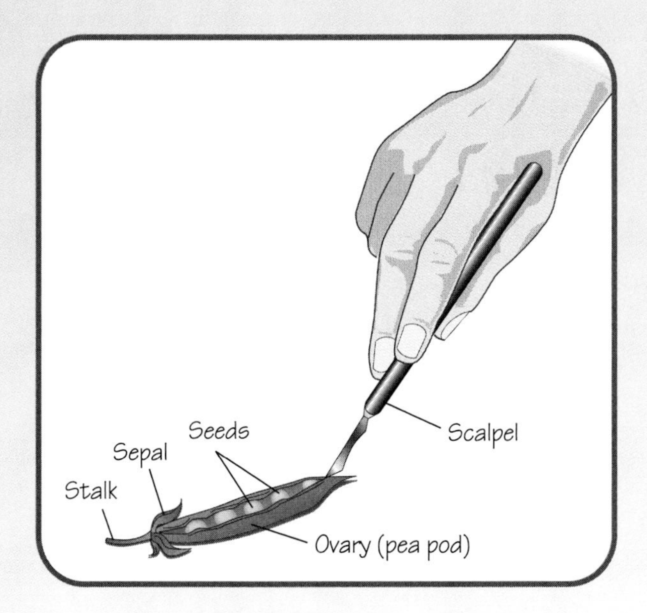

3. At the opposite end of the pod, find the remains of the style.

4. Carefully open the pod along the curved edge using the razor blade or scalpel. CAUTION: *Be very careful when using sharp instruments.*

5. Count the number of peas in the pod and note their characteristics. Record the number of peas as well as your observations of their appearance.

6. Notice the fibers to which the peas are attached by a short stalk. Record the number of fibers you see.

7. Obtain a soaked pea and examine it. Locate the scar that shows where the pea was attached to the pod.

8. Using the dissecting needle, carefully remove the seed coat from the pea.

9. Separate the cotyledons. Use a hand lens to observe the embryo plant.

Observations

1. Describe the external appearance of the pea pod.
2. How many sepals does the pea pod have?
3. Are all the pea seeds attached to the same side of the pod?
4. Are all the peas alike?
5. Draw a sketch of the embryo plant of the soaked pea and label its parts.

Analysis and Conclusions

1. Based on your observations of the number of sepals and cotyledons present, classify the pea as a monocot or a dicot.
2. How do you explain any differences among the peas?
3. What is the function of the stalk that attaches the pea to the pod?
4. What reproductive structure does the pod represent?
5. Based on the color of the pea pod, can you identify one of the processes it carries out?

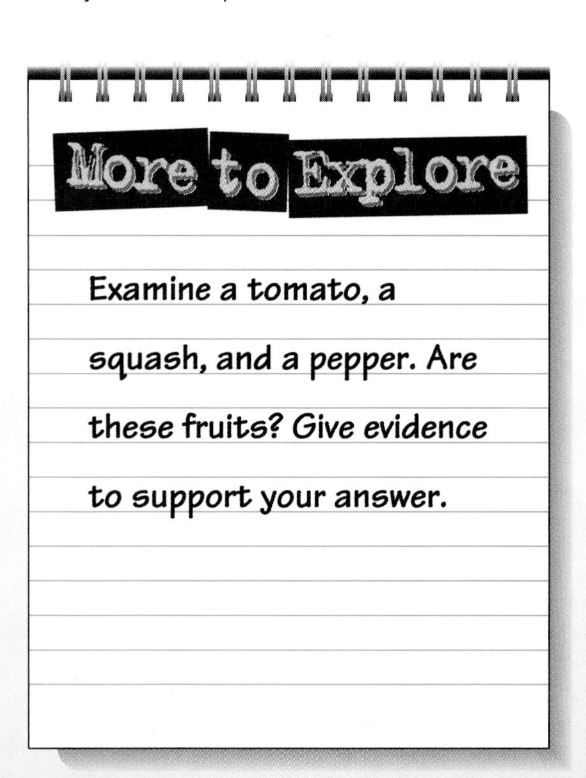

Study Guide

Summarizing Key Concepts

The key concepts in each section of this chapter are listed below to help you review the chapter content. Make sure you understand each concept and its relationship to other concepts and to the theme of this chapter.

25–1 Seed-Bearing Plants

- A seed is a reproductive package that contains a plant embryo and a supply of stored food inside a protective coating.
- Seed-bearing plants do not require standing water for reproduction.

25–2 Gymnosperms

- There are five divisions of living seed-bearing plants. Four are known as the gymnosperms—cycads, ginkgoes, gnetophytes, and conifers.
- The life cycle of a gymnosperm displays alternation of generations of sporophytes and gametophytes within the sporophytes.

25–3 Angiosperms

- Angiosperms produce seeds encased in protective tissue of the sporophyte, the ovary.
- Flowers are the reproductive organs of angiosperms, formed from four types of leaves—sepals, petals, stamens, and carpels.
- Reproduction in angiosperms also follows alternation of generations of sporophytes and gametophytes. Production of gametophytes, fertilization of eggs, and development of seeds take place within the structures of flowers.
- A fruit is a combination of seed and ovary.

25–4 What's in a Fruit?

- Fruits vary as a result of changes that take place in the ovary wall as the seeds develop inside.
- Variation in fruit type is tied to the plant's strategy for producing offspring.

Reviewing Key Terms

Review the following vocabulary terms and their meaning. Then use each term in a complete sentence.

25–1 Seed-Bearing Plants

seed
egg
pollen

25–2 Gymnosperms

scale
female cone
gymnosperm
seed cone
male cone

25–3 Angiosperms

angiosperm
flower
ovary
sepal
fruit
petal
stamen
filament
anther
carpel
ovule
style
stigma
pistil
endosperm
double fertilization
cotyledon
monocot
dicot

25–4 What's in a Fruit?

nut
legume
drupe
annual
berry
perennial

CHAPTER 25

Recalling Main Ideas

Choose the letter of the answer that best completes the statement or answers the question.

1. The sporophyte of seed-bearing plants contains the
   a. gametophyte. **c.** gymnosperm.
   b. angiosperm. **d.** antheridia.

2. Plants that bear their seeds on the surface of reproductive structures are called
   a. gymnosperms. **c.** angiosperms.
   b. bryophytes. **d.** lichens.

3. Scales are grouped into clusters known as
   a. crowns. **c.** bunches.
   b. cones. **d.** plates.

4. Spruce, pine, cedar, redwood, and cypress trees are all examples of
   a. bryophytes. **c.** monocots.
   b. conifers. **d.** flowering plants.

5. The specialized reproductive structure that distinguishes angiosperms from other seed plants is the
   a. egg. **c.** root.
   b. pistil. **d.** flower.

6. The four types of specialized leaves from which flowers are formed are sepals, petals,
   a. stamens, and carpels.
   b. tassels, and anthers.
   c. scales, and tissues.
   d. pollen grains, and ovules.

7. The food-rich tissue that surrounds the embryo is known as the
   a. nutrient cell. **c.** placenta.
   b. endosperm. **d.** female tissue.

8. In plant reproduction, the ovary and the seeds it contains are called the
   a. endosperm. **c.** fruit.
   b. spore. **d.** cone.

9. A fleshy fruit such as an apple is well-adapted for
   a. wind dispersal.
   b. animal dispersal.
   c. water dispersal.
   d. pollination.

Putting It All Together

Using the information on pages xxx to xxxi, complete the following concept map.

SEED PLANTS

consist of the

four living divisions of

and the division of

bear their seeds in structures called

1

bear their seeds in structures called

Assessment

Reviewing What You Learned

Answer each of the following in a complete sentence.

1. What is the importance of a seed?
2. How do spores differ from seeds?
3. What are the two groups of seed-bearing plants called?
4. What are gymnosperms? Give examples of these plants.
5. **Observe** a flower and list the structures you see.
6. What are the male and female parts of a flower?
7. What is double fertilization?
8. Compare gymnosperms and angiosperms.
9. Compare features of monocots and dicots.
10. What is a nut? A drupe?
11. Give examples of the dispersal of seeds by wind, water, and animals.

Expanding the Concepts

Discuss each of the following in a brief paragraph.

1. Why are the cone-bearing plants—the conifers—so unique and successful?
2. Are peppers classified as fruits? Give evidence to support your answer.
3. Are seed-bearing plants basically sporophytes or gametophytes? Explain.
4. What changes in Earth's geography and climate paralleled the development of seed plants?
5. Compare the male and female cones of a typical gymnosperm.
6. Explain how pollination and fertilization occur in angiosperms.
7. Describe the flower in terms of its structure and function.
8. How does pollen deposited on the stigma reach the egg?
9. What function is served by the fruit in terms of survival of the plant species?
10. Describe a method of seed dispersal involving a mutually beneficial relationship between organisms.
11. Explain why monocots and dicots are placed in two different groups.

Extending Your Thinking

Use the skills you have developed in this chapter to answer the following.

1. **Sequencing** Describe the events in the life cycles of an angiosperm and a gymnosperm.
2. **Relating** Lumber for home construction and industry essentially comes from gymnosperm forests. Because these forests are considered to be renewable resources, extensive programs have been developed for reforestation. These plans include forest and wildlife management. Why is it essential to include wildlife management in maintaining a healthy environment for trees?
3. **Drawing conclusions** Most conifers in temperate regions on the Earth keep their leaves through the winter, whereas angiosperm trees shed their leaves each autumn. What are the advantages and disadvantages of an evergreen tree? Of a tree shedding and regrowing its leaves?
4. **Designing an experiment** Are fruit-eating animals attracted to the color of a fruit or to its fragrance? Formulate a hypothesis and then design an experiment to find out the answer. Be sure to include a control.
5. **Evaluating** Seed plants have moved into every habitat since their origin. It has been estimated that of the 3 million species of plants, only 20 or 30 major species have been cultivated for food. Considering that only a few species are so important, why do we frequently hear or read about the need to maintain genetic diversity of plants?

Applying Your Skills

Can Seeds Fly?

The survival of plants depends on their ability to reproduce. In seed-bearing plants, the ability to reproduce is enhanced by mechanisms that protect and disperse the seeds and fruits. This allows less competition with the parent plants for nutrients, light, and water. How have some plants adapted to different methods of seed dispersal?

1. Working in a group, choose three seeds from the variety of seeds provided by your teacher.
2. Observe each type of seed. Draw and list the characteristics of each seed.
3. Place one seed on a flat surface and gently blow on the seed. What happens? Repeat this procedure with each of the other seeds.
4. What parts of the seeds are important for their dispersal?

• GOING FURTHER •

5. Construct a classification key using the seeds observed in this activity.

CHAPTER 26

Plant Structure, Function, and Growth

FOCUSING THE CHAPTER THEME: *Systems and Interactions*

26–1 Plant Structure and Function

- **Identify** the main parts of a plant and **explain** their functions.
- **Explain** how water uptake and fluid transport take place in plants.
- **Discuss** some specializations of plant structures that help plants adapt to their environment.

26–2 Plant Growth

- **Describe** plant tropisms.
- **Examine** the role of hormones in plant growth.

BRANCHING OUT *In Depth*

26–3 Plant Propagation

- **Describe** vegetative reproduction in nature and in agriculture.

LABORATORY INVESTIGATION

- **Observe** the water-transport tissues in certain types of vegetables.

Biology and Your World

BIO JOURNAL

Have you ever wondered why the leaves of plants display an amazing variety of sizes and shapes? In your journal, make a note of the leaf types found in and around your home and school. Then formulate hypotheses to explain the connection between shape and function.

A peach orchard in blossom

Plant Structure and Function

SECTION 26–1

GUIDE FOR READING

- **Explain** the functions of roots, stems, and leaves in plants.
- **Describe** the process of fluid transport in xylem and phloem tissues in plants.

MINI LAB

- **Design an experiment** to determine how the number of leaves affects water uptake in a plant.

THE GREAT SAND DUNES OF Cape Cod National Seashore in Massachusetts are battered by storms and laced with saltwater sprays. In the winter, they endure bitter cold and in the summer, blinding heat. Yet, every year, plants poke through the sand to claim a portion of this difficult landscape. Thousands of kilometers to the west, a bristlecone pine ekes out a living on the stony mountainsides of the Sierra Nevada. Neither the lack of soil nor the strong winds of the high country keep these rugged plants from growing.

How do plants manage to live just about anywhere? The best place to start to answer this question is with the plants themselves—their structure and the ways in which they are adapted to their specific environments.

The Structure of a Plant

Plants provide nearly all the food that makes life on land possible. They release the oxygen that animals breathe and even fashion the places in which animals live. They trap the energy of sunlight and convert it into forms that other living things can use. In this section, you will see how the different tissues within

Figure 26–1
Plants manage to thrive in some of the most difficult places imaginable. (a) Joshua trees and (b) birdcage evening primroses growing in the desert soil of southern California and (c) evergreen trees rising up from the nearly vertical rock cliffs of the Oregon coast are just a few examples of these plants.

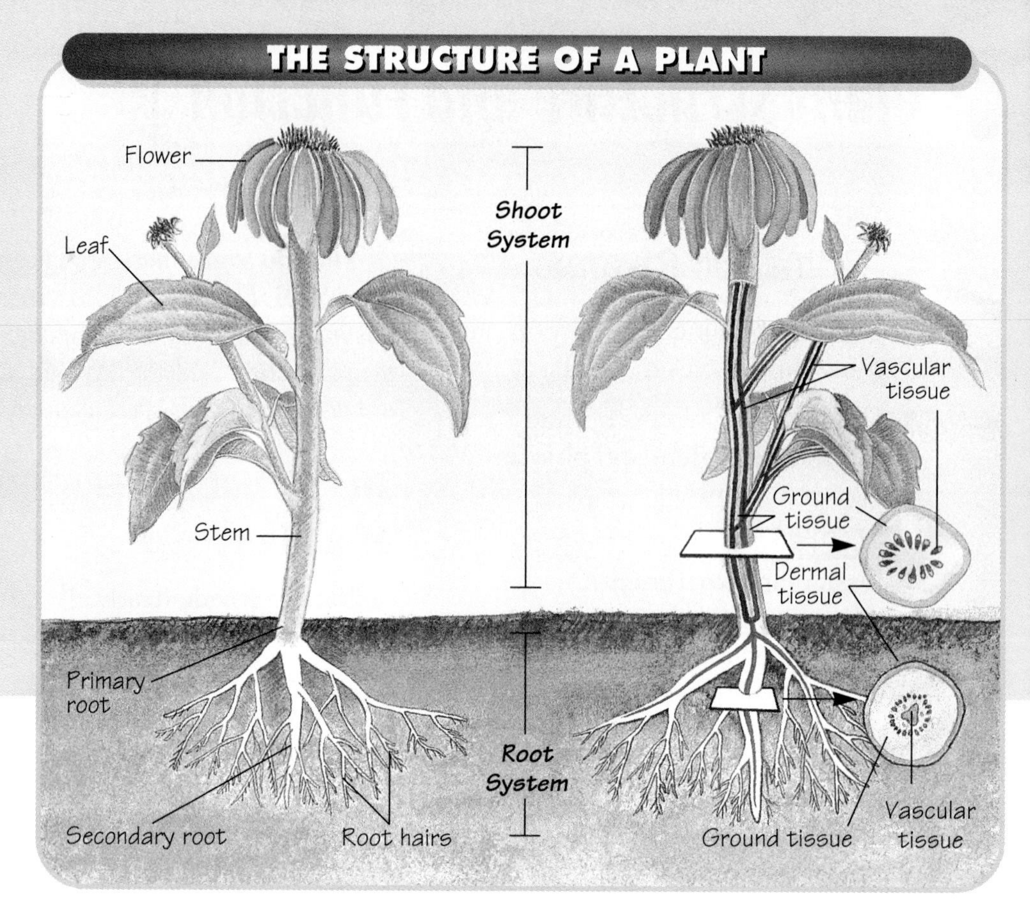

Figure 26–2
A flowering plant consists of a root system below the soil and a shoot system above the soil. Stems, leaves, flowers, and fruits comprise the shoot system, while primary and secondary roots along with root hairs make up the root system. The root system and the shoot system of a flowering plant are made up of three types of tissues: dermal tissue, vascular tissue, and ground tissue.

a plant work together to make these organisms so successful.

Roots, Stems, and Leaves

The body of a plant consists of three distinct regions known as roots, stems, and leaves. What are their primary functions? **Roots** anchor a plant in the ground, drawing water and minerals from the soil. **Stems** typically rise above the ground, support the body of the plant, and carry water and nutrients from one end of the plant to the other. **Leaves** are the main organs of photosynthesis—the process by which plants convert energy from sunlight into chemical energy—so they are usually arranged to capture as much sunlight as possible.

Plant Tissue

Plants generally contain three kinds of tissue. Dermal tissue is the outer covering of the plant. Vascular tissue makes up the fluid-conducting system of the organism. And the rest of the plant is ground tissue, which provides most of a plant's supporting strength and also contains most of the cells that are active in photosynthesis. If you were to compare a plant to a typical mammal, you could say that the dermal tissue is its skin, the vascular tissue is its bloodstream, and ground tissue is everything in between.

☑ **Checkpoint** What are the three types of plant tissue?

Plant Cells

The three types of plant tissue contain many cell types specialized to perform certain functions. Dermal tissue, as you might expect, has to protect the plant from its environment while allowing gases such as oxygen and carbon dioxide to flow between the plant and the atmosphere.

In many plants, the ground tissue consists mainly of parenchyma (puh-REHN-kih-muh) cells. These thin-walled cells usually form the bulk of tissue in roots, stems, and leaves. Parenchyma cells in leaves are very active in photosynthesis. The plant roots that we use as

Figure 26–3
(a) A dandelion's root system shows the thicker primary root and the thinner secondary roots that branch out from the primary roots. (b) A young radish seedling's primary root has put out hundreds of tiny root hairs to help it absorb water and other nutrients from its environment.

food, such as carrots and radishes, are mostly parenchyma cells. The ground tissue may also contain collenchyma (kuh-LEHN-kih-muh) cells and sclerenchyma (sklih-REHN-kuh-muh) cells. The thick walls of these cells provide strong support for the rest of the plant.

The vascular tissues carry fluids from one part of the plant to another. The principal vascular cell types are **xylem** (ZIGH-luhm), which carries water, and **phloem** (FLOH-ehm), which carries sugars and other foods throughout the plant.

Roots 

As a plant grows, it sends roots into the soil to collect nutrients and water, and also to help provide support for the portion of the plant that is above ground. A growing seedling first sends a single primary root into the soil. ***Figure 26–3*** shows secondary roots that branch off the primary root as further growth takes place. As the secondary roots grow, their surface area enlarges dramatically. The surface area of the roots, even in a small plant, may be 50 or 100 times the surface area of its leaves!

Epidermis

The **epidermis** is the outer covering of a root. Epidermal cells grow tiny thin-walled projections, known as root hairs, that make direct contact with the soil. These root hairs, which even in a small plant may number in the billions, are responsible for most of the root's surface area. Plants absorb nearly all the water and nutrients directly through root hairs.

☑ **Checkpoint** What are root hairs?

Cortex

Just beneath the epidermis is a layer of spongy cells known as the **cortex.** The parenchyma cells of the root cortex are important in moving water from the epidermis to the vascular tissue near the center of the root.

Vascular Cylinder

In most roots, a **vascular cylinder**—a central region of xylem and phloem cells—carries water and nutrients between the roots and the rest of the plant. Water and minerals that have passed through the cortex enter the vascular cylinder, where they are transported upward into the rest of the plant.

How Roots Work

During its growing season, a single corn plant may take as much as 3 liters of water a day from the soil. How does it do

THE STRUCTURE OF A ROOT

Epidermis
Root hair
Endodermis
Cortex
Xylem
Phloem
Vascular cylinder
Epidermis
Cortex
Endodermis
Vascular cylinder
Casparian strip

Figure 26–4
A section of a typical dicot root shows concentric layers of epidermis, cortex, and endodermis surrounding the vascular cylinder in the center. The Casparian strip, or endodermis, separates the cortex from the vascular cylinder. Like police officers directing traffic, these cells ensure that water travels into the vascular cylinder and does not leave it.

INTEGRATING CHEMISTRY

What is the significance of the concentration of a solution? What units do chemists use to measure solution concentration?

this? The answer is osmosis. Osmosis is the movement of water across a membrane. If the concentration of dissolved material is higher on one side of a membrane, then water will flow across the membrane toward the side of higher concentration. ● This means that water will move out of damp soil into root hairs, which contain high concentrations of dissolved salts and sugars.

When there is plenty of water in the soil, osmosis rapidly draws water into the root epidermis. As the epidermal cells are diluted by the incoming water, osmosis causes water to move out of the epidermis into the cells of the cortex and then into the cells of the vascular cylinder.

This process will not work if the concentration of dissolved material in the soil is greater than the concentration inside the root cells. Could this happen? It certainly could. When plants are flooded with salty water, osmosis may cause water to flow out of the roots back into the soil. The rapid loss of water from roots, known as "root burn," may weaken or kill a plant. Applying too much fertilizer to the soil can also cause root burn.

☑ **Checkpoint** What is osmosis?

Active Transport

Besides water, root hairs must take in minerals that plants need to survive. Unlike water, nearly all the minerals are brought across the cell membrane of the root hair by active transport. Protein molecules in the cell membrane use energy in the form of ATP to "pump" mineral ions across the cell membrane.

Once they enter the root, these ions continue to move by active transport. They are pumped from the epidermis to the cortex and then into the vascular cylinder. By using energy to move mineral ions into the vascular cylinder, the plant also draws water right along with the ions. How does this happen? As mineral ions are moved toward the center of the root, their concentration in these cells increases. Water then moves from the outer cortex toward the center of the root, drawn by osmosis.

The Casparian Strip

In most roots, the inner boundary of the cortex is formed by a layer of cells

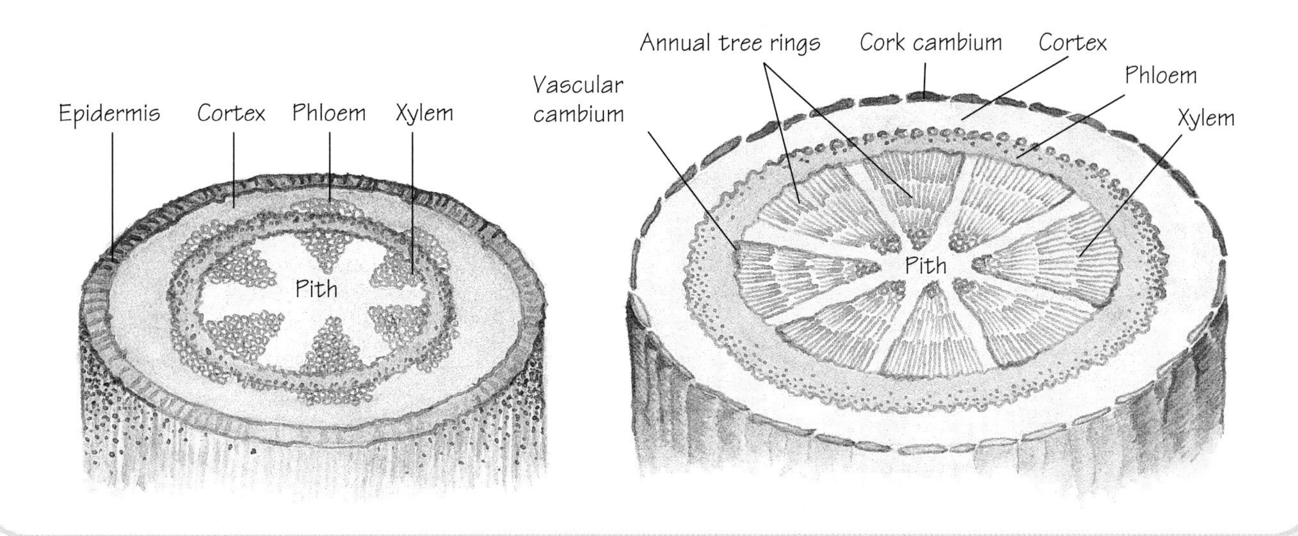

Figure 26–5
This diagram shows the cross sections of a young woody dicot stem (left) and an older woody dicot stem (right). In the older stem, notice the location of the vascular cambium, which, over the years, produces wood and the cork cambium. The cork cambium forms the bark of a tree.

called the **endodermis.** These cells form a tight layer that separates the cortex from the vascular cylinder. In fact, the seal is so tight that each cell in the endodermis is set—almost like a brick in mortar—in a waxy layer known as the Casparian strip. Neither water nor mineral ions can move through this waxy layer. Therefore, the endodermal cells are able to control entry to the vascular cylinder.

☑ **Checkpoint** What is the role of the Casparian strip?

Stems

Stems connect the roots that gather water and nutrients with the leaves that carry out photosynthesis. Stems may be as short as a few centimeters or they may rise tens of meters into the air.

Stems are surrounded by a layer of epidermal cells. Like the roots, they also contain ground tissue and vascular tissue, but these are arranged differently in stems than they are in roots. In monocots, vascular bundles containing xylem and phloem cells are scattered through the ground tissue. In dicots, these vascular bundles are arranged in a ring. The ground tissue inside the ring is known as the **pith.** The ground tissue outside the ring is called the cortex.

Wood

You've probably noticed that the stems of many plants not only get longer as they grow, they also get thicker. Some plants get thicker stems every year by producing a tough material called wood.

Vascular bundles of xylem and phloem surround the pith in the center of the stem. As the stem grows, new cells are produced at the boundary between the xylem and the phloem. These cells push outward, adding more xylem and phloem and increasing the diameter of the stem. This layer of rapidly dividing cells is called the vascular cambium because it produces more and more vascular tissue.

Most of what we call wood is xylem. As more and more xylem tissue is added, the woody stem gets larger and larger. But the phloem tissue, which is outside the vascular cambium, has a problem. As the xylem layer gets larger, the older layers of phloem crack open, leaving gaps between them. To solve this problem,

Figure 26–6
Found in approximately concentric circles, annual tree rings form due to a seasonal variation in the production of xylem. As a result, this tree trunk provides a record of the weather conditions of years past.

woody plants have another layer of cells, called the cork cambium. This layer produces cork, a tough layer with thick cell walls loaded with waxes and oils that form the bark covering the stem.

The Stem of a Tree

As ***Figure 26–5*** on page 605 shows, nearly all the wood of a tree is actually xylem. In an older tree, the innermost xylem layers no longer conduct water and are known as heartwood. The outer layers, which are active in water transport, are called sapwood. In temperate climates, the cambium produces much more xylem in the summer than it does in the fall and winter. This leads to a seasonal variation in the texture of wood, producing **annual tree rings.** Tree rings often provide important historical information. Thick rings indicate that weather conditions were favorable and much new wood was added to the tree. Thin rings may mean that a drought or other problems allowed little growth.

INTEGRATING HISTORY

What historical information has been provided by the study of the growth rings in old trees?

As you have seen, outside the vascular cambium is a thin layer of phloem, which carries nutrients, and then a layer of cork cambium, which produces the bark. This means that trees carry most of their nutrients and do most of their growing in thin layers of cells just under the bark.

These tissues just under the bark are delicate and easily damaged. If a strip of bark is removed from the base of a tree, often these layers will be pulled off with the bark. The loss of cambium and phloem means that the tree will stop growing and will lose the ability to carry nutrients to its roots. Eventually, the roots of the tree may starve and the tree will die.

☑ **Checkpoint** What are annual tree rings?

Leaves

Leaves are the main organs in which plants carry out photosynthesis. One way to understand plants is to think of them as organisms that "eat" sunlight. If you consider sunlight as "food" for plants, then the structure of a typical plant makes perfect sense. Most leaves are thin, broad structures that have a lot of surface area with little mass—most leaves aren't heavy. That means they are efficient collectors of solar energy—the primary job of a leaf.

Leaves come in an incredible variety of shapes and sizes, adapted to the environment in which a plant lives, as illustrated by the photographs on page 609. Like roots and stems, leaves have an outer covering of epidermal cells, fluid-carrying vascular tissue, and ground tissue consisting of parenchyma cells. Leaves are attached to stems by a thin structure called a petiole (PEHT-ee-ohl). The petiole is where a leaf separates from the stem when a plant drops its leaves in the fall.

The Epidermis

Leaf epidermal cells are covered with a waterproof waxy layer called the cuticle. This layer protects the leaf against water loss and insect invasion. But leaves cannot be sealed off completely from the air around them. Leaves need to "breathe" just as you do. The reason is photosynthesis. In sunlight, plants take in enormous amounts of carbon dioxide and give off oxygen. How do these gases

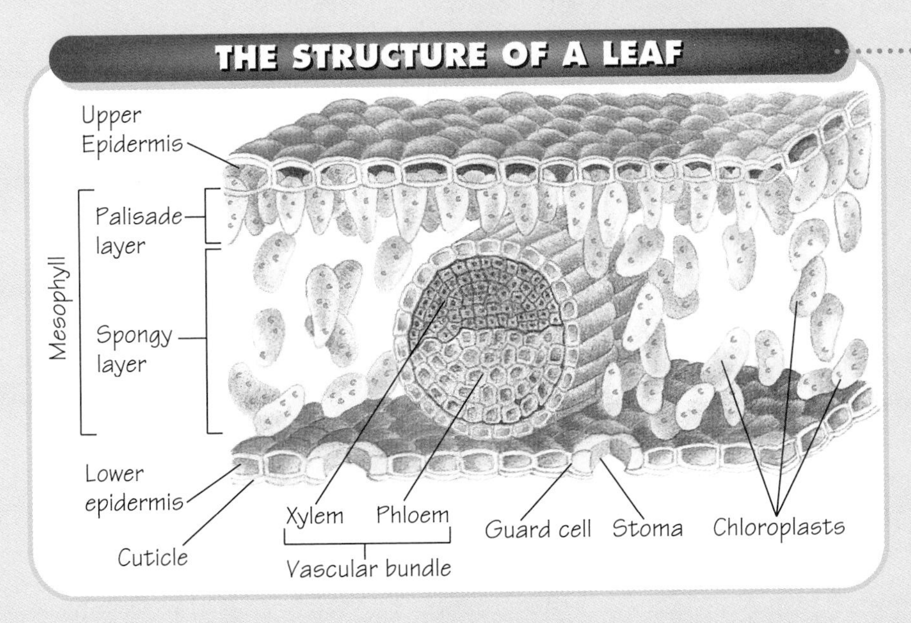

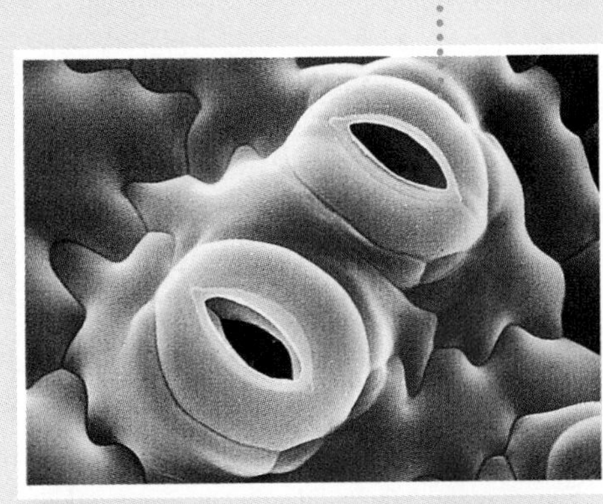

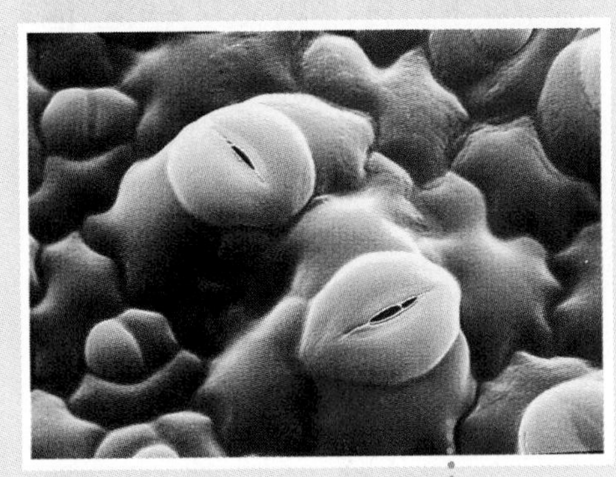

Figure 26–7
(a) In a dicot leaf, a mesophyll layer containing vascular bundles and chloroplasts is sandwiched between the upper epidermis and the lower epidermis. Located in the lower epidermis are stomata. (b) Stomata open (magnification: 257X) and (c) close (magnification: 343X) to allow gases in and out of the leaves.

get in and out of the leaf? The undersides of leaves have small openings known as **stomata** (STOH-muh-tuh; singular: stoma). Each stoma is surrounded by guard cells that control the passage of gases by opening and closing the stomata.

At first glance, it might seem to make sense for a plant to keep its stomata open all the time. But that would lead to a problem. In order for a gas to enter the cells inside a leaf, it must first dissolve in water. That means the cell surfaces inside the leaf must always be kept wet. What happens to a wet surface exposed to air? It dries out. So if a plant always kept its stomata open, it would lose water rapidly.

Plants solve this problem by balancing their need for carbon dioxide against their need to conserve water. Stomata generally open up during periods of rapid photosynthesis. This allows carbon dioxide to enter the leaf. However, if a plant begins to lose too much water, the thick cell walls of the guard cells cause the cells to push together, closing the opening. If more water becomes available, the guard cells swell as they take in fluid, causing them to expand and opening the stomata to allow gas exchange to take place.

☑ **Checkpoint** What are stomata?

Mesophyll Tissue

The ground tissue of most leaves consists of a tissue known as mesophyll. Mesophyll cells are packed with chloroplasts, which are cells that perform most of a plant's photosynthesis. A typical leaf consists of two types of mesophyll—a layer of tall palisade cells just beneath the upper epidermis and a layer of irregularly shaped spongy cells just above the lower epidermis. Spongy mesophyll has plenty of air spaces, and these cells are the ones that carbon dioxide enters first.

Leaf Veins—Vascular Tissue

Leaf tissue needs plenty of water, and the vascular system of a plant makes sure that it gets that water. Xylem cells carry water into the leaf, and osmosis carries the water from cell to cell within the leaf. Phloem tissue carries the products

MINI LAB .... Experimenting ...

Leaf Me Alone

PROBLEM *How does the number of leaves affect water uptake in a plant?* ***Design an experiment*** *to answer the question.*

SUGGESTED PROCEDURE 

1. Using three stalks of celery with all their leaves and three beakers containing red food coloring, design an experiment that will determine the effect the number of leaves on a plant has on its ability to transport water up the stem.
2. Formulate a hypothesis and be sure to include a control. You may want to use a metric ruler to measure your results.
3. Have your teacher approve your experimental design and then perform the investigation. Record your observations in a data table or diagram.

ANALYZE AND CONCLUDE

1. How did the data differ among the three setups? Explain your answer.
2. How did the number of leaves affect the plant's water uptake? Give evidence to support your answer.
3. Explain how transpirational "pull" is involved in your results.

INTEGRATING CHEMISTRY

What are hydrogen bonds? How do they determine the properties of water?

of photosynthesis from the leaf into the rest of the plant. In most plants, xylem and phloem cells are found together in vascular bundles, known as the "veins" of the leaf. In most monocots, the veins run parallel to each other. In dicots, they usually form a branched network.

Checkpoint Why do leaves contain vascular tissue?

Fluid Transport in Plants

You have already seen how xylem and phloem form a vascular network that carries water and nutrients throughout a plant. Vascular systems are a bit like the circulatory systems of animals, which carry fluids within the body. However, plants don't have a muscular "heart" to push fluid through their "veins." How, then, are plants able to move fluid from one end of the plant body to the other?

Xylem Transport

Recall that water enters the cells of root tissue by osmosis. But osmosis cannot generate enough force to lift the water more than a few centimeters above the ground. This is one of the reasons why nonvascular plants, such as mosses, are so small. In vascular plants, individual xylem cells are joined end to end—like stiff-walled drinking straws—to form continuous tubes that reach from one end of the plant to the other. Two powerful forces draw water up into these tubes.

Water molecules are strongly attracted to each other, a property known as cohesion. Because of their ability to form hydrogen bonds, water molecules can also be strongly attracted to other substances, a property called adhesion. If a thin tube is placed in a bowl of water, these properties will cause the water level to rise inside the tube, a process known as capillary action. Capillary action is one of the forces that draw water upward from the roots into the stems of vascular plants.

A second, and stronger, force is produced by the plant's own use of water. The leaves of a plant use water for photosynthesis, and they lose large amounts of water vapor through their stomata, a process known as **transpiration.** A large tree can lose as much as 100 liters of water a day through transpiration. As leaves lose water, they draw water out of xylem vessels through osmosis. Then, as a locomotive pulls a train hundreds of cars long, the movement of water into the leaves creates an upward "pull" through the xylem. The strong cohesion

Visualizing Plant Adaptations

The forces of natural selection affect all organisms, and plants are no exception. Facing some of the most extreme conditions on Earth—from parched deserts to the frozen Arctic—plants have adapted to meet the demands of these environments.

1 Desert Plants

The cactuses have extensive root systems that are able to quickly soak up the rare desert rainfall. Their leaves have been reduced to sharp spines, and they do most of their photosynthesis in thick, water-conserving stems.

2 Water Plants

Water lilies have stomata on the upper surfaces of their leaves. This important adaptation allows them to take carbon dioxide directly from the air, rather than absorbing it from water.

3 Salt-Tolerant Plants

A saltbush is a halophyte, meaning "salt-tolerant plant," that takes up sodium into its roots by active transport, drawing water along with it by osmosis. The extra sodium is then transported to the leaves, where it is pumped into expandable bladder cells on the surface of the leaf. The delicate bladder cell walls eventually burst, and the excess salt is washed away by rain or a high tide.

4 Carnivorous Plants

Carnivorous plants turn the tables on the animal kingdom! Many of these plants are found in nitrogen-poor soil, so capturing small organisms is an important way of meeting their nutritional needs. When an insect lands on a leaf of the Venus flytrap (right), the flytrap's leaves slam together. The insect is digested by enzymes released from the leaves. The annual sundew (left) is a tiny plant with tentacles on its leaves. The tentacles secrete a sticky liquid that attracts insects. When an insect lands on a tentacle, other tentacles bend over to surround the insect, secreting enzymes that digest the insect.

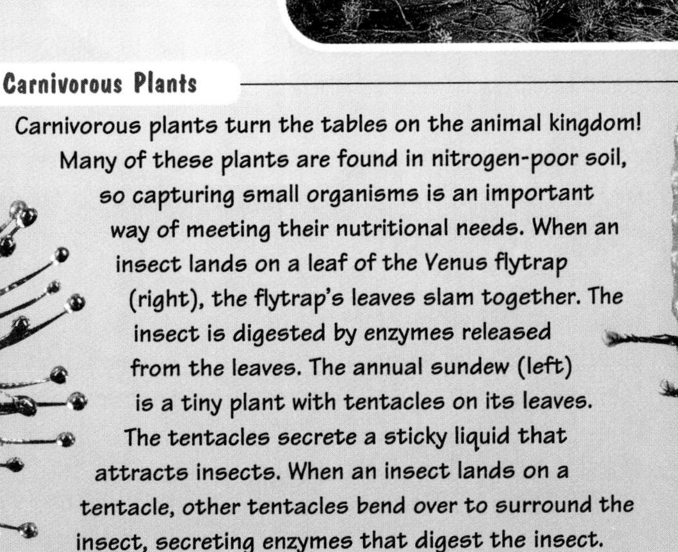

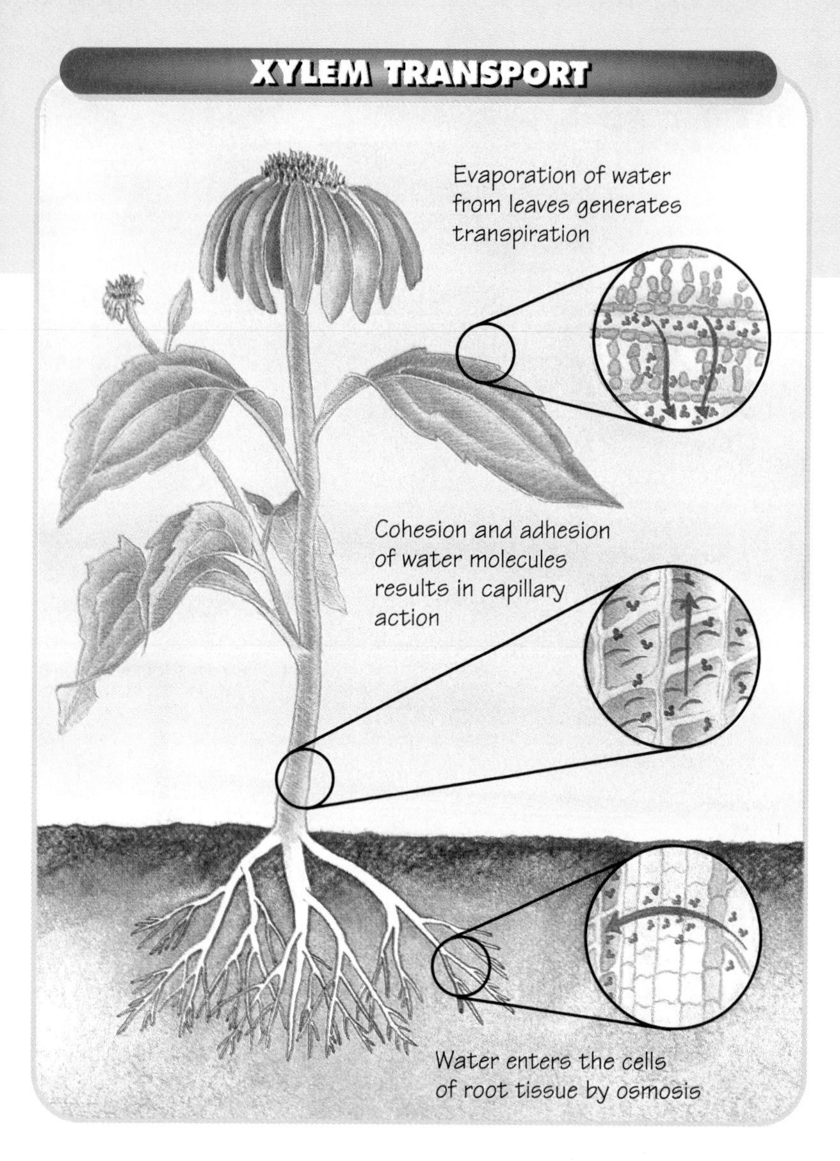

Figure 26–8
Xylem tissues are capable of transporting water from the tip of a plant's roots to the tops of its shoots. This action is the result of a combination of factors, illustrated in the diagram.

of water molecules produces a powerful force—called transpiration pull—that draws water upward into the leaves. **The combination of capillary action, osmosis, and transpiration pull is capable of lifting water to the tops of the tallest trees—such as the giant redwoods.**

☑ *Checkpoint* What is transpiration?

Phloem Transport

Like xylem, phloem forms a continuous network that reaches from the roots to the leaves. The combination of sugars and other nutrients dissolved in water forms phloem sap. Naturally, phloem sap carries the sugars produced in the leaves by photosynthesis down into the stems and roots of the plant. However, phloem also moves sugars into developing fruits. And if one part of a plant is kept in darkness or shade, phloem will carry nutrients to wherever they are needed.

The mechanism of phloem transport is not completely understood. **A combination of active transport and osmosis, known as the pressure-flow hypothesis, is believed to transport sugars through a plant's phloem tissue.** Suppose that leaf tissue is actively producing sugars and transporting them into the phloem sap. Osmosis will cause a flow of water into the sap at the same point, producing pressure. If sugars are needed at another point, the cells absorb the sugar from the phloem, and osmosis causes water to follow the sugars. The combination of these effects produces a steady flow of phloem sap between the parts of the plant that produce sugars and the parts that use them.

Section Review 26–1

1. **Explain** the functions of roots, stems, and leaves in plants.
2. **Describe** the process of fluid transport in xylem and phloem tissues in plants.
3. **Identify** the adaptations of plants that enable them to thrive in their environment.
4. **Critical Thinking—Relating** Describe the structure and function of a leaf.
5. **MINI LAB** How can you **design an experiment** to find out how the number of leaves affect water uptake in a plant?

Plant Growth

SECTION 26–2

Guide for Reading

- **Define** tropism.
- **Discuss** the role of auxins in plant growth.
- **Explain** why timing is important in plant life cycles.

MINI LAB

- **Predict** the effect of various factors on root growth.

PLANTS ARE DIFFERENT FROM animals in many ways, but one of the most important is the way in which they grow. Unlike most animals, plants continue to grow and increase in size throughout their lives. Even the oldest plants, including trees in the Pacific Northwest that are thousands of years old, continue to grow and produce new tissue.

Tropisms

Plants grow in response to cues from their environment. These responses are known as tropisms. The term **tropism** is derived from a Greek word that means "to turn." There are several common tropisms, each of which demonstrates the ability of plants to respond to changes in their environment.

Geotropism

Every seedling has the ability to sense and respond to the force of gravity. **Geotropism,** as this response is called, helps seedlings find their way out of the soil and into the sunlight. It affects roots and stems differently. Roots turn toward the force of gravity, while stems grow away from it.

Phototropism

The ability of plants to grow in response to light is known as **phototropism.** When a plant is grown near a window or other source of light, it will generally grow in the direction of the light source. The phototropic response can be so quick that young seedlings turn toward the light in a matter of a few hours!

Figure 26–9
(a) A pine tree, a gymnosperm, begins as a tiny seed from a pine cone. And a (b) coconut palm, an angiosperm, grows from one of the largest seeds known, a coconut. Yet these plants perform many common functions—such as growth and development—and exhibit characteristic responses to various conditions.

Figure 26–10
The sensitive Briar leguminosae *plant from Texas is shown here with its leaves* (a) *expanded and* (b) *folded to illustrate the plant's response to touch—a response called thigmotropism.*

Thigmotropism

Many plants respond to touch, an ability known as **thigmotropism** (thihg-MAH-truh-pihz-uhm). Climbing plants, such as ivy and pole beans, use thigmotropism to regulate their growth patterns in a way that enables them to wind around solid objects for support.

Plants can respond to other environmental factors, too, including the length of day and the time of year. Some flowers, for example, open only during the daytime. Others, including the cereus cactus of the American Southwest, open only at night.

☑ **Checkpoint** What is phototropism?

Plant Hormones

How do plants control their growth to respond to cues such as light and gravity? In many cases they do this by using chemical messengers known as hormones. A hormone is a substance produced in one part of an organism that affects activities in another part. You may know that animals produce hormones, too.

Phototropism was first explored by Charles Darwin and his son Francis in the 1880s. Their simple experiment making use of dark bands around the tips of seedlings is illustrated in ***Figure 26–11.***

Figure 26–11
(a) *Charles and Francis Darwin investigated the effect of covering the tip of a seedling with opaque and clear bands. Their experiments showed that the tip of a seedling releases a substance that causes the plant to bend toward the light.* (b) *These young plants are growing toward light that is coming from the right.*

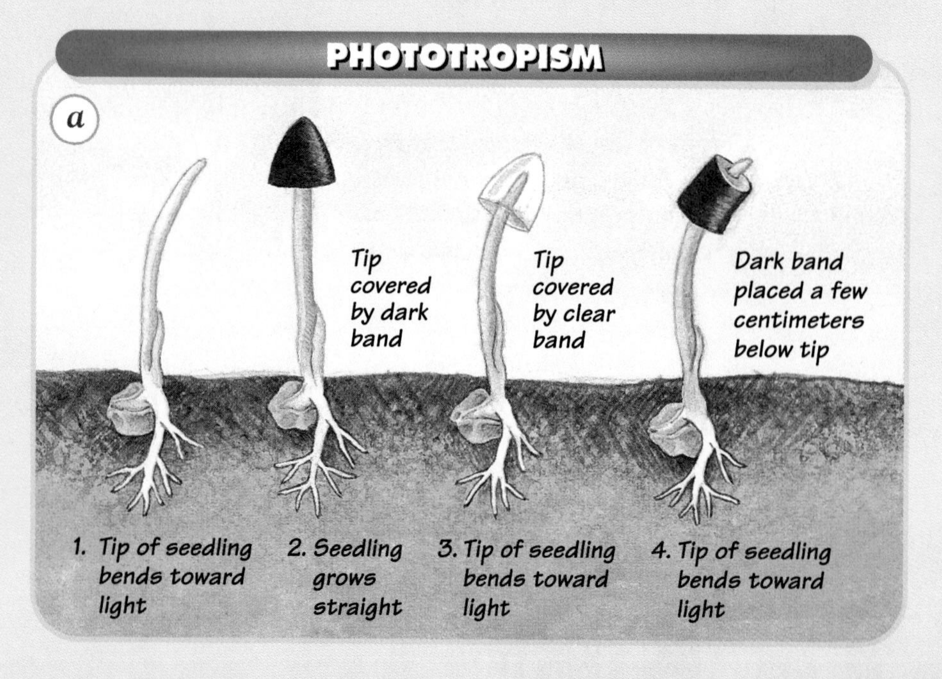

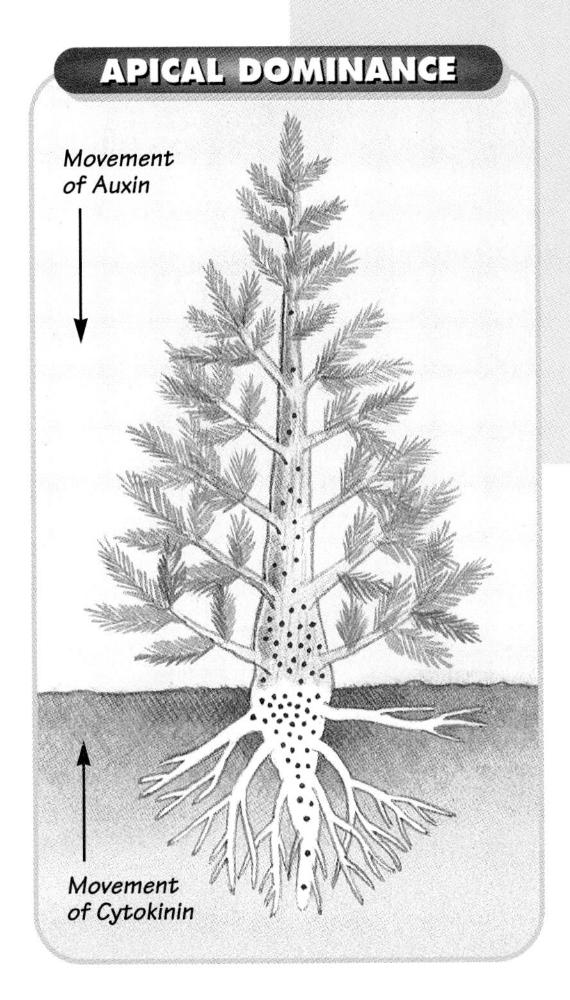

Figure 26–12
As a plant gets larger, side branches sprout from the main body of the plant. These side branches, however, grow more slowly near the tip of the plant than they do near the base, giving many plants their characteristic shapes. This growth pattern, called apical dominance, is the result of an auxin released from the apical meristem. This auxin inhibits the growth of meristems in the side branches closest to the apex.

Auxins

The Darwins suggested that the tip of a growing plant releases a substance that slows down growth on the side of the plant facing the light and speeds up growth on the other side. This, they reasoned, causes the plant to bend. Forty years later, that substance was discovered. It was a compound called indoleacetic (IHN-dohl-uh-seet-ihk) acid, and it was given the name **auxin,** from a Greek word that means "to increase."

Auxins stimulate cell growth and are produced by cells in the apical meristem, which is the rapidly growing region near the tip of a root or stem. Auxin is the hormone that produces phototropism. When light hits just one side of the plant, auxin builds up in the side away from the light, causing an enhanced growth of cells on that side. This causes the plant to bend toward the light. In a similar way, if a tree is knocked on its side by a storm, the production of auxins will cause the tree to begin to grow upward. Released from the meristems (regions of rapid growth), auxins redirect new growth in the direction of sunlight.

An auxin is also responsible for geotropism. By a mechanism that is still not understood, auxins build up on the lower sides of roots and stems. In stem tissue, auxin stimulates cell growth, turning the plant upright. In roots, auxin inhibits growth, causing roots to grow downward.

Cytokinins

Cytokinins (sigh-toh-KIGH-nihnz) are plant hormones that, like auxins, affect the rates of plant growth and cell division. However, many of the effects of cytokinins are opposite those of auxins. For example, auxins inhibit growth in lateral branches, but cytokinins stimulate it. Cytokinins are produced by cells throughout the plant, including those of the root tissue. Chemically, cytokinins are very much like adenine, one of the bases found in DNA and RNA.

Cytokinins seem to act in tandem with auxins. Recent experiments show that the ratio between auxin and cytokinin concentrations determines cell growth, rather than the level of either hormone by itself.

☑ **Checkpoint** What are auxins? Cytokinins?

Growing My Way?

PROBLEM *How can you **predict** the effect of various factors on root growth?*

PROCEDURE  

1. Obtain six corn seeds that have been soaked overnight. Arrange the seeds across the bottom of a Petri dish with their pointed ends facing the same direction. Lay clear plastic tape across the seeds and attach the ends of the tape to the sides of the dish.
2. Cover the seeds with six layers of moistened paper towels trimmed to fit inside the dish. Be sure there is no water collecting in the dish. Cover and tape the dish shut.
3. Stand the dish on its edge in some modeling clay with the pointed ends of the seeds pointing downward. Place the upright dish in a dark place for about two days.
4. When the roots are about 2 cm long, remove the cover of the Petri dish and moistened paper. Using a single-edge blade and a ruler, trim the last 2 mm from the root tips of three seedlings. Record the length of all the roots. Replace the moistened paper and tape the lid back on.
5. Rotate the dish so that the roots are pointing upward, replace it in the clay, and return the dish to a dark spot. Predict what will happen.
6. Examine the dish the following day.

ANALYZE AND CONCLUDE

1. Did all the roots grow during the second part of the experiment? If not, explain why or why not.
2. In which direction did the roots grow?
3. Was your prediction correct? If not, what information would have been helpful?

Gibberellin

Gibberellin is a hormone that was first discovered in rice plants. Japanese rice farmers knew that some of their plants were affected by a disease that caused them to grow so quickly that the plants became tall and spindly. Eventually, this caused the tall, thin plants to fall over and die. The farmers' name for this disorder was "foolish seedling" disease.

Gibberellin regulates the rate at which stems elongate. It can cause dramatic size increases in plants, even causing dwarf varieties to reach the size of normal plants. Certain tissues in seeds release large amounts of gibberellin, which serves as a signal that it is time for the seed to sprout.

Ethylene

When natural gas was first used for lighting and heating in the nineteenth century, people noticed that fruit on indoor plants seemed to ripen quickly in rooms where gas was present. The effect was traced to ethylene, one of the minor components of natural gas. Surprisingly, fruit tissues, in response to an auxin, release small amounts of ethylene that stimulate the ripening process.

Commercial producers of fruit sometimes take advantage of this hormone. Many crops, including lemons and tomatoes, are picked before they ripen so they can be handled more easily. Then, just before they are delivered to market, they are treated with synthetic ethylene to quickly produce a ripe color. Unfortunately, this trick doesn't always produce a ripe flavor—one reason why naturally ripened fruits often taste much better.

☑ ***Checkpoint*** What is the role of gibberellin in plant growth?

Controlling Plant Life Cycles

Some plants, known as **annuals,** live for just a single year. Annuals such as marigolds, corn, and peas grow from seed to maturity, flower, and produce new seeds in just a single growing season. A few plants, including carrots and sugar beets, live for two years and are called

biennials. Biennials usually flower and produce seeds in the second year of their life. Plants that live for more than two years are called **perennials.** Most common trees and shrubs are perennials.

Whether annual or perennial, timing is everything to a plant. **Because plants cannot search for mates as animals do, plants must time their reproductive cycles so their reproductive cells will be ready at the same time as those of other members of their species.** Also, because plants cannot migrate when the weather changes, they must react to seasonal changes—such as storing food for the long winters in temperate regions, preparing to grow again when spring arrives.

Figure 26–13
Compare the leaves of Oxalis stricta, *commonly known as prayer plant, during (a) day and (b) night. The folding and unfolding of the leaves is caused by the action of phytochrome.*

Plants seem to know what time of year it is. In the 1950s, two scientists at the U.S. Department of Agriculture discovered that a red pigment, called **phytochrome,** was used by plants to sense day and night. Phytochrome enables plants to sense the changing seasons by changes in the length of light and dark periods each day. Phytochrome then acts as a master timing switch to coordinate other events in the plant life cycle.

Abscisic acid is a hormone regulated by phytochrome. As the nights become longer and cooler, synthesis of the green pigment chlorophyll stops in many plants. Nutrients are drawn from the leaves into the body of the plant, and, eventually, the leaves fall off.

The appearance of abscisic acid and the drop in auxin production from these leaves has other effects on the plant. The tips of branches grow thick, forming waxy bud scales that will protect the apical meristem from winter weather. Finally, xylem and phloem cells pump salts and organic molecules into the fluids of the plant, producing a thick sap that, like the antifreeze in a car's radiator, will resist cold temperatures.

Section Review 26-2

1. **Define** tropism.
2. **Discuss** the role of auxins in plant growth.
3. **Explain** why timing is important in plant life cycles.
4. **Critical Thinking—Comparing** What features do geotropism, phototropism, and thigmotropism have in common? How do they differ?
5. **MINI LAB** **Predict** the effect that gravity has on the growth of root tips.

Plant Propagation

GUIDE FOR READING

- **Define** the term vegetative reproduction.

MINI LAB

- **Compare** the way vegetative reproduction occurs in different plants.

MANY PLANTS ARE ABLE TO grow from a detached stem or leaf, so with a little care it is possible to grow an entire plant from just a small cutting! Why is it so easy to regenerate a plant from such small pieces? Earlier you explored many of the ways in which various types of plants reproduce sexually. However, one of the most important attributes of plants is their ability to reproduce asexually, and the results are familiar to any gardener.

Vegetative Reproduction

Plants grow from regions known as meristems, which contain actively dividing cells capable of producing any cell type in the mature plant. This means that the meristematic cells in even a small cutting are capable of replacing any of the cell types that were present in the original organism from which the cutting was taken. **The propagation of plants by asexual reproduction, in which offspring are produced from the division of cells of the parent plant, is known as vegetative reproduction.**

Vegetative Reproduction in Nature

A well-known example of **vegetative reproduction** occurs in the spider plant, *Chlorophytum*, shown in **Figure 26–14.** This plant produces slender lateral shoots, called runners, that can produce buds of their own. If they find soil, these new buds can put down roots

Figure 26–14
In addition to reproducing sexually, many plants are capable of reproducing asexually, through vegetative reproduction. In one type of vegetative reproduction, a small part of the plant—such as a stem or a leaf—gives rise to a completely new plant. Examples of plants that commonly reproduce in this manner are the (a) strawberry, (b) spider plant, and (c) Japanese iris.

Figure 26–15
(a) In silverweed plants, vegetative reproduction occurs when horizontal stems, called runners, grow over the ground surface and develop new plants at the tips. (b) In lily-of-the-valley plants, vegetative reproduction occurs when horizontal stems, called rhizomes, grow at or below the ground's surface and produce new shoots and roots at the nodes.

and grow into completely independent plants. Strawberries grow in much the same way, and gardeners know that a thick bed of delicious fruit can be produced from just a few plants by allowing the runners to spread out and take root.

Incidentally, the new plants that are produced in this way are genetically identical to their "parent" plants. This means that a group of such plants is a **clone,** a group of organisms produced by cell division from a single cell. In this case, that single cell was the fertilized zygote from which the original plant was first produced. Vegetative reproduction is an important method of propagation for many plants, such as those shown in **Figure 26–15.**

☑ **Checkpoint** What is vegetative reproduction?

Vegetative Reproduction in Agriculture

Not surprisingly, humans have taken advantage of vegetative reproduction for thousands of years. Many common houseplants, including African violets, can be propagated from cuttings.

Grafting is sometimes used by farmers who wish to combine the best characteristics of two different plants. A bud or stem from one plant is carefully sliced off and inserted onto the stem of another plant. To minimize injury, this is usually done when the plants are dormant.

Nearly all the wine grapes grown in the world are the product of careful grafting. In the late nineteenth century, a series of diseases all but wiped out grape vines across Europe. Luckily, the Concord grape, a native of North America, was resistant to these diseases.

Figure 26–16
Humans have put their understanding of vegetative reproduction to use, as in (a) grafting to produce desirable characteristics in an orange tree. (b) CAREER TRACK This plant propagator is carefully transferring pine shoots to Petri dishes to monitor their early growth.

Bio FRONTIER Connections

Genetically Engineered Food

Recently, the owner of a small chain of grocery stores in a midwestern state was approached by a vegetable distributor who had a new tomato he wanted to sell. As the result of genetic engineering, this tomato can stay on the vine much longer before it softens, allowing it to ripen more fully on the vine. This means that the tomato will have a better flavor than many other tomatoes and yet will still be firm enough to withstand the trip from farm to market.

Engineering a Better Tomato

Until now, farmers have relied on selective breeding to take advantage of the natural mutations that occur in their crops. With selective breeding, it may take 10 to 15 years to create a successful new tomato variety. Using genetic engineering, in which scientists isolate individual genes from one kind of organism and transfer them into a different kind of organism, it is possible to create a new variety in a matter of months.

Vine-ripened tomatoes

So What's the Problem?

The U.S. Food and Drug Administration (FDA) expects more than 100 genetically engineered foods to appear in supermarkets in the next few years. Some people are concerned about the potential dangers of genetically engineered foods. In some cases, the gene transfer might cause allergic reactions in people who are allergic to the organism from which the gene came. Other genetic-engineering methods involve creating plants that produce their own natural insecticides. These chemicals might be harmful to insects that farmers find beneficial.

If animal genes are placed in plants, then vegetarians, who don't eat animal products, might accidentally ingest them. For example, in one experiment, the "antifreeze" gene from arctic flounder, a fish, was placed into strawberries to prevent them from freezing on cold nights.

Getting back to the new tomato, the FDA has said it is as safe as traditionally bred tomatoes and does not require special labels to alert consumers that it is genetically altered. Yet the food-chain owner is concerned because she has heard that a number of organizations have complained that genetically engineered foods are not safe. On the other hand, the grower will sell the tomatoes to her at a very reasonable price because she will be one of the first to try the new tomato.

Making the Connection

Although the food-chain owner and her company are fictitious, the situation is real. What should she do? What reasons does she have to buy the tomato? What are some reasons not to buy it? Should the owner inform her customers that the tomatoes are genetically altered? Why or why not?

The fruit of the Concord grape, however, does not make a very good wine. So vintners grafted European wine stems onto American roots. The result was successful, and even today the best wine grapes of Europe are grown on American roots.

☑ *Checkpoint* What is grafting?

The New Technology of Plant Cloning

For years, plant biologists worked to find ways to take a single cell from one plant and stimulate it to grow into a complete organism. Today, for many species of plants, that dream is a reality.

Isolated cells from many plants can be grown in laboratory culture in nutrient-rich broths. Under the right conditions, these cells can be kept as protoplasts—cells that do not grow cell walls. Why would anyone want to grow protoplasts? These cells can be very useful in genetic engineering. DNA from other sources can be injected into protoplasts, and in some cases that DNA finds its way into the cell nucleus and becomes part of the plant cell genome.

Using a carefully controlled mix of hormones, protoplasts will begin to make cell walls and will gradually form a small clump of cells known as a callus. Each callus is transferred to a sterile dish or tube, where it grows into a small plantlet. As the plant increases in size, it can be planted in ordinary soil and grown to maturity.

The ability to grow and manipulate plant cells in this way gives biologists a new opportunity to study the genetics of plants and to produce more productive varieties for agriculture.

Growing Plant Parts

PROBLEM *How does vegetative reproduction in various plants compare?*

PROCEDURE 

1. From a coleus plant, cut a stem that has several leaves above the cut. Place the cutting in a container of water.
2. Partially insert four toothpicks around the middle of an onion bulb. Place the onion, base down, in a container of water so the toothpicks sit on the edge of the container and allow only the bottom of the onion to touch the water.
3. Wash a potato and, using toothpicks, suspend it in a container of water.
4. Observe and compare the growth of each plant.

ANALYZE AND CONCLUDE

1. How did the three plants compare in terms of the structures that grew?
2. How did they compare in terms of time, length, or other factors you observed?

Section Review 26–3

1. **Define** vegetative reproduction.
2. **Describe** ways in which growers take advantage of vegetative reproduction.
3. **MINI LAB** How does vegetative reproduction in different plants **compare?**
4. **BRANCHING OUT ACTIVITY** Visit a plant nursery and **identify** the plants that are grown by vegetative reproduction. Consult a plant grower if necessary.

Laboratory Investigation

Plant Tissues and Their Functions

A plant, like an animal, must transport water, minerals, and nutrients through its body. It must also transport the products of photosynthesis—such as sugars—to all parts of its body. In this investigation, you will examine tissues and organs that transport water through a plant.

Problem

What water-transport tissues can you **observe** in certain types of vegetables?

Materials (per group)

- Petri dish with radish seedlings
- hand lens
- whole carrot
- scalpel or single-edged razor blade
- round slice of carrot
- lettuce leaf
- microscope slide
- coverslip
- medicine dropper
- distilled water
- dissecting needle
- compound microscope
- forceps

Procedure  

1. **Using a hand lens, examine the radish seedlings in the Petri dish. Sketch and label one seedling.**
2. **Obtain a whole carrot and use the scalpel to cut it in half lengthwise. Make a second cut along the length of one of the carrot halves so that you have a very thin long slice. CAUTION:** ***Be careful when using sharp instruments.***
3. **Hold the thin slice up to the light and examine it. Sketch and label this lengthwise section.**
4. **Using the hand lens, observe the round carrot slice. Sketch and label what you observe.**
5. **Force the inner core out of the round carrot slice by pushing it firmly. Examine the surface of the core.**
6. **Obtain a lettuce leaf that has had its lower surface soaking in distilled water. Bend the leaf so that it breaks and peels away the lower surface.**

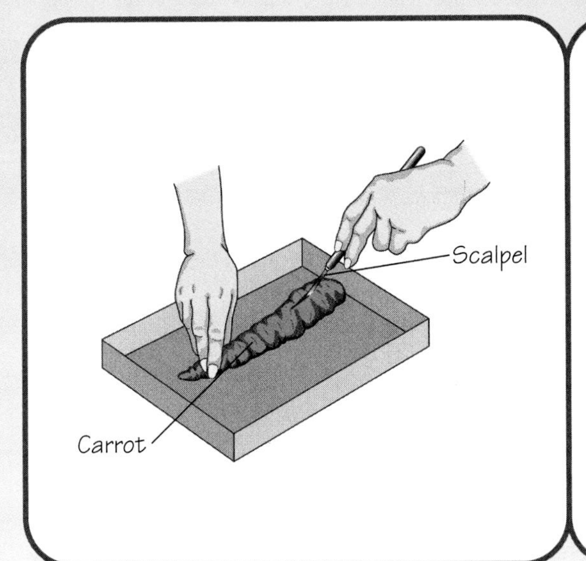

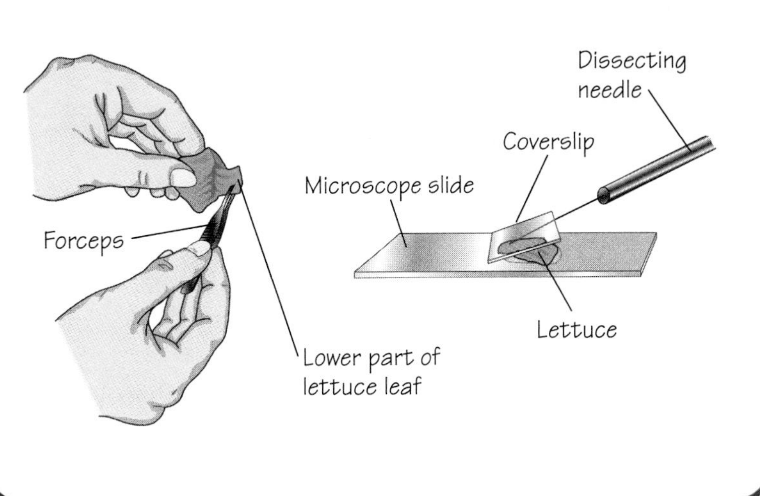

7. With the forceps, peel off a small piece of the leaf tissue. Put this piece of tissue on a microscope slide. Using a medicine dropper, add a drop of water to the slide. With a dissecting needle, slowly lower one edge of the coverslip and then the other onto the lettuce leaf.

8. Observe the leaf tissue under the low-power objective of the microscope. Then switch to the high-power objective. Look for jigsaw-puzzle-shaped cells and pairs of rounded cells.

Observations

1. Describe the outer surface of the radish seedling root. What structures did you observe? Did the entire surface of the root have the same appearance?
2. What was the shape of the inner core of the carrot?
3. Can you find extensions growing from the surface of the carrot core? What are they?
4. What did you observe on the lettuce leaf between the two rounded cells in each pair?

Analysis and Conclusions

1. Are there root hairs at the tip of the radish root? Explain your answer.
2. What kind of tissue is found in the center of the carrot root? What is its function?
3. What kind of tissue makes up the thick, outer part of the carrot? What is its function?
4. What kind of tissue did you peel from the lettuce?
5. What are the rounded cells? What is their function?

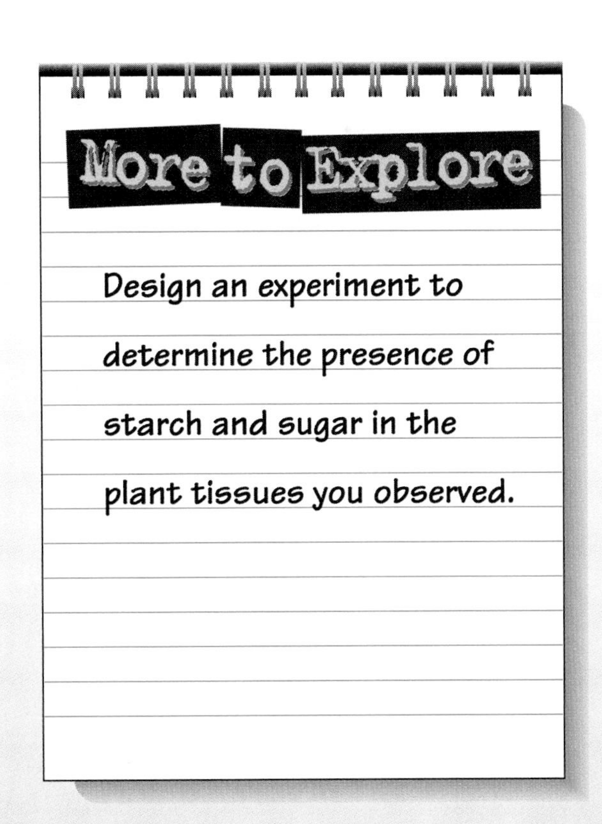

Summarizing Key Concepts

The key concepts in each section of this chapter are listed below to help you review the chapter content. Make sure you understand each concept and its relationship to other concepts and to the theme of this chapter.

26–1 Plant Structure and Function

- The body of a plant consists of three distinct regions known as roots, stems, and leaves.
- Roots anchor a plant in the ground, drawing water and minerals from the soil. They consist of the epidermis, the cortex, the endodermis, and the vascular cylinder. Roots take up water and minerals from soil by osmosis and active transport.
- Stems rise above the ground, support the body of the plant, and carry water and nutrients from one end of the plant to the other. They consist of the epidermis, the cortex, the pith, and vascular bundles.
- Leaves, the organs of photosynthesis, have an epidermis, fluid-carrying vascular tissue, and ground tissue. The undersides of leaves have small openings known as stomata through which gases enter and exit.
- A combination of capillary action, osmosis, and transpiration pull enables water to rise to the tops of trees in xylem tissue. A combination of active transport and osmosis is probably responsible for phloem transport.

26–2 Plant Growth

- A plant's response to its environment is known as a tropism. Geotropism is a response to gravity, phototropism is a response to light, and thigmotropism is a response to touch.
- Auxins stimulate cell growth and are produced by cells in the apical meristem.

26–3 Plant Propagation

- The process of asexual reproduction, in which offspring are produced from the division of cells of the parent plant, is known as vegetative reproduction.
- A clone is an organism produced by the division of a single cell.

Reviewing Key Terms

Review the following vocabulary terms and their meaning. Then use each term in a complete sentence.

26–1 Plant Structure and Function

root
stem
leaf
xylem
phloem
epidermis
cortex
vascular cylinder
endodermis
pith
annual tree ring
stoma
transpiration

26–2 Plant Growth

tropism
geotropism
phototropism
thigmotropism
auxin
annual
biennial
perennial
phytochrome

26–3 Plant Propagation

vegetative reproduction
clone
grafting

Recalling Main Ideas

Choose the letter of the answer that best completes the statement or answers the question.

1. In plants, the main organs of photosynthesis are the

a. roots. **c.** leaves.
b. stems. **d.** petioles.

2. The function of a plant's ground tissue is to

a. support and carry out photosynthesis.
b. protect the plant.
c. conduct fluids through the plant.
d. prevent water evaporation.

3. Plants take in almost all their water and nutrients through structures called

a. guard cells. **c.** osmosis.
b. vascular cylinders. **d.** root hairs.

4. Most of the wood in a tree is made up of

a. cork cambium cells. **c.** xylem tissue.
b. phloem tissue. **d.** bark cells.

5. Phototropism results in

a. seedling stems growing toward the light.
b. seedling stems growing away from the light.
c. seedling roots growing toward the light.
d. seedling roots growing toward the force of gravity.

6. The Darwins' hypothesis about the cause of phototropism was confirmed when

a. geotropism was discovered.
b. auxin was discovered.
c. gibberellin was discovered.
d. ethylene was discovered.

7. What determines whether a plant is classified as an annual or a perennial?

a. the kind of flowers it has
b. its age
c. whether or not it will produce fruit
d. its ability to sense day and night

8. Vegetative reproduction results in offspring that

a. are grafts of two plants.
b. inherit genetic traits from both parent plants.
c. have some traits unlike the parent plant.
d. are genetically identical to the parent plant.

9. Farmers use the technique called grafting to

a. combine the best characteristics of two different plants.
b. produce protoplasts.
c. grow new kinds of potatoes.
d. produce genetically identical plants.

Putting It All Together

Using the information on pages xxx to xxxi, complete the following concept map.

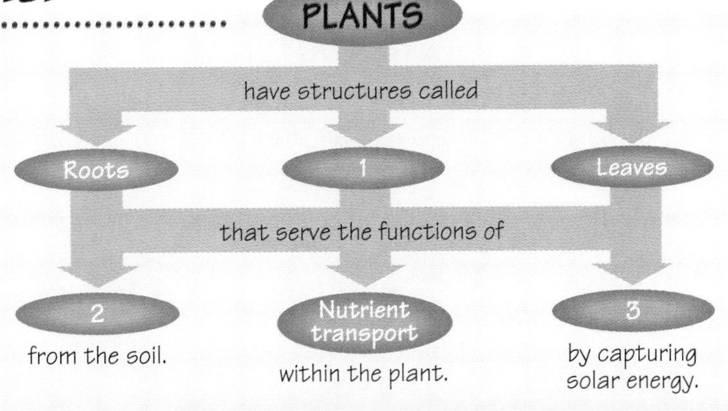

Reviewing What You Learned

Answer each of the following in a complete sentence.

1. What are the three main kinds of plant tissue?
2. Explain the two functions of the guard cells on the underside of most plant leaves.
3. What is the difference between a primary root and a secondary root?
4. Describe the role that osmosis plays in root cells.
5. What is the difference between heartwood and sapwood?
6. What do carnivorous plants gain from the insects they trap?
7. What are stomata?
8. How are roots and stems affected by geotropism?
9. What is a hormone?
10. Explain the effect ethylene has on plants.
11. What roles does phytochrome play in plants?
12. Which characteristic of meristematic cells enables cuttings to develop into complete plants?
13. What is the definition of a clone?
14. How do spider plants or strawberry plants reproduce vegetatively?
15. How are protoplasts useful in genetic engineering?

Expanding the Concepts

Discuss each of the following in a brief paragraph.

1. What are some ways that a plant uses the energy produced by photosynthesis?
2. How do guard cells help to regulate the rate of photosynthesis?
3. Describe how wood is produced in a tree stem.
4. Why don't some desert plants have leaves?
5. Describe how a single plant can exhibit geotropism, phototropism, and thigmotropism.
6. What effect would paving the ground around a tree's trunk have on the tree?
7. **Compare** osmosis and active transport.
8. What information about ancient people can archaeologists infer from tree rings?
9. Why do people turn the containers of plants grown on windowsills?

Extending Your Thinking

Use the skills you have developed in this chapter to answer the following.

1. **Designing an experiment** How does capillary action vary with the diameter of a tube? Design an experiment to answer the question. Make sure that your experiment has only one variable. Plan a control, too. Draw a labeled diagram to explain your design. Have your teacher check your proposed experiment.
2. **Predicting** In the early 1700s, Stephen Hales experimented with plants. In one experiment, he attached a tube of water to the root of a tree and measured the amount of water pulled up the tube into the tree. Based on what you have learned, predict the conditions under which the tree will exert the strongest pull: a sunny day, a cloudy day, or at night. Explain your choice.
3. **Observing** Find two logs or tree stumps that display the cross sections of two tree trunks. Describe the differences you see in the annual tree rings.
4. **Hypothesizing** When digging up and transplanting a tree from one site to another, the branches must be cut back severely. Form a hypothesis to explain why this is necessary to help the plant become established in its new location.

Applying Your Skills

Modeling Plant Parts in 3-D

A plant has three major kinds of organs—roots, stems, and leaves. Even though these structures have many differences, they share three common tissue types: dermal tissue, vascular tissue, and ground tissue.

1. Working in groups of three, have each member of a group choose a different plant organ: root, stem, or leaf.
2. Discuss the materials your group will use to build three-dimensional models of these plant organs. Then collect the materials.
3. Establishing a common scale, build a model of each plant organ. Include in each model a representation of the tissues or cells.
4. Write labels for your model that will demonstrate your understanding of the relationship between structures and functions.

• GOING FURTHER •

5. With members of your group, propose a plan to show how you could use easily available materials to make a working model of a plant.

UNIT 7

Animals

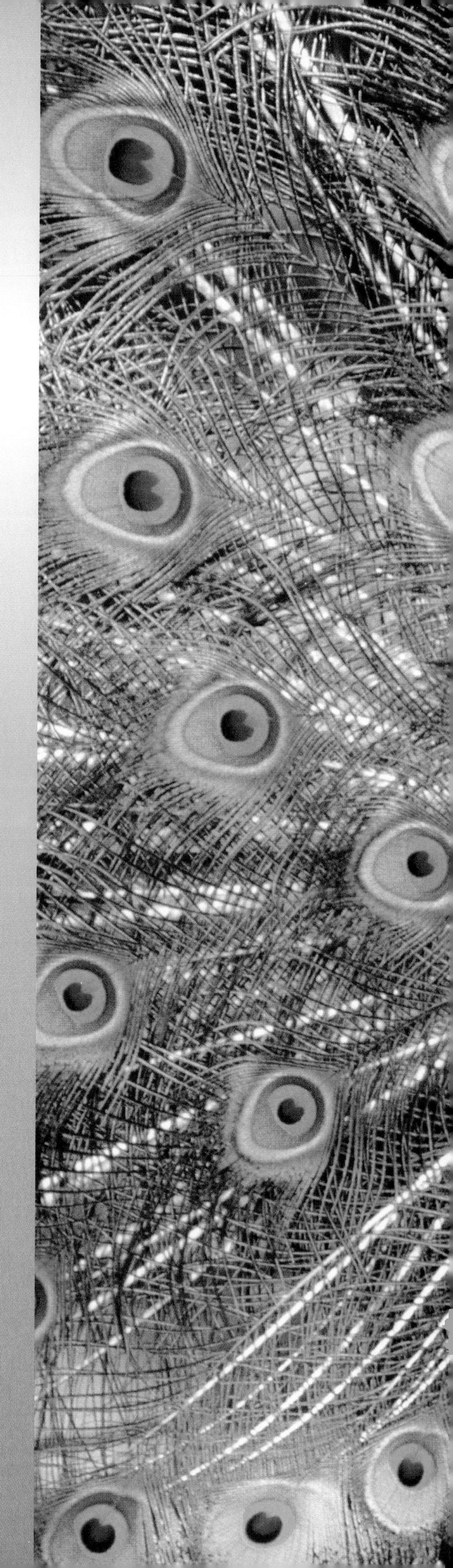

CHAPTERS

"Biological diversity is the key to the maintenance of the world as we know it."

— Edward O. Wilson

A magnificently colored peacock showing the diversity of animals

CHAPTER 27

Sponges, Cnidarians, and Unsegmented Worms

FOCUSING THE CHAPTER THEME: Unity and Diversity

27–1 Sponges and Cnidarians
- **Describe** the characteristics of sponges and cnidarians.

27–2 Unsegmented Worms
- **Compare** the free-living and parasitic unsegmented worms.

BRANCHING OUT *In Depth*

27–3 Nematodes—A Help or a Hindrance?
- **Describe** some examples of nematode parasites.

LABORATORY INVESTIGATION
- **Design an experiment** to determine how a hydra and a planarian respond to stimuli.

Biology and Your World

BIO JOURNAL

In your journal, describe the organisms shown in the photographs on this page and the next. Have you ever seen these organisms before? If so, where have you encountered them?

Sea fan on a coral reef

Sponges and Cnidarians

SECTION 27–1

GUIDE FOR READING

- **Explain** how sponges differ from other animals.
- **Describe** the organization of tissues in cnidarians.

MINI LAB

- **Compare** different sponges.

THE SEA IS HOME TO AN incredibly diverse set of animals, some of which are shown on this page. Jellyfishes are nearly transparent, and they drift and swim with the currents in the open sea. Corals can be as hard as rocks, and they are the only animals—other than humans—that create entire ecosystems. Many sea anemones are among both the most beautiful and the most harmless animals, while others carry venom powerful enough to kill a human. And sponges are some of the most unusual animals of all.

Let's take a look at each of these fascinating groups of organisms. You may never look at sea life the same way again!

Sponges

Adult sponges are sessile, which means they live attached to one spot. For this reason, early naturalists thought sponges were unusual aquatic plants. However, sponges are animals, not plants. In fact, they are among the most ancient of the multicellular animals, although they have less in common with the rest of us than any other phylum!

Classifying Sponges

Sponges are members of the phylum Porifera, meaning "pore bearing." Tiny pores, or openings, penetrate a sponge's body, allowing sponges to filter food from the water that passes through these pores.

Why are sponges classified as animals and not as plants? First, sponge cells lack cell walls—the tough outer boundaries that surround plant cells. And second, sponges are heterotrophic, which means they obtain energy from the food they take in. Plants

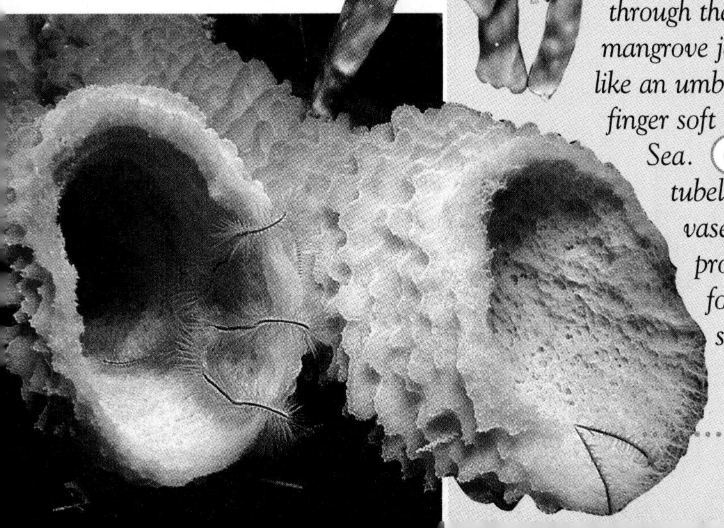

Figure 27–1 *Sponges and cnidarians are some of the strangest and most interesting animals in the sea. (a) To move through the water, the body of this mangrove jellyfish opens and closes like an umbrella. (b) These lady finger soft corals live in the Red Sea. (c) This tubelike azula vase sponge is providing shelter for a brittle starfish.*

a

c b

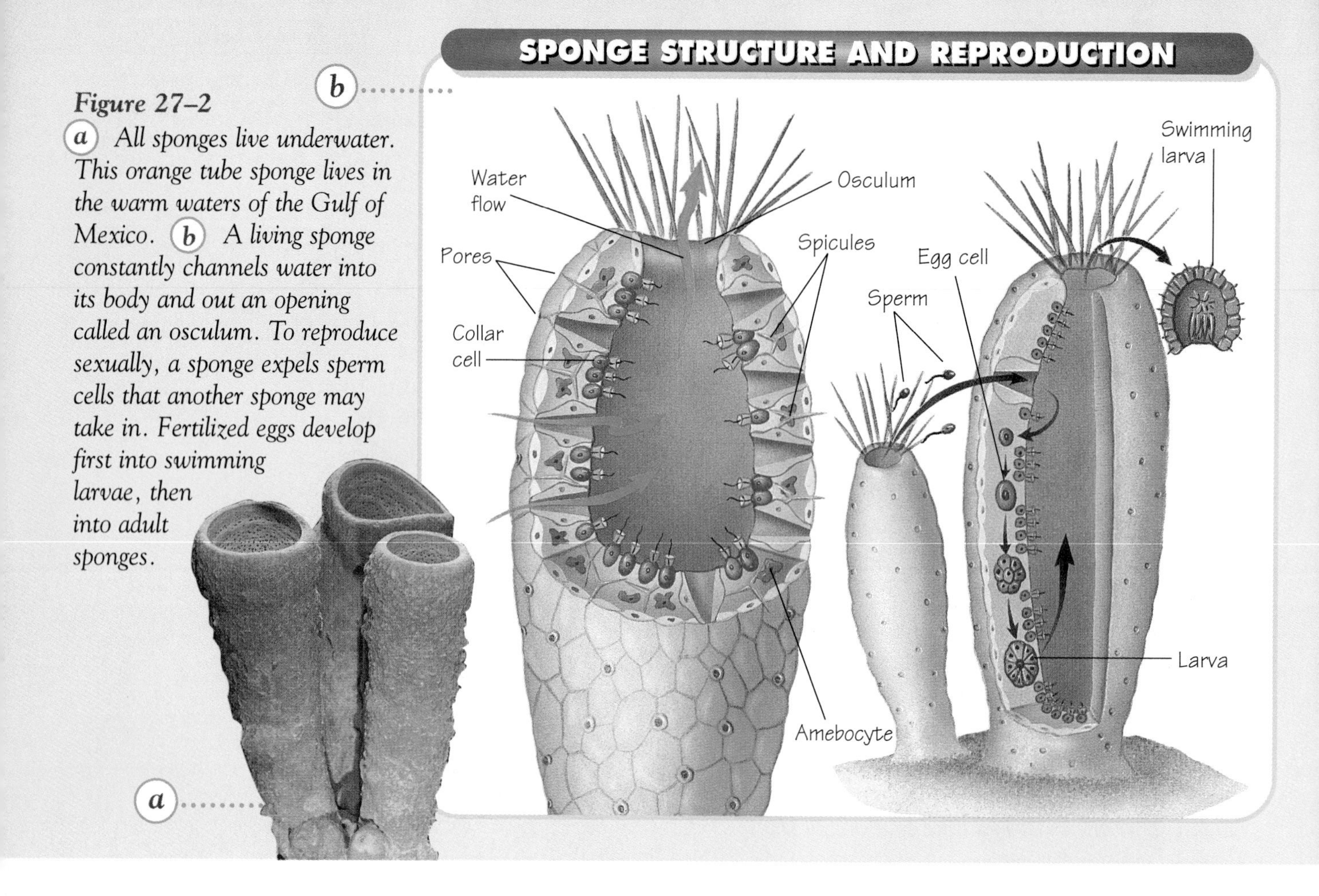

Figure 27–2 (a) *All sponges live underwater. This orange tube sponge lives in the warm waters of the Gulf of Mexico.* (b) *A living sponge constantly channels water into its body and out an opening called an osculum. To reproduce sexually, a sponge expels sperm cells that another sponge may take in. Fertilized eggs develop first into swimming larvae, then into adult sponges.*

are autotrophic, meaning they manufacture their own food from raw materials.

The differences between sponges and other animals are even greater than you might expect by looking at them. **Unlike the cells of other animals, a sponge's cells are not organized into recognizable organs—or even recognizable tissues.** For this reason, most biologists believe that sponges evolved separately from other animals.

☑ ***Checkpoint*** How do sponges differ from other animals?

Sponge Structure

As shown in ***Figure 27–2,*** sponges use different types of cells for different purposes. Their **collar cells,** for example, have whiplike flagella that beat water through the body wall and into the central cavity. These cells also snag food particles from the water moving past them—particles which they then engulf and digest. Sponges also use wandering cells called **amebocytes** (uh-MEE-boh-sights) to digest and distribute food, as well as to produce components of the sponge skeleton.

The sponge's body plan allows it to perform many different tasks efficiently. Respiration, feeding, and the elimination of wastes are each accomplished by exchanging materials with the water that flows through the sponge's body.

INTEGRATING CHEMISTRY

What are the physical and chemical properties of calcium carbonate and silica?

Some sponges support their bodies with small crystallike spikes called **spicules.** Spicules are made of either calcium carbonate ($CaCO_3$) or silica (SiO_2), and they can interlock to form rigid yet delicate networks, as shown in ***Figure 27–3.*** ●

Sponges also support their body with a protein called **spongin.** The skeletons of sponges that contain this protein are soft, flexible, and elastic—qualities that give them a "spongy" texture.

Sponge Life Cycles

Sponges reproduce sexually by producing sperm and eggs, as shown in ***Figure 27–2***. However, like many animals, sponges can reproduce asexually, too. One way sponges do this is by producing ball-shaped structures called **gemmules** (JEHM-yoolz). Gemmules are clusters of amebocytes that are protected by a tough outer covering of spicules. This allows gemmules to survive long periods of harsh conditions that would kill adult sponges. When conditions improve, gemmules grow into complete sponges.

A process called **budding** is another way in which sponges reproduce asexually. In budding, a piece of the sponge falls off and grows into a new sponge.

☑ **Checkpoint** What are gemmules?

Cnidarians

The phylum Cnidaria (nigh-DEHR-ee-uh) contains the corals, jellyfishes, and sea anemones. All cnidarians live underwater, and nearly all live in the sea.

Cnidarians represent an important step up the evolutionary ladder of complexity. Why? **Unlike sponges, the cnidarians have layers of differentiated cells that are organized into three specialized layers of tissues.** The innermost tissue layer is called the **endoderm,** and the outermost layer is called the **ectoderm.** In between these layers is a poorly developed jellylike layer called the **mesoglea** (mehs-oh-GLEE-uh).

☑ **Checkpoint** What are the cnidarians?

Cnidarian Structure

A cnidarian's body plan is radially symmetric, meaning that its body parts are arranged in circles around a central mouth. For this reason, you could say that cnidarians have a top and a bottom but no front or back.

Spongy Skeletons

PROBLEM *How do different sponges **compare?***

PROCEDURE    

1. Add a few drops of bleach to a microscope slide. **CAUTION:** *Avoid direct contact with bleach.*
2. Place a small piece of a *Grantia* sponge on the slide. The sponge's cells will deteriorate in the bleach, but the spicules will not be affected.
3. Add two drops of water, place a coverslip on the slide, and study the sponge under the low-power objective of a compound microscope.
4. Repeat steps 1 to 3 for a *Spongilla* sponge and a *Hexactinellida* sponge.

ANALYZE AND CONCLUDE

1. Describe the sponge structures you observed.
2. What is the role of spicules in a living sponge?
3. Compare the different sponges you studied. How do living sponges compare with artificial sponges used for cleaning?

Figure 27–3
The skeleton of a Venus' flower basket sponge consists of an intricate network of spicules.

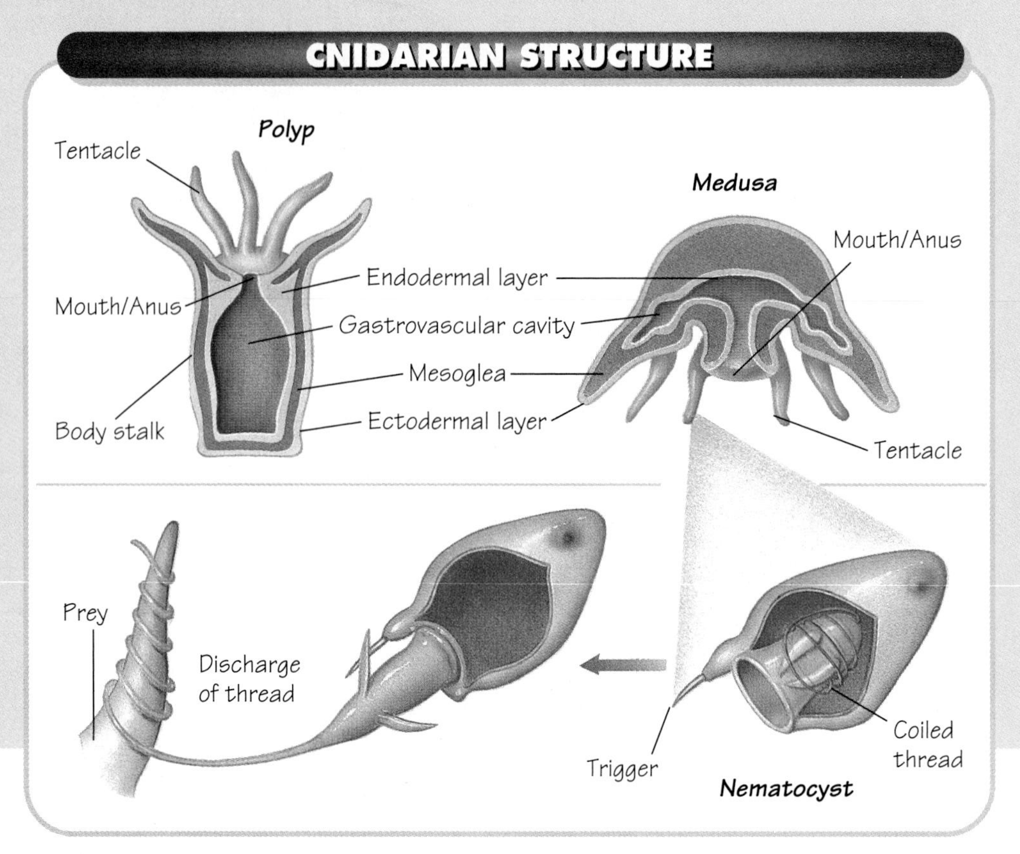

Figure 27–4
Cnidarians have two forms—a sedentary polyp and a free-swimming medusa. In both, the jellylike mesoglea separates the layers of endoderm and ectoderm, forming a gastrovascular cavity with a single opening. Both a polyp and a medusa can release poison-filled nematocysts to attack prey. Some nematocysts penetrate a prey's skin, while others, such as the nematocyst shown here, entangle a prey.

To send information around its body, a cnidarian uses a simple nervous system called a nerve net. The nerve net shows no centralization or cephalization—meaning "located in a head"—so a cnidarian has nothing that you could describe as a brain. Some species, however, have specialized nerve cells that function as sensory organs. Organs called **ocelli** (oh-SEHL-igh; singular: ocellus) are simple eyespots that detect the presence or absence of light. And organs called **statocysts** provide information on which way is up, helping the cnidarian to balance itself.

Although cnidarians do not have recognizable muscle tissue, they do have musclelike cells that contract when stimulated by the nerve net. These cells help cnidarians to move and to feed.

Cnidarians also have specialized structures called **nematocysts,** which are used for defense and for catching prey. A nematocyst is a sac containing poison and a tiny springlike harpoon. When triggered, the spring uncoils, flinging the poison-tipped harpoon toward the cnidarian's prey. Once the prey is paralyzed, the cnidarian uses its tentacles to push the prey through its mouth and into the **gastrovascular cavity,** where most digestion takes place.

The nematocysts of most cnidarians threaten only fishes and other small animals. But some species, such as the tiny sea wasp, carry powerful stings that can seriously injure—or even kill—a human.

Cnidarian Life Cycles

Typically, cnidarians alternate between two stages: a free-swimming stage called a **medusa** and a sessile stage called a **polyp.** In some groups, such as jellyfishes, the cnidarian spends most of its life as a medusa. In other groups, the dominant stage is the polyp. And in still other groups, such as corals, the medusa stage is left out altogether. In addition, many cnidarians also alternate between sexual and asexual reproduction.

☑ **Checkpoint** What is a polyp? A medusa?

Problem Solving

INTERPRETING GRAPHS

The Hydra Test

Like all organisms that live in water, hydras are affected by the concentration of many substances in the water surrounding them. The following experiment was designed to show how two substances—sodium ions (Na^+) and potassium ions (K^+)—affect the growth of a hydra's tentacles.

Hypothesis: The growth of a hydra's tentacles is equally affected by Na^+ and K^+.

Five solutions of growth medium were prepared, with different combinations of sodium and potassium ions added to each. Hydras were placed in each solution, then the lengths of their tentacles were measured each day for six days. The trends in the data and the contents of the five solutions are presented in the diagram below.

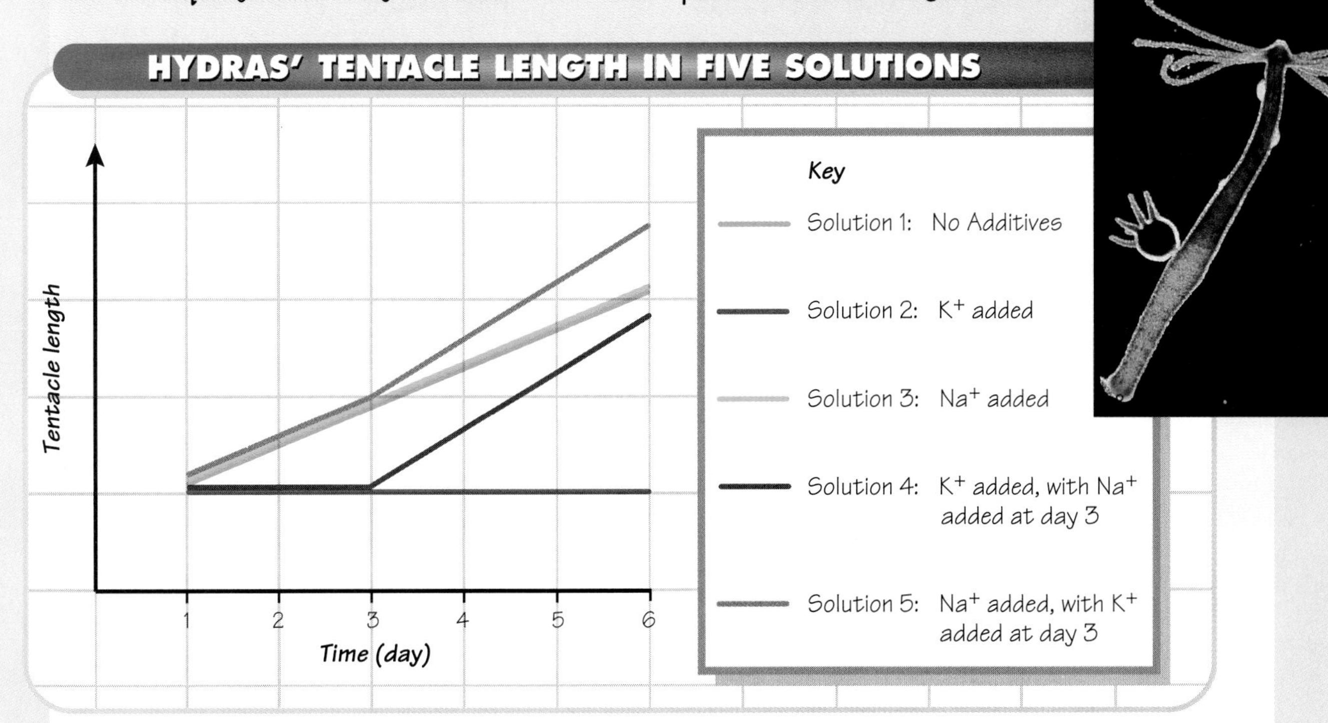

A polyp of Hydra, *with a small bud at its side (magnification: 10X)*

THINK ABOUT IT

1. Identify the control group and the variables in this experiment.
2. Describe the growth of the hydras' tentacles in each of the five solutions. In which solution did the tentacles grow the most? The least?
3. Did the results of the experiment support or contradict the hypothesis? Explain your answer.
4. What can you conclude about the growth of a hydra's tentacles from this experiment? Identify any assumptions that you used to reach this conclusion.

Figure 27–5
These hydrozoans live in the Gulf of Mexico. (a) *The Portuguese man-of-war is a floating hydrozoan colony that contains several different polyps.* (b) *Beware the fire coral if you encounter it—its stings are painful!*

Hydrozoans

There are several important classes of cnidarians, one of which is the class Hydrozoa (high-droh-ZOH-uh). The hydrozoans spend most of their lives as polyps, although many species have a brief medusa stage. Most hydrozoans grow as branching collections of polyps called colonies. As shown in **Figure 27–6,** a colony can range in width from a centimeter to more than a meter.

In many cases, the polyps in a colony are specialized to perform different tasks, including feeding and defense. The colony can also contain reproductive polyps, which produce small free-swimming medusae. The medusae mature rapidly, reproduce sexually by producing sperm and eggs, and then die.

The best-known hydrozoans are members of the genus *Hydra*—which just happen to be among the least typical members of the group! Unlike most hydrozoans, hydra live in fresh water, lack a medusa stage, and spend most of their lives as solitary polyps. Hydras also move from place to place with a peculiar somersaulting motion.

Jellyfishes

True jellyfishes are members of the class Scyphozoa (sigh-fuh-ZOH-uh). In most respects, these animals have typical

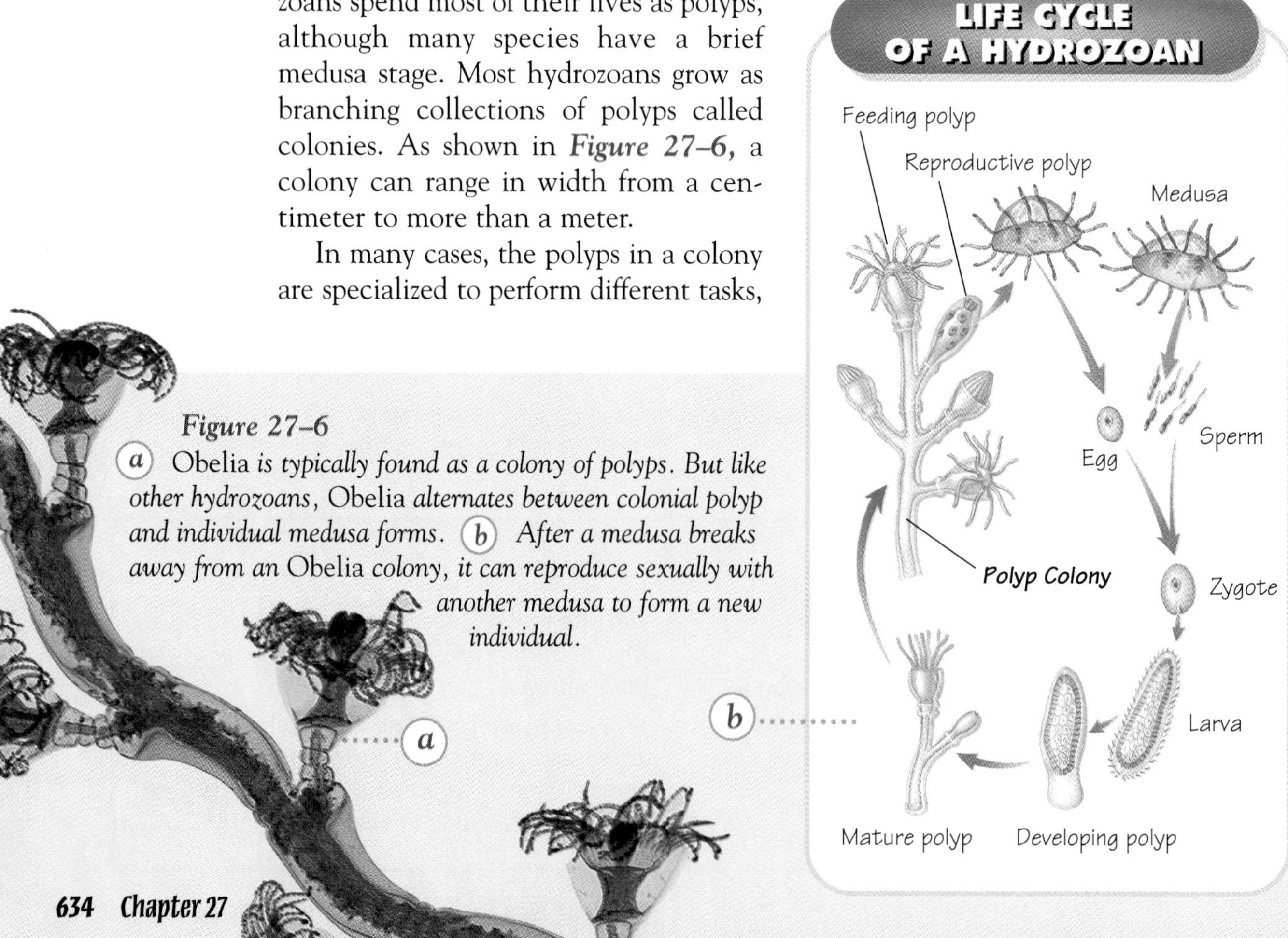

Figure 27–6
(a) Obelia *is typically found as a colony of polyps. But like other hydrozoans,* Obelia *alternates between colonial polyp and individual medusa forms.* (b) *After a medusa breaks away from an* Obelia *colony, it can reproduce sexually with another medusa to form a new individual.*

Figure 27–7
This large sea anemone has nematocysts on its tentacles that can paralyze fishes. However, it dines not on the small clownfish shown here, but on larger fishes that the clownfish attract.

cnidarian life cycles. The polyp stage is small and usually short-lived, however, while the medusa stage can grow to more than 3 meters in diameter and live a long time.

Most jellyfishes are harmless to humans, but several have stings that can cause allergic reactions in some people. A few species that live mostly in the South Pacific have venom powerful enough to kill a human in minutes.

Corals and Sea Anemones

The corals and sea anemones—members of the class Anthozoa (an-thuh-ZOH-uh)—include many of the most beautiful animals in the sea, as well as some of the most ecologically important. These animals grow either as solitary or colonial polyps and have no medusa stage.

Corals and sea anemones reproduce sexually when mature polyps produce eggs and sperm. Fertilization produces free-swimming larvae that attach themselves to rocks and grow into polyps. In addition, many anthozoans reproduce asexually by budding. And pieces of coral that break off colonies often survive and begin new colonies.

Coral Reefs

Hard corals, soft corals, and sea anemones are the best-known inhabitants of coral reefs, which are large living structures found near the coastlines in many tropical regions. As shown in the photograph on page 628, coral reefs can be as intricate as an Oriental carpet.

How do coral reefs form? With the help of the algae living inside them, hard coral colonies form layers of skeleton made of calcium carbonate. Over thousands of years, this calcium carbonate—together with carbonate rocks produced by algae—create the reef's foundation.

Section Review 27-1

1. **Explain** how sponges differ from other animals.
2. **Describe** the organization of tissues in cnidarians.
3. **Describe** the two stages of a cnidarian's life cycle.
4. **Critical Thinking—Analyzing** All sponges and cnidarians live underwater. Why are they ill-equipped to live on land?
5. **MINI LAB** How do living sponges **compare** with artificial household sponges?

SECTION 27–2

Unsegmented Worms

Guide for Reading

- **Describe** the characteristics of flatworms and nematodes.

MINI LAB

- **Identify** the structures of a tapeworm.

WHEN YOU HEAR THE WORD worm, do you think only of earthworms—the long, squirmy creatures that crawl out of the ground after it rains? If so, look at the worms shown on this page. In fact, worms come in different colors, shapes, and sizes. They are found in soil and in water. And they fill many different roles in nature—including the roles of pests and parasites. As you can see, there is more to worms than meets the eye!

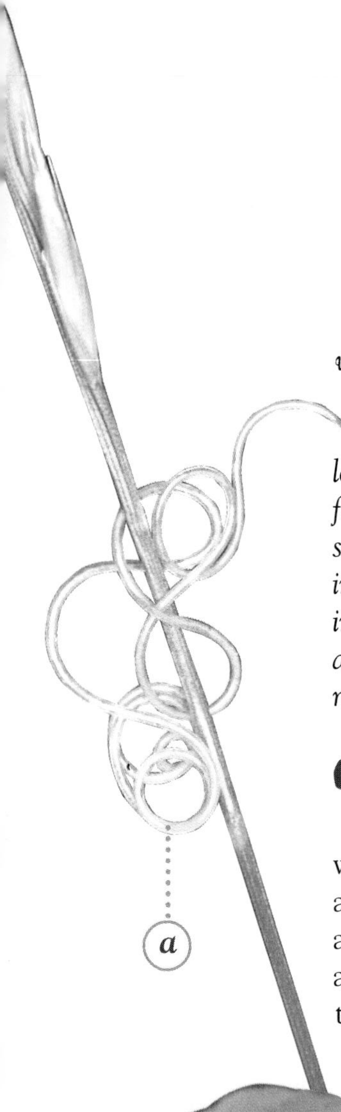

Classifying Worms

In everyday language, the word worm applies to a wide range of animals. In fact, many of the animals called worms, such as inchworms, are not true worms but insects!

To classify the real worms, biologists first look at a worm's body plan. Worms that have bodies divided into parts are called the segmented worms, and worms that are not divided in this way are called the unsegmented worms. There are two phyla of unsegmented worms, and we'll present both phyla in this section. Segmented worms will be discussed in the next chapter.

Platyhelminths

The members of phylum Platyhelminthes (plat-ih-hehl-MIHN-theez) are called platyhelminths or, more commonly, flatworms. **Flatworms are the simplest worms. They are also the simplest animals to show bilateral symmetry—meaning they have symmetric sides that can be identified as left and right.**

The name flatworm is appropriate because these worms are flat from top to bottom. Although flatworms can be several meters long, they are usually no more than a few millimeters thick. Otherwise, the flatworms are an enormously

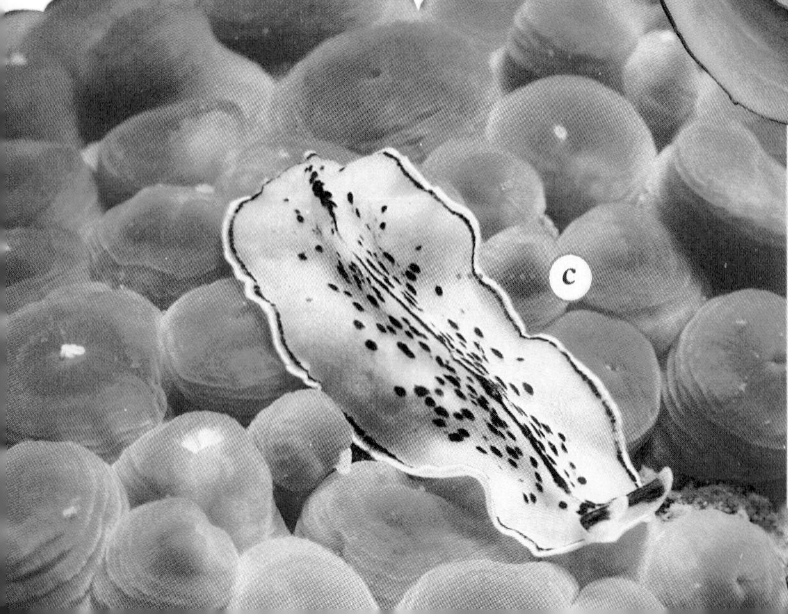

a

b

c

Figure 27–8
Worms can be aquatic or terrestrial, harmless or dangerous, and relatively large or very small. (a) This tiny nematode, often called a threadworm or eelworm, is coiled around a blade of grass. It is barely 10 centimeters long. (b) The parasitic nematode Toxocara canis *normally infects dogs, but sometimes it infects humans (magnification: 450X). (c) This marine flatworm is swimming over a bed of club anemones.*

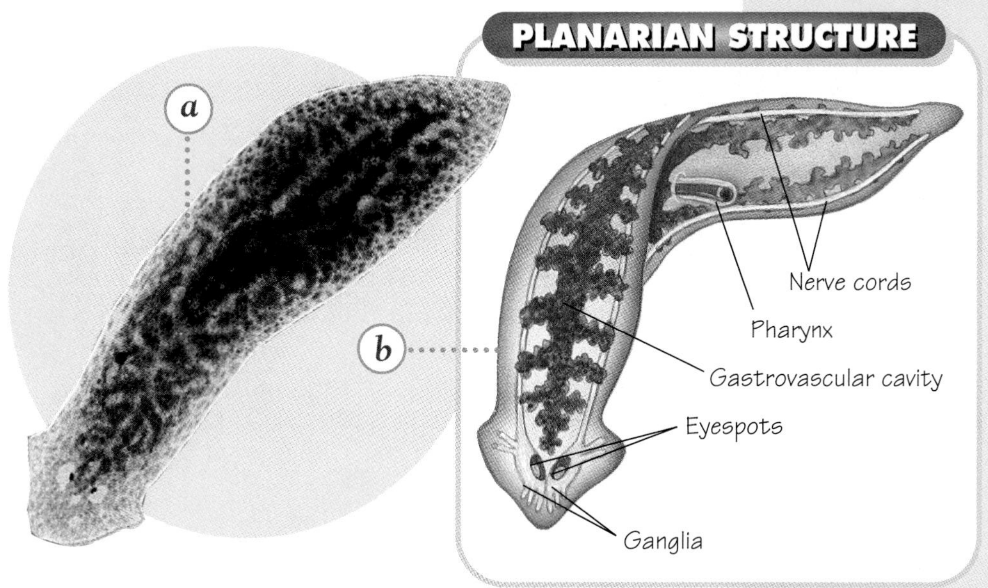

Figure 27–9
(a) Planarians are common in lakes, ponds, and streams (magnification: 10X).
(b) To take in food, a planarian secretes digestive juices on a prey, then sucks in food particles through its pharynx. The planarian's nervous system contains two eyespots connected to collections of nerve cells called ganglia. The ganglia act as a very simple brain.

diverse group. For example, some flatworms are harmless and free-living, while others are destructive parasites of plants, animals, and humans.

Along with bilateral symmetry, most flatworms have a few basic features in common. Typically, a flatworm's nervous and sensory systems show enough cephalization to reasonably label one end as a head. And because their bodies are so thin, they do not need respiratory or circulatory systems. Beyond these features, however, free-living and parasitic flatworms have very different anatomies and body plans. Let's take a closer look at three examples of free-living and parasitic flatworms.

☑ **Checkpoint** What is a flatworm?

Planarians

The free-living flatworms include the members of the class Turbellaria, commonly called planarians. Most planarians are either carnivores that feed on much smaller aquatic animals or scavengers that feed on dead or decaying organic matter.

Planarians are common throughout the world. Many species live in lakes, streams, and oceans, while other species live on land in wet parts of the tropics. In the United States, the most common planarians live in fresh water.

Flukes

The members of the class Trematoda (trehm-uh-TOH-duh) are called flukes. Flukes are parasites that live in the blood and tissues of various hosts.

Although flukes are usually less than a centimeter long, their actions as parasites cause serious, painful, and even life-threatening damage to millions of humans and countless other animals around the world. Some of the most destructive fluke parasites are native to Africa and Asia, and many were spread to South America and Central America during the 1800s, where they are now well-established. Flukes of the genus *Schistosoma* (shihs-tuh-SOHM-uh) infect about 200 million people around the world.

INTEGRATING HEALTH

What are the symptoms of a Schistosoma *infection? How can it be treated?*

Few of the flukes that cause serious human diseases are common in the United States, so most Americans know little about them. However, some species of *Schistosoma* are parasites of freshwater fishes and water birds, and they occasionally infect humans. These flukes are not adapted to be human parasites, however, so they don't live long. Instead, they produce a short-lived rash known as "swimmer's itch" before the body's defense mechanisms eliminate them.

Tapering Tapeworms

PROBLEM *How can you **identify** the structures of a tapeworm?*

PROCEDURE

1. Study a slide of a tapeworm under the low-power and high-power objectives of a compound microscope.
2. Sketch the tapeworm's body. Label the structures that you can identify.

ANALYZE AND CONCLUDE

1. What structures does the scolex contain? What is the role of the scolex?
2. How much of the tapeworm's body consists of proglottids? Why are proglottids important for the tapeworm?
3. Why does a tapeworm lack elaborate systems for digestion and elimination?

Tapeworms

The members of the class Cestoda (sehs-TOHD-uh)—the tapeworms—are another class of parasitic flatworms. Tapeworms live in the intestines of their hosts. With this arrangement, the tapeworms' food is not only collected for them, it is also digested by the host's enzymes! For this reason, tapeworms have no gut or mouth of their own, and they have no use for a nervous system or sense organs.

As shown in ***Figure 27–10,*** the front end of a tapeworm contains several suckers and a ring of hooks that attach to the intestinal wall of the host. Because the front end is not really a proper head, it is called a **scolex** (SKOH-lehks). A tapeworm hangs in the intestine from its scolex, absorbing digested food as it passes by.

The rest of the tapeworm's body consists of segments called **proglottids** (proh-GLAHT-ihdz). Proglottids consist of little more than male and female reproductive organs, which constantly produce eggs and sperm. One proglottid can contain more than 100,000 eggs, and a single tapeworm can produce as many as half a billion eggs in a single year!

Tapeworms rarely cause death directly. But because they "steal" a lot of food, they can cause their hosts to lose weight and become weak.

Flatworm Life Cycles

Like their body plans, the life cycles of flatworms vary enormously between free-living and parasitic species. The free-living flatworms usually have simple life cycles. However, many have the unusual feature of being simultaneous hermaphrodites, which means that they carry functioning male and female reproductive organs at the same time. These flatworms almost never fertilize their own eggs, however. Instead, they pair with another member of their species and the two cross-fertilize each other.

Figure 27–10
The scolex—the front end of a tapeworm—contains hooks and suckers that attach to a host's intestines. The rest of a tapeworm's body consists of segments called proglottids, each of which constantly produce eggs and sperm.

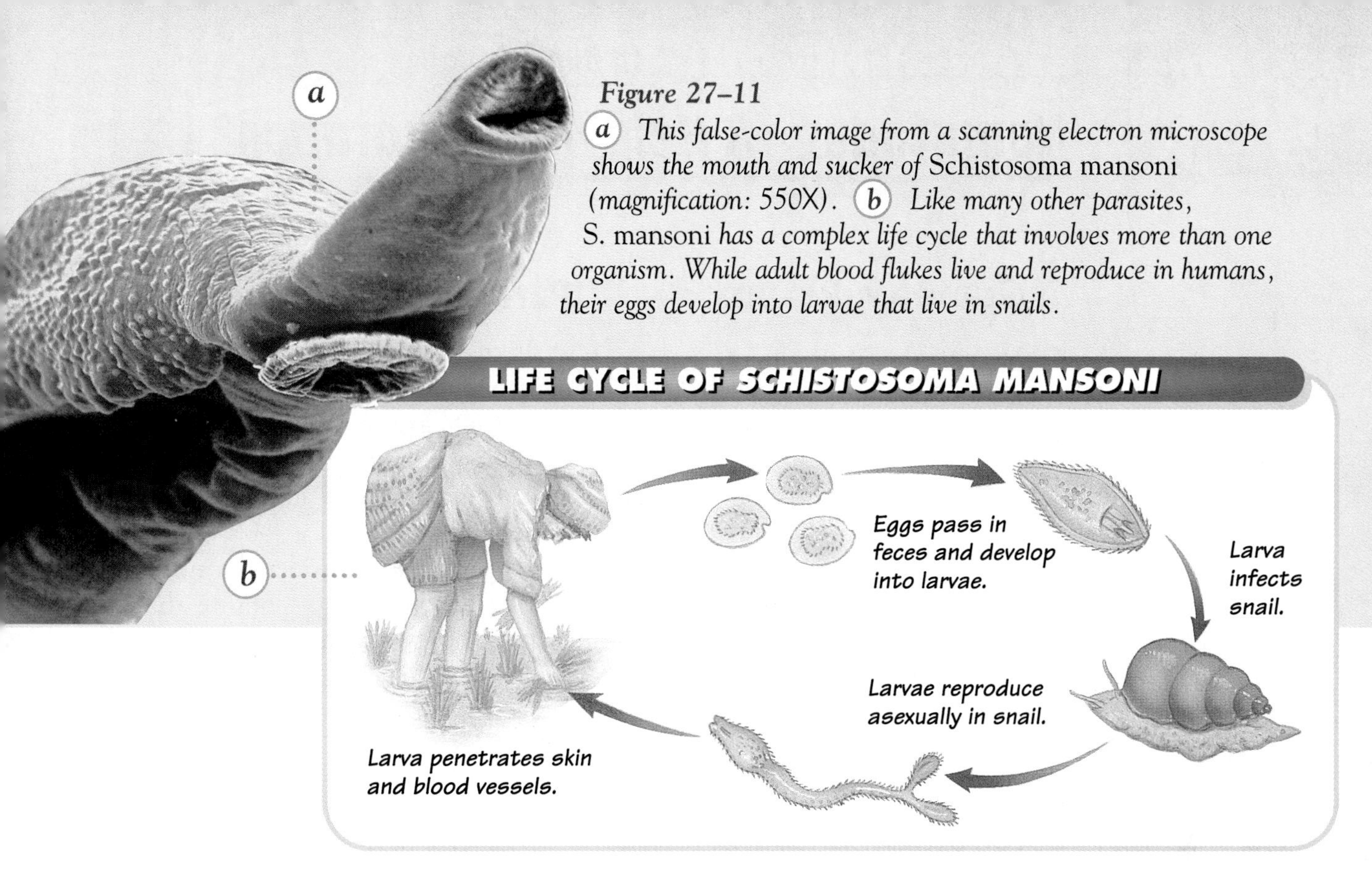

Figure 27–11
(a) *This false-color image from a scanning electron microscope shows the mouth and sucker of* Schistosoma mansoni *(magnification: 550X).* (b) *Like many other parasites,* S. mansoni *has a complex life cycle that involves more than one organism. While adult blood flukes live and reproduce in humans, their eggs develop into larvae that live in snails.*

Many parasitic flatworms have complicated life cycles, often involving two host species or even three or more. Blood flukes, for example, have several stages in a life cycle that involves two separate hosts—humans and snails. Tapeworms also switch back and forth between two different hosts.

Nematodes

The members of the phylum Nematoda—the nematodes—are also called roundworms. You may never have heard of these worms, but they play important roles in the biosphere—and in the study of biology. **The nematodes are the simplest animals with a digestive tract similar to a human's, with a mouth at one end and an anus at the other.** Although most nematodes are free-living, others are parasites of humans, other animals, and plants. In fact, nematodes are parasites of almost every kind of plant and animal.

Nematodes range in length from 1 millimeter to more than 1 meter. Although nematodes are often inconspicuous, they may be the most numerous of all multicellular animals. The free-living nematodes live in soil, water, sand, and on or around the bodies of other organisms. Just one rotting apple may contain nearly 100,000 of them!

Section Review 27–2

1. **Describe** the characteristics of flatworms and nematodes.
2. **Critical Thinking—Comparing** Compare the life cycles and body structures of a planarian and a tapeworm. Why does a tapeworm produce a far greater number of eggs than a planarian produces?
3. **MINI LAB** How can you **identify** the structures of a tapeworm?

Nematodes—A Help or a Hindrance?

Guide for Reading

- **Describe** the life cycles of parasitic nematodes.
- **Explain** why researchers study *Caenorhabditis elegans*.

IN OUR MODERN, INDUSTRIAL society, it is often easy to ignore animals that are too small to see or that live in natural habitats that we don't often visit. But that doesn't mean these animals are unimportant.

Nematodes arrived on the Earth hundreds of millions of years ago and have been evolving relationships with other organisms ever since. Unfortunately, this process has produced worms that cause a great deal of human pain and suffering. But not all nematodes are harmful to humans. In fact, one nematode species just might hold the key to some very useful information about all life on Earth.

Parasitic Nematodes

Like the parasitic flatworms, parasitic nematodes typically have life cycles that are quite complex, often involving more than one host. In addition, some of these life cycles can be quite gruesome!

For example, consider the nematode *Ascaris lumbricoides*, a parasite that infects millions of humans every year. Like the eggs of flukes, the eggs of *A. lumbricoides* are carried in feces, and they can spread by improper disposal of human wastes. If those eggs are eaten on food that has not been properly washed and cooked, the life cycle shown in ***Figure 27–13*** could be the result.

Trichinella, another parasitic nematode, causes a terrible disease called trichinosis. Although adult *Trichinella* worms live in the intestines, much of the damage of trichinosis comes from the worms' larval forms. The larvae travel through the bloodstream and burrow into muscles and other organs, which is extremely painful.

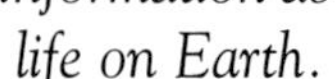

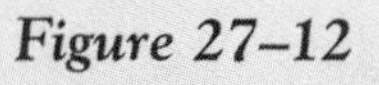

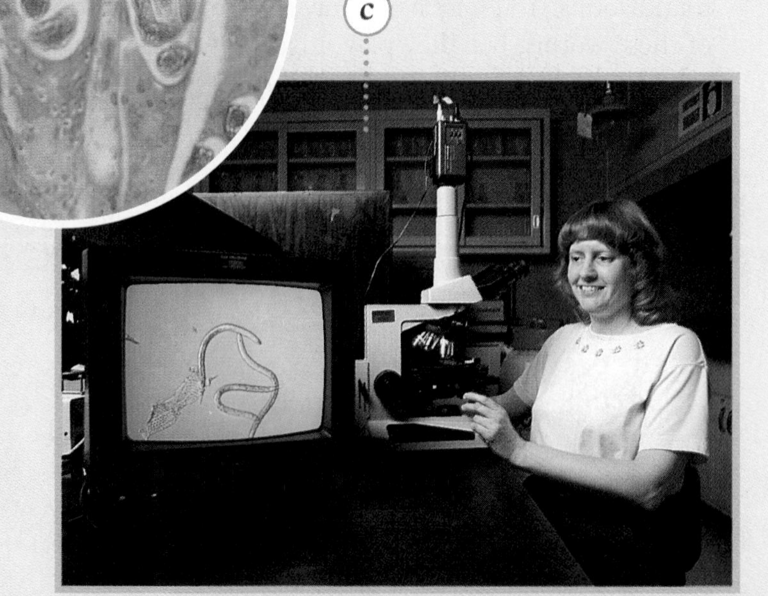

Figure 27–12
Although many nematodes cause a great deal of human suffering, others are beneficial—and in some surprising ways. (a) Caenorhabditis elegans, *a free-living nematode, has proved valuable in genetic research (magnification: 32X).* (b) *Larvae of the nematode* Trichinella *can wall themselves inside human muscle cells, a condition that is both painful and difficult to treat (magnification: 35X).* (c) CAREER TRACK *Nematologists work to protect plants and animals from nematode pests.*

Like other nematode parasites, *Trichinella* has a complex life cycle that involves more than one host. Almost always, humans acquire trichinosis by eating undercooked pork that contains *Trichinella* cysts.

☑ **Checkpoint** What causes the disease trichinosis?

Caenorhabditis elegans

Caenorhabditis elegans, a tiny nematode, is providing incredibly important information to a wide variety of different researchers. **Because *Caenorhabditis elegans* is such a simple animal, it contains relatively few genetic instructions. This means that researchers have a chance to identify and study its entire genome.** What's more, these animals are easy to culture, and they grow to maturity in only three-and-a-half days. This means that their characteristics can be tracked quickly.

C. *elegans* has another unusual feature. Curiously, every adult worm of this species has exactly 959 cells. And because the worms are transparent, researchers can watch each cell develop. As a result, scientists now know the precise location and complete history of each C. *elegans* cell from the moment the egg is fertilized! This information helps researchers to study how genes direct growth and development.

Research on C. *elegans* is yielding clues to all sorts of mysteries, such as why animals age. Recently, researchers uncovered several genes that enable a worm to live as long as two months, which is five times longer than its usual life span. It may take years to determine just how these genes work, but the results may provide clues to determining why all animals—including humans—live to the ages that they do.

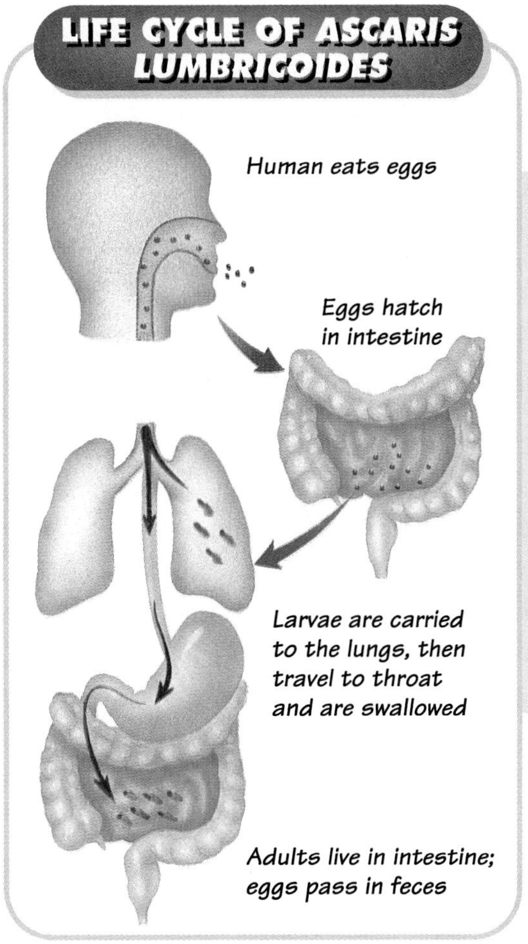

LIFE CYCLE OF ASCARIS LUMBRICOIDES

Figure 27–13 Ascaris lumbricoides *has an unusually complex life cycle that involves a human's digestive tract, bloodstream, and lungs. Although most* Ascaris *infections are not serious, severe infections can block a segment of the gut or spread to the appendix or other organs.*

Section Review 27–3

1. **Describe** the life cycles of parasitic nematodes.
2. **Explain** why researchers study *Caenorhabditis elegans.*
3. **BRANCHING OUT ACTIVITY** Nematodes are found all over the Earth. **Research** different nematodes and the roles they fill.

CHAPTER 27

Laboratory Investigation

DESIGNING AN EXPERIMENT

Observing a Hydra and a Planarian

Like all living things, hydras and planarians respond to stimuli, or changes in their environment. In hydras and planarians, these responses depend on a nervous system. The hydra's nervous system consists of a nerve net, whereas the planarian's nervous system has a small brain, eyespot, and nerve cord. In this investigation, you will observe a hydra and a planarian and see how they respond to different stimuli.

Problem

How do a hydra and a planarian respond to different stimuli? **Design an experiment** to answer this question.

Suggested Materials

hydra culture
planarian culture
medicine droppers
culture dishes
colored brine shrimp
compound microscope
toothpicks
stereomicroscope
blunt metal probe
coverslips
petroleum jelly
depression slides
dilute acid solution

Suggested Procedure

1. **Using a medicine dropper, transfer a hydra to a small culture dish half filled with water.**
2. **Observe the hydra under the stereomicroscope. Sketch and label its structures.**
3. **Add a few brine shrimp to the culture dish. Use the stereomicroscope to observe how the hydra reacts to brine shrimp.**
4. **To observe the hydra under the compound microscope, prepare a "hanging drop." First, use a toothpick to dab a small amount of petroleum jelly on the corners of the upper side of a coverslip.**

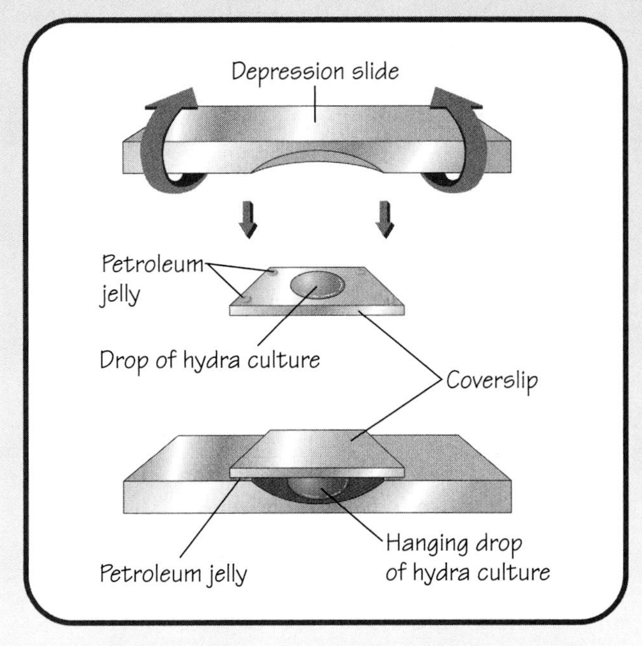

5. **Use a medicine dropper to transfer the hydra to the middle of the coverslip. Hold a depression slide over the coverslip, depression side down, and lower it onto the coverslip, as shown in the illustration. Turn the slide over so the coverslip faces up.**
6. **Using the low-power objective of a compound microscope, observe the hydra and brine shrimp. Record your observations. When you have finished, return the hydra to your teacher.**
7. **Using a similar procedure, design an experiment to determine how a hydra responds to the following stimuli:**
   - **light touch from a toothpick**
   - **dilute acid solution**
8. **Using steps 1 to 3, design an experiment to determine how a planarian responds to the following stimuli:**
   - **light touch from a metal probe**
   - **brine shrimp**
9. **Write your hypotheses. With your teacher's approval, carry out the experiments you designed. Record your observations.**

Observations

1. Draw and label the parts of the hydra and the planarian.
2. Describe the hydra's responses to brine shrimp, touch, and the dilute acid solution.
3. Describe the planarian's responses to touch and the brine shrimp.

Analysis and Conclusions

1. Did the hydra detect the shrimp at a distance or by physical contact?
2. What reaction did the hydra have to touch? To the dilute acid solution? Why might such reactions be useful?
3. Based on your observations, how does the planarian respond to brine shrimp? What can you infer about the structures in the planarian that are involved in this response?

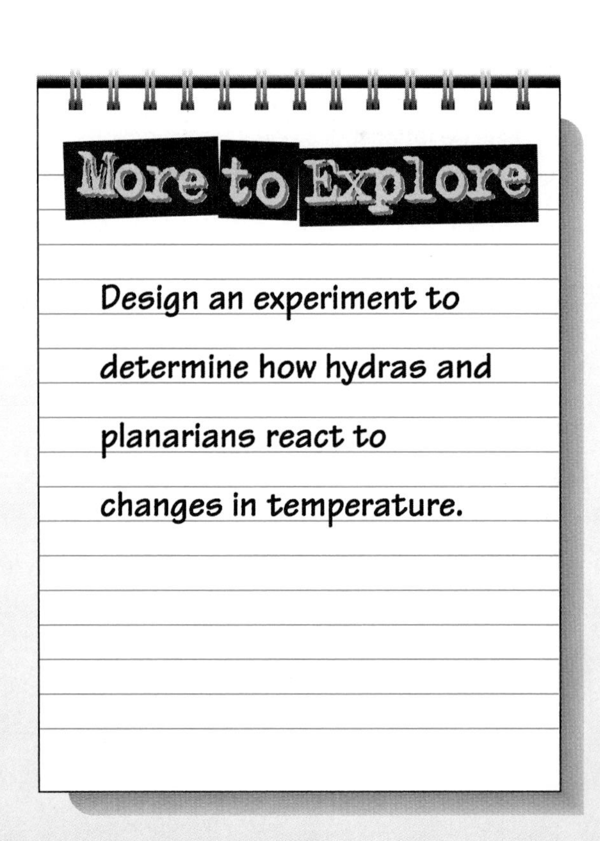

More to Explore

Design an experiment to determine how hydras and planarians react to changes in temperature.

Study Guide

Summarizing Key Concepts

The key concepts in each section of this chapter are listed below to help you review the chapter content. Make sure you understand each concept and its relationship to other concepts and to the theme of this chapter.

27–1 Sponges and Cnidarians

- Unlike the cells of other animals, a sponge's cells are not organized into recognizable organs or tissues.
- Sponges can reproduce sexually with sperm and egg cells. They can also reproduce asexually by making ball-shaped structures called gemmules or through a process called budding.
- Cnidarians are the corals, jellyfishes, and sea anemones. All cnidarians live underwater.
- Unlike sponges, cnidarians have layers of differentiated cells that are organized into three specialized layers of tissues.
- Cnidarians alternate between two stages—a free-living stage called a medusa and a sessile (attached) stage called a polyp. Many cnidarians also alternate between sexual and asexual reproduction.

27–2 Unsegmented Worms

- Flatworms are the simplest worms. They are also the simplest animals to show bilateral symmetry—identifiable left and right sides.
- Planarians are common free-living flatworms. Parasitic flatworms include flukes, which live in blood and other tissues, and tapeworms, which live in intestines.
- Nematodes are the simplest animals with a digestive tract similar to a human's, with a mouth at one end and an anus at the other.
- Free-living nematodes include carnivores, herbivores, and detritus feeders such as *Caenorhabditis elegans.*

27–3 Nematodes—A Help or a Hindrance?

- Like the parasitic flatworms, parasitic nematodes typically have life cycles that are quite complex, often involving more than one host. Nematodes cause trichinosis and other diseases.
- *Caenorhabditis elegans* contains few genetic instructions, allowing researchers the chance to identify and study all of them.

Reviewing Key Terms

Review the following vocabulary terms and their meaning. Then use each term in a complete sentence.

27–1 Sponges and Cnidarians

collar cell
amebocyte
spicule
spongin
gemmule
budding
endoderm
ectoderm
mesoglea
ocellus
statocyst
nematocyst
gastrovascular cavity
medusa
polyp

27–2 Unsegmented Worms

scolex
proglottid

Recalling Main Ideas

Choose the letter of the answer that best completes the statement or answers the question.

1. Which animal group has the least in common with the other three?

   a. sponges **c.** flatworms
   b. cnidarians **d.** roundworms

2. Sponges produce ball-shaped structures called gemmules for

   a. asexual reproduction.
   b. sexual reproduction.
   c. food digestion.
   d. structural support.

3. In cnidarians, the endoderm, ectoderm, and mesoglea are the names of

   a. reproductive cells. **c.** nerve cells.
   b. tissue layers. **d.** sensory organs.

4. Cnidarians balance themselves with information from

   a. ocelli. **c.** statocysts.
   b. collar cells. **d.** amebocytes.

5. A typical jellyfish spends most of its life as a

   a. sessile organism. **c.** solitary polyp.
   b. medusa. **d.** colonial polyp.

6. Flatworms are the simplest animals to have

   a. radial symmetry. **c.** a mouth and anus.
   b. bilateral symmetry. **d.** a respiratory tract.

7. Which are free-living flatworms?

   a. planarians **c.** tapeworms
   b. flukes **d.** nematodes

8. Nematodes are parasites of

   a. humans only.
   b. dogs and cats only.
   c. plants only.
   d. many animals and plants.

9. *Ascaris lumbricoides* is an example of a

   a. free-living flatworm. **c.** free-living nematode.
   b. parasitic flatworm. **d.** parasitic nematode.

Putting It All Together

Using the information on pages xxx to xxxi, complete the following concept map.

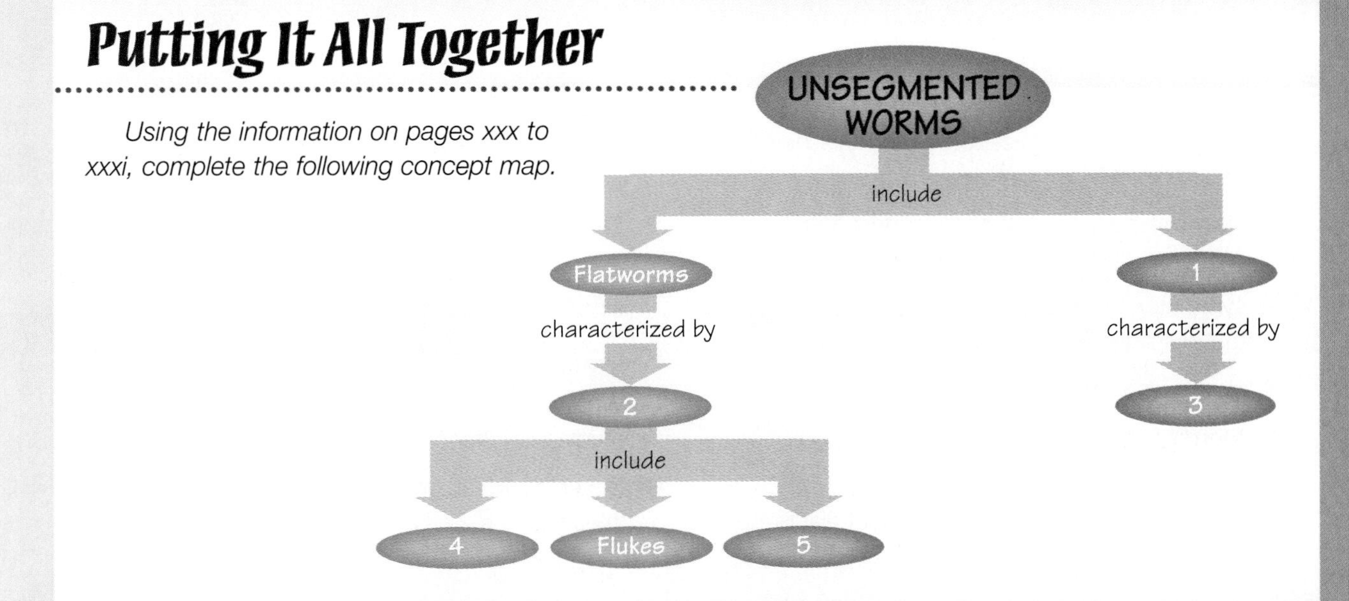

Reviewing What You Learned

Answer each of the following in a complete sentence.

1. How do sponges differ from other animals?
2. How does a sponge support its body?
3. What are collar cells?
4. What are ocelli?
5. What is budding?
6. What is a gastrovascular cavity? Name two organisms that have a gastrovascular cavity.
7. Identify three different hydrozoans.
8. Which flatworms are free-living? Which are parasites?
9. How does a tapeworm attach to its host's intestines?
10. What are proglottids?
11. What is a simultaneous hermaphrodite?
12. How is a nematode's digestive tract similar to a human's?
13. What is trichinosis?
14. List the names of two nematode species.

Expanding the Concepts

Discuss each of the following in a brief paragraph.

1. Why are sponges classified as animals instead of as plants?
2. How do sponges take in and digest food?
3. **Compare** radial symmetry with bilateral symmetry. Give an example of an organism that displays each type of symmetry.
4. How do coral reefs form?
5. **Identify** the features that all flatworms have in common.
6. Why can tapeworms survive without elaborate digestive and circulatory systems?
7. What is "swimmer's itch"? How is it transmitted?
8. Describe the life cycle of a blood fluke.
9. Describe reproduction in flatworms that are simultaneous hermaphrodites.
10. How does *Trichinella* infect humans?
11. What useful information has been provided by studies of *Caenorhabditis elegans*?
12. Describe the stages of an *Ascaris lumbricoides* infection.

CHAPTER 27

Extending Your Thinking

Use the skills you have developed in this chapter to answer the following.

1. **Designing an experiment** Suppose that while transplanting a tomato plant, you discover several small white worms attached to the plant's roots. Design an experiment to determine whether these worms damage tomato plants.
2. **Problem solving** A medical student suspects that a patient has contracted an intestinal disease caused by flatworms or nematodes. To make an accurate diagnosis, what questions should the medical student ask the patient?
3. **Hypothesizing** On a certain tropical island, the wind typically blows in one direction only. Offshore from the island, coral reefs grow on the side that faces the wind but not on the side away from the wind. Formulate a hypothesis to explain why coral reefs would grow in this way.
4. **Analyzing data** In its medusa stage, a cnidarian moves freely in the water. In its polyp stage, it is sessile—attached to a rock or other structure. What advantages does this two-stage life cycle provide?
5. **Constructing a model** Construct a model of a sponge, planarian, tapeworm, or other organism presented in this chapter. Label the organism's important structures.

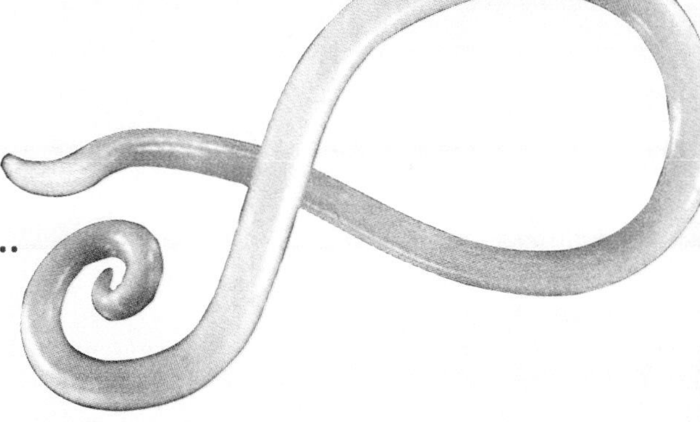

Applying Your Skills

How Are They Alike?

In this chapter, you studied a wide variety of animals. Although these animals are quite different from one another, they are similar in several basic ways.

1. Working in groups, observe a piece of coral. Record the structures you observe, then label the structures you can identify.
2. Compare the coral's anatomy with the anatomies of a hydra and a planarian. How do these organisms obtain food? How do they reproduce? Why are they classified as animals?
3. Compare corals, hydras, and planarians with other animals you have studied in this chapter, such as sponges, jellyfishes, flukes, tapeworms, and nematodes. Describe their different feeding and reproductive strategies.

• GOING FURTHER •

4. Construct a chart that presents the important information about these animals.
5. Research one of the diseases caused by unsegmented worms. Prepare a report on your findings.

CHAPTER 28

Mollusks, Annelids, and Echinoderms

FOCUSING THE CHAPTER
THEME: Unity and Diversity

28–1 Mollusks
- **Compare** the different classes of mollusks.

28–2 Annelids
- **Identify** the main characteristics of annelids.

28–3 Echinoderms
- **Describe** the features of echinoderms.

BRANCHING OUT *In Depth*

28–4 Adaptations of Cephalopods
- **Explain** why some cephalopods grow larger than other mollusks.

LABORATORY INVESTIGATION
- **Design an experiment** to discover how earthworms respond to stimuli.

Biology and Your World
BIO JOURNAL

In your journal, describe the starfish shown on this page. How is it similar to other organisms you have studied? How is it different?

Red sunstar—a ten-armed echinoderm

Mollusks

Guide for Reading

- **Identify** the characteristics of mollusks.
- **Describe** the different classes of mollusks.

MINI LAB

- **Classify** a collection of mollusk shells.

THEY CLIMB TREES IN TROPICAL rain forests and float over coral reefs. They crawl through garbage cans, eat their way through farm crops, and speed through the deep ocean. Some are so small that you can hardly see them with the unaided eye, while others are 20 meters long! They are the mollusks—one of the oldest and most diverse animal phyla.

Mollusks come in so many sizes, shapes, and forms that you might wonder why they are classified in the same phylum. To learn the answer, read on.

Mollusk Structures

All mollusks have similar body plans that typically include a foot, gut, mantle, and shell. They also pass through similar stages in their early development. The **foot** is a soft, muscular structure that usually contains the mouth of the mollusk. The mouth typically contains other feeding structures, such as a structure called the **radula** (RAJ-oo-luh). The mollusk's body contains its digestive tract—its **gut**—and other internal organs. The **mantle** is a thin, delicate layer of tissue that surrounds the mollusk's body, much like a cloak. Glands in the mantle secrete calcium carbonate, which forms the mollusk's **shell.** Although not all mollusks have shells, those that do not are thought to have evolved from shelled ancestors.

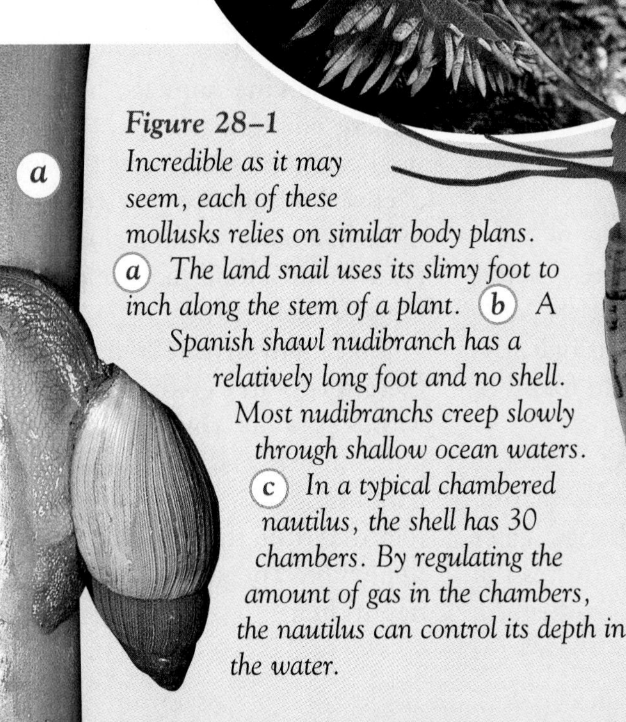

Figure 28–1 *Incredible as it may seem, each of these mollusks relies on similar body plans. (a) The land snail uses its slimy foot to inch along the stem of a plant. (b) A Spanish shawl nudibranch has a relatively long foot and no shell. Most nudibranchs creep slowly through shallow ocean waters. (c) In a typical chambered nautilus, the shell has 30 chambers. By regulating the amount of gas in the chambers, the nautilus can control its depth in the water.*

MOLLUSK BODY PLANS

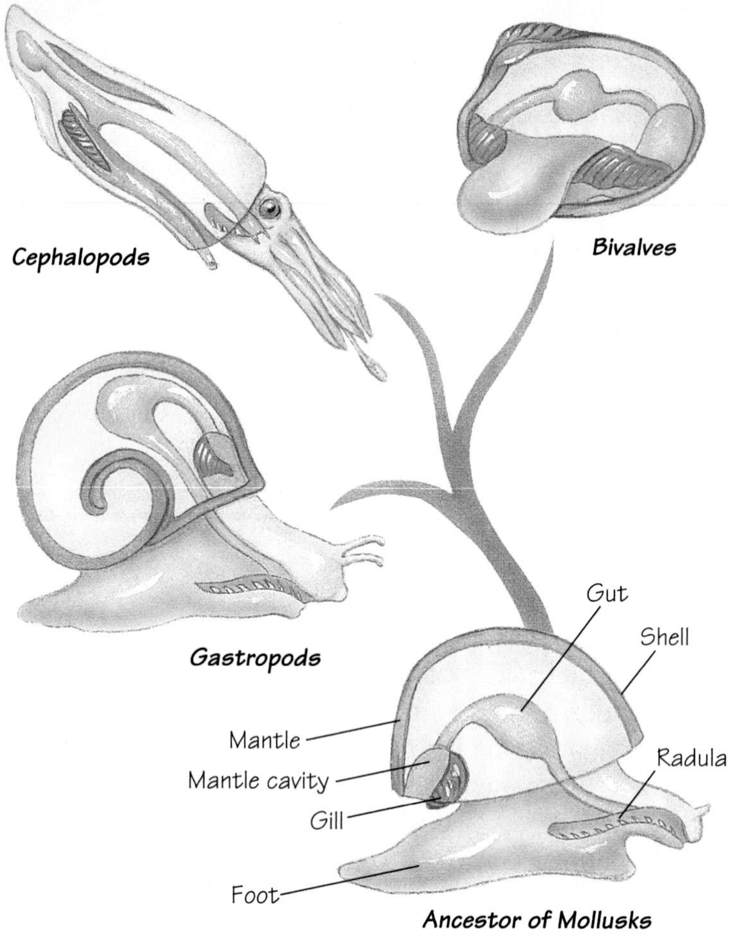

Figure 28–2
The first mollusks lived in water and were protected by a hard external shell. As they evolved, their body parts changed and adapted for different purposes, producing the diversity of mollusks that live today.

Over the past 600 million years, the members of the phylum Mollusca have evolved to live in nearly every habitat and to eat almost anything. You can think of the mollusk's basic parts as a "tool kit" that adapted for different uses by changing shape, size, and position.

☑ **Checkpoint** What is a mollusk?

Feeding and Respiration

Mollusks may be carnivores, herbivores, filter feeders, detritus feeders, scavengers, or parasites. Not surprisingly, the mollusks' feeding structures are found in a wide variety of sizes and shapes.

Most aquatic mollusks breathe through gills located inside the mantle cavity, and they also exchange gases through exposed, wet skin. Terrestrial mollusks, such as slugs and land snails, usually breathe through a specially adapted mantle cavity. This mantle is well supplied with blood vessels and is folded to increase its surface area.

The mantle of terrestrial mollusks regularly loses moisture to dry air, so it must be kept moist. That's one reason why most terrestrial mollusks prefer to move around at night, during rainstorms, and in places where the air is moist.

Internal Transport and Excretion

Internal transport in mollusks is managed by a circulatory system with a simple heart. In sessile and slow-moving mollusks, blood is pumped from the heart through open spaces called sinuses. The blood then enters vessels that pass through gills and, eventually, return to the heart. This type of circulatory system is called an **open circulatory system.** While this system is not especially efficient, it suffices for "stick-in-the-mud" mollusks like clams and oysters.

Fast-moving animals, however, need a more efficient transport system. Mollusks such as squids and octopuses use a **closed circulatory system,** which is more like our own. In a closed circulatory system, blood is pumped through blood vessels only.

Like many invertebrates, mollusks excrete wastes in two different ways. Solid wastes pass out through the end of the gut, known as the anus. Nitrogen-containing wastes, such as ammonia, are removed from the blood by organs called **nephridia** (nee-FRIHD-ee-uh; singular: nephridium).

Figure 28–3
(a) Like other bivalves, this giant clam filters small food particles from the water. (b) A pteropod—also known as a sea butterfly—has a poison-filled radula that shoots like a dart at unwary prey.

Response

Mollusks have almost as many different kinds of nervous systems as they have habitats. Clams and their close relatives have a few ganglia and nerve cords, as well as scattered sensory organs used for balance, touch, taste, and detection of light. Other mollusks evolved more advanced nervous systems. Octopuses and their kin have well-developed brains, excellent vision, and sophisticated senses of touch and taste.

Mollusk Life Cycles

Most mollusks have straightforward life cycles that involve separate males and females. In almost all species, the fertilized egg develops into a type of free-swimming larva called a **trochophore** (TRAHK-oh-for). Eventually, this free-swimming larva transforms into a miniature adult, which in some species settles down and becomes sessile and in other species continues swimming.

Many aquatic mollusks release sperm and egg cells into the open water, where external fertilization occurs more or less by chance. Other species, such as octopuses and squids, reproduce with internal fertilization, in which the male uses a tentacle to deliver sperm to the female.

Quite a number of mollusks are hermaphrodites—animals that have both male and female reproductive organs. Certain hermaphroditic snails often pair together and fertilize each other's eggs. Sometimes they line up in long chains, each fertilizing the individual in front while being fertilized by the one behind.

☑ **Checkpoint** What is a trochophore?

Classifying Mollusks

As you can see, there is more to mollusks than meets the eye! Let's take a closer look at the different classes in this fascinating phylum of animals.

Gastropods

The members of the class Gastropoda (gas-TRAHP-oh-duh) include the familiar snails and slugs found across North America and elsewhere. The gastropods also include such exotic and brilliantly colored species as abalones, nudibranchs, and sea hares.

Figure 28–4
This photograph shows a marine snail in an advanced larval stage. The snail's transparent shell reveals its stomach, liver, heart, and other internal organs.

Problem Solving

The Invasion of the Zebra Mussels

In 1986, a ship from Europe emptied its ballast water in the Great Lakes region of North America. Unfortunately, the ballast water contained a type of mollusk called a zebra mussel. Zebra mussels have been a nuisance in the Great Lakes ever since!

Zebra mussels attach to almost anything underwater, then reproduce in great numbers. In the Great Lakes, dense mats of zebra mussel shells are clogging water intake pipes and other structures. The zebra mussels also consume huge quantities of plankton, a food source for many other marine organisms.

Biologists hope that ducks that live on the Great Lakes will start eating more zebra mussels. In one experiment that took place in Lake Erie, researchers monitored several rocks on which zebra mussels were growing. On half the rocks, they placed cages that ducks could not penetrate. After a few weeks, they captured several ducks and measured the mass of zebra mussel shells inside them. The results are shown in the graphs.

RESULTS OF ZEBRA MUSSEL EXPERIMENT

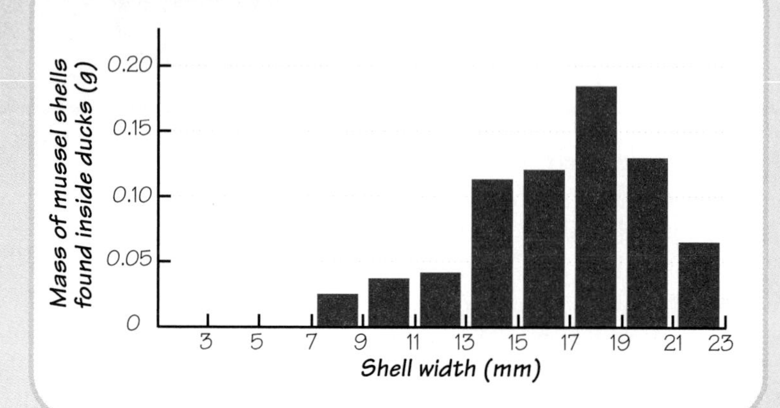

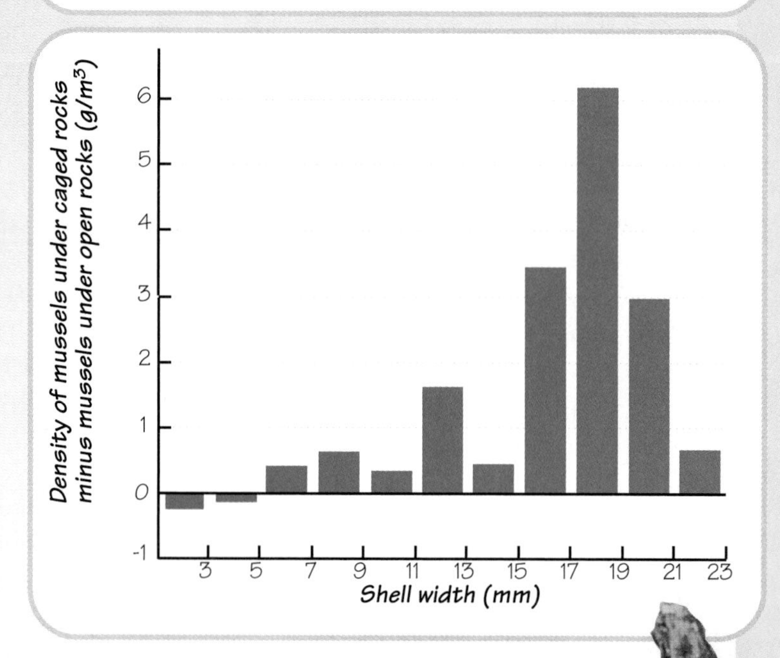

THINK ABOUT IT

1. Interpret the data presented in the graphs.
2. Explain why the researchers placed cages on half the rocks they monitored.
3. From the information presented, what can you conclude about the ducks' ability to control the zebra mussel population of Lake Erie?

Seashells by the Seashore

PROBLEM *How can you **classify** different mollusks?*

PROCEDURE

1. Obtain a sample of mollusk shells from your teacher. Study the outer and inner surfaces of the shells.
2. Group the shells according to their similarities.

ANALYZE AND CONCLUDE

1. Compare the outer and inner surfaces of the shells.
2. Explain how you classified the shells into groups. Discuss any alternative classification scheme you considered.
3. From studying these shells, what can you infer about the organisms that produced them?

Most gastropods have a one-piece shell into which they can withdraw for defense. Others have only small internal shells, while still others have lost their shells altogether. Many shell-less species protect themselves by producing powerful poisons or by having a bad taste to predators.

Bivalves

You may have dined on members of the class Bivalvia, which include clams, oysters, mussels, and scallops. Most bivalves live in shallow waters near the shore. They typically are filter feeders, which means they feed by filtering organisms from huge volumes of water.

Bivalves have two shells that are hinged at the back and held together by strong muscles. The mantle of the bivalve secretes the shell that encloses the organism, as well as a substance called nacre (NAY-ker), which provides the shell with a smooth, shiny inner coating.

Nacre also produces a highly valued product of oysters and other bivalves. Because a bivalve opens its valves so often, it sometimes takes in a pebble or a sand grain. The mantle responds by coating such objects with nacre, which over many years creates a pearl. For this reason, nacre is also called mother-of-pearl.

☑ *Checkpoint* What is a bivalve?

Cephalopods

The members of the class Cephalopoda (sehf-uh-LAHP-oh-duh) are the largest, most active, and most intelligent mollusks. **Most living cephalopods have shells that are either small, internal, or missing altogether.** The only living cephalopod with a large external shell is the chambered nautilus, shown in *Figure 28–1* on page 649. The chambered nautilus resembles the cephalopods from the beginning of the Cambrian Period, more than 500 million years ago.

Most cephalopods have long tentacles armed with sucker disks. These tentacles can grab and handle prey ranging from other mollusks to crabs, shrimps, and fishes.

Section Review 28–1

1. **Identify** the main characteristics of mollusks.
2. **Describe** the different classes of mollusks.
3. **Critical Thinking—Analyzing** Why can bivalves live without an elaborate brain and nervous system, such as those found in cephalopods?
4. **MINI LAB** How can you **classify** mollusk shells?

SECTION

28–2 Annelids

GUIDE FOR READING

- **Describe** the annelids.
- **Identify** the three classes of annelids.

YOU PROBABLY ARE FAMILIAR with one annelid—the common earthworm. But annelids include a huge number of species, including those pictured on this page.

Annelids live in many different habitats—including oceans, fresh water, and land. Some are less than a millimeter long, while others grow to 3 meters. Some are herbivores, others are harmless filter feeders, while still others are fearsome predators—at least for their size!

Annelid Body Plan

Segmented worms are members of the phylum Annelida. **All annelids have bodies that are divided into individual segments.** The segments are filled with fluid and are tightly sealed by body walls called **septa.**

Because of its fluid-filled segments and muscle organization, an annelid moves in an interesting way. Running up and down the annelid are **longitudinal muscles.** When these muscles contract, they shorten the segments. Surrounding these longitudinal muscles are ring-shaped **circular muscles.** When these muscles contract, they lengthen the segments. By coordinating these two sets of muscles, the annelid can move—albeit slowly—in almost any direction.

One common annelid is the earthworm. To learn more about earthworms, study the illustration on the next page.

Classifying Annelids

The annelids are very numerous and live in all sorts of habitats on Earth. **Annelids are divided into these three classes: the Oligochaeta, the Polychaeta, and the Hirudinea.**

Figure 28–5 *Different annelids evolved different uses for their setae, or bristles. (a) This bristleworm uses its setae to swim, as well as to sting potential predators. (b) In plume worms, the setae are adapted for filtering food from the water surrounding them. This plume worm lives in the ocean off the coast of Florida. (c) Sandworms use spikelike setae to move through desert sands.*

Visualizing an Earthworm

Most segments of an earthworm are nearly identical—both on the inside and on the outside. A few segments, however, contain specialized structures for such functions as digestion, circulation, and control of the nervous system.

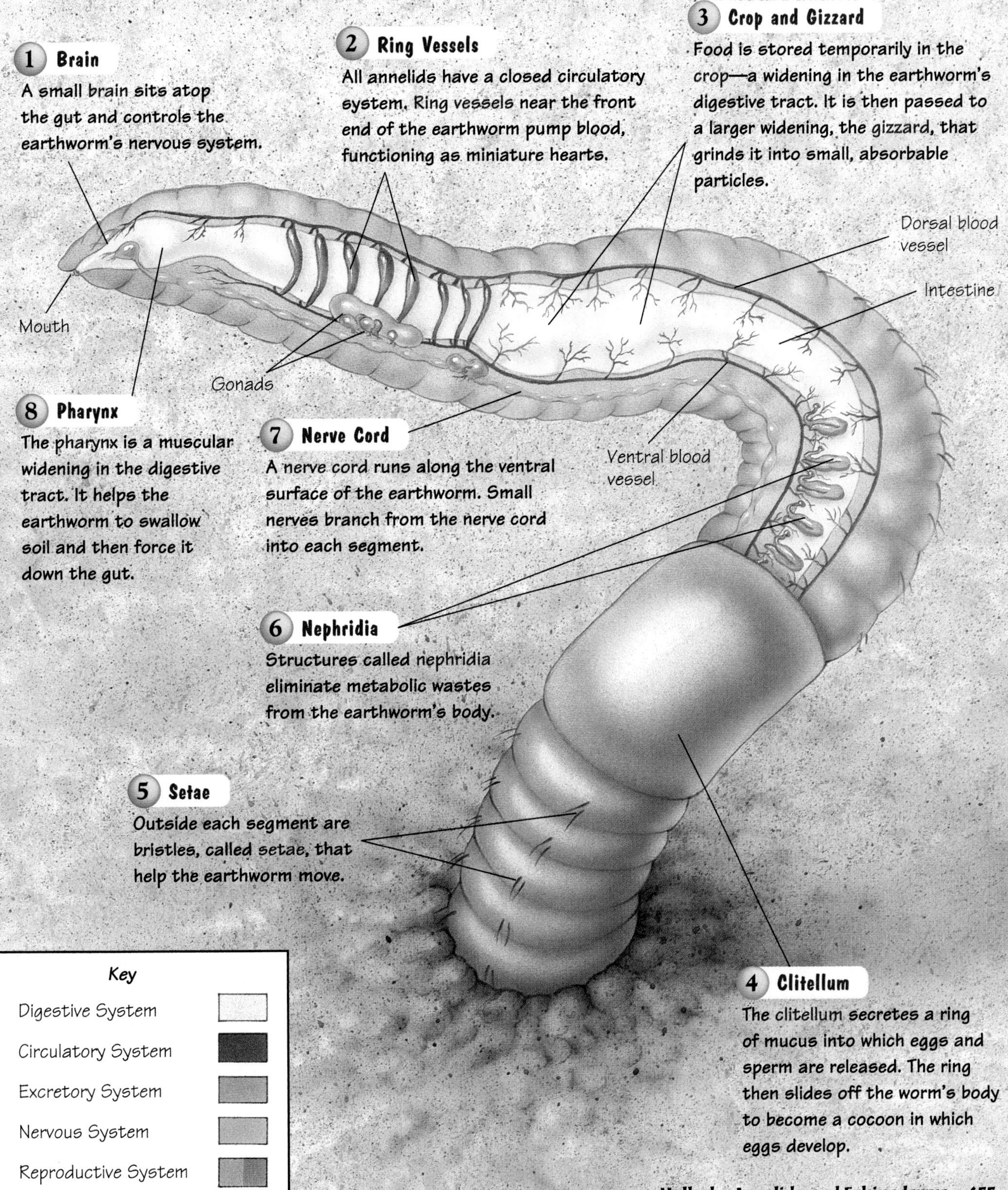

Figure 28–6
(a) *Earthworms are hermaphrodites. These two earthworms are exchanging sperm, which eventually will be used to fertilize egg cells.* (b) *Leeches produce a chemical that prevents blood from clotting. This helps them to tap the blood supply of their hosts.* (c) CAREER TRACK *These researchers at the U.S. Department of Agriculture are studying earthworms.*

Oligochaetes

The name Oligochaeta (ahl-ih-goh-KEE-tuh) means "few bristles," and an oligochaete—such as an earthworm—has a few short bristles on its body. These bristles, called setae, are arranged in rows along the worm's underside. As the earthworm moves, these bristles dig into the soil, providing traction.

Although oligochaetes may not be beautiful or attention grabbing, some are ecologically important. As the English naturalist Charles Darwin noted, earthworms swallow and grind up incredible amounts of soil and organic matter. This action aerates soil and recycles many nutrients, including nitrogen.

☑ **Checkpoint** How do earthworms use their bristles?

Polychaetes

The class Polychaeta (pahl-ih-KEE-tuh) includes the worms shown in **Figure 28–5** on page 654. These worms swarm throughout the world's oceans—from the polar regions to the tropics.

The name Polychaeta means "many bristles." Typically, each body segment of a polychaete includes a pair of paddlelike structures tipped with bristles. These structures are short and barely visible in some polychaetes but are long and brightly colored in others.

Leeches

The class Hirudinea (hir-yoo-DIHN-ee-uh) contains the leeches, a group of familiar but not-so-popular animals. Many leeches are blood-drinking parasites of aquatic animals, and some will feed on humans if given the chance. Most leech species live in the tropics, including leeches so large and powerful they can penetrate human skin. In temperate climates, leeches that usually feed on fishes or other aquatic animals will sometimes attach themselves to swimmers.

Section Review 28–2

1. **Describe** the body plan of annelids.
2. **Identify** the three classes of annelids.
3. **Critical Thinking—Comparing** Besides the annelids, what other animals show segmentation? Do humans show segmentation of any kind?

SECTION 28–3

Echinoderms

GUIDE FOR READING

- **Describe** the characteristics of echinoderms.

MINI LAB

- **Observe** the structures of a starfish.

SOME OF THE ECHINODERMS are delicate, brightly colored, feathery armed creatures. Others look like mud-brown half-rotten cucumbers! Echinoderms live only in the sea, where they seldom affect humans. But their body plans and life history combine some unique features found in no other animals—living or extinct. For this reason alone, echinoderms are well worth studying.

Echinoderm Structures

Each of the animals shown in *Figure 28–7* is a member of the phylum Echinodermata (ee-kigh-noh-DER-muh-tuh), meaning "spiny skin." **Echinoderms have radial symmetry, an internal skeleton, spiny skin, and a network of tubes and appendages known as a water vascular system.**

Echinoderms are radially symmetric, meaning their bodies include units repeated around a central axis. Their internal skeleton, called an endoskeleton, is made of stiff plates of calcium carbonate that may be hinged to one another or fused together into a solid casing.

However, the most unusual feature of echinoderms is their **water vascular system.** Physically, the water vascular system contains a network of fluid-filled canals connected to countless **tube feet,** each of which looks like a medicine dropper attached to a suction cup. In many species, the water vascular system is tied to all sorts of life functions, including feeding, movement, internal transport, respiration, and excretion.

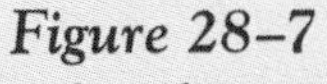

Figure 28–7
Echinoderms include the unusual organisms shown here. (a) *Sphere-shaped sea urchins are protected by sharp spikes, which in many species contain poison. This photograph shows small white sea urchins surrounding a larger sea urchin.* (b) *Some sea cucumbers look like warty pickles. What do you think this specimen looks like?* (c) *This starfish is attacking a bivalve. After it pries open the shell, it will extend its stomach and secrete digestive enzymes into the bivalve's body.*

Figure 28–8
(a) Starfishes are able to regenerate lost arms. This starfish seems to be growing several arms at once! (b) A starfish's skin is lined with spines. Inside its body, the water vascular system consists of radial canals down each arm, joined by a central ring canal. The system opens to the outside through a hole called a madreporite.

(a)

Like cnidarians, echinoderms have no front or back end, no head, and no brain. They do have top and bottom sides, however. The bottom side is called the oral surface because it contains the mouth, whereas the top side is called the aboral surface.

Checkpoint What is an echinoderm?

Feeding, Respiration, and Elimination

Starfishes are predators of clams, oysters, and other bivalves. Other echinoderms are filter feeders, using individual tube feet to snag passing plankton. And still others use tube feet to pick up a mixture of sand and detritus, which they then shove into their mouth. Once the food is inside, digestive glands distributed throughout the body process it further.

Thanks to their water vascular system, many echinoderms have an enormous surface area exposed to the surrounding water. As a result, the echinoderm's cells can exchange gases directly with the water, as well as eliminate wastes into it. Because the water vascular system so efficiently handles these tasks, echinoderms do not need specialized respiratory, elimination, or circulatory systems. However, some species have feathery tufts called skin gills that aid in gas exchange.

Response, Movement, and Reproduction

Like cnidarians, echinoderms lack any structure that even resembles a brain. However, echinoderms do have a few specialized nerve cells. Sensory cells provide an adequate sense of taste and smell. Some species have eyespots that can distinguish light from dark, although they cannot detect objects. And some

(a)

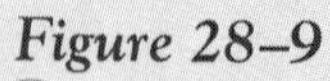

(b)

Figure 28–9
(a) Disk-shaped sand dollars often settle on the ocean floor. (b) Crinoids use their feathery arms to filter plankton from the passing water.

echinoderms have **statocysts,** specialized structures that tell which way is up.

An echinoderm moves with its tube feet. However, an echinoderm has thousands of these appendages, yet only a simple nervous system. How it coordinates its movement remains a mystery!

Most echinoderms have separate sexes that release eggs and sperm into the water. This is less of a hit-or-miss process than it might seem, however, because an echinoderm can detect the egg and sperm cells of its species in the water and then releases its own reproductive cells in response. Once fertilized, echinoderm larvae swim in the open water before changing to the adult form.

Starfish Hunt

PROBLEM *What structures can you **observe** in a starfish?*

PROCEDURE

1. Obtain a starfish specimen from your teacher.
2. Sketch the starfish, including both its top and bottom sides. Label the structures you can identify.

ANALYZE AND CONCLUDE

1. Compare the two sides of the starfish. With which side does it attack and eat prey?
2. Describe the tube feet of the starfish. What is the purpose of these structures?
3. Why can starfishes live only in water?

Classifying Echinoderms

Starfishes are members of the class Asteroidea. Although the most familiar starfishes have five arms, other species grow many more. Most starfishes are carnivores that feed on bivalve mollusks, corals, and other sedentary prey.

Brittle stars belong to the class Ophiuroidea (ahf-ih-yoo-ROI-dee-uh). These animals look like starfishes whose arms have been stretched until they are long, thin, and flexible. Typically, they hide during the daylight hours, wandering about only under cover of darkness.

Sand dollars and sea urchins are members of the class Echinoidea (ehk-ih-NOI-dee-uh). Most sea urchins will wedge themselves into cracks and crevices by day, coming out only after dusk.

Sea cucumbers are members of the class Holothuroidea (hahl-oh-thoo-ROI-dee-uh). Most sea cucumbers are detritus feeders, picking up the food they encounter as they bulldoze their way across the sandy ocean bottom.

Members of the class Crinoidea (krigh-NOI-dee-uh) include sea lilies and feather stars. These delicate-looking animals may be the most ancient members of their phylum.

Section Review 28-3

1. **Describe** the characteristics of echinoderms.
2. For each of the five classes of echinoderms, **give an example** of an organism in that class and **describe** its characteristics.
3. **Critical Thinking—Comparing** How do the feeding strategies of different echinoderms compare?
4. **MINI LAB** What structures can you **observe** in a starfish?

Adaptations of Cephalopods

Guide for Reading

- **Explain** how squids and octopuses increase the speed of nervous transmissions.

EARTHWORMS, CLAMS, SNAILS, and starfishes move from place to place fairly slowly, if they move at all. And they are small creatures, typically growing no longer than about 30 centimeters. But cephalopods are different. Octopuses and squids travel with speed and agility, and some grow to incredible sizes. Including its tentacles, a giant squid can grow over 16 meters in length—longer than 8 humans placed end to end!

Why can cephalopods move so quickly and grow to such huge sizes? The answers to these questions begin with a very important body system—the nervous system.

The Nervous System

Usually, humans take their nervous system for granted. When you decide to wiggle your big toes, for example, your nervous system takes only a few milliseconds to transmit a message from your brain to muscles in your foot.

How do messages travel so quickly through nerves? In part, the answer lies with a white fatty substance called **myelin.** Myelin surrounds many nerve cells, and it acts like the insulation around an electrical wire. With myelin, nerves can carry transmissions as fast as 200 meters per second!

Living Without Myelin

Humans and other vertebrates—animals with backbones—all make myelin to increase the speed of nervous transmissions. But invertebrates are not able to make myelin. And without myelin, a small nerve can transmit impulses at a rate of only a few millimeters per

Figure 28–10 *Few invertebrates show the sophisticated adaptations of cephalopods.* (a) *This deep-sea octopus uses a well-developed nervous system to control its long, flexible tentacles.* (b) *To surprise a prey, a cuttlefish can change color to blend into the background. When the unsuspecting prey is within reach, the cuttlefish quickly extends its tentacles to seize it.* (c) *Cephalopods evolved unusually wide nerves, including the squid giant axon shown between tweezers in this photograph. Wide nerves transmit impulses faster than narrow nerves do.*

second—too slow to coordinate distant body parts. As a result, most invertebrates have small, compact bodies.

Cephalopods, however, evolved a different way to increase the speed and efficiency of nervous transmissions. It turns out that wide nerve cells carry transmissions faster than narrow nerve cells. **Squids and octopuses evolved unusually wide nerve cells, thus increasing the speed of their nervous transmissions without the use of myelin.** In a nerve called the squid giant axon, the cells are over a millimeter in diameter—a significant width for a single cell of any kind.

Squids and octopuses can grow to large sizes because wide nerves allow their brain to communicate quickly with distant body parts. Without such communication, octopuses and squids could not grow to the sizes they are today.

☑ ***Checkpoint*** What is myelin?

Research on Nerves

Aside from their size, the nerves of cephalopods are very similar to the nerves of other animals, including humans. And in a laboratory experiment, large, wide nerves are much easier to handle and manipulate than small, narrow nerves. For these reasons, researchers investigating nerve function typically study the squid giant axon and other nerves of cephalopods. In fact, much of what scientists have learned about nerves has come from these studies.

Figure 28–11 *An octopus may discharge a cloud of ink in response to a threatening presence—in this case, a deep-sea diver.*

Other Adaptations

The nervous system is only one of the fascinating features of cephalopods. For example, these animals can move by a very unusual method—jet propulsion! Cephalopods typically draw water into their flexible mantle cavity, then rapidly force out the water through a tube called a **siphon.** Most cephalopods have several movable siphons, allowing for quick movements in different directions.

To protect themselves, many cephalopods release a cloud of dark-colored, strong-tasting ink just before they make their retreat. And some use specialized cells called **chromatophores** to change color, allowing the cephalopod to blend into its background. Cephalopods are also very intelligent. In fact, an octopus can learn to perform simple tasks, such as unscrewing a lid from a jar to reach food inside it.

Section Review 28–4

1. **Explain** how squids and octopuses increased the speed of nervous transmissions without the use of myelin.
2. **Identify** some adaptations in cephalopods.
3. **BRANCHING OUT ACTIVITY** Investigate the research performed on the large nerves of cephalopods. **Describe** how messages are transmitted through nerves.

Laboratory Investigation

DESIGNING AN EXPERIMENT

How Earthworms Respond to Stimuli

As earthworms burrow their way through the soil, they respond to stimuli. Their responses to certain stimuli affect their chances of survival in their environment. To detect stimuli, the earthworm uses sensory cells in its skin rather than specialized sense organs such as eyes. In this investigation, you will observe an earthworm and see how it responds to different stimuli.

Problem

How does an earthworm respond to different stimuli? **Design an experiment** to answer this question.

Suggested Materials

live earthworm in a storage container
tray
paper towels
medicine dropper
hand lens
lamp
cold water

Suggested Procedure

1. **With a hand lens, observe the skin, mouth, setae, and segments of the earthworm. CAUTION:** ***Be careful not to harm the earthworm when handling it.*** **Make a sketch of the earthworm and label its external features.**
2. **Fill the medicine dropper with water and use it to wet the earthworm. Make sure you keep the earthworm moist. It will die if its skin dries out.**

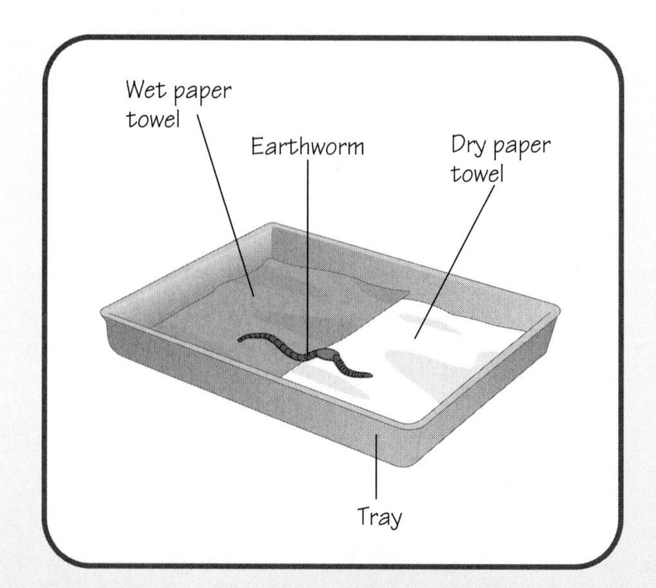

3. **Fold a dry paper towel in half and place it on one side of your tray. Fold a dampened paper towel in half and place it on the other side of the tray. See the diagram.**
4. **Place the earthworm in the center of the tray, between the dry paper towel and the moist paper towel.**
5. **Place the tray in an area where the earthworm will remain undisturbed for 5 minutes.**
6. **After 5 minutes, observe the location of the earthworm. Record your observations.**
7. **Using a similar procedure, design an experiment to determine how earthworms respond to the following stimuli:**
   - **light**
   - **cold**
8. **Write your hypotheses. With your teacher's approval, carry out the experiments you designed. Record your results.**
9. **When you have completed your experiments, return the earthworm to your teacher.**

Observations

1. Describe the earthworm's external features, including color and texture.
2. How did the earthworm respond when placed between the moist and dry environments?
3. What was the earthworm's response to light? To cold?

Analysis and Conclusions

1. How is an earthworm's body adapted for movement into and through soil?
2. What is the function of the earthworm's slimy skin?
3. How does an earthworm's response to moisture help it to survive?
4. Do an earthworm's responses to light and cold have any protective value? Explain your answer.
5. Would you expect to find earthworms in hard soil? Explain your answer.

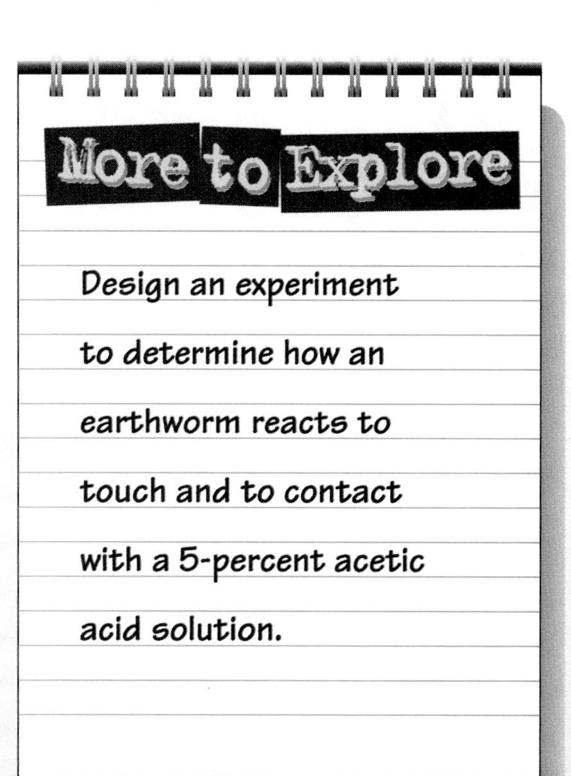

More to Explore

Design an experiment to determine how an earthworm reacts to touch and to contact with a 5-percent acetic acid solution.

Study Guide

Summarizing Key Concepts

The key concepts in each section of this chapter are listed below to help you review the chapter content. Make sure you understand each concept and its relationship to other concepts and to the theme of this chapter.

28–1 Mollusks

- All mollusks have similar body plans that typically include a foot, gut, mantle, and shell. They also pass through similar stages in their early development.
- Gastropods include snails and slugs. Bivalves include clams and oysters, which have two shells joined at a hinge. The largest mollusks are the cephalopods, including octopuses and squids.

28–2 Annelids

- All annelids have a body divided into segments separated by walls called septa.
- Annelids have a tube-within-a-tube digestive tract, a small brain, and a closed circulatory system. Nephridia remove metabolic wastes.
- Earthworms are oligochaetes. Sandworms and bloodworms are polychaetes. Leeches are hirudineans.

28–3 Echinoderms

- Echinoderms have radial symmetry, an internal skeleton, spiny skin, and a water vascular system—a network of tubes and appendages used for many purposes.
- Echinoderms include starfishes, brittle stars, sand urchins, sea urchins, sea cucumbers, sea lilies, and feather stars.

28–4 Adaptations of Cephalopods

- To increase the speed of their nervous transmissions without myelin, squids and octopuses evolved wide nerve cells.

Reviewing Key Terms

Review the following vocabulary terms and their meaning. Then use each term in a complete sentence.

28–1 Mollusks

foot
radula
gut
mantle
shell
open circulatory system
closed circulatory system
nephridium
trochophore

28–2 Annelids

septum
longitudinal muscle
circular muscle
brain
ring vessel
crop
gizzard
clitellum
seta
nephridium
nerve cord
pharynx

28–3 Echinoderms

water vascular system
tube foot
statocyst

28–4 Adaptations of Cephalopods

myelin
siphon
chromatophore

CHAPTER 28

Recalling Main Ideas

Choose the letter of the answer that best completes the statement or answers the question.

1. The shell of a mollusk is secreted by glands in the
   a. mantle. **c.** gut.
   b. foot. **d.** radula.
2. Mollusks excrete nitrogen-containing wastes through organs called
   a. gills. **c.** trochophores.
   b. nephridia. **d.** radulas.
3. The circulatory system of a clam or an oyster is described as
   a. hermaphroditic. **c.** open.
   b. water vascular. **d.** closed.
4. Snails and slugs are members of the class
   a. Gastropoda. **c.** Cephalopoda.
   b. Bivalvia. **d.** Hirudinea.
5. The segments of an annelid are divided by body walls called
   a. trochophores. **c.** nephridia.
   b. ring vessels. **d.** septa.
6. Earthworms are members of the class
   a. Oligochaeta. **c.** Hirudinea.
   b. Polychaeta. **d.** Echinoidea.
7. Which type of body plan do echinoderms have?
   a. radial symmetry **c.** spherical symmetry
   b. bilateral symmetry **d.** no symmetry
8. The name echinoderm means
   a. segmented body. **c.** spiny skin.
   b. large brain. **d.** one-piece shell.
9. For feeding, respiration, excretion, and other life functions, an echinoderm uses its
   a. nephridia.
   b. gills.
   c. water vascular system.
   d. closed circulatory system.
10. The fatty substance that insulates many nerve cells in vertebrates is
    a. myelin. **c.** chromatophore.
    b. nacre. **d.** radula.

Putting It All Together

Using the information on pages xxx to xxxi, complete the following concept map.

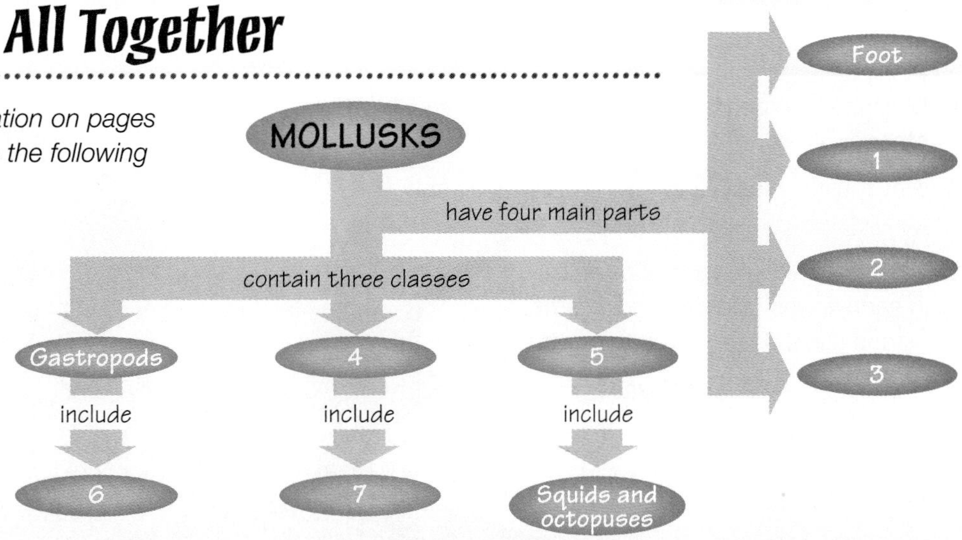

MOLLUSKS
have four main parts
Foot
1
2
3
contain three classes
Gastropods
4
5
include
include
include
6
7
Squids and octopuses

Reviewing What You Learned

Answer each of the following in a complete sentence.

1. Describe the different ways in which mollusks obtain food.
2. List the basic body parts found in almost all mollusks.
3. What is the purpose of nephridia?
4. What is a trochophore?
5. Some gastropods lack shells. How do they protect themselves?
6. What is nacre?
7. What are cephalopods? Describe the characteristics they have in common.
8. Describe an annelid's digestive tract.
9. Why are earthworms important for fertile soil?
10. How do earthworms use the bristles on their ventral surface?
11. What is the function of the clitellum in earthworms?
12. What are the characteristics of echinoderms?
13. Describe the different ways in which echinoderms feed.
14. Why do most invertebrates have small, compact bodies?

Expanding the Concepts

Discuss each of the following in a brief paragraph.

1. Describe the roles of the four basic parts of a mollusk.
2. Compare an open circulatory system with a closed circulatory system. Give examples of organisms that use each.
3. Discuss the different reproductive strategies in mollusks.
4. Describe how pearls are formed in oysters and other bivalves.
5. Why must mollusks and annelids keep their skin moist?
6. What are the two types of muscles in earthworms? Describe how earthworms use these muscles to move.
7. Compare the different classes of annelids.
8. Describe the structure and function of the water vascular system in echinoderms.
9. How do starfishes feed on clams, oysters, and other bivalves?
10. Discuss the adaptations that allow cephalopods to move faster and grow larger than other mollusks.

Extending Your Thinking

Use the skills you have developed in this chapter to answer the following.

1. **Analyzing** Researchers have identified a chemical in leeches that suppresses blood clotting. Why is this chemical important in leeches?
2. **Designing an experiment** Which foods can earthworms eat? Design an experiment to discover whether earthworms will eat bread, vegetables, meat, or other foods. Conduct the experiment only with the permission of your teacher.
3. **Observing** Suppose that you are asked to make daily observations of an octopus that lives in an aquarium. What characteristics or behaviors might indicate the intelligence of the octopus?
4. **Classifying** Describe the characteristics of the animal shown in the photograph on page 666. Based only on the photograph, into which phylum and class does the animal belong? What further information would help you classify the animal? Explain your answer.
5. **Predicting events** Suppose that pollutants contaminated a patch of farmland and killed the earthworm population in the soil. Predict how this would affect the farmland.

Applying Your Skills

The Great Snail Pull

Land snails are famous for moving slowly. But because of their muscular foot, these animals are unusually strong for their size. Perform this activity to find out how strong snails actually are.

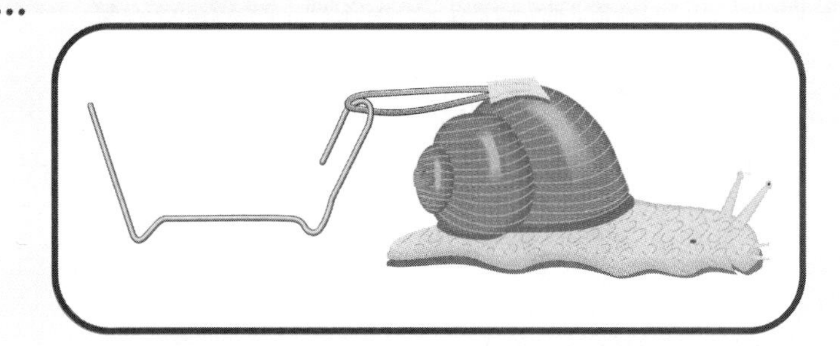

1. Obtain a land snail from your teacher. On a piece of paper, sketch its body and label the parts you can identify.
2. Gently tape a loop of thread to the snail's back. Bend open a paper clip and link it to the loop of thread, as shown in the illustration. Make sure you do not damage the snail's shell or harm the snail in any way.
3. Add washers or other weights to the paper clip until the snail can no longer move forward.
4. Measure the mass of the snail and the mass of the weights it was able to pull, then return the snail to your teacher. How many times its own mass did the snail pull?
5. Describe the way in which a snail moves. How is the snail able to pull the added weight?

• GOING FURTHER •

6. Design an experiment to determine how water, light, aromas, or other factors influence a snail's behavior.

CHAPTER 29

Arthropods

FOCUSING THE CHAPTER
THEME: Unity and Diversity

29–1 Arthropod Form and Function
- **Describe** an arthropod's structure and life cycle.

29–2 A Tour of the Arthropods
- **Identify** the subphyla of arthropods.

BRANCHING OUT *In Depth*

29–3 The Importance of Arthropods
- **Discuss** the importance of arthropods to humans.

LABORATORY INVESTIGATION
- **Design an experiment** to determine the foods that fruit fly larvae prefer.

Biology and Your World

BIO JOURNAL

Do you think that arthropods are interesting? Beautiful? Bothersome? In your journal, describe your thoughts and feelings about arthropods.

Lobster in the Indian Ocean

SECTION 29–1

Arthropod Form and Function

GUIDE FOR READING

- **Describe** an arthropod's body.
- **Explain** the purpose of molting.

MINI LAB

- **Design an experiment** to determine how fruit flies respond to light.

IF YOU HAVE EVER ADMIRED A lady bug, been chased from a picnic by wasps, or enjoyed eating a shrimp dinner, then you've had close encounters with arthropods. From mites so small that they ride on dust particles to king crabs that scuttle on meter-long legs, arthropods are the most successful animals on Earth. There may be more than 15 million species of insects alone. What's more, insects outnumber humans by about 200 million to 1!

What are these animals? Why are they so successful? Let's explore the answers to these questions.

Arthropod Body Plan

Arthropods have segmented bodies, jointed appendages, and are surrounded by a tough exoskeleton. You can think of the arthropods' original design as a biological version of the first mass-produced automobiles. The first automobiles had an internal-combustion engine, gas tank, steering wheel, transmission, and other familiar parts. Over time, however, automobiles changed. Engineers designed electric starters to replace hand cranks, fuel injectors to replace carburetors, and body shapes that range from small sports cars to large semitrailers. But the changes have been little more than variations on the original theme.

Like automobiles, the arthropods' original body plan and basic parts remained the same, while their size, shape, and complexity evolved along different paths.

Checkpoint What is an arthropod?

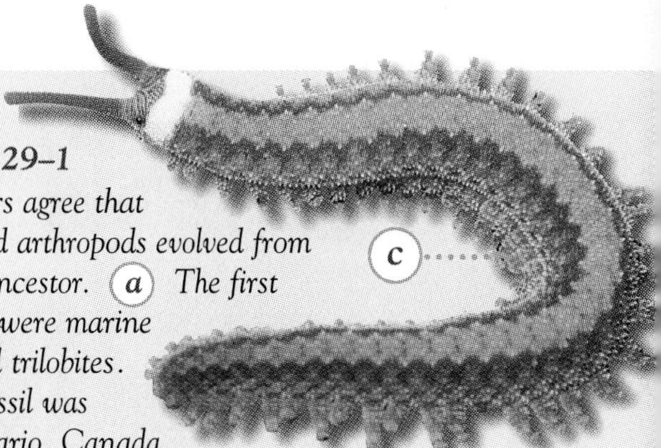

Figure 29–1 Researchers agree that annelids and arthropods evolved from a common ancestor. (a) *The first true arthropods were marine organisms called trilobites. This trilobite fossil was found in Ontario, Canada, and is over 400 million years old.* (b) *Today, arthropods vary greatly in size and shape. This man-faced beetle from Malaysia gets its name from its unusual markings.* (c) Peripatus, *the velvet worm, is neither an annelid nor an arthropod but has characteristics of both.*

MINI LAB .... Experimenting ...

Flying to the Light

PROBLEM *How does a fly react to light? **Design an experiment** to answer this question.*

SUGGESTED PROCEDURE  

1. Obtain a light source and a culture tube containing fruit flies.
2. Position the culture tube at different distances in front of the light source. Observe the behavior of the fruit flies at each position.
3. Formulate a hypothesis and design an experiment to find the farthest distance at which the light affects the behavior of the fruit flies. Check with your teacher before beginning the experiment.

ANALYZE AND CONCLUDE

1. What conclusion can you draw from your results? Explain how the data you collected support this conclusion.
2. What other variables might have affected your experiment? Explain.
3. Design an experiment to determine how the color of light affects the behavior of fruit flies.

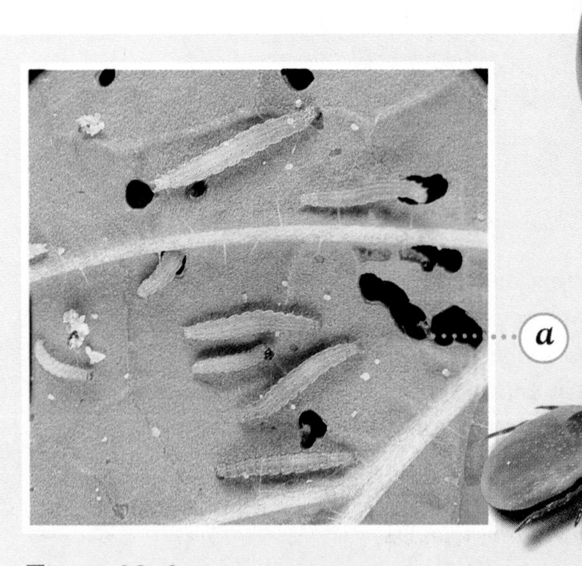

Figure 29–2
(a) These diamondback moth larvae are feeding on a cabbage leaf. (b) The larger of these two Eastern wood ticks has filled its body with a blood meal.

Arthropod Structures

Arthropods have a tough exoskeleton made mostly of a carbohydrate called **chitin** (KIGH-tihn). Some exoskeletons, like those of caterpillars, are firm yet leathery. Others, especially those of ticks and some crabs, are so tough and hard that they are almost impossible to crush by hand. In many terrestrial arthropods, the exoskeleton has a waxy covering that helps prevent loss of body fluids.

In addition, an arthropod's body is divided into segments. Centipedes and millipedes can have dozens of segments, while other arthropods—such as ants—have only three. Because the exoskeleton is rigid, it contains joints between segments and parts of appendages. The joints enable body parts to flex and extend. To learn more about arthropods, study the illustration on the next page.

Feeding and Respiration

Feeding in arthropods is very diverse. The arthropods include run-of-the-mill herbivores and carnivores, as well as filter feeders, detritus feeders, bloodsuckers, and a host of specialized parasites.

To breathe, arthropods use a variety of different structures. Many aquatic species—such as lobsters and crabs—breathe through **gills,** which are featherlike organs located in a chamber beneath the exoskeleton. Horseshoe crabs breathe through **book gills,** and spiders breathe through **book lungs**—both of which are made of layers of respiratory tissue that resemble the pages in a book.

Most terrestrial arthropods, however, breathe through a branching, air-filled network of structures called **tracheal** (TRAY-kee-uhl) **tubes.** Tracheal tubes connect the arthropod's tissues with the atmosphere, and oxygen passes through the tracheal tubes by diffusion.

Visualizing a Grasshopper

The bodies of all insects have three distinct parts—head, thorax, and abdomen. Most insects, including grasshoppers, have three pairs of legs.

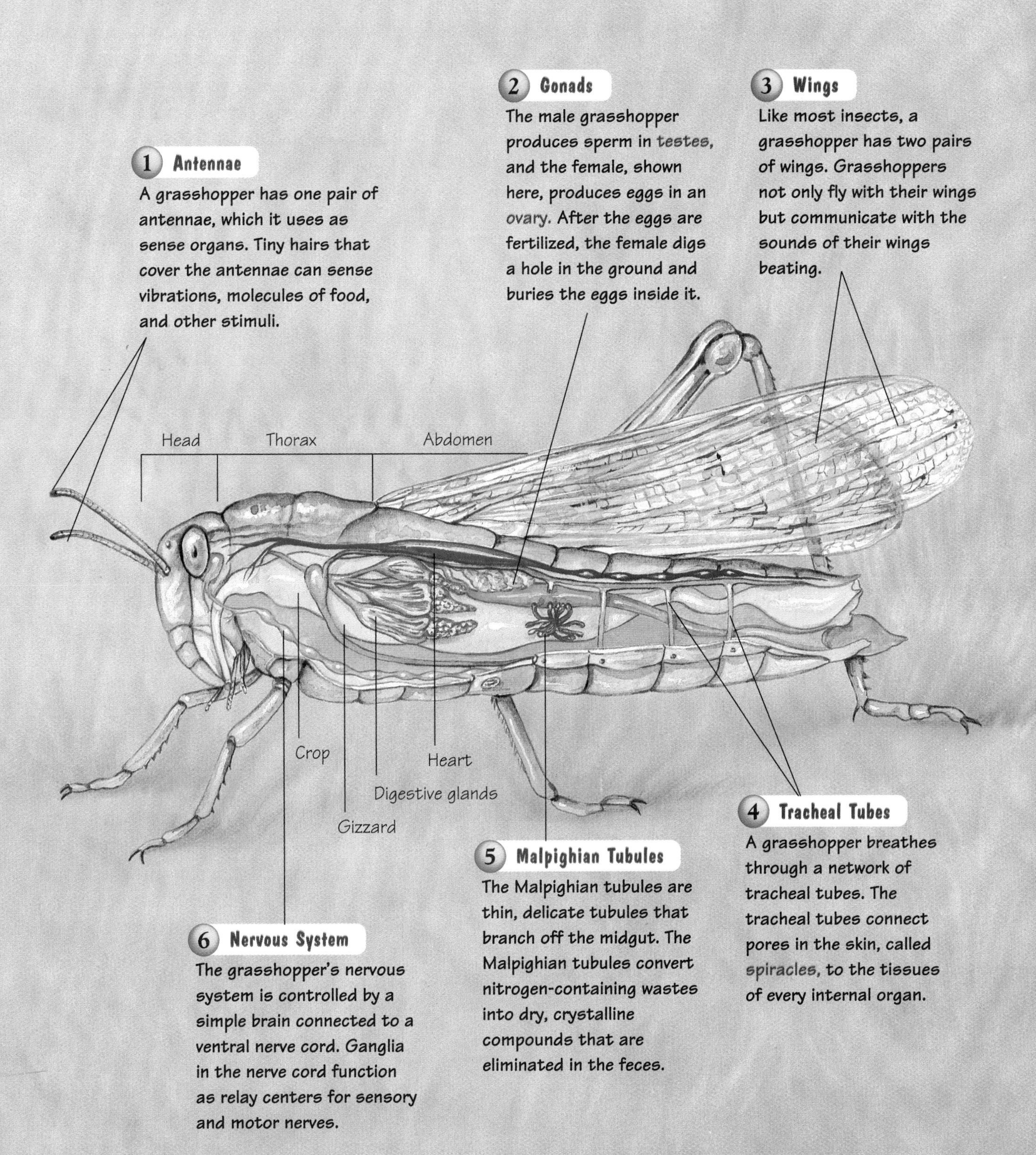

1 Antennae
A grasshopper has one pair of antennae, which it uses as sense organs. Tiny hairs that cover the antennae can sense vibrations, molecules of food, and other stimuli.

2 Gonads
The male grasshopper produces sperm in testes, and the female, shown here, produces eggs in an ovary. After the eggs are fertilized, the female digs a hole in the ground and buries the eggs inside it.

3 Wings
Like most insects, a grasshopper has two pairs of wings. Grasshoppers not only fly with their wings but communicate with the sounds of their wings beating.

4 Tracheal Tubes
A grasshopper breathes through a network of tracheal tubes. The tracheal tubes connect pores in the skin, called spiracles, to the tissues of every internal organ.

5 Malpighian Tubules
The Malpighian tubules are thin, delicate tubules that branch off the midgut. The Malpighian tubules convert nitrogen-containing wastes into dry, crystalline compounds that are eliminated in the feces.

6 Nervous System
The grasshopper's nervous system is controlled by a simple brain connected to a ventral nerve cord. Ganglia in the nerve cord function as relay centers for sensory and motor nerves.

Figure 29–3
This image from a scanning electron microscope shows the head of a housefly (magnification: 24X). Each compound eye contains about 4000 separate image-forming units.

Transport and Excretion

Like clams and other mollusks, arthropods have an open circulatory system. A well-developed heart pumps blood through arteries into smaller vessels, from which it flows into spaces called sinuses. There, muscles slosh the blood to bathe body tissues. Eventually, the blood collects in a large sinus surrounding the heart. It then re-enters the heart and begins its journey again.

Arthropods may excrete nitrogen-containing wastes in several ways. In most terrestrial species, these wastes are excreted by dead-end sacs called **Malpighian** (mal-PIHG-ee-uhn) **tubules.** Malpighian tubules extract nitrogenous wastes from blood in body sinuses. The tubules concentrate those wastes and add them to feces moving through the gut.

Most aquatic arthropods excrete nitrogenous wastes by allowing ammonia to diffuse across gill surfaces. Marine species may also have a pair of antennal glands that control the balance of water and solutes in body fluids.

☑ ***Checkpoint*** What are Malpighian tubules?

Response and Movement

Did you ever try to sneak up on a fly? Flies and most other arthropods have a well-developed nervous system and sense organs. Large compound eyes detect color and movement. Many arthropods "smell" with their feathery antennae and "taste" with sensory hairs on their legs. And although insects' ears are located in what we would consider to be odd places, they can often detect sounds far above the range that humans can hear.

Arthropods move using their well-developed muscular system, which works well with the exoskeleton. Muscles are arranged around each body joint so that some muscles flex the joint while others extend it, allowing arthropods to walk, jump, swim, or fly.

Arthropod Growth and Development

Suppose that you lived inside a suit of armor. In fact, suppose this armor enclosed every part of your body, was tailored exactly to your measurements, and was actually a part of your skin. What would happen? In a few months you would be in bad shape, because the suit of armor couldn't stretch to accommodate your growing body!

Arthropods are faced with exactly this problem because an exoskeleton is just like a suit of armor. Thus, arthropods evolved a process called **molting**—a complex, dangerous, and physiologically expensive process in which the arthropod literally changes its skin. **In molting, an arthropod sheds its entire exoskeleton and manufactures a larger one, thus allowing its body to grow.**

Molting involves many steps. As the time for molting approaches, skin glands digest the inner part of the exoskeleton. Meanwhile, other glands secrete a new

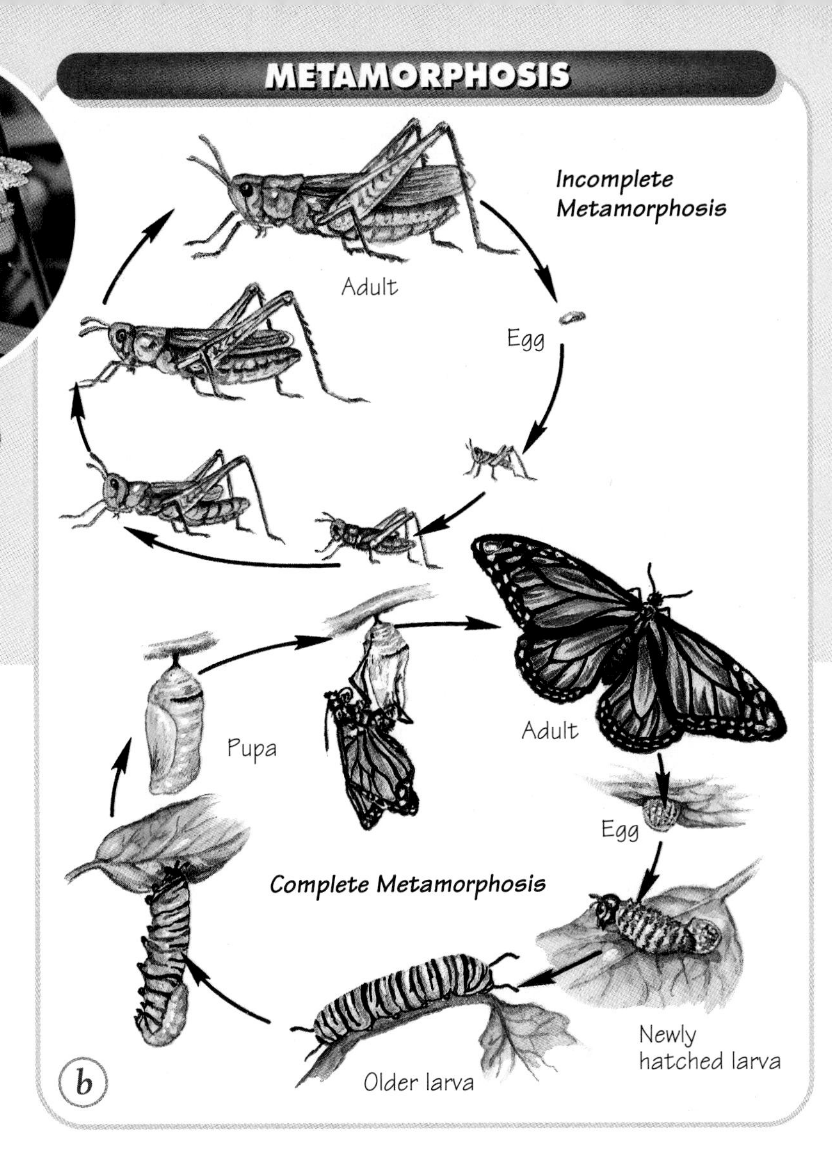

Figure 29–4
(a) Molting allows arthropods to grow and develop. This photograph shows a Libellula depressa *dragonfly emerging from its old exoskeleton.*
(b) Arthropods that undergo incomplete metamorphosis do not change as drastically as those that undergo complete metamorphosis. While a grasshopper retains the same body plan throughout its life, a butterfly changes from a wormlike larva to a winged adult.

skeleton, which at first is quite soft. When the arthropod is ready, it pulls itself out of what remains of the original skeleton—a process so difficult that it can take several hours. As soon as the arthropod is free, it usually eats its old skin, thus recycling chitin and minerals.

Before the new exoskeleton hardens, the arthropod pumps itself with air or fluids to enlarge itself. This makes the new exoskeleton as large as possible before molting is necessary again.

Molting not only allows arthropods to change size, it allows them to change their form and shape, a process called **metamorphosis.** As shown in ***Figure 29–4,*** some arthropods undergo incomplete metamorphosis, which involves minimal changes. Others undergo complete metamorphosis, which produces distinct larval, pupal, and adult stages.

You probably know all about caterpillars turning into butterflies. But that familiarity should not hide the wonder of this change. Just try to imagine a baby mouse changing into an adult bird. Even in aquatic arthropods that don't have a pupal stage, metamorphosis could be compared to a mouse changing into a guinea pig that grows into an elephant!

Section Review 29-1

1. **Describe** an arthropod's body.
2. **Explain** the purpose of molting.
3. **Critical Thinking—Analyzing Events** The first arthropods—the trilobites—lived in the sea. Describe the features that help terrestrial arthropods adapt to life on land.
4. **MINI LAB** How can you **design an experiment** to determine a fly's response to light?

SECTION

29–2 A Tour of the Arthropods

Guide for Reading

- **Describe** each subphylum of arthropods.
- **Identify** the structure of insects.

MINI LAB

- **Infer** how crustaceans move by studying a shrimp.

YOU COULD SPEND A LIFETIME touring the arthropods—the largest phylum of the animal kingdom. Millions of species live all over the Earth, and their ways of making a living are diverse enough to boggle the imagination.

Living arthropods are divided into three groups called subphyla, each of which contains several classes. One of these classes—insects—includes more species than any other class of multicellular organisms.

Chelicerates

Spiders are the most familiar members of the subphylum Chelicerata (kuh-lihs-er-AT-uh), which also includes mites, scorpions, and horseshoe crabs. **A chelicerate has a body that is divided into two parts—a cephalothorax and an abdomen.** The **cephalothorax** includes the head and carries the legs. The **abdomen** contains most internal organs.

Chelicerates also have two pairs of unique, specialized mouthparts. The first pair are called **chelicerae** (kuh-LIHS-er-ee), from which the subphylum gets its name. The second pair are known as **pedipalps.** Chelicerae and pedipalps serve different purposes in different species.

☑ *Checkpoint* What are the chelicerates?

Horseshoe Crabs

The horseshoe crabs—which aren't true crabs at all—are among the most ancient living arthropods. They first appeared in the Ordovician Period, more than 430 million years ago, and have changed little since that time. The larvae of horseshoe

Figure 29–5
Arthropods live almost everywhere on Earth. (a) Horseshoe crabs live in shallow waters and sandy beaches. (b) This female scorpion is carrying her offspring on her back. Scorpions are common in deserts. (c) Tiny bedbugs live in bedsheets, where they feed on dead skin cells (magnification: 10X).

crabs resemble even more ancient arthropods—so much so that they are called trilobite larvae.

Arachnids

Spiders, mites, ticks, and scorpions are grouped in the class Arachnida (uh-RAK-nih-duh). All arachnids have four pairs of walking legs attached to their cephalothorax.

Spiders are predators, usually of other arthropods. Some spiders lie in ambush and pounce on their prey. But more commonly, spiders spin intricate webs, such as the one shown in **Figure 29–6.**

Mites and ticks, which usually live as parasites of plants and animals, are the only arachnids that typically target humans and domesticated plants and animals. Their chelicerae and pedipalps are specialized for digging into host tissues and sucking out blood or plant fluids.

In scorpions, the pedipalps evolved into large claws, while their elongated abdomens carry a venomous stinger. Although a scorpion's sting is painful to humans—and occasionally fatal—its normal prey are insects or other small invertebrates.

Figure 29–6
(a) This spider, Argiope aurantia, *builds round webs in open areas. The silk of a spider web is incredibly strong and thin. (b) Although crustaceans are very diverse, the crayfish is a good representative. As in an arachnid, the head and thorax of a crayfish are fused into a rigid cephalothorax, which carries mouthparts, claws, and walking legs.*

Crustaceans

The subphylum Crustacea (kruhs-TAY-shee-uh) includes crabs, shrimps, and crayfishes. Although most of the 35,000 crustacean species live in the ocean, some live in fresh water and a few live in moist places on land. Crustaceans range in size from *Daphnia,* a tiny water flea, to spider crabs that have a mass of up to 20 kilograms.

Nearly all crustaceans have two pairs of antennae, several pairs of mouthparts, and appendages with two branches. One specialized pair of mouthparts is the **mandibles,** which in different crustaceans are adapted for a variety of uses. The appendages typically contain a leg, claw, or mouthpart on the outer branch and a gill on the inner branch. In addition, crustaceans often incorporate calcium carbonate into their exoskeleton, making it especially strong.

Checkpoint What are crustaceans?

Uniramians 

Centipedes, millipedes, and insects belong to the subphylum Uniramia (yoo-nih-RAH-mee-uh). **All uniramians have a single pair of antennae and appendages with only a single branch.** The uniramians are by far the most diverse, widespread, and successful of the three arthropod subphyla.

What's a Shrimp?

PROBLEM *What can you **infer** about the way in which crustaceans move by studying a shrimp?*

PROCEDURE 

1. Obtain a shrimp from your teacher. Sketch the shrimp's body, and note any appendages. Label the parts of the shrimp that you can identify.
2. Open and close each joint in the shrimp's exoskeleton. In your sketch, identify the locations of the joints.
3. Compare the dorsal (top) and ventral (bottom) sides of the shrimp.

ANALYZE AND CONCLUDE

1. From your data, what can you infer about a shrimp's range of motion and the way it moves from place to place? Explain your answer.
2. Do you suspect that a shrimp is more easily attacked from its dorsal or its ventral side? Explain.

Centipedes and Millipedes

Because the exoskeletons of centipedes and millipedes lack a waterproof covering, these uniramians live in moist places. Centipedes are typically carnivores that eat other arthropods, earthworms, small amphibians, or even small snakes and mice. All body segments except the first and last carry a single pair of legs. Millipedes feed on dead and decaying plant material, and they carry two pairs of legs on each segment.

Insects

Estimates of the total number of insect species vary by several million! Despite their diversity, members of the class Insecta have certain features in common. **All insects have a three-part body consisting of head, thorax, and abdomen, and three pairs of legs.** Many insects also have two pairs of wings attached to the thorax and a single pair of antennae.

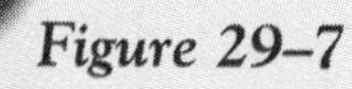

Figure 29–7 *These photographs show three of the nearly countless number of uniramians. (a) This colorful long-horned beetle lives in Texas. (b) Millipedes scurry under logs or around tree bark. (c) This multiple-exposure photograph shows the intricate body of a green lacewing fly.*

Section Review 29–2

1. **Describe** the characteristics of chelicerates, crustaceans, and uniramians.
2. **Identify** the structure of insects.
3. **Critical Thinking—Inferring** Spiders use two kinds of silk in their web—one that is sticky and one that is not. Infer why they need both types of silk.
4. **MINI LAB** What can you **infer** about how crustaceans move by studying a shrimp?

The Importance of Arthropods

GUIDE FOR READING

- **Discuss** the importance of arthropods.

WHAT ANIMALS MAKE THE world go round? You could make a case for the Earth's tiny armor-plated creatures—arthropods. From bees that flit from flower to flower to insect pests that feed on crops, arthropods seem to be everywhere—influencing life on Earth in all sorts of ways. To get an idea of the arthropods' impact, consider this astonishing fact: In the Amazon basin, the total mass of living ants and termites alone accounts for nearly one third of all animal biomass there!

Arthropod Interactions

Because arthropods are so varied and numerous, their roles in food webs, nutrient cycles, and partnerships with other animals are enormously important in nearly every habitat. Arthropods eat plants, animals, decomposing organic matter—and each other. In the process, they form a variety of symbiotic relationships with other organisms. Some are mutualistic partnerships, such as the partnership between bees and flowers, as illustrated in ***Figure 29–8.***

However, other arthropods live in relationships that are not so pleasant. Several arthropods sting or bite humans—as you probably already know! Bees and wasps typically sting to defend themselves, and some arthropods bite as part of their life cycle. A female mosquito, for example, needs a blood meal from a human or other animal to develop her eggs. Although bites from arthropods are usually only irritating, they can transmit organisms that cause serious diseases, such as malaria, encephalitis, and Lyme disease.

Figure 29–8
Arthropods interact with almost every other kind of organism. (a) Cleaner shrimps eat small parasites that live on marine animals, such as the moray eel shown here. (b) Termites digest wood, which helps to recycle it. Unfortunately, termites also destroy wooden buildings all over the world. (c) Bumblebees visit flowers for a sweet fluid called nectar. In the process, they transfer pollen from plant to plant.

INTERPRETING GRAPHS

Do Pesticides Kill Soil-Dwelling Arthropods?

Suppose you are trying to grow zucchini in your garden but have been discouraged because insects are gobbling up your plants. Your neighbor loans you three pesticide sprays—each deadly to a certain arthropod species. However, you are concerned that after a pesticide is applied, it could find its way into the soil and harm arthropods there, too.

To compare the pesticides, you test them on three small plots of soil. You also monitor a fourth plot of soil as a control.

In your experiment, you first collect a ten-liter sample of soil from the center of each plot and count the number of individual arthropods living in the sample. Then you apply a different pesticide to three of the plots but apply no pesticide to the fourth plot. Every day for the next three weeks, you collect ten-liter samples of soil from each plot and count the arthropods in the samples. The graph shows the results of this experiment.

RESULTS OF TESTS ON THREE PESTICIDES

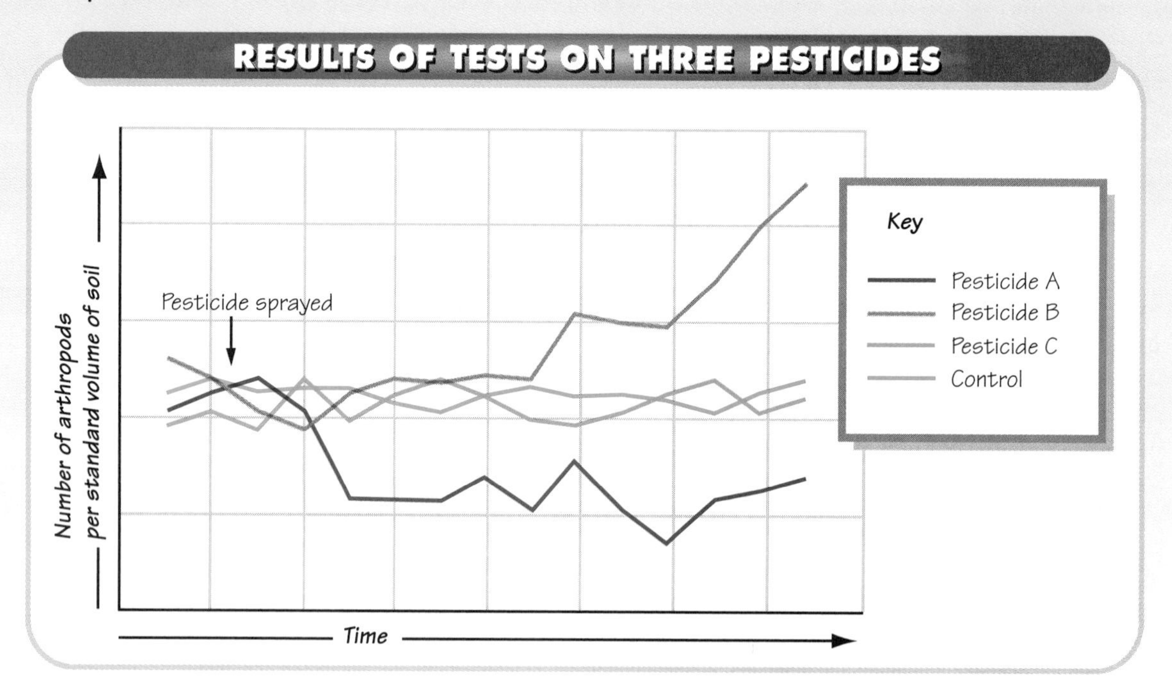

THINK ABOUT IT

1. **Compare** the number of arthropods in each plot before the pesticides were applied.
2. For each of the three pesticides, what can you **conclude** about the pesticide's effects?
3. **Identify** any assumptions behind the conclusions you reached.
4. **Discuss** the importance of arthropods in the soil.

Not surprisingly, arthropods inflict damage on plants and plant products, too.  Insects are common pests of nearly every agricultural crop, and farmers spend huge amounts of time and money trying to kill or control them. ●

Figure 29–9
CAREER TRACK
Entomologists study insects and the ways in which insects affect life on Earth.

The Pollination Crisis

To reproduce, most flowering plants depend on pollination—the transfer of pollen from one flower to another. Plants have evolved all sorts of mechanisms for pollination, but by far the most significant rely on animals. In fact, out of every three mouthfuls of food you eat, one mouthful depends on plants pollinated by animals!

Most people think only of honeybees and bumblebees as pollinators. But mosquitoes pollinate orchids in peat bogs, scarab beetles pollinate Amazonian water lilies, and there are many other examples. Biologists estimate that animal pollinators include between 130,000 and 200,000 animal species—most of which are insects.

Unfortunately, pollinators in the United States are in trouble. Their problems began almost 400 years ago, when colonists in Jamestown, Virginia, introduced European honeybees to North America. These bees spread into every habitat in the continent, competing with native pollinators and reducing their numbers. Native pollinators suffered another blow in the 1950s, when the widespread use of pesticides cut their population by more than 20 percent. As a result, many plants in the United States now rely on European honeybees as their principal pollinators.

This might not be so bad—except now honeybees are also in trouble! In the past 15 years, mites and other parasites have infected honeybee colonies, in some cases killing up to half the members of a hive. In addition, honeybees in South America have interbred with aggressive African bees. These Africanized honeybees reached Texas in 1990, and they are sure to spread farther in the years ahead. These bees carry diseases and attack humans.

Researchers estimate that if the pollinators' problems continue, up to a third of the United States alfalfa crop could be lost, and damage to other crops could cost more than $1.25 billion. Researchers have just begun to study this growing problem. Its solution is vital for both agriculture and conservation.

INTEGRATING AGRICULTURE

Which insects are pests of the crops raised in your state? How do farmers control these pests?

Section Review 29–3

1. **Discuss** the importance of arthropods.
2. **BRANCHING OUT ACTIVITY** Select a flowering plant species that is common where you live. **Design an experiment** to determine whether insects pollinate this species. Conduct the experiment only with your teacher's supervision.

Laboratory Investigation

DESIGNING AN EXPERIMENT

Feeding Fly Larvae

Fruit fly larvae are particularly fond of yeast and other microorganisms present in overripe, fermenting fruit. In this investigation, you will determine which foods fruit fly larvae prefer.

Problem

Which foods do fruit fly larvae prefer? **Design an experiment** to answer this question.

Suggested Materials

- drawing paper
- pencil
- drawing compass
- metric ruler
- various foods
- yeast
- water
- fruit fly larvae

Suggested Procedure

1. Using a compass, draw a circle with a diameter of 10 cm on a sheet of paper. You may also want to use the pencil and ruler to draw more lines at any time during the investigation.
2. Working with a partner, choose four different food samples you will test to find which one the fruit fly larvae prefer. You might try cooked oatmeal, canned pumpkin, or pieces of crushed ripe fruit.

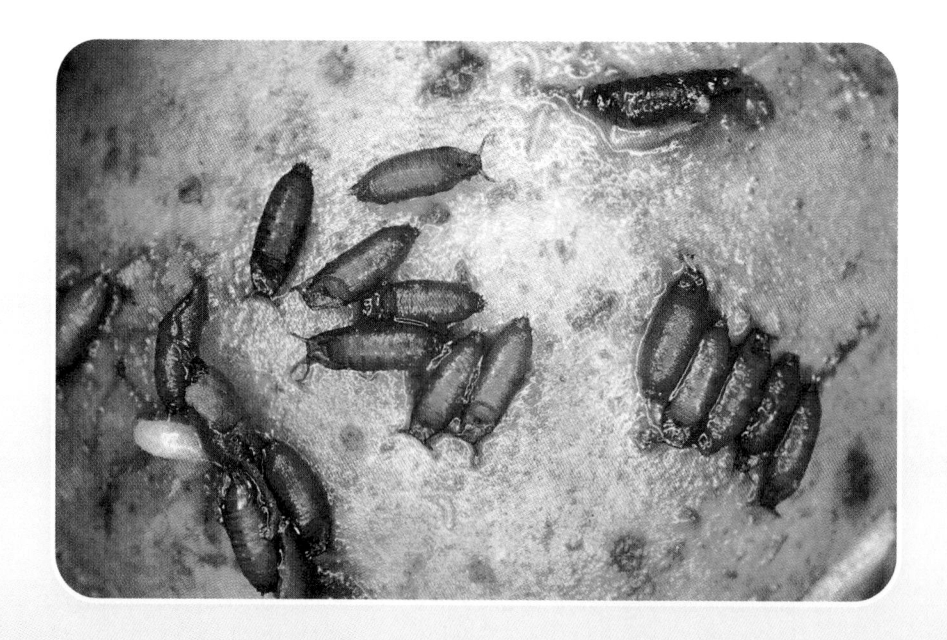

3. Decide how to arrange the food and larvae on the paper circle. Remember to include a control setup and to verify your experiment by repeating it.

4. Discuss with your partner what behaviors you will look for during each test.

5. Check with your teacher before carrying out the experiment. Make sure you keep a record of your procedure and observations at each step.

Observations

1. Prepare a data table to record the types of food and the reactions of the fruit fly larvae.
2. Describe the behavior of the larvae during each part of your experiment.

Analysis and Conclusions

1. Which of the four foods did the fruit fly larvae prefer? What evidence in your data table supports your conclusion?
2. How do you think fruit fly larvae sense food?
3. What sources of error could be present in your experiment?
4. How would you improve your experiment if you were to do it again?
5. Examine the laboratory investigations done by other pairs of students. What did you learn that gave you ideas about how you might change your own investigation?

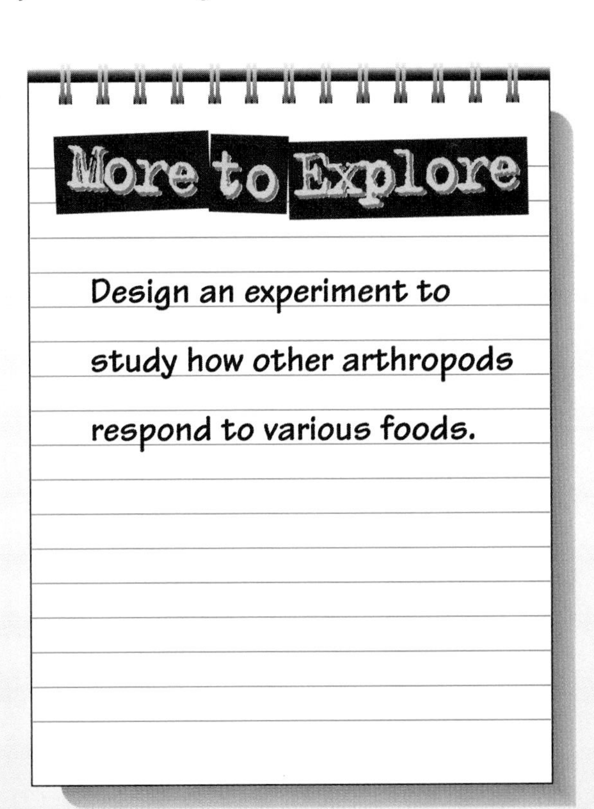

Study Guide

Summarizing Key Concepts

The key concepts in each section of this chapter are listed below to help you review the chapter content. Make sure you understand each concept and its relationship to other concepts and to the theme of this chapter.

29–1 Arthropod Form and Function

- Arthropods have segmented bodies, jointed appendages, and are surrounded by a tough exoskeleton.
- Most terrestrial arthropods breathe through a network of tracheal tubes. Nitrogen-containing wastes are eliminated by Malpighian tubules.
- In molting, an arthropod sheds its entire exoskeleton and manufactures a larger one, thus allowing the arthropod's body to grow.
- Some arthropods—such as grasshoppers—undergo incomplete metamorphosis, in which they grow larger but do not change drastically as they molt. Other arthropods—such as butterflies—undergo complete metamorphosis, involving distinct larval and pupal stages.

29–2 A Tour of the Arthropods

- A chelicerate has a body that is divided into two parts—a cephalothorax and an abdomen. The chelicerates include horseshoe crabs, spiders, mites, ticks, and scorpions.
- Nearly all crustaceans have two pairs of antennae, a pair of specialized mouthparts, and appendages with two branches. The crustaceans include crabs, shrimps, and crayfishes.
- Uniramians have a single pair of antennae and appendages with only a single branch. Uniramians include centipedes, millipedes, and insects.
- All insects have a three-part body consisting of head, thorax, and abdomen, and three pairs of legs.

29–3 The Importance of Arthropods

- Because arthropods are so varied and numerous, their roles in food webs, nutrient cycles, and partnerships with other animals are important in nearly every habitat.
- The United States is facing a pollination crisis because of threats to bees and to native pollinators.

Reviewing Key Terms

Review the following vocabulary terms and their meaning. Then use each term in a complete sentence.

29–1 Arthropod Form and Function

chitin
gill
book gill
book lung
tracheal tube
testis
ovary
spiracle
Malpighian tubule
molting
metamorphosis

29–2 A Tour of the Arthropods

cephalothorax
abdomen
chelicera
pedipalp
mandible

Recalling Main Ideas

Choose the letter of the answer that best completes the statement or answers the question.

1. The first arthropods to appear in the fossil record are the

a. mosquitoes. **c.** velvet worms.
b. trilobites. **d.** annelid worms.

2. An arthropod's exoskeleton is made mostly of

a. calcium carbonate. **c.** chitin.
b. protein. **d.** wax.

3. Lobsters, crabs, and other aquatic arthropods breathe through

a. gills. **c.** book lungs.
b. tracheal tubes. **d.** Malpighian tubules.

4. Because grasshoppers do not change their body shape drastically as they molt, their pattern of growth is called

a. exoskeleton growth.
b. pupal growth.
c. complete metamorphosis.
d. incomplete metamorphosis.

5. Which of these organisms is an arachnid?

a. centipede **c.** grasshopper
b. spider **d.** horseshoe crab

6. Water fleas, spider crabs, and crayfishes are members of the subphylum

a. Chelicerata. **c.** Crustacea.
b. Uniramia. **d.** *Macrotermes.*

7. The exoskeletons of centipedes and millipedes lack

a. a waterproof covering. **c.** spiracles.
b. body segments. **d.** chitin.

8. Termites are significant to humans because they

a. feed on dead skin cells.
b. cause allergies.
c. spread disease among livestock.
d. destroy wood and wooden structures.

9. Biologists estimate that animal pollinators include

a. about 10 different species.
b. about 1000 different species.
c. over 130,000 different species.
d. only honeybees and bumblebees.

Putting It All Together

Using the information on pages xxx to xxxi, complete the following concept map.

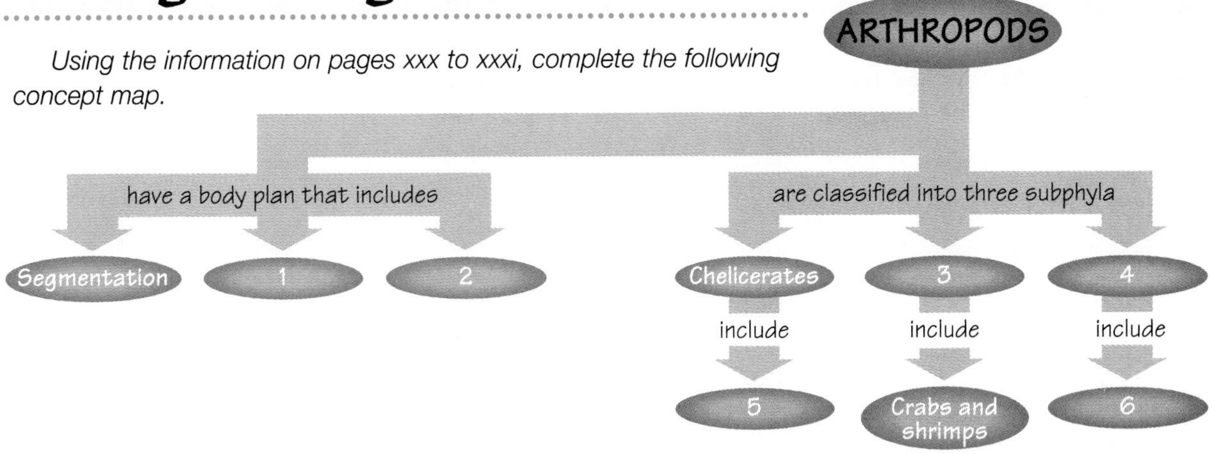

ARTHROPODS
have a body plan that includes
Segmentation
1
2
are classified into three subphyla
Chelicerates
3
4
include
include
include
5
Crabs and shrimps
6

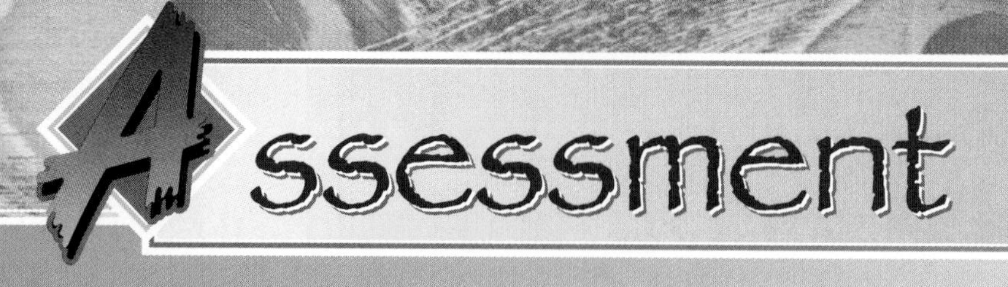

Reviewing What You Learned

Answer each of the following in a complete sentence.

1. What are the characteristics of arthropods?
2. What is a trilobite?
3. What is the function of the waxy covering on the exoskeleton of some arthropods?
4. Which structures do most land-dwelling arthropods use for breathing?
5. What are Malpighian tubules?
6. What is molting?
7. Compare incomplete metamorphosis and complete metamorphosis.
8. How do most spiders capture prey?
9. How do mites and ticks feed?
10. Describe the characteristics of a centipede and a millipede.
11. What characteristics do all insects have in common?
12. Why are insects important to many flowering plants?
13. Describe the threat that Africanized bees pose in the United States.

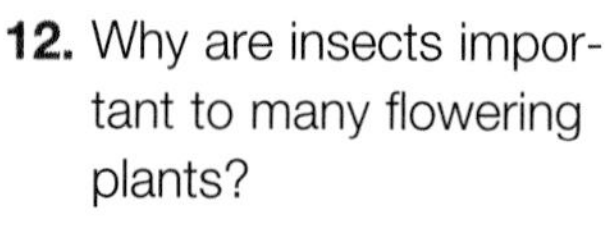

Expanding the Concepts

Discuss each of the following in a brief paragraph.

1. Describe the diversity of arthropods.
2. What characteristics do biologists use to divide arthropods into three subphyla?
3. List the different ways in which arthropods feed.
4. Describe an arthropod's circulatory system.
5. Why do arthropods molt? Why is molting a dangerous and physiologically expensive process for arthropods?
6. Compare the structures used for breathing in terrestrial arthropods and aquatic arthropods.
7. Describe the advantages and disadvantages that a hard exoskeleton provides an arthropod.
8. Compare the body plans of a chelicerate and a uniramian.
9. Discuss some of the problems that insects cause humans.
10. Describe the pollination crisis in the United States and identify its causes.

Extending Your Thinking

Use the skills you have developed in this chapter to answer the following.

1. **Designing an experiment** Many factors can influence an insect's molting cycle. Design an experiment to determine whether atmospheric temperature influences when a butterfly larva—a caterpillar—forms a cocoon and enters the pupa stage.
2. **Inferring** Arthropods are rarely found in sizes longer than half a meter, and most are less than a few centimeters in length. From your knowledge of the arthropod body plan, infer which structures limit an arthropod's growth.
3. **Evaluating arguments** Your friend argues that the pollination crisis in the United States is not a serious problem because new species of pollinators will evolve to replace those that are in decline. Do you agree with this assertion? Explain your answer.
4. **Making judgments** Arthropods are often the subjects of scientific research, and they sometimes die during the course of an experiment. Prepare a list of arguments both for and against the use of arthropods in research. How do you feel about this issue?
5. **Formulating hypotheses** At a park, you notice a red can and a green can resting at opposite ends of a picnic table. Several bees are swarming around the red can, but they seem to be avoiding the green can. Formulate a hypothesis to explain this observation.

Applying Your Skills

Creating a Dichotomous Key

To classify organisms, biologists use a type of flowchart called a dichotomous key. A dichotomous key is an ordered series of yes-or-no questions and direction arrows. Each question is followed by two direction arrows—one for yes and one for no—that lead to the next questions in the series or to the name of the organism.

1. Obtain a collection of eight photographs of different arthropods from your teacher. Research the characteristics of these arthropods and how they are classified.
2. Write a dichotomous key that can be used to classify the eight arthropods.
3. Exhange the eight photographs and your dichotomous key with another group of classmates. Follow the dichotomous key that the other group prepared to classify the organisms that they were given.
4. With the members of the other group, discuss the strengths and weaknesses of the dichotomous keys that your groups prepared.

• GOING FURTHER •

5. Prepare a poster to illustrate the way in which arthropods are classified.

CHAPTER 30

Fishes and Amphibians

FOCUSING THE CHAPTER THEME: Unity and Diversity

30–1 Fishes
- **Describe** the body plan of fishes.

30–2 Amphibians
- **Explain** how amphibians are adapted to live in water and on land.

BRANCHING OUT *In Depth*

30–3 The World of Fishes
- **Discuss** the importance of fishes in ecosystems.

LABORATORY INVESTIGATION
- **Design an experiment** to determine the effect of colored light on tadpoles.

Biology and Your World

BIO JOURNAL

In your journal, list the external features of a fish using the photograph on this page. How are fishes different from other animals you have encountered? How are they similar?

School of snappers near the Great Barrier Reef, Australia

Fishes

SECTION 30–1

GUIDE FOR READING

- **Describe** the chordates.
- **Discuss** the characteristics of fishes.
- **Identify** the three classes of living fishes.

MINI LAB

- **Classify** a fish.

FROM FRESHWATER SPRINGS to salty seas and from sunlit mountain streams to dark, frigid ocean depths, fishes are the masters of the underwater world. Some fish species have changed little since the dinosaurs prowled the Earth, while scores of other species have evolved in just the last few thousand years. Some are beautiful, some are frightening, but all are fascinating.

To biologists, fishes are important for another reason: They provide clues about the development of the vertebrates—animals with backbones. Indeed, certain fishes may be the closest living examples of the common ancestor of all vertebrates.

The First Chordates

Fishes, frogs, birds, reptiles, and humans—all are members of the phylum Chordata. You will learn a great deal about the chordates as you read this chapter and the following chapters.

At some stage of its life, a chordate has a notochord, a hollow dorsal nerve cord, and pharyngeal slits. A structure called the **notochord** is a long, flexible rod that typically develops in the embryonic stage. In fishes and most other chordates, the notochord is replaced by a vertebral column—a backbone. The **hollow dorsal nerve cord** lies next to the notochord. Typically, it develops into the main nerve pathway from the body to the brain. And the **pharyngeal** (fuh-RIHN-jee-uhl) **slits** are slits in the **pharynx,** or throat region of the body. In fishes, these structures develop into featherlike structures called **gills,** which are used for breathing.

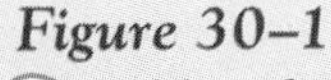

Figure 30–1
(a) Although this tunicate may look like an underwater flower, it is one of the few examples of an invertebrate chordate. (b) Fishes, such as these brook trout, are true vertebrates—animals with backbones. The vertebral column helps support a fish's body and is flexible enough to allow a variety of movements. (c) Different fishes evolved a variety of adaptations to help them survive. The unusual shape and coloring of this frogfish help it to blend into its surroundings.

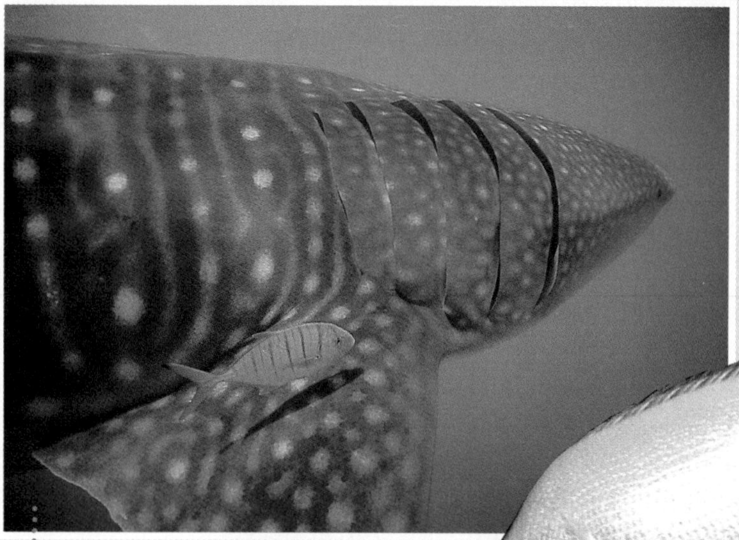

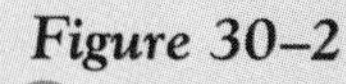

Figure 30–2
(a) *This whale shark has several gill slits on each side of its head. Aside from sharks and lampreys, most other fishes—such as the golden trevally swimming next to the shark—have only one pair of gills.* (b) *Some fishes breathe in unusual ways. These kissing gouramis exchange gases with air bubbles they pick up during trips to the surface.*

A few chordates are invertebrates, meaning they lack a backbone. The invertebrate chordates include lancelets and the tunicates shown in *Figure 30–1* on page 687. However, the vast majority of chordates—including the fishes—do have backbones. These chordates are the vertebrates.

☑ **Checkpoint** What are chordates? Vertebrates?

Fish Form and Function

All fishes have fins, and almost all fishes breathe through gills and have scales. In other respects, fishes are incredibly diverse. However, most fishes have the basic characteristics shown in the illustration on the next page.

Respiration

To breathe, most fishes use gills to exchange gases with the surrounding water. Gills provide a great amount of surface area for gas exchange and are richly supplied with blood vessels. In many fishes, gills are protected by a bony structure called an **operculum,** which also forms a single gill slit on each side of the head.

How does water reach the gills? The answer lies with the coordinated movements of the mouth and pharynx, which together act as a two-stage water pump. First, the gill slits close, the mouth opens, and the fish's "cheeks" are pulled sideways. This action increases the volume of the mouth and pharynx, and water is sucked in. Next, the gill slits open, the mouth closes, and the volume of the mouth decreases. This pushes the water over the gills and out through the gill slits.

☑ **Checkpoint** How do fishes breathe through gills?

Feeding

No other vertebrates come close to fishes in variety of feeding strategies. In part, this is because a fish's head is made of many bones, joints, and muscles. In different groups of fishes, these structures evolved in various ways, producing different mouth sizes and shapes, as well as a great variety in size, shape, and number of teeth. As a result, different fishes have evolved ways to eat almost every conceivable food.

Part of the bony fishes' success comes from an interesting link between eating and breathing. By combining their ability to suck in water with their specialized mouthparts, many fishes can literally suck in their prey.

Visualizing a Fish

The body plan of fishes has allowed them to thrive in waters throughout the Earth.

1 Gills

At a fish's gills, oxygen diffuses from the water to the blood, and carbon dioxide diffuses from the blood to the water. Fishes eliminate salts and nitrogenous wastes across their gills, as well.

2 Brain, Spinal Cord, and Vertebral Column

A fish's brain lies at the front end of a spinal cord—a hollow nerve cord that runs along the dorsal side of the body. Surrounding and protecting the spinal cord is a backbone—the vertebral column.

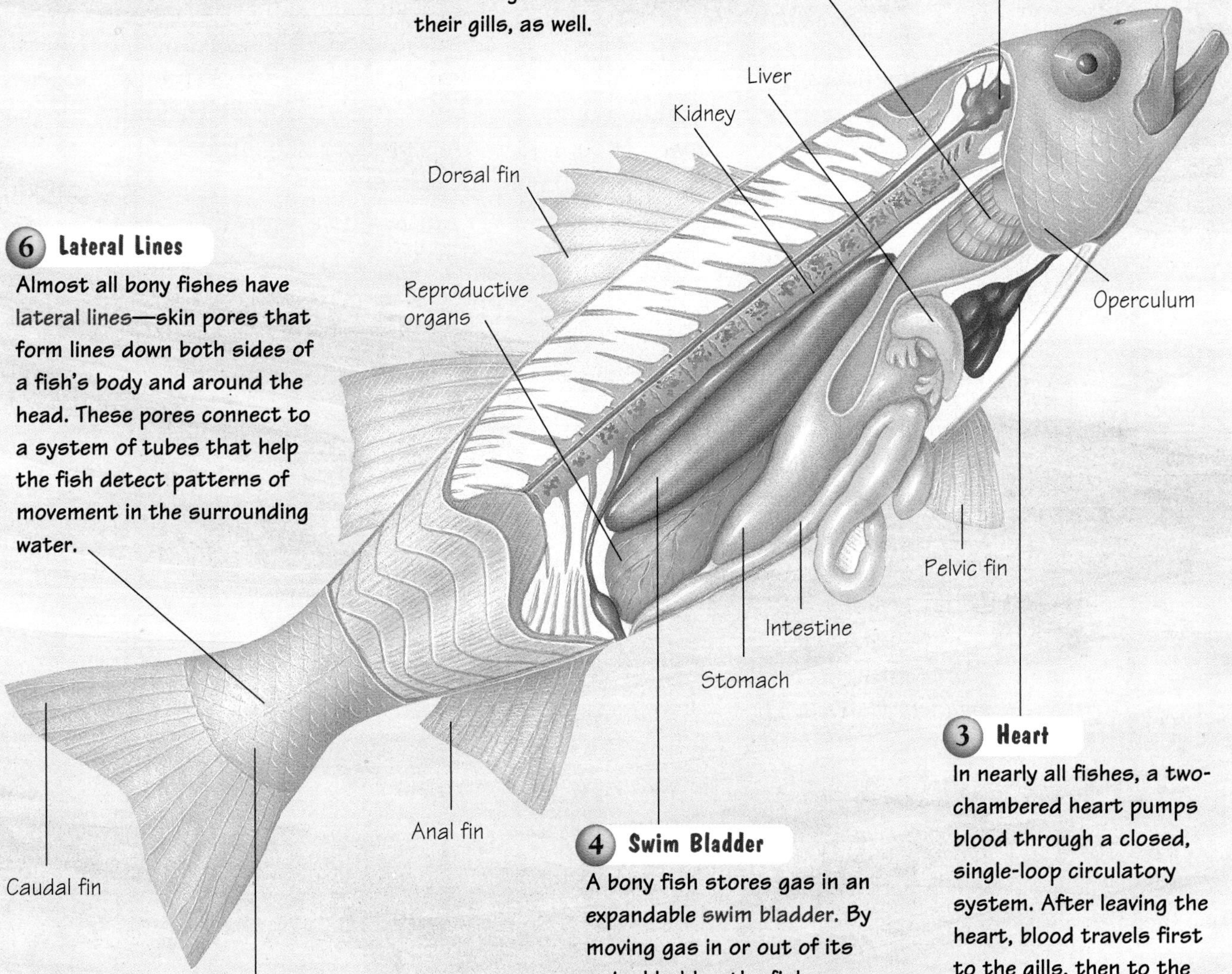

Liver
Kidney
Dorsal fin
Reproductive organs
Operculum
Pelvic fin
Intestine
Stomach
Anal fin
Caudal fin

6 Lateral Lines

Almost all bony fishes have lateral lines—skin pores that form lines down both sides of a fish's body and around the head. These pores connect to a system of tubes that help the fish detect patterns of movement in the surrounding water.

3 Heart

In nearly all fishes, a two-chambered heart pumps blood through a closed, single-loop circulatory system. After leaving the heart, blood travels first to the gills, then to the rest of the body.

4 Swim Bladder

A bony fish stores gas in an expandable swim bladder. By moving gas in or out of its swim bladder, the fish can change its internal density, thus changing its depth in the water.

5 Scales

A bony fish coats its scales with a slimy mucus, thus reducing friction with the water and helping the fish to escape predators.

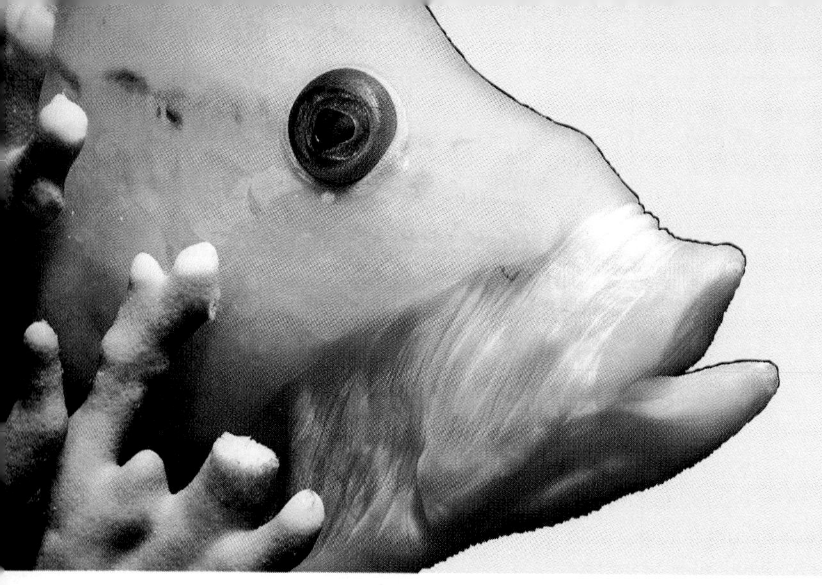

Figure 30–3
The mouth of this slingjaw wrasse expands like a telescope. Together, its mouth and pharynx act as a powerful aquatic vacuum cleaner to suck in prey.

Excretion

Many fishes get rid of much of their nitrogenous wastes in the same way that many other aquatic animals do—by allowing ammonia to diffuse across their gills and into the water. Fishes also use a simple **kidney** to filter nitrogenous wastes from the blood. In addition, fishes can actively pump salts in or out of their body fluids by using specialized cells in their gills.

Do you think that fishes drink a lot of water? In fact, freshwater fishes don't swallow water at all! Why? The blood and body fluids of freshwater fishes contain more dissolved substances than the water around them. As a result, water constantly enters the body through osmosis. The fishes eliminate this water by passing lots of dilute urine through their kidneys.

Saltwater fishes, on the other hand, constantly lose water through osmosis. These fishes do drink continuously, but they produce very little urine.

☑ **Checkpoint** How do fishes get rid of nitrogenous wastes?

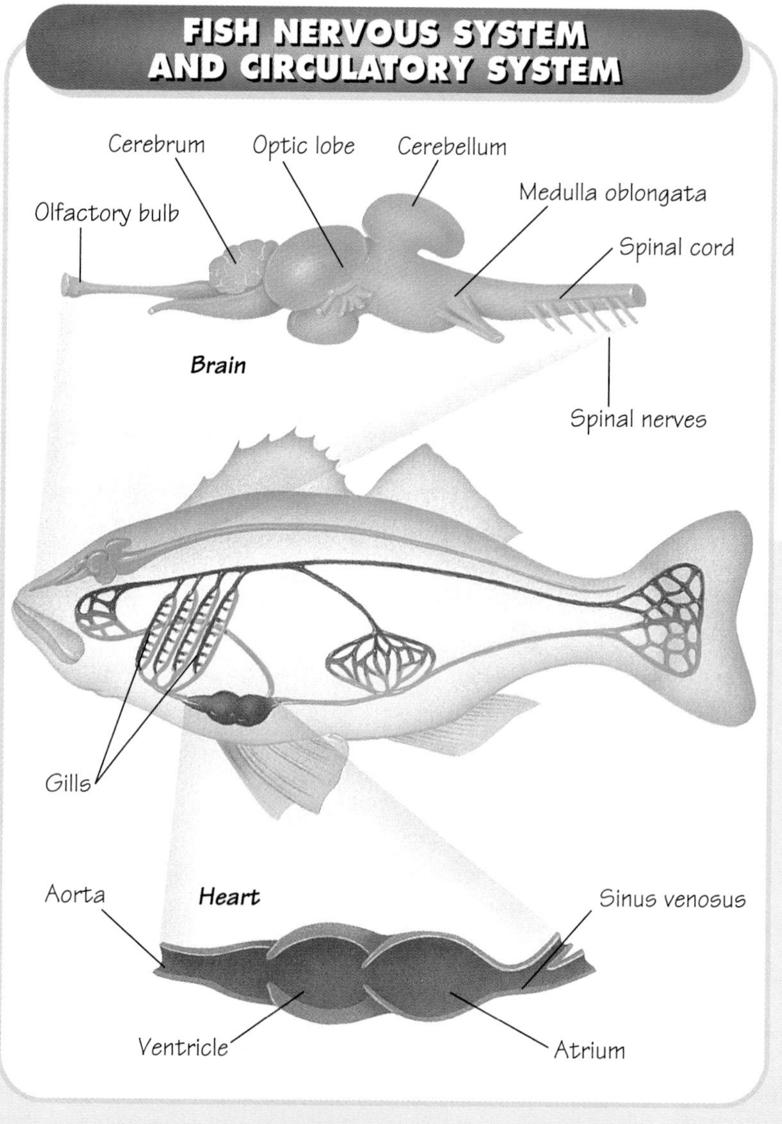

FISH NERVOUS SYSTEM AND CIRCULATORY SYSTEM

Cerebrum
Optic lobe
Cerebellum
Medulla oblongata
Olfactory bulb
Spinal cord
Brain
Spinal nerves
Gills
Aorta
Heart
Sinus venosus
Ventricle
Atrium

Figure 30–4
Like other vertebrates, a fish has a heart located near its respiratory organs and a brain located at the front of the spinal cord. The heart has two chambers—an atrium and a ventricle. Blood travels from the heart first to the gills, then to the rest of the body. The brain contains many of the same parts as the brains of other vertebrates, including the cerebrum, cerebellum, and medulla oblongata.

Nervous System

Although the nervous system in fishes is far simpler than in humans, it is set up in much the same way. Up front are **olfactory bulbs,** which are specialized for the sense of smell. The olfactory bulbs are directly connected to the two lobes of the **cerebrum.** In humans, the cerebrum is where thinking occurs. But in many fishes, the cerebrum seems to be mostly involved with the sense of smell and basic behaviors. Next come the **optic lobes,** which process visual information. Then comes the **cerebellum,** where body movements are coordinated. Finally, connecting the brain with the spinal cord, is the **medulla oblongata,** which regulates the function of many internal organs and helps maintain balance.

Sense Organs

As different species of fishes evolved in different environments, their sensory abilities adapted to provide them with information about their surroundings. Fishes that live in clear water and are active in the daytime, for example, have well-developed eyes and color vision. In some cases, this vision is better than our own. On the other hand, species that are active at night have enormous eyes that see only in black and white but function well in what to us is pitch darkness.

Many fishes, especially those that live in murky water, carry large numbers of chemical receptors—in what we might think are strange places. Some fishes can "smell" with their lips and "taste" with their barbels—a kind of whiskers. These chemical receptors can be very sensitive. Some predatory sharks respond to a single drop of blood in 115 liters of water!

Some fishes have other senses that humans can only imagine. Sharks, for example, have specialized cells that detect tiny electric currents. Why might this sense be useful?

Figure 30–5
(a) The large, lively eyes of these longspine squirrelfish help them to prowl for food at night, when light is scarce. (b) This polka-dot catfish is able to "taste" with chemical receptors along its barbels, or whiskers. (c) Electric eels have the amazing ability to generate an electric current, which they use mainly to shock would-be predators.

(a)

(b)

(c)

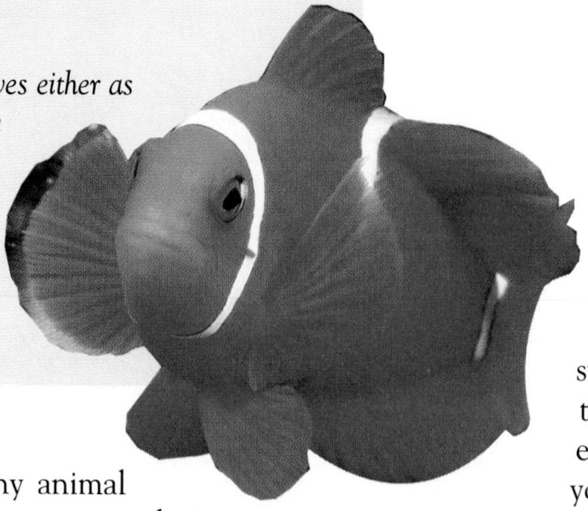

Figure 30–6
Most fishes spend all of their lives either as males or females. But there are exceptions! This clownfish was born male, then changed into a female as it got older. Fishes of other species can change from female to male.

Whenever any animal moves, its nerves and muscle cells produce electric currents. By detecting these currents, sharks can find prey that are otherwise hidden. Sharks also seem to use their electric sense to "zero in" on prey in the final moments before a strike.

Certain fishes not only detect electric currents but produce their own. Several species from the murky waters of the Amazon basin are known to navigate and communicate with one another using this electric sense.

Reproduction

Fishes reproduce in several different ways. Many are **oviparous,** meaning they lay eggs just before or soon after the eggs are fertilized. These species may or may not take care of the eggs. In either case, the developing embryos are nourished by a yolk sac attached to their gut.

Other species are **ovoviviparous** (oh-voh-vigh-VIHP-uh-ruhs), meaning the female holds eggs inside its body while the eggs grow into free-swimming babies. However, the embryos are nourished entirely by the original yolk. Still other fishes are **viviparous** (vigh-VIHP-er-uhs), meaning that the female provides additional nourishment to developing young inside its body. Oviparous species may have either internal or external fertilization, while ovoviviparous and viviparous forms have internal fertilization.

Although most fishes are either male or female, a few have the interesting ability to change genders, as illustrated in ***Figure 30–6.*** And fishes of a few species may be both male and female at the same time.

Classifying Fishes

We lump together all sorts of aquatic vertebrates under the name fishes. However, fishes are classified not in a single class but in three classes of the superclass Pisces. **The three classes of living fishes are jawless fishes, cartilaginous fishes, and bony fishes.**

Figure 30–7
Hagfishes and lampreys are the only living jawless fishes, a class once very common and diverse. (a) Adult lampreys are eellike fishes that live as parasites of other aquatic animals. These two lampreys are living off a carp. (b) A lamprey attaches to its host with the aid of strong teeth around its mouth.

Figure 30–8
The teeth of this sand tiger shark can rip and tear almost any prey. Unlike most vertebrates, a shark regularly replaces its teeth throughout its lifetime.

Jawless Fishes

For millions of years, jawless fishes ruled the sea. However, the jawless fishes of today—hagfishes and lampreys—bear little resemblance to their ancient ancestors or to any other fishes! Unlike other vertebrates, these animals have lost their bony skeletons. They also have no scales.

Lampreys spend the first seven years of their life buried in sand, where they live as filter-feeding larvae. Eventually, they mature into parasitic adults.

Hagfishes are even more peculiar. These parasitic fishes have six hearts, use an open circulatory system, and produce huge amounts of slime, which helps them to escape predators. Their bodies are so flexible that they can literally tie themselves in knots!

Cartilaginous Fishes

Members of the class Chondrichthyes (kahn-DRIHK-theez) are called cartilaginous fishes. They include sharks, rays, and skates. This ancient and successful group of fishes lives predominantly in the sea. Only a few species adapted to fresh water.

Although bones had evolved in their ancestors, living cartilaginous fishes have a skeleton made entirely of cartilage, as their name implies. The body is covered with platelike scales, each armed with toothlike structures that make the scales rough to the touch. In addition, scales around the mouth have evolved into teeth.

There are only about 750 living species of cartilaginous fishes. However, they are diverse and important in marine ecosystems. Most are carnivores, but their diets range from small worms and mussels to large fast-swimming fishes, marine mammals—and an occasional human. Other species are harmless filter feeders that feed on plankton.

☑ **Checkpoint** What are the cartilaginous fishes?

Figure 30–9
Bony fishes have evolved unique adaptations. (a) Archerfish shoot water at unwary prey—here, an insect on a reed. When the insect falls in the water, the archerfish eats it. (b) When threatened, a puffer expands its body like a balloon. If it expands rapidly while in a predator's mouth, the predator could suffocate.

Is That a Fish on Your Pizza?

PROBLEM *How can you **classify** a fish?*

PROCEDURE

1. Obtain an anchovy from your teacher. Sketch the anchovy, including close-ups of a dorsal fin, a pectoral fin, and the tail (caudal fin).
2. Count the number of scales along the lateral line, which runs from the gill slit to the caudal fin.
3. Using a scalpel, cut open the underside of the fish. Describe the structures you observe.

ANALYZE AND CONCLUDE

1. Why are anchovies included in the class Osteichthyes?
2. How can you further classify this fish based on the shape of its fins?
3. How does the number of scales you counted compare with the numbers counted by your classmates? How is this significant?

Bony Fishes

The members of the class Osteichthyes (AHS-tuh-ihk-theez)—the bony fishes—have skeletons made of strong, lightweight bone. Biologists have identified more than 30,000 species of bony fishes, and more species remain to be discovered. While some bony fishes are rare, others are quite abundant. For example, researchers estimate that there are as many as a billion billion individuals of various species of herring!

Biologists divide the bony fishes into two subclasses. Ray-finned fishes include nearly every bony fish you can think of. These fishes range from guppies to groupers, from bluefish to flounders, and from anchovies to eels and salmon. Many perform feats you would never dream fishes could manage. Archerfish can use their mouth like a water pistol, as shown in ***Figure 30–9*** on page 693. Flying fish leap out of the water and use large pectoral fins to glide through the air. And some species of catfish can actually climb a tree!

The other subclass of bony fishes are fleshy-finned fishes, which include lungfishes and lobe-finned fishes. Lungfishes have retained their ancestors' primitive lung, which is an air-filled sac connected to the gut. A lungfish fills its lung by swallowing air and empties it by belching.

Lobe-finned fishes were common in the Devonian Era. But today, the only living species is *Latimeria,* also known as a coelacanth (SEE-luh-kanth). Coelacanths live in the deep sea, far from the shallow freshwater swamps of their ancestors. Still, they offer a fascinating look at ancient fishes. Of all living animals, they are among the closest to the common ancestor of all four-limbed vertebrates.

Section Review 30–1

1. **Describe** the chordates.
2. **Discuss** the characteristics of fishes.
3. **Identify** the three classes of living fishes.
4. **Critical Thinking—Explaining Facts** For fishes to survive in an aquarium, the water must be kept clean and well-ventilated. Explain why water quality is so important to a fish's survival.
5. **MINI LAB** How can you **classify** a fish?

Amphibians

Guide for Reading

- **Describe** the characteristics of amphibians.
- **Identify** two orders of amphibians.

MINI LAB

- **Design an experiment** to find out how temperature affects frogs.

AMPHIBIANS HAVE SURVIVED for millions of years, typically in places where fresh water is plentiful. Amphibians also provide a link to the past. The 2500 species of living amphibians are the only surviving descendants of an ancient group that gave rise to all other land vertebrates.

Amphibian Form and Function

The name amphibian means "double life," emphasizing that these animals live both in water and on land. **With some exceptions, amphibians lay eggs in water, live in water as larvae and on land as adults, and have moist skin that lacks scales.**

Although the class Amphibia is relatively small, it is diverse enough to make it difficult to identify a typical species. Even so, frogs are good representatives. To learn more about the anatomy of frogs, study the illustration on the next page.

☑ *Checkpoint* What is an amphibian?

Feeding and Respiration

The double life of amphibians is reflected in their feeding habits. Amphibian larvae—commonly called tadpoles—are typically filter feeders or herbivores that graze on algae. Like other herbivores, tadpoles eat almost constantly, and their long, coiled intestines are usually filled with food. However, when tadpoles change into adults, their feeding apparatus and digestive tract are transformed to a strictly meat-eating design, complete with a much shorter intestine.

In most larval amphibians, gas exchange occurs through both skin and gills. Lungs typically replace gills when an amphibian becomes an adult, although some gas exchange continues across the skin.

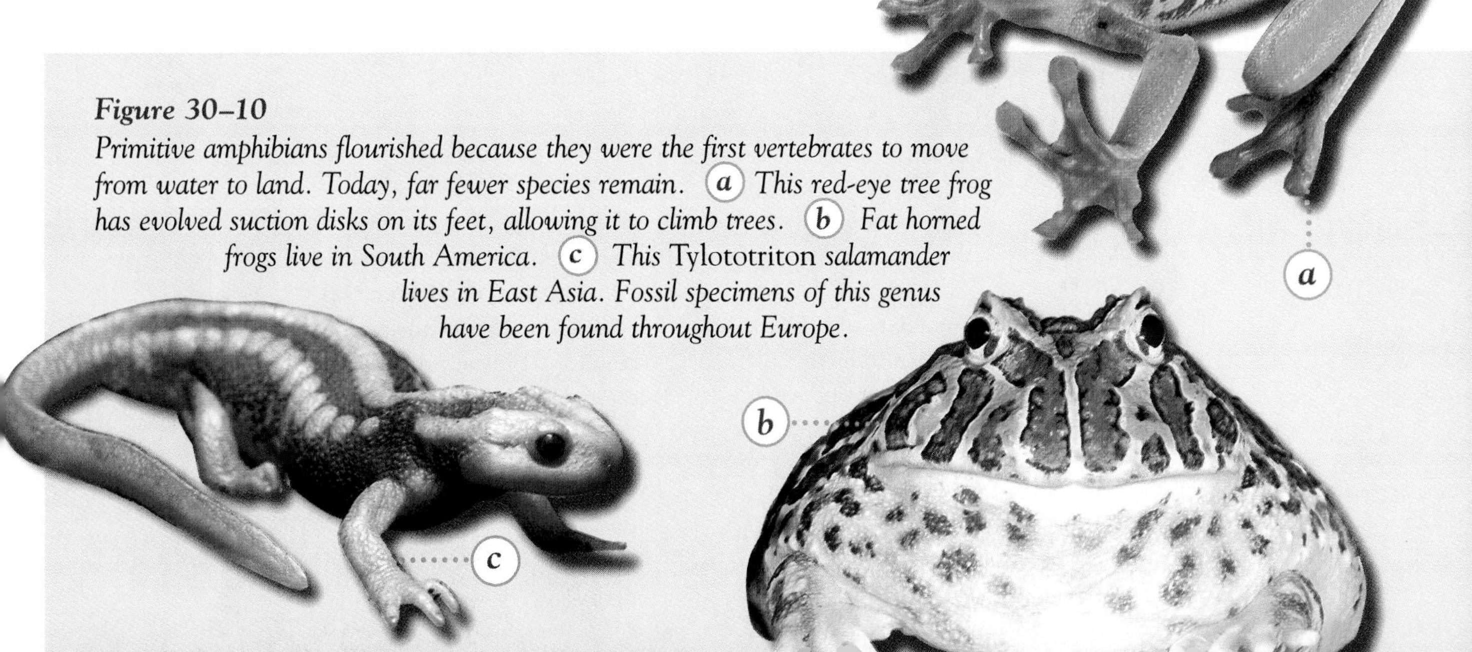

Figure 30–10
Primitive amphibians flourished because they were the first vertebrates to move from water to land. Today, far fewer species remain. (a) This red-eye tree frog has evolved suction disks on its feet, allowing it to climb trees. (b) Fat horned frogs live in South America. (c) This Tylototriton *salamander lives in East Asia. Fossil specimens of this genus have been found throughout Europe.*

Visualizing a Frog

Frogs and other amphibians represent an important link in the evolution of terrestrial vertebrates. While a larval frog lives only in water, an adult frog has adaptations for life on land.

1 Lungs

Although a frog has functional lungs, it lacks the specialized chest and stomach muscles that other vertebrates use to move air in and out of their lungs. Instead, a frog fills its mouth with air, forces air into the lungs, and holds it there for some time.

2 Tympanic Membrane

As in a human ear, a frog's **tympanic membrane** vibrates in response to sound—the first step in the process of hearing. Some frogs hear well both on land and in water.

3 Eyes

A frog's eyes are specially "tuned" to detect objects resembling flying insects, the frog's main prey. In addition, transparent eyelids called **nictitating membranes** protect the eyes underwater and keep them moist in air.

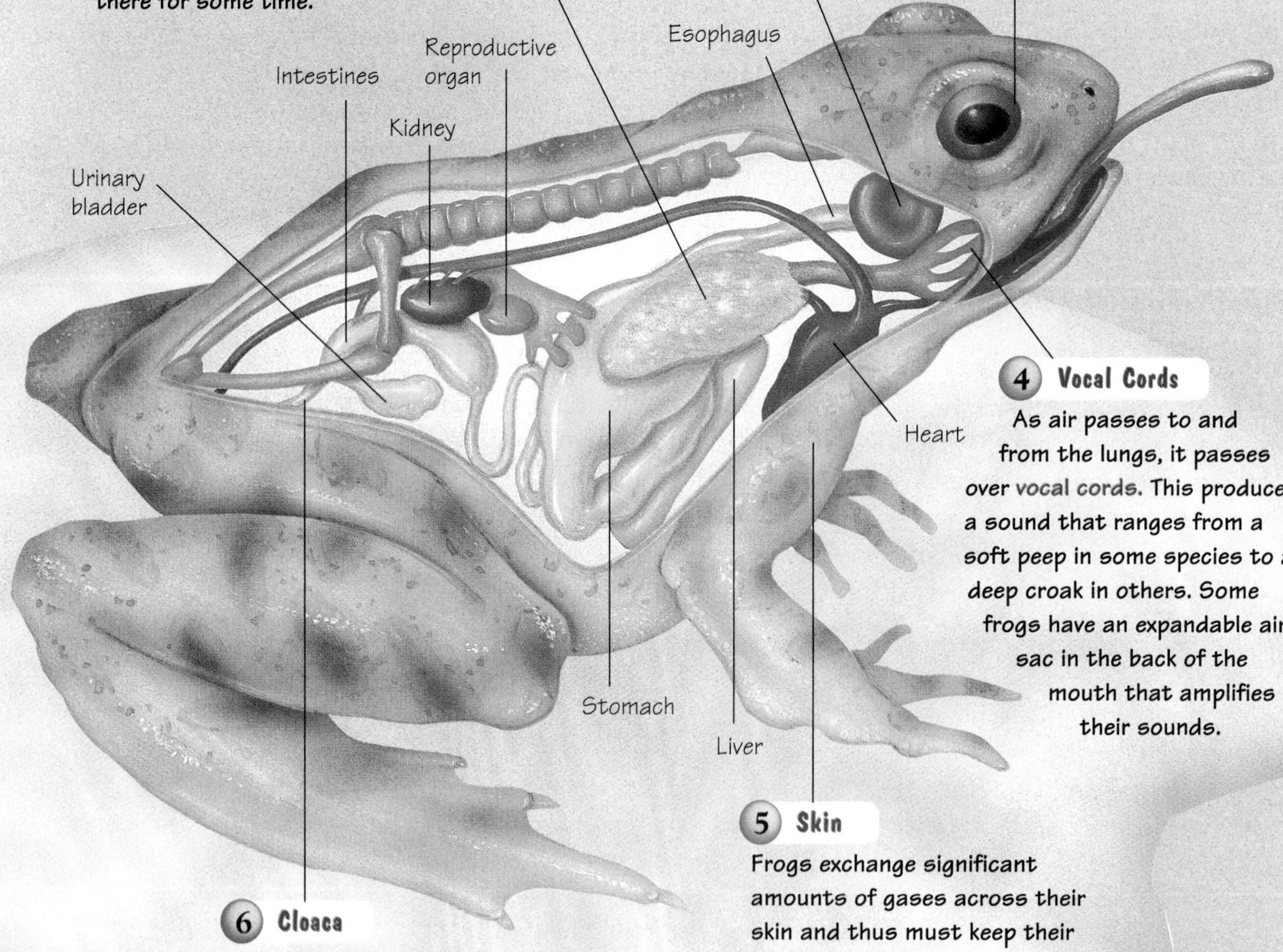

4 Vocal Cords

As air passes to and from the lungs, it passes over **vocal cords**. This produces a sound that ranges from a soft peep in some species to a deep croak in others. Some frogs have an expandable air sac in the back of the mouth that amplifies their sounds.

5 Skin

Frogs exchange significant amounts of gases across their skin and thus must keep their skin moist. This is one reason why frogs typically stay close to bodies of water.

6 Cloaca

The **cloaca** receives undigested food from the intestines, urine from the kidneys, and eggs or sperm from the reproductive organs. All these materials exit through a single opening.

Figure 30–11
An amphibian typically begins its life in water, then moves to land. (a) Frog eggs hatch into aquatic larvae called tadpoles. Like most fishes, a tadpole has gills, fins, and a two-chambered heart. As a tadpole changes into a terrestrial adult, gills and fins shrink away as lungs, heart, and legs develop. (b) This midwife toad is protecting a long string of eggs around its hind feet.

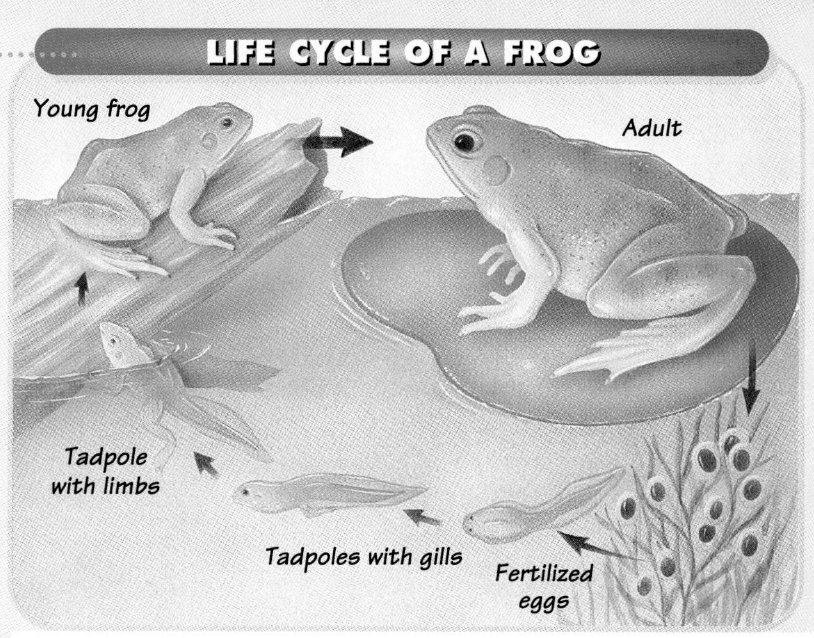

Internal Transport and Excretion

In the heart of an adult amphibian, oxygenated and deoxygenated blood are received separately in two chambers called atria, then mix together in a single pumping chamber called a ventricle. Although this mixing would cause all sorts of problems for a reptile or mammal, an amphibian compensates for this problem by exchanging significant amounts of gases through its skin.

Like fishes, amphibians have little problem excreting nitrogenous wastes. Some wastes diffuse across the skin, while others are removed from the bloodstream by kidneys. Amphibians need kidneys because they constantly gain water through osmosis, just as freshwater fishes do. To get rid of all that water, amphibians produce large amounts of dilute urine.

☑ **Checkpoint** How do amphibians eliminate wastes?

Movement

Although most adult amphibians possess the same four limbs found in most terrestrial vertebrates, they use these limbs differently. As shown in **Figure 30–12** on page 698, an amphibian's limbs stick out sideways from its body, making it difficult for it to walk on four limbs as mammals do.

Instead, amphibians move in other ways. Many salamanders throw their body into an s-shaped curve, then use their legs to push themselves along the ground. Land-dwelling frogs and toads use strong hind legs to hop from place to place. And tree frogs have small "suction cups" on the ends of their fingers and toes that help them to climb.

Reproduction

In most species of amphibians, the female lays eggs in water, then the male fertilizes them externally. In a few species, including salamanders, eggs are fertilized internally. These species may be oviparous, ovoviviparous, or even viviparous. Most amphibians abandon eggs after they lay them, but some take great care of both eggs and young.

Classifying Amphibians

Newts and salamanders belong to the order Urodela (yoor-oh-DEE-luh). Although the ancestors of these amphibians grew up to 3 meters long, living salamanders rarely grow longer than about

MINI LAB .... Experimenting ....

Frogs in Cold Water

PROBLEM *How does temperature affect a frog? Design an experiment to find out.*

PROCEDURE   

1. Obtain from your teacher a frog in a small jar containing water.
2. Count the number of times the frog breathes in 1 minute by watching the movements of its throat. Record this information.
3. Design an experiment to determine how water temperature affects a frog's breathing rate. Be sure to include a control. Have your teacher approve your procedure before you begin the experiment. Your experiment must not harm the frog in any way. Return the frog to your teacher.

ANALYZE AND CONCLUDE

1. How did water temperature affect the frog's breathing rate? How is this an advantage to the frog?
2. Formulate a hypothesis to explain your results.

15 centimeters. Their larvae are usually fully aquatic, having gills and a tail. Although most adults lose their gills as they move to live in moist woods, not all species develop lungs. And some species, such as the axolotl (AK-suh-laht-'l) and mud puppy, keep their gills as adults and live in water throughout their lives. Still other species switch back and forth between water and land.

Frogs and toads are members of the order Anura (uh-NOOR-uh). These amphibians live in many regions of the world and are common throughout the continental United States. One difference between frogs and toads is that frogs are more closely tied to water as adults. In fact, some toads have developed remarkable ways to adapt to environments with little or no surface water. These toads dig burrows deep into moist soil, from which they absorb water through their skin.

Figure 30–12
Amphibians are not especially numerous, but they are diverse. (a) Newts thrive in a variety of habitats. This alpine newt lives in western Europe. (b) To moisten its skin, this eastern spadefoot toad is burying itself in wet mud.

Section Review 30–2

1. **Describe** the characteristics of amphibians.
2. **Identify** two orders of amphibians.
3. **Critical Thinking—Formulating Hypotheses** Suppose that you discover the population of frogs and toads at a small lake is down by 25 percent from a year ago. Formulate a hypothesis to explain this observation.
4. **MINI LAB** How can you **design an experiment** to find out how temperature affects frogs?

The World of Fishes

GUIDE FOR READING

- **Explain** the importance of fishes in ecosystems.

WATER COVERS NEARLY 70 percent of Earth's surface, so most of our planet is home turf for fishes. In fact, fishes are Earth's most numerous vertebrates and can be found in all kinds of water. You shouldn't be surprised to discover that your life is affected by all these fishes—from the largest sharks to the smallest herring and anchovies.

Fishes in Ecosystems

Fishes live in almost every aquatic habitat on Earth, where they fill many different roles in natural ecosystems. Why are fishes so numerous and diverse? First, the fish body plan has proved successful in fishes of many different sizes, ranging from guppies and minnows—which are less than a centimeter long—to giant whale sharks that span over 15 meters. Second, fishes feed on nearly every type of food you could imagine, including some substances that you probably would never consider as fish food! Thus, fishes are part of many different food webs in a wide range of ecosystems.

In addition, many fish eggs and larvae are much smaller than adult forms. Thus, many species fit into different places of food webs at different times in their life cycle. The larvae of some fishes, for example, may be eaten by the same animals that are prey for the adults!

Fishes as Food

When you think of a good fish to eat, you probably think of a large fish, such as a salmon, trout, or tuna. But in nature, arguably the most important food fishes are small fishes, such as the herring shown in ***Figure 30–13***.

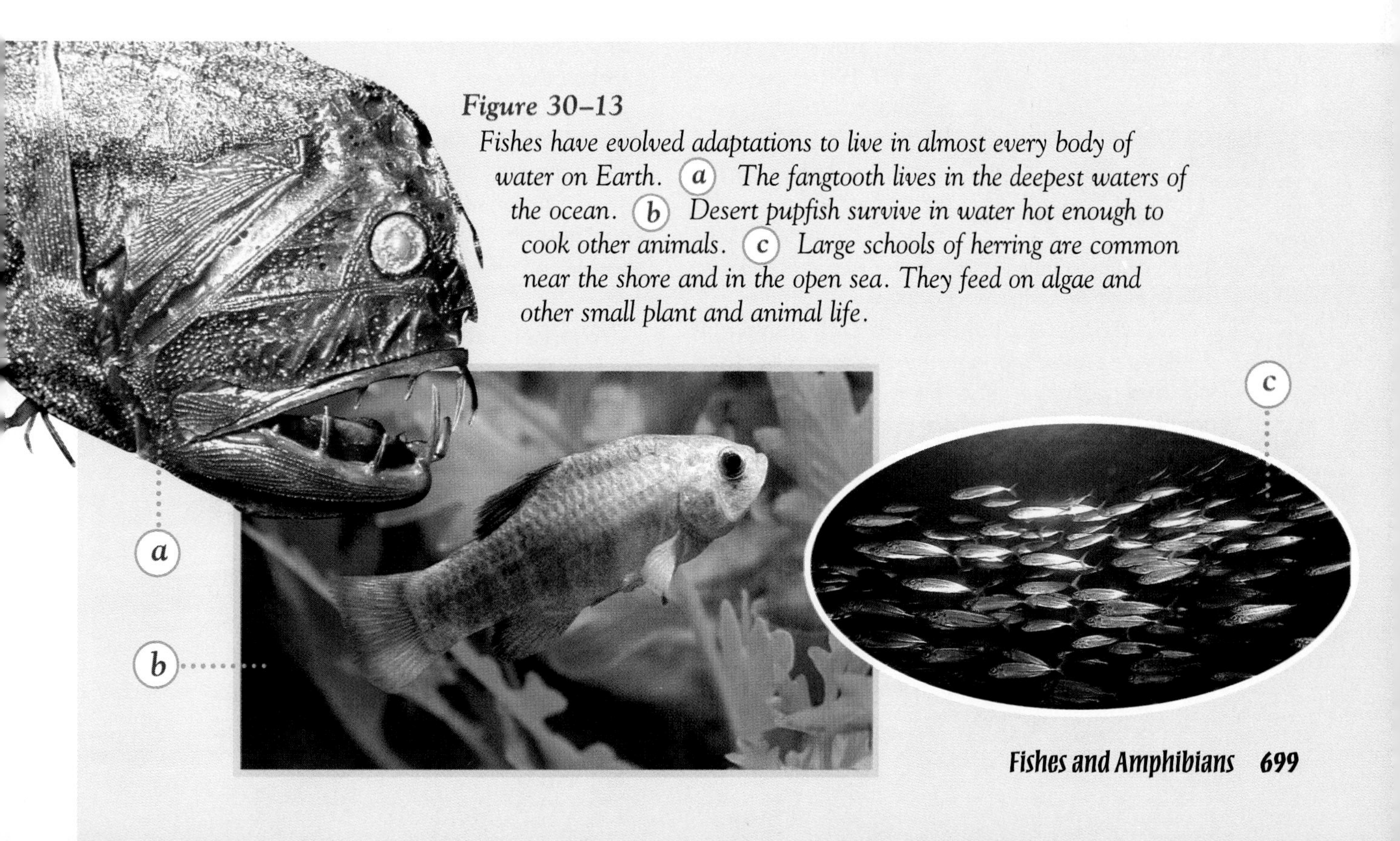

Figure 30–13
Fishes have evolved adaptations to live in almost every body of water on Earth. (a) The fangtooth lives in the deepest waters of the ocean. (b) Desert pupfish survive in water hot enough to cook other animals. (c) Large schools of herring are common near the shore and in the open sea. They feed on algae and other small plant and animal life.

Biology AND YOU .... Connections

Fish Farming

For many years, the commercial fishing industry has used huge nets to harvest fishes from oceans and large lakes. However, fishes can also be raised on aquatic farms—just as cows and pigs are raised on land.

Fish farming is not a new idea—humans have been raising and breeding fishes for thousands of years. People in China, Japan, and Thailand have a long tradition of raising and breeding fishes as pets. These fishes include goldfish, koi, and Siamese fighting fish. Today, the best koi sell for more than $10,000 per fish!

Koi fish are valued for their gracefulness and color.

Fish Farming in the United States

Fish farming is common in many countries. However, there are relatively few large commercial fish farms in the United States. Many of these farms raise catfish, a popular food. Unfortunately, most of the other fishes that Americans like to eat cannot be cultivated easily.

In recent years, however, the U.S. fishing industry has undergone many changes. Overfishing and pollution has been threatening the populations of many fish species in the wild, making these fishes more expensive to catch and bring to market. As a result, biologists are working to develop better and more economical techniques for raising popular fishes on farms.

Techniques of Fish Farming

Some species, such as catfish, grow well in crowded conditions, so they can be raised in shallow ponds. Other species, such as trout, require steady supplies of clean, cool, running water. These fishes may be farmed in streams, natural ponds, or cages set out in open water.

To raise salmon—a very popular food—fish farmers take advantage of the salmon's unusual life cycle. A salmon begins its life in a freshwater stream. It then migrates to the sea, grows to adulthood, and returns to the place it hatched in order to spawn. So, in a technique called salmon ranching, salmon eggs are placed in an aquaculture facility that is set apart from a stream. The salmon hatch and swim downstream to the sea. When they mature, they instinctively swim back to the facility, where they can be captured.

Although salmon ranching has proved successful, it is also controversial. Some researchers fear that mass releases of artificially raised salmon will reduce the genetic diversity—and thus the evolutionary fitness—of salmon species in the wild.

Making the Connection

What are the advantages and disadvantages of fish farming? Do you think that fishing is adequately regulated in the bodies of water near your community?

CAREER TRACK *This fish farmer is testing the water quality of a fish tank.*

Figure 30–14
Fishes are a vital part of the ecosystem of the Amazon basin—and not just the part underwater. When the river floods, fishes wander into the surrounding rain forest.

In the ocean, huge schools of herrings, anchovies, and other small fishes provide food for predators ranging from larger fishes to squids, sea birds, sea lions, dolphins, whales—and humans. You may or may not eat small fishes, but they are an important part of the food supply in many countries. What's more, enormous quantities of fish meal are used to supplement livestock feed in the United States and in Europe.

Unfortunately, several stocks of small fishes have been overharvested. More than 30 years ago, overfishing nearly wiped out the enormous anchovy schools off the coast of South America. Almost overnight, flocks of sea birds and schools of dolphins and porpoises also disappeared. Why? Because these carnivores depend heavily on anchovies for food.

People in the fishing industry have been slow to learn the lessons of overharvesting. In fact, many delicious fishes that once were very common—including cod, halibut, and several other species—have been overfished almost to extinction. Hope for fishes may lie with laws that protect their spawning grounds and nurseries, as well as a new emphasis on fish farming, which is described on the opposite page.

Checkpoint Why are anchovies and other small fishes important?

Fishes in the Rain Forest 

Fishes also play important roles in places you might normally consider to be terrestrial ecosystems. In the rain forest of the Amazon basin, for example, each year's rainy season brings torrential downpours that can raise river levels as much as 11 meters. When this happens, the rivers overflow into thousands of square kilometers of rain forest, creating what Brazilians call *varzea,* or flooded forest, as shown in **Figure 30–14.**

During the rainy season, many fishes wander from the riverbed into the flooded forest. There, they feed on ants and other insects that fall into the water, as well as fruits and seeds of rain forest trees. In fact, fish that are relatives of the ferocious piranhas use their huge teeth and strong jaws to crack open Brazil nuts and other hard-shelled seeds. Biologists suspect that certain rain forest plants depend on these types of fishes, rather than birds or mammals, to spread their seeds.

The vital interactions between the fishes and the forest offer another reason to preserve tropical forests. If the *varzea* disappears, so too will the river fishes that depend on the flooded forest for food and breeding places.

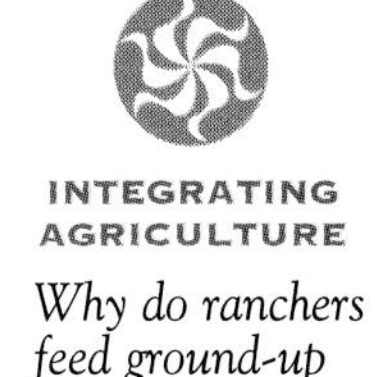

INTEGRATING AGRICULTURE

Why do ranchers feed ground-up fish to livestock?

Section Review 30–3

1. **Explain** the importance of fishes.
2. **BRANCHING OUT ACTIVITY** Conduct a survey to determine which fishes are popular foods with your classmates. **Research** how these fishes are raised or harvested.

Laboratory Investigation

DESIGNING AN EXPERIMENT

Raising Tadpoles

Frogs lay their eggs in a variety of watery habitats—from woodland ponds shaded by the leaves of overhanging trees to the edges of lakes in the full sun. In this investigation, you will determine whether tadpoles are affected by different kinds of light.

Problem

How can you determine the effect of colored light on tadpoles? **Design an experiment** to answer this question.

Suggested Materials

3 large plastic containers
pond water or dechlorinated tap water
12 tadpoles of the same species and age
tadpole food
sheets of cellophane, in two colors
tape

Suggested Procedure

1. **Use the large plastic containers to create three identical aquariums. Fill each aquarium with water from the same source—either pond water or dechlorinated tap water. You may choose to add clean rocks or other decorations to the aquariums. Place 4 tadpoles in each aquarium.**
2. **Design a procedure to determine how colored light affects tadpoles. To expose the tadpoles to colored light, tape sheets of colored cellophane**

around all sides of the aquariums. Do not tape cellophane directly to a light source. Be sure that your experiment includes a control group.

3. With your teacher's permission, carry out the experiment you designed. Prepare a data table to record the observations you will be making.

4. During the course of your experiment, follow your teacher's instructions for the care and upkeep of tadpoles. As a general rule, tadpoles should be fed twice a week. The water in the aquariums should be changed every 3 days and be kept at a temperature near 20°C. Avoid handling the tadpoles directly, which can affect their health.

5. Observe the tadpoles every day for 2 months. Note their appearance, general behavior, and feeding habits. Record all observations in your data table. **CAUTION:** ***Always wash your hands after handling tadpoles, their food, or the aquariums.*** At the end of the experiment, return the tadpoles to your teacher.

Observations

1. How did the tadpoles change over the course of the experiment?
2. Compare the development of the three groups of tadpoles. How did they develop similarly? Differently?
3. In each aquarium, did the living conditions remain the same throughout the experiment? If not, describe any differences that you observed.

Analysis and Conclusions

1. How does the color of light that a tadpole receives affect its development? Use the data you collected to support your answer.
2. Identify any assumptions you made in your answer to question 1. How could you change the design of your experiment to account for these assumptions?
3. Suppose that tadpoles were submitted to colored light in their natural environment. Do you think they would develop as they did in your experiment? Explain.
4. Formulate a hypothesis to explain the tadpoles' response to colored light. Discuss how you could test this hypothesis.

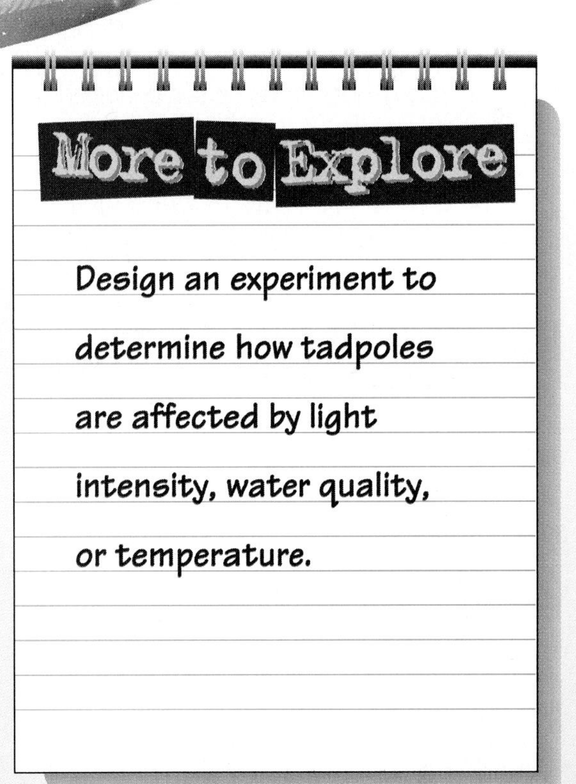

Study Guide

Summarizing Key Concepts

The key concepts in each section of this chapter are listed below to help you review the chapter content. Make sure you understand each concept and its relationship to other concepts and to the theme of this chapter.

30–1 Fishes

- At some stage of its life, a chordate has a notochord, a hollow dorsal nerve cord, and pharyngeal slits. Invertebrate chordates lack a backbone. Most chordates are vertebrates.
- All fishes have fins, and almost all fishes breathe through gills and have scales. In other respects, fishes are very diverse.
- A fish has a much simpler nervous system than a human has, but it is organized in a similar way. In addition, most fishes have a closed, single-loop circulatory system, a kidney to eliminate wastes, and well-developed sense organs.
- Fishes reproduce in several different ways and may be oviparous, ovoviviparous, or viviparous. Individuals of some species can change gender under certain conditions.
- The three classes of fishes are jawless fishes, cartilaginous fishes, and bony fishes. The bony fishes include two subclasses: ray-finned fishes and fleshy-finned fishes.

30–2 Amphibians

- With a few exceptions, amphibians lay eggs in water, live in water as larvae and on land as adults, and have moist skin that lacks scales.
- Frogs have a well-developed nervous system and sense organs. A frog's eyes are specially "tuned" to detect objects resembling flying insects, the frog's main prey.
- Newts and salamanders are members of the order Urodela. Frogs and toads are members of the order Anura.

30–3 The World of Fishes

- Fishes live in almost every aquatic habitat on Earth, where they fill many different roles in natural ecosystems.
- Herrings, anchovies, and other small fishes are important foods for larger animals—including sea birds and dolphins. Over-harvesting these fishes can decrease the populations of their predators.

Reviewing Key Terms

Review the following vocabulary terms and their meaning. Then use each term in a complete sentence.

30–1 Fishes

notochord
hollow dorsal nerve cord
pharyngeal slit
pharynx
gill
operculum
spinal cord
vertebral column
swim bladder
scale
lateral line
kidney
olfactory bulb
cerebrum
optic lobe
cerebellum
medulla oblongata
oviparous
ovoviviparous
viviparous

30–2 Amphibians

tympanic membrane
nictitating membrane
vocal cord
cloaca

Recalling Main Ideas

Choose the letter of the answer that best completes the statement or answers the question.

1. The vertebrate chordates include
   - **a.** arthropods.
   - **b.** tunicates.
   - **c.** lancelets.
   - **d.** amphibians.
2. Gills are located in the area of a fish's body called the
   - **a.** lungs.
   - **b.** sinus venosus.
   - **c.** pharynx.
   - **d.** cerebrum.
3. To move up and down in the water, a fish moves gas in and out of its
   - **a.** kidney.
   - **b.** swim bladder.
   - **c.** pharynx
   - **d.** gills.
4. Some fishes lay eggs just before or soon after the eggs are fertilized. This type of reproduction is described as
   - **a.** ovoviviparous.
   - **b.** external fertilization.
   - **c.** oviparous.
   - **d.** viviparous.
5. Which of these organisms is a cartilaginous fish?
   - **a.** coelacanth
   - **b.** herring
   - **c.** dolphin
   - **d.** shark
6. Frogs eggs typically develop into larvae
   - **a.** on land.
   - **b.** in water.
   - **c.** inside the female's body.
   - **d.** inside the male's body.
7. In adult amphibians, gas exchange occurs through
   - **a.** lungs only.
   - **b.** gills only.
   - **c.** skin only.
   - **d.** lungs and skin.
8. Nictitating membranes cover and protect a frog's
   - **a.** eyes.
   - **b.** gills.
   - **c.** lungs.
   - **d.** kidneys.
9. In natural ecosystems, fishes feed on
   - **a.** plants only.
   - **b.** other fishes only.
   - **c.** plants and small animals only.
   - **d.** many different foods.

Putting It All Together

Using the information on pages xxx to xxxi, complete the following concept map.

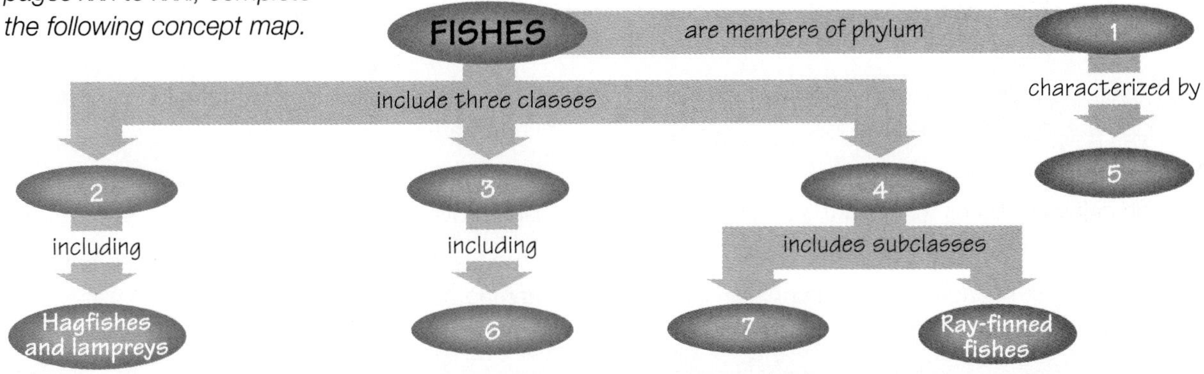

FISHES
are members of phylum
1
characterized by
5
include three classes
2
3
4
including
including
includes subclasses
Hagfishes and lampreys
6
7
Ray-finned fishes

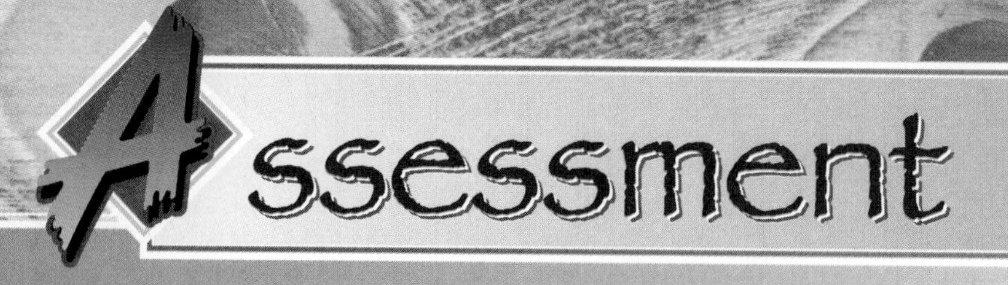

Reviewing What You Learned

Answer each of the following in a complete sentence.

1. Identify three structures that all chordates have at some time during their life cycle.
2. In fishes, what is the function of the operculum?
3. Identify the chambers of a fish's heart.
4. What are the olfactory bulbs?
5. What is a coelacanth? Why is it significant?
6. Identify the three classes of fishes. Which class is most numerous?
7. What is the purpose of the lateral line in fishes?
8. List three characteristics common to almost all amphibians.
9. Describe the heart of an adult amphibian.
10. Why must amphibians keep their skin moist?
11. What is the main difference between frogs and toads?
12. Identify the different orders of amphibians.
13. How do fishes help to disperse seeds in the rain forest of the Amazon basin?

Expanding the Concepts

Discuss each of the following in a brief paragraph.

1. Compare the nervous systems of fishes and humans.
2. Describe the function of scales in bony fishes and cartilaginous fishes.
3. How are sharks able to "zero in" on prey buried in the sand?
4. Describe the three general ways in which fish eggs are fertilized and developed.
5. Compare the ways in which freshwater fishes and saltwater fishes maintain the amount of water in their body.
6. Describe the changes that an amphibian larva undergoes when it becomes an adult.
7. In which stage of an amphibian's life—larva or adult—does it more closely resemble a fish? Explain.
8. Describe the three classes of fishes.
9. How do frogs produce their characteristic sounds?
10. Compare a frog's lungs with the lungs of other vertebrates.
11. Describe the different ways in which amphibians move.
12. Discuss the importance of small fishes, such as herring and anchovies.

Extending Your Thinking

Use the skills you have developed in this chapter to answer the following.

1. **Classifying** Examine the bony fish shown in the photograph on the opposite page. What structures can you identify from this photograph? List some other structures you would expect to find in this fish.
2. **Designing an experiment** A biologist hypothesizes that water pollution is causing the frog population to decline at a local lake. Design an experiment to provide evidence for or against this hypothesis.
3. **Analyzing concepts** Discuss the factors that may have contributed to the decline in the number of fishes caught off the coastal waters in many regions of the United States.
4. **Making judgments** Currently, laws in the United States protect endangered animals of any species—including fishes and amphibians. Research an instance in which the government halted a commercial development in order to protect a fish or amphibian species. Do you think that the government's action was appropriate? Discuss both sides of the issue.
5. **Communicating** Research the structure and life cycle of an interesting fish or amphibian, such as a shark, sea horse, or salamander. Present a report on your findings to the class.

Applying Your Skills

An Incredible Journey

A salmon is born in a freshwater stream, migrates to the sea, then returns years later to the same stream where its life began. Although biologists have learned a great deal about salmon, much about a salmon's behavior remains a mystery.

1. Research the structures or processes that allow salmon to survive in both salt water and fresh water.
2. Formulate a hypothesis to explain how a salmon is able to find the stream where it hatched.
3. As a salmon swims up a river to its home stream, it swims against strong currents and overcomes obstacles such as waterfalls. Do salmon have stronger muscles than other fishes? Research the answer to this question.

4. A female salmon lays about 8000 eggs, but only about 1 percent of the eggs survive to adulthood. Discuss the significance of this fact.

• GOING FURTHER •

5. Create a poster to illustrate a salmon's life cycle. You may also include information on salmon fishing, salmon farming, and factors that influence the salmon population in the wild.

CHAPTER 31

Reptiles and Birds

FOCUSING THE CHAPTER
THEME: Unity and Diversity

31–1 Reptiles
- **Describe** the characteristics of reptiles.

31–2 Birds
- **Explain** the adaptations that allow birds to fly.

BRANCHING OUT *In Depth*

31–3 Evolution of Birds
- **Discuss** the evolution of birds and flying reptiles.

LABORATORY INVESTIGATION
- **Design an experiment** to show how a chameleon changes color in response to its environment.

Biology and Your World

BIO JOURNAL

Do you have a favorite bird? A favorite reptile? What roles do birds and reptiles play in your life? Answer these questions in your journal.

Lanner falcon of eastern Africa

Reptiles

GUIDE FOR READING

- **Describe** the characteristics of reptiles.
- **Explain** why reptiles were able to adapt to land.

MINI LAB

- **Infer** how a snake moves by examining its skeleton and scales.

HUMANS HAVE ALWAYS BEEN *fascinated by reptiles—as well as frightened by many of them. Many people fear snakes for their venomous bites or slithery movements. And explorers' encounters with lizards and crocodiles inspired images of dragons in European folk tales.*

Yet reptiles are as astonishing as any creatures of human imagination. Dinosaurs grew to incredible sizes on land, and other extinct reptiles conquered sea and air. In fact, birds evolved from one line of small dinosaurs! Although living reptiles are smaller and less powerful than their ancestors, they are wondrous animals all the same.

The Reptile Body Plan

Reptiles are scaly-skinned vertebrates that breathe with lungs and lay eggs that hatch on land. A reptile's scales may be smooth or rough. Folklore to the contrary, reptile skins are never slimy.

The basic reptilian body plan is typical for terrestrial vertebrates, with a well-developed skull, a backbone and tail, two limb girdles, and four limbs. However, two types of reptiles have slightly different body plans: Turtles have a hard shell that is fused to their vertebrae, and snakes have lost both sets of legs.

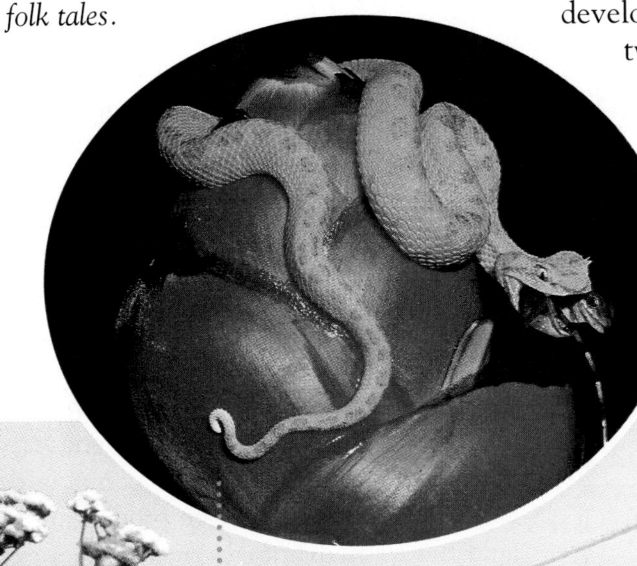

Figure 31–1
Reptiles have evolved a wide range of unusual adaptations. (a) *A snake's mouth can open much wider than the mouths of other animals, allowing snakes to swallow relatively large prey. This eyelash viper is eating a small lizard.* (b) *This Parson's chameleon captures prey with a long, sticky tongue.* (c) *Turtles and tortoises, like this radiated tortoise, are protected by a hard shell called a carapace.*

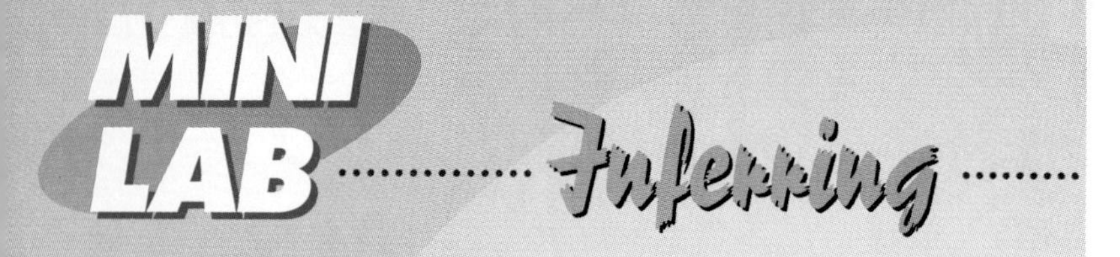

Ready to Slither

PROBLEM *What can you **infer** about the movement of snakes by observing their scales and skeletons?*

PROCEDURE

1. From your teacher, obtain a mounted snake skeleton or a photograph of a snake skeleton. Study the skeleton, then count the number of vertebrae and ribs.
2. Obtain the shed skin of a snake. Examine the shed skin, then count the number of belly scales.

ANALYZE AND CONCLUDE

1. How many vertebrae and ribs does the skeleton have? How is it adapted for a snake's type of movement?
2. How many belly scales are on the snake's skin? Compare this number to the number of vertebrae and ribs.
3. Compare the scales on the dorsal and ventral sides of the snake. How does a snake use its scales to move?

Reptiles are **ectotherms,** meaning "heat from the outside." Because a reptile's body does not generate much heat and lacks effective insulation, reptiles need to gain heat from their environment—typically by basking in the sun.

Feeding

Reptiles eat a wide range of foods. Certain iguanas and other reptiles are herbivores, while most lizards and snakes are carnivores. Some snakes paralyze prey with a powerful venom, while others suffocate prey with the force of their body, which they wrap into tight coils. Other reptiles, such as crocodiles, grab prey with powerful jaws filled with sharp teeth.

Interestingly, most reptiles cannot chew very well. Instead, their digestive tracts are specialized for processing large pieces of food. In addition, many reptiles have developed unusual ways of eating, as shown in ***Figure 31–1*** on page 709.

Respiration and Internal Transport

Because gases cannot diffuse across scaly skin, reptiles need well-developed and efficient lungs. Unlike amphibians, reptiles have muscles attached to their rib cage that enable their lungs to inflate and deflate.

Reptiles have a double-loop circulatory system that pumps blood first through the lungs and then through the rest of the body. A few reptiles, including crocodiles and alligators, have a four-chambered heart, similar to the hearts of birds and mammals. However, most reptiles have what might be called a "not-quite-four-chambered" heart, as shown in ***Figure 31–2.***

Excretion

Reptiles that live mostly in water, such as crocodiles and alligators, excrete most of their nitrogenous wastes in the form of ammonia—a compound toxic to most animals. For this reason, crocodiles and alligators drink large amounts of water, which dilutes the ammonia and helps to carry it out of the body.

Terrestrial reptiles, however, don't always live near water. Thus, most convert nitrogenous wastes into a compound called uric acid. As a waste product, uric acid has two advantages over ammonia. First, it is much less toxic. Second, it is not very soluble in water. Why is that an advantage? In many reptiles, urine from kidneys flows into an organ called the cloaca. As water is absorbed from the cloaca, crystals of uric acid precipitate, forming a white, semisolid paste. By eliminating dry wastes, reptiles conserve water.

☑ ***Checkpoint*** How do reptiles eliminate nitrogenous wastes?

Figure 31–2
Many features of the reptile body plan can be seen in other terrestrial animals, including humans. One unusual structure in a reptile is the heart, in which oxygenated and deoxygenated blood mix slightly as they leave the ventricles.

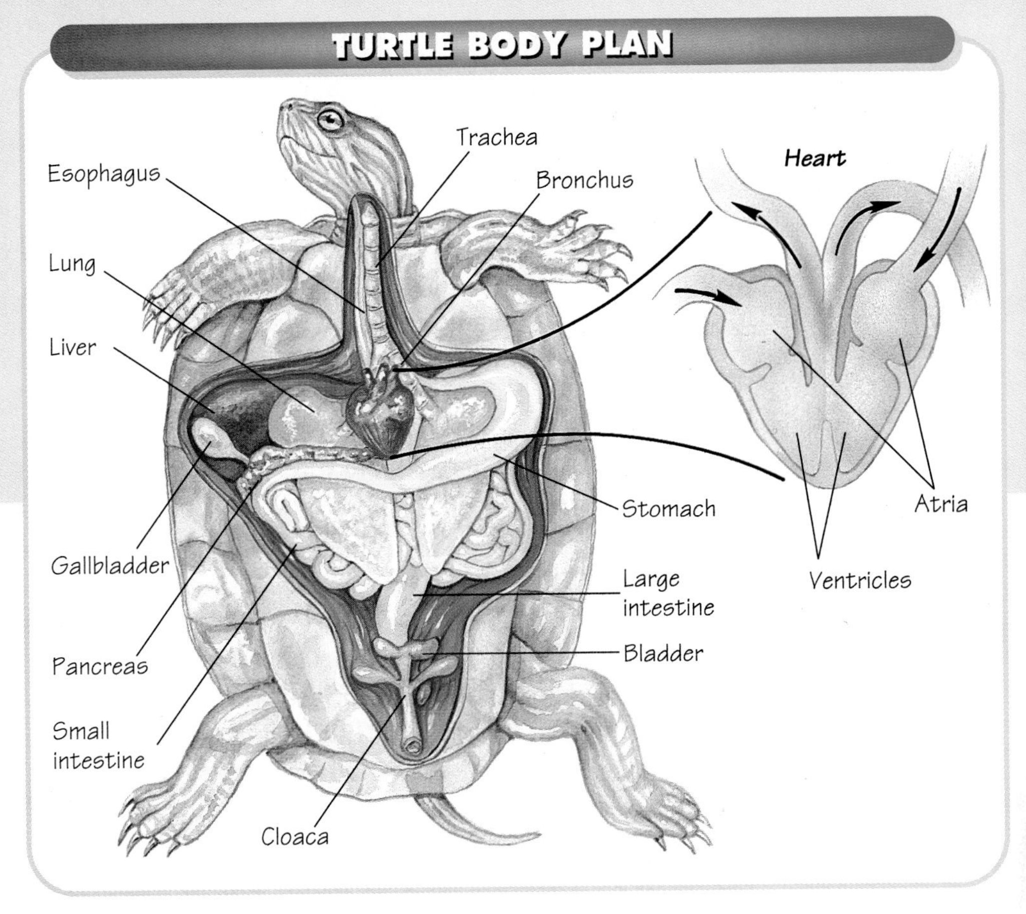

Nervous System and Sense Organs

Reptile brains are similar to the brains of amphibians, although reptiles have a larger cerebrum and cerebellum. As a group, reptiles are not known for complex behaviors. Many seem to do little more than sit, stalk prey, eat, and sleep.

Reptiles also have well-developed sense organs, some of which they use in unusual ways. For example, have you ever watched a snake flicking its tongue? If so, you have seen it "tasting" its environment. The tips of the tongue pick up molecules from the air, then deposit them on taste receptors in the mouth.

Although reptiles typically do not have a well-developed sense of hearing, many reptiles have excellent vision. In fact, some turtles and snakes probably see more colors than you do.

Several types of sense organs are found only in reptiles. For example, snakes known as pit vipers have an extraordinary ability to detect the body heat of prey, as illustrated in ***Figure 31–3.***

☑ **Checkpoint** How do snakes taste the air?

Movement

Compared to amphibians, most reptiles have larger and stronger legs that support their body weight easily. In different reptiles, legs are adapted for running, swimming, burrowing, or climbing.

Figure 31–3
Reptiles use information from their sense organs in many unusual ways. (a) *Like other chameleons, this flap-necked chameleon changes color in response to its environment.* (b) *Near the eye of this Chinese green tree viper is a depression called a pit organ. The pit organ contains heat receptors, allowing the viper to sense any prey that maintains a high body temperature.*

Many reptiles can stay absolutely still for a long period of time, then dash away with incredible speed.

The ancestors of snakes lost their legs while adapting to a burrowing existence. In the process, they evolved large bandlike scales on their ventral surface. To move, a snake uses its ventral scales for traction as it pushes its body into long, curving waves.

Reproduction

Early reptiles were the first vertebrates to evolve a life cycle that does not rely on standing water. **Reptiles were able to adapt to terrestrial life because they fertilized eggs internally, developed an amniotic egg, and typically cared for their eggs in some way.**

An **amniotic egg** gets its name from the amnion, a membrane that surrounds and protects the developing embryo. In fact, an amniotic egg contains several membranes and a leathery external shell. Inside, the embryo is bathed in a watery liquid—an environment that in some ways resembles the ocean in which life originally developed. The egg holds ample food in the form of a yolk and provides a place to store wastes until hatching. In addition, the shell and internal membranes "breathe" by allowing oxygen and carbon dioxide to diffuse across them.

Because amniotic eggs are produced and packaged inside the female's reproductive tract, they must be fertilized internally. Almost all male reptiles have an external reproductive organ—a penis—that delivers sperm into the female's cloaca. Lizards and snakes have two such structures, often called hemipenes.

Most reptiles are oviparous, laying eggs that hatch outside the body. Creating these eggs requires a lot of energy, so it is not surprising that many reptiles take special precautions to help their eggs develop. Sea turtles, for example, make a difficult journey to their nesting beaches. There, they bury their eggs at just the right depth and distance from the sea for the eggs to incubate properly.

☑ **Checkpoint** Why were reptiles able to adapt to land?

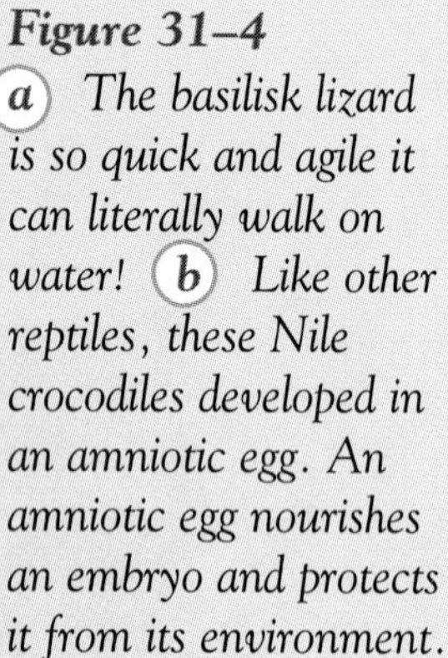

Figure 31–4 (a) *The basilisk lizard is so quick and agile it can literally walk on water!* (b) *Like other reptiles, these Nile crocodiles developed in an amniotic egg. An amniotic egg nourishes an embryo and protects it from its environment.*

Classifying Reptiles

Almost all ancient reptiles became extinct millions of years ago, but two surviving species still live on small islands near New Zealand. These reptiles are tuataras (too-uh-TAH-ruhz), the last survivors of the order Rhynchocephalia. Tuataras show many features of the ancient reptiles from which they evolved.

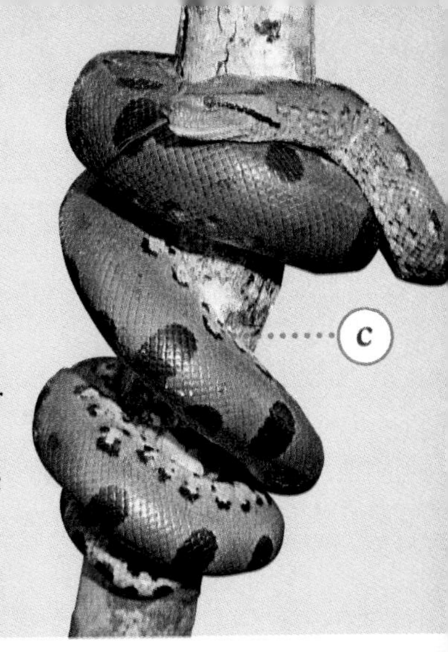

Figure 31–5
Although reptiles are not as numerous or diverse as they were millions of years ago, many fascinating species still live today. (a) *The American alligator lives in swampy regions of the southeastern United States.* (b) *Tuataras are found only on islands near New Zealand. This tuatara is poised to eat a tree weta, a large insect.* (c) *Snakes are common in many areas of the world. This anaconda lives in Peru.*

Tortoises and turtles are members of the order Chelonia. The first of these reptiles evolved over 200 million years ago, and they have changed little since then. Turtles usually live in or near freshwater ponds and streams or in the open ocean. Tortoises tend to be more terrestrial, living in dry places such as deserts and the rocky Galapagos Islands.

Members of the order Crocodilia—which includes crocodiles and alligators—are about as old as tortoises and turtles. Crocodilians live only in places that are at least as warm as southern Florida and Mississippi.

What's the difference between an alligator and a crocodile? Alligators live only in fresh water and are found almost exclusively in North America and China. Crocodiles, on the other hand, may live in fresh or salt water and are native to Southeast Asia, India, and Africa. In the United States, an alligator has a shorter, blunter snout and its teeth do not protrude outside its closed mouth.

Lizards, snakes, and most other living reptiles belong to the order Squamata. Lizards range in size from insect-eating geckos, which are only a few centimeters long, to the reptile commonly called the Komodo dragon. A Komodo dragon can grow up to 3 meters in length and can kill animals as large as a full-grown goat!

Different snake species live on land, in water, and even in trees. Poisonous snakes include cobras in India, rattlesnakes in North America, and the deadly fer-de-lance in Central America. Common fears to the contrary, however, poisonous snakes don't normally hunt humans. They usually bite humans only when they feel threatened or cornered.

Section Review 31–1

1. **Describe** the characteristics of reptiles.
2. **Explain** why reptiles were able to adapt to land.
3. **Critical Thinking—Analyzing Concepts** Explain why crocodilians live only in warm, shallow waters.
4. **MINI LAB** By studying the skeleton and scales of a snake, what can you **infer** about how it moves from place to place?

SECTION

31–2 Birds

Guide for Reading

- **Describe** the characteristics of birds.
- **Explain** why a bird can fly.

MINI LAB

- **Observe** the characteristics of a down feather and a contour feather.

WHETHER THEY ARE GREETING the dawn with song or coloring the air with brilliant feathers, birds are among the most prevalent and welcome of all animals. From common house sparrows to the spectacular and rare quetzal of Central America and from tiny hummingbirds to huge ostriches, the approximately 8700 living bird species seem to live everywhere. Not surprisingly, one secret to their success lies in the adaptations that allow them to fly.

The Bird Body Plan 

Birds are descendants of small dinosaurs. Thus, you shouldn't be surprised to discover that birds are much like reptiles. Unlike reptiles, however, birds have **feathers**—an adaptation that has allowed them to become very successful. Nothing like feathers is seen elsewhere in the animal kingdom, and no one knows how or why they first evolved.

All birds have feathers. They also have hind limbs that they use for walking, running, or perching, as well as front limbs that are modified into wings. Most birds use their feathers and wings to fly, although many birds—including ostriches and emus—cannot fly. Some flightless birds, such as penguins, use their wings as flippers for swimming.

Compared to reptiles, birds have a high rate of metabolism—the total chemical and physical processes that go on inside the body. In addition, a bird's body is insulated enough to conserve most of its metabolic energy and can thus maintain a constant high body temperature. For this reason, birds are described as **endotherms,** which literally means "heat from within." Some birds keep a body temperature near 40°C even on a cold, winter day.

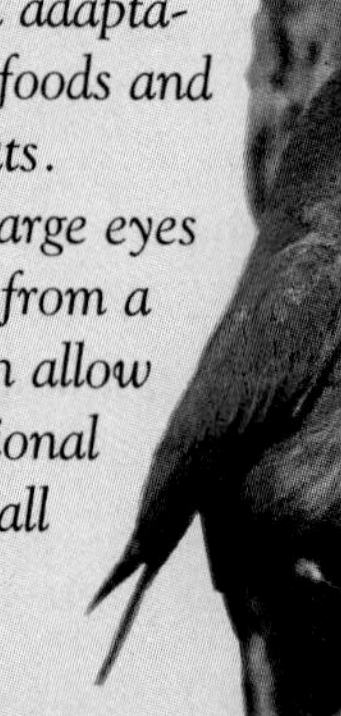

Figure 31–6 Birds have evolved adaptations to eat different foods and to live in different habitats. (a) The great horned owl has large eyes set on the front of its face, allowing it to spot small prey from a great distance. (b) The long legs of this great blue heron allow it to stand tall in the waters of the Florida Everglades National Park. (c) This eastern bluebird can squeeze through a small hole in a tree trunk, where it makes its nest.

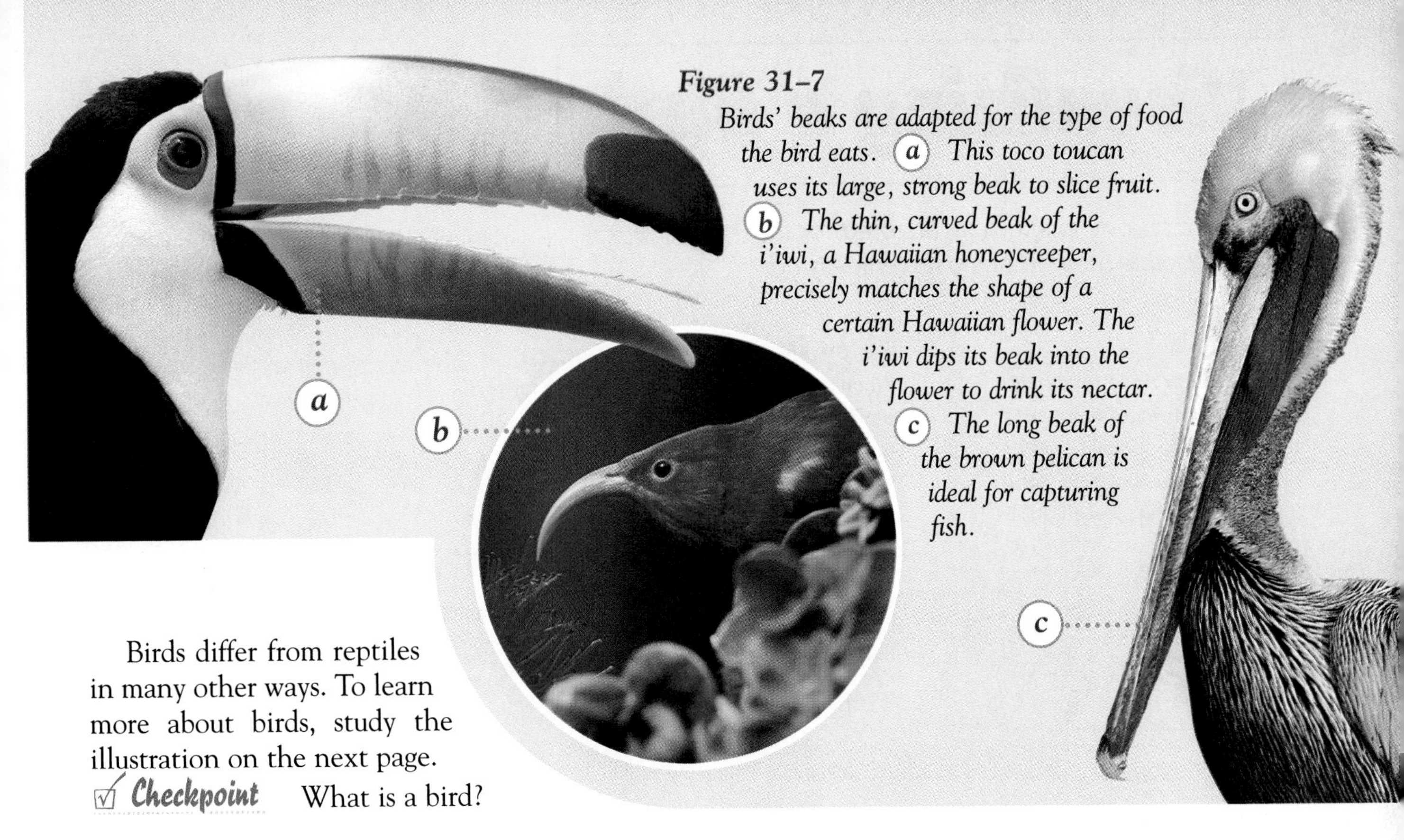

Figure 31–7
Birds' beaks are adapted for the type of food the bird eats. (a) *This toco toucan uses its large, strong beak to slice fruit.* (b) *The thin, curved beak of the i'iwi, a Hawaiian honeycreeper, precisely matches the shape of a certain Hawaiian flower. The i'iwi dips its beak into the flower to drink its nectar.* (c) *The long beak of the brown pelican is ideal for capturing fish.*

Birds differ from reptiles in many other ways. To learn more about birds, study the illustration on the next page.

Checkpoint What is a bird?

Beaks

You probably know that birds have beaks, but you may not realize the many different ways in which birds use beaks. From flamingoes that strain plankton out of water to eagles that rip small mammals apart, beaks perform scores of different functions. As a result, different birds have beaks of different shapes, sizes, and strengths, as shown in ***Figure 31–7.***

Respiration and Internal Transport

For several good reasons, birds need efficient respiratory and circulatory systems. First, birds must maintain a high rate of metabolism. Second, although birds are so graceful that they make flying look easy, they actually use a great deal of energy to fly. For a human, the equivalent to a bird's flying would be long-distance running, with a slow, steady pace alternating with sprints.

To keep up that pace, birds must take in a steady stream of oxygen and get rid of large amounts of carbon dioxide. Thus, birds have a remarkable respiratory system. When fresh air is inhaled, it bypasses the lungs and enters large air sacs located around them. When air is exhaled, it travels through the lungs in a series of small tubes. These tubes are lined with tiny air sacs called **alveoli** (al-VEE-uh-ligh; singular: alveolus), which is where gas exchange takes place.

This one-way flow means that alveoli are constantly exposed to fresh, oxygen-rich air. At the same time, air lower in oxygen and higher in carbon dioxide is constantly exhaled. This is a much more efficient system than the lungs that operate like inflatable balloons—the type of lungs found in humans and most other vertebrates.

To carry food, oxygen, and wastes, birds have a double-loop circulatory system. The pump is a four-chambered heart consisting of a pair of two-part halves divided by a structure called a **septum.** Because of a bird's generally small body size, high metabolic rate, and high-energy activity, a bird's heart pumps rapidly—at rates that range from 150 to 1000 beats per minute!

Checkpoint Why do birds need an efficient respiratory system?

Visualizing a Bird

A bird's body is specially adapted for its remarkable method of movement—flying!

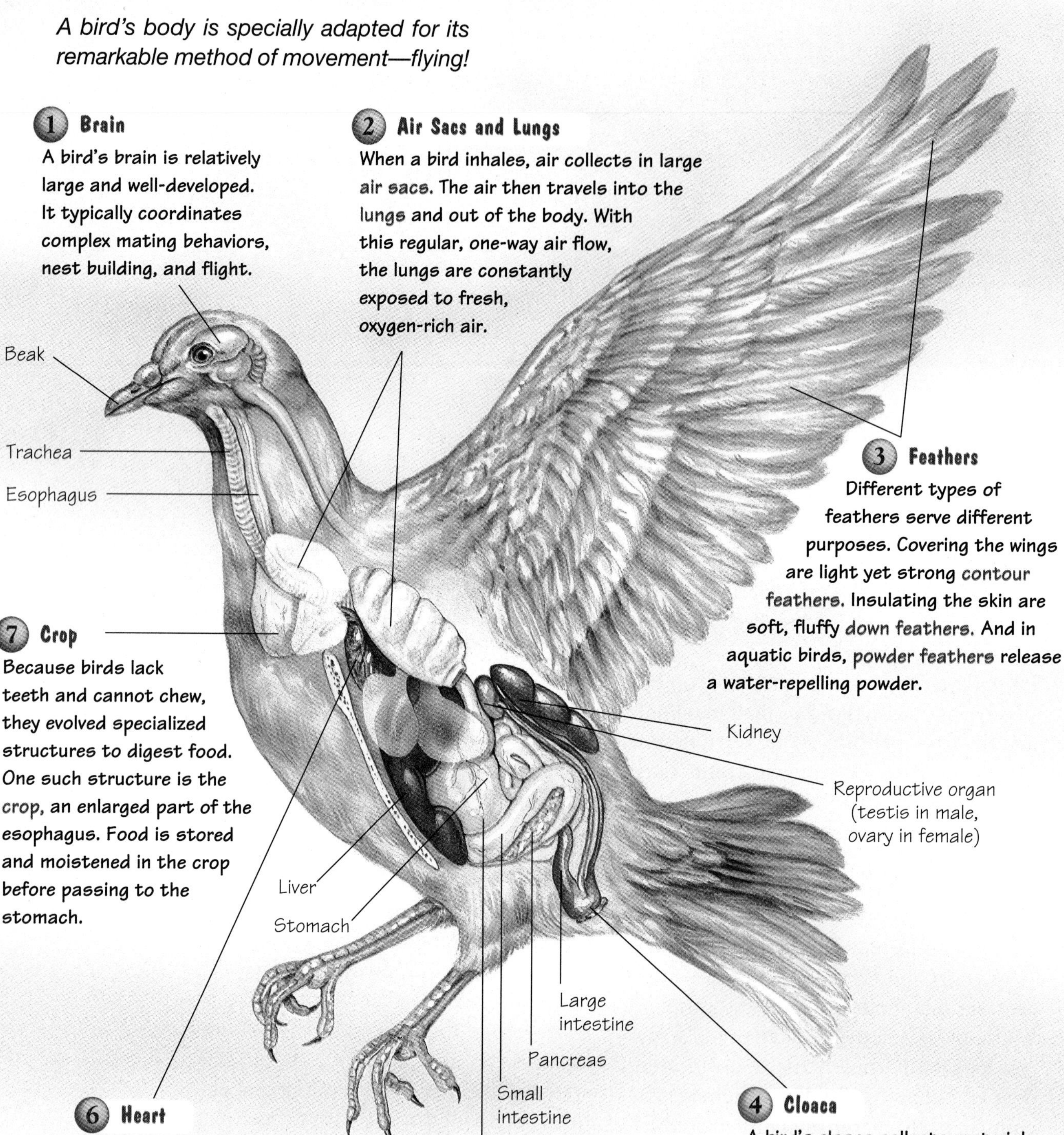

1 Brain
A bird's brain is relatively large and well-developed. It typically coordinates complex mating behaviors, nest building, and flight.

2 Air Sacs and Lungs
When a bird inhales, air collects in large air sacs. The air then travels into the lungs and out of the body. With this regular, one-way air flow, the lungs are constantly exposed to fresh, oxygen-rich air.

3 Feathers
Different types of feathers serve different purposes. Covering the wings are light yet strong contour feathers. Insulating the skin are soft, fluffy down feathers. And in aquatic birds, powder feathers release a water-repelling powder.

4 Cloaca
A bird's cloaca collects material from the digestive, excretory, and reproductive systems. To reproduce, male and female birds firmly press the lips of their cloacas together, which allows sperm to transfer.

5 Gizzard
The gizzard is a section of the stomach with thick, muscular walls. The gizzard holds gravel, which a bird usually swallows intentionally. The muscular walls grind gravel against the food, crushing it and making it easier to digest.

6 Heart
Like humans, birds have a four-chambered heart. The two chambers on one side receive deoxygenated blood and pump it to the lungs. The two chambers on the other side receive oxygenated blood from the lungs and pump it to the body.

7 Crop
Because birds lack teeth and cannot chew, they evolved specialized structures to digest food. One such structure is the crop, an enlarged part of the esophagus. Food is stored and moistened in the crop before passing to the stomach.

Excretion

Birds handle wastes in much the same way that reptiles do. Nitrogenous wastes are removed from blood by the kidney, converted to uric acid, and deposited in the cloaca. There, water is reabsorbed, causing uric acid crystals to precipitate and form the whitish, semisolid paste you recognize as bird droppings.

Many birds that live in marine habitats take in more salt than they can keep. Because vertebrate kidneys cannot excrete salt efficiently, these birds have evolved specialized salt glands near the base of the beak. These glands excrete enough concentrated salt solution to enable the bird to maintain a salt and water balance.

Nervous System and Sense Organs

Although you might use the term "bird brain" to refer to someone foolish, birds are in fact relatively intelligent animals. Although a bird's medulla and spinal cord are much like those of its reptilian ancestors, its cerebrum and cerebellum are both large and well-developed.

Would you guess that birds have well-developed sense organs? Before you answer, think about the demands of bird life. Many species fly at high speeds, catch prey from great distances, and migrate thousands of kilometers between summer and winter nesting grounds. All these tasks demand the gathering and processing of large amounts of information from the environment. So it is not surprising that bird senses equal or surpass our own.

Most birds have excellent vision, and birds that stay awake in the daytime often see color far better than humans do. Hawks and their relatives can see incredibly fine detail, which lets them spot small prey on the ground while flying high overhead. Although birds lack external ears, many species have an acute sense of hearing. Owls, for example, from quite a distance can detect mice rustling along the forest floor.

Of the five human senses, only taste and smell are not well-developed in birds. However, many birds have a sense that has long puzzled and amazed humans. This sense is the ability of migratory birds to navigate across continents and oceans. As we now know, some of these birds have a magnetic sense that works like an internal compass. Other birds use a combination of acute eyesight and a built-in clock to navigate by the stars.

☑ **Checkpoint** Why do birds need well-developed senses?

Fine Feathered Friends

PROBLEM *What characteristics can you **observe** in a contour feather and a down feather?*

PROCEDURE 

1. Use a hand lens to observe a contour feather from a bird's wing or tail. Sketch and label your observations.
2. Observe and sketch a bird's down feather.
3. Dip each feather into water for 2 seconds. Record your observations of how the feathers react to water.

ANALYZE AND CONCLUDE

1. How are the contour feather and down feather different in appearance? How are they similar?
2. What is the function of barbs and hooks on contour feathers?
3. From the feathers' reactions to water, infer how birds respond to rain.

Biology AND YOU .... Connections

Secrets in a Bird's Brain

Although medical scientists have learned a great deal about the nervous system, many serious diseases and injuries that affect this system remain incurable. As a rule, dead nerve cells do not regenerate in the human brain and spinal cord. Thus, physicians have limited treatments for severe injuries to the nervous system and diseases such as Alzheimer's disease, in which the nerves in the brain are gradually destroyed.

Can human nerves be stimulated to regenerate? Incredibly enough, the answer may come from studies on the brains of another class of vertebrates—birds!

Song Centers

At one time, biologists thought that a vertebrate brain never formed new nerve cells after the brain was fully developed. In fact, they thought that the brain steadily lost nerve cells over time.

However, studies on canaries have provided evidence against this idea. In these studies, researchers concentrated on a part of the canary's brain called the song center—the part used in learning and performing the canary's familiar songs. A canary's song center expands during breeding season—the time when canaries sing frequently. The song center shrinks at other times, when canaries are more quiet.

How do these changes occur? At Rockefeller University in New York, one group of scientists showed that canaries develop new nerves in the song center, and they do so quite regularly. What's more, the scientists discovered that canaries periodically develop new cells throughout the brain! In male canaries, at least some of these changes are triggered by changing levels of sex hormones. However, scientists are still looking for the genes that put these hormones into action.

A canary's brain regularly develops new nerves—a feat that scientists once thought was impossible!

Help for Humans?

Do the genes that cause brain cells to reproduce and grow in birds also exist in humans? If humans have these genes, is there a way to turn them on? If the answers to these questions are yes, then scientists could someday develop treatments to repair human brains damaged by injury or disease.

Research on bird brains may someday help us to understand the human brain.

Making the Connection

Which animals are able to regenerate lost body parts? Which human organs are able to regenerate?

BIRD FLIGHT

Downstroke

Upstroke

Airflow

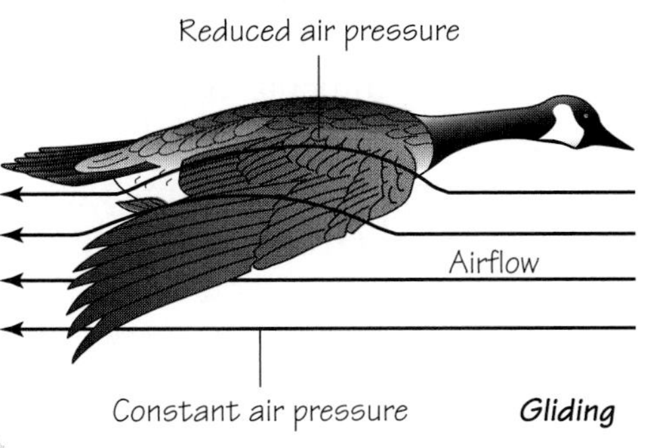

Figure 31–8
To fly, birds need to coordinate the movements of their feathers and wings. (a) *During the downstroke, feathers cover the wings' surface. This pushes air downward and provides lift. During the upstroke, the feathers are spread apart, allowing air to pass through and making the wings easier to lift. Birds can glide because the wings' shape reduces air pressure above the wings.* (b) *Both downstrokes and upstrokes can be seen in these Canada geese.* (c) CAREER TRACK *Veterinary assistants help veterinarians treat pets, farm animals, or wildlife.*

Birds and Flight

Various birds evolved ways in which to run and swim. However, unlike almost all other vertebrates, most birds can fly. **Birds can fly because of their relatively light bodies, powerful breast muscles, and aerodynamic feathers and wings.**

First, birds are lighter than other vertebrates mainly because of their bones, which have spaces within them. In addition, air sacs used in breathing extend inside several bones, making these bones even lighter. Birds minimize their body mass in other ways, too. Sex organs, for example, are small and light most of the year, then increase to working size during breeding season.

Second, birds have large, strong breast muscles, which they use to flap their wings. These huge muscles provide the force that lifts a bird into the air.

Third, a bird's wings and feathers are shaped in just the right way for flight, as shown in ***Figure 31–8***. In fact, birds' wings served as the model for the wings that humans created—the wings of airplanes.

INTEGRATING PHYSICS

How does an airplane fly? How are an airplane's wings similar to a bird's wings?

Section Review 31–2

1. **Describe** the characteristics of birds.
2. **Explain** why a bird can fly.
3. **Critical Thinking—Analyzing Concepts** Explain why crops and gizzards are especially common in birds that eat seeds, while they are less common in carnivorous birds.
4. **MINI LAB** What characteristics can you **observe** in a down feather and a contour feather?

Evolution of Birds

Guide for Reading

- **Discuss** the evidence that shows that birds evolved from dinosaurs.

ROUGHLY 150 MILLION YEARS ago, what is now the Solnhofen region of Germany was a very different place from what it is today. Sharks, fishes, and turtles swam in a lagoon. Small dinosaurs and lizards scurried on land. And overhead soared a surprising number of vertebrates. Some were the earliest birds—extinct relatives of the birds that live today.

How do we know that birds flew over the Solnhofen lagoon? Where did they come from? How did they evolve the ability to fly? The fossil record provides answers for each of these questions.

Birds From Dinosaurs?

When a dead animal sank to the bottom of the Solnhofen lagoon, it was eventually covered with a fine mud that isolated its body from oxygen. Over millions of years, the mud turned into fine-grained limestone. As a result, the Solnhofen limestone today contains some of the most beautiful and complete fossils ever discovered. And the star of this collection is one of the first birds—*Archaeopteryx*, meaning "ancient wing."

At first, proper identification of *Archaeopteryx* was difficult because its skeleton looks almost exactly like the skeleton of a small two-legged dinosaur. An *Archaeopteryx* skeleton has teeth, a heavy skull, a jointed tail, and other dinosaurlike features. But the exceptional specimens of the Solnhofen region included structures that almost never fossilize well—feathers!

Because feathers are found only in birds, their presence in the dinosaurlike *Archaeopteryx* provides evidence that birds evolved from dinosaurs. *Archaeopteryx* is a transitional species, a relic of a time when early birds had a combination of dinosaurlike and birdlike characteristics.

Checkpoint What evidence shows that birds evolved from dinosaurs?

Figure 31–9 *Quetzalcoatlus, an ancient reptile, had enormous wings and no tail. Could it fly? (a) To answer this question, a team of biologists and engineers built a mechanical model. (b) To the surprise of many scientists, the model flew! (c) Among living birds, ostriches and other ostrich-like birds may most resemble their dinosaur ancestors.*

Did Archaeopteryx Fly?

For many years, researchers debated whether *Archaeopteryx* could fly. Although *Archaeopteryx* had plenty of feathers, it had heavy bones and lacked an enlarged breastbone, the structure to which large flight muscles attach in birds today. These facts suggested that the feathers were important only for insulation and that *Archaeopteryx* moved by running or climbing.

However, the wings and flight feathers of *Archaeopteryx* look a great deal like those of modern species. In addition, the discovery of other flying reptiles has caused researchers to rethink their ideas about what sorts of animals can fly.

In Texas, paleontologists discovered fossils of a very unlikely reptile. As they assembled the bones, they found wings larger than those of any other known animal—living or extinct. Some specimens had wingspans of 12 meters, the width of a small hang glider! What's more, these creatures had extremely large heads and no tail. This unusual species was named *Quetzalcoatlus*, after a feathered serpent deity in Aztec mythology.

At first, scientists questioned whether *Quetzalcoatlus* could ever have gotten off the ground. They argued that its body was too large, that it had no tail to stabilize it, and that its flight muscles could not possibly have been strong enough. But some determined biologists and engineers built a mechanical model of this improbable beast, using its fossils as their guide. To their delight and amazement, the model flew!

Quetzalcoatlus clearly would have spent more time gliding like an albatross than flying like a bluejay. But it could fly! And so, it seems, could other pterosaurs and early birds such as *Archaeopteryx*.

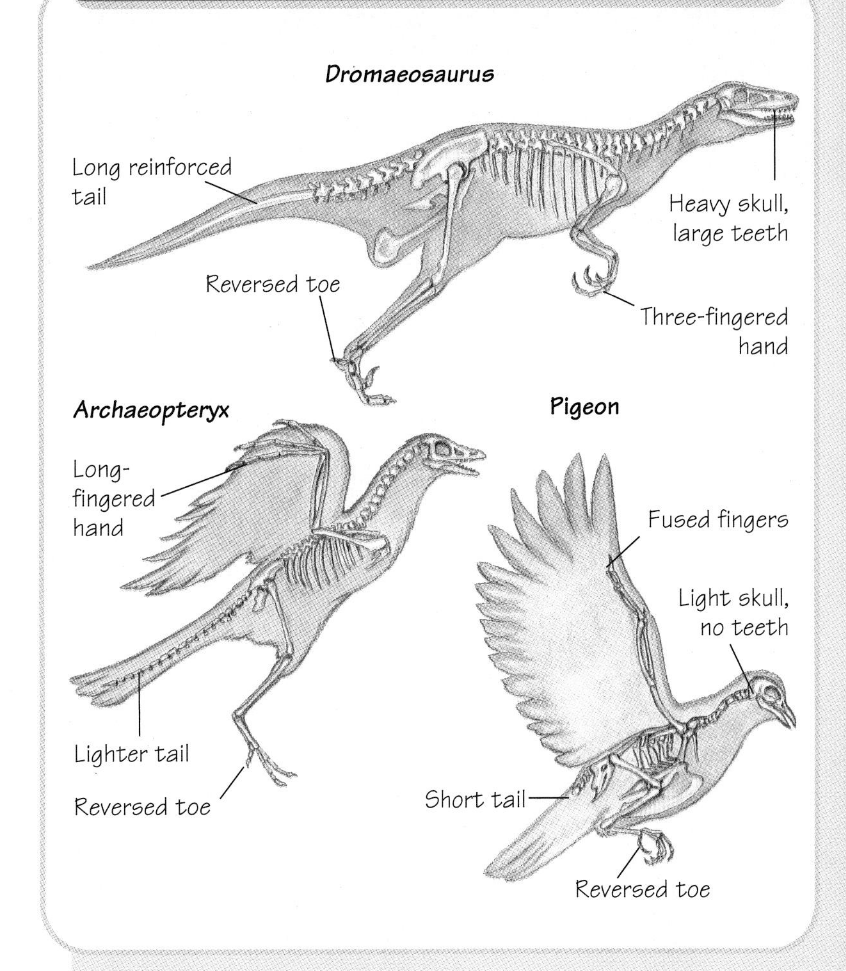

Figure 31–10
The skeleton of Archaeopteryx *has characteristics in common with dinosaurs, such as* Dromaeosaurus, *as well as with modern birds.*

Section Review 31–3

1. **Discuss** the evidence that shows that birds evolved from dinosaurs.
2. **BRANCHING OUT ACTIVITY** Research the body plan of a flying dinosaur or early bird. Under your teacher's supervision, **construct a model** of one of these animals.

Laboratory Investigation

DESIGNING AN EXPERIMENT

Adaptation for Survival

American chameleons, or anoles, are native to the southeastern United States, where they can be seen on fences, trees, and wooden buildings. In this investigation, you will explore an unusual adaptation of chameleons—their ability to change color rapidly.

Problem

How can you **design an experiment** to determine how chameleons change color with their environment?

Suggested Materials

American chameleon
20-gallon terrarium
screen terrarium lid
sand or gravel
dried branches
rocks
moistened sponge, with a surface area of about 4 cm × 4 cm
mealworms or other small live insects
heat source
thermometer
colored pencils
colored paper
light source
watch or clock

Suggested Procedure

1. To create housing for the chameleon, construct a screen top that fits tightly over the 20-gallon terrarium. Cover the bottom of the terrarium with sand or gravel. Add rocks and small branches to create an environment in which the chameleon can climb and hide. Suspend the moistened sponge from the screen to maintain proper humidity.
2. Place the thermometer inside the terrarium. If necessary, use a heat source to maintain the temperature between 22°C and 27°C.
3. Obtain a chameleon from your teacher. **CAUTION:** *Follow your teacher's instructions for handling chameleons. Be especially careful not to grab a chameleon by the tail. This could trigger a response in which the tail falls off!* Place the chameleon in the terrarium. Cover the terrarium with the screen top.

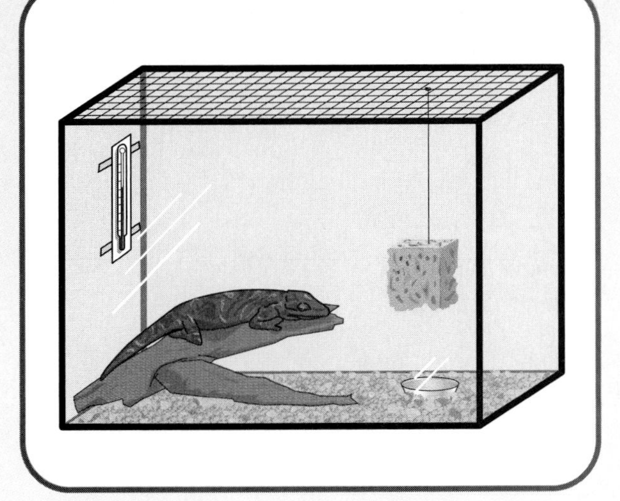

4. **Feed live insects to the chameleon three to four times a week. CAUTION:** ***Always wash your hands after handling live animals.***
5. **Observe the chameleon twice a day. Note its coloration and behavior. Use colored pencils to sketch the chameleon's appearance.**
6. **Design an experiment to test the effect of an environmental factor on color changes in chameleons. Choose one of these variables to investigate:**
   - **background color**
   - **light intensity**
   - **temperature (within 22°C to 27°C)**

   Your experiment should test only one variable and should test the chameleon's long-term and short-term response to its environment. Your experiment should not involve handling the chameleon and should not harm the chameleon in any way.
7. **With your teacher's approval, perform the experiment you designed.**

Observations

1. Describe any color changes you observed in the chameleon.
2. If there was a color change, was it the same on all parts of the chameleon's body?
3. How much time did it take for you to notice a change? For the change to be complete?

Analysis and Conclusions

1. Did the chameleon change color in response to the environmental factor you investigated? Explain.
2. How quickly was the chameleon able to change color? For how long was the chameleon able to maintain a new color?
3. How does the chameleon's ability to change color help it to survive?
4. Formulate a hypothesis to explain how a chameleon is able to change color. How might you test this hypothesis?

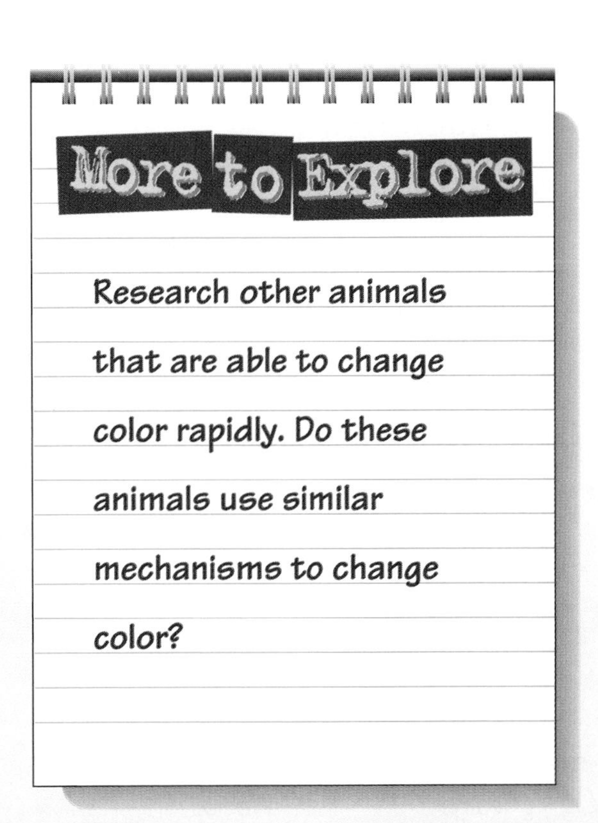

More to Explore

Research other animals that are able to change color rapidly. Do these animals use similar mechanisms to change color?

Summarizing Key Concepts

The key concepts in each section of this chapter are listed below to help you review the chapter content. Make sure you understand each concept and its relationship to other concepts and to the theme of this chapter.

31–1 Reptiles

- Reptiles are scaly-skinned vertebrates that breathe with lungs and lay eggs that hatch on land. They are also ectotherms, meaning they gain body heat from their environment.
- Reptiles have an efficient respiratory system and a double-loop circulatory system. Crocodiles and alligators excrete ammonia, which they dilute with large amounts of water. Other reptiles convert nitrogenous wastes into a dry paste of uric acid.
- Reptiles were able to adapt to terrestrial life because they fertilized eggs internally, developed an amniotic egg, and typically cared for eggs after the eggs were laid.
- Tortoises and turtles have remained almost unchanged for 200 million years, as have crocodiles and alligators. Other reptiles include snakes and lizards.

31–2 Birds

- All birds have feathers. They also have hind limbs that they use for walking, running, or perching, as well as front limbs that are modified into wings.
- A bird's body is insulated enough to conserve its metabolic energy. Thus, birds are endotherms—meaning "heat from within."
- Birds have an efficient respiratory system in which air is stored in air sacs, then passed through the lungs and out of the body. This one-way flow constantly provides the lungs with fresh, oxygen-rich air.
- Birds use different feathers for different purposes. Contour feathers line and shape the wing. Down feathers line the skin and insulate the bird. And powder feathers help repel water in aquatic birds.
- Birds can fly because of their relatively light bodies, powerful breast muscles, and aerodynamic feathers and wings.

31–3 Evolution of Birds

- Because feathers have been found only in birds, their presence in the dinosaurlike *Archaeopteryx* established that birds evolved from dinosaurs.

Reviewing Key Terms

Review the following vocabulary terms and their meaning. Then use each term in a complete sentence.

31–1 Reptiles

ectotherm
amniotic egg

31–2 Birds

feather
endotherm
alveolus
septum
air sac
lung
contour feather
down feather
powder feather
cloaca
gizzard
crop

Recalling Main Ideas

Choose the letter of the answer that best completes the statement or answers the question.

1. Which of these animals is an endotherm?

a. snake
b. turtle
c. alligator
d. hummingbird

2. Snakes and lizards excrete nitrogenous wastes in the form of

a. carbon dioxide.
b. uric acid.
c. toxic venom.
d. ammonia.

3. In pit vipers, the pit organs sense

a. taste.
b. smell.
c. sound.
d. heat.

4. In most reptiles, reproduction is

a. viviparous.
b. ovoviviparous.
c. oviparous.
d. asexual.

5. In birds and reptiles, materials from the reproductive, digestive, and excretory tracts collect in the

a. cloaca.
b. crop.
c. kidney.
d. gizzard.

6. Which type of feathers enable birds to fly?

a. down feathers
b. smooth feathers
c. powder feathers
d. contour feathers

7. Birds swallow gravel, which is stored in the

a. crop.
b. gizzard.
c. stomach.
d. cloaca.

8. Small sex organs, hollow bones, and air sacs are all adaptations in birds for

a. internal fertilization.
b. internal transport.
c. flight.
d. movement.

9. Although *Archaeopteryx* is classified as a bird, it resembled a

a. large four-legged dinosaur.
b. small two-legged dinosaur.
c. pterosaur.
d. snake.

Putting It All Together

Using the information on pages xxx to xxxi, complete the following concept map.

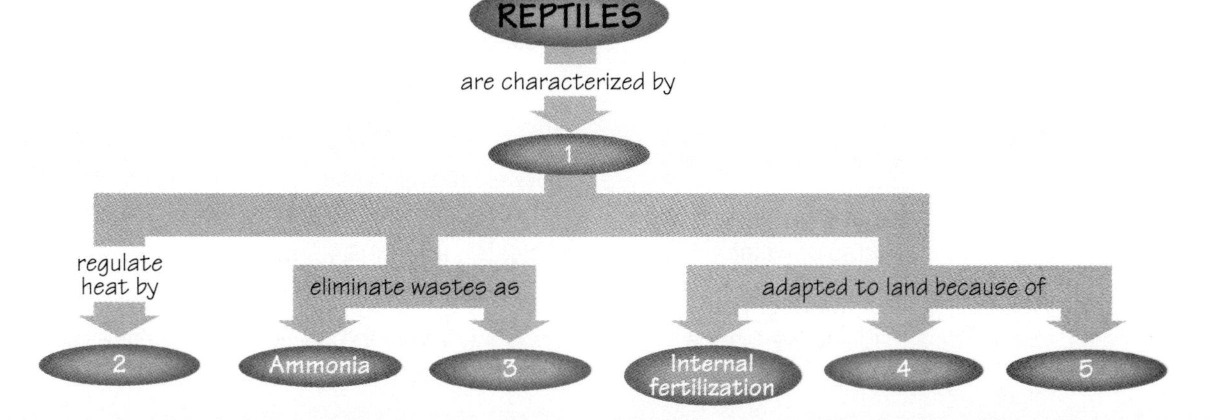

Assessment

Reviewing What You Learned

Answer each of the following in a complete sentence.

1. Compare the ways in which reptiles and birds regulate their body heat.
2. Why does a snake flick its tongue in the air?
3. What is an amniotic egg?
4. How does excretion differ in terrestrial and aquatic reptiles?
5. How do snakes move without legs?
6. Identify four orders of reptiles. Give an example of a reptile in each order.
7. What are alveoli?
8. Why did beaks evolve different sizes and shapes in different bird species?
9. Identify three different types of feathers. Describe the purpose of each type.
10. Compare a bird's heart with a reptile's heart.
11. How are birds able to migrate over great distances?
12. How do birds reproduce?
13. Although *Archaeopteryx* resembled dinosaurs in many ways, why is it classified as a bird?
14. What is *Quetzalcoatlus*?

Expanding the Concepts

Discuss each of the following in a brief paragraph.

1. Compare the basic body plans of reptiles and birds.
2. Describe a bird's respiratory system. Why do birds have a more efficient respiratory system than does an animal that lives only on land?
3. Discuss the adaptations that allow reptiles and birds to live away from bodies of water.
4. Why do most reptiles live in warm or hot climates, while birds are able to live in other climates?
5. Describe the different ways in which reptiles capture prey.
6. What adaptations allow birds to fly?
7. Discuss the different sense organs in birds.
8. Why are well-preserved fossils especially prevalent in the Solnhofen region of Germany?
9. Explain the evidence that birds evolved from dinosaurs.
10. Could *Archaeopteryx* fly? Discuss the evidence that applies to this question.

Extending Your Thinking

Use the skills you have developed in this chapter to answer the following.

1. **Inferring** Some species of birds are highly social and live in huge flocks. Based on what you know about birds, infer the survival advantages of group living.
2. **Observing** Examine the photograph of the animal shown on this page. Describe the characteristics that you observe. How would you classify this animal?
3. **Developing a hypothesis** On average, reptiles that live in water are larger than reptiles that live on land. Formulate a hypothesis to explain this difference.
4. **Applying concepts** As the seasons change, many birds migrate over great distances. What advantages does migration offer birds?
5. **Designing an experiment** Parakeets, parrots, and other birds can mimic the sounds they hear. Design an experiment to determine whether a parakeet or parrot can be trained to repeat a phrase when prompted.
6. **Using the writing process** In reference to penguins, Herman Melville asked, "What outlandish being are these?" Kurt Vonnegut wrote, "They were skinny things underneath their head waiters' costumes." Write a creative description of a penguin or another bird of your choice.

Applying Your Skills

Take a Bird Census

Wildlife biologists use many methods for collecting information about birds. At least one method, called a census, requires your help. A census counts the number of individuals of each species in a particular place and at a particular time.

1. Work in three-member teams. Each team should select a leader, an expert, and a recorder. The leader does the actual counting, the expert consults references to help identify species, and the recorder develops and uses a data sheet.
2. As a team, select an open outdoor plot and assemble in the middle of the plot. For 10 minutes, count the number of birds you see and identify the species of each bird. If you cannot identify a bird, write a description of it.
3. Analyze your data. Include the total number of individuals of each species.
4. Compare your census data with the data from other teams. Prepare a report of the bird population that lives in the regions you surveyed.

• GOING FURTHER •

5. Contact an environmental organization active in your community. Request information on how to participate in bird counts that they sponsor.

CHAPTER 32

Mammals

FOCUSING THE CHAPTER
THEME: Unity and Diversity

32–1 Characteristics of Mammals
- **Describe** the characteristics of mammals.
- **Identify** different orders of mammals.

BRANCHING OUT *In Depth*

32–2 The Influence of Humans
- **Discuss** the ways in which humans have changed the populations of other mammals.

LABORATORY INVESTIGATION
- **Infer** the characteristics of a mammal by studying its skull.

Biology and Your World

BIO JOURNAL

Study the mountain goats shown on this page. In your journal, describe their characteristics. How do these characteristics help the mountain goats survive in their environment?

Mountain goats in Olympic National Park, Washington

Characteristics of Mammals

Guide for Reading

- **List** the characteristics of mammals.
- **Explain** how mammals are classified.

MINI LAB

- **Classify** mammals from their footprints.

GROUNDHOGS AND FIELD MICE burrow holes in prairies, apes and lemurs swing from tree to tree in rain forests, and whales and dolphins swim majestically through oceans. What do these animals have in common? All are mammals. As you will discover, mammals evolved all sorts of shapes, sizes, and lifestyles and live in a great variety of habitats. And mammals include humans—a typical mammal species in some ways, but very different in others.

Mammalian Body Plan

Like the bodies of other terrestrial vertebrates, the mammalian body plan evolved from that of ancient reptiles. For this reason, the body parts and internal organs of mammals are similar in appearance and function to those of today's reptiles. Yet several important characteristics set mammals apart. **Mammals are endotherms, have external body hair, and have a layer of fat beneath the skin. Female mammals produce milk to feed their offspring.**

To generate their body heat, all mammals—especially small ones—have a much higher metabolic rate than most other vertebrates. Two other mammalian traits—external body hair and a layer of fat beneath the skin—evolved as insulation to store that important heat.

Mammals have several other characteristics, many of which help paleontologists to identify mammalian bones and fossils. These characteristics include a simpler lower jaw than reptiles, complex teeth that are replaced just once in a lifetime, a unique set of bones in the middle ear, and several features of limb bones, limb girdles, and the vertebral column.

☑ **Checkpoint** What is a mammal?

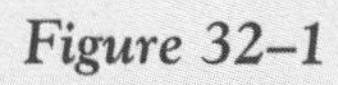

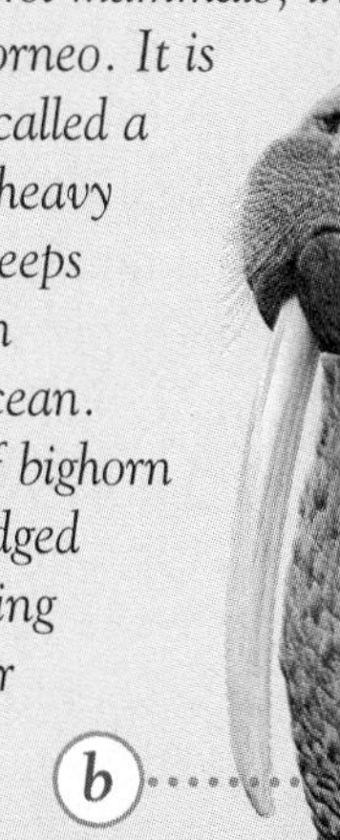

Figure 32–1
Mammals live almost everywhere on Earth. (a) This tree shrew, which is thought to resemble the first mammals, lives on the island of Borneo. It is eating an insect called a katydid. (b) A heavy layer of blubber keeps a walrus warm on land and in the ocean. (c) The hooves of bighorn sheep are sharp-edged and elastic, allowing them to roam over rocky terrain.

complex design of these teeth allows top and bottom teeth to interlock during chewing, like the blades of scissors. Interlocking teeth are important because they can crush, slice, or tear food more quickly than simple teeth can. The more thoroughly food is chewed, the more quickly and efficiently the digestive tract can break it down. And the more efficiently an animal can process and digest its food, the more food it can eat!

The rest of the digestive tract evolved to digest the type of food the mammal eats. Because digestive enzymes can quickly break down meat, carnivores have a relatively short intestine. And because tough plant tissues take much more time to digest, most leaf-eating herbivores have a much longer intestine. Cows and their relatives, which are among the most successful grazing animals, evolved ways to digest plant foods that are nearly useless to other animals.

Figure 32–2
Different mammals evolved special adaptations for their different environments and food supplies. (a) The long hair of these llamas provides insulation from the cold. (b) Like other meat-eating mammals, this leopard has sharp front teeth called canines that it uses to bite and tear prey. (c) Cows can digest grass because of their elaborate digestive tract, which includes a four-chambered stomach.

Feeding

Because of its high metabolic rate, a mammal must eat nearly ten times as much food as a reptile of the same size! As a result, mammals evolved a variety of specialized jaws and teeth, some of which are shown in ***Figure 32–3.*** In mammals, the joint between the skull and lower jaw is simpler, stronger, and more versatile than it is in reptiles. This joint allowed mammals to evolve larger, more powerful jaw muscles and different ways of chewing.

Mammals also evolved specialized teeth called **molars** and **premolars.** The

Internal Transport and Respiration

Like birds, mammals have a double-loop circulatory system powered by a four-chambered heart. Blood travels between the heart and lungs in one loop, then between the heart and the rest of the body in the other loop.

A mammal's lungs inflate in two ways at once. Muscles in the chest lift the rib cage up and outward, increasing the volume of the chest cavity. At the same time, a powerful muscle called the **diaphragm** pulls the bottom of the chest cavity downward, which further increases its volume. As a result, air enters the

EVOLUTION OF SKULLS AND TEETH

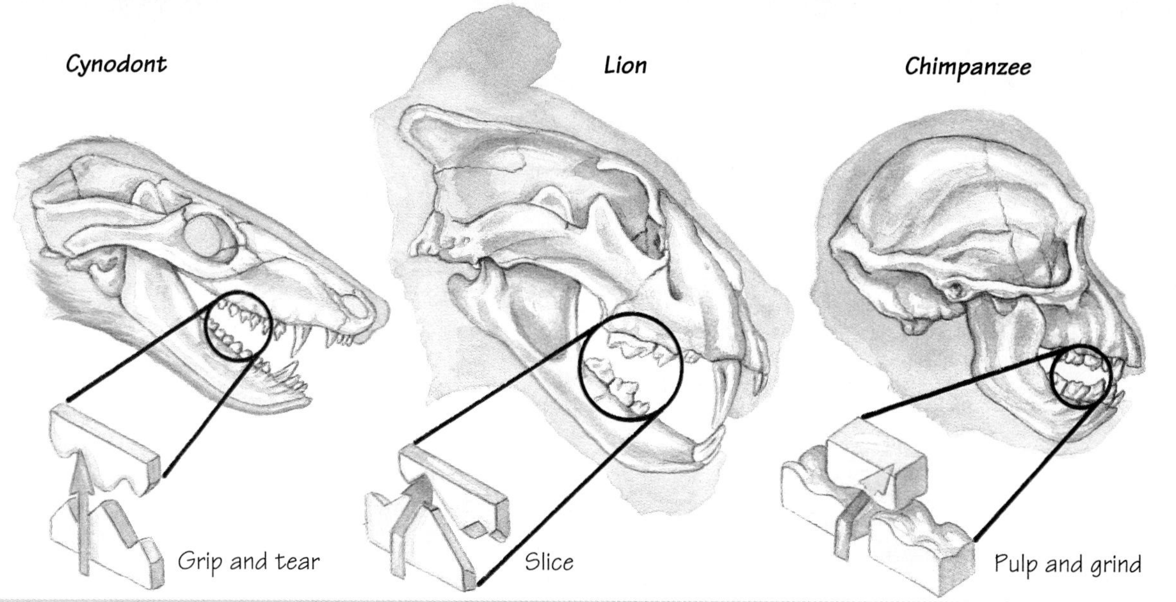

Cynodont

Lion

Chimpanzee

Grip and tear

Slice

Pulp and grind

Figure 32–3
An ancient reptile called a cynodont had teeth that did not precisely interlock but could grip and tear food. In true mammals, such as lions, the lower jaw is slightly narrower than the upper jaw. Thus, when the jaws close, the back teeth slice food as they shear past each other. Plant-eating mammals, such as chimpanzees, evolved back teeth that were square and flat, ideal for grinding food.

lungs. When these muscles reverse their activity, air rushes out of the lungs as the chest cavity shrinks.

Even marine mammals, such as seals and whales, breathe air with this type of lungs. However, these animals have evolved both physical and physiological adaptations for life in the water. They are able to store oxygen in their tissues, withstand the buildup of carbon dioxide, and hold their breath during long dives.

☑ **Checkpoint** How do a mammal's lungs inflate?

Excretion

Mammals have a complex, well-developed excretory system. The liver transforms nitrogenous wastes into urea, which the **kidneys** filter from the blood and combine with other waste products to form urine.

Mammalian kidneys not only eliminate wastes, they do an excellent job of conserving important compounds, including water, salts, and sugars. This enables some mammals to inhabit deserts, where water is scarce, and others to live in rain forests, where salt is rare. However, mammalian kidneys cannot excrete salt without eliminating water at the same time. That's why most mammals can't survive by drinking sea water.

Nervous System

As in the brains of other vertebrates, the medulla oblongata of the mammalian brain regulates breathing, heart rate, and other bodily functions not normally under conscious control. The brain also includes a cerebellum, which coordinates movement. In the cerebellum, the conscious intention to move is translated into a series of muscle contractions.

However, a mammal's brain differs from the brains of other animals because of its well-developed cerebral cortex—the center of thinking and other complex behaviors. Some activities, such as reading a textbook, are possible only with the human cerebral cortex. But many other complex behaviors—ranging from social behaviors to the use of tools—may be more widespread among other mammals than was previously thought.

Figure 32–4
(a) *Because a cat's eye is adapted for low-intensity light, it can tolerate only small amounts of bright light. Its pupils close into thin slits during the day.* (b) *A human eye is adapted for bright light. Its round pupil allows a relatively large amount of light into the eye.*

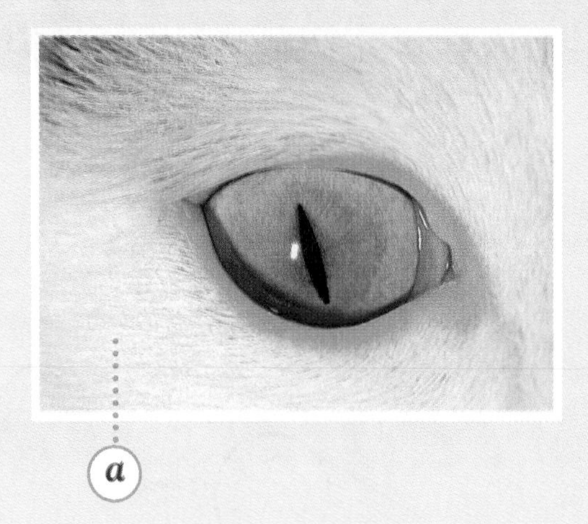

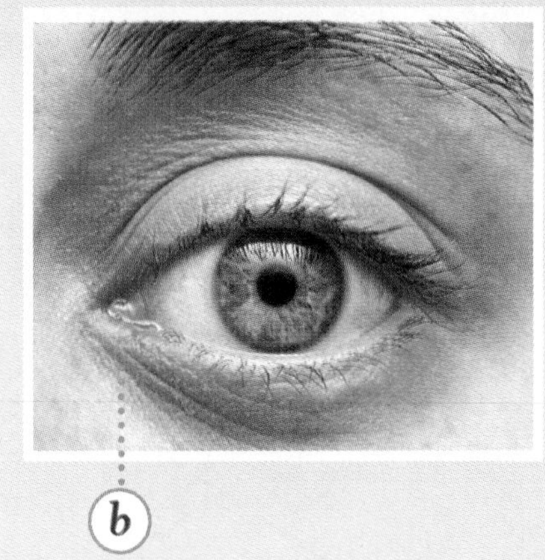

Sense Organs

According to current theory, the first mammals were small and nocturnal, meaning "active at night." If this is true, it helps to explain the range of mammalian senses.

Nocturnal animals must see in dim light, which means their eyes adapted for low-light vision. Like certain black-and-white photographic film, low-light eyes are very sensitive to differences in light intensity but cannot detect differences in color. If the first mammals had this type of eye, it would help to explain why most of their descendants do not see color well.

The ancestors of humans and some other primates, on the other hand, were active during the day. Human ancestors evolved eyes that work more like color film—less sensitive in dim light but able to distinguish color well.

Nocturnal mammals also benefit from sharp senses of smell and taste. This is still the rule among mammals, although humans again are an exception. Although our senses of smell and taste serve us well, they are pitiful in comparison with other mammals. Dogs and cats, for example, can identify individual humans on the basis of subtle differences in body odor that we cannot recognize.

In addition, a sharp sense of hearing is enormously important to nocturnal animals. When ancestral mammals evolved a simplified jaw joint, three bones that were once part of the jaw evolved into the bones of the middle ear. These bones form a delicate mechanism that transfers vibrations from the eardrum to the inner ear. There, sensory cells are stimulated in an organ called the cochlea, allowing the brain to interpret the vibrations as sounds.

Movement

The first mammals evolved several changes in their skeleton, each of which was passed on to their descendants. Among these changes was a backbone that flexed vertically as well as side to side. This flexibility allows mammals to move with a bouncing, leaping stride. Shoulder and pelvic girdles also became more streamlined and flexible, permitting both front and hind limbs to move in a great variety of ways.

Figure 32–5
Unlike most mammals, this mole is blind. Most species of moles lost their eyesight as they adapted to an underground existence. These animals are members of the order Insectivora, meaning "insect-eating."

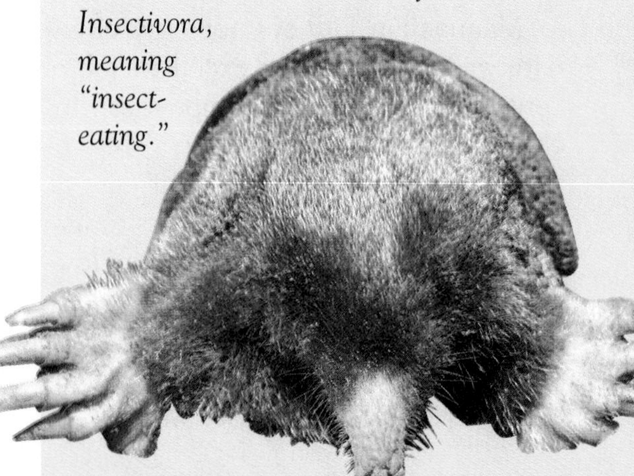

Reproduction and Development

Compared with other animals, mammals put more energy and effort into nourishing and protecting their young, both before and after birth. Most mammals reproduce in the same way that humans do. The male deposits sperm inside the reproductive tract of the female, where fertilization occurs. Typically, the female carries the developing embryo for a significant length of time.

In many mammals—with humans an important exception—newborn young can stand up and move around on their own a short time after birth. Newborn mammals feed for varying lengths of time on their mother's milk.

☑ **Checkpoint** How do mammals reproduce?

Figure 32–6 This female impala is nursing its offspring. Milk is produced in the female's mammary glands—the structures that provide the class Mammalia with its name.

Classifying Mammals

Compared with other classes of animals, there are fewer species of mammals—even mollusks are more diverse! Nor are there many individual mammals alive today, especially compared with insects. However, mammals have evolved a variety of fascinating physical and behavioral traits. They include some of the Earth's largest and most familiar organisms.

Mammals are classified into three groups based on the way in which they reproduce and develop. These groups are monotremes, marsupials, and placental mammals. To learn more about mammals, study the photographs on the next two pages.

Monotremes

Members of the order Monotremata are descendants of the most primitive mammals and seem to be part reptile and part mammal. They are represented today only by the duck-billed platypus and two species of spiny anteaters, also called echidnas. These living relics live only in Australia and New Guinea.

In monotremes—a name that means "one hole"—both the reproductive system and the urinary system open into a cloaca, like the cloaca of reptiles. Development in monotremes is also very different from development in other mammals. As in reptiles, the female lays leathery-shelled eggs that hatch outside the body. And although newly hatched young are nourished with milk, the young do not suckle at their mother's breast as the young of other mammals do. Instead, milk trickles from pores onto the surface of the mother's abdomen, from which the young lap it up.

☑ **Checkpoint** What is a monotreme?

Visualizing Mammals

Of the approximately 4500 living species of mammals, approximately 90 percent are classified in the orders discussed on these two pages.

1 Monotremata

Reptilelike monotremes lay eggs, which contain enough yolk to nourish the developing embryo. Monotremes include this spiny, short-beaked echidna.

2 Marsupialia

Marsupial offspring develop in the external pouch of the female. This order includes this kangaroo, as well as wombats and koalas.

5 Edentata

Edentates either lack teeth or have small, simple teeth. They include this anteater, as well as armadillos and sloths.

7 Dermoptera

Often called flying lemurs, mammals of this order are native to Southeast Asia. They do not actually fly but glide on skin stretched between their legs.

8 Primates

Primates include apes and monkeys, such as this mandrill baboon. The most primitive living primates are small tree dwellers called lemurs. The most advanced primates are humans.

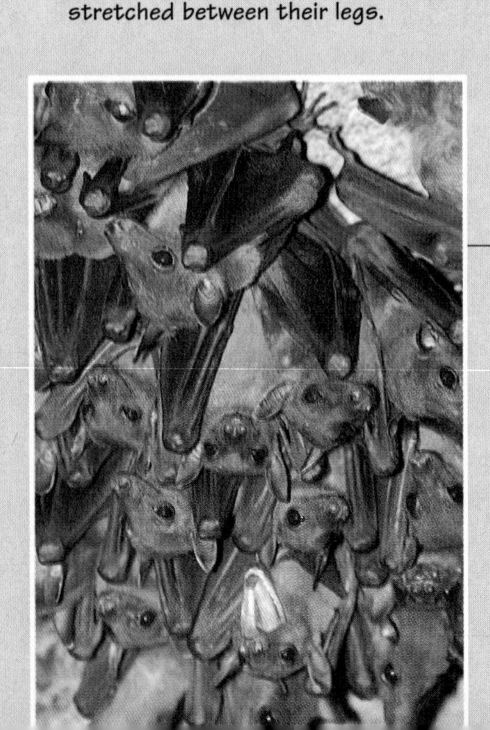

12 Chiroptera

This order includes bats. Most bats fly at night and navigate using a sophisticated echolocation system.

13 Lagomorpha

In many ways, rabbits and hares look and act like rodents. They use sharp front teeth to devour a wide range of plants.

3 Rodentia

Rodents live all over the world and are the most numerous order of mammals. They include this beaver and rats and mice.

4 Carnivora

Carnivores—meaning meat-eaters—include this hyena, as well as dogs, cats, wolves, bears, weasels, and seals.

6 Artiodactyla

The name artiodactyl means "even-toed"—most artiodactyls have two functional toes on each foot. The order includes many grazing animals, such as cows, pigs, camels, antelopes, deer, hippopotamuses, and the giraffe shown here.

10 Cetacea

Cetaceans had terrestrial ancestors that adapted to a totally aquatic existence. This order includes whales and the dolphins shown here.

9 Perissodactyla

Perissodactyls, meaning "odd-toed," have a single functional toe on each foot. In many cases, the toe forms a hoof. The order includes horses, rhinoceroses, tapirs, and these zebras.

11 Sirenia

Sirenians include these manatees, also known as sea cows. These animals live in waters off Africa, South America, and Florida.

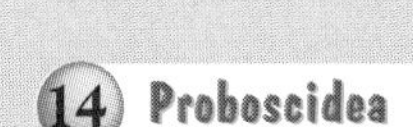

14 Proboscidea

Mammals with trunks survive today only in two endangered species, the Indian elephant and this African elephant.

Step on It!

PROBLEM *How can you **classify** mammals from their footprints?*

PROCEDURE

1. Obtain a set of photographs of mammal footprints from your teacher. Study the relative sizes and shapes of the footprints. Record your observations.
2. Devise a classification scheme for the footprints. Infer the characteristics of each category.
3. Obtain a list of the mammals represented in the photographs from your teacher. Revise your classification scheme as necessary. Try to match each mammal with its footprint.

ANALYZE AND CONCLUDE

1. Which mammal was the easiest to identify from its footprint? Which was the most difficult? Discuss the usefulness of footprints in identifying a mammal's characteristics.
2. Compare the classification scheme you devised with the way in which biologists classify these mammals.

Marsupials

Members of the order Marsupialia once roamed in great numbers across what is now South America, Antarctica, Australia, and New Guinea. Today, only one species, the common opossum, survives in North America. The rest of the living marsupials—including kangaroos, wombats, koalas, and a number of other unusual species—live only in Australia and New Guinea. Unfortunately, most are in danger of extinction because of the human introduction of rats, sheep, rabbits, and other animals.

In marsupials, young are born alive, but at an incredibly early stage in development. When they are only a little more advanced than embryos, newborn marsupials crawl across their mother's belly and into a pouch. There, the babies attach to nipples and feed on their mother's milk as they develop.

Placental Mammals

Almost all mammals—including humans—are classified as placental mammals. In these mammals, the developing embryo attaches to the wall of the mother's uterus. There, tissues from both embryo and mother grow into a structure called the **placenta.** The placenta provides an interface for the circulatory systems of embryo and mother. Through it, oxygen and nutrients from the mother are exchanged for carbon dioxide and waste products from the embryo. The developing young are kept inside the mother—attached to the placenta—until they reach an advanced stage of development.

☑ *Checkpoint* What is a placental mammal?

Section Review 32-1

1. **List** the characteristics of mammals.
2. **Explain** how mammals are classified.
3. **Critical Thinking—Making Inferences** Some early mammals were nocturnal, and others were active during the day. Compare the structures and adaptations of these two types of mammals.
4. **MINI LAB** How can you **classify** mammals from their footprints?

The Influence of Humans

GUIDE FOR READING

- **Identify** the common ancestor of dogs.
- **Discuss** how humans changed the Earth's animal populations.

WILDEBEEST, ANTELOPES, AND zebras once migrated in enormous herds across Africa. Bison roamed the great plains of North America in herds large enough to defy counting. These animals ate some plants and ignored others, churned the ground with their hooves and fertilized it with their dung. Today, however, humans have replaced wild plants with fields of corn, wheat, and other agricultural crops, and they have replaced wild animals with cows, goats, pigs, and sheep.

As human civilization has advanced, it has dramatically altered the populations of nearly every species it has encountered. In the future, decisions you make may further influence the course of life on Earth.

Domestication

As humans evolved, they learned to hunt and capture wild animals for food. Over time, however, they learned techniques of **domestication**—ways to tame, raise, and breed animals for human purposes.

What was the first domesticated animal? Evidence indicates that over 10,000 years ago, humans began interacting with *Canis lupus*, the gray wolf. **Today, descendants of the gray wolf include the many breeds of the first domesticated animal—*Canis familiaris*, the dog.**

Modern dogs look and act not like adult wolves, but like wolf puppies. A dog's head, for example, develops much like the head of a wolf puppy, but it does not change further to resemble an adult wolf. And typical dog activities—such as tail-wagging, face-licking, forming social bonds, and playing games—are all behaviors that juvenile wolves outgrow.

Figure 32–7
Human actions have changed the population of almost every mammal species on Earth. (a) Up to 60 million bison once roamed North America. By 1900, however, bison had been hunted to near extinction! (b) The Earth is now home to huge numbers of domesticated animals, such as these pigs. (c) CAREER TRACK This animal breeder is holding the reins of a Père David's deer, a species that has been kept alive in captivity.

Biology AND YOU .... Connections

Wildlife and Disease

The Masai Mara and Serengeti parks of East Africa are among the world's most spectacular wildlife reserves. But the important goal of these parks—to provide a safe home for endangered wildlife—is not always easy to achieve.

Male lions in the Serengeti National Park in Tanzania

In 1993, for example, many of the lions that lived in these parks became sick, eventually dying in violent convulsions. By the time the epidemic reached its peak, more than 1000 lions were dead.

What killed these lions? Tissues from infected lions revealed an unexpected culprit: canine distemper virus. This virus is most often found in dogs. But it mutated in recent years, adapting to infect other species. At the Masai Mara and Serengeti parks, the virus first infected dogs that lived near the parks. It then spread to wild hyenas that shared food with the dogs, then to the lions that shared food with the hyenas.

To stop this disease, international wildlife experts and veterinarians decided on an interesting course of action—they vaccinated all the domestic dogs in and around the parks. Today, the disease seems to be under control.

Deer, Ticks, and Lyme Disease

While some diseases have spread from domestic animals to wildlife, other diseases have spread in a different direction. In New England, for example, hunting and conservation laws stimulated the recovery of deer populations. But with increased deer came increased numbers of *Ixodes* ticks, parasites of deer. *Ixodes* ticks also bite humans, to whom they sometimes transmit the microorganisms that cause Lyme disease. This infection affects the skin, joints, and nerves.

Complex Relationships

The outbreaks of canine distemper in lions and Lyme disease in humans illustrate the complex relationships that can form among humans, domestic animals, and wildlife. We share the Earth with a huge variety of other forms of life. When we deliberately change one aspect of our relationship with other living things, we can easily change an unforeseen aspect of this relationship as well.

As deer populations have increased, so too have incidents of Lyme disease.

Making the Connection

What changes have humans made to the environment in your community? How has wildlife changed in response to these changes?

Researchers hypothesize that dog breeders may have unintentionally selected for mutations in "master control genes." Such genes affect how other genes are switched on and off and thus control how an animal develops. Conceivably, mutations in master control genes could halt an animal's development at some point before adulthood.

Figure 32–9 (a) *Researchers have identified* Canis lupus, *the gray wolf, as the common ancestor of all dogs.* (b) *Today, breeders have created dogs of all sorts of shapes, sizes, and colors.*

Changing Life on Earth

In the centuries that followed the domestication of the dog, humans learned to domesticate goats, sheep, camels, horses, and other familiar animals. Humans also learned to make new and better weapons, thus becoming more efficient at hunting wild animals. These abilities undoubtedly helped human communities and civilizations to develop.

However, the success of humans has dramatically altered the populations of the Earth's animals. **By overhunting and by using huge tracts of land for farming and ranching, humans have greatly reduced the populations of many species of wildlife—including most large mammals.**

In the 1800s, for example, the huge buffalo herds of the Great Plains of the United States were hunted to near extinction. Today, overhunting in many countries is threatening elephants, rhinoceroses, tigers, and many other animals. In addition, humans typically claim large lands for farming or ranching wherever they settle, often by draining swamps, chopping down rain forests, or destroying other natural ecosystems.

To protect wildlife, people in many countries have passed special laws, set aside tracts of land as wildlife reserves, and tried to reintroduce some species into their former habitats. However, humans have made the Earth a far different place from the days when wild animals roamed it freely. Buffaloes, elephants, gazelles, and lions may play roles in nature in the future, but those roles are sure to differ from their roles of the past.

Section Review 32–2

1. **Identify** the common ancestor of dogs.
2. **Discuss** how humans have changed the Earth's ecosystems.
3. **BRANCHING OUT ACTIVITY** In the United States and elsewhere, zoos and wildlife conservation societies run programs to maximize the biodiversity of many species in their care. **Research** the effects of these programs.

CHAPTER 32

Laboratory Investigation

Investigating Mammalian Skulls

Paleontologists have inferred all sorts of information about ancient animals from the bones they left behind. One of the most useful bones to study is the skull. The skull provides clues about an animal's feeding habits, the way it used its senses, and its relative intelligence. In this investigation, you will explore some of the ways in which this information can be revealed.

Problem

What can you **infer** about a mammal by studying its skull?

Materials (per group)

mammal skull
metric ruler
balance
uncooked rice
graduated cylinder

Procedure

1. **Working with a partner, measure the overall dimensions of the skull in centimeters. Use the balance to measure the mass of the skull in grams. Record this information.**
2. **Locate the lower jaw, which is called the mandible. The rest of the skull is called the cranium, and it is formed by several bones fused together. Study the lines on top of the cranium, which indicate where bones joined. The older the animal, the more fused the bones will be.**
3. **Holding the skull on its side, measure the length and width of the snout. Look for any indication of antlers or horns.**
4. **Locate the large opening at the bottom of the skull, which is where the skull attached to the vertebral column. The farther back the opening, the more forward the skull projected from the body. In humans and some primates, the opening is close to the middle of the skull. Thus, a primate's head is balanced when it stands upright. Predict whether your specimen is from a primate.**

5. **The brain is stored in the braincase, the large cavity just beneath the top of the skull. Cover all openings to the braincase except the large one at the bottom. Fill the braincase with rice, then empty the rice into a graduated cylinder. Measure and record its volume.**
6. **Study any teeth present in the skull. Identify the incisors, which are sharp, flat teeth located in front of the mouth. Next to the incisors are rounded canine teeth. And in back of the mouth are premolars and molars. Carnivores have large canine teeth, whereas rodents have large incisors. In addition, a gap between the incisors and premolars indicates that it is the skull of a rodent, horse, deer, or sheep. Study the arrangement of teeth in your specimen, then record your observations.**
7. **Study the shape and structure of the molars. Broad, flat molars were used for chewing plant matter, whereas ridges on these teeth indicate they were used for tearing meat. Infer which type of food the animal ate.**
8. **Study the eye sockets of the skull. Determine whether the animal's eyes pointed forward or lay on either side of the head.**

Observations

1. Sketch and describe the general shape of the skull and the shape of the braincase. Include the measurements of the skull on the drawing.
2. Label all the parts of the skull that you can identify.

Analysis and Conclusions

1. From your data and observations, what can you infer about the age of your specimen?
2. What can you infer about the animal's diet?
3. Classify the animal as best you can. Describe the evidence that supports this classification.

4. Describe any evidence that does not support how you classified the animal. What additional information about the skull would help you to classify it?
5. With your classmates, compare the braincase volumes of the different skulls you investigated. Which skull had the greatest braincase volume? The smallest braincase volume? How useful is braincase volume in predicting the intelligence of an animal?

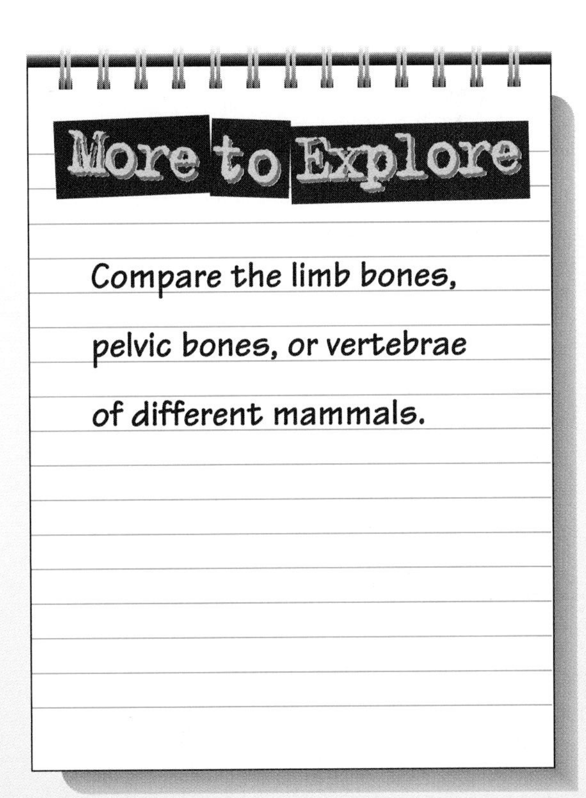

More to Explore

Compare the limb bones, pelvic bones, or vertebrae of different mammals.

Study Guide

Summarizing Key Concepts

The key concepts in each section of this chapter are listed below to help you review the chapter content. Make sure you understand each concept and its relationship to other concepts and to the theme of this chapter.

32–1 Characteristics of Mammals

- Mammals are endotherms, have external body hair, and have a layer of fat beneath the skin. Female mammals produce milk to feed their offspring.
- To help them eat great quantities of food, mammals evolved specialized jaws and interlocking teeth. Because digestive enzymes can quickly break down meat, carnivores have a relatively short intestine. Because plant matter takes more time to digest, herbivores have a longer intestine.
- Like birds, mammals have a double-loop circulatory system powered by a four-chambered heart. Mammals use a diaphragm to expand and contract the chest cavity, which forces air in and out of the lungs.
- Unlike the brains of other animals, a mammal's brain has a well-developed cerebral cortex—the center of thinking and other complex behaviors.
- Nocturnal mammals must see in dim light, which means their eyes adapted for low-light vision. They also have well-developed senses of hearing, taste, and smell. Because humans evolved from animals active during the day, they evolved excellent color vision, but other senses are relatively less developed.
- Most mammals reproduce in the same way that humans do. The male deposits sperm inside the reproductive tract of the female, where fertilization occurs.
- Monotremes, such as the duck-billed platypus and echidna, seem to be part reptile and part mammal. Marsupials include kangaroos, wombats, and koalas. Most mammals are placental mammals—a placenta joins developing offspring with the mother.

32–2 The Influence of Humans

- The descendants of the gray wolf include the many breeds of the first domesticated animal—*Canis familiaris,* the dog.
- By overhunting and by using huge tracts of land for farming and ranching, humans have greatly reduced the populations of many species of wildlife—including most large mammals.

Reviewing Key Terms

Review the following vocabulary terms and their meaning. Then use each term in a complete sentence.

32–1 Characteristics of Mammals

molar
premolar
diaphragm
kidney
placenta

32–2 The Influence of Humans

domestication

Recalling Main Ideas

Choose the letter of the answer that best completes the statement or answers the question.

1. Mammals are classified as endotherms because they

a. conserve body heat.
b. lose body heat.
c. produce milk for offspring.
d. are ovoviviparous.

2. In mammals, the diaphragm is used to

a. move from place to place.
b. encase the heart.
c. help move air in and out of the lungs.
d. move food through the digestive tract.

3. Compared with other vertebrate brains, the primate brain contains a proportionately larger

a. medulla oblongata. **c.** cerebellum.
b. hindbrain. **d.** cerebral cortex.

4. Compared with nocturnal animals, animals that are active during the daytime are more sensitive to

a. low-intensity light. **c.** sounds.
b. colors. **d.** smells.

5. Mammals that lay eggs and lack a placenta are classified in the order

a. Dermoptera. **c.** Edentata.
b. Monotremata. **d.** Marsupialia.

6. Kangaroos, wombats, and koalas are members of the order

a. Chiroptera. **c.** Rodentia.
b. Monotremata. **d.** Marsupialia.

7. Which of these mammal classifications contains the most species?

a. monotremes **c.** placental mammals
b. marsupials **d.** rodents

8. The first domesticated animals were

a. sheep. **c.** horses.
b. rabbits. **d.** dogs.

Putting It All Together

Using the information on pages xxx to xxxi, complete the following concept map.

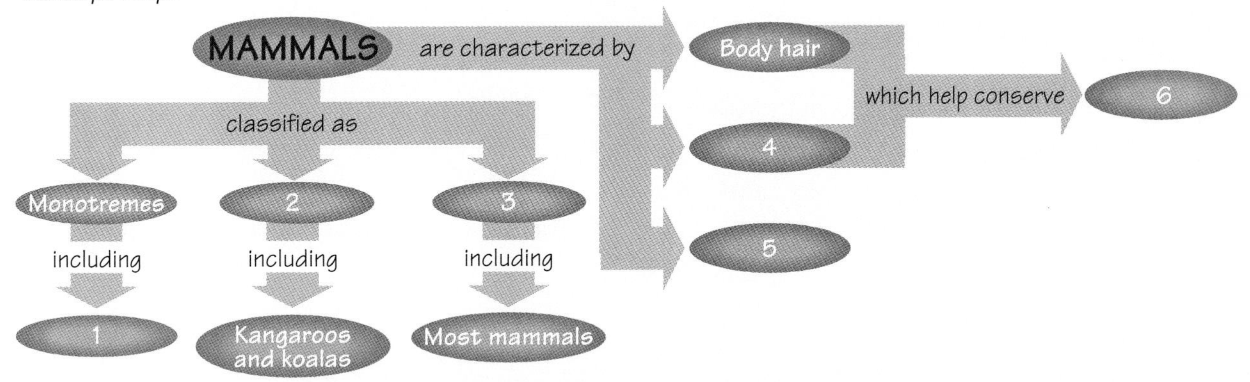

Reviewing What You Learned

Answer each of the following in a complete sentence.

1. How do mammals differ from other vertebrates?
2. How do mammals conserve their body heat?
3. Why must mammals take in and digest a greater amount of food than reptiles or amphibians do?
4. What is the diaphragm?
5. What is the function of a mammal's kidneys?
6. Compare the mammal's brain with the brains of other vertebrates.
7. How did adaptations of the backbone, shoulder, and pelvic girdle affect the way mammals move?
8. What distinguishes marsupials from other mammals?
9. How do monotremes differ from other mammals?
10. What are the placental mammals?
11. List three orders of placental mammals.
12. List three examples of domesticated animals.
13. Give an example of a mammal that humans have overhunted.

Expanding the Concepts

Discuss each of the following in a brief paragraph.

1. Compare a mammal's body plan with the body plans of reptiles.
2. Discuss the characteristics that paleontologists use to identify mammalian bones and fossils.
3. Describe the importance of interlocking teeth to mammals.
4. Explain why most mammals cannot survive by drinking sea water.
5. Compare the sensory organs in nocturnal mammals with those in mammals that are active during the day.
6. Describe the ways in which monotremes resemble reptiles.
7. Mammals typically produce fewer offspring than other vertebrates do. Discuss the adaptations that help mammalian offspring survive.
8. Describe the structure of a mammal's ear.
9. Compare reproduction and development among monotremes, marsupials, and placental mammals.
10. Compare modern dogs with their ancestor, the gray wolf.
11. Explain how human actions have changed the animal populations of the Earth.

Extending Your Thinking

Use the skills you have developed in this chapter to answer the following.

1. **Applying concepts** Mammalian carnivores tend to eat large meals separated by long intervals of time, whereas herbivores tend to eat continuously throughout their waking hours. Describe how these behaviors are supported by the digestive systems of these animals.
2. **Inferring** Small mammals have a greater surface-to-volume ratio than large mammals have. Infer how this difference affects their metabolic rates.
3. **Classifying** Study the mammal shown in the photograph on the opposite page. To which order does this mammal belong? Explain your answer.
4. **Making predictions** In the 1800s, human females typically gave birth to many more offspring than they do today. Many of these babies did not survive infancy. Predict what might happen to the human birth rate in the future.
5. **Using the writing process** Research how the mammal populations where you live have changed over the past few hundred years. Present your findings in a report or incorporate them into a play, short story, or poem.

Applying Your Skills

Mammal Inventory

Mammals can populate almost any habitat, including the one where you live. In this activity, you'll learn more about the roles that mammals play in your community.

1. Select an area the size of a city block or a neighborhood park. Draw a map of this area and describe how the area is used.
2. With a partner, walk through the area you selected. Record the names of any mammals you observe and note their activities. If you cannot identify an animal, draw a rough sketch of it and list its characteristics. Do not approach any of the animals.
3. Repeat step 2 at least two times. Make your walks at different times of the day.
4. If necessary, consult field guides or other references to identify an animal.
5. Prepare a report on the mammals that live in the area you studied. Describe the roles that these mammals play.

• GOING FURTHER •

6. Perform this activity in a different location. Compare the mammal populations in the two areas you studied.

CHAPTER 33

Animal Behavior

FOCUSING THE CHAPTER THEME: Systems and Interactions

33–1 Behaviors and Societies

- **Discuss** different animal behaviors.

BRANCHING OUT *In Depth*

33–2 Insect and Primate Societies

- **Compare** the societies that insects and primates form.

LABORATORY INVESTIGATION

- **Design** an experiment to determine the instinctive behaviors of a planarian.

Biology and Your World

BIO JOURNAL

Look at the meerkats shown in the photograph on this page. In your journal, describe the meerkats' behavior. What purpose do you think this behavior serves?

Meerkats emerging from their den

SECTION 33–1

Behaviors and Societies

GUIDE FOR READING

- **Explain** the purpose of animal behaviors and **identify** different types of behaviors.
- **Explain** why animals live in societies.

MINI LAB

- **Classify** a learned behavior.

OF THE SIX KINGDOMS OF LIFE on Earth, the animals have evolved the most complex set of behaviors. From earthworms that emerge above ground after rainstorms to peacocks that strut their colorful tail feathers to attract a mate to humans who constantly ask questions to learn about the world around them, animals interact with their environment in an impressive and diverse set of ways. To truly understand animals, you need to understand why they behave in the ways they do.

Perception and Response

Why do animals behave in certain ways? The answers begin with their sense organs, the organs that allow animals to perceive the world around them.

We humans know the world only through our five senses—vision, smell, touch, taste, and feel—each of which operates within certain limits. However, these limits differ in other animals. A fly on your pant leg, for example, can see colors in your clothes that you cannot see, use its legs to taste soda you spilled

Figure 33–1
Animals perceive and respond to stimuli in vastly different ways—some of which are quite amazing. (a) A spider has a relatively simple brain, yet it is able to build a complex, intricate web. (b) The spotted bowerbird needs no special training to build its unusual nest. (c) In the compound eyes of this black fly, thousands of lenses each form individual images (magnification: 30X). This creates a very different view of the world from the one your eyes provide.

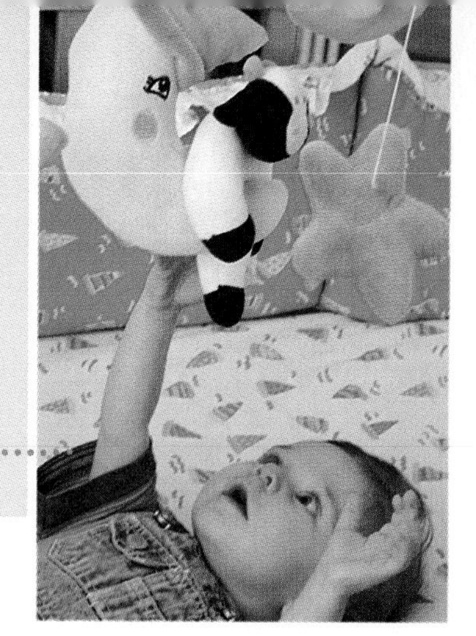

Figure 33–2
(a) Ticks can detect few stimuli and have a relatively limited nervous system. Thus, their responses are simple, genetically programmed behaviors. (b) Humans, in contrast, have a complex nervous system, which they use to learn a wide variety of behaviors throughout their lifetime.

on your jeans, and see your fast-moving hand much better than a human can—which is why flies are so hard to swat! Meanwhile, a dog sitting at your feet recognizes you by your smell—even if you've taken a shower—and hears your parents walking down the hall long before you can. On the other hand, the dog probably can't detect the differences in the colors of the pencils on your desk.

Among the animal species, sensory abilities differ in these sorts of ways. Because of these differences, animals respond to their environment in different ways, too.

Behavior

Biologists use the word **behavior** to describe an animal's response to its environment. Obviously, different animal species behave very differently, as shown in ***Figure 33–1*** on page 747.

Despite the enormous differences among the ways in which animals behave, the typical reasons for their behavior are very similar. **All animals have evolved behaviors that help them survive. Some behaviors are instinctive, or inborn in the animal, and others are learned as the animal develops.**

Instincts

Some behaviors are genetically programmed into every member of a species. These behaviors are often called innate behaviors, or **instincts.** The simplest instincts are fixed and cannot be changed.

For example, consider a female tick waiting on a blade of grass. This tick may have lived 20 years or longer without ever encountering the odor of butyric acid, a compound that is part of mammalian sweat. When the tick does "smell" this compound, however, it lets go of the grass and—if it's lucky—drops onto a passing mammal, such as a mouse. It then follows temperature cues to a place where blood is near the surface. After sinking its mouthparts into the mouse's skin, the tick drinks the blood until it is full. At that point, it lets go once again, drops to the ground, lays its eggs, and dies. ●

INTEGRATING CHEMISTRY

What is the chemical formula for butyric acid? What are its properties?

This sequence of behaviors is genetically programmed into the tick's nervous system. Instincts control most of the behaviors of ticks and most other invertebrates.

☑ ***Checkpoint*** What are instincts?

Advantages of Instincts

Instincts are inflexible behaviors that may often be a drawback for an animal. However, instincts also provide enormous advantages. They permit animals to perform tasks essential to their survival, with no learning period required.

All sorts of animals rely on instinctive behaviors, including large animals with a complex nervous system. A female graylag goose, for example, will instinctively retrieve an egg that rolls out of its nest. Spiders build webs instinctively, birds construct nests without ever having done so before, and newborn mammals—from mice to humans—will instinctively suckle at their mother's breast.

Problem Solving

INTERPRETING DIAGRAMS

A Fishy Investigation

Many animals have instincts for movement or for orienting their body. For example, when an insect's environment changes suddenly, it typically will move rapidly in a random direction. This type of movement is called kinesis. With an instinct for kinesis, insects increase their chances of avoiding danger.

Another type of instinctive movement is called a taxis (plural: taxes). Taxes are movements toward or away from a stimulus. Examples of taxes include the instinctive movements of moths toward light, earthworms toward moisture, and reptiles toward heat.

To orient their body in water, fishes rely on one taxis that responds to light and another taxis that responds to gravity. In one experiment, researchers removed the gravity-detecting organ from several fishes, then observed their response to light. The results of this experiment are illustrated below.

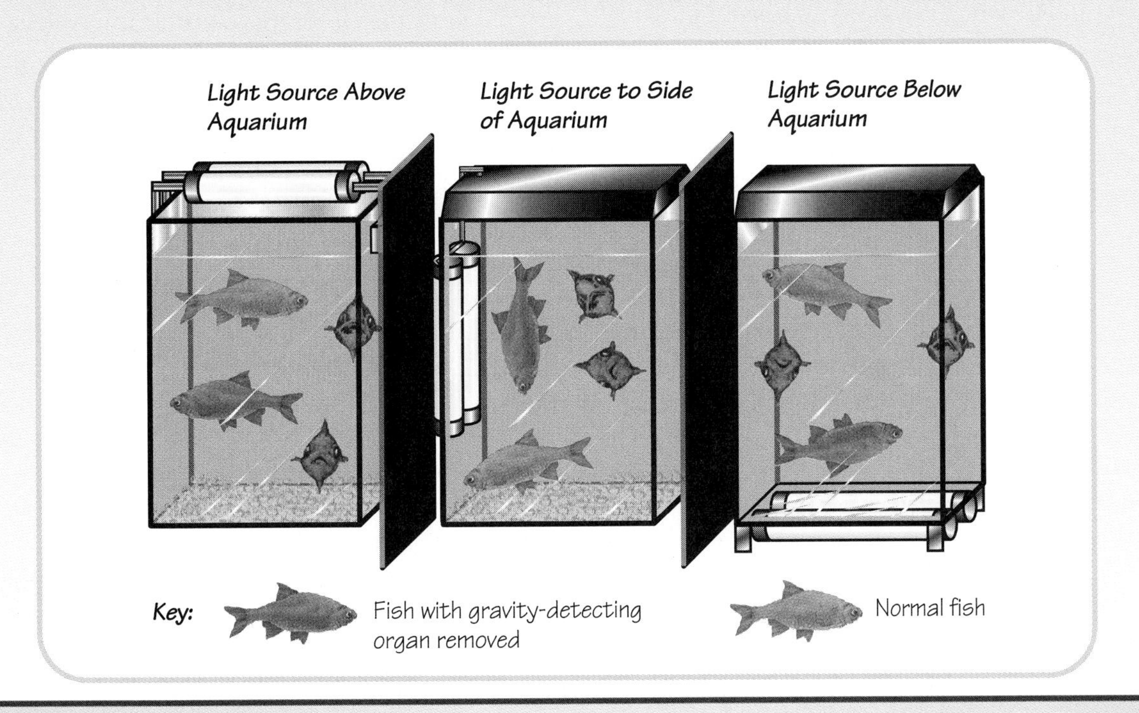

THINK ABOUT IT

1. Explain how taxes for gravity and light help fishes to survive.
2. Describe the results of the experiment illustrated above.
3. From these results, what can you conclude about the taxes toward light and gravity in fishes?
4. Do you suspect that a fish could learn to compensate for the loss of its gravity-detecting organ? Design an experiment to answer the question.

Figure 33–3
Most animals are capable of one or more types of learning. (a) *Ivan Pavlov taught dogs to expect food whenever he turned on a light—an example of classical conditioning.* (b) *In nature, predators learn that brightly colored animals, such as this tomato frog, are often poisonous. This is an example of operant conditioning.* (c) *This chimpanzee is signing the word "me"—a skill only a few species can master.*

Solid evidence indicates that genes control many instinctive behaviors. However, no one yet understands just how these genetic instructions are carried out. And, as you might expect, efforts to prove genetic control become much more complicated when dealing with complex behaviors, such as those that humans perform.

Learning

While animals often rely on instincts, many species also have the ability to change their behaviors as a result of experience. This is called **learning.** You may be surprised to discover the wide range of animals that learn behaviors of one sort or another. These animals range from humans to insects to cnidarians!

The simplest form of learning is **habituation,** in which an animal decreases or stops its response to a stimulus that neither helps nor harms the animal. If you give pieces of plain paper to a sea anemone, for example, the sea anemone will swallow the paper at first. But after a few trials, the sea anemone learns that paper is neither a food nor a danger and it simply stops responding to it.

In **classical conditioning**—a form of learning also called associative learning—an animal learns to associate a stimulus with either a reward or a punishment. One of the first scientists to study classical conditioning was the Russian biologist Ivan Pavlov, shown in *Figure 33–3.*

If you have a pet, you may have helped it learn a behavior through classical conditioning. If you always ring a bell or perform a certain kitchen chore before meal time, for example, your pet may run eagerly to its food dish before you even get out the pet food!

In **operant conditioning**—often called trial-and-error learning—animals learn to perform some sort of task in order to get a reward or to avoid punishment. Laboratory rats, for example, can be trained to run through a maze or to push a lever in order to obtain food or to avoid a mild electric shock. In nature, birds and other predators can learn to associate the bright colors of some butterflies and tree frogs with an unpleasant taste.

Insight learning occurs when an animal applies past experiences to a new situation without any trial and error. This kind of learning is the rarest and most complicated kind of learning, and it is common only in primates. Even otherwise intelligent animals such as dogs and cats rarely show insight learning. The typical dog, for example, will repeatedly wrap its leash around a tree and gain no insight on this behavior—as many dog owners know all too well! Chimpanzees, on the other hand, will use an object as a tool in a new situation, particularly to obtain food.

☑ **Checkpoint** What is insight learning?

Complex Behaviors

Years ago, people studying animal behavior argued to great lengths about whether behavior was instinctive or learned. This debate was sometimes called the "nature versus nurture" controversy.

Today, this debate has largely ended. Why? Because it is clear that many behaviors are produced by a combination of instinct and learning. This combination can be called instinctively guided learning, and it provides many advantages to an animal. Because learned behaviors depend in part on heredity, they evolve over time to adapt the species to its environment. Yet because they are also somewhat flexible, they allow individual animals to adjust to local conditions.

Birds and fishes provide several examples of instinctively guided learning. For example, geese are born with an instinct to follow the first large moving object that they see during a critical time early in their lives. Normally, that object is their mother, whom they follow in search of food or shelter. This phenomenon is known as **imprinting.** Interestingly, if researchers remove eggs from the mother and expose goslings to another moving object—even another kind of animal—the birds will imprint on that object! See ***Figure 33–5*** on page 752.

In a similar way, newly hatched salmon imprint on the particular odor of

MINI LAB .... Classifying ....

Pieces of the Puzzle

PROBLEM *How can you **classify** a learned behavior?*

PROCEDURE 

1. Working in groups, cut an unlined index card into 5 pieces of various sizes. Make each cut as straight as possible.
2. Shuffle the pieces, then have one member of the group time you as you try to reassemble the pieces.
3. Repeat step 2 three times. Record your time for each trial.

ANALYZE AND CONCLUDE

1. Compare how quickly you reassembled the index card in each of the four trials.
2. Classify the types of learning that you displayed.
3. Compare your ability to reassemble an index card with other behaviors you have learned.

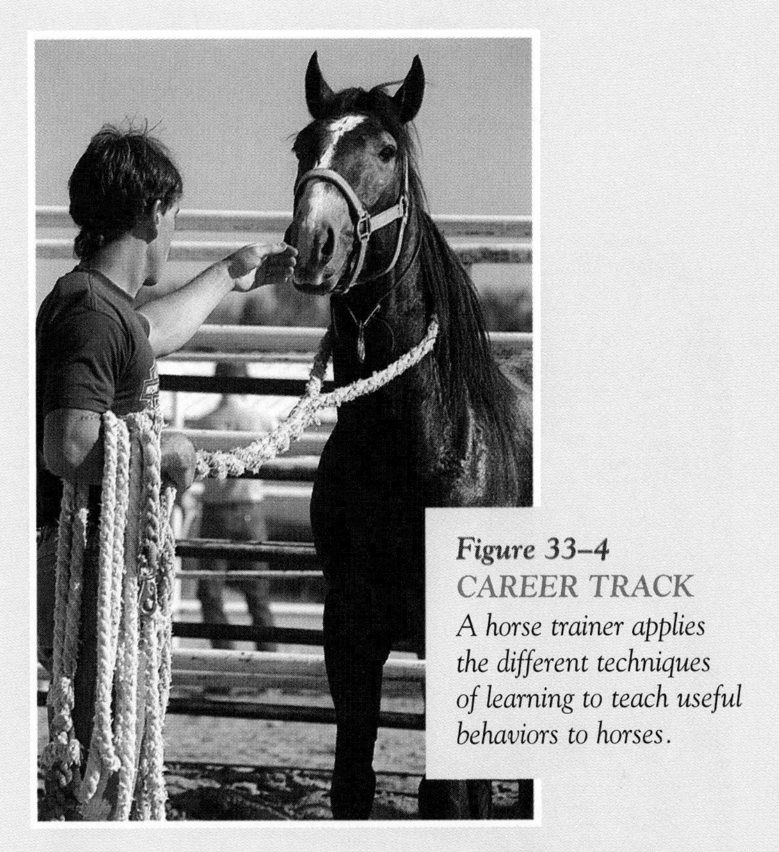

Figure 33–4
CAREER TRACK
A horse trainer applies the different techniques of learning to teach useful behaviors to horses.

the stream in which they hatch. Young salmon head out to sea, where they spend several years feeding and growing. When they mature and prepare to mate, they perform an amazing feat. They remember the odor of their home stream and return there to spawn.

Animal Societies

Any animal that reproduces sexually needs to locate and mate with another of its species—even if the two animals spend no time together afterward. Thus, an animal lives at least part of its life with a member of its own kind. In the broadest sense, an **animal society** is any group of animals living together. **Animal societies help the species to survive and provide benefits for the individual animals.**

☑ **Checkpoint** What is an animal society?

Types of Animal Societies

Animals live in a wide variety of different societies. For example, songbirds and other species live as **mated pairs,** in which a male and a female live together and jointly raise their offspring. As mated pairs, songbirds can defend their home and feed their young more successfully than a single parent could alone.

Other animals live in **family groups,** such as a pride of lions, a herd of elephants, and various groups of apes and monkeys. Family groups use strength in numbers to improve their ability to hunt, to protect their territory from competitors, to guard their young from predators, and to fight with rivals if necessary.

Still other animals rely on highly structured living groups. Ants, termites, and bees live in **social insect colonies,** as shown in ***Figure 33–6.***

Communication

In order for individual animals to form and operate a society—from a mated pair of bluebirds to the largest beehive—they must communicate with one another. In this sense, biologists define **communication** as the passing of information from one animal to another in a way that influences both animals from that point on.

Figure 33–5
In imprinting, a stimulus early in an animal's life establishes an irreversible behavior. (a) *In a famous series of experiments, baby geese imprinted on biologist Konrad Lorenz. The geese behaved as if Lorenz was their mother!* (b) *Salmon return to their home stream—often overcoming all sorts of obstacles—because they imprinted on the stream's scent.*

Figure 33–6
Animal societies range from two birds living together in a nest to hundreds of thousands of ants or bees living in a colony. (a) As a mated pair, these adult crimson chats together raise their offspring. (b) These crescent-tail bigeyes live in a large group called a school. Swimming in schools helps to protect fishes from predators. (c) By living in a highly organized colony, these honeybees accomplish tasks that they never could accomplish individually.

Animals communicate to find and select mates, to pass on information about food or danger, to assert dominance, to claim territory, or to threaten to fight. In fact, many of the most intriguing, entertaining, and even frightening activities in the animal kingdom are behaviors used by animals to communicate with one another.

Different species communicate by using one or more of their senses. Birds, for example, have well-developed hearing and vision. Thus, it is not surprising that male birds attract mates with sound signals, such as songs or calls, and visual signals, such as brightly colored feathers.

Animals that have a highly developed sense of smell often use chemical signals to transmit information. Social insects such as ants, for example, often give off a variety of chemical signals known as **pheromones** (FER-uh-mohnz). Pheromones tell nest mates about food, the condition of the nest, or danger. Many mammals use pheromones to mark their territory or to signal that they are ready to mate.

Section Review 33–1

1. **Explain** the purpose of animal behaviors and **identify** different types of behaviors.
2. **Explain** why animals live in societies.
3. **Critical Thinking—Applying Concepts** Do humans learn in each of the different ways described in this section? Give examples to justify your answer.
4. **MINI LAB** How can you **classify** a learned behavior?

SECTION 33–2 BRANCHING OUT In Depth

Insect and Primate Societies

Guide for Reading

- **Compare** individual ants and the superorganism they form.
- **Describe** the typical nonhuman primate society.

YOU MIGHT THINK THAT ANTS and primates have very little in common. Ants are tiny insects with minuscule collections of nerve cells that can barely be called brains. Primates, on the other hand, are large, intelligent mammals—among them, the most intelligent animals of all.

Yet despite their differences, ants and primates have each developed the most sophisticated societies on Earth. These societies are well worth studying—their secrets may apply to humans in some surprising ways.

Ant Societies

Place up to 100 army ants or leaf-cutter ants on a table, and they will probably walk around in circles until they die. But watch a colony of either of these ants in action, and you will be amazed at what 500,000 ants can do!

Army ants send raiding parties up to 200 meters from the nest, form living bridges of their bodies to cover difficult terrain, and form teams to kill and transport large prey. Leaf-cutter ants cut leaves into small pieces and carry the pieces home. There, they chew the leaves into a pulp, which they use to grow a special fungus that they weed, water, harvest, and eat!

How do ant colonies accomplish such amazing feats? In a colony, ants constantly click and rub each others' antennae. Ants pass information back and forth with these actions. In a way that researchers do not yet understand, this constant communication creates what is called a **superorganism**—a colony that is more than the sum of its parts.

Figure 33–7
In social insect colonies, each individual is born to fill a specific role. (a) This soldier army ant has razor-sharp mouthparts, making it a fierce predator. (b) This queen fire ant is surrounded by smaller workers. In all ant colonies, only the queen bears offspring. (c) Among leaf-cutter ants, worker ants have the job of carrying leaves back to the nest.

Figure 33–8
Primate societies are composed of unique, intelligent individuals that develop close relationships with one another. (a) To learn the rules of their society, young primates—such as this infant olive baboon—typically stay close to their mother. (b) The kneeling Japanese macaque is cleaning the fur of another macaque—an activity called grooming.

Although ants individually are helpless and unintelligent, as a superorganism they are able to process information and accomplish complicated tasks. As a whole, an ant colony "knows" how to build a nest, how to raid territory, how to raise offspring, and how to protect the colony. The colony also can "learn" where to find food and where enemies are located.

Some researchers compare ant colonies to the human brain, which is formed from huge numbers of individual neurons. Like ants, the neurons of the brain are "stupid" individually, but somehow connect and communicate with one another to create intelligence.

Primate Societies

Nonhuman primates don't build much, do not farm crops, and only occasionally use tools. But their societies may offer insights into the evolution of human societies—and human intelligence.

Nonhuman primates typically live in groups of closely related individuals, with females as central members. In many species, most males periodically switch from group to group. Most females, on the other hand, spend their lives with the groups into which they were born.

As researchers have shown, primate social groups are built on a complex web of relationships. Individuals build friendships and make enemies, do favors for one another, and even have been seen to trick or manipulate one another for their own ends. All these factors come into play when primates hunt, travel, or encounter neighboring groups.

Of course, it takes a good memory and lots of intelligence to keep track of all these relationships. Perhaps in part to develop this intelligence, primate infants depend on their parents for relatively long periods of time. Some researchers suggest that the growth of the primate social system was the major reason why intelligence developed to such a great extent in ancestral humans.

Section Review 33–2

1. **Compare** individual ants and the superorganism they form.
2. **Describe** the typical nonhuman primate society.
3. **BRANCHING OUT ACTIVITY** Find an ant colony in your neighborhood and **observe** the colony for 15 minutes. Make a list of all your observations.

CHAPTER 33

Laboratory Investigation

DESIGNING AN EXPERIMENT

Investigating the Instinctive Behaviors of Planarians

Many animals rely on simple instincts called taxes (singular: taxis). A taxis is a movement either toward or away from a stimulus, such as light, gravity, or chemical changes in the environment. In this investigation, you will explore the taxes of a planarian, a type of free-living flatworm found in ponds and streams.

Problem

How can you determine the taxes of a planarian? **Design an experiment** to answer the question.

Suggested Materials

- glass-marking pencil
- test tube, with cork or stopper
- planarian
- pond water
- test-tube rack
- Petri dish
- small piece of liver
- index card
- dilute acetic acid
- salt
- forceps

Suggested Procedure

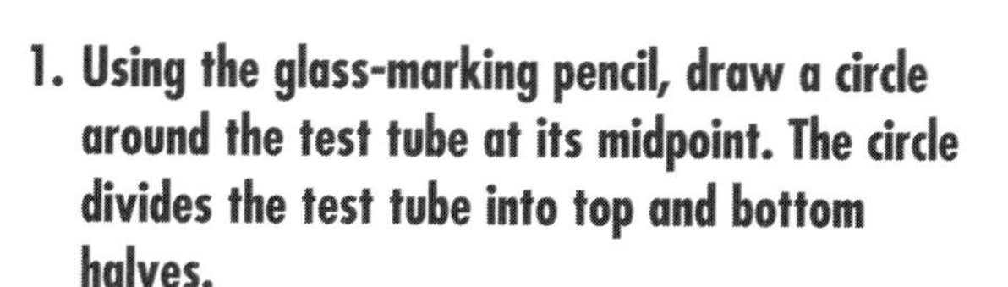

1. **Using the glass-marking pencil, draw a circle around the test tube at its midpoint. The circle divides the test tube into top and bottom halves.**
2. **Obtain a planarian from your teacher. Gently transfer the planarian to the test tube. CAUTION:** ***Be careful when handling live animals.*** **Add pond water as necessary to fill the test tube. Seal the test tube with the stopper or cork.**

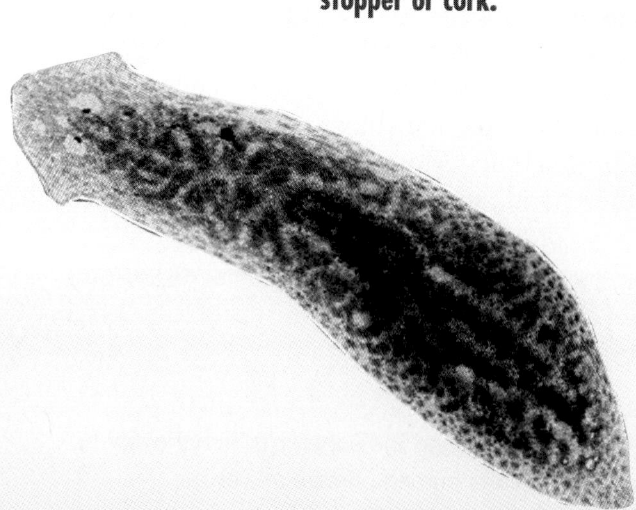

3. **Gently move the test tube back and forth until the planarian is in the center of the test tube.**
4. **Place the test tube in a test-tube rack. For 3 minutes, measure the amount of time the planarian spends in the top half of the test tube and the amount of time it spends in the bottom half of the test tube.**
5. **Turn the test tube upside down, then repeat steps 3 and 4.**
6. **Carefully transfer the planarian and pond water to a Petri dish. Add more pond water, if necessary, to cover the bottom of the dish. Using forceps, place a piece of liver in the Petri dish. Observe the planarian's response.**
7. **Design an experiment to determine whether the planarian has a taxis for one of the following stimuli:**
   - **light**
   - **change in pH**
   - **change in salinity**

   Make sure your experiment does not harm the planarian in any way. Your experiment should include a control group.
8. **With your teacher's permission, perform the experiment you designed.**

Observations

1. Compare the time the planarian spent in the top half of the test tube with the time it spent in the bottom half of the test tube.
2. Describe the planarian's response to liver.
3. Describe the planarian's response to the stimulus you tested in your experiment.

Analysis and Conclusions

1. What is a taxis? Why are taxes important to animals with a limited nervous system?
2. Do planarians have a taxis for gravity? Explain your answer.
3. Why was it necessary to turn the test tube upside down in step 5?
4. Can planarians detect food in nearby water, then move toward the food? Or do they encounter food only by chance? Explain your answer.
5. Do planarians have a taxis for the stimulus you tested in your experiment? Use the data you generated to justify your answer.
6. Explain how a planarian's taxes help it to survive in its environment.

More to Explore

Try to train a planarian to navigate through a simple maze.

Study Guide

Summarizing Key Concepts

The key concepts in each section of this chapter are listed below to help you review the chapter content. Make sure you understand each concept and its relationship to other concepts and to the theme of this chapter.

33–1 Behaviors and Societies

- An animal's sense organs shape the ways in which it perceives and responds to its environment. Because sense organs differ among animals, their behaviors differ as well.
- All animals have evolved behaviors that help them to survive. Some behaviors are instinctive, or inborn in the animal, whereas others are learned as the animal develops.
- Instincts are genetically programmed behaviors. The simplest instincts are fixed and cannot be changed.
- Many animals can change their behavior as a result of experience—a process called learning. The simplest form of learning is habituation, in which an animal decreases or stops its response to a stimulus that is neither harmful nor helpful.
- In classical conditioning, an animal learns to associate a stimulus with either a reward or a punishment. In operant conditioning, an animal learns to perform some sort of task in order to get a reward or to avoid punishment.
- Insight learning occurs when an animal applies past experiences to a new situation without any trial and error. Instinctively guided learning is a combination of instinct and learning.
- An animal society is any group of animals living together. Animal societies help the species to survive and provide benefits for the individual animals.
- Types of animal societies include mated pairs, family groups, and social insect colonies. All animal societies rely on communication among members.

33–2 Insect and Primate Societies

- Although ants individually are helpless and unintelligent, in a social insect colony—a superorganism—they are able to process information and accomplish complicated tasks.
- Nonhuman primates typically live in groups of closely related individuals, with females as central members.

Reviewing Key Terms

Review the following vocabulary terms and their meaning. Then use each term in a complete sentence.

33–1 Behaviors and Societies

behavior
instinct
learning
habituation
classical conditioning
operant conditioning
insight learning
imprinting
animal society
mated pair
family group
social insect colony
communication
pheromone

33–2 Insect and Primate Societies

superorganism

Recalling Main Ideas

Choose the letter of the answer that best completes the statement or answers the question.

1. Another name for an innate behavior is a(an)

a. sensory behavior. **c.** learned behavior.
b. instinct. **d.** insight behavior.

2. An animal stops responding to a stimulus that is neither helpful nor harmful. This is called

a. habituation. **c.** classical conditioning.
b. instinctive learning. **d.** operant conditioning.

3. Pavlov trained dogs to expect food whenever he turned on a light. This is an example of

a. habituation. **c.** classical conditioning.
b. instinctive learning. **d.** operant conditioning.

4. Another name for trial-and-error learning is

a. habituation. **c.** classical conditioning.
b. insight learning. **d.** operant conditioning.

5. Insight learning is common only in

a. birds and reptiles. **c.** dogs and cats.
b. primates. **d.** humans.

6. Baby geese follow the first large object they encounter during a critical period in their development. This phenomenon is an example of

a. insight learning. **c.** imprinting.
b. habituation. **d.** operant conditioning.

7. Many animals use pheromones as

a. chemical signals. **c.** sound signals.
b. visual signals. **d.** temperature signals.

8. By clicking and rubbing their antennae together, ants are able to

a. communicate. **c.** mate.
b. attack each other. **d.** groom each other.

9. Unlike male primates, female primates typically

a. build friendships and make enemies.
b. rely on a complex web of relationships.
c. move from group to group.
d. remain in the same group.

Putting It All Together

Using the information on pages xxx to xxxi, complete the following concept map.

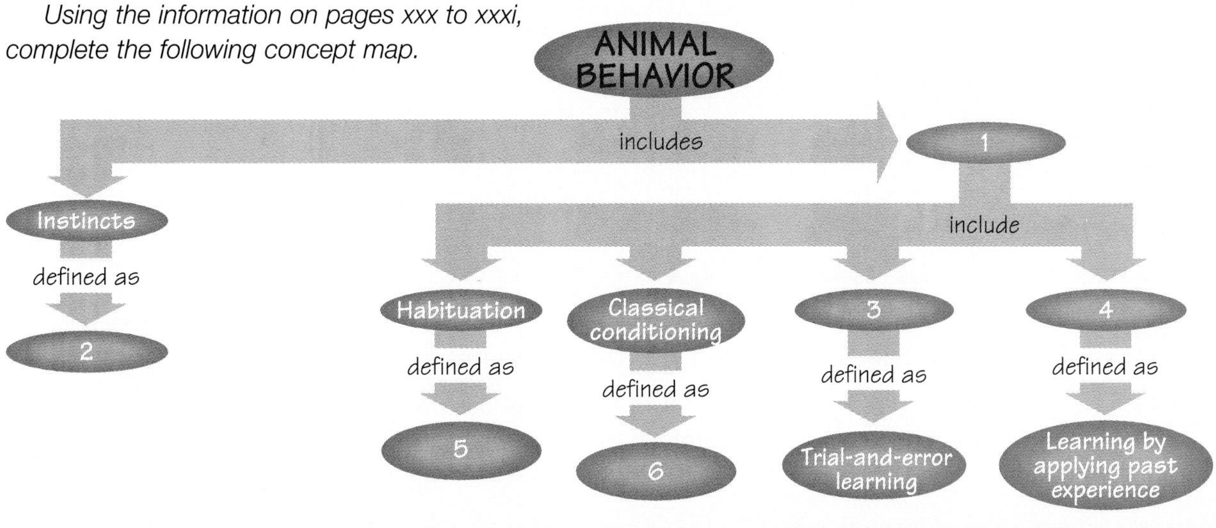

Assessment

Reviewing What You Learned

Answer each of the following in a complete sentence.

1. What is a behavior?
2. List three examples of instinctive behavior.
3. Describe a human behavior controlled by an instinct.
4. Give an example of habituation.
5. What is classical conditioning?
6. Explain operant conditioning.
7. Describe the once-controversial issue of "nature versus nurture."
8. What is instinctively guided learning?
9. Give an example of imprinting.
10. Give an example of an animal that lives in mated pairs.
11. List three family groups of animals.
12. Identify three ways in which animals communicate.
13. What is a superorganism?
14. In what way is a social insect colony similar to the human brain?
15. List three behaviors typical of primate groups.

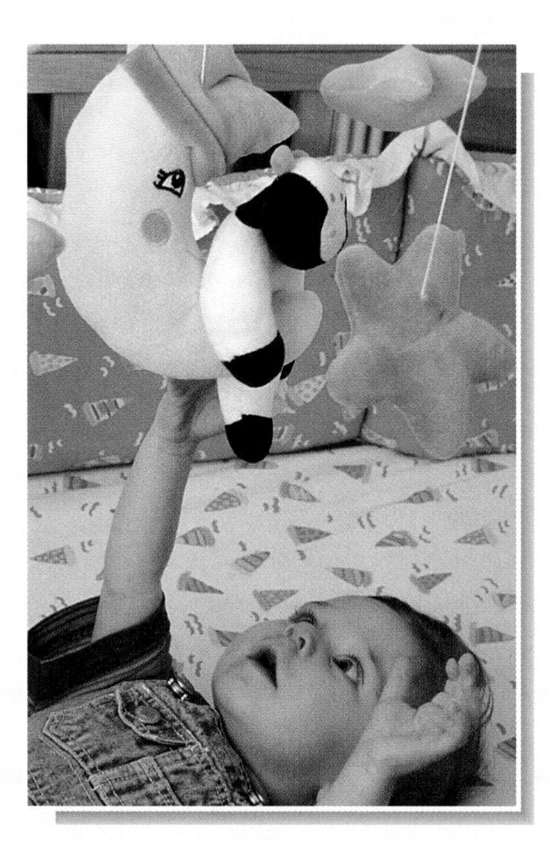

Expanding the Concepts

Discuss each of the following in a brief paragraph.

1. Explain why animals of different species perceive the world in different ways.
2. In animals with a primitive nervous system, are most behaviors instinctive or learned? Explain your answer.
3. Compare the advantages and disadvantages that instincts provide an animal.
4. Compare instinctive behaviors with learned behaviors.
5. Why has the debate ended over the issue of "nature versus nurture"?
6. Why do animals live in societies?
7. Why is communication essential in any animal society?
8. Describe some of the complex behaviors that insect societies carry out that the individual insects cannot.
9. Compare social insect colonies with primate societies.
10. Discuss how primate societies may have helped primate intelligence to evolve.

Extending Your Thinking

Use the skills you have developed in this chapter to answer the following.

1. **Classifying** Give examples of ways in which humans use sound communication, visual communication, or chemical communication.
2. **Predicting** A salmon imprints on the stream in which it hatches and has an instinct to return there to spawn. Suppose that over the course of a year, a stream becomes polluted with industrial wastes. Predict the effects on the salmon population that hatched in the stream.
3. **Designing an experiment** Young herring gulls peck at a brightly colored spot on their parent's beak whenever the parent returns to the nest. The pecking stimulates the parent to regurgitate food for the young gull to eat. Design an experiment to determine whether the size, shape, or color of the spot affects the pecking response of the young gull.
4. **Analyzing concepts** To what extent do instincts affect human behavior? Explain your answer.
5. **Using the writing process** Write a short story, play, or poem in which a character learns a new behavior. Classify the type of learning that your writing illustrates.

Applying Your Skills

Classifying Animal Behavior

Although some animal behaviors are easy to understand and classify, other behaviors are not so straightforward. In this activity, you will apply what you have learned about animal behavior by studying a real animal.

1. Select an animal to study. The animal may be a pet, a farm animal, or an animal in nature. Observe the animal for about 1 hour. Do not touch or approach the animal. Record all the behaviors that you observe.
2. Infer which behaviors are forms of communication. Classify the communication as sound signals, visual signals, or chemical signals. Record any responses to the communications from other animals.
3. Classify other behaviors that you can identify. Typical behaviors include feeding, courtship, care of young, shelter-seeking, claiming territory, and grooming.
4. Analyze your data and present your conclusions in a report.

• GOING FURTHER •

5. In 1973, biologists Konrad Lorenz, Niko Tinbergen, and Karl von Frisch were awarded a Nobel prize for their studies of animal behavior. Research their work.

The Human Body

CHAPTERS

"Science is part of the reality of living; it is the what, the how, and the why for everything in our experience."

— Rachel Carson

CAREER TRACK

As you explore the topics in this unit, you will discover many different types of careers associated with biology. Here are a few of these careers:

- Biomedical Equipment Technician
- Electroneurodiagnostic Technologist
- Physical Therapist
- Emergency Medical Technician
- Dietician
- Physician Assistant
- Immunologist

Two dancers showing the various ranges of motion possible to the human body

CHAPTER 34

Introduction to Your Body

FOCUSING THE CHAPTER THEME: *Scale and Structure*

34–1 Organization of the Human Body
- **Describe** the basic organization of the human body.

34–2 Communication and Control
- **Explain** the relationship between the nervous system and the endocrine system.

BRANCHING OUT *In Action*

34–3 Physiology in Action
- **Explain** how the systems of the body work together.

LABORATORY INVESTIGATION
- **Observe** the effect of adrenaline on *Daphnia*.

Biology and Your World

BIO JOURNAL

The runners in this photograph are just starting a race. Imagine that you are one of these runners. In your journal, make a list of all the body systems that you will be using in this race. After you have read the chapter, refer back to your list and modify it if necessary.

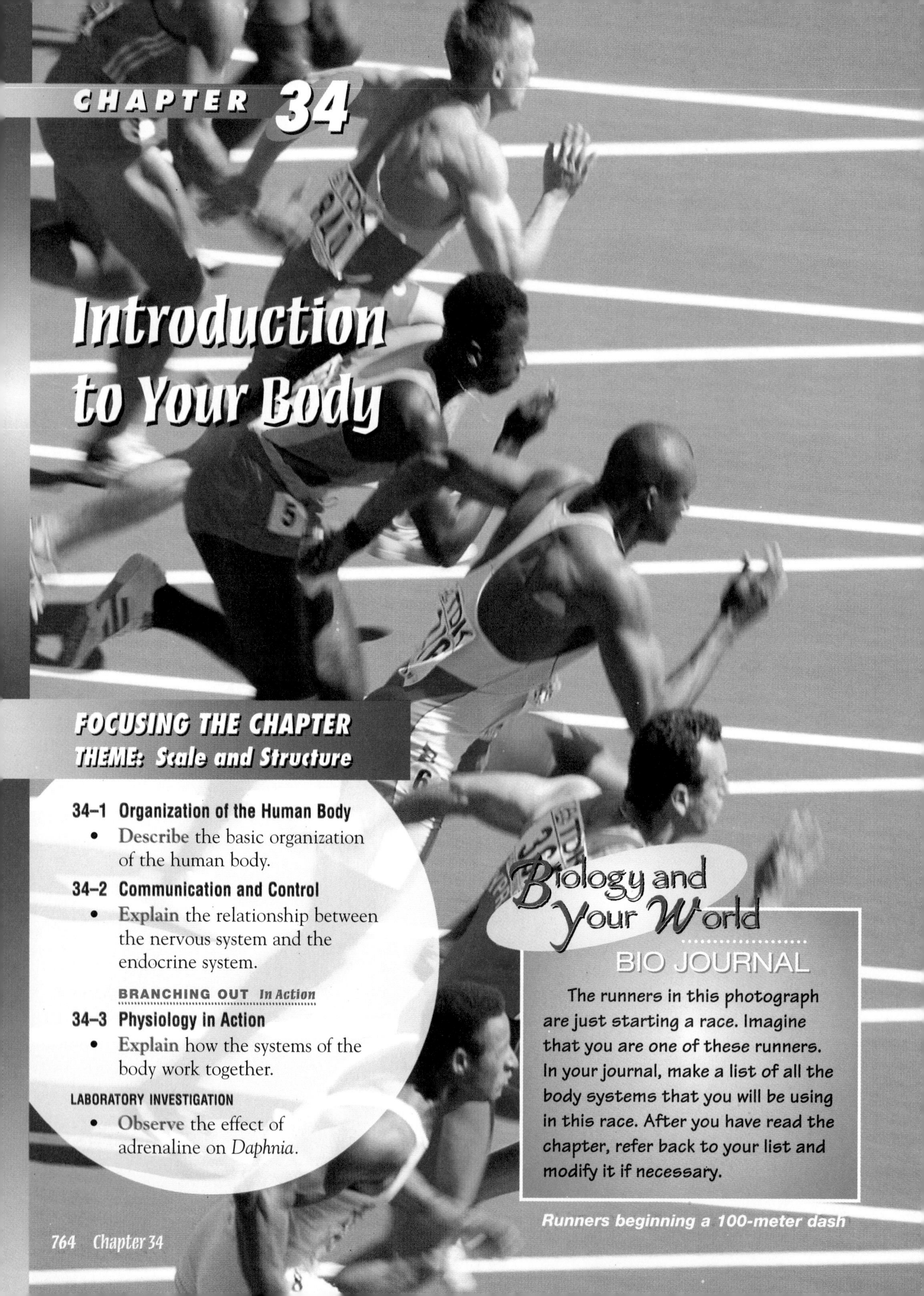

Runners beginning a 100-meter dash

Organization of the Human Body

SECTION 34–1

GUIDE FOR READING

- **Name** the four basic types of tissues.
- **List** the eleven organ systems of the body and **describe** their functions.

MINI LAB

- **Design an experiment** to observe the different types of tissue.

IT'S THE LAST INNING AND the score is tied as the batter looks over at her coach. The coach gives a series of signals, one of which is the signal to bunt. The batter's heart beats faster, because she knows that a successful bunt will score the winning run from third base. As the pitcher winds up, the batter takes a shallow breath. Her eyes follow the release of the ball, which becomes a blur as it spins toward the plate. The batter drops the bat to waist level, catches the ball squarely, and bunts it down the base line, just inside fair territory.

She drops the bat and runs toward first base. The runner from third sprints down the base line and slides easily across home plate. The winning run is scored!

100 Trillion Cells 

A spectator watching a softball game might see this winning play as a remarkable example of teamwork between batter, base runner, and coach. As impressive as it might be, the real teamwork involves a much larger number of players—the 100 trillion cells that make up the human body. Each of these cells has a purpose of its own, but each is also part of something larger. How can so many individual cells work together? How are their activities controlled and coordinated? And how is the human body organized to use the activities of these cells to handle the everyday business of life? These are some of the questions that we will try to answer in this chapter and in those that follow.

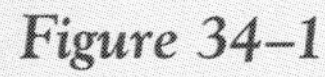

Figure 34–1
Just as the members of a softball team work together for one common goal, so too do the organ systems in this batter's body, as she prepares to hit the ball.

Figure 34–2
The human body contains eleven major organ systems. This chart lists each of these systems, along with its function.

SOME MAJOR BODY SYSTEMS

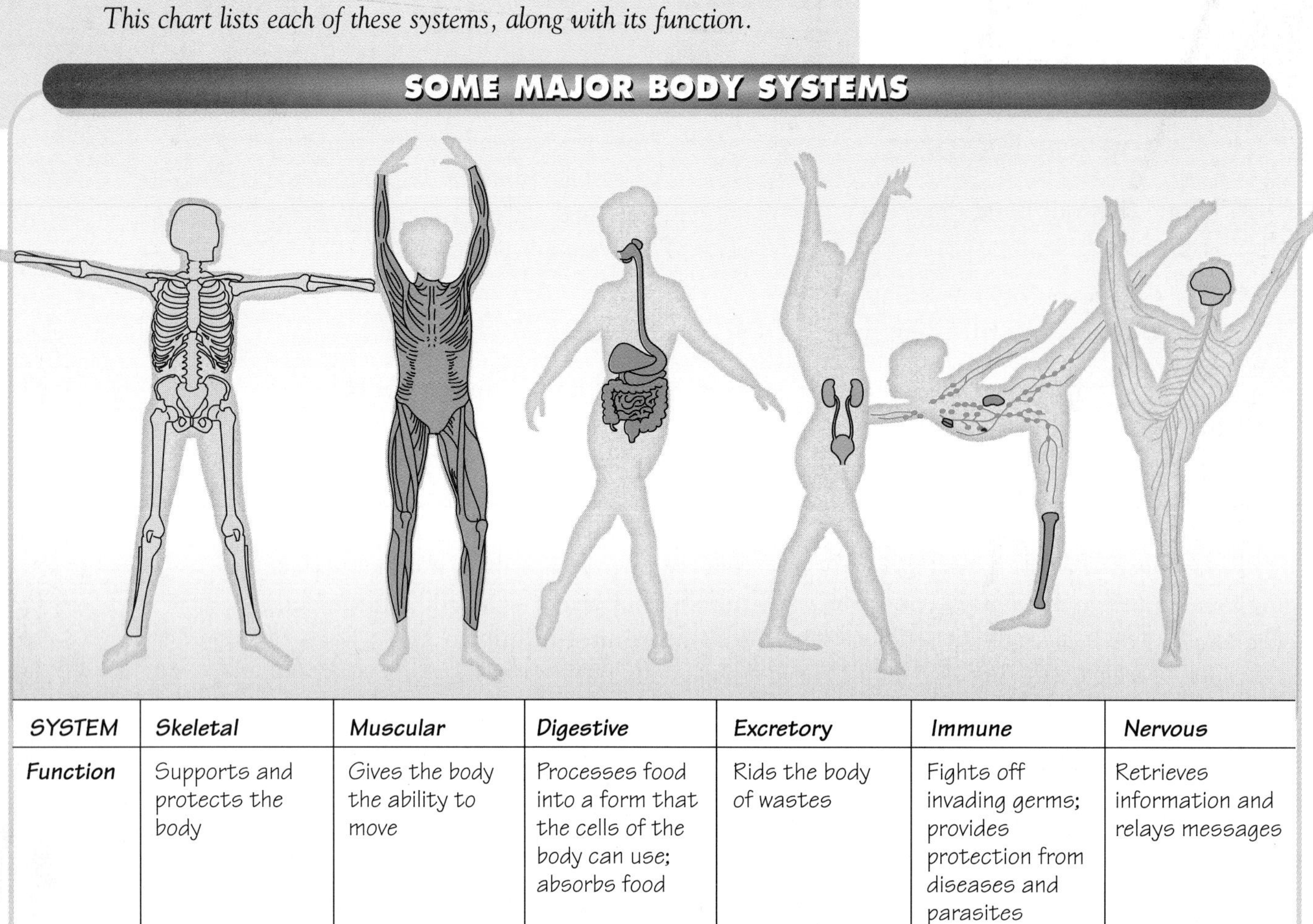

SYSTEM	Skeletal	Muscular	Digestive	Excretory	Immune	Nervous
Function	Supports and protects the body	Gives the body the ability to move	Processes food into a form that the cells of the body can use; absorbs food	Rids the body of wastes	Fights off invading germs; provides protection from diseases and parasites	Retrieves information and relays messages

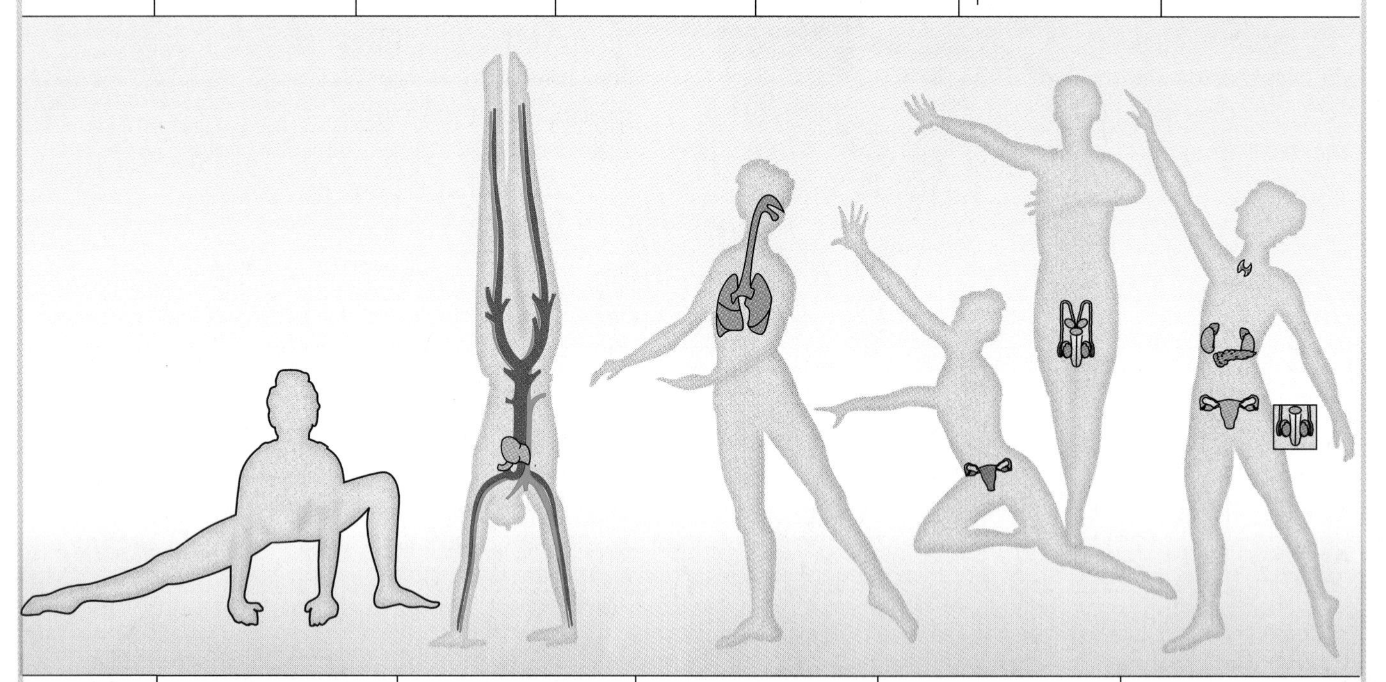

SYSTEM	Integumentary	Circulatory	Respiratory	Reproductive	Endocrine
Function	Protects the body from injury, infection, and dehydration	Brings oxygen, food, and chemical messages to cells	Brings oxygen to the body and rids the body of carbon dioxide	Produces reproductive cells; in females, nurtures and protects developing embryo	Helps to regulate and control the body's functions

Levels of Organization

In a country with millions of people, every person is an individual. However, people sometimes associate in groups. The most basic of these groups is the family, but there are larger groups as well. Communities, counties, and states are groups of people who work together to organize and choose governments.

Tissues

As in a human society, it's possible to classify the cells of the human body into groups. A group of similar cells that perform a single function is called a **tissue.** Your body contains many tissues. **There are four basic types—epithelial** (ehp-ih-THEE-lee-uhl)**, connective, nerve, and muscle tissues.** Epithelial tissue covers interior and exterior body surfaces. Connective tissue provides support for the body and connects all its parts. Nerve tissue transmits nerve impulses through the body. And muscle tissue, along with bones, enables the body to move.

Organs

A group of tissues that work together to perform a single function is called an **organ.** The eye is an organ made up of epithelial tissue, nerve tissue, muscle tissue, and connective tissue. As different as these tissues are, they all work together for a single function—sight.

MINI LAB .... Experimenting

Bones, Muscles, and Skin—Oh My!

PROBLEM *What types of tissue can you observe in a chicken leg? **Design an experiment** to answer this question.*

SUGGESTED PROCEDURE  

1. Using a hand lens, forceps, and scalpel, design an experiment to observe the different types of tissues in a cooked chicken leg with skin.
2. Formulate a hypothesis and have your teacher approve your planned experiment.
3. Carry out your experiment. Identify the different tissues you observed.

ANALYZE AND CONCLUDE

1. From your observations, what tissues are present in the chicken leg?
2. Which tissue seems to be most plentiful? Explain the reason.

Organ Systems

An **organ system** is a group of organs that perform closely related functions. For example, the eye is one of the organs of the nervous system, which gathers information about the outside world and uses it to control many of the body's functions. ***Figure 34–2*** shows the eleven organ systems in the body.

Section Review 34–1

1. **Name** the four basic types of tissues.
2. **List** the eleven organ systems in the body and **describe** their functions.
3. **Critical Thinking—Relating Facts** The eleven systems of the human body are all dependent upon one another. Explain how a virus, which affects the respiratory system, could affect the other systems of the body.
4. **MINI LAB** What different types of tissue can you observe? **Design an experiment** to find out.

SECTION

34–2 Communication and Control

Guide for Reading

- **Describe** the endocrine system.
- **Explain** how negative feedback works.

IN A HUMAN SOCIETY, THERE are many forms of communication. Every day, you communicate with your teacher and with other students by writing or speaking to them. You use this kind of communication to express your thoughts and ideas, to ask questions, and to exchange greetings.

Communication is even more important in a large society. Think of the ways in which newspapers and magazines affect your view of the world. Remember the images that you may have seen on television last night. In many ways, these mass communications bind the people of a society together with common experiences and expose them to a range of diverse ideas. The same principle applies to the cells, tissues, organs, and organ systems of the body. Unless they communicate, they cannot act together.

The Nervous System

Similar to a telephone network that reaches every home and office in a large city, the body contains a cellular network that carries messages from cell to cell. That network is the **nervous system.** The message-carrying cells of the nervous system are called **neurons.** Neurons can relay signals from one end of a cell to the other. They can also pass these impulses from cell to cell.

You can think of neurons as tiny wires in a vast communications network. They gather information from every region of the body and relay it to central locations. Here the information can be analyzed, then instructions and commands can be carried to organs throughout the body.

The center of this network is the brain, where impulses arrive from every part of the body. In the brain they are analyzed and compared, past information is stored and retrieved, and appropriate responses are sent back out through the network.

Figure 34–3
Communication is an important means of exchanging information. In the body, the nervous system and endocrine system are responsible for communication. (a) Talking with your friends, (b) speaking on the telephone, and (c) reading newspapers are three common ways in which information is passed from person to person.

The Endocrine System

If you wanted to send a message to just one or two people, you could call them on the telephone. But if you had to reach thousands of people at the same time, you might broadcast your message on radio or television.

It might not have radio or television, but your body does have a chemical "broadcasting" system that can send messages to millions of cells at the same time. This broadcasting system is the **endocrine system.** ***Figure 34–4*** illustrates the endocrine system. **The endocrine system is made up of a series of glands located throughout the body. Glands are organs that produce and release chemicals, and endocrine glands generally release their chemicals into the bloodstream.**

Hormones

Unlike the organs of the nervous system, the endocrine glands are not connected to one another. The endocrine glands send signals between each other and to other cells in the body.

These signals are produced by the endocrine glands in the form of **hormones**—the chemicals that travel through the bloodstream and affect the behavior of other cells. Only those cells that have **receptors**—specific chemical binding sites—for a particular hormone can respond to it. Cells that have receptors for a particular hormone are called **target cells.** Hormone receptors are similar to radios tuned to just one station. A cell that does not have receptors for a hormone cannot respond to it, just as you would miss an announcement on a jazz station if you were listening to a rock station.

☑ **Checkpoint** What are hormones?

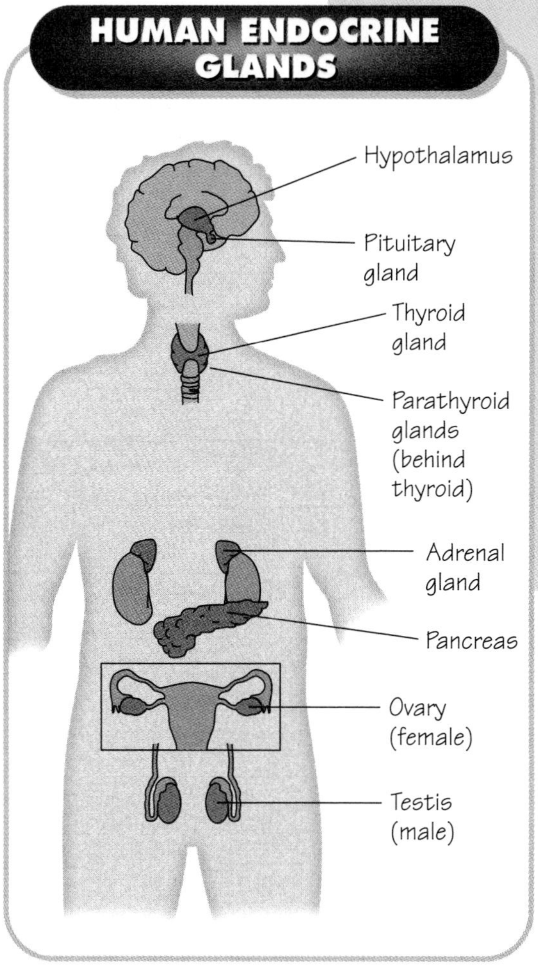

HUMAN ENDOCRINE GLANDS

Hypothalamus
Pituitary gland
Thyroid gland
Parathyroid glands (behind thyroid)
Adrenal gland
Pancreas
Ovary (female)
Testis (male)

Figure 34–4 *The endocrine system consists of glands located throughout the body. Notice that in females the sex glands are called ovaries and in males they are called testes.*

Pituitary Gland

A network of broadcasting stations usually has a place where the activities of individual stations are coordinated—and the endocrine system is no exception. Its headquarters is a tiny structure at the base of the brain known as the

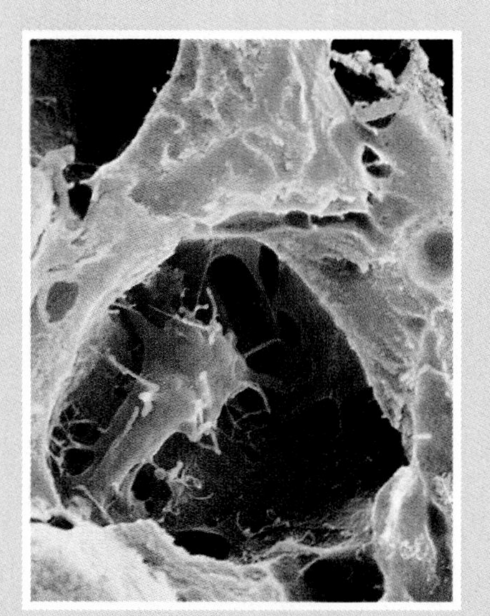

Figure 34–5 *The pituitary gland, located in the brain, is responsible for secreting nine different hormones. In this scanning electron micrograph, the pituitary gland is colored a pale red-brown. The yellow objects in the center and top of the gland are large white blood cells (magnification: 2100X).*

HORMONES AND THEIR ACTIONS

Gland	Hormone	Action
Thyroid	Thyroxine	Increases metabolic rate and body temperature; regulates growth and development
	Calcitonin	Inhibits release of calcium from bone
Parathyroid	Parathyroid hormone (PTH)	Stimulates release of calcium from bone
Pituitary	Antidiuretic hormone (ADH)	Stimulates reabsorption of water
	Oxytocin	Stimulates uterine contractions and release of milk
	Follicle-stimulating hormone (FSH)	Stimulates follicle maturation in females and sperm production in males
	Luteinizing hormone (LH)	Stimulates ovulation and growth of corpus luteum in females and testosterone secretion in males
	Thyroid-stimulating hormone (TSH)	Stimulates thyroid to release thyroxine
	Adrenocorticotropic hormone (ACTH)	Stimulates adrenal cortex to release hormones
	Growth hormone (GH) or somatropin	Stimulates growth; synthesizes protein; inhibits glucose oxidation
	Prolactin	Stimulates milk production
Adrenal	Aldosterone	Controls salt and water balance
	Cortisol, other corticosteroids	Regulate carbohydrate, protein, and fat metabolism
	Adrenaline and noradrenaline	Initiate the body's response to stress; increase blood glucose level; dilate blood vessels; increase rate and strength of heartbeat; increase metabolic rate
Pancreas	Glucagon	Stimulates conversion of glycogen to glucose, raising the blood glucose level
	Insulin	Stimulates conversion of glucose to glycogen, lowering the blood glucose level
Ovary	Estrogen	Develops and maintains female sex characteristics; initiates buildup of uterine lining
	Progesterone	Promotes continued growth of uterine lining and formation of placenta
Testis	Testosterone	Develops and maintains male sex characteristics; stimulates sperm development

Figure 34–6
Each gland of the endocrine system releases a different hormone. This chart lists the endocrine gland, the hormone or hormones it produces, as well as the action of each hormone.

pituitary gland. The hormones produced by the pituitary gland are important because many of them regulate the other endocrine glands.

Hypothalamus

What controls the pituitary gland? The pituitary gland is attached to a region of the brain known as the **hypothalamus** (high-poh-THAL-uh-muhs). Either directly or indirectly, the hypothalamus controls the release of hormones from the pituitary gland.

Some hormones released from the pituitary gland are actually made in the hypothalamus. Others are made in the pituitary gland, then released into the bloodstream only when chemical signals are received from the hypothalamus.

The link between the hypothalamus and the pituitary gland is important for two reasons. First, it explains how the pituitary gland works, controlling the activities of other endocrine glands in response to signals that it receives from the hypothalamus. Second, it shows that the nervous system and the endocrine system are interconnected. That means that input into the nervous system can influence the endocrine system by way of the connection between the hypothalamus and the pituitary gland.

☑ **Checkpoint** Where in the body is the hypothalamus found?

Figure 34–7
CAREER TRACK
A biomedical equipment technician is responsible for the maintenance and repair of medical equipment. In this photograph, a technician is running an annual validation test for temperature and temperature uniformity on a sterilizing oven in a pharmaceutical research facility.

Endocrine System Control

To better understand how the endocrine system works, let's take a look at one of the body's most important hormones, thyroxine. Thyroxine is made by the **thyroid gland.** To make thyroxine, the thyroid gland needs the amino acid tyrosine and a small amount of iodine, which must be obtained from food.

Thyroxine affects nearly all the body's cells, increasing their metabolic rate—the rate at which they use food and oxygen. It also increases the rate at which cells grow.

Too much thyroxine in the bloodstream results in increased blood pressure, nervousness, increased pulse rate, and dangerous weight loss. Too little thyroxine results in a lowered pulse rate, lowered blood pressure, excessive sleepiness, and weight gain. How does the thyroid gland usually manage to get the level of this important hormone just right?

Control in the Hypothalamus

Like most cells in the body, the cells of the hypothalamus also have receptors for thyroxine. When levels of thyroxine drop, most cells slow down their activity, but some cells of the hypothalamus increase their activity. These cells now produce more of a thyroid-releasing hormone (TRH).

Remember how close the hypothalamus is to the pituitary? TRH travels from the hypothalamus to the pituitary gland through a network of tiny blood vessels. It then causes cells in the pituitary gland to release increased amounts of another hormone, thyroid-stimulating hormone (TSH), into the bloodstream. The target cells for TSH are the cells of the thyroid gland itself. TSH causes the thyroid gland to release more thyroxine.

Feedback Regulation

As you can see, the combination of the hypothalamus, pituitary gland, and thyroid gland automatically regulates the level of thyroxine in the bloodstream. When there is too little thyroxine in the blood, the thyroid gland is stimulated by TSH to make more. When there is too much thyroxine, TSH is not released, and the thyroid gland makes less. You could say that the activity of the thyroid gland "feeds back" to the hypothalamus. When the thyroid gland is working too fast, its own product—thyroxine—causes the hypothalamus to signal the thyroid gland to slow down. When it is working too slow, low levels of thyroxine cause the hypothalamus to signal the thyroid gland to speed up.

The relationship between the thyroid gland and the hypothalamus is an example of **negative feedback.** In humans, the release of hormones is regulated through negative feedback. **Through negative feedback, the secretion of a hormone inhibits further production of another hormone.** Negative feedback enables the conditions within the body to remain relatively constant over time.

Biology AND YOU .... Connections

The Salt Connection

As you have just read, thyroxine is a hormone produced by the thyroid gland and needed by the body for its cells to function properly. Thyroxine not only stimulates nerve cell growth, it is also important for brain development.

In order to produce thyroxine, the thyroid gland requires a small amount of the mineral iodine. Like all minerals, the iodine that is needed to make thyroxine must be obtained from the foods we eat. Because iodine is rarely found in soil, most foods contain very little. Fortunately, seafood is rich in iodine, and eating a small amount of seafood every now and then provides the iodine your body needs.

Iodine Deficiency

What happens if your diet does not contain the small amount of iodine your body needs? The thyroid gland attempts to compensate for the lack of iodine. It does this by increasing in size so that it can absorb as many atoms of iodine as possible. This increase in size produces a noticeable swelling of the thyroid gland, a condition called goiter.

Iodized salt

At one time, goiter was common in parts of the midwestern United States, where seafood was hard to get. Today, however, goiter is almost unheard of in the United States. Why have incidences of goiter decreased? Because in many countries, table salt is now iodized. Trace amounts of iodine are added to much of the table salt in the United States and other industrial nations. A few sprinkles of iodized salt can supply enough iodine for the body to make all the thyroxine it needs.

Severe Problems

For children, the consequences of iodine deficiency are much more severe than for adults. Because the developing brain requires thyroxine to stimulate nerve cell growth, lack of thyroxine can result in mental retardation.

Unfortunately, some parts of the world have been slow to initiate production of iodized salt to supplement the diets of people who live inland and do not eat seafood. One such place is China, and that country is experiencing a major health crisis because the diets of many Chinese people do not contain enough iodine. The Public Health Ministry of China estimates that as many as 10 million Chinese now suffer mental retardation because of iodine deficiencies during childhood.

Recently, public health authorities in China have realized the severity of the health crisis in their country. They're now rushing to ensure that iodized salt becomes available in all parts of China as soon as possible.

Making the Connection

Can you think of places other than inland China where people might not be able to get the iodine they need? Iodized salt is more expensive in China than regular salt. What solutions would you offer to solve the problem of cost versus health?

HORMONE ACTION

Polypeptide hormone
Receptor
Target cell membrane
Second messenger
ATP
cAMP
Enzyme activities
Nucleus
Altered cellular function

Polypeptide and Amino Acid Hormones

Steroid hormone
Target cell membrane
Receptor
Hormone-receptor complex
Altered cellular function
Nucleus
DNA
mRNA
Protein synthesis

Steroid Hormones

Figure 34–8
Polypeptide and amino acid hormones do not enter the cell, but bind to receptors on the cell membrane where they activate second messengers, such as cAMP. Steroid hormones enter the cell and bind to receptors inside it.

How Do Hormones Work?

Hormones come in many different shapes and sizes. Some are proteins, others are lipids, and still others are chemically modified amino acids. How do hormones affect their target cells?

There are two basic patterns of hormone action. Polypeptide and amino acid hormones bind to receptors on the cell surface. As *Figure 34–8* shows, when a hormone binds to a receptor on the cell surface, it activates enzymes on the inner side of the cell membrane to produce a large number of second messenger molecules. These second messengers may be ions such as Ca^{2+} or small molecules such as cyclic adenosine monophosphate, cAMP. Once released, these second messengers can activate or inhibit a wide range of other cell activities.

Unlike the polypeptide hormones, steroid hormones easily pass across the cell membrane. Steroid hormones are lipids produced from cholesterol. Target cells for these hormones have protein receptors that tightly bind to a specific steroid hormone. This hormone-receptor complex, as it is called, enters the nucleus, where it binds to a specific DNA sequence. Hormone receptor complexes work as regulators of gene expression—they can turn on or turn off whole sets of genes. Steroid hormones, which directly affect gene expression, can produce dramatic changes in cellular function.

INTEGRATING CHEMISTRY

What are polypeptides? What are amino acids? Use a chemistry textbook to find out.

Section Review 34–2

1. **Describe** the function of the endocrine system.
2. **Explain** how negative feedback works.
3. **Critical Thinking—Comparing** How do polypeptide and steroid hormones compare in terms of hormone action?

Physiology in Action

Guide for Reading

- Define human physiology.

MINI LAB
- Construct a model of negative feedback in the endocrine system.

AS COMPLICATED AS IT IS, IT sometimes seems remarkable to think that all the cells of the human body are capable of working together for a common purpose. How does this happen? To take a case in point, how did the body of the ballplayer we described at the beginning of the chapter make it possible for her to drive in the winning run?

Human Physiology

When you think about the variety of things your body is capable of doing—including the simple things you take for granted—you realize how incredible the human body is. **The study of how the body works is called human physiology** (fihz-ee-AHL-uh-jee). Let's use the example of the softball player to see how the systems of the body work together.

Getting Ready

Two days ago, this player missed practice because she had chills and a fever. Today she feels better because her immune system fought off a cold virus.

At lunch time, she ate two slices of pizza, a salad, a glass of milk, and an apple. Her digestive system quickly broke down the complex molecules in this food so they could be absorbed into the blood of the circulatory system. The sugars from that meal were carried throughout the body, providing her with energy.

Preparing to Play

As she waited in the on-deck circle, the pressure was on her to bring home the third-base runner. Her nervous

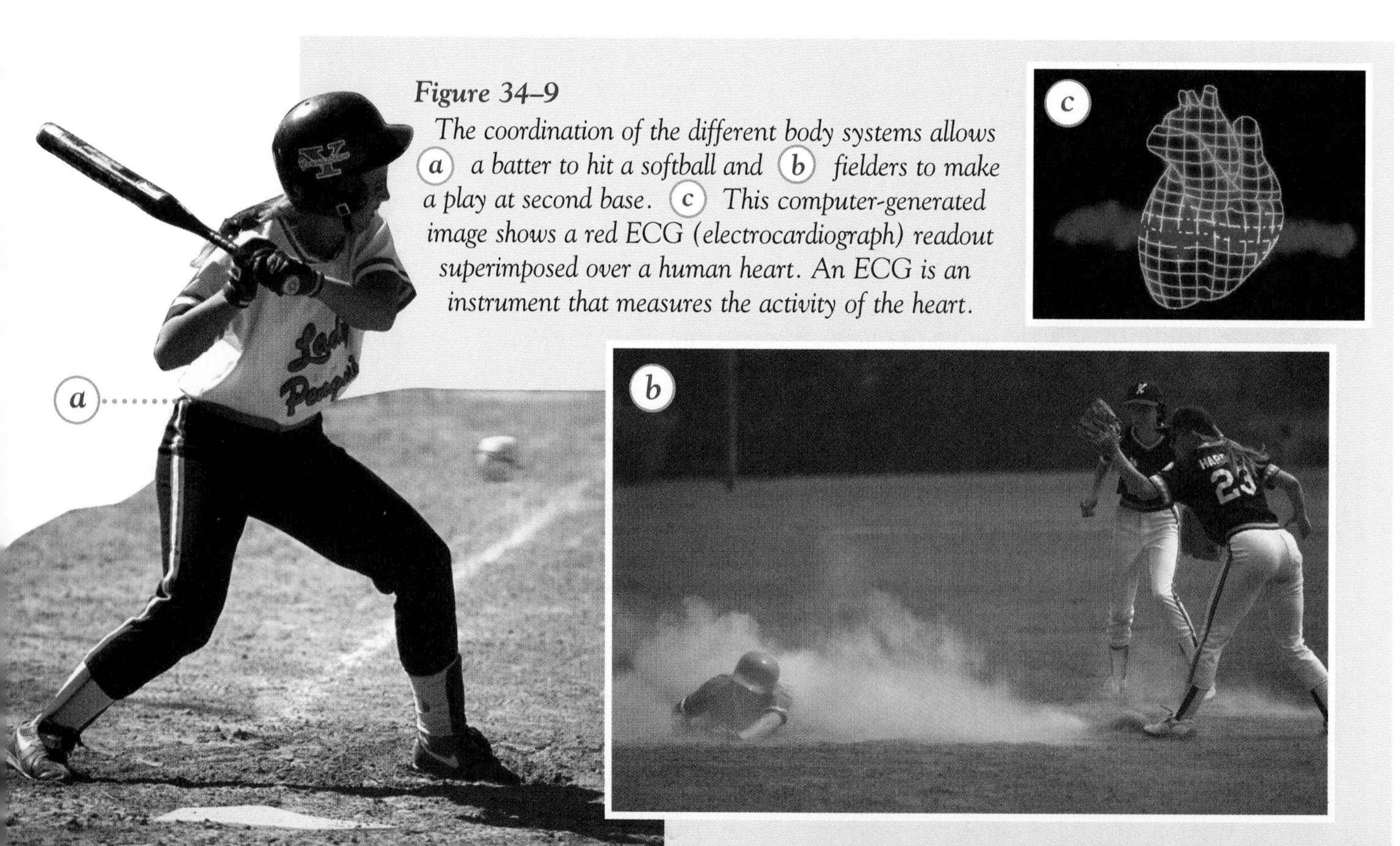

Figure 34–9 The coordination of the different body systems allows (a) a batter to hit a softball and (b) fielders to make a play at second base. (c) This computer-generated image shows a red ECG (electrocardiograph) readout superimposed over a human heart. An ECG is an instrument that measures the activity of the heart.

system buzzed with anxiety, and that message was passed to the endocrine system. Nerve impulses stimulated the adrenal cortex to release adrenaline into the bloodstream. Within seconds, her heart rate and blood pressure rose, and the blood supply to her muscles increased.

With a few deep breaths, her respiratory system cleared the carbon dioxide from her lungs and filled her blood with the oxygen it needed. As her coach flashed a series of signals, her nervous system searched her stored memory of pre-game plans to recall the signs.

Time for Action

As the pitcher threw the ball to the plate, the batter's eyes watched the ball. Her nervous system took less than three tenths of a second to realize that the ball was headed for the strike zone. Instantly, nerves carried the message to key muscles in her arms and legs.

After she bunted the ball, her nervous system issued a new set of commands to the muscular system. Now muscles in her legs contracted in a pattern that placed tremendous force on the long bones of the skeletal system, enabling her to run.

As she continued to run, her respiratory system provided more oxygen to the bloodstream, and her muscles quickly used the energy stored in them. Finally, as she crossed the base safely, she relaxed.

Dropping the Ball

PROBLEM *How can you **construct a model** of negative feedback in the endocrine system?*

PROCEDURE

1. Working with three other students, stand in a line. The first person in line will be the hypothalamus. The second person will be the pituitary gland, the third person will be the thyroid gland, and the fourth person will be the body.
2. Give the *hypothalamus* a red ball labeled TRH, the *pituitary gland* a green ball labeled TSH, and the *thyroid gland* 3 yellow balls labeled thyroxine, to be held in one hand.
3. At a signal from the *body,* the *hypothalamus* passes the TRH ball to the *pituitary gland.* The *pituitary gland* passes the TSH ball to the *thyroid gland.* The *thyroid gland* passes one thyroxine ball to the right hand of the *body* and a second thyroxine ball to the left hand of the *body.*
4. The *body* is now saturated with thyroxine. In order to accept the last thyroxine, the *body* must pass one thyroxine to the *hypothalamus.* The *body* is now able to accept the third thyroxine.
5. When the *hypothalamus* receives the thyroxine, the *thyroid gland* should drop the TSH ball.

ANALYZE AND CONCLUDE

1. What is the function of TRH? Of TSH?
2. Why did the *body* pass the thyroxine to the *hypothalamus?* What effect did this have?
3. What would happen if the *body* kept accepting thyroxine?

Section Review 34-3

1. **Define** human physiology.
2. **MINI LAB** How did you **construct a model** of negative feedback in the endocrine system?
3. **BRANCHING OUT ACTIVITY** Imagine that you are about to take a final exam. In a one-page essay, **summarize** how the systems of your body will work together to prepare you for this task and to carry it through.

Laboratory Investigation

DESIGNING AN EXPERIMENT

Daphnia and Adrenaline

Although hormones are very small molecules that are released in minute amounts, they cause very noticeable effects in an organism. Adrenaline is a hormone that prepares the body to deal with stress. It increases the heart rate and the metabolic rate. In this investigation, you will observe the heart of a small crustacean called *Daphnia*, or water flea, and you will determine the effects adrenaline has on its heart.

Problem

What can you observe about the effect adrenaline has on a *Daphnia*? **Design an experiment** to answer this question.

Suggested Materials

Daphnia culture
0.01% adrenaline solution
depression slides
medicine droppers
microscope

DAPHNIA

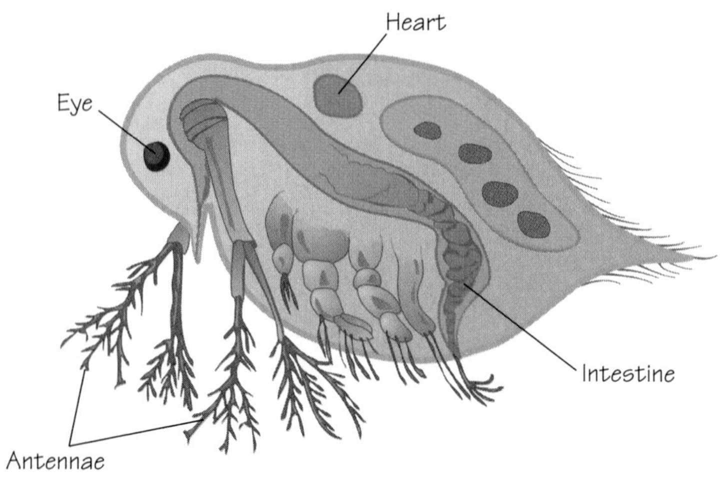

Suggested Procedure

1. Use the medicine dropper to transfer a single *Daphnia* from the culture to the center of the depression slide. The *Daphnia* will look like a small white dot.
2. Place the slide under the low-power objective of a microscope and observe the *Daphnia*. Locate the heart.
3. Count the number of times the heart beats in one minute. Record this information in a data table similar to Data Table 1.
4. Repeat step 3 two more times. Then average the three measurements. Record your data. Return the *Daphnia* to the culture dish.
5. Using a procedure similar to the one given in steps 1 to 4, determine the effect a 0.01% solution of adrenaline has on the *Daphnia*'s heart.

DATA TABLE 1

Normal Heart Rate	
Count 1	
Count 2	
Count 3	
Average	

DATA TABLE 2

Heart Rate With Adrenaline

Time (minutes)	Heart Rate
1	
3	
5	
7	
9	

6. **Formulate a hypothesis. Make sure that you have your teacher's approval for the experiment.**
7. **Be sure to return the *Daphnia*'s heart beat to its normal rate. To do so, remove some of the adrenaline solution from the slide with a medicine dropper. Using a clean medicine dropper, replace the volume of the liquid with water from the culture. Continue observing and counting the heart rate every other minute until the heart rate returns to normal.**
8. **Carry out your experiment and record your data in a data table similar to Data Table 2.**

Observations

1. Share your observations and measurements with the rest of the class. As a class, find the average number of heartbeats for the *Daphnia* before the adrenaline solution was added and just after it was added.
2. What was the effect on the heart rate when you added the adrenaline solution?
3. How long did it take the *Daphnia* to return to a normal heart rate?

Analysis and Conclusions

1. Formulate a hypothesis to explain the results of your experiment. Could a different hypothesis also explain the results? Discuss this possibility.
2. How do you think the *Daphnia* adjusted its heart rate to the adrenaline solution?
3. Why might it be a good idea to take an average of the class data in this experiment?
4. Why do you think that *Daphnia* are useful organisms in an experiment such as this? Explain your answer.

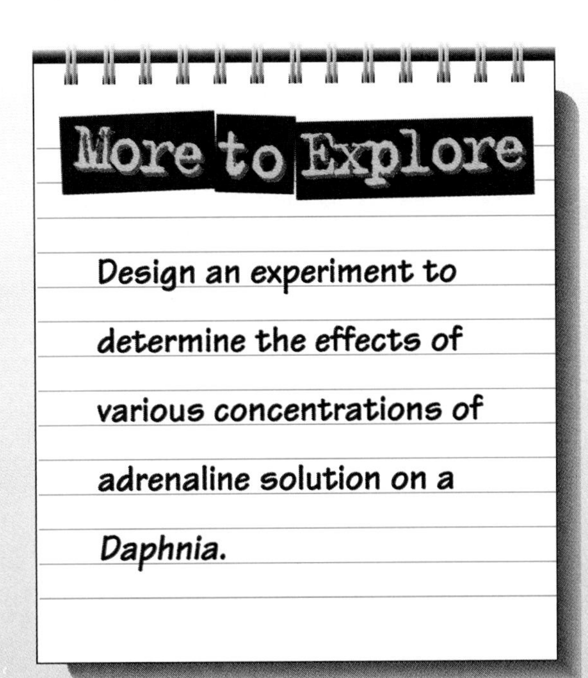

Study Guide

Summarizing Key Concepts

The key concepts in each section of this chapter are listed below to help you review the chapter content. Make sure you understand each concept and its relationship to other concepts and to the theme of this chapter.

34–1 Organization of the Human Body

- There are four basic types of tissue—epithelial, connective, nerve, and muscle.
- The eleven systems of the body include the skeletal system, muscular system, digestive system, excretory system, immune system, nervous system, integumentary system, circulatory system, respiratory system, reproductive system, and endocrine system.

34–2 Communication and Control

- The endocrine system is made up of a series of glands located throughout the body. Glands are organs that produce and release chemicals, and endocrine glands generally release their chemicals into the bloodstream.
- Hormones are chemicals that travel through the bloodstream and affect the behavior of other cells.
- During negative feedback, the secretion of a hormone inhibits further production of another hormone.

34–3 Physiology in Action

- The study of how the body works is called human physiology.

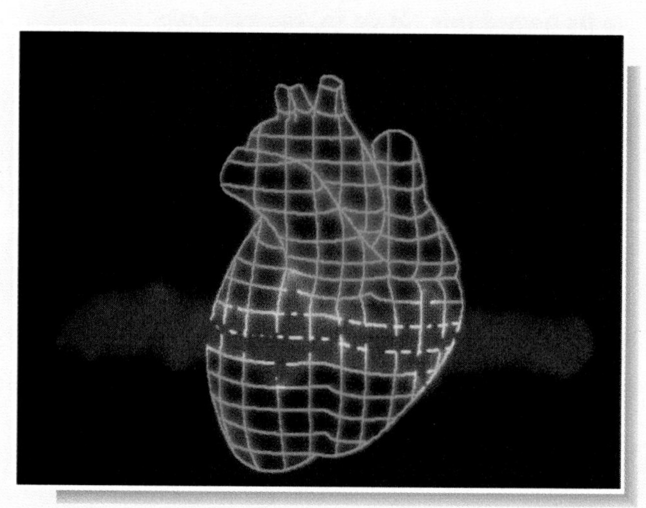

Reviewing Key Terms

Review the following vocabulary terms and their meaning. Then use each term in a complete sentence.

34–1 Organization of the Human Body

tissue
organ
organ system

34–2 Communication and Control

nervous system
neuron
endocrine system
hormone
receptor
target cell
pituitary gland
hypothalamus
thyroid gland
negative feedback

Recalling Main Ideas

Choose the letter of the answer that best completes the statement or answers the question.

1. A group of similar cells that perform a similar function is called a(an)

   a. organism. **c.** tissue.
   b. organ. **d.** organ system.

2. Which type of tissue covers interior and exterior surfaces?

   a. epithelial **c.** connective
   b. muscle **d.** nerve

3. The body system that relays messages from one part of the body to another is the

   a. immune system. **c.** excretory system.
   b. nervous system. **d.** digestive system.

4. Organs that produce and release substances are called

   a. hormones. **c.** glands.
   b. receptors. **d.** target cells.

5. What controls the release of hormones from the pituitary gland?

   a. hypothalamus **c.** thyroid
   b. adrenal glands **d.** pancreas

6. Which endocrine gland produces estrogen?

   a. testes **c.** adrenal gland
   b. ovaries **d.** pituitary gland

7. Which system is responsible for clearing the blood of carbon dioxide and replacing it with oxygen?

   a. circulatory **c.** excretory
   b. digestive **d.** respiratory

Putting It All Together

Using the information on pages xxx to xxxi, complete the following concept map.

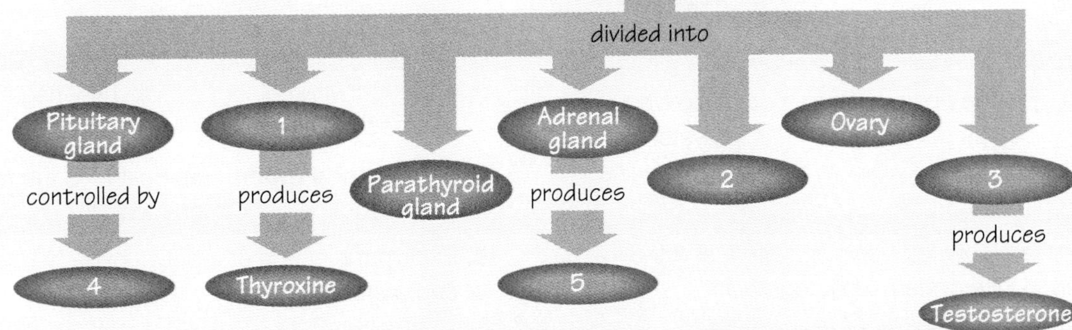

Assessment

Reviewing What You Learned

Answer each of the following in a complete sentence.

1. Explain the function of muscle tissue.
2. What is a tissue?
3. What is an organ system?
4. Explain the function of the excretory system.
5. What are neurons?
6. How does the endocrine system work?
7. What are glands?
8. What are hormones? In which system are they produced?
9. How does the integumentary system work?
10. What are receptors?
11. What are target cells?
12. Where is the pituitary gland located?
13. Where is the hypothalamus located?
14. Which gland produces thyroxine?
15. What kinds of chemicals are hormones usually made of?
16. Explain the purpose of the second messenger hormone.
17. Why is negative feedback important?

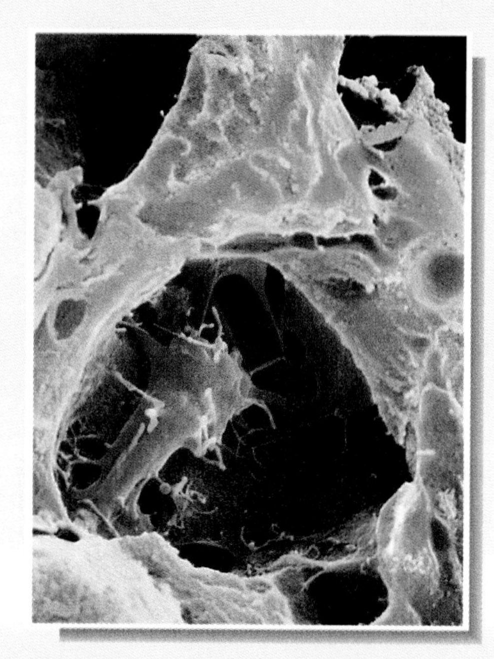

Expanding the Concepts

Discuss each of the following in a brief paragraph.

1. What is the function of each of the four types of tissue?
2. Compare tissues with organs.
3. Can any of the eleven body systems work in isolation of another? Explain your answer.
4. Why is the nervous system often referred to as the coarse control of the human system while the endocrine system is considered the fine control?
5. Explain how the nervous system and the endocrine system work together.
6. Explain how glucagon and insulin have opposite effects on the body.
7. What is the relationship between the pituitary gland and the hypothalamus?
8. Using the hypothalamus, the pituitary gland, and the thyroid gland, explain how negative feedback works.
9. **Construct a model** that illustrates the two basic patterns of hormone action.
10. Explain the action of a polypeptide hormone and a steroid hormone.

Extending Your Thinking

Use the skills you have developed in this chapter to answer the following.

1. **Analyzing** Explain how the negative-feedback mechanism works in a way similar to a thermostat that maintains the temperature of a room.
2. **Making judgments** With the advent of genetic engineering, many techniques have led to bacterially produced hormones. One of these hormones—human growth hormone (GH)—is now readily available. Consider the advantages and disadvantages of administering this hormone over an extended period of time to adjust the size of a patient.
3. **Designing an experiment** A person driving a car is maintaining a constant speed at the speed limit. How could you design an experiment to illustrate how this is an example of negative feedback?
4. **Hypothesizing** You have probably heard stories of people who had incredible strength when placed in an extremely dangerous situation. How might a person be able to summon such strength for a short period of time?
5. **Making judgments** Anabolic steroids are synthetic drugs produced from the hormone testosterone. Although anabolic steroids are used for medical purposes to reduce swelling and promote healing, they are sometimes misused to develop muscle mass and muscle strength. Whatever the case, the long-term effects of steroid use are serious and dangerous. Should anabolic steroids be reclassified as an illegal substance? Why or why not?

Applying Your Skills

The Squeeze Play

In order to do things effectively, the systems of the body must all work together. As a class, conduct the following activity, which involves reaction time.

1. In groups of 8 to 10 students, stand in line holding hands. Identify the first person in line as the timekeeper. He or she will need a stopwatch or clock with a second hand. All students but the timekeeper should close their eyes.
2. When everyone in the group is ready, the timekeeper begins timing and, at the same time, squeezes his or her neighbor's hand. As soon as the neighbor feels the squeeze, he or she squeezes the hand of the next person. Each person does this in turn. When the last person feels the squeeze, he or she yells "stop." This is the signal to stop timing and note the time. Record your group's reaction time.
3. Repeat steps 1 and 2 two more times, recording your data each time. Calculate a class average for reaction time.

• GOING FURTHER •

4. Do you think you react faster earlier in the day than you do later in the day?
5. Will practice doing this activity decrease your reaction time? Explain your answer.

CHAPTER 35

Nervous System

FOCUSING THE CHAPTER
THEME: Stability

35–1 The Human Nervous System
- **Describe** the basic structures of the nervous system.

35–2 Organization of the Nervous System
- **Explain** how the nervous system is organized.

35–3 The Senses
- **Describe** each of the five senses.

BRANCHING OUT *In Depth*

35–4 Nerve Impulses and Drugs
- **Explain** the effects various drugs have on nerve impulses.

LABORATORY INVESTIGATION
- **Identify** the parts of a sheep brain.

Biology and Your World

BIO JOURNAL

The gymnast in this photograph relies on her nervous system for balance and coordination. Pick a favorite sport or activity. In your journal, explain how you rely on your nervous system to perform that sport or activity. Remember to include your five senses as well.

Gymnast on balance beam

The Human Nervous System

SECTION
35–1

Guide for Reading

- **Name** the three parts of a neuron.
- **Describe** the way in which a nerve impulse begins.

A COMPUTER CAN DELIVER A *page of text, graphics, and even a video in a few seconds. However, there is another "machine" less than a meter away from the computer that can perform equally astonishing feats. In a fraction of a second, this machine can take in everything on the computer screen; scan the area around the computer; monitor the sound, light, and heat in the room; and automatically regulate the activities of hundreds of devices running at the same time. What kind of information-processing system can do all this? The answer is inside you—the human nervous system.*

Neurons

To understand a computer, you start with the basics—with electricity, wires, and switches. Gradually, you see how thousands of individual components are wired together to form a functional unit. We can do almost the same thing with the nervous system. The basic units of the nervous system are cells called **neurons.** These cells carry messages in the form of electrical signals known as **impulses.**

Some features of a typical neuron are shown in ***Figure 35–2*** on the next page. **A neuron is made up of a cell body, dendrites, and an axon.** The largest part of the neuron is its **cell body.** Most of the metabolic activity of the cell takes place there. In addition, the cell body collects information from the **dendrites**—small branched extensions that spread out from the cell body. Dendrites carry impulses toward the cell body. And the long

Figure 35–1
(a) Just as a computer is able to process large amounts of information in a short time, so too can the human nervous system. (b) Neurons (magnification: 4600X) act as (c) a computer circuit board to carry electrical signals through the body.

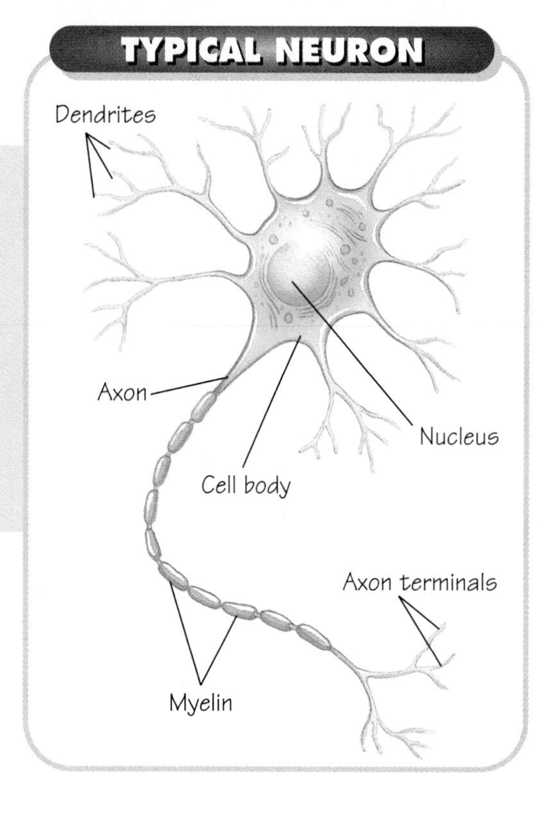

Figure 35–2 *The main structures of a typical neuron include the dendrites, the cell body, and the axon.*

branch that carries impulses away from the cell body is called an **axon.** A neuron may have many dendrites, but it usually has only one axon.

In most animals, neurons are clustered into bundles of fibers called **nerves.** Some nerves contain only a few neurons, but many others have hundreds or even thousands of neurons.

There are three general types of neurons, distinguished by the directions in which they carry impulses. The **sensory neurons** carry impulses from the sense organs to the brain and the spinal cord. **Motor neurons** carry impulses in the opposite direction—from the brain or spinal cord to muscles or other organs. **Interneurons** connect sensory and motor neurons, and carry impulses between them.

☑ **Checkpoint** What are the three types of neurons?

The Nerve Impulse

A neuron not carrying an impulse is said to be at rest. The resting neuron has an electrical potential—a charge difference—across its cell membrane. The inside of the cell is negatively charged and the outside is positively charged.

This difference in electrical charges is called the **resting potential.** Although negative and positive ions are found on both sides of the membrane, there is a net excess of negative charges on the inside of the membrane, and that's what produces the resting potential.

An impulse begins when a neuron is stimulated by another neuron or by the environment. Once it begins, the impulse travels rapidly down the axon away from the cell body. As ***Fig. 35–3*** shows, the impulse is a sudden reversal of the membrane potential. For a few milliseconds, positive ions rush across the cell membrane, reversing the charge difference. For that brief instant, the inside of the membrane is more positive than the outside. In less than 10 milliseconds, the potential reverses itself again, and the resting potential is restored.

This rapid change in voltage on the inside of the axon—negative to positive and back to negative—is called an **action potential.** A nerve impulse is an action potential traveling down an axon.

How does the action potential move? Imagine a row of dominoes. When one domino falls, it causes the next one to fall, causing the next to fall, and so on. That's almost what happens in a neuron. Like a domino toppling, the flow of positive charges into one region of the axon causes the membrane just ahead of it to open up and let positive charges flow across the membrane there, too. This happens again and again, until the impulse moves along the length of the axon. In a typical axon, the impulse can move as quickly as one meter per second. Once the impulse passes, the resting potential is restored, and the neuron is ready to conduct another impulse.

☑ **Checkpoint** What is an action potential?

Myelin

As you know, most electrical wires are insulated—that is, they are covered with rubber or plastic to prevent a short circuit. The nervous system has a kind of insulation, too. In some nerve cells, Schwann cells surround the axons of certain neurons. As Schwann cells grow around an axon, they wrap it in layers of their own cell membrane, forming a material known as **myelin** (MIGH-uh-lihn). The Schwann cells that surround a single long axon leave many gaps—called nodes—between themselves where the axon membrane is exposed.

When an impulse moves down an axon covered with myelin, the action potential jumps from one node to the next. This happens because electrical current flows from one node to the next, which greatly speeds up the rate at which the impulse moves. A large axon with myelin can carry messages at speeds as great as 200 meters per second!

Checkpoint What are Schwann cells?

NERVE IMPULSE

Resting Potential

Action Potential

Impulse travels along axon

Figure 35–3 At rest, the outside of the neuron's membrane is more positively charged than is the inside. If a stimulus is applied, causing an impulse, the electrical charges become reversed. The reversal of electrical charges continues as the action potential moves down the axon.

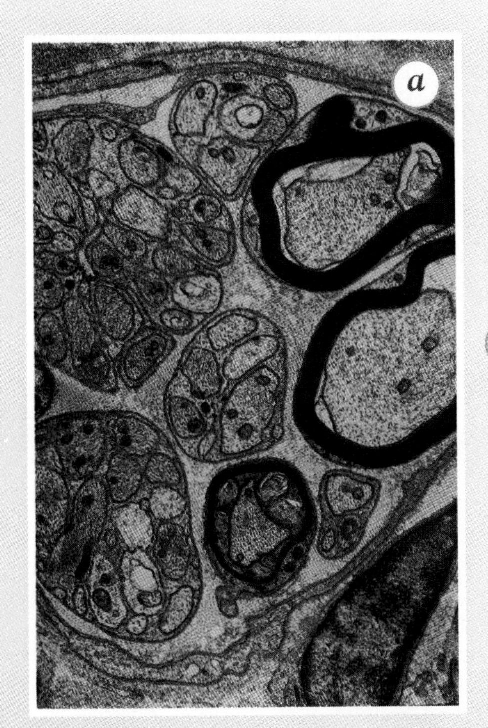

Figure 35–4 (a) This transmission electron micrograph of the cross section of nerve tissue shows both myelinated and unmyelinated axons. The axons that are covered with myelin are those that seem to have black rings around them (magnification: 47,000X). (b) An action potential can move much faster along a myelinated axon because it can jump from node to node, rather than moving continuously along the membrane.

b

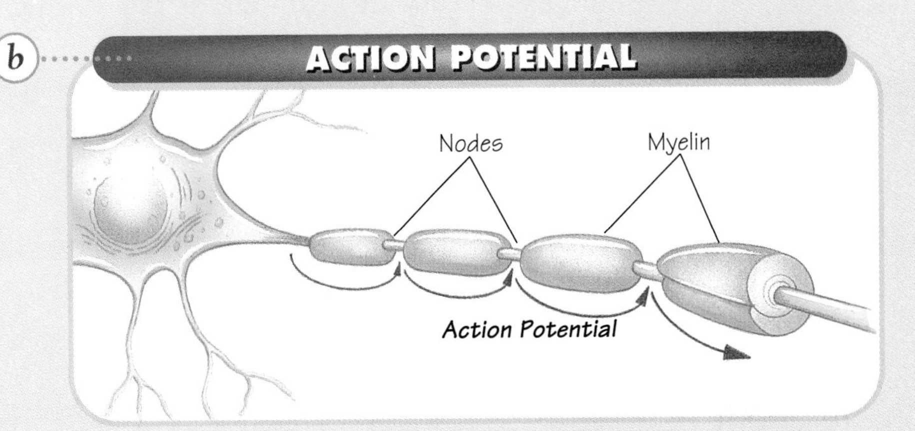

Figure 35–5
When an impulse reaches the end of the axon of one neuron, neurotransmitters are released into the synaptic gap. The neurotransmitters bind to the receptors on the membrane of an adjacent neuron. As a result, the nerve impulse continues to the next neuron.

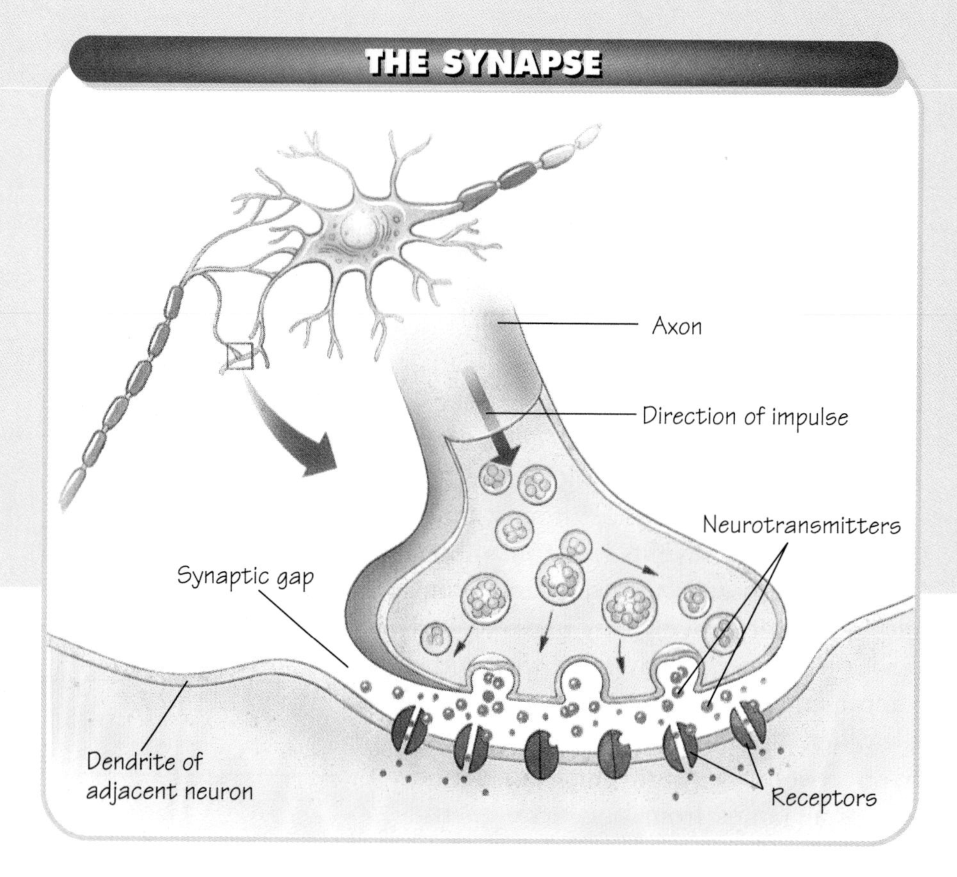

The Synapse

Even though some neurons are among the longest cells in your body, there comes a point when an impulse reaches the end of an axon. Usually the neuron then makes contact with another cell, and it may even pass the impulse to this cell. Motor neurons, for example, pass their impulses to muscle cells.

The point at which a neuron can transfer an impulse to another cell is called the **synapse** (SIHN-aps). The synapse is a small space between the axon of one neuron and the dendrites of the next neuron. The synapse contains tiny sacs filled with **neurotransmitters** (NOO-roh-trans-miht-erz). Neurotransmitters are chemicals used by one neuron to signal another cell.

When an action potential arrives at the end of an axon, the sacs release the neurotransmitters into the synapse between the two cells. The neurotransmitter molecules attach to receptors on the neighboring cell. This causes positive ions to rush across the cell membrane, stimulating that cell. If the stimulation is great enough, a new impulse begins.

A fraction of a second after binding to their receptor, the neurotransmitter molecules are released from the cell surface. They may then be broken down by enzymes or recycled by the axon.

Section Review 35–1

1. **Name** the three parts of a neuron.
2. **Describe** the way in which a nerve impulse begins.
3. **Critical Thinking—Relating Concepts** Your "funny bone" is actually a nerve in your arm. How does this explain your reaction when you "hit" it?

SECTION 35–2

Organization of the Nervous System

Guide for Reading

- **Describe** the two major divisions of the nervous system.

THE HUMAN NERVOUS SYSTEM is similar to a complex telephone network in a large city. In a telephone network, telephone lines and wires connect homes, businesses, and schools through a central telephone switching station. The nervous system has connecting wires as well—the neurons. The nervous system itself works like the central switching station of the body. It receives, compares, and analyzes information, then it sends messages and commands to the rest of the body.

The Central Nervous System

The human nervous system is divided into two main parts—the central nervous system and the peripheral nervous system. **The central nervous system consists of the brain and the spinal cord.** The brain and spinal cord share many structural similarities. The brain is protected by the bones of the skull, and the spinal cord is protected by the vertebrae of the backbone. Both are cushioned by three layers of tough, elastic tissue called **meninges** (muh-NIHN-jeez). Between the meninges is a space filled with **cerebrospinal** (ser-uh-broh-SPIGH-nuhl) **fluid,** which cushions the brain and spinal cord. These three means of protection help to prevent many injuries to the central nervous system.

The Brain

The brain contains about 100 billion cells—a far greater number than the number of people in the world! That fact alone should prepare you for how complicated this organ is. Although the brain represents only about 2 percent of the mass of the body, it is so active that it may use as much as 25 percent of the body's energy. To supply the food and oxygen needed to support that activity, the brain has a rich blood supply. If that blood supply is interrupted—even for just a few minutes—the brain may suffer damage serious enough to cause death.

INTEGRATING HEALTH

Refer to a health book to find out what causes a stroke.

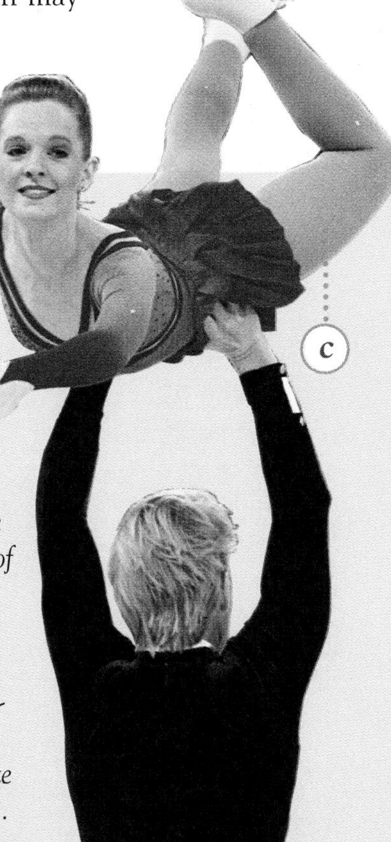

Figure 35–6 *The human nervous system is often compared to a telephone network. Just as (a) these telephone wires connect homes to a central station, these (b) nerves connect the different parts of the nervous system. The individual nerve fibers, which are falsely colored blue, are motor nerves (magnification: 500X). (c) Without a properly functioning nervous sytem, these skaters would not have the ability to coordinate and perform these graceful movements.*

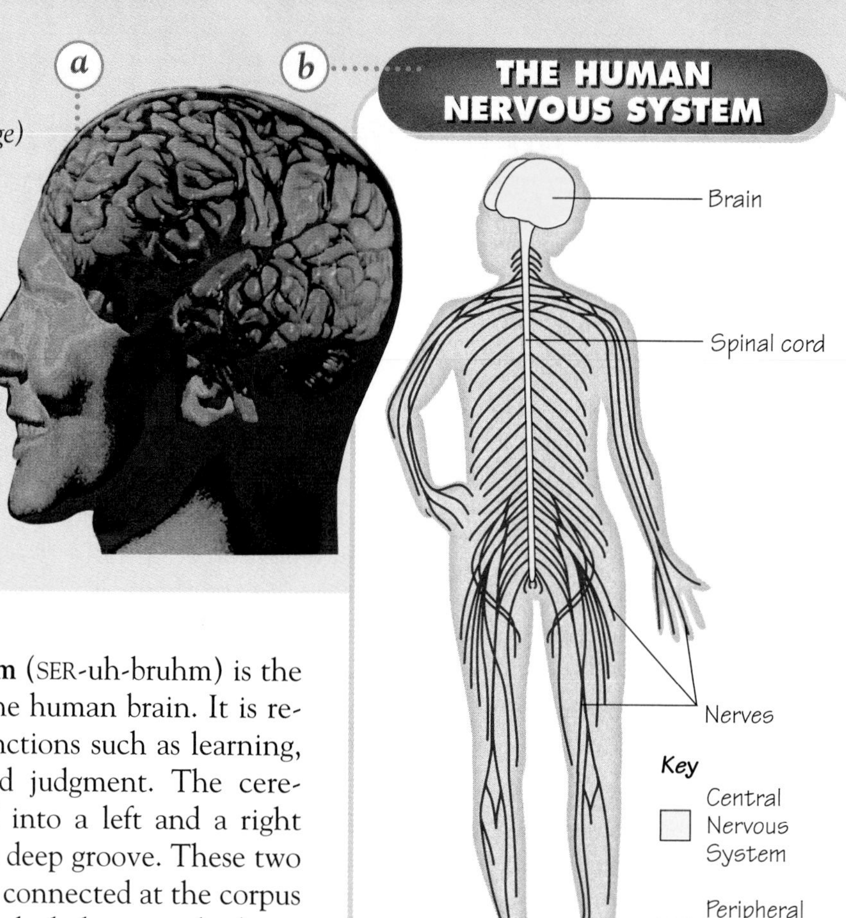

Figure 35–7
(a) An MRI (magnetic resonance image) of the brain—the control center of the human body—has been superimposed over an image of the head. (b) The human nervous system consists of two main divisions. The central nervous system, colored yellow, consists of the brain and the spinal cord. The peripheral nervous system, colored red, consists of all the nerves that carry information to and from the spinal cord.

The **cerebrum** (SER-uh-bruhm) is the largest part of the human brain. It is responsible for functions such as learning, intelligence, and judgment. The cerebrum is divided into a left and a right hemisphere by a deep groove. These two hemispheres are connected at the corpus callosum (KOR-puhs kuh-LOW-suhm).

The surface of the cerebrum, the **cerebral cortex,** is deeply creased. The creasing enlarges its surface area and increases the number of cells that can be packed into this layer. The cerebral cortex processes information from the senses and controls body movements.

The **cerebellum** (ser-uh-BEHL-uhm), the second-largest part of the brain, is located just below the cerebrum at the base of the skull. When the cerebral cortex commands a muscle group to move, that message is routed through the cerebellum. The cerebellum coordinates and balances the actions of muscles so the body moves gracefully and efficiently.

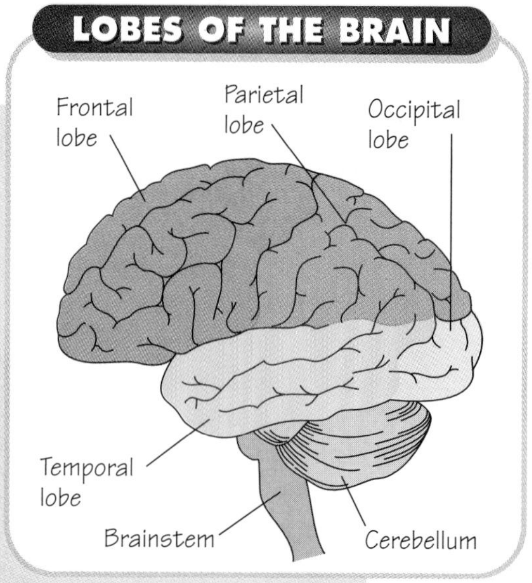

The **brainstem** connects the brain to the spinal cord. The brainstem includes regions called the **medulla oblongata** (mih-DUHL-uh ahb-lahn-GAHT-uh) and the **pons.** Some of the body's most important functions—including blood pressure, heart rate, breathing, and swallowing—are controlled in this part of the brain.

The thalamus is located just beneath the cerebrum. The thalamus receives messages from sense organs, including the eyes and the nose, before they are relayed to the cerebral cortex. Just beneath the thalamus is the hypothalamus, a small region that is linked to the pituitary gland.

Figure 35–8
Each hemisphere, or half, of the cerebrum contains the same four lobes—frontal lobe, parietal lobe, occipital lobe, and temporal lobe.

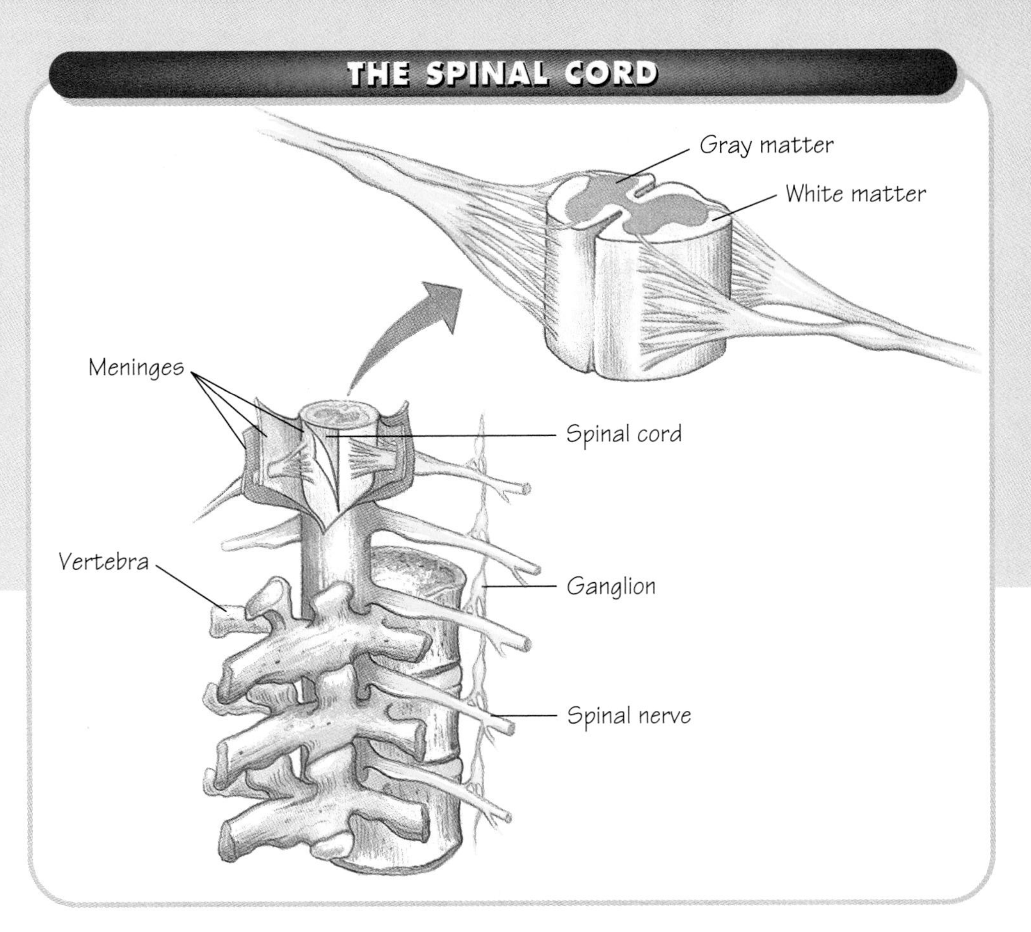

Figure 35–9
The spinal cord is surrounded by three layers of meninges. In the upper right, the cross section of the spinal cord shows both the gray matter and the white matter. The gray matter gets its color from the color of the cell bodies, of which it is mostly made. The white matter consists mainly of long axons, which get their color from myelin sheaths.

The Spinal Cord

The spinal cord is the primary link between the brain and the rest of the body. Thirty-one pairs of spinal nerves branch out from the spinal cord, connecting the brain to all parts of the body. Certain kinds of information are processed directly in the spinal cord. One example is the well-known knee-jerk reflex. A **reflex** is a quick, automatic response to a stimulus. As shown in ***Figure 35–10,*** a tap on the knee stimulates a reflex. The nerve impulses travel through sensory neurons to the spinal cord. There the impulse synapses with motor neurons, which sends the message back to the leg muscle to contract.

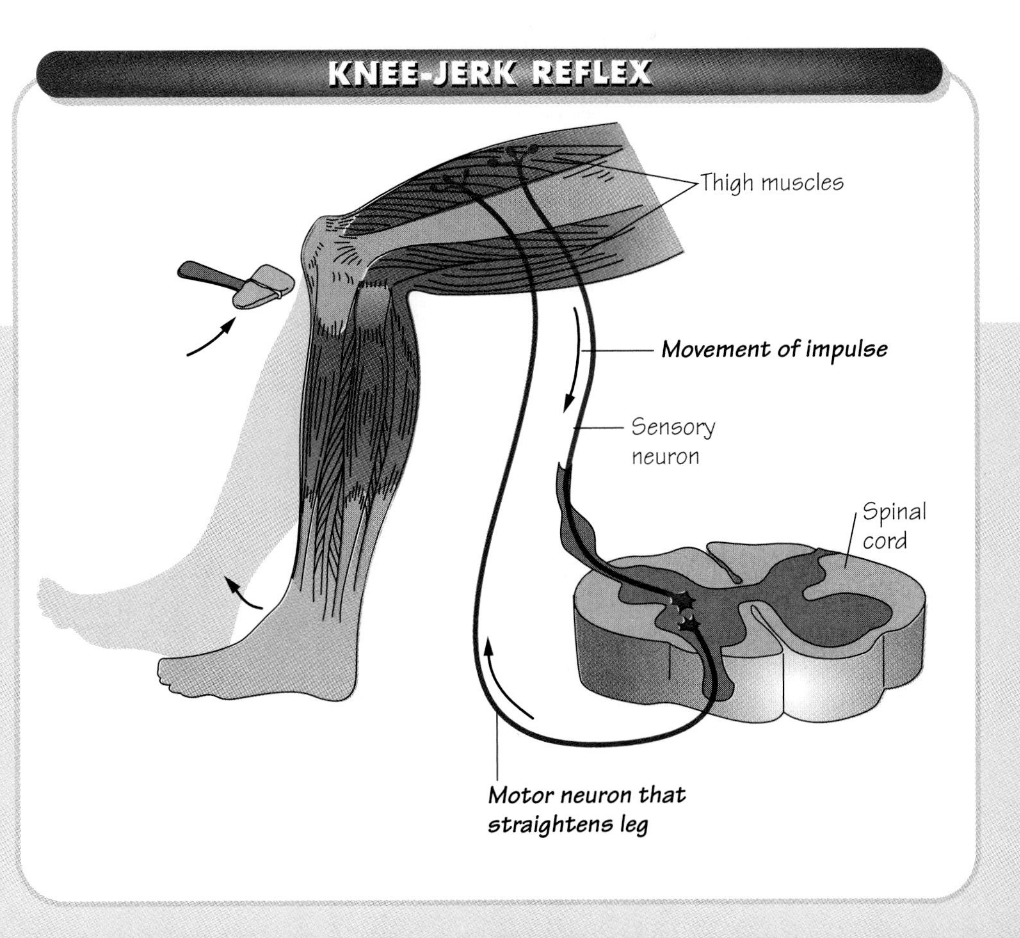

Figure 35–10
The knee-jerk reflex is one of the simplest neural circuits in the body. Tapping the kneecap causes a sensory neuron to send a signal to the spinal cord, where it stimulates an interneuron. The interneuron stimulates a motor neuron, which sends the signal back to the leg "telling" the leg muscles to contract, thereby straightening the leg.

Bio FRONTIER Connections

Informed Consent

Whenever Kim came home from work lately, all she wanted to do was lie down and take a nap. At first, she didn't think too much of it, even though she usually liked to go for a run before dinner.

When her husband became concerned and asked her about her lack of energy, she would say, "It's nothing. I just have a headache."

Strange Symptoms

Then one day, Kim noticed a strange metallic taste in her mouth. She also began smelling perfume, even though she did not usually wear it. Finally, Kim decided to go to see her doctor.

The doctor suggested that Kim undergo a few tests. One test was called an MRI, or Magnetic Resonance Imaging, which was able to take a "snapshot" of Kim's brain.

Test Results

Unfortunately, the results of Kim's MRI showed a small brain tumor. And although the tumor was not cancerous, Kim would probably need surgery to avoid further, more serious, complications. Specifically, the tumor might eventually lead to blindness if left untreated. Kim agreed to the operation.

Possible Side Effects

The night before Kim was to have surgery, her doctor visited her in her hospital room. Before proceeding with the operation, the doctor wanted to discuss all the possible side effects—which might include infection, paralysis, or loss of speech—so that Kim could make a final decision based on the best possible information. This is called informed consent.

One of the problems doctors often face is just how much (or how little) to tell a patient. On the one hand, patients like Kim need to know certain information before they can consent to surgery or other procedures that might have serious side effects. In fact, there are laws that make it mandatory for doctors to keep their patients informed of all possible outcomes. On the other hand, how much information is too much information, which might only cause more fear in the patient? How much knowledge does a patient need in order to give informed consent? There is a fine line between knowing enough and knowing too much.

An MRI scan of the back of the head showing a noncancerous tumor (green area)

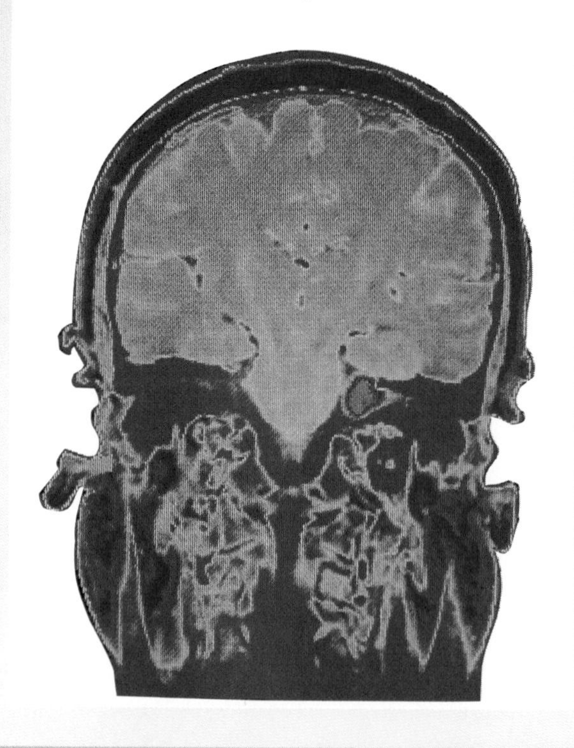

Making the Connection

If you were facing a serious operation, what kinds of questions would you ask your doctor? How would the answers to those questions help you to make an informed decision?

Figure 35–11
CAREER TRACK
An electroneurodiagnostic technologist is responsible for taking electroencephalogram (EEG) images of the brain. EEGs show the average electrical activity of the brain and are useful in studying the process of sleep and in diagnosing some brain abnormalities.

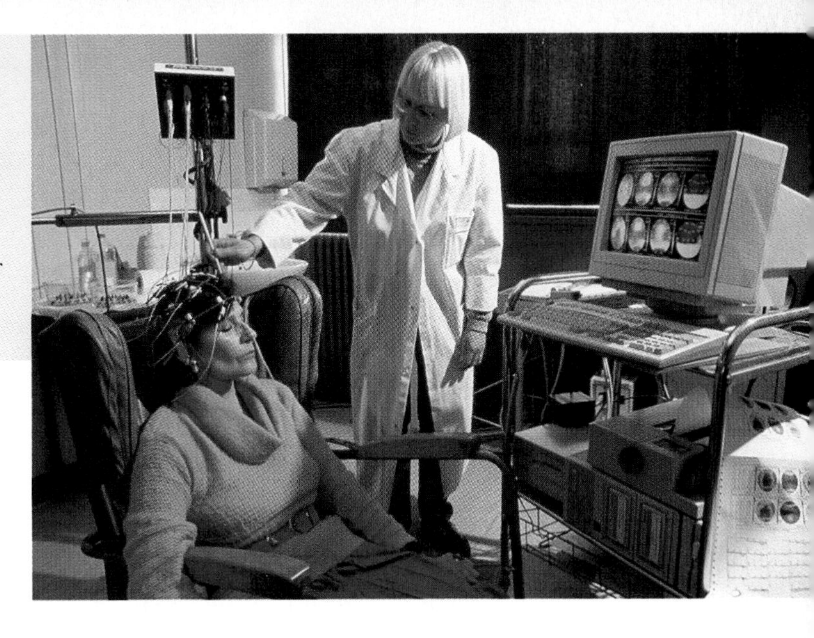

The Peripheral Nervous System

The peripheral nervous system includes all the nerves and associated cells that connect the brain and spinal cord to the rest of the body. The peripheral nervous system receives information from the environment and relays commands from the central nervous system to organs throughout the body.

There are two divisions in the peripheral nervous system—the sensory division and the motor division. The sensory division carries information from the sense organs to the central nervous system. The motor division transmits messages in the other direction—from the central nervous system to the rest of the body. The motor division is further divided into the **somatic nervous system** and the **autonomic nervous system.**

The Somatic Nervous System

The somatic nervous system controls voluntary movements. Every time you turn a page of this book, you are using the somatic nervous system to command the movements of your hand. Some somatic nerves are also part of reflexes and act with or without conscious control.

The Autonomic Nervous System

The autonomic nervous system regulates activities that are not under conscious control, including the beating of the heart and the contraction of muscles surrounding the digestive system. This system regulates the activities of many organs throughout the body. The autonomic nervous system consists of two distinct parts—the sympathetic nervous system and the parasympathetic nervous system. In nearly every case, nerves from both systems regulate each organ.

Why is each organ regulated by both nervous systems? The answer is that the effects of each system are different. For example, sympathetic nerves cause the heart rate to speed up, but parasympathetic nerves cause it to slow down—like the gas pedal and the brake of a car. Therefore, the autonomic nervous system can quickly speed up the activities of important organs or slam on the brakes—whichever is necessary.

Section Review 35–2

1. **Describe** the two major divisions of the nervous system.
2. **Critical Thinking—Making Inferences** Each hemisphere of the cerebrum receives sensory information and controls movement on the opposite side of the body. In a right-handed person, which hemisphere is dominant? In a left-handed person?

SECTION 35–3

The Senses

Guide for Reading

- List the five senses.

MINI LAB

- **Design an experiment** to determine your threshold for taste.

HOW DO YOU KNOW THAT there is a world around you? That might seem to be an obvious question, but philosophers have wondered about this for thousands of years. Many of them have come to an interesting conclusion—that what we know of the world depends entirely on our senses. In this very important way, we are at the mercy of our sensory systems.

The Five Senses

Our five senses—vision, hearing, smell, taste, and touch—each begins with specialized sense organs that respond to the environment. Sensory neurons carry impulses from these sense organs back to the central nervous system, where they form the basis for our understanding of the world.

Vision

Similar to other primates, humans have exceptionally good eyesight. In a world that is filled with sunlight, our vision, more than any other sense, shapes our understanding of everything around us.

Light enters the eye through the **cornea,** a tough transparent layer at the surface of the eye. The cornea focuses the entering light, which then passes through a fluid-filled chamber. At the back of this chamber is a disk of tissue called the **iris.** Tiny muscles adjust the size of the opening in the iris, called the **pupil,** to regulate the amount of light that enters the eye. Pigments in the iris give your eye its color, making it appear blue, brown, or green.

Just behind the pupil is the **lens,** a flexible structure filled with a transparent protein. Small muscles attached to the lens change its shape to help you adjust your eyes' focus to see near or distant objects. Behind the lens is a large chamber filled with a transparent fluid called the **vitreous** (VIH-tree-uhs) **humor.**

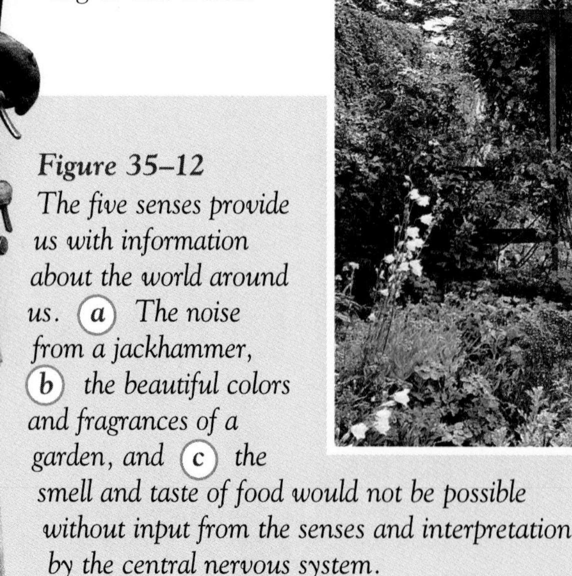

Figure 35–12 *The five senses provide us with information about the world around us. (a) The noise from a jackhammer, (b) the beautiful colors and fragrances of a garden, and (c) the smell and taste of food would not be possible without input from the senses and interpretation by the central nervous system.*

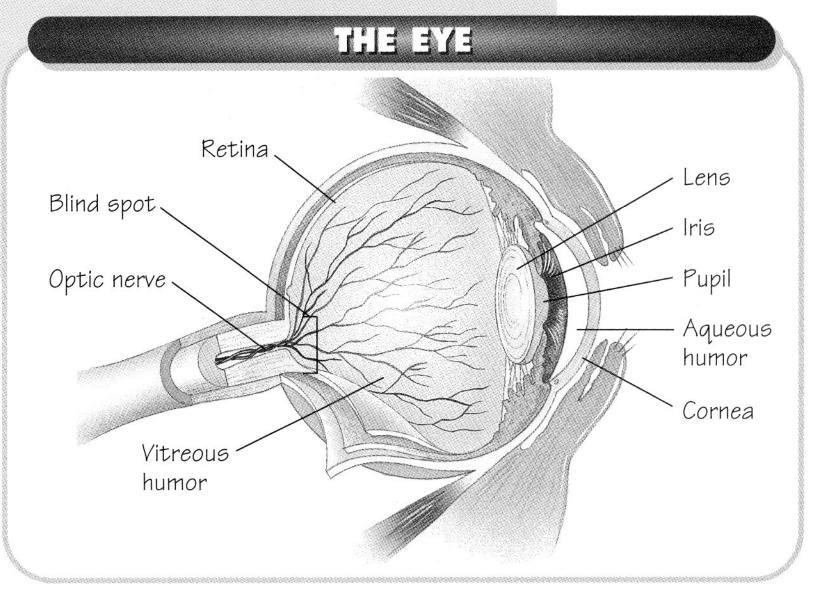

Figure 35–13
The human eye is a complicated organ responsible for the sense of vision. The area on the retina that contains no rods or cones is called the blind spot.

Similar to a miniature slide projector, the lens and cornea project an image of the scene in front of your eyes directly onto the **retina,** the layer of cells at the back of the eye. The light-sensitive photoreceptor cells in the retina come in two kinds—rods and cones. Rods are extremely sensitive to light, but they cannot detect color. Cones can detect color, but they need more light than rods to work properly.

Rods and cones are wired directly to a series of interneurons that are part of the retina itself. These cells help to analyze the pattern of cells stimulated by light and then relay that information to the brain through the optic nerve. A major portion of the human brain is devoted to receiving and analyzing signals from the optic nerve.

Checkpoint What are rods and cones?

Hearing and Balance

Sound is nothing more than vibrations in the air around us. Slow vibrations—those that shake the air 100 to 500 times a second—produce deep, low-pitched sounds. Higher pitches result from faster vibrations—1000 to 5000 times per second. Our ears allow us to sense the pitch and determine its loudness—how strong the vibration is.

Vibrations enter the ear through the **auditory canal,** causing the **tympanum,** the eardrum, to vibrate. The vibrations are picked up by three tiny bones, the

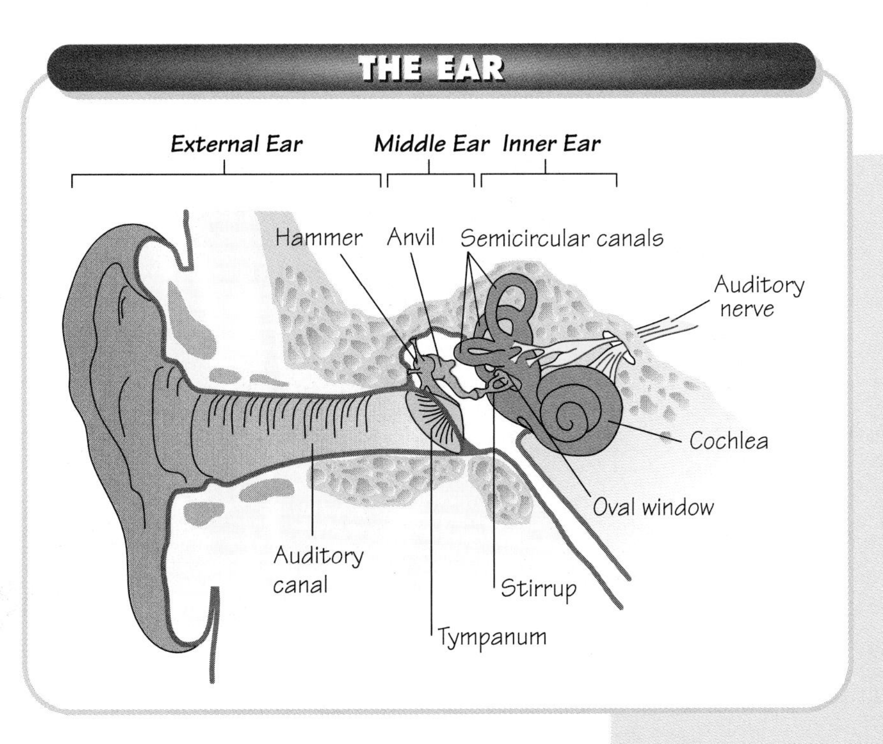

Figure 35–14
The human ear is divided into three parts—the outer ear, the middle ear, and the inner ear. Each of the structures within the ear is responsible for either hearing or maintaining balance.

Figure 35–15
Nerve cells located in tissues lining the upper part of the nasal cavity are responsible for the sense of smell.

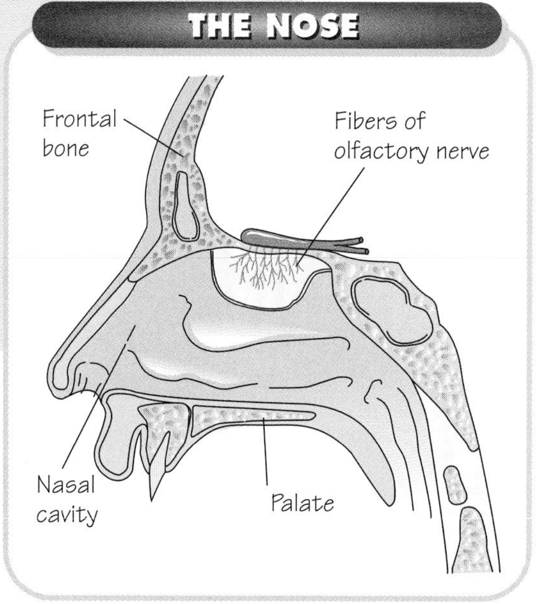

hammer, the anvil, and the stirrup. These bones transmit the vibrations to a thin membrane called the oval window. The oval window covers the opening for the inner ear. Vibrations of the oval window create pressure waves in the fluid-filled **cochlea** (KAHK-lee-uh) of the inner ear.

The cochlea is lined with tiny hair cells that are pushed back and forth by these pressure waves. In response to these movements, the hair cells produce nerve impulses that are sent to the brain through the auditory nerve.

The ear also contains a tiny organ that helps you to sense your position in space and maintain your balance. Three fluid-filled **semicircular canals** enable the nervous system to sense changes in the position of the head. When the head is moved quickly in any direction, pressure is developed in one or more of the canals, and tiny hairs near the ends of the canals sense these pressures.

Two tiny sacs located below the semicircular canals are embedded in a gelatinlike substance that contains tiny grains of calcium carbonate. The downward pressure produced by these tiny grains enables you to sense the pull of gravity. Together, the semicircular canals and these sacs enable you to sense your body's position and keep your balance steady.

☑ **Checkpoint** What is the function of the semicircular canals?

(a) THE TONGUE

Bitter
Sour
Salty
Sweet

Figure 35–16
(a) Although the tongue is covered with taste buds, the taste buds are grouped into four regions, depending on the "taste" they perceive. (b) This scanning electron micrograph shows what the surface of the tongue really looks like. The large disk-shaped objects are the taste buds (magnification: 240X).

(b)
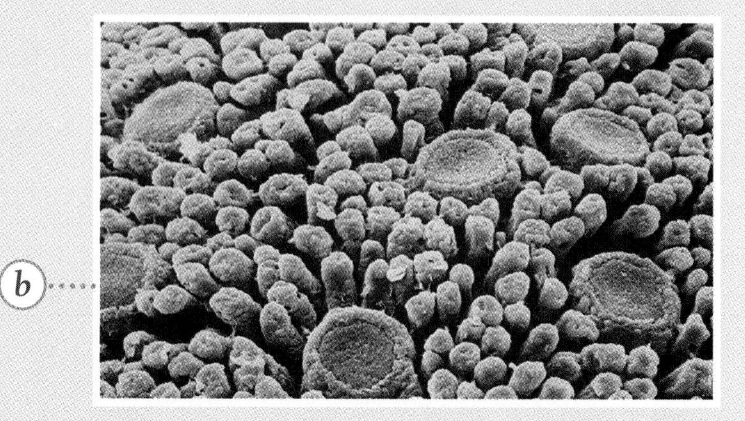

Smell

You may never have thought of it this way, but your sense of smell is actually the ability to detect chemicals. Special cells in the upper part of the nasal passageway act as receptors for a variety of chemicals. When stimulated, these cells produce the nerve impulses that travel to the central nervous system.

Your sense of smell is capable of producing thousands of different sensations. In fact, much of what is commonly called the "taste" of food actually depends on your sense of smell. To prove this, eat a few bites of food while holding your nose. You'll discover that the food doesn't have much taste until you open your nose and breathe freely.

Taste

Your mouth contains chemical receptors called **taste buds,** which are located on the tongue. Although you are able to perceive hundreds of tastes, there are only four types of taste receptors—sweet, sour, salty, and bitter. Sensitivity to these different tastes varies on different parts of the tongue, as shown by the drawing in *Figure 35–16.*

How Tasty Is It?

PROBLEM *How can you determine your threshold for taste?* ***Design an experiment*** *to answer this question.*

SUGGESTED PROCEDURE

1. Rinse your mouth with water.
2. Using a clean cotton swab, place a drop of sugar solution on the tip of your tongue. Record your observations. Rinse your mouth with water.
3. Using a procedure similar to the one used in steps 1 and 2, design an experiment to determine the effects of two different concentrations of sugar solution on the tip of your tongue. Have your teacher check your procedure before carrying out your experiment.

ANALYZE AND CONCLUDE

1. Which sugar solutions were you able to taste? Explain why.
2. What do you think the word "threshold" means with regard to taste?
3. Why was it important for you to rinse your mouth between tastings?

Touch and Related Senses

The largest sense organ in your body is the skin, which contains different receptors for touch, pain, heat, and cold. Each receptor responds to its particular stimulus and produces nerve impulses that signal the central nervous system.

Not all parts of the body are equally sensitive to touch, because not all parts have the same number of receptors. The greatest densities of touch receptors, for example, are found on your fingers, toes, and lips.

Section Review 35–3

1. **List** the five senses.
2. **Critical Thinking—Drawing Conclusions** Why do you think you feel dizzy after spinning around for a few seconds?
3. **MINI LAB Design an experiment** to determine your threshold for taste.

SECTION 35–4 BRANCHING OUT In Depth

Nerve Impulses and Drugs

GUIDE FOR READING

- **Describe** the function of the sodium-potassium pump.
- **Describe** the effects of drugs on the nervous system.

MINI LAB
- **Predict** how the nervous system will respond to a stimulus.

AS YOU HAVE READ, IN THE nervous system, neurons are similar to telephone wires carrying coded messages and passing important signals from one cell to the next. How, exactly, do neurons do this? What produces the electrical potential across the neuron cell membrane? What happens when an impulse moves along an axon? And what happens when chemical substances alter the nervous system? To answer these questions in depth, we have to look closely at the cell membrane and at the point of synapse between two nerves.

The Resting Potential

As you have learned, there is a difference in electrical charge in a neuron's membrane, called the resting potential. There are two forces that produce this potential, both of which depend upon the special properties of the neuron cell membrane.

The first force is an active-transport protein built into the membrane, the **sodium-potassium pump.** What does this pump do? **The sodium-potassium pump uses the energy from ATP to pump sodium ions (Na^+) out of the cell while at the same time pumping potassium ions (K^+) into the cell.**

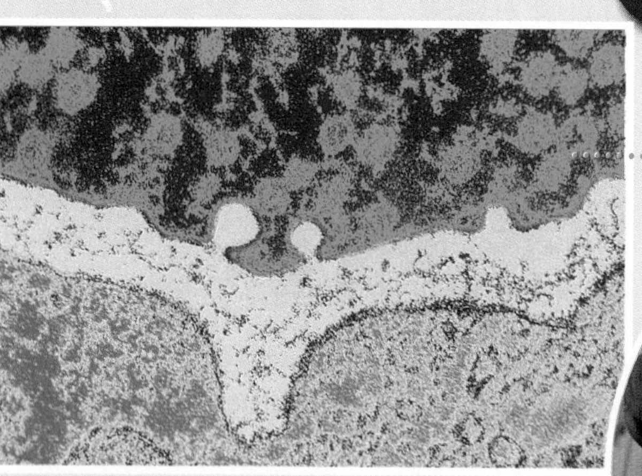

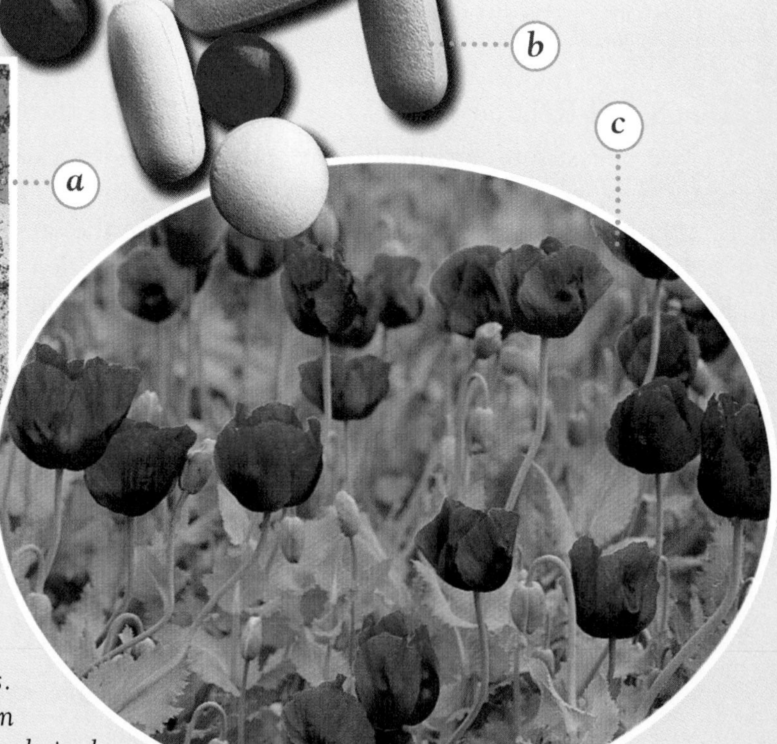

Figure 35–17
(a) The color-enhanced transmission electron micrograph shows the synapse between two neurons. (b) Drugs can cause changes in the synapse, which, in turn, can affect nerve impulses. (c) The centers of the opium poppy flowers contain pods from which opiates, the pain-killing drugs, are derived.

Figure 35–18
At resting potential, sodium ions (Na^+) are outside the cell membrane. As the action potential begins, the sodium gates open and Na^+ ions rush across the membrane. In a few milliseconds, the sodium gates close and the potassium gates open. This allows potassium ions (K^+) to cross the membrane. So many K^+ ions rush across the membrane that the outside of the membrane becomes more positive than the inside, thus restoring the resting potential.

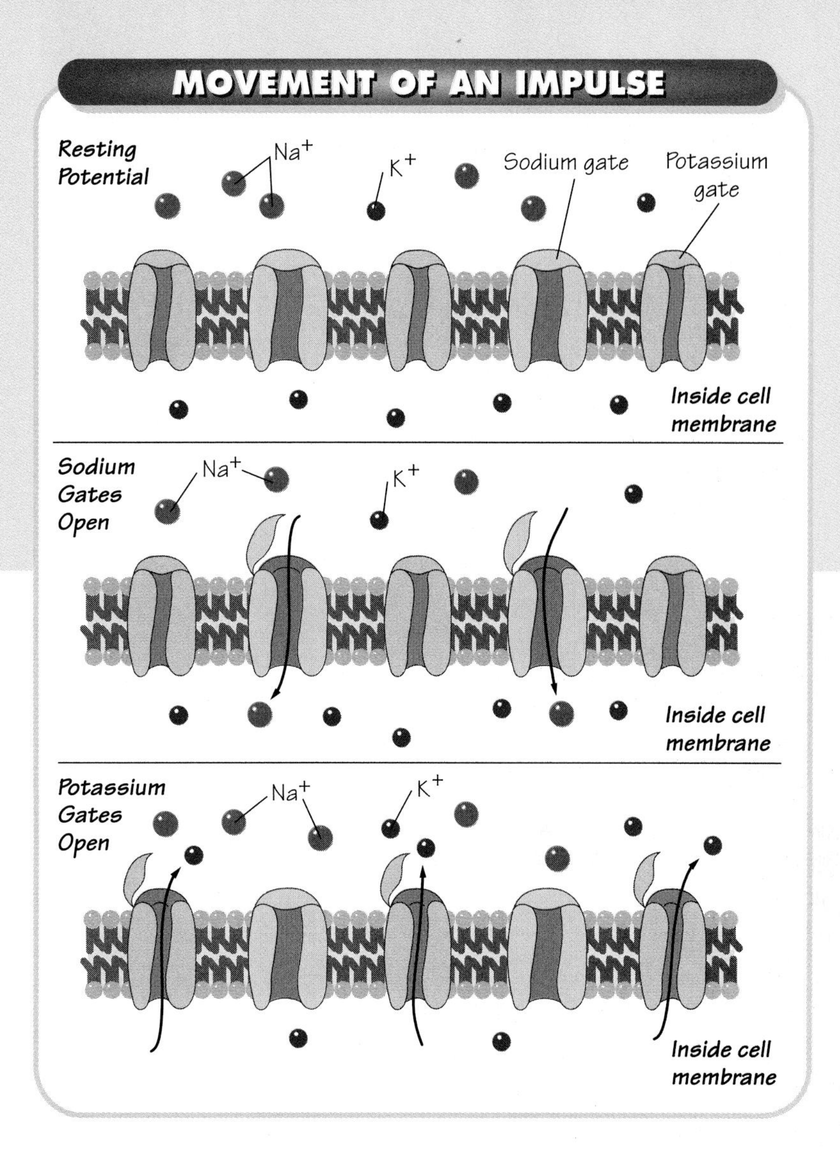

The difference in the concentrations of sodium and potassium ions on the two sides of the membrane is important because both of these ions are positive. Ions do not move across most cell membranes very easily, and once the sodium ions are pumped out of the cell, they tend to stay out. However, significant numbers of potassium ions do manage to leak across the membrane. Remember, potassium ions are located mostly inside the cell. This means that large numbers of positive ions leak out of the cell but few leak into it.

What is the result of positive charges leaking out of the cell? As you might expect, this leaves fewer positive ions inside the cell, meaning that it is negatively charged compared with the outside. The great difference in charges between the two sides of the membrane produces the resting potential.

The Action Potential

A nerve impulse consists of an action potential that rapidly moves down the axon. How does this happen? Remember that an impulse can be started by the environment, by another neuron, or by an electrical stimulus.

The cell membrane of a neuron contains thousands of tiny protein channels known as **voltage-sensitive gates,** which can allow either sodium or potassium to pass through. Generally, the gates are closed. However, when the voltage across the membrane changes—as it does when a stimulus is present—the sodium gates open, allowing Na^+ ions to rush across the membrane.

So many Na^+ ions rush across the membrane that for a few milliseconds, the inside of the membrane becomes more positive than the outside! As you might expect, the inward rush of positive charges in one region of the membrane causes the sodium gates just ahead of it to respond to the voltage change, and these gates open, too. Very quickly, the impulse spreads along the axon.

However, almost as rapidly as they open, the sodium gates close, and the

Face to Face

PROBLEM ***Predict*** *how the nervous system will respond to a stimulus.*

PROCEDURE

1. Have two members of your group stand a few meters away from each other. They should be facing each other.
2. Have one group member hold a piece of screening or a sheet of clear plastic in front of his or her face.
3. Have the second group member toss a crumpled piece of paper at the screening or plastic. Observe the eyes of the group member behind the screening.
4. Repeat steps 1 to 3, but this time have the group member behind the screening try not to respond to the tossed paper. Predict what will happen.

ANALYZE AND CONCLUDE

1. What response did the group member have when the paper was tossed?
2. Was your prediction correct about the group member being successful in stopping the response to the stimulus (tossed paper)? Why or why not?
3. What is the advantage of having this quick response?
4. Which part of the central nervous system is responsible for this response? What is this response called?

movement of sodium stops. A few milliseconds after the sodium gates open, the potassium gates open, allowing K^+ ions to rush out of the cell. Within a few milliseconds, the inside of the membrane is negative once again.

The rapid opening and closing of sodium and potassium gates makes the impulse possible. The small amounts of sodium and potassium that cross the cell membrane during an impulse are quickly pumped back by the sodium-potassium pump, and the neuron is ready for another impulse.

When an action potential reaches the synapse, it triggers the release of a neurotransmitter. The neurotransmitter molecules diffuse across the gap and bind to receptors in the dendrites of the next neuron. The receptors cause the ion gates to open, and the impulse continues.

☑ **Checkpoint** What are voltage-sensitive gates?

Drugs and the Synapse

The nervous system depends on neurotransmitters to relay information about the world from cell to cell. This means that the synapse—the connection from one neuron to the next—is one of the body's most important relay stations.

What might happen if a chemical such as a **drug** that affected the synapse was introduced into the body? The nervous system could malfunction. A drug is any substance that causes a change in the body. **Drugs can affect the body in a variety of ways, causing changes in the brain, the nervous system, and the synapses between nerves.**

Stimulants

Some drugs, such as **stimulants,** increase the release of neurotransmitters at some synapses in the brain. This speeds up the nervous system, leading to a feeling of energy and well-being. When the effects of stimulants wear off, however, the brain's supply of neurotransmitters has been depleted. The user quickly falls into fatigue and depression. Long-term use causes hallucinations, circulatory problems, and psychological depression.

Even stronger effects are produced by drugs, such as **cocaine,** that act on neurons in what are known as the pleasure centers of the brain. The effects of cocaine are so strong that they produce **addiction**—an uncontrollable craving for more of the drug.

Cocaine causes the sudden release of a neurotransmitter called dopamine. Normally, dopamine is released when a basic need, such as hunger or thirst, is fulfilled. By fooling the brain into releasing dopamine, cocaine produces intense feelings of pleasure and satisfaction.

Cocaine is a powerful stimulant that increases the heart rate and blood pressure. For many first-time users, this stimulation is just too much—cocaine can damage the heart and has produced heart attacks, even in young people. In the United States, an inexpensive form of cocaine called crack has become one of the most dangerous drugs on the street. The intense high produced by crack wears off quickly and leaves the brain with too little dopamine. As a result, the user suddenly feels sad and depressed and quickly seeks another dose.

Depressants

Other drugs, called **depressants,** decrease the rate of brain activity. Some depressants, such as alcohol, enhance the effects of neurotransmitters that prevent some nerve cells from starting action potentials. This calms some parts of the brain that sense fear and relaxes the individual. However, long-term use of this type of drug can also cause problems. Depressant drugs reduce the effects of natural inhibitors of these neurons. As a result, the user comes to depend on the drug to relieve the anxieties of everyday life, which may seem unbearable without the drug.

Figure 35–19 *Cocaine is made from the* Erythroxylum coca *plant, which grows mainly in South America.*

☑ **Checkpoint** What are depressants?

Opiates

The opium poppy produces a powerful class of pain-killing drugs called **opiates.** These include chemical derivatives of opium, such as morphine and heroin. Opiates mimic natural chemicals in the brain known as endorphins, which normally help to overcome sensations of pain. The first few doses of these drugs produce strong feelings of pleasure and security, but the body quickly adjusts to the higher levels of endorphins. Once this happens, the body literally cannot do without them. If the user attempts to stop taking these drugs, the body cannot produce enough of the natural endorphins that are needed to prevent the user from the uncontrollable pain and sickness that accompany withdrawal from the drug.

Section Review 35–4

1. **Describe** the function of the sodium-potassium pump.
2. **Describe** the effects of drugs on the nervous system.
3. **MINI LAB** **Predict** how the nervous system will respond to a stimulus.
4. **BRANCHING OUT ACTIVITY** Use library references to make a chart of the following groups of drugs—stimulants, depressants, hallucinogens, opiates, cocaine, marijuana, alcohol, and tobacco. **Summarize** their long- and short-term effects on the nervous system.

Laboratory Investigation

Observing the Structures of the Brain

In structure, the brain of a sheep is similar to the brain of a human. Most of the parts are located in identical areas. In this investigation, you will use the sheep brain to illustrate the anatomy of the human brain.

Problem

How can you **identify** the parts of the sheep brain?

Materials (per group)

dissecting pan
scalpel
probe
gloves
sheep brain

Procedure

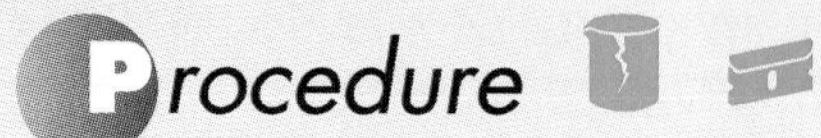

1. **Put on the pair of gloves provided. Place the sheep brain in a dissecting pan so that the raised side (dorsal) is up and the flat side (ventral) is resting on the pan.**
2. **Notice the large anterior section, the cerebrum. The cerebral cortex has many convolutions, or gyri (singular: gyrus), and grooves, called sulci (singular: sulcus). Each sulcus separates the cerebrum into lobes.**

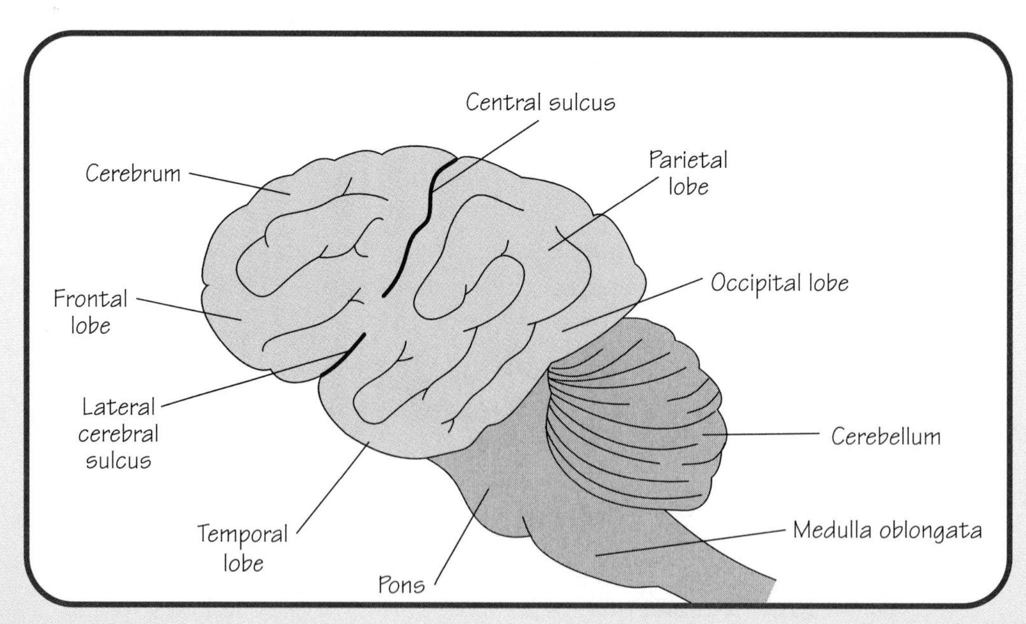

3. Locate the central sulcus that runs down from the top to the bottom of the cerebrum, midway between the front (anterior) and the back (posterior). In front of the central sulcus, you will find the frontal lobe. Behind the central sulcus is the parietal lobe.

4. Locate the lateral cerebral sulcus, a groove that runs horizontally along the cerebrum, separating the frontal lobe (above) from the temporal lobe (below).

5. Locate the occipital lobe, which is found posterior to the parietal lobe.

6. Posterior to the cerebrum is a smaller lobed structure called the cerebellum. Holding the brain in your hand, gently bend the cerebellum down. Although in the sheep the cerebellum is directly behind the cerebrum, in humans it is positioned similarly to where it is located when you bend the sheep's brain downward.

7. Posterior to the cerebellum, you will find the medulla oblongata, which forms a triangular shape. This part of the brain narrows into and becomes the spinal cord.

8. The midbrain section is more visible on the ventral (underneath) side. Locate the pons—a bridgelike, raised area in the cerebellum region. The pons is in front of the medulla oblongata.

9. With a scalpel, carefully cut the brain in half lengthwise. CAUTION: ***Be very careful when using a sharp instrument.***

10. Distinguish between the outer gray cortex area of the cerebellum and its inner white matter. The gray matter is slightly darker than the white and not really gray at all.

11. Dispose of the brain as directed by your teacher. Clean your dissecting pan and scalpel and probe. Remove your gloves and wash your hands thoroughly.

Observations

1. Compare the sheep brain to the human brain.
2. Compare the size of the cerebrum with the other parts of the brain. What significance does the size of the cerebrum have?
3. Describe the location of the "gray" matter and the "white" matter in the brain. Other than by color, how do they differ?

Analysis and Conclusions

1. Were you able to identify the four lobes of the brain? Why or why not?
2. The hypothalamus is connected to the pituitary gland. What is the relationship between the hypothalamus and the pituitary gland?
3. What is the function of the medulla oblongata? Of the cerebellum?

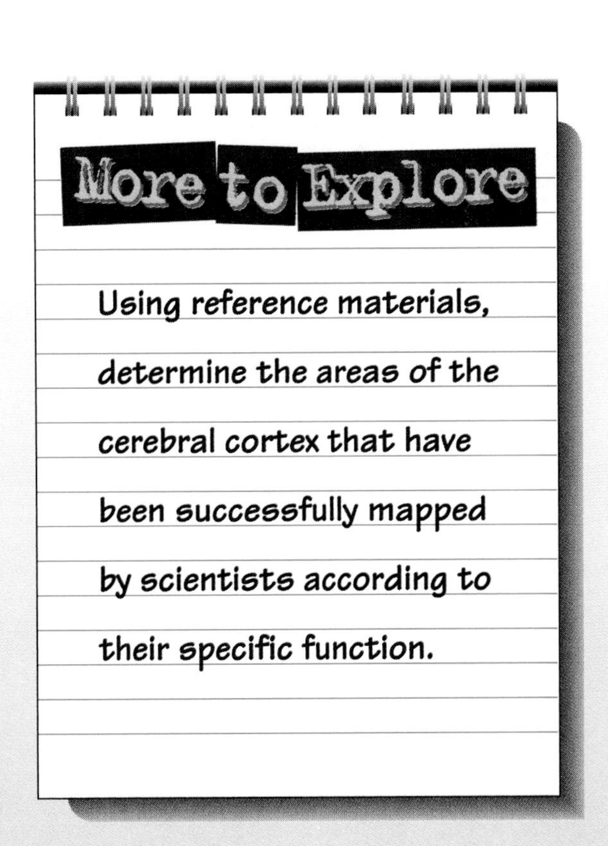

Study Guide

Summarizing Key Concepts

The key concepts in each section of this chapter are listed below to help you review the chapter content. Make sure you understand each concept and its relationship to other concepts and to the theme of this chapter.

35–1 The Human Nervous System

- A neuron is made up of a cell body, dendrites, and an axon.
- An impulse begins when a neuron is stimulated by another neuron or by the environment. Once it begins, the impulse travels rapidly down the axon away from the cell body.

35–2 Organization of the Nervous System

- The central nervous system consists of the brain and the spinal cord.
- The peripheral nervous system includes all the nerves and associated cells that connect the brain and spinal cord to the rest of the body.

35–3 The Senses

- Our five senses—vision, hearing, smell, taste, and touch—each begins with specialized sense organs that respond to the environment.

35–4 Nerve Impulses and Drugs

- The sodium-potassium pump uses the energy from ATP to pump sodium ions out of the cell while at the same time pumping potassium ions into the cell.
- Drugs can affect the body in a variety of ways, causing changes in the brain, the nervous system, and the synapses between nerves.

Reviewing Key Terms

Review the following vocabulary terms and their meaning. Then use each term in a complete sentence.

35–1 The Human Nervous System

neuron
impulse
cell body
dendrite
axon
nerve
sensory neuron
motor neuron
interneuron
resting potential
action potential
myelin
synapse
neurotransmitter

35–2 Organization of the Nervous System

meninges
cerebrospinal fluid
cerebrum
cerebral cortex
cerebellum
brainstem
medulla oblongata
pons
reflex
somatic nervous system
autonomic nervous system

35–3 The Senses

cornea
iris
pupil
lens
vitreous humor
retina
auditory canal
tympanum
cochlea
semicircular canal
taste bud

35–4 Nerve Impulses and Drugs

sodium-potassium pump
voltage-sensitive gate
drug
stimulant
cocaine
addiction
depressant
opiate

CHAPTER 35

Recalling Main Ideas

Choose the letter of the answer that best completes the statement or answers the question.

1. Axons connect with other nerve cells at

a. synapses. **c.** the brainstem.
b. nodes. **d.** Schwann cells.

2. Dendrites

a. transmit impulses away from the cell body.
b. contain nuclei.
c. are specialized to receive impulses from other cells.
d. are surrounded by Schwann cells.

3. Which type of neuron is responsible for transmitting impulses to a muscle?

a. sensory **c.** interneuron
b. myelinated **d.** motor

4. What is the space between two neurons called?

a. dendrite **c.** synapse
b. neurotransmitter **d.** axon

5. In which function is the cerebral cortex not involved?

a. sensory **c.** associative
b. motor **d.** involuntary

6. A loss of balance might result from a disease of the

a. cerebellum. **c.** pons.
b. cerebrum. **d.** cranium.

7. The brain and the spinal cord are cushioned by layers of tough, elastic tissue called

a. myelins. **c.** corneas.
b. meninges. **d.** synapses.

8. Which of these is an active-transport protein located in a neuron's membrane?

a. neurotransmitter
b. myelin
c. sodium-potassium pump
d. vitreous humor

9. Substances that cause a change in the body are known as

a. myelins. **c.** neurotransmitters.
b. drugs. **d.** meninges.

Putting It All Together

Using the information on pages xxx to xxxi, complete the following concept map.

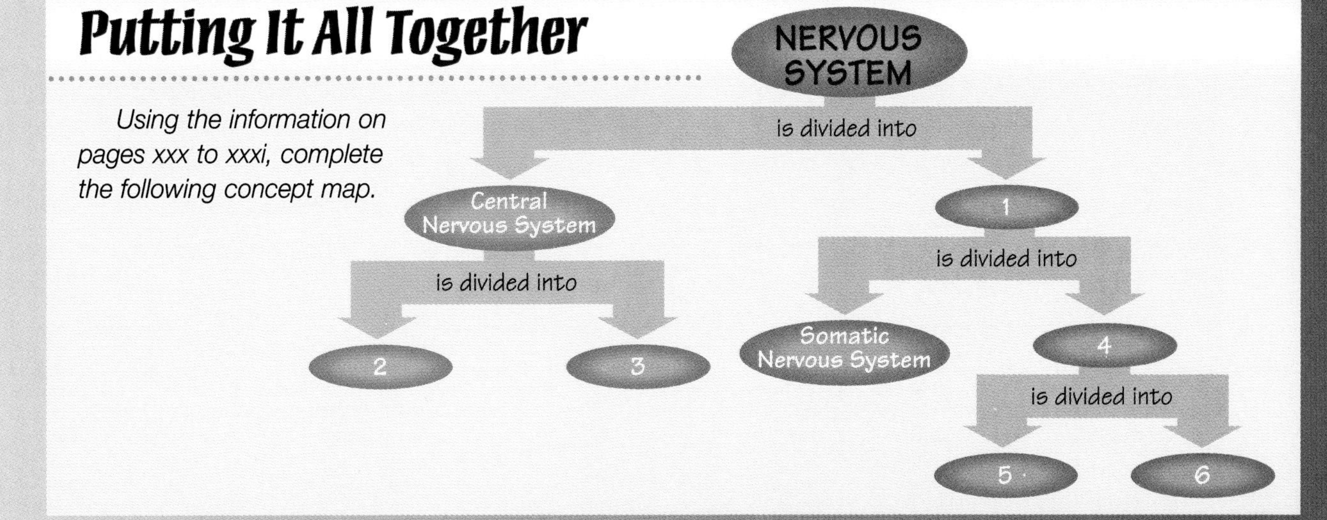

Assessment

Reviewing What You Learned

Answer each of the following in a complete sentence.

1. What is a neuron?
2. List the three types of neurons.
3. What is a resting potential? An action potential?
4. What is myelin?
5. What is a synapse?
6. Explain the difference between a sensory neuron and a motor neuron.
7. Describe the two parts of the human nervous system.
8. Give the function of the meninges and the cerebrospinal fluid.
9. List the functions of the three parts of the brain.
10. Which area of the brain is linked to the pituitary gland?
11. What is a reflex?
12. How many divisions are there in the peripheral nervous system? Name them.
13. Describe the functions of the sympathetic and parasympathetic nervous systems.
14. Which organs are associated with the five senses?
15. What is the sodium-potassium pump?
16. List three types of drugs and their effects on the body.

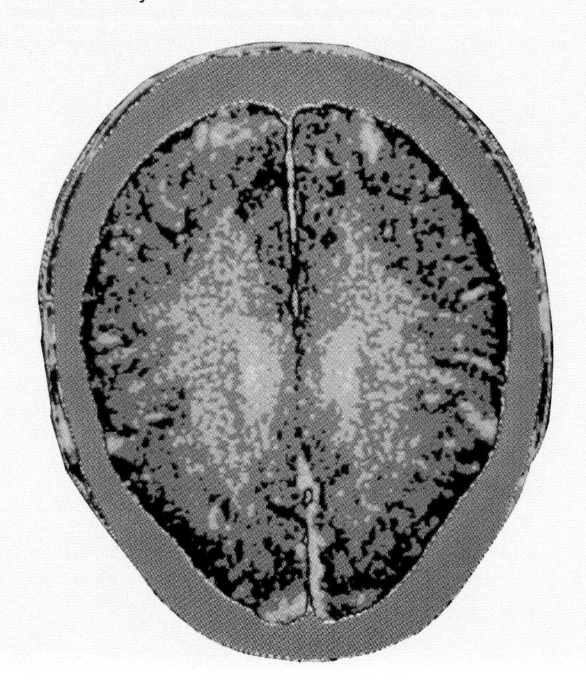

Expanding the Concepts

Discuss each of the following in a brief paragraph.

1. Describe the advantage of having an axon covered with myelin.
2. Describe how an action potential is carried from one neuron to the next.
3. Compare the central nervous system and the peripheral nervous system.
4. Describe the electrical state of the resting neuron.
5. Explain the sequences of changes associated with the passage of an impulse along the axon.
6. What are two factors that affect the rate at which impulses are transmitted along an axon?
7. Which parts of your nervous system are involved in reading and answering this question?
8. What changes in ion distribution occur in the area of an impulse?
9. Describe the function of the sodium-potassium pump.
10. How do the different types of drugs affect the nervous system?

Extending Your Thinking

Use the skills you have developed in this chapter to answer the following.

1. **Relating concepts** The effects of the two divisions of the autonomic nervous system are said to be antagonistic. What does this mean? What is the advantage of this relationship?
2. **Applying concepts** When you go snorkeling with a face mask, you can see very clearly in the water. But if you remove the face mask, things become blurry. Why is this so?
3. **Predicting** Multiple sclerosis (MS) is characterized by the patchy destruction of myelin. Predict the symptoms that might be produced.
4. **Interpreting** Heat receptors of mammals are particularly concentrated on the tongue. These receptors keep humans from burning the mouth with hot food. What advantage is it for a wild mammal that doesn't cook its food to have so many heat receptors on its tongue?
5. **Designing an experiment** Design an experiment to determine the effects of fatigue on reaction time. Formulate a hypothesis and write up a procedure. Have your teacher check your experimental plan before you begin.

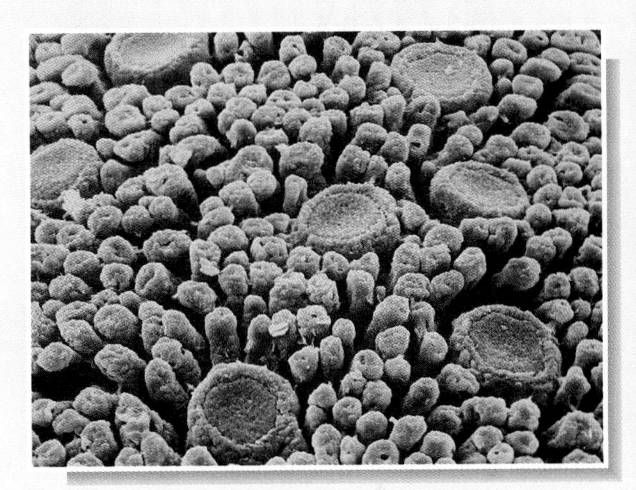

Applying Your Skills

Sensing Trouble

We respond to many stimuli in our environment every day. Through the use of our sense organs, we react to the stimuli, process the information, and respond very quickly. How does this happen?

1. Choose any two of the following situations:
   a. Seeing an approaching ambulance in the rear-view mirror of a car, then pulling over to the side to get out of the way
   b. Hearing a telephone ringing, then running to answer it
   c. Eating a vegetable you don't like, then deciding to take a drink of water
   d. Feeling chilly in a swimming pool, then deciding to get out of the water
2. Diagram the path taken when the sense organ first reacts to the stimulus in each situation to when the impulse is processed as a response.
3. Label the various neurons as well as the parts of one neuron.

• GOING FURTHER •

4. Determine the time it takes you to react to smelling a pizza burning and getting up to turn off the oven.

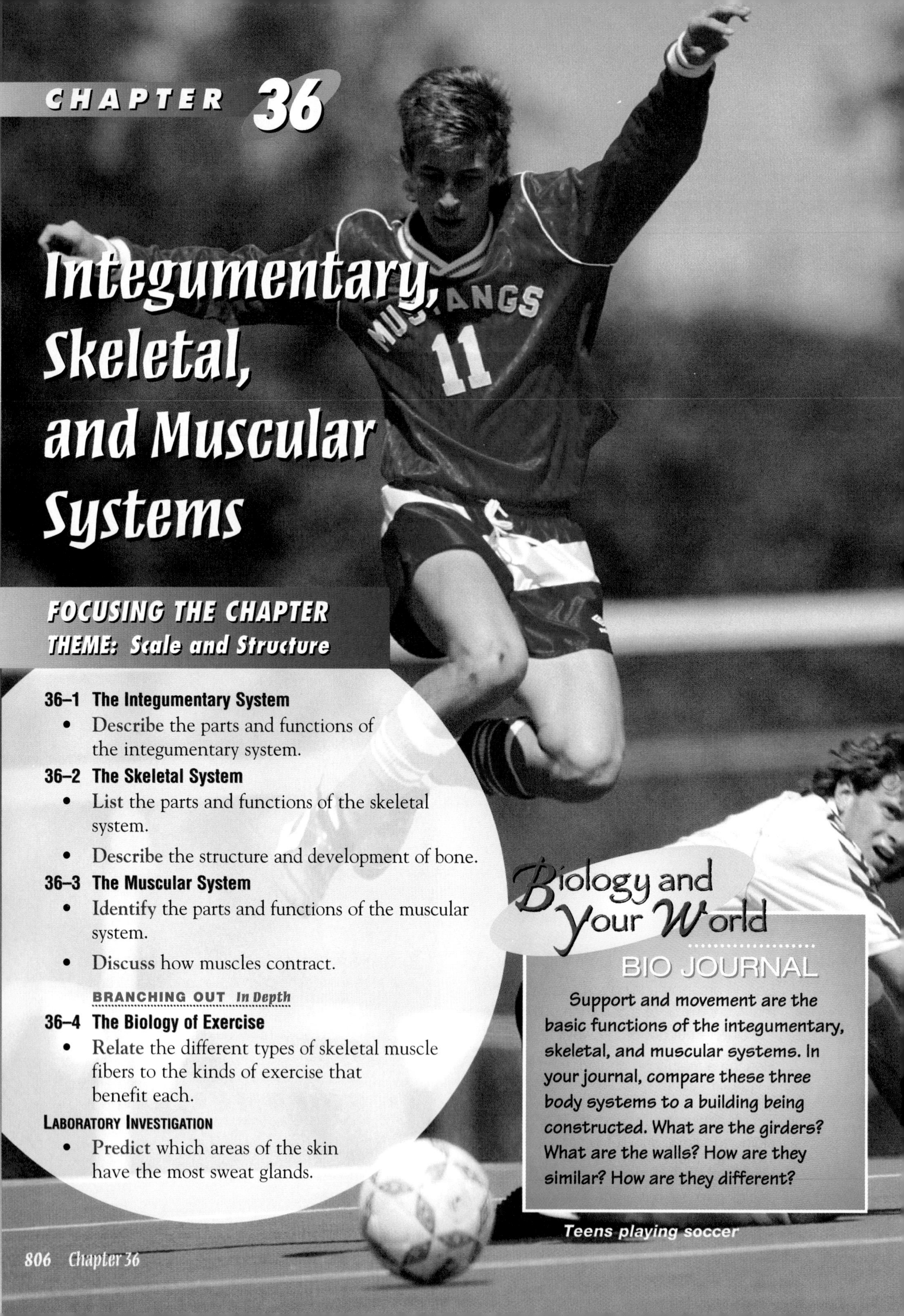

CHAPTER 36

Integumentary, Skeletal, and Muscular Systems

FOCUSING THE CHAPTER

THEME: *Scale and Structure*

36–1 The Integumentary System

- **Describe** the parts and functions of the integumentary system.

36–2 The Skeletal System

- **List** the parts and functions of the skeletal system.
- **Describe** the structure and development of bone.

36–3 The Muscular System

- **Identify** the parts and functions of the muscular system.
- **Discuss** how muscles contract.

BRANCHING OUT *In Depth*

36–4 The Biology of Exercise

- **Relate** the different types of skeletal muscle fibers to the kinds of exercise that benefit each.

LABORATORY INVESTIGATION

- **Predict** which areas of the skin have the most sweat glands.

Biology and Your World

BIO JOURNAL

Support and movement are the basic functions of the integumentary, skeletal, and muscular systems. In your journal, compare these three body systems to a building being constructed. What are the girders? What are the walls? How are they similar? How are they different?

Teens playing soccer

The Integumentary System

SECTION 36–1

Guide for Reading

- **Identify** the basic structures of the integumentary system.

WHEN YOU LOOK AT YOURSELF in the mirror, what do you notice first? Maybe you notice your hair or your skin—or another freckle. Did you know that one system of the body is responsible for the shade of your skin, your hair, and even your nails? This body system is the integumentary system. Your integumentary system acts like a protective covering for the rest of your body. In this section, we will examine the structures of the integumentary system.

Layers of Skin

The largest organ of the body—the **skin**—is part of the integumentary (ihn-tehg-yoo-MEHN-ter-ee) system. **The integumentary system includes your skin, hair, nails, and a number of important glands in the skin.** The skin itself is made up of two layers. The outer layer is called the **epidermis,** whereas the inner layer is called the **dermis.**

INTEGRATING LANGUAGE ARTS

The word integere *is a Latin verb meaning "to cover." How does this explain how the integumentary system got its name?*

Epidermis

The epidermis is made up of layers of epithelial cells. Deep in the epidermis, these cells grow and divide rapidly, producing new cells that are gradually pushed toward the surface of the skin. As they move upward, the cells begin making **keratin,** a tough, flexible protein. In humans, keratin is the major protein found in hair and fingernails.

Figure 36–1
The skin—the single largest organ of the body—is part of the integumentary system. (a) The small dark granules within this melanocyte are responsible for protecting the skin from the harmful ultraviolet rays of the sun. (b) Sweating is one way the body regulates temperature. (c) This sweat gland, which is one of about 3 million in the skin, helps to regulate body temperature. The green spheres inside the sweat gland are bacteria (magnification: 10,000X).

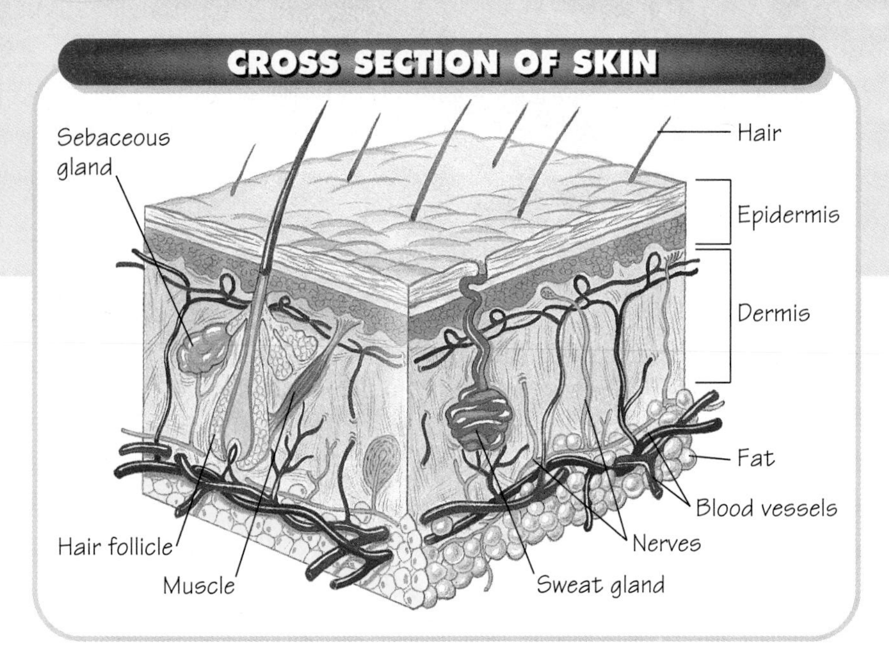

Figure 36–2
The skin consists of two layers—the epidermis and the dermis. The layer of fat in the dermis provides insulation.

As they near the surface, these cells die, but the keratin within remains, forming a tough, waterproof layer on the surface of the skin. As you walk, work, and bathe, you wash and scrape off millions of these dead cells every day. But don't worry, your skin produces more than enough new cells to take their place.

Cells that produce pigments, called melanocytes, are also found in the epidermis. They contain granules of **melanin,** a dark-brown pigment that gives skin its color. Most people have the same number of melanocytes in their skin, but dark-skinned people produce more melanin than people with light skin.

Dermis

The dermis supports the epidermis and contains cells such as nerve endings, blood vessels, and smooth muscles. When you feel something by touching it with your fingers, touch receptors in your dermis are picking up the sensation.

The dermis reacts to the body's needs in a number of ways. For example, on cold days—when you need to conserve heat—blood vessels in the dermis narrow, limiting the loss of heat. When the body is hot, the same vessels widen, bringing more blood to the skin. This warms the skin, causing the loss of heat.

The dermis also contains two types of glands—sweat glands and sebaceous (suh-BAY-shuhs), or oil, glands. Sweat glands produce a watery secretion that contains ammonia, salt, and other compounds. These glands are controlled by the nervous system, which activates them to cool the body when it gets too hot. Sebaceous glands produce an oily secretion that helps keep the epidermis flexible and waterproof.

Hair and Nails

Hair is produced from columns of cells that are filled with keratin and then die. Clusters of such cells make up **hair follicles,** which are anchored in the dermis. Cells multiply in the base of the follicle, causing the hair to grow longer.

Toenails and fingernails are formed in almost the same way, except that the keratin-forming cells in these tissues form a flattened plate. Nails cover and protect the tips of your fingers and toes.

Section Review 36–1

1. **Identify** the basic structures of the integumentary system.
2. **Critical Thinking—Inferring** Some scientists are concerned about the destruction of the ozone layer, which prevents the sun's ultraviolet radiation from reaching the Earth. Why might this be of concern for humans?

The Skeletal System

GUIDE FOR READING

- **List** the functions of an internal skeleton.
- **Compare** the three main kinds of joints.

MINI LAB

- **Classify** your joints based on the motions they make.

LIKE THE FRAMEWORK OF A tall building, the human skeleton contains important clues as to the kind of organisms we are. The shape of our hip bones shows that we walk on two legs. The structure of the bones in our hands, especially our thumbs, gives us the ability to hold and grasp objects. And the size and shape of our skull indicate that we have a well-developed nervous system.

Bones

All vertebrates, humans included, have an internal skeletal system. **An internal skeleton provides support for the body, attachment sites for muscles, and protection for internal organs.** The skull protects the brain, and the ribs protect the heart and lungs. Bones store supplies of calcium and phosphorus that can be used by other tissues, and they also produce blood cells.

Bone Structure

Bones are surrounded by a tough membrane called the **periosteum** (per-ee-AHS-tee-uhm). Just inside the periosteum is a dense layer of **compact bone.** Compact bone appears to be solid, but a series of **Haversian** (huh-VER-zhuhn) **canals** containing nerves and blood vessels runs through it. A region of **spongy bone** is usually located just inside the compact bone. Although it is less dense than compact bone, spongy bone is strong and resilient.

Figure 36–3
The human skeleton is made up of four types of connective tissue—bone, cartilage, ligament, and tendon. (a) *This electron micrograph shows a mature bone cell, or osteocyte, that makes up normal bone tissue.* (b) *Like the skeletal system,* (c) *the framework of this building provides structure and support.*

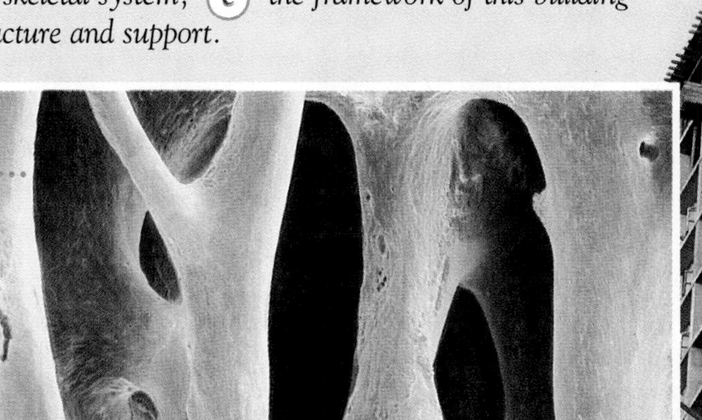

Figure 36–4
The human skeleton is made up of 206 bones. The axial skeleton, colored blue, includes the skull, the vertebral column, and the rib cage. The appendicular skeleton, colored beige, includes the bones of the arms, legs, hands, and feet.

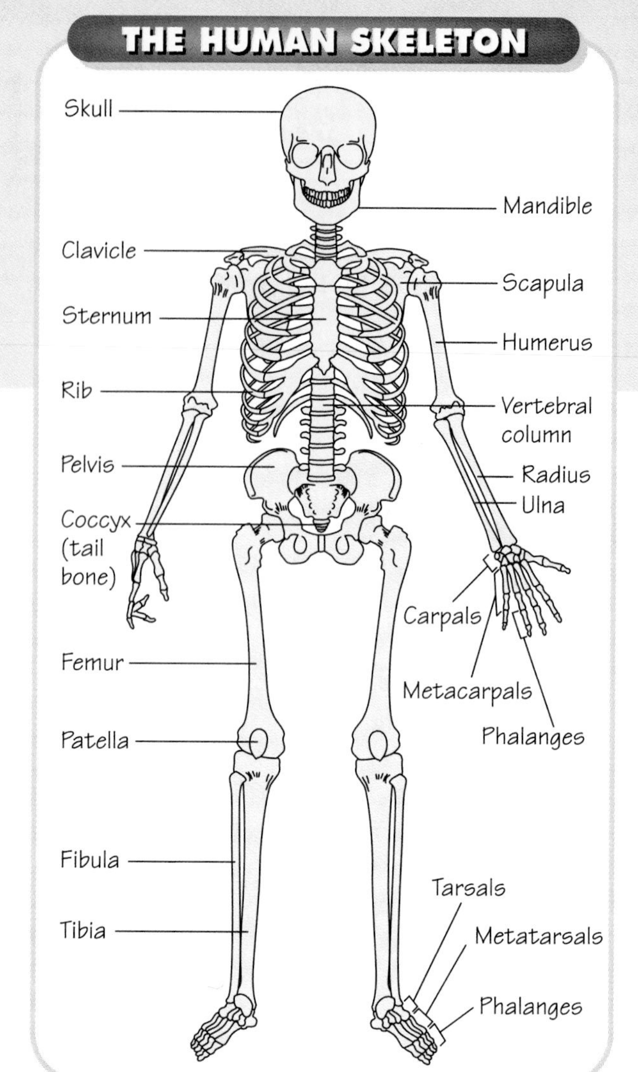

Embedded in both compact and spongy bone are cells called **osteocytes** (AHS-tee-oh-sights). Osteocytes help to build and maintain bones. They deposit the minerals that make up bone and can reabsorb them when the body needs them elsewhere.

As you can see in **Figure 36–6,** inside many larger bones is a region of blood-forming tissue called **bone marrow.** White and red blood cells are produced in the bone marrow. The marrow also plays an important role in the immune response.

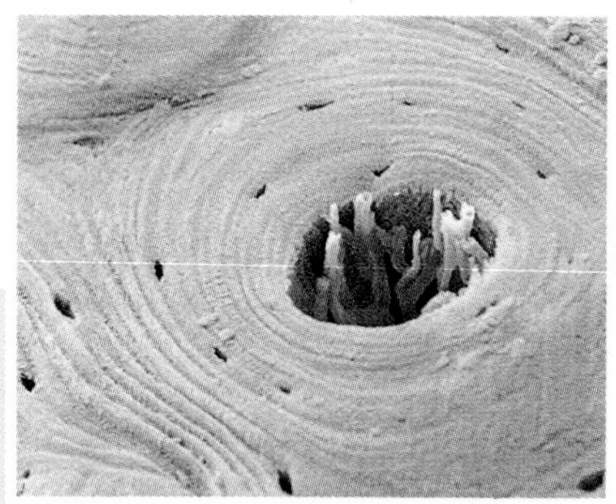

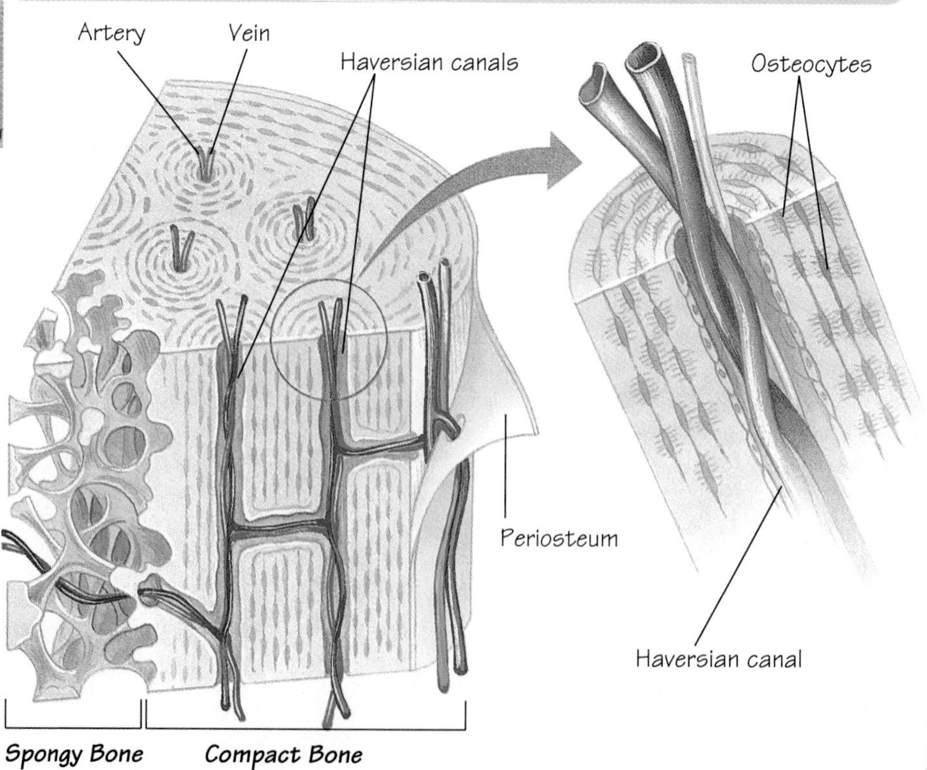

Figure 36–5
(a) This illustration shows the structures of compact and spongy bone. Running through compact bone is a network of tubes called Haversian canals, which contain blood vessels and nerves. (b) In this scanning electron micrograph of compact bone, some blood vessels can be seen in a Haversian canal (magnification: 315X).

All the Right Moves

PROBLEM *How can the motions you make be used to classify your joints?*

PROCEDURE

1. Try each of the actions listed below. Notice which freely movable joints are used and the kind of motion in each joint:
   - waving a paper fan
   - looking behind yourself
   - shrugging your shoulders
   - rotating your index finger
   - pushing open a door
   - lifting a backpack
2. Compare the motion in each joint with the illustration on page 812.

ANALYZE AND CONCLUDE

1. Which type of freely movable joint did you use in each action? Construct a table to record your answer.
2. Which joint was the easiest to classify? Which was the hardest? Why do you think this is so?

Bone Growth

Bones are produced from cartilage. During embryonic development, the human skeleton first appears almost as a cartilage "scale model." Gradually, this cartilage "model" is replaced by bone.

Osteocytes near the surface of the bone begin this process, gradually moving inward, depositing minerals that replace the cartilage with bone. The long bones of the arms and legs, for example, have growing points at either end called growth plates. Cartilage is produced at these plates and then is gradually replaced by bone as the skeleton enlarges. By the time you have stopped increasing in height—usually between the ages of 18 and 20—the growth plates have disappeared and the bone has reached its final size and shape.

Joints

Joints are the places where two bones meet. Your ability to move depends not only on your muscles, but also on the joints that must allow bones to move smoothly past each other. There are three kinds of joints—**fixed, slightly movable,** and **freely movable.**

Fixed joints allow little or no movement between bones. Some of the most important fixed joints are those located in the skull. Skull

THE STRUCTURE OF BONE

Figure 36–6
In a typical bone, such as the femur, blood vessels pass through the periosteum, carrying oxygen and nutrients to the bone. In addition to red marrow, most bones also contain yellow marrow, which is made up of blood vessels, nerve cells, and fat.

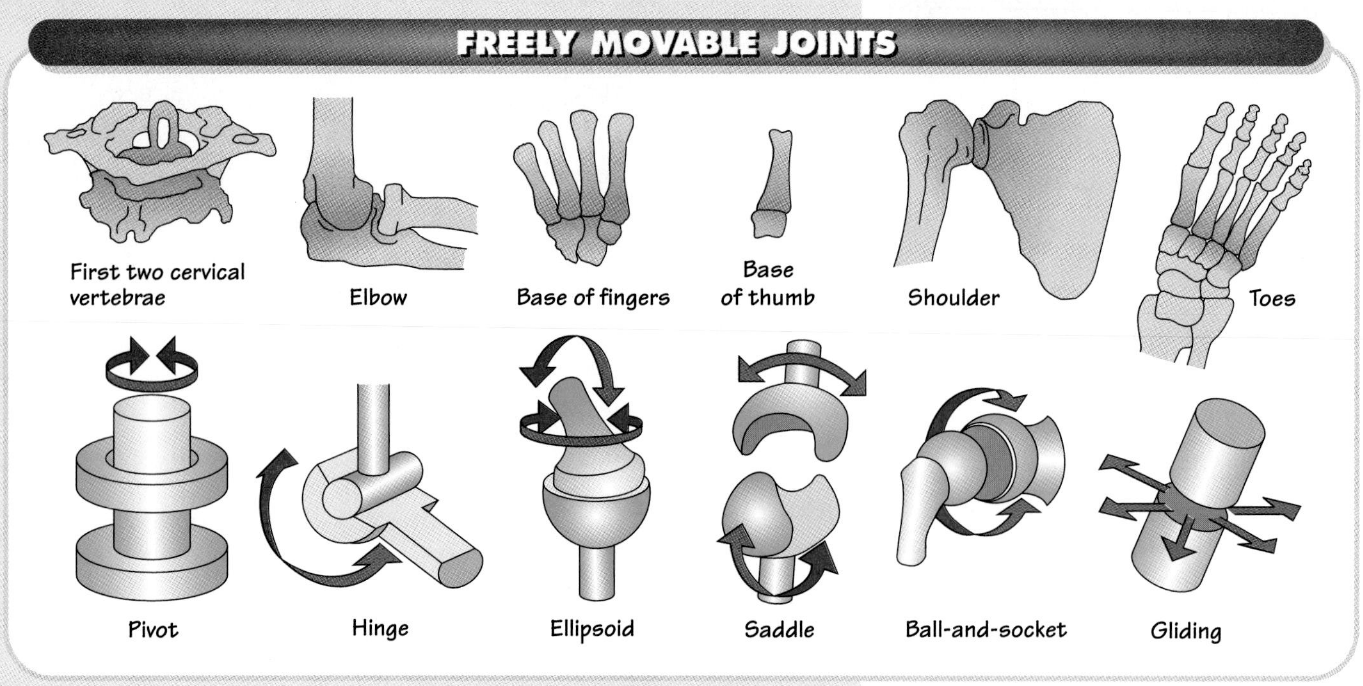

FREELY MOVABLE JOINTS

First two cervical vertebrae
Elbow
Base of fingers
Base of thumb
Shoulder
Toes
Pivot
Hinge
Ellipsoid
Saddle
Ball-and-socket
Gliding

Figure 36–7
The six types of freely movable joints allow a wide range of motion. A pivot joint allows for rotation of one joint around another; a hinge joint permits back-and-forth motion; an ellipsoid joint allows for a hinge-type movement in two directions; a saddle joint allows for movement in two planes; a ball-and-socket joint permits circular movement; and a gliding joint permits a sliding motion of one bone over another.

bones do not move because their purpose is to protect the brain and sense organs in the head.

Slightly movable joints allow a small amount of movement. The bones of the spinal column as well as the ribs are slightly movable joints. The bones of the spinal column are separated from each other by pads of cartilage, called disks. Cartilage is extremely flexible and absorbs most of the shocks and strains of everyday activity.

Freely movable joints allow a wide range of movement. Ball-and-socket joints, located in the shoulders and hips, allow the widest range of movement. The other types of freely movable joints are shown in ***Figure 36–7***.

The ends of bones in freely movable joints are covered with layers of cartilage, providing a smooth surface at the point of contact. The joint itself is enclosed by a joint capsule. The capsule may include **ligaments**—tissues that connect the bones of the joint—and **tendons,** which connect muscles and bones. Inside the capsule is **synovial** (sih-NOH-vee-uhl) **fluid,** a natural lubricant that reduces friction and allows the cartilage-coated bones to slip past each other easily.

Section Review 36–2

1. **List** the functions of an internal skeleton.
2. **Compare** the three main kinds of joints.
3. **Critical Thinking—Inferring** Why do you think that the amount of cartilage as compared to bone decreases as a person develops?
4. **MINI LAB** How can your motions be used to **classify** your joints?

The Muscular System

SECTION 36–3

GUIDE FOR READING

- **Compare** the three types of muscle tissue.
- **Describe** the process of muscle contraction.

MINI LAB

- **Compare** the actions of the muscles of your left hand with those of your right hand.

DESPITE THE FANTASIES OF Hollywood horror films, the skeleton cannot move by itself. Muscles provide the forces that put the body into motion. More than 40 percent of the mass of the human body is muscle—making it the most common tissue in the body. The muscular system includes the large muscles, which athletes proudly display as signs of physical development. But it also includes thousands of tiny muscles throughout the body that regulate blood pressure, move food through the digestive system, and power every movement of the body—from the blink of an eye to the hint of a smile.

Muscle Tissue

There are three types of muscle tissue—**skeletal, cardiac,** and **smooth**—that are specialized for a different job in the body. Each of these three types of muscle has a different cellular structure.

Skeletal Muscle Tissue

Skeletal muscle tissue is generally attached to the bones of the skeleton and is usually under voluntary control.

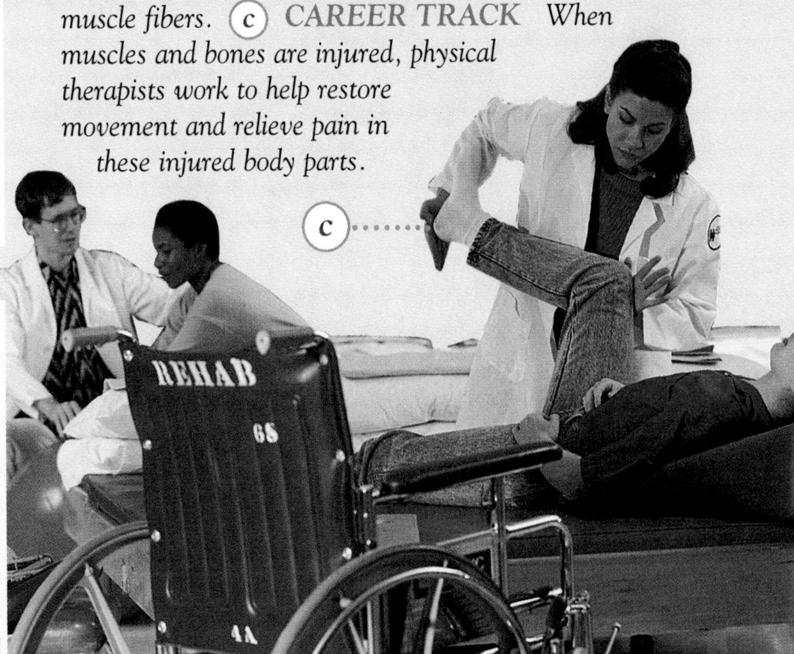

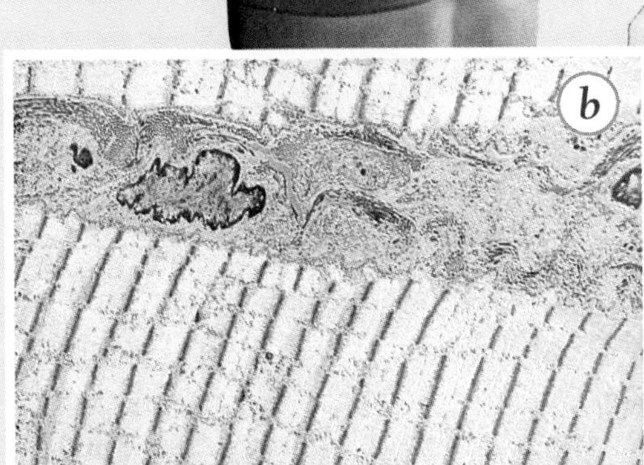

Figure 36–8 *Muscle tissue is found everywhere within the body—from beneath the skin to deep within the body. (a) Skeletal muscles are generally connected to bones and are at work every time we move. (b) This transmission electron micrograph of skeletal muscle shows the banding pattern of the myofibrils—the units that make up muscle fibers. (c) CAREER TRACK When muscles and bones are injured, physical therapists work to help restore movement and relieve pain in these injured body parts.*

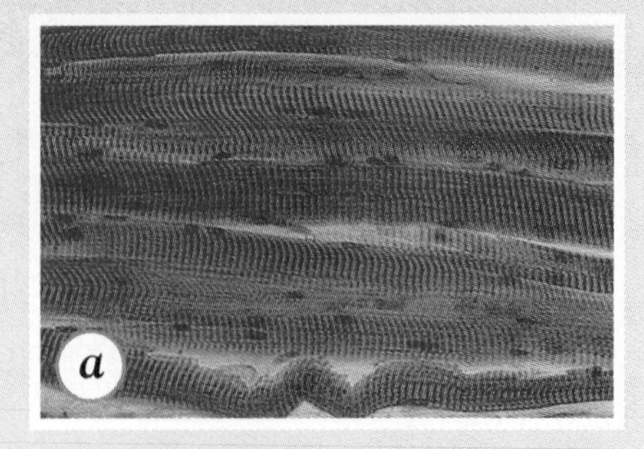

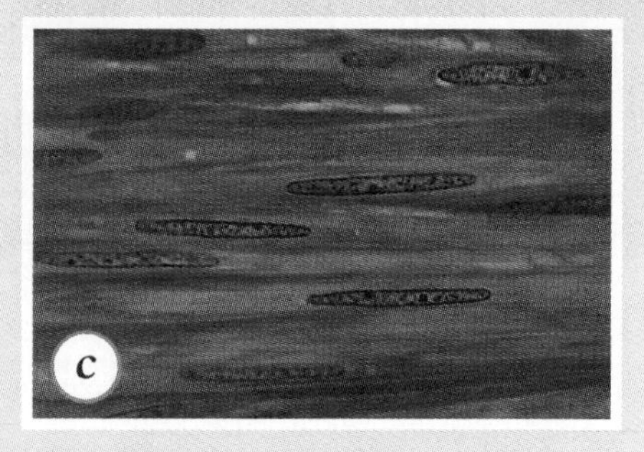

Figure 36–9
There are three types of muscle tissue—skeletal, cardiac, and smooth. (a) Because skeletal, or striated, muscle cells are long and slender, they are often called muscle fibers rather than muscle cells (magnification: 140X). (b) Unlike skeletal muscle tissue, cardiac muscle tissue contracts without direct stimulation by the nervous system (magnification: 200X). (c) Smooth muscle tissue is found in many internal organs and in the walls of many blood vessels. Their contractions move food through the digestive system and control the flow of blood through the circulatory system (magnification: 360X).

Skeletal muscle tissue is behind every conscious movement you make, whether you are lifting a weight or tying your shoelaces. This is because most skeletal muscle tissue is controlled directly by the nervous system. Skeletal muscle cells have many nuclei because they form during development from the fusion of scores of individual cells.

Skeletal muscle cells can be very large—as long as 60 centimeters in the cells in the large muscles of your arms and legs! Under the light microscope, skeletal muscle cells appear striated, or striped. For this reason, skeletal muscle tissue is often called striated muscle tissue.

Figure 36–10
Skeletal muscles are made up of densely packed muscle fibers. Each muscle fiber is made up of many thin fibers called myofibrils. Each myofibril, in turn, is made up of both thick contractile filaments—myosin—and thin contractile filaments—actin.

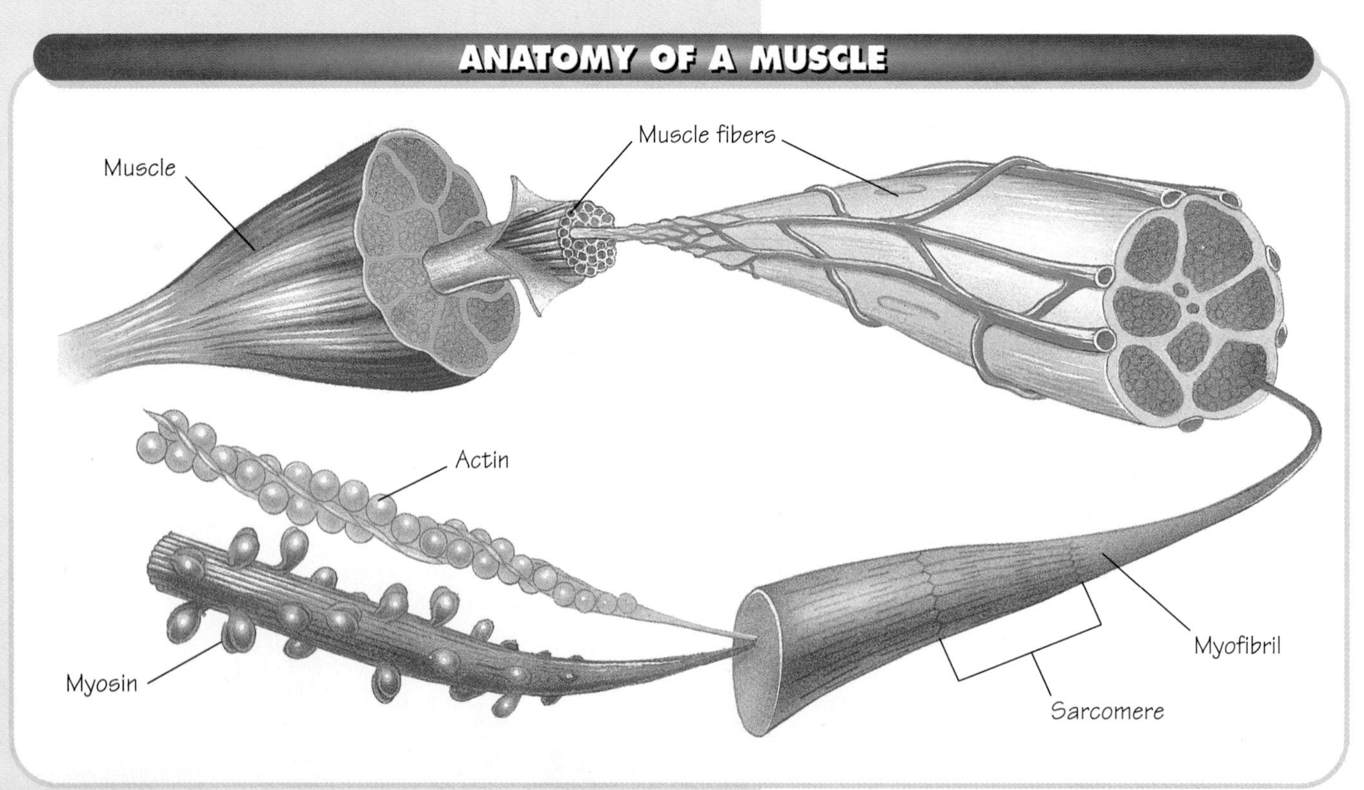

A Stimulating Situation

Your class is discussing the muscular system and muscular contraction. One of the students asks what causes a muscle twitch. The teacher responds by saying that a muscle twitch is a quick contraction of the muscle followed by immediate relaxation.

Another student wonders how you can maintain a muscle contraction over a length of time. The teacher explains that when there is a continuous stream of nerve impulses, the muscle fibers receive new stimulation before they are able to relax. The individual muscle twitches together form a smooth, continuous contraction called tetanization. If the muscle is continuously stimulated for a long period of time, however, it would be unable to respond to further stimulation and fatigue would set in.

The teacher shows the class the following graph to further explain muscular contraction. Look at the graph and answer the questions that follow.

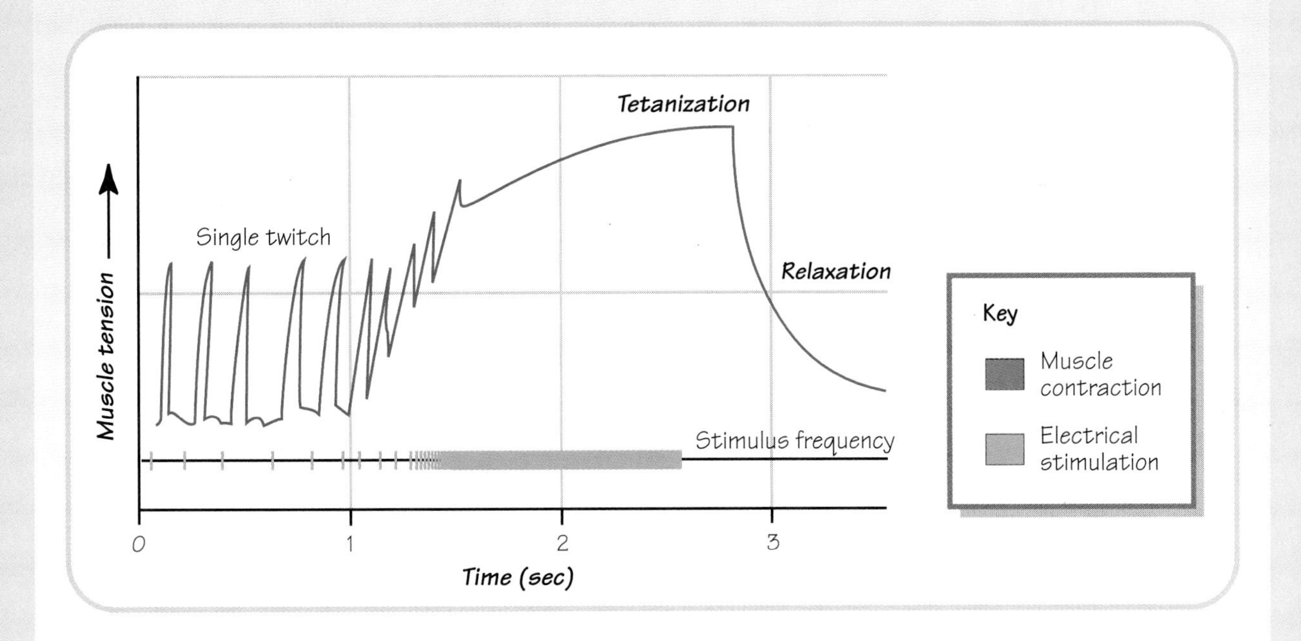

THINK ABOUT IT

1. Describe what happens when the stimulus frequency is slow.
2. Describe what happens when the stimulus frequency increases.
3. According to this graph, when does tetanization occur?
4. Fatigue can occur only in skeletal muscles. Can you explain why this is so?

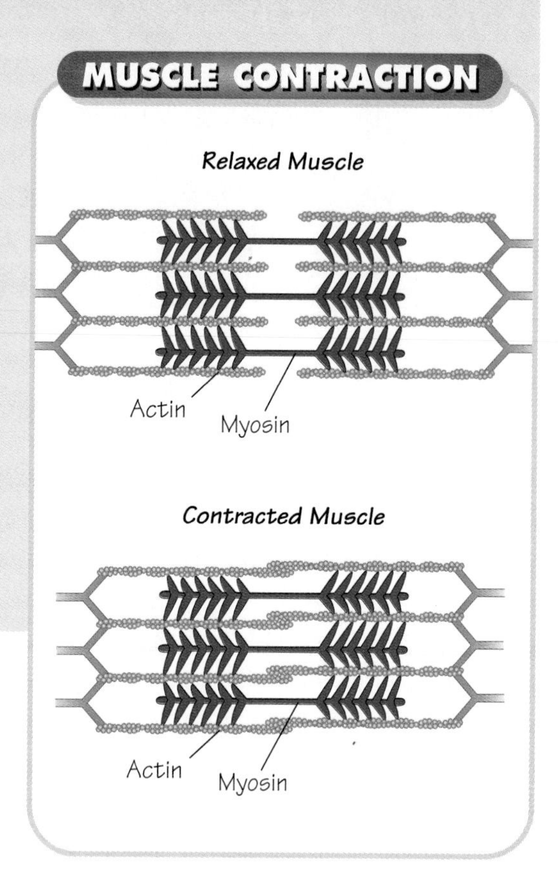

Figure 36–11 When a muscle is stimulated to contract, the myosin and actin filaments slide past each other, causing the muscle cells to shorten and the muscle to contract.

Cardiac Muscle Tissue

Cardiac muscle tissue is found in just one place in the body—the heart. **Cardiac muscle tissue is striated, but the smaller cardiac muscle cells have just one nucleus, and they are not under the direct control of the central nervous system.** Adjacent cardiac muscle cells form branching fibers that allow nerve impulses to move from cell to cell.

Smooth Muscle Tissue

The cells of smooth muscle tissue are spindle-shaped, have a single nucleus, and are not striated. Smooth muscle tissue is generally not under the conscious control of the nervous system. Smooth muscle tissue is found in the walls of many internal organs, except the heart. It is responsible for actions not under voluntary control, such as digestion. Although smooth muscle tissue contracts much more slowly than skeletal muscle tissue does, it can maintain its contraction for a much longer period of time.

Muscle Structure

To understand how a muscle works, we can look inside the kind of muscle cell that we understand best—the skeletal muscle cell. Scientists have learned a great deal about skeletal muscle, partly because the regular structure of these cells has made them easy to study.

Under the electron microscope, we can see that the striations in these cells are actually formed by an alternating pattern of thick and thin filaments. The thick filaments are made of a protein called **myosin.** Thin filaments contain another protein, called **actin.** These thick and thin filaments overlap in a regular pattern, producing the striations that appear under the light microscope.

These tiny actin and myosin filaments are the force-producing engines that cause a muscle to contract. One end of the myosin molecule sticks out just enough from the thick filament to form a cross-bridge that makes firm contact with the thin filament. Using the energy supplied by the splitting of ATP, the cross-bridge changes shape, pulling the thin filament along. The cross-bridge then releases the thin filament, snaps back to its original position, and "grabs" the thin filament again to start another cycle. The cross-bridges repeatedly bend, release, and reattach farther along the thin filament.

When hundreds of thousands of actin-myosin cross-bridges go through their cycle in a fraction of a second, the muscle cell contracts with considerable force. This process is known as the **sliding filament theory** of muscle contraction because the thick and thin filaments slide past each other as the muscle shortens. One molecule of ATP supplies the energy for each cycle.

☑ **Checkpoint** What is myosin? Actin?

Muscle Contraction

To make well-coordinated movements, muscle contractions must be carefully controlled. In most skeletal muscles, this is the job of motor neurons.

A single motor neuron may form synapses to one or several muscle cells. An impulse in the motor neuron causes the release of a neurotransmitter, **acetylcholine** (as-ih-tihl-KOH-leen), which causes a new action potential. This, in turn, causes the release of calcium ions into the cytoplasm of muscle cells. When calcium flows into the cytoplasm, cross-bridges form, and the muscle contracts.

The contraction stops when enzymes break down the neurotransmitter. Calcium is then removed from the cytoplasm, and the contraction stops.

Whose Side Are You On?

PROBLEM *How do the actions of the muscles of your left hand* ***compare*** *with those of your right hand?*

PROCEDURE

1. Count the number of times you can fully open a spring clothes pin with your right thumb and right index finger for 2 minutes. Have your partner time you. Record the results.
2. Rest for 1 minute and repeat step 1. Record the results.
3. Repeat steps 1 and 2 with your left hand. Record the results.

ANALYZE AND CONCLUDE

1. Compare the two performances of your right-hand muscles.
2. How did the performance of your right hand compare with that of your left hand?
3. What conclusions can you draw from these results?

Muscles and Movement

Muscles produce force by contracting. Attached to bones by tendons, a muscle can pull two bones together, using the joint between them as a lever. When you lift a heavy object or pull something close to you, it's easy to see how muscle contraction produces the movement.

An individual muscle can pull by contracting, but it cannot push. If that is true, then how can you push a door open or do a pushup in gym class? The answer is that most skeletal muscles are arranged in pairs. These pairs oppose each other and produce forceful movements in either direction. Your upper arm, for example, contains one muscle—the biceps—that flexes, or bends, the arm. On the other side of the upper arm is another muscle—the triceps—that extends the arm. When you do a pushup, the triceps contracts, forcing the arm to extend and push down on the gym floor.

Section Review 36-3

1. **Compare** the three types of muscle tissue.
2. **Describe** the process of muscle contraction.
3. **Critical Thinking—Relating Concepts** Why are skeletal muscles arranged in pairs?
4. **MINI LAB** How do the actions of the muscles of your left hand **compare** with those of your right hand?

SECTION 36–4 BRANCHING OUT In Depth

The Biology of Exercise

Guide for Reading

- **List** the two main types of muscle fibers in skeletal muscles.
- **Compare** aerobic exercises and resistance exercises.

AN ATHLETE MAY DEVOTE hours every day to developing and conditioning the muscular system. To a runner, a swimmer, or a weight lifter, the strength, flexibility, and endurance of the muscular system is the difference between winning and losing. But a well-conditioned muscular system is important even to those of us who are not competitive athletes. We depend on our skeletal muscles—whether walking, running, working, or playing—and we can make many of life's everyday challenges easier by keeping our muscles in good condition.

Specialized Skeletal Muscle Fibers

If you look at any large office that performs many different jobs, you will find specialists. One person might sell the product, another might handle the bookkeeping, and another the scheduling. What's the advantage of an arrangement like this? By allowing a few people to specialize at different tasks, the jobs get done more quickly and efficiently.

Even though the job description for a muscle is simple—contract on command—a large skeletal muscle contains specialists of its own. **Skeletal muscles contain two main types of muscle fibers—red and white—whose properties make them specialists at different kinds of exercise.**

Red muscle fibers contain large amounts of the reddish oxygen-storing protein **myoglobin.** Red fibers also have rich blood supplies and plenty of mitochondria to produce ATP through aerobic

Figure 36–12
Different types of exercise affect different types of skeletal muscle fibers. (a) Aerobic exercises, such as bicycling, affect red muscle fibers, whereas (b) resistance exercises, such as weight lifting, affect white muscle fibers. (c) When planning an exercise program, both aerobic and resistance-type exercises should be included.

(oxygen-dependent) respiration. Red fibers are also called slow-twitch muscle fibers because they contract slowly after being stimulated by a motor neuron.

Red muscle fibers are usually able to meet their ATP needs from their own mitochondria. Red muscle fibers enable muscles to contract again and again, against slight resistance, without fatigue.

White muscle fibers contain little or no myoglobin, giving them a pale color. They have few mitochondria, but they store large reserves of **glycogen.** Glycogen is a compound that stores excess glucose in the body. When white muscle fibers need extra ATP, that glycogen is broken down to glucose, which is then used anaerobically (without oxygen) to generate ATP by fermentation.

White fibers are also called fast-twitch muscle fibers and can generate powerful contractions. Having few mitochondria, these fibers contain greater densities of contractile proteins than red fibers do. These powerful fibers fatigue easily, however, which means that they can produce maximum contractions for only a few seconds at a time.

Exercise and Muscle Cells

As you probably know, exercising a muscle causes it to get stronger. There are different kinds of exercise, of course, and each kind has different effects on each kind of skeletal muscle cell.

Aerobic exercises—such as running, swimming, and bicycling—cause your body systems to become more efficient. For example, your circulatory system benefits because the number of capillaries in the muscle increases, which increases its blood supply. More myoglobin is synthesized, especially in red fibers. Aerobic exercises also benefit the heart and lungs, helping to increase their capacity. All these effects increase physical endurance—the ability to perform an activity without fatigue. However, aerobic exercises do not result in large increases in muscle size.

Resistance exercises, such as weight lifting, increase muscle size. This effect is most pronounced in the white fibers, which grow in size and add more contractile proteins after such exercises.

Resistance exercises have little effect on red fibers, which means that they do not improve endurance. However, they do have a dramatic effect on maximum muscle strength. Because most skeletal muscles function in opposing pairs, weight training should include exercises that develop both muscles in each pair. This helps an individual to maintain coordination and flexibility.

Different forms of exercise develop different types of muscle fibers. This means that the best exercise programs should blend resistance and aerobic training, helping to develop muscles, heart, and lungs.

Section Review 36–4

1. **List** the two main types of muscle fibers found in skeletal muscles.
2. **Compare** aerobic exercises and resistance exercises.
3. **BRANCHING OUT ACTIVITY** Use a health book as a guide and **design** and **write** an exercise program for yourself that includes both aerobic and resistance exercises. Do not perform your exercise program unless it has been checked and approved by your doctor.

Laboratory Investigation

Mapping Your Sweat Glands

A solution of iodine reacts with starch by turning a blue-black color. When dried iodine comes in contact with the water given off by sweat glands, it becomes a solution again and will react with the starch in a piece of blotting paper, causing it to turn a blue-black color. In this investigation, you will take advantage of these changes to locate sweat glands in your skin.

Problem

Which areas of your skin have the most sweat glands? **Make a prediction** to answer this question.

Materials (per group)

4 squares of blotting paper, 1 cm × 1 cm
medical adhesive tape
iodine solution
cotton swabs

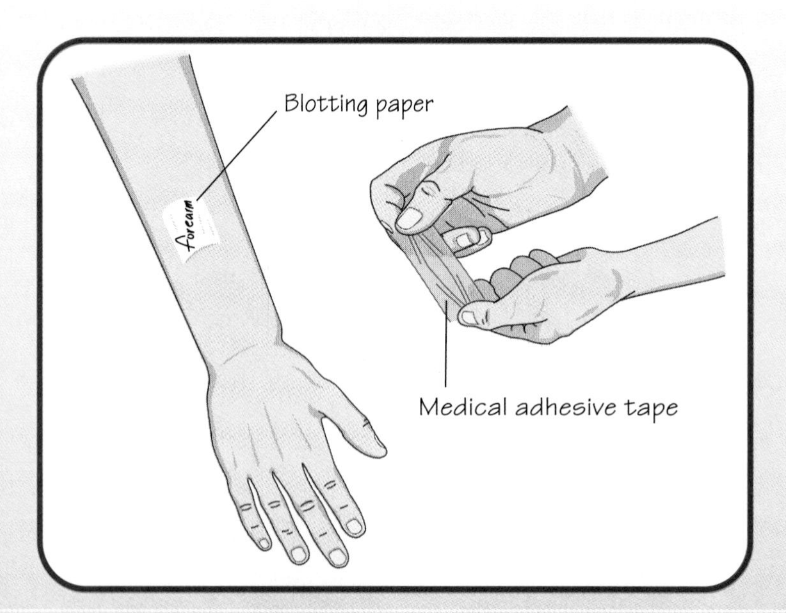

Procedure

1. Wash and dry your hands and forearms thoroughly. Roll your sleeves up to above the elbow. Carefully dip the cotton swab into the iodine solution. **CAUTION:** *Be careful when using iodine because it stains clothing.*
2. Find an area in the center of the palm of one of your hands that is free of creases. Using the iodine-soaked cotton swab, paint a 2-cm square on your palm and let it dry.
3. Repeat step 2 on the inside of one of your forearms.
4. Using a pencil, label one of the blotting-paper squares "Palm" and the other one "Forearm." Have your partner tape the blotting paper square labeled "Palm" over the iodine on your palm and the square labeled "Forearm" over the iodine on your forearm. Allow the blotting paper to remain in place for 20 minutes.

5. **While you are waiting, write a prediction about whether the palm of your hand or your forearm has more sweat glands. Also predict whether all members of the class will have the same results.**

6. **After 20 minutes, have your lab partner remove the two paper squares. Count the number of blue-black dots on each one. Each dot indicates the presence of an active sweat gland.**

7. **Construct a data table similar to the one shown. Record the number of dots on the paper square for your palm. In the space marked "Distribution of Sweat Glands," draw or write a description of the pattern of dots.**

Observations

1. What difference did you observe in the number of sweat glands per square centimeter on your palm and forearm?
2. What differences were there in the patterns of sweat glands?

DATA TABLE

Location of Sweat Glands	Number of Active Sweat Glands	Distribution of Sweat Glands
Middle surface of palm		
Inside of forearm		

Analysis and Conclusions

1. Which area tested—palm or forearm—had the greatest density of active sweat glands?
2. Were your predictions correct? Explain your answer.
3. How did the information in your data table compare with the information your classmates compiled?
4. What conclusion can you draw about the distribution of sweat glands on the skin?
5. Do all sweat glands produce the same amount of sweat? What evidence did you observe to support your answer?

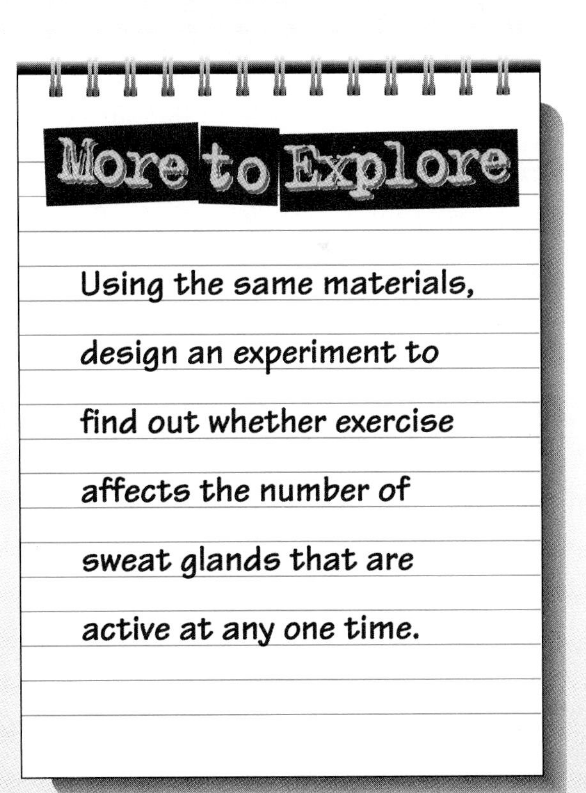

Study Guide

Summarizing Key Concepts

The key concepts in each section of this chapter are listed below to help you review the chapter content. Make sure you understand each concept and its relationship to other concepts and to the theme of this chapter.

36–1 The Integumentary System

- The integumentary system includes your skin, hair, nails, and a number of important glands in the skin.

36–2 The Skeletal System

- An internal skeleton provides support for the body, attachment sites for muscles, and protection for internal organs.
- There are three kinds of joints in the body. Fixed joints allow little or no movement between bones. Slightly movable joints allow a small amount of movement. Freely movable joints allow a wide range of movement.

36–3 The Muscular System

- Skeletal muscle tissue is generally attached to the bones of the skeleton and is usually under voluntary control. Cardiac muscle tissue is striated, but the smaller cardiac muscle cells have one nucleus and are not under the direct control of the central nervous system. Smooth muscle tissue is generally not under the conscious control of the nervous system.
- When hundreds of thousands of actin-myosin cross-bridges go through their cycle in a fraction of a second, the muscle cell contracts with considerable force.

36–4 The Biology of Exercise

- Skeletal muscles contain two main types of muscle fibers—red and white—whose properties make them specialists at different kinds of exercise.
- Aerobic exercises—such as running, swimming, and bicycling—cause your body systems to become more efficient. Resistance exercises, such as weight lifting, produce an increase in muscle size.

Reviewing Key Terms

Review the following vocabulary terms and their meaning. Then use each term in a complete sentence.

36–1 The Integumentary System

skin
epidermis
dermis
keratin
melanin
hair follicle

36–2 The Skeletal System

periosteum
compact bone
Haversian canal
spongy bone
osteocyte
bone marrow
fixed joint
slightly movable joint
freely movable joint
ligament
tendon
synovial fluid

36–3 The Muscular System

skeletal muscle tissue
cardiac muscle tissue
smooth muscle tissue
myosin
actin
sliding filament theory
acetylcholine

36–4 The Biology of Exercise

myoglobin
glycogen

Recalling Main Ideas

Choose the letter of the answer that best completes the statement or answers the question.

1. The cells that form keratin are found in the

a. ligaments. **c.** bone marrow.
b. myoglobin. **d.** epidermis.

2. The dermis helps to regulate body temperature by responses in the

a. oil glands.
b. blood vessels.
c. sweat glands and blood vessels.
d. sweat glands.

3. The pigment that gives skin its color is called

a. melanocyte. **c.** melanin.
b. keratin. **d.** osteocyte.

4. Cartilage is part of the

a. nervous system.
b. integumentary system.
c. muscular system.
d. skeletal system.

5. Haversian canals contain

a. bone cells.
b. nerves and blood vessels.
c. periosteum.
d. calcium phosphate crystals.

6. Fixed joints are most flexible

a. at birth. **c.** in childhood.
b. in adolescence. **d.** in adulthood.

7. In freely movable joints, the ends of bones are covered with

a. spongy bone. **c.** bone marrow.
b. cartilage. **d.** blood vessels.

8. Which contains the greatest amount of skeletal muscle tissue?

a. cerebrum **c.** small intestine
b. kidney **d.** foot

9. The tissues that connect muscles to bones are called

a. ligaments. **c.** sliding filaments.
b. tendons. **d.** cartilage.

10. Which of the following is true of resistance exercises?

a. increase the capacity of the heart
b. increase the number of capillaries
c. increase muscle size
d. increase endurance

Putting It All Together

Using the information on pages xxx to xxxi, complete the following concept map.

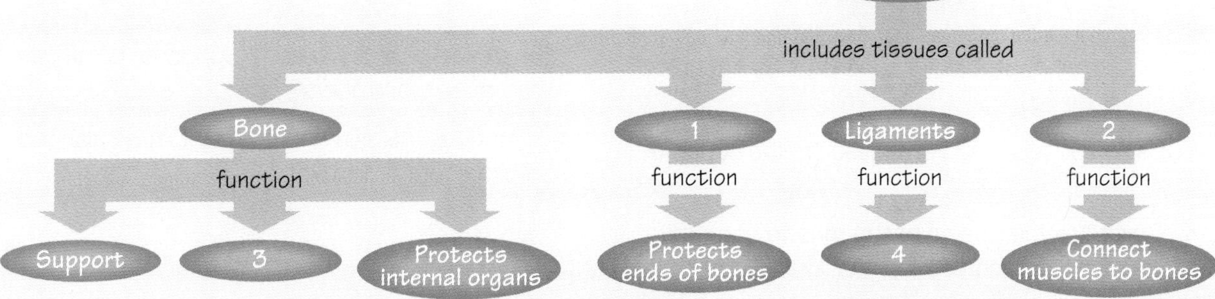

Assessment

Reviewing What You Learned

Answer each of the following in a complete sentence.

1. What is the integumentary system?
2. Which structures does keratin form in humans and other animals?
3. Which glands are found in the dermis?
4. How are the formation of hair and nails related?
5. What kinds of tissue make up an internal skeletal system?
6. How does compact bone differ from spongy bone?
7. What is the function of osteocytes?
8. List three examples of freely movable joints.
9. Explain the purpose of synovial fluid.
10. How do the tissues that make up the skeleton change during human development?
11. What are the three types of muscles and where are they found?
12. What is acetylcholine?
13. How can muscles, which only contract, enable the skeleton to push as well as pull?
14. What is myoglobin?
15. Explain the benefits of aerobic exercises on the body.

Expanding the Concepts

Discuss each of the following in a brief paragraph.

1. What is meant when a biologist says that "we are alive because our surface is dead"?
2. How does the dermis differ from the epidermis?
3. Name three main classes of joints and their differences.
4. Tendinitis is an inflammation of the tendon. Explain what probably happens when a tennis player develops tendinitis.
5. Using a labeled diagram, describe the structure of a joint.
6. **Compare** the three types of muscle tissue by their function and cellular structure.
7. Describe the sliding filament theory of muscle contraction.
8. Describe the role of calcium in the skeletal system.
9. If an athlete has a greater number of red muscle fibers, for what kinds of events would he or she be more suited? Suppose the athlete had more white fibers?
10. How does the skin help to maintain a balanced environment in humans?

Extending Your Thinking

Use the skills you have developed in this chapter to answer the following.

1. **Classifying** Two athletes have had their muscle tissue analyzed. One athlete's leg muscles contained between 80 and 90 percent red muscle fibers, while the other athlete's leg muscles had almost 70 percent white muscle fiber. Which athlete would you classify as a sprinter, and which one as a marathon runner?
2. **Using the writing process** At birth, the joints in an infant are flexible and not yet fixed. As the child develops, the bones become more rigid and grow together. Use reference materials to find out why the skull bones are flexible in a newborn, yet rigid as an adult. Then write a brief summary of your findings.
3. **Drawing conclusions** Although exercising can increase your strength and endurance, overexercising can have some adverse effects on the body. What are some adverse effects?
4. **Constructing a model** Using cardboard for bones and string for muscles, construct a working model of the biceps and triceps muscles of the upper arm. Be sure to show the relationship of these muscles to the humerus, ulna, and radius bones.
5. **Interpreting data** The line graph below plots the force of contraction of three different skeletal muscles. Using the data, describe the differences among the three muscles.

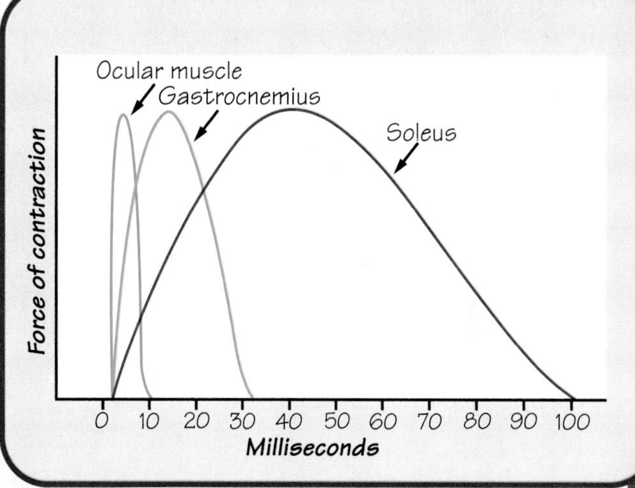

Applying Your Skills

Up Close

Biologists use a microscope to better understand the workings of the human body on a cellular level. In this activity, you'll use a microscope to differentiate among the following muscle tissue: smooth, skeletal, and cardiac.

1. Working with a partner, set up a compound microscope and obtain three "mystery slides," labeled A, B, and C, from your teacher.
2. Construct a data table to record the letter of each slide, the identity of the tissue, and characteristics that support your identification.
3. Place slide A on the stage of the microscope and examine it closely.
4. Discuss with your partner the identity of the cells or tissues. You may have to examine all the slides and identify some by a process of elimination.
5. Complete the data table for slide A and repeat the procedure using slides B and C.

• GOING FURTHER •

6. In your journal, think about the ease or difficulty of this activity. Which tissue was the easiest to identify? Which was the most difficult? Explain your answers.

CHAPTER 37

Circulatory and Respiratory Systems

37–1 The Circulatory System
- Describe the structure and function of the circulatory system.
- Describe the components of blood.

37–2 The Respiratory System
- Describe the structure and function of the respiratory system.

BRANCHING OUT *In Depth*

37–3 The Hazards of Smoking
- Describe the effects smoking has on the body.

LABORATORY INVESTIGATION
- Measure the products of burning tobacco.

Biology and Your World

BIO JOURNAL

This swimmer knows how important exercise is for his circulatory and respiratory systems. In your journal, make a list of things you do that affect your circulatory and respiratory systems. After completing your list, place a check mark next to those that are harmful. Pick one harmful habit and write a paragraph explaining how you will change or eliminate that habit.

A swimmer taking a deep breath

The Circulatory System

SECTION 37–1

GUIDE FOR READING

- **List** the structures of the circulatory system.
- **Identify** the three types of blood vessels.
- **Compare** the functions of red blood cells, white blood cells, and platelets.

MINI LAB

- **Relate** your pulse rate to your activity level.

ONE OF THE SIGNS OF LIFE itself is your heartbeat. Even when you drift off to sleep, your heart beats out a steady rhythm. Why is this process so important that it must be kept going even while you sleep? Does it meet some great need of the trillions of cells that live inside you? It certainly does.

Each breath you take brings air into your body. Oxygen in that air is needed by every one of your cells. Not surprisingly, that oxygen needs to be delivered, and that's where the heart comes in. Its beating produces the force to move oxygen-carrying blood through the circulatory system to every part of your body. The circulatory system supplies cells throughout your body with substances they need to stay alive.

Functions of the Circulatory System

If an organism is composed of a small number of cells, it doesn't really need a circulatory system. Most cells in such organisms are in direct contact with the environment so that oxygen, nutrients, and wastes can easily diffuse across cell membranes from the outside.

Larger organisms, however, don't have this advantage. They need a circulatory system. Most of their cells are not in direct contact with the environment, and the substances made in one part of the organism may be needed in another part. In a way, this is the same problem faced by people who live in a large city. What is needed, of course, is a transportation system that moves people, goods, and waste materials from one place to another. The transportation

Figure 37–1
The circulatory system consists of the heart, the blood vessels, and the blood. (a) A measure of the pressure produced by the contractions of the heart and the muscles surrounding the heart is known as blood pressure. (b) The heart, which is made up almost entirely of cardiac muscle tissue, contracts at regular intervals, forcing blood through the circulatory system. (c) The instrument that measures blood pressure is called a sphygmomanometer.

Figure 37–2 In this diagram, the arrows show the path of blood through the body. The path of oxygen-poor blood is shown in blue. The path of oxygen-rich blood is shown in red.

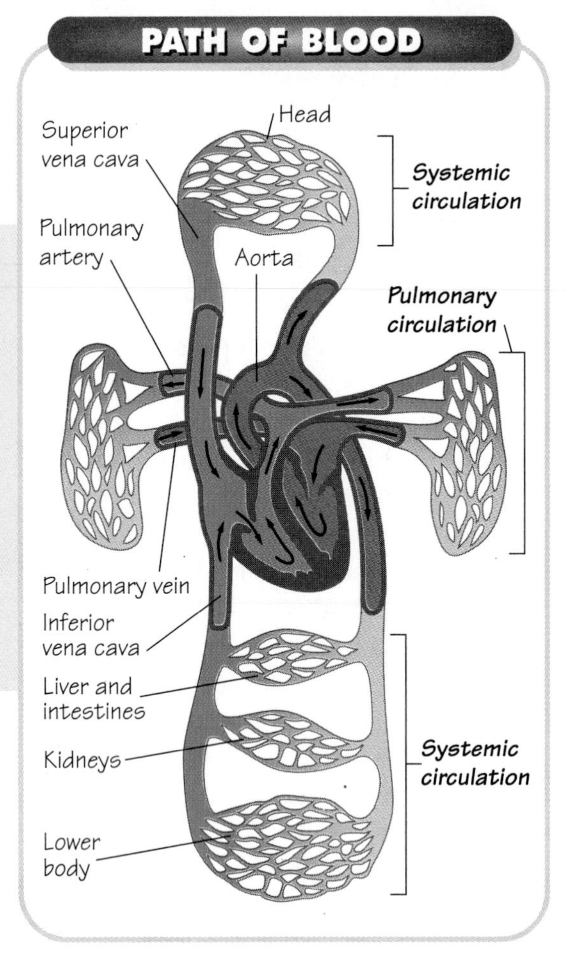

system of a city is its streets, highways, and rail lines. The transportation system of a living organism is its circulatory system.

Humans and other vertebrates have closed circulatory systems. This means that a circulating fluid—called **blood**—is pumped through a system of vessels. **The human circulatory system consists of the heart, a series of blood vessels, and the blood itself.**

The Heart 

The heart is a hollow organ near the center of the chest composed almost entirely of muscle. The human heart is really like two separate pumps sitting side by side. Each side has two chambers: an **atrium** (AY-tree-uhm; plural: atria), the upper chamber that receives blood, and a **ventricle,** the lower chamber that pumps blood out of the heart. This means that the heart has a total of four chambers—two chambers on each side.

Each side of the heart pumps blood to a different part of the circulatory system. The right side of the heart pumps blood from the heart to the lungs. This pathway is called **pulmonary circulation.** Oxygen-poor blood is pumped by the right side of the heart through the lungs, where it gives off carbon dioxide and picks up oxygen. This oxygen-rich blood is then returned to the left side of the heart by the pulmonary veins.

The left side of the heart pumps blood to the rest of the body. This pathway is called **systemic circulation.** Oxygen-rich blood leaves the heart and supplies the body with oxygen-rich blood. By the time that blood returns from systemic circulation, cells throughout the body have picked up much of its oxygen and loaded it with carbon dioxide. In short, it is ready for another trip to the lungs. ***Figure 37–2*** shows the path of blood through the body.

☑ **Checkpoint** What is pulmonary circulation? Systemic circulation?

Blood Flow Through the Heart

Blood enters the heart through the right and left atria. As the heart contracts, blood is forced first into the ventricles, then out from the ventricles into circulation. When it contracts, why doesn't some blood flow backward from the ventricles into the atria? Special flaps of tissue called **valves** reach across the passageways between the atria and the ventricles. Blood moving from the atria easily forces these valves open. But when the ventricles contract, the valves slam shut, preventing any backflow. There are four valves in the heart. Each valve ensures that blood moves through the heart in a one-way direction and increases the pumping efficiency of the heart.

Visualizing the Heart

The heart is a powerful muscle that continuously pumps blood throughout the body. It is about the size of a fist and is located near the center of the chest. It beats an average of 70 times per minute at rest.

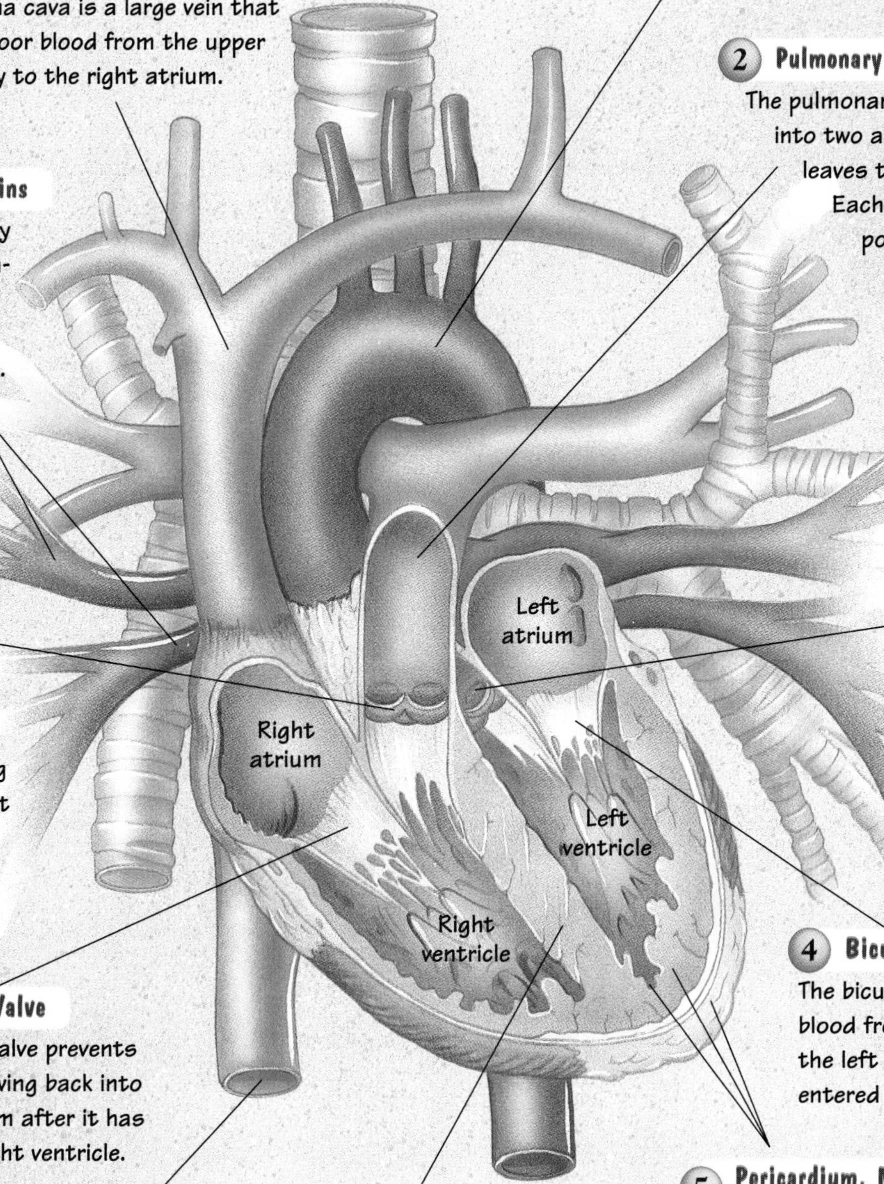

1 Aorta

The aorta is the largest artery in the body. It brings oxygen-rich blood from the left ventricle to the rest of the body.

2 Pulmonary Artery

The pulmonary artery branches into two arteries after it leaves the right ventricle. Each branch brings oxygen-poor blood to one lung.

3 Aortic Valve

The aortic valve prevents blood from flowing back into the left ventricle after it has entered the aorta.

4 Bicuspid Valve

The bicuspid valve prevents blood from flowing back into the left atrium after it has entered the left ventricle.

5 Pericardium, Myocardium, and Endocardium

There are three layers that make up the walls of the heart. The pericardium and endocardium cover and protect the heart tissues. The middle layer, or myocardium, is responsible for pumping the blood.

6 Septum

The septum is a thick muscular wall that divides the heart into two halves and prevents the mixing of oxygen-rich and oxygen-poor blood.

7 Inferior Vena Cava

The inferior vena cava is a vein that brings oxygen-poor blood from the lower part of the body to the right atrium.

8 Tricuspid Valve

The tricuspid valve prevents blood from flowing back into the right atrium after it has entered the right ventricle.

9 Pulmonary Valve

The pulmonary valve prevents blood from flowing back into the right ventricle after it has entered the pulmonary artery.

10 Pulmonary Veins

The two pulmonary veins bring oxygen-rich blood from each of the lungs to the left atrium.

11 Superior Vena Cava

The superior vena cava is a large vein that brings oxygen-poor blood from the upper part of the body to the right atrium.

MINI LAB ········ Relating ········

Feel the Beat

PROBLEM *How does your pulse rate **relate** to your activity level?*

PROCEDURE

CAUTION: *If you have any respiratory or circulatory conditions, do not perform this activity.*

1. Using the first two fingers of one hand, locate the pulse point on the inside of your wrist. It is next to the tendon near your thumb.
2. Lightly place the same two fingers against the same side of your neck near the corner of your jawbone.
3. At each location, count and record the number of pulses in 1 minute.
4. Repeat steps 1 to 3 after exercising in place for 1 minute.

ANALYZE AND CONCLUDE

1. How did your resting pulse rate compare in the wrist and the neck?
2. What effect did activity have on your pulse rate?

The Heartbeat

Although the heart is a single muscle, all of its cells do not all contract at the same time. Instead, cells of the atria contract first, and a wave of contraction spreads from the right atrium over the rest of the heart. This pattern of contraction makes the heart a more efficient pump, squeezing blood from one chamber to the next. How does this wavelike contraction happen?

Each contraction begins in a small group of cells in the right atrium. Because these cells "set the pace" for the heart as a whole, they are called the **pacemaker.** From the pacemaker, the contraction impulse is spread from cell to cell, producing a wave of contractions that reach all four chambers.

Blood Vessels

Blood leaving through the heart travels through a series of blood vessels that will carry it on its round trip through the body and back to the heart. **The three types of blood vessels that blood**

INTEGRATING TECHNOLOGY AND SOCIETY

Some people need an artificial pacemaker to help maintain a steady heart rate. Use reference materials to find out how an artificial pacemaker works.

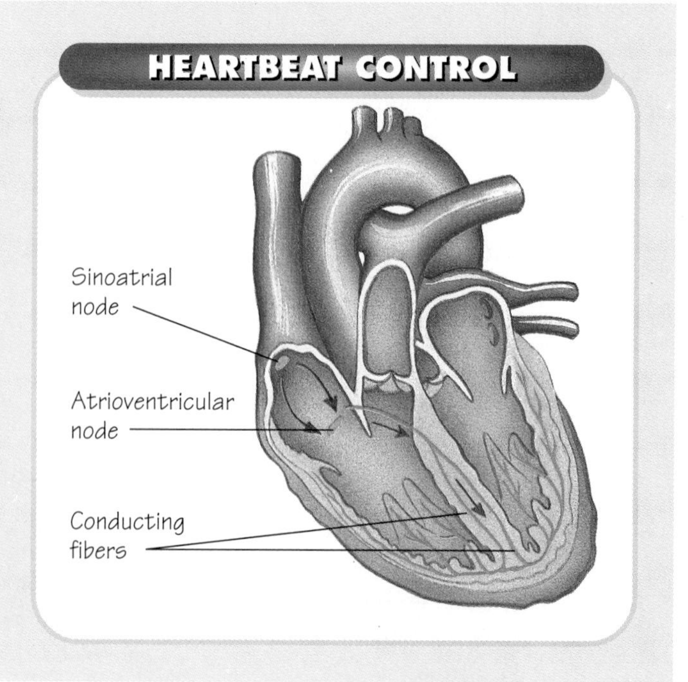

Figure 37–3
The rhythmic beating of the heart is maintained by the sinoatrial node—the pacemaker—located in the right atrium. The signal to contract spreads from the pacemaker through the cardiac muscle cells, causing the atria to contract. Then the impulse is picked up by the atrioventricular node, which is a bundle of fibers that carry the impulse to the ventricles, causing them to contract.

moves through are the arteries, capillaries, and veins.**

Arteries—the superhighways of the circulatory system—are blood vessels that carry blood from the heart to the body. Except for the pulmonary arteries, all arteries carry oxygen-rich blood. Arteries have thick elastic walls that help them withstand the powerful spurts of blood produced when the heart contracts. The lining of an artery is surrounded by layers of elastic tissue and smooth muscle cells that allow an artery to expand under pressure.

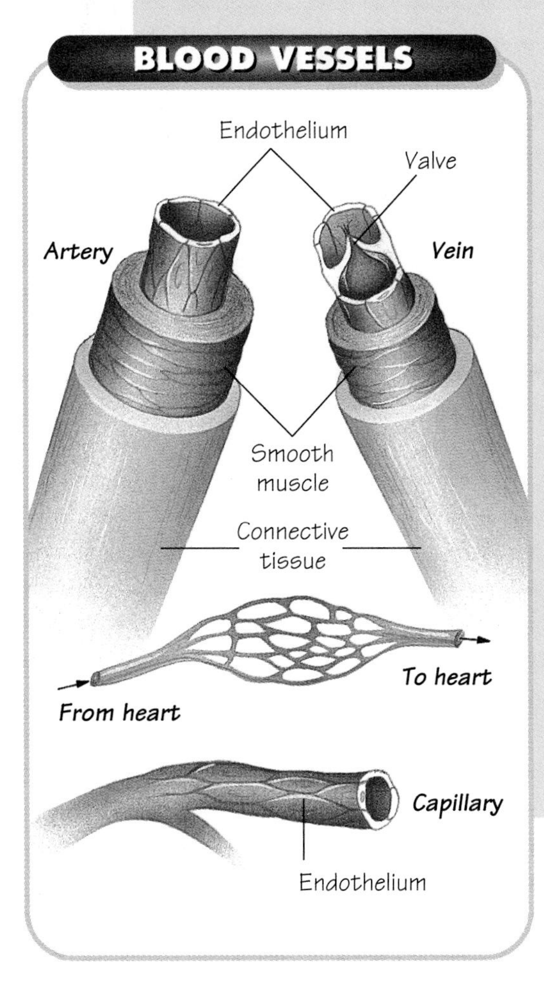

Figure 37–4
The layer of smooth muscle in arteries is much thicker than that of veins because the arteries have to withstand the high pressure of blood as it is pumped from the heart. Capillaries, on the other hand, contain only epithelial tissue and are only one cell thick.

Capillaries, the smallest of the blood vessels, are the side streets and alleys of the circulatory system. The exchange of nutrients and wastes takes place in the capillaries. Their walls are only one cell thick and may be so narrow that blood cells must pass through in single file.

Veins collect blood after it has passed through the capillary system. Like arteries, the walls of veins are lined with elastic tissue and smooth muscle. Veins, however, have thin walls and are less elastic than arteries. The largest veins contain one-way valves that keep blood flowing toward the heart. Many veins are located near skeletal muscles, and the contractions of those muscles help to push blood along to the heart. This is one reason why it is important to exercise regularly. Exercise helps to keep blood from accumulating in the limbs and from stretching the veins out of shape.

Blood Pressure

Any pump produces pressure, and the heart is no exception. When the heart contracts, it produces a wave of fluid pressure in the arteries. Although the force of blood on the walls of the arteries, known as **blood pressure,** falls when the heart relaxes, the system still remains under pressure. This is good, too, because without that pressure, blood would not flow through the arteries and into the capillaries.

The body regulates blood pressure in two different ways. Sensory neurons at several places in the body detect the level of blood pressure and send impulses to the brainstem. When blood pressure is too high, the autonomic nervous system releases neurotransmitters that cause the smooth muscles around blood vessels to relax, lowering blood pressure. When blood pressure is too low, neurotransmitters are released that elevate blood pressure by causing these muscles to contract.

The kidneys also help to regulate blood pressure. Hormones produced by the heart and other organs cause the kidneys to remove more water from the

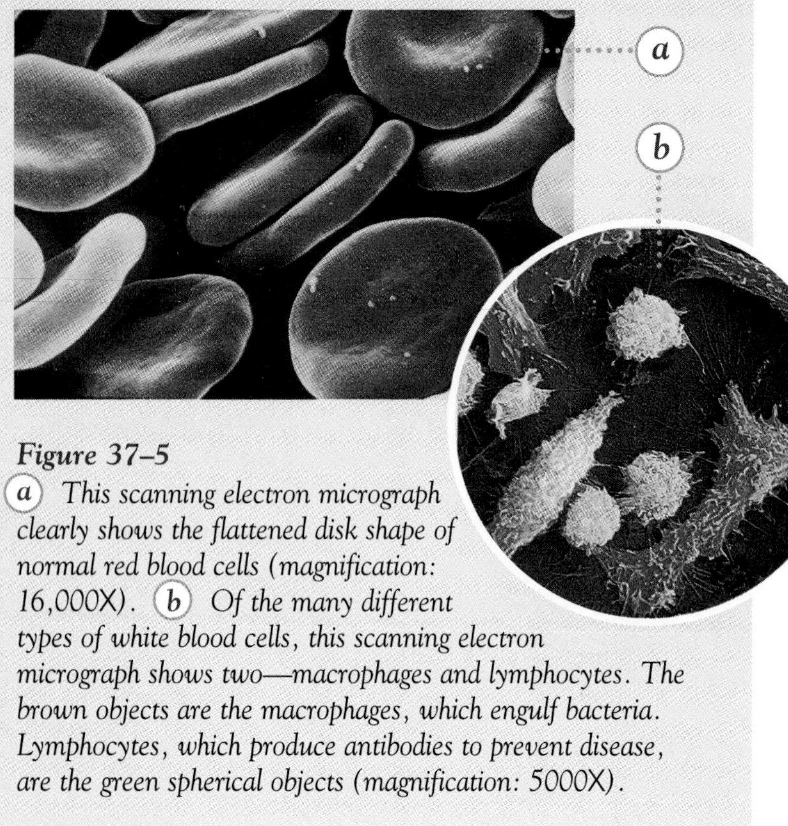

Figure 37–5
ⓐ *This scanning electron micrograph clearly shows the flattened disk shape of normal red blood cells (magnification: 16,000X).* ⓑ *Of the many different types of white blood cells, this scanning electron micrograph shows two—macrophages and lymphocytes. The brown objects are the macrophages, which engulf bacteria. Lymphocytes, which produce antibodies to prevent disease, are the green spherical objects (magnification: 5000X).*

blood when blood pressure is high. This reduces blood volume and lowers blood pressure.

Medical problems may result if blood pressure is either too high or too low. High blood pressure forces the heart to work harder, which may weaken or damage the heart muscle. People with high blood pressure are more likely to develop heart disease and to suffer from other diseases of the circulatory system.

The causes of high blood pressure are complex, but one of them is well understood—obesity. Although scientists have developed a number of drugs that can lower blood pressure, high blood pressure is easier to prevent than to cure. Exercise, weight control, and a sensible diet seem to be the keys to avoiding high blood pressure.

Blood

Roughly 8 percent of the mass of the human body is blood. For most of us, that means we contain anywhere from 4 to 6 liters of blood. About 45 percent of the volume of blood consists of living cells—**red blood cells, white blood cells,** and **platelets.** The remaining 55 percent is a fluid called **plasma.**

Plasma itself is 90 percent water. The remaining 10 percent consists of salts, sugars, and three groups of plasma proteins. The first, called serum albumin, helps to regulate osmotic pressure. Other plasma proteins, called globulins, are produced by the immune system and help to protect against infection. And the last of the plasma proteins is fibrinogen, which regulates blood clotting.

Red Blood Cells

Red blood cells are the most numerous cells in the blood—1 milliliter of blood contains nearly 5 million of them. Their scientific name—erythrocyte, means "red cell" and comes from the bright-red **hemoglobin** inside them. Hemoglobin is an iron-containing protein that dramatically increases the ability of blood to carry oxygen. **Red blood cells are oxygen carriers.** As red blood cells pass through the lungs, the hemoglobin within them quickly absorbs dissolved oxygen. When the same cells pass through capillaries in oxygen-poor regions of the body, oxygen is released into the surrounding tissues.

Red blood cells are produced in the bone marrow. As red blood cells develop, they lose their nuclei and their ability to divide. A typical red blood cell has a life span of roughly 120 days, which means that nearly 1 percent of your red blood cells must be replaced each day.

☑ **Checkpoint** What is hemoglobin?

White Blood Cells

White blood cells are blood cells that do not contain hemoglobin. White blood cells are also called leukocytes, which means "white cells." They are much less common than red cells, which outnumber them almost 500 to 1. Like red blood cells, white blood cells are produced in the bone marrow and are released into the blood as cells with nuclei. Unlike red blood cells, white blood cells may live for many months and possibly even for years.

More than 20 different types of white blood cells are known. **White blood cells guard against infection, fight parasites, and attack bacteria.** Some actually engulf and digest these foreign cells. Others attack invading organisms in the tissues of the body. Still others produce chemical signals that activate the body's immune system to help fight infection.

Like an army with units in reserve, the body is able to increase the number of white blood cells dramatically when a "battle" is underway. In fact, a sudden increase in the number of white cells is one way in which physicians can tell that the body is fighting a serious infection.

Platelets and Blood Clotting

Try as we might to protect ourselves against injury, sooner or later just about everyone receives a cut or a scrape. Most of these injuries aren't serious. A minor cut or scrape may bleed for a few minutes, then stop. Have you ever wondered why bleeding stops so quickly?

The answer is that blood has the ability to form a clot, a tangle of microscopic fibers that block the flow of blood. **Blood clotting is made possible by cell fragments, called platelets, and a number of plasma proteins.**

When platelets come into contact with the broken edges of a blood vessel, their surfaces become sticky, and a cluster of platelets develops around the wound. These platelets then release a series of chemicals that help to start a complicated clotting reaction. This reaction results in the formation of **fibrin,** which forms a netlike trap that covers the wound and traps the red blood cells. If the wound is small, a network of platelets and fibrin seals the leak within a few minutes, and bleeding stops.

Figure 37–6
(a) In response to an injury to a blood vessel, proteins called fibrin form in the blood. Fibrin produces a netlike web that traps the red blood cells, thus forming a clot that constricts the wound and promotes healing. (b) Platelets, which are responsible for blood clotting, are actually fragments of larger cells. This color-enhanced scanning electron micrograph shows some inactive platelets (magnification: 9200X).

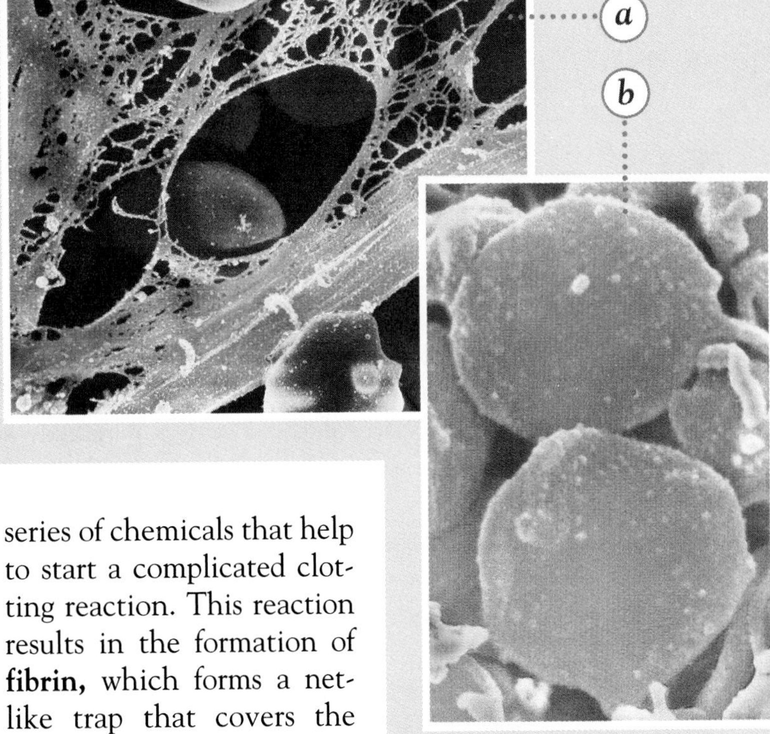

Checkpoint What is fibrin?

Diseases of the Circulatory System

Unfortunately, diseases of the circulatory system are common. ● Many of them stem from a condition known as **atherosclerosis** (ath-er-oh-skluh-ROH-sihs), in which fatty deposits build up on the inner surfaces of arteries. If these deposits get too large, they obstruct the flow of blood.

Atherosclerosis is particularly dangerous in the coronary arteries, a set of small

INTEGRATING CAREERS

Use reference materials to find out the special training needed by a cardiologist, a doctor who treats heart disease.

Figure 37–7
The lymphatic system is made up of lymphatic vessels and lymph nodes. The lymphatic vessels collect lymph—fluid that leaves the circulatory system—and returns it to veins in the neck. The lymph nodes act as filters that prevent harmful materials from entering body cells.

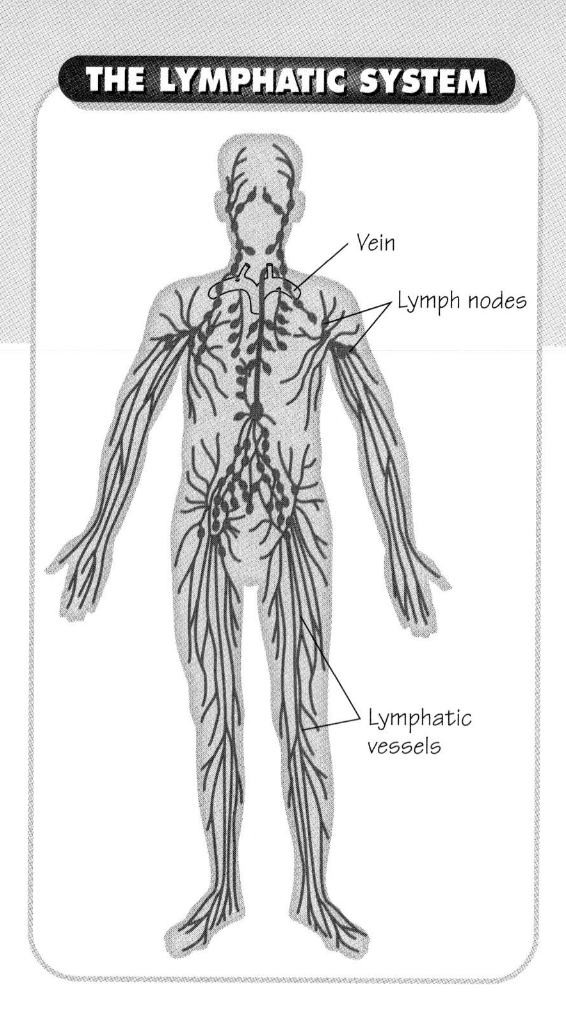

arteries that bring oxygen and nutrients to the heart muscle itself. If one of these becomes blocked, part of the heart muscle may begin to die from a lack of oxygen, a condition called a heart attack. The symptoms of a heart attack include nausea, shortness of breath, radiating pain down the left arm, and severe, crushing chest pain. People who show symptoms of a heart attack should be given medical attention immediately.

When one of the blood vessels leading to part of the brain is blocked, a stroke results. Brain cells served by that blood vessel gradually die from a lack of oxygen, and brain function in that region may be lost. Depending on the part of the brain that is affected, a stroke may cause paralysis, loss of the ability to speak, and even death.

The Lymphatic System

Fluid from the bloodstream is constantly leaking from the capillaries into the surrounding tissues. The leaking fluid helps to bring salts and nutrients into tissues where they are needed. More than 3 liters of fluid leak from the circulatory system into surrounding tissues every day.

What happens to all this fluid? A network of vessels called the **lymphatic** (lihm-FAT-ihk) **system** collects the fluid—**lymph** (LIHMF)—and returns it to the circulatory system. As ***Figure 37–7*** shows, these vessels empty the lymph back into general circulation through a vein under the left shoulder. Lymph vessels contain one-way valves, like veins, to keep lymph flowing in one direction.

Section Review 37–1

1. **List** the structures of the circulatory system.
2. **Identify** the three types of blood vessels.
3. **Compare** the functions of red blood cells, white blood cells, and platelets.
4. **Critical Thinking—Applying Concepts** Hemophilia is a genetic disorder in which the gene for normal blood clotting is missing. How would injections of normal clotting proteins help a hemophiliac?
5. **MINI LAB** How does your pulse rate **relate** to your activity level?

The Respiratory System

SECTION 37–2

GUIDE FOR READING

- **Describe** the function of the respiratory system.

MINI LAB

- **Measure** your lung capacity.

WHEN PARAMEDICS RUSH TO the aid of an injured person, one of the first things they usually do is to check to see whether the victim is breathing. If the person is not breathing, paramedics will ignore other injuries—even broken bones and serious wounds—to get the person breathing again. There's no time to lose! If breathing stops for more than a few minutes, the person may die!

A well-trained paramedic understands how important the respiratory system is to life. This system provides nearly every cell in the body with oxygen—the same oxygen that is carried throughout the body by the circulatory system.

What Is Respiration?

In biology, the word respiration is used in two slightly different ways. At the cellular level, respiration is defined as the release of energy from the breakdown of food molecules in the presence of oxygen. Without oxygen, cells lose much of their ability to produce ATP, and that means that they cannot synthesize new molecules, pump ions, carry nerve impulses, or even move.

Because trillions of cells have a pressing need for oxygen, the human body must find a way to get that oxygen to them. It must also dispose of the carbon dioxide produced when food molecules are broken down. At the level of the organism, respiration is the exchange of gases—oxygen and carbon dioxide—between the organism and its environment.

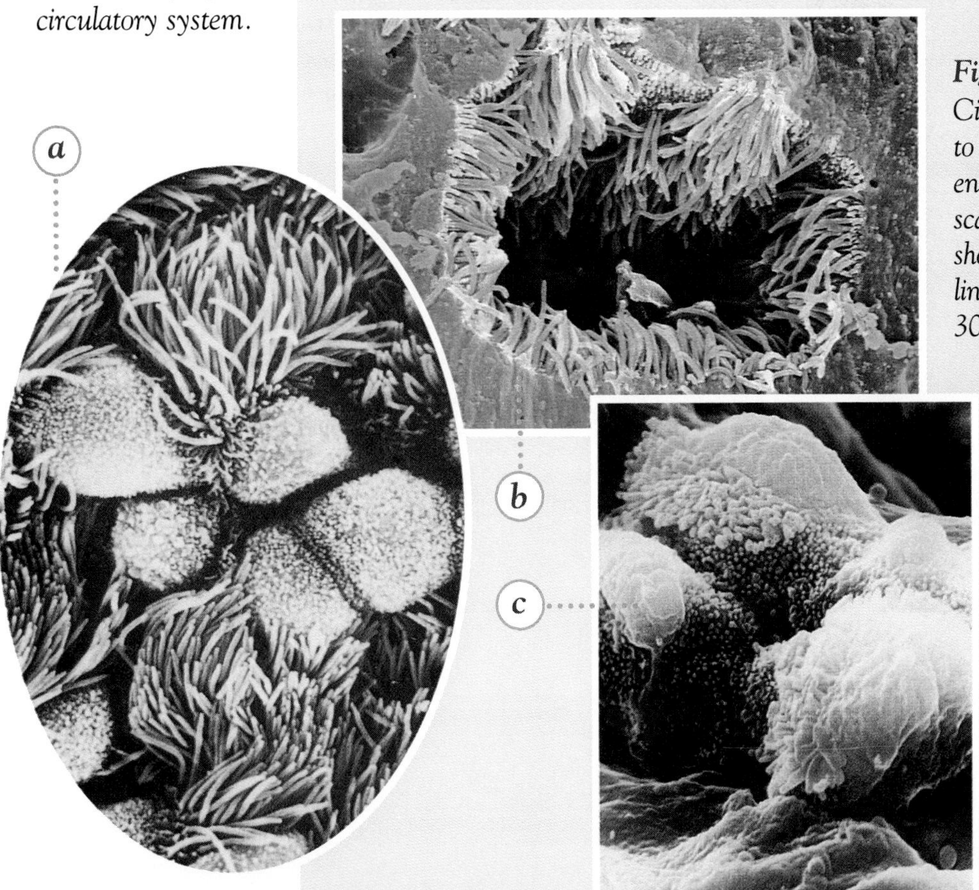

Figure 37–8
Cilia and goblet cells work together to prevent foreign particles from entering the lungs. (a) This scanning electron micrograph shows a close-up of the cilia that line the trachea (magnification: 3000X). (b) In this cross section of the trachea, the cilia have been colored green (magnification: 3570X). The yellow structures at the base of the cilia are goblet cells, which produce mucus. (c) Once the incoming air has been cleaned and filtered, it enters the alveoli of the lungs, where actual gas exchange takes place. (magnification: 5100X).

Visualizing Human Respiration

The human respiratory system is responsible for bringing oxygen to the blood so that it can be distributed to the body cells. In addition, it removes carbon dioxide from the body.

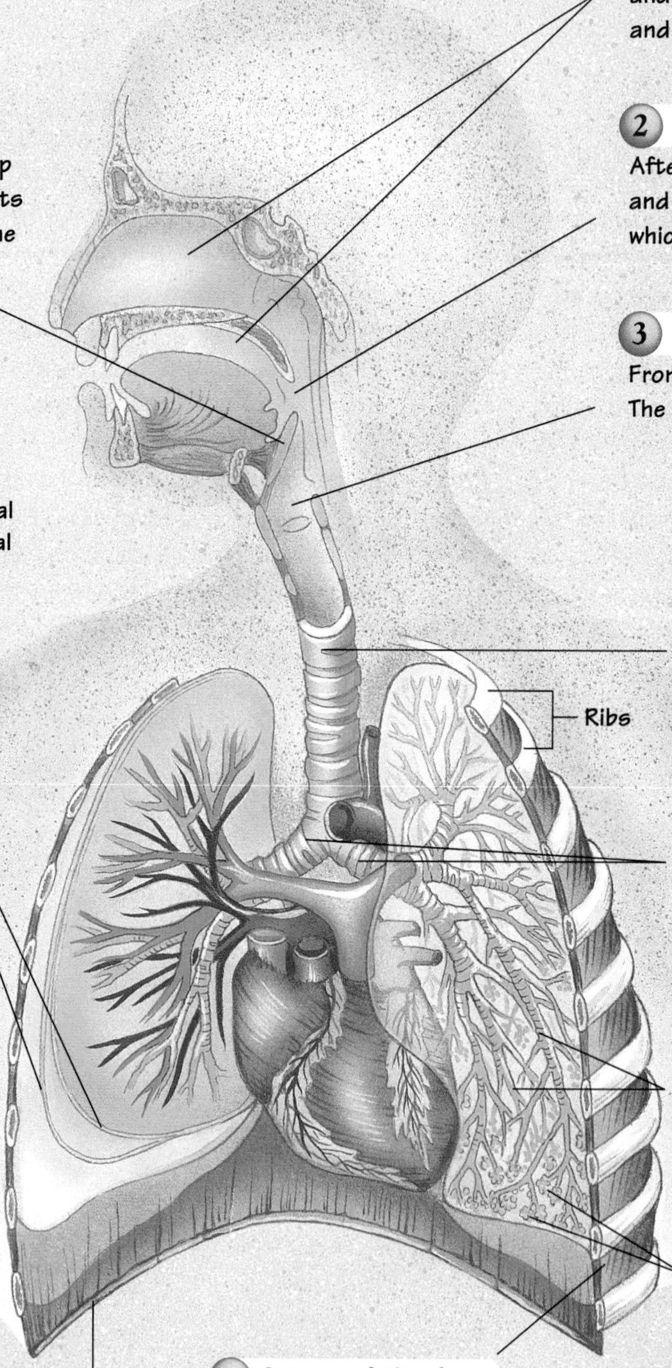

1 Nose and Mouth

Air enters the body through the nose and mouth, where it is filtered, warmed, and moistened.

2 Pharynx

After the air has been filtered, warmed, and moistened, it enters the pharynx, which is also a passageway for food.

3 Larynx

From the pharynx, air enters the larynx. The larynx also contains the vocal cords.

4 Trachea (Windpipe)

After the larynx, air enters the trachea—the main airway to the lungs. The trachea divides into two bronchi.

Ribs

5 Bronchi

The bronchi (singular: bronchus) are the tubes that bring air to the lungs from the trachea. Each bronchus leads to one lung.

6 Bronchioles

Once inside the lungs, the bronchi branch into smaller and smaller air passageways called bronchioles. The bronchioles continue to divide until they finally end in clusters of tiny air sacs.

7 Alveoli

The alveoli (singular: alveolus) are tiny air sacs that appear in grape-like clusters. Surrounding each alveolus is a network of capillaries, where gas exchange takes place.

8 Intercostal Muscles

These muscles cause the chest cavity to expand when air is inhaled and get smaller when air is exhaled.

9 Diaphragm

A dome-shaped muscle called the diaphragm separates the chest cavity from the abdominal cavity. Breathing is directed by contracting and relaxing the diaphragm.

10 Pleural Membranes

Each lung is enclosed by two membrane layers called the pleural membranes. The pleural membranes form an airtight seal between the lungs and the chest wall.

11 Epiglottis

The epiglottis is a flap of tissue that prevents food from entering the trachea.

The Human Respiratory System

The function of the respiratory system is to bring about the exchange of oxygen and carbon dioxide. With each breath, air enters the body through the air passageways and fills the lungs, where gas exchange takes place.

As air moves through the respiratory system, it is warmed, moistened, and filtered. Many of the cells lining the respiratory system produce a thin layer of protective mucus. This layer also traps inhaled particles of dust or smoke. Cilia lining the passageways then sweep such materials away from the lungs, keeping them clean and open for the important work of gas exchange. The human respiratory system is illustrated on page 836.

Checkpoint What do cilia do?

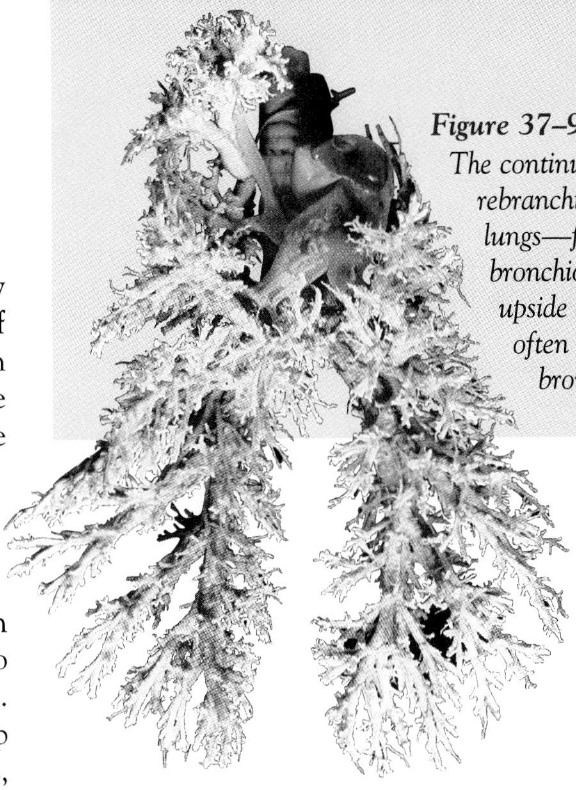

Gas Exchange and Hemoglobin

There are nearly 300 million alveoli in a healthy lung, providing an enormous surface for gas exchange. Oxygen dissolves in the moisture on the inner surface of the alveoli and then diffuses across the thin capillary walls into the blood. Carbon dioxide in the bloodstream diffuses in the opposite direction—across the wall of the alveolus and into the air within it.

The process of gas exchange in the lungs is very efficient. The air that you

Figure 37–9 *The continuous branching and rebranching of tubes in the lungs—from bronchi to bronchioles—resembles an upside down tree, so it is often given the name the bronchial tree.*

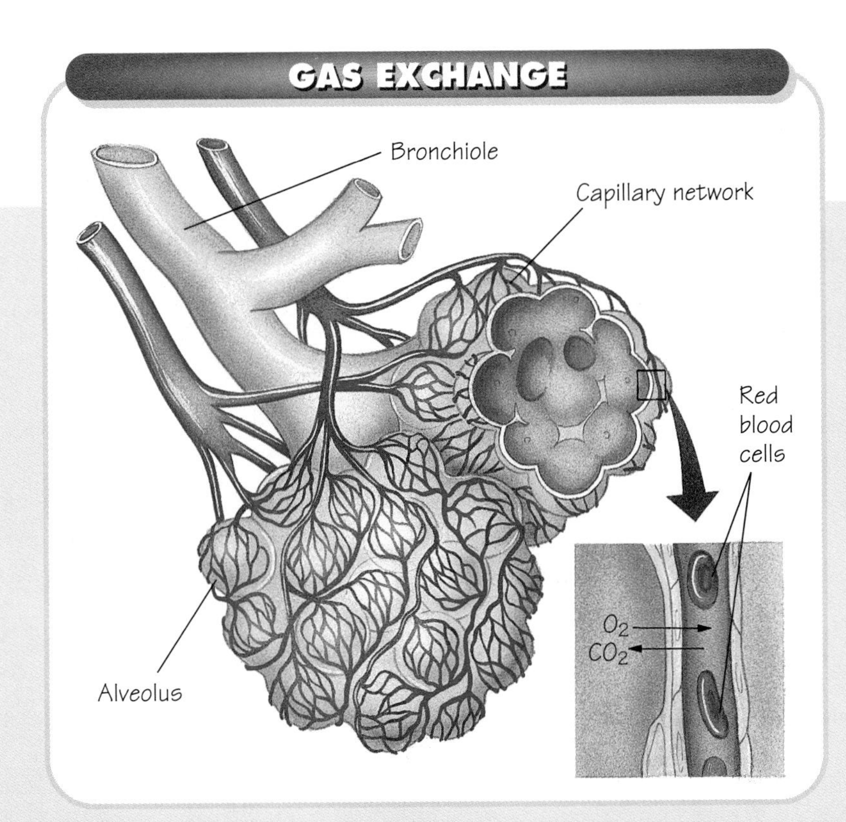

Figure 37–10 *Gas exchange takes place in the alveoli of the lungs. Oxygen enters the blood by diffusing through the alveolus into the capillaries. Carbon dioxide, on the other hand, diffuses from the blood into the alveoli, where it will be exhaled out of the body.*

MINI LAB .... Measuring ......

A Ballooning Effect

PROBLEM *How can you* ***measure*** *your lung capacity?*

PROCEDURE

CAUTION: *If you have any respiratory or circulatory conditions, do not perform this activity.*

1. Take two normal breaths. On the next breath, inhale as much air as you can. Then exhale into an empty round balloon, trying to empty your lungs as much as possible.
2. Hold the balloon closed while your partner uses a string to measure the circumference of the balloon at its widest part. Record the measurement.
3. Repeat steps 1 and 2.
4. Properly dispose of the balloons when you are finished.

ANALYZE AND CONCLUDE

1. How did your measurements compare? Compare them with those of other members of your class.
2. How did the measurements of males and females compare?
3. Do you think people who exercise regularly would have a larger lung capacity? Explain why or why not.

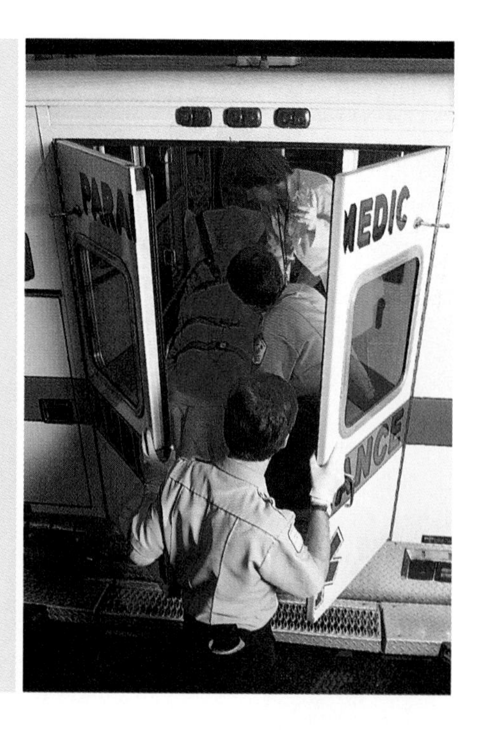

Figure 37–11 CAREER TRACK *Emergency Medical Technicians are trained to evaluate injuries and provide appropriate first-aid care. Certification in cardiopulmonary resuscitation (CPR) and advanced first aid is required for EMT certification.*

inhale contains 21 percent oxygen and 0.04 percent carbon dioxide. Exhaled air is usually less than 15 percent oxygen and 4 percent carbon dioxide. This means that the lungs remove about one third of the oxygen in the air that you inhale and increase the carbon dioxide content of that air by a factor of 100!

As you may recall, oxygen dissolves easily. You may therefore wonder why hemoglobin is needed at all. The reason is efficiency. Hemoglobin binds with so much oxygen that it increases the oxygen-carrying capacity of the blood more than 60 times.

Breathing

Breathing is the movement of air into and out of the lungs. Surprisingly, there are no muscles connected to the lungs. The force that drives air into the lungs comes from ordinary air pressure. The lungs are sealed in two sacs, called the pleural membranes, inside the chest cavity. At the bottom of the cavity is the diaphragm. When you inhale, or breathe in, the diaphragm contracts and expands the volume of the chest cavity. Because the chest cavity is tightly sealed, this creates a partial vacuum inside the cavity. Atmospheric pressure does the rest, filling the lungs as air rushes through the breathing passages.

Most of the time, exhaling is a passive event. When the diaphragm muscle relaxes, elastic tissues surrounding the chest cavity return to their original positions, placing pressure on the lungs. As a result of that pressure, air rushes back out of the lungs. As you know, sometimes you exhale with much greater force, as when you blow out a candle. Muscles surrounding the chest cavity provide that extra force, contracting vigorously just as the diaphragm relaxes.

☑ **Checkpoint** What is breathing?

Bio FRONTIER Connections

The Price of an Organ Donation

Heart and lung diseases strike millions of people each year. Those people whose organs cannot perform at a level that can keep them alive are candidates for heart or lung transplants. In order to receive a heart or lung transplant, these people must meet strict guidelines. After qualifying for a transplant, their names are placed on national waiting lists that match them with organs as they become available.

The Problems of Organ Donations

The problem with waiting for a heart or a lung is that there are not enough donors to keep up with the demand. In fact, twice as many people are on waiting lists as there are healthy organs available.

Unfortunately, organs of this type are donated only when other people die. That is, if they have given written permission or if their families give permission upon their death. Some states have a check-off box on driver's licenses that authorize organ donation in the event of death.

Although most people say they would be willing to donate their organs, few actually fill out a donor card. As a result, there are not enough available organs to meet the needs.

The Debate

Recently, someone suggested that families should be paid for the organs. This plan would require that a 1984 federal law making it illegal to receive payment for organs be changed. Some people believe that donating organs is a moral duty and that payment would merely encourage people to do the right thing.

Others believe that it is unethical to pay for organs. They fear that families might be too eager to receive payment and go against the wishes of the potential donor.

Critics also say that the idea of paying for organs would increase the costs of organ transplants. At present, the cost of obtaining an organ for transplant is nearly $50,000.

Organ donor registration card

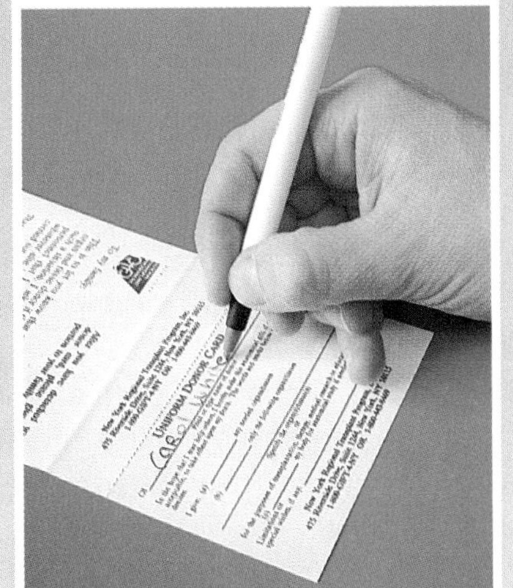

Heart specialist explaining procedure to a patient

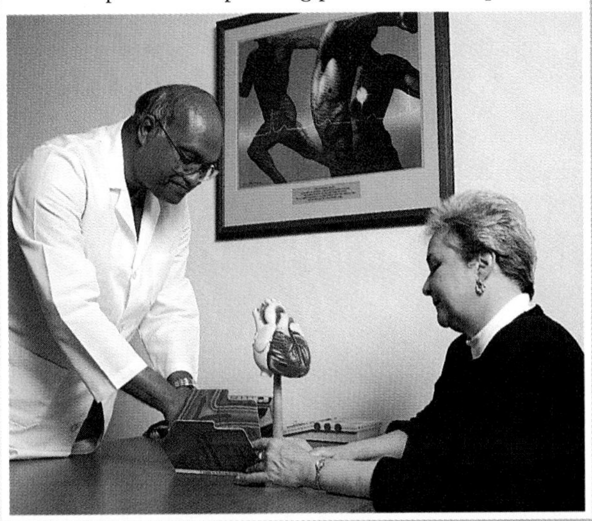

Making the Connection

There is also concern over who would be able to pay for the organs. Would only those patients who could afford them get organs because they could pay the price? Or would insurance companies be responsible for payment? What other reasons for or against this plan are there? Would you support this plan? Explain why or why not.

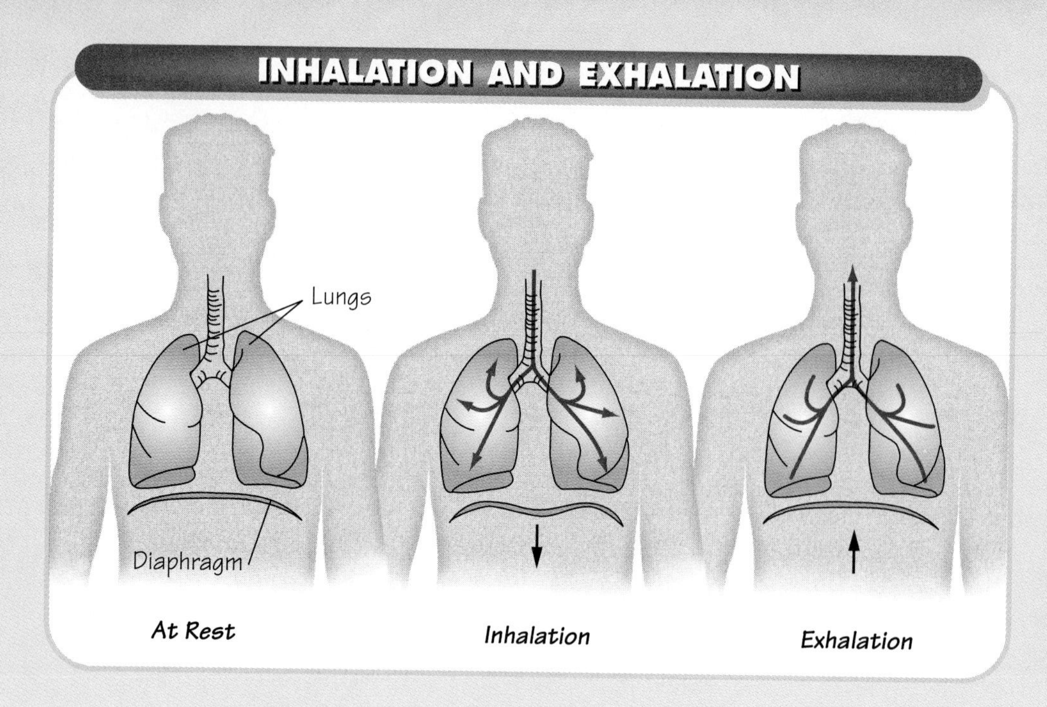

Figure 37–12
At rest, the pressure inside the lungs is equal to the atmospheric pressure, or pressure outside the lungs. During inhalation, the diaphragm contracts, increasing the size of the chest cavity. This action causes the pressure inside the lungs to decrease and air to enter. As the diaphragm relaxes, the chest cavity gets smaller, which increases the pressure in the lungs. To equalize the pressure again, air is exhaled.

How Breathing Is Controlled

As you know, you can control your breathing almost anytime you want—whether it's to blow up a balloon or to play a musical instrument. But this does not mean that breathing is purely voluntary. If you hold your breath for a minute or so, you'll see what we mean. Your chest begins to feel tight, your throat begins to burn, the muscles in your mouth and throat struggle to keep from breathing, and eventually your body takes over. It "forces" you to breathe!

Breathing is such an important function that your nervous system simply will not let you have complete control over it. The brain controls this process in a breathing center located in the medulla oblongata—part of the brain just above the spinal cord. Autonomic nerves from the medulla oblongata to the diaphragm and chest muscles produce the cycles of contraction that bring air into the lungs.

How does the medulla know when it's time to breathe? Cells in the breathing center monitor the amount of carbon dioxide in the blood. As the carbon dioxide level rises, nerve impulses from the center cause the diaphragm to contract, bringing air into the lungs. The higher the carbon dioxide level, the stronger these impulses. If the carbon dioxide level reaches a critical point, the impulses become so powerful that you cannot stop your breathing.

The breathing center responds to high carbon dioxide levels, and not to a lack of oxygen. As a result, when the air is thin, people sometimes do not sense a problem, and must be told to begin breathing pressurized air.

Section Review 37–2

1. **Describe** the function of the respiratory system.
2. **Critical Thinking—Inferring** As you have read, the breathing center in the brain responds to the level of carbon dioxide in the blood and not to the oxygen level. What consequences could this have on people at high altitudes?
3. **MINI LAB** How can you **measure** your lung capacity?

The Hazards of Smoking

GUIDE FOR READING

- **List** the substances in cigarette smoke.
- **Describe** some health problems caused by smoking tobacco.

WHEN IT IS FUNCTIONING well, the respiratory system is simple and efficient. First, air enters through the breathing passages. Next, that air inflates the millions of alveoli in the lungs. And finally, oxygen from the air diffuses across thin membranes to enter the bloodstream, where it is quickly absorbed by hemoglobin. The circulatory system does the rest, taking this oxygen-rich blood to the rest of the body.

When people smoke, things can go wrong. The passageways can be blocked, the lungs themselves can be damaged, and other gases can interfere with hemoglobin. Understanding the biology behind the dangers of smoking is one of the best ways to avoid them.

Tobacco Use

Tobacco is a plant that was cultivated and smoked by Native Americans long before Europeans emigrated to the North American continent. The dried leaves of the plant are chewed or smoked in pipes, cigars, and cigarettes.

Tobacco contains many different chemical compounds. When it burns, many harmful compounds are produced. **Three of the most dangerous substances in tobacco are tar, carbon monoxide, and nicotine.** Tar is a brown sticky mixture of chemicals, nicotine is a stimulant drug, and carbon monoxide is a poisonous gas. When tobacco smoke is inhaled, these compounds quickly enter the airways and blood, affecting the body.

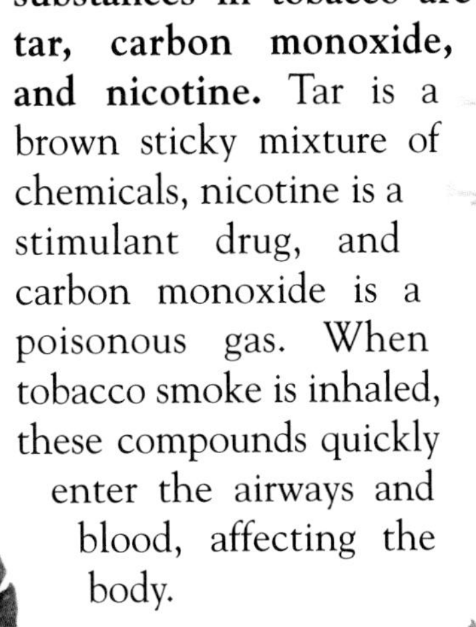

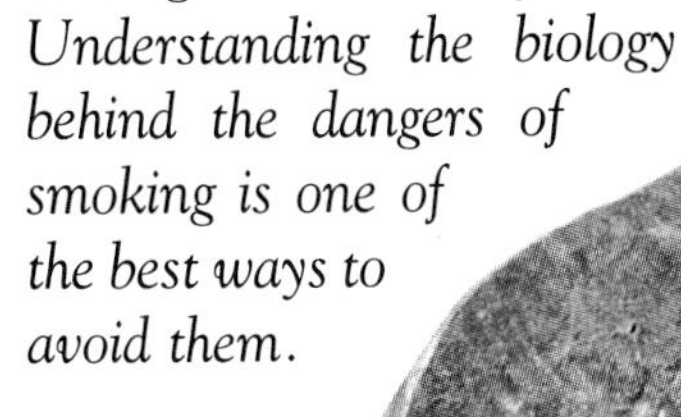

Figure 37–13 *Smoking causes many different disorders. Notice how (a) the lung of a nonsmoker differs from (b) that of a smoker. The holes in the smoker's lung are caused by ruptured alveoli. (c) Parts of this tobacco plant are used to make cigarettes, cigars, pipe tobacco, and chewing tobacco. Smoking any of these products has the greatest impact on the lungs.*

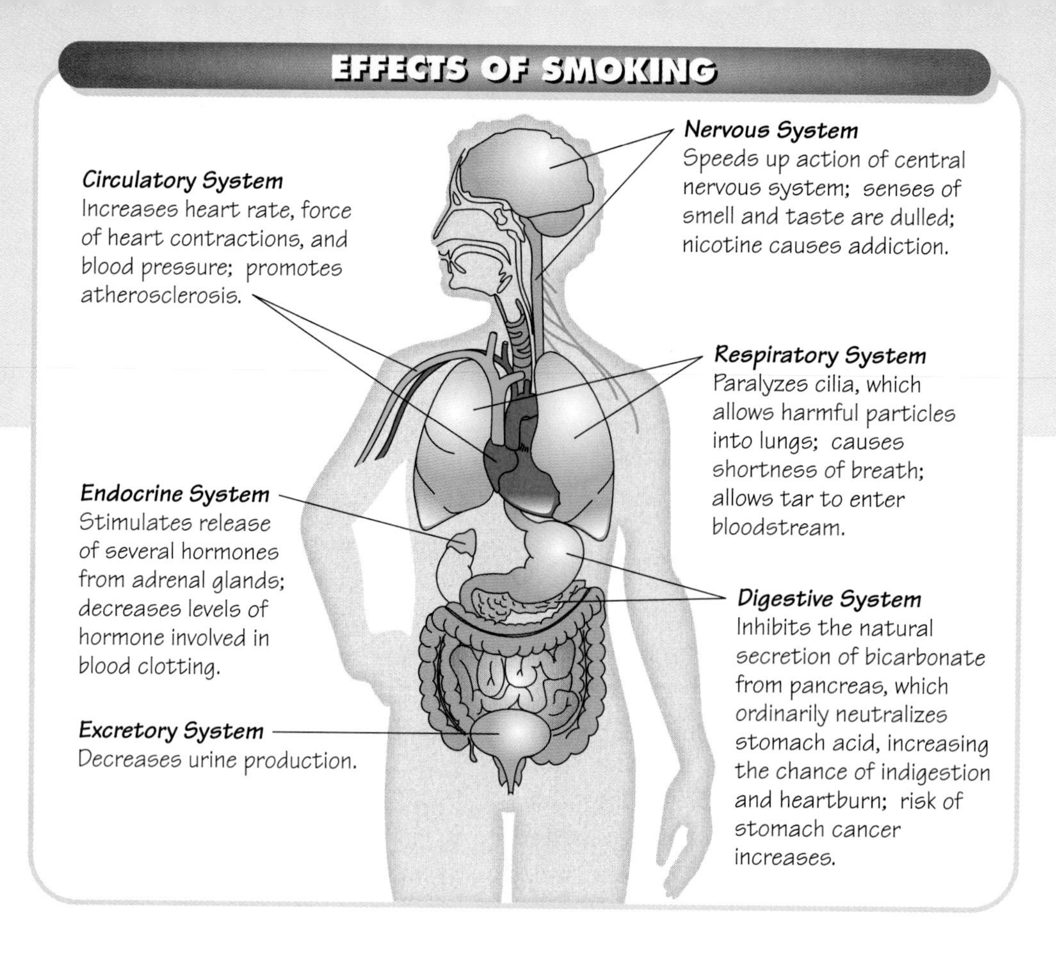

Figure 37–14
This illustration shows the effects of smoking on different systems of the body.

Effects on the Respiratory System

INTEGRATING CAREERS

A respiratory therapist works with people who have respiratory problems. What are some responsibilities of a respiratory therapist?

As you have read earlier, the upper part of the respiratory system is generally able to filter out dust and foreign particles that might otherwise damage the lungs. ● Incredibly, millions of people engage in a habit—smoking tobacco—that damages and eventually destroys this protective system.

The respiratory system helps to protect the body from impurities and disease. Cilia and mucus keep dust and other foreign particles away from the lungs. Nicotine and carbon monoxide paralyze the cilia. With the cilia out of action, the inhaled particles stick to the walls of the respiratory tract or enter the lungs. The paralyzing effects of just one cigarette can last up to an hour!

In response to the irritation of cigarette smoke, the respiratory system increases mucus production. Without cilia to sweep it along, mucus builds up and obstructs the airways. This explains why smokers often cough—they have to clear their airways. Smoking also results in the swelling of the lining of the respiratory tract, which results in less air flow to the alveoli.

The tar in cigarettes affects the respiratory system, too. The tar accumulates in the lungs, where it can pass directly into the bloodstream. And a number of compounds in tar can cause cancer.

Checkpoint What happens to the cilia as a result of carbon monoxide and nicotine?

Respiratory Disorders Caused by Smoking

Smoking can cause such respiratory diseases as bronchitis, emphysema, and lung cancer. In **bronchitis,** the bronchi become swollen and clogged with mucus. Even smoking a moderate number of cigarettes can produce bronchitis. People with bronchitis often find simple activities, such as climbing stairs, difficult.

Long-term smoking can also cause **emphysema.** Emphysema is a loss of elasticity in the tissue of the lungs. This makes breathing difficult. People with emphysema cannot get enough oxygen to the body tissues or rid the body of carbon dioxide.

The most serious consequence of smoking is **lung cancer.** Nearly 180,000 people in the United States develop lung cancer each year, and very few survive it. Lung cancer is particularly deadly because it spreads easily—small groups of cancer cells from the lungs break off and spread to other places in the body. This is called metastasis. By the time the cancer is detected, it usually has spread to dozens of other places in the body, causing a painful death. Lung cancer claims 87 percent of its victims in the first five years after it is detected.

☑ ***Checkpoint*** What is bronchitis?

Effects on the Circulatory System

Smoking affects the circulatory system, too. Every part of the circulatory system—the heart, the blood vessels, and the blood—is affected by smoking. **People who smoke have twice the rate of heart disease of nonsmokers. Besides an increased chance of a heart attack or stroke, smokers often have high blood pressure.** The circulatory system is most affected by the nicotine and carbon monoxide found in the smoke of tobacco products.

Carbon monoxide is an invisible, odorless, and highly poisonous gas. ● This gas can attach to the oxygen-binding site of hemoglobin. In fact, carbon monoxide binds more tightly than oxygen itself! Recall that hemoglobin is the oxygen-carrying agent in the blood. As more carbon monoxide combines with it, it has less room for oxygen. This continues until the oxygen-carrying ability of the blood is almost gone. As a result, the heart must work harder in order to deliver oxygen to the cells of the body.

INTEGRATING CHEMISTRY

What are some other sources of carbon monoxide?

The nicotine in cigarette smoke causes blood vessels to constrict, thus inhibiting blood flow. It also causes a rise in resting heart rate—an added burden for the heart. Nicotine also causes blood pressure to rise.

In addition, smoking tends to increase the buildup of fatty materials on the walls of blood vessels. This buildup leads to the development of atherosclerosis.

Effects on Other Body Systems

As you can see in ***Figure 37–14,*** smoking affects other body systems as well. Because nicotine is a stimulant drug, it has effects similar to those of other stimulant drugs, especially on the nervous system. In addition, smoking affects the digestive system, the endocrine system, and the excretory system.

Section Review 37–3

1. **List** the main components of cigarette smoke.
2. **Describe** some health problems caused by smoking tobacco.
3. **BRANCHING OUT ACTIVITY** **Design** and **construct** a poster for a middle school classroom that discourages students from smoking. Have your teacher approve your poster before it is displayed.

CHAPTER 37

Laboratory Investigation

Burning Tobacco

Scientists have identified many harmful substances that are produced when tobacco is burned in a cigarette, pipe, or cigar. Of these, the most dangerous are nicotine, tar, and carbon monoxide. In this investigation, you will see how much tar is produced when different tobacco products are burned.

Problem

How can you **measure** the products of burning tobacco?

Materials (per group)

triple-beam balance
filter paper
tobacco from a cigarette, pipe, and cigar
cotton
test tube
test-tube holder
test-tube rack
Bunsen burner
matches

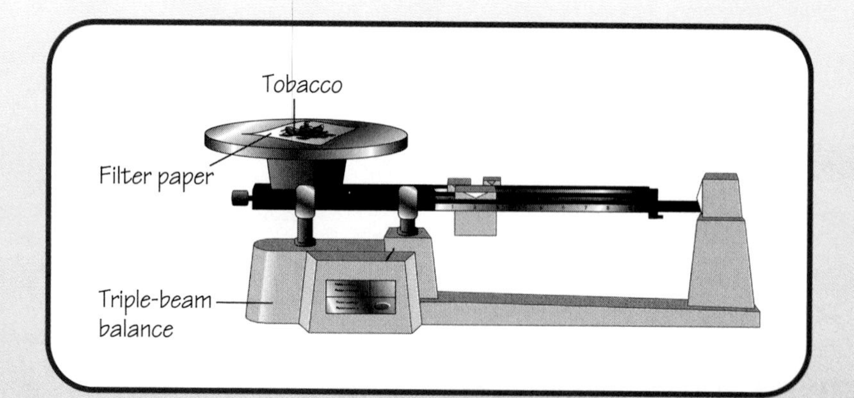

Procedure

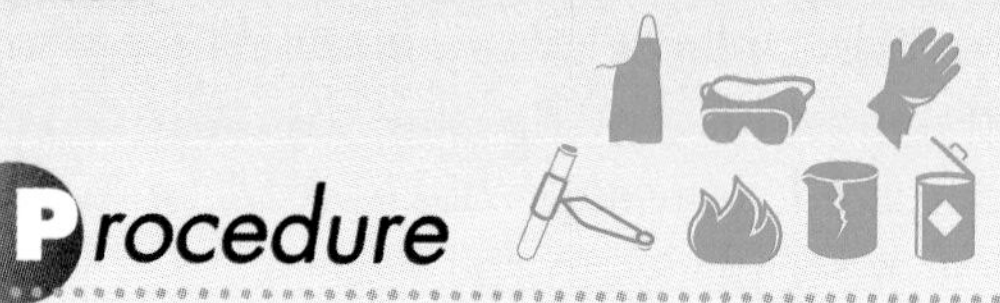

1. **Place a piece of filter paper on a triple-beam balance. Using the balance, measure 2 g of cigarette tobacco. Place the tobacco in a test tube.**
2. **Copy the data table shown on a separate sheet of paper. Record your measurements in your data table.**
3. **Find the mass of a wad of cotton large enough to fill the opening of the test tube. Record its mass.**
4. **Put the cotton wad into the open end of the test tube.**
5. **Using a test-tube holder, heat the bottom of the test tube over a Bunsen burner flame. CAUTION:** ***Be careful with open flames. Keep the cotton pointed away from the flame, and keep the opening of the test tube away from others.***

DATA TABLE

Type of Tobacco	Mass of Cotton Before Heating	Mass of Cotton After Heating	Difference in Mass

6. **After heating the tobacco for 3 minutes, turn off the Bunsen burner and place the test tube in a rack to cool.**
7. **Remove the cotton wad and measure and record its mass.**
8. **Follow the same procedure—steps 1 to 7—with the pipe tobacco and cigar tobacco.**

Observations

1. What did you observe on the inside of the test tube while the tobacco was burning?
2. Did the appearance of the cotton change after it was heated?
3. Describe the appearance of tar.

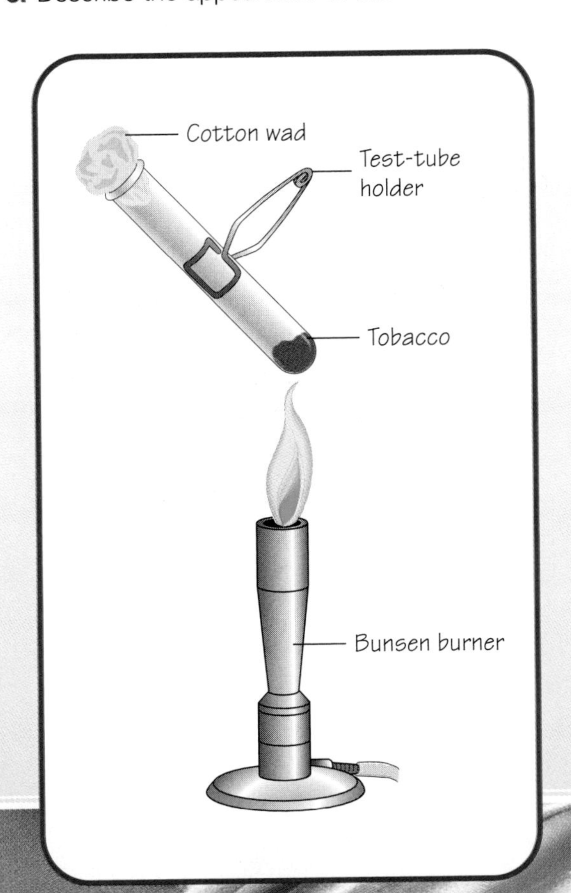

Analysis and Conclusions

1. Based on its appearance, what effect would tar have on the respiratory system?
2. Which of the three types of tobacco produced the most tar? How do you know?
3. Which type produced the least tar?
4. What could be done to make these measurements more precise?

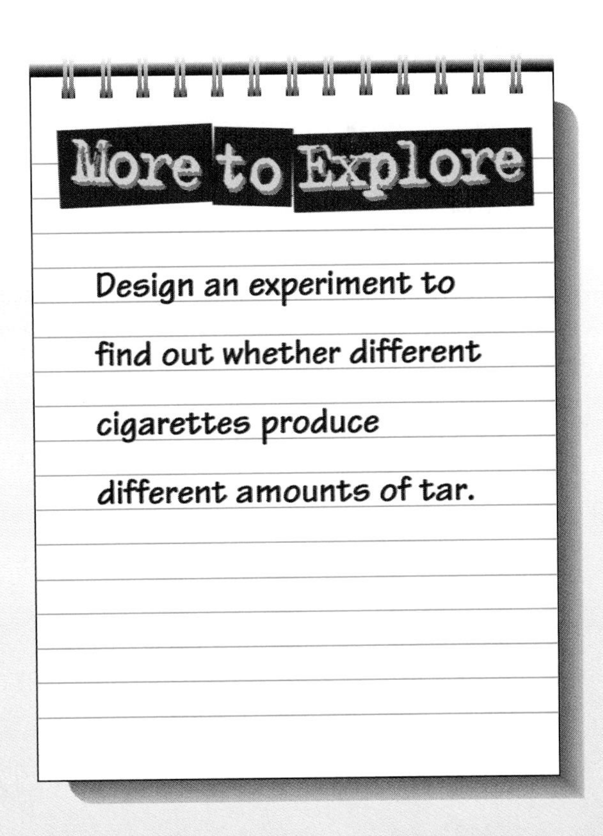

Study Guide

Summarizing Key Concepts

The key concepts in each section of this chapter are listed below to help you review the chapter content. Make sure you understand each concept and its relationship to other concepts and to the theme of this chapter.

37–1 The Circulatory System

- The human circulatory system consists of the heart, a series of blood vessels, and the blood.
- The three kinds of blood vessels are arteries, capillaries, and veins.
- Red blood cells are oxygen carriers. White blood cells guard against infection, fight parasites, and attack bacteria. Platelets are responsible for blood clotting.

37–2 The Respiratory System

- The function of the respiratory system is to bring about the exchange of oxygen and carbon dioxide.

37–3 The Hazards of Smoking

- Three of the most dangerous substances in tobacco are tar, carbon monoxide, and nicotine.
- Smoking can cause such respiratory diseases as bronchitis, emphysema, and lung cancer.
- People who smoke have twice the rate of heart disease than nonsmokers. Besides an increased chance of a heart attack or stroke, smokers often have high blood pressure.

Reviewing Key Terms

Review the following vocabulary terms and their meaning. Then use each term in a complete sentence.

37–1 The Circulatory System

blood
atrium
ventricle
pulmonary circulation
systemic circulation
valve
aorta
pacemaker
artery
capillary
vein
blood pressure
red blood cell
white blood cell
platelet
plasma
hemoglobin
fibrin
atherosclerosis
lymphatic system
lymph

37–2 The Respiratory System

pharynx
larynx
trachea
bronchus
alveolus
diaphragm
epiglottis

37–3 The Hazards of Smoking

bronchitis
emphysema
lung cancer

Recalling Main Ideas

Choose the letter of the answer that best completes the statement or answers the question.

1. The upper chambers of the heart are called the

a. atria. **c.** ventricles.
b. myocardium. **d.** septum.

2. Oxygen-rich blood returns from the lungs to the heart's

a. left atrium. **c.** left ventricle.
b. right atrium. **d.** right ventricle.

3. Oxygen-poor blood enters the lungs from the

a. left atrium. **c.** right atrium.
b. left ventricle. **d.** right ventricle.

4. The largest vessel that transports oxygen-rich blood away from the heart is the

a. pulmonary artery. **c.** vena cava.
b. pulmonary vein. **d.** aorta.

5. Red blood cells contain an oxygen-absorbing protein called

a. fibrinogen. **c.** hemoglobin.
b. serum albumin. **d.** plasma.

6. Oxygen and carbon dioxide are exchanged with the circulatory system at the

a. pharynx. **c.** alveoli.
b. bronchi. **d.** bronchioles.

7. The medulla oblongata regulates breathing by monitoring the blood's level of

a. carbon dioxide. **c.** hemoglobin.
b. oxygen. **d.** carbon monoxide.

8. The condition that results from a loss of elasticity in the lungs is

a. bronchitis. **c.** emphysema.
b. lung cancer. **d.** stroke.

9. Which stimulant drug is found in cigarette smoke?

a. globulin **c.** tar
b. nicotine **d.** carbon dioxide

Putting It All Together

Using the information on pages xxx to xxxi, complete the following concept map.

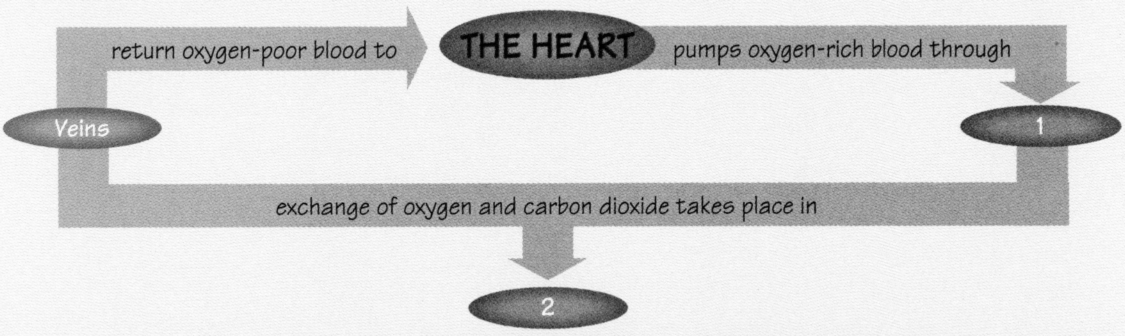

THE HEART
pumps oxygen-rich blood through
1
exchange of oxygen and carbon dioxide takes place in
2
Veins
return oxygen-poor blood to

Reviewing What You Learned

Answer each of the following in a complete sentence.

1. Explain the function of the circulatory system.
2. What is the difference between pulmonary circulation and systemic circulation?
3. What are the functions of the valves located in the heart and veins?
4. Explain the function of the pacemaker.
5. List the parts of the blood.
6. What is the name and function of the substance in red blood cells that gives them their red color?
7. How do platelets aid in blood clotting?
8. What is a function of the lymphatic system?
9. List, in order, the organs through which air passes on its way to the lungs.
10. Trace the path of oxygen as it moves from an alveolus to the capillaries.
11. Describe what happens when the diaphragm muscle relaxes.
12. What are three components of cigarette smoke?

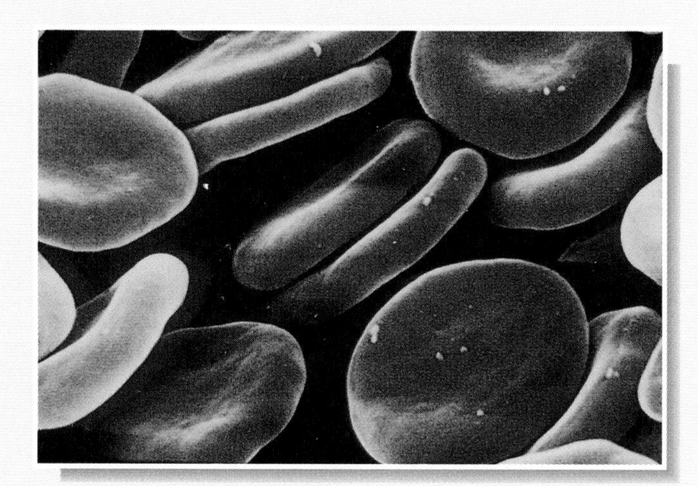

Expanding the Concepts

Discuss each of the following in a brief paragraph.

1. What is the difference between a closed circulatory system and an open circulatory system?
2. Draw a labeled diagram to trace the flow of blood from your "big" toe to your fingers.
3. Compare red blood cells and white blood cells.
4. How would you **relate** an elevated white blood cell count to the presence of an infection?
5. What is atherosclerosis? How does it affect blood pressure?
6. How are a stroke and a heart attack similar? How are they different?
7. How does respiration at the cellular level compare with respiration at the level of the organism?
8. If your body didn't contain hemoglobin, how might it look and work?
9. Children born at high altitudes develop more alveoli and more blood vessels in their lungs than children born at low altitudes. Is this an advantage or a disadvantage? Explain your answer.
10. Explain how smoking affects the circulatory system.

Extending Your Thinking

Use the skills you have developed in this chapter to answer the following.

1. **Applying concepts** Even if you don't smoke, you may be exposed to secondhand smoke. This "passive smoking" exposes you to carbon monoxide. What effect does passive smoking have on your ability to perform in sports?
2. **Giving examples** *Pneumo-* or *pneum-* are prefixes taken from Greek words meaning "related to air." For example, pneumonia is an inflammation of the lungs. Find other examples of words that have these prefixes and write their definitions.
3. **Designing an experiment** Design an experiment to find out how breathing rate varies in different age groups after short-term exercise. Exercise is considered short term if it is minimal and a person recovers in a brief amount of time.
4. **Using the writing process** Studies have shown that smoking tobacco harms the circulatory and respiratory systems of the body. As a result, many people believe that tobacco should be classified as an illegal drug. However, in some parts of the country, farmers depend upon their tobacco crops for their income and would face financial problems if tobacco use decreased. Choose one side of this issue and write a persuasive argument supporting your stand.
5. **Measuring** People who smoke often do not have the physical stamina that nonsmokers have. How could you measure the lung capacity of a smoker? Would you expect to find the same capacity as in a nonsmoker?

Applying Your Skills

Have a Heart

Place your right hand on your chest as if you were going to salute the flag. Can you feel the beating of your heart? Your heart is responsible for pumping blood to all parts of your body. Does your heart rate change if you are sitting, standing, or lying down? Try this activity to find out.

1. Sit down for 2 minutes. After 2 minutes, take your pulse for 1 minute. Record your pulse in your journal.
2. Stand for 2 minutes. Then take your pulse for 1 minute and record it in your journal.
3. Now lie down on the floor for 2 minutes. Then take your pulse for 1 minute and record it in your journal.

• GOING FURTHER •

4. Compare your standing, sitting, and lying-down pulse rates.
5. Did you observe any difference in the rates per minute? In your journal, explain your reasons for any differences.

CHAPTER 38

Digestive and Excretory Systems

FOCUSING THE CHAPTER
THEME: Systems and Interactions

38–1 Nutrition
- **Identify** the role played by food in the body.
- **Discuss** the function of the major nutrients in the body.

38–2 The Digestive System
- **Discuss** the function of the digestive system.
- **Outline** the steps in the process of digestion.

38–3 The Excretory System
- **Describe** how the excretory system functions.

BRANCHING OUT *In Action*

38–4 A Cure for Ulcers?
- **Explain** how ulcers are caused and how they can be treated.

LABORATORY INVESTIGATION
- **Design an experiment** to determine the effect of air exposure on the vitamin C content of foods.

Biology and Your World

BIO JOURNAL

Perhaps you know someone who undergoes regular dialysis treatments or who has a transplanted kidney. Find out why renal patients require these treatments and how the lives of these individuals change as a result of having to undergo the treatments. Describe your findings in your journal.

False-color X-ray image of the abdomen showing the large intestine

Nutrition

GUIDE FOR READING

- **Explain** the function of food in the body.
- **List** the various classes of nutrients in the body.

MINI LAB

- **Observe** the presence of iron in cereal.

HOW IMPORTANT IS FOOD IN your life? Before you answer, think of some of the most important holidays or occasions you celebrate. What pictures come to mind? No matter where you live, chances are that a meal was the centerpiece of that special day. To most of us food is more than just nourishment—it is an important part of our culture. Human societies around the globe organize meetings and family gatherings around certain kinds of food.

The need to eat is one of your body's first priorities. In this section, you will see why your body needs food and what kinds of food are required to keep it healthy and strong.

Energy and Materials

Why do you need food? The most obvious answer is energy—the ability to do work. You need energy to climb stairs, lift books, run, and even to think. Like a car that needs gasoline, your body needs fuel for all that work, and food is your fuel.

But there is another reason that you need food. There's an old saying that "you are what you eat." And, like many sayings, it teaches us a valuable lesson. **The foods that we eat not only provide us with the energy to perform various types of actions, they also contain the materials from which our body cells and tissues are made.**

The energy available in food can be measured in a surprisingly simple way—by burning it! The amount of heat given off is measured and expressed in terms of calories. A calorie is the amount of heat energy needed to raise the temperature of 1 gram of water by 1 degree Celsius. On packaged foods, nutrition labels use the unit **Calorie,** which

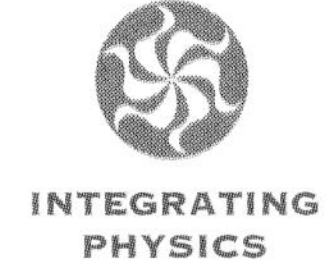

INTEGRATING PHYSICS

What is energy? Work? How are they related? Measured?

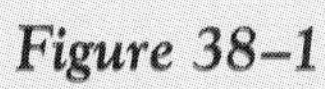

Figure 38–1 People of all cultures devote a great deal of time and effort to the (a) preparation and presentation of food. By doing so, eating, which is necessary to maintain life, is made more enjoyable. Food provides (b) this gymnast with the muscular strength he needs to support himself on the parallel bars and (c) these soccer players with the energy they need to run and kick.

Magnetism in Your Cereal?

PROBLEM *How can you **observe** the iron in your cereal?*

PROCEDURE 

1. In a sealable plastic bag, crush 50 grams of a dry breakfast cereal that contains 100 percent of the U.S. Required Daily Allowance for iron.
2. Shake all the crushed cereal to one corner of the bag. Pass the magnet under the cereal and observe what happens. Use the magnet to drag any particles that are attracted to it to the empty part of the bag.
3. Remove the magnetized cereal and find its mass.

ANALYZE AND CONCLUDE

1. What percentage of the cereal was attracted by the magnet?
2. What evidence do you have that iron was in the cereal?
3. Describe a procedure you could use to compare the amount of iron in the cereal at the top of a cereal box and at the bottom.

Figure 38–2
CAREER TRACK
One of the jobs of a dietician is to help people choose a diet that will best meet that person's bodily needs.

represents 1000 calories, or 1 kilocalorie. Thus, a teaspoon of sugar contains 16 Calories, or 16,000 calories.

The basic energy needs of an average-sized teenager are between 1800 and 2800 kcal per day. If you engage in vigorous physical activity, however, your energy needs may be higher. What kinds of foods can meet those energy needs? Just about any kind. Chemical pathways in the body's cells can extract energy from almost any type of food.

☑ ***Checkpoint*** What is the function of food in the body?

Nutrients

Although the body is able to manufacture many of the molecules it needs, it must still obtain the materials for this from the food it takes in. **The classes of nutrients that are part of any healthy diet are water, carbohydrates, fats, proteins, vitamins, and minerals.**

Water

Water is the most important of all nutrients. Water is needed by every cell in our body, and it makes up the bulk of blood, lymph, and other body fluids. Water dissolves food taken into the digestive system. Water in the form of sweat cools the body. Water is lost from the body as vapor in every breath we exhale and as the primary component of urine.

If enough water—at least a liter a day—is not taken in to replace what is lost, dehydration can result. This leads to problems with the circulatory, respiratory, and nervous systems. Drinking plenty of pure water is one of the best things you can do to help keep your body healthy.

Carbohydrates

Carbohydrates are major sources of food energy. Sugars—found in fruits, honey, sugar cane and sugar beets—are examples of simple carbohydrates.

Problem Solving

INTERPRETING DATA

It's All in the Analysis

You and a friend have just read an article in a sports magazine that says that diets high in fats increase an individual's chances of getting heart disease and certain cancers when the individual gets older. The article also points out that sufficient fiber—found in fruits, vegetables, beans, and whole grain breads—helps to lower the chance of getting these diseases.

Your friend wonders whether his diet is helping to prevent disease or causing it. He has decided to find out just how his diet compares with a balanced diet recommended by nutritionists. Using the list of all the foods he ate the day before, answer the questions that follow.

Breakfast	*Dinner*
2 scrambled eggs 2 slices white bread 1 teaspoon butter orange juice ($\frac{3}{4}$ cup)	fried chicken (6 oz) roasted potatoes (1 cup) asparagus ($\frac{1}{2}$ cup) 1 slice white bread 1 glass soda
Lunch	***Snacks***
fast-food cheeseburger (bun, one-quarter pound beef, 1 slice cheese, mayonnaise, lettuce) French-fried potatoes (1 cup) milkshake (12 oz)	potato chips (one small bag) pretzels (1 cup)

Serving size   One serving from the bread group is equal to 1 slice of bread or $\frac{1}{2}$ cup cooked cereal, rice, or pasta. One serving from the vegetable group is equal to $\frac{1}{2}$ cup cooked or raw vegetables, 1 cup leafy raw vegetables, or $\frac{1}{2}$ cup cooked beans. A serving of fruit is the size of a medium apple or banana, $\frac{1}{2}$ cup diced or cooked fruit, or $\frac{3}{4}$ cup juice. One cup of milk or yogurt or $1\frac{1}{2}$ ounces of cheese are equal to one serving from the milk group. One egg or 3 ounces of meat make up a serving from the meat group.

• THINK ABOUT IT •

1. Using the Food Guide Pyramid on page 856, classify the foods your friend ate by food group.
2. Determine how many servings of each food group your friend consumed.
3. Did your friend meet the minimum number of servings recommended in the guide? Explain your answer.
4. Based on the guide, what would you recommend to your friend regarding his diet?

VITAMINS

VITAMIN	FOOD SOURCES	FUNCTION	RESULTS OF VITAMIN DEFICIENCY
Water-Soluble Vitamins			
B_1 (thiamine)	Yeast, liver, grains, legumes	Coenzyme for carboxylase	Beriberi, general sluggishness, heart damage
B_2 (riboflavin)	Milk products, eggs, vegetables	Coenzyme in electron transport (FAD)	Sores in mouth, sluggishness
Niacin	Red meat, poultry, liver	Coenzyme in electron transport (NAD)	Pellagra, skin and intestinal disorders, mental disorders
B_6 (pyridoxine)	Dairy products, liver, whole grains	Amino acid metabolism	Anemia, stunted growth, muscle twitches and spasms
Pantothenic acid	Liver, meats, eggs, whole grains, and other foods	Forms part of coenzyme A, needed in Krebs cycle	Reproductive problems, hormone insufficiencies
Folic acid	Whole grains and legumes, eggs, liver	Coenzyme in biosynthetic pathways	Anemia, stunted growth, inhibition of white cell formation
B_{12}	Meats, milk products, eggs	Required for enzymes in red cell formation	Pernicious anemia, nervous disorders
Biotin	Liver and yeast, vegetables, provided in small amounts by intestinal bacteria	Coenzyme in a variety of pathways	Skin and hair disorders, nervous problems, muscle pains
C (ascorbic acid)	Citrus fruits, tomatoes, potatoes, leafy vegetables	Required for collagen synthesis	Scurvy: lesions in skin and mouth, hemorrhaging near skin
Choline	Beans, grains, liver, egg yolks	Required for phospholipids and neurotransmitters	Not reported in humans
Fat-Soluble Vitamins			
A (retinol)	Fruits and vegetables, milk products, liver	Needed to produce visual pigment	Poor eyesight and night blindness
D (calciferol)	Dairy products, fish oils, eggs (also sunlight on skin)	Required for cellular absorption of calcium	Rickets: bone malformations
E (tocopherol)	Meats, leafy vegetables, seeds	Prevents oxidation of lipids in cell membranes	Slight anemia
K (phylloquinone)	Intestinal bacteria, leafy vegetables	Required for synthesis of blood-clotting factors	Problems with blood clotting, internal hemorrhaging

Figure 38–3
The table lists the food sources and the functions of the fourteen essential vitamins. Notice that the vitamins are categorized as water soluble or fat soluble. Because fat-soluble vitamins can be stored in the fatty tissues of the body, these vitamins should not be taken in excess of daily required amounts or they can become toxic.

Starches—found in grains, potatoes, and vegetables—are examples of complex carbohydrates. Starches are easily broken down by the digestive system into simple sugars and passed into the bloodstream, where they can provide energy to cells throughout the body.

Many carbohydrate-rich foods also contain cellulose, commonly referred to as "fiber." Unlike starch, the sugars in cellulose are linked together in such a way that humans and many other animals cannot break them apart. Even though we cannot digest cellulose, it is still important in the diet. Cellulose, or fiber, adds bulk to the material moving through the digestive system. This bulk helps the muscles of the digestive system to process foods more effectively. Foods such as lettuce, whole grain breads, and bran are rich in fiber.

MINERALS

MINERAL	FOOD SOURCES	FUNCTION	RESULTS OF MINERAL DEFICIENCY
Calcium	Milk, cheese, legumes, dark-green vegetables	Bone formation, blood-clotting reactions, nerve and muscle function	Stunted growth, weakened bones, muscle spasms
Phosphorus	Milk products, eggs, meats	Bones and teeth, ATP and related nucleotides	Loss of bone minerals
Potassium	Most foods	Acid-base balance, nerve and muscle function	Muscular weakness, heart problems, death
Chlorine	Salt	Acid-base balance, nerve and muscle function, water balance	Intestinal problems, vomiting
Sodium	Salt	Acid-base balance, nerve and muscle function, water balance	Weakness, diarrhea, muscle spasms
Magnesium	Green vegetables	Enzyme cofactors, protein synthesis	Muscle spasms, stunted growth, irregular heartbeat
Iron	Eggs, leafy vegetables, meats, whole grains	Hemoglobin, electron-transport enzymes	Anemia, skin lesions
Fluorine	Drinking water, seafood	Structural maintenance of bones and teeth	Tooth decay, bone weakness
Iodine	Seafood, milk products, iodized salt	Thyroid hormone	Goiter (enlarged thyroid)

Figure 38–4
Nine of the body's important minerals are listed in this table, which also identifies some of the foods in which the mineral is found and the function the mineral has in the body.

Fats

Fats, or lipids, may have a bad reputation in today's society, but in proper amounts they are still an important part of a healthy diet. Your body needs certain essential fatty acids to manufacture the lipids in cell membranes and to produce certain hormones. Fats that can meet these requirements are found in most foods. In fact, only 60 grams of vegetable oil—about 2 tablespoons—meet the daily fatty acid needs of an average person.

Despite the fact that little fat is actually required in the diet, about 40 percent of the Calories in the diet of a typical American come from fat! Most nutritionists believe that this is far too much fat and may lead to serious health consequences, such as high blood pressure, heart disease, obesity, and diabetes.

☑ **Checkpoint** What are fats?

Proteins

Proteins are important nutrients because of the amino acids they contain. The body uses the amino acids in proteins to build new cells and tissues. Rapidly growing tissues, including those in the skin and the lining of the digestive system, must constantly replace dead and dying cells with new ones. Of the 22 most common amino acids, your body is able to make 14—the other 8 must be obtained from food. Diets that are deficient in protein interfere with cell growth and can cause serious health problems.

☑ **Checkpoint** What is the function of proteins in the body?

Vitamins and Minerals

If you think of proteins, fats, and carbohydrates as the building blocks of the body, you might think of vitamins as the tools that help to put them together. Vitamins are organic molecules that the body needs to help perform important chemical reactions.

Although vitamins are needed in very small amounts, vitamin deficiencies can have serious, even fatal, consequences.

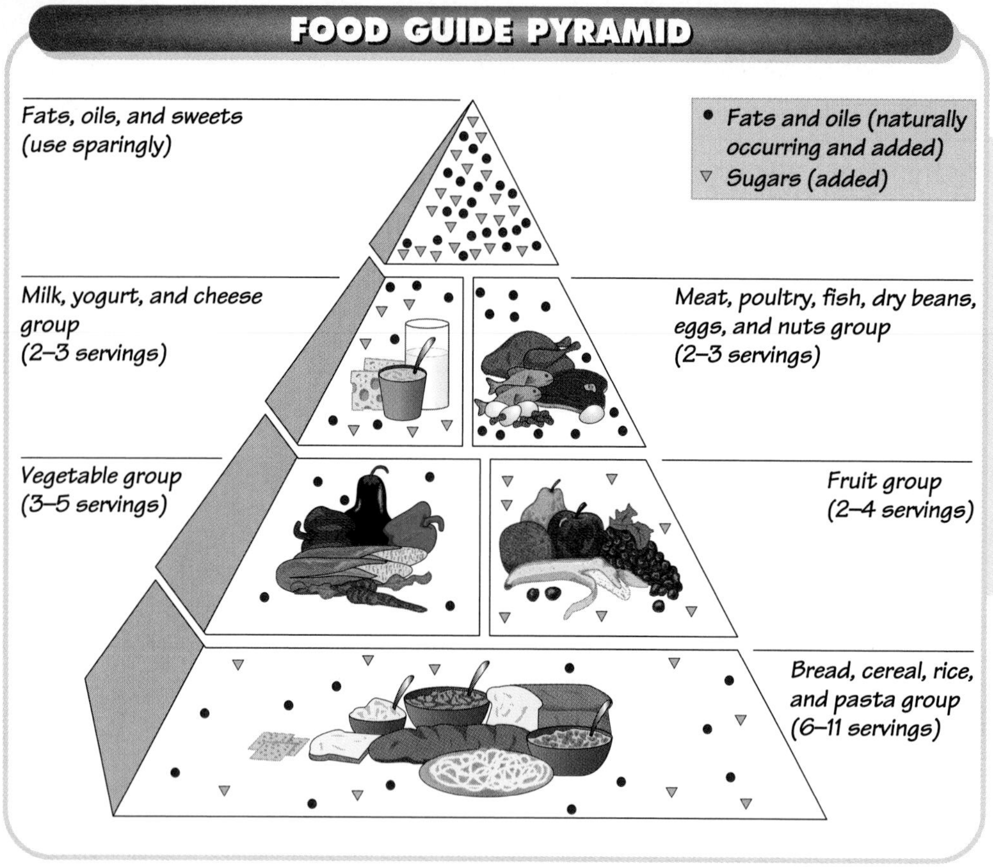

FOOD GUIDE PYRAMID

Fats, oils, and sweets (use sparingly)

• Fats and oils (naturally occurring and added)
▽ Sugars (added)

Milk, yogurt, and cheese group (2–3 servings)

Meat, poultry, fish, dry beans, eggs, and nuts group (2–3 servings)

Vegetable group (3–5 servings)

Fruit group (2–4 servings)

Bread, cereal, rice, and pasta group (6–11 servings)

Figure 38–5
The food guide pyramid illustrates the main characteristics of a balanced diet. Carbohydrate-rich foods should make up the major portion of the diet, while foods containing fats and sugars should be eaten sparingly.

Fourteen vitamins, listed in ***Figure 38–3*** on page 854, are generally recognized as essential to human health—and there may be more. Eating a diet containing a variety of foods will supply the daily vitamin needs of nearly everyone.

As you know, food stores and pharmacies sell vitamin supplements containing the minimum daily requirement of each important vitamin. Unfortunately, taking extra-large doses of vitamins does not benefit the body and in some cases may cause it harm.

Minerals are inorganic substances that are needed in small amounts by the body. ***Figure 38–4*** on page 855 lists nine of the most important minerals. Calcium, for example, is required to produce the calcium phosphate that goes into bones and teeth. Iron is needed to make hemoglobin, the oxygen-carrying protein in red blood cells. And thyroxine, a hormone required for normal growth and development, cannot be synthesized without iodine.

☑ **Checkpoint** What roles do vitamins and minerals play in the body?

Balancing the Diet

It's no easy matter to figure out the best balance of nutrients for the human diet, but nutritional scientists have done their best to do exactly that. The result is the food pyramid shown in ***Figure 38–5.*** The basic idea behind the pyramid is sound and simple—you should eat a variety of fresh foods each day, and you should limit your intake of fatty foods.

Section Review 38–1

1. **Explain** the function of food in the body.
2. **List** the various classes of nutrients in the body.
3. **Critical Thinking—Evaluating** Why should you eat a balanced diet?
4. **MINI LAB** How can you **observe** the presence of iron in cereal?

The Digestive System

Guide for Reading

- **Explain** the function of the digestive system in the body.
- **Describe** the action of insulin and glucagon.

FOOD PRESENTS EVERY ANIMAL with at least two challenges. The first, of course, is how to obtain it. Then, when an animal has caught, gathered, or engulfed its food, it faces a new challenge—how to break that food down into small molecules that can be passed to the cells that need them. In this section, we will focus our attention on this challenge.

To meet this challenge, the body is equipped with a remarkable organ system that begins to process food as soon as it is placed in the mouth. As the food passes through these organs, it gets disassembled, contributing its value to the body along the way.

The Digestive Tract

The human digestive system, like those of other vertebrates, is built around an alimentary canal—a one-way tube that passes through the body. **The function of the digestive system is to convert foods into simple molecules that can be absorbed and used by the cells of the body.**

The Mouth

Food enters the **mouth,** where the work of the digestive system begins. Chewing, which takes place in the mouth, seems simple enough—teeth tear and crush the moistened food to a fine paste until it is ready to be swallowed. But there is a great deal more to it than that.

Teeth are anchored in the bones of the jaw and are connected to the jaw by a network of blood vessels and nerves that enter through the roots of the teeth. The surfaces of the teeth, which are much tougher than ordinary bone, are protected by a coating of mineralized enamel. Teeth do much of the mechanical work

Figure 38–6
The foods we eat—such as (a) pizza, (b) meats and vegetables, and (c) strawberries—must be broken down into molecules that are further broken down to release the energy and provide the nutrients they contain. The process by which foods are broken down is known as digestion.

Figure 38–7
Digestion begins in the mouth, where teeth cut, grind, and crush food into a paste with the aid of saliva. Saliva, a fluid that is produced by salivary glands, not only moistens the food for easy swallowing, it also contains enzymes that begin to break large, complex starch molecules into simple glucose molecules.

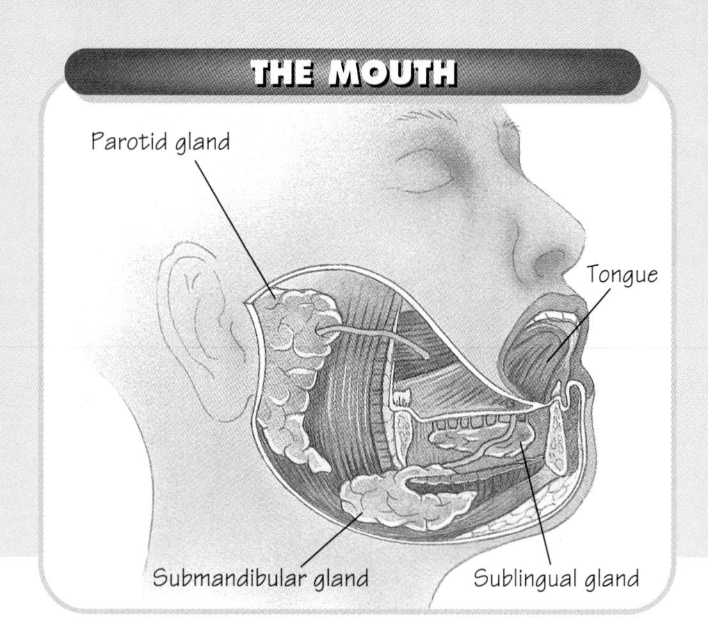

of digestion by cutting, tearing, and crushing food into small fragments.

Human teeth include incisors, which are sharp enough to cut directly through meat; cuspids and bicuspids, which grasp and tear food; and molars, which have large, flat surfaces ideal for grinding food. Human tooth structure is intermediate between that of a plant-eating herbivore, in which molars are most common, and a carnivore, in which incisors are prominent. Our tooth structure reflects a mixed diet of meats and plants.

In the mouth, saliva, a fluid produced by the mouth's three pairs of **salivary glands,** helps to moisten the food and make it easier to chew. The release of saliva is under the control of the nervous system, and it can be triggered by the scent of food—especially when you are hungry!

Saliva not only helps to moisten food, it also helps to ease its passage through the digestive tract. Saliva also contains an enzyme called **amylase,** which breaks the chemical bonds in starches, releasing sugars. If you chew on starchy foods like crackers long enough, they will begin to taste sweet—a taste that comes from the chemical action of amylase on starch. Saliva also contains **lysozyme,** an enzyme that fights infection by digesting the cell walls of many bacteria.

Food is passed from the mouth into the rest of the digestive system by swallowing. Pushed by the tongue and muscles of the throat, the chewed clump of food, called a bolus, is forced down the throat. Just as this happens, a flap of tissue, known as the epiglottis (ehp-uh-GLAHT-ihs), is forced over the opening to the air passageways. This prevents food from clogging the air passageways—most of the time, anyway!

☑ **Checkpoint** What is amylase?

The Esophagus

After the bolus is swallowed, it passes through the **esophagus,** or food tube, into the stomach. Did you know that food can travel through the esophagus whether you're sitting up, lying down, or standing on your head? Even in astronauts, food passes through the esophagus in the weightlessness of space. The reason is that food is moved along by contractions of smooth muscle surrounding the esophagus. Known as **peristalsis** (per-uh-STAL-sihs), these contractions, which occur throughout the alimentary canal, squeeze the food through the 25 centimeters of the esophagus.

The Stomach

Food from the esophagus empties into a large muscular sac called the **stomach.** A thick ring of muscle, known as the cardiac sphincter, closes the esophagus after food has passed into the stomach,

Figure 38–8
Once food is swallowed, it is forced through the alimentary canal—esophagus, stomach, small intestine, and large intestine—by the action of peristalsis, a series of periodic muscular contractions. Just below the entry to the large intestine is a small pouch known as the appendix. In humans the appendix is of little importance except when it becomes infected. This condition, which is known as appendicitis, usually requires the surgical removal of the appendix.

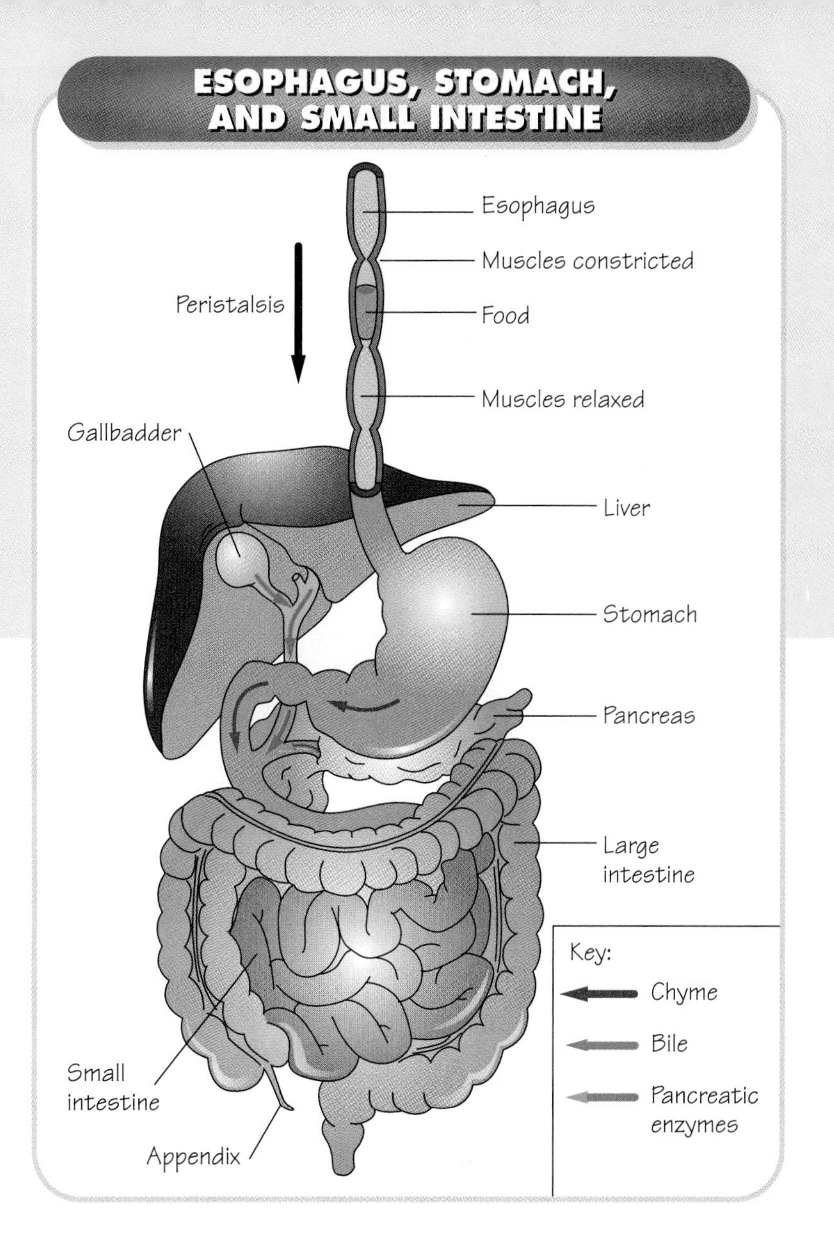

preventing the contents of the stomach from moving back into the esophagus. The size of the stomach enables you to eat a few large meals a day, rather than having to nibble all the time. Its walls produce a powerful combination of enzymes and strong acids, and contractions of its smooth muscles thoroughly mix the food you swallow.

The lining of the stomach contains millions of microscopic **gastric glands** that release a number of substances into the stomach. Some of these glands produce mucus, a fluid that lubricates and protects the stomach wall. Other glands produce hydrochloric acid, and still others produce **pepsin.** Pepsin, an enzyme that digests proteins, works best at low pH. Remember that low pH corresponds to high acidity. ● The combination of pepsin and hydrochloric acid in the stomach unfolds large proteins in foods and breaks them into smaller polypeptide fragments.

As digestion proceeds, stomach muscles contract to churn and mix stomach fluids and food, gradually producing a mixture known as chyme (KIGHM). After a few hours in the stomach, the pyloric valve—located at the junction of the stomach and small intestine—opens, and chyme is forced into the duodenum.

Checkpoint What is chyme?

The Small Intestine

The duodenum is the first of three parts of the **small intestine,** and it is the place where most of the chemical work of digestion takes place. As chyme enters from the stomach, it is mixed with enzymes and digestive fluids from the other accessory digestive organs and even from the lining of the duodenum itself.

INTEGRATING CHEMISTRY

What is pH? How does pH characterize acids? Bases?

Located just below the stomach is a gland called the **pancreas,** which serves two important functions. One function is to produce hormones that regulate blood sugar, which we will examine later. Within the digestive system, the pancreas plays the role of master chemist, producing enzymes that break down carbohydrates, proteins, lipids, and nucleic acids. The pancreas also produces sodium

THE SMALL INTESTINE

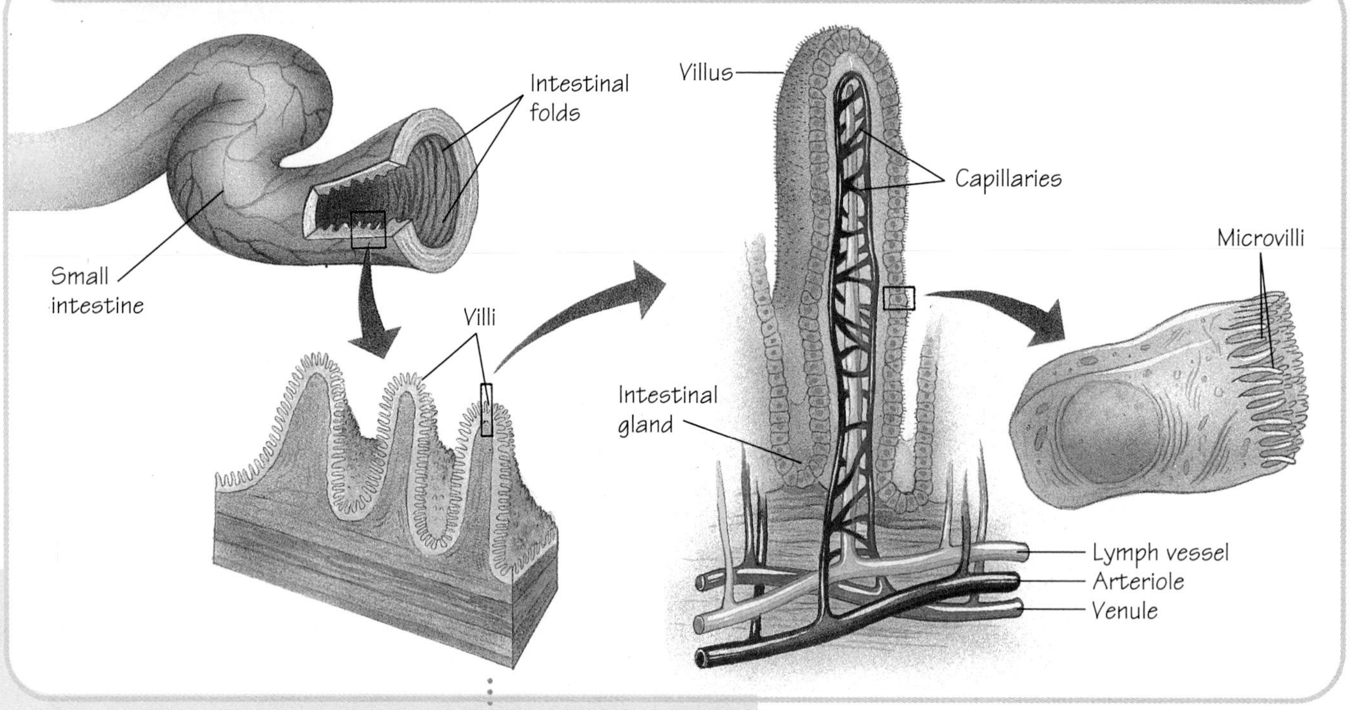

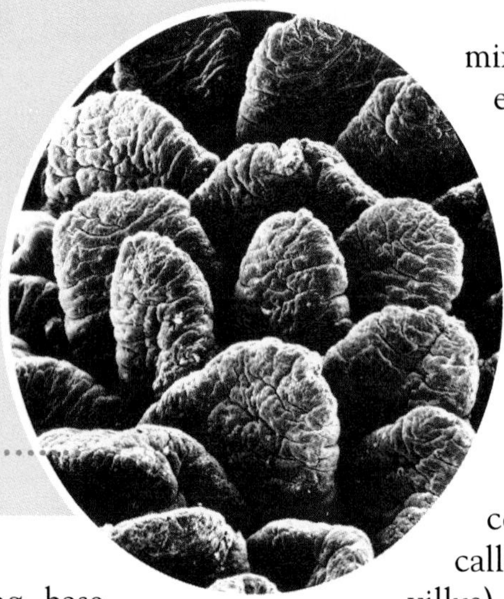

Figure 38–9
(a) *The lining of the small intestine consists of folds that are covered with tiny projections called villi. Within each villus there is a network of blood capillaries and lymph vessels that absorb and carry away nutrients.*
(b) *This electron micrograph of villi shows how they greatly increase the surface area of the small intestine (magnification: 300X).*

bicarbonate, a strong base that neutralizes stomach acid so that these enzymes can go to work.

To aid in this process, the **liver,** a large gland just above the stomach, produces **bile,** a fluid loaded with lipids and salts. Bile acts almost like a detergent, dissolving and dispersing the droplets of fat found in fatty foods. This makes it possible for enzymes to reach these fat molecules and break them down.

By the time chyme enters the remaining parts of the small intestine, nearly all the chemical work of digestion has been completed. The chyme is now a rich mixture of small nutrient molecules ready to be absorbed by the body. The remaining parts of the small intestine—the jejunum and the ileum—are specially adapted to absorb these nutrients. Nearly 7 meters in length, the folded surfaces of the jejunum and the ileum are covered with projections called **villi** (VIHL-igh; singular: villus). The surfaces of the cells of the villi are themselves covered with thousands of fingerlike projections known as microvilli. Slow, wavelike contractions of smooth muscles move the chyme along this absorptive surface.

Nutrient molecules are rapidly absorbed into the cells lining the small intestine. Sugars, amino acids, and other nutrients are then passed directly into the bloodstream. Fats take a different route. They are transported into tiny lymph vessels before being delivered to the bloodstream.

As you might expect, blood leaving the capillaries of the small intestine is nutrient-rich after a meal. Does this mean that levels of sugars and other nutrients in the blood rise and fall dramatically during the day? Not at all. All the blood leaving the small intestine passes through the liver before it flows to the rest of the body. This unusual arrangement places the liver in a position to monitor the blood before it enters the general circulation.

Not long after you eat something sweet, blood from the small intestine is loaded with energy-producing sugar. The liver efficiently removes most of the extra sugar from the blood and stores it in the form of glycogen, a polysaccharide. Liver cells keep this nutrient in reserve, and when blood sugar levels begin to fall, they release the stored sugars.

☑ **Checkpoint** What are villi?

The Large Intestine

Nearly all the available nutrients have been removed from the chyme that enters the **large intestine,** or colon. The primary job of the colon is to remove water from the undigested material that is left, producing a concentrated waste material known as feces. This material passes through the rectum and is eliminated from the body.

Water is moved quickly across the colon wall while rich colonies of bacteria grow on the undigested material. These intestinal bacteria are helpful to the digestive process—some even produce compounds that the body is able to use, including vitamin K.

Regulating Nutrient Levels

Regulating the level of blood sugar is one of the body's most important jobs. If the blood has too little sugar, organs—including the brain—that depend upon it as a source of energy will begin to suffer. Two hormones produced by the pancreas help to regulate blood sugar. Inside the pancreas there are tiny "islands" of cells known as the **islets of Langerhans.** Some of these cells produce **insulin,** a polypeptide hormone.

When the blood sugar level rises, as it might after a meal, the islets of Langerhans in the pancreas detect this and release insulin into the bloodstream. **Insulin stimulates cells in the liver, muscles, and fatty tissues to remove sugar from the bloodstream and store it in the form of glycogen and fat.**

When blood sugar levels fall, as they might after several hours without eating, another hormone, called **glucagon,** is released by the islets of Langerhans. **Glucagon stimulates the liver, muscles, and fatty tissues to break down glycogen and fats and to release sugars into the blood.** This raises blood sugar back to a safe level.

When the body cannot produce enough insulin, blood sugar cannot be regulated and a disorder known as diabetes mellitus results. Most forms of juvenile-onset diabetes, a type of diabetes mellitus, can be treated with a carefuly controlled diet and daily injections of insulin.

Section Review 38-2

1. **Explain** the function of the digestive system in your body.
2. **Describe** the action of insulin and glucagon in your body.
3. **Critical Thinking—Relating** What functions do the enzymes amylase and pepsin serve in digestion?

SECTION

38–3 The Excretory System

GUIDE FOR READING

- **Explain** the function the kidneys perform in the body.
- **Describe** the processes of filtration and reabsorption.

MINI LAB

- **Relate** the structure of a kidney to its function.

INTEGRATING CHEMISTRY

How does protein metabolism produce urea?

THE CHEMISTRY OF THE human body is a marvelous thing. An intricate system of checks and balances controls everything from your blood pressure to your body temperature. Nutrients are absorbed, stored, and carefully released just as they are needed. But every living system, including the human body, produces chemical wastes—byproducts of chemical reactions that are no longer useful. Some are even toxic. Therefore, the body occasionally needs to throw things away. The body relies on a group of organs that work together to remove the unwanted, and often toxic, wastes.

Chemical Wastes

The elimination of chemical wastes from the body is known as excretion. The lungs, for example, excrete carbon dioxide, a chemical waste produced when energy is captured from food compounds. The skin excretes excess water and salt in sweat. This makes the lungs and the skin part of the excretory system, a system of organs that remove chemical wastes from the body. In this section, however, we will focus our attention on a pair of organs whose main function is excretion—the **kidneys.**

Why does the body need organs specialized for excretion? Part of the answer has to do with the chemistry of proteins. When the body uses the amino acids from proteins for food, it sometimes must remove their amino ($-NH_2$) groups.

Figure 38–10
(a) As food is utilized by the body to perform work—such as in this track event—waste products, which the body must get rid of, are generated and released into the bloodstream. As (b) the runner and (c) the cyclist can testify, water plays a key role in the excretion of these wastes by the kidneys.

Figure 38–11
(a) *The breakdown of an amino acid such as alanine yields pyruvic acid and ammonia. The pyruvic acid is used for energy, while the ammonia is converted to urea by a complex series of reactions.*
(b) *The kidneys function in the removal of urea as well as the regulation of water in the bloodstream.*

(a) **PROTEIN METABOLISM**

$$H_2N-\underset{COOH}{\overset{CH_3}{C}}-H + \frac{1}{2}O_2 \longrightarrow \underset{COOH}{\overset{CH_3}{C}}=O + NH_3$$

Alanine — Pyruvic acid — Ammonia

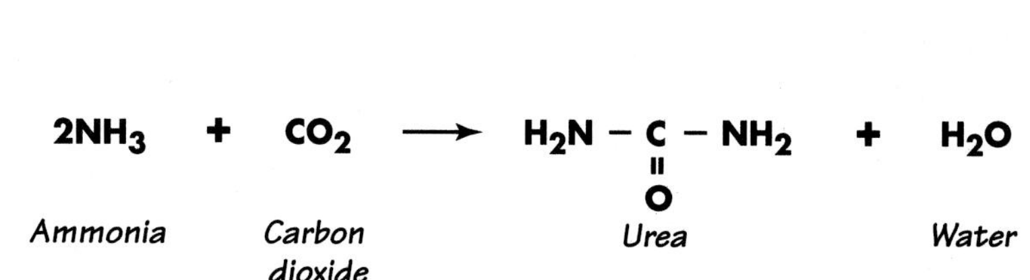

(b) **THE EXCRETORY SYSTEM**

Vena cava
Aorta
Renal artery
Kidney
Renal vein
Ureter
Urinary bladder
Urethra

This produces ammonia, NH_3, a poisonous compound that the body quickly converts to urea. ● Urea is less toxic than ammonia and is very soluble in water, but it still must be removed from the bloodstream. **The removal of urea—a substance that is the result of the metabolism of proteins in body cells—along with the regulation of water in the bloodstream, is the principal job of the kidneys.**

☑ **Checkpoint** What is the function of the kidneys?

The Kidneys

The kidneys, each about the size of a fist, are located on either side of the spinal column in the lower back. Blood flows into each kidney through a **renal artery** and leaves through a **renal vein.** A third vessel, called the **ureter,** leaves each kidney, carrying fluid to the **urinary bladder.**

Within each kidney is a complex filtration system that removes urea, excess water, and other wastes from the blood. As purified blood is returned to the body, these wastes are collected to form a fluid known as urine. Urine is stored in the bladder and then is eliminated from the body.

The Nephrons

Each kidney contains about a million tiny functional units known as **nephrons.** Refer to the illustration on page 865 as you read about the workings of a nephron. Blood enters each nephron through a single arteriole (small artery) and leaves through a venule (small vein). Each nephron also contains a collecting duct, which leads to the renal pelvis. Wastes and excess water are

Examining a Kidney

PROBLEM *How does the structure of a kidney **relate** to its function?*

PROCEDURE 

1. Obtain a mammal kidney and five different-colored pushpins.
2. Using reference materials, place a pin in or next to the renal artery, renal vein, and ureter. Slice the kidney lengthwise and identify as many structures as you can.
3. Use colored pencils and a file card to make a key to identify each part.

ANALYZE AND CONCLUDE

1. How does the size of the kidney you labeled compare with the size of a human kidney?
2. What structures and functions of the kidney are easier to understand when a real organ is examined?

removed from the bloodstream and gathered in the collecting duct. When blood enters the nephron, it flows into a spherical meshwork of thin-walled capillaries called a **glomerulus**—the filtering unit of the nephron. The blood is under such pressure in these small vessels that much of the blood plasma seeps out, almost like water from a leaky garden hose.

Filtration and Reabsorption

Blood cells, platelets, and plasma proteins are too large to leave the blood as it passes through the glomerulus. **In the glomerulus, blood plasma—containing water, salts, sugars, and nutrients—filters out of the bloodstream and is collected in Bowman's capsule.** This plasma, called the primary filtrate, is then passed from **Bowman's capsule** to the proximal tubule. It looks as though most of this plasma is about to be thrown away, but appearances can be deceiving.

More than 180 liters of blood plasma pass into the tubules of the kidneys each day. Needless to say, if you excreted this much fluid, life would be very difficult! Fortunately, this is not what happens. **As fluid moves through the proximal tubule, nearly all the material first removed from the blood is put back into the blood.** This process is called **reabsorption.**

Why should the nephron first throw nearly everything away, only to reabsorb most of the water and dissolved materials? This is a bit like cleaning a room by first carrying everything out into the hallway, then bringing back only the things you want to keep. The likely answer is that this is exactly what makes the kidney such an excellent filter.

By first removing nearly everything from the bloodstream, the kidneys ensure that drugs, poisons, and other dangerous compounds will be taken out of the bloodstream. As they then reabsorb most of what they have filtered out, these toxic compounds are left in the proximal tubule, where they will be eliminated in the urine.

Concentration and Water Balance

As the kidneys reabsorb material into the bloodstream, they draw back nearly 99 percent of the water that had first been filtered out. Urine, which now moves through the nephron toward the ureter, is further concentrated in a region of the nephron known as the **loop of Henle.** Depending on the demands of the body, the nephron can produce urine that is very concentrated, which conserves water, or very dilute, which eliminates water.

☑ ***Checkpoint*** How do kidneys filter and reabsorb plasma?

Visualizing Kidney Structure and Function

The most important organs of the excretory system are the kidneys, which consist of numerous tiny fluid-filtering structures called nephrons.

1 Renal Artery and Renal Vein

The renal artery brings blood to the kidney, while the renal vein carries the filtered blood away.

2 Renal Cortex and Renal Medulla

The human kidney consists of an outer renal cortex and an inner renal medulla, in which nephrons perform the task of filtration.

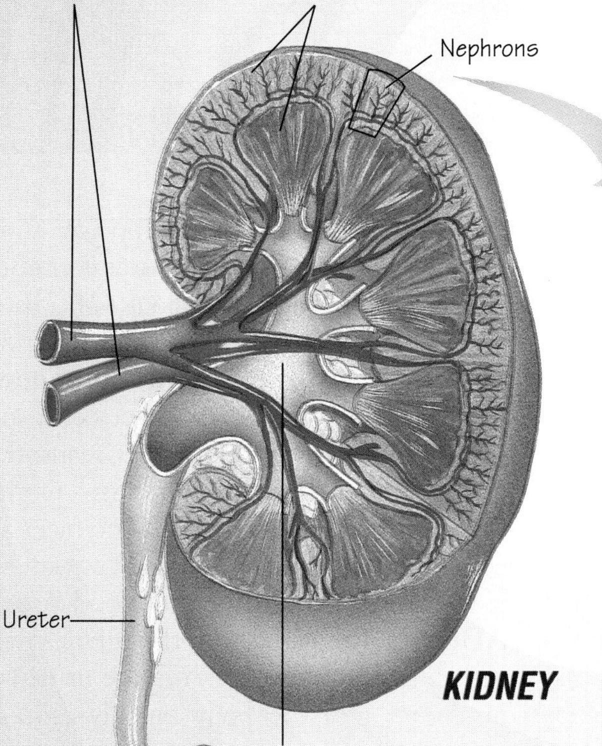

3 Renal Pelvis

The collecting ducts at the ends of nephrons merge and empty urine into the inner region of the kidney, called the renal pelvis, from which the urine is carried to the urinary bladder by way of the ureter.

4 Bowman's Capsule

Water, nutrients, and wastes are filtered from the blood that enters the glomerulus. This filtrate is collected in Bowman's capsule.

5 Proximal Tubule

Most of the water and nutrients are reabsorbed into the blood as the filtrate passes through the proximal tubule of the nephron.

6 Distal Tubule

Wastes that still remain in the blood are secreted into the distal tubule of the nephron and become part of the urine.

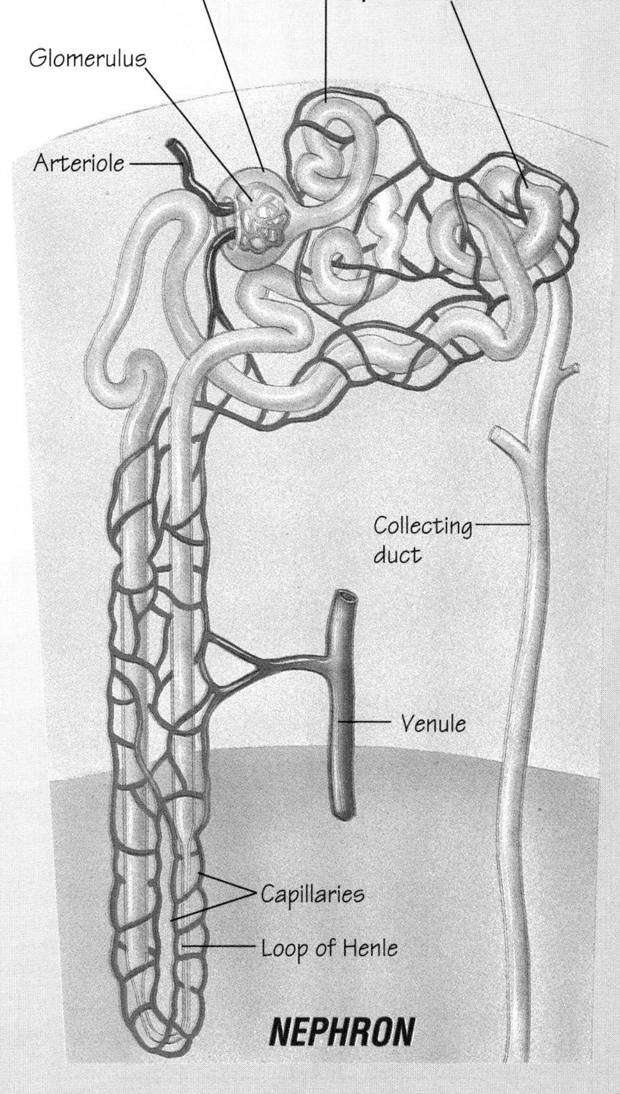

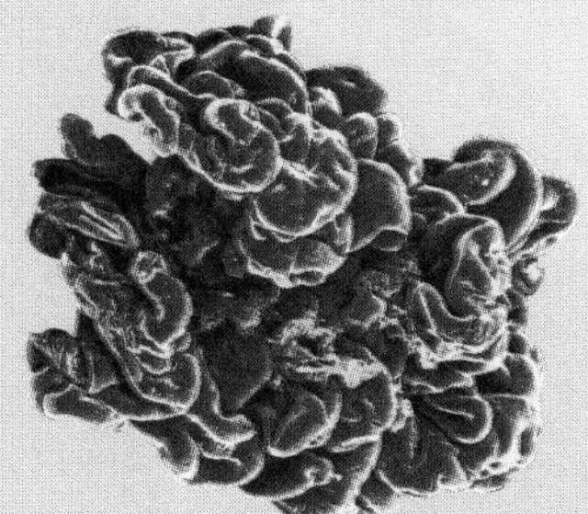

Scanning electron micrograph of a glomerulus (magnification: 400X)

Figure 38–12
When disease or injury reduces kidney function, a kidney dialysis machine is used to filter the blood—much like the kidney itself. These young patients are undergoing kidney dialysis on a beach, using portable kidney dialysis machines. As their blood flows through porous tubes inside the machine, it is bathed in a solution whose composition is close to that of blood plasma. Waste products diffuse into this solution, returning purified blood to the patient. Kidney dialysis makes it possible for people to lead relatively normal lives while waiting for the transplant of a healthy kidney, which will set them free from dialysis machines.

Control of Water Balance

When you drink a glass of water, it is quickly absorbed by the digestive system and passed into the bloodstream. If that's all that happened, every time you took in fluids your blood would become more dilute. Osmosis would cause water to flow into the tissues of the body, and many of the tissues would swell. Needless to say, this doesn't usually happen. The reason, of course, is that the kidneys remove that excess water.

The kidneys don't accomplish this by removing any more water in filtration—they do it by putting less water back into the bloodstream during reabsorption. The kidneys, however, don't make this "decision" on their own—they have help from the brain.

The hypothalamus region of the brain closely monitors the water content of the blood. When it is about right, the hypothalamus signals the pituitary gland to release an **antidiuretic hormone** (ADH). ADH causes its target cells, in the tubules of the kidneys, to reabsorb more water. This action returns more water to the bloodstream and produces a concentrated urine that conserves water. If the water content of the blood rises, ADH is not released, and the tubules reabsorb much less water. A dilute urine is produced, and the body quickly eliminates the excess water.

Section Review 38–3

1. **Explain** the function the kidneys perform in your body.
2. **Describe** the processes of filtration and reabsorption.
3. **Critical Thinking—Comparing** How does the process that takes place within the glomerular capillaries differ from the one that occurs in the capillaries surrounding the collecting tubule?
4. **MINI LAB** How does the structure of a kidney **relate** to its function?

BRANCHING OUT In Action

A Cure for Ulcers?

SECTION 38–4

GUIDE FOR READING

- **Explain** how ulcers are caused and **describe** how they can be treated.

FOR MANY YEARS, DR. JOHN Warren, a pathologist at Royal Perth Hospital in Western Australia, had examined tissue taken from ulcer patients. In nearly all of them, he found a curious corkscrew-shaped bacterium. In 1982, he brought these bacteria to the attention of Barry Marshall, a 30-year-old physician at the hospital.

Marshall then examined his own ulcer patients and discovered that nearly all of them had this bacterium in their stomach linings. Could this bacterium be the cause of ulcers? And if so, what would be the implications of this discovery? Before we discuss the answer to these important questions, you need to learn what ulcers are and how they are treated today.

Peptic Ulcers

As you have read earlier, powerful acids and enzymes are released into the stomach during the process of digestion. They are capable of breaking down just about anything you eat—animal or vegetable. If that's the case, you might wonder why the stomach doesn't digest itself! Why don't these acids eat holes right through the lining of the stomach itself? The answer is that sometimes they do.

Normally, the lining of the stomach is protected by a thick layer of mucus. In addition, the cells lining the stomach are tightly connected to each other, which prevents acid from leaking between them. Finally, millions of damaged cells in the lining are discarded every day and replaced by new ones. Most of the time, these three mechanisms keep the lining strong and healthy.

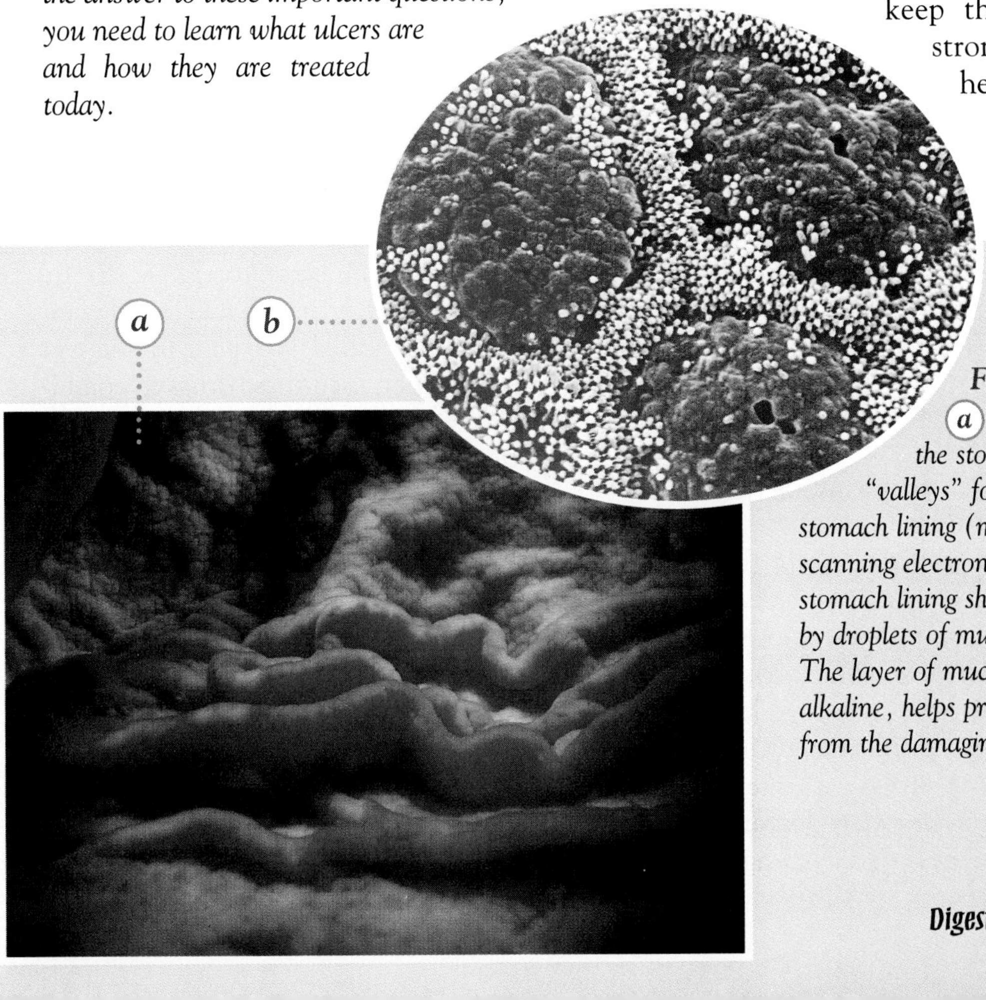

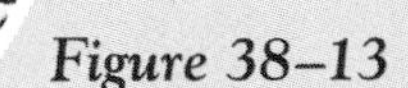

Figure 38–13

(a) A view of the inner surface of the stomach shows the "hills" and "valleys" formed due to folds in the stomach lining (magnification: 15X). (b) A scanning electron micrograph of a portion of the stomach lining shows epithelial cells surrounded by droplets of mucus (magnification: 3550X). The layer of mucus, which is viscous and alkaline, helps protect the lining of the stomach from the damaging effects of acidic secretions.

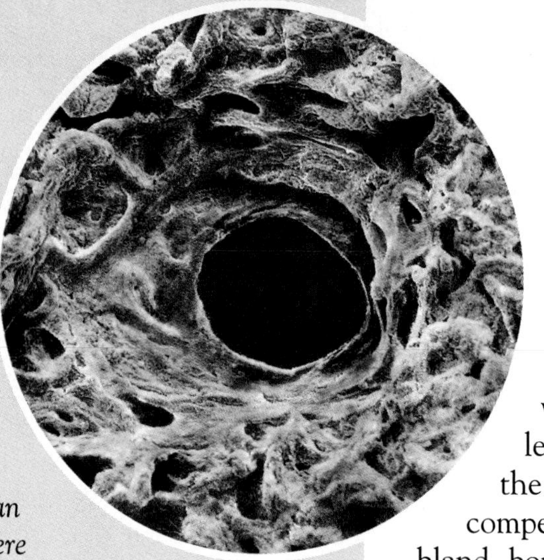

Figure 38–14
Because acid can damage the lining of the stomach, physicians assumed that a peptic ulcer, such as one shown in this scanning electron micrograph, was caused by the production of too much stomach acid. And because stress, overeating, and spicy foods stimulate acid production, each of these was assumed to play a role in causing an ulcer. As a result, ulcer victims were often told to relax, avoid certain foods, and take medication that suppressed the release of stomach acid.

Sometimes, however, stomach acids make direct contact with the cells of the stomach lining. This damages these cells in the same way that a strong acid would burn the skin of your hand. The pain and inflammation that results is known as **gastritis**—inflammation of the stomach.

If acid continues to attack these cells, it may literally eat its way through the stomach lining to produce a deep crater or even a hole in the stomach wall. This kind of damage is known as a **peptic ulcer.** Because the ulcer is actually an open wound, each release of acid into the stomach causes intense pain.

Peptic ulcers present serious medical problems for people around the world. As much as 10 percent of the adult population of the United States suffers from ulcers at some point in their lives. In some countries, the numbers are greater.

Because the release of stomach acid causes further damage to the ulcer, physicians have tried to prevent the release of stomach acid as a way of treating ulcers. A number of drugs can block the release of stomach acids, and physicians have prescribed these drugs to help ulcers heal themselves. Unfortunately, the success rate for this kind of treatment has been very low. Even when doctors were able to heal ulcers using such drugs, the ulcers usually reappeared when the drugs were stopped.

What was the result? There were millions of people taking expensive, ineffective drugs. To make matters worse, by blocking acid release, these drugs interfere with the normal digestive process. To compensate, ulcer patients must eat bland, boring foods that do not require stomach acid for their digestion.

☑ **Checkpoint** What is a peptic ulcer?

What Causes Ulcers?

With so many treatments that didn't work and so much guesswork as to what caused ulcers, could it be that the real cause of this problem was something else? Something hidden in the bodies of ulcer patients? In the early 1980s, two young Australian medical scientists, Dr. John Warren and Dr. Barry Marshall, began to wonder. Could the bacterium—now known as *Helicobacter pylori*—that they both found in their ulcer patients be the cause of ulcers? After months of unsuccessful attempts, Marshall finally found the right conditions to grow this bacterium in the laboratory.

Before long, Marshall had found compounds that killed *H. pylori* in the test tube. He then used these drugs in combination with some antibiotics on his ulcer patients. Seven out of ten patients were completely cured of ulcers. He was sure that he had found the answer to this terrible affliction.

A Daring Experiment

You might think that the scientific community would have jumped at Marshall's exciting results. But his initial reports were met with skepticism. Most scientists believed that acid and stress caused ulcers, and they pointed out that

Marshall had never proved that this bacterium could actually cause an ulcer.

In June 1984, Marshall decided to try the ultimate experiment. He walked into his lab, mixed up a strong dose of *H. pylori* bacteria in a test tube, and swallowed it. Three days later, Marshall awoke with the excruciating pain of gastritis—the first sign that acid has attacked the wall of the stomach and an ulcer is developing. Vomiting and stomach pain kept on for days. He had proven—at least to his own satisfaction—that this bacterium was the cause of gastritis and ulcers.

Figure 38–15
Today, peptic ulcers are thought to be caused by a bacterium called Helicobacter pylori, *which means "screw-shaped bacterium from the stomach" (magnification: 5100X). Approximately 50 percent of Americans over the age of 50 are infected with* H. pylori. *In some developing countries, nearly all adults are infected with* H. pylori. *Infections by this bacterium seem to run in the family, which suggests that it is spread from person-to-person by contaminated drinking water, eating utensils, and food. Public health officials around the world are now devising ways to deal with this bacterium.*

Rigorous Testing

Marshall then conducted a full-scale experiment, with 100 ulcer patients, to test his ideas. He divided his patients into four groups. One of the groups received no medication, one received just antibiotics, one received the exact mixture of *H. pylori*-killing drugs and antibiotics that Marshall now thought best, and one just the *H. pylori*-killing drugs. After a year, the results were very clear.

Ulcers persisted in three of the groups. But in one group—the one receiving Marshall's combination of *H. pylori*-killing drugs and antibiotics—the *H. pylori* bacteria and the ulcers were gone in 70 percent of the patients. **Dr. Marshall had successfully shown that the *H. pylori* bacteria could cause ulcers and that ulcers could be cured by destroying the bacteria.** In the last few years, one medical panel after another has endorsed Marshall's work, and other scientists have duplicated his findings. By using newer, more powerful antibiotics, cure rates as high as 90 percent have now been reported. Doctors around the world are now getting used to a new bit of knowledge: Ulcers can be cured!

Barry Marshall's work is a perfect example of the scientific method in action. Part luck, part intuition, and part Marshall's hard work had produced the kind of overwhelming evidence that was needed to overcome the doubts of other scientists—and, ultimately, to save thousands of human lives.

Section Review 38–4

1. **Explain** how ulcers are caused and describe how they can be treated.
2. **BRANCHING OUT ACTIVITY** Many drugs originally marketed as anti-ulcer medications because they suppress acid release are now sold over the counter for antacid relief and the prevention of heartburn. **Formulate a hypothesis** to explain why this has happened.

Laboratory Investigation

DESIGNING AN EXPERIMENT

Vitamin C in Fruit Juice

Although many vertebrate animals can synthesize their own vitamin C, humans cannot. For this reason, you need to consume foods that will supply you with adequate amounts of this essential nutrient. Vitamin C is found in many vegetables, such as potatoes and broccoli, but orange juice is probably the best source for vitamin C. In this investigation, you will design an experiment to determine the effect exposure to air has on the vitamin C content of selected foods.

Problem

How does exposure to air affect the vitamin C content of certain foods? **Design an experiment** to answer the question.

Suggested Materials

- small beakers
- 100 percent orange juice, refrigerated
- water
- burette
- indophenol solution
- other kinds of fruit drinks
- glass stirring rod
- funnel

Suggested Procedure

1. **Set up your equipment as shown in the diagram.**
2. **While carefully standing on a chair, pour orange juice into the funnel at the top of the burette. Record the initial level of the juice in milliliters.**
3. **Pour 10 mL of indophenol indicator solution into a beaker. CAUTION:** ***Be careful using indophenol because it stains. Do not drink any of the substances used in the lab. Wash your hands thoroughly after cleaning up.***
4. **Place the beaker below the burette and turn the stop valve to the open position so that only one drop of orange juice falls into the indophenol.**
5. **As each drop of juice falls into the indophenol, stir the mixture and note the color.**
6. **Keep adding drops until the blue indicator color becomes colorless when you stir. Turn the burette valve off.**

7. **Read the level of the juice in the burette. Record this measurement.**
8. **Using the same technique as in steps 1 to 7, design an experiment to determine the effect that exposure to air has on the vitamin C content in orange juice. You may also wish to do a similar experiment with powdered drinks that are fortified with vitamin C.**
9. **Propose a hypothesis and write up your procedure. Make sure the procedure has only one variable and includes a control.**
10. **Obtain the approval of your teacher before carrying out your experiment.**

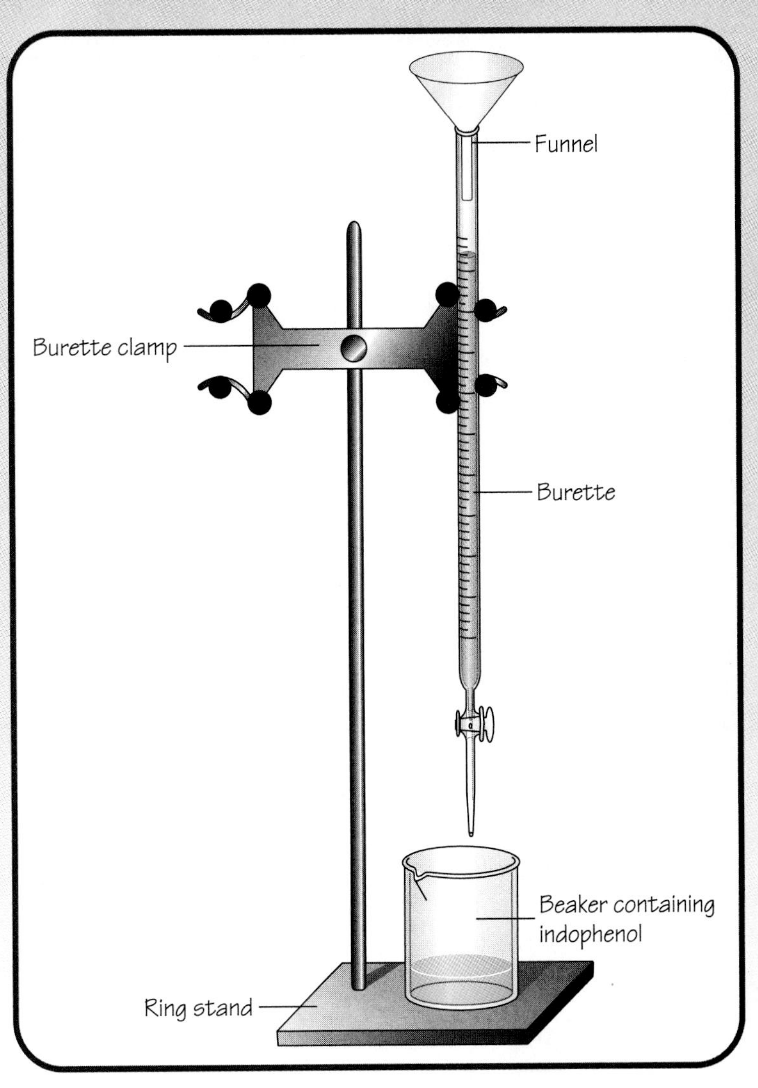

Observations

1. How many milliliters of the orange juice were needed to change the color of the indophenol?
2. Construct a bar graph to compare the number of milliliters of juice required in each trial of your experiment.
3. How did your data compare with that of other groups?

Analysis and Conclusions

1. What is the relationship between the volume of juice required to reach the end point of the titration and the vitamin C content of the juice?
2. Under what conditions did the orange juice have the highest vitamin C content? The lowest? Did your results support your hypothesis?
3. Do the results of your experiment suggest how to handle foods with vitamin C? Explain your answer.

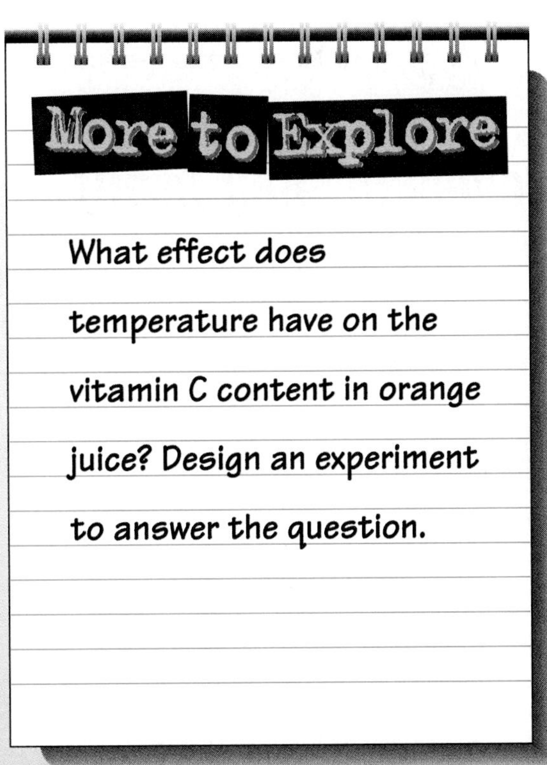

More to Explore

What effect does temperature have on the vitamin C content in orange juice? Design an experiment to answer the question.

Study Guide

Summarizing Key Concepts

The key concepts in each section of this chapter are listed below to help you review the chapter content. Make sure you understand each concept and its relationship to other concepts and to the theme of this chapter.

38–1 Nutrition

- Foods provide energy to perform work and raw materials to build body tissues.
- Food energy is measured in units of the Calorie, which is equivalent to 1000 calories.
- A human body needs water, carbohydrates, fats, proteins, vitamins, and minerals.
- Carbohydrates, fats, and proteins provide energy and the material to build body tissues. Vitamins and minerals are needed in small amounts for the proper functioning of the body.

38–2 The Digestive System

- The digestive system converts food into simple molecules that the body can use.
- The process of digestion begins in the mouth by the action of enzymes such as amylase and pepsin. Food is digested in the stomach and is absorbed by the cells of the villi, which make up the lining of the small intestine.
- Insulin and glucagon help in regulating nutrient levels in the blood.

38–3 The Excretory System

- The kidneys remove wastes, such as urea, and regulate the water in the bloodstream.
- The kidneys, consisting of nephrons, carry out the filtration of blood plasma.
- The urine produced passes through the ureter to the bladder.

38–4 A Cure for Ulcers?

- Recent experiments indicate that the *H. pylori* bacterium can cause ulcers—craters in the stomach wall—by reducing the stomach's layer of protective mucus.

Reviewing Key Terms

Review the following vocabulary terms and their meaning. Then use each term in a complete sentence.

38–1 Nutrition

Calorie

38–2 The Digestive System

mouth
salivary gland
amylase
lysozyme
esophagus
peristalsis
stomach
gastric gland
pepsin
small intestine
pancreas
liver
bile
villus
large intestine
islet of Langerhans
insulin
glucagon

38–3 The Excretory System

kidney
renal artery
renal vein
ureter
urinary bladder
nephron
glomerulus
Bowman's capsule
reabsorption
loop of Henle
antidiuretic hormone

38–4 A Cure for Ulcers?

gastritis
peptic ulcer

Recalling Main Ideas

Choose the letter of the answer that best completes the statement or answers the question.

1. The amount of heat energy given off by food is measured in
   a. ATP. **c.** glucose molecules.
   b. calories. **d.** carbohydrates.

2. Examples of carbohydrates include
   a. vegetable oils. **c.** sugars and starches.
   b. amino acids. **d.** lipids.

3. The foods at the top of the Food Guide Pyramid should be eaten
   a. only sparingly.
   b. at every meal.
   c. four to six servings a day.
   d. six to eleven servings a day.

4. The enzyme amylase breaks down the chemical bonds in
   a. fats. **c.** sugars.
   b. proteins. **d.** starches.

5. Food moves through the alimentary canal by
   a. the cardiac sphincter. **c.** air pressure.
   b. gravity. **d.** peristalsis.

6. Insulin is produced by the
   a. stomach. **c.** pancreas.
   b. duodenum. **d.** liver.

7. Where in the nephron is the blood plasma filtered out of the bloodstream?
   a. glomerulus **c.** collecting tubule
   b. Bowman's capsule **d.** loop of Henle

8. Which hormone returns more water to the bloodstream?
   a. insulin **c.** ADH
   b. glucagon **d.** pepsin

9. A peptic ulcer is an inflammation of the
   a. gums. **c.** colon.
   b. stomach wall. **d.** kidneys.

10. Dr. Barry Marshall's research showed that peptic ulcers are caused by
   a. acid-blocking drugs. **c.** a virus.
   b. stress and anxiety. **d.** a bacterium.

Putting It All Together

Using the information on pages xxx to xxxi, complete the following concept map.

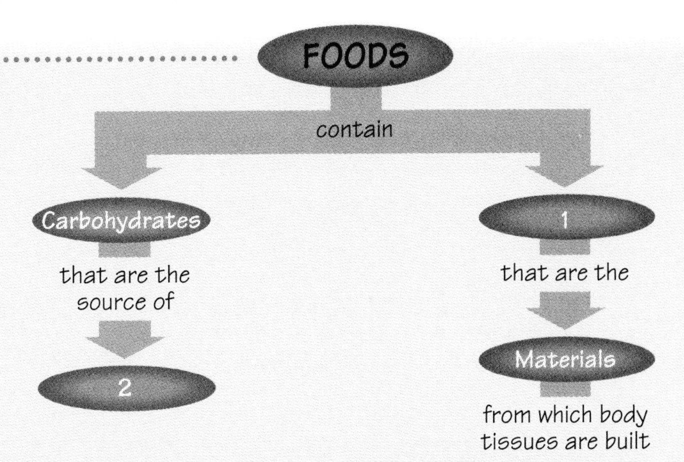

Assessment

Reviewing What You Learned

Answer each of the following in a complete sentence.

1. Why must humans consume nutrients?
2. How much energy does a calorie represent?
3. What are six categories of essential nutrients?
4. Why is cellulose an important part of the diet?
5. How are vitamins and minerals similar? How are they different?
6. What are some food sources of vitamin K? What is the function of this vitamin in the body?
7. What is the Food Guide Pyramid? Explain the significance of this shape.
8. Why is it important to chew food thoroughly before swallowing?
9. Explain the function of peristalsis.
10. List the organs that make up the excretory system.
11. What essential nutrients are the primary sources of wastes that the kidneys excrete?
12. What is urine?
13. Explain the function of a kidney dialysis machine.
14. What is gastritis?

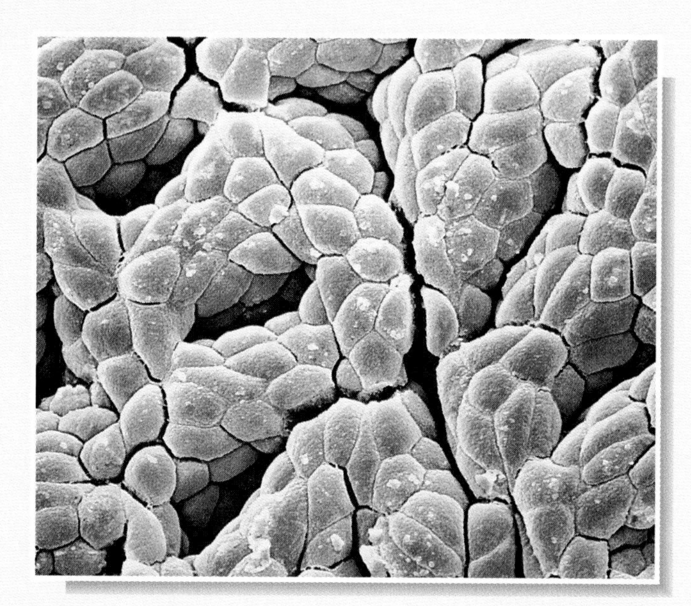

Expanding the Concepts

Discuss each of the following in a brief paragraph.

1. Plants are autotrophs and humans are heterotrophs. Explain what makes them different.
2. How might sipping water with a meal aid in digestion?
3. How does the liver contribute to digestion before and after absorption?
4. In what order do the fats, carbohydrates, and proteins in a mouthful of food begin chemical digestion? Explain your answer.
5. What enzymes and hormones does the pancreas produce? Explain their functions.
6. Explain how kidneys remove wastes. Relate the structure of a kidney to its function.
7. How does insulin affect the liver, muscles, and fatty tissues?
8. How might the kidneys respond to dehydration?
9. Describe the "feedback" mechanism that helps the kidneys maintain the body's water balance.
10. How could washing your hands before preparing a meal reduce the incidence of stomach ulcers?

Extending Your Thinking

Use the skills you have developed in this chapter to answer the following.

1. **Observing** Obtain a pocket-sized mirror. Hold it about 3 cm from your mouth and exhale onto it. What substance do you observe on the mirror? Explain how your observation illustrates a function of the excretory system.
2. **Relating** Create a poster describing a "Specials of the Day" menu. Include selections for a breakfast, lunch, and dinner that will appeal to your classmates and that also follow the dietary guidelines in the Food Guide Pyramid.
3. **Predicting** If you had your gallbladder removed, would you have to change your diet? Explain your answer.
4. **Applying concepts** Explain how the work of the Australian physicians John Warren and Barry Marshall on *Helicobacter pylori* involved each of the following parts of the scientific method:
   - Developing a question
   - Stating a problem
   - Gathering and interpreting data
   - Formulating a hypothesis
   - Experimenting
5. **Hypothesizing** Some scientists predict that more minerals than the ones already known will be discovered to be essential nutrients. For example, garlic may contain minerals that help fight infection. Formulate a hypothesis to explain how this may happen.

Applying Your Skills

A Recipe for Tenderness

The molecules in protein-rich foods—such as meats, eggs, and cheese—are often complex and enormous in size. Each protein molecule is coiled and folded so that the bonds between its atoms must be broken down bit by bit to reach its center. The stomach has seven different protein-digesting enzymes. In this activity, you will observe the effects of several enzymes.

1. Obtain four test tubes containing solidified gelatin.
2. Add 1 mL of each of the following three solutions to the top of the gelatin: fresh pineapple, canned pineapple, meat tenderizer, and water. Why is water a part of this investigation? Why is gelatin used?
3. Put the test tubes in a refrigerator for 24 hours. After 24 hours, measure the amount of gelatin that has dissolved in each of the test tubes. Record your results in a bar graph.
4. Was there any difference between the actions of the fresh and the canned pineapple? How can you explain your results?
5. Carefully clean out the test tubes using hot water and properly discard the contents.

• GOING FURTHER •

6. Design an experiment to compare the effects of two brands of meat tenderizers containing different enzymes.

CHAPTER 39

Reproductive System

FOCUSING THE CHAPTER THEME: *Patterns of Change*

39–1 The Human Reproductive System
- **Identify** the organs of the male reproductive system and **describe** their functions.
- **Identify** the organs of the female reproductive system and **describe** their functions.

39–2 Fertilization and Development
- **Sequence** the events from the fertilization of an egg through childbirth.

BRANCHING OUT *In Depth*

39–3 Sexually Transmitted Diseases
- **Describe** some of the most common sexually transmitted diseases.

LABORATORY INVESTIGATION
- **Observe** the microscopic structures of reproductive organs and gametes.

Biology and Your World

BIO JOURNAL

The infants in this photograph may look contented and easy to care for; however, raising a child is harder than you may think. Have you ever baby-sat for a younger sibling or another person's child? Was this easier or harder than you expected? In your journal, describe your experience. If you have never cared for a younger child, describe what you think the experience would be.

A group of babies

The Human Reproductive System

GUIDE FOR READING

- **Describe** the functions of the male and female reproductive systems.
- **List** the four phases of the menstrual cycle.

WHAT IS THE SINGLE MOST important day in your life? If you said "my birthday," you are not alone. Not only is your birthday important to you, it's also important to those closest to you. Think about how many things depend on your birthday. Your birthday determines when you are allowed to go to school, when you qualify for a driver's license, and when you may vote.

To the rest of society, your birthday may indeed be the day your life began. To a biologist, however, birth is as much the end of a process—the process of reproduction—as it is a beginning. The trillions of cells in a newborn baby all come from the fusion of two reproductive cells. Human reproduction is the story of how these cells are produced and come together to form a single cell, from which a new life develops.

Human Sexual Development

For the first six weeks of development, male and female human embryos are nearly identical. Then, in the seventh week, the primary reproductive organs, which will eventually produce reproductive cells, or **gametes** (GAM-eets), begin to develop into either **ovaries** in females or **testes** (TEHS-teez; singular: testis) in males.

The ovaries and testes are endocrine glands. The ovaries produce estrogens—a group of steroid hormones that cause development of the female reproductive organs. And the testes produce androgens—steroid hormones that cause the male reproductive organs to develop. Although the male and female reproductive organs develop from the same tissues in the embryo, these hormones determine whether they will develop in the male or female pattern.

Figure 39–1
Many changes in your life—both social and physical—are dependent on your age, including (a) getting your driver's license, (b) graduating from high school, and (c) shaving.

During childhood, the reproductive glands, or gonads, continue to produce small amounts of estrogen and androgen, which help to shape development. However, the gonads are not capable of producing reproductive cells until puberty, a period of rapid sexual development that usually occurs in the early teenage years. During puberty, the gonads complete their development and the reproductive system becomes fully functional. Puberty may occur at any time between 9 and 15 years of age, and on average occurs about a year earlier in females than in males.

Puberty begins when hormone levels begin to increase. The hypothalamus releases substances that signal the pituitary gland to begin producing two hormones—**follicle-stimulating hormone** (FSH) and **luteinizing hormone** (LH).

In males, LH causes certain cells in the testes to produce **testosterone** (tehs-TAHS-ter-ohn), the principal androgen, or male hormone. FSH and testosterone are necessary for other cells in the testes to develop into **sperm,** the male gametes. These diploid (2n) cells undergo meiosis, which results in 4 haploid (n) cells, each of which contains 23 chromosomes. After the completion of meiosis, these cells develop into mature sperm cells with long flagella.

In addition to sperm development, testosterone is responsible for the development of a male's secondary sex characteristics. In puberty, the male's voice deepens, he begins to grow body hair, and he finds it easier to develop large muscles.

INTEGRATING HEALTH

Testosterone is a natural anabolic steroid. What are some of the risks involved in taking a synthetic anabolic steroid?

In females, FSH and LH stimulate the ovaries to produce the female sex hormones, **progesterone** and **estrogen,** responsible for the development of female secondary sex characteristics, such as breast development and widening of the hips. In addition, these hormones work with FSH and LH to produce female gametes called **ova** (singular: ovum), or eggs.

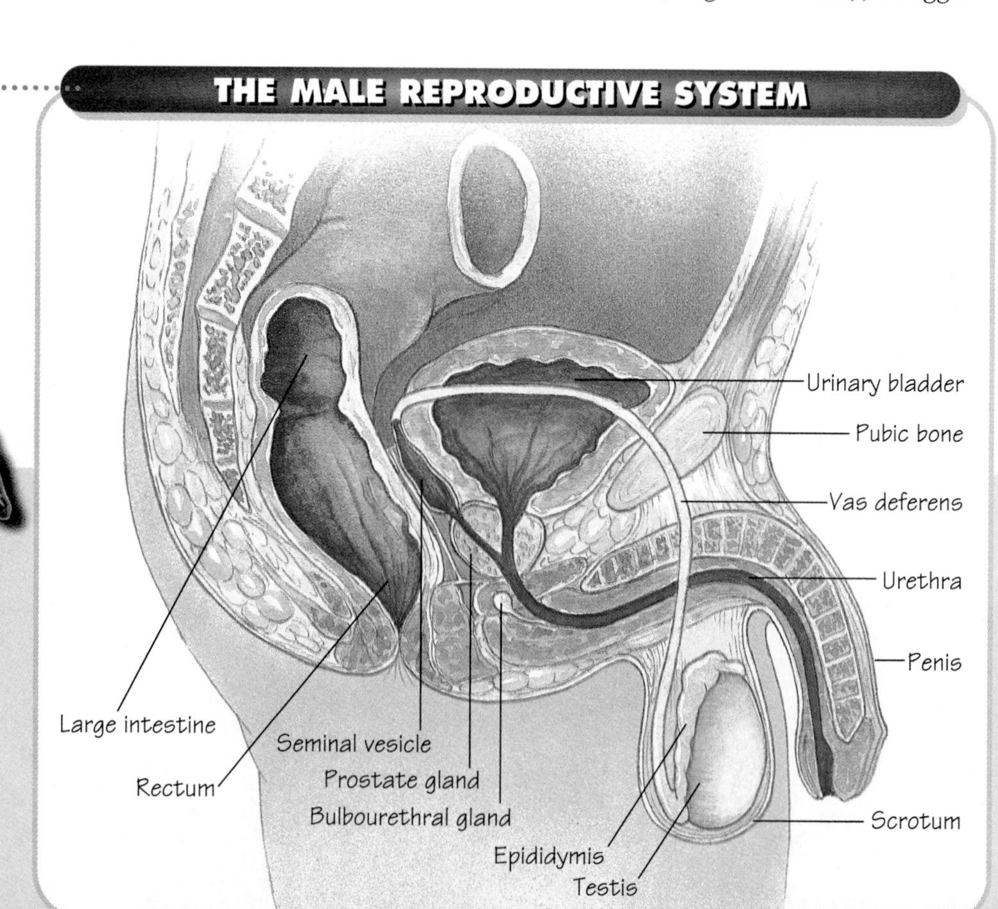

Figure 39–2
(a) The male reproductive system is a collection of organs and glands responsible for the production and delivery of (b) sperm, one of which is shown in this electron micrograph (magnification: 87,413X).

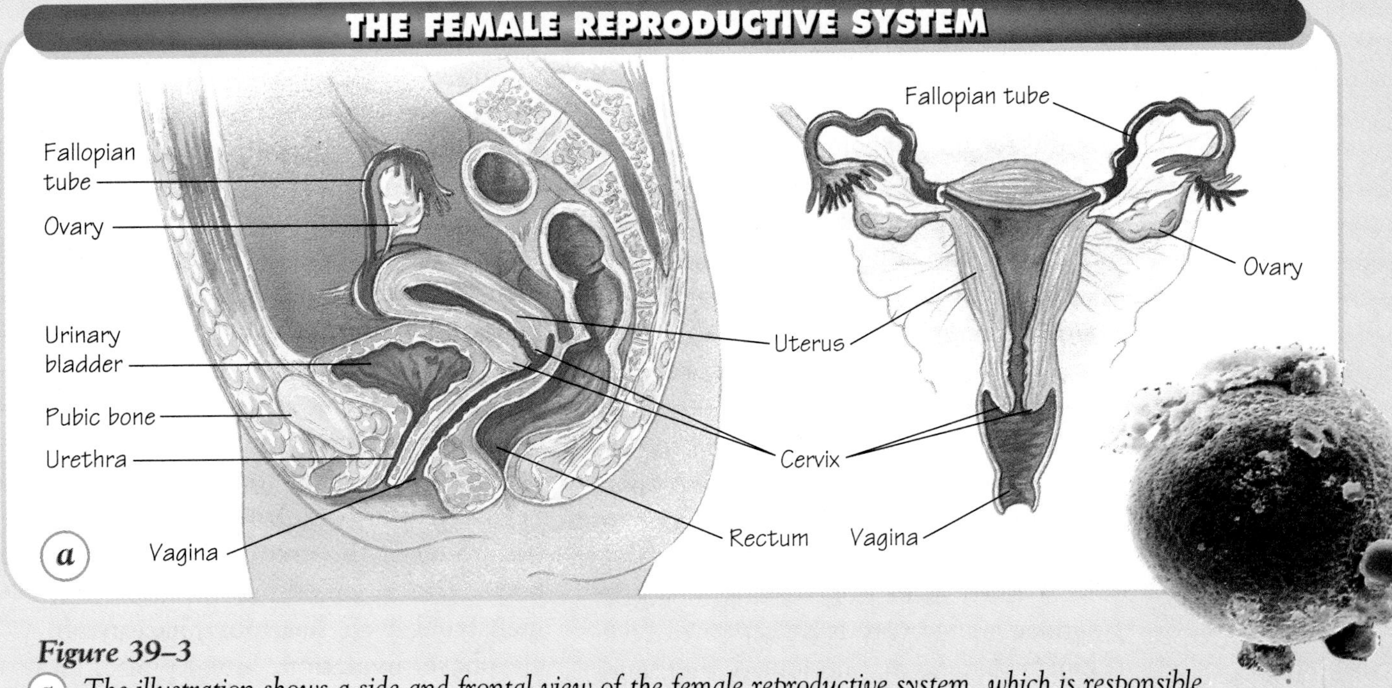

Figure 39–3
(a) *The illustration shows a side and frontal view of the female reproductive system, which is responsible for the production of eggs and for providing a place for fertilization and development.* (b) *This color-enhanced electron micrograph of a single human egg shows its protective covering (colored red) as well as the remnants from the follicle (colored yellow) in which the egg developed (magnification: 470X).*

The Male Reproductive System

The male reproductive system has the task of producing and delivering sperm. Sperm are produced in the testes, which descend from the abdominal cavity just before birth into a sac called the **scrotum,** located outside the body cavity. This location keeps the temperature of the testes about 2°C cooler than the rest of the body, for sperm development.

The testes are made up of tightly coiled tubules called **seminiferous** (sehm-uh-NIHF-er-uhs) **tubules.** More than 100 million sperm cells are produced in the seminiferous tubules every day. When mature, the sperm travel into the **epididymis** (ehp-uh-DIHD-ih-mihs), where they are stored. After a brief period, they are released into the **vas deferens** (VAS DEHF-uh-rehnz). The vas deferens extends upward from the scrotum into the abdominal cavity.

Glands lining the reproductive tract—including the seminal vesicle, the prostate, and the bulbourethral gland—release a nutrient-rich fluid called semen, in which sperm are suspended.

The Female Reproductive System

The female reproductive system produces gametes, too. However, the fact that fertilization and development of a baby both take place inside the female's body places special demands on the female reproductive system. In contrast to the millions of sperm produced each day, the ovaries produce, on average, just one egg every 28 days or so. Each time an egg is released from one of the ovaries, the rest of the reproductive tract must be prepared to nourish and support a developing embryo.

FSH stimulates the development of a **follicle**—a cluster of cells that contains a developing egg. As the follicle gets larger and larger, the egg passes through the early stages of meiosis. When meiosis is complete, a single large haploid (n) egg and three smaller cells called polar bodies are produced. Because the polar bodies have little cytoplasm, they disintegrate.

While the follicle is developing, cells surrounding the egg produce larger and larger amounts of estrogen. When the level of these hormones reaches a critical point, hormones from the pituitary gland cause the follicle to rupture, and the egg is released into one of the two **Fallopian tubes.** The release of an egg is known as **ovulation.**

As the mature egg breaks through the surface of the ovary, it is swept into a Fallopian tube by the motions of thousands of microscopic cilia. The egg then passes into a muscular chamber called the **uterus.** The outer end of the uterus is called the cervix. Beyond the cervix is a canal known as the vagina, which is the passageway that leads to the outside of the body.

The Menstrual Cycle

Ovulation is just one of a series of events that occur in a pattern called the **menstrual cycle.** The menstrual cycle takes, on average, about 28 days. As an egg develops, the lining of the uterus is prepared to receive it. If the egg is fertilized after ovulation, it is implanted in the uterus and embryonic development begins. If it is not fertilized, the egg and the lining of the uterus pass out of the body. This is called menstruation. **The menstrual cycle has four phases: follicle phase, ovulation, luteal phase, and menstruation.**

Follicle Phase

The follicle phase begins when the level of estrogen in the bloodstream is low. The hypothalamus senses the low level of estrogen and causes the pituitary

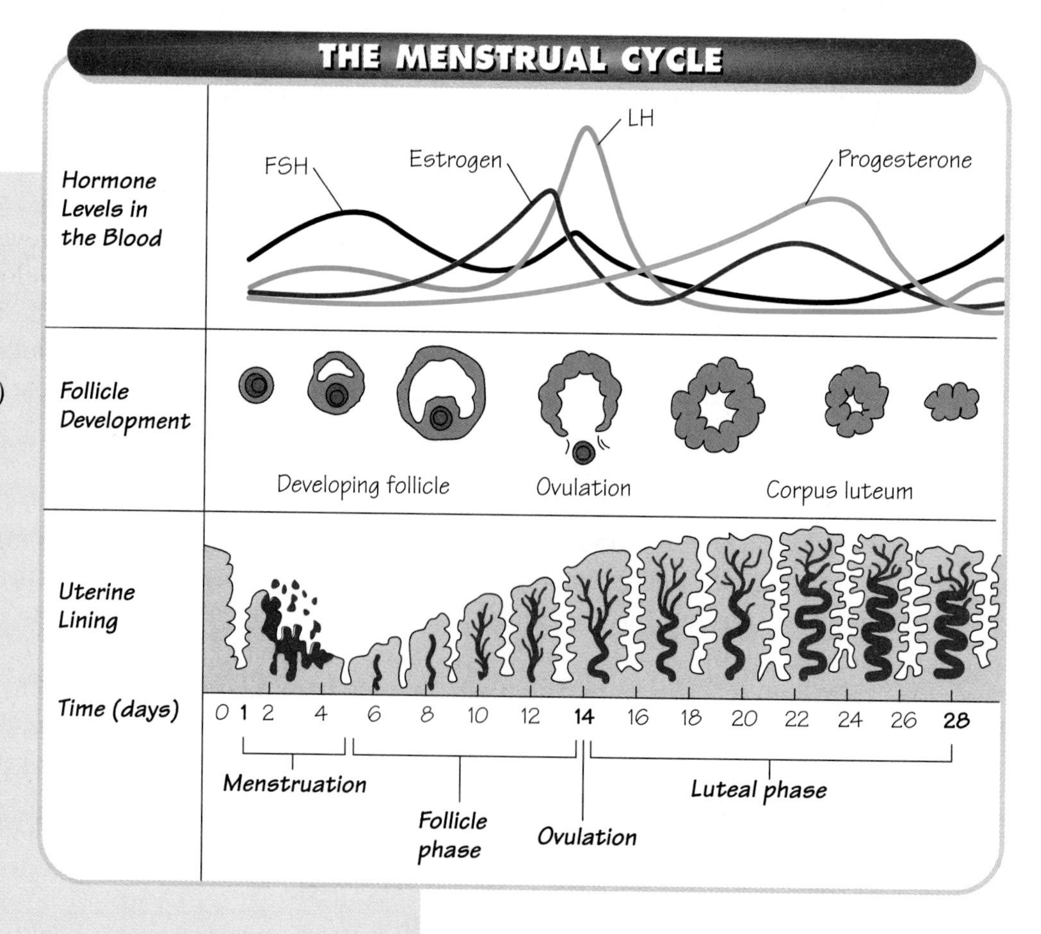

Figure 39–4
The menstrual cycle, which is controlled by hormones produced by the ovary (progesterone and estrogens) and the pituitary gland (FSH and LH), occurs on average about every 28 days. Notice the changes in hormone levels in the blood, the development of the follicle, and the changes in the uterine lining during the cycle, illustrated in the diagram.

gland to release FSH and LH into the bloodstream. These hormones act on the ovary, stimulating a follicle to develop to maturity. As the egg develops, the cells of the follicle produce more and more estrogen. Estrogen causes the lining of the uterus to thicken in preparation for a fertilized egg.

Ovulation

Ovulation is the shortest phase of the cycle. When the egg is mature, the pituitary gland sends out a burst of LH. LH causes the wall of the follicle to break open and the egg is released into one of the Fallopian tubes.

Luteal Phase

The luteal phase of the cycle begins after the egg is released. As the egg moves through the Fallopian tube, the cells of the ruptured follicle undergo a transformation. The follicle turns yellow and is now known as the **corpus luteum** (KOR-puhs LOOT-ee-uhm), which means "yellow body" in Latin. The corpus luteum continues to produce estrogen, but now produces large amounts of progesterone too. Progesterone stimulates cell and tissue growth in the lining of the uterus, preparing it for pregnancy. Progesterone also inhibits the release of LH and FSH.

The chances of an egg being successfully fertilized are greatest around the time of ovulation. If the egg is not fertilized, the corpus luteum breaks down and estrogen levels decrease.

Menstruation

Menstruation begins when the level of estrogen in the blood becomes so low that the lining of the uterus cannot be maintained. Tissues detach from the uterine wall and are discharged through the vagina along with blood and the unfertilized egg. Menstruation usually lasts 3 to 7 days. When it ends, a new cycle begins.

What starts a new cycle? The drop in estrogen that causes menstruation is sensed by the hypothalamus. It signals the pituitary gland to release FSH, which starts the development of a new follicle—and the cycle begins all over again.

Figure 39–5
When hormone levels reach a certain point, an egg bursts from one of the ovaries into a Fallopian tube. The halo around the egg in this photograph is a protective layer of follicle cells called the corona radiata.

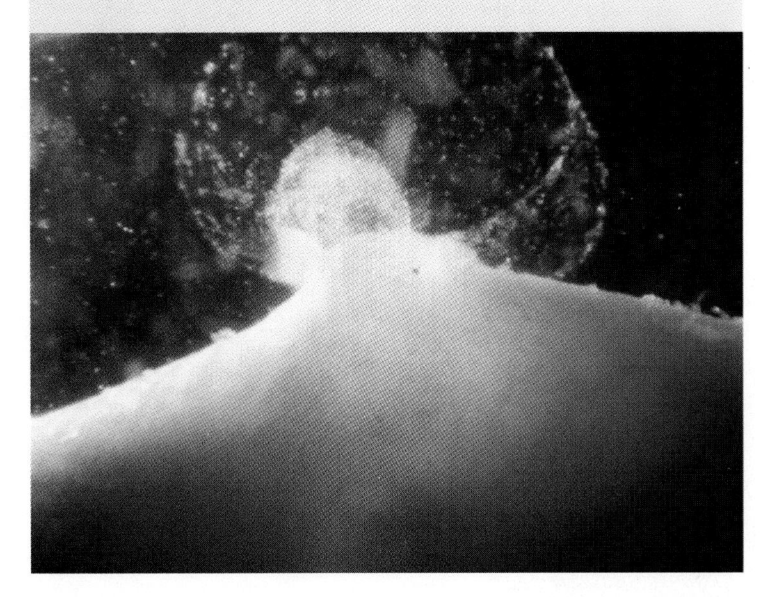

Section Review 39-1

1. **Describe** the functions of the male and female reproductive systems.
2. **List** the four phases of the menstrual cycle.
3. **Critical Thinking—Interpreting Diagrams** Which hormone is at its highest level during ovulation?

SECTION 39–2

Fertilization and Development

Guide for Reading

- **Describe** fertilization.
- **Describe** the importance of the placenta.

MINI LAB

- **Construct a model** of early human development.

A SINGLE CELL FORMED BY THE fusion of sperm and egg begins the process of development. To produce this cell, the reproductive system must bring these two gametes together. In humans and most other mammals, the fertilized egg develops within the body of its mother. This means that the human reproductive system must not only produce reproductive cells, it must also protect and nourish the developing embryo from fertilization to birth.

Fertilization

During sexual intercourse, the penis is inserted into the vagina to a point just below the cervix. Involuntary smooth muscle contractions of the penis cause an ejaculation, which is the release of several milliliters of semen into the vagina.

Semen contains as many as 100 million sperm per milliliter. Therefore, it's no exaggeration to say that after they are released, millions of sperm swim through the uterus toward the Fallopian tubes. Of the millions of sperm released, only a relatively small number will reach the upper region of the Fallopian tube. If an egg is present in one of the Fallopian tubes, its chances of being fertilized by one of these sperm are very good.

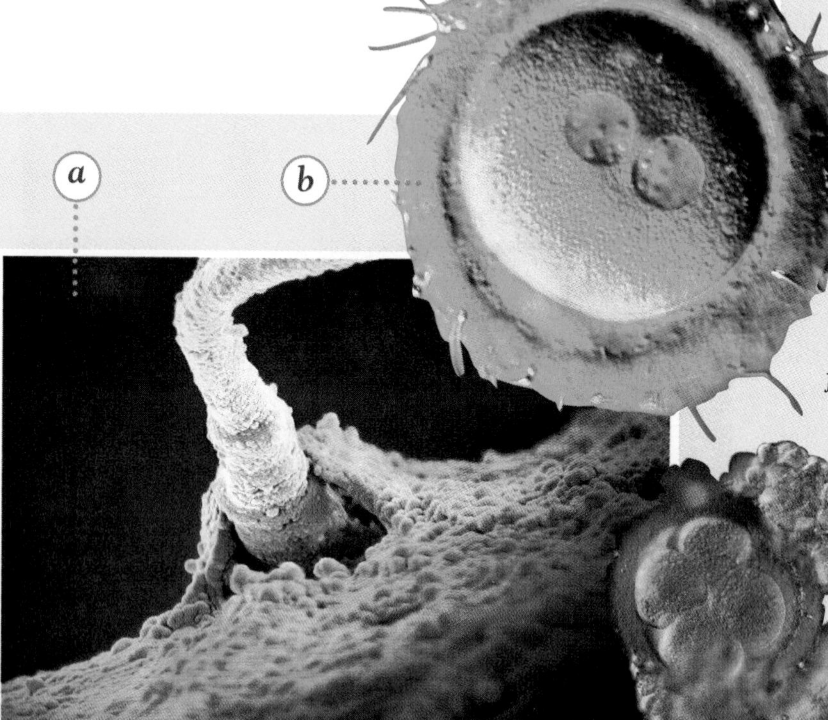

Figure 39–6 *Fertilization of an egg by a sperm is the first step in human development. In this series of photographs, (a) the head of a sperm has just entered an egg, (b) a newly fertilized egg that has just undergone one cell division is surrounded by sperm, and (c) a clump of cells has undergone several cell divisions. Surrounding these cells are smaller cells that provide nutrients for the embryo as it makes its way from the Fallopian tube to the uterus.*

The Fusion of Egg and Sperm

It might seem impossible for such a tiny sperm to fuse with the much larger egg. The egg is surrounded by a thick, protective layer that also includes cells from the follicle in which the egg developed. However, the outermost protective layer also contains binding sites to which sperm cells can attach. When a sperm attaches to these cells, a sac at the end of the sperm breaks open, releasing powerful enzymes that help to dissolve the protective layers around the egg.

Very often, more than one sperm attaches to the egg, and each begins to work its way through to the egg cell membrane. **As soon as one of the sperm makes direct contact with the egg cell membrane, the two cells fuse, the tail of the sperm breaks away, and the sperm nucleus enters the egg's cytoplasm. This process is known as fertilization.** Then, immediately following **fertilization,** rapid electrical and chemical changes take place in the egg cell membrane that prevent any other sperm cells from entering. A **zygote** (ZIGH-goht), or a fertilized egg, has been formed, and development begins.

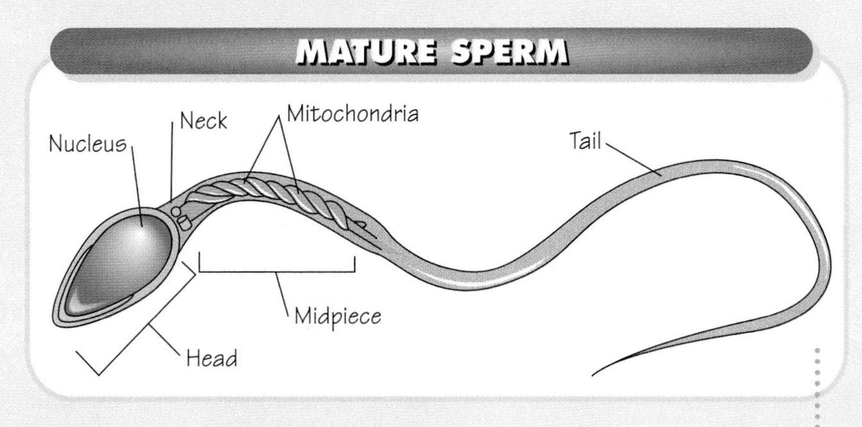

Figure 39–7
(a) A single human sperm consists of a head, a midpiece, and a tail. The head contains the nucleus, with its genetic material. The midpiece is packed with mitochondria that are needed for energy production. And the tail is used to propel the sperm forward. (b) Once inside a female's body, sperm are attracted by chemicals produced by an egg, causing all sperm to swim in the same direction toward the egg.

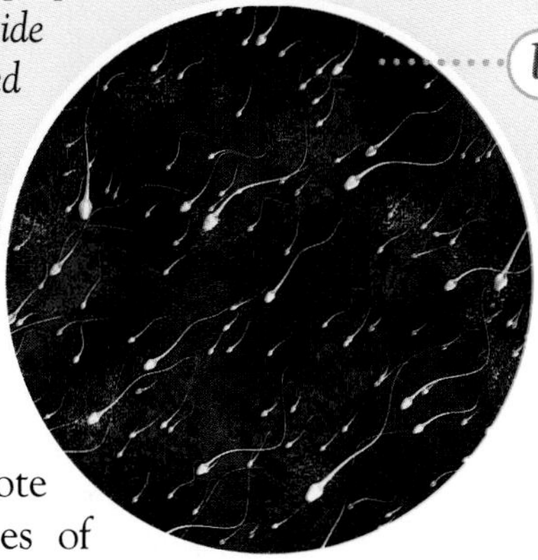

Cell Division

The newly formed zygote goes through a rapid series of cell divisions, gradually forming a solid ball of cells known as a **morula** (MOR-yoo-luh). As cell division in the zygote continues, it gradually develops a hollow cavity and becomes known as a blastocyst. Three to four days after ovulation, the blastocyst attaches itself to the wall of the uterus. The blastocyst may not look like much, but within a few months it will develop into the trillions of cells in a human baby. The precision and intricacy of this process is wonderful to behold—no builder ever executed a plan more exacting, yet the blueprints for this remarkable process are written on the smallest possible scale, the DNA sequence in the nucleus of a single human cell.

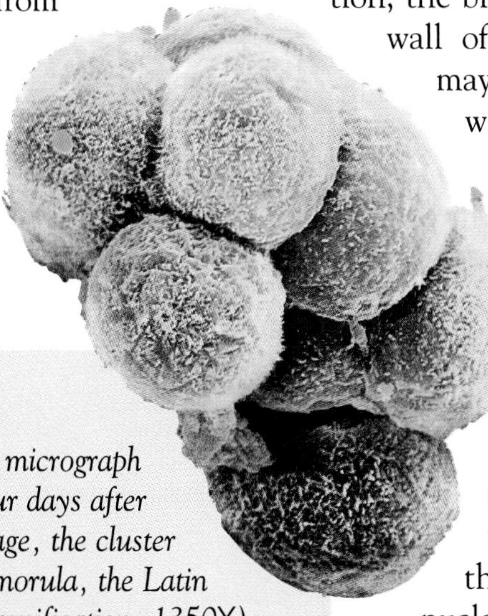

Figure 39–8
This scanning electron micrograph shows an egg about four days after fertilization. At this stage, the cluster of cells is known as a morula, the Latin word for mulberry (magnification: 1350X).

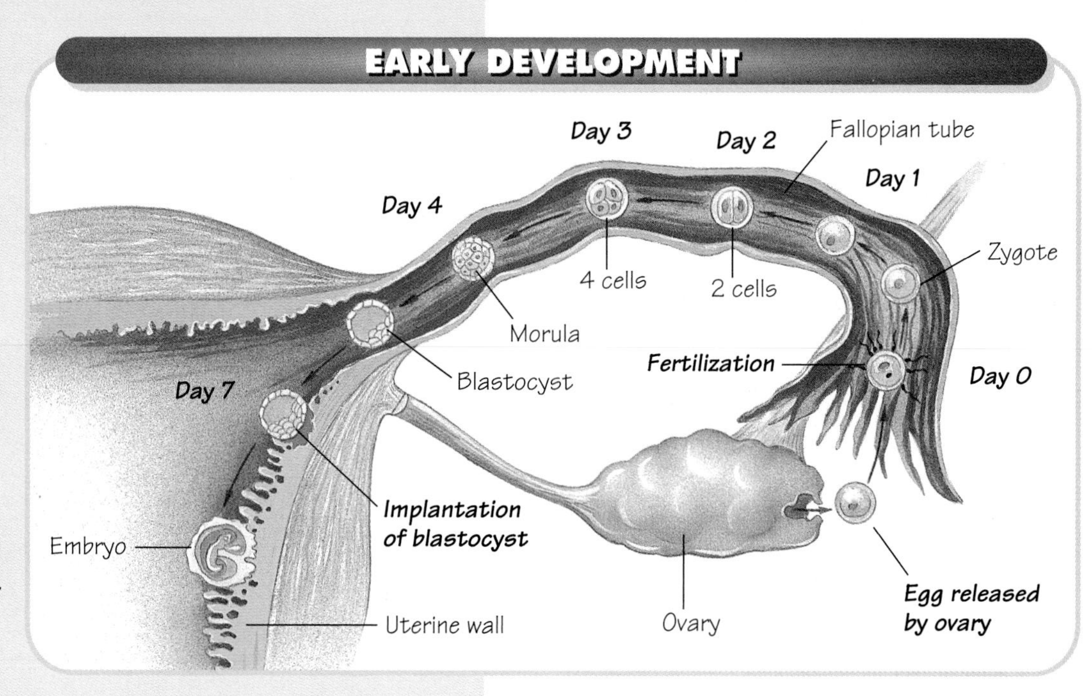

Figure 39–9
Once an egg is released from an ovary, it may become fertilized in a Fallopian tube. If fertilization occurs, the fertilized egg becomes known as a zygote. As the zygote continues its trip toward the uterus, it undergoes many cell divisions, becoming a morula and then a hollow ball of cells called a blastocyst. The blastocyst implants itself in the uterine wall and development begins.

Implantation

The cells of the zygote then grow into the uterine wall in a process known as **implantation.** Implantation is critical to the zygote for two reasons. First, the outermost cells of the zygote and the cells in the uterine wall will grow together to form structures that will nourish and support the developing **embryo.** Without that support, the zygote would not have enough stored food to develop on its own. Second, and most importantly, implantation produces chemical signals that preserve the uterine lining.

Here's how those chemical signals work. Recall that the corpus luteum, which produces progesterone, remains active for about two weeks after ovulation. Then it breaks down, progesterone synthesis stops, and the uterine wall is discarded in menstruation. However, as soon as they implant in the uterine wall, the outermost cells of the zygote produce a substance called **human chorionic gonadotropin** (HCG). HCG is a powerful chemical signal that keeps the corpus luteum alive and active. As a result, the uterine lining is preserved and the embryo can continue to develop.

PRIMARY GERM LAYERS

Ectoderm
Mesoderm
Endoderm

Primary Germ Layer	Develops Into
Ectoderm	Epidermis of skin; lining of mouth and nose; hair; nervous system; lens of eye
Mesoderm	Dermis of skin; muscle; skeleton; circulatory system; gonads
Endoderm	Inner lining of digestive and respiratory systems; liver and pancreas

Figure 39–10
During gastrulation, the three primary germ layers—ectoderm, mesoderm, and endoderm—are formed. All embryonic structures are derived from one of these three layers.

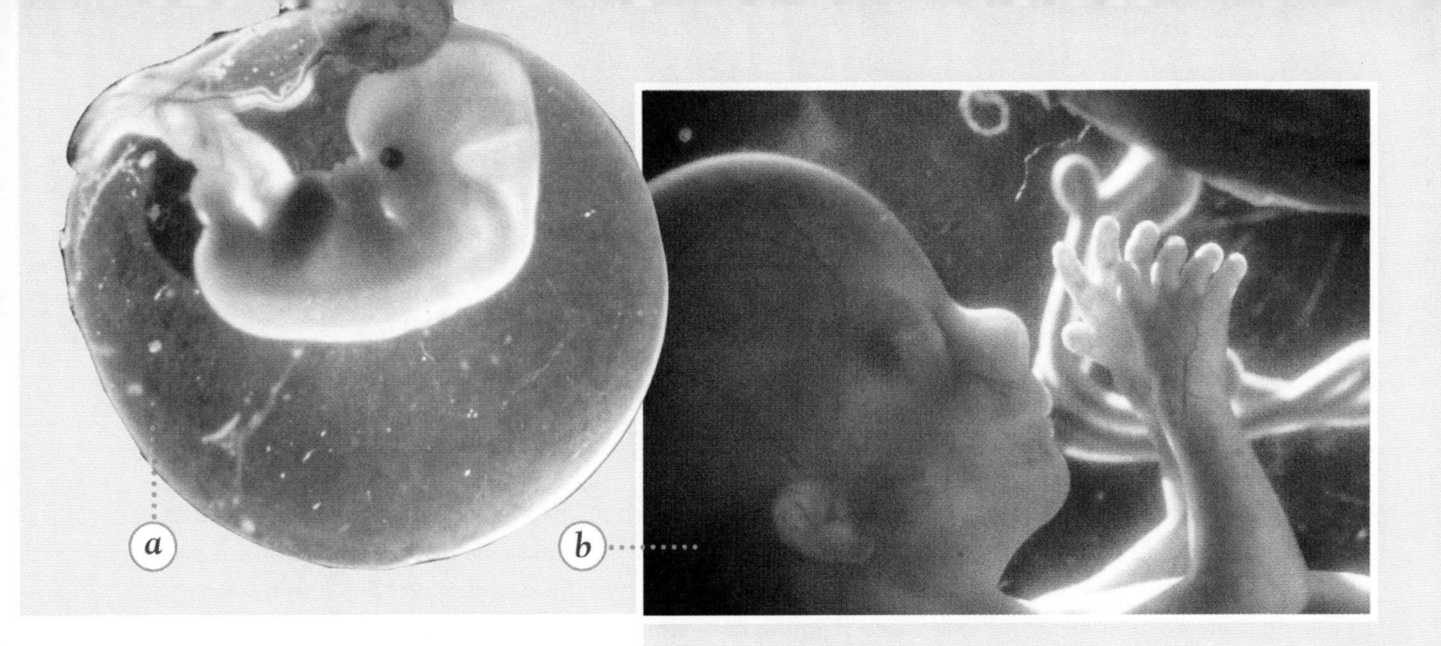

Figure 39–11
Up until the eighth week of development, a developing human is called an embryo. After this time, it is referred to as a fetus. (a) *At six to seven weeks after fertilization, the embryo is 13- to 18-mm long, has the beginnings of eyes, a beating heart, fingers, and a brain, which is nearly as large as the rest of the body.* (b) *By the fourth month, the ears, nose, and mouth have formed. Notice the umbilical cord, which connects the fetus to the placenta.*

The Developing Embryo

The early events of embryonic development are the same for humans as they are for most other mammals. The outer layer of the blastocyst grows into the uterus, surrounding the tissues of the developing embryo. A mass of cells that forms on one side of the blastocyst will become the embryo itself. In the first two weeks after fertilization, the cells in this mass sort themselves into two distinct layers.

Gastrulation

In a process known as **gastrulation** (gas-troo-LAY-shuhn), cells from the upper layer migrate inward to form a third layer between the first two. These three layers of cells—the ectoderm, mesoderm, and endoderm—are known as the primary germ layers of the embryo. In ***Figure 39–10,*** you can see that all the other organs and tissues of the embryo will be formed from these three primary germ layers.

The outer layers of the blastocyst form two important membranes that surround the embryo. These extra-embryonic membranes, located outside the embryo, are called the amnion and the chorion. Both of these membranes fill with fluid that helps to cushion the embryo and protect it from injury.

Checkpoint What is gastrulation?

Neurulation

Near the end of the third week of development, an especially critical event—neurulation—takes place. At about the seventeenth day after fertilization, the ectoderm begins to thicken into a line pointing to the head of the embryo. Gradually, this thickening produces a groove, and the raised edges of the groove move toward each other, forming a tube. By the end of the twenty-third day, the tube has sealed and then sunk beneath the ectoderm. What is this tube, and why is its formation so important? This neural tube, as it is called, becomes the nervous system. One end of the tube will eventually develop into the brain, and the remainder of the tube will become the spinal cord. The remaining ectoderm will then begin to form the skin of the embryo.

Checkpoint What is neurulation?

Problem Solving

INTERPRETING DIAGRAMS

Picturing Pregnancy

During a class discussion of human reproduction, one of your classmates says, "I have always wondered how the fetus 'knows' when it is time to be born. And how does it know to stay inside the uterus for nearly nine months?"

Your teacher shows the class an illustration that may help answer these questions. Look at the diagram and line graph below, then answer the questions that follow.

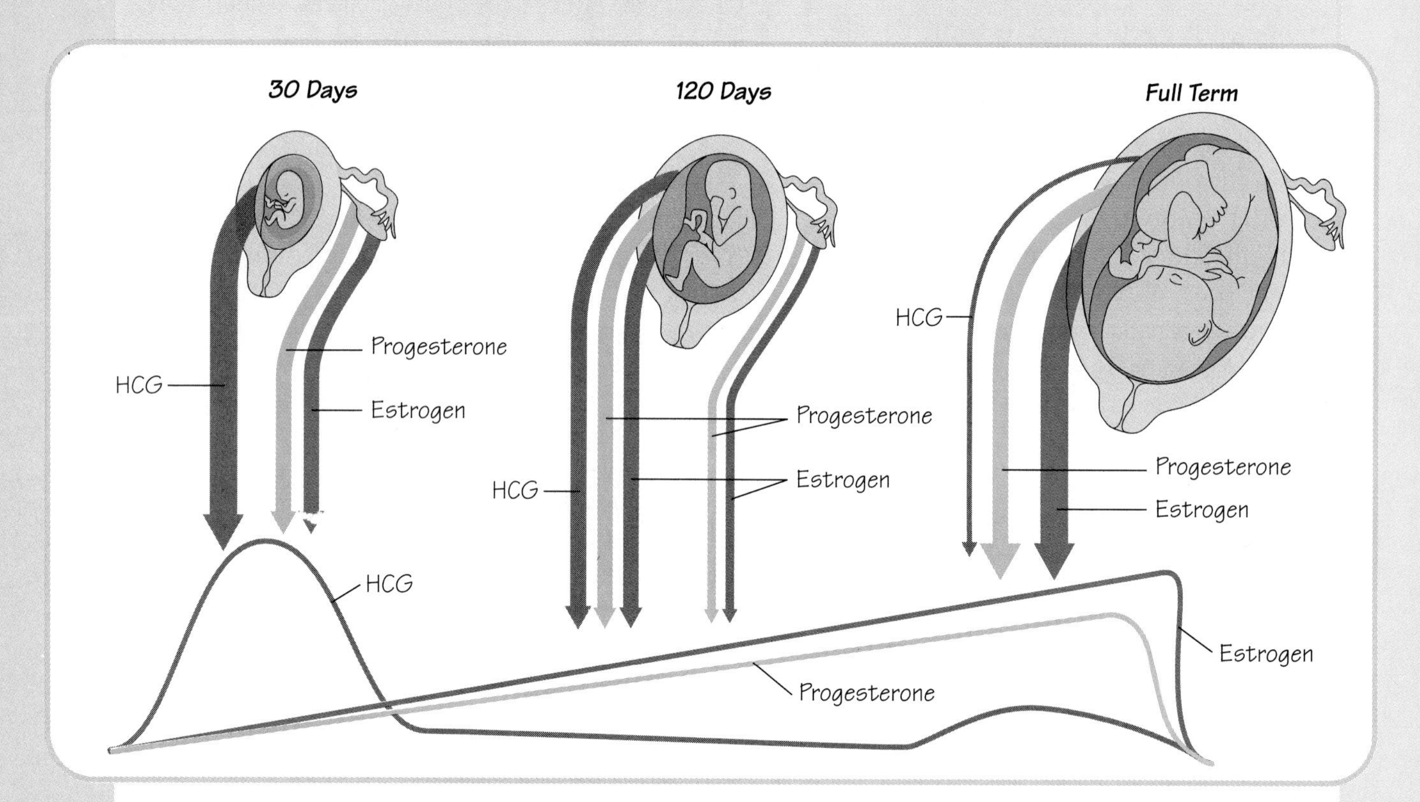

THINK ABOUT IT

1. What organs produce progesterone and estrogen at 30 days and 120 days in the pregnancy? How does this change at full term?
2. Which hormone is most abundant at 30 days? Give evidence to support your answer.
3. Which hormones are increasing at 120 days? Which hormone is decreasing?
4. Describe what happens to the levels of the hormones at the end of the pregnancy.
5. Based on what you know about the roles of these hormones, predict the "birth" day.

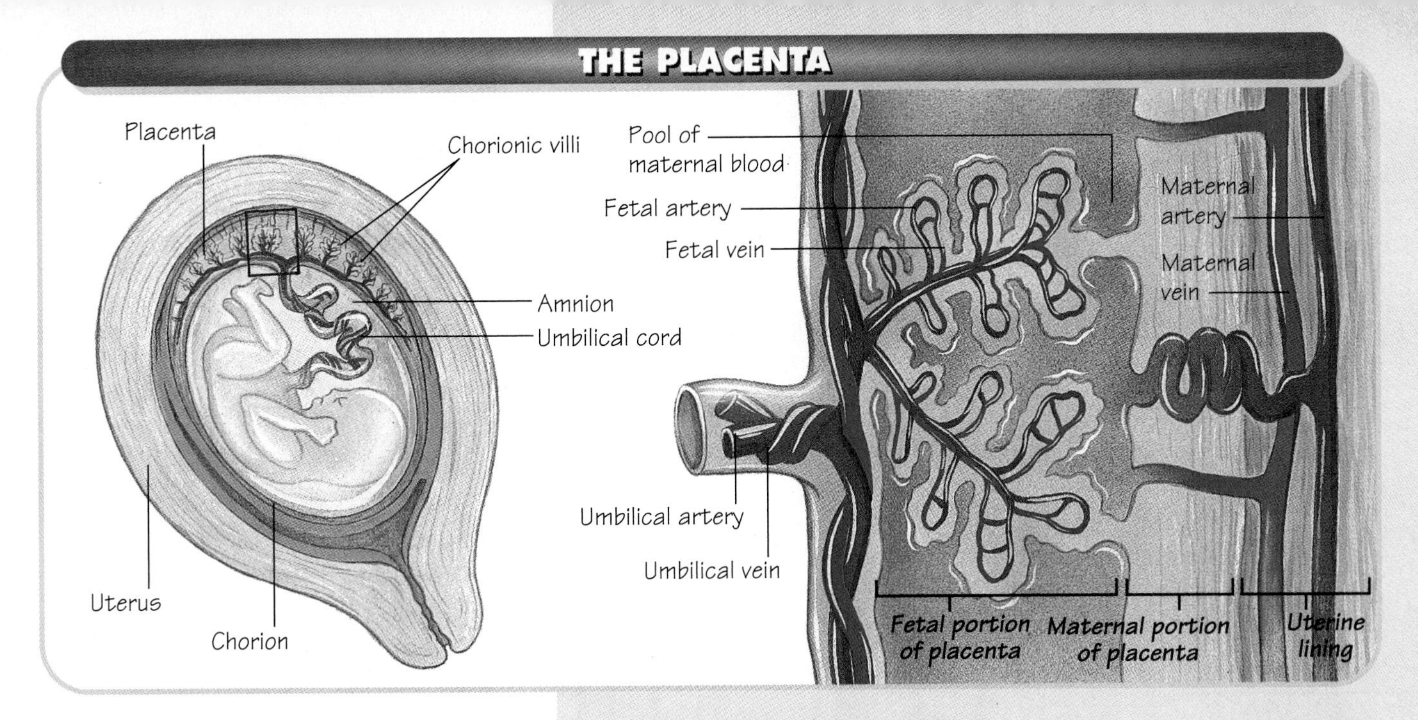

Figure 39–12
The placenta, formed from the chorion layer of the embryo and the uterine lining of the mother, is the connection between the mother and the developing embryo. Although fetal blood vessels branch extensively into pools of maternal blood, there is no mixing of blood. The placenta provides a large surface area for diffusion of oxygen, carbon dioxide, nutrients, and wastes between fetus and mother.

The Placenta

During the first few weeks of development, the cells of the embryo simply absorb nourishment from the surrounding cells of the mother. But within a few days, the embryo is too large for this to continue to meet its needs. Shortly after gastrulation is complete, the embryo forms an organ that will supply its needs for the rest of its development—the **placenta.**

In the fourth week of development, the amniotic sac expands to the point where it completely surrounds the embryo. This expansion squeezes the other extra-embryonic tissue into a thin stalk, connecting the embryo to the uterus. This stalk is the beginning of the **umbilical cord.**

You might think that the best way to bring food and oxygen to the embryo would be to link its blood directly to its mother through the umbilical cord. However, if this were the case, diseases could easily spread from mother to child. Serious problems could also result if the embryo and mother have different blood groups. ● So it should not be surprising that the blood supplies of mother and embryo do not mix directly.

Instead, they come into very close contact in the placenta. The placenta forms at the base of the umbilical cord, where it meets the uterine wall. Tiny blood vessels from the embryo pass through pools of maternal blood. Although the blood supplies do not mix, food, oxygen, carbon dioxide, and metabolic wastes pass back and forth between mother and embryo. **As the embryo grows, the placenta serves as its main organ of respiration, nourishment, and excretion.** Anything that the mother takes into her body, including drugs, passes through the placenta to the embryo.

Checkpoint What is the placenta?

INTEGRATING HEALTH

Use reference materials to find out what happens when an embryo and mother have different Rh factors in their blood.

Fetal Development

In the next few weeks, the skeletal system and the limbs of the embryo take shape. This is a particularly important time for the embryo because a number of

MINI LAB ········ Modeling ········

Early Human Development

PROBLEM *How can you* ***construct a model*** *of human development?*

PROCEDURE

1. Use modeling clay and reference materials to create a three-dimensional representation of each of the following stages in early human development: zygote, first cell division, morula, blastocyst, five-week embryo, and eleven-week fetus.
2. Label each model with its name and an estimation of its actual size.

ANALYZE AND CONCLUDE

1. At which stage can you begin to recognize human structures?
2. What changes have occurred at each stage of development?

external factors can disrupt development at this point. Although the placenta does act as a barrier to some disease-causing organisms, others—such as rubella (German measles) and most drugs, including alcohol, tobacco, and medications—can penetrate the placenta and prevent normal development.

By the end of eight weeks, the embryo is called a **fetus.** Its muscular system enables it to move around a bit, and its sexual organs have developed to the point where it is possible to tell a male from a female. Its heartbeat becomes loud enough to be heard through a stethoscope. It now begins to grow rapidly, and by the end of six months, it is nearly 35 centimeters long and has a mass of between 500 and 800 grams.

As the fetus enters its seventh month of growth, reflexes in its nervous system enable it to respond to sudden noises. Its lungs are enlarging rapidly, and its digestive system is almost ready to function. In many respects, after six months of growth, the fetus is almost ready to lead an independent existence.

Pregnancy usually continues for three more months, however, and there is good reason for that. A baby born early in the last three months of development may have serious problems adjusting to life outside. In particular, the respiratory system is not fully matured. Many babies born at this time are unable to regulate their body temperature. For this reason, premature babies are always at risk and must be given special care to increase their chances of survival.

☑ ***Checkpoint*** At what stage is an embryo called a fetus?

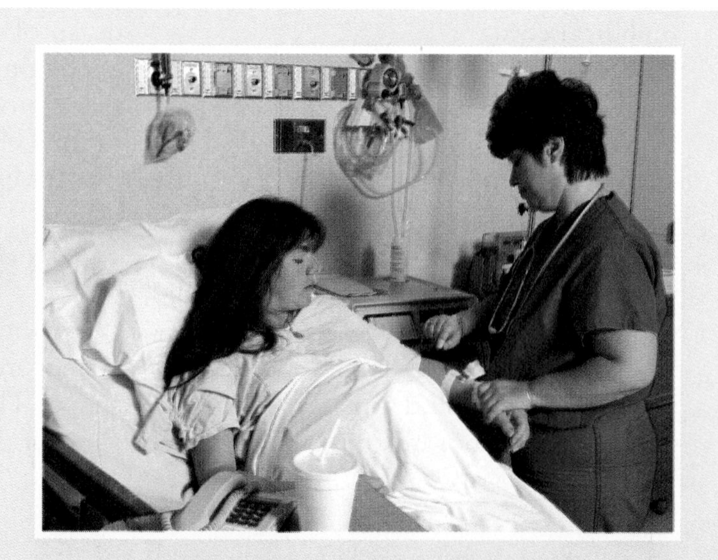

Figure 39–13
CAREER TRACK
A physician assistant performs routine medical care—physical examinations, administering injections and immunizations, administering or ordering certain medical procedures such as X-rays—and generally instructs and counsels patients.

Labor and Birth

Childbirth in humans takes place, on average, about nine months after fertilization. It is not known exactly how the body "decides" when it is time for a child to be born. The immediate signal is a pituitary hormone called **oxytocin.** This hormone stimulates contractions of the smooth muscles surrounding the uterus. Slowly at first, these muscles begin a series of contractions known as labor. As the contractions become more frequent and more powerful, the opening of the cervix expands until it reaches a diameter of nearly 10 centimeters—large enough for the head of the baby to pass through. At some point during this time, the amniotic sac bursts and the fluid in the sac passes out of the mother's body.

Childbirth

When the cervix has fully expanded and contractions come at intervals of 2 to 3 minutes, childbirth is about to begin. Contractions of the uterus force the baby, usually head first, out through the cervix, the vagina, and into the world.

As the baby leaves its mother's body, it may cough or cry to open up its fluid-filled lungs and take its first breath. Blood flow to the placenta dries up, and the placenta itself detaches from the wall of the uterus. It will follow the baby out through the birth canal. The umbilical cord is tied and cut, leaving only a small piece attached to the baby. The baby is now ready to begin life on its own.

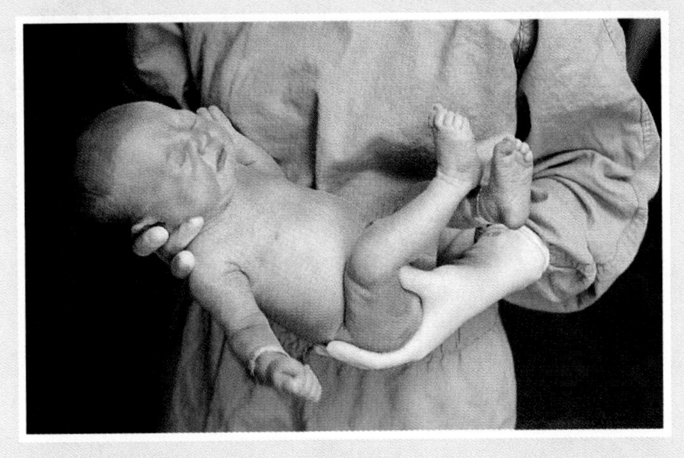

Figure 39–14
About nine months after fertilization, a baby is born. Identification bands, which match those of the mother, are placed around the baby's wrist and ankle.

After Childbirth

Just a few hours after childbirth, a pituitary hormone known as **prolactin** stimulates the production of milk in the breast tissues of the mother. By nursing her child soon after it is born, the mother stimulates the release of more oxytocin, which also helps to stop uterine bleeding. The mother also passes an important secretion to the baby in her first breast milk. This fluid, called colostrum, contains a remarkable mixture of special antibodies produced by the immune system. Colostrum helps to protect the baby from infections for many weeks. Colostrum is gradually replaced by mature milk. The milk that humans and other mammals produce for their offspring is a complete food. It contains all the vitamins, minerals, and other nutrients that the baby needs for the first few months of life.

Section Review 39–2

1. **Describe** fertilization.
2. **Describe** the importance of the placenta.
3. **Critical Thinking—Hypothesizing** What might happen to a fetus if the placenta became detached from the uterine wall?
4. **MINI LAB** How can you **construct a model** of early human development?

SECTION 39–3

BRANCHING OUT In Depth

Sexually Transmitted Diseases

Guide for Reading

- **List** three common sexually transmitted diseases caused by bacteria.
- **List** three common sexually transmitted diseases caused by viruses.

MINI LAB

- **Interpret data** from a graph to learn about an STD.

INTEGRATING HEALTH

Use reference materials to find out about STDs caused by protozoans, fungi, and arthropods.

IT IS A BIOLOGICAL FACT THAT human reproduction can take place only when living sperm cells come in contact with an egg. As you have read, these reproductive cells are sheltered from contact with the outside world, surrounded with nutrient-rich fluids to support their activities, and pass directly from the reproductive tract of the male into that of the female. Internal fertilization is an effective method of reproduction. However, it also creates a perfect opportunity for disease-causing organisms to exploit the reproductive system.

STDs—A Growing Problem

Diseases that are spread from one person to another by sexual contact are known as **sexually transmitted diseases,** or STDs. Sexually transmitted diseases are a serious health problem in the United States, infecting millions of people each year and accounting for thousands of deaths. Both bacteria and viruses can cause STDs. Because each sexual contact carries with it the risk of an STD, it is important to know and understand the most serious of these diseases.

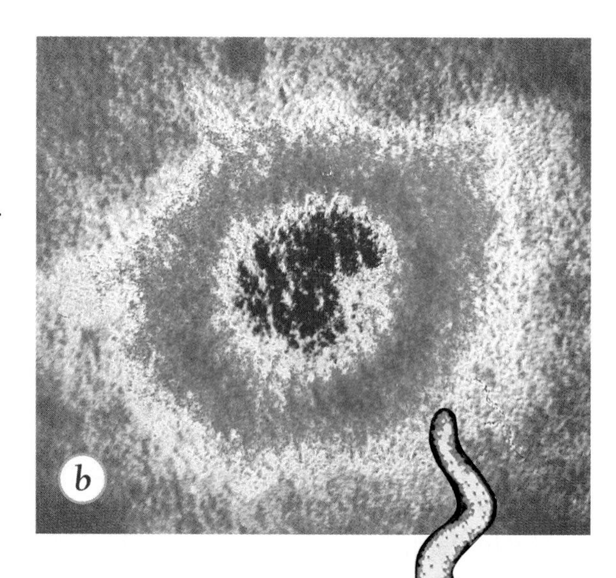

Figure 39–15 Most sexually transmitted diseases are caused by bacteria and viruses. (a) Gonorrhea is caused by the bacterium called Neisseria gonorrhoeae, which is kidney-shaped and often found in pairs (magnification: 29,000X). (b) There are two forms of herpes virus, called Herpes simplex Type 1 and Type 2. Herpes simplex Type 1 causes sores around the mouth. Herpes simplex Type 2 causes sores on the reproductive organs and is sexually transmitted. Although the two forms are identical in appearance, they are different in their chemical makeup. This electron micrograph shows Herpes simplex Type 1 (magnification: 200,000X). (c) The bacterium that causes syphilis, Treponema pallidum, is a threadlike, spiral-shaped bacterium (magnification: 4000X).

Bacterial Diseases

Although many of the most serious STDs are caused by bacteria, they can usually be treated and cured with antibiotics. Early detection and treatment are necessary to prevent any serious consequences of infection. **Three common STDs caused by bacteria are syphilis, gonorrhea, and chlamydia.**

Syphilis

One of the most serious and dangerous bacterial STDs is **syphilis.** This disease is caused by a corkscrew-shaped bacterium called a spirochete. The syphilis spirochete is a delicate organism that dies quickly when exposed to the air. However, sexual contact provides it with the perfect opportunity to pass from one person to another.

The first sign of this disease appears from 10 to 90 days after infection—a hard sore called a chancre (SHAN-ker) forms at the point where the organism passes through the skin. The sore disappears after a few weeks, but the infection does not. The spirochete spreads throughout the body, gradually infecting the circulatory and nervous systems.

In its early stages, syphilis can be cured with large doses of antibiotics. If left untreated, however, syphilis can cause serious physical damage and may even be fatal.

☑ **Checkpoint** What is syphilis?

Gonorrhea

Another serious STD is **gonorrhea.** Gonorrhea infects the urinary and reproductive tracts, and it is easily passed from person to person through sexual contact. Males are usually aware of being infected because the bacterium causes pain during urination and may result in the discharge of blood and pus. In females, the symptoms are much milder. Many females do not even know they are infected. This is particularly dangerous, because gonorrhea can produce pelvic inflammatory disease (PID), which may make it impossible to ever have children.

☑ **Checkpoint** What is gonorrhea?

Chlamydia

The single most common STD in the United States is one you may never have heard of—**chlamydia.** The bacterium that causes this disease thrives in the moist reproductive tracts of both males and females. It produces only a few symptoms, which may include a mild burning sensation during urination. Because of this, most people with chlamydia are unaware of the infection and continue to spread it to others without being treated. This is a cause for serious concern because long-term infection with this bacterium is now the leading cause of preventable infertility among women in their thirties.

☑ **Checkpoint** What is chlamydia?

Figure 39–16
Chlamydia trachomatis, *the bacterium that causes chlamydia, is a viruslike bacterium. The two smaller spheres in this color-enhanced electron micrograph show two* Chlamydia *bacteria (magnification: 40,000X).*

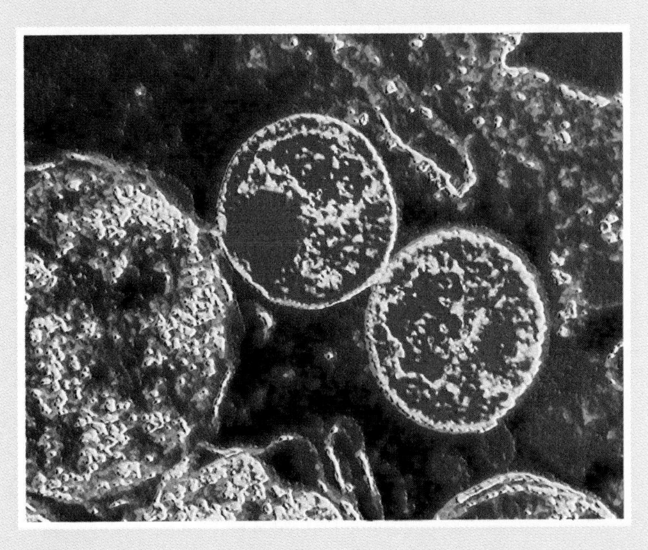

Number Sense

PROBLEM *What can you learn about an STD by interpreting data on a graph?*

Year	Number of Cases of Syphilis (per 100,000 people)	Year	Number of Cases of Syphilis (per 100,000 people)
1985	11.5	1990	20.0
1986	11.5	1991	17.0
1987	14.5	1992	13.5
1988	17.0	1993	10.0
1989	19.0	1994	7.5

Source: Centers for Disease Control and Prevention, September 1995

PROCEDURE

1. Use the data in the table to construct a line graph.
2. List some generalizations you can make from the graph.

ANALYZE AND CONCLUDE

1. In which year was the number of reported cases of syphilis the highest? The lowest?
2. Based on the line graph you constructed, what can you say about the general trend in the number of cases between 1985 and 1994?
3. From this data, what trends could you predict for the next 10 years?

Viral Diseases

Unlike bacteria-caused STDs, which can usually be treated with antibiotics if discovered early enough, there are currently no drugs that will cure viral STD infections. Therefore, STDs caused by viruses are particularly serious. **Three common STDs caused by viruses are genital herpes, hepatitis B, and AIDS.**

Genital Herpes

One of the most common STDs is **genital herpes.** A virus known as Herpes Type II infects the genital areas of males and females, causing small reddish blisters. Visible evidence of the virus may disappear for weeks or months at a time, but the infection stays in the body. Visible herpes blisters may appear without warning, causing pain and shedding virus particles that may be passed on to sexual partners. There are some drugs available that can help to reduce the severity of genital herpes outbreaks, but none that can permanently cure the infection.

☑ **Checkpoint** What is genital herpes?

Hepatitis B

Hepatitis B is a virus that infects the liver and kills several thousand people each year. Unlike other STDs, the hepatitis B virus can survive on objects outside the body for a long period of time. Although there is now an effective vaccine, there is no cure for hepatitis B once an individual has been infected. Hepatitis B is particularly common among people with many sexual partners and presents a serious health risk.

☑ **Checkpoint** How does hepatitis B differ from other STDs?

AIDS

In many respects, the most dangerous STD of all is **AIDS** (acquired immune deficiency syndrome). The virus that causes AIDS attacks the immune system, leaving the person vulnerable to a variety of infections. Virus particles are present in the blood, semen, and vaginal fluids of people infected with AIDS. Therefore, sexual contact is one of the principal ways in which the disease is spread from person to person. Currently, there is no cure for AIDS.

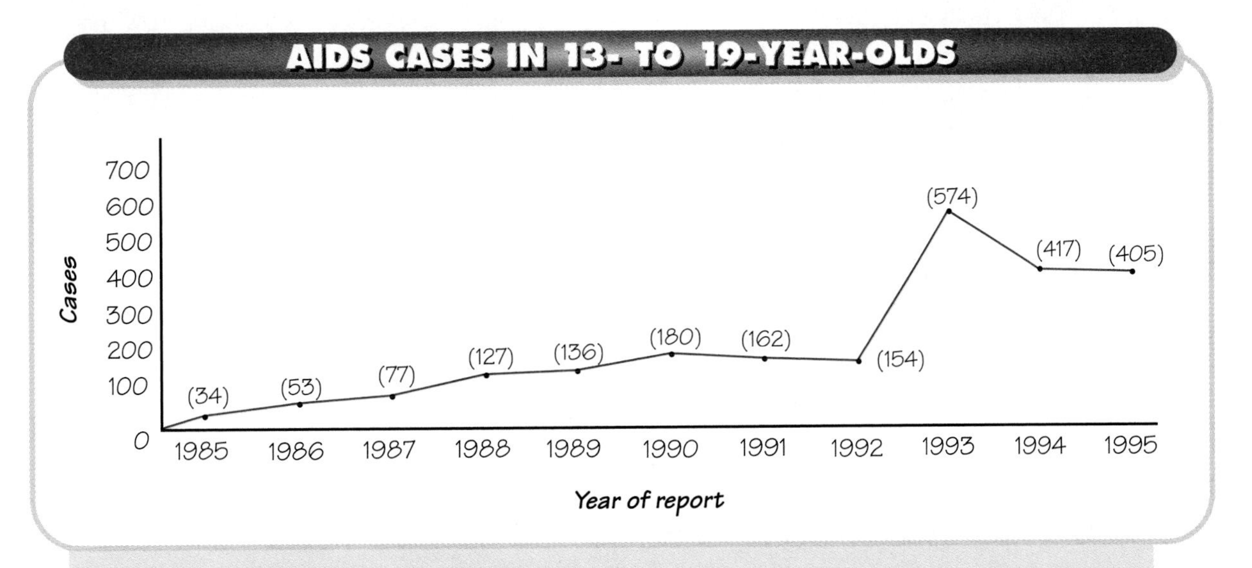

Figure 39–17
This graph shows the number of AIDS cases in 13- to 19-year-olds in the United States from 1985 through 1995. Notice the sharp increase in 1993, followed by a gradual decline.

Avoiding STDs

As you have seen, STDs are serious diseases that are spread by sexual contact. Are there any ways to avoid these diseases, which can ruin your health and even threaten your life? Yes, there are. The great irony of STDs is that although many of them cannot be cured, all of them can be prevented by understanding what conditions they require to spread from person to person.

Nearly all STDs infect the reproductive tracts of males and females. This usually means that bacteria or infectious virus particles are present in semen or vaginal fluids of infected individuals. Because these fluids are exchanged during intercourse, infectious particles from one person can easily be transmitted to another.

STDs require intimate contact between people for infections to be transmitted. Therefore, individuals who engage in sexual intercourse risk not only an unintended pregnancy but serious diseases as well. In contrast, abstaining from sexual contact provides complete protection against all STDs and pregnancy as well.

Section Review 39–3

1. **List** three common sexually transmitted bacterial diseases.
2. **List** three common sexually transmitted viral diseases.
3. **MINI LAB** What can you learn about an STD by **interpreting data** from a graph?
4. **BRANCHING OUT ACTIVITY** Choose one of the sexually transmitted diseases discussed in this chapter and research the effects it would have on a pregnant woman or her developing fetus. **Summarize** your findings in a brief report.

Laboratory Investigation

Reproductive Organs and Gametes

The egg and sperm are microscopic structures that can best be observed in slides that have been professionally sectioned, stained, and preserved. In this investigation, you will examine some prepared slides and use the knowledge you have gained from this chapter to interpret what you see.

Problem

What can you **observe** in the microscopic structure of reproductive organs and gametes?

Materials (per group)

compound microscope
prepared slides of the following:
- human testis
- ovary
- sperm

Procedure 

1. **Obtain a prepared slide of the cross section of a human testis. Using the low-power objective of the microscope, examine the slide and draw what you see. Find an area that has well-defined circular structures. These are the seminiferous tubules.**
2. **Switch to the high-power objective on the microscope. Select one seminiferous tubule and draw the structures inside it.**

Cross section of a human testis (magnification: 250X)

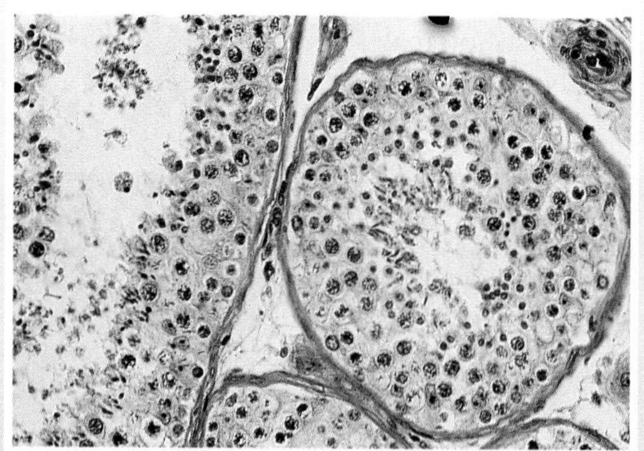

3. **Locate and identify the cells at various stages of development. The cells located just inside the walls of the tubule are the cells in which meiosis begins. The cells closest to the center of the tubule are at later stages of meiosis, and the cells closest to the center of the tubule are the sperm.**
4. **Obtain a prepared slide of human sperm. Using the low-power objective of the microscope, examine the slide. Locate one sperm and then switch the microscope to high power. Draw what you observe.**
5. **Obtain a prepared slide of a human ovary and observe it under the low-power objective of the microscope. Locate a large cell that has a distinct nucleus surrounded by a lightly stained area of cytoplasm. This is an egg.**

Cross section of a human ovary and egg (magnification: 250X)

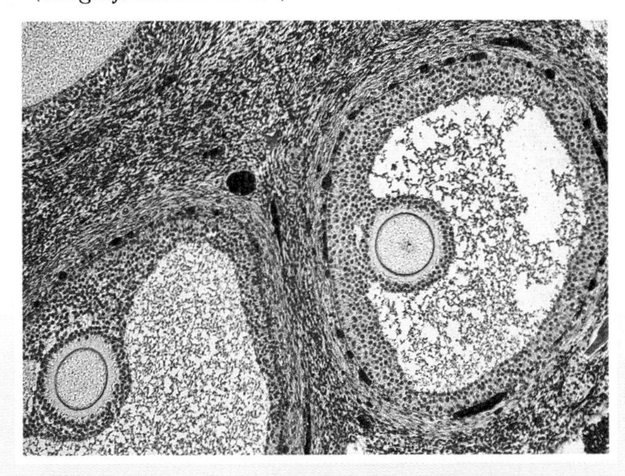

6. **Find the largest follicle, a fluid-filled cavity that surrounds an egg. Switch the microscope to high power and draw what you observe. Try to locate the fluid-filled space, the egg itself, the egg's nucleus, and the layers of follicle cells surrounding the egg.**

Observations

1. How many enlarged follicles with mature eggs did you observe in the cross section of the ovary?
2. How many mature sperm did you observe in the seminiferous tubule?

Analysis and Conclusions

1. How much of the testis did you see?
2. Explain how a human testis is able to produce millions of sperm each day.
3. How does the volume of cytoplasm in the egg compare with the volume of cytoplasm in the sperm?
4. Which structures were easiest to identify? Which were the hardest?

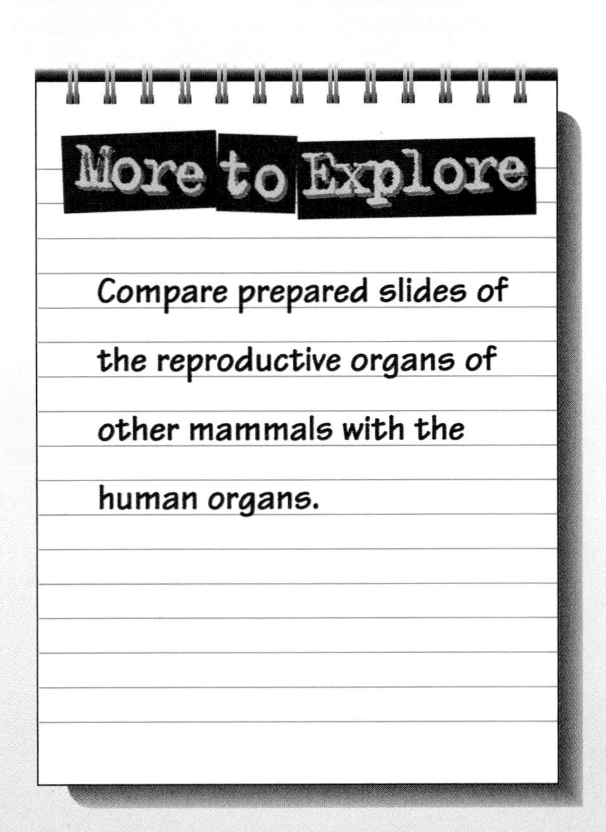

Study Guide

Summarizing Key Concepts

The key concepts in each section of this chapter are listed below to help you review the chapter content. Make sure you understand each concept and its relationship to other concepts and to the theme of this chapter.

39–1 The Human Reproductive System

- The male reproductive system has the task of producing and delivering sperm. The female reproductive system produces gametes, too. Because fertilization and the development of a baby both take place inside the female's body, this puts special demands on the female reproductive system.
- The menstrual cycle has four phases: follicle phase, ovulation, luteal phase, and menstruation.

39–2 Fertilization and Development

- As soon as one sperm makes direct contact with the egg cell membrane, the two cells fuse, the tail of the sperm breaks away, and the sperm nucleus enters the egg's cytoplasm. This process is called fertilization.
- As the embryo grows, the placenta serves as its main organ of respiration, nourishment, and excretion.

39–3 Sexually Transmitted Diseases

- Three common STDs caused by bacteria are syphilis, gonorrhea, and chlamydia.
- Three common STDs caused by viruses are genital herpes, hepatitis B, and AIDS.

Reviewing Key Terms

Review the following vocabulary terms and their meaning. Then use each term in a complete sentence.

39–1 The Human Reproductive System

gamete
ovary
testis
follicle-stimulating hormone
luteinizing hormone
testosterone
sperm
progesterone
estrogen
ovum
scrotum
seminiferous tubule
epididymis
vas deferens
follicle
Fallopian tube
ovulation
uterus
menstrual cycle
corpus luteum

39–2 Fertilization and Development

fertilization
zygote
morula
implantation
embryo
human chorionic gonadotropin
gastrulation
placenta
umbilical cord
fetus
oxytocin
prolactin

39–3 Sexually Transmitted Diseases

sexually transmitted disease
syphilis
gonorrhea
chlamydia
genital herpes
hepatitis B
AIDS

Recalling Main Ideas

Choose the letter of the answer that best completes the statement or answers the question.

1. Among other changes, human gonads begin to produce reproductive cells during

a. puberty.
b. the first six weeks of development.
c. formation of the zygote.
d. childhood.

2. Testosterone controls the development of

a. secondary sex characteristics.
b. secondary sex characteristics and sperm.
c. sperm.
d. eggs.

3. Fertilization generally takes place in the

a. ovary. **c.** vagina.
b. uterus. **d.** Fallopian tube.

4. During the follicle phase, cell and tissue growth in the uterus is stimulated by

a. progesterone. **c.** estrogen.
b. FSH. **d.** LH.

5. In the developing embryo, the neural tube forms from the

a. ectoderm. **c.** mesoderm.
b. endoderm. **d.** amnion.

6. As the embryo develops, the placenta supplies it with

a. metabolic wastes. **c.** food and blood.
b. blood and oxygen. **d.** food and oxygen.

7. Childbirth is initiated by

a. the cervix. **c.** oxytocin.
b. prolactin. **d.** colostrum.

8. Genital herpes is caused by a

a. bacterium. **c.** protozoan.
b. virus. **d.** fungus.

Putting It All Together

Using the information on pages xxx to xxxi, complete the following concept map.

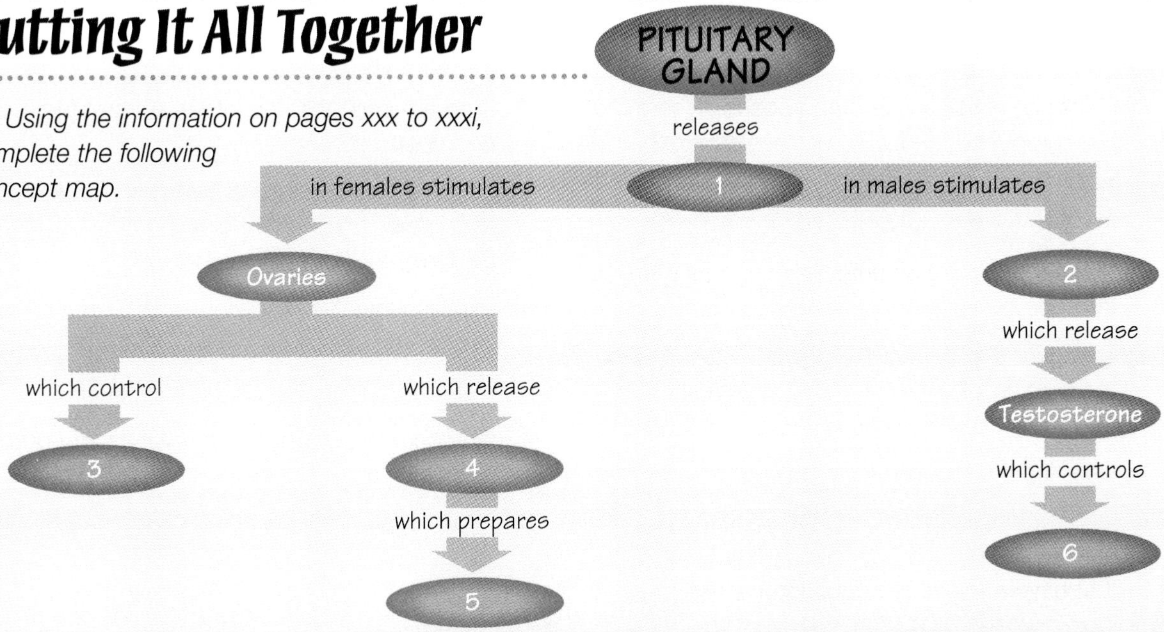

PITUITARY GLAND
releases
1
in females stimulates
Ovaries
which control
3
which release
4
which prepares
5
in males stimulates
2
which release
Testosterone
which controls
6

Reviewing What You Learned

Answer each of the following in a complete sentence.

1. What are the two human gametes called?
2. What is puberty?
3. Identify two secondary sex characteristics of human males.
4. Where is progesterone produced, and what is its function?
5. Explain ovulation.
6. What change in the blood triggers menstruation?
7. Where does fertilization usually take place? Where does the fertilized egg go next?
8. How does a zygote form?
9. Describe a blastocyst.
10. What is gastrulation?
11. From what tissue does the umbilical cord form?
12. When does an embryo become a fetus?
13. What hormone stimulates contractions of the uterus?
14. What is the first symptom of a syphilis infection?
15. Name three sexually transmitted diseases caused by a virus.

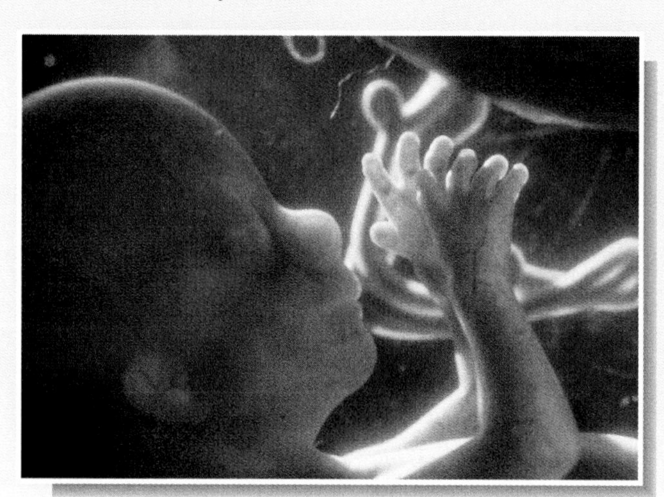

Expanding the Concepts

Discuss each of the following in a brief paragraph.

1. Compare estrogens and androgens. Where are they produced, and what do they control?
2. Trace the path of a sperm as it travels from a testis to the outside of the body.
3. Trace the path of an unfertilized egg as it travels from the ovary to the outside of the body.
4. Compare the number of sperm produced with the number of eggs produced. Suggest reasons for the difference.
5. Describe the four phases of the menstrual cycle.
6. Identify and describe the function of the hormones produced by the pituitary gland that have a role in reproduction.
7. Diagram and label the interactions of the egg and sperm that result in fertilization.
8. Describe the changes a zygote goes through as it travels through a Fallopian tube until it implants itself on the uterine wall.
9. Explain what the presence of HCG in the blood of a female indicates.
10. Describe the process of neurulation.
11. Compare the symptoms of syphilis and genital herpes.
12. Why is AIDS considered to be the most dangerous STD?

Extending Your Thinking

Use the skills you have developed in this chapter to answer the following.

1. **Modeling** Construct a model placenta. Explain how it acts as a lifeline between mother and baby.
2. **Developing hypotheses** The World Health Organization (WHO) states that the percentage of adults in the United States who are reported to have at least one STD is much higher than the percentage in other countries. Develop a hypothesis to account for this difference.
3. **Making judgments** You have learned that drugs can be passed from the mother to the embryo or fetus through the placenta. Drinking alcohol during pregnancy can cause irreversible problems for a baby, including slowed growth, later behavioral problems, and mental retardation. Should drinking alcohol during pregnancy be made illegal? Why or why not?
4. **Interpreting data** Examine the graph showing the reported rates of gonorrhea. In what year was the rate of gonorrhea highest in the United States? Lowest? What changes have taken place between 1970 and 1994? Predict how many people might contract gonorrhea in a city of one million.

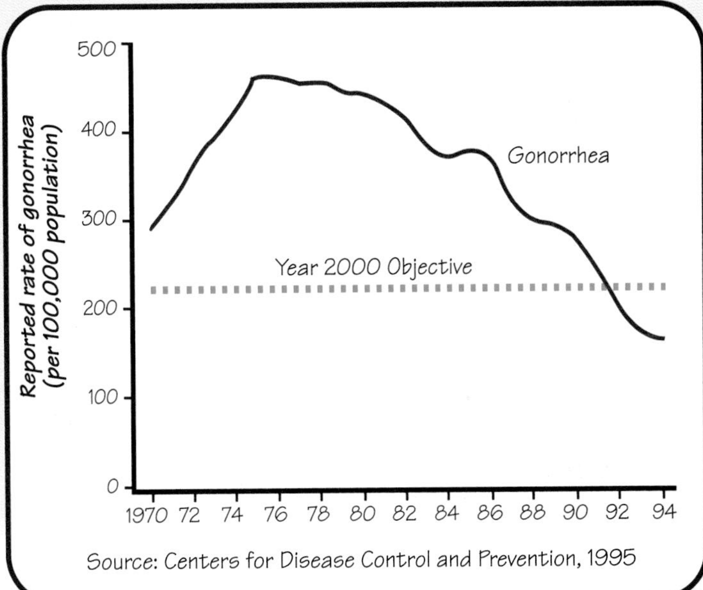

Source: Centers for Disease Control and Prevention, 1995

Applying Your Skills

In Vitro Fertilization

Since the birth of Louise Brown in 1979, in vitro *fertilization has enabled some couples with reproductive problems to conceive a child from their own egg and sperm. In this research project, you'll apply the information in this chapter to explain how* in vitro *fertilization works.*

1. Obtain reference materials on *in vitro* fertilization from an encyclopedia, library book, magazine articles, or other sources.
2. Describe the reproductive problems that *in vitro* fertilization can overcome.
3. List the steps in the *in vitro* fertilization process.
4. Explain each step, comparing it with the process of fertilization described in this chapter.

• GOING FURTHER •

5. In your journal, discuss reasons for and against this means of reproduction.

CHAPTER 40

Immune System

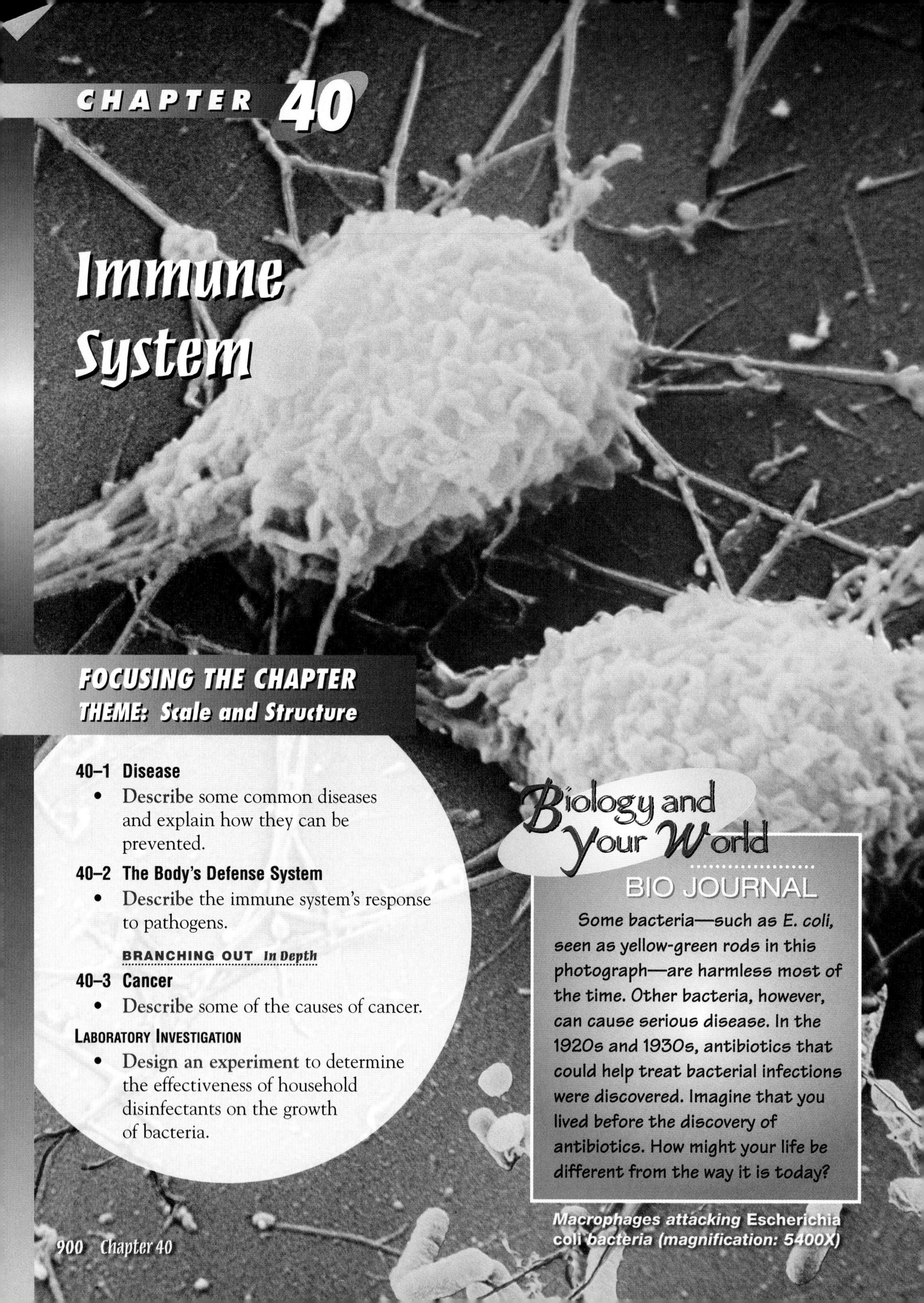

FOCUSING THE CHAPTER

THEME: *Scale and Structure*

40–1 Disease
- **Describe** some common diseases and explain how they can be prevented.

40–2 The Body's Defense System
- **Describe** the immune system's response to pathogens.

BRANCHING OUT *In Depth*

40–3 Cancer
- **Describe** some of the causes of cancer.

LABORATORY INVESTIGATION
- **Design an experiment** to determine the effectiveness of household disinfectants on the growth of bacteria.

Biology and Your World

BIO JOURNAL

Some bacteria—such as *E. coli*, seen as yellow-green rods in this photograph—are harmless most of the time. Other bacteria, however, can cause serious disease. In the 1920s and 1930s, antibiotics that could help treat bacterial infections were discovered. Imagine that you lived before the discovery of antibiotics. How might your life be different from the way it is today?

Macrophages attacking Escherichia coli *bacteria (magnification: 5400X)*

Disease

SECTION 40–1

GUIDE FOR READING

- **Define** disease.
- **State** the germ theory of disease.
- **List** Koch's postulates.

MINI LAB

- **Predict** the growth of bacteria by using a model.

"IF YOU'VE GOT YOUR HEALTH, *you've got everything." There's a lot of truth to this old saying. Good health makes it possible for you to work, study, and play. Unfortunately, most of us take our health for granted—until we catch a cold or develop a disease, something that interferes with the body's normal activities. When that happens, the value of good health becomes all too obvious. Why do we get sick? What is the best way to avoid disease? In this section, we will try to answer these questions.*

Defining Disease

A disease is any change that disrupts the normal functions of the body. Some diseases, such as Huntington disease or cystic fibrosis, are genetic. Others, such as brown lung, a disease of coal miners, are produced by materials in the environment. But many diseases are produced by other organisms, including bacteria, fungi, protozoans, and viruses. A disease-causing organism is called a **pathogen**—literally, a "sickness-maker." The diseases caused by pathogens are generally known as **infectious diseases** because the organisms that cause them usually enter, or infect, the body of the person they make ill.

☑ **Checkpoint** What is an infectious disease?

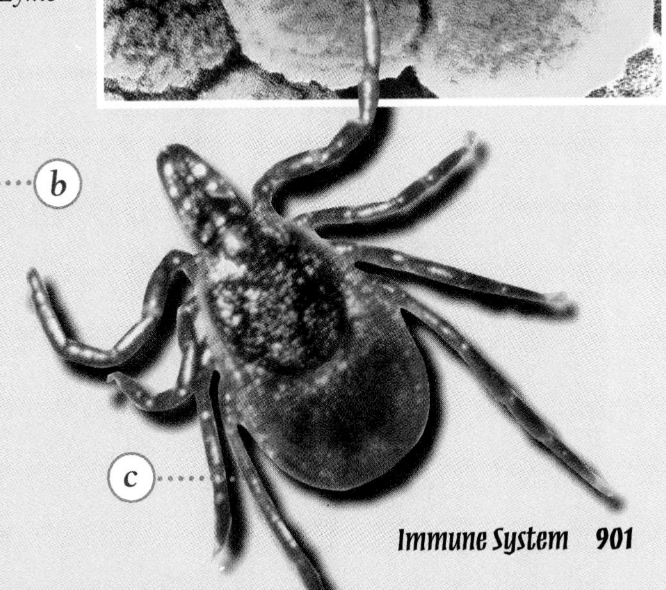

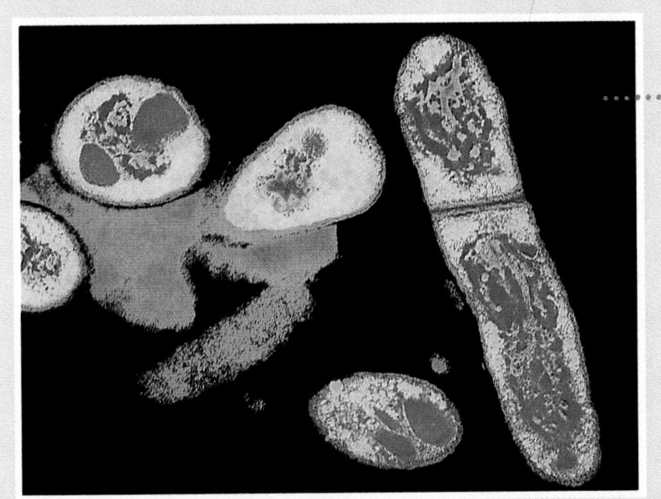

Figure 40–1
Diseases are caused by a variety of pathogens, including (a) Neisseria meningitis, *the bacterium that causes meningitis (magnification: 20,580X),* (b) Mycobacterium tuberculosis, *the bacterium that causes tuberculosis (magnification: 30,000X), and* (c) Borrelia burgdorferi, *the bacterium that causes Lyme disease, which is carried by the deer tick shown here (magnification: 12X).*

Another Generation

PROBLEM *How can you use a model to **predict** the growth of bacteria?*

PROCEDURE

1. Place a kidney bean (or other small object), representing a bacterium, on the table. Wait 30 seconds—which represents the actual 30 minutes that it takes a bacterium to divide—and then put another kidney bean down next to the first.
2. After another 30 seconds, add one kidney bean for each kidney bean that is already on the table.
3. Repeat step 2 until there are 128 kidney beans on the table. Construct a data table to record each generation and the number of bacteria produced.

ANALYZE AND CONCLUDE

1. How much time does it take to produce 128 bacteria?
2. Predict how many bacteria would be produced in 24 hours.
3. What can you infer from this activity about why it is necessary to treat wounds quickly?

The Germ Theory of Disease

At one time, many people believed that infectious diseases were caused by evil spirits. In the mid-nineteenth century, however, a new theory of the way in which disease spread gained acceptance. **Pioneered by the French chemist Louis Pasteur and the German bacteriologist Robert Koch, the germ theory of disease suggested that infectious diseases were caused by microorganisms, which many people call "germs."**

As you know, the world around us is filled with microorganisms of various shapes and descriptions. How can we be sure which of these scores of possible organisms actually causes a particular disease? Koch provided the answer in a series of rules, called **Koch's postulates,** that are used to identify the microorganism that causes a specific disease. **Koch's postulates include the following:**

- **The pathogen should always be found in the body of a sick organism (the host) and should not be found in a healthy organism.**
- **The pathogen must be isolated and grown in a pure culture.**
- **When purified pathogens are injected into a new host, they cause the disease.**
- **The same pathogen should be reisolated from the second host and grown in a pure culture. The pathogen should still be the same as the original pathogen.**

By focusing attention on the biological causes of disease, Pasteur, Koch, and many other scientists produced a revolution in human thinking. For the first time, disease did not seem to be an unavoidable consequence of being alive. And, if a particular pathogen could be identified, there was hope that the disease it caused could be prevented or cured.

Pathogens

Many of the microorganisms that fill the world around also live in and around the human body. The large intestine, for example, harbors dense colonies of bacteria. Bacteria and yeast are found in the mouth, throat, and excretory system. Fortunately, most of these organisms are harmless, and a few of them are actually beneficial.

This being the case, why are some organisms considered pathogens? In some cases, the pathogens actually destroy cells as they grow. Many pathogens, including viruses and many bacteria, grow directly inside cells of the human body, eating their food and ultimately destroying the cell. Other pathogens, including many bacteria, produce **toxins,** or poisons that disrupt bodily functions and produce illness. Finally, many pathogens, especially parasitic worms, produce sickness when they block the flow of blood, remove nutrients from the digestive system, and disrupt other bodily functions. ***Figure 40–2*** lists some of the most common pathogens, including viruses and bacteria, and the diseases they cause.

Viruses are tiny particles that invade and replicate within living cells. Viruses attach to the surface of a cell, insert their genetic material in the form of RNA or DNA, and take over many of the functions of the host cell. Nearly all living organisms—including plants, insects, mammals, and even bacteria—can be infected by viruses.

Although bacteria are often helpful, certain bacteria cause some of the most serious diseases of all. Bacteria are one of the main reasons why it is important to handle food carefully. Bacteria grow quickly in warm, partially-cooked food

Figure 40–2 *This chart lists some of the viruses, bacteria, protozoans, worms, and fungi that cause disease. Notice how each disease is spread.*

PATHOGENS AND DISEASE

TYPE OF PATHOGEN	DISEASE	ORGANISM THAT CAUSES THE DISEASE	METHODS OF SPREADING THE DISEASE
Viruses	Smallpox	Variola	Airborne; personal contact
	Common cold	Rhinovirus	Airborne; direct contact with infected person
	Influenza	Two types (A, B), plus many subtypes	Airborne; droplet infection; direct contact with infected person
	Measles	Measles	Droplets in air; direct contact with secretions of infected person
	AIDS	HIV	Sexual contact; contaminated blood products or hypodermic needles
	Chickenpox	Varicella	Airborne; direct contact with infected person
Bacteria	Tuberculosis	*Mycobacterium tuberculosis*	Droplets in air; contaminated milk and dairy products
	Meningitis	*Neisseria meningitis*	Direct exposure to organism
	Diphtheria	*Corynebacterium diphtheriae*	Contact with infected person or carrier; contaminated raw milk
	Rocky Mountain spotted fever	*Rickettsia rickettsii*	Bite of infected tick
	Lyme disease	*Borrelia burgdorferi*	Bite of infected tick
	Cholera	*Vibrio cholerae*	Contaminated drinking water
	Tetanus	*Clostridium tetani*	Dirty wound; usually a puncture wound
Protozoans	African sleeping sickness	*Trypanosoma*	Spread by tsetse fly
	Malaria	*Plasmodium*	Spread by mosquitoes
Worms	Schistosomiasis	*Schistosoma mansoni*	Freshwater streams and rice paddies
	Beef tapeworm	*Taenia saginata*	Contaminated meat
Fungi	Athlete's foot	Imperfect fungi	Contact with infected person; shower stalls
	Ringworm	Imperfect fungi	Exchange of hats, combs, and athletic headgear with infected person

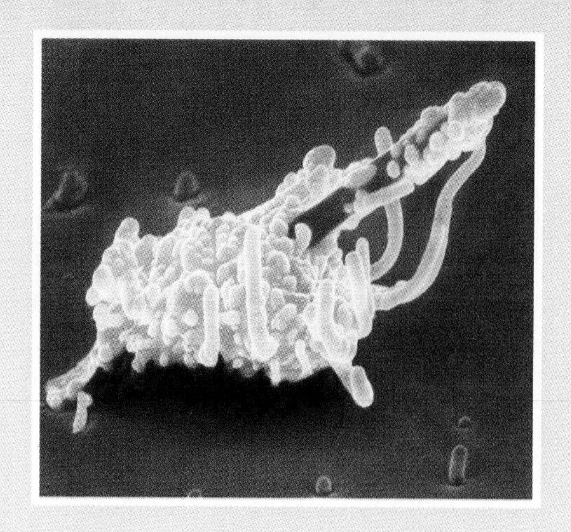

Figure 40–3 *Although white blood cells normally trap and digest pathogens, this white blood cell is engulfing a cancer-causing asbestos fiber. Unfortunately, it cannot be digested.*

INTEGRATING SOCIAL STUDIES

What effect did the bubonic plague have on the populations of Europe in the Middle Ages?

and are always present in uncooked meat. Many types of bacteria, especially the species known as *Salmonella,* can grow quickly enough to produce high levels of the toxins that produce food poisoning. The only way to make food completely safe is to cook it completely.

Although many pathogens are either viruses or bacteria, there are other pathogens as well. Fungi, protozoans, insects, and parasitic worms also cause disease.

Fighting Disease

Once a pathogen has been identified, biologists search for clues as to how it is passed from person to person. If a disease is thoroughly understood, preventing it usually becomes clear. Then, the best way to fight the disease is to avoid it.

Many diseases are spread by animals. The **vectors**—animals that carry disease-causing organisms from host to host—are the key to stopping such diseases. Malaria and yellow fever are fought by controlling the mosquito population. Lyme disease can be prevented by avoiding deer ticks, which carry it. Bubonic plague, the "black death" that killed millions in Europe in the Middle Ages, is spread by fleas that live on rats and mice. Aggressive measures to control these rodents have made serious outbreaks of the plague rare.

Diseases that are spread from one person to another can be controlled by simple habits of personal hygiene. Washing one's hands thoroughly helps to prevent the spread of many pathogens. The common cold is spread by coughing, sneezing, and hand-to-hand contact. Such measures as covering your mouth with a tissue can limit infection.

If prevention fails, drugs have been developed for use against all sorts of pathogens. Perhaps the most useful single class of infection-fighting drugs are **antibiotics**—compounds that kill bacteria without harming cells of humans or animals. Many of these compounds are produced naturally. Antibiotics work by interfering with the cellular processes of microorganisms. For example, penicillin, the first antibiotic to be discovered, interferes with cell wall synthesis, disabling and killing bacteria. Streptomycin and tetracycline interfere with protein synthesis on bacterial ribosomes. Because viruses use the ribosomes of the infected cell to make their proteins, antibiotics are not effective against viruses.

Section Review 40–1

1. **Define** disease.
2. **State** the germ theory of disease.
3. **List** Koch's postulates.
4. **Critical Thinking—Drawing Conclusions** Why might Koch's postulates be difficult to apply to a viral infection?
5. **MINI LAB** How can you **predict** the growth of bacteria by using a model?

The Body's Defense System

SECTION 40–2

GUIDE FOR READING

- **Describe** the function of the immune system.
- **Define** immunity.

SOMETIMES IT MAY SEEM THAT we are surrounded by a world of pathogens that threaten our existence every moment of the day. In such a hostile environment, how does the body cope with the threat of infection? The answer is that the body has a protective system, a series of defenses that guard against disease.

The body's defense against infection is provided by the immune system. Unlike many of the body's other systems, the work of the immune system goes on behind the scenes. In fact, if it is working properly, you may never notice it. But when something does go wrong with the immune system, life itself is threatened by the pathogens it normally holds at bay.

Nonspecific Defenses

The immune system is the body's primary defense against pathogens. It consists of nonspecific and specific defenses against infection. The skin is the body's most important nonspecific defense. Few pathogens can penetrate the tough layers of keratin protein at the skin's outer surface. The importance of these layers becomes obvious as soon as the skin is broken. As you know, even a small cut or scrape quickly becomes infected if it is not taken care of. Why does this happen? Because even the smallest cut breaks the protective barrier, giving microorganisms

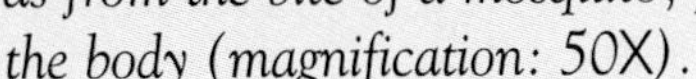

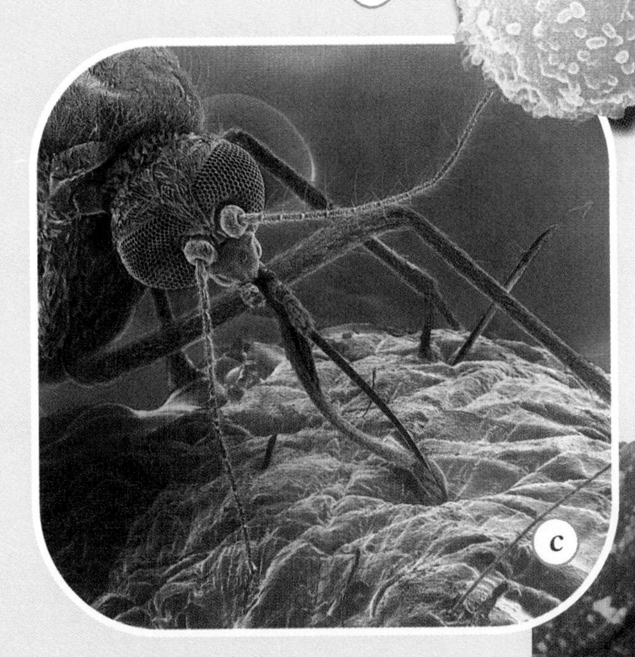

Figure 40–4 The body responds to infection with specific defenses and nonspecific defenses. (a) Specialized white blood cells such as lymphocytes are the body's main specific defense (magnification: 1500X). (b) The skin provides the body with its most important nonspecific defense. It keeps bacteria, seen as green rods in this electron micrograph, from breaking through (magnification: 8000X). (c) When there is a break in the skin, such as from the bite of a mosquito, pathogens may enter the body (magnification: 50X).

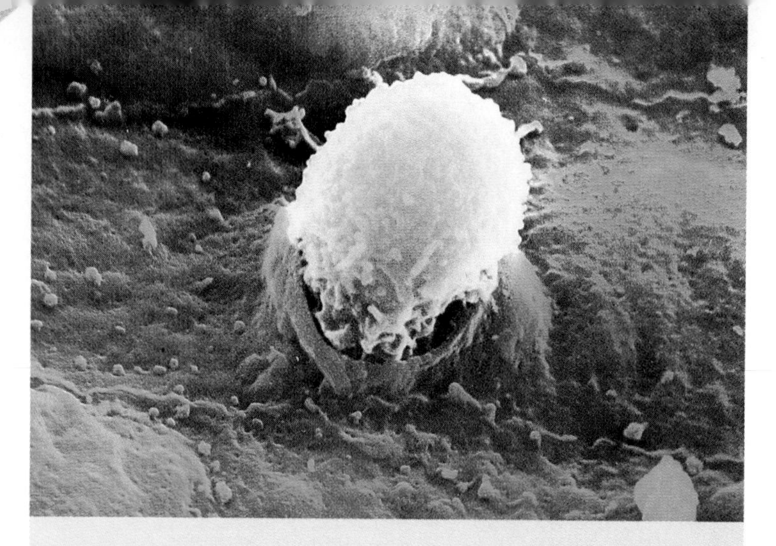

Figure 40–5
During the inflammatory response, white blood cells leak out of the blood vessels to attack a pathogen. In this scanning electron micrograph, a white blood cell is squeezing through the endothelium of a vein to join the attack (magnification: 3450X).

on the skin's surface a chance to enter the body.

The skin is considered a nonspecific defense because it is not directed against any one pathogen. Instead, a nonspecific defense guards against all infections, regardless of their cause. Other nonspecific defenses are found in the mouth and respiratory passages, where millions of microorganisms enter every day. Passages leading to the lungs are coated with mucus that traps airborne pathogens. Cells lining these passages sweep these trapped cells into the digestive system, where they are destroyed by digestive enzymes. Other pathogens that enter the mouth, eyes, and excretory system may be destroyed by lysozyme, an enzyme that breaks down cell walls of some bacteria.

☑ **Checkpoint** What is a nonspecific defense?

Inflammation

If pathogens do enter the body through a cut, they grow quickly, spreading and releasing toxins into the tissues. When this happens, a second nonspecific defense is activated. This is called the **inflammatory response** because it can cause the skin to turn a "flaming" red color. Blood vessels near the wound expand, and white blood cells leak from the vessels to invade the infected tissues. Many of these white cells, which are called **phagocytes,** engulf and destroy bacteria. The infected tissue may become swollen and painful as the battle between the pathogens and white blood cells rages.

☑ **Checkpoint** What is the inflammatory response?

Fever

When a serious infection causes pathogens to spread, the body responds in two ways. First, the immune system releases chemicals that increase the body's temperature. You have probably experienced this elevated body temperature, called a fever. Second, the immune system produces millions of white blood cells to fight the infection. An increased number of white cells in the blood is a sign that the body is dealing with a serious infection.

Specific Defenses

The immune system is also capable of powerful specific defenses that produce **immunity** against particular diseases. **Immunity is the ability of the body to resist a specific pathogen.** There are two kinds of immunity—antibody immunity and cell-mediated immunity.

Ever since ancient times, people have observed that individuals who recover from certain diseases, such as mumps and measles, never again become sick with the same disease. In other words, most people who recover from the mumps or measles are permanently immune—they will never get the mumps or measles again.

Antibody Immunity

Exposure to certain diseases produces permanent immunity because it stimulates cells in the immune system to make proteins called **antibodies.** The molecules that stimulate the production of antibodies are known as **antigens.** An antigen is any substance that triggers the specific defense of the immune system. The antibody is the basic functional unit of the specific immune response.

As ***Figure 40–6*** shows, an antibody is shaped like the letter Y and has two identical antigen-binding sites. This means that an antibody can attach to two antigens. Why is this good? If the antigen is on the surface of a virus particle, it means that antibodies can link the virus particles into a clump, preventing them from entering a cell. The clump of viruses and antibodies attract phagocytes, which engulf and destroy the whole mass. If the immune system produces enough antibodies to a particular virus, it can prevent that virus from causing disease.

Antibodies can prevent bacterial infections, too. When antibodies bind to the surfaces of bacteria, they mark the cells for destruction by phagocytes and other white blood cells.

Lymphocytes

The immune system also includes a special class of white blood cells called **lymphocytes.** One group of lymphocytes, the B lymphocytes, or **B cells,** matures in bone marrow. These cells produce antibodies.

How do B cells produce antibody proteins that fit their antigens so precisely? In a sense, they are custom-made. As B cells develop, the antibody genes within each of them rearrange themselves in a slightly different way. When their development is complete, the immune system contains literally millions of B cells—each capable of producing a slightly different antibody.

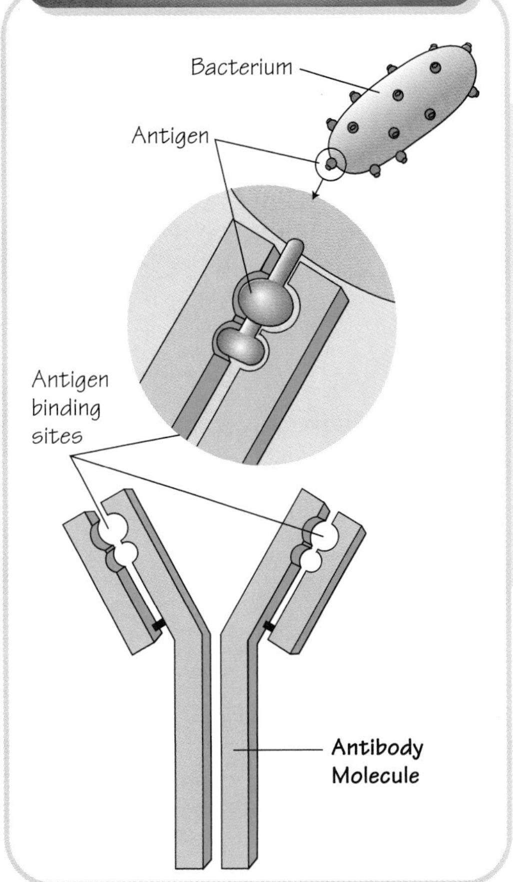

Figure 40–6 *The antibody molecule is the basic functional unit of an immune response. Each antibody molecule has two antigen-binding sites, which are specific for a particular antigen.*

When a pathogen enters the body, molecules on its surface act as antigens to stimulate the immune system. This is helped along by **T cells,** lymphocytes that have matured in the thymus gland. In a few days, B cells whose antibodies closely match the shape of the antigen have started to grow and divide rapidly. This produces a large number of specialized B cells called **plasma cells,** which release antibodies into the bloodstream to fight the infection. Millions of plasma cells may be produced from a handful of B cells stimulated by antigen.

Although it may take many days for enough plasma cells to form and help to overcome an infection, these cells have another important function—they produce permanent immunity. When a person contracts measles, for example, antigens on the surface of the virus stimulate the production of millions of plasma cells to fight the infection. As a result, if the person survives the infection, his or her immune system retains those cells and is always ready to respond with a massive supply of the measles antibody.

Vaccines

Vaccines take advantage of this fact. A **vaccine** is a weakened or mild form of a pathogen that causes permanent immunity when injected into the body. For example, when you were vaccinated against polio, weakened polio viruses were used to stimulate the B cells in your body that are capable of making anti-polio antibodies. As a result, if you are ever exposed to the polio virus, your body is prepared to fight this virus with millions of plasma cells ready to make polio antibodies.

☑ **Checkpoint** What is a vaccine?

Figure 40–7
If an infection occurs, the body responds by either of two immune responses—the production of antibodies or the production of killer T cells. In antibody immunity, antigens cause B cells to multiply. Some B cells develop into plasma cells that secrete antibodies. Other B cells develop into memory B cells, which store information about the pathogen in order to provide future immunity. In cell-mediated immunity, T cells multiply, producing three types of T cells—killer T cells, helper T cells, and suppressor T cells.

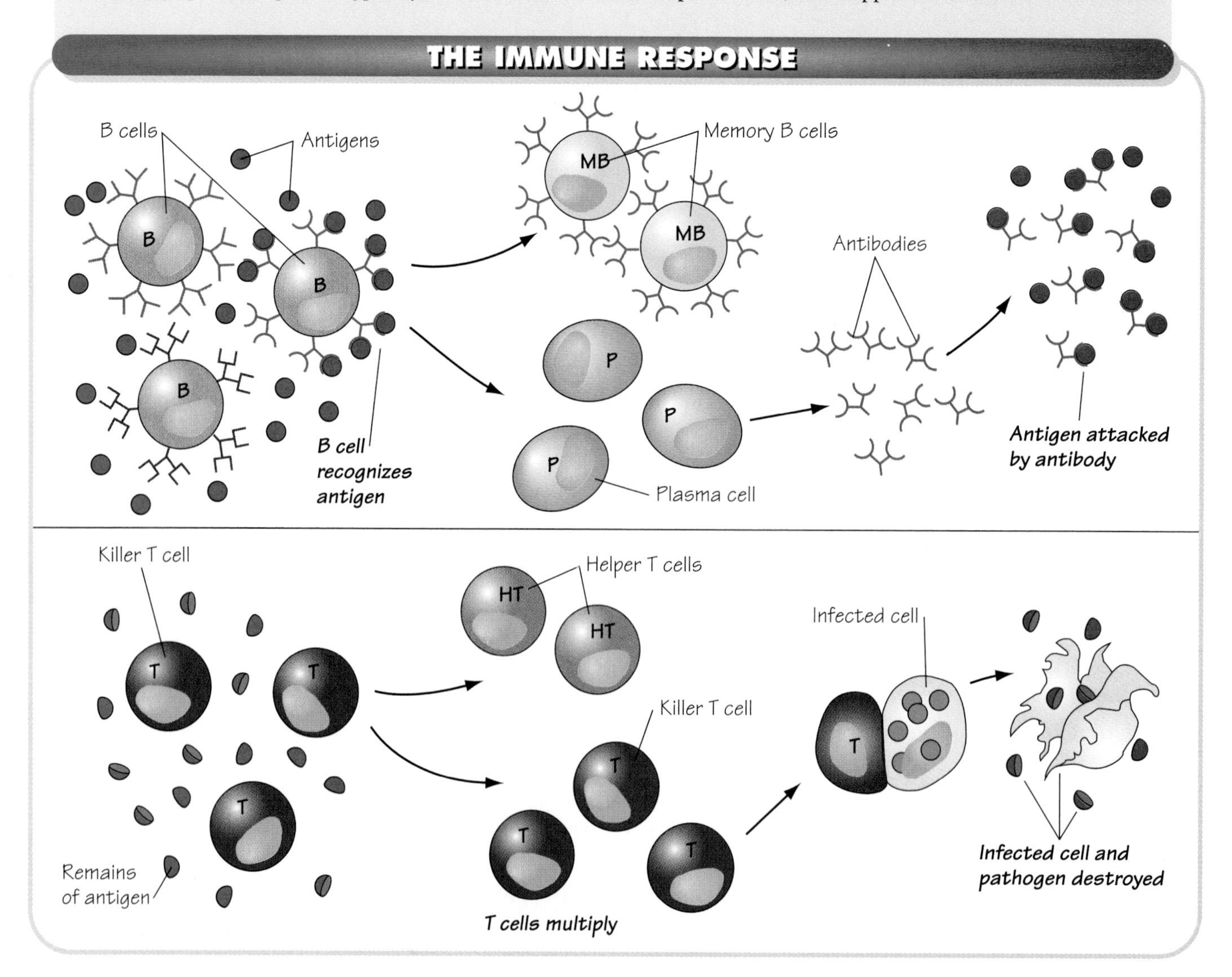

Cell-Mediated Immunity

Antibodies alone are not enough to protect the body against some pathogens. In these cases, the immune system has a powerful weapon—the **killer T cell.** Killer T cells make direct contact with antigen-bearing cells, disrupting their cell membranes and destroying them. This process is known as **cell-mediated immunity.** The actions of killer T cells are extremely important in fighting certain types of infections. Killer T cells attack and destroy virus-infected cells, and they are also important in fighting protozoans and multicellular parasites.

Unfortunately, killer T cells are also responsible for the rejection of tissue transplants. The cells of your body have a special set of protein markers on their surfaces that enable the immune system to recognize them as your own. If tissues from another person are transplanted into the body, the immune system checks these markers, recognizes them as foreign, and attacks them. To prevent tissue rejection, physicians search for an organ donor whose cell markers are nearly identical to those of the recipient. They may also use drugs, such as cyclosporine, that suppress the cell-mediated immune response.

In addition to killer T cells, there are other types of T cells that are involved in the immune response. **Helper T cells** are the first cells to identify the specific pathogen in the body. The helper T cells then send a message to the B cells "telling" them to produce antibodies for that particular pathogen.

A third kind of T cell is a **suppressor T cell.** After an infection has been fought off successfully, suppressor T cells shut off the immune response in both B cells and killer T cells.

☑ ***Checkpoint*** What is the function of helper T cells?

Figure 40–8
In this photograph, the killer T cells, which become elongated when active, are attacking a cancer cell's membrane (magnification: 6000X).

Immune Disorders 

Although the immune system protects the body against many pathogens, sometimes disorders occur. There are two main types of disorders. In the first type, the body overreacts to harmless substances. Allergies and autoimmune diseases are examples of this type of disorder. In the second type, the immune system becomes too weak to fight infection. AIDS is an example of this type of disorder.

Allergies

When the immune system overreacts to an antigen in the environment, an **allergy** results. An allergic reaction occurs when antigens trigger a type of immune cell, called a mast cell, to release chemicals called **histamines.** Histamines produce an inflammatory response that includes sneezing, runny nose, and itchy eyes, which allergy sufferers know all too well. ● The offending antigens may be part of a pollen grain, a mold spore, the toxin from a bee sting, or even household dust.

INTEGRATING HEALTH

What are antihistamines? How do they help allergy sufferers?

Biology AND YOU .... **Connections**

AIDS

Although scientists do not have enough information to cure AIDS, they do know more than enough to prevent it. HIV, the virus that causes AIDS, is not easily spread. People who live with, care for, work with, or go to school with people with HIV are not at risk of contracting the virus by casual contact. The virus that causes AIDS can be spread only by coming into direct contact with an infected person's body fluids, such as blood, semen, and vaginal secretions.

Each square of this quilt is dedicated to a person who has died of AIDS.

Blood Transfusions

Prior to 1985, donated blood used for transfusions and other medical procedures, such as treatments for hemophilia, was not tested for the presence of HIV. As a result, several thousand people became infected with HIV and eventually died of AIDS. Today, blood supplies are carefully screened for HIV and other viruses.

Injected Drug Use

HIV infection among users of illegal drugs, on the other hand, has skyrocketed. Sharing needles to inject drugs may pass on HIV-laden blood directly from one person's bloodstream into the bloodstream of another. This action is one of the most dangerous and irresponsible things a person can do. Not only are drugs such as heroin and cocaine dangerous in their own right, but the injection of these drugs also carries with them the added risk of contracting AIDS.

A Sexually Transmitted Disease

AIDS can also be spread by sexual contact. The presence of the virus in the semen and vaginal fluids of an infected person means that every sexual contact that person has carries with it the risk of passing on AIDS. The most effective way to protect yourself from HIV infection is to abstain from sexual intercourse. By doing so, you help to prevent the spread of this deadly disease.

Making the Connection

How does information about AIDS transmission help you to make decisions about your behavior? How does it help you when you encounter someone who has HIV? Do you think employers should have the right to require all their employees to be tested for HIV? Why or why not?

Figure 40–9
The helper T cell, colored green in this scanning electron micrograph, has been infected with HIV. The red objects are HIV, the virus responsible for causing AIDS (magnification: 6000X).

Allergic reactions can create a dangerous condition called **asthma,** in which smooth muscle contractions reduce the size of air passageways in the lungs, making breathing difficult. Asthma attacks are usually triggered by a particular antigen, and so the best response to the condition is to avoid that antigen. New drugs make it possible to provide immediate relief from asthma attacks, relaxing the smooth muscles to make breathing easier.

Autoimmune Diseases

Sometimes the immune system makes mistakes and attacks its own cells, producing what is called an **autoimmune disease.** Myasthenia gravis and multiple sclerosis are two autoimmune diseases. In myasthenia gravis, antibodies attack the nerve receptors of the muscles. In multiple sclerosis, the immune system attacks the myelin sheath that surrounds nerve fibers. Multiple sclerosis, which may be triggered by a viral infection, usually strikes people between the ages of 20 and 40.

AIDS

The importance of a healthy, functioning immune system has been underscored by a disease that was first recognized in the early 1980s. At that time, physicians began to see an increase in the number of unusual infections. Some of these infections included protozoans in the lungs, severe fungal infections in the mouth and throat, and a rare form of skin cancer. Normally, such infections are prevented by the immune system, so doctors immediately realized that the immune systems of their patients had been weakened. They called the disease **AIDS** (acquired immune deficiency syndrome).

The spread of the disease made scientists suspect that it was caused by a virus. In 1984, that virus—now known as **HIV** (human immunodeficiency virus)—was discovered. HIV infects, weakens, and gradually destroys the helper T cells. As the disease progresses, which may take many years, HIV-infected individuals suffer one infection after another from organisms that the immune system normally controls with ease.

To date, there is no cure for AIDS, although some progress has been made in developing drugs that make it difficult for HIV to infect cells and to reproduce. Fortunately, HIV does not spread easily from person to person, and the way in which the virus is transmitted—in blood and other body fluids—is now well understood. The best way to stop this dangerous disease is to avoid contact with the virus.

Section Review 40–2

1. **Describe** the function of the immune system.
2. **Define** immunity.
3. **Critical Thinking—Comparing** Compare the role of B cells and T cells in the immune response.

SECTION 40–3 BRANCHING OUT In Depth

Cancer

Guide for Reading

- **Define** cancer.
- **Describe** some of the causes of cancer.

MINI LAB

- **Interpret** data regarding cancer.

CANCER! FEW WORDS IN THE English language conjure up the dread and fear that this word does. Cancers are the second leading cause of death in the United States, claiming more than half a million lives each year. And cancers are a worldwide threat, affecting people of every nationality and culture.

What is this disease that takes such a toll on human life? Where does it come from, and why is it so difficult to treat? As you will see, one of the things that makes cancer so different from other diseases is that the cells that cause it are not invading pathogens—they are the body's own cells, which have, in a sense, turned against it.

A Cellular Disease

The growth of the many trillions of cells throughout the body is closely regulated. Controls on cell growth are necessary in a complex organism to keep tissues and organs together. **Cancer begins when a cell or a group of cells escapes the body's normal growth controls.** When this happens, the cells grow into a **tumor,** a mass of growing, unregulated cells.

If a tumor forms in a sensitive area of the body, such as the brain or spinal cord, its very presence can be dangerous.

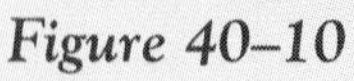

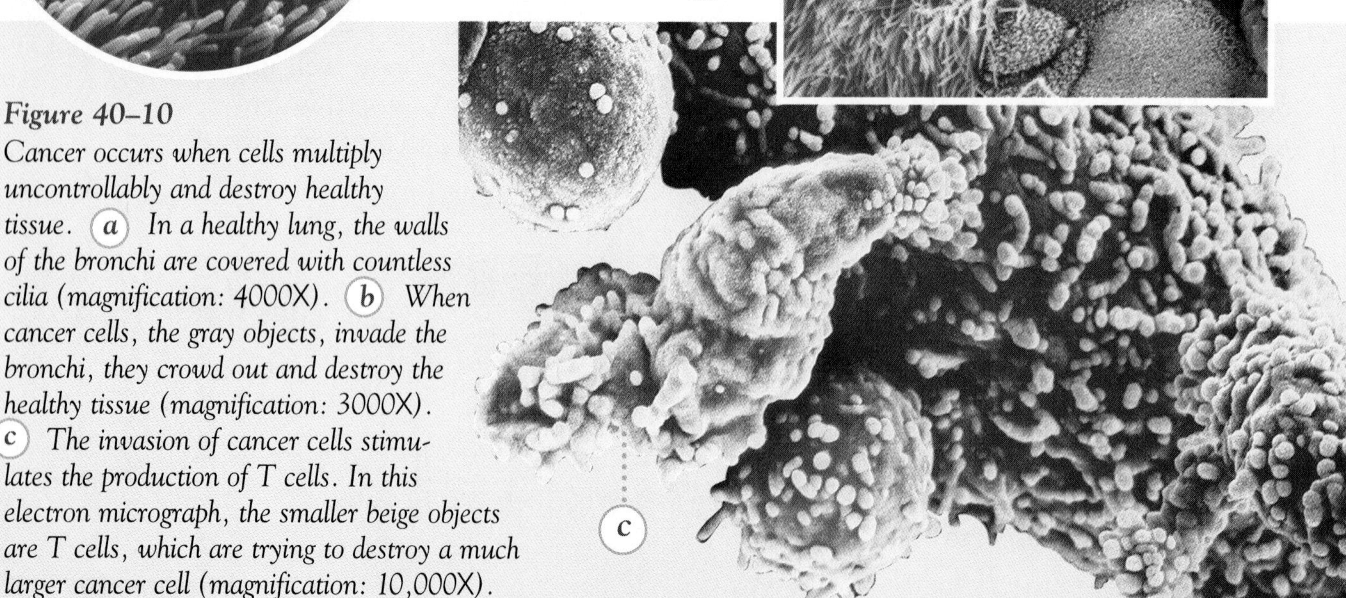

Figure 40–10
Cancer occurs when cells multiply uncontrollably and destroy healthy tissue. (a) *In a healthy lung, the walls of the bronchi are covered with countless cilia (magnification: 4000X).* (b) *When cancer cells, the gray objects, invade the bronchi, they crowd out and destroy the healthy tissue (magnification: 3000X).* (c) *The invasion of cancer cells stimulates the production of T cells. In this electron micrograph, the smaller beige objects are T cells, which are trying to destroy a much larger cancer cell (magnification: 10,000X).*

However, the most dangerous tumors are those that are capable of spreading from their places of origin to invade other parts of the body.

Physicians classify tumors according to their ability to spread. A tumor that does not spread to surrounding tissue is called a benign tumor. A tumor that spreads and destroys healthy tissue is called a malignant tumor.

It is often said that cancer is not one disease but many. Cancers differ from each other according to the tissues in which they originate. Lung cancers originate in the tissues of the lung, skin cancers are produced when cells in the skin begin to grow out of control, and breast cancers begin in the tissues of the breast. Each type of cancer has its own pattern of growth and invasion, and each presents unique problems.

☑ **Checkpoint** What is a tumor?

Figure 40–11
Immunofluorescence is a technique that uses antibodies to attach fluorescent dyes to specific structures within the cell. This immunofluorescent light micrograph shows a malignant form of skin cancer (magnification: 320X).

Causes of Cancer

One of the most baffling elements of the cancer puzzle has been the many different causes of cancer. **Some cancers are caused by biological agents, such as viruses. Others are caused by chemicals in the environment or in the foods we eat. And others have a physical cause—radiation.**

Viruses

Cancer-causing viruses were first discovered in animals. In the early 1900s, Peyton Rous discovered that a type of cancer in chickens could be caused by viruses passed from one animal to the next. Although many similar viruses have been discovered in animals, only a few human cancers seem to be caused by viruses. One of these is the human papilloma virus (HPV), which is sexually transmitted. Chronic HPV infections can lead to cancer of the reproductive organs. Hepatitis, which infects the liver, may lead to liver cancer. And the Epstein-Barr virus, in very rare instances, can produce a cancer known as Burkitt's lymphoma.

Radiation

Radiation has been recognized as a cause of cancer ever since the first experiments with radioactive substances. Exposures to high levels of radioactivity can cause leukemia, a cancer of the blood, and several other cancers. However, the most common and most dangerous form of cancer-causing radiation is all around us—sunlight. Exposure to the ultraviolet radiation in sunlight brings with it a risk of skin cancer. Although most forms of skin cancer can be successfully treated with surgery, one often fatal form of skin cancer—malignant melanoma—is increasing in frequency at an alarming rate.

Chemicals

Chemicals can also cause cancer. A cancer-causing chemical is known as a **carcinogen,** and scientists have identified hundreds of them. Carcinogens

Going by the Numbers

PROBLEM *How can you **interpret** data regarding cancer?*

PROCEDURE

LUNG CANCER DEATH RATES *

Gender	*1960–1962*	*1990–1992*
Male	40.2	74.4
Female	6.0	32.2

* per 100,000 U.S. population

1. Based on the data table, construct a bar graph.
2. After examining the data table and bar graph with your partner, discuss some reasons to explain the changes that occurred in lung cancer death rates over the 30-year span.
3. Discuss some reasons to explain why there are differences in lung cancer death rates between males and females.

ANALYZE AND CONCLUDE

1. Determine the percentage change in lung cancer death rates in males and females during the 30-year period.
2. Because 80 percent of lung cancers are caused by smoking tobacco, what can you infer about the use of tobacco in men over the past 30 years? In women?

cause cells to lose control of their normal functions, thus causing cancer. Some carcinogens are naturally occurring compounds, such as aflatoxin, a compound produced by fungi. Others, such as benzene, are synthetic compounds. Health researchers have worked hard to identify the most dangerous carcinogens so they can be removed from the workplace and the food supply.

Ironically, the single most damaging source of carcinogens in most countries is not a contaminant of food, water, or air. Some of the most powerful carcinogens are found in tobacco smoke. Smoking tobacco or inhaling tobacco smoke causes lung cancer, which is responsible for more than 30 percent of all cancer deaths in the United States.

Cancer Genes

For many years, researchers were puzzled as to how causes as different as sunlight, chemicals, and viruses could produce cancer. However, all the causes of cancer have one thing in common—they all produce mutations, or changes in DNA. Carcinogenic chemicals cause errors in DNA replication; sunlight and other forms of radiation damage nucleic acids directly; and cancer-causing viruses introduce new genes into the cells they infect.

Why should mutations be so important? In recent years, biologists have discovered that a series of important proteins regulate the rate at which cells pass through the phases of the cell cycle. These proteins, and the genes that encode them, are responsible for the controlled cell growth that occurs normally in the body. If one of these key genes is damaged or mutated, controls over cell growth would be lost. And a group of cells would begin to grow out of control, just like a tumor.

One of these key genes is a protein called p53. Although p53 has many functions, its most important function is to check the condition of a cell's genetic information before it is allowed to divide. If damage to DNA is detected, p53 prevents the cell from entering mitosis. In that way, it ensures that DNA damage is repaired before the cell is allowed to divide. How important is p53 in avoiding cancer? More than half of all human

Figure 40–12
CAREER TRACK
An immunologist studies how organisms defend themselves against infection by pathogens. Here, an immunologist is using a technique that analyzes and measures the concentration of antibodies in order to test for a parasitic disease.

tumors have a damaged or inactive p53 gene! Damage to p53 or the many other genes that help to regulate cell growth seems to be the source of nearly all human cancers.

Treating Cancer

In their early stages, most cancers produce few symptoms. This is unfortunate, because cancers are easier to treat if they are found early. Once a tumor is discovered, it should receive immediate medical attention. In some cases, surgery is the best treatment. In their early stages, all the skin cancers, as well as cancers of the breast, the intestine, and the bladder, can be completely removed by surgery. This often results in a complete cure.

If the cancer has spread or if the tumor cannot be completely removed, a different approach is needed. Some cancers can be treated with radiation, to which fast-growing cells are especially sensitive. Radiation is used for localized tumors. Other types of cancers are treated with **chemotherapy**—mixtures of drugs that interfere with important processes such as DNA replication and cell division.

Chemotherapy has proven especially useful against certain forms of childhood leukemia, a cancer of the white blood cells. However, the powerful drugs used in chemotherapy to kill cancer cells have serious side effects—they kill many normal cells as well, causing nausea, headaches, and sometimes a temporary loss of hair.

Because no drug has proven completely effective against all cancers, the search goes on for compounds that will do a better job of killing tumor cells with fewer side effects. Medical science is closer than ever before in pinpointing the exact genetic causes of cancer, and there is good reason to hope that new technologies may conquer this disease. ●

INTEGRATING BIOLOGY AND SOCIETY

Find out from the American Cancer Society the seven warning signs of cancer.

Section Review 40–3

1. **Define** cancer.
2. **Describe** some of the causes of cancer.
3. **MINI LAB** How can you **interpret** data regarding cancer?
4. **BRANCHING OUT ACTIVITY** Using reference materials, find out the ten leading causes of death from cancer in 1960 to 1962 for both men and women. Then find out the ten leading causes of death from cancer in 1990 to 1992 for both men and women. **Compare** the lists and try to explain any differences you see.

Laboratory Investigation

DESIGNING AN EXPERIMENT

Inhibiting Bacterial Growth

Disinfectants are products used to destroy bacteria and viruses. How do you know which type of disinfectant is effective against which type of bacteria? In this investigation, you will compare the bacteria-inhibiting powers of several kinds of disinfectants.

Problem

Which household disinfectant is the most effective in preventing the growth of bacteria? **Design an experiment** to find the answer.

Suggested Materials

- 2 Petri dishes containing sterile nutrient agar
- glass-marking pencil
- sterile medicine dropper
- transparent tape
- hydrogen peroxide
- metric ruler
- 2 other types of household disinfectants

Suggested Procedure

1. **Obtain 2 Petri dishes. Open each Petri dish and have your partner rub his or her fingers over the entire surface of the sterile agar. Replace the covers.**
2. **Open the cover of one Petri dish and, using the medicine dropper, place 2 drops of hydrogen peroxide in the middle of the dish.**

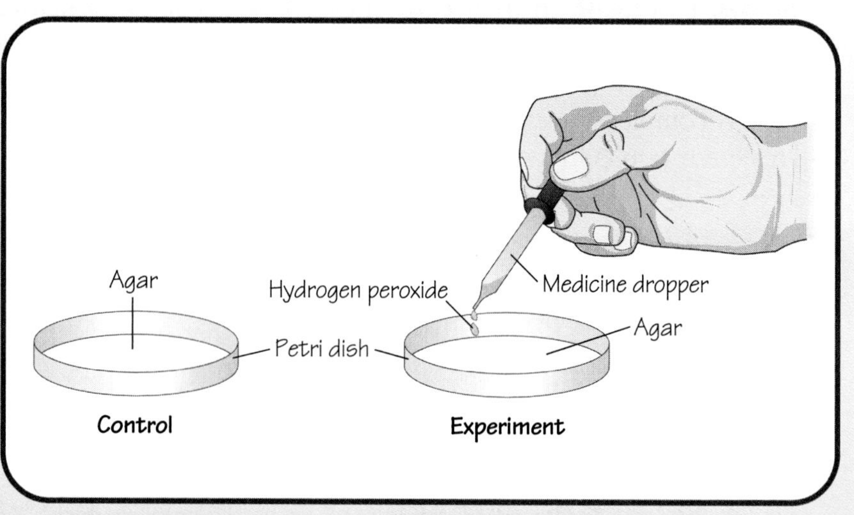

3. **Turn the dish over and, with the glass-marking pencil, label it Hydrogen Peroxide. Using transparent tape, tape the cover closed.**
4. **Turn over the other dish and label it Control.**
5. **Place the Petri dishes in an area where they will remain undisturbed and at room temperature for 24 hours. CAUTION:** ***Do not open the Petri dishes again.***
6. **After 24 hours, without opening the Petri dishes, observe the surfaces of the nutrient agar in each dish. Record your observations in a data table similar to the one shown.**
7. **Return the Petri dishes to their storage place and observe them after 48 hours and after 72 hours. Record your observations.**
8. **Return the unopened Petri dishes to your teacher for sterilization and proper disposal. CAUTION:** ***Always wash your hands after handling a used Petri dish.***
9. **Using a procedure similar to the one given in steps 1 to 8, design an experiment to determine the effectiveness of various disinfectants on the prevention of the growth of bacteria.**
10. **Be sure to include a control. Obtain the approval of your teacher before beginning your procedure.**

Observations

1. What effect did hydrogen peroxide have on the growth of bacteria after 24, 48, and 72 hours?
2. What effect did each of the other disinfectants have on the growth of bacteria after 24, 48, and 72 hours?

DATA TABLE

Disinfectant	After 24 Hours	After 48 Hours	After 72 Hours
Hydrogen peroxide			

Analysis and Conclusions

1. Which disinfectant was the most effective against the growth of bacteria? Give evidence to support your answer.
2. Did every group in the class reach the same conclusion? How can you account for any differences?
3. What conclusion can you draw from your experiment about choosing disinfectants for your home?

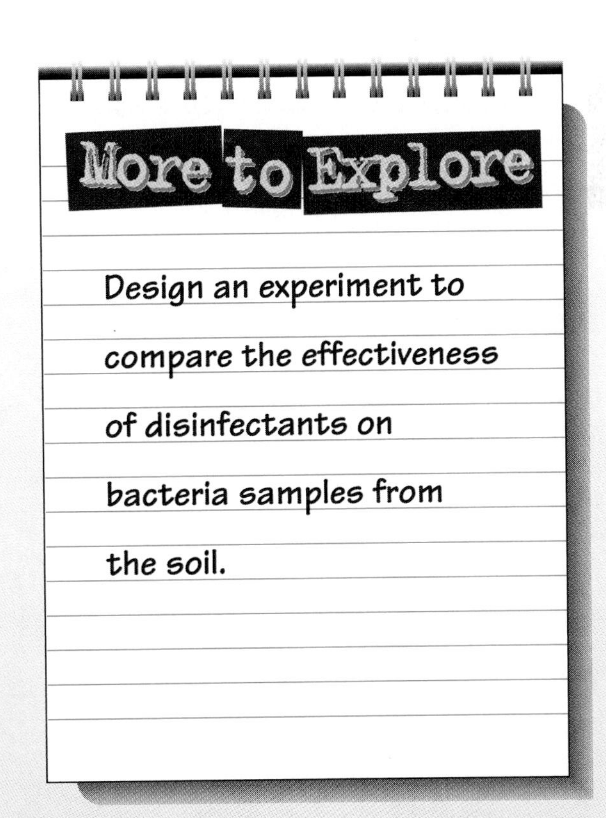

Study Guide

Summarizing Key Concepts

The key concepts in each section of this chapter are listed below to help you review the chapter content. Make sure you understand each concept and its relationship to other concepts and to the theme of this chapter.

40–1 Disease

- A disease is any change that disrupts the normal functions of the body.
- Pioneered by the French chemist Louis Pasteur and the German bacteriologist Robert Koch, the germ theory of disease suggested that infectious diseases were caused by microorganisms, which many people call germs.
- Koch's postulates include the following: The pathogen should always be found in the body of a sick organism and should not be found in a healthy organism. The pathogen must be isolated and grown in a pure culture. When purified pathogens are injected into a new host, they cause the disease. The same pathogen should be re-isolated from the second host and grown in a pure culture. The pathogen should still be the same as the original pathogen.

40–2 The Body's Defense System

- The immune system is the body's primary defense against pathogens. It consists of nonspecific and specific defenses against infection.
- Immunity is the ability of the body to resist a specific pathogen.

40–3 Cancer

- Cancer begins when a cell or a group of cells somehow escapes the body's normal growth controls.
- Some cancers are caused by biological agents such as viruses. Others are caused by chemicals in the environment or in the food we eat. And others have a physical cause—radiation.

Reviewing Key Terms

Review the following vocabulary terms and their meaning. Then use each term in a complete sentence.

40–1 Disease

pathogen
infectious disease
Koch's postulates
toxin
vector
antibiotic

40–2 The Body's Defense System

inflammatory response
phagocyte
immunity
antibody
antigen
lymphocyte
B cell
T cell
plasma cell
vaccine
killer T cell
cell-mediated immunity
helper T cell
suppressor T cell
allergy
histamine
asthma
autoimmune disease
AIDS
HIV

40–3 Cancer

tumor
carcinogen
chemotherapy

Recalling Main Ideas

Choose the letter of the answer that best completes the statement or answers the question.

1. Disease-causing organisms are called
   a. toxins. **c.** pathogens.
   b. bacteria. **d.** antigens.

2. Measles, smallpox, the common cold, and AIDS are caused by
   a. viruses. **c.** fungi.
   b. bacteria. **d.** parasitic worms.

3. Cholera, a disease caused by a bacterium, is spread by
   a. mosquitoes.
   b. contaminated water.
   c. fleas on rats and mice.
   d. hand-to-hand contact.

4. Antibiotics are drugs used against
   a. viruses. **c.** fungi.
   b. bacteria. **d.** AIDS.

5. Nonspecific defense mechanisms of the immune system include
   a. B lymphocytes. **c.** vaccination.
   b. antibodies. **d.** the skin and lysozymes.

6. Antibody production is stimulated by the presence of
   a. antigens. **c.** killer T cells.
   b. inflammation. **d.** phagocytes.

7. When antigens trigger mast cells,
   a. histamines are released.
   b. an allergic reaction occurs.
   c. the immune system overreacts.
   d. all of the above.

8. HIV weakens and gradually destroys
   a. mast cells. **c.** helper T cells.
   b. B cells. **d.** plasma cells.

9. Tumors capable of spreading from one part of the body to another are said to be
   a. malignant. **c.** benign.
   b. cellular. **d.** lung cancer.

10. All cancers seem to be caused by
    a. chemicals. **c.** radiation.
    b. mutations. **d.** viruses.

Putting It All Together

Using the information on pages xxx to xxxi, complete the following concept map.

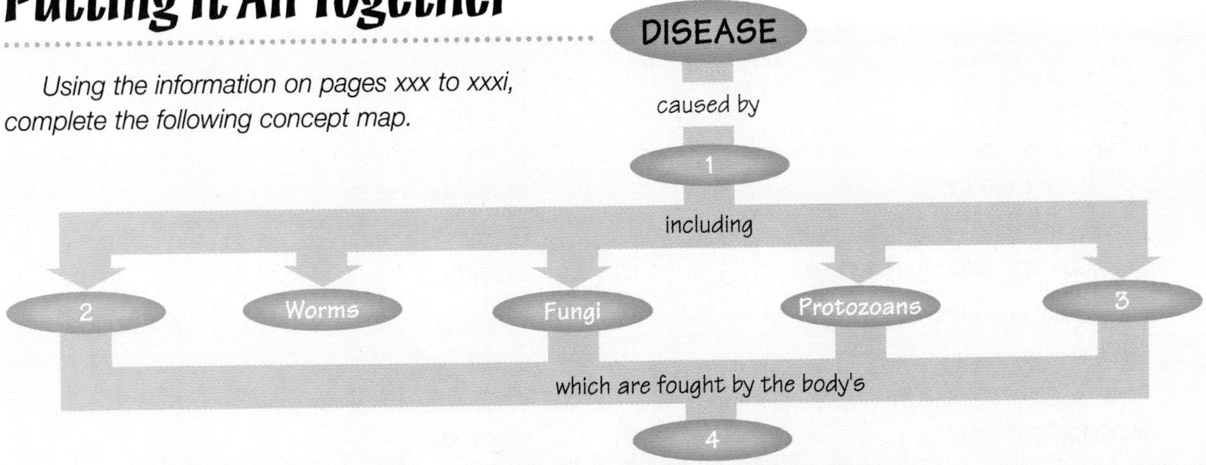

DISEASE
caused by
1
including
2
Worms
Fungi
Protozoans
3
which are fought by the body's
4

Reviewing What You Learned

Answer each of the following in a complete sentence.

1. What is a pathogen?
2. Explain the germ theory of disease.
3. List two diseases caused by protozoans.
4. What kind of pathogen causes tuberculosis?
5. Give some examples of the body's nonspecific defense against pathogens.
6. How does the inflammatory response fight infection?
7. What is immunity?
8. What are B cells? T cells?
9. Name the two types of immunity.
10. Describe the function of killer T cells.
11. What are autoimmune diseases?
12. What is HIV?
13. What is a benign tumor? A malignant tumor?
14. Name three ways in which cancer can be treated.
15. What is a carcinogen?

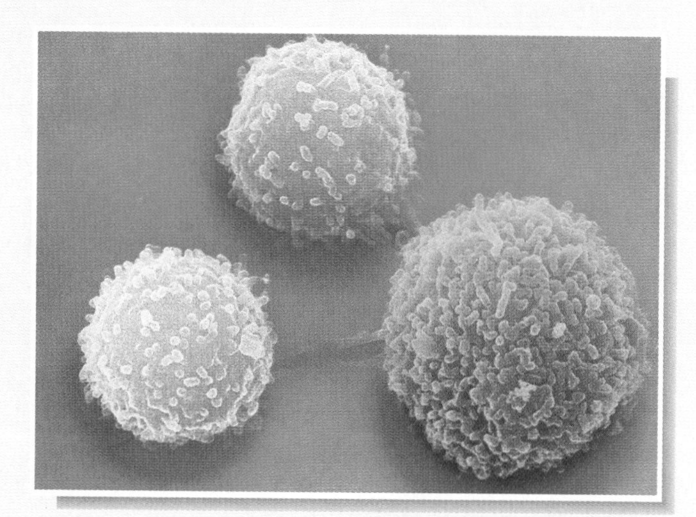

Expanding the Concepts

Discuss each of the following in a brief paragraph.

1. Explain some of the ways in which viruses differ from bacteria.
2. What are some ways in which humans encounter pathogens?
3. Explain the difference between nonspecific defenses and specific defenses.
4. How is inflammation of an injury part of the healing process?
5. How is cell-mediated immunity different from antibody immunity?
6. How does a vaccination protect you from a disease?
7. Explain why good personal hygiene is important to human health.
8. Explain the effect HIV has on the immune system.
9. Why is the cancer caused by smoking the most preventable type of cancer?
10. Why is cancer considered not one but many diseases?

Extending Your Thinking

Use the skills you have developed in this chapter to answer the following.

1. **Predicting** *E. coli* bacteria, which normally live in your large intestine, sometimes contaminate meat and cause food poisoning. Under ideal conditions, these bacteria can divide every 15 minutes. Starting with one bacterium, predict how many bacteria would be produced in 4 hours.
2. **Making judgments** The first vaccine was developed by Edward Jenner to fight smallpox. Jenner tested his theory of immunity on an eight-year-old boy. Do you think Jenner was right in doing so? Defend your answer.
3. **Applying concepts** Some scientists say that HIV has not been proven to cause AIDS. By referring to Koch's postulates, explain what they mean.
4. **Inferring** Because the immune system generally protects you with active immunity after an infection, why is it possible to get "the flu" year after year?
5. **Interpreting data** Examine the statistics for stomach cancer in the data table. How would you interpret the information found there?

STOMACH CANCER DEATH RATES*

Gender	1960–1962	1990–1992	Change
Male	16.2	6.7	−59%
Female	8.2	3.0	−63%

*per 100,000 U.S. population

Applying Your Skills

"This Will Only Hurt a Little"

When vaccinations are involved, a visit to the doctor's office is not always pleasant for a child. However, vaccinations are necessary in preventing many diseases.

1. Using library references, find out what vaccinations for children are recommended by the American Academy of Pediatrics.
2. Describe the pathogen and symptoms that each vaccine prevents.
3. Describe how each vaccine is made.
4. List any side effects associated with the vaccine.
5. Organize your data into a table.

• GOING FURTHER •

6. In your journal, discuss whether parents should be required to have their children vaccinated and who should pay for the vaccinations if the parents are unable to pay.

Periodic Table of the Elements

Key

6	Atomic number
C	Element symbol
Carbon	Element name
12.011	Atomic mass

Period	1 1A	2 2A	3 3B	4 4B	5 5B	6 6B	7 7B	8 8B	9 8B
1	1 H Hydrogen 1.00794								
2	3 Li Lithium 6.941	4 Be Beryllium 9.0122							
3	11 Na Sodium 22.990	12 Mg Magnesium 24.305							
4	19 K Potassium 39.098	20 Ca Calcium 40.08	21 Sc Scandium 44.956	22 Ti Titanium 47.88	23 V Vanadium 50.94	24 Cr Chromium 51.996	25 Mn Manganese 54.938	26 Fe Iron 55.847	27 Co Cobalt 58.9332
5	37 Rb Rubidium 85.468	38 Sr Strontium 87.62	39 Y Yttrium 88.9059	40 Zr Zirconium 91.224	41 Nb Niobium 92.91	42 Mo Molybdenum 95.94	43 Tc Technetium (98)	44 Ru Ruthenium 101.07	45 Rh Rhodium 102.906
6	55 Cs Cesium 132.91	56 Ba Barium 137.33	57 to 71	72 Hf Hafnium 178.49	73 Ta Tantalum 180.95	74 W Tungsten 183.85	75 Re Rhenium 186.207	76 Os Osmium 190.2	77 Ir Iridium 192.22
7	87 Fr Francium (223)	88 Ra Radium 226.025	89 to 103	104 Rf Rutherfordium (261)	105 Db Dubnium (262)	106 Sg Seaborgium (263)	107 Bh Bohrium (262)	108 Hs Hassium (265)	109 Mt Meitnerium (266)

57 La Lanthanum 138.906	58 Ce Cerium 140.12	59 Pr Praseodymium 140.908	60 Nd Neodymium 144.24	61 Pm Promethium (145)	62 Sm Samarium 150.36
89 Ac Actinium 227.028	90 Th Thorium 232.038	91 Pa Protactinium 231.036	92 U Uranium 238.029	93 Np Neptunium 237.048	94 Pu Plutonium (244)

Phase at 20°C		Metallic Properties	
C	Solid	Li	Metal
Br	Liquid	B	Semimetal
H	Gas	C	Nonmetal

10	11 1B	12 2B	13 3A	14 4A	15 5A	16 6A	17 7A	18 8A
								2 He Helium 4.003
			5 B Boron 10.81	6 C Carbon 12.011	7 N Nitrogen 14.007	8 O Oxygen 15.999	9 F Fluorine 18.998	10 Ne Neon 20.179
			13 Al Aluminum 26.98	14 Si Silicon 28.086	15 P Phosphorus 30.974	16 S Sulfur 32.06	17 Cl Chlorine 35.453	18 Ar Argon 39.948
28 Ni Nickel 58.69	29 Cu Copper 63.546	30 Zn Zinc 65.39	31 Ga Gallium 69.72	32 Ge Germanium 72.59	33 As Arsenic 74.922	34 Se Selenium 78.96	35 Br Bromine 79.904	36 Kr Krypton 83.80
46 Pd Palladium 106.42	47 Ag Silver 107.868	48 Cd Cadmium 112.41	49 In Indium 114.82	50 Sn Tin 118.71	51 Sb Antimony 121.75	52 Te Tellurium 127.60	53 I Iodine 126.905	54 Xe Xenon 131.29
78 Pt Platinum 195.08	79 Au Gold 196.967	80 Hg Mercury 200.59	81 Tl Thallium 204.383	82 Pb Lead 207.2	83 Bi Bismuth 208.98	84 Po Polonium (209)	85 At Astatine (210)	86 Rn Radon (222)
110 Uun Ununnilium (269)	111 Uuu Unununium (272)	112 Uub Ununbium (277)						

The names of elements 104-108 are under dispute. This table provides proposed names from the International Union of Pure and Applied Chemistry (IUPAC).

Mass numbers in parentheses are those of the most stable or common isotope.

63 Eu Europium 151.96	64 Gd Gadolinium 157.25	65 Tb Terbium 158.925	66 Dy Dysprosium 162.50	67 Ho Holmium 164.93	68 Er Erbium 167.26	69 Tm Thulium 168.934	70 Yb Ytterbium 173.04	71 Lu Lutetium 174.967
95 Am Americium (243)	96 Cm Curium (247)	97 Bk Berkelium (247)	98 Cf Californium (251)	99 Es Einsteinium (252)	100 Fm Fermium (257)	101 Md Mendelevium (258)	102 No Nobelium (259)	103 Lr Lawrencium (260)

Care and Use of the Microscope

THE COMPOUND MICROSCOPE

One of the most essential tools in the study of biology is the microscope. With the help of different types of microscopes, biologists have developed detailed concepts of cell structure and function. The type of microscope used in most biology classes is the compound microscope. It contains a combination of lenses and can magnify objects normally unseen with the unaided eye.

The eyepiece lens is located in the top portion of the microscope. This lens usually has a magnification of 10×. A compound microscope usually has two other interchangeable lenses. These lenses, called objective lenses, are at the bottom of the body tube on the revolving nosepiece. By revolving the nosepiece, either of the objectives can be brought into direct line with the body of the tube.

The shorter objective is of low power in its magnification, usually 10×. The longer one is of high power, usually 40× or 43×. The magnification is always marked on the objective. To determine the total magnification of a microscope, multiply the magnifying power of the eyepiece by the magnifying power of the objective being used. For example, the eyepiece magnifying power, 10×, multiplied by the low-power objective, 10×, equals 100×. The total magnification is 100×.

A microscope also produces clear contrasts to enable the viewer to distinguish between objects that lie very close together. Under a microscope the detail of objects is very sharp. The ability of a microscope to produce contrast and detail is called resolution, or resolving power. Although microscopes can have the same magnifying power, they can differ in resolving power.

Learning the name, function, and location of each of the microscope's parts is necessary for proper use. Use the following procedures when working the microscope:

1. Remove the microscope from its storage area by placing one hand beneath the base and grasping the arm of the microscope with the other hand.
2. Gently place the microscope on the lab table with the arm facing you. The microscope's base should be resting evenly on the table, approximately 10 centimeters from the table's edge.
3. Raise the body tube by turning the coarse adjustment knob until the objective lens is about 2 centimeters above the opening of the stage.
4. Revolve the nosepiece so that the low-power objective (10×) is directly in line with the body tube. A click indicates that the lens is in line with the opening of the stage.
5. Look through the eyepiece and switch on the lamp or adjust the mirror so that a circle of light can be seen. This is the field of view. Moving the lever of the diaphragm permits a greater or smaller amount of light to come through the opening of the stage.
6. Place a prepared slide on the stage. Place the specimen over the center of the opening of the stage. Fasten the stage clip to hold the slide in position.
7. Look at the microscope from the side. Carefully turn the coarse adjustment knob to lower the body tube until the low-power objective almost touches

MICROSCOPE PARTS AND THEIR FUNCTION

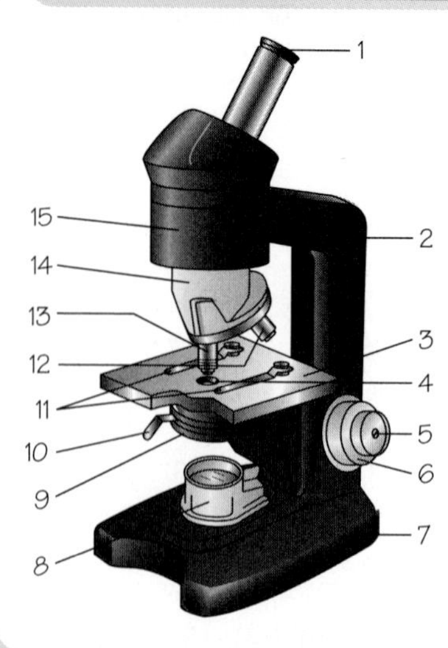

1. ***Eyepiece*** Contains a magnifying lens
2. ***Arm*** Supports the body tube
3. ***Stage*** Supports the slide being observed
4. ***Opening of the stage*** Permits light to travel up to the eyepiece
5. ***Fine adjustment*** Moves the body tube slightly to sharpen the focus
6. ***Coarse adjustment*** Moves the body tube up and down for focusing
7. ***Base*** Supports the microscope
8. ***Illuminator*** Produces light or reflects light up through the body tube
9. ***Diaphragm*** Regulates the amount of light entering the body tube
10. ***Diaphragm lever*** Opens and closes the diaphragm
11. ***Stage clips*** Hold the slide in position
12. ***Low-power objective*** Provides a magnification of 10× and is the shorter of the objectives
13. ***High-power objective*** Provides a magnification of 43× and is the longer of the objectives
14. ***Revolving nosepiece*** Contains the low- and high-power objectives and can be rotated to change magnification
15. ***Body tube*** Maintains a proper distance between the eyepiece and the objective lenses

the slide or until the body tube can no longer be moved. Do not allow the objective to touch the slide.

8. Look through the eyepiece and observe the specimen. If the field of view is out of focus, use the coarse adjustment knob to raise the body tube while looking through the eyepiece. When the specimen comes into view, use the fine adjustment knob to focus the specimen. Be sure to keep both eyes open when viewing a specimen. This helps prevent eyestrain.

9. Adjust the lever of the diaphragm to allow the right amount of light to enter.

10. To view the specimen under high power (43×), revolve the nosepiece until the high-power objective is in line with the body tube and clicks into place.

11. Look through the eyepiece and use the fine adjustment knob to bring the specimen into focus.

12. After every use remove the slide. Clean the stage of the microscope and the lenses with lens paper. Do not use other types of paper to clean the lenses, as they may scratch the lenses.

PREPARING A WET-MOUNT SLIDE

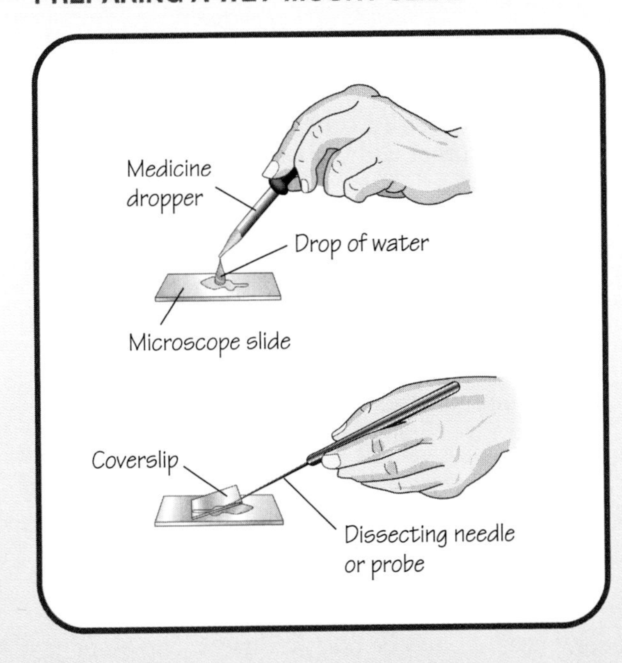

1. Obtain a clean microscope slide and a coverslip. A coverslip is very thin, permitting the objective lens to be lowered close to the specimen.

2. Place the specimen in the middle of the microscope slide. The specimen must be thin enough for light to pass through it.

3. Using a medicine dropper, place a drop of water on the specimen.

4. Lower one edge of the coverslip so that it touches the side of the drop of water at a 45-degree angle. The water will spread evenly along the edge of the coverslip. Using a dissecting needle or probe, slowly lower the coverslip over the specimen and water. Try not to trap any air bubbles under the coverslip because they will interfere with the view of the specimen. If air bubbles are present, gently tap the surface of the coverslip over the air bubble with a pencil eraser.

5. Remove any excess water at the edge of the coverslip with a paper towel. If the specimen begins to dry out, add a drop of water at the edge of the coverslip.

STAINING TECHNIQUES

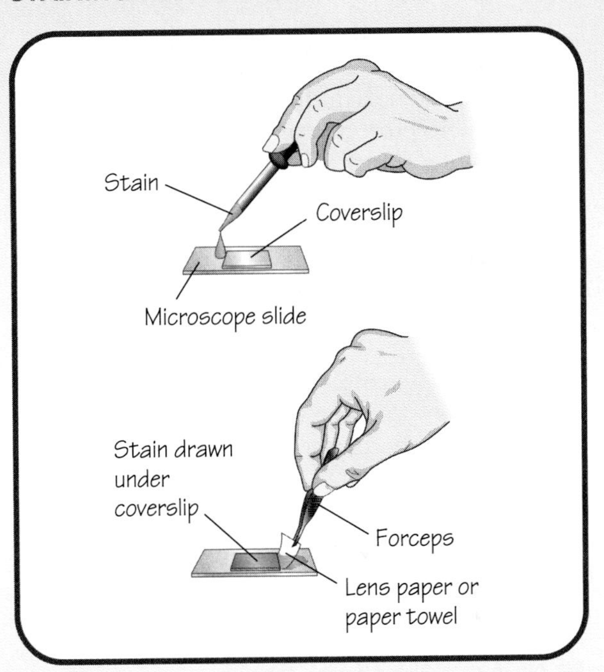

1. Obtain a clean microscope slide and coverslip.

2. Place the specimen in the middle of the microscope slide.

3. Using a medicine dropper, place a drop of water on the specimen.

4. Place one edge of the coverslip so that it touches the side of the drop of water at a 45-degree angle. After the water spreads along the edge of the coverslip, use a dissecting needle or probe to lower the coverslip over the specimen.

5. Add a drop of stain at the edge of the coverslip. Using forceps, touch a small piece of lens paper or paper towel to the opposite edge of the coverslip. The paper causes the stain to be drawn under the coverslip and stain the cells. Some common stains are methylene blue, iodine, fuchsin, and Wright's.

Science Safety Rules

ONE OF THE FIRST THINGS A SCIENTIST *learns is that working in the laboratory can be an exciting experience. But the laboratory can also be quite dangerous if proper safety rules are not followed at all times. To prepare yourself for a safe year in the laboratory, read over the following safety rules. Then read them a second time. Make sure you understand each rule. Ask your teacher to explain any rules you don't understand.*

DRESS CODE

1. Many materials in the laboratory can cause eye injury. To protect yourself from possible injury, wear safety goggles whenever you are working with chemicals, burners, or any substance that might get into your eyes. Never wear contact lenses in the laboratory.
2. Wear a laboratory apron or coat whenever you are working with any chemicals or heated substances.
3. Tie back long hair to keep it away from any chemicals, burners, candles, or any other laboratory equipment.
4. Before working in the laboratory, remove or tie back any article of clothing or jewelry that could hang down and touch chemicals and flames.

GENERAL SAFETY RULES

5. Read all directions for an experiment several times. Then follow the directions exactly as they are written. If you are in doubt about any part of the experiment, ask your teacher for assistance.
6. Never perform investigations that are not authorized by your teacher. Obtain permission before "experimenting" on your own.
7. Never handle any equipment unless you have specific permission.
8. Take extreme care not to spill any material in the laboratory. If spills occur, ask your teacher immediately about the proper cleanup procedure. Never simply pour chemicals or other substances into the sink or trash container.
9. Never eat in the laboratory.

FIRST AID

10. Report all accidents, no matter how minor, to your teacher immediately.
11. Learn what to do in case of specific accidents, such as getting acid in your eyes or on your skin. (Rinse any acids that may splash on your body with lots of water.)
12. Know the location of the first-aid kit. Your teacher should administer any required first aid due to injury. Or your teacher may send you to the school nurse or call a physician.
13. Know where and how to report an accident or fire. Find out the location of the fire extinguisher, phone, and fire alarm. Keep a list of important phone numbers, such as those for the fire department and school nurse, near the phone. Report any fires to your teacher at once.

HEATING AND FIRE SAFETY

14. Again, never use a heat source such as a candle or burner without wearing safety goggles.
15. Never heat a chemical you are not instructed to heat. A chemical that is harmless when cool can be dangerous when heated.
16. Maintain a clean work area and keep all materials away from flames.
17. Do not plug too many devices into one socket. Never touch electrical devices with wet hands.
18. Make sure you know how to light a Bunsen burner. (Your teacher will demonstrate the proper procedure for lighting a burner.) If the flame leaps out of a burner toward you, turn the gas off immediately. Do not touch the burner. It may be hot. And never leave a lighted burner unattended!
19. When heating a test tube or bottle, point it away from yourself and others. Chemicals can splash or boil out of a heated test tube.
20. Never heat a liquid in a closed container. The expanding gases produced may blow the container apart, causing injury to yourself or others. Never reach across a flame.

21. Never pick up a container that has been heated without first holding the back of your hand near it. If you can feel the heat on the back of your hand, the container may be too hot to handle. Use tongs or heat-proof gloves when handling hot containers.

USING CHEMICALS SAFELY

22. Never mix chemicals "for the fun of it." You might produce a dangerous, possibly explosive substance.

23. Never touch, taste, or smell any chemicals in the laboratory. Many chemicals are poisonous. If you are instructed to note the fumes in an experiment, gently wave your hand over the opening of a container and direct the fumes toward your nose. Do not inhale the fumes directly from the container.

24. Use only those chemicals needed in the investigation. Keep all lids closed when a chemical is not being used. Notify your teacher whenever chemicals are spilled.

25. Dispose of all chemicals as instructed by your teacher. To avoid contamination, never return chemicals to their original containers.

26. Be extra careful when working with acids or bases. Pour such chemicals over the sink, not over your workbench.

27. When diluting an acid, pour the acid into water. Never pour water into the acid.

28. If any acids get on your skin or clothing, rinse them off with water. Immediately notify your teacher of any acid spill.

USING GLASSWARE SAFELY 

29. Never force glass tubing into a rubber stopper. A turning motion and lubricant will be helpful when inserting glass tubing into rubber stoppers or rubber tubing. Your teacher will demonstrate the proper way to insert glass tubing.

30. Never heat glassware that is not thoroughly dry. Use a wire screen to protect glassware from any flame.

31. Keep in mind that hot glassware will not appear hot. Never pick up glassware that is not thoroughly cooled.

32. If you are instructed to cut glass tubing, fire-polish the ends immediately to remove sharp edges.

33. Never use broken or chipped glassware. If any glassware breaks, notify your teacher and dispose of the glassware in the proper trash container.

34. Never eat or drink from laboratory glassware. Thoroughly clean glassware before putting it away.

USING SHARP INSTRUMENTS 

35. Handle scalpels or razor blades with extreme care. Never cut material toward you; cut away from you.

36. Notify your teacher immediately if you cut yourself when in the laboratory.

ANIMAL SAFETY 

37. No experiments that will cause pain, discomfort, or harm to mammals, birds, reptiles, fishes, and amphibians should be done in the classroom or at home.

38. Animals should be handled only if necessary. If an animal is excited or frightened, pregnant, feeding, or with its young, special handling is required.

39. Your teacher will instruct you as to how to handle each animal species that may be brought into the classroom.

40. Clean your hands thoroughly after handling animals or the cage containing animals.

END-OF-EXPERIMENT RULES 

41. When an experiment is completed, clean up your work area and return all equipment to its proper place.

42. Wash your hands before and after you perform every experiment.

43. Turn off all burners before leaving the laboratory. Check that the gas line leading to the burner is off as well.

Metric System

THE METRIC SYSTEM OF MEASUREMENT *is used by scientists throughout the world. It is based on units of ten. Each unit is ten times larger or ten times smaller than the next unit. The most commonly used units of the metric system are given below. After you have finished reading about the metric system, try to put it to use. How tall are you in meters? What is your mass? What is your normal body temperature in degrees Celsius?*

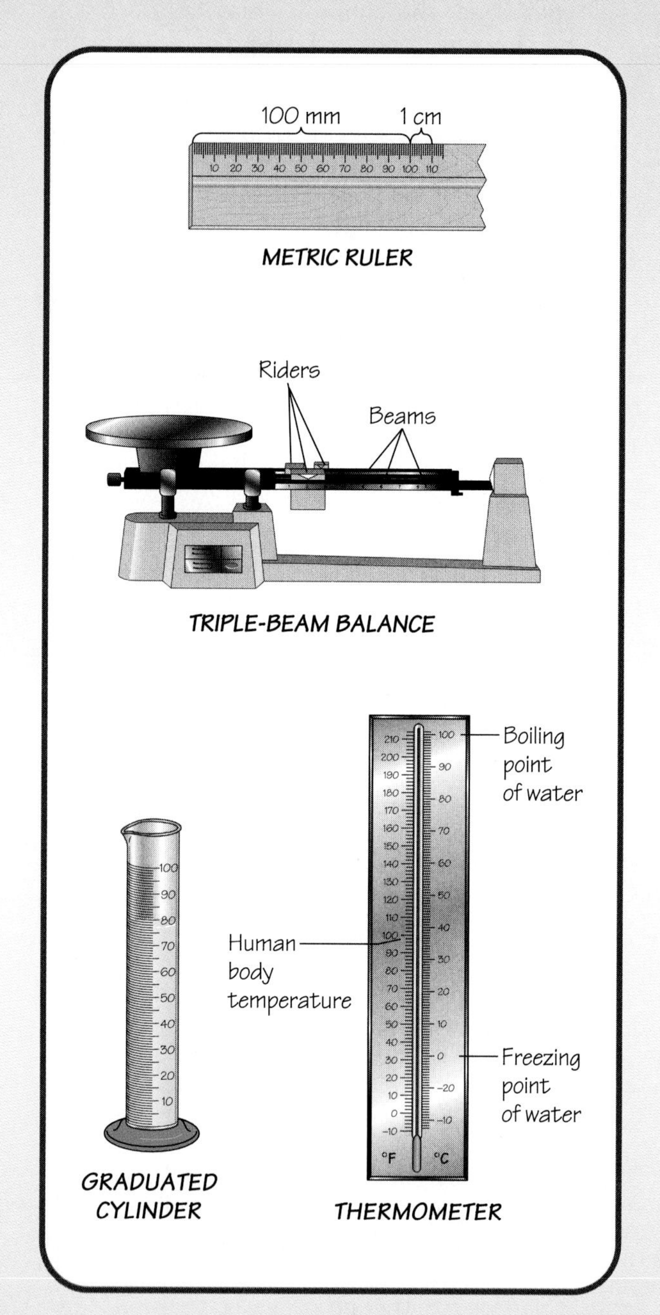

METRIC RULER

TRIPLE-BEAM BALANCE

GRADUATED CYLINDER

THERMOMETER

COMMONLY USED METRIC UNITS

Length The distance from one point to another

meter (m)	A meter is slightly longer than a yard. 1 meter = 1000 millimeters (mm) 1 meter = 100 centimeters (cm) 1000 meters = 1 kilometer (km)

Volume The amount of space an object takes up

liter (L)	A liter is slightly more than a quart. 1 liter = 1000 milliliters (mL)

Mass The amount of matter in an object

gram (g)	A gram has a mass equal to about one paper clip. 1000 grams = 1 kilogram (kg)

Temperature The measure of hotness or coldness

degrees Celsius (°C)	0°C = freezing point of water 100°C = boiling point of water

METRIC—ENGLISH SYSTEM EQUIVALENTS

2.54 centimeters (cm) = 1 inch (in.)
1 meter (m) = 39.37 inches (in.)
1 kilometer (km) = 0.62 miles (mi)
1 liter (L) = 1.06 quarts (qt)
250 milliliters (mL) = 1 cup (c)
1 kilogram (kg) = 2.2 pounds (lb)
28.3 grams (g) = 1 ounce (oz)
$°C = 5/9 \times (°F - 32)$

Six-Kingdom Classification System

Kingdom Eubacteria

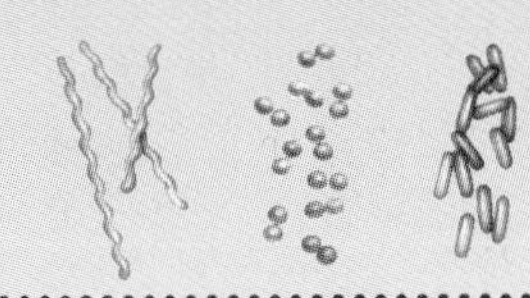

Outer cell wall contains complex carbohydrates; all species have at least one inner cell membrane; live in diverse environments.

PHYLUM CYANOBACTERIA (blue-green bacteria)
Photosynthetic autotrophs, once called blue-green algae; contain pigments phycocyanin and chlorophyll *a*; some fix atmospheric nitrogen. Examples: *Anabaena*, *Nostoc*.

PHYLUM PROCHLOROBACTERIA
Photosynthetic autotrophs containing chlorophylls *a* and *b*; strikingly similar to chloroplasts; few species identified to date. Example: *Prochloron*.

Kingdom Archaebacteria

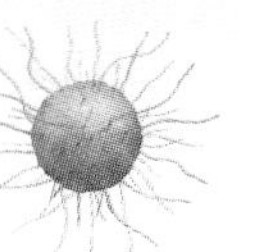

Single-celled prokaryotic organism; cell membrane contains lipids not found in any other organism; many survive in absence of oxygen; known as methanogens and produce methane gas; live in harsh environments. Examples: *Thermoplasma*, *Sulfolobus*.

Kingdom Protista

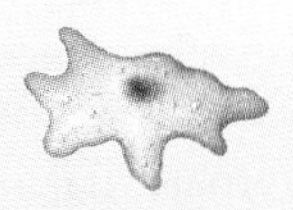

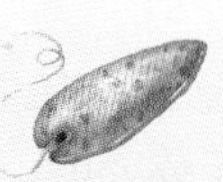

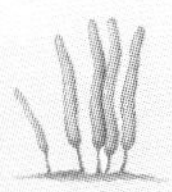

Eukaryotic; usually unicellular; some multicellular or colonial; heterotrophic or autotrophic organisms.

ANIMALLIKE PROTISTS Unicellular; heterotrophic; usually motile; also known as protozoa.

PHYLUM CILIOPHORA (ciliates)
All have cilia at some point in development; almost all use cilia to move; characterized by two types of nuclei: macronuclei and micronuclei; most have a sexual process known as conjugation. Examples: *Paramecium*, *Didinium*, *Stentor*.

PHYLUM ZOOMASTIGINA (animallike flagellates)
Possess one or more flagella (some have thousands); some are internal symbionts of wood-eating animals. Examples: *Trypanosoma*, *Trichonympha*.

PHYLUM SPOROZOA
Nonmotile parasites; produce small infective cells called spores; life cycles usually complex, involving more than one host species; cause a number of diseases, including malaria. Example: *Plasmodium*.

PHYLUM SARCODINA
Use pseudopods for feeding and movement; some produce elaborate shells that contain silica or calcium carbonate; most free-living; a few parasitic. Examples: *Ameba*, *foraminifers*.

PLANTLIKE PROTISTS Mostly unicellular photosynthetic autotrophs that have characteristics similar to green plants or fungi. A few species are multicellular or heterotrophic.

PHYLUM EUGLENOPHYTA (plantlike flagellates)
Primarily photosynthetic; most live in fresh water; possess two unequal flagella; lack a cell wall. Example: *Euglena*.

PHYLUM PYRROPHYTA (fire algae)
Two flagella; most live in salt water, are photosynthetic, and have a rigid cell wall that contains cellulose; some are luminescent; many are symbiotic. Examples: *Gonyaulax*, *Noctiluca*.

PHYLUM CHRYSOPHYTA (golden algae)
Photosynthetic; aquatic; mostly unicellular; contain yellow-brown pigments; most are diatoms, which build a two-part cell covering that contains silica. Examples: *Thallasiosira*, *Orchromonas*, *Navicula*.

PHYLUM CHLOROPHYTA (green algae)
Live in fresh and salt water; unicellular or multicellular; chlorophylls and accessory pigments similar to those in vascular plants; food stored as starch. Examples: *Ulva* (sea lettuce), *Chlamydomonas*, *Spirogyra*, *Acetabularia*.

PHYLUM PHAEOPHYTA (brown algae)
Live almost entirely in salt water; multicellular; contain brown pigment fucoxanthin; food stored as oils and carbohydrates. Examples: *Fucus* (rockweed), kelps, *Sargassum*.

PHYLUM RHODOPHYTA (red algae)
Live almost entirely in salt water; multicellular; contain red pigment phycoerythrin; food stored as carbohydrates. Examples: *Chondrus* (Irish moss), coralline algae.

FUNGUSLIKE PROTISTS Lack chlorophyll and absorb food through cell walls. Unlike true fungi, the funguslike protists contain centrioles and lack chitin in their cell walls.

PHYLUM ACRASIOMYCOTA (cellular slime molds)
Spores develop into independent, free-living amebalike cells that may come together to form a multicellular structure; this structure, which behaves much like a single organism, forms a fruiting body that produces spores. Example: *Dictyostelium*.

PHYLUM MYXOMYCOTA (acellular slime molds)
Spores develop into haploid cells that can switch between flagellated and amebalike forms; these haploid cells fuse to form a zygote that grows into a plasmodium, which ultimately forms spore-producing fruiting bodies. Example: *Physarum*.

PHYLUM OOMYCOTA (water molds)
Unicellular or multicellular; mostly aquatic; cell walls contain cellulose or a polysaccharide similar to cellulose; form zoospores asexually and eggs and sperms sexually. Example: *Saprolegnia* (freshwater molds).

KINGDOM FUNGI

Eukaryotic; unicellular or multicellular; cell walls typically contain chitin; mostly decomposers; some parasites; some commensal or mutualistic symbionts; asexual reproduction by spore formation, budding, or fragmentation; sexual reproduction involving mating types; classified according to type of fruiting body and style of spore formation; heterotrophic.

PHYLUM ZYGOMYCOTA (conjugation fungi)
Cell walls of chitin; hyphae lack cross walls; sexual reproduction by conjugation produces diploid zygospores; asexual reproduction produces haploid spores; most parasites; some decomposers. Example: *Rhizopus stolonifer* (black bread mold).

PHYLUM ASCOMYCOTA (sac fungi)
Cell walls of chitin; hyphae have perforated cross walls; most multicellular; yeasts unicellular; sexual reproduction produces ascospores; asexual reproduction by spore formation or budding; some cause plant diseases such as chestnut blight and Dutch elm disease. Examples: *Neurospora* (red bread mold), baker's yeast, morels, truffles.

PHYLUM BASIDIOMYCOTA (club fungi)
Cell walls of chitin; hyphae have cross walls; sexual reproduction involves basidiospores, which are borne on club-shaped basidia; asexual reproduction by spore formation. Examples: most mushrooms, puffballs, shelf fungi, rusts.

PHYLUM DEUTEROMYCOTA (imperfect fungi)
Cell walls of chitin; sexual reproduction never observed; members resemble ascomycetes, basidiomycetes, or zygomycetes; most are thought to be ascomycetes that have lost the ability to form asci. Examples: *Penicillium*, athlete's foot fungus.

KINGDOM PLANTAE

Eukaryotic; overwhelmingly multicellular and nonmotile; photosynthetic autotrophs; possess chlorophylls *a* and *b* and other pigments in organelles called chloroplasts; cell walls contain cellulose; food stored as starch; reproduce sexually; alternate haploid (gametophyte) and diploid (sporophyte) generations; botanists typically use the term division rather than phylum.

DIVISION BRYOPHYTA (bryophytes)
Generally small, multicellular green plants; live on land in moist habitats; lack vascular tissue; lack true roots, leaves, and stems; gametophyte dominant; water required for reproduction. Examples: mosses, liverworts, hornworts.

DIVISION PSILOPHYTA (whisk ferns)
Primitive vascular plants; no differentiation between root and shoot; produce only one kind of spore; motile sperm must swim in water.

DIVISION LYCOPHYTA (lycopods)

Primitive vascular plants; usually small; sporophyte dominant; possess roots, stems, and leaves; water required for reproduction. Examples: club mosses, quillworts.

DIVISION SPHENOPHYTA (horsetails)

Primitive vascular plants; stem comprises most of mature plant and contains silica; produce only one kind of spore; motile sperm must swim in water. Only one living genus: *Equisetum*.

DIVISION PTEROPHYTA (ferns)

Vascular plants well-adapted to live in predominantly damp or seasonally wet environments; sporophyte dominant and well-adapted to terrestrial life; gametophyte inconspicuous; reproduction still dependent on water for free-swimming gametes. Examples: cinnamon ferns, Boston ferns, tree ferns, maidenhair ferns.

GYMNOSPERMS Four divisions of seed plants—Cycadophyta, Ginkgophyta, Coniferophyta, and Gnetophyta. Gymnosperms are characterized by seeds that develop exposed, or "naked," on fertile leaves—there is no ovary wall (fruit) surrounding the seeds. Gymnosperms lack flowers.

DIVISION CYCADOPHYTA (cycads)

Evergreen, slow-growing, tropical and subtropical shrubs; many resemble small palm trees; palmlike or fernlike compound leaves; possess symbiotic cyanobacteria in special roots; sexes are separate—individuals have either male pollen-producing cones or female seed-producing cones.

DIVISION GINKGOPHYTA

Deciduous trees with fan-shaped leaves; sexes separate; outer skin of ovule develops into a fleshy, fruitlike covering. Only one living species: *Ginkgo biloba* (ginkgo).

DIVISION GNETOPHYTA

Few species; live mostly in deserts; functional xylem cells are alive; sexes are separate. Examples: *Welwitschia*, *Ephedra* (joint fir).

DIVISION CONIFEROPHYTA (conifers)

Cones predominantly wind-pollinated; most are evergreen; most temperate and subarctic shrubs and trees; many have needlelike leaves; in most species, sexes are not separate. Examples: pines, spruces, cedars, firs.

DIVISION ANTHOPHYTA

Members of this division are commonly called angiosperms, or flowering plants. Nearly all familiar trees, shrubs, and garden plants are angiosperms. Seeds develop enclosed within ovaries; fertile leaves modified into flowers; flowers pollinated by wind or by animals, including insects, birds, and bats; occur in many different forms; found in most land and freshwater habitats; a few species found in shallow saltwater and estuarine areas.

Class Monocotyledonae (monocots) Embryo with a single cotyledon; leaves with predominantly parallel venation; flower parts in threes or multiples of three; vascular bundles scattered throughout stem. Examples: lilies, corns, grasses, irises, palms, tulips.

Class Dicotyledonae (dicots) Embryo with two cotyledons; leaves with variation in netlike patterns; flower parts in fours or fives (or multiples thereof); vascular bundles arranged in rings in stem. Examples: roses, maples, oaks, daisies, apples.

KINGDOM ANIMALIA

Multicellular; eukaryotic; typical heterotrophs that ingest their food; lack cell walls; approximately 35 phyla; in most phyla, cells are organized into tissues that make up organs; most reproduce sexually; motile sperm have flagella; nonmotile egg is much larger than sperm; development involves formation of a hollow ball of cells called a blastula.

SUBKINGDOM PARAZOA

Animals that possess neither tissues nor organs; most asymmetrical.

PHYLUM PORIFERA (sponges)

Aquatic; lack true tissues and organs; motile larvae and sessile adults; filter feeders; internal skeleton made up of spongin and/or spicules of calcium carbonate or silica. Examples: Venus' flower baskets, bath sponges, tube sponges.

SUBKINGDOM METAZOA

Animals with definite symmetry; definite tissues; most possess organs.

PHYLUM CNIDARIA

Previously known as coelenterates; aquatic; mostly carnivorous; two layers of true tissue; radial symmetry; tentacles bear stinging nematocysts; many alternate between polyp and medusa body forms; gastrovascular cavity.

Class Hydrozoa Polyp form dominant; colonial or solitary; life cycle typically includes a medusa generation that reproduces sexually and a polyp generation that reproduces asexually. Examples: hydras, Portuguese man-of-wars.

Class Scyphozoa Medusa form dominant; some species bypass polyp stage. Examples: lion's mane jellyfishes, moon jellies, sea wasps.

Class Anthozoa Colonial or solitary polyps; no medusa stage. Examples: reef corals, sea anemones, sea pens, sea fans.

PHYLUM PLATYHELMINTHES (flatworms)

Three layers of tissue (endoderm, mesoderm, ectoderm); bilateral symmetry; some cephalization; acoelomate; free-living or parasitic.

Class Turbellaria Free-living carnivores and scavengers, live in fresh water, in salt water, or on land; move with cilia. Example: planarians.

Class Trematoda (flukes) Parasites; life cycle typically involves more than one host. Examples: *Schistosoma*, liver flukes.

Class Cestoda (tapeworms) Internal parasites; lack digestive tract; body composed of many repeating sections (proglottids).

PHYLUM NEMATODA (roundworms)

Digestive system has two openings—a mouth and an anus; pseudocoelomates. Examples: *Ascaris lumbricoides* (human ascarid), hookworms, *Trichinella*.

PHYLUM MOLLUSCA (mollusks)

Soft-bodied; usually (but not always) possess a hard, calcified shell secreted by a mantle; most adults have bilateral symmetry; muscular foot; divided into seven classes; digestive system with two openings; coelomates.

Class Pelecypoda (bivalves) Two-part hinged shell; wedge-shaped foot; typically sessile as adults; primarily aquatic; some burrow in mud or sand. Examples: clams, oysters, scallops, mussels.

Class Gastropoda (gastropods) Use broad, muscular foot in movement; most have spiral, chambered shell; some lack shell; distinct head; some terrestrial, others aquatic; many are cross-fertilizing hermaphrodites. Examples: snails, slugs, nudibranchs, sea hares, sea butterflies.

Class Cephalopoda (cephalopods) Foot divided into tentacles; live in salt water; closed circulatory system; sexes separate. Examples: octopuses, squids, nautiluses, cuttlefishes.

PHYLUM ANNELIDA (segmented worms)

Body composed of segments separated by internal partitions; digestive system has two openings; coelomate; closed circulatory system.

Class Polychaeta (polychaetes) Live in salt water; pair of bristly, fleshy appendages on each segment; some live in tubes. Examples: sandworms, bloodworms, fanworms, feather-duster worms, plume worms.

Class Oligochaeta (oligochaetes) Lack appendages; few bristles; terrestrial or aquatic. Examples: *Tubifex*, earthworms.

Class Hirudinea (leeches) Lack appendages; carnivores or blood-sucking external parasites; most live in fresh water. Example: *Hirudo medicinalis* (medicinal leech).

PHYLUM ARTHROPODA (arthropods)

Exoskeleton of chitin; jointed appendages; segmented body; many undergo metamorphosis during development; open circulatory system; ventral nerve cord; largest animal phylum.

Subphylum Trilobita (trilobites) Two furrows running from head to tail divide body into three lobes; one pair of unspecialized appendages on each body segment; each appendage divided into two branches—a gill and a walking leg; all extinct.

Subphylum Chelicerata (chelicerates) First pair of appendages specialized as feeding structures called chelicerae; body composed of two parts—cephalothorax and abdomen; lack antennae; most terrestrial. Examples: horseshoe crabs, ticks, mites, spiders, scorpions.

Subphylum Crustacea (crustaceans) Most aquatic; most live in salt water; two pairs of antennae; mouthparts called mandibles; appendages consist of two branches; many have a carapace that covers part or all of the body. Examples: crabs, crayfishes, pill bugs, water fleas, barnacles.

Subphylum Uniramia Almost all terrestrial; one pair of antennae; mandibles; unbranched appendages; generally divided into five classes.

Class Chilopoda (centipedes) Long body consisting of many segments; one pair of legs per segment; poison claws for feeding; carnivorous.

Class Diplopoda (millipedes) Long body consisting of many segments; two pairs of legs per segment; mostly herbivorous.

Class Insecta (insects) Body divided into three parts—head, thorax, and abdomen; three pairs of legs and usually two pairs of wings attached to thorax; some undergo complete metamorphosis; approximately 25 orders. Examples: termites, ants, beetles, flies, moths, grasshoppers.

PHYLUM ECHINODERMATA (echinoderms)

Live in salt water; larvae have bilateral symmetry; adults typically have five-part radial symmetry; endoskeleton; tube feet; water vascular system used in respiration, excretion, feeding, and locomotion.

Class Crinoidea (crinoids) Filter feeders; feathery arms; mouth and anus on upper surface of body disk; some sessile. Examples: sea lilies, feather stars.

Class Asteroidea (starfishes) Star-shaped; carnivorous; bottom dwellers; mouth on lower surface. Examples: crown-of-thorns starfishes, sunstars.

Class Ophiuroidea Small body disk; long armored arms; most have only five arms; lack an anus; most are filter feeders or detritus feeders. Examples: brittle stars, basket stars.

Class Echinoidea Lack arms; body encased in rigid, boxlike covering; covered with spines; most grazing herbivores or detritus feeders. Examples: sea urchins, sand dollars.

Class Holothuroidea (sea cucumbers) Cylindrical body with feeding tentacles on one end; lie on their side; mostly detritus or filter feeders; endoskeleton greatly reduced.

PHYLUM CHORDATA (chordates)

Notochord and pharyngeal gill slits during at least part of development; hollow dorsal nerve cord.

Subphylum Urochordata (tunicates) Live in salt water; tough outer covering (tunic); display chordate features during larval stages; many adults sessile, some free-swimming. Examples: sea squirts, sea peaches.

Subphylum Cephalochordata (lancelets) Fishlike; live in salt water; filter feeders; no internal skeleton. Example: *Branchiostoma*.

Subphylum Vertebrata Most possess a vertebral column (backbone) that supports and protects dorsal nerve cord; endoskeleton; distinct head with a skull and brain.

Jawless Fishes Characterized by long eellike body and a circular mouth; two-chambered heart; lack scales, paired fins, jaws, and bones; ectothermic; possess a notochord as adults. Once considered a single class—Agnatha—jawless fishes are now divided into two classes: Myxini and Cephalaspidomorphi. Although the term agnatha no longer refers to a true taxonomic group, it is still used informally.

Class Myxini (hagfishes) Mostly scavengers; live in salt water; short tentacles around mouth; rasping tongue; extremely slimy; open circulatory system.

Class Cephalaspidomorphi (lampreys) Larvae filter feeders; adults are parasites whose circular mouth is lined with rasping, toothlike structures; many live in both salt water and fresh water during the course of their lives.

Class Chondrichthyes (cartilaginous fishes) Jaw; fins; endoskeleton of cartilage; most live in salt water; typically several gill slits; tough small scales with spines; ectothermic; two-chambered heart; males possess structures for internal fertilization. Examples: sharks, rays, skates, sawfishes.

Class Osteichthyes (bony fishes) Bony endoskeleton; aquatic; ectothermic; well-developed respiratory system, usually involving gills; possess swim bladder; paired fins; divided into two groups—ray-finned fishes (Actinopterygii), which include most living species, and fleshy-finned fishes (Sarcopterygii), which include lungfishes and the coelacanth. Examples: salmons, perches, sturgeons, tunas, goldfishes, eels.

Class Amphibia (amphibians) Adapted primarily to life in wet places; ectothermic; most carnivorous; smooth, moist skin; typically lay eggs that develop in water; usually have gilled larvae; most have three-chambered heart; adults either aquatic or terrestrial; terrestrial forms respire using lungs, skin, and/or lining of the mouth.

ORDER URODELA (newts and salamanders) Possess tail as adults; carnivorous; usually have four legs; usually aquatic as larvae and terrestrial as adults.

ORDER ANURA (frogs and toads) Adults in almost all species lack tail; aquatic larvae called tadpoles; well-developed hind legs adapted for jumping.

ORDER APODA (legless amphibians) Wormlike; lack legs; carnivorous; terrestrial burrowers; some undergo direct development; some are viviparous.

Class Reptilia (reptiles) As a group, adapted to fully terrestrial life, although some live in water; dry, scale-covered skin; ectothermic; most have three-chambered heart; internal fertilization; amniotic eggs typically laid on land; extinct forms include dinosaurs and flying reptiles.

ORDER RHYNCHOCEPHALIA (tuatara) "Teeth" formed by serrations of jawbone; found only in New Zealand; carnivorous. One species: *Sphenodon punctatus*.

ORDER SQUAMATA (lizards and snakes) Most carnivorous; majority terrestrial; lizards typically have legs; snakes lack legs. Examples: iguanas, geckos, skinks, cobras, pythons, boas.

ORDER CROCODILIA (crocodilians) Carnivorous; aquatic or semiaquatic; four-chambered heart. Examples: alligators, crocodiles, caimans, gharials.

ORDER CHELONIA (turtles) Bony shell; ribs and vertebrae fused to upper part of shell; some terrestrial, others semiaquatic or aquatic; all lay eggs on land. Examples: snapping turtles, tortoises, hawksbill turtles, box turtles.

Class Aves (birds) Endothermic; feathered over much of body surface; scales on legs and feet; bones hollow and lightweight in flying species; four-chambered heart; well-developed lungs and air sacs for efficient air exchange; about 27 orders. Examples: owls, eagles, ducks, chickens, pigeons, penguins, sparrows, storks.

Class Mammalia (mammals) Endothermic; subcutaneous fat; hair; most viviparous; suckle young with milk produced in mammary glands; four-chambered heart; four legs; use lungs for respiration.

Monotremes (egg-laying mammals)

ORDER MONOTREMATA (monotremes) Exhibit features of both mammals and reptiles; possess a cloaca; lay eggs that hatch externally; produce milk from primitive nipplelike structures. Examples: duck-billed platypuses, spiny anteaters, short-beaked echidnas.

Marsupials (pouched mammals)

ORDER MARSUPIALIA (marsupials) Young develop in the female's uterus but emerge at very early stage of development; development completed in mother's pouch. Examples: opossums, kangaroos, koalas.

Placentals Young develop to term in uterus; nourished through placenta; some born helpless, others able to walk within hours of birth; about 16 orders.

ORDER INSECTIVORA (insectivores) Among the most primitive of living placental mammals; feed primarily on small arthropods. Examples: shrews, moles, hedgehogs.

ORDER CHIROPTERA (bats) Flying mammals, with forelimbs adapted for flight; most nocturnal; most navigate by echolocation; most species feed on insects, nectar, or fruits; some species feed on blood. Examples: fruit bats, flying foxes, vampire bats.

ORDER PRIMATES (primates) Highly developed brain and complex social behavior; excellent binocular vision; quadrupedal or bipedal locomotion; five digits on hands and feet. Examples: lemurs, monkeys, chimpanzees, humans.

ORDER EDENTATA (edentates) Teeth reduced or absent; feed primarily on social insects such as termites and ants. Examples: anteaters, armadillos.

ORDER LAGOMORPHA (lagomorphs) Small herbivores with chisel-shaped front teeth; generally adapted to running and jumping. Examples: rabbits, pikas, hares.

ORDER RODENTIA (rodents) Mammalian order with largest number of species; mostly herbivorous, but some omnivorous; sharp front teeth. Examples: rats, beavers, guinea pigs, hamsters, gerbils, squirrels.

ORDER CETACEA (cetaceans) Fully adapted to aquatic existence; feed, breed, and give birth in water; forelimbs specialized as flippers; external hindlimbs absent; many species capable of long, deep dives; some use echolocation to navigate; communicate using complex auditory signals. Examples: whales, porpoises, dolphins.

ORDER CARNIVORA (carnivores) Mostly carnivorous; live in salt water or on land; aquatic species must return to land to breed. Examples: dogs, seals, cats, bears, raccoons, weasels, skunks, pandas.

ORDER PROBOSCIDEA (elephants) Herbivorous; largest land animal; long, flexible trunk.

ORDER SIRENIA (sirenians) Aquatic herbivores; slow-moving; forelimbs modified as flippers; hindlimbs absent; little body hair. Examples: manatees, sea cows.

ORDER PERISSODACTYLA (odd-toed ungulates) Hooved herbivores; odd number of hooves; one hoof generally derived from middle digit on each foot; teeth, jaw, and digestive system adapted to plant material. Examples: horses, donkeys, rhinoceroses, tapirs.

ORDER ARTIODACTYLA (even-toed ungulates) Hooved herbivores; hooves derived from two digits on each foot; digestive system adapted to thoroughly process tough plant material. Examples: sheep, cows, hippopotamuses, antelopes, camels, giraffes, pigs.

Glossary

Pronunciation Key

When difficult names or terms first appear in the text, a pronunciation key follows in parentheses. A syllable in small capital letters receives the most stress. The key below lists the letters used in the pronunciations. It includes examples of words using each sound and shows how those words would be written.

Symbol	Example	Respelling
a	hat	(HAT)
ay	pay; late	(PAY); (LAYT)
ah	star; hot	(STAHR); (HAHT)
ai	air; dare	(AIR); (DAIR)
aw	law; all	(LAW); (AWL)
eh	met	(MEHT)
ee	bee; eat	(BEE); (EET)
er	learn; sir; fur	(LERN); (SER); (FER)
ih	fit	(FIHT)
igh	mile; sigh	(MIGHL); (SIGH)
oh	no	(NOH)
oi	soil; boy	(SOIL); (BOI)
oo	root; rule	(ROOT); (ROOL)
or	born; door	(BORN); (DOR)
ow	plow; out	(PLOW); (OWT)

Symbol	Example	Respelling
u	put; book	(PUT); (BUK)
uh	fun	(FUHN)
yoo	few; use	(FYOO); (YOOZ)
ch	chill; reach	(CHIHL); (REECH)
g	go; dig	(GOH); (DIHG)
j	jet; gently; bridge	(JEHT); (JEHNT-lee); (BRIHJ)
k	kite; cup	(KIGHT); (KUHP)
ks	mix	(MIHKS)
kw	quick	(KWIHK)
ng	bring	(BRIHNG)
s	say; cent	(SAY); (SEHNT)
sh	she; crash	(SHEE); (KRASH)
th	three	(THREE)
y	yet; onion	(YEHT); (UHN-yuhn)
z	zip; always	(ZIHP); (AWL-wayz)
zh	treasure	(TREH-zher)

A

abdomen: in arthropods, the body segment containing most of the internal organs, including the reproductive organs; in mammals, the cavity region between the diaphragm and the pelvis; in vertebrates, the part of the body cavity containing the digestive and reproductive organs

abiotic factor: physical environmental factor, such as climate, type of soil and its acidity, or availability of nutrients

acid: compound that donates H^+ ions

actin: protein found in the thin filaments of muscle fiber

action potential: rapid change in voltage on the inside of the axon from negative to positive and then back to negative

active site: region of an enzyme where a substrate binds

active transport: movement of a substance against a concentration difference; a process that requires energy

adaptation: evolution of physical and behavioral traits that make organisms better suited to survive in their environment

adaptive radiation: pattern of evolution in which selection and adaptation lead to the formation of a new species in a relatively short period of time

addiction: uncontrollable craving for a substance

age-structure diagram: diagram that illustrates the percentages of individuals within different age groups

AIDS (acquired immune deficiency syndrome): fatal disease of the immune system caused by HIV infection

air sac: in birds, an extension of the respiratory system where air collects

alcoholic fermentation: in yeasts, the process by which pyruvic acid is converted to alcohol and carbon dioxide

alga; pl. **algae**: single-celled photosynthetic organism classified as a protist

alimentary canal: one-way digestive track that begins at the mouth; includes the esophagus, stomach, small intestine, and large intestine; ends at the anus

allele (uh-LEEL): different form of a gene for a specific trait

allergy: overreaction of the immune system to an antigen in the environment

alternation of generations: variation in a life cycle that switches back and forth between the production of diploid (2n) and haploid (n) cells; in plants, pattern of reproduction in which the organism alternates between sporophyte and gametophyte phases

alveolus (al-VEE-uh-lus; pl. **alveoli**, al-VEE-uh-ligh): tiny air sac in the lungs appearing in a grapelike cluster surrounding a network of capillaries where gas exchange takes place

amebocyte (uh-MEE-boh-sight): wandering cell in sponges

amino acid: molecule of which proteins are made, containing a central carbon bonded to an amino group, a carboxyl group, a hydrogen atom, and another group called R group

amniocentesis (am-nee-oh-sehn-TEE-sihs): prenatal technique that involves withdrawing a small amount of fluid from the sac surrounding the fetus

amnion: membrane of the sac that envelops the embryo

amniotic (am-nee-AHT-ik) **egg**: self-sufficient developing egg encased in a shell that allows for gas exchange but keeps liquid inside

amylase: enzyme in saliva that breaks down the chemical bonds in starches, releasing sugars

anal pore: waste-discharging region on a paramecium

analogous structure: similar in appearance and function but dissimilar in anatomical development and origin

anaphase: third phase of mitosis during which duplicated chromosomes separate from each other

angiosperm: flowering plant whose seeds develop within a matured ovary (fruit)

animal society: any group of animals living together

annual: plant that completes its life cycle (from seed to maturity, flower, and seed production) within a single growing season

annual tree ring: layer of xylem cells produced in a tree stem in one year, with growth affected by seasonal variation; in cross section, seen as a concentric circle

anther: sac at the tip of a filament where pollen is produced and released

anthropoid (AN-thruh-poid): any of a higher order of primates; includes humans, apes, and monkeys

antibiotic: drug or compound that can destroy bacteria

antibody: large protein that is the basic functional unit of a specific immune response

anticodon: three nucleotides in transfer RNA that bind to a codon in mRNA

antidiuretic hormone (ADH): hormone that helps to regulate the fluid balance of the body

antigen: molecule that stimulates the production of an antibody

aorta: largest artery in the human body

artery: blood vessel that carries blood from the heart to the body

artificial selection: method of selective breeding of organisms to produce offspring with desirable characteristics

ascus: tough sac of an ascomycete that contains the spores produced by sexual reproduction

asthma: condition in which smooth muscle contractions reduce the size of air passageways in the lungs, making breathing difficult

atherosclerosis (ath-er-oh-skluh-ROH-sihs): condition in which fatty deposits build up on the inner surfaces of arteries, obstructing the flow of blood

atom: smallest unit of a chemical element

ATP (adenosine triphosphate): energy-storage compound in cells

atrium (AY-tree-uhm; pl. **atria**, AY-tree-ah): upper chamber of the heart that receives the blood

auditory canal: region where vibrations enter the ear

autoimmune disease: condition that results when the immune system attacks its own cells

autonomic nervous system: part of the motor division of the peripheral nervous system that regulates activities not under conscious control

autosome: any chromosome other than a sex chromosome

autotroph: organism that uses energy from the sun to change simple nonliving chemical nutrients in its environment into living tissue

auxin: plant hormone that produces phototropism, stimulating cell growth near the tip of a root or stem

axon: single long, branched extension of the cell body of a neuron that carries impulses away from the cell body

B

B cell: lymphocyte that matures in bone marrow and produces antibodies

bacillus (buh-SIHL-uhs; pl. **bacilli**, buh-SIHL-igh): rod-shaped bacteria

bacteriophage: any virus that infects bacteria

bacterium; pl. **bacteria:** prokaryote with a cell membrane and genetic material not surrounded by a nuclear envelope

base: compound that donates OH^- ions or accepts H^+ ions

behavior: animal's response to its environment

berry: soft ovary wall that encloses many seeds

biennial: plant that flowers and produces seeds in the second year of its life cycle

bilateral symmetry: body form of an organism that has identical left and right sides, specialized front and back ends, and upper and lower sides

bile: lipid and salt fluid produced by the liver to aid digestion

binary fission: asexual form of reproduction in which a cell divides in half to produce two identical daughter cells

binocular vision: ability to perceive depth using both eyes at the same time

biodegradable: capable of being broken down into unharmful products by the life processes of living things

biodiversity: variety of organisms, their genetic information, and biological communities in which they live

biological magnification: process whereby substances such as toxic metals and chemicals are passed up the trophic levels of the food web at increasing concentrations

biology: science of life

biome: ecosystem identified by its climax community

biotic factor: biological environmental factor, such as the living things with which an organism might interact

blood: fluid medium of transport of the circulatory system in vertebrates

blood pressure: measure of the force exerted by blood against the walls of arteries

bone marrow: blood-forming tissue that produces white and red blood cells

book gill: respiratory organ of some marine arthropods

book lung: respiratory organ of most spiders made of numerous layers of tissue that resemble the pages of a book

brain: bundle of nerves and neural connections that controls the nervous system

brainstem: structure that connects the brain to the spinal cord; includes the medulla oblongata and the pons

bronchitis: respiratory disease caused by the bronchi becoming swollen and clogged with mucus

bronchus (BRAHN-kus; pl. **bronchi**, BRAHN-kigh): one of many air tubes that enter the lungs and branch off, for gas exchange

budding: type of asexual reproduction in which an outgrowth of an organism breaks away and develops into a new organism

C

calorie: amount of heat energy needed to raise the temperature of 1 gram of water by 1°C

Calvin cycle: chemical pathway used to convert energy from ATP and $NADP^+$ into sugars; also called the light-independent reactions

capillary: small blood vessel in which the exchange of nutrients and wastes takes place

carbohydrate: class of macromolecules that includes sugars and starches; source of chemical energy

carcinogen: cancer-causing chemical

cardiac muscle tissue: muscle tissue found in the heart; not under direct control of the central nervous system

carpel: female leaf of a flower
carrying capacity: largest number of individuals of a particular species that can survive over long periods of time in a given environment
cartilage: strong, resilient connective tissue
catalyst: substance that speeds up a chemical reaction without itself being used up in the reaction
cell: smallest working unit of living things
cell body: largest part of the neuron in which metabolic activity takes place
cell cycle: period of time from the beginning of one cell division to the beginning of the next
cell division: process in which a cell divides into two independent daughter cells
cell-mediated immunity: immune response in which foreign cells are destroyed by contact with T cells
cell membrane: part of the cell's outer boundary; contains a lipid bilayer
cell theory: principle stating that all living things are composed of cells, cells are the smallest working units of living things, and all cells come from preexisting cells by cell division
cell transformation: the changing of a cell's genetic makeup by the insertion of DNA
cell wall: tough, porous boundary that lies outside the cell membrane; found in plant cells and in some bacteria but not in animal cells
centralization: concentration of nerve cells that form nerve cords or nerve rings around the mouth of most primitive invertebrates
centriole: small structure in animal cells that helps to organize microtubules
centromere: the part of a chromosome in which the chromatids are attached
cephalization (sehf-uh-lih-ZAY-shun): concentration of nerve cells and sensory cells in the head of an organism
cephalothorax: in chelicerates, body segment formed from fused head and thorax; carries the legs
cerebellum (ser-uh-BEHL-uhm): second-largest part of the brain; coordinates movement
cerebral cortex: deeply creased surface of the cerebrum
cerebrospinal (ser-uh-broh-SPIGH-nuhl) **fluid**: fluid that fills the space between the meninges to cushion the brain and spinal cord
cerebrum (SER-uh-bruhm): largest and most complex structure of the nervous system; consists of two lobes or hemispheres; controls voluntary activities
chelicera (kuh-LIHS-er-uh; pl. **chelicerae**, kuh-LIHS-er-ee): unique, specialized mouthpart of chelicerates used for grasping and crushing
chemical compound: substance formed by the bonding of atoms of different elements in definite proportions
chemical reaction: process that changes one set of substances into a new set of substances
chemiosmosis: process of ATP formation in chloroplasts and mitochondria
chemotherapy: mixture of drugs that interfere with important cell processes; used for treating cancer
chitin (KIGH-tihn): structural carbohydrate that is the main component of arthropod exoskeletons; reinforces the cell walls of fungi; also found in insect skeletons
chlamydia: sexually transmitted disease caused by a bacterium that can bring about infertility in females
chlorophyll: principal pigment of green plants
chloroplast: organelle found in plants and certain types of algae; harvests the energy of sunlight
chorionic villus (kor-ee-AHN-ihk VIHL-uhs) **sampling**: technique that involves the removal and examination of tissue surrounding the fetus
chromatid: strand of a chromosome that occurs in identical pairs; combined with its sister chromatid, constitutes a chromosome
chromatin: material of chromosomes that consists of DNA and proteins
chromatophore: specialized cell capable of changing color
chromosomal mutation: change in the number or structure of a cell's chromosomes; a mutation that affects the entire chromosome
chromosome: structure in the nucleus of a cell that contains DNA bound to proteins
chromosome deletion: phenomenon in which a broken-off piece of chromosome is left out during meiosis, which usually results in a genetic disorder
chromosome translocation: phenomenon in which a broken-off piece of chromosome becomes reattached to another chromosome

cilium; pl. **cilia**: short, hairlike projection of some cells; in ciliates, used to pull the organism through water with a coordinated rowing movement

circular muscle: in annelids, a muscle that surrounds the longitudinal muscle, lengthening the segment as it contracts

class: classification category of related orders

classical conditioning: associative learning; learning to associate a stimulus with either a reward or a punishment

climate: temperature range, average annual precipitation, humidity, and amount of sunshine that a region typically experiences

climax community: collection of plants and animals that results when an ecosystem reaches a relatively stable state in the interaction between organisms and their environment

clitellum: in an annelid, the structure that secretes a ring of mucus into which eggs and sperm are released

cloaca: body cavity into which the intestinal and genitourinary tracts empty

clone: group of genetically identical organisms produced by the division of a single cell

closed circulatory system: system in which blood moves only through blood vessels

cocaine: addictive stimulant drug that causes the brain to release dopamine and acts on the neurons of the brain

coccus (KAHK-uhs; pl. **cocci**, KAHK-sigh): spherical bacterium

cochlea (KAHK-lee-uh): fluid-filled portion of the inner ear that creates pressure waves

codominance: genetic condition in which both dominant and recessive alleles are expressed

codon: group of three nucleotides in mRNA that specifies an amino acid

coelom (SEE-lohm): mesoderm-lined cavity that provides an open space inside the body within which organs can grow and function

coevolution: simultaneous, progressive change of structures and behaviors in two different organisms in response to changes in each organism over time

collar cell: sponge cell made of fused cilia surrounding a whiplike flagella; used for beating water and filtering food

common descent: principle that species have descended from common ancestors

communication: passing information from one organism to another

compact bone: dense layer of hard bone with minuscule spaces for Haversian canals

compound light microscope: microscope that uses lenses and light to magnify an image

cone: reproductive structure of a conifer; male cone produces pollen and female cone produces ovules

conjugation: form of sexual reproduction that results in new combinations of genes

conservation: managing of natural resources in a way that maintains biodiversity

consumer: organism that eats other organisms to obtain energy and nutrients

continental drift: geological theory suggesting that continents move slowly over hundreds of millions of years

contour feather: strong, lightweight feather that provides a large surface area for a bird's flight

contractile vacuole: specialized structure in some protists; used to collect and expel water

control: in an experiment, the test group in which the variable is not altered; used as a benchmark to measure the variable's effect

convergent evolution: process by which unrelated species independently evolve superficial similarities when adapting to similar environments

cornea: tough transparent layer at the surface of the eye through which light enters

corpus luteum (KOR-puhs LOOT-ee-uhm): name given to the follicle after ovulation because of its yellow color

cortex: layer of spongy cells beneath the epidermis of a plant

cotyledon (kaht-uh-LEED-uhn): tiny seed leaf found in a plant embryo inside a seed

covalent bond: attraction between two atoms in which electrons are shared between two atoms

crop: widening in the digestive tract for the temporary storage of food

crossing-over: exchange between homologous chromosomes

cyclin: protein that regulates the timing of the cell cycle

cytokinesis: division of the cytoplasm that takes place during the anaphase and telophase phases of mitosis in most cells

cytoplasm: portion of the cell outside the nucleus

cytoskeleton: supporting framework of a eukaryotic cell

D

decomposer: organism that feeds on the dead bodies of animals and plants or on their waste products

demographic transition: change in growth rate resulting from changes in birth rate

dendrite: any of many small branched extensions of the cell body of a neuron that carry impulses toward the cell body

density-dependent limiting factor: population-limiting factor that operates more strongly on large, dense populations than on small, less crowded ones

density-independent limiting factor: population-limiting factor that acts on organisms regardless of the size of the population

depressant: one of a group of drugs that decreases the rate of brain activity and slows down the actions of the nervous system

dermis: inner layer of skin; supports the epidermis and contains nerve endings, blood vessels, smooth muscle, and glands

diaphragm: dome-shaped muscle that pulls the bottom of the chest cavity downward, increasing its volume; separates the chest cavity from the abdominal cavity

dicot: angiosperm that produces seeds with two cotyledons

diffusion: process by which substances spread through a liquid or gas from regions of high concentration to regions of low concentration

diploid: description of a cell that contains a double set of chromosomes; represented by the term 2n

DNA (deoxyribonucleic acid): nucleic acid that transmits genetic information from one generation to the next and codes for the production of proteins

DNA fingerprinting: technique used to identify an individual from the unique pattern of DNA

DNA replication: process in which DNA is copied

domesticate: to tame, raise, and breed animals for human purposes

dominant: form of gene that is expressed when present and excludes the recessive form; represented with a capital letter

dorsal hollow nerve cord: develops into the main nerve pathway from the body to the brain; characteristic of chordates

double fertilization: process of two fertilization events taking place inside an embryo sac

down feather: soft, fluffy feather that insulates the skin of a bird

drug: substance that causes a change in the body

drug-resistant bacteria: bacteria not susceptible to one or more antibodies

drupe: soft, fleshy ovary wall that encloses a single tough, stony seed of a fruit

E

ecological pyramid: diagram showing the decreasing amounts of energy, living tissue, or number of organisms at successive trophic levels

ecological succession: process by which an existing ecosystem is gradually and progressively replaced by another ecosystem

ecology: scientific study of interactions between different kinds of living things and the environments in which they live

ecosystem: collection of organisms—producers, consumers, and decomposers—interacting with each other and with their physical environment

ecosystem diversity: variety of habitats, living communities, and ecological processes in the living world

ectoderm: outer embryonic cell layer from which the nervous system, the skin, and other associated body coverings are derived in animal development

ectotherm: organism—such as a fish, amphibian, or reptile—that relies on interactions with the environment to control body temperature; cold-blooded organism

egg: female reproductive cell

electron: subatomic particle that carries a negative charge

electron microscope: microscope that uses a beam of electrons to examine a sample

electron transport chain: series of molecules located in the inner membrane of the mitochondrion that receive high-energy electrons from electron carriers

embryo: early stage of development of an organism resulting from fertilization

emphysema: respiratory disease in which the alveoli lose their elasticity and breathing becomes difficult

endocrine system: body system made up of a series of glands that produce and release chemicals into the bloodstream

endoderm: inner embryonic cell layer from which the tissues and organs of the digestive tract and other internal organs are developed

endodermis: layer of cells forming the inner boundary of the cortex of most roots and controlling entry into the vascular cylinder

endoplasmic reticulum: network of membranes within a cell that process and transport proteins and other macromolecules

endoskeleton: skeletal system located within the body

endosperm: food-rich tissue that surrounds the embryo of a plant

endospore: type of asexual spore formed inside a bacterial cell that develops a thick wall enclosing part of the cytoplasm and DNA, enabling the bacteria to survive for years

endosymbiont hypothesis: hypothesis that billions of years ago, eukaryotic cells arose as a combination of different prokaryotic cells

endotherm: organism, such as a mammal or a bird, that generates and maintains body heat through chemical reactions in the body; warmblooded organism

enhanced greenhouse effect: increased retention of heat in the Earth's atmosphere as a result of increased levels of carbon dioxide and other greenhouse gases to the atmosphere

environment: combination of physical and biological factors that influence life

enzyme: molecule that serves as a catalyst in organic reactions

epidemic: uncontrolled spread of disease

epidermis: outer layer of skin made up of layers of epithelial cells

epididymis (ehp-uh-DIHD-ih-mihs): tube in males that stores sperm

epiglottis: flap of tissue that prevents food from entering the trachea

esophagus: tube through which food passes from the pharynx to the stomach

estrogen: hormone in females involved in the development of the reproductive organs

eukaryote: organism made of cells that contain nuclei

evolution: process of change over a period of time

exon: expressed sequence of mRNA; a region that remains in mRNA after the introns are removed

exoskeleton: external skeletal system

experiment: controlled test to determine the validity of a hypothesis

exponential growth: rapid growth of a population whose living conditions are ideal

external fertilization: method of fertilization in which eggs and sperm meet outside the organism's body

extracellular digestion: process in which food is digested outside the cells

F

Fallopian tube: one of the two tubes in a female's reproductive system that receives the ovum from the ovary and passes it to the uterus

family: classification of a group of closely related genera

family group: group of related animals living together

feather: lightweight covering of a bird's body

fermentation: regeneration of NAD^+ to keep glycolysis running in the absence of oxygen

fertilization: fusion of egg and sperm to form a zygote

fetus: name given to an embryo after eight weeks of development

fibrin: netlike trap of plasma protein that forms a blood clot

filament: stamen that emerges from a flower

filtration: transport process that removes urea, excess water, and other wastes from the blood

fin: winglike structure on a fish used in swimming

fitness: an organism's ability to successfully pass on its genes to its offspring

fixed joint: joining place of two bones where there is little or no movement, such as in the skull

flagellum; pl. **flagella**: whiplike projection found on some cells; typically used for movement

flame cell: specialized cell with a tuft of cilia that conducts water and wastes through the branching tubes that serve as an excretory system in flatworms

flower: reproductive structure of an angiosperm

follicle: cluster of cells that contains a developing egg

follicle-stimulating hormone (FSH): hormone secreted by the pituitary gland that stimulates the development of sperm in males and the follicle in females

food chain: sequence of organisms related to one another as food and consumer

food web: interconnecting food chains in an ecological community

foot: in a mollusk, the muscular structure that usually contains the mouth

fossil: preserved bone or other trace of an ancient organism

frameshift mutation: gene mutation that involves the insertion or deletion of a nucleotide, thus changing the grouping of codons

freely movable joint: joining place of two bones where there is a wide range of movements, such as in the shoulders, hips, elbows, and knees

frond: large leaf of a fern

fruit: ripened ovary that contains angiosperm seeds

fruiting body: reproductive structure that produces spores

G

gamete (GAM-eet): haploid reproductive cell that can unite with another haploid cell to form a new individual

gametophyte: in plants, the haploid gamete-bearing generation that reproduces by fertilization

gastric gland: microscopic gland that appears in great numbers in the lining of the stomach and produces mucus, acid, and pepsin to aid digestion

gastritis: inflammation of the stomach caused by stomach acids making direct contact with the cells of the stomach lining

gastrovascular cavity: digestive sac with a single opening to the outside from which food enters and wastes leave

gastrulation (gas-troo-LAY-shuhn): process of cell migration during which the primary germ layers of the embryo are formed

gemmule (JEHM-yool): in sponges, a cluster of ball-shaped amebocytes that can grow into a new individual

gene: segment of DNA that codes for a specific protein; the unit by which hereditary characteristics are transmitted

gene mutation: mutation that involves only a single gene

gene pool: all the alleles of all the genes of the members of a population that interbreed

genetic code: language of the instructions in DNA and RNA that code for the amino acid sequence of a polypeptide

genetic diversity: genetic material in the gene pool of a species; variety of different forms of genes present in a population

genetic drift: random change in allele frequency, often producing offspring that will be different from the original population by chance

genetic engineering: manipulation and insertion of genes and DNA from different sources into an organism

genetics: study of heredity

genital herpes: sexually transmitted disease caused by a virus that affects the genital areas of males and females

genotype: genetic composition of an organism

genus; pl. **genera**: classification of a major group of closely related organisms

geologic time scale: unit of time—such as eons, eras, periods, and epochs—based on information contained in rocks

geotropism: response of an organism to the force of gravity, usually by turning downward or upward

gill: featherlike respiratory organ of many aquatic species; used to obtain oxygen from the water

gizzard: portion of digestive tract where food is ground into small, absorbable particles

global warming: prediction that the enhanced greenhouse effect will cause a significant rise in Earth's average temperature

glomerulus: ball of capillaries in a nephron that filters blood

glucagon: hormone that breaks down glycogen and fats and releases sugars into the blood

glycogen: compound that stores excess glucose in the body

glycolysis: series of reactions in which a molecule of glucose is broken down

Golgi apparatus: network of membranes within a cell that, in conjunction with the endoplasmic reticulum, processes and transports proteins and other macromolecules; contains special enzymes that attach carbohydrates or lipids to a protein

gonorrhea: sexually transmitted disease caused by a bacterium that infects the urinary and reproductive tracts

gradualism: theory that evolutionary change occurs slowly and steadily over long periods of time

grafting: artificial method of propagation

greenhouse effect: retention of heat in the Earth's atmosphere due to the presence of greenhouse gases

green revolution: substantial increase in crop yields that resulted from the introduction of modern agricultural practices

growth rate: change in the size of a population

gullet: depression on a paramecium used for the intake of food particles

gut: digestive tract

gymnosperm: seed plant in which the seeds are exposed to the air, usually in a cone-shaped structure

H

habitat: surroundings in which a species lives and thrives; defined in terms of the plant community and the abiotic factors

habituation: decreased response to a repeated stimulus

hair follicle: clusters of columns of epidermal cells anchored in the dermis; produce hair

half-life: length of time required for half of a radioactive element to decay to its more stable form

haploid: description of a cell that contains a single set of chromosomes; often represented by the letter n

Haversian (huh-VER-zhuhn) **canal**: small channel in compact bone through which nerve and blood vessels run

helper T cell: type of lymphocyte in the immune system that identifies a specific pathogen in the body

hemoglobin: iron-containing protein found in red blood cells that helps to transport oxygen

hepatitis B: disease caused by a virus that can cause inflammation of the liver; can be transmitted sexually

heredity: biological inheritance of traits from parent to offspring

hermaphrodite (her-MAF-ruh-dight): organism that has both male and female reproductive organs and produces both sperm and eggs

heterotroph: organism that cannot manufacture its own food

heterozygous (heht-er-oh-ZIGH-guhs): description of an organism that has a mixed pair of alleles for a trait

histamine: chemical released by a mast cell to produce an inflammatory response to an allergy

HIV (human immunodeficiency virus): virus that infects cells in the immune system and destroys helper T cells

homeostasis: process by which organisms respond to stimuli to keep conditions in their bodies suitable for life

hominid: any of a family of two-legged primates; includes humans and closely related primates but not apes

hominoid: any of a superfamily of primates; includes humans and apes

homologous structures: structures that have a common origin but not necessarily a common function

homozygous (hoh-moh-ZIGH-guhs): description of an organism that has an identical pair of alleles for a trait

hormone: chemical that travels throughout the bloodstream and affects the behavior of other cells

human chorionic gonadotropin (HCG): hormone produced by the cells' zygote that keeps the corpus luteum alive and preserves the uterine lining

hybrid: offspring of parents with different characteristics

hybridization: mating of two organisms with dissimilar genetic characteristics

hydrostatic (high-droh-STAT-ihk) **skeleton**: skeletal system in which the muscles surround and are supported by a water-filled body cavity

hypha (HIGH-fuh; pl. **hyphae**, HIGH-fee): threadlike, branching filament that is the most basic structure in a fungus

hypothalamus (high-poh-THAL-uh-muhs): region of the brain that directly or indirectly controls the release of hormones from the pituitary gland

hypothesis; pl. **hypotheses**: possible explanation of, preliminary conclusion about, or guess at the solution to a problem

I

immunity: ability of the body to resist a specific pathogen

implantation: process during which the cells of a zygote grow into the uterine wall

imprinting: learning mechanism in the early life of an animal in which a stimulus establishes a characteristic behavior

impulse: electrical signal carried by neurons in the nervous system

inbreeding: mating of organisms with similar genetic characteristics

incomplete dominance: genetic condition in which neither allele is completely dominant or recessive

independent assortment: process by which different genes do not influence each other's segregation into gametes

infectious disease: disease caused by a pathogen

inflammatory response: activation of a nonspecific defense to a pathogen

inorganic compound: very generally, a compound that does not contain carbon chains

insight learning: application of past experience to a new situation

instinct: genetically programmed, innate behavior

insulin: polypeptide hormone that removes sugar from the bloodstream

internal fertilization: method of fertilization in which the eggs and sperm meet inside the body of the egg-producing individual

interneuron: neuron that connects the sensory and motor neurons and carries impulses between them

interphase: G_1, S, and G_2 phases; occurs between cell divisions

intracellular digestion: process by which food is digested inside the internal cells of a simple multicellular animal

intron: intervening sequence that is removed from mRNA and thus not expressed

ion: charged particle formed by an atom that has gained or lost one or more of its electrons

ionic bond: attraction between oppositely charged ions

iris: disk of tissue located at the back of the cornea

islets of Langerhans: clusters of endocrine cells in the pancreas

K

karyotype (KAR-ee-uh-tighp): diagrammatic representation of individual chromosomes cut out from a photograph and grouped together

keratin: tough, flexible structural protein found in hair, fingernails, and epidermis

kidney: one of a pair of specialized excretion organs that removes nitrogen and other nonsolid wastes from the body and regulates water in the bloodstream

killer T cell: type of lymphocyte in the immune system that attacks and destroys a virus-infected cell

kingdom: highest ranking classification of living organisms that falls into one of the six major groups

Koch's postulates: series of rules used to identify the microorganism that causes a specific disease

Krebs cycle: series of reactions in which the chemical bonds in pyruvic acid are broken apart; also called citric acid cycle

L

lactic acid fermentation: in animals, the conversion of pyruvic acid to lactic acid

large intestine: colon; part of the digestive system that produces a waste material known as feces

larynx: structure in which air enters the respiratory system and the vocal cords are located

lateral line: collection of skin pores that connect to a system of tubes that help a fish detect patterns of movement in the surrounding water

leaf: specific outgrowth of a vascular plant that captures sunlight for photosynthesis

learning: ability to change behavior as a result of experience; acquisition of knowledge or a skill

legume: fruit with a pod that splits open on two sides

lens: flexible structure behind the pupil filled with a transparent protein that helps adjust eyes to focus to see near or far objects

ligament: band of tissue that connects the bones of a joint

light-dependent reaction: response that requires the direct involvement of light; produces NADPH

light-independent reaction: response that does not directly involve light; *see also* Calvin cycle

lipid: waxy, fatty, or oily compound used to store and release energy; one of the classes of macromolecules

lipid bilayer: double-layered pattern formed by phospholipids in water; the principal component of cell membranes

liver: large gland situated above the stomach that produces bile to aid digestion

longitudinal muscle: in annelids, a muscle that runs along the length of the body, shortening the segments as it contracts

loop of Henle: region of the nephron of the kidney where concentrated urine is produced

lung: respiratory organ in which gas exchange takes place

lung cancer: disease caused by small groups of cancer in the lungs; usually the consequence of smoking

luteinizing hormone (LH): hormone secreted by the pituitary gland that stimulates testosterone production in males and the development of the follicle, ovulation, and production of the corpus luteum in females

lymph (LIHMF): fluid found in intracellular spaces and in the lymphatic vessels of vertebrates

lymphatic (lihm-FAT-ihk) **system**: network of vessels that collects fluid that leaks from the capillaries and returns it to the circulatory system

lymphocyte: white blood cell that produces antibodies to assist the immune system

lysogenic infection: process in which viral genes combine with the host cell's DNA, produce viral mRNA, and gradually make new viruses

lysosome: saclike membrane filled with chemicals and enzymes that can break down almost any substance within a cell

lysozyme: enzyme in saliva that fights infection by digesting the cell walls of bacteria

lytic infection: process in which viral enzymes destroy a host cell's DNA, ribosomes, and resources to reproduce

M

macromolecule: large organic polymer, such as a carbohydrate, lipid, protein, or nucleic acid

macronucleus: the larger of the two nuclei of a ciliate; stores multiple copies of commonly used genes

Malpighian (mal-PIHG-ee-uhn) **tubule**: structure that excretes nitrogenous wastes; found in many arthropods

mammary gland: gland that enables a female to nourish her young with milk

mandible: in arthropods, specialized mouthpart used for biting

mantle: thin, delicate layer of tissue that covers the internal organs of a mollusk

mated pair: male and female living together

medulla oblongata (mih-DUHL-uh ahb-lahn-GAHT-uh): part of the brainstem that regulates the flow of information between the brain and the rest of the body; controls involuntary functions

medusa: free-swimming stage in the life cycle of a cnidarian

meiosis (migh-OH-sihs): process of cell division that reduces the number of chromosomes in the cell by half, from diploid to haploid; creates gametes used for sexual reproduction

melanin: dark-brown pigment that gives skin its color

meninges (muh-NIHN-jeez): three layers of tough, elastic tissues that cushion the brain and spinal cord

menstrual cycle: pattern of events in females that involves the development and release of an egg for fertilization and the preparation of the uterus to receive a fertilized egg

mesoderm: middle embryonic cell layer from which the skeletal, muscular, and other tissues are developed

mesoglea (mehs-oh-GLEE-uh): in cnidarians, the jellylike layer between the endoderm and ectoderm

messenger RNA (mRNA): form of RNA that carries genetic information from the DNA in the nucleus to the ribosomes in the cytoplasm

metamorphosis: process of changing form and shape

metaphase: second phase of mitosis during which the chromosomes complete their attachment to the spindle and line up across the center of the cell

microclimate: climate conditions of a particular area that vary over small distances

micronucleus: the smaller of the two nuclei of a ciliate; stores copies of all the cell's genes

mitochondrion; pl. **mitochondria**: an organelle found in the cells of most plants and animals; produces energy from a chemical fuel and oxygen

mitosis: process of cell division in eukaryotic cells

molar: tooth in the rear of the mouth adapted for grinding

molecular clock: theory that mutations in DNA occur at a constant rate; used to estimate time frames of departure from common ancestry

molecule: group of atoms united by covalent bonds

molting: process of shedding an exterior layer or exoskeleton to allow for renewal or growth

monocot: angiosperm that produces seeds with one cotyledon

monoculture: farming strategy whereby a single highly productive crop is planted in a large field

monomer: small, individual molecule that forms a polymer

morula (MOR-yoo-luh): solid ball of cells produced by cell division of a zygote

motor neuron: specialized neuron that carries impulses from the brain or spinal cord to muscles or other organs

mouth: opening through which food is taken in for digestion

multiple allele: type of gene that is determined by more than two alleles for a single trait

mutation: abrupt alteration in the genetic information of a cell

mutualism: reciprocal relationship in which two organisms benefit each other

mycelium (migh-SEE-lee-uhm): mass of hyphae that grows into the food source and forms the body of the fungus

myelin (MIGH-uh-lihn): material that forms a protective sheath around an axon; white fatty substance that surrounds many vertebrate nerve cells

myoglobin: reddish oxygen-storing protein found in skeletal muscles

myosin: protein that makes up the thick filaments of muscle fiber

N

natural selection: process in nature that over time results in the survival of the fittest

negative feedback: regulatory system that enables conditions within the body to remain constant

nematocyst: poisonous stinger used in cnidarians for defense and catching prey

nephridium (nee-FRIHD-ee-uhm; pl. **nephridia**, nee-FRIHD-ee-uh): excretory organ that removes nitrogen-containing wastes from the blood; found in many invertebrates

nephron: any of the numerous blood-filtering units of a kidney

nerve: bundle of nerve fibers

nerve cord: group of nerve cells that extend along the length of the body

nervous system: network of nerve cells and nervous tissue that receives and relays information about activities within the body and monitors and responds to internal and external changes

neuron: cell that carries impulses throughout the nervous system; consists of a cell body, dendrites, and an axon

neurotransmitter (NOO-roh-trans-miht-er): a chemical used by one neuron to signal another cell

neutron: subatomic particle that carries no charge

niche: full range of physical and biological conditions in which the organisms in a species can live and the way in which the organisms use those conditions

nictitating membrane: transparent eyelid that protects the eye underwater and keeps it moist in air

nonbiodegradable: incapable of being broken down into unharmful products by the life processes of living things

nondisjunction: failure of chromosome pair to separate correctly during meiosis

notochord: flexible, rodlike structure that provides body support; unique to chordates

nucleic acid: DNA or RNA that consists of nucleotides and genetic information

nucleotide: compound made of a phosphate group, a nitrogenous base, and a 5-carbon sugar; forms the basic structural unit of DNA

nucleus; pl. **nuclei:** in an atom, the compact core that contains the neutron and protons; in a cell, the structure that contains nearly all of the cell's DNA

nut: fruit with a hard ovary wall forming a protective shell around the seed

nutrient cycle: path along which nutrients that are available in fixed quantities on Earth are passed from one organism to another and from one part of the biosphere to another

O

ocellus (oh-SEHL-uhs; pl. **ocelli**, oh-SEHL-igh): simple eyespot that detects the presence or absence of light

olfactory bulb: part of the brain specialized for the sense of smell

open circulatory system: system in which blood is pumped from the heart through vessels and open spaces

operant conditioning: trial-and-error learning

operator: special region of DNA to which the repressor binds

operculum: bony structure that protects the gills; found in bony fishes

opiate: one of a group of drugs derived from the opium poppy; plant that mimics endorphins in the brain; often used as a pain-killing drug

optic lobe: structure in the brain that processes visual information

order: classification of several families of similar organisms

organ: group of tissues that work together to perform a specific function

organelle: small structure that performs a specialized function within a cell

organic compound: very generally, a substance that contains a chain of a least two carbon atoms

organism: individual living thing

organ system: group of organs that perform several closely related functions

osmosis: diffusion of water through a selectively permeable membrane

osteocyte (AHS-tee-oh-sight): cell embedded in both compact and spongy bone that helps build and maintain bones

ovary: in animals, the female reproductive gland that produces eggs and female hormones; in flowering plants, the structure that contains the egg cells of a flower

oviparous (oh-VIHP-uh-ruhs): producing eggs that develop and hatch outside the female's body

ovoviviparous (oh-voh-vigh-VIHP-uh-ruhs): producing eggs that develop and hatch within the female's body and are born alive

ovulation: process in which an egg is released from the ovary

ovule: in a seed plant, the place where female gametophytes are produced

ovum; pl. **ova:** female gamete or egg

oxytocin: pituitary hormone that stimulates contractions of the smooth muscles surrounding the uterus

P

pacemaker: area of the heart that regulates the heartbeat; the sinoatrial node

pancreas: gland situated below the stomach that produces digestive enzymes and regulates blood sugar

parasite: organism that takes nourishment from and lives at the expense of its host

passive transport: movement of substances across the cell membrane from regions of high concentration to regions of low concentration; occurs without the cell expending energy

pathogen: disease-causing organism

pedigree: diagram that tracks the inheritance of a single gene through several generations of a family

pedipalp: specialized leglike appendage of chelicerates that are used for grasping, sensing, and fertilizing

pepsin: protein-digesting enzyme produced by gastric glands in the stomach

peptic ulcer: lesion in the stomach wall caused by stomach acid

perennial: plant that lives for more than two years

periosteum (per-ee-AHS-tee-uhm): tough membrane that covers the bones

peristalsis: muscular contractions that pass food through the alimentary canal or digestive tract

petal: one of the white or colorful leaflike parts of a flower

phagocyte: white blood cell that engulfs and destroys bacteria

pharyngeal (fuh-RIHN-jee-uhl) **slit**: a narrow opening in the throat region of chordates

pharynx: muscular structure in the back of the mouth; connects the mouth with the rest of the digestive tract

phenotype: form of a genetic trait displayed by an organism

pheromone (FAIR-uh-mohn): chemical signal produced by an organism to influence the behavior or development of another organism of the same species

phloem (FLOH-ehm): vascular tissue that transports products of photosynthesis and substances from one part of the plant to another

photosynthesis: process by which green plants use the energy of sunlight to produce carbohydrates

phototropism: growth response of a plant to light, usually by turning toward or away from light

pH scale: measurement system that indicates the acidity or basicity of a solution

phylum (FIGH-luhm; pl. **phyla**, FIGH-luh): category made up of several classes of different organisms that share important characteristics

phytochrome: red-light-sensitive pigment that enables plants to sense day and night, changing seasons, and developmental processes

pigment: colored substances that reflect or absorb light

pistil: female reproductive organ of a flowering plant, composed of ovary, style, and stigma

pith: in dicots, the ground tissue inside the ring of a vascular bundle

pituitary gland: tiny endocrine gland at the base of the brain that secretes hormones that regulate the activity of other endocrine glands

placenta: organ that connects a mother with her developing embryo and provides a place for the exchange of nutrients, oxygen, wastes, and carbon dioxide

plasma: fluid part of the blood; constitutes about 55 percent of the total volume of blood

plasma cell: specialized B cell that releases antibodies into the bloodstream to fight infection

plasmid: small, circular DNA molecule in some bacteria that can be used for cell transformation

platelet: cell fragments in the blood that aid in blood clotting

point mutation: gene mutation that involves a single nucleotide

pollen: tiny spore that contains the male reproductive cells of a plant

pollination: transfer of pollen that precedes fertilization

polygenic trait: inherited characteristic controlled by more than one gene

polymer: large molecule assembled from small, individual molecules

polyp: sessile stage in the life cycle of a cnidarian

pons: region of connecting tissue at the base of the brainstem

population: group of organisms of a single species that live in a given area

powder feather: feather in aquatic birds that releases a water-repelling powder

premolar: teeth directly in front of the molar teeth

primary producer: organism that uses energy from the sun to change simple nonliving chemical nutrients in its environment into living tissue

progesterone: female hormone that promotes development of the uterine wall

proglottid (proh-GLAHT-ihd): segment of a tapeworm's body that contains reproductive structures

prokaryote: organism that does not contain nuclei; typically is small and single celled

prolactin: pituitary hormone that stimulates the production of milk in the breast tissue of the mother

promoter: special region of DNA to which RNA polymerase binds at the beginning of the process of transcription

prophase: first phase of mitosis during which each chromosome consists of two chromatids

prosimian (proh-SIHM-ee-uhn): any of a suborder of primates

protein: polymer of amino acids used for building cells, catalyzing reactions, and other purposes

protist: eukaryotic organism that does not share a unique set of characteristics

proton: subatomic particle that carries a positive charge

protonema (proht-oh-NEE-muh): tangle of thin filaments germinating from a moss spore and developing into a leafy moss plant

provirus: viral DNA that has become part of the host cell's DNA

pseudopod (SOO-doh-pahd): temporary projection from an ameboid cell used for movement and feeding; cytoplasm streams into the pseudopod and the rest of the cell follows

pulmonary circulation: pathway of blood vessels on the right side of the heart that carries blood between the heart and lungs

punctuated equilibrium: pattern of long periods of stability that are interrupted by episodes of rapid change

pupil: opening in the iris that regulates the amount of light that enters the eye

R

radial symmetry: arrangement of body parts that repeat around an imaginary line drawn through the center of an organism; shown in cnidarians and some adult echinoderms

radula (RAJ-oo-luh): the feeding structure in the mouth of many mollusks

reabsorption: process by which the material that was removed from the blood is put back into the blood without the toxic compounds

receptor: specific chemical binding site for a particular hormone

recessive: form of gene that is not expressed in the presence of the dominant form; represented with a lowercase letter

recombinant DNA: pieces of DNA from two or more sources that are reassembled to act as a single DNA molecule

red blood cell: blood cell that contains hemoglobin and constitutes almost half the total volume of blood; also called an erythrocyte

reflex: quick, automatic response to a stimulus

renal artery: artery through which blood flows into the kidney

renal vein: vein through which blood leaves the kidney

repressor: protein that blocks a gene's transcription by binding to the operator

reproductive isolation: separation of different species that cannot interbreed

respiration: release of energy from the breakdown of food molecules in the presence of oxygen

resting potential: product of a net excess of negative charges on the inside of the membrane

restriction enzyme: protein that cuts DNA at a specific sequence of nucleotides

retina: layer of cells at the back of the eye

retrovirus: virus containing an enzyme that copies its genetic information from RNA to DNA

RFLP (RIHF-lihp) **(restriction fragment length polymorphism):** dark band revealed when pieces of DNA are probed; can be used to identify and classify an individual's unique DNA pattern

rhizoid (RIGH-zoid): rootlike anchoring structure of moss plants that absorbs water and nutrients from the soil

rhizome: underground stem of a vascular plant

ribosomal RNA (rRNA): form of RNA that is an important component of ribosomes

ribosome: small particles in a cell that are made of RNA and protein; sites of protein assembly

ring vessel: in an anneid, structure that pumps blood, thus functioning as a miniature heart

RNA (ribonucleic acid): principal molecule that carries out the instructions coded in DNA

root: descending structure of a plant that branches into the soil, anchors the plant, and absorbs water and nutrients

S

salivary gland: gland in the mouth that produces saliva, a fluid that moistens food and makes food easier to chew

scale: in fish, one of many overlapping rigid plates that form a protective covering; in seed-bearing plants, the surface of a reproductive structure on which the seeds are exposed

scanning probe microscope: microscope that traces the surface of a sample with a small tip called a probe

science: process of thinking and learning about the world

scientific method: system of asking questions, developing explanations, and testing those explanations against the reality of the natural world

scolex (SKOH-lehks): front end of a tapeworm; contains suckers and hooks

scrotum: sac located outside a male's body cavity that contains the testes

sedimentary rock: kind of rock formed when silt, sand, or clay builds up on the bottom of a river, lake, or ocean

seed: reproductive structure that includes a developing plant and a food reserve enclosed in a resistant outer covering

seed cone: female cone that contains mature seeds

segment: one of several body compartments that allows an animal to increase in body size with minimal new genetic material

segregation: process that separates the two alleles of a gene during gamete formation

selective breeding: producing a new generation by mating individuals with desired characteristics

semicircular canal: one of three fluid-filled organs that help sense position in space and maintain balance

seminiferous (sehm-uh-NIHF-er-uhs) **tubule**: tightly coiled tubules in males in which sperm cells are produced

sensory neuron: specialized neuron that carries impulses from the sense organs to the brain and the spinal cord

sepal: structure that encloses and protects the developing flower bud and opens as the flower blooms

septum; pl. **septa**: dividing wall or membrane

seta; pl. **setae**: external bristles

sex chromosome: X and Y chromosomes that determine the sex of an individual

sex-linked gene: gene located on the sex chromosome

sexually transmitted disease (STD): disease spread from one person to another by sexual contact

shell: in mollusks, protective structure formed by glands in the mantle

siphon: tube through which water is forced out

skeletal muscle tissue: muscle tissue generally attached to bones; can be contracted voluntarily

skin: outer protective covering of the body; largest organ of the body

sliding filament theory: concept that thick and thin filaments slide past each other and cause the muscle to contract

slightly movable joint: joining place of two bones where there is a small amount of movement and flexibility, such as in the spinal column or ribs

small intestine: portion of the digestive tract in which most of the chemical work of digestion takes place

smooth muscle tissue: spindle-shaped, unstriated muscle tissue found in internal organs and blood vessels; not under conscious control of the nervous system

social insect colony: highly structured living group performing tasks that no single insect could accomplish

sodium-potassium pump: protein in nerve cell that moves sodium ions out of the cell and potassium ions into the cell

solution: uniform mixture of substances

somatic nervous system: part of the motor division of the peripheral nervous system that controls voluntary movements

speciation: formation of a new species brought about by genetic changes that prevent breeding between the new, genetically different groups

species: smallest group in the classification system of organisms that share similar characteristics and interbreed in nature

species diversity: number and variety of different life forms

sperm: male gamete

spicule: one of many small, spikelike structures that form the skeleton of a sponge

spinal cord: collection of nerve fibers that extends from the brain; part of the central nervous system of a vertebrate

spindle: cluster of microtubules that span the cell nucleus

spirillum (spigh-RIHL-uhm; pl. **spirilla**, spigh-RIHL-uh): spiral-shaped bacteria

spongin: protein that makes up the tough but flexible skeleton of some sponges

spongy bone: region of resilient, supportive bone tissue within the compact bone with an interlaced pattern that withstands stress

spore: small, typically single-celled structure capable of producing a new individual, either immediately or after a period of dormancy

sporophyte: in plants, the diploid spore-bearing generation that reproduces by spores

stamen: male leaf that produces pollen

statocyst: organ of balance found in many invertebrates

stem: main, upward-growing part of a vascular plant that provid-s support and conducts water and nutrients

stereoisomer: molecule that has the same atoms and bonds of another molecule but has atoms oriented differently in space

stigma: sticky tip of the style of a plant

stimulant: any one of a group of drugs that increase the release of neurotransmitters at some synapses in the brain to speed up the nervous system

stoma (STOH-muh; pl. **stomata**, STOH-muh-tuh): in the epidermis of a plant, one of many small openings that can open and close to allow gas exchange and to prevent water loss

stomach: large muscular sac where contractions mix food and enzymes and acids digest food

style: in a flower plant, the stemlike narrow part of the carpel

substrate: in a chemical reaction, the component that binds to an enzyme

superorganism: colony of interdependent organisms that act as a unit, able to achieve far more than individuals acting separately

suppressor T cell: type of lymphocyte in the immune system that shuts off the immune response in killer T cells and in B cells

sustainability: degree to which a human activity is in harmony with the biosphere and does not deteriorate the biosphere's living and nonliving parts

swim bladder: expandable structure that holds gas to change a fish's internal density and depth in the water

symbiosis (sihm-bigh-OH-sihs): beneficial relationship between two organisms that live together

synapse (SIHN-aps): place where a neuron can transfer an impulse to another cell

synovial (sih-NOH-vee-uhl) **fluid**: lubricant found in a joint that reduces friction and allows bones to slip past each other easily

syphilis: sexually transmitted disease caused by a bacterium that can result in death

systemic circulation: pathway of blood vessels on the left side of the heart that supplies the body with oxygen-rich blood and returns oxygen-poor blood to the heart

T

target cell: cell that has a receptor for a particular hormone

taste bud: one of many chemical receptors located on the tongue

taxonomy: science of naming organisms and assigning them to groups

T cell: white blood cell that matures in the thymus gland and regulates other cells of the immune system

telophase: fourth and final phase of mitosis during which two distinct nuclei form within the cell

tendon: cord of tissue that connects muscles and bones

testis (TEHS-tihs; pl. **testes**, TEHS-teez): male reproductive gland that produces sperm and male hormones

testosterone (tehs-TAHS-ter-ohn): male hormone produced in the testes that stimulates sperm production and the development of male sex organs and secondary sex characteristics

tetrapod: body plan that includes four limbs or legs

theory: logical explanation for a broad range of observations

thigmotropism (thihg-MAH-truh-pihz-uhm): response to touch
thyroid gland: an endocrine gland that produces the hormone thyroxine
tissue: mass of similar cells that performs a specific function
toxin: poisonous substance
trachea: tube that carries air from the larynx to the lungs; also called the windpipe
tracheal (TRAY-kee-uhl) **tube**: air-conducting passage for the diffusion of oxygen
tracheid (TRAY-kee-ihd): specialized water-conducting thick-walled tubelike cell of a vascular plant
trait: inherited characteristic that distinguishes one organism from another
transcription: process in which the nucleotide sequence of a DNA molecule is copied into RNA
transfer RNA (tRNA): form of RNA that carries an amino acid to the ribosome during the assembly of a protein
transformation: process of reproduction in which genetic material is added to or replaces portions of a bacteria's DNA
transgenic: description of an organism that has been transformed or altered with genes from another organism
translation: process by which the nucleotides in mRNA are decoded into a sequence of amino acids in a polypeptide
transpiration: loss of water vapor though the stomata of a vascular plant
trichocyst: tiny bottle-shaped structure embedded in the pellicle of a paramecium and discharged for purposes of defense
trisomy: condition caused by cells that contain three copies of a chromosome rather than two
trochophore (TRAHK-oh-for): free-swimming larva stage of a mollusk
trophic level: feeding level in the flow of food energy and nutrients from primary producers to highest level consumers
tropism: response of an organism to an environmental stimulus
tube foot: suction-cuplike structure connected to the water vascular system of an echinoderm
tumor: mass of cells
tympanic membrane: portion of the ear that vibrates in response to sound; eardrum
tympanum: eardrum

U

umbilical cord: thin tube of embryonic tissue that connects the embryo to the uterus
ureter: vessel that carries urine from the kidney to the urinary bladder
urinary bladder: sac that stores urine before it is eliminated from the body
uterus: muscular chamber in a female's reproductive system in which a fertilized egg can develop

V

vaccine: weakened or mild form of a pathogen that causes permanent immunity when injected into the body
vacuole: saclike structure in a cell that stores materials—such as proteins, fats, and carbohydrates—in animal cells, and water and dissolved salts in plant cells
valve: specialized flap of tissue that prevents a backflow of blood
variable: factor that differs among test groups in an experiment and is measured against a control
vascular cylinder: central region of xylem and phloem cells carrying water and nutrients between the roots and the rest of the plant
vascular plant: plant with tracheids that draw water upward
vascular tissue: specialized tissue that transports water and nutrients throughout a land plant
vas deferens (VAS DEHF-uh-rehnz): duct that extends from the scrotum to the ejaculatory duct
vector: animal that carries a disease-causing organism from host to host
vector pollination: spread of pollen from one plant to another by an insect or animal
vegetative reproduction: process of asexual reproduction in which offspring are produced from the division of cells of the parent plant

vein: blood vessel that returns blood to the heart

ventricle: lower chamber of the heart that pumps the blood out of the heart

vertebra; pl. **vertebrae:** any of the individual segments of bone that make up the backbone, or vertebral column

vertebral column: backbone that encloses and protects the spinal cord

vestigial organ: structure in an organism that seems to have little or no obvious purpose

villus; pl., **villi:** any of the numerous projections on the folded surfaces of the small intestine that increase the surface area for the absorption of food molecules

virus: nonliving particle that contains DNA or RNA and that can infect a living cell

vitreous (VIH-tree-uhs) **humor:** transparent fluid that fills the large chamber behind the lens of the eye

viviparous (vigh-VIHP-er-uhs): retaining the developing embryo inside the female's body and bearing offspring alive

vocal cord: elastic fold of tissue that vibrates and produces sound when exhaled air is passed by it

voltage-sensitive gate: one of thousands of tiny protein channels in the cell membrane of a neuron through which sodium or potassium passes

W

water vascular system: in echinoderms, a network of fluid-filled tubes and appendages, used for many purposes

white blood cell: blood cell that fights infection, parasites, and bacterial disease; also called a leukocyte

X

xylem (ZIGH-luhm): vascular tissue that carries water and nutrients from the roots to the branches and leaves of a plant

Z

zero population growth: lack of population growth due to equality of a population's birth rate and death rate

zoospore (ZOH-oh-spor): reproductive cell that produces a new individual by cell division

zygote (ZIGH-goht): fertilized egg

Index

A

B

C

D

E

F

G

H

I

J

K

L

M

N

O

P

T

U

Credits

Photo Research: Natalie Goldstein

Photo Credits

Cover Frans Lanting/Minden Pictures, Inc.; Borders Corel Professional Photos CD-ROM™; **iv** t; Patricia Agre/Photo Researchers, Inc.; **iv** c; David Scharf/Peter Arnold, Inc.; **iv** bl; ©Philippe Plailly/Science Photo Library/Photo Researchers, Inc.; **iv** br; Leonard Lessin/ Peter Arnold, Inc.; **v** t; Corel Professional Photos CD-ROM™; **v** bl; ©Alfred Pasieka/Scince Photo Library/ Photo Researchers, Inc.; **v** br; Rich Cane/Sports Chrome East/West ; **vi** tl; Corel Professional Photos CD-ROM™; **vi** t;r; Corel Professional Photos CD-ROM™; **vi** bl; Corbis-Bettmann; **vi** br; Corel Professional Photos CD-ROM™; **vii** t; ©Biophoto Associates/Photo Researchers, Inc.; **vii** c; Dr. Dennis Kenkel/Phototake; **vii** b; Gopal Morti/CNRI/ Phototake; **viii** tr; K. G. Murti/Visuals Unlimited; **viii** cl; ©Oliver Meckes/Photo Researchers, Inc.; **viii** cr; ©Michael Fairchild/Peter Arnold, Inc.; **viii** b; Corel Professional Photos CD-ROM™; **ix** tl; Tui De Roy/ Bruce Coleman, Inc.; **ix** tr; Frans Lanting/Minden Pictures, Inc.; **ix** bl; Gerard Lacz/Peter Arnold, Inc.; **ix** br; Corel Professional Photos CD-ROM™; **x** t; Corel Professional Photos CD-ROM™; **x** bl; Dan Budnik/ Woodfin Camp & Associates; **x** br; Michael Fogden/ DRK Photo; **xi** tl; Art Wolfe Incorporated; **xi** tr; Lionel Isy Schwart/The Image Bank; **xi** c; D. Cavagnaro/DRK Photo; **xi** b; Bullaty/Lomeo/The Image Bank; **xii** tl; ©Holt Studios International (Miss P. Peackock)/Photo Researchers, Inc.; **xii** tr; T.E. Adams/Visuals Unlimited; **xii** bl; David M. Phillips/Visuals Unlimited; **xii** b; r; ©Phil A. Dotson/Photo Researchers, Inc.; **xiii** tl; Visuals Unlimited; **xiii** tr; Larry Lipsky/DRK Photo; **xiii** c; Corel Professional Photos CD-ROM™; **xiii** b; Frans Lanting/Minden Pictures, Inc.; **xiv** tl; ©David Scharf/ Peter Arnold, Inc.; **xiv** tr; Arthur J. Olson, The Scripps Research Institute, La Jolla, California, Copyr; 1988; **xiv** c; David Phillips/Visuals Unlimited; **xiv** b; ©Roger HartRainbow; **xv** t; Jeff Foot/DRK Photo; **xv** c; Robert & Linda Mitchell Photography; **xv** b; D. Cavagnaro/ DRK Photo; **xvi** tl; Jeffrey L. Rotman; **xvi** tr; ©Jackie Lewin, EM Unit Royal Free Hospital/Science Photo Library/Photo Researchers, Inc.; **xvi** bl; Larry Lipsky/ DRK Photo; **xvi** br; ©Photo Researchers, Inc.; **xvii** tl; photographer/DRK Photo; **xvii** tr; Art Wolfe Incorporated; **xvii** bl; ©Andrew Syred/Science Photo Library/Photo Researchers, Inc.; **xvii** br; S. Nielsen/ DRK Photo; **xviii** t; ©Sophie de Wilde Jacana/Jacana Scientific Control/Photo Researchers, Inc.; **xviii** c; Runk/Schoenberger/Grant Heilman Photography; **xviii** bl; Martim Harvey/The Wildlife Collection; **xix** t; John Callanan/The Image Bank; **xix** c; Johnny Johnson/DRK Photo; **xix** bl; Mark Moffett/Minden Pictures, Inc.; **xix** br; Art Wolfe Incorporated; **xx** tl; ©Professors P.M. Motta and S. Correr/Science Photo Library/Photo Researchers, Inc.; **xx** tr; Rob Tringali, Jr./Sports Chrome East/West ; **xx** c; GJLP/CNRI/ Phototake; **xx** b; ©Dr. Morley Read/Science Photo Library/Photo Researchers, Inc.; **xxi** tl; Paul J. Sutton/ Duomo Photography, Inc.; **xxi** tr; ©Prof. P. Motta/ Dept. of Anatomy/ University "La Sapienza", Rome/ Science Photo Library/Photo Researchers, Inc.; **xxi** c; ©Prof. Arnold Brody/Science Photo Library/Photo Researchers, Inc.; **xxi** b; David Phillips/Visuals Unlimited; **xxii** tl; William Sallaz/The Image Bank; **xxii** tr; Lennart Nilsson ©Boehringer Ingelheim International GmbH; **xxii** c; ©Professors P.M. Motta and J. Van Blerkom/Science Photo Library/Photo Researchers, Inc.; **xxii** b; Electra/Phototake; **xxiii** t; ©Don Fawcett/Photo Researchers, Inc.; **xxiii** cr; ©Lennart Nilsson, THE INCREDIBLE MACHINE; **xxiii** cl; ©Boehringer Ingelheim International GmbH, Photo by Lennart Nilsson, THE INCREDIBLE MACHINE; **xxiii** b; ©Boehringer Ingelheim International GmbH, Photo by Lennart Nilsson; **1** Stephen Wilkes/The Image Bank; **2** Thomas D. Mangelsen/Peter Arnold, Inc.; **3** t; Jeff Hunter/The Image Bank; c; Johnny Johnson/Animals Animals/ Earth Scenes; b; C & M Denis-Huot/C & M Denis-Huot; **4** l; David Scharf/Peter Arnold, Inc.; r; Jim Brandenburg/Minden Pictures, Inc.; **5** tl; Robert & Linda Mitchell Photography; tc; Dwight Kuhn Photography; tr; Luiz C. Marigo/Peter Arnold, Inc.; c; Frans Lanting/Minden Pictures, Inc.; bl; Tom and Pat Leeson/DRK Photo; br; A. Cosmos Blank/Photo Researchers, Inc.; **6** t; Vic Verlinder/The Image Bank; b; ©Norbert Wu; **7** Corel Professional Photos CD-ROM™; **8** MC. Chamberlain/DRK Photo; **9** t; David Scharf/Peter Arnold, Inc.; c; Corel Professional Photos CD-ROM™; b; David Scharf/Peter Arnold, Inc.; **10** NASA; **11** l; Photo Researchers, Inc.; r; Frans Lanting/ Minden Pictures, Inc.; **13** l; Patricia Agre/Photo Researchers, Inc.; r; F. Ruggeri/The Image Bank; **15** l; Coco McCoy/Rainbow; c; Penny Tweedie/Woodfin Camp & Associates; r; Bios (Klein-Hubert)/Peter Arnold, Inc.; **16** l; D. Cornwell/The Granger Collection Ltd.; r; Courtesy National Archives; **19** l; Robert Frerck/Woodfin Camp & Associates; r; USDA/ Science Source/Photo Researchers, Inc.; **24** Thomas D. Mangelsen/Peter Arnold, Inc.; **26** ©Philippe Plailly/ Science Photo Library/Photo Researchers, Inc.; **27** t; Corel Professional Photos CD-ROM™; b; Harald Sund/The Image Bank; **30** t; ©Hermann Eisenbeiss/ Photo Researchers, Inc.; b; ©Jerry Mason/Science Photo Library/Photo Researchers, Inc.; **31** ©Will and Deni McIntyre/Photo Researchers, Inc.; **33** Michael Fogden/DRK Photo; inset; Joe Van Os/The Image Bank; **35** David Young-Wolff/PhotoEdit; **36** Leonard Lessin/Peter Arnold, Inc.; **37** t; Peter L. Chapman/ Stock, Boston ; b; ICM Production/The Image Bank; **38** ©Bachmann/Photo Researchers, Inc.; **39** ©Leonard Lessin/Peter Arnold, Inc.; **40** © Tom & Pat Leeson; **46** Joe Van Os/The Image Bank; **48** ©Cecil Fox/Science SourcePhoto Researchers, Inc.; **49** l; The Bettmann Archive; c; ©Leonard Lessin/Peter Arnold, Inc.; r; Corbis-Bettmann; **50** tl; Corel Professional Photos CD-ROM™; b; ©Mark Burnett/Photo Researchers, Inc.; b inset; George J. Wilder/Visuals Unlimited; tr; Joe Devenney/The Image Bank; **52** tl; ©CNRI/Science Photo Library/Photo Researchers, Inc.; tr; ©Manfred Kage/Peter Arnold, Inc.; b; ©Philippe Plailly/Science Photo LibraryPhoto Researchers, Inc.; **53** l; ©Manfred Kage/Peter Arnold, Inc.; r; ©Manfred Kage/Peter Arnold, Inc.; **54** Antonio M. Rosario/The Image Bank; **55** t; ©Dan McCoy/Rainbow; b; Paul Silverman/ Fundamental Photographs; **57** t; Dr. Dennis Kunkel/ Phototake; c; Dr. Dennis Kunkel/Phototake; b; Dr. Dennis Kenkel/Phototake; **58** t; Visuals Unlimited/ David M. Phillips; b; ©Andrew McClenaghan/Science Photo Library/Photo Researchers, Inc.; **59** ©Dr. Arnold Brody/Science Photo Library/Photo Researchers, Inc.; **60** t; M. Eichelberger/Visuals Unlimited; b; Michael Abramson/Woodfin Camp & Associates; **61** ©Ed Reschke/Peter Arnold, Inc.; **63** tl; ©Don Fawcett/ Science Source/Photo Researchers, Inc.; tr; ©Biophoto Assoc./Science SourcePhoto Researchers, Inc.; b; Don W. Fawcett/Visuals Unlimited; **64** ©Dr. Gopal Hurti/ Science Photo Library/Photo Researchers, Inc.; **65** t; K.R. Porter/Photo Researchers, Inc.; b; ©Biophoto Association/Science SourcePhoto Researchers, Inc.; **66** l; ©Dan McCoy/Rainbow; c; James Dennis/CNRI/ Phototake; r; ©Don W. Fawcett/Rainbow; **72** Dr. Gopal Hurti/Science Photo Library/Photo Researchers, Inc.; **73** David M. Phillips/Visuals Unlimited; **74** Lindsay Hebberd/Woodfin Camp & Associates; **75** t; F. M. Whitney/The Image Bank; b; Corel Professional Photos CD-ROM™; **77** l; Runk/Scoenberger/Grant Heilman Photography; r; ©Alfred Pasieka/Science Photo Library/Photo Researchers, Inc.; **78** The Image Bank; inset; Ross M. Horowitz/The Image Bank; **79** t; Rich Cane/Sports Chrome East/West; b; Alan PitcairnGrant Heilman Photography; **81** l; Richard Jackson/RO-MA Stock©; r; Flip Nicklin/Minden Pictures, Inc.; **85** t; Kuhn, Inc./The Image Bank; b; Martin Rogers/Stock, Boston; **86** l; ©Biophoto Assoc./Photo Researchers, Inc.; r; C. Bradley Simmons/Bruce Coleman, Inc.; **92** ©Photo Researchers, Inc.; **94** t; Alan Pitcairn/Grant Heilman Photography; b; David M. Phillips/Visuals Unlimited; **98** Runk/Scoenberger/Grant Heilman Photography; **99** Corel Professional Photos CD-ROM™; **100** K.G. Murti/Visuals Unlimited; **101** t; ©Clay Myers/Photo Researchers, Inc.; c; ©Tim Davis/Photo Researchers, Inc.; b; ©Photo Researchers, Inc.; **105** l; David M. Phillips/Visuals Unlimited; r; R. Celentine/Visuals Unlimited; **106** Lennart Nilsson/ Bonnier Alba; **108** t; ©Dr. Gopal Murti/Photo Researchers, Inc.; b; R. Calentine/Visuals Unlimited; **109** l; ©CNRI/Science Phtoto Library/Photo Researchers, Inc.; r; Lennart Nilsson/Bonnier Alba; **111** ©Dr. Brain Eyden/Science Photo LibraryPhoto Researchers, Inc.; **112** t; Corel Professional Photos CD-ROM™; bl; ©Hans Pfletschinger/Peter Arnold, Inc.; br; ©Hans Pfletschinger/Peter Arnold, Inc.; **113** ©Will and Deni McIntyre/Science Source/Photo Researchers, Inc.; **114** Corel Professional Photos CD-ROM™; **117** ©Dr. Gopal Murti/Photo Researchers, Inc.; **118** David M. Phillips/Visuals Unlimited; **119** Corel Professional Photos CD-ROM™; **121** ©The Stock Market/William Roy; **122** ©Archive Photos/Lambert Photography, 1993/PNI; **123** l; John Eastcott/Yva Momatiuk/DRK Photo; c; Johnny Johnson/DRK Photo; r; ©Bill BacHman/Photo Researchers, Inc.; **124** t; ©Tom and Pat Leeson/Photo Researchers, Inc.; b; Corbis-Bettmann; b inset; Larry Lefever/Grant Heilman Photography; **125** ©Philippe Plailly/Science Photo LibraryPhoto Researchers, Inc.; **131** l; Corel Professional Photos CD-ROM™; r; ©David M. Phillips/Photo Researchers, Inc.; **135** l; Cliff Riedinger/ Natural Selection Stock Photography, Inc.; r; Al Hamdan/The Image Bank; **138** l; ©IFA/Peter Arnold, Inc.; r; Corel Professional Photos CD-ROM™; **139** bl; Hans Reinhard/Bruce Coleman, Inc.; br; Larry Lefever/ Grant Heilman Photography; **144** ©Jerome Wexler/ Photo Researchers, Inc.; **146** Chip Henderson/Tony Stone Images; **147** l; Runk/Schoenberger/Grant Heilman Photography; r; ©Oliver Meckes/Photo Researchers, Inc.; **148** l; ©Martin Cooper/Peter Arnold, Inc.; r; ©CNRI/Science Photo Library/Photo Researchers, Inc.; **150** ©Ed Reschke/Peter Arnold, Inc.; **151** l; Carnegie Institution of Washington; r; ©Stephen Collins/National Audobon Society/Photo Researchers, Inc.; **153** ©Biophoto Associates/Photo Researchers, Inc.; **154** ©Gunn & Stewart/Mary EvansPhoto Researchers, Inc.; **155** l; Dr. Dennis Kunkel/Phototake; r; Stanley Flegler/Visuals Unlimited; **156** ©Reinhard Kunkel/Peter Arnold, Inc.; **158** l; Kunkel/Phototake; r; A. Berliner/Liaison International; **160** t; Cabisco/Visuals Unlimited; b; ©Robert Maier/ Animals Animals; **161** ©Will & Deni McIntyre/Photo Researchers, Inc.; **162** ©Applied Biosystems/Peter Arnold, Inc.; **165** FBI; **168** ©Biophoto Associates/ Photo Researchers, Inc.; **170** ©Kenneth Eward/ BioGrafx—Science SourcePhoto Researchers, Inc.; **171** tl; Ian Yedmans/Woodfin Camp & Associates; tr; ©Dr. Gopal Murti/Science Photo LibraryPhoto Researchers, Inc.; b; Lynn Saville; **173** Cold Spring Harbor Laboratory; **175** ©R. Langridge/D. McCoy/ Rainbow; **176** l; Cold Spring Harbor Laboratory; r; ©Science SourcePhoto Researchers, Inc.; 177 The Hulton-Deutsch Collection/Woodfin Camp & Associates; **178** t; ©James Holmes/Cellmark Diagnostic/ Science Photo Library/Photo Researchers, Inc.; b; Superstock; **181** l; National Museum of American History, Smithsonian Institution; c; Professor Oscar Miller/Science Photo Library/Photo Researchers, Inc.; r; ©Ken Eward/Biografx/Photo Researchers, Inc.; **187** l; Larry Lefever/Grant Heilman Photography; r; Gopal Morti/CNRI/Phototake; **194** Claude Revy, Jean/ Phototake; **196** ©Mitsuaki Iwago/Photo Researchers, Inc.; **197** t; Photograph courtesy of Appaloosa Museum & Heritage C, Moscow, ID; b; ©Fritz Prenzel/Peter Arnold, Inc.; **198** Corbis-Bettman; **201** l; ©Philippe Plailly/Science Photo Library/Photo Researchers, Inc.; **201** r; ©Matt Meadows/Peter Arnold, Inc.; **203** K. G. Murti/Visuals Unlimited; **204** ©Leonard Lessin/Peter Arnold, Inc.; **205** l; Wackson Lab/Visuals Unlimited; r; Museum of Israel/Jerusalem/Giraudon, Paris, Superstock; **206** Keith Wood/Visuals Unlimited; **207** l; Gerard R. Lazo; r; Stewart Cohen/Tony Stone Images; **208** Grant Heilman Photography; **209** ©Philippe Plailly/Science Photo Library/Photo Researchers, Inc.; **214** Wackson Lab/Visuals Unlimited; **215** ©Fritz Prenzel/Peter Arnold, Inc.; **217** Henry Holdsworth/Minden Pictures, Inc.; **218** Frans Lanting/Minden Pictures, Inc.; **219** l; N. H. (Dan) Cheatham/DRK Photo; rt; Clyde H. Smith/Peter Arnold, Inc.; rb; ©Oliver Meckes/Photo Researchers, Inc.; **220** Charles Darwin Museum, Down

House & The Royal College of Surgeons; **221** tl; Syndics of Cambridge University Library; tr; Joe McDonald/Bruce Coleman, Inc./PNI; b; ©Fritz Polking/ Peter Arnold, Inc.; **222** Barbara Gerlach/DRK Photo; **224** t; ©Jeffrey L. Rotman/Peter Arnold, Inc.; bl; ©Soames Summerhays/Photo Researchers, Inc.; br; Carr Clifton/Minden Pictures, Inc.; **225** l; Stanley Breeden/DRK Photo; r; T. Wiewandt /DRK Photo; **226** l; Frans Lanting/Minden Pictures, Inc.; r; Frans Lanting/Minden Pictures, Inc.; inset; David Cavagnaro/ DRK Photo; **227** Derek Witty; **228** l; Lewis Kemper/ DRK Photo; r; Frans Lanting/Minden Pictures, Inc.; **230** l; Runk/Schoenberger/Grant Heilman Photography; r; ©M.E. Warren/Photo Researchers, Inc.; **231** l; ©Biophoto Associates/Photo Researchers, Inc.; r; ©David M. Phillips/Science Source/Photo Researchers, Inc.; **232** t; Robert & Linda Mitchell Photography; c; ©Michael Fairchild/Peter Arnold, Inc.; b; Mitsuaki Iwago/Minden Pictures, Inc.; **233** Breck P. Kent; **234** Robert & Linda Mitchell Photography; **235** t; Runk/Schoenberger/Grant Heilman Photography; b; Breck P. Kent; **236** Corel Professional Photos CD-ROM™; **238** Corel Professional Photos CD-ROM™; **239** Robert & Linda Mitchell Photography; **240** Mark Moffett/Minden Pictures, Inc.; **241** t; Jeremy Woodhouse/DRK Photo; bl; Benn Mitchell/The Image Bank; br; S. Nielsen/DRK Photo; **242** l; Frans Lanting/ Minden Pictures, Inc.; rt; Perry Conway/Corbis; rb; Ron Kimball Studios; **246** tr; Darell Gulin/DRK Photo; tl; Tom Bean/DRK Photo; b; Dietrich Gehring/DRK Photo; **249** ©Hank Morgan/Rainbow; **253** tl; Tui De Roy/Bruce Coleman, Inc.; tr; Tui De Roy/Bruce Coleman, Inc.; bl; Tui De Roy/Bruce Coleman, Inc.; br; Norman Owen Tomalin/Bruce Coleman, Inc.; **260** Darell Gulin/DRK Photo; **261** Corel Professional Photos CD-ROM™; **262** ©The Stock Market/ZEFA Germany; **263** tl; Corel Professional Photos CD-ROM™; tr; Corel Professional Photos CD-ROM™; b; Corel Professional Photos CD-ROM™; **265** l; Corel Professional Photos CD-ROM™; c; ©Stephen Dalton/ Photo Researchers, Inc.; r; Robert & Linda Mitchell Photography; **268** l; Corel Professional Photos CD-ROM™; c; Art Wolfe Incorporated; r; ©M.I. Waler/ Science SourcePhoto Researchers, Inc.; **269** Wolfgang Kaehler; **271** l; D. Cavagnaro/DRK Photo; c; G. Prance/Visuals Unlimited; r; Robert Burke; **272** t; ©Ken Edward/Science Source/Photo Researchers, Inc.; b; Jeff Spielman/The Image Bank; **274** ©Ken Edward/ Science SourcePhoto Researchers, Inc.; **278** Frans Lanting/Minden Pictures, Inc.; **279** Tim Laman/The Wildlife Collection; **281** Charles Gurche/The Wildlife Collection; **282** Holt Studios International (Nigel Cattlin)/Photo Researchers, Inc.; **283** t; Art Wolfe/Art Wolfe Incorporated; bl; New England Aquarium/Paul Erickson/Animals Animals/Earth Scenes; br; Breck P. Kent; **286** l; Earth Satellite Corporation/Science Photo Library/Photo Researchers, Inc.; r; ©Mark Burnett/ Photo Researchers, Inc.; **287** tl; Wolfgang Kaehler; tr; ©Johnny Johnson/Animals Animals; b; Robert Ferck/ Odyssey Productions; **288** Robert Ferck/Odyssey Productions; l inset; Peter Miller/The Image Bank; r inset; Tom Brakefield/DRK Photo; **291** t; Gary Gray/ DRK Photo; bl; Kennan Ward /DRK Photo; br; Tom Bean /DRK Photo; **297** t; John Gerlach/DRK Photo; b; Doug Milner Photography/DRK Photo; **300** t; Marty Cordano/DRK Photo; bl; Stephen J. Krasemann/DRK Photo; br; Frans Lanting/Minden Pictures, Inc.; **304** Robert & Linda Mitchell Photography; **308** Art Wolfe Incorporated; **309** Charlie Palek/Animals Animals; **310** ©Clyde H. Smith/Peter Arnold, Inc.; **311** l; Photograph courtesy of Dr. George Bowes, Department of Botany and the C for Aquatic Plants, University of Florida, Gainesville, Florida; r; Momatiuk/Eastcott/ Woodfin Camp & Associates; **315** l; ©Tui De Roy/ Oxford Scientific Films/Animals Animals; c; Dan Budnik/Woodfin Camp & Associates; r; Dan Budnik/ Woodfin Camp & Associates; **318** Jeffrey L. Rotman; **319** l; ©1994, Art Montes De Oca/FPG International Corp.; r; Michael Quackenbush/The Image Bank; **320** ©Norbert Wu; **322** l; Corel Professional Photos CD-ROM™; r; Guido Alberto Rossi/The Image Bank; **326** David M. Phillips/Visuals Unlimited; **330** Dan Budnik/ Woodfin Camp & Associates; **332** Kevin Schafer/Tom Stack & Associates; **333** t; Bullaty/Lomeo/The Image Bank; b; Steve Krongard/The Image Bank; **336** l; ©Michael Sewell/Peter Arnold, Inc.; r; ©Allan Morgan/Peter Arnold, Inc.; **337** t; Lionel Isy Schwart/ The Image Bank; bl; Anthony Johnson/The Image Bank; **337** br; John Banagan/The Image Bank; **340** tl; Wolfgang Kaehler; tr; Frans Lanting/Minden Pictures, Inc.; b; ©Francois Gohier/Photo Researchers, Inc.; **341** The Image Bank; **342** t; Lynn M. Stone/DRK Photo; b; Carr Clifton/Minden Pictures, Inc.; **343** t; Tom Bean/ DRK Photo; c; Frans Lanting/Minden Pictures, Inc.; b; Art Wolfe Incorporated; **342-343** Background; Rand McNally; **344** tl; Jeffrey L. Rotman; tr; Carr Clifton/ Minden Pictures, Inc.; bl; ©Norbert Wu; br; C.C. Lockwood/DRK Photo; Background; Rand McNally; **346** Jim Brandenburg/Minden Pictures, Inc.; r inset; Frank S. Balthis/Natural Selection Stock Photography, Inc.; l inset; Bullaty/Lomeo/The Image Bank; **348** l; Charles Mauzy/Natural Selection Stock Photography, Inc.; r; Geoffrey Clifford/Woodfin Camp & Associates; **350** l; ©NASA/Science Photo Library/Photo Researchers, Inc.; r; Ira Block/The Image Bank; **356** Tom Bean/DRK Photo; **357** Frans Lanting/Minden Pictures, Inc.; **358** Andre Gallant/The Image Bank; **359** l; American Museum of Natural History, Photo by P. Hollembeak/J. Beckett; c; Bill Ross/Woodfin Camp & Associates; r; Corel Professional Photos CD-ROM™; **360** Bob Harrington/Michigan Department of Natural Resources; **363** l inset; Giraudon/Art Resource, NY; r inset; Grant V. Faint/The Image Bank; A. Baccaccio/ The Image Bank; **364** l; Miao Wang/The Image Bank; r; A. Upitis/The Image Bank; **365** Photo provided by the National Cattlemen's Association; **367** Jim Brandenburg/Minden Pictures, Inc.; **368** t; Guido A. Rossi/The Image Bank; c; Guido Rossi/The Image Bank; bl; ©Uniphoto, Inc.; br; ©Lowell Georgia/Photo Researchers, Inc.; Background; Corel Professional Photos CD-ROM™; **370** Paul McCormick/The Image Bank; **371** t; A. Ramey/Woodfin Camp & Associates; inset; ©Photo Researchers, Inc.; b; Jeffrey D. Smith/Woodfin Camp & Associates; **372** l; Robert Phillips/The Image Bank; r; ©John Mead/Photo Researchers, Inc.; **373** t; Kevon Schafer/Natural Selection Stock Photography, Inc.; c; G. Brimacombe/ The Image Bank; b; Mark Moffett/Minden Pictures, Inc.; **374** ©E.R. Degginger/Animals Animals; **375** ©Bernard Giani/Photo Researchers, Inc.; inset; Robert Cameron/Cameron & Company; **377** Jeff Foott Productions; **382** Wolfgang Kaehler; **383** D. Cavagnaro/DRK Photo; **385** John Trager/Tom Stack & Associates; **386** ©Francois GohierPhoto Researchers, Inc.; **387** t; Corel Professional Photos CD-ROM™; b; NASA; **388** l; Tom Bean/DRK Photo; r; David M. Dennis /Tom Stack & Associates; **391** Jonathan Blair/ Woodfin Camp & Associates; **392** Joe Englander/ Natural Selection Stock Photography, Inc.; **393** t; Corel Professional Photos CD-ROM™; bl; Steve Proehl/The Image Bank; br; Corel Professional Photos CD-ROM™; **398** t; T.E. Adams/Visuals Unlimited; bl; ©CNRI/Science Photo Library/Photo Researchers, Inc.; br; ©Dr. L. Caro/Science Photo Library/Photo Researchers, Inc.; **399** l; Corel Professional Photos CD-ROM™; r; ©Bonnie Sue Rauch/Photo Researchers, Inc.; **401** l; ©Holt Studios International (Miss P. Peackock)/Photo Researchers, Inc.; r; ©NIBSC/Science Phot Library/Photo Researchers, Inc.; **402** ©CNRI/ Science Photo Library/Photo Researchers, Inc.; **403** ©NIBSC/Science Phot LibraryPhoto Researchers, Inc.; **404** t; ©Soames Summerhays/Photo Researchers, Inc.; b; T.A. Wiewandt/DRK Photo; **405** Roger Ressmeyer/ Corbis; **406** Sidney Fox/Visuals Unlimited; **408** Don & Pat Valenti/DRK Photo; **412** ©Karen Tweedy-Holmes Animals Animals; **413** ©Lee D. Simon/Science Source/Photo Researchers, Inc.; **414** Chuck Place/The Image Bank; **415** t; David M. Phillips/Visuals Unlimited; bl; A.M. Siegelman/Visuals Unlimited; br; ©Alfred Pasieka/Peter Arnold, Inc.; **418** l; John D. Cunningham/Visuals Unlimited; r; ©Norbert Wu; **419** l; William E. Ferguson; r; Mike Abbey/Visuals Unlimited; **420** l; ©Phil A. Dotson/Photo Researchers, Inc.; c; Tom Bean/DRK Photo; r; Corel Professional Photos CD-ROM™; **422** l; ©Noble Proctor/Science Source/Photo Researchers, Inc.; r; John D. Cunningham/Visuals Unlimited; **423** tl; Jack M. Bostrack/Visuals Unlimited; bl; ©Viard/Jacana/Photo Researchers, Inc.; r; S. Nielsen/DRK Photo; **424** Michael Fogden/DRK Photo; **425** tl; Corel Professional Photos CD-ROM™; tr; Corel Professional Photos CD-ROM™; b; ©Ed Reschke/Peter Arnold, Inc.; **427** r; Walt Anderson/Visuals Unlimited; l; William E. Ferguson; inset; Jeff Foott/DRK Photo; **428** Anne Rippy/The Image Bank; **429** Antonio M. Rosario/The Image Bank; **431** Gary Braasch/Woodfin Camp & Associates; **432** l; Wayne Lankinen/DRK Photo; r; ©Hans Pfletschinger/Peter Arnold, Inc.; **433** ©Merlin D. Tuttle/Bat Conservation Photo Researchers, Inc.; **434** Carolina Biological Supply Company / Phototake NYC; **435** t; William E. Ferguson; b; ©Jeff Lepore/ Photo Researchers, Inc.; **438** ©Phil A. Dotson/Photo Researchers, Inc.; **440** John Eastcott/Yva Momatiuk/ DRK Photo; **441** ©Merlin D. Tuttle/Bat Conservation International/Photo Researchers, Inc.; **442** Stephen Frink/Water House Stock Photography; **443** l; O. Louis Mazzatenta/National Geographic Image Collection; c; K. Push/Visuals Unlimited; r; ©Kevin Schafer/Peter Arnold, Inc.; **444** Ken Lucas/Visuals Unlimited; **446** t; Jeffrey L. Rotman; bl; Doug Sokell/Visuals Unlimited; br; Robert Parks/The Wildlife Collection; **447** Dave B. Fleetham/Visuals Unlimited; **448** Daniel W. Gotshall/ Visuals Unlimited; **449** t; ©Zig Leszczynski/Animals Animals; bl; Visuals Unlimited; br; ©CNRI/Science Photo Library/Photo Researchers, Inc.; **450** t; ©Robert Dunne/Photo Researchers, Inc.; b; A. Kerstitch/Visuals Unlimited; **451** tl; Henry Holdsworth/The Wildlife Collection; tr; Mark W. Moffett/Minden Pictures, Inc.; b; Hal Beral/Visuals Unlimited; **452** t; ©Norbert Wu; b; Hal Beral/Visuals Unlimited; **457** Jeffrey L. Rotman; **459** Flip Nicklin/Minden Pictures, Inc.; **460** l; ©Bruce Watkins/Animals Animals; r; Carolina Biological Supply/Phototake; **461** ©Kevin Aitke/Peter Arnold, Inc.; **462** Carolina Biological Supply/Phototake; **466** Hal Beral/Visuals Unlimited; **468** Stuart Westmorland/ Tony Stone Images; **469** t; Corel Professional Photos CD-ROM™; bl; Corel Professional Photos CD-ROM™; br; Art Wolfe/Tony Stone Images; **470** From WONDERFUL LIFE: The Burgess Shale and the Nature of History by Stephen Jay Gould. Copyr ©1989 by Stephen Jay Gould. Reprinted by permission of W.W. Norton & Company, Inc.; **471** t; ©Steinhart Aquarium/Tom McHughPhoto Researchers, Inc.; b; ©Estate of Dr. J. Metzner/Peter Arnold, Inc.; **473** Davd M. Dennis/Tom Stack & Associates; **475** tl; ©BIOS (Pu Tao)/Peter Arnold, Inc.; tr; Corel Professional Photos CD-ROM™; b; Superstock ; **476** tl; Corel Professional Photos CD-ROM™ ; tr; Corel Professional Photos CD-ROM™; bl; Art Wolfe/Tony Stone Images; br; Larry LipskyDRK Photo; **477** Art Wolfe/Tony Stone Images; **478** tl; Frans Lanting/Minden Pictures, Inc.; tr; Hans Reinhard/Bruce Coleman, Inc.; b; ©Gerard Lacz/Peter Arnold, Inc.; **479** t; ©The Stock Market/ZEFA Germany; c; ©The Stock Market/ZEFA Germany; b; ©The Stock Market/ZEFA Germany; **480** Peter Arnold, Inc.; **484** l; John Cancalosi/DRK Photo; r; Kim Heacox/DRK Photo; **486** t; ©Michael Fogden/ Animals Animals; bl; Visuals Unlimited; br; Corel Professional Photos CD-ROM™; **492** Larry Lipsky/ DRK Photo; **493** ©Phil A. DotsonPhoto Researchers, Inc.; **494** Norman Owen Tomalin/Bruce Coleman, Inc.; **495** t; Douglas Mazonowicz/Bruce Coleman, Inc.; bl; ©De Sazo/Photo Researchers, Inc.; br; Michael Holford Photographs; **496** tl; Rod Williams/Bruce Coleman, Inc.; tr; Frans Lanting/Minden Pictures, Inc.; bl; Chamberlain, M.C./DRK Photo; br; ©BIOS (Compost/Visage)/Peter Arnold, Inc.; **498** l; Cleveland Museum of Natural History, c; Phil Schermeister, 1989/ All Stock/PNI, rt, rb; David L. Brill Photograph; **501** ©Jean-Loup Charmet/Science Photo Library/Photo Researchers, Inc.; **502** l; The Metropolitan Museum of Art, Bashford Dean Memorial Collection, Purchase, 1929.(29.158.142); c; The Metropolitan Museum of Art, Gift of Jean Jacques Reubell, in memory of his mother, Julia C. Coster, and of his wife, Adeline E. Post, both of New York City, 1926. (26.145.104); r; Museo America, Madrid/Bridgeman Art Library, London/Superstock; **505** ©Hank Morgan/Photo Researchers, Inc.; **508** ©BIOS (Compost/Visage)/Peter Arnold, Inc.; **510** ©John Reader/Science Photo Library/Photo Researchers, Inc.; **511** ©Photo Researchers, Inc.; **513** Tammy Peluso/Tom Stack & Associates; **514** ©Manfred Kage/Peter Arnold, Inc.; **515** t; ©NIBSC/Science Photo Library/Photo Researchers, Inc.; bl; ©CNRI/Science Photo

LibraryPhoto Researchers, Inc.; br; David M. Phillips/Visuals Unlimited; **516** l; CNRI/Phototake; r; Martin Rotker/Phototake; **517** tl; David Phillips/Visuals Unlimited; tc; David M. Phillips/Visuals Unlimited; tr; Mike Gabridge/Visuals Unlimited; b; Richard Nowitz/Phototake; **519** l; David M. Phillips/Visuals Unlimited; c; ©David Scharf/Peter Arnold, Inc.; r; Cabisco/Visuals Unlimited; **520** l; ©Dr. Jeremy Burgess/Science Photo Labrary/Photo Researchers, Inc.; r; ©R. Knanft/Photo Researchers, Inc.; **521** ©Philippe Plailly/Eurelios/Science Photo Library/Photo Researchers, Inc.; **522** Hans Gelderblom/Visuals Unlimited; **524** Arthur J. Olson, The Scripps Research Institute, La Jolla, California, Copyr 1988; **525** l; Corel Professional Photos CD-ROM™; rt; Corel Professional Photos CD-ROM™; rb; Corel Professional Photos CD-ROM™; **527** l; John Edwards/Tony Stone Images; r; John Colwell/Grant Heilman Photography; **528** Jeffrey Hutcherson/DRK Photo; **532** ©Oliver Meckes/Photo Researchers, Inc.; **533** CDC/Phototake; **534** Frans Lanting/Minden Pictures, Inc.; **535** tl; Philip Sze/Visuals Unlimited; tr; Cabisco/Visuals Unlimited; b; Stanley Flegler/Visuals Unlimited; **536** ©Scott Camazine/Photo Researchers, Inc.; inset; ©Eric Grave/Science SourcePhoto Researchers, Inc.; **537** t; A.M Siegelman/Visuals Unlimited; c; A.M. Siegelman/Visuals Unlimited; b; A.M. Siegelman/Visuals Unlimited; **538** t; M. Abbey/Visuals Unlimited; c; A.M. Siegelman/Visuals Unlimited; b; Ken Wagner/Visuals Unlimited; **539** ©Tom E. Adams/Peter Arnold, Inc.; **540** l; ©Cabisco/Peter Arnold, Inc.; r; David Phillips/Visuals Unlimited; **542** l; ©Ed Reschke/Peter Arnold, Inc.; r; ©Stanley Flegler/Peter Arnold, Inc.; inset; ©James W. Richardson/Peter Arnold, Inc.; **543** l; ©Ralph Lee Hopkins/Peter Arnold, Inc.; r; ©Ed Reschke/Peter Arnold, Inc.; inset; ©Stanley Flegler/Photo Researchers, Inc.; **544** Don & Pat Valenti/DRK Photo; **545** t; D. Cavagnaro/DRK Photo; b; ©Ed Reschke/Peter Arnold, Inc.; **546** ©CNRI/Science Photo LibraryPhoto Researchers, Inc.; **547** t; John D. Cunningham/Visuals Unlimited; bl; Mitch Kezar/Phototake; br; ©Nigel Cattlin/Holt Studios Int. Photo Researchers, Inc.; **548** t; W.E. Ferry/Visuals Unlimited; b; ©Breck Kent/Animals Animals; **549** ©Rod Planck/Photo Researchers, Inc.; **550** M. Abbey/Visuals Unlimited; **554** Michael Melford/The Image Bank; **556** D. Cavagnaro/DRK Photo; **557** t; ©Michel Viard/Peter Arnold, Inc.; bl; ©Linda Sims/Photo Researchers, Inc.; br; ©Manfred Kage/Peter Arnold, Inc.; **558** ©Bud Lehnhausen/Photo Researchers, Inc.; inset; ©Nancy Sefton/Photo Researchers, Inc.; **560** ©Philip Szer/Photo Researchers, Inc.; **561** ©Daniel W. Gotshall/Photo Researchers, Inc.; **562** l; Ken Wagner/Phototake; c; ©E.R. Degginger/Photo Researchers, Inc.; r; John D. Cunningham/Phototake; **565** t; ©Scott Camazine/Photo Researchers, Inc.; bl; ©Walter H. Hodge/Peter Arnold, Inc.; bc; ©Henry Holdsworth/Peter Arnold, Inc.; br; ©Roger HartRainbow; **566** ©Ed Reschke/Peter Arnold, Inc.; **568** t; Jeff Foot/DRK Photo; bl; Garry Gay/The Image Bank; br; Michael Newman/PhotoEdit; **569** t; Bob Elsdale/The Image Bank; b; Michal Heron/Woodfin Camp & Associates; **570** ©John D. Cunningham/Photo Researchers, Inc.; **572** ©Stephen Dalton/Photo Researchers, Inc.; **574** Frans Lanting/Minden Pictures, Inc.; **575** ©Ed Reschke/Peter Arnold, Inc.; **576** ©Riccardo Marcialis/Photo Researchers, Inc.; **577** t; Alan Pitcairn/Grant Heilman Photography; bl; Corel Professional Photos CD-ROM™; br; Corel Professional Photos CD-ROM™; **578** Runk/Schoenberger/Grant Heilman Photography; **580** l; Robert & Linda Mitchell Photography; c; Holt Confer/Grant Heilman Photography; r; E.F. Anderson/Visuals Unlimited; **581** l; Robert & Linda Mitchell Photography; r; ©Larry Brownstein/Rainbow; **583** S. Nielsen/DRK Photo; **584** l; Michael Fogden/DRK Photo; c; Dwight Kuhn/DRK Photo; r; Pascal Perret/The Image Bank; **585** Thomas Hovland/Grant Heilman Photography; **588** l; Robert & Linda Mitchell Photography; c; Runk/Schoenberger/Grant Heilman Photography; r; Paul E. Loven/The Image Bank; **589** l; ©Ed Reschke/Peter Arnold, Inc.; r; ©Ed Reschke/Peter Arnold, Inc.; **590** l; Christi Carter/Grant Heilman Photography; c; Alan Pitcairn/Grant Heilman Photography; r; Alan Pitcairn/Grant Heilman Photography; **591** t; ©Herminia Dosal/Photo Researchers, Inc.; c; ©G. Buttner/Naturbild/Okapia/Photo Researchers, Inc.; b; Steve Kaufman/DRK Photo; **593** D. Cavagnaro/DRK Photo; **594** ©Adam Hart-Davis/Science Photo Library/Photo Researchers, Inc.; **598** Jim Brandenburg/ Minden Pictures, Inc.; **599** John Eastcott/Yva Momatiuk/DRK Photo; **600** Larry Lefever/Grant Heilman Photography; **601** t; Carr Clifton/Minden Pictures, Inc.; b; Joseph Van Os/The Image Bank; b inset; ©S.E. Cornelius /Photo Researchers, Inc.; **603** l; Dwight R. Kuhn/DRK Photo; r; Robert & Linda Mitchell Photography; **606** Stephen J. Krasemann/ DRK Photo; **607** t; ©Dr. Jeremy Burgess/Science Photo Library/Photo Researchers, Inc.; b; ©Dr. Jeremy Burgess/Science Photo Library/Photo Researchers, Inc.; **609** t; John W. Banagan/The Image Bank; c-1; Robert & Linda Mitchell Photography; c-2; ©Jim Steinberg/ Photo Researchers, Inc.; b l; Robert & Linda Mitchell Photography; b r; Robert & Linda Mitchell Photography; **611** l; Carr Clifton/Minden Pictures, Inc.; c; John D. Cunningham/Visuals Unlimited; rt; Corel Professional Photos CD-ROM™; rb; Tim Laman/Minden Pictures, Inc.; **612** tl; Robert & Linda Mitchell Photography; tr; Robert & Linda Mitchell Photography; b; ©Jerome Wexler/Photo Researchers, Inc.; **615** t; Ken Wagner/Phototake; b; Ken Wagner/ Phototake; **616** l; ©Coco McCoy/Rainbow; c; ©Kevin Schafer/Peter Arnold, Inc.; r; D. Cavagnaro/DRK Photo; **617** tl; ©Ed Reschke/Peter Arnold, Inc.; tr; Alan & Linda Detrich/Photo Researchers, Inc.; bl; ©Holt Studios Int./Photo Researchers, Inc.; br; Brownie Harris/Tony Stone Images; **618** Patricia Agre/Photo Researchers, Inc.; **624** Robert & Linda Mitchell Photography; **625** ©Jerome Wexler /Photo Researchers, Inc.; **627** ©Terry Qing/FPG International Corp.; **628** Hall Baral/Visuals Unlimited; **629** t; ©Norbert Wu; bl; ©Stephen Frink/Photo Researchers, Inc.; br; Jeffrey L. Rotman; **630** ©Louisa PrestonPhoto Researchers, Inc.; **631** ©Dr. E.R. Degginger/Photo Researchers, Inc.; **633** ©M.I. Walker/Science Source/ Photo Researchers, Inc.; **634** tl; ©Oxford Scientific Films/Animals Animals; tr; ©Norbert Wu; b; John D. Cunningham/Visuals Unlimited; **635** ©Norbert Wu; **636** t; ©Dr. Jeremy Burgess/Science Photo Library/ Photo Researchers, Inc.; c; ©Jackie Lewin, EM Unit Royal Free Hospital/Science Photo Library/Photo Researchers, Inc.; b; ©Milton Love/Peter Arnold, Inc.; **637** ©M.I. Walker/Science Sourcer/Photo Researchers, Inc.; **638** Dwight Kuhn Photography; **639** ©CNRI/ Science Photo Library/Photo Researchers, Inc.; **640** t; Sinclair Stammers/Science Photo Library/Photo Researchers, Inc.; c; ©L. Jensen/Photo Researchers, Inc.; b; Photo courtesy of USDA Nematology Laboratory, Beltsville, Maryland; **642** ©M.I. Walker/Science Sourcer/Photo Researchers, Inc.; **644** Cabisco/Visuals Unlimited; **646** ©Norbert Wu; **647** Arthur M. Siegelman/Visuals Unlimited; **648** ©Steve Krasemann/Peter Arnold, Inc.; **649** t; Jeff Foott Productions; bl; ©M.H. Sharp/Photo Researchers, Inc.; br; ©Tom McHugh/Photo Researchers, Inc.; **651** tl; ©Lynn Funkhouser/Peter Arnold, Inc.; tr; ©Norbert Wu/DRK Photo; b; ©Peter Parks/Mo Young Productions; **652** ©Ed Reschke/Peter Arnold, Inc.; **654** t; Larry Lipsky/DRK Photo; c; ©B.B. Cadbury/Photo Researchers, Inc.; b; ©Norbert Wu/DRK Photo; **656** l; ©David Thompson/Oxford Scientific Films/Animals Animals; c; ©Oxford Scientific Films/Animals Animals; r; Richard Nowitz/PNI; **657** t; ©Kjell B. Sandved/Photo Researchers, Inc.; c; ©Norbert Wu/DRK Photo; b; ©Fred Winner/Jacana/Photo Researchers, Inc.; **658** t; ©Fred McConnaughey/Photo Researchers, Inc.; bl; ©Fred Bavendam/Peter Arnold, Inc.; br; ©Norbert Wu; **660** t; ©Norbert Wu; c; ©Dan McCoy/Rainbow; b; ©Fred Bavendam/Peter Arnold, Inc.; **661** F. Stuart Westmorland/Mo Young Productions; **663** ©Jeanne White/Photo Researchers, Inc.; **666** ©Seitre/Peter Arnold, Inc.; **668** Marty Snyderman; **669** t; Art Wolfe Incorporated; b; Robert & Linda Mitchell Photography; **670** l; ©Holt Studios International (Nigel Cattlin)/Photo Researchers, Inc.; r; ©J H Robinson/Photo Researchers, Inc.; **672** ©Andrew Syred/Science Photo Library/Photo Researchers, Inc.; **673** ©Hans Pfletschinger/Peter Arnold, Inc.; **674** t; Robert & Linda Mitchell Photography; bl; ©Mark Burnett/Photo Researchers, Inc.; br; ©Manfred Kage/Peter Arnold, Inc.; **675** t; ©Michael Lustbader/Photo Researchers, Inc.; b; ©Tom McHugh/Photo Researchers, Inc.; **676** lt; Robert & Linda Mitchell Photography; lb; ©John R. McGregor/Peter Arnold, Inc.; r; ©Stephen Dalton/Oxford Scientific Films/Animals Animals; **677** tl; Larry Lipsky/DRK Photo; tr; Robert & Linda Mitchell Photography; b; Dwight R. Kuhn/DRK Photo; **679** Robert F. Sisson, ©National Geographic Society; **680** ©Dr. Jeremy Burgess/Science Photo Library/Photo Researchers, Inc.; **681** ©Oxford Scientific Films/Animals Animals; **684** S. Nielsen/DRK Photo; **685** Robert & Linda Mitchell Photography; **686** ©Norbert Wu; **687** l; Stephen Frink/Water House Stock Photography; rt; Stephen Frink/Water House Stock Photography; rb; ©Gerard Lacz/Peter Arnold, Inc.; **688** t; ©Kelvin Aitken/Peter Arnold, Inc.; b; ©A. W. Ambler/Photo Researchers, Inc.; **690** Jeffrey L. Rotman; **691** t; ©Norbert Wu; c; ©Steinhart Aquarium/Tom McHugh/Photo Researchers, Inc.; b; ©Nicole Duplaix/Peter Arnold, Inc.; **692** t; ©Sophie de Wilde Jacana/Jacana Scientific Control/Photo Researchers, Inc.; b inset; Runk/Schoenberger/Grant Heilman Photography; b; Tom Stack/Tom Stack & Associates; **693** t; Jeffrey L. Rotman; bl; ©Stephen Dalton/Photo Researchers, Inc.; br; ©J. W. Mowbray/Photo Researchers, Inc.; **695** t; ©Dr. Paul A. Zahl/Photo Researchers, Inc.; bl; ©Tom McHugh/Steinhart Aquarium/Photo Researchers, Inc.; br; James H. Carmichael Jr./The Image Bank; **697** ©Nuridsany et Perennou/Photo Researchers, Inc.; **698** l; ©J. L. Lepore /Photo Researchers, Inc.; r; ©Hans Pfletschinger/Peter Arnold, Inc.; **699** l; ©Gregory Ochocki/Photo Researchers, Inc.; c; ©Steinhart Aquarium/Tom McHugh/Photo Researchers, Inc.; r; ©Jeffrey Rotman/Peter Arnold, Inc.; **700** t; ©Peter Skinner/Photo Researchers, Inc.; b; Robert Frerck/Woodfin Camp & Associates; **701** Corel Professional Photos CD-ROM™; **702** ©Oxford Scientific Films/Animals Animals; ©Oxford Scientific Films/Animals Animals; **706** ©Norbert Wu; **707** ©Francois Gohier/Photo Researchers, Inc.; **708** Art Wolfe Incorporated; **709** t; Michael Fogden/DRK Photo; c; Frans Lanting/Minden Pictures, Inc.; b; Frans Lanting/Minden Pictures, Inc.; **711** l; Martim Harvey/The Wildlife Collection; r; ©Ed Reschke/Peter Arnold, Inc.; **712** t; ©Stephen Dalton/Photo Researchers, Inc.; b; Martin Harvey/The Wildlife Collection; **713** lt; Robert Lankinen/The Wildlife Collection; lb; Mark Moffett/Minden Pictures, Inc.; r; ©Martha Cooper/Peter Arnold, Inc.; **714** l; John Giustina/The Wildlife Collection; c; ©Jim Zipp /Photo Researchers, Inc.; r; ©Adam Jones/Photo Researchers, Inc.; **715** l; Frans Lanting/Minden Pictures, Inc.; c; Frans Lanting/Minden Pictures, Inc.; r; ©C. K. Lorenz/Photo Researchers, Inc.; **718** l; ©Breck P. Kent/Animals Animals; r; ©Robert Holmgren/Peter Arnold, Inc.; **719** t; Art Wolfe Incorporated; b; Penny Tweedie/Woodfin Camp & Associates; **720** t; Martyn Cowley; bl; Martyn Cowley; br; Frans Lanting/Minden Pictures, Inc.; **722** David Norhcott/DRK Photo; **726** Frans Lanting/Minden Pictures, Inc.; **727** ©Stephen Dalton/Photo Researchers, Inc.; **728** David Fritts/Tony Stone Images; **729** l; ©Stephen J. Krasemann/Peter Arnold, Inc.; r; Thomas Kitchin/Tom Stack & Associates; inset; Frans Lanting/Minden Pictures, Inc.; **730** tl; Corel Professional Photos CD-ROM™; tr; Gerard Lacz/Peter Arnold, Inc.; b; John Callanan/The Image Bank; **732** tl; ©Michael P. Gadomski /Photo Researchers, Inc.; tr; ©Martin Dohrn/Photo Researchers, Inc.; b; Chris Huss/The Wildlife Collection; **733** Frans Lanting/Minden Pictures, Inc.; **734** tl; Martin Harvey/The Wildlife Collection; tr; ©The Stock Market/ZEFA Germany; cl; Michael Fogden/DRK Photo; cc; Ivan Polunin/Bruce Coleman, Inc; cr; Corel Professional Photos CD-ROM™; bl; Robert & Linda Mitchell Photography; br; ©Jim Sipp/Photo Researchers, Inc.; **735** tl; Johnny Johnson/DRK Photo; tr; Corel Professional Photos CD-ROM™; cl; Corel Professional Photos CD-ROM™; cr; Doug Perrine/DRK Photo; bl; Corel Professional Photos CD-ROM™; bc; Doug Perrine/DRK Photo; br; Stephen KrasemannDRK Photo; **737** t; Kevin R. Morris/Corbis; bl; ©C. Prescott-Allen/Animals Animals; br; Bates Littlehales/National Geographic Image Collection; **738** t; Johnny Johnson/DRK Photo; b; Corel Professional Photos CD-ROM™; **739** t; ©Tim Davis/Photo Researchers, Inc.; b; Superstock; **740**

Darrel Gulin/PNI; 741 David Smart/DRK Photo; 744 Gerard Lacz/Peter Arnold, Inc.; 745 Corel Professional Photos CD-ROM™; 746 ©David McDonald/Oxford Scientific Films/Animals Animals; 747 t; Michael Fogden/DRK Photo; bl; Frans Lanting/Minden Pictures, Inc.; br; ©Manfred Kage/Peter Arnold, Inc.; 748 l; Robert & Linda Mitchell Photography; r; Laura Dwight/PhotoEdit; 750 tl; Sovfoto/Eastfoto/PNI; tr; ©John Mitchell/Photo Researchers, Inc.; b; ©Susan Kuklin/Photo Researchers, Inc.; 751 John Eastcott/Yva Momatiuk/DRK Photo; 752 l; Otis Imboden/National Geographic Image Collection; r; ©Francois Gohier/ Photo Researchers, Inc.; 753 lt; Stanley Breeden/DRK Photo; lb; Fred Bavendam ; r; ©Scott Camazine/Photo Researchers, Inc.; 754 t; Mark Moffett/Minden Pictures, Inc.; bl; ©Gary Retherford/Photo Researchers, Inc.; br; Mark Moffett/Minden Pictures, Inc.; 755 l; Manoh Shah/DRK Photo; r; Art Wolfe Incorporated; 756 ©M.I. Walker/Science Sourcer/Photo Researchers, Inc.; 760 Laura Dwight/PhotoEdit; 761 Mark Moffett/ Minden Pictures, Inc.; 763 Photo © Lois Greenfield, 1996; 764 Leo Mason/The Image Bank; 765 l; Mel Di Giacomo/The Image Bank; c; Rob Tringali, Jr./Sports Chrome East/West; r; Silver Burdett Ginn; 768 t; Kagen/Monkmeyer; bl; Bill Truslow/Tony Stone Images; br; Kopstein/Monkmeyer; 769 ©Professors P.M. Motta and S. Correr/Science Photo Library/Photo Researchers, Inc.; 771 ©Geoff Tompkinson/Science Photo Library/Photo Researchers, Inc.; 772 Ken Karp Photography; 774 t; David Wagner/Phototake; bl; Mark C. Burnett/Stock, Boston/PNI; br; Stock, Boston/ PNI; 778 David Wagner/Phototake; 779 Kagen/ Monkmeyer; 780 ©Professors P.M. Motta and S. Correr/Science Photo Library/Photo Researchers, Inc.; 782 Robert Tringali/Sports Chrome East/West; 783 l; Ken Karp Photography; c; Dr. David Scott/CNRI/ Phototake; r; Jeff Spielman/The Image Bank; 785 David M. Phillips/Visuals Unlimited; 787 l; Gary Cralle/The Image Bank; c; ©Prof. P. Motta/Dept. of Anatomy/ University of " LaSapienza", Rome/Science Photo Library./Photo Researchers, Inc.; r; ©Allsport/ Simon Bruty; 788 GJLP/CNRI/Phototake; 790 ©Mehau Kulyk/Science Photo Library/Photo Researchers, Inc.; 791 ©Catherine Poudras/Science Photo Library/Photo Researchers, Inc.; 792 l; Robert Brenner/PhotoEdit; c; Val Corbett/Tony Stone Images; r; Phillip Kretchmar/The Image Bank; 794 ©Omikron/ Science Source/Photo Researchers, Inc.; 796 t; John Ramey/The Image Bank; bl; ©Don Fawcett/Heuser, Reese/Science Source/Photo Researchers, Inc.; br; Robert Freck/Woodfin Camp & Associates; 799 ©Dr. Morley Read/Science Photo Library/Photo Researchers, Inc.; 804 GJLP/CNRI /Phototake; 805 ©Omikron/ Science Source/Photo Researchers, Inc.; 806 ©Bob Daemmrich; 807 c; Paul J. Sutton/Duomo Photography, Inc.; r; © Lennart Nilsson, Behold Man, Little Brown and Company; 809 t; Dr. Gobal Murti/ Phototake; bl; ©Prof. P. Motta/Dept. of Anatomy/ University "La Sapienza", Rome/Science Photo Library/ Photo Researchers, Inc.; br; Brett Froomer/The Image Bank; 810 ©Andrew Syred/Science Photo Library/ Photo Researchers, Inc.; 813 bl; ©CNRI/Science Photo Library/Photo Researchers, Inc.; br; Ydav Levy/ Phototake; t; William R. Sallaz/Duomo Photography, Inc.; 814 l; John D. Cunningham/Visuals Unlimited; c; Don W. Fawcett/Visuals Unlimited; r; Carolina Biological Supply Company/Phototake; 818 t; William R. Sallaz/Duomo Photography, Inc.; bl; ©Allsport/Gary Newkirk, 1993); br; William R. Sallaz/Duomo Photography, Inc.; 824 ©Allsport/Gary Newkirk; 826 Zao-Grimberg/The Image Bank; 827 t; ©Will and Deni McIntyre/Photo Researchers, Inc.; c; ©Eric Grave/ Science Source/Photo Researchers, Inc.; b; T. Rosenthal/Superstock; 832 l; Visuals Unlimited; r; Microworks/Phototake; 833 l; Lennart Nilsson, THE INCREDIBLE MACHINE, National Geographic Society, @ Boehringer Ingelheim International GmbH; r; ©Biophoto Assoc. Science Source/Photo Researchers, Inc.; 835 t; ©Proff. Motta, Correr & Nottola/University "LaSapienza", Rome/Science Photo Library/Photo Researchers, Inc.; bl; Michael Gabridge/ Visuals Unlimited; br; ©Prof. Arnold Brody/Science Photo Library/Photo Researchers, Inc.; 837 Art Siegel; 838 ©Will & Deni McIntyrePhoto Researchers, Inc.; 839 t; Ydav Levy/Phototake; b; David M. Grossman/ Phototake; 841 l; ©A. Glauberman/Photo Researchers, Inc.; r; Peter Holden/Visuals Unlimited; 848 David Phillips/Visuals Unlimited; 850 ©CNRI/Science Photo Library/Photo Researchers, Inc.; 851 lt; Phillip Kretchmar/The Image Bank; lb; ©Bob Daemmrich; r; William Sallaz/The Image Bank; 852 Nathan Benn/ Stock, Boston/PNI; 853 l; Corel Professional Photos CD-ROM™; rb; Corel Professional Photos CD-ROM™; rt; Corel Professional Photos CD-ROM™; 857 lt; R. Hutchings/PhotoEdit; lb; Corel Professional Photos CD-ROM™; r; Mary Kate Denny/PhotoEdit; 860 Lennart Nilsson ©Boehringer Ingelheim International GmbH; 862 t; Walter Iooss, Jr./The Image Bank; bl; ©Allsport/Al Bello; br; ©Allsport/Billy Strickland; 865 Lennart Nilsson ©Boehringer Ingelheim International GmbH; 866 ©Dan McCoy/ Rainbow; 867 t; ©CNRI/Science Photo Library/Photo Researchers, Inc.; b; Lennart Nilsson ©Boehringer Ingelheim International GmbH; 868 ©Prof. J. James/ Science Photo Library/Photo Researchers, Inc.; 869 ©Prof. R. Motta/Dept. of Anatomy/University "La Sapienza", Rome/Science Photo Library/Photo Researchers, Inc.; 874 ©Prof. P. Motta/Dept. of Anatomy/University "La Sapienza", Rome/Science Photo Library/Photo Researchers, Inc.; 876 Penny Gentieu/Tony Stone Images; 877 t; ©Mr. Richard Hutchings/Photo Researchers, Inc.; bl; P. Rivera/ Superstock ; br; ©Arthur Tilley/FPG International Corp.; 878 Electra/Phototake; 879 ©Motta & Familiari/Anatomy Dept./Univerisity "La Sapienza", Rome/Science Photo Library/Photo Researchers, Inc.; 881 ©C. Edelmann/La Villette/Photo Researchers, Inc.; 882 t; Tony Stone Images; bl; ©Lennart Nilsson, A CHILD IS BORN; br; ©Lennart Nilsson, A CHILD IS BORN; 883 t; Douglas Struthers/Tony Stone Images; b; ©Professors P.M. Motta and J. Van Blerkom/ Science Photo Library/Photo Researchers, Inc.; 885 l; ©Petit Format/Nestle/Science SourcePhoto Researchers, Inc.; r; ©Petit Format/Nestle/Photo Researchers, Inc.; 888 T. Rosenthal/Superstock ; 889 ©Suzanne Szasz/ Photo Researchers, Inc.; 890 t; ©Dr. Kar Lounatmaa/ Science Photo Libray/Photo Researchers, Inc.; bl; ©A.B. Dowsett/Science Photo Library/Photo Researchers, Inc.; br; ©Alfred Pasieka/Science Photo Library/Photo Researchers, Inc.; 891 ©Alfred Pasieka/ Science Photo Library/Photo Researchers, Inc.; 894 ©Manfred Kage/Peter Arnold, Inc.; 895 ©Manfred Kage/Peter Arnold, Inc.; 898 ©Petit Format/Nestle/ Photo Researchers, Inc.; 900 ©Manfred Kage/Peter Arnold, Inc.; 901 rt; ©Tektoff-Merieux, CNRI/Science Photo Library/Photo Researchers, Inc.; rb; National Institutes of Health, Rocky Mountain Lab; l; ©Dr. Kari Lounatmaa/Science Photo Library/Photo Researchers, Inc.; 904 ©Boehringer Ingelheim International GmbH, Photo by Lennart Nilsson, THE INCREDIBLE MACHINE; 905 t; ©Don Fawcett/Photo Researchers, Inc.; bl; ©Lennart Nilsson, THE INCREDIBLE MACHINE; br; ©Lennart Nilsson, THE INCREDIBLE MACHINE, National Geographic Society; 906 ©NIBSC/Science Photo Library/Photo Researchers, Inc.; 909 Science VU/©Boehringer-Ingelheim/Visuals Unlimited; 910 ©R. Ellis/Sygma; 911 ©NIBSC/ Science Photo Library/Photo Researchers, Inc.; 912 tl; ©Boehringer Ingelheim International GmbH, Photo by Lennart Nilsson, THE INCREDIBLE MACHINE, National Geographic Society; tr; ©Boehringer Ingelheim International GmbH, Photo by Lennart Nilsson, THE INCREDIBLE MACHINE; b; ©Boehringer Ingelheim International GmbH, Photo by Lennart Nilsson; 913 ©Nancy Kedersha/Science Photo Library/Photo Researchers, Inc.; 915 ©Sinclair Stammers/Science Photo Library/Photo Researchers, Inc.

Illustration

Academy Artworks Inc. 90; 95; 139 top; 159; 257; 353; 436 right; 437 top; 463 top; 488; 489; 500; 550; 571 top; 595 top; 621 top; 643 top left; 662; 667; 719 top left; 723 top left; 749; 773; 776; 820; 844; 845 bottom left; 856; 871 top; 916; 922 - 923; 924; 925 left, right; 928. Ernest Albanese 35 top, bottom; 79 top; 82; 83 top; 88; 91; 816; 863 top. Glory Bechtold 125; 126 spot art; 127; 128; 129 spot art; 137 spot art; 421 top; 426; 429 spot art; 430; 560; 561; 564; 567; 579 spot art; 586; 602; 604; 605; 607 top left; 610; 612 bottom left; 613. Scott Bodell 108 top, middle. Dartmouth Publishing, Inc. 27 inset; 39; 172; 189; 190; 199; 203 bottom; 210; 220 top; 223; 254; 323; 388 bottom; 389 top, bottom. Electra-Graphics, Inc. 36 top left; 40 bottom; 367 bottom; 379; 405; 529 top. Carlyn Iverson 14 top left; 106 top left; 134; 152 spot art; 174; 175 bottom left; 177 bottom; 179; 182; 183; 188; 191; 203 top; 285; 288; 290; 292; 298; 316 spot art; 317; 348 bottom; 349; 630; 632; 634 bottom right; 637; 639; 641; 690 bottom; 697 top right; 929 spot art; 930 spot art; 931 spot art; 302; 362; 371. Thomas LaPadula for S.I. International 18. Laura Maggio 8 top; 28; 29; 30 top inset, bottom; 33 inset. Elizabeth Morales 54 top, bottom; 56 top, bottom; 59 top; 61; 62; 63 insets; 65 insets; 67; 92; 93; 470 bottom; 472 top, bottom; 474; 481; 482; 483; 485; 487; 497; 516 top; 522 top right; 769 top; 784; 785 bottom right; 786; 788 top, bottom; 789 top, bottom; 793 top, bottom; 794 top, bottom; 797; 800; 808; 810 top, bottom; 811; 812; 814 bottom; 828; 830; 831; 834; 837 bottom; 840; 842; 858; 859; 860; 863 middle; 878; 879; 880; 883 top right; 884 top, bottom; 886; 887; 907; 908. Hilda Muinos 68; 69; 115; 270. Laurie O'Keefe 111insets; 228 top; 245 top; 264; 299; 301; 334; 335 top; 338 top, bottom; 394; 417; 445; 447 top; 448 bottom; 453; 454 top, bottom; 455; 456; 458; 462 bottom right; 537 top right; 538 top right; 539 top right; 544; 545 top, bottom; 549; 650; 658 top right; 673; 711 top; 721; 731. Rose Sievers 36 top right; 47; 76.

Charts and Graphs

Ernest Albanese 341; 504; 505; 507 top; 579; 592; 652; 678; 770; 777 top; 785 top; 815; 825; 845 top; 853; 855; 893. Dartmouth Publishing, Inc. 312; 314; 320; 321; 324 top; 361 bottom. ElectraGraphics, Inc. 347. Joe Galka 390-391; 396-397. Marnie Ingman 10 top right; 14 top right; 21 bottom left; 102 bottom; 113; 284; 303; 305. Michelle LoGerfo 821 left; 854; 903; 914; 917 left; 921. Laura Maggio 899. Hilda Muinos 102 top. Evelyn O'Shea 152; 395; 555. Rose Sievers 32; 42; 43; 83 bottom; 87; 103; 126; 128 top, bottom; 136; 137; 141 top; 149 top, bottom; 150; 184; 195; 211 top; 230; 243; 245 middle; 248; 250; 267; 275; 313; 316; 327; 335; 351; 372 top; 376; 377; 407; 409 top; 416; 418 top; 421 bottom; 429; 436 left; 559; 633.

Visual Essays

Glory Bechtold 582; 587. Scott Bodell 107. Andy Graziano 251. Carlyn Iverson 84; 89; 133; 185; 523; 689; 696. Elizabeth Morales 499; 766; 829; 836; 865. J/B Woolsey Associates 293; 294; 295. Laurie O'Keefe 368 spot art; 655; 671; 716.

Editorial Support

Lillian Duggan, Holly Gordon, Mary Hicks, Kim Merlino, Allan Sison